THE HOME & FARM MANUAL

A PICTORIAL ENCYCLOPEDIA OF
FARM, GARDEN, HOUSEHOLD, ARCHITECTURAL, LEGAL
MEDICAL AND SOCIAL INFORMATION

With More Than Twelve Hundred Engravings

CLASSIC 1884 EDITION

By JONATHAN PERIAM

GREENWICH HOUSE
Distributed by Crown Publishers, Inc.
New York

Publisher's Note: This book was originally published in 1884. Much
of the technical and medical information is now obsolete. This
book should be used as a historical reference work and as an insight
into the social history of the late 1800s, not as a guide applicable
to modern times.

Copyright © 1984 by Greenwich House, Inc. a division of
Arlington House Inc. All rights reserved.

This 1984 edition is published by Greenwich House, a division of
Arlington House, Inc., distributed by Crown Publishers, Inc.,
One Park Avenue, New York, New York 10016.

Manufactured in the United States of America

Library of Congress Cataloging in Publication Data

Periam, Jonathan.
   The home and farm manual.

   Includes index.
   1. Home economics—United States—Handbooks, manuals, etc.
2. Agriculture—United States—Handbooks, manuals, etc.
I. Title.
TX153.P43   1984   640   84-6155

ISBN: 0-517-445301

h g f e d c b a

FOREWORD

This edition of *The Home and Farm Manual* is a reprint of a book first published in 1884. It was Jonathan Periam's masterpiece, the culmination of years of learning and research. His plan was ambitious—to compile virtually all information necessary for farmers and householders into one concise, clear, and easy-to-understand volume.

When *The Home and Farm Manual* was written, information was difficult to obtain and often its accuracy was questionable. It was a time prior to the advent of radio, television, and easy travel. A book of the sort Periam envisioned was meant to fill a very real void, providing information that people all across America truly needed.

Periam's *The Home and Farm Manual* was a terrific success. His tone was confident and helpful, his information accurate and modern. The topics ranged from highly technical discussions of crop yield and specific farming techniques to questions of etiquette, the metric system, and advice on how to raise children. Accompanying the text were extremely detailed and lovely drawings that added greatly to the book's practicality.

Now, exactly one hundred years after the book was first published, this classic reprint allows the reader to experience the range of Periam's knowledge and wisdom. Of course, much of the information that was then so modern and up-to-date is now useful only as a historical record of previous methods and techniques. But there is still much material that the reader will find enjoyable and profitable—exquisite botanical drawings, ideas and techniques for landscaping, simple and economical recipes, general principles of etiquette, and Periam's sensible, rational approach to solving the problems of everyday life.

New York
1984

A.F.

AUTHOR'S PREFACE

THIS book is intended for the average American citizen—for the Household and the Farm. It is intended for the man of work and business who has not the time nor disposition to plod through scores of volumes of elaborate dissertations on the subjects touching the affairs of his every-day life. It is dedicated to the household and the library of the poor. It is designed for the use of the father, the mother, the son and the daughter of the American family. It is hoped that by its teachings and use they will all fulfill the purposes of their lives and labor better—more successfully—more happily. If so, and by its directions and simple teachings the man and woman of work will be led to labor with more intelligent thought, and more pleasure and profit, the author will be abundantly repaid.

The principles which underlie success or failure in the affairs of business or society are so momentous as to deter any one from undertaking the task of outlining them. But, encouraged by the flattering and really unexpected success and favor with which former works by the author have been received, the labor of preparing this work was undertaken.

The information can not fail to be of general, even universal value. Because one has lived on the farm, or managed a household, it is no reason that either has been done to the best advantage. So long as we see the many instances which occur of men, long experienced in business, failing, we may properly conclude that all is not yet known that needs to be known by even the experienced and the fairly successful; and it is quite probable that many, whom, by reason of the length of time they have been in business, the world regards as wise and skillful in managing their affairs, have yet much to learn.

The aim of the author has been to produce a book which will give to all, who may honor it with their attention, advantages of information and knowledge not possessed heretofore in any ordinary library, and never in one volume, and which fathers may place in the hands of their wives, and their families, with the commendation that it contains the essence of the combined experience of generations of practical, thinking men; and that wisdom to which Solomon alluded when he said, "The merchandise of it is better than the merchandise of silver, and the gain thereof than fine gold."

In the preparation of the work the object has been to preserve a clear and systematic arrangement of the several subjects, giving to every fact such plain statement as would make it of easy understanding, and its application made alike easy. To this end the work has been divided into *eleven* distinct departments, according to general subjects—these departments into chapters, which are again carefully subdivided, so as to make the whole of easy reference and consultation. In furtherance of this end an elaborate alphabetical index is appended so that in matters of frequent and general reference it will be most convenient.

Its Objective Presentation by means of *Charts*, *Maps*, *Drawings* and *Diagrams* of all the more important features, adds, it is hoped, force and strength to the written word to materially aid the understanding.

As a combination of Object-Teaching with simply written instructions—the methods of imparting knowledge so successful in our best schools—it is certainly more full than any book or combination of books yet published concerning the many important subjects embraced. The value of this double method of conveying fact to the mind through the eye by accurate and elegant illustration—as well as by simple word—cannot well be over-estimated. The perusal of the following pages and the effort to reduce its teachings to practice will, however, impress one with the admirable utility of this illustrative method of imparting practical fact.

The prodigal liberality and broad enterprise of my publishers have, at an expense truly enormous, enabled me to utilize this superb method to an extent, which, I think I am safe in saying, has never been attempted in a book of like practical character in this or any other country. For this I am pleased to acknowledge my indebtedness, and believe I am in the bounds of truth when I say that every reader will be disposed to recognize a similar obligation.

To preserve in the written text a style and method of easy comprehension, which will be in keeping with the accuracy, simplicity and elegance of the illustration, has been the constant effort. Whether this important end has been attained it is the province of the reader to decide.

One idea that has sustained the author in the immense labor involved in the preparation of this volume is, that he may thus be useful to those living in the country districts, many of whom have no access to large libraries, and by inadequacy of income are prevented from largely increasing their own. The hope has been to put within their reach a volume which, because of its wealth, variety and suggestiveness, should be a *practical every-day library* in itself.

AUTHOR'S PREFACE

Of matters concerning *Live Stock* no mention has been made except those of Farm Building, and Agricultural Law as connected therewith. In a previous work in connection with Dr. A. H. Baker, V. S., the author prepared and published a large book on the animals of the farm, the extraordinary success of which encouraged the preparation of this work, which is designed as a fit companion volume. Combined they cover, practically, it is believed, the entire subjects of Household and Farm interests—the one appropriately supplementing the other. In both it has been the aim to *sift* from the great mass of information, statistics and suggestions, in contemporary and newspaper literature, only those matters of actual interest and every-day usefulness, and condense into space the most convenient for a hasty moment's reference; and yet to retain all necessary details to reward the leisurely perusal and study of the book with interest and great profit.

It is believed that, whether read by the man of business, the farmer, the mechanic, the housewife, or the child, it will be a source of profitable knowledge, entertainment and pleasure.

For special assistance in my work I desire to acknowledge my obligations to Messrs. FURST & BRADLEY, of Chicago, who have made Farm Machinery a special study; to Mr. J. C. VAUGHAN, of Chicago, widely known as an adept in all that pertains to general Floriculture; to Messrs. ELLWANGER & BARRY, of Rochester, N. Y., who have a world-wide reputation in what pertains to the Nursery, Ornamental and other Trees; to Mr. H. DE VRY, Superintendent of Lincoln Park, Chicago, for many ideas in Landscape Work and Ornamental Planting; and also to Dr. LORING, Commissioner of Agriculture, at Washington, than whom no one has a truer practical knowledge of the manifold necessities of progressive Agriculture. The assistance asked has in each case been promptly and cheerfully accorded.

To Mr. E. C. SIMMONS, of St. Louis, President of Simmons Hardware Company, the brilliantly successful man of business, I am likewise indebted for the use of many valuable cuts, illustrating, practically, the text in *Third* and *Ninth* Departments.

Trusting that it will be favorably received, I commit my book into the hands of that critic whose judgment has been so partial to my previous efforts, and which is deemed the best, most important and final—the practical reading public.

JONATHAN PERIAM.

CONTENTS

PART I.

THE HOME AND FARM.

MAKING CONVENIENT, COMFORTABLE AND HAPPY HOMES. ANCIENT AND MODERN AGRICULTURE ILLUSTRATED AND COMPARED. IMPROVED FARM IMPLEMENTS AND MACHINERY. PRINCIPLES AND PRACTICE.

CHAPTER I.

BUILDING HAPPY HOMES.

I—Contrasted Pictures. II—The Ideal Country Home. III—The Farmer's Wife. IV—The Overworked Wife. V—Sons and Daughters on the Farm. VI—Youthful Activity VII.—Adorning the Home. VIII—Improving the Homestead. IX—Sports of Childhood. X—Lessons from the Garden. XI—What we Hope to Teach. 33

CHAPTER II.

ANCIENT AND MODERN AGRICULTURE.

I—The Value of Books. II—Agriculture Among the Savages. III—The Arizona Indians. IV—Mound Builders. V—Agriculture Defined. VI—Its Divisions. VII—Its History. VIII—The Books of Mago. IX—Mago on Working Cattle. X—Rome's Agricultural Writers. XI—Chronicles of Columella. XII—Ancient Farms and Implements. XIII—Cultivate Little, Cultivate Well. XIV—A Fancy Farmer. XV—Arable Lands and Pastures. XVI—Water Meadows. XVII—A Rich Meadow. XVIII—Roman Rotation. XIX—Roman Small Grains. XX—Some Ancient Methods. XXI—Antique Crops. XXII—Crops Pulled by Hand. XXIII—Fallow Crops. XXIV—Ancient Harvesting. XXV—Roman Fertilizers. XXVI—A Question not yet Settled. XXVII—Little and Often. XXVIII—Commercial Fertilizers. XXIX—Ancient Plowing. XXX—Plows. XXXI—Seeding. XXXII—Yield Per Acre. XXXIII—Mediæval and Modern Agriculture. 44

CHAPTER III.

MODERN FARM IMPLEMENTS.

I—A Revolution in Fifty Years. II—The Pioneer's Plow. III—The First Steel Plow. IV—The Webster Plow. V—The Plow of To-day. VI—The Art of Plowing. VII—Laying Out the Land. VIII—Turning the First Furrow. IX—The Back Furrow. X—Re-Plowing. XI—Subsoil Plowing. XII—French Plowing. XIII—Implements for Smoothing and Disintegrating. XIV—Leveling, Compacting and Grinding. XV—The Plank Soil Grinder. XVI—The Leveler. XVII—Implements of Cultivation. XVIII—History of the Cultivator. XIX—One-Horse Cultivators. XX—Seeding Machines. XXI—The Grain Drill. XXII—Corn Planters. XXIII—Harvesting Machinery. XXIV—Use Only the Best. XXV—Plowing Irregular Areas. 81

CONTENTS

CHAPTER IV.

PRINCIPLES AND PRACTICE.

I—Study Your Farm. II—Analysis of the Soil Unnecessary. III—Soil Does Not Wear Out. IV—Organic and Inorganic Matter. V—Economy of Fertilizers. VI—Practical Test of Fertility. VII—Rotation and Crops. VIII—A Simple Rotation. IX—Effect of Bad Seasons. X—Elaborate Rotation. XI—Grass-Seed and Meadows. XII—An Eastern Man on Rotation. XIII—A Southern Planter's Testimony. XIV—Rotation in Europe. XV—Substitution in Rotation. XVI—Potash and Phosphate Crops. XVII—Soft and Hard Ground Crops. XVIII—Science in Agriculture. XIX—Ignorance vs. Intelligence. XX—Soils and their Capabilities. XXI—Percentage of Sand in Soils. XXII—Absorbing Power of Soils. XXIII—Absorption of Oxygen by the Soil. 75

PART II.

PRACTICAL AND SYSTEMATIC HUSBANDRY.

CEREAL CROPS AND THEIR CULTIVATION. GRASSES, FODDER AND ROOT CROPS. SILK CULTURE—SPECIAL CROPS. CROPS FOR SUGAR MAKING. VARIETIES ILLUSTRATED AND COMPARED.

CHAPTER I.

CEREALS AND THEIR CULTIVATION.

I—The Cereals Described. II—Wheat and Corn Belts. III—Corn in the United States. IV—Different Kinds of Wheat. V—Variations Illustrated. VI—Proper Wheat Soils. VII—Preparing the Soil. VIII—Drilling Gives the Best Results. IX—Depth of Covering for Wheat. X—Time to Seed and Harvest. XI—Harvesting Wheat. XII—How to Shock the Grain. XIII—Importance of Good Seed. XIV—Pedigree Grain. XV—General Conclusions. XVI—Artificial Cross Fertilization. XVII—Reputable Old Varieties in the United States. XVIII—Rye and its Cultivation. XIX—Barley and its Cultivation. XX—Time for Sowing Barley. XXI—Harvesting and Threshing Barley. XXII—New Varieties of Barley. XXIII—Oats and their Cultivation. XXIV—Export of Food Crops. XXV—Species of Oats—their Latitude. XXVI—Soil and Cultivation of Oats. XXVII—Harvesting and Threshing Oats. XXVIII—Varieties of Oats to be Cultivated. XXIX—Buckwheat. XXX—Seeding and Harvesting Buckwheat. 89

CHAPTER II.

INDIAN CORN, RICE, AND SPECIAL CROPS.

I—The Crop in the United States. II—How to Increase the Average. III—Proper Manures for Corn. IV—The Cultivation of Corn—Plowing. V—Preparing the Soil. VI—Planting the Crop. VII—Harrowing the Young Corn. VIII—After Cultivation of Corn. IX—How Often to Cultivate. X—Depth of Cultivation. XI—Harvesting the Crop. XII—Cutting and Shocking. XIII—Seed Corn. XIV—Cost of a Corn Crop. XV—Varieties of Corn. XVI—Rice and its Cultivation. XVII—True Water Rice, or Commercial Rice. XVIII—Cultivation of Rice in Carolina. XIX—Management of Rice Fields. XX—Cultivating the Crop. XXI—Flooding the Crop. XXII—Harvesting and Threshing. XXIII—Hulling for Market. XXIV—Rice in the Mississippi Delta. XXV—Some Special Crops. 109

CHAPTER III.

MEADOW AND PASTURE GRASSES.

PAGE

I—The Value of Grass. II—What is Grass? III—How to Know Grass. IV—Testing the Value of Species. V—The Value of Accurate Knowledge. VI—Well-known Cultivated Grasses. VII—Grasses for Hay and Pasture. VIII—A list of Good Grasses. IX—Valuable Native Western Grasses. X—Disappearance of Native Grasses. XI—Valuable Introduced Grasses, South. XII—Bermuda Grass. XIII—Guinea Grass. XIV—Brome or Rescue Grass. XV—Seeding Meadows. XVI—The Alphabet of Agriculture. XVII—Sowing for Hay and for Pasture. XVIII—The Celebrated Woburn Experiments. XIX—A Summary of Meadow Grasses. XX—About Pastures. XXI—Genera, Species and Varieties. XXII—Favorite Pasture Grasses. XXIII—Bent Grasses. XXIV—Orchard Grass. XXV—Grasses for Various Regions. XXVI—Clover in its Relation to Husbandry. XXVII—The Seed Crop. XXVIII—Valuable Varieties of Clover. XXIX—Dutch, or White Clover. XXX—Alsike, or Swedish Clover. XXXI—Clovers for the South—Alfalfa. XXXII—Japan Clover. XXXIII—Mexican Clover. XXXIV—Importance of the Pulse Family. XXXV—Interchange of Grasses Between Nations..... 132

CHAPTER IV.

SOILING, FODDER AND ROOT CROPS.

I—Soiling Compared with Pasturing. II—Soiling Indispensable in Dairy Districts. III—Soiling as Against Fencing. IV—How to Raise a Soiling Crop. V—Corn and Sorghum for Soiling. VI—The Clovers as Soiling Crops. VII—Millet and Hungarian Grass. VIII—Prickly Comfrey. IX—The Advantages of Soiling. X—Results of Soiling in Scotland. XI—Root Crops for Forage. XII—Things to Remember in Root Culture. XIII—Preparing for the Root Crop. XIV—Sowing and Cultivating. XV—Harvesting Root Crops. XVI—Pitting and Cellaring the Roots. XVII—The Artichoke.................. 167

CHAPTER V.

SILOS AND ENSILAGE.

I—What is Ensilage? II—Silos and Ensilage Long Known in Europe. III—Two Methods Illustrated. IV—The Father of Ensilage. V—Fermentation Should be Avoided. VI—What Ensilage May do. VII—The History of Ensilage. VIII—Feeding Value of Ensilage. IX—Ensilage in the United States. X—Effects of Fermentation in the Silo. XI—Size of Silos for Certain Number of Stock. XII—How to Build a Silo. XIII—Practical Experience and Results. XIV—Perfect Food and Rations. XV—Some Statements of the Quantity Fed. XVI—Cost of Ensilage in Massachusetts. XVII—Building a Model Silo. XVIII—Practical Conclusions from Careful Experiment................. 178

CHAPTER VI.

TEXTILE CROPS AND FIBERS.

I—Cotton: Its History and Cultivation. II—The Family to which Cotton Belongs. III—The Soils for Cotton. IV—History of Cotton Cultivation in the United States. V—Increasing Importance of Cotton. VI—Cotton by States. VII—The Climate for Cotton. VIII—The Best Cotton States. IX—The Cultivation of Cotton. X—Preparation of the Soil. XI—Tending the Growing Crops. XII—Flax and its Cultivation. XIII—Proper Soil for Flax—Seeding.

CONTENTS

XIV—Harvesting Flax. XV—Hemp and its Cultivation. XVI—Raising a Crop of Hemp-Seed. XVII—Raising Hemp for Lint. XVIII—The Time to Harvest Hemp. XIX—Rotting and Breaking for Market. XX—Conclusions on Flax and Hemp. XXI—Jute and its Cultivation. XXII—Growth and Harvesting of Jute. XXIII—Preparing Jute Fiber. XXIV—The Ramie Plant in the United States. XXV—Soil and Planting. XXVI—Ramie is a Perennial Plant. 194

CHAPTER VII.

SILK AND SILK-WORMS

I—Silk Culture in America. II—Silk Producing Insects. III—From the Egg to the Moult. IV—Varieties of the Silk-worm. V—Keeping and Hatching the Eggs. VI—Preparing to Feed the Worms. VII—Feeding and Care of Silk-worms. VIII—Moulting or Casting the Skin. IX—Winding Frames on Which the Worms Spin. X—Killing the Worms. XI—Reeling the Silk. XII—Marketing Cocoons and Eggs. XIII—Food of the Silk-worm. XIV—Raising Mulberry Trees. 211

CHAPTER VIII.

SPECIAL CROPS—HOPS, TOBACCO, PEANUTS AND SWEET POTATOES.

I—Hop Growing in America. II—Cost of Raising. III—Establishing a Hop Yard. IV—The Proper Situation and Soil. V—Preparing for the Crop. VI—Trenching the Soil. VII—Setting the Plants. VIII—Care of the Hop Yard. IX—Cultivation in Crop Years. X—Picking the Hops. XI—Drying the Hops. XII—Management in the Kiln. XIII—The Cultivation of Tobacco. XIV—Soils and Situation for Tobacco. XV—The True Tobacco Belt. XVI—Raising the Crop South. XVII—Transplanting, Cultivating and Worming. XVIII—The Seed Bed. XIX—Raising Plants North. XX—Preparing the Land. XXI—Planting at the North. XXII—Proper Way to Transplant Tobacco. XXIII—Cultivation. XXIV—Cutting and Curing Tobacco. XXV—The Tobacco House. XXVI—Twelve Rules for Tobacco Growers. XXVII—Peanuts, or Goubers. XXVIII—The Cultivation for Peanuts. XXIX—Gathering the Nuts. XXX—After Management and Care of Seed. XXXI—Sweet Potatoes. XXXII—Field Culture of Sweet Potatoes. XXXIII—Keeping Sweet Potatoes in Winter. XXXIV—Garden Cultivation. 223

CHAPTER IX.

CROPS FOR SUGAR-MAKING.

I—Sugar and its Manufacture. II—Cane and Other Sugars Compared. III—History of Beet Sugar in the United States. IV—Our Two Great Sugar Plants. V—The Various Saccharine Products. VI—The Three Sugars Compared. VII—Cultivation of Sugar-Cane. VIII—Cultivation of Sorghum. IX—When to Cut Sorghum Cane. X—Cutting and Handling the Cane. XI—Specific Gravity as a Basis of Value. XII—Specific Gravity and Composition of Juices. XIII—Table of Juices. XIV—Value of Sorghum During Working Period. XV—Four Important Points. XVI—Valuable Canes South. XVII—The Real Test of Value. XVIII—Table of Comparative Values During Working Period. XIX—The Manufacture of Sorghum. XX—Making Sugar on the Farm. XXI—General Conclusions. XXII—Maple Sugar. XXIII—Tapping the Trees. XXIV—Boiling and Sugaring. XXV—Sugaring Off. XXVI—To Tell When Sugar is Done. 246

PART III.

ARRANGEMENT OF FARMS.

HOW TO SECURE COMFORT AND PROFIT IN THE HOMESTEAD. FENCING AND DRAINAGE ART. FARM IMPROVEMENTS ILLUSTRATED AND EXPLAINED.

CHAPTER I.

COMFORT AND PROFIT IN THE HOMESTEAD.

I—Pioneer Farming. II—Improving the Farm. III—True Success in Farming. IV—Look to the Details. V—Thrift and Unthrift Illustrated. VI—How to Select a Claim of Land. VII—Commencing the Farm. VIII—The Crops to Raise. IX—The Second Year's Crops. X—The Third Year's Work. XI—Wind-Breaks and Groves. XII—Starting the Orchard. XIII—How to Clear a Timbered Farm. XIV—Making a Clearing and Building the House. XV—Carrying up the Sides. XVI—Putting on the Roof. XVII—Building the Fireplace. XVIII—Chinking the House. XIX—Deadening Timber. XX—The Work of Improvement. 271

CHAPTER II.

FARMS AND THEIR IMPROVEMENT.—LEASING.

I—Soils Indicating Variety of Crops. II—Adaptation of Soils to Crops. III—Adaptation of Crops to Localities. IV—Starting a Dairy. V—When to Sell the Crop. VI—Study the Probabilities. VII—When to Hold the Crop. VIII—How to Select a Farm. IX—Important Things to Consider. X—Situation of the Farm. XI—Some Things to be Remembered. XII—Leasing a Farm. XIII—Forms of Lease and Certificate. XIV—Plan for Laying Out a Farm. XV—A Garden Farm..286

CHAPTER III.

FENCES, HEDGES AND GATES.

I—Relative Cost of Fences and Buildings. II—Cost of Farm Fences in the United States. III—The Cost of Fence Per Rod. IV—Worm, or Virginia Fences. V—Staking and Ridering the Fence. VI—Post and Rail Fence. VII—Preparing the Timber. VIII—Mortising the Posts and Sharpening the Rails. IX—Setting the Posts. X—Fastening the Rails and Finishing. XI—How to Build a Board Fence. XII—Stringing a Wire Fence. XIII—Sod-and-Ditch Fence. XIV—Compound Fences. XV—Portable Fences. XVI—Fencing Steep Hillsides. XVII—Bars and Gates. XVIII—The Slide and Swinging Gate. XIX—Swing Gates and Slide Gates Explained. XX—Self-Closing Slide Gates. XXI—Southern Strap-Hinge Farm Gate. XXII—Double-Braced Gate. XXIII—Adjustable Swing Gate. XXIV—How to Prevent Posts from Sagging. XXV—Ornamental Gates. XXVI—Flood and Water Gates. XXVII—Stream Gate and Footway...............................298

CHAPTER IV.

FARM AND ORNAMENTAL HEDGES.

I—The Poetry of Hedges. II—Advantages and Disadvantages of Hedges. III—How to Prepare the Hedge-Row. IV—Setting the Hedge. V—Finishing the Planting—Cultivation. VI—Trimming the Hedge. VII—Ornamental Hedges. VIII—Ornamental Plants for Hedges. IX—How to Plant the Hedge. X—Care of Deciduous Hedges. XI—Trees for Barriers and Protection...316

CHAPTER V.

DRAINAGE AND THE DRAINER'S ART.

I—The Importance of Draining. II—The Antiquity of Drainage. III—Ancient Writers on Drainage. IV—Drainage Among the Greeks. V—Drainage Defined. VI—Drainage Among the Romans. VII—Drainage by French Monks. VIII—Some Fathers of Modern Drainage. IX—The Origin of Tile. X—Practical Men on Tile Drainage. XI—A Dry Surface May Need Drainage. XII—What an Ohio Farmer Says. XIII—Draining in Indiana. XIV—Draining in Michigan. XV—Illinois Experience. XVI—A Right and a Wrong Way for Open Drains. XVII—Stock Water from Drains. XVIII—How to Excavate the Pond. XIX—Drainage and Fences. XX—The Formation of Underdrains. XXI—Various Means of Drainage. XXII—Stone-Laid Drains. 324

CHAPTER VI.

DRAINAGE AND THE DRAINER'S ART—Continued.

I—Slab and Pole Drains. II—Tile Drains. III—Laying Out the Work. IV—Draining Tools. V—Grading the Ditch. VI—Leveling the Bottom. VII—Challoner's Level. VIII—Leveling from the Surface. IX—Altering the Grade—Silt Wells. X—The Water Carried by Tile. XI—Capacity of Soils for Water. XII—Velocity of Water in Tiles. XIII—Connecting Laterals with Mains. XIV—Draining a Field. XV—When it Pays to Drain a Farm. XVI—Sinks and Wallows. XVII—Springs, Soaks and Sloughs. XVIII—Draining Large Areas. XIX—Lands Requiring Drainage. XX—Wet Weather Plants. XXI—How to Know Lands Requiring Drainage. XXII—Importance of Drainage to Stockmen. 340

PART IV.

RURAL ARCHITECTURE.

ILLUSTRATED PLANS AND DIRECTIONS FOR VILLAGE AND COUNTRY HOUSES. BUILDING MATERIAL AND THE BUILDER'S ART. INCLUDING EVERY GRADE OF RESIDENCE, OUT-HOUSES, GARDEN AND ORNAMENTAL STRUCTURES. MECHANICS AS APPLIED TO THE FARM.

CHAPTER I.

PLANS AND DIRECTIONS FOR COUNTRY HOUSES.

I—Building According to Means. II—Improving the Old Homestead. III—An Elegant Country Home. IV—Farm and Suburban Cottage. V—When to Build. VI—The Provident Farmer's Marriage Settlements. VII—How to Build. VIII—What to Build. IX—Taste and Judgment in the Details. X—Where to Build. XI—A Hillside Cottage. XII—Ice-House and Preservatory. XIII—The Water Supply. XIV—House Drainage. XV—Ventilation. 361

CHAPTER II.

BUILDING MATERIAL AND THE BUILDER'S ART.

I—Building Material. II—How to Make Unburned Brick III—Specifications of Farm and Other Buildings. IV—Outline of Specifications for House of Wood with Stone or Brick Foundations. V—Masonry and Mason's Work. VI—Carpentry and Carpenter's Work. VII—

Painter's Work. VIII—Tinner's and Plumber's Work. IX—Contract for Performance of Obligations. X—How to Consult an Architect. XI—Glossary of Scientific Terms Used in Architecture... 376

CHAPTER III.

RURAL BUILDINGS, OUT-HOUSES AND GARDEN STRUCTURES.

I—Farm Houses and Cottages. II—Cottage for Farm Hand. III—Square Cottage. IV—Suburban or Farm Cottage. V—A Pretty Rural Home. VI—A Convenient Cottage. VII—Farm House in the Italian Style. VIII—English Gothic Cottage. IX—Plan of Rural Grounds. X—School-House and Church Architecture. XI—Children's Wigwam. XII—Rustic Seats and Summer Houses. XIII—Some Rural Out-Buildings. XIV—Poultry Houses and Chicken Coops. XV—Glass Structures. XVI—Smoke-Houses. XVII—The Farm Ice-House. XVIII—Privies and their Arrangement........................... 393

CHAPTER IV.

BARNS, STABLES AND CORN-CRIBS.

I—Grouping Farm Buildings. II—A Complete Cattle-Feeding Barn. III—Horse and Cow Barn with Shed. IV—Suburban Carriage-House and Stable. V—Sheep Barns and their Arrangement. VI—Hog Barns. VII—Granaries, Corn-Houses and Corn-Cribs. VIII—Rat-Proof Granary and Corn-Crib. IX—Corn-Cribs with Driveway. X—Section of Western Corn-Crib. 414

CHAPTER V.

MECHANICS AS APPLIED TO THE FARM.

I—The Farm Workshop. II—Mechanics' Tools on the Farm. III—Arrangement and Care of Tools. IV—How to Keep Farm Implements. V—Sharpening Tools. VI—Proper Way to File an Implement. VII—Repairing Common Implements. VIII—The Farm Paint Shop. IX—Putting up Rough Buildings. X—Shingling a Roof. XI—Making a Hay Rack. XII—Stone Fences. XIII—Moving Heavy Stones. XIV—For and Against Stone Walls. XV—How to Build the Wall. XVI—The Balloon Frame in Building. XVII—How to Build the Frame.. 426

PART V.

HORTICULTURE.

VARIETIES AND CULTIVATION OF FRUITS, FLOWERS AND SHRUBS. THE PRACTICAL ART OF GRAFTING AND BUDDING. LANDSCAPE GARDENING AND LANDSCAPE TREES. COMMON SENSE TIMBER PLANTING. INCLUDING FISH AND FISH CULTURE.

CHAPTER I.

ORCHARD, VINEYARD AND SMALL-FRUIT GARDEN.

I—The Farm Orchard and Garden. II—Arrangement of the Home Orchard. III—How to Prepare for an Orchard. IV—Laying Out the Orchard and Planting. V—When to Buy Trees and When to Plant Them. VI—What Varieties to Plant. VII—Apples, their Cultivation and

CONTENTS

Varieties. VIII—Pears, their Varieties and Cultivation. IX—The Forms of Fruit Explained. X—Peaches. XI—Nectarines. XII—The Cherry XIII—Picking and Packing Orchard Fruits. XIV—The Small Fruits. XV—The Vineyard. XVI—The Grapes for Farmers. XVII—Cultivation of the Cranberry. 445

CHAPTER II.

GRAFTING AND BUDDING.

I—Grafts, Cuttings and Seedlings. II—The Grafter's Art. III—How to Graft. IV—Tools for Grafting. V—Grafting by Approach. VI—Grafting Old Orchards. VII—Cutting and Saving Scions. VIII—Grafting Wax. IX—Budding. X—When to Bud. XI—How to Prepare the Buds. XII—How to Bud. XIII—Spring Budding. XIV—Time to Cut Scions. XV—Grafting the Grape. 481

CHAPTER III.

VEGETABLE GARDENING.

I—Economy of the Garden. II—How One Man Became a Gardener. III—Starting a Market Garden. IV—Troughs for Forcing Plants. V—The Number of Plants to Raise. VI—The Hot-Bed. VII—Laying up the Hot-Bed. VIII—Market and Kitchen Gardening. IX—Water and Ventilation. X—How to Have Early Rhubarb. XI—"Take Time by the Forelock." XII—What to Raise for Market. XIII—Economy in Cultivation. XIV—Preparing Vegetables for Market. XV—How to Raise Potatoes. XVI—"Planting in the Moon." XVII—Potatoes Illustrated. 491

CHAPTER IV.

GARDEN FLOWERS AND SHRUBS.

I—The Flower Garden. II—How to Cultivate Flowers. III—Select List of Flowers for General Cultivation. IV—Biennial and Perennial Flowers. V—Summer Flowering Bulbs. VI—Flowering Plants and Vines—Roses. VII—Flowering Shrubs. VIII—Climbing and Trailing Shrubs. IX—Flowering Trees. X—Everlasting Flowers and Ornamental Grasses. XI—Water Plants. XII—Trellises. 507

CHAPTER V.

LANDSCAPE GARDENING AND LANDSCAPE TREES.

I—The Landscape Gardener's Art. II—Studying Effects. III—Design for a Village Lot. IV—Design for Secluded Grounds. V—Trees and Terraces—Tree Protectors. VI—Laying Out Curves of Walks and Drives. VII—Laying Out and Planting Flower Beds. VIII—Landscape Effects. IX—Trees for Landscape Planting. X—Tropical Plants. 535

CHAPTER VI.

FISH AND FISH PONDS.

I—Fish on the Farm. II—Fishes for Cultivation. III—River and Pond Fish and their Time for Spawning. IV—The Families of River and Pond Fish. V—Rules for the Transportation of Fish. VI—Artificial Fish Breeding. VII—Hatching the Fish. VIII—Fish-Hatching Boxes. IX—Breeding Fish in Ponds. X—Carp Breeding. XI—How to Form the Pond. 557

CHAPTER VII.

COMMON SENSE TIMBER PLANTING.

I—The Economy of Timber. II—What Timber Really Does for a Country. III—What Timber to Plant. IV—Our Experience in Tree Planting. V—The Poetry of the Forest. 570

PART VI.

INSECTS AND BIRDS IN THEIR RELATION TO THE FARM.

INSECTS INJURIOUS AND BENEFICIAL. ILLUSTRATED CLASSIFICATION OF INSECTS. REMEDIES AND PREVENTIVES AGAINST DAMAGE. BIRDS TO BE FOSTERED OR DESTROYED.

CHAPTER I.

ENTOMOLOGY ON THE FARM.

I—Practical Value of Entomology. II—Destroying Insects on Nursery Trees. III—Orchard Culture in Relation to Insects. IV—Care of Trees in Relation to Insects. V—Predatory Birds and Insects. VI—The Study of Insects. VII—The Classification and Anatomy of Insects. VIII—Divisions of Insects According to their Food. IX—Noxious and Injurious Insects. 579

CHAPTER II.

INSECTS INJURIOUS AND BENEFICIAL.

I—Plant-Lice. II—Scale Insects. III—Plant Bugs. IV—General Means for Destroying Bugs. V—Remedies for Chinch-Bugs. 595

CHAPTER III.

INSECTS DESTROYING GRASS AND GRAIN.

I—Insects that Prey Upon Grass. II—Insects Injuring Clover. III—Clover-Leaf Beetle. IV—The Army Worm. V—Vagabond Crumbus. VI—Insects Injuring Grain. VII—The Sorghum Web-Worm. VIII—Sugar Cane Beetle. The Smaller Corn-Stalk Borer. X—The Rice-Stalk Borer. XI—Grass-Worm of the South. XII—Corn Bill-Bug. XIII—The Corn or Cotton-Boll Worm. XIV—Remedy for the Cotton-Worm, South. XV—Poisons for Worms. XVI—The Hateful Grasshopper or Locust. XVII—Remedies Against the Grasshopper. 617

CHAPTER IV.

OTHER DESTRUCTIVE INSECTS AND THEIR ENEMIES.

I—Insects Injurious to Trees. II—Insects Injurious to Coniferous Trees. III—Insects Injuring the Grape. IV—Insects Injuring Fruit Trees. V—Leaf Rollers. VI—Apple Tree Case Bearer. VII—The Orange Leaf-Notcher. VIII—Fuller's Rose Beetle. IX—Insects Injuring Plants. X—Snout Beetles. XI—The White Grub or May Beetle. XII—The Spanish Fly or Blister Beetle. XIII—Beneficial Insects—Lady Birds. XIV—Soldier Beetles. XV—Tiger Beetles. XVI—Other Beetles and Parasites. XVII—Conclusions. 637

CHAPTER V.

BIRDS IN THEIR RELATION TO AGRICULTURE.

I—Birds in the Economy of Nature. II—What Birds Shall We Kill? III—Food of Some Common Birds. IV—Birds Classified by their Food. V—Birds the Natural Enemy of

Insects. VI—Birds to be Carefully Fostered. VII—Birds of Doubtful Utility. VIII—Birds to be Exterminated. IX—Destroying Insects. 651

PART VII.

FARM LAW AND ITS PRINCIPLES.

LEGAL FORMS AND OBLIGATIONS. AGRICULTURAL LAW. STOCK, GAME AND FENCE LAW. SECURING A HOMESTEAD. HIRING HELP, ETC.

CHAPTER I.

PRINCIPLES IN RURAL LAW.

I—Law Governing Farmers' Animals. II—Liability for Injury by Dogs. III—Trespassing Upon Property. IV—Division Fences. V—Railway Fences and Trespass. VI—Railways Running Through Farms. VII—Public Roadways. VIII—The Rights of the Public in the Road. IX—Avoiding Obstructions in the Road. X—Right of Way Over Lands of Others. XI—Liability of the Farmer for his Servants. XII—Rights Relative to Water and Drainage. XIII—Liability of Dealers. XIV—Hiring Help—Specific Wages. XV—What Is a Farm? XVI—Getting a Free Farm. XVII—The Public Land System. XVIII. Pre-Emption, Homestead and Timber-Culture Acts. XIX—Land Taken Under the Three Acts. XX—The Desert Land Act. XXI—Land Yet Open to Settlers. . . . 661

CHAPTER II.

LAWS RELATING TO AGRICULTURE.

I—Needed Reforms in Farm Laws. II—Laws that Every Farmer Should Know. III—Fish and Game Laws. IV—Game Laws in Old and New States. V—Laws Relating to Dogs. VI—Stock and Estray Laws. VII—Stock Laws of the New England States. VIII—Stock Laws of the Middle States. IX—Stock Laws of the Southern States. X—Stock Laws of the Western States. XI—State Laws Relating to Fences. XII—Fence Laws in General. XIII—Fence Laws in New England. XIV—Fence Laws in the Middle States. XV—Fence Laws in the South. XVI—Fence Laws in the Western States. XVII—Fence Laws of the Pacific Slope. 674

CHAPTER III.

LAW FORMS RELATING TO BUSINESS TRANSACTIONS.

I—Guarding Against Swindlers. II—Rules of Guidance in Business. III—Rules Relating to Banking. IV—Indorsements. V—Forms of Notes. VI—Judgment Notes. VII—Due-Bills, Receipts, Orders, Etc. VIII—Some Defenses Which May Defeat Payment of Negotiable Paper. IX—Remarks Concerning Notes. X—Drafts Explained. XI—Remittances. XII—Obligation for Married Women. XIII—Drawing up Important Papers. XIV—Short Form of Lease for Farm and Buildings. XV—Agreements Between Landlord and Tenant. XVI—Wills. XVII—Power of Attorney. XVIII—Mortgage—Short Form. XIX—Warranty Deeds. XX—Bills of Sale. XXI—Bonds. XXII—Arbitration. XXIII—Award of Arbitrators. XXIV—Counterfeit Money. XXV—Good Business Maxims. XXVI—Some Points on Business Law. XXVII—Definitions of Mercantile Terms. XXVIII—Business Characters. 703

PART VIII.

HOUSEHOLD ART AND TASTE.

BEAUTIFYING THE HOME. DRESS AND TOILET ART. THE NURSERY AND SICK ROOM. RULES FOR THE PRESERVATION OF HEALTH. REMEDIES AND PREVENTIVES OF DISEASE. COOKING FOR THE SICK, ETC., ETC.

CHAPTER I.

HOUSEHOLD ART AND TASTE.

I—Beautifying the Home. II.—Furnishing the House. III—The Parlor Furniture. IV—The Dining-Room. V—The Kitchen. VI—The Bed-Rooms. VII—The Cellar. VIII—The Water Supply. IX—Soft-Water Cisterns. X—Laying Down Carpets. XI—Painting and Kalsomining. XII—Arrangement of Furniture. XIII—House Cleaning. XIV—Sweeping and Dusting—Renovating Carpets. 725

CHAPTER II.

THE PARLOR AND LIBRARY.

I—The Rooms for Company. II—Guests of the House. III—Etiquette of the Parlor. IV—Entertaining Visitors and Guests. V—Daily Duties Not Interrupted by Guests. VI—Going to Bed. VII—Servants and Parlor Service. VIII—Duty to Children. IX—What Constitutes Vulgarity. X—Parlor Decoration. XI—Decoration Not Necessarily Costly. XII—A Rocking Chair. XIII—A Practical Family. XIV—Ingenious and Useful. . . . 737

CHAPTER III.

THE DINING-ROOM AND ITS SERVICE.

I—Dining-Room Furniture and Decoration. II—Table Etiquette. III—Carving at Table. IV—Carving Four-Footed Game. V—Carving Birds and Fowls. VI—Carving Fish. VII—The Service of the Table. VIII—Some Dishes for Epicures. IX—Queer Facts About Vegetables. X—The Use of Napkins. 746

CHAPTER IV.

DRESS, AND TOILET ART.

I—Dress, Ancient and Modern. II—The Real Purposes of Dress. III—Clothe According to Circumstances. IV—Mending Clothes. V—Altering Clothes. VI—The Kind of Clothes to Wear. VII—Taste in Ladies' Dress. VIII—Something About Color. IX—Toilet-Room and Bath. X—Garments Next the Skin. XI—The Care of Clothes. XII—The Care of Brushes and Combs. 754

CHAPTER V.

THE NURSERY AND SICK-ROOM.

I—To Preserve Health and Save Doctors' Bills. II—The Care of Children. III—Nursery Bathing. IV—Duration of and Proper Time for Bathing. V—Exercise of Children. VI—Study and Relaxation. VII—The Sick-Room. VIII—Cookery for Invalids. IX—Table of Foods and Time of Digestion. X—Some Animal Foods in their Order of Digestibility. XI—The Time Required to Cook Various Articles. XII—Cooking for Convalescents—Recipes and Directions. XIII—Jelly of Meat. XIV—Other Simple Dishes.

XV—Gruels. XVI—Teas and Other Refreshing Drinks. XVII—Remedies for the Sick. XVIII—Doses and their Graduation. XIX—Disinfection. XX—Tests for Impurities in Water. XXI—Simple Poisons and their Antidotes. XXII—Virulent Poisons and their Antidotes. XXIII—Health-Board Disinfectants. XXIV—How to Use Disinfectants. . . 761

CHAPTER VI.

CONTRIBUTIONS FROM FRIENDS ON HOUSEHOLD ECONOMY.

I—Value of Condensed Information. II—Origin of Our Household Recipes. III—Economy in the Kitchen—Washing Dishes. IV—The Damper in the Stove. V—Regulating Coal Fires. VI—The Use of Waste Paper. VII—Cleaning Soiled Marble, etc. VIII—Verminous Insects. IX—Cloth and Fur Moths. X—Book-Destroying Insects. XI—Kerosene. XII—The Laundry—Some Helps in Washing. XIII—Starching and Ironing. XIV. Bleaching Linens, etc. XV—Home-Made Soap and Candles. XVI—To Clean Silver. XVII—Sweeping. XVIII—Papering, Kalsomining and Painting. XIX—Kalsomining. XX.—Painting. XXI—Spring House-Cleaning. XXII—Household Hints. XXIII—Toilet Recipes. XXIV—Home-Made Wines. XXV—Home-Made Inks. XXVI—Recipes for Glue. XXVII—The Dyer's Art. XXVIII—Coloring Dress and Other Fabrics. XXIX—Coloring Yellow, Blue and Green. XXX—Scarlet and Pink. XXXI—Coloring Black, Brown and Slate. XXXII—Walnut Coloring—Black Walnut. XXXIII—Coloring Carpet Rags. . 780

PART IX.

PRACTICAL, COMMON SENSE HOME COOKING.

KITCHEN ECONOMY AND KITCHEN ART. OUR EVERY-DAY EATING AND DRINKING. RECIPES FOR ALL STYLES OF COOKING. EXCELLENT DISHES CHEAPLY MADE. ECONOMY OF A VARIED DIET.

CHAPTER I.

THE LARDER AND KITCHEN.

I—The Meat-Room. II—Hanging, Testing and Preserving Pork, etc. III—Mutton and Lamb. IV—Calves and their Edible Parts. V—Beef on the Farm. VI—The Kitchen. VII—The Floor, Walls and Furniture. VIII—Cleanliness Indispensable. IX—Kitchen Utensils. X—Chemistry of the Kitchen. XI—The Component Parts of Meat. XII—A Famous Cook on Boiling. XIII—Boiled and Stewed Dishes. XIV—How to Stew. . . . 809

CHAPTER II.

SOME USEFUL RECIPES.

I—Vegetable Soup. II—Clear Beef Soup. III—Soups of Various Meats. IV—Fish Soup. V—Boiled Dishes. VI—Stewing. VII—How to Make Stock. VIII—To Clarify Stock or Soup. IX—To Color Soups. X—Roasted and Baked Meats. XI—Beef a la Mode. XII—Preparing the Roast. XIII—Roast Saddle of Venison. XIV—Fowl and Turkey. XV—Baked Ham. XVI—Baked Beans. XVII—Broiling and Frying. XVIII—Prepared Dishes Baked. XIX—Pastry for Meat Pies. XX—Ingredients for Meat Pies. XXI—Dishes of Eggs. XXII—Steamed Dishes. 821

CHAPTER III.

SAUCES, SALADS, PICKLES AND CONDIMENTS.

I—Sauces and Gravies. II—Salads and their Dressing. III—Various Made Dishes. IV—Pickles, Catsups and Condiments. V—Leaves for Flavoring. VI—Sour Pickles—Cucumbers. VII—Chow-Chow. VIII—Piccalilli. IX—Sweet Pickles. X—Catsups. XI—Condiments. XII—Flavored Vinegar. XIII—Strawberry Acid. 838

CHAPTER IV.

BREAD-MAKING.

I—Selecting the Flour. II—Some Things to be Remembered. III—Yeast and Yeast-Making. IV—Bread of Fine Flour. V—Heating the Oven. VI—Milk Bread, Potato Bread and Cream Bread. VII—Rye Bread. VIII—Graham Bread. IX—Boston Brown Bread. X—Various Recipes for Bread. XI—Biscuits, Rolls, Gems, etc. XII—Oatmeal Breakfast Cakes. XIII—Rusks and Rolls. 854

CHAPTER V.

PASTRY AND PUDDINGS.

I—Digestible Pastry. II—Pies for Dyspeptics. III—Mince Pies. IV—Rhubarb Pie. V—Some Every-Day Pies. VI—Tarts and Tart Crusts. VII—Fruit Short-Cake. VIII—Puddings and their Sauces. IX—Devonshire Cream. X—English Plum Pudding. XI—Oatmeal Pudding or Porridge. XII—Four Puddings of Potatoes. XIII—Brown Betty. XIV—Some Good Puddings. XV—Dumplings. XVI—A Hen's Nest and the Sauce. XVII—Fruit Puddings. XVIII—Puddings of Grain. XIX—Miscellaneous Puddings. XX—Custards and Creams—Frozen Custard. 862

CHAPTER VI.

CAKE-MAKING.

I—Cake an Economical Food. II—General Rules for Making Cake. III—Icing, Glazing and Ornamenting. IV—Recipes for Frosting. V—Ornamenting Cake. VI—Special Preparations. VII—Fruit Cake, Dark. VIII—Rich Pound-Cake. IX—Miscellaneous Cakes. X—More Good Cakes. XI—Gingerbread and Other "Homely" Cakes. XII—The Housewife's Table of Equivalents. 875

CHAPTER VII.

BEVERAGES, ICES AND CANDIES.

I—Pure Water as a Beverage. II—Tea and Coffee. III—How to Make Tea. IV—The Tea-Making of Various Peoples. V—A Cup of Coffee. VI—Chocolate. VII—Refreshing Drinks. VIII—Summer Drinks. IX—Tomato Beer. X—Ice Cream and Water Ices. XI—Candy-Making. XII—Candied Fruit. 884

CHAPTER VIII.

PRESERVING, DRYING AND CANNING FRUIT.

I—Old and New Ways of Preserving. II—Canning Fruit. III—How to Preserve Fruit. IV—Canning Whole Fruit—Peaches. V—Canning Tomatoes. VI—Canning Vegetables. VII—Preserving in Sugar. VIII—Marmalade. IX—Jam of Apples and Other Fruits. X—Jellies. XI—Syrups—Blackberry, etc. XII—Drying Fruits. XIII—Miscellaneous Recipes for Preserving. XIV—Brandied Peaches and Other Brandied Fruits. 892

PART X.

DEPORTMENT AND SOCIETY.

SOCIAL FORMS AND CUSTOMS. SELF-HELP, RULES OF ETIQUETTE, ETC. DIRECTIONS FOR LETTER-WRITING, ETC. COMPLETE SOCIAL GUIDE.

CHAPTER I.

PHILOSOPHY AND PRECEPTS OF ETIQUETTE.

I—The Philosophy of Etiquette. II—Etiquette an Aid to Success. III—What it Inculcates. IV—Etiquette of Dining—How Many to Invite. V—Dinner Costumes. VI—Informal Dinners. VII—How to Receive Guests. VIII—At the Table. IX—How to Serve a Dinner. X—Family Dinners. XI—A Few Useful Hints. XII—Table Usages; What to Do and What to Avoid. XIII—Wines at Formal and Official Dinners. XIV—Sensible Hints for Dinner-Givers. XV—After Dinner. XVI—Breakfast and Supper. XVII—Luncheon—Invitation and Service. XVIII—Etiquette of Dress and Conversation. XIX—The Golden Rule. XX—Things to Avoid. XXI—Calls. XXII—General Etiquette of Calls. XXIII—Evening Calls. XXIV—Visiting Cards. XXV—New Year's Calls. 903

CHAPTER II.

ETIQUETTE OF THE STREET, BALL, CHURCH, ETC.

I—Street Deportment. II—General Rules of Street Deportment. III—Special Rules of Street Deportment. IV—Etiquette of Introductions. V—Salutations. VI—Riding and Driving. VII—Ball and Party Etiquette. VIII—The Supper, Dressing-Room, etc. IX—Some General Rules of Party Etiquette. X—Evening Parties—The Conversazione. XI—Concerts, Theatricals, etc. XII—Parlor Lectures. XIII—Church Etiquette. XIV—Etiquette of Visits. XV—Rules for General Guidance. XVI—Etiquette of the Funeral. XVII—Etiquette of the Christening—Godfather and Godmother—Presents, etc. 919

CHAPTER III.

ETIQUETTE OF THE WEDDING, THE ROAD AND THE CAPITAL.

Etiquette of Wedding Engagements. II—The Wedding. III—The Ceremony in Church. IV—Wedding Receptions. V—Etiquette of the Road—Traveling. VI—Ladies Traveling—The Escort. VII—General Rules for Traveling. VIII—Etiquette in Washington. IX—Etiquette of Shopping. X—Special Rules of Deportment. XI—George Washington's One Hundred Rules of Life Government. 932

CHAPTER IV.

FORMS, LETTERS, FRENCH PHRASES, ETC.

I—Written Invitations to Dinner and Social Parties. II—Other Invitations—Evening Party. III—Acceptances and Regrets. IV—Friendly Invitations. V—Friendly Acceptances and Regrets. VI—Letters of Introduction. VII—Letters of Recommendation. VIII—Asking a Loan and the Reply. IX—Directing a Letter. X—Suggestions for Letter-Writers. XI—Styles of Cards XII—French Words and Phrases in General Use. XIII—Treatment of Children. XIV—Seventy-five Cardinal Rules of Etiquette. XV—Alphabet of Etiquette. 945

CONTENTS

xxi

PART XI.

MISCELLANEOUS.

VALUABLE TABLES AND RECIPES. FOODS, SPICES AND CONDIMENTS. WEIGHTS, MEASURES, LEGAL FORMS, ETC.

CHAPTER I.

FOOD PRODUCTS OF COMMERCE.

PAGE

I—Flour and its Manufacture. II—Rye and its Products. III—Barley and its Products. IV—Oats and their Products. V—Maize and its Products. VI—Beans and Peas and their Products. VII—Potatoes and Potato Products. VIII—Sage and Tapioca. IX—Chocolate and Cocoa. X—Coffee. XI—Tea. XII—Cotton-Seed Oil. XIII—Spices and their Adulteration—Pepper. XIV—Cinnamon; How to Know it Pure. XV—Cloves and Allspice. XVI—Nutmegs and Mace. XVII—Ginger and its Preparation. XVIII—Capers—True and Spurious Kinds. XIX—The Tamarind. 961

CHAPTER II.

LAW, COMMERCIAL AND OTHER FORMS.

I—Indenture of Apprenticeship or for Service. II—Arrears of Pay and Bounty. III—Forms for Bounty Land. IV—Agreements and Contracts. V—Warranty Deed. VI—Mortgage of Personal Property. VII—Bills of Sale. VIII—Certificates, Releases and Discharges. IX—Powers of Attorney. X—Revocation of Power of Attorney. XI—Proxy Revoking all Previous Proxies. 971

CHAPTER III.

TABLES OF WEIGHTS, DIMENSIONS, STRENGTH, GRAVITY, ETC.

I—Tables of Weights. II—The Metric System. III—The Metric System Compared with Our Own. IV—Tables Relating to Money. V—Foreign Exchange. VI—Specific Gravity. VII—Earths and Soils. VIII—Cohesion of Materials. IX—Strength of Common Ropes. X—Human Force. XI—Heat and its Effects. XII—Capacity of Soils for Heat. XIII—Radiating Power, Absorption and Evaporation. XIV—Temperatures Required by Plants. XV—Temperatures of Germination. XVI—Contrasts between Animal and Plant Life. XVII—Thermometers. XVIII—Dimensions and Contents of Fields, Granaries, Corn-Cribs, etc. XIX—Rainfall in the United States. XX—Force and Velocity. XXI—Weight of Agricultural Products. 982

CHAPTER IV.

TABLES AND DIAGRAMS OF PRACTICAL VALUE.

I—Seeds and Plants to Crop an Acre. II—Vegetable Seeds to Sow 100 Yards of Drills. III—Plants per Acre at Various Distances. IV—Vitality of Seeds. V—Plants per Square Rod of Ground. VI—Foretelling the Weather. VII—Comparison of Crops in Great Britain and

CONTENTS

the United States. VIII—Improved and Unimproved Lands in the States and Territories. IX—Forest Areas—Europe and United States. X—Surveyed and Appropriated Lands in States and Territories. XI—Tables of Nutritive Equivalents, etc. XII—Table Showing Prices per Pound. XIII—Table of Interest at Six per Cent. XIV—Growth of Money at Interest. XV—Mean Duration of Life. XVI—Mortality Rates. XVII—How to Calculate Salaries and Wages. XVIII—The Earth's Area and Population. XIX—The World's Commerce. XX—Pay of the Principal Officers of the United States. XXI—Public Debt of the United States. XXII—The United States and Territories. XXIII—Diagrams giving Valuable Statistics, ... 1006

ILLUSTRATIONS

	PAGE.
A village (rural) home,	34
Rude home of the pioneer,	38
Nature's classic halls (waterfall),	39
Window gardening,	40
The settler's first home—log cabin,	41
Indian corn, ancient and modern,	45
Early Greek implements of agriculture,	48
Ancient Chinese plow,	48
Ancient Roman agriculture (*five illustrations*),	51
Old Moorish plow,	58
Norman farm tools (*six figures*),	59
Plowing in the Orkney Islands,	59
English steam plow at work,	60
Daniel Webster's plow,	62
Plow with chain for turning under trash,	63
Plow with attachment for clearing trash,	63
Gang stubble plow,	64
Stirring and stubble plow,	66
Skim and trench furrow in trench plowing,	67
Trench plowing ten inches deep,	67
The furrow moved back,	67
Deep trench plowing,	68
Double harrow, slanting teeth,	68
Harrow folded,	68
Sectional field roller,	69
Soil grinder,	69
Walking cultivator,	71
Five-tooth cultivator,	71
Seed sower,	72
Diagrams for plowing fields (*four figures*),	74
Wheat without fertilizers,	77
Wheat with fertilizers,	78
Illustrations of wheat (*four figures*),	93
Wheat planted at different depths,	94
Caps for shocks (*two figures*),	96
The shock finished,	96
Hallet pedigree wheat (*two sections*),	97
Heads of wheat (*three figures*),	100
Montana spring rye,	101
Winter barley—plant and head,	102
Annat barley,	104
Chevalier-barley,	104
English barley,	104
Horse-mane oats,	105
White Russian oats,	106
Yellow dent corn,	111
White dent corn,	112
Michigan yellow dent corn,	113
Yellow flint corn,	114
A field of shocked corn,	115
Corn horse for shocking,	116
Corn-shock binder,	117
Varieties of Indian corn (*twenty illustrations*),	119
Michigan yellow dent corn,	120
White dent or Parrish corn,	120
Mammoth yellow dent,	121
North star corn,	121
Eight-rowed flint corn,	122
Waushakum corn,	122
Silver white flint corn,	122
Wild rice of the Northwest,	124
Grasses and clover,	134
Flowering of grasses (*thirteen illustrations*),	136
Flowering of grasses (*thirteen illustrations*),	137
Flowering of grasses (*fourteen illustrations*),	138
Flowering of grasses (*thirteen illustrations*),	139
Flowering of grasses (*twelve illustrations*),	140
Flowering of grasses (*ten illustrations*),	141
Timothy or cat's-tail grass,	143
The proper form of stack,	144
Prairie blue joint (broom grass),	145
Buffalo grass (*Buchloe*),	146
Gama grass of the South,	147
Indian grass (*sorghum nutans*),	148
Mesquit grass,	148
Sweet-scented vernal grass,	150
Rye grass,	153
Tall oat grass,	153
Blue grass (Kentucky),	158
Red-top (*agrostis*),	159
Orchard grass,	160
Mammoth red clover,	161
White (Dutch) clover,	162
Alsike (Swedish) clover.	163
Alfalfa or lucerne,	164
Japan clover (*lespedeza*),	164
Sorghum,	169
Hungarian grass,	170
Pearl millet,	171
Prickly comfrey,	172
Belgian carrot,	175
Mangel-wurzel,	175
Parsnips (*two cuts*),	176
Carrots (*two cuts*),	176

ILLUSTRATIONS

	PAGE.
Red altringham carrot,	177
Jerusalem artichoke,	177
Tall corn close shocked,	179
Dwarf corn in two tiers,	179
Dwarf corn in three tiers,	179
Ensilage illustrated—before covering,	179
Ensilage illustrated—after covering,	179
Ensilage illustrated—final compression,	179
One of the earlier silos,	181
Section of double silo,	185
Spraying with poisoned water to destroy cotton worms,	202
Ramie roots and stem,	210
Silk-worm larva, full grown,	213
Silk-worm moth,	213
Silk-worm cocoon,	213
Piedmontese silk-reel,	217
Lath frame,	218
French silk-reeling machine,	219
Plain view of old French reel,	220
Hop kiln or dry house,	228
Tobacco plant in blossom,	233
Tobacco moth worm,	234
Tobacco worm (larva),	234
Pupa of tobacco worm,	235
Tobacco plant properly set,	237
Tobacco house,	239
Hand of tobacco,	239
Moth of tomato worm—destroys tobacco in the North,	240
Peanuts—plants and tubers,	242
Head of Liberian cane,	257
Head of Neeanzana cane,	257
Head of wolf-tail cane,	258
Head of black-top cane,	258
Head of rice, or Egyptian corn,	259
Head of hybrid cane,	259
Head of white mammoth cane,	260
Head of early amber cane,	261
Head of gray-top cane,	261
Head of Oomseeana cane,	262
Head of goose-neck cane,	262
Head of Honduras cane,	263
Modern prairie breaker,	271
A pioneer's cottage,	273
Ground plan of cottage, with lean-to,	273
Shall I move the barn, or the manure pile,	274
The successful farmer's model barn,	275
Reins for three horses abreast,	278
Three-horse draft,	278
Vertical breaking plow,	279
Flat furrows breaking,	276
Orchard and wind break,	281

	PAGE.
Adze-eyed mattock,	284
Dairy-house, elevations,	288
Butter worker,	289
Diagram of farm,	295
A garden farm,	296
Panels of Virginia fence,	300
Locked Virginia fence,	300
Perpendicular staking and capping,	301
Post-and-rail fence,	301
Post-hole auger,	302
Cast post maul,	303
Portable board fence,	304
End posts and braces for wire fence or trellis,	305
Fence for a hillside,	307
Slide and swing gate,	308
Gate swinging on rings,	308
Gate with strut,	308
Gate with tie latch,	308
Balance gate,	309
Sliding gate (*two views*),	309
Rising gates (*two views*),	309
Self-shutting upper hinge,	310
Gate latch (half size),	310
Self-closing slide gate,	310
Roller hangers,	311
Heavy strap-hinge,	311
Southern strap-hinge farm gate (figured),	312
Double-braced gate,	312
Adjustable swing gate,	313
Ornamental gate and fence,	313
Incorrect form of water gate,	314
Correct form of water gate,	314
Stream gate and footway (*two views*),	315
Corn knife,	318
Bill hook,	319
Norway spruce and arbor-vitæ hedge,	319
White or evergreen thorn,	320
Hedge clipper,	321
Osage orange as a tree,	323
Drainage of sloughs,	333
Deep-tiller plow for working ditches,	334
Watering-box from under-drain,	334
Drainage-map—space for water pond,	336
Round-pointed shovel,	337
Cross section of under-drain,	338
Flat-stone drain,	339
Round-stone drain,	339
Slab and pole drains (*four cuts*),	340
Tile drain,	341
Drainer's level,	342
Finding the level,	343
Protecting the bank,	343
German spade,	344

ILLUSTRATIONS

	PAGE
Drainage tools (*eight cuts*),	345
Silt well,	347
Ice or tile pick,	350
Connection of lateral with main drain,	350
Diagram of drained field,	351
Draining water halls,	352
Draining 160 acres,	354
Ashland, the home of Henry Clay,	362
Farm-house built from increasing profits,	363
Ground plans of farm-house (*two figures*),	364
The old house remodeled,	364
Ground plan of house,	365
Farm or suburban cottage,	366
First floor, farm cottage,	367
Upper floor, farm cottage,	367
Park of the farm cottage,	368
Farm stable and carriage-house,	368
Hillside cottage,	371
Ground plan,	372
Plan of attic,	372
Ground plan of horse barn,	373
Main floor of farm barn,	373
Ice-house and preservatory,	373
Poor reservoir for fountain,	374
A plain farm-house,	393
Ground plan, No. 1,	394
Ground plan, No. 2,	394
Cottage for farm hand,	394
First floor, square cottage,	395
Second floor, square cottage,	395
Suburban or farm cottage,	396
Ground plan,	396
A pretty rural home,	397
Ground plan of rural home,	397
Ground plan of convenient cottage,	398
Second floor of same,	398
Farm house, Italian style,	398
A convenient cottage,	399
English gothic cottage,	400
Ground plan of English gothic cottage,	401
Plan of rural grounds,	402
School and meeting house combined,	403
Neighborhood primary school-house,	404
Interior of primary school-house,	404
Scholar's wigwam,	405
A rustic seat,	406
Summer-house of bark,	406
Square summer-house,	406
An elegant summer-house,	407
Drinking fountain,	408
A wicket coop,	408
Barrel coop,	408
Poultry house,	409
Chicken and duck enclosure,	409
Lean-to propagating pit,	410
Propagating and dry house,	410
Farm ice-house,	411
Outline drawings of earth-closets (*two cuts*),	411
Reservoir earth-closet,	412
Brick smoke-house,	413
Framed smoke-house,	413
Suburban carriage-house and stable,	414
Complete dairy barn elevation,	415
The stable floor,	416
Horse and cow barn,	416
Ground plan of barns and sheds,	417
Barn basement,	417
Feed box,	418
Sheep barn and sheds,	418
Wagon jack,	419
Improved wagon jack,	419
Plan of sheep barn and yards,	419
Sheep dipping box,	419
Ground plan of sheep barn and yards,	420
Sheep rack for open yard,	420
Square hog barn with extended wings,	421
Corn crib of poles,	422
Ventilated granary,	423
Ground plan of granary,	424
Corn crib and granary,	424
Skeleton of crib,	424
Crib extended inwards,	424
Western corn crib,	425
A family set of tools (*twenty-one figures*),	427
One end of tool-house (*thirty figures*),	428
Second end of tool-house (*thirty figures*),	429
View of one side of tool-house and workshop (*123 figures*),	430
The other side of tool-house and workshop (*fifty-one figures*),	431
Improved saw-set,	432
A saw clamp,	433
Newly wooded singletree,	433
Open link,	434
Arm-chair turned back,	434
The arm-chair closed,	434
Mole trap,	434
Paint brush—best,	435
Sash brush,	435
Section of adjustable plumb and level,	435
Simplest form of stone boat,	437
One-story frame (balloon),	439
Diagonal lining, inside and out,	440
Isothermal view, balloon frame,	440
Canada reinette apple,	446
Tetafsky apple,	447

ILLUSTRATIONS

	PAGE.
Early Joe apple,	450
Higby sweet apple,	451
Nursery trees (*four figures*),	452
Summer rose apple,	454
Grimes' golden apple,	455
Bonne du Puits ansault pear,	457
Frederic Clapp pear,	458
Howell pear,	459
Dix pear,	460
Paradise D'Automne pear,	461
Little Marguerite pear,	462
George IV. peach,	463
Noblesse peach,	464
Nectarine or smooth peach,	465
Early Richmond and late morello cherry,	468
Black eagle cherry,	469
Governor Wood cherry,	470
Knight's early black cherry,	471
Fruit-picking ladder,	472
Grapevine trellis,	473
View of two canes trained to stake,	475
Trellis and vine, renewal system,	476
Short vine, cultivated cranberry,	478
Varieties of the cranberry (*seven figures*),	479
Saddle grafting,	482
Budding, pruning and grafting knife,	483
Grafting by approach (*five figures*),	484
Stock and grafts (*three figures*),	484
Grafting chisel and wedge,	484
Budding knife,	487
Budding illustrated (*four cuts*),	488
The wealth of the garden,	491
Hand cultivator,	492
Early dwarf peas,	492
Phinney's early melon,	492
Long scarlet radish,	492
Martynia,	494
Sweet potato and vine,	495
Tripoli onion,	496
Danvers yellow globe onion,	496
Peppers,	496
Carrots (*three figures*),	496
Celery,	496
Parsnips (*three figures*),	496
French breakfast radish,	496
Summer golden crook-neck squash,	496
Egg plant,	496
The farm hot-bed,	497
Kohl rabi,	499
Growing cucumbers in green-house,	500
Okra,	501
How to cut potatoes for seed (*two diagrams*),	504
Potatoes as they should grow,	505
Potatoes, illustrated (*eight cuts*),	506
Crested moss rose (half size),	507
Pansies,	508
Moss pink,	508
Perennial daisy,	509
Caladium,	509
Hybrid tea rose—La France,	512
Charles Lefebvre rose,	513
Countess of Serene rose,	514
Louis Van Houtte rose,	515
Flowers of white-flowering dogwood,	516
Flowers of Japan quince,	517
Rose-colored weigela,	518
Variegated cornelian cherry,	518
Dentzia blossoms,	519
Fortune's forsythia,	520
Double-flowering plum,	521
Hydrangea otaksa,	522
Syringa or mock orange,	522
Silver bell (*halesia*),	523
Meadow sweet (*spirea*),	523
Lance-leaved spirea,	524
Japanese spirea,	525
Guelder rose (*viburnum*),	525
Clematis jackmanni,	526
Hall's Japan honeysuckle,	526
Chinese wistaria,	527
Flowers of magnolia speciosa,	528
Racemes of double-flowering horse chestnut,	529
Double-flowering cherry,	530
Flowers of double flowering thorn,	531
Flowers of catalpa speciosa,	531
Chinese double-flowering crab,	532
Statice latifolia,	533
Ornamental grasses,	533
Water lilies,	534
Ornamental trellises (three illustrations),	534
English oak,	535
Mulberry tree,	536
Design for village lot,	537
Plan for secluded grounds,	538
Trees massed for effect in height,	539
Sodding terraces,	539
Tree protector,	539
Transplanting trees,	540
Road scraper,	540
Flower beds (three diagrams),	540
Planting flower beds (three diagrams),	541
The great flower garden in Lincoln park, Chicago,	542
A landscape effect in Lincoln park, Chicago	543
Heavy wooded pine,	544
Lamson's cypress,	545

ILLUSTRATIONS

	PAGE.
Magnolia glauca,	546
Siberian arbor-vitæ,	546
White spruce,	547
Leaves of crisp-leaved maple,	548
Leaves of Wier's cut-leaved maple,	548
Leaves of Acutia-leaved ash,	549
Leaves of tricolored-leaved sycamore,	549
Leaf of fern-leaved beech,	549
Fern-leaved beech,	550
Weeping birch,	551
Cut-leaved weeping birch,	552
White-leaved weeping linden,	553
Yellow wood (vergelea),	554
Scarlet maple,	555
Leaf of maple (variety tripartitum),	555
English elm,	555
The ivory-nut plant,	556
Deciduous cypress of the South,	556
Persimmon tree,	556
Climbing fish,	557
A fish nursery,	558
Pond and fish-way,	559
Black bass of the West,	560
Striped or brassy bass of the Mississippi,	560
Brook trout,	561
Artificial spawning of fishes' eggs,	563
Out-door hatching-box,	565
In-door hatching-box (*four figures*),	566
Succession of hatching-boxes,	567
Artificial spawning bed,	568
Illustrations of common insects (*eleven figures*),	586
Beneficial insects (*sixteen figures*),	587
Pear-tree lice (*five illustrations*),	595
Plant lice magnified (*ten illustrations*),	596
Lice and scale insects magnified (*seventeen illustrations*),	597
Orange scale insect magnified (*four illustrations*),	598
Glover's orange scale magnified (*four illustrations*),	599
Plant bugs (*seven cuts*),	602
Plant bugs (*seven figures*),	603
Land bugs (*six figures*),	604
Land bugs (*six figures*),	605
Plant bugs (*four figures*),	606
Chinch and other bugs (*six figures*),	607
Plant bugs (*seven figures*),	608
Plant bugs (*four figures*),	609
Plant bugs and bed bug (*eleven figures*)	610
Corsair and reduvius (*three figures*),	611
Predatory bugs (*six figures*),	612
Predatory bugs (*six figures*),	613
Water bugs (*three figures*),	614

	PAGE.
Water bugs (*two illustrations*),	615
Device for destroying chinch bugs, army worms, etc.,	616
Clover-stem borer (*ten figures*),	618
Clover-root borer (*four figures*),	618
Clover-leaf midge, (*four figures*),	618
Army worm, moth, pupa, and eggs (*four figures*)	620
Army worm, larva,	620
Clover-leaf beetle (*thirteen figures*),	621
Lamp for killing night-flying moths,	622
The vagabond crambus (*five figures*),	622
Wheat isosoma (*ten figures*),	623
Wheat hopper,	623
Sorghum web worm (*thirteen figures*),	624
The smaller corn-stalk borers (*thirteen figures*),	624
Sugar-cane beetle,	625
Rice-stalk borer,	626
Grass worm of the South,	626
Moth of grass worm,	626
Corn bill bug,	627
Corn, cotton-boll or tomato worm,	628
Spraying cotton from below,	630
Map of regions infested with locusts,	632, 633
The Riley locust gatherer,	635
The catalpa sphinx,	638
Osage orange sphinx,	639
Pine-tree borer,	640
Resin-inhabitating deplosis,	640
Pine-leaf miner,	641
Juniper web-worm,	641
Lime-tree winter moth,	642
Flea beetle,	642
Apple-leaf sewer,	643
Orange-leaf notcher,	643
Apple-tree case bearer,	643
Fuller's rose beetle (*nine figures*),	644
Melon worm and moth,	644
Thacina parasite on melon worm,	644
Asparagus beetles (*three figures*),	645
Snout beetle,	645
Sweet-potato borer,	645
Distended May beetle,	645
Lady birds of California (*twenty-one cuts*),	646
Nuttall's blister beetle,	647
Lady birds (*six cuts*),	647
Soldier beetles (*ten figures*),	648
Tiger beetles (*two cuts*),	649
Soldier bug,	649
Ground beetles (*three cuts*),	649
Chalcis fly (*seven figures*),	649
Epax apicaulis,	649
Lebia grandis (*two cuts*),	650

ILLUSTRATIONS

	PAGE.
Tachina fly (*four figures*),	650
Beautifying the home,	725
Glass case for house plants,	727
Aquarium,	728
Walking fern,	729
Pineapple and fruit,	729
Case of ferns,	729
Window plants in dining-room,	730
Tub filter,	731
Towel rack,	731
Carpet stretcher,	733
Kalsomine brush,	734
Floor brush,	736
An oriel window,	740
Dragon-like ornament,	741
The living-room window,	741
Design for chair cover,	742
The dragon chair,	742
Work-box and seat,	742
Scissors case,	742
A plant case,	743
A plant fumigator,	743
Plant-case bottom,	743
Sarah's what-not,	744
Moss water cooler,	744
Aleck's quilting frame,	744
Completely arranged dinner-table,	747
A dressed ham,	748
Sirloin of beef,	748
Fillet of veal,	748
Leg of mutton,	748
Roast turkey,	749
Roast pig,	749
Trussed fowl (breast),	750
Trussed fowl (back),	750
Pheasant,	750
Partridge,	750
Pigeon (breast),	750
Pigeon (back),	750
Goose roasted,	750
Codfish head and shoulders,	751
Pan fish,	751
A piece of salmon,	751
Napkins about a decanter,	753
Folded napkins (*three cuts*),	753
Infant bath tub,	763
Oval jelly mould,	768
Jelly sieve,	768
Milk, porridge or rice boiler,	770
Earth closets for invalids,	776
Adjustable stove damper,	781
Hog figured for cutting up,	810
Dressed carcase of mutton,	811

	PAGE.
Dressed lamb,	812
Carcase of veal,	813
Dressed ox,	814
Towel rack,	815
Roller towel,	815
Clothes bars,	815
Refrigerator,	816
Meat cutter,	816
Family meat cleaver,	816
Soup digester,	816
Mortars and pestles,	816
Potato masher,	818
Tinned skewers,	818
Wooden steak maul,	819
Meat block,	819
Round-bottom pot,	820
Granite-ware stew kettle.	820
Stew pot and lid,	821
Skillet and lid,	821
Porcelain-lined fish kettle,	823
Brass kettle,	823
Convex stew-pan,	827
Flesh fork,	830
Broiler and cover,	832
Oyster broiler,	833
Improved frying pan,	833
Vegetable or egg boiler,	836
Omelet pan,	836
Steamer,	836
Soup or sauce strainer,	838
Gravy strainer,	838
Salad washer,	840
Toaster and light broiler,	842
Pudding or timbale pan,	844
Mushrooms,	845
Colander,	847
Preserving and pickling kettle,	850
Kitchen sieve,	852
Kneading pan,	858
Corn-cake pans,	859
Wood rolling pin.	859
Bake-pans for rolls,	861
Scalloped pie-plate,	863
Oblong pie-plate,	864
Scalloped patty-pan,	865
Farina and porridge boiler,	868
Deep pudding-pan,	868
Apple corer,	870
Oval pudding pan,	872
Charlotte Russe pan,	873
Beating bowl,	875
Octagon cake-mould,	877
Turk's-head cake-mould,	877

ILLUSTRATIONS

	PAGE.
Sponge-cake pan,	878
Deep jelly pans,	878
Cake cutter,	880
Cookie pans,	881
Water filter and cooler,	884
Tea leaves, natural size,	885
Coffee roaster,	887
Ice-cream freezer,	889
Fruit and jelly press,	896
Illustrations showing forms of cards,	951, 952
Germination of the bean,	963
Germination of the pea,	963
Wild potato of New Mexico,	964
Young plant of Arabian coffee,	965
Tea in the various stages of manufacture,	967
Diagram, production of corn,	1024
Diagram, production of wheat,	1025
Diagram, export and consumption of cotton,	1026
Diagram, average rate of wages, groups of States,	1027
Diagram, number of miles of railroad built annually,	1028
Diagram, aggregate tons of freight moved, Erie canal,	1029

Part I.

THE HOME AND FARM.

MAKING CONVENIENT, COMFORTABLE AND HAPPY HOMES.

ANCIENT AND MODERN AGRICULTURE ILLUSTRATED AND COMPARED.

IMPROVED FARM IMPLEMENTS AND MACHINERY.

PRINCIPLES AND PRACTICE.

THE HOME AND FARM.

"Thy free, fair homes, my country!
Long, long, in hut and hall,
May hearts of native proof be reared
To guard each hallow'd wall!
And green forever be the groves,
And bright the flowery sod,
Where first the child's glad spirit loves
Its country and its God!"—FELICIA HEMANS.

CHAPTER I.

BUILDING HAPPY HOMES.

I. CONTRASTED PICTURES.—II. THE IDEAL COUNTRY HOME.—III. THE FARMER'S WIFE.—IV. THE OVERWORKED WIFE.—V. SONS AND DAUGHTERS ON THE FARM.—VI. YOUTHFUL ACTIVITY.—VII. ADORNING THE HOME.—VIII. IMPROVING THE HOMESTEAD.—IX. SPORTS OF CHILDHOOD.—X. LESSONS FROM THE GARDEN.—XI. WHAT WE HOPE TO TEACH.

I. Contrasted Pictures.

THERE is no sweeter word than Home! Around the fireside cluster all that makes life beautiful,—Love, Trust, Charity, Truth and Beauty. There husband and wife prove the loveliness of unselfish union. There the youth gains aspiration and the training for a noble life. There the maiden learns the sweetness of unsullied purity and gentle deeds.

Much lies upon the man before he can be worthy of a happy home, much upon the woman. Some examples teach by warning, as others by furnishing models for imitation. Let us take a common case. A girl marries. She has been reared by an unwise though fond mother, whose slavish devotion to her children has made her an unlovely household drudge. She has been brought up to be that wretched thing, a gaudy slattern; she is unkempt at breakfast and elsewhere at home, but gay beyond the household means for others; is ashamed of, and discontented with, her surroundings. City life to such an one is a cheerless, if not fatal, thing. After her marriage, the young couple live with her parents, and what the wretched home-education has taught grows into life-habit. Or, perhaps, they board in some house where idleness and gossip grow like noxious weeds, choking the possibilities of good. There is no wholesome work of head or hand; a wretched life of complaint ensues; the girl becomes the mother of children she is all unfit to rear: a querulous, discontented wife doing nothing—often unable to see anything to do—to aid in building a HOME. In the end, when her husband has won a house of his own, this woman drifts into a likeness of her mother. She is shrill-voiced, careless of raiment, old before her time, with no sign of the fair, calm matronly beauty, that second blossoming after seed-time, which should come to replace the young charm — the Indian summer, almost as fair as wakening spring-time. Her very love for her children works their hurt because there is no guidance.

A VILLAGE HOME.

The man is as often to blame, seeming to live for business alone, or vastly worse, only for boon companions. It is true that the wife, if she be one of those exceptional beings who can answer harsh words or the more bitter neglect with a smile, who will make home sweet even when her own life is as ashes within her lips, will, in the end, win any man to home and duty. Of such women there are a few,—martyrs as worthy of our highest homage as any that ever perished at the stake. But such a husband has no right to expect his wife to prove one of them. God's law is that "whatsoever a man sows that shall he also reap." The enforcement of this law is nearly always as speedy and obvious as it is ultimately certain. The man sows indifference, neglect, unfaithfulness, and he reaps bitter recriminations, domestic broils and jealousies, a full and hideous crop.

II. The Ideal Country Home.

In country life, the way is more smooth for both, though far from easy. The life of a farm is hard, especially so for the woman, but there is work for willing hands to do. There is the home to be made a haven of rest and sweet content. The wife should never forget her high ideal of home life, and that she must be the centre of its beauty. Though they have but a log hut amidst the wilderness, she may make that wilderness blossom as the rose. She may have small share of beauty, yet she may still be exceeding fair in her husband's eyes. Let them both remember it is not the harsh word that heals the breach.

When the children come, let there be order, but remember children are not machines. Train them as you would a vine by daily, hourly care and thought; by example, not by hard rules. A child needs play, air, sunlight, and above all, love and sympathy. He needs a gentle mother-breast wherein to pour the little griefs which, though quickly flown, are at the moment all as poignant as the weightier woes of later years. Teach by love. Teach by example. Be chary of stern precepts for which the child can see no reason but your arbitrary will. If you would have your child respect you—to say nothing of his love—never punish him in a spirit of anger. Never make him a promise without performing it. Remember, there are nearly always other and better modes of punishment, than beating him. Remove some present, or deny some expected pleasure, instead. Let him, if the fault be grave, feel your grave displeasure. Never scold. Govern firmly, but don't govern too much. Threaten seldom—never idly. The parent who tells a child, "If you do so-and-so, I will do so-and-so with you," and then weakly forgets both the broken command and the assigned penalty for it, merits and receives the child's contempt. In all things, remember the tremendous force of parental example. Long before he learns his letters, your toddling one has read your daily life through and through. He moulds his little life by the pattern you present him. For your child's sake, no less than your own, see to it, then, that your life is upright, true and pure. Oh, the tender grace and sweetness of the home where love and duty reign supreme; where the husband and father may cast off his load of daily care; where the wife and mother, grown lovelier by her self-restraint and thought for others, shines beside the hearth the dearest and the sacredest of all created things. In many a home, even in these degenerate days, may such a wife and mother be found:

> A woman, not too pure and good
> For human nature's daily food,
> For wholesome pleasures, simple wiles,
> Praise, blame, love, kisses tears and smiles.

III. The Farmer's Wife.

From the day, when a bride, she has entered that house wherein the Home lies as does the sculptor's dream of genius within the marble block, needing the patient, loving toil to bring forth its lines of beauty; through days or years of sorrow or of sunshine; with many a rebellious thought, fancy or longing to be trodden down in the path of duty; amid griefs and heart-aches not merely to be endured, but to be made stepping-stones to a yet higher and nobler life. Through child-birth pain and weary illness—still guided by the light of Love and Truth—the true woman moves on, blessing all who come within her influence. "Her children rise up and call her blessed."

Woe to the man who shall mar the happiness of the home life. And how many a farmer unthinkingly does this! He amuses himself; he goes to town to buy and sell; he hires labor when there is much to do, but he habitually neglects his fellow-toiler and helpmeet in the house. At the busy season the work heaped upon the "women folks" almost crushes the life out of them. All this is to his own future infinite loss. The life of too many farmers' wives is what no man could bear, and no woman should be made to suffer. It would be a standing shame to the men of America—a disgrace to our nation—if anywhere the women should become slaves without even the slave's holidays, as brutally sacrificed to the chase for the almighty dollar, as ever victim dragged before the throne of Moloch. As a child needs play, so men and women need some form of innocent pleasure. If "all work and no play makes Jack a dull boy," so from Jill it either crushes all her brightness and beauty, or else almost forces her to rebel against social and domestic law in search of a less intolerable lot. Work, the wife of a farmer must, but he should make the burden as light as possible.

IV. The Overworked Wife.

Does any reader recognize this picture of the overworked wife, drawn by Ella Wheeler, one of the most sympathetic poets of the West? If so, let him have a care, lest he, too, become such a tyrant to such a slave:

> "Up with the birds in the early morning—
> The dew-drop glows like a precious gem;
> Beautiful tints in the skies are dawning,
> But she's never a moment to look at them.
> The men are wanting their breakfast early;
> She must not linger, she must not wait;
> For words that are sharp and looks that are surly,
> Are what the men give when the meals are late.

> "Oh, glorious colors that clouds are turning,
> If she would but look over hills and trees;
> But here are the dishes, and here is the churning—
> Those things must always yield to these.

The world is filled with the wine of beauty,
  If she could but pause and drink it in;
But pleasure, she says, must wait for duty—
  Neglected work is committed sin.

"The day grows hot, and her hands grow weary;
  Oh, for an hour to cool her head,
Out with the birds and winds so cheery!
  So she must rise in the morning and make her bread.
The busy men in the hay-field working,
  If they saw her sitting with idle hands,
Would think her lazy, and call it shirking,
  And she never could make them understand.

"They do not know that the heart within her
  Hungers for beauty and things sublime;
They only know that they want their dinner,
  Plenty of it, and just 'on time.'
And after the sweeping and churning and baking,
  And dinner dishes are all put by,
She sits and sews, though her head is aching,
  Till time for supper and 'chores' draws nigh.

"Her boys at school must look like others,
  She says, as she patches their frocks and hose,
For the world is quick to censure mothers
  For the least neglect of their children's clothes.
Her husband comes from the field of labor;
  He gives no praise to his weary wife;
She's done no more than has her neighbor,
  'Tis the lot of all in country life.

"But after the strife and weary tussle
  With life is done, and she lies at rest,
The nation's brain and heart and muscle—
  Her sons and daughters—shall call her blest;
And I think the sweetest joy of Heaven,
  The rarest bliss of eternal life,
And the fairest crown of all will be given
  Unto the wayworn farmer's wife."

V. Sons and Daughters on the Farm.

It is not necessary that the boy reared in the country should be a farmer. Farmers' sons often become leaders in trade, commerce, the arts, science, politics or letters. In fact, it is from the country that the vigor of the city is constantly recruited. Hence the necessity of educating every boy to fit him, not for some single groove in life, but to occupy any plane his talents and industry may enable him to reach. But does he choose the farm? There is here as high an ideal — as great a field for action as anywhere in the wide world.

Nor are the daughters of the household, because they are the children of farmers, all, of necessity, to become farmers' wives. It may be happy for them if they do, for there is no condition in life where more true enjoyment may be had than in the tillage of the soil, in the rearing of stock, a well-kept garden, an orchard dropping luscious and healthful fruits, a comfortable dwelling, and well-kept grounds. These, every industrious family may have, however few the acres.

We can no more control the affections of the daughters than the talents of the sons. But much may be accomplished by so directing education that these talents and affections may be carried in natural channels. The boy who is the mere

RUDE HOME OF THE PIONEER.

drudge of the farm, and the girl that of the kitchen, will always be looking afar for that happiness denied them at home. It is the instinct of all young animals to play. By both his physical and mental constitution, the child requires exercise, to promote growth, harden the bones, strengthen the muscles and sinews, and recreate the brain. This must be found outside the daily routine of labor, whether it be of the farm, the workshop or the school. In directing these matters, nature must be counseled and co-operated with. She cannot be rudely over-ridden and disregarded, without exacting a heavy penalty in a stunted and misshapen life.

VI. Youthful Activity.

The idler is the product of bad training. He is peculiarly in danger of becoming vicious. Hence, when a child is inclined to be idle, otherwise than as the result of grinding overwork, care should be taken to arouse the natural activity. Some natures develop slowly, yet bear noble fruit. These need a stimulus; others, unduly precocious, should be checked. If the child becomes too early absorbed in study, the life may be brilliant, indeed, but is likely to be short. The tree that too soon puts on fruitage is the tree that prematurely decays.

There is no better place for the precocious youth than the farm. Let such watch the squirrels darting here and there in the groves, gather flowers in warm nooks in the spring, play in the new-mown hay in the summer meadows, fish or swim in brook or pond, go nutting in the autumn, and coast, skate or snare rabbits in the winter. It will round out and freshen the growth, and, when time again comes for study, renewed health will enable the brain to carry its load. Thus the slow child should be led, and the too quick-witted one held back; but the lash in the one case, or too sharp a curb in the other, might be fatal. We must, in every case, try to wisely guide; to be able to understand the nature of a child; to diagnose the mind, even as a good physician would an illness.

NATURE'S CLASSIC HALLS

The duties of parents do not end when the children are fed and clothed. The moral is higher than the physical. They must have a pleasant home, must be interested in all that is going on, and help in their small way to create beauty. Thus they will learn to love labor for what it brings, and to love beauty for what it gives. In after life, however successful one may be, the old homestead, even though it be but the simplest cottage, should be looked back to, as the place where the happiest days of life were spent; the remembrance of father and mother be cherished as those to whom the mistakes and successes of life might always be carried, as to careful counselors; the sisters and brothers, ever ready to assist with word or deed.

VII. Adorning the Home.

WHEREVER the home may be, whether in city or country, it is little things that make up the household comforts. In cities little can be done, except to keep the surroundings, small though they may be, neat and tidy. The small yard, if any, may have a little green grass and a few plants; the windows, in any event, should have a few pots of choice flowers. In the ordinary city home, great display should not be thought of. A single plant well grown is better than a window crowded full of ill-looking and untidy starvelings. The village home presents greater capabilities. A few handsome trees for shade, a smooth green lawn, with here and there a bed of

WINDOW GARDENING.

flowers, running roses trained to the veranda, a clinging vine over the porch, and a path winding gracefully with gentle curves to the door, will speak eloquently of taste and contentment in the owner. It will be a suggestion of happy, smiling children, a careful father, a fond and earnest mother. Inside you are sure to find neatness, order, and reliance one on another. The walls will not be bare of pictures, nor the windows of flowers, nor will there be wanting those little elegancies of feminine work that tell of taste and refinement in every department of the household. There may not be wealth, but there will be something better—comfort. The husband may be at work all day in his shop, the wife perhaps working at home, but it will be cheerful labor.

VIII. Improving the Homestead.

The workman in city or village, may not own his own home; the majority do not. The farmer usually owns the farm he works. He may be in debt, and, of course, his first endeavor must be to make himself and family free. Yet, even while doing this, there is many a labor of love that will make the place increase yearly in value and beauty. An orchard may be planted, a vegetable garden cultivated, and trees set out to shade the lawn between the house and the road. Fences may be repaired, and vines and trellis-work made to beautify the home. Such labor is scarcely felt, and as the years roll by, the cattle and horses, sheep and other stock will be increasing and growing in numbers, as the home increases in value and attractiveness.

THE SETTLER'S FIRST HOME.

IX. Sports of Childhood.

There is no aristocracy among children. If we see a child sneering at one not so well dressed as herself, or bragging about his parent's riches, be sure something is wrong at home. It is after we grow up, that we really look down on those not so favored as ourselves. But if the proper training has been given in youth, the man or woman will have only kindly feelings, and a pleasant word for all, where the person is not bad at heart, and average human nature is not so. The well-bred child is as happy in sport, with one cleanly dressed child as another. Childhood is a true republic, where all contribute to the general weal. It is the duty, then, of the parents, to provide such amusements as may lie within their means. Skipping-ropes, swings, dolls, and other feminine articles for the girls,—the coveted knife or hatchet, the little wagon or wheelbarrow for the boys, and the jolly ride behind the farm team that is pleasure always.

If there is water near, both boys and girls should be taught to manage a boat, and, as a matter of precaution, both should learn to swim. Bathing dresses are cheap, and danger of accident will be lessened. Athletic sports should never be denied to boys, and girls should be allowed to race to their hearts' content. Dresses may be soiled, and clothes be torn. Such things inhere among the necessities of childhood. Indeed, we would give but little for the girl who never soiled her dress, or the boy who never had a rent in coat or trousers. Far better these annoyances, when supplemented with the glow of health, the strength of muscle, and the innocent cheerfulness that come with them, than that children should always look as though they had "just come out of a band-box."

As children grow, their sport may be directed in practical channels. Both boys and girls should be taught to gather plants. These may be studied, and thus the first lessons in botany taught. Let them learn to distinguish noxious plants from innocent ones; plants of use from what we call weeds (for weeds are simply plants out of place — many of them being valuable for their medicinal virtues). Even the chores about the farm may come, with a little instruction, to be regarded as near of kin to play. The calves, and colts, and lambs are to be conciliated while being fed, the older animals taught that, although boys are sometimes rough, they are nevertheless kindly. Even the village boy and girl may thus be trained to love rural life, in the little attentions they bestow on the pet calf or lamb, and the thriving pig, reared although these may be, for the butcher.

X. Lessons from the Garden.

The garden everywhere may be made a never-ceasing source of pleasure, until even its labors will be eagerly sought. The preparation of the tiny seeds, the careful planting, the wonders of germination and growth, the blossoming and the ripe fruits, all will be enjoyed when we come to understand something of the mysteries of vegetable life.

Why the South ripens the pineapple, the banana and the pomegranate; the Middle region the grape, the pear, the peach, and the ever-welcome apple. From whence we get the tomato, the melon, okra, egg-plant, the potato, and other exotics, not known beyond their native homes until civilization and commerce brought the products of the four quarters of the globe even to our doors; how fruits, vegetables and brilliant flowers have become possible about every home, even in lands but a few years ago supposed to be almost uninhabitable,—these and a thousand other entertaining questions may be asked and answered in connection with the boy's work in the garden. Thus you train him in habits of thought as well as of industry. It is a great thing for your boy to rise to the conception that work is more than sweat and muscle—more even than the greasy dollars received for the crop.

Why does the farmer and mechanic of to-day live more comfortably, and really better than the nobles of two hundred years ago? Why have we a broader and wider intelligence to-day than in the old feudal times? Education has been different among the masses. Every man is his own master, and head-work directs the labor of his hands. Why are we, as a people, more prosperous and happy than others? It is the feeling that all honest labor is alike honorable, and that *agriculture is the groundwork of permanent wealth.*

XI. What we Hope to Teach.

WHEN reliance on agriculture has ceased, nations have invariably receded from their high position, and sunk into oblivion. The wonderful prosperity of the United States is due, almost solely, to the immense agricultural resources of the country, its wealth in grain, grass, stock, cotton, fruits, and other products; and this prosperity has continued in the face of enormous taxes and tariffs, partly rendered necessary by our great debt. The power of France to pay her huge war debt, and yet recuperate so quickly, was due to the fact, first, that the peasantry in a great measure own the soil, small fields though most of them be, and to the excellence of her agriculture; second, to the fact, that nearly all of the population are engaged in some productive industry.

In this volume, the aim is practical and helpful. Our mission is to the homes of the great working classes, farmers, artizans, laborers, and all who work with hand and brain. We hope to show the farmer, for example, how he can better himself by better tillage, to explain and illustrate the value of grass, grain, drainage, textile and other special crops, and how to improve them; to exhibit, in its true light, the wealth of the orchard and garden, and how he may cheaply enjoy them. In short, we hope to make this work valuable, nay, indispensable, to all who own a rood of ground, by hints as to its cultivation and improvement with a view to founding there a happy and prosperous Home.

CHAPTER II.

ANCIENT AND MODERN AGRICULTURE.

I. THE VALUE OF BOOKS.—II. AGRICULTURE AMONG THE SAVAGES.—III. THE ARIZONA INDIANS.—IV. MOUND BUILDERS.—V. AGRICULTURE DEFINED.—VI. ITS DIVISIONS.—VII. ITS HISTORY.—VIII. THE BOOKS OF MAGO.—IX. MAGO ON WORKING CATTLE.—X. ROME'S AGRICULTURAL WRITERS.—XI. CHRONICLES OF COLUMELLA.—XII. ANCIENT FARMS AND IMPLEMENTS.—XIII. CULTIVATE LITTLE, CULTIVATE WELL.—XIV. A FANCY FARMER.—XV. ARABLE LANDS AND PASTURES.—XVI. WATER MEADOWS.—XVII. A RICH MEADOW.—XVIII. ROMAN ROTATION.—XIX. ROMAN SMALL GRAINS.—XX. SOME ANCIENT METHODS.—XXI. ANTIQUE CROPS.—XXII. CROPS PULLED BY HAND.—XXIII. FALLOW CROPS.—XXIV. ANCIENT HARVESTING.—XXV. ROMAN FERTILIZERS.—XXVI. A QUESTION NOT YET SETTLED.—XXVII. LITTLE AND OFTEN.—XXVIII. COMMERCIAL FERTILIZERS.—XXIX. ANCIENT PLOWING.—XXX. PLOWS.—XXXI. SEEDING.—XXII. YIELD PER ACRE.—XXXIII. MEDIÆVAL AND MODERN AGRICULTURE.

I. The Value of Books.

SOME of the processes, maxims and data that have come down to us from farmers of the remote past, are interesting as well as instructive, and none more so than those which show the high estimation and honor in which this, the most important industry of mankind, was held by the ancients. It will also be instructive to observe how slow was the growth of agriculture among the moderns, after the advanced husbandry of the ancients was lost in the darkness of the Middle Ages. It was not until the invention of printing, and the universal dissemination of knowledge, through books, giving the practical experience of the best minds in the profession, that improved methods of farming became possible to the masses. And yet there are fogies to-day, who sneer at what they call "BOOK FARMING." But for books, the great majority of farmers and farm-laborers would still be mere serfs and beasts of burden, as they were three hundred years ago. But for the information published in good books the West could not have become the Granary of the World, and the South could not have supplied, as it now does, cheap cotton to clothe the people of all lands.

II. Agriculture Among the Savages.

AGRICULTURE, as practiced by barbarous tribes of the present day, for instance, by some of our American Indians, consists simply of gathering grass-seeds, digging wild roots, and storing acorns for the winter's bread. This is the agriculture of the Digger Indians of the Pacific slope. In the country east of the Mississippi river, the aborigines were one grade above this. Their squaws planted corn, beans and pumpkins with the rudest hand implements. They also gathered nuts for winter use, but they were improvident to the last degree. The instinct or experience of the squaws taught them to hide the seed for the next crop in

INDIAN CORN—ANCIENT AND MODERN.

holes in the ground, in order to save it from the rapacity of the "braves," when their gluttony had consumed the general store. The seed-corn was buried in dry soil, the ears, with the husk on, standing on end. It is a curious fact that this method of keeping the seed corn is the best known for preserving its germinating quality. The reason is obvious. The grain is not dried down to extreme hardness, and in the partly-confined atmosphere, the germs slowly undergo the preliminary change necessary to prompt sprouting, just as do the seeds of forest trees under their natural winter covering of leaves.

III. The Arizona Indians.

The Zuni Indians, the Moquis, and some other tribes of Arizona and New Mexico, were far in advance of the more warlike savages, and yet far below respectable husbandry. Still, their methods are perhaps better than those practiced by the patriarch, Abraham. They raise a variety of crops, including fruit, and grind their grain in mills. They live in permanent dwelling-places, raise, spin and weave wool, and had, when the Spaniards first came among them, three hundred years ago, advanced beyond the condition of wandering barbarians. Driven from other homes by the savage Comanches and Apaches, they had fled to those arid regions, there to follow more peaceful avocations, far from their brutal conquerors. Are they an effete offshoot of the people whom the Spaniards found practicing a still higher art of husbandry in Mexico, Peru and Chili? That prehistoric people whose forefathers were saved from being engulfed in "Lost Atlantis," may have given civilization to the ancient Chinese, and, perhaps, through them to the ancient people of Asia, the so-called cradle of the human race. There is good evidence tending that way, even from the Chinese, who acknowledge that their civilization came from the eastward.

The three representations of Indian corn, ancient and modern, are an interesting study, first, as showing the constancy of species to the original type; and second, as illustrating the cultivation of maize in America, by a people, perhaps, more ancient than those of the so-called Old World. Fig. 1 represents a cob of Indian corn from one of the rock caves of Arizona. Fig. 2 shows the corn now cultivated by the Pueblo Indians of Arizona. Fig. 3 exhibits a nubbin, found with a mummy of ancient Peru, the date so remote that more than its geological age cannot be guessed. That nubbin was probably coeval with, if not anterior to the Mound Builders, and they belong to so remote a period that there is not even a tradition of them left.

IV. Mound Builders.

The Mound Builders, who once occupied the whole Mississippi Valley, or, at least, its Eastern and Central portions, were undoubtedly far advanced in civilization. They were undoubtedly an agricultural people. They understood mining and the working of metals. They certainly had commercial intercourse with peoples, many hundreds of miles from their homes. But no tradition of them remained among the

Indians found by the first settlers in America. Nothing is left of their civilization but the mounds, covered in many cases by gigantic forests. The skulls exhumed seem to show that they consisted of a superior and an inferior race, or caste, as was anciently the case in Egypt.

V. Agriculture Defined.

THE word Agriculture is derived from two Latin words, *ager*, a field, and *cultura*, cultivation. From these words we get our word acre, originally a field, and our word culture. In its broad sense the word Agriculture is now used to express all that pertains to the farm and garden. The preparation of the soil, sowing, cultivating, reaping, and preparing the crop for market. It also includes breeding, feeding, fattening, care and training of farm stock, as given in the companion volume to this, the "CYCLOPEDIA OF LIVE STOCK, AND COMPLETE LIVE STOCK DOCTOR."

VI. Its Divisions.

MODERN Agriculture proper is embraced in two grand subdivisions: 1. Husbandry, or, as the latter term is now generally used, Agriculture, and in which sense it will be used in this work; 2. Horticulture. The first relates to the farm proper; the second to the garden. The word Horticulture, again, is derived from two Latin words, *hortus*, a garden, and *cultura*, from *colo*, to till—literally, the art of tilling a garden. This, again, in its broad sense, embraces all that relates to the orchard, the garden, the forest, and even landscape adornment. Agriculture, or husbandry, again, is subdivided into, 1, Field Husbandry, and, 2, Animal Husbandry.

Field Husbandry.—Field Husbandry is divided into grain husbandry, grass husbandry, seed husbandry, cotton planting, sugar-cane planting, sugar-beet husbandry, tobacco husbandry, and mixed husbandry. The latter is that pursued by the majority of farmers North, and by the planters South. It must embrace two or more of the subdivisions named, as, for instance, hemp and tobacco in the North, and textile crops, as cotton and jute, in the South. But mixed farming usually embraces a variety of crops, and also the breeding and care of farm animals.

Animal Husbandry.—This branch of agriculture is divided into horse breeding and rearing, mule breeding, cattle breeding and rearing, sheep husbandry, swine husbandry, poultry raising, stock feeding and fattening, and mixed animal husbandry.

Horticulture.—This, the second grand division of Agriculture, may be divided as follows: Pomology, or Fruit Husbandry; Aboriculture, or all that pertains to trees; Vegetable Gardening; Floriculture; Landscape Gardening; the Nursery, or the cultivation of young trees and plants; Seed Husbandry, and Mixed Horticulture, or a combination of two or more of the foregoing.

VII. Its History.

THE history of agriculture is by no means perfect. Of the ancient agriculture of Southern Europe, Asia and Africa, we have the written record of the Bible, and the writings of the Greeks, Romans and Carthaginians. The writings of the Greeks on the subject are meagre; their taste lay in the direction of art, and the

EARLY GREEK IMPLEMENTS.

agriculture of the country was left entirely to slaves. Their implements of tillage never reached the perfection of the Romans. Fig. 1 is the form of an implement sculptured on an ancient tombstone at Athens; 2, shows the ancient Grecian plow; 3, the spade; 4 and 5, hoes.

Among the Romans and the Carthaginians, the highest officers did not disdain to practice agriculture, when not engaged in battle or on the forum. By them, agriculture was considered the most honorable of professions, and the most illustrious men of Rome and Carthage prided themselves on their skill in farming.

Mago, the Carthaginian, was the father of agricultural literature. This was the great Mago, who is supposed to have lived in the time of Darius, and to have been the founder of the great Punic family from which came Hannibal.

VIII. The Books of Mago

The books of Mago on agriculture were twenty-eight in number. At the final destruction of Carthage, when the whole literature of the conquered nation was given over by the Romans to their African allies, these twenty-eight treatises were con-

ANCIENT CHINESE PLOW.

sidered so valuable, that they were translated at the public expense. Hereen says there are thirty-one distinct passages in which the maxims of the Carthaginian author are handed down to us. Curiously enough, none of these passages refers to the cultivation of grain. One gives directions for grinding or pounding grain, lentils, vetches and sesame. Another recommends the proprietor to reside on his farm, for,

"He to whom an abode in the city lies close at heart, has no need of a country estate." Mago condemned the practice of destroying bees when the honey was taken.

He seems to have been well informed upon horticulture, for he gives directions for the cultivation of vines, nut trees, poplars and reeds. He asserted that in Africa female mules were nearly as prolific as mares, and Cato makes the same assertion. This is very different from modern experience, except in occasional instances in warm climates.

IX. Mago on Working Cattle.

VARRO borrowed much from the writings of Mago, "which I make my herdsmen carefully read." Mago's directions for buying working oxen hold good, even to-day. He says: "The young oxen which we buy should be square in form, large-limbed, with strong, lofty, and dark-colored horns, broad and curly fronts, rough ears, black eyes and lips, prominent and expanded nostrils. Long and brawny neck, ample dewlaps pendant nearly to the knees, a wide chest and large shoulders, roomy-bellied, with well-bowed ribs, broad on the loin, with a straight, level, or even slightly depressed back. Round buttocks, straight and firm legs, by no means weak in the knee, large hoofs, very long and bushy tails, the body covered with thick short hair of a tawny color." There is no doubt that he understood "judging by the touch;" they should, he says, be very soft handlers.

X. Rome's Agricultural Writers.

Cato.—Cato claims precedence as first in time, and first in honor, among Roman writers on agriculture. He died in 150 B. C., at the age of eighty-eight. He was a practical farmer, and recommended precise, if not high, farming. He wrote more, however, in favor of economy than of improvement. Plutarch lashes him for heartlessly recommending the sale of worn-out slaves and oxen.

Varro the Valiant.—Varro lived during nearly the whole of the century preceding the Christian era. A general and an admiral, he was a voluminous writer, but only two of his works have come down to us. In his *De Re Rustica* he frequently refers to operations on his own farm, but relies principally upon the authority of Mago and some Greek writers.

An Orator's Testimony.—The writings of Cicero show some practical knowledge, but relate more to the pleasures than to the labors of agriculture. He regarded agriculture as an honor to princes, and the ornament and solace of old age.

The Pastoral Poet.—The poet Virgil was born seventy years before Christ. His "Georgics" may be called a hand-book of agriculture. Its directions are ample, precise and practical. In it the imagination of the poet never clashes with the art of the farmer, however smooth the verse and elegant the diction. He has been accused of borrowing more than a foundation upon which to rear a structure possessing all the charms of originality. No one who has read this didactic poem will quarrel either with structure or foundation. The first book treats of the proper cultivation of the soil. The second tells how to manage fruit trees. The third is on horses and cattle; the fourth on bees. If original genius did not belong to Virgil, his taste, skill and powers of versification have made his name remembered as the greatest pastoral poet of ancient or modern times.

XI. Chronicles of Columella.

COLUMELLA, who was born at about the Christian era, wrote twelve books on agriculture, and is supposed to have established the Merino sheep as a distinct breed in Spain. He was a native of what is now Spain, and had a farm in the Pyrennees. He mentions an uncle of his name, who greatly improved his flock by introducing African rams. Columella wrote more largely of his success in vineyarding than in any other department of agriculture. Of his twelve books, two are on farming, three treat of the vine, the olive and orchard fruits, two are devoted to farm and domestic animals, one to poultry, one to bees, and three to the bailiff, his wife, and their respective duties — wine, vinegar and other domestic preparations, and the kitchen garden. From the fact that he excludes the sporting dog from domestic animals, the inference is plain, that he was well aware of their sheep-killing pro-propensities. The curs of our day are no less fond of mutton.

Pliny, who died A. D. 79, was a diligent compiler, and not always discriminating.

Palladius, who wrote A. D. 355, was voluminous, but a copyist of well-known writers.

XII. Ancient Farms and Implements.

THAT the agriculture of the Romans in their best days was in every way superior to that of a century ago, there is every reason to believe. Nor were their implements so crude and inefficient as they have been represented. They had no expensive machinery, but neither did America or other countries even so late as a hundred years ago. Because Attilius Regulus, hearing, while he was pursuing his conquests in Africa, that the bailiff of his estate of four acres was dead and his slave had run away, sent to the Senate a catalogue of his spades, rakes, hoes and spuds, with the threat that if they did not take care of his tools, and replace his bailiff and slave, he would return immediately home, we are not to take him as the average farmer, but rather as a suburban cultivator. The fact that sixty-two and a half acres were allotted to a plow, indicates superior cultivation and large estates, though estates were not generally so large as now.

A Prophet's Patrimony.—The Book of Kings speaks of the Prophet Elisha plowing with twelve yoke of oxen, the prophet himself guiding one of the plows. This, according to the Roman allotment, would show him to have had some eight hundred acres of plow-land. Ancient plowing must have been superficial, as a rule, for, with our superior implements, sixty acres to the team of horses is good work. As the ancients were given to caste, perhaps the labors of the farm were subdivided.

It was the policy of the Romans to limit the size of farms, and this was the genius of Carthaginian cultivation. The smaller farms, then as now, were the best cultivated. Columella is the author of the maxim, that the farm should be weaker than the farmer, denoting careful tillage.

XIII. Cultivate Little, Cultivate Well.

PLINY is the author of an oft-told and much-garbled story: C. Furius Cresinus, a freedman, became the object of much ill-will on the part of his neighbors, because he gathered from a small field much greater crops than they could from larger fields. They accused him of attracting the fertility of their fields through charms. A court day was appointed to hear the accusation. Cresinus fearing the result, when the tribe were about to bring in their verdict, brought his live stock and implements into the forum, and with him came a stout serving-woman, who, Pisa says, was well-fed and well-clad. His iron implements were of excellent manufacture; the spades strong, the plowshares powerful, and his oxen in high condition. "Here, Romans," he exclaimed, "are my charms! But I cannot show you, or bring into the forum, my labor of thought, my vigils, nor the sweat of my brow." Thus, "a little farm well tilled," is older than Christianity itself. The cuts shows the more ancient implements of Roman agriculture and successive stages in crude improvements in the plow. 1, 2 and 3 are ancient Roman plows, of successive ages. 4 is the Roman yoke and means of attachments. 5 is the reaping-hook, and 6 the scythe, and they all belong to an age anterior to the Christian era.

ANCIENT ROMAN TILLAGE.

XIV. A Fancy Farmer.

PLINY tells of a rich man who ruined himself by his prodigally expensive outlay in farming, and cites as the general opinion, "*nihil minus expedire quam agram optime colere*,"—nothing pays worse than fancy farming. A good many people are still learning this truth. By his expression, "*bonis malis*," is meant that a thing should be done well and done cheaply. He holds to a middle course, and says a tenant working himself, and having a family to support, may do some things profitably, that a master living at a distance and hiring his labor could not do without ruin. True enough, again, to-day.

XV. Arable Lands and Pastures.

PASTURING was recommended on the score that it required little outlay. Cato being asked, how can a man get rich quickest by farming, replied, "by being a good grazier." How next? "By being a middling grazier." The question being put a third time, the answer was, "by being a bad grazier." Columella admits that there can be no doubt that middling grazing is more profitable than the best management in any other line of agriculture. It would not be so accepted now. Indeed, Pliny, while admitting the truth of the first two answers, accuses Columella of "drawing a long bow" in relating the third, and says, Cato's inculcation was that we should use those means which bring the largest returns at the least expense.

XVI. Water Meadows.

MEADOWS were called by the Romans, *prata, quasi parata*—always ready to yield without culture. We all know the value of meadows, where they may be flooded at will, or watered. Cato's advice is to make them in preference to anything else, but says, if you cannot get water, nevertheless have meadows. Minute directions are given how to pass the water slowly and evenly over the land, so as to guard against stagnation; and says that too much water is as bad as too little. Columella admits that the grass from rich upland is better than that from watered land, but holds that from thin land, whether light or heavy, watering is the only means to bring a good crop.

Precise instructions were given for renewing old and mossy meadows, by breaking them up. One plan was to plow and thoroughly work one summer, and sow in the autumn with turnips or beans, and the next year with grain. The third year the land must be kept entirely clean of weeds of every kind, and then sown with vetches and hay-seeds. And special directions were given that the soil be made quite fine and smooth, so that there may be no impediment to the scythe. New-sown grass must not be fed to cattle, but the second year small cattle might be allowed to graze on it if the soil was dry and firm. Pigs were not allowed on meadows, on account of their rooting. The ancients did not have patent ringers. It was advised that early and weak meadows should be cleared of stock in January, if a full crop of hay was desired, but better lands might be pastured until February or March. Upon such meadows as could not be watered, a dressing of the greenest manure was advised; and the pasturage of the aftermath was accounted as of nearly as much value as the crop of hay—not bad precepts to follow to-day.

XVII. A Rich Meadow.

The Campus Rosea was deemed the richest plat of land in Italy, and it is this field to which may be traced the English story of the meadow-grass that covered by its growth the walking-stick of the owner—while he ate his dinner and smoked his pipe—so he could not find it. The Roman grass, however, required a whole night to accomplish the feat. Dickson thinks that a single mowing of a Roman meadow (it must have been a good one) would yield two and a half tons an acre. It is said that the mowing of a *jugurum*—three-fifths of an acre—was a day's work. The grass must have been heavy, or the Roman scythes dull. Grass was cut before the seed was ripe, and the most minute directions were given for curing and storing the hay. It must not be under-sweated nor overheated. Pliny thought when hay was stored too green, the sun set the ricks on fire. We know better. It burns from accumulated heat within. Scythes were sharpened with oil-stones, and a horn of oil was fastened to the leg of the mower. Pliny first mentions stones that would whet a scythe with the aid of water. In Italy short scythes were used, and in Gaul long ones. The Romans certainly understood haymaking in its minutest details.

XVIII. Roman Rotation.

The usual rotation in Roman agriculture was most simple. It was a crop and a fallow. One-half of the arable land in grain and one-half fallow. One-third of the fallow land was sown to some crop to be cut green for feeding, and this alone was manured. Hence, three grain crops and one green crop were taken from the land for each manuring. In other words, the whole soil was manured once in six years — better manuring than the average American farmer gives to-day. The naked fallow received three or four plowings in the season, besides the seed furrow. Wheat and barley were the principal bread grains.

XIX. Roman Small Grains.

Wheat.—White, red, black bearded, and smooth wheat are mentioned among the principal varieties. The Romans understood the varieties adapted to different soils. *Siligo, triticum* and *far adoreum* are named as the best sorts, and Pliny held that *siligo* sown in certain soils for three years in succession, would turn to *triticum*. *Triticum* is our botanical name for the wheat tribe. If *siligo* was what we now term a pedigree grain, there is no doubt that Pliny was correct. Choice seed will degenerate, whatever the species. Wheat was generally sown in the autumn, but, since it was also sown in spring—though this was condemned—they must have had spring varieties.

Barley.—Barley was sown in September and October, and also in January and March. The spring sowing of barley was less condemned than spring sowing of wheat. The ancients doubtless thought wheat would turn to chess, for Pliny says that if oats are sown on a certain day of the moon, it will come up barley. Pliny probably, sometimes " said things sarcastic." He believed in books, for he thought no book was so bad that something good might not be learned from it.

XX. Some Ancient Methods.

SMALL grain was covered with a double mold-board plow, and it was considered a clumsy plowman who required an *occator* (a person to cover in the seed) to follow. Another plan was to sow on the plowed ridges, and cover with rakes, so lightly that the ridges were not obliterated. The crop that was covered by the double mold-board plow was sometimes plowed once, after it had attained some height, but generally it received two hoeings, one in the autumn and another in the spring. This seems to have been universal, and sometimes three or four hoeings were given. Hand-weeding was also practiced, if necessary. The Romans were clean farmers.

XXI. Antique Crops.

BESIDES the cereal grains, pasture and meadow grasses and millet, the ancients cultivated green crops for soiling. *Cicer* (pulse) is mentioned. Vetch, lentil, lupin, fenugreek, peas and beans were commonly grown. Lucerne seems to have been much raised, and one sowing is said to have lasted ten, or, as Pliny says, thirty years. Hemp, flax and poppy were also cultivated. Pliny tells how flax was used for fine linen, sail-cloth, wicks, fish-nets, snares for wild beasts, and says that each thread in a breast-plate, then extant, of Amasis, King of Egypt, consisted of three hundred and sixty-five ply. Flax, then as now, was considered an exhausting crop.

XXII. Crops Pulled by Hand.

THE word *legum,* or legume, seems to have been used to denote all plants or crops that were pulled by hand; for not only peas and beans, but flax, hemp, turnips, rape and radish were so gathered. Turnips were sown after five plowings, on dry friable soil, in manured rows, and thinned to eight inches apart. Pliny records a turnip weighing forty pounds, and says they were valuable in ornamental cookery, because capable of receiving six colors besides their own, one of them purple,—a fact that may be valuable to fancy cooks of the present day. The crop was subject to the fly, then as now. These pests were fought by steeping, by the use of soot, etc. Columella says that in Gaul turnips were fed to cattle and sheep; and so England is not the only country where turnips may have been the sheet-anchor of agriculture.

XXIII. Fallow Crops.

THE belief was prevalent among the ignorant that certain crops, even if gathered and carried away, improved the soil—a belief now disproved. Intelligent people knew better. Columella says: "Some tell us that a crop of beans stand in the place of a manuring of the land, which opinion I would interpret thus: not that one can make the land richer by sowing them, but that this crop will exhaust the land less than some others. For of this I am certain, that land that has had nothing on it will produce more wheat than that which has produced pulse the preceding year." We of to-day know well that a covered fallow—a crop sown and plowed under—is more enriching than a naked fallow.

The ancients seem to have been aware of this, for they did sow, for plowing under, beans, vetches, and especially lupines. So it is quite probable that they used the naked fallow, as we do, for cleaning the soil of weeds.

XXIV. Ancient Harvesting.

The ancient mode of harvesting was by means of the sickle, not unlike that practiced one hundred years ago. They also harvested grain with a machine that gathered the heads in a comb-like arrangement, the straw being cut afterwards, probably with a scythe.

The Romans, however, usually cut with the sickle, the straw being taken about the middle. This upper part was fed to cattle, after threshing, and the butts were used as bedding for stock. Some nations, according to Pliny, pulled the grain by the roots, imagining that disturbing the surface was beneficial to the soil. The shield of Achilles shows reaping by means of the sickle, and with much division of labor. Homer says the practice of rich men was to start a gang of reapers at each end of the field; and he likens their approach towards each other to that of the Greek and Trojan hosts in conflict.

Reaping Machines.—Reaping machines were known in Rome. They are mentioned both by Pliny and Palladius, the latter saying that they were used on the large farms in Gaul. They were as primitive as they were unwieldy.

Threshing.—Threshing was performed both with the flail and by tramping by cattle and horses; also by machines drawn by cattle, one having teeth and the other rollers. The grain was generally threshed immediately after harvest, and often in the fields where it was cut. Where permanent threshing-floors were made, much care and skill were used. They were made of rolled and hard-rammed earth, mixed with lees of oil. Columella says the floor is better if straw is used with the mixture. Finely broken flint, such as macadam roads are now composed of, is also spoken of; also flagged earth, or a surface of flag-stones. Virgil is both precise and poetical in his description of a threshing-floor.

Winnowing.—The grain was cleaned by throwing it from shovels against the wind. Some sort of fanning-mill must also have been used, for Columella recommends it, and Virgil mentions among the "*duris agrestibus arma*" this improvement, as "*mistica vannus Jacchi.*"

XXV. Roman Fertilizers.

The ancients would seem to have been as careful savers of manure as the Chinese now are. The value of all animals on the farm as manure-makers is minutely estimated. Curiously enough, they considered the manure of water-fowl as of very little value. Pliny sneers at Columella for considering the manure of pigs not worth much. Possibly, Columella's pigs were poorly fed. The ancients, at least, understood that, "the richer the feed the better the manure." Every available thing was used to increase the compost heap—leaves, litter-road scraping, etc. Both Pliny and Columella enjoin that the heaps shall be covered, so as to suffer neither by drying in the wind nor parching in the sun. Also, that the bottom be water-tight, so that the liquid parts cannot escape. "It is an idle farmer," says Columella, "who does not get together some manure, even though he keeps no cattle." He also delicately hints that the manure-pile need not necessarily occupy the front yard, so to speak. Oak leaves are recommended to be mixed in the pile, or else a stake driven through, "to prevent serpents breeding in the manure." This is no worse than some of the superstitions that prevail nowadays.

XXVI. A Question not yet Settled.

THE value of fermented as compared with unfermented manure was much discussed by Roman writers. Cato and Varro held that manure heaped, turned and rotted down was stronger than green manure. Columella and Palladius held that the only good of fermentation was to destroy the seeds of weeds, and that it weakened the manure. Those who have prepared manure in water-tight yards, will probably agree with Cato and Varro. Palladius was a mere copyist at best. From where he got the idea that sea-weed should be washed before it is applied as manure, he does not say. Neither has a list of sidewalk farmers, other than himself, come down to us. Was Palladius the originator of the guild?

XXVII. Little and Often.

A LITTLE manure often applied was considered the best, and it was known then, as now, that manure does not produce so great an effect on wet land as on dry. Why should it? Manure that is dissolved and escapes over the surface is lost.

Dickson says that about eight hundred bushels of manure, well prepared, was an average Roman cast. Not too much, certainly, when the land was only manured once in six years. It is not more than good modern gardeners apply annually, but, perhaps, more than the average farmer now gives his land even once in six years.

XXVIII. Commercial Fertilizers.

LIME and marl were much used for manure. Six varieties of marl are recorded. There was clayey for light lands, and sandy marl for stiff soils. A rock-like marl is mentioned, which did not become reduced for many years, and one kind brought from Britain, where it was dug from pits thirty yards (?) deep, is said to have held the fertility of the land for eighty years. The ancients seem to have known how to advertise their wares fully as well as the modern fertilizer man.

XXIX. Ancient Plowing.

The number of plowings the ancients gave their fallows prove that they were no mean plowmen. Dickson would have us believe that their ordinary plowing was nine inches deep, and, from the fact that Pliny says that, once in the fallow course, it was not unusual to attach six or eight oxen to a plow, the probability is that at such times the soil was deeply stirred. Their word *sulcus* meant a certain breadth stirred up, and the object of their fallow plowing was thoroughly to stir all the soil to an equal depth. They plowed narrow lines of equal breadth. A good Roman plowman would leave the surface so even that it would be difficult to tell the lines of soil moved by the plow. Each line must be perfectly straight. The characteristic of Roman plowing was precision.

Because the plows had no mold-boards, as we understand the term, the plowman returned upon his own furrow, plowing back and forth.

Plowing uneven breadths, *sulco vario*, was condemned. If the ground was left lumpy it was called *scamna*, and this was thought to bring a bad name to the land. Crooked plowing was called *prævaricare*, and Pliny says this term was afterwards used in the law courts to designate those who testified falsely or told crooked stories. The word has come down to our time, to denote crookedness. The ridge upon which the grain was sown was called *lira* and *porca*, implying a ridge-backed sow. There are not a few plowmen nowadays who "prevaricate" and make "ridge-backed" sows of their two-back furrows.

Cato forbade his bailiff to plow land when wet, to cart over it, or even to allow the treading of cattle, because it would not recover in three years. Various ancient authors say if you plow wet land you lose the whole season. How many farmers of to-day know this? Very few it seems, for they go on plowing, even stiff clay, year after year, when so wet that the fields dry into great lumps and clods somewhat softer only than half-burned bricks.

XXX. Plows.

It is generally supposed that ancient plows were pretty much after one pattern, and all of them of the crudest make. The more ancient ones, certainly, were mere pointed sticks, as we have shown. There is no proof that any of them turned a furrow completely over, and pulverized it. Neither did the plows of one hundred years ago. It is only within the last forty years that plows have been brought to moderate perfection. Adam Dickson, who wrote in the last half of the last century (in 1764, he published the first volume of a "Treatise of Agriculture") exhaustively investigated the construction of Roman plows. Here are his conclusions:

The ancients had all the different kinds of plows that we have at present in Europe, though perhaps not so exactly constructed. They had plows without mold-boards, and plows with mold-boards; plows without coulters, and plows with coulters; plows without wheels, and plows with wheels. They had broad-pointed shares and narrow-pointed shares. They even had, what he says, "I have not, as yet, met with among moderns—shares not only with sharp sides and points, but also with high-raised cutting tops."

Were we well acquainted with the construction of all these, perhaps it would be found that the improvements made by the moderns up to one hundred years ago, in this implement of cultivation, are not so great as many persons believe.

XXXI. Seeding.

The Romans had many good notions about sowing. One was: "Early sowing sometimes deceives the husbandman; late sowing, never—because the crop is always bad." Their usual mode of sowing was broadcast, or, more properly, overcast, like that of the ancient Egyptians. The sowing was always single-handed. Two-handed casting does not seem to have been known. The Romans paid great attention to the seed. It must be sound, plump and well-formed. It was carefully selected in the field, while the crop was still uncut. The standard quantity of wheat sown per acre would seem to have been somewhat less than two and a quarter bushels. Two bushels seem to have been the least quantity, and two and a half bushels the maximum quantity per acre.

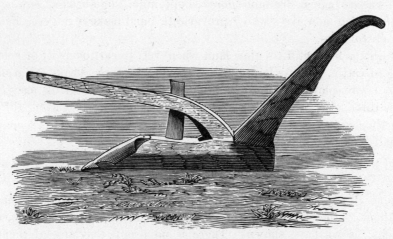

OLD MOORISH PLOW.

XXXII. Yield Per Acre.

Varro claims ten bushels of crop for one of seed, in the average, and fifteen for one in very rich land. This was on the basis of about two bushels sown per acre. Cicero quotes the rich lands of Sicily as yielding eight for one, on the basis of two and a half bushels of seed per acre. He records that in Columella's time, over the greater part of Italy, the usual return was not more than four to one. Hence we see that in ancient agriculture, as in modern, the continued sowing of wheat always caused deterioration of the soil. The rule will apply, and without exception, to all crops, unless due provision is made for re-fertilizing, or resting the soil.

XXXIII. Mediæval and Modern Agriculture.

The Dark Ages carried the world almost back to barbarism. Then there was no progress. The Moorish plow was as good as the best. One much like it was used in Mexico fifty years ago, and probably is to-day in some portions of that country. If the implements shown by drawings preserved in the British Museum are an index to the agriculture of the early Norman period, it was crude, indeed, even making allowance for the art of drawing in those days.

NORMAN FARM TOOLS.

The illustration here shows the state of British agriculture soon after the Norman conquest of England. 1 represents the plow and a rough hatchet or maul carried by the plowman for breaking the clods. 2 is a sower casting the grain. 3 is a reaper with reaping-hook. 4, threshing. 5, the scythe, and mode of sharpening the blade. 6, beating and breaking hemp. To show how slow was agricultural progress in some regions, we illustrate a plow in actual use in the Orkney Islands not more than forty years ago. The yoke is not a bad copy of the most ancient plow and yoke (Chinese) known, and supposed to have been common in the days of the Patriarch Abraham. An interesting companion picture to these is the modern English steam plow, worked by a stationary engine on each side of the field. The engines move forward from time to time, and anchor themselves as the plowing progresses.

PLOWING IN THE ORKNEY ISLANDS.

Steam plows moved by the traction power of the engine have so far been found impractical, though many have been invented in England, and more in the United States, only to prove failures. In fact, steam plowing is not a success in America generally. Horses and cattle, and their food, are too cheap. Besides, the necessary teams for the other farm work, and the marketing of the crops, will also do the cultivating. In the great corn region of the country, it requires about the same teams and hands to do the cultivating that are required for the plowing. Besides, wood for fuel is becoming scarcer in the

ENGLISH STEAM PLOW AT WORK.

West every year, and it may long remain cheaper to raise corn for work-stock than to buy coal for steam plowing.

CHAPTER III.

MODERN FARM IMPLEMENTS.

I. A REVOLUTION IN FIFTY YEARS.—II. THE PIONEER'S PLOW.—III. THE FIRST STEEL PLOW.—IV. THE WEBSTER PLOW.—V. THE PLOW OF TO-DAY.—VI. THE ART OF PLOWING.—VII. LAYING OUT THE LAND.—VIII. TURNING THE FIRST FURROW.—IX. THE BACK FURROW.—X. RE-PLOWING.—XI. SUBSOIL PLOWING.—XII. FRENCH PLOWING.—XIII. IMPLEMENTS FOR SMOOTHING AND DISINTEGRATING.—XIV. LEVELING, COMPACTING AND GRINDING.—XV. THE PLANK SOIL GRINDER.—XVI. THE LEVELER.—XVII. IMPLEMENTS OF CULTIVATION.—XVIII. HISTORY OF THE CULTIVATOR.—XIX. ONE-HORSE CULTIVATORS.—XX. SEEDING MACHINES.—XXI. THE GRAIN DRILL.—XXII. CORN PLANTERS.—XXIII. HARVESTING MACHINERY.—XXIV. USE ONLY THE BEST.—XXV. PLOWING IRREGULAR AREAS.

I. A Revolution in Fifty Years.

THE farm implements of fifty years ago were of the rudest construction, by comparison with those of to-day. In Europe, and especially in England, some progress was made from time to time, but until the investigations of Jefferson and others in the United States started the era of improvement in iron plows, there was no radical or permanent advance. The iron plow of Jethro Wood in 1819, marks the beginning of the revolution.

II. The Pioneer's Plow.

IN a report to the New York State Agricultural Society in 1856, Mr. A. B. Allen thus describes the plows of the early part of the century: "A winding tree was cut down, and a mold-board hewed from it, with the grain of the timber running as nearly along its shape as it could well be obtained. On to this mold-board, to prevent its wearing out too rapidly, were nailed the blade of an old hoe, thin straps of iron, or worn-out horseshoes. The land-side was of wood, its base and sides shod with thin plates of iron. The share was of iron, with a hardened steel point. The coulter was tolerably well made of iron, steel edged, and locked into the share nearly as it does in the improved lock coulter plow of the (then) present day. The beam was usually a straight stick. The handles, like the mold-board, split from the crooked trunk of a tree, or as often cut from its branches; the crooked roots of the white ash were the most favorite timber for plow handles in the Northern States. The beam was set at any pitch that fancy might dictate, with the handles fastened on almost at right angles with it, leaving the plowman little control over his implement, which did its work in a very slow and imperfect manner."

III. The First Steel Plow.

WE have seen plows such as Mr. Allen describes in use in the West considerably less than fifty years ago. In fact, up to forty years ago, most of the ground was broken with plows with wooden mold-boards covered with strips of iron. Middle-aged men may easily remember when their fathers first brought home the steel plow. It is a Western invention, without which our mucky soils could not have been successfully cultivated. It is an open question, whether Deere, of Moline, or Lane, of Yankee Settlement, near Lockport, Illinois, made the first steel plow. They both made the first mold-boards out of saw-plates. Then came cast plows, cast-steel plows, and next, plows of silver steel, so that now we have implements that will scour perfectly, if properly kept, even in the worst soils.

IV. The Webster Plow.

DANIEL WEBSTER'S PLOW.

The plow that the great "Expounder of the Constitution" helped to make, and which he held, in breaking a piece of bad, grubby land, at his home in Marshfield, was always a source of pride to the statesman. It was shown at the Centennial Exhibition, and attracted much attention. It was an immense, clumsy and heavy affair, drawn by a long string of oxen; and yet, forty years ago, it was not a bad plow for stumpy and grubby land. The engraving is an exact representation of this historic implement. The plow of to-day will do better work with half the team.

V. The Plow of To-day.

The plows of to-day may be said to have reached perfection. They combine lightness of draft, with great excellence of work. The notion is now discarded, that one plow can be adapted to all the uses of the farm. The different kinds of plows are now counted by hundreds. We have stirring plows, stubble plows, deep tillers and subsoil plows. There are trenching plows, plows for turning meadow, pastures, broken sward, low-land and upland prairie. For laying flat inverted furrows, there is a special description of plow, and also for lap furrows, and there are even plows for setting the furrows on the edge.

The old cast plow is a thing of the past. For sandy or other soils that scour fairly the chilled plow is much used. The polished steel share cleaves its way in all difficult soils, doing its work perfectly; and in all soils not unusually trashy the debris

PLOW WITH CHAIN FOR TURNING UNDER TRASH.

is effectually turned under if the plowman attends to his business. For turning under growing crops and weeds there are so many appliances that it would be difficult to enumerate them. The chain drag and the trash cleaning attachment, shown in the

PLOW WITH ATTACHMENT FOR CLEARING TRASH.

illustrations, explain themselves. The object with every form of chain or hook is simply to catch the material to be plowed under and drag it forward into the position where the furrow in closing down will cover it. The longer the trash the less does it need pressing. Uneven or short material is the hardest to cover. The favorite plow in the great prairie regions of the West is the gang stubble plow, worked by three or four horses, as shown in the engraving on next page.

VI. The Art of Plowing.

The furrows must be straight and even; whether the furrows be shallow or deep, the depth should be uniform. The lands once laid out straight and of uniform widths, there is no trouble in keeping the furrows straight. Even in stumpy land and land with large rocks that must be plowed around, there is no need that the furrows remain permanently crooked. A little calculation at the proper points in narrowing the furrows or running them completely out, will soon bring the furrows straight again. There is nothing lost by this. There will be no more furrows to plow and the furrows are turned equal, to say nothing of the shocking appearance of a field plowed in a hap-hazard way.

VII. Laying Out the Lands.

SUPPOSE the field is to be plowed in regular lands, two rods wide. Select four poles eight feet three inches long, and perfectly straight. Fasten a short flag to each pole,

GANG STUBBLE PLOW.

and provide two baskets filled with inch-square, sharpened pegs, twelve inches long. Let an assistant take two poles, one of the baskets of pegs, and a hammer, reserving a similar outfit for yourself. Measure on opposite sides of the field twice the length of a pole, and place it securely, one at each side of the field. Sighting along these, let the assistant advance one-quarter across the field, and set his second pole, moving it one way and another until you find it in a direct line with the other two. If the lands are very long, still another pole may be necessary. Drive a peg securely in each of these marks. Then the assistant measures four lengths of the pole, or two rods, from the next land at the end. You do the same, and then he sights for you to fix the second pole. Drive pegs as before, and so proceed until you have the whole field laid off. Then fix the flag-poles at the peg marks.

VIII. Turning the First Furrows.

THE next thing is to turn the back furrows, or two first furrows of each land. This is especially necessary when several plowmen are employed. Many who can do fair work, after a land is laid out, fail utterly in laying out or plowing the first furrows. They either plow crooked or make balks, or both. Others, again, will lay out lands correctly, and yet become inattentive to their work in plowing. The merit of a field of plowing is the poorest furrow laid—not the best—and the test of the work in laying out lands is to be able to drive a "fresh team" straight across a forty-acre field. Have the reins so they will carry the horses' heads rather wide apart, in order that the plowman can see between the team and to the stakes beyond. Arrange the lines so they will have about six inches or more slack when the team is pulling, and hang on the left handle of the plow. Take the lines in one, two or three fingers of the left hand, in such a manner that a pull straight back will bring the team to the left. If you wish to turn to the right, carry the hand to the left and pull the line around the handle of the plow. Have the plow sharp and perfectly bright. A little practice will, by keeping your eye constantly on the stakes ahead, enable you to drive a straight furrow, however long the distance. But if the land is irregular, stakes enough must be set for you to have two always in line. If the team are fractious, plow them upon the land first laid out until they become steady.

IX. The Back Furrow.

It is usual, in drawing the first furrow, to turn the plow somewhat over, so the share will cut less at the edge than at the heel—especially if the back furrow is to be turned into the first furrow made, thus cutting all the soil. Then in coming back, the plow should be so set as to do this, plowing a flat furrow-slice, with the off horse walking in the furrow. Twice up and back ought to prepare any land so the plowman following, will have clean furrows to begin with. Then make him keep them so. The lands as you leave them for the plowman should show two furrows, one on top of the others, and the other two sharply lapped against them. This is the plan that should be followed for all plowing, unless the ground is simply to be stirred, or in re-plowing land that was plowed in the fall. In the latter case, it is sometimes admissible to lap the first two furrows, one against the other. Then the merit of the work is that the first two furrow-slices turned, shall leave each furrow clean for the plowmen.

Drainage Furrows.—Having laid out one land, proceed in the same manner with the others, until you have the whole field laid out. This often serves the double purpose of regularity in the lands, and drainage, if wet weather ensues. If the eye is good, one may often save a week's time in drying the land, by plowing furrows through the field along the lines of natural drainage, as early as possible in the season, and previous to laying out the regular lands.

Prairie or Sward.—In breaking prairie or sward land, the practice of laying the two first furrows together, so they just meet, is usually followed. We prefer cutting the two first furrows thin, and laying one on top of the other, thus cutting the whole of the soil.

X. Re-plowing.

IF the land has been fall-plowed, and is to be re-plowed in the spring, we prefer in laying out lands to throw two light furrows out to start with and then cover them back. This is the best plan unless it is desired to still further ridge up the soil, in which case the first furrow may be laid as previously directed. This plowing, however, should never be deep. The deep plowing, if necessary, should have been done in the fall, and the spring plowing should be done with a stirring plow, the notion being simply to loosen up the surface. Many persons do this with a corn cultivator. We

STIRRING AND STUBBLE PLOW.

have never found any saving in this, since the ground must be gone over several times in order to make the work equal to plowing. The stirring and stubble plow shown is much liked in the South and Southwest, and especially in Texas, as a plow for general use. It is used to best advantage in loams and other soils containing a fair proportion of sand

XI. Subsoil Plowing.

MANY persons use the words subsoil plowing when trench plowing is meant. Trench plowing is inverting one furrow over another by one turning plow following in the track of another. Subsoil plowing is simply loosening up soil in the bottom of the furrow to a depth of four or six or more inches, as may be desired, and leaving the mellowed subsoil lying in the furrow to be again covered by the next furrow, the upper soil remaining on top. Subsoil plowing is not advantageous except on naturally dry soils or artificially drained ones. On undrained soils, when they become wet and sodden after subsoiling, the value of the subsoiling is destroyed. On drained soils, for special crops, it is of great value in increasing the depth of the tilth, and this will remain for years from a single subsoiling.

XII. Trench Plowing.

TRENCH plowing, as before said, is the inverting of one furrow on another, or the mixing the two furrow-slices more or less together. The Michigan subsoil plow is, in reality, a trench plow, inverting one furrow upon another. Many of the best gang plows are now fitted not only to perform this operation, but also to mix the soil. They are much used now for covering under sward, for special crops, by which a deep and cultivable tilth is left on the surface. The illustration shows the skin, or sod-slice, *a*, turning into the bottom of the furrow, *c*; and *b* shows the thick sub-furrow and its twist, in the process of being inverted on the sod-slice. Trench plowing, in connection with manure, is practiced in mellow soil, where deep tillage is necessary, as for root and other special crops. As a rule, deep trench plowing should never be done except where plenty of manure is used. And trench plowing should not be given to thin, poor soils. It often pays, however, say to a depth of eight inches, when it is necessary carefully to cover under trash, leaving a clean surface for cultivation.

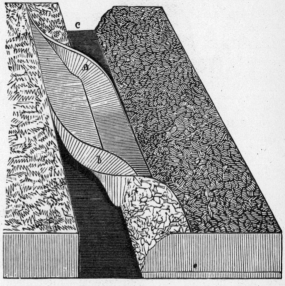

SKIN AND TRENCH FURROW, IN TRENCH PLOWING.

Fig. 1.
TRENCH PLOWING, TEN INCHES DEEP.

Figure 1 shows trench plowing ten inches deep, the capacity of the plow for lifting having been exhausted at this depth. Now, if it is desired to go still deeper, the bank of earth must be removed as in Figure 2. In theory it seems easy. In practice it does not work. It may be partly done with the plow, but must be finished with shovels. By successive removals, and with proper plows, a depth of twenty to twenty-two inches may be gotten with eighteen inches of clean furrow, as shown in Figure 3, provided the soil is a pretty firm loam. In lighter soils the furrow will be more largely filled, as shown by the dark line below, in Figure 3. The cuts are used mainly to show the impracticable nature of very deep trenching.

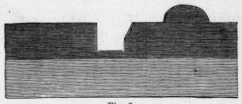

Fig. 2.
THE FURROW MOVED BACK.

Fortunately, the theory which advocated it some years ago is exploded. Deep trenching is now never done except for some special crop, and then only in connection with large quantities of manure. Subsoiling has taken its place.

The Subsoil.—When a subsoil plow is run in the ten-inch furrow, the subsoil is

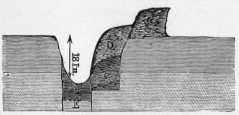

Fig. 3.
DEEP TRENCH PLOWING.

left pulverized in the furrow, and will fill the furrow quite half full, by reason of its lying lighter than before. A depth of sixteen inches may thus be obtained, and the subsoil will lie in the furrow up to the top of the dark line shown running to a point in Fig. 3. But it will be uniformly, and you have the subsoil where it ought to be, in the bottom of the furrow. In Chapter I, Part III, the subject will be again referred to.

XIII. Implements for Smoothing and Disintegrating.

THE principal smoothing and fining implement is the harrow. There are now many varieties of harrows, but only two principal forms are used, the one with teeth placed perpendicularly, the other with the teeth inclined more or less back. Some of them are arranged so they may be used at will, with the teeth slanting or straight. There are also many modifications of the principle involved in the harrow, by which curved and cutting teeth, or shares, are fixed in a frame for slicing the soil. There are also discs or hollowed wheels which slice and raise the soil. This latter device has also been applied to the cultivation of crops with more or less success, and also to plowing, the concave surface of the wheel throwing the earth outward as it cuts its way through the soil.

The Rotary Harrow.—One of the most perfect forms of the harrow, for simply stirring and disintegrating the soil, is the rotary harrow, which revolves as it passes forward, and which is adjusted by loading a box on top, more or less, with earth or stones. It has this advantage that, in revolving, the teeth free themselves of trash. It is useful on trashy and especially on compacted soils.

The Double Harrow.—The form of harrow best adapted to the general uses of the farm is the square, double harrow, jointed in the middle for ease in lifting and freeing it from trash. Whatever the frame adopted, it is indispensable that the teeth may be easily removed for sharpening, and that, when replaced, they remain firmly fixed. The teeth should be of well-tempered steel, and always sharp. A dull harrow is costly in its work, because inefficient. It is a hindrance that no good farmer will tolerate. The illustrations show, in two positions, one of the best forms of square harrows, the teeth of which may be used slanting or upright, single or double, and which will fold compactly for transportation.

FULL-SIZED HARROW WITH TEETH SLANTING.

HARROW FOLDED.

XIV. Leveling, Compacting and Grinding.

There are various implements, of more or less intricate construction, for leveling, grinding or compacting the surface of the soil. Except the roller, the necessary implements may be made on the farm. The use of the smooth roller is now almost entirely abandoned, except when it is necessary to compact the surface. It has little of the grinding action that is necessary to break down lumps and clods. It simply presses them into the surface, to be again dragged out intact if the land is subsequently harrowed. The corrugated roller does grind, as well as compress, and hence is superior. For settling and compressing roads, the smooth roller is valuable, and its use is now pretty much confined to this purpose.

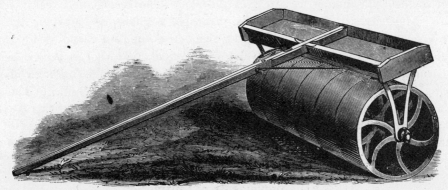

SECTIONAL IRON FIELD ROLLER.

On some soils, especially light, fluffy ones, requiring strong compression, the roller is indispensable. For breaking down clods, it has almost entirely gone into disuse. It simply presses them into the earth, and does not disintegrate them. It is, nevertheless, valuable for compacting meadows when they heave. It presses the soil firmly about grass and other seeds, and for a variety of uses on the farm, the sectional iron roller is best.

XV. The Plank Soil Grinder.

This has come to be one of the most efficient implements on the farm, for grinding down lumps and clods, and reducing the surface to a firm tilth. It has sufficient direct pressure to firm the surface, so as to bring the earth into direct and close contact with small seeds. To make this implement, select five two-inch hardwood planks, eight feet long and eight inches wide, and one of the same length for the forward piece ten inches wide. Spike or pin these together, lapping one edge on the other two inches, as in weather-boarding a house. Turn it over, and pin two strips from front to rear to hold it solid. This is drawn over the field so the laps will not dig into the soil, but slip over it. A chain fastened to hooks two feet from each end, will form a triangular

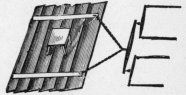

SOIL GRINDER.

hitching point, and if not heavy enough to do the work, a boy may ride when driving, being located at a suitable point so the implement will run straight. The cut fully explains itself, and shows the riding stool.

XVI. The Leveler.

This is another important and easily-made implement for the farm. Its use is to bring a field comparatively even and level, when flax, roots, or other special field crops are to be raised. It is made of four pieces, eight feet long, of 4 x 4 hardwood scantling, pinned together, sixteen inches apart, and so braced that each piece is firmly held. The front lower edge of the first and third scantlings is rounded so they will incline to slip over the soil. The second and rear pieces are left with sharp edges. It is drawn by means of a chain, as shown in the plank grinder. The operator rides on a plank thrown across the top, changing his position as may be necessary to make the leveler run straight. The soil, where it projects, will be caught and deposited in the lower places. Its action is both grinding and smoothing, and, when necessary, the earth may be thrown to one side or the other, by a change of position of the driver. Its use, however, should be entrusted only to a good driver, and one of sound judgment, who will take advantage of anything that will assist in leveling the field. With this implement, a sharp harrow, and the grinding planks, a field may be put in condition for almost any garden crop.

XVII. Implements of Cultivation.

The implements already described are those principally used in the preparation of the soil. The first two are indispensable to every farm. The two last named are indispensable for all crops that need a smooth, fine surface, as, for instance, grass. As an implement of tillage, the harrow plays an important part, especially the smoothing and other slender-toothed harrows, for stirring and loosening the surface of the soil. A sharp harrow is often useful for scarifying an old meadow, and especially a pasture when more seed is to be sown. If a field of wheat, or other grain, becomes badly crusted, a light harrow is needed. In the first cultivation of corn, and other hoed crops, its use is well known. The implement of cultivation, however, that has revolutionized the production of hoed crops, is that originally known as the straddle-row cultivator, and now as the walking cultivator, which completely finishes a row of corn each time through.

XVIII. History of the Cultivator.

Horse cultivators began to be used in the eighteenth century. Jethro Tull was the father of drill husbandry. From time to time, horse hoes were improved, and were at last developed into the straddle-row implement of the present day. It is said that the original idea was applied in England as long ago as 1701, but, while the principle—the cultivation of one or more rows at a time—is the same, the successive improvements have been so numerous and important, that now eight to twelve acres may be cultivated to a man and team, in a day, much better than a single acre could have been done thirty years ago with the implements then in use. Not only is level cultivation accomplished, but hilling and ridging are much better done with this

WALKING CULTIVATOR.

implement than with the plow. The cut shows a cultivator with plows that may be reversed, to throw earth to or from the hill, or to press the earth equally to each side.

XIX. One-Horse Cultivators.

NOTWITHSTANDING the general use of two-horse machines, there are many farmers who still use the single five-tooth cultivator or some modification of it. The Eastern farmer whose corn planting in a single year would be between five and ten acres would have little use for the two-horse machine. The home of the latter implement is on the prairies, North and South, or where more crops are raised in large fields. Still the double or single shovel plow and the one-horse cultivator will not go out of fashion. In the smaller fields perhaps the most important single-horse implement that combines level and hill cultivation is the horse hoe.

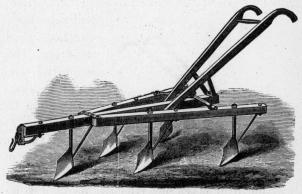

FIVE-TOOTHED CULTIVATOR.

The Horse Hoe.—This is simply an improved shovel plow or bull tongue. From the top of the shares there is expanded a wing on each side. These are so adjustable that they may be set off at any required angle. They catch the earth as it is pressed outward, and move it slowly back into a ridge, lower or higher, as may be desired. Thus, excellent ridges may be raised for sweet potatoes, or crops requiring ridges; or, with it, any crop, such as potatoes, celery, etc., may be hilled to the required height. During the early cultivation of the crop, the wings are not used, and the soil is simply stirred. The implement is not recommended for flat cultivation, but except this, it is most thorough in its work. A form known as the double-shovel plow gives better satisfaction in working corn than the five-toothed cultivator, and the latter is now mostly used for garden crops.

XX. Seeding Machines.

SEEDING machines stand next to cultivating implements in the rank of useful farm tools. They are as indispensable in the garden as in the field, and not only save seed by distributing it more equally and regularly, but enable the cultivators, in drill husbandry, to cover the seed to the exact depth wanted. They are not, however, confined to drills. There are broadcast sowers, with or without covering attachments. Probably the most valuable to the small farmer is that form which turns with a crank, and casts the seed in spiral circles from a cone-shaped orifice, distributing the seed evenly and perfectly. One man will sow from ten to fifteen acres a day, and with the large machine—which may be attached to any farm wagon—we have sown one hundred and ten acres of wheat in a day of ten hours

XXI. The Grain Drill.

No person who makes wheat raising an important part of his farm work can afford to be without an improved force-feed grain drill. They will seed and cover at one operation, with a span of horses, from eight to twelve acres a day, and in seeding one hundred acres will save their cost in a single season. For sowing all kinds of garden crops, or for field cultivation of root crops, the single-row drill will be found

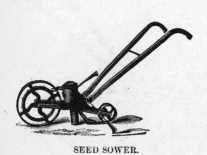

SEED SOWER.

indispensable. The best of these now have markers attached, which mark the next row while seeding the first. They open the furrow, deposit the seed as desired—much or little—and cover as fast as a man can walk. For field work, a larger machine, drawn by a horse, is used. In fact, any of these larger ones may be so drawn, by hitching the horse to the end of a light rope, six feet long, fastened to the machine. In this manner it is easily controlled, and if the horse swerves from the line, the drill may, nevertheless, be kept true.

XXII. Corn Planters.

NOT less important to the farmer is the check-row corn planter. Without it, the great cornfields of the West and South could not be planted. The check-row attachment is an important part of the machine. The corn is dropped with much more precision than can be done by a boy sitting on the machine, and besides, the marking of the land is entirely done away with. It will plant from eight to twelve acres a day much better than can be done by hand, unless great care is taken. The seed is dropped in narrow and exact lines, a matter of no small importance in the subsequent cultivation, especially in harrowing the young corn, as it is less likely to be torn out by the harrow.

XXIII. Harvesting Machinery.

NEXT in importance to implements of cultivation, are harvesting implements. These are now of so many kinds, that a somewhat critical knowledge is necessary to determine the relative merits of one to another. With modern machinery on a Western or Southern farm, the master may ride and do perfectly almost all the work required to prepare the land, seed, cultivate, harvest and store away the crop. In pitching hay and grain on the wagon, some hand-work must be done. In pitching hay from the wagon to the stack or mow, machinery alone may be used; so also in gathering the hay from the windrow, and automatically carrying it on to the wagon. Hay is taken from the cock, and with a sweep is carried to the stack, deposited on a carrier, which raises it on the stack without touching it with the hand. It is mowed, raked, cocked, carried and stacked by machinery. Grain is cut, bound and delivered in lots of a dozen bundles, from the machine, ready for shocking. The steam thresher then separates the grain from the straw, separates the foul seeds from the good, bags it up, and keeps tally of the bushels.

XXIV. Use Only the Best.

THE great improvements in farm machinery, that cause all these operations to be so perfectly performed, from plowing the field to cleaning the grain, have been made within the last thirty years. Any farmer who does not make use of them is far behind the times, and is working at such a disadvantage that he will surely be distanced in the struggle for wealth. It is in the great agricultural region of the Mississippi Valley, lying between the slopes of the Alleghanies and the Rocky Mountains, and between the great Northern Lakes and the Gulf, and on the Pacific slope, that agricultural machinery may most economically be employed. The influx of population yearly pouring into this region from more eastern States, and from foreign countries, attests the ease with which farms may be made and competence secured to the in-comer. How to do this most easily and economically, will be unfolded as this volume progresses.

XXV. Plowing Irregular Areas.

THE New York State Agricultural Society had diagrams prepared, some years ago, showing how to plow irregular fields. It is often economical, especially in small fields, to plow without dead furrows. We here give these diagrams, and also the text fully explaining them.

Dead furrows are a nuisance, especially where hoed crops are cultivated; and when land is stocked down for meadows, deep dead furrows make an uneven surface for the mowers and horse-rakes to work over. When a field is plowed in lands beginning on the outside, turning all the furrows outward, and finishing the plowing in the middle of the field, there will be a dead furrow from every corner to the middle dead furrow of each land, and a strip of ground eight or ten feet wide on one side of every dead furrow will be trodden down firmly by the teams when turning around. Plowing a field without dead furrows is simply commencing at the middle and turning the furrow slices all inward. If the plowing be done with a right-hand plow the teams will "gee around," always turning on the unplowed ground. When a field is plowed in this manner there are no ridges or dead furrows, and the surface is even, so that the operation of any machine is never hindered. When sod ground is plowed in lands there is always a strip of ground beneath the first two furrow slices at every ridge that is not broken up. This is to a great extent avoided when the whole field is plowed as one land, and may be entirely avoided if back-furrowed. The accompanying diagrams will show how to plow a square field, or one of irregular boundary, commencing in the middle and finishing at the outsides. Fig. 1 shows a rectangular field. The plowman finds a point equally distant from three sides, measuring of course at right angles to the sides, and sets a stake. Then he finds the point equally distant from the three sides at the other end, and sets another stake. From these two stakes to the corners of the field he turns two furrow slices together, and then plows the field, being guided by them, and occasionally measuring to the outside to see if he is keeping his furrows of equal width at setting in and running out, and on each side. In Fig. 2, a four-sided lot, where the angles are not right angles, precisely the same rule is followed. In the case of the triangular field, the plowman begins by plowing about a single point, which, though awkward at first, may be executed with ease after a few trials. In the case of the irregular five-sided lot, represented by Fig. 4, it is a little more difficult to start exactly right, but the ruling gives a clear idea of how the furrows run, and it is always well to pace off frequently to the outside of the lot—or rather from the fence starting at right angles to it—to be sure that the portion remaining unplowed on each side, and at each end of each side, remains always of a corresponding width as the plowing progresses.

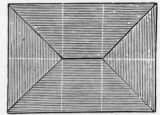

Fig. 1.

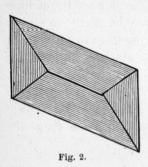

Fig. 2.

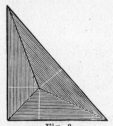

Fig. 3.

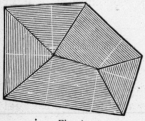

Fig. 4.

CHAPTER IV.

PRINCIPLES AND PRACTICE.

I. STUDY YOUR FARM.—II. ANALYSIS OF THE SOIL UNNECESSARY.—III. SOIL DOES NOT WEAR OUT.—IV. ORGANIC AND INORGANIC MATTER.—V. ECONOMY OF FERTILIZERS.—VI. PRACTICAL TEST OF FERTILITY.—VII. ROTATION AND CROPS.—VIII. A SIMPLE ROTATION.—IX. EFFECT OF BAD SEASONS.—X. ELABORATE ROTATION.—XI. GRASS-SEED AND MEADOWS.—XII. AN EASTERN MAN ON ROTATION.—XIII. A SOUTHERN PLANTER'S TESTIMONY.—XIV. ROTATION IN EUROPE.—XV. SUBSTITUTION IN ROTATION.—XVI. POTASH AND PHOSPHATE CROPS.—XVII. SOFT AND HARD GROUND CROPS.—XVIII. SCIENCE IN AGRICULTURE.—XIX. IGNORANCE VS. INTELLIGENCE.—XX. SOILS AND THEIR CAPABILITIES.—XXI. PERCENTAGE OF SAND IN SOILS.—XXII. ABSORBING POWER OF SOILS.—XXIII. ABSORPTION OF OXYGEN BY THE SOIL.

I. Study Your Farm.

THE best farmers are those who study the capabilities of their farms, with a view to the selection of the most remunerative crops. The first question to be decided in settling in new regions distant from markets, is, What crops will bear the farthest carriage without consuming their value? These are wheat, flax, and grass-seeds. And these crops, on new lands, are raised with the least outlay of labor. This sort of cultivation is, of course, ruinous to the land, and, if long persisted in, will certainly end in so reducing the fertility of the soil that even other crops cannot be profitably raised. The soil must not only possess all the elementary substances necessary to the production of a crop, but it must, to yield the greatest return, have all these elements in excess of the requirements. And if only one of these elements is lacking or deficient, the crop is subject to such changes as not only to cease to be profitable, but often to become impossible to be grown.

II. Analysis of the Soil Unnecessary.

It used to be said by visionaries, years ago, that an analysis of the soil would give true indications of what it was capable of, so that if the lacking constituents were supplied, success was certain. The difficulty in all this is, that while the analysis of a soil will show correctly the constituents of that particular cube, it never shows the capabilities of a field. Even if it did show the constituents of a field, these constituents might be so locked up as to be valueless, or if not absolutely locked up (that is, insoluble, or not in a state in which they can be assimilated by the plant), the mechanical nature of the soil might be such as to render them inoperative. The fertility of a soil may be locked up in its gravel, sand or clay. The soil might not be able to give up its constituents from saturation of the soil by water, or its adhesion may be too great. Other reasons might be given to prove that analysis of the soil is often of little value as showing its fertility. Those mentioned are sufficient, aside from the constant fact, that the analysis of a six-inch cube of soil can never, even approximately, show the agricultural value of a field.

III. Soil Does Not Wear Out.

SOILS do not wear out, as is generally supposed, from cropping. Soils vary in their capabilities of converting plant food as they do in their inherent qualities. A blowing sand lacks not only the inorganic elements of fertility, but is also incapable of arresting and holding the organic elements. And yet, sand is one of the most important constituents of the soil. The strongest arable clays contain from fifty to sixty per cent of sand, fertile loams sixty to seventy-five per cent of sand, and fertile sandy soils from seventy-five to eighty-five per cent of sand. As a rule, the more clay a soil contains, the greater its capability of taking and holding organic matter, and also the greater its composition of inorganic or mineral matter.

IV. Organic and Inorganic Matter.

THE inorganic matter of soils is that produced by the breaking-down of the rocks of which all soils are originally composed. As a rule, the surface soil contains more organic, and the subsoil more of inorganic matter, though this rule is often reversed. The organic matter of soils is that formed by the decay of vegetable matter from generation to generation in the soil. Both are necessary to a fertile soil, for if either is lacking, plants will not produce seed.

The organic matter produces woody fiber, starch, sugar, gum, gluten, albumen, etc. They may all be resolved into four elements: carbon, oxygen, hydrogen and nitrogen, and are essential as assisting in the formation of inorganic compounds.

The inorganic elements of soils are much more numerous, and it is important for the farmer to know those which are considered essential to the composition of all plants. They are given in tabular form as follows:

NAME.	FORMING.	IN COMBINATION WITH.
Chlorine,	Chlorides,	Metals.
Sulphur,	Sulphurets,	"
"	Sulphuric Acid,	Oxygen.
"	Sulphureted Hydrogen,	Hydrogen.
Phosphorus,	Phosphoric Acid,	Oxygen.
Potassium,	Potash,	"
"	Chloride of Potassium,	Chlorine.
Sodium,	Soda,	Oxygen.
"	Common Salt,	Chlorine.
Calcium,	Chloride of Lime,	"
"	Lime,	Oxygen.
Magnesium,	Magnesia,	"
Aluminum,	Alumina,	"
Silicon,	Silica,	"
Iron,	Oxide of Iron,	"
"	Sulphuret of Iron,	Sulphur.
Manganese,	Oxide of Manganese,	Oxygen.
"	Sulphuret of Manganese,	Sulphur.

Iodine is also another element which in combination with metals produces iodides of the metals; but this element is not considered essential. The most important of the above-named are sulphur, phosphorus, potassium, calcium, aluminum (the base of clay), silicon (the base of sand) and iron. Neither the organic nor the inorganic elements exist in plants in their simple state, but in combination with other substances, and to be taken up, must be in some form soluble in water.

A soil to be fertile, we have said, must contain an excess of organic and inorganic matter, in a state capable of being taken up by a plant and converted into sap. To be capable of producing all the crops cultivated on the farm, the soil must consist of three earths — sand, clay and lime — intimately mixed; four gases, forming organic matter; and eleven or twelve inorganic elements, chemically combined.

V. Economy of Fertilizers.

The results of fertility cannot be better illustrated than by the specimens of ripe wheat shown in Fig. 1 and Fig. 2. They are from the results of careful experiments in France. Fig. 1 shows wheat grown on good soil, without fertilizers; Fig. 2 (next page), wheat grown in the same soil, contiguous, the same year, and under the same conditions, except that the patch of No. 2 was enriched to the maximum with special fertilizers. The squares show the height and bulk of each. The product is given in the French terms, hectare — 2.471 acres (two and a half acres nearly) — and in kilogram — 2.2046 pounds (two and a quarter pounds nearly). The products were: No. 1, 2,640 kilograms of straw, and 903.11 kilograms of grain; No. 2, 6,931 kilograms of straw, and 3,790.46 kilograms of grain. Reduced to our measure, the fertilized portion produced 3,307 pounds of grain, or fifty-five (55) bushels and six (6) pounds per acre, and the unfertilized portion only thirteen bushels and six pounds per acre — a difference of nearly forty-two bushels per acre in favor of fertilization. Analogous cases might be multiplied indefinitely. It would seem to be unnecessary.

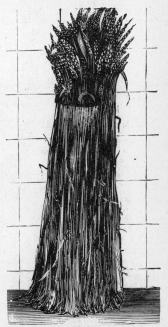

Fig. 1.
WHEAT, WITHOUT FERTILIZER.

VI. Practical Test of Fertility.

The practical test is that the soil shall uniformly produce plants from the seed sown that will make healthy and vigorous growth in average seasons, and ripen perfectly. If the soil will not do this, something is wrong with it. It either lacks the proper organic or inorganic elements, or their mechanical mixture is at fault. The farmer must find out what the trouble is. A well-drained sandy loam, or clayey loam, will produce all the crops generally raised on the farm, if the rotation is such that one kind of crop follows another kind at the proper intervals. And as no two species contain the same quantity or quality of organic or inorganic matter, it follows that the more widely plants differ in character, the better are they suited for rotation. Hence, turnips and wheat being widely different, and both naturally adapted to the climate of Great Britain, turnips are the great fallow, or cleaning, crop of that country. In the West, Indian corn is the great cleaning crop, and clover may be called our principal renovating crop.

VII. Rotation and Crops.

In Europe especially, and in the older portions of the United States in a more limited degree, an elaborate rotation, with liberal application of manures, is found necessary to bring back the soil to a state of full fertility and keep it so. In the West, and in some portions of the South, a more simple rotation, with or without manure, is practiced. In newly settled districts little attention is paid to rotation, and less to manuring, except by the more sagacious settlers. The farmers raise wheat and flax until the soil begins to show signs of exhaustion, and then alternate with corn, or else seed down the land for mowing and pasture, making corn the principal grain-crop, and thus naturally gliding into stock husbandry, in the place of grain husbandry. The better-informed acquire stock as quickly as possible, and before their soil refuses to raise wheat and flax. Those who do this early, make the most money; for thus all but a small portion of that taken from the soil may be returned to it. The soil simply loses the phosphates of the bones and the nitrogen of the flesh of the animals sold.

VIII. A Simple Rotation.

The rotation in mixed farming is of the simplest kind. One-quarter of the farm in small grain, three-eighths in corn, and three-eighths in pasture and meadow is a natural rotation. It is evident here that one-quarter of the pasture must be broken every year. It would be inconvenient, but let us see how this may

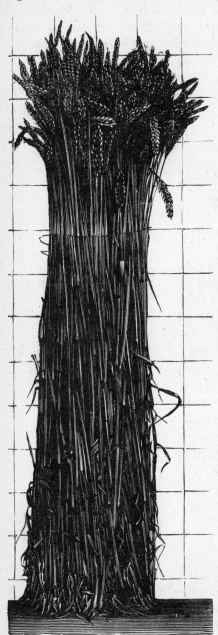

Fig. 2.
WHEAT, WITH FERTILIZER

be accomplished by dividing the farm into six fields. Take one hundred and sixty acres: A section of each field is shown below; the figures at the top show the fields:

	1	2	3	4	5	6
First Year,	Corn,	Wheat,	Corn,	Wheat,	Grass,	Grass.
Second Year,	Wheat,	Corn,	Wheat,	Corn,	Grass,	Grass.
Third Year,	Grass,	Wheat,	Corn,	Wheat,	Corn,	Grass.
Fourth Year,	Grass,	Corn,	Wheat,	Corn,	Wheat,	Grass.
Fifth Year,	Grass,	Grass,	Corn,	Wheat,	Corn,	Wheat.

This rotation will give a cleaning crop of corn the year before every seeding of wheat or grass, and every third year one field of grass is to be plowed. This may be called a three-course crop, and it will be seen that it will require a long time to bring the fields into their original order again. In the third year, field six will be field one, and the entire rotation will not be complete so the fields occupy their original place until the twelfth year, a far better plan than the usual hap-hazard plan generally adopted.

IX. Effect of Bad Seasons.

WHATEVER the rotation, whether simple or elaborate, an unfavorable season may frustrate the best-laid plans. Winter wheat is liable to be destroyed by freezing. Winter and spring wheat are both subject to the depredations of the fly and other insects, often ruining the crop. We have known the corn crop destroyed by rain and flood to such a degree in one of the great corn counties in Illinois (Livingston), that the feed of the farm teams was not made, and all through June, and into July, the land could not be entered on for re-planting. We have seen whole meadows in Northern Illinois (Cook county), destroyed by the white grub (larvæ of the May beetle), which eats the roots of meadow grass below the surface. We have seen, we repeat, such meadows, when the turf might be rolled up like a carpet. All these, and other contingencies, will interfere with the regular rotation, and often destroy the sequence. In such cases, the meadow must often be plowed up and lost entirely as a meadow, and cannot be recovered in less than two years. The loss of an annual crop, however, need not seriously interfere. The land may be fallowed, or some temporary crop put in for that year, or a fallow crop may be sown and plowed under. Hence the loss is light, and the regular crop of the next season may come in its turn. In no rotation can more than the general idea be followed.

X. Elaborate Rotation.

IN the foregoing, we have given the most simple rotation possible, as an example. Few farmers raise so simple a list of crops. Oats, barley and flax are generally raised. These all come under the same category as wheat. The cereal grains form one year's crop, as a whole, and may be divided up to suit; but flax should not follow the cereals, nor should one cereal follow another. Sorghum, potatoes, roots, and all that class of crops, must have a place in the rotation. They should be allotted to the corn land. They are cleaning crops. Hungarian grass, millet, and other special forage crops may encroach either upon the small grain crops, or corn. Flax, hemp, and other fiber crops are exhausting, and must also follow the cleaning or fallow crop.

No Rigid Rule.—The idea in all this, is not to lay down a system of elaborate rotation to be rigidly followed. This notion has long since been exploded. Every farmer must figure it out for himself, and select his own system as best adapted to his particular needs. But to reach the best success, a system must be adopted and adhered to, so far as possible. It is the want of system that costs money, or, what is the same thing, time. It is the knowledge as to the crops best adapted to a soil and climate, or the lack of that knowledge, that marks the successful farmer from the unsuccessful one. It is not the intention of this work to speculate upon what crops pay best, but to point out that which shall be of value to every reader to know, in the management of the farm.

XI. Grass-seed and Meadows.

In the simple rotation already described, there may seem too little meadow and pasture. It is one adapted to new countries where the corn raised is supplementary to the grass crop. This brings the feeding resources of the farm, in proportion to that of grain raised to sell, as three-fourths of the first to one-fourth of the latter. As stock increases, the pasturage may be increased. A seed crop of grass and clover may be taken the first year, but when this is to be done, the grass—timothy, blue-grass, red-top, orchard-grass, fowl-meadow, etc., must be sown separate. Clover, whatever the variety, must be sown separate, until stock can be obtained. Then seed crops may be made profitable, since the seed will bear transportation long distances, and still yield a profit. The seed crop taken, the aftermath may be plowed under, and the straw converted into rough fodder and manure. It will always pay to seed grass with any cereal crop, with a view to turning it under out of its rotation, and independently of the meadow and pasture in their regular rotation. The notion is, not only to prevent exhaustion and keep the soil rich, but to make it richer. This is the true secret in all cultivation.

XII. An Eastern Man on Rotation.

A SIMPLE and excellent rotation is given in one of the United States Agricultural Reports, as adopted by a gentleman in Vermont. This was on a one-hundred-acre farm, of which twenty acres were woodland. The farm had eight lots, of ten acres each. Labor was high, and since hay paid well, as much grass was raised as possible. In going through these eight fields in eight years, one ten-acre field would be in corn or roots; the second year in wheat, barley, oats, or some other grain-crop seeded to grass; the next two years mowed for hay, and the next four years in pasture, and about equally divided for the keeping of stock summer and winter, the owner to feed all the crops on his farm. By this system of rotation and feeding all the produce on the farm, it was estimated he could keep two-thirds more stock than the majority of farmers, and the land would be all under cultivation. He estimated ten acres in hoed crops, ten acres in grain, and the roots at three hundred bushels of potatoes, or one thousand bushels of rutabagas or beets per acre, besides four or five hundred bushels of grain annually.

XIII. A Southern Planter's Testimony.

Some years ago, a wide-awake Southern planter gave his plan of rotation, adapted to the cotton region, the farm containing five hundred acres of open land under fence, two hundred and fifty acres being devoted to arable purposes, and the rest to grazing. The writer held that the rotation might be as follows: 1, Cotton and corn in the same field in suitable proportions. 2, Oats sown in August on the cotton and corn land. 3, Rye, or rye and wheat, sown in September, the land having been twice plowed in order to kill the germinant oats. 4 and 5, Clover, if the land is in sufficient heart to produce it; if not, the fourth year rest ungrazed, and the fifth year sheep and cattle penned upon it every night during the year, using a portable fence. An ordinary farm of five hundred acres, it was held, would support five hundred sheep, besides the crops in the above rotation. The oats and rye should feed them during the winter nearly or entirely, without injury to the grain. Five hands would be sufficient to work such a farm and take care of the live stock. During the first year, the following results might be expected from an ordinary farm, without manure:

25 acres in cotton, 12 bales, at 15 cents,	$ 900
25 acres in corn, 250 bushels, at $1,	250
50 acres in oats, 500 bushels, at 80 cents,	400
25 acres in rye, 200 bushels, at $1,	200
25 acres in wheat, 150 bushels, at $1.50,	225
Increase and mutton sales of 500 sheep,	500
Wool, 3 pounds per head, at 33 cents per pound,	495
Manure, at $1 per head,	500
	$3,470

The Second Year.—This gives an average of six hundred dollars per hand for the first year, fully three times the average per hand in the Cotton States then. The next year the writer holds that the cotton and corn would be more than double by penning five hundred sheep at night on fifty acres, and says that ten sheep regularly penned will manure well one acre in a year. Five hundred would, therefore, manure well fifty acres. He acknowledges that the appearance of the ground would not indicate this high manuring; but says it should be remembered that the liquid manure, which is equal in value to the solid, is not visible. If, in addition, a stock of cattle were kept and penned on the same fifty acres, then fertility would be increased in proportion.

The experience of the last few years of those in the South who have applied themselves to diversified crops, where stock forms a prominent feature, shows that this is not overdrawn, and that the enrichment of worn farms by natural means is no more difficult there than in other sections of the country.

XIV. Rotation in Europe.

In Great Britain, and on the continent, the most elaborate systems of rotation and manuring are followed. Careful cultivation and liquid manure have made Flanders the agricultural model of many writers. In France the system of culture for field crops is equally elaborate and careful. The table below will give a rotation proposed in the early part of the century by order of the French government, and will be interesting as showing the diversity of crops then cultivated. It is given simply as showing how a diversity of crops may be cultivated in localities that furnish a suitable market, as, for instance, the neighborhood of cities. Vetches are the only crop in the list not used in American agriculture. They may be substituted by any fallow crop, to be cut green. This rotation is as follows:

First Year. Acres.	Second Year. Acres.	Third Year. Acres.	Fourth Year. Acres.	Fifth Year. Acres.	Sixth Year. Acres.	Seventh Year. Acres.
30 Wheat	5 Turnips 5 Cabbages 2½ Field Beet 2½ Carrots	10 Oats 5 Barley	15 Clover	15 Wheat	10 Potatoes 3 Vetches 2 Beans	30 Wheat.
15 Clover	10 Potatoes 3 Vetches 2 Beans	15 Wheat	5 Turnips 5 Cabbages 2½ Field Beet 2½ Carrots	10 Oats 5 Barley	15 Clover	15 Clover.
5 Turnips 5 Cabbages 2½ Field Beet 2½ Carrots	10 Oats 5 Barley	15 Clover	30 Wheat	5 Turnips 5 Cabbages 2½ Field Beet 2½ Carrots	10 Oats 5 Barley	5 Turnips. 5 Cabbages. 2½ Field Beet. 2½ Carrots.
10 Oats 5 Barley	15 Clover	15 Wheat	10 Potatoes 3 Vetches 2 Beans	10 Potatoes 3 Vetches 2 Beans 15 Wheat	5 Turnips 5 Cabbages 2½ Field Beet 2½ Carrots	10 Oats. 5 Barley.
10 Potatoes 3 Vetches 2 Beans	15 Wheat	5 Turnips 5 Cabbages 2½ Field Beet 2½ Carrots	10 Oats 5 Barley	15 Clover	15 Wheat	10 Potatoes. 3 Vetches. 2 Beans.

XV. Substitution in Rotation.

However simple a rotation may be, or however elaborate, the crops must be selected with judgment, that is, with a view to the money they will bring. This is what marks the successful from the unsuccessful farmer. Nor need the rotation be a cast-iron one. A rotation, to be perfect, must be flexible—capable of having one crop substituted for another. But the general character must be stuck to; and the main idea, enrichment of the soil and paying crops, must always be kept in mind. The more stock that can be fed, the easier will this quality of enrichment be retained. Hence the value of grass and clover in the rotation.

Ten Ideals.—Ten ideals in agriculture may be stated as follows:
 1.—Grass, hay and corn make fat cattle.
 2.—Cattle furnish manure.
 3.—Manure ripens heavy crops of grain.
 4.—A judicious rotation gives the greatest average yield.
 5.—Cleaning crops prevent the growth of weeds.
 6.—Weedy crops never gave large yields.
 7.—Barren fields leave empty granaries.
 8.—Good tillage is manuring with brains.
 9.—Follow potash crops with phosphorous crops.
 10.—And let the master's eye be vigilant over all that concerns the farm.

XVI. Potash and Phosphate Crops.

The crops that do well in a soil containing potash are corn, the grasses, clover and potatoes. Rye, barley, oats, and nearly all the garden vegetables, also, are fairly natural to such a soil. Wheat, sorghum, sugar-cane, the beet, where cultivated for sugar, and in fact all sugar crops, require a soil rich in phosphates. Barley, oats and flax like a soil rich in both phosphate and potash, and the same is true of the grasses and clover. Potatoes require phosphate for their tops and seed, and potash for their tubers. Sweet potatoes require plenty of phosphate. Soils rich in potash are generally rich, also, in humus or vegetable matter, and, if they also abound in phosphate, will produce large crops of all the cereals, including winter wheat. But such soils are usually liable to heave; drainage will in a great measure obviate this.

XVII. Soft and Hard Ground Crops.

There are crops that require hard or firm soils, and others that must have soft soils, to give the best results. Among the crops natural to soft, or humus soils, are corn, Irish potatoes, flax, hemp, most textile crops, and most garden crops. The crops natural to non-humus soils, or firm soils, are wheat especially, the cereal grains generally, grass, clover, sweet potatoes, sorghum and all saccharine crops, including sugar-cane, melons and squashes, and among garden plants, flat turnips and onions. Hence the necessity of compacting soils rich in potash, and also those rich in the phosphates, on which such crops are grown. The nearest approach to bringing such soils into proper condition for these crops, is to cultivate upon a newly turned sward or upon a second year's plowing of sward. All such soils are assisted by tramping, rolling, etc., before sowing, only requiring a shallow harrowing on top. Sometimes it is difficult to get grass-seed to take on these light, fluffy soils; but once growing, it soon becomes firm and luxuriant. The prairie soils of the West, as a rule, are rich in potash, and fairly so in the phosphates. The difficulty with wheat on such soils is heaving out, rust and lodging. They are better adapted to spring wheat than to winter wheat. Timber soils and all loess (sandy-alluvial) soils are adapted to winter wheat. As we go West, after crossing the Mississippi, the soil becomes better and better adapted to the cultivation of this crop.

XVIII. Science in Agriculture.

The practical farmer has little or nothing to do with abstruse science. And yet, science has done much for agriculture within the last fifty years. Chemistry especially has lent its aid in all directions, even in mechanical appliances. It has been called the corner-stone of agriculture, and in fact it is, since the growth of all crops is due largely to chemical action. The natural sciences are largely connected with agriculture. Botany teaches the structure and physiology of plants. Entomology treats of insects beneficial and injurious to crops. Zoölogy describes animals valuable or destructive. Mechanical science has perfected implements of tillage and use. Veterinary science has rendered it possible for any intelligent farmer to treat the ordinary diseases of domestic animals without the aid of the professional veterinarian. Such works as the CYCLOPEDIA OF LIVE STOCK AND COMPLETE STOCK DOCTOR give minute directions concerning all the animals of the farm, and most diseases incident to farm stock, and this one is so fully illustrated that everything is presented in the clearest manner.

Manures requisite for any crop may now be bought or made on the farm. The soils adapted to particular crops may now be easily and accurately studied. Agricultural invention has fairly kept pace with other inventive progress within the last ten years, and they who read and reflect have profited thereby. Agriculture can no more stand still than can any other art. The conditions requisite to success are constantly changing, and must ever change. The middle-aged man who would be content with the agriculture of his father, must go back to the heavy eyeletted hoe, the scythe and reaping-hook, the old team-killing plow, the one-horse plow for cultivaitng, the flail or tramping-floor for threshing, and a spade-edge over a half-bushel for shelling corn.

XIX. Ignorance vs. Intelligence.

They who do not believe in books, and in improved agriculture,—and there are not a few such,—are toiling from twelve to fifteen hours a day to scratch a hard-earned pittance from an unwilling soil, while their better-informed neighbors are working less hours, reading more, using improved seed, implements and processes, and gaining a competence. Not by studying books a quarter of a century or a hundred years old, not works of theory and dry detail, but paragraphs and condensed and illustrated reading, that give ideas to be elaborated and made to fit, by each individual, his own particular wants. In other words, the application of agricultural truths, new and old, to the every-day labors of the farm.

XX. Soils and their Capabilities.

Among other things, every farmer ought to understand soils, their characteristics and capabilities. We are, for instance, in the habit of using the terms, "light" and "heavy" soils, just as we are in the habit of saying the air is "heavy," the air is "light," etc. When we say the air is "light," it is probably heavy, but it is bracing.

When we say it is heavy, it really is light. That is, the pressure on the barometer is not heavy. We say a soil is "light" because it is sandy and easily worked; heavy, when it is clayey; "hard," when it is dry, and "sticky" when it is wet. Better terms would be, "open" and "close." Only humus soils are really light. Sandy soils are the heaviest in gravity, and clay soils are lighter in weight than any others except those wholly or largely composed of vegetable matter. The following table from Schubler will be valuable. The first column shows the kind of soil, the second the weight per cubic foot, and the third the weight of one foot in depth per acre:

KIND OF SOIL.	WEIGHT PER CUBIC FOOT.	WEIGHT ONE FOOT DEEP PER ACRE.
Dry silicious or calcareous sand,	about 110 pounds,	6,792,000 pounds.
*Half sand, half clay,	" 96 "	4,182,000 "
†Common arable soil,	" 80 to 90 "	3,485,000 to 3,920,000 "
Stiff clay,	" 75 "	3,267,000 "
‡Garden mold, rich in vegetable matter,	" 70 "	3,049,000 "
Peat earth,	" 30 to 50 "	1,307,000 to 2,178,000 "

XXI. Per Cent of Sand in Soils.

THE composition of soils is important. They are designated as light, heavy, warm, cold, dry, wet, compact, porous, fine, coarse, hungry, leachy, loamy, sour, sweet, clayey, sandy, limy, marly. In fact, no two soils are precisely alike, and each acre of a field may differ essentially from the rest. Common sand, flint sand, alumina, lime, magnesia, potash, and various salts and metalloid compounds unite in various combinations to make up these soils. The humus, which gives richness and blackness of color, is chiefly derived from successive growths and decays of the vegetation for untold generations. The following statement shows the percentages of sand and clay, from pure clay (alumina and sand) to humus and peaty soils.

1.—Pure clay or pipe clay is sixty per cent silica and forty per cent alumina.

2.—Strongest clay soil (brick clay), pure clay, with five to ten per cent of sand that can be separated by washing.

3.—Clay loam, pure clay, with fifteen to thirty per cent of sand.

4.—Loamy clay, pure clay, with thirty to sixty per cent of sand added.

5.—Sandy loam, pure clay, with sixty to ninety per cent of sand.

6.—Sandy soil contains ten per cent or less of clay.

7.—Marly soils, from five to twenty per cent of marl, by weight, of the dry soil.

8.—Calcareous soils, twenty per cent or more of lime.

9.—Humus soils, from five to fifteen per cent of vegetable matter.

10.—Peat soils have sixty per cent or more of vegetable or organic matter

11.—Gravelly soil, in which gravel is a distinct constituent.

12.—Rocky soil, in which ledges appear, or which consists largely of boulders or other rock.

---

* This soil would correspond to what we call a sandy loam.

† This soil would correspond to what we call a clay loam.

‡ This soil would correspond to our strong, rich prairie soils.

XXII. Absorbing Power of Soils.

The power of a soil to absorb water indicates its quality, since a soil that will absorb water and hold it, is generally fertile. Schubler presents this absorbing power as follows, for the soils named:

KINDS OF EARTH.	1,000 GRAINS OF EARTH ON A SURFACE OF FIFTY SQUARE INCHES ABSORBED IN			
	12 HOURS.	24 HOURS.	48 HOURS.	72 HOURS.
	Grains.	Grains.	Grains.	Grains.
Silicious Sand,	0 Water.	0 Water.	0 Water.	0 Water.
Arable Soil,	16 "	22 "	23 "	23 "
Sandy Clay,	21 "	26 "	28 "	28 "
Loamy Clay,	25 "	30 "	34 "	32 "
Brick Clay,	30 "	36 "	40 "	41 "
Gray Pure Clay,	37 "	42 "	48 "	49 "
Garden Mold,	38 "	45 "	50 "	52 "
Humus,	80 "	97 "	110 "	120 "

Thus we see why dry pure sands are hungry soils. They cannot hold moisture, nor the soluble portions of manure, which are the only fertilizing elements.

XXIII. Absorption of Oxygen by the Soil.

GRAINS.	KIND OF EARTH.	CUBIC INCHES ABSORBED.
1,000	Silicious Sand, in a wet state, absorbed oxygen,	0.24
1,000	Sandy Clay,	1.39
1,000	Loamy Clay,	1.65
1,000	Brick Clay,	2.04
1,000	Gray Pure Clay,	2.29
1,000	Garden Mold,	2.60
1,000	Arable Soil,	2.43
1,000	Humus,	3.04

If the oxygen of the air is absorbed, as shown by this table, the carbon, hydrogen and nitrogen of the air may also enter into combination if the conditions are right. The table from Schubler will show the power of one thousand grains of soil for absorbing oxygen, from fifteen cubic inches of air, containing twenty-one per cent of oxygen.

Part II.

PRACTICAL AND SYSTEMATIC HUSBANDRY.

CEREAL CROPS AND THEIR CULTIVATION.

GRASSES, FODDER AND ROOT CROPS.

SILK CULTURE—SPECIAL CROPS.

CROPS FOR SUGAR MAKING.

VARIETIES ILLUSTRATED AND COMPARED.

PRACTICAL AND SYSTEMATIC HUSBANDRY.

CHAPTER I.

CEREALS AND THEIR CULTIVATION.

I. THE CEREALS DESCRIBED.—II. WHEAT AND CORN BELTS.—III. CORN IN THE UNITED STATES.—IV. DIFFERENT KINDS OF WHEAT.—V. VARIATIONS ILLUSTRATED.—VI. PROPER WHEAT SOILS.—VII. PREPARING THE SOIL.—VIII. DRILLING GIVES THE BEST RESULTS.—IX. DEPTH OF COVERING FOR WHEAT.—X. TIME TO SEED AND HARVEST.—XI. HARVESTING WHEAT.—XII. HOW TO SHOCK THE GRAIN.—XIII. IMPORTANCE OF GOOD SEED —XIV. PEDIGREE GRAIN.—XV. GENERAL CONCLUSIONS.—XVI. ARTIFICIAL CROSS FERTILIZATION.—XVII. REPUTABLE OLD VARIETIES IN THE UNITED STATES.—XVIII. RYE AND ITS CULTIVATION.—XIX. BARLEY AND ITS CULTIVATION.—XX. TIME FOR SOWING BARLEY.—XXI. HARVESTING AND THRESHING BARLEY.—XXII. NEW VARIETIES OF BARLEY.—XXIII. OATS AND THEIR CULTIVATION.—XXIV. EXPORT OF FOOD CROPS.—XXV. SPECIES OF OATS—THEIR LATITUDE.—XXVI. SOIL AND CULTIVATION OF OATS—XXVII. HARVESTING AND THRESHING OATS.—XXVIII. VARIETIES OF OATS TO BE CULTIVATED.—XXIX. BUCKWHEAT.—XXX. SEEDING AND HARVESTING BUCKWHEAT.

I. The Cereals Described.

THE cereals are the edible seeds of the grasses, or those cultivated for food. In the American usage, the cereals include wheat, rye, Indian corn, rice, barley and oats. In its broader sense, the word also includes sorghum, doura corn, some varieties of millet which are used as food by oriental nations and tribes, besides the seeds of the bene-plant (*sesamum*), a grain from which oil is expressed. The seeds of the bene-plant are eaten by some tribes, and were once used to a limited extent for food by the negroes in the South. In this work we shall not have occasion to notice any of the cereals except wheat, Indian corn, rye, barley, oats, buckwheat, rice and millet. Of these, wheat, rice and Indian corn are the most important food-plants of the world. In the United States, Indian corn is the most important food-crop, if we take into consideration its use for stock; wheat coming second. Of the food-crops of the world, as a whole, wheat stands first, rice second, Indian corn third, rye fourth; buckwheat, oats and barley coming last among civilized nations. Oats are coming more into use year by year as a staple article of food, in the shape of grits and meal, and are among the most nutritious of the cereals. Barley is becoming more important every year, being the chief ingredient in the manufacture of beer. All the cereals produce alcohol, by fermentation and distillation, but Indian corn is the great staple, and rye the next, for this purpose. For the manufacture of grape-sugar, or glucose (a saccharine product about forty per cent of the strength of cane sugar), Indian corn has within the last few years assumed great importance, and now employs immense capital in its production.

II. Wheat and Corn Belts.

WITH reference to wheat and Indian corn, the United States may be divided into three great belts,—the Atlantic, the Central, and the trans-Mississippi. The Atlantic belt produces about nineteen per cent of the wheat and corn; the Central belt, forty per cent, and the trans-Mississippi about forty-one per cent. In 1850 the percentage stood: Atlantic belt about fifty-one per cent., the Central belt forty-three, and the trans-Mississippi, six per cent of the whole crop—a wonderful exhibit of Western growth, which also forcibly shows how exhausted fertility of the soil and natural causes operate to change relative importance and values of crops cultivated.

III. Corn in the United States.

AN idea of the great importance and value of the corn crop of the United States is given by the immense crop, averaging since 1878 about 1,500,000,000 bushels a year, and this notwithstanding the crop failure of 1882. The following table, prepared by the Department of Agriculture, gives in a compact form all the facts about the corn crop of the United States for a period of sixteen years, during which time the production increased more than threefold. Since 1878 the quantity raised and the percentage exported have steadily increased.

YEARS.	ACREAGE.	YIELD PER ACRE.	TOTAL PRODUCT.	PRICE PER BUSHEL.	TOTAL VALUE OF PRODUCT.	TOTAL VALUE PER ACRE.	Corn and Corn-Meal exported in the fiscal year closing June 30, following.	Proportion of Crops exported.
		Bush.	Bushels.				Bushels.	P. ct.
1863	15,312,441	25.98	397,839,212	$0.69.9	$278,089,609	$18 16	5,146,192	1.29
1864	17,438,752	30.42	530,451,403	99.5	527,718,183	30 26	3,610,402	.68
1865	18,990,180	37.09	704,427,853	46.0	324,168,698	17 07	14,465,751	2.05
1866	33,306,538	25.30	867,946,295	68.2	591,666,295	17 21	16,026,947	1.85
1867	32,520,249	23.63	768,320,000	79.5	610,948,390	18 49	12,493,522	1.62
1868	34,887,246	25.9	906,527,000	62.8	569,512,460	16 32	8,286,665	.91
1869	37,103,245	23.5	874,320,000	75.3	658,532,700	17 74	2,140,487	.24
1870	38,646,977	28.3	1,094,255,000	54.9	601,839,030	15 57	10,676,873	.98
1871	34,091,137	29.1	991,898,000	48.2	478,275,900	14 02	35,727,010	3.60
1872	35,526,836	30.7	1,092,719,000	39.8	435,149,290	12 24	40,154,274	3.68
1873	39,197,148	23.8	932,274,000	48.0	447,183,020	11 41	35,985,834	3.86
1874	41,036,918	20.7	850,148,500	64.7	550,043,080	13 40	30,025,036	3.53
1875	44,841,371	29.4	1,321,069,000	42.0	555,445,930	12 38	50,910,532	3.85
1876	49,033,364	26.1	1,283,827,000	37.0	475,491,210	9 69	72,652,611	5.66
1877	50,369,113	26.6	1,342,518,000	35.8	480,643,400	9 54	87,172,110	6.59
1878	51,585,000	26.9	1,388,218,750	31.8	441,153,405	8 55	. . . . .	. .
Average of whole period,	35,930,407	26.7	959,174,938	52.3	501,616,287	13 96	. . . . .	. .
Average 1863–'70.	28,650,703	26.8	768,010,845	67.7	520,309,421	18 16	. . . . .	. .
Average 1871–'78.	43,210,111	26.6	1,150,341,531	42.0	482,923,154	11 18	. . . . .	. .

Specially noticeable is the rapid increase in the corn product west of the Mississippi River and in the Southern States. West of the great river, the settlement of new lands is rapid, and in the South the notion is constantly gaining ground, that it is cheaper to raise corn than to buy it. The distribution of the corn crop is shown by the following table, except that the acreage has greatly increased in the South and in the country west of the Mississippi since 1878:

STATES.	CORN CROP 1878.			STATES.	CORN CROP 1878.		
	BUSHELS.	ACRES.	VALUE.		BUSHELS.	ACRES.	VALUE.
Maine,	2,180,000	54,500	$ 1,417,000	Tennessee,	37,422,700	1,939,000	$15,343,307
New Hampshire,	2,207,400	56,600	1,346,514	West Virginia,	10,118,400	372,000	4,249,728
Vermont,	2,275,500	55,500	1,319,790	Kentucky,	45,922,100	2,023,000	18,368,840
Massachusetts,	1,260,000	35,000	781,200	Ohio,	108,643,700	3,113,000	35,852,421
Rhode Island,	268,800	8,400	142,464	Michigan,	31,247,700	835,500	11,874,126
Connecticut,	2,220,000	75,000	1,376,400	Indiana,	138,252,000	4,215,000	37,328,040
New York,	25,020,000	695,000	12,510,000	Illinois,	225,932,700	8,337,000	56,483,175
New Jersey,	9,792,000	272,000	4,406,400	Wisconsin,	36,900,000	984,000	10,701,000
Pennsylvania,	44,065,000	1,259,000	21,151,200	Minnesota,	17,106,900	449,000	4,961,001
Delaware,	4,500,000	180,000	1,755,000	Iowa,	175,256,400	4,686,000	28,041,024
Maryland,	11,209,500	477,000	5,044,275	Missouri,	93,062,400	3,552,000	24,196,224
Virginia,	18,200,400	1,040,000	7,826,000	Kansas,	81,563,000	2,406,000	15,497,046
North Carolina,	22,603,200	1,662,000	10,171,440	Nebraska,	54,222,000	1,291,000	8,675,520
South Carolina,	12,276,000	1,320,000	6,629,040	California,	3,467,250	100,500	2,080,350
Georgia,	24,398,000	2,218,000	14,882,780	Oregon,	166,500	5,000	153,180
Florida,	2,124,000	236,000	1,550,520	Nevada, Colorado and the Territories,	2,670,000	89,000	1,602,000
Alabama,	23,928,000	1,994,000	14,117,520				
Mississippi,	19,474,000	1,498,000	12,463,360				
Louisiana,	16,875,200	848,000	10,125,120				
Texas,	58,396,000	2,246,000	25,694,240	Total,	1,388,218,750	51,585,000	$441,153,405
Arkansas,	22,992,000	958,000	11,036,160				

IV. Different Kinds of Wheat.

THE many varieties of wheat cultivated may be divided into two principal classes: hard wheats and soft wheats. The hard wheats are natives of warm or semi-tropical climates, and the soft wheats of cold climates. These are true wheats—that is, the seeds are not attached to the chaff. An inferior variety, but very hardy, is spelt wheat, also a hard wheat, but with the seed adhering to the chaff like barley. Another division is into bearded and smooth wheat, and still another, into red and white wheat. Polish wheat resembles rye; it is a hard wheat. St. Peter's corn, or one-grained wheat, is a variety in which the seeds adhere to the chaff, a whitish-seeded, flinty wheat, which makes a sweetish bread, and is sparingly raised in some portions of Southern Europe. Another variety, Emmer or Amel corn, is raised in the Alpine valleys; it is a vigorous, hardy and productive variety, used for bread, for cattle, and for making starch. The seeds are broadly furrowed, pointed at both ends, the upper end woolly, and the color grayish red, and very glassy.

Favorite Varieties.—The varieties of wheat are so numerous, and so many new ones are coming to the front every year, that a list of them would be of little value. The practical farmer must experiment in a small way for himself before adopting any new variety. Of spring wheats, club varieties have long been noted for early ripening on high, dry lands. Fife wheat does well on moister rich soils. Both these varieties are beardless. Of winter varieties, the May or amber wheats have the best general reputation. The only true way is to experiment each for one's-self, or accept the experience of those whose location and soil are similar.

V. Variations Illustrated.

To show variation in wheat, we give a series of four cuts, two of bearded and two of smooth or beardless wheat. Fig. 1 represents strongly bearded wheat (Mediterranean hybrid); Fig. 2, lightly bearded wheat (black bearded centennial; Fig. 3, a bald white wheat (Clauson), and Fig. 4, a red bald wheat, red blue stem.

These are given, not with a view of showing varieties, but to illustrate some principal variations. [See next page.]

VI. Proper Wheat Soils.

THE best soils for winter wheat are those that are compact, and not liable to shrink and swell (heave) in freezing and thawing weather—soils rich in phosphates, lime and potash. The same soils suit spring wheat, except that spring wheat may be raised on soils that do heave somewhat. Very soft (fluffy) soils containing large amounts of humus are not at all adapted to wheat, since all such soils are liable to rust, mildew and smut, especially in moist seasons. If a soil is wet, it may be improved by under-draining. If it is a rich humus, as much of the prairie land east of the Mississippi is, it is worth more for other crops than for wheat. Well-drained sandy loams are the best wheat soils, since these lands are compact, and generally rich. The best wheat soils of the West and South lie in the undulating regions, and on the plains of Minnesota, Dakota, Nebraska, Kansas, and in the valleys of the Rocky Mountains and the Pacific slope.

VII. Preparing the Soil.

IN new soils, wheat may be sown among corn, in the latter part of August or in September. Or the corn may be cut and shocked, the land plowed, and the seed drilled in, in September in the North, and later in the South. On soils more worn, wheat may follow in the rotation, either on a clover by being turned five or six inches deep, or it may follow the seed crop of corn. Where sown on fallow land, the soil may be plowed three times, once as early in the spring as possible, and deep. Upon this at the proper time sow some crop to be plowed under. The last plowing should be shallow, only sufficient for the tilth, and not deep enough to disturb the crop turned under. After turning the fallow crop under, roll the ground to compact it thoroughly. The more solid the under-soil, and the better the surface tilth, the greater the probability of a crop. If, from any cause, the wheat is destroyed, spring wheat may be drilled in or sown broadcast, unless it be frozen after the wheat has begun to "shoot" in the spring. If so, there may yet be time

Fig. 1. MEDITERRANEAN HYBRID. Fig. 2. BLACK BEARDED. Fig. 3. CLAUSON WHEAT. Fig. 4. RED BLUE STEM.

to plant to corn or to sow to some of the annual grasses—Hungarian or millet—for forage.

VIII. Drilling Gives the Best Results.

THE quantity of seed must be determined by the nature of the soil and its conditions. As a rule, poor land requires the most seed, since the plant does not tiller so readily. Two bushels is heavy seeding broadcast; one and a half bushels per acre is the usual quantity. If the seed is drilled, one-quarter less may be used. Almost every one who has carefully noticed results, will admit that drilling the seed in gives the best crops. There is no use in going into an argument to prove this. Careful scientific experiment, as well as the experience of practical farmers in every part of the country, shows that drilling effects a saving of from ten to thirty per cent in seed, and gives an increase of five to twenty per cent in the crop gathered.

IX. Depth of Covering for Wheat.

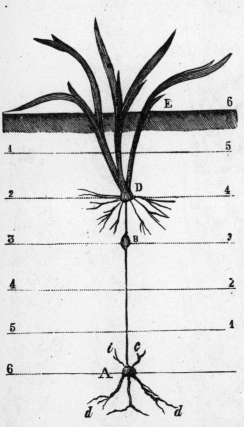

WHEAT PLANTED AT DIFFERENT DEPTHS.

HERE we give an object-lesson illustrating germination when planted too deep. The cut shows that when planted three inches deep the plumule throws out roots at a depth of two inches below the surface. In making that extra growth of one inch, the constitutional vigor of the plant is impaired, and this loss of vigor is in proportion to the increased depth below two inches from the surface, until planted at six inches in depth, the vitality of the seed is exhausted in reaching the surface.

The same progression will apply to all seeds. Of barley sown twelve inches deep not a grain germinated. As a rule, the smaller the seed the less should be its depth of covering. For any one who has not experience with any particular seed, four times the diameter of the seed is a good depth to insure prompt germination in soil of average texture, always supposing that the earth is in full contact with the seed, and that moisture is present. The cereal grains do not require a high temperature for germination; between forty-two and fifty degrees is about right.

The Proper Depth.—From one to two inches is the proper depth for wheat—deepest in loose soils. The earth should be firmly pressed about the seed. The

CEREALS AND THEIR CULTIVATION.

observation of intelligent farmers shows this, and the experiments of Petri prove it. Here are the results of his experiments, one inch being the best depth for germination. The seed sown being of a given quantity in each case:

SEED SOWN TO THE DEPTH OF—	CAME ABOVE GROUND IN—	PROPORTION OF PLANTS WHICH CAME UP.
Half inch	Eleven days	Seven-eighths.
One inch	Twelve days	All.
Two inches	Eighteen days	Seven-eighths.
Three inches	Twenty days	Three-fourths.
Four inches	Twenty-one days	One-half.
Five inches	Twenty-two days	Three-eighths.
Six inches	Twenty-three days	One-eighth.

X. Time to Seed and Harvest.

THE average time to seed and harvest wheat in different parts of the United States is given in the following table. It is compiled by the Department of Agriculture from answers from the different States, and also gives the average quantity of seed and the best wheat soils for the localities named:

STATES.	TIME OF SOWING.	AVERAGE BUSHELS OF SEED PER ACRE.	TIME OF HARVEST.	BEST SOIL.
Maine,	May 15 to June,	1½ bush.,	August 20 to 30,	Sward corn stubble; high ridges; dry pasture.
New Hampshire,	April to May 20,	1½ to 2 bush.,	August 1 to 20,	Clay loam; new upland; diluvial; black loam.
Massachusetts,	April 10 to 25,	1¼ to 2 bush.,	June 25 to Aug. 10,	
Vermont,	May 1 to Sept.,	2 to 2½ bush.,	Last Aug. to Sept. 1,	Loam clay: clay loam.
New York,	May 10 to Sept 1.,	1¼ to 2 bush.,	July 2 to Aug. 10,	Sandy loam; clay loam; loam mixed with gravel.
New Jersey,	Sept. 1 to Oct. 15,	1⅓ to 2 bush.,	June 28 to July 7,	Friable loam; loam; clay loam; sandy loam, rather stiff.
Pennsylvania,	Sept. 1 to Oct. 15,	1¼ to 2 bush.,	June 15 to July 15,	Light sandy; clay soil; sandy loam; limestone; do. clay, mixed with gravel; clay; do.; clay and gravel.
Delaware,		1⅙ to 2 bush.,		
Maryland,	October,	1½ bush.,	June,	Rich loam; clay.
Virginia,	Sept. 15 to Nov. 30,	1 to 2 bush.,	June 15 to July 15,	Clay; do. do.; clay and lime.
South Carolina,	Oct. and Nov.,	50 lbs.,	June 1,	Clay.
Georgia,	Sept. 15 to Nov.,	¾ to 1 bush.,	June 1,	Red mulatto.
Alabama,	Sept. to Dec.,	½ to 2 bush.,	June to July,	Loam; oak and hickory.
Tennessee,	Oct. 12,	1 to 1½ bush.,	June 15,	Dark loam; all kinds.
Kentucky,	Sept. and Oct.,	75 lbs.,	July,	Clay.
Ohio,	Sept. 1 to Oct. 25,	1 to 1½ bush.,	June 28 to July 20,	Oak and maple land; clay; do.; very warm; limestone; clay loam; yellow clay; clay; sandy.
Indiana,	Sept. to Oct ,	1 to 2 bush.,	June 15 to July 20,	Sand and loam; clay loam; clay; improved clay; loam do.; clay; sandy loam.
Illinois,	Aug. to Sept. 30,	1 to 1½ bush.,	May to July 1,	Sandy loam; clay; oat or clover stubble; clover; rich loam.
Michigan,	Sept. 3 to Oct. 1,	1¼ to 1½ bush.,	June to July 30,	Marl clay; clay and sand; oak; clay loam.
Iowa,	Aug. 20 to Sept. 15,	90 lbs. to 1½ bu.,	July 5 to 20,	
Texas,	Oct. 1 to Dec. 15,	¾ bush.,	May 1 to June 10,	Lime soil.

Cultivating.—There can be very little cultivation of the growing wheat under our system of tillage. We must have a much denser population, great division of farms, and very much cheaper labor first. A light harrowing in the spring if the ground becomes crusted, or a rolling if the land heaves, is almost all that can be done. Hence, the advantage of clean land and a thorough preparation of the soil. No crop that is largely composed of weeds ever yet paid the cultivator.

XI. Harvesting Wheat.

The proper time for harvesting wheat is still open to discussion. The miller insists that wheat cut just in the dough state, and carefully cured, makes the most and the best flour per bushel. There is no doubt of it, but it costs more to harvest the crop, and the yield per acre is not so large. Will the miller, or rather the buyer in the market, pay more for wheat harvested in this condition than for wheat harvested when fully ripe? No! Then, the farmer will let his crop stand until it is so nearly ripe that there will be no difficulty in curing it in the shock and stack. In this, the farmer must so calculate, that the whole harvest shall be cut before the grain crinkles down or shells from the head. If harvest facilities are not just what you desire, cut the first of the harvest rather green; the grain is good for milling, that is, it will ripen for milling, as soon as the grain, squeezed between the fingers, shows a fairly firm, pasty consistency. But that intended for seed should not only be fully ripe; it must be taken from the very best part of the field and stacked separately from the rest.

XII. How to Shock the Grain.

The importance of careful shocking is almost always underestimated. Whenever there is danger of rain during harvest, this work should be done in the most careful manner. To do this, the sheaves must be properly placed, and the cap sheaves properly broken over to turn rain. If well done, it takes a long-continued storm to injure the grain. To illustrate the whole fully we have prepared these figures as a guide. Stand four sheaves in a row, the two end sheaves slanting somewhat, and pressing against the others. Strike the butts firmly into the stubble, then place sheaves firmly against these, three on a side. This makes a round shock. For the first cap take a long, smooth sheaf, break all the heads back toward the band, to one side and the other as shown in the cut. Lay it on the shock, throwing the heads each way, and with the butt to the east. Then break the second cap down at the band, spreading somewhat, and lay it on with the butt to the west. The shock will then look like the third figure, entitled "The Shock Finished," and, when fairly settled, will not be liable to be blown over by any ordinary wind. Let your shocks

SECOND CAP.   FIRST CAP.

THE SHOCK FINISHED.

run perfectly straight across the field, however long it may be, and at equal distances apart. The cost of shocking is hardly an appreciable quantity in the cost of a crop, and whatever is worth doing at all is worth doing well. It is a great comfort for the farmer to know, in bad weather, that his crop is safe.

XIII. Importance of Good Seed.

HALLETT'S PEDIGREE WHEAT.

HALLETT'S PEDIGREE WHEAT.

WE have already shown that the ancients thought it worth while to select the best ears directly from the field. The best cultivators of to-day do this, and thus produce pedigree grain that sells for large prices. Every farmer should do this. If it pays the seedsmen it will pay the farmer. If in your examination you find a head distinct in its characteristics, save it and plant it in an experimental plat. The success of Major F. F. Hallett, of Kemptown, England, in this line has been famous for many years. In 1874, he delivered, before the Midland Farmer's Club, of Birmingham, an address, in which he stated his whole plan of operations. His mode of procedure, results and general conclusions are given in the next two articles. But what is pedigree grain? It is—and the same is true of all grains—seed selected and cultivated for years under the best possible conditions. The engraving shows an accurate and life-size representation of the wheat grown by Maj. Hallett. Compare this with the head of red bluestem wheat, on page 93, also shown in its natural size, as taken from fields—the others represented being select heads. The comparison will convince any intelligent reader of the importance of selection in seed.

XIV. Pedigree Grain.

"A GRAIN produces a plant consisting of many ears. I plant the grain from these ears in such a manner that each ear occupies a row by itself, each of its grains occupying a hole in this row, the holes being twelve inches apart every way. At harvest, after the most careful study and comparison of the plants from all these grains, I select the finest one, which I accept as a proof that its parent grain was the best of all, under the peculiar circumstances of that season. This process is repeated annually, starting every year with the proved best grain, although the verification of this superiority is not obtained until the following harvest."

Table showing the importance of each additional generation of selection.

YEAR.	SELECTED EARS.	LENGTH.	CONTAINING.	NO. OF EARS ON FINEST STOOL.
		Inches.	Grains.	
1857	Original ear,	4⅜	47	....
1858	Finest ear,	6¼	79	10
1859	Finest ear,	7¾	91	22
1860	Ears imperfect from wet season,	....	....	39
1861	Finest ear,	8¾	123	52

Thus, by means of repeated selection alone, the length of the ears was doubled, their contents nearly trebled, and the "tillering" power of the seed increased fivefold.

The following table gives similar increased contents of ear obtained in three other varieties of wheat:

VARIETIES OF WHEAT	GRAINS IN ORIGINAL EAR.	GRAINS IN IMPROVED EAR.
Original red, commenced 1857,	45	123
Hunter's white, commenced 1861,	60	124
Victoria white, commenced 1862,	60	114
Golden drop, commenced 1864,	32	96

XV. General Conclusions.

1.—Every fully developed plant, whether of wheat, oats or barley, presents an ear superior in productive power to any of the rest on that plant.

2.—Every such plant contains one grain which, upon trial, proves more productive than any other.

3.—The best grain in a given plant is found in its best ear.

4.—The superior vigor of this grain is transmissible in different degrees to its progeny.

5.—By repeated careful selection the superiority is accumulated.

6.—The improvement, which is at first rapid, gradually, after a long series of years, is diminished in amount, and eventually so far arrested that, practically speaking, a limit to improvement in the desired quality is reached.

7.—By still continuing to select, the improvement is maintained, and practically a fixed type is the result.

XVI. Artificial Cross Fertilization.

The wheat plant, and the grasses generally, have perfect blossoms — that is, stamens or male organs; and the female organs, or stigma and ovary. Perfect seeds of two races are selected and planted in separate patches of alternate rows of male and female, the individual seeds ten inches apart each way, as shown in the diagram:

```
No. 1.  *M*M*M*M*M*M         No. 2.  *M*M*M*M*M*M
    2.  *F*F*F*F*F*F             1.  *F*F*F*F*F*F
    1.   M*M*M*M*M*M             2.  *M*M*M*M*M*M
    2.   F*F*F*F*F*F             1.  *F*F*F*F*F*F
    1.   M*M*M*M*M*M             2.  *M*M*M*M*M*M
    2.   F*F*F*F*F*F             1.  *F*F*F*F*F*F
```

The notion is that No. 1 of the first plat shall fertilize No. 2 of the second, and *vice versa*. As soon as the anthers show, clip off all carefully from the lines marked 1 throughout one of the patches, 1 being supposed to be of one variety and 2 of another. From the other cut the anthers from all the lines marked 2, cover with gauze to keep off insects for three days, or until the anthers have lost all their pollen and shrunk up. You will have the product of two races; the male of No. 1 with female of No. 2, and the male of No. 2 with the female of No. 1. From selections of the produce of these experiments you may preserve heads with distinct and valuable characteristics.

The three cuts show some results in this direction. Fig. 1, White Russian, is a beardless, white-chaff amber wheat. Fig. 2, Defiance, is also a beardless spring wheat, a cross of club wheat upon a variety from California. Fig. 3, Martin Amber, is a cross on Red Mediterranean, having the quality in the young plant of lying close to the ground; beardless, an amber berry, with a thin hull. [See next page.]

XVII. Reputable Old Varieties of Wheat.

A REPORT from the Department of Agriculture on samples of wheat from various States at the Centennial Exhibition, shows the general estimation of well-tested varieties, and is valuable for reference. In relation to new varieties, yearly appearing, every person must be guided by experiment, or by the advice of those who have tried them. The statement in relation to well-tested sorts, most of which will continue to hold their popularity, is as follows:

In the New England States we find the Lost Nation, Tappahannock and Lancaster Red Chaff, the most commonly cultivated; while samples of Arnautka, Canada, Hybrid, White Laisette and White Italian occur. New York adds Diehl, Treadwell, China Tea and other varieties. In the remaining Middle States and Maryland, Virginia and North Carolina we find Fultz and Mediterranean grown; Tappahannock, White Canada and Golden Chaff are also represented. Ohio has sent nearly the same wheats as are grown in Pennsylvania, only one name not previously occurring, that of "Todd" wheat, being observed. Indiana and Illinois grow Lancaster, Michigan, Amber, Tappahannock, Odessa Red, Fultz, China, Missouri, Velvet, Early, Oran, Scotts, Egyptian, and two or three other varieties. In the Missouri collection we still find Fultz and Odessa, together with New York Flint, Independence Spring, etc. Iowa contributes Rio Grande, Canada, Fife and White Chili. Among the varieties grown in Minnesota are Scotch Fife, Rio Grande and China Tea, before mentioned, with the addition of Eureka, Early Sherman and White Hamburg. Michigan sends Diehl, Gold Medal and White Mountain.

The wheats of Kansas and Colorado, approaching in appearance those of California, are White Colorado, White California, Turkey and Colorado Red Chaff, while Nebraska gives the names of Priest Spring, Otoe and Russian Club. Among the wheats of the Pacific coast, principally white wheats, the White Australian appears to be the general favorite. The White Chili is also grown, and such varieties as Canada Club, Jones, Propo, Pride of Butte and Nonpareil are represented in the collection. From Texas and New Mexico we have Sonora and Zaragoza. From the remaining Southern States the collection of wheats is very meager.

The varieties of wheat that have originated apparently by accident or from peculiar culture, do not enjoy the surroundings necessary for continuous product. The care of man is necessary to preserve or to render perfect the already improved varieties. That cross fecundation may be practiced was proved by Maund and Raynbird as early as 1851. In that year their specimens took the prize in the London exhibition.

Fig. 1. WHITE RUSSIAN WHEAT.
Fig. 2 DEFIANCE WHEAT.
Fig. 3. MARTIN AMBER WHEAT.

XVIII. Rye and Its Cultivation.

RYE is comparatively little raised in the United States. It succeeds on thin, sandy land, not strong enough for wheat. As a crop for pasturing, or for plowing under, it is valuable if sown among corn in August, and should really find a place on every farm. It is not so liable to freeze out as wheat, and stands the winter better in almost every way. As the soil becomes less adapted to wheat, rye will gain more and more in importance. There are few varieties. The white winter varieties should be sown, if the grain is to be harvested; but for pasture and plowing under, the so-called black varieties are hardier. The rye raised in the United States is almost in the proportion of one bushel to fifteen of wheat, and as one to eighty of corn. Our export of rye is from 200,000 to 500,000 bushels a year. The cultivation and care are precisely like that of wheat; but it is generally sown on a single plowing.

Some attention has been paid of late years to improved varieties. The cut shows the so-called Montana Spring Rye, natural size. We think more attention should be paid to the spring varieties of this grain, in all that great portion of the West devoted to the raising of spring wheat.

For a seed crop rye is sown the last of September in the North, as late as the last of October in the latitude of St. Louis, and even later farther south. The best rye flour is made from that raised on sandy land; but rye is adapted to a great variety of soils not really wet.

Rye is almost unknown for bread in the United States, though in the northern parts of Europe, notably in Russia and Germany, it is one of the principal bread grains. In the extreme northern portions of the United States, and on some sandy soils not well adapted to other grain, more or less rye is raised. But the proportion to other grains is very small, being not more than one to seventy-five of Indian corn, which has usurped its place for distilling and as food for horses. In this country it is principally cultivated to furnish green forage for horses, and is usually sown broadcast in October, at the rate of a bushel and a half to the acre, or if drilled at the rate of one bushel to the acre. Of course the time of sowing and harvesting, as well as the quantity of seed to the acre, and soil best adapted to this grain, will vary with the latitude and climatic conditions.

MONTANA SPRING RYE.

The following table, carefully compiled, will give more information as to the time of sowing, etc., than could be gathered into pages of description:

STATES.	TIME OF SOWING.	AVERAGE BUSHELS OF SEED PER ACRE.	TIME OF HARVEST.	BEST SOIL
Maine,	Fall and Spring,	1¼ bush.,		New, burnt land.
New Hampshire,	Sept. and April,	1 to 2 bush.,	July and Aug.,	Sandy; silicious; newly cleared land,
Massachusetts,	Aug. to Sept.,	1 to 1½ bush.,		High, warm, light land.
Vermont,	September,	1¼ to 1½ bush.,	Last July to Aug.	Light.
New York,	Sept. 1 to Nov.,	1 to 2 bush.,	July 10 to 25,	Sandy and slate; sand; sandy loam. or gravel; gravelly loam.
New Jersey,	September 1,	1 bush.,	July 1,	Light, sandy loam.
Pennsylvania,	Sept. 1 to 15,	1 to 1½ bush.,	Last June to July	Gravel; serpentine; stubble.
Maryland,	October,	1 bush.,	June,	Light.
South Carolina,	October,	⅝ bush.,	June,	Gray land.
Georgia,	September,	¾ bush.,	July,	Mulatto.
Alabama,	October,	½ bush.,	May,	Light.
Mississippi,	September,	½ bush.,	May and June,	Rich loam.
Tennessee,	September 1,	1 bush.,	June 15 to July,	Black, thin loam.
Ohio,	Sept. and Oct.,	1 to 1½ bush.,	June to July,	Clay; clay, light; sandy and warm oak and hickory clay.
Indiana,	Sept. to Oct.,	1 to 1¼ bush.,	June 20 to July,	Dry; just cleared.
Illinois,	Oct. and Nov.,	2 to 2½ bush.,	June 20 to July,	Clay; or sandy loam.
Michigan,	October 10,	1½ bush.,	July 15,	Clay or loam.
Iowa,	September,	2 bush.,	July,	

XIX. Barley and Its Cultivation.

THERE is no farm crop liable to so many vicissitudes, which reduce its market value, as barley. Were it not used in such immense quantities by brewers, in the manufacture of beer, its cultivation would be almost abandoned. Except in localities where the grain can be harvested bright and in good condition, it is now little cultivated. Barley is less nutritious than wheat, but contains one-seventh more feeding value than oats, pound for pound. Strong dry loams and sandy soils give the best quality of grain, and very considerable care should be given to the preparation of the soil to have it light and mellow; the prairie region of the far West, the plains, the valleys of the Rocky Mountain region, and the Pacific slope, where rain does not interfere with curing it properly, are the principal sources from which the brewers get their malting supplies.

XX. Time for Sowing Barley.

THERE is winter and spring barley, but the spring varieties are generally sown. There are two principal classes, the two-rowed and the six-rowed, and many varieties, some of them pedigree sorts. In the South, winter barley is usually sown; in the North, spring barley almost universally. The usual yield on good land is from twenty to twenty-five bushels an acre, but under exceptional

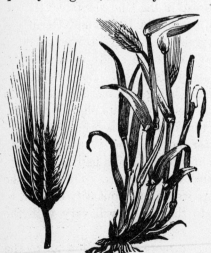

WINTER BARLEY—PLANT AND HEAD.

CEREALS AND THEIR CULTIVATION. 103

circumstances as high as sixty bushels have been harvested. The time for sowing winter barley is about the same as for winter wheat. Spring barley should not be sown until the soil has acquired some warmth, or about three weeks before the usual time of planting Indian corn. The following table gives precise information, compiled from many sources in various States where barley is grown.

STATES.	WHEN SOWN.	AVERAGE SEED SOWN PER ACRE.	TIME OF HARVEST.	BEST SOIL.
Maine,	Last of May,	2 bush.,		
New Hampshire,	April; May,	2½ to 4 bush.,	August,	Black loam.
Massachusetts,	May,	2½ to 3 bush,	July 30,	High, warm land.
Vermont,	July 1,	2 bush.,	August 1,	Dry.
New York,	April 10 to May 10	2 to 3 bush.,	July 1 to Aug.,	Loam; warm loam; loam and muck; sandy loam; black sandy loam.
Pennsylvania,	March 15,	1½ to 2 bush.,	June and July,	Heavy clay; sandy loam.
South Carolina,	September,	2 bush.,	May,	Clay.
Tennessee,	March 1,	1 bush.,	July 1,	
Ohio,	April 1 to May 1,	1½ to 2 bush.,	July 1 to 25,	Clay, mixed with sand; clay loam; loose do.
Indiana,	April to Sept.,	1¼ to 2½ bush.,	June 25 to Aug.,	Clay; do. loam; dry, sandy loam.
Michigan,	April 15 to May 1	½ to 2 bush.,	July 7 to Aug. 1,	Sandy loam; rich loam.
Iowa,	March to April 1	1¼ bush.,	July 1,	

XXI. Harvesting and Threshing Barley.

HARVESTING is the most difficult thing in making a crop of barley; but the price of first-class grain will always pay for the trouble. Barley is never harvested until it is ripe, so the seed will germinate evenly; but it should not be allowed to get dead ripe, else the grain will be dark-colored. When the red streaks, which run lengthwise in the ripening grain, disappear, and the head begins to hang down, and the straw takes on a yellow hue, it is ready for the harvester. Where cured without binding, the grain sooner comes into condition, sweats more uniformly in the shock, and is somewhat better in color; but in all regions subject to rain, it is more easily protected from staining when bound. Sheaves should be small and of even size. This is easily regulated with self-binders.

Threshing.—The threshing should never be done with a spike-thresher. A beater-machine is better, and the flail, or tramping out by horses, best of all. The reason of this is, that if the germ is injured—which is almost sure to be the case when it is threshed with the ordinary thresher—the value of the grain for malting is greatly diminished, for upon its germinating powers depends its value. After threshing, care must be taken that the barley does not sweat in the heap or bin. Unless quite dry, it must be moved often until it is wholly cured.

XXII. New Varieties of Barley.

AMONG the better kinds of barley for malting, is the Chevalier, a pedigree variety that is in repute in England and France. In the United States, it has given great satisfaction wherever tried. It is a two-rowed variety. The Manshury, a six-rowed sort, originated in Canada, and was new in 1882. It is recommended as standing up well in the richest soils, and superior for malting. Sibley's Imperial

ANNAT BARLEY.   CHEVALIER BARLEY.   ENGLISH BARLEY.

barley originated in Vermont. It is a six-rowed variety, tillering freely, long in the straw and head, and with medium beards. It is hardy and prolific. Adams's heavy barley originated in Western New York. It is a six-rowed variety, of stiff straw, medium height, and is especially free from crinkling down when ripe. Annat barley is a variety that has given good satisfaction. The cuts show the common English barley, Chevalier barley, and Annat barley, natural sizes.

XXIII. Oats and Their Cultivation.

THE value of oats in the agriculture of the United States and Canada is enormous. However small the farm, oats are an important factor in the crop, and for feeding horses are considered a necessity in spring and summer. This crop really stands next in importance after wheat in the cereal crops of the country. For feeding young animals oats are coming more and more into favor every year, on account of their bone and muscle making properties. They are adapted to nearly all soils not really sandy or wet, and the straw is more useful on the farm than that of any other grain. Taking the year 1879, a fairly productive year throughout the country, we find the values of the principal crops of the United States to be as follows: Corn, $580,486,217; wheat, $497,030,142; hay, $330,804,494; cotton, $242,140,987; oats, $120,533,294; potatoes, $79,153,673; barley, $23,714,444; tobacco, $22,727,524; rye, $15,507,431; and buckwheat, $7,856,191.

XXIV. Export of Food Crops.

HORSE-MANE OATS.

THE following table of exports shows that oats are one of the crops consumed at home. Both the years 1878 and 1879 were of fair average export.

PRODUCTS.	1878.		1879.	
BREADSTUFFS AND OTHER PREPARATIONS.	QUANTITY.	VALUE.	QUANTITY.	VALUE.
Barley, bushels,	3,921,501	$2,565,736	715,536	$401,180
Bread and biscuit, pounds,	14,392,231	730,317	15,565,190	682,471
Corn, bushels,	85,461,098	48,030,358	86,296,252	40,655,120
Corn-meal, barrels,	432,753	1,336,187	397,160	1,052,231
Oats, bushels,	3,715,479	1,277,920	5,452,136	1,618,644
Rye, bushels,	4,207,912	3,051,739	4,851,715	3,103,970
Rye-flour, barrels,	6,962	30,775	4,351	15,113
Wheat, bushels,	72,404,961	96,872,016	122,353,936	130,701,079
Wheat-flour, barrels,	3,947,333	25,095,721	5,629,714	29,567,713
Other small grain and pulse,		1,077,433		817,536
Other preparations of grain,		1,709,639		1,740,471
Rice, pounds,	631,105	33,953	740,136	35,538
Total value of breadstuffs, etc.,		$181,811,794		$210,391,066

XXV. Species of Oats—Their Latitude.

The following are acknowledged species of oats, the botanical name of which is *Avena Sativa:*

Avena trevis, or short oat, which ripens early; it is raised in some parts of France as forage.

Avena Fatua, or California oats, which is thought to be identical with White Tartarian oats.

WHITE RUSSIAN OATS.

Avena Nuda, or hulless oats, an old variety, probably produced by cultivation. It has been known in England for more than three hundred years, and comes up again and again under new names. It is not valuable for general cultivation, on account of shelling so easily.

Avena Orientalis, or Tartarian oats, probably brought into Europe in the latter part of the seventeenth century.

Avena Strigosa, or bristle-pointed oat. The seeds are small, it is not productive, and may be called worthless.

As a rule white oats are more salable, but the black and brown varieties are thought to be hardier and more prolific. The cut shows a panicle of White Russian oats reduced in size. The oat is essentially a grain adapted to cultivation in temperate and northern latitudes, and has both winter and spring varieties. The winter varieties are raised in the latitude of Kentucky and in the South. In Europe oats are cultivated as far north as latitude 64° to 65°. It is grown successfully up to the northern limits of the United States, and in Canada. The limit of successful culture reaches farther and farther northward, as we pass west to the Pacific coast.

XXVI. Soil and Cultivation of Oats.

Oats require rich, moist land. Any good Indian corn land will produce uniformly heavy crops of oats, except in seasons of early drought. Strong, fairly drained clays, and strong loams are adapted to this crop. Oats will not stand drought, nor hot suns. The best crops are raised in cool, rather moist seasons. The table following is useful for reference.

How and When to Plant.—As a rule, especially on strong and tenacious clay, the land should be fall-plowed, and as deep as the soil will admit. The sowing should not be too early, since the young crop is easily killed by freezing. A fair rule is to sow about a month before the time for corn-planting; but if the season is untoward, the sowing may be delayed up to the time for planting corn. But in all late sowings the crop is apt to suffer from heat and drought, so it is important to get the seeding done as early as the season will admit. Oats are almost always sown broadcast, two and a half bushels to the acre. Three bushels an acre on land in good heart would be better, since the crop does not tiller much, and thick sowings give a more equal ripening of the grain. The seed should in no case be covered more than two inches, and the soil must be in firm contact with the seed to ensure prompt germination.

STATES.	TIME OF SOWING.	SEED PER ACRE.	TIME OF HARVEST.	BEST SOIL.
Maine, ...	April to May, . .	2½ to 3 bush., . .	August 10 to 20, .	Dry; gravelly.
New Hampshire,	April and May, . .	3 to 4 bush., . .	August 1, ...	Clay; sandy; free.
Massachusetts, .	April 10 to May 10,	2½ to 3 bush., . .	July 15 to Aug. 20,	High, warm land.
Vermont, . . .	April to May 15, .	3 bush., ...	August, ...	Light; sandy.
New York, . .	March 15 to May 25	1½ to 3 bush., .	July 10 to Aug. 15,	Loam and muck; loam; deep, black muck; rich, sandy loam.
New Jersey, . .	April 1 to 15, . .	2 to 2½ bush., .	July 20 to Aug. 1,	Sandy loam; clay.
Pennsylvania, .	March 15 to April 15	1 to 3 bush., . .	July 10 to Aug. 1,	Sandy loam; light; sandy; sandy loam; limestone do.
Delaware, . . .	. . . .	. . . .	. . . .	Rich; moist.
Maryland, . .	April, ...	2 bush., . . .	July, ...	Dry loam; clay and lime; sandy loam; do.; do.
Virginia, . . .	Feb. to April 1, .	1½ to 2½ bush., .	July 10 to Aug. 10,	
South Carolina, .	Dec. to Feb., . .	1 bush., ...	June 1 to last June,	Moist; sandy.
Georgia, . . .	Jan. to March 1, .	1 to 1½ bush., .	June to July 1, .	Slate loam.
Alabama, . . .	Nov. to April, . .	¾ to 1 bush., .	May and June, .	Sandy loam.
Mississippi, . .	Oct. to Feb., . .	1 bush., . . .	July, ...	Light do.; clay.
Tennessee, . .	Feb. 15 to March, .	1½ bush., . . .	July 10, ...	Black loam, thin.
Kentucky, . .	March and April, .	48 lbs., ...	July, ...	
Ohio, ...	March and April, ,	2 bush., . . .	June to Aug. 1, .	Loose loam; do. do.; clay loam; sandy loam; oak and hickory loam.
Indiana, . . .	March to May 1, .	1¼ to 2 bush., .	July 1 to August, .	Sandy loam; loam; clay; do.; do.; do.; sandy.
Illinois, . . .	March 20 to April 4	1½ to 2 bush., .	June to Aug. 1, .	Sandy loam; light loam; sandy do.
Michigan, . . .	April 10 to 30, . .	2 to 3 bush., .	July 7 to Aug. 1, .	Clay or sand; rich loam; sandy.
Iowa, ...	April, ...	2 to 4 bush., .	July 15 to Aug., .	
Texas, ...	February, . . .	1 bush., . . .	May, ...	

XXVII. Harvesting and Threshing Oats.

OATS are usually harvested before they are fully ripe. When fully ripe they shell easily, the hull becomes hard and glassy, and the straw is much reduced in value. Oats cure readily, and are seldom put in close, capped shocks. They are not easily injured by rain, and hence they are generally shocked by setting a row of sheaves leaning together, uncapped. Whenever the straw is found valuable, it will always pay to shock like wheat and barley, with two cap-sheaves to each shock.

XXVIII. Varieties of Oats to Cultivate.

IN the South, where oats are subject to rust and blight, the winter varieties are sown. The Red Rust-proof is usually preferred. The White Winter oats is growing in favor on rich uplands and drained bottom lands. It stands the winter as far north as Virginia; is said to be rust-proof, and improved by moderate grazing.

In the North, the Black Tartarian is the most universally grown. It is the best of black oats. It is probably adapted to a greater variety of soil and climate than any other one variety. Among white oats, the Schoenen and Probsteier are generally liked. Among newer varieties, White Zealand, said to be rust-proof; White Australian and White Russian (see cut) are yearly gaining in favor.

XXIX. Buckwheat.

THE cultivation of buckwheat receives little attention in the West, and in the East it is sown principally as a secondary crop, where others have failed. It is sometimes difficult to eradicate it the second season, since the seeds shelled out in harvesting, germinate the next season, producing a volunteer crop. Sandy soils, and indeed, rather poor sands, produce the best buckwheat for flouring. It is one of the best fallow crops for turning under green, or just at the time of blossoming.

For a seed crop, the sowing should be so timed, that it will be in full seed at the time of the first frosts, since it requires cool nights to fruit properly. Sow about the fifteenth to the twentieth of June in the latitude of Maine and Minnesota, and later, even to the first of July, in the latitude of Philadelphia and Central Illinois.

XXX. Seeding and Harvesting Buckwheat.

THE plant is tropical and killed by the slightest frost. It was introduced originally from Persia, and its name buckwheat is a corruption of beech-wheat, from the resemblance of its three-cornered seeds to those of the beech. The blossoms are eagerly sought by bees, from the abundance of honey they contain.

The seed is sown on fresh-plowed land, at the rate of two or three pecks, or if very late, one bushel per acre. The seed should be covered lightly (one-half inch). At the time of the first frost the crop is cut, laid in gavels, and set together, without binding until dry. Then it is threshed with the flail, or tramped out by horses. The yield varies according to the seasons from ten to forty bushels per acre, and the seed approaches, in price, about that of wheat. There are few varieties. The Silver-hull is considered the best, but the common buckwheat is generally sown.

Time of Seeding, etc.—The following table gives the desired information:

STATES.	TIME OF SOWING.	SEED PER ACRE.	TIME OF HARVEST.	BEST SOIL.
Maine,	Middle of June,	½ bush.,		
New Hampshire,	June,	½ bush.,	August; Sept.,	Silicious.
Vermont,	July 1,	¼ to ½ bush.,	September.	
New York,	June 10 to July 20,	½ to 1 bush,	Sept. 15 to Nov,	Rich, sandy loam; sand and loam; deep black muck; light sandy.
New Jersey,	July 13 to last June	½ to 1 bush.,	September 15,	Sandy.
Pennsylvania,	June 1 to last July,	¼ to 1 bush.,	Sept. 1 to Oct. 15,	Slate; sandy loam; gravel or slate.
Tennessee,	May 20,	1 bush.,	October 15,	Mountain.
Kentucky,				
Ohio,	Mid. June to July 1	¼ to 1 bush.,	Sept. 20 to Oct. 1.	Loose loam; sandy; black, thin and compact loam.
Indiana,	July 1 to Aug. 1,	½ to 1 bush.,	Sept. to Oct 1,	Clay; black loam.
Illinois,	June to July 1,	15 to 25 lbs.,	September,	Sandy loam; wheat stubble; black muck.
Michigan,	June 15 to July 5,	½ to 2 bush.,	Aug. 10 to Oct. 10,	Light loam; light sandy do.
Iowa,	June 20,	½ to 1½ bush.,	September 20,	

CHAPTER II.

INDIAN CORN, RICE, AND SPECIAL CROPS.

I. THE CROP IN THE UNITED STATES.—II. HOW TO INCREASE THE AVERAGE.—III. PROPER MANURES FOR CORN.—IV. THE CULTIVATION OF CORN—PLOWING.—V. PREPARING THE SOIL.—VI. PLANTING THE CROP.—VII. HARROWING THE YOUNG CORN.—VIII. AFTER CULTIVATION OF CORN.—IX. HOW OFTEN TO CULTIVATE.—X. DEPTH OF CULTIVATION.—XI. HARVESTING THE CROP.—XII. CUTTING AND SHOCKING.—XIII. SEED CORN.—XIV. COST OF A CORN CROP.—XV. VARIETIES OF CORN.—XVI. RICE AND ITS CULTIVATION.—XVII. TRUE WATER RICE, OR COMMERCIAL RICE.—XVIII. CULTIVATION OF RICE IN CAROLINA.—XIX. MANAGEMENT OF RICE FIELDS.—XX. CULTIVATING THE CROP.—XXI. FLOODING THE CROP.—XXII. HARVESTING AND THRESHING.—XXIII. HULLING FOR MARKET.—XXIV. RICE IN THE MISSISSIPPI DELTA.—XXV. SOME SPECIAL CROPS.

I. The Crop in the United States.

THE United States now raise as the average 1,600,000,000 bushels of corn yearly, an increase of 100,000,000 bushels per year, for the present decade, as compared with the last few years of the last decade. The American crop is seventy-eight per cent of the Indian corn crop of the world, the total production outside the United States being only 360,000,000 bushels yearly. And yet the general average yield per acre in this country is only about twenty-three bushels per acre, and the best average yield in the great corn year of 1880, only twenty-seven and a half bushels per acre, while authenticated yields of one hundred bushels per acre could be cited on one-hundred-acre fields, and special yields of one hundred and forty bushels to the acre on smaller areas. Whole counties have averaged sixty bushels, and some States forty bushels per acre in particular years. Taking all these facts, and remembering that no good farmer is satisfied with less than forty to sixty bushels in ordinary seasons, and it seems certain that a majority of farmers must be wofully negligent in their cultivation and recklessly inattentive to their best interests.

Comparing Results.—Let us estimate the loss from ignorance, or bad cultivation of the corn crop, as shown by the best average of the State and the general average of the country, remembering that the light averages are not in hilly, rocky, worn New England, but in countries of so-called virgin soil. Take the average annual yield at 1,500,000,000 bushels for the country, the general average of twenty-six bushels per acre, and the best State average of forty bushels per acre. Suppose the general average brought up to forty bushels, and the corn crop of the country would be increased fifty per cent, making a total of 2,250,000,000 bushels. At the average price of fifty-two cents, this would increase the annual value of the crop by over $390,000,000. Would this pay for the better cultivation of the crop? Even those who do not believe in advanced farming, must admit that it would.

II. How to Increase the Average.

There are only three reasons why the average yield of corn is not forty bushels per acre, as the minimum crop. In no ordinary season should it go below that, over any large area. The causes which keep down the average are: 1, want of drainage; 2, want of manure; and 3, bad cultivation. There are also these three causes, which may reduce the crop locally: 1, destruction by insects; 2, an excessively wet season, preventing proper cultivation; and 3, excessive drought. Untimely frosts can hardly be taken into account, since they occur so seldom that drainage, by allowing earlier planting and steady growth, would entirely throw this out of the calculation, and it would also practically do away with danger of severe loss from wet and dry seasons. As it is, through all the great corn region of the West, our dry seasons are our best ones. "Drought scares the farmer, but water utterly destroys his hopes."

Therefore, if the land needs drainage, attend to it at once. (See Chapter V, Part III, on Drainage.) It is the best investment, because it is a permanent one. If the soil lacks fertility, improve it by manure, a proper rotation, and by plowing under suitable crops. (See articles on Manure, Rotation and Fallow Crops.)

III. Proper Manures for Corn.

Corn is one of the gross feeders. That is, it is a humus-loving crop, and the roots feed upon any decaying substance. Barn-yard manure produces the best results. Ashes are valuable, since corn is one of the potash-loving plants; plaster (gypsum) is good on sandy soils. Phosphate, guano and other commercial manures are always valuable, where they may be cheaply obtained. Barn-yard manure should be spread in the autumn, plowed under rather deep, and the land plowed again in the spring, but not so deep as to bring the manure to the surface; then some special fertilizer should be added. Ashes, plaster, and other commercial fertilizers should be applied in the spring broadcast, and lightly covered, say with a harrow. The roots will be certain to find it. The soil between the rows will become a perfect net-work of roots before the plant attains its full growth.

Prize Crops.—The prize crops of corn are raised upon rich soil, heavily manured with green barn-yard manure, plowed under deep in the autumn before planting—not less than forty good loads per acre. In the winter cover the land with twenty loads of rotten manure and plow this in, say, four inches deep. Then use some commercial fertilizer for a surface dressing. We have raised one hundred bushels of field corn to the acre by this plan, using plaster and home-made manure of night-soil for surface dressings, and this without hoeing with the hand-hoe. By this plan we have also raised one thousand dozen of green corn per acre for the market. Farther on we will show how about forty bushels per acre were raised on a field of over twelve hundred acres—on land that had never been manured—and the cost.

IV. The Cultivation of Corn—Plowing.

It is a great mistake to suppose that fall plowing is not advantageous in the cultivation of corn. We speak of clay loams and clay soils, of course, such as raise our best crops. On sandy soils, fall plowing is not necessary, since such soils do not require the action of frost to mellow them. It is true, the spring plowing should not be omitted, but this is always superficial, never more than four inches deep, and may be done at the rate of three acres per day to the single plow. Besides the better disintegration of the soil, by freezing and thawing, when fall-plowed, the soil comes into condition earlier in the spring, absorbs warmth quicker, and as a rule may be worked much earlier than land not fall-plowed.

V. Preparing the Soil.

In the introduction to the chapter on cereal grains, we advised to follow with the planter immediately after plowing. This advice is general. It is by no means true, however, that letting the soil lie for a week or ten days is injurious. On the contrary, it may be decidedly beneficial. In this case the land must again be made fine at the surface, and this kills one crop of weeds. In the case of fall plowing, there is nothing gained by letting the ground lie after plowing, and before planting. In the case of

YELLOW DENT CORN. Two-thirds Natural Size. See page 119.

spring plowing, without fall plowing, the planter must be guided by circumstances. Never neglect plowing for corn, when the land is in good condition, because the season is too early for planting; and never let the land lie unplanted, after plowing, when the season for planting has arrived. Loosening the soil, or smoothing it after fresh plowing, is best done by a sharp harrow, but harrowing must never be delayed until the weeds are up in a field to show green. The work of the harrow, to be effective, must be done just as the weeds are coming up, and this will be in about a week or ten days after spring plowing, according to the season. If you have not underdrained wet places, turn to the chapter on Artificial Drainage, and see how easily and cheaply the surplus water may be carried off. Only injury will result from plowing, harrowing or cultivating the soil when it is not in a thoroughly friable state. Then please remember that fully half the cost of raising a crop of corn, in the West at least lies in the proper plowing and fining the soil before planting.

VI. Planting the Crop.

THERE are four principal things to be remembered in planting a field of corn: 1, The rows should be perfectly straight. 2, The seed must be strong in its germ. 3, The planter must drop the seed accurately, and rather closely together, rather than widely spread. 4, Whatever the number of grains planted, more than four stalks should never be allowed to the hill; three is better.

If your farm is too small to allow you to own a check-row planter, hire your planting done by all means. There is no doubt but corn may be planted as well by hand as by a machine; but children and hired men will not do it, and, at best, it is tedious and costly. Do not plant too close. Three feet and a half is close enough between rows for the dwarf varieties, three feet eight or ten inches for the medium varieties, and four feet is not too much for the Mammoth Southern varieties. If you have been careful to lay out your first row straight, as in the directions for plowing, you may with care keep every other row straight, by means of a re-marker attached

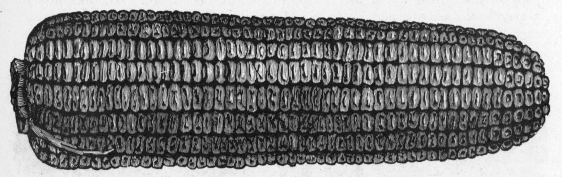

WHITE DENT CORN. Two-thirds Natural Size. See page 120.

to the planter and the check-rower. Perfectly straight rows should add five bushels an acre to the crop, through the better cultivation possible thereby. There is no doubt that more corn may be raised per acre by accurately drilling so the stalks will stand twelve inches apart in the row, but, except in small fields, or where the intention is to make a premium crop, the extra cost will not allow drill-cultivation to become profitable, especially in the great corn regions of the United States.

VII. Harrowing the Young Corn.

THE harrowing is the best cultivation young corn ever receives. Of course, the ground must be measurably free from trash, and no sensible farmer plants on trashy ground. With the present perfection in plows, trash may all be so deeply turned under that the harrow will not find it, and as the corn gains size, the trash will be so decayed as not to interfere seriously with good work. The harrowing should be given with a sharp, light harrow, at the first indication of weeds, whether the corn is up or not. If the corn is just pushing through the ground, care must be taken. The germ is thus easily broken. Otherwise harrow the field without reference to anything, except to destroy the weeds. If no weeds appear, and the top soil is not crusted, the harrowing may be delayed until the rows of corn can be seen. It will often pay to harrow both ways, once before the corn is up and once after. After the corn is up we have always found it pay to have careful hands uncover such as may have been covered with trash and lumps. Two rows may be attended to each time, going across the field, using a forked stick, or better, one crooked at the end. The back of a hoe or rake is also useful for this purpose.

VIII. After Cultivation of Corn.

THE hand-hoe finds no place in the cultivation of corn, except in very small fields; in those so rocky or stumpy that horse implements cannot work to advantage,

MICHIGAN YELLOW DENT CORN. Two-thirds Natural Size. See page 119.

or in fields where the weeds have got the full start of the crop. In all fields of this kind, the cultivation is attended with such disadvantages as often to bring the balance on the wrong side of the ledger. That is, it will be cheaper to buy corn than to raise it, unless the special purpose be to clean the land for other crops. A roller may be used with success in some cases, after the harrow, if the ground is very lumpy. We have rolled corn eight inches high, and had it rise again all right; but the land should not be lumpy, nor need it be if the directions in relation to fall plowing have been followed. We repeat, never stir the land in the spring or summer unless it will work friably.

A Busy Time.—From the time the corn is up four inches high the cultivator must be kept moving. In catching weather every hour must be improved when the soil is in condition. If rainy weather has interfered with cultivation, and weeds begin to show unduly, pay no attention to regular hours, work the men from daylight until it is too dark for them to see the rows at night, changing teams and paying for extra time. This kind of work often saves a crop, for if once the weeds get a full start it is difficult to overcome them. Remember always: the time to kill weeds is while they are young.

Clean Crops.—No man ever raised a good crop who waited for the weeds to grow before cultivating his land. The primary object of cultivation is to keep the soil in such condition that it will admit air properly through its pores. Killing weeds is only a secondary consideration. A weedy crop never pays its cost. A rich soil always grows weeds. They are easy to kill when young; when their roots get strong it is difficult. The Chinese, who have cultivated the same soil for over four thousand years, have a saying that, "a clean crop is always good." Their fields are kept as clean as a garden.

IX. How Often to Cultivate.

The cultivator should be kept going until the crop is so large that the stalks cannot be pressed under the arch of the implement. Whenever the surface is crusted from rain, moving the soil will be beneficial. Two harrowings and two to three plowings are what the average crop should get. The operation should be guided by the farmer's own observation of the necessities of the case. Wet, rainy weather interferes with cultivation, and the farmer who calculates on the basis of fifty acres to the hand, will, in bad seasons, not be able to do full justice to the crop; while in dry seasons sixty-five acres to the hand may be well plowed. Why? Because the team can work every day, and an average of eight acres a day will get over this area once in about eight working days. A field ought to be plowed over once every ten working days.

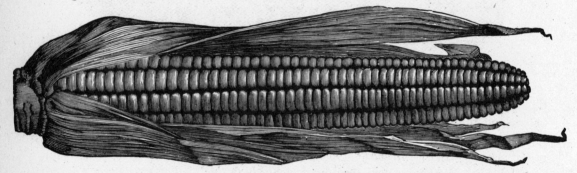

YELLOW FLINT CORN.—Two-thirds Natural Size. See page 122.

X. Depth of Cultivation.

There is a diversity of opinion as to the proper depth of cultivation. Our experience is that the cultivation should be to the depth of about three inches while the corn is young, but after it has made good root, the cultivation should be superficial. By the time the corn is knee-high the soil becomes pretty well filled with roots. In moist weather if the roots are torn they will quickly recuperate; if they are torn in dry weather a decided injury ensues. After the corn begins to shoot, that is, to joint, and prepare for blossoming, cutting the roots is a decided injury. After this time the cultivation, if any be necessary, should be simply surface stirring, not more than an inch and a half deep. Roots do not penetrate the soil by forcing their way through solid earth; that is impossible. They find their way between the minute interstices which are always present, however compact the soil. If the surface is kept mellow, the sub-surface never becomes so compact but the pores are amply sufficient for the roots. Corn is a fast-growing crop under heat and moisture. It is a crop that must have mellow soil to give the best results. Other crops, such as wheat, onions, etc., require compact—not hard—sub-surface. Deep cultivation is not required for what are known as hard or compact soil crops. It is decidedly injurious to the soft-soil crops after the ground becomes filled with roots. A safe rule for corn, is to give deep and clean cultivation while the crop is young; deep cultivation in the middle of the rows, while the corn is eighteen inches to two feet high. After that the cultivation should be shallow—simply sufficient to keep the surface fine and mellow. When the corn fully shades the soil, the earth will no longer be beaten down by the rain. It will not be liable to crust, nor will it become impacted or lose much moisture by evaporation at the surface. The roots will arrest all this.

XI. Harvesting the Crop.

There are two ways of securing the crop of corn—by husking on the hill, and by cutting and shocking, and husking from the shock. There are only three conditions under which corn should be husked and shocked: 1, when the fodder will pay for the

A FIELD OF SHOCKED CORN.

extra cost of cutting and shocking, and the extra cost of husking from the shock; 2, when the corn is to be fed to cattle directly from the shock; 3, when, from danger of early frosts, it becomes necessary to shock the corn to assist it in ripening.

It costs about as much to cut and shock an acre of corn as to husk an acre on the hill, or when the corn stands in the field as it grew. It takes twice as long to husk an acre of corn from the shock, and tie up and re-shock the corn, as it does to husk it standing in the field.

Husking from the Hill.—A team should be allowed to every two men. The wagon should be provided with one wide extra side-board, with cleats on each side, so they will slip down easily over the ordinary side-board. This is to prevent the corn from flying over when thrown into the wagon. The wagon should always be to the right of the huskers if possible, and two or four rows may be husked at a time. A short board, ten inches wide, should slant into the rear of the wagon, for ease in shoveling out the corn. When the wagon is filled and goes to the crib, the remaining hand husks and throws the corn in piles on the ground, to be picked up on returning. By this means, if the corn is dry, about one acre may be husked a day by each good hand. We have known one man thus to husk seventy-five bushels in a day, and it is said that one hundred bushels have been husked in a day by one man. It is certain that a man will husk an acre of heavy, sound, dry, standing corn, easier than an acre of soft and inferior corn, even when the yield in the first case is double what it is in the latter.

XII. Cutting and Shocking.

CUTTING, shocking and tying corn, like any other work on the farm, must be done systematically. The rows of shocks must run continuously through the field, and at regular intervals. If set around a single hill, a corn-horse should be used to support the stalks until ready for binding. This consists of a strong, smooth paling twelve feet long, sharpened at one end, with an inch and a quarter hole two feet from the end, and a pair of feet three feet beyond, as shown in the cut. A rod five feet long, sharpened at one end and fitting loosely in the hole, holds the corn until bound, when it is withdrawn and carried to the next shock.

CORN-HORSE FOR SHOCKING.

Cutting.—The diagram shows the manner of cutting, when forty-nine hills are placed in a shock, to save travel in carrying the corn. Begin at a, and cut three hills as shown on the dotted lines, and carry to the shock. Then walk to b, and cut as designated, and so in succession until you get around back, and cut the three hills, beginning at i. Then bind firmly with a twisted hay-band.

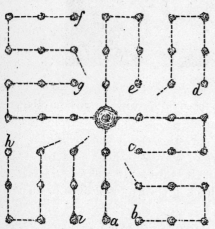
CUTTING AND SHOCKING CORN.

Shocking around Tables.—In the West, where the fields are exposed to the full force of the winds, shocks are often made sixty-four hills square. By this plan the shocking places may be formed by twisting four hills diagonally together, so that what is known as a table is formed for each shock. Formed in this way, and securely tied, they will stand upright and secure against the strongest winds. One man goes ahead of the cutters, and forms these tables at regular intervals, by twisting two hills diagonally together, and then another two. Afterwards he returns and ties the shocks.

Making the Bands.—Never trust to binding with corn-stalks or other material found in the fields. Get your blacksmith to make you an iron crank with a simple hook at one end. Before the hook is turned, slip a section of a rod suitable for gripping, and properly bored, upon the shank. To twist the bands, throw down a lot of hay upon the barn floor, shaking it up very light, moisten it, catch a little with the hook, and begin walking backward, twisting as you go, while a boy feeds the hay to the hook. When you have gone the whole length of the barn floor, while the boy holds his end securely, slip your end off the hook and roll the whole into a ball, and pass a skewer through the end of the hay rope and ball, first tucking the end under so it will not pull out. Proceed in this way until you have bands enough for your field. You will be surprised to find how many you can make in a day.

A Binder.—One difficulty in binding large shocks of corn is, that one man cannot put the hay-band around the top and draw tight enough. The illustration of implement to draw the tops of the shocks securely for binding, almost explains itself. A is a piece of hardwood inch board, two feet long, or more, and five inches wide, bored with three holes, the two outside ones to receive a ten-foot cord, and the center one to take the shank of the crank, C. The board is laid against the shock, the spindle-shank is thrust through the center hole, one end of the rope is passed through an outside hole and fastened to the crank, as shown at C. The rope is passed around the shock, and the other end fastened to a hook at the hole at the other end of the board. Then the crank is turned until the whole is drawn together, the hay-band is fastened above, and, the crank being let go, the spring of the shock holds all secure. It is, perhaps, unnecessary to say that the balls of band should be dipped in water before carrying to the field. By placing them out of the sun and wind, they will remain moist until used, and one ball may tie three or four shocks, the cut end every time being held by the skewer passing through the ball.

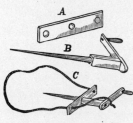
CORN-SHOCK BINDER.

XIII. Seed-corn.

SEED-CORN that will germinate surely is indispensable. Have you carefully selected at husking-time, or before, the soundest and most perfect ears, and attended to their careful curing? If not, lose no time early in the spring in selecting the best you have, again carefully sorting this over. From that you think is pretty sure to grow, shell a small quantity from a number of ears selected as they run, mix all well together, count out fifty grains, place them between folds of flannel cloth, kept constantly moist and at a temperature of fifty-five to sixty degrees, not more. Corn does not germinate at a temperature much below fifty degrees. Note the time it takes to sprout. If it does so in seven or eight days it is good. Ascertain the number of grains that come up promptly, and you can decide how much to drop in a hill. If your corn proves bad, buy good seed, whatever may be the cost. You cannot afford to risk uncertain seed. There are contingencies enough, even with the best seed. Never neglect carefully to select and save seed-corn in the autumn.

XIV. Cost of a Corn Crop.

There is no crop that varies more in its cost than corn. Manure, rough land, hand-hoeing and small fields are expensive. In the great corn region of the West the cost is reduced to a minimum. Some years ago, while engaged extensively in general farming and stock-feeding in Central Illinois, every crop was itemized and correctly kept. Actual figures on a crop of corn from 1,225 acres were as follows. It must, however, be remembered that the smaller the area the more it costs per acre. Nevertheless, the smaller the field the greater the average yield. The field yielded a little over thirty-nine bushels an acre, and the tillage was at the rate of sixty-five acres to the man and team for cultivating. But the season was an exceptionally good one for working. Here are the figures:

Fall plowing,	$ 600 00
275 bushels seed-corn @ 80c.,	220 00
50 bushels seed-corn @ $1.50,	75 00
Manual labor,	1,958 13
Team labor,	1,174 25
	$4,027 38

This brings the corn ready to husk. The husking cost:

1,470 days manual labor,	$1,837 50
735 days team work,	918 75
Thus the corn cost in the crib,	$6,773 63

The crop was 48,225 bushels. That portion not fed on the farm brought 42½c. in the crib, making a total for 48,225 bushels of $20,495.63. Deduct from this the cost of producing the crop, and the balance is $13,691 for the crop, or $11.09 per acre for the use of the land.

The men were all paid at the rate of $1.25 per day, and the teams were estimated at the same price for each double team. Every individual item was correctly charged, as plowing, harrowing, rolling, planting, cultivating, uncovering corn, etc., and there was even a charge of $13.20 for cutting and pulling weeds. Looking at the matter in another light, it will be seen that the whole expense of making the crop ready to husk, for man and team, was, counting man and team at $2.50 per day, at the rate of one and one-fifth days work per acre, or, in other words, counting the value of seed-corn, the cost of raising an acre of corn was $3.29 per acre. The cost of husking was $2.25 per acre, or, per bushel, nearly six cents; the whole cost of raising and cribbing the corn was $5.54 per acre, and the corn cost, in the crib, fifteen cents per bushel to raise, not counting ground rent. There is no reason why it should ever cost more for labor, in any clean, rich soil, free from stumps, stones, or other obstructions.

A well-kept account book is always useful in enabling the farmer to tell exactly what any crop has cost, but it is a curious fact that hardly one farmer in a dozen ever keeps a record of the debit and credit on a farm. Book-keeping takes time, of course, but without it the farmer never knows on what crops he is making or losing money. A single book of, say, two hundred pages, with lines ruled for dollars and cents on the right, and one line on the left for the date of each entry, will be sufficient for most practical purposes.

XV. Varieties of Corn.

The well-defined species of corn, true to type, are comparatively few. The varieties tolerably pure are many. The six principal species may be defined as "Oregon corn," Figs. 1 and 2, in which each grain is enveloped in a separate husk; flint corn, Fig. 20, with hard, rounded grains; horse-tooth corn, Figs. 7, 13 and 19, with thin broad soft grains, roughened and scaly at the top; sweet or sugar corn, Fig. 16, containing much gluten; rice or Guinea corn (popping corn), Fig. 3, and Tuscarora or flour corn, Fig. 5, the substance of the kernel being peculiarly soft and farinaceous. The illustration shows twenty varieties, giving the forms of the grains. These are: Figs. 1 and 2, Oregon or so-called wild corn; Fig. 3, rice corn; Fig. 4, Jersey white flint; Fig. 5, Tuscarora; Fig. 6, Ohio white flint; Fig. 7, Kentucky white; Fig. 8, Virginia golden; Fig. 9, King Philip; Fig. 10, Yankee or eight-rowed yellow; Fig. 11, Samassoit; Fig. 12, improved Dutton; Fig. 13, Ohio dent; Fig. 14, small eight-rowed yellow; Fig. 15, blood red; Fig. 16, Mexican black; Fig. 17, Oregon shoe-peg; Fig. 18, Canada pop-corn (flint corn); Fig. 19, white gourd-seal; Fig. 20, golden Sioux (very dwarf).

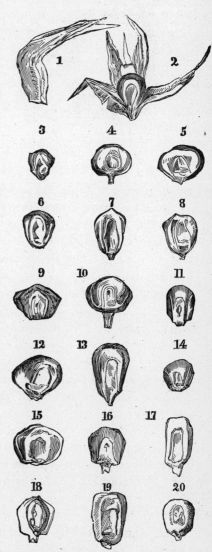

VARIETIES OF INDIAN CORN.

All the varieties of dent corn are probably crosses from the species named. All dent corn is either white, yellow—or rather orange—red and speckled. Flint corn is white, yellow, dusky and red. Sweet corn is white, cream-colored or blue-black, but varieties of all the classes vary infinitely between the colors named. The varieties of field corn in best repute are either white or yellow.

Dent Corn.—One of the oldest varieties of dent corn cultivated North is early dent, known also as Reynolds, Murdock and ninety-day dent. Varieties of this corn are known by many names. They will ripen perfectly even in the latitude of Minnesota. Another variety, larger than the above, the Hathaway or Michigan yellow dent, will ripen usually in one hundred days. It ripens up to forty-three and one-half degrees north. The illustrations on this page show the grain perfectly; other cuts in this chapter show the ears of the several varieties, longitudinally and sectionally.

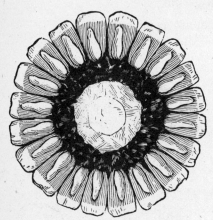

MICHIGAN YELLOW DENT.

Matchless.—A White dent corn, adapted to the Middle region and South. Originated by selection from the so-called Mammoth dent, in Pulaski county, Virginia. Ears about nine inches long by two and a fourth inches in diameter; uniform in shape; fourteen to sixteen rowed. Kernels five-eighths of an inch long and of the "horse-tooth" shape; hard and of excellent texture, closely set upon the cob. Cob small for so large an ear, and white.

Adams' Early (*Burlington*).—White dent. Ear about eight inches long, two in diameter; twelve to fourteen rowed. Cob white, small. Kernel white, deeper than broad. This variety is much used for table purposes, by those who do not like the flavor of sweet corn. It is the earliest dent corn known.

Of the mammoth varieties of white corn, the improved White dent, or Parrish corn, is well known and liked in the central corn zone. The illustration will show its characteristics.

Blount's Prolific.—This is a white half dent variety. Originated by A. E. Blount, in Tennessee, who bred it especially to develop the tendency to produce several ears on each stalk. The result is that usually two, frequently six, and even eight ears are produced on a stalk. The ears are about eight or nine inches long, eight to ten rowed, uniform in shape. Kernel white, hard, as broad as long, closely packed on the cob in straight rows. It is adapted only to the middle and southern latitudes. Stalks above the average height; they sucker freely, hence the plant is well adapted for ensilage and fodder purposes.

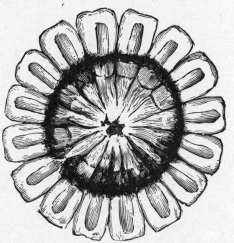

WHITE DENT OR PARRISH CORN.

Maryland Prolific.—This is a White dent variety. Ears nine inches long and upwards, two and one-fourth to two and one-half inches in diameter, slightly tapering, well filled at both ends. Cob medium size, white. This is an improvement by selection from the ordinary Horse-tooth variety of the South. Kernel one-half inch long, narrow and thin, hard, white and glassy, closely set upon the cob. A popular variety in Maryland and Virginia. Adapted to middle and southern sections, where it is principally grown for use upon the farm—wheat, cotton or tobacco being the money crop.

Horse Tooth (*Southern White and Yellow*).—Dent, South. The original type of the large-eared dent varieties. The ears are ten inches and upward in length, two and a half in diameter, and nominally sixteen-rowed, but varying from fourteen to thirty-two. The kernels are half-inch long and longer, broad and thick, rather soft in texture. The cob is large; color both red and white. This variety is extensively grown in the South, and is well adapted for ensilage.

MAMMOTH YELLOW DENT.

Among the better known of the improved varieties of Mammoth Yellow Dent corn, adapted to growth in the middle corn region of the United States, are the Mammoth, or Chester County Yellow Dent, large, prolific, rows sixteen to twenty-four in number, and the stalks leafy, making desirable fodder.

Southern Prolific.—This is a red and yellow dent variety, sixteen-rowed, adapted to the Middle region and South. It originated by selection and breeding for twelve years on the Missouri Agricultural College Farm. The ears are ten inches long and upwards, and a little over two inches in diameter; sixteen to eighteen rowed. The kernels are five-eighths of an inch long, closely set upon the cob, narrow and thick, dark to light red in color, yellow at the outer ends. Cob small and white.

Little Red Cob.—A white dent, eighteen-rowed variety. Ears seven to nine inches long, two and a half inches through. Cob small and pale-red in color. Kernels white, five-eighths of an inch long, thick and narrow, quite hard and heavy. Stalk six feet high by one and a half inches through; very leafy and yielding a large amount of fodder. It is largely cultivated in Georgia; is an excellent stock corn, and adapted to the lower middle and southern corn-growing sections of the country.

North Star.—Another variety, adapted to the North; ripening in Southern Minnesota and Dakota; is a red cob, twelve to sixteen rowed; variety known as the "North Star Yellow Dent," prolific and ripening in one hundred days. The cut shows its characteristics of cob and grain.

Flint Corn—The flint varieties of corn are little grown in the West, except pretty far north, and even there the newer, extra-early varieties of dent corn are favorites. East of the Alleghanies and in Canada, the flint varieties are almost universally grown. For meal, the flint corn is superior, and the large round-grained varieties are used for hominy.

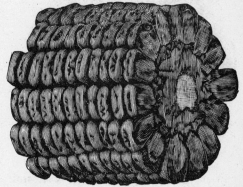

NORTH STAR.

The flint varieties are also said

to be weevil-proof; but this is not so, as all will testify who examined the foreign samples of flint corn at the Centennial Exposition.

White Pearl or Hominy Corn.—This is the variety usually grown for hominy and samp. It is also the best variety for making hulled corn. The stalks are large and tall, the grains pearly white and flinty, ears nine inches and upward in length by one and a half inches in diameter. It is a southern variety, ripening always up to forty degrees of latitude.

Yankee Corn.—The Early Eight-rowed Flint or Yankee Corn may be regarded as the type of the several varieties of yellow flint corn. The improved yellow flint has a small cob, a deep yellow grain, and among the earliest of any of the varieties. Its characteristics are shown in the cut, natural size.

EIGHT-ROWED FLINT CORN.

Waushakum Corn.—This is without doubt one of the best of eight-rowed flint varieties, and is strictly a pedigree corn, originating by selection, and having been improved by careful cultivation by Dr. E. L. Sturtevant, of Massachusetts, well known throughout the country from his contributions to practical, scientific agriculture. The ear is perfect in shape; about nine inches long, eight-rowed; about forty-five kernels to the row; rows straight and even, full from tip to stem; the ears uniform in size from end to end. The kernels are compactly set upon the small cob, are flinty, dense and heavy. The section of ear shows its characteristics.

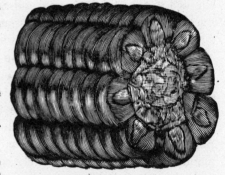

WAUSHAKUM CORN.

Silver White.—Of white flint varieties the silver white flint, originated by H. E. Alvord, of New York, is also a pedigreed variety, exceedingly early, hardy and prolific.

Following are brief descriptions of some of the more reputable varieties of flint corn:

Compton's Early (*Yellow Flint*).—Ten-rowed. Ears ten to twelve inches long, well filled, often two on a stalk. Kernels bright yellow, medium size. Stalk eight to ten feet high.

Dutton (*Yellow Dutton*).—Yellow Flint. Ears nine to ten inches long, ten to twelve rowed; rows close together; ears uniform, symmetrical, tapering, well filled at both ends. Kernel as broad as deep, bright yellow color, flinty; of superior quality for meal. Cob above the average, white.

SILVER WHITE FLINT CORN.

Early Canada (*Canada Yellow*).—Eight-rowed. Ear small, symmetrical, seven to eight inches long, tapering from butt to tip; rows separated into pairs at the butt. Cob small, white. Kernel as broad as deep, compact, flinty, smooth, of a deep yellow color; of superior quality for meal. Stalk five or six feet high, slender, leaves not plentiful, bearing one or two ears near the ground. Matures very early. The yield is light on account of the small size of the ear, but its extreme earliness makes it very desirable for the more northern sections of the corn belt.

Improved King Philip.—Copper-colored Flint; eight-rowed. Ears ten to twelve inches long, uniformly eight-rowed when pure. Cob below the average, pinkish white. Kernel copper-red, or brown, varying to yellow. Kernel large, somewhat broader than deep, smooth, glossy and hard. Stalk six feet high and upwards. In favorable seasons ripens in about ninety days.

Longfellow (*Yellow Flint*).—Eight-rowed. Ears ten inches long and upwards, some of them fifteen inches; one and one-half to one and three-quarter inches in diameter; uniform, cylindrical shape, well filled at both ends. Cob small. Kernel very large, broader than deep.

Red Blazed (*Yellow Flint*).—Blazed, or striped, with red. Ears large, well filled at both ends. Eight-rowed; early.

Rural Thoroughbred Flint.—Dingy white Flint. Introduced by the *Rural New Yorker* in 1882. Ears eight-rowed, ten inches long, often fifteen, and occasionally sixteen and seventeen inches, slightly tapering. Cob large and white. The kernels are broad and short, a dull white in color, hard and flinty. The stalks have a habit of suckering to an unusual degree, making it of great value as a fodder plant. But one plant is grown in a hill. The main stalks frequently bear two ears.

Sanford.—Dingy white Flint, eight-rowed. Ear eight to ten inches long, one and one-half inch in diameter; slightly tapering; rows separated in pairs by the rather large white cob; kernel broader than deep, hard. Stalks large and leafy.

XVI. Rice and Its Cultivation.

RICE is probably one of the earliest cultivated plants in tropical and sub-tropical countries. It was already known in China three thousand years before Christ, and in India it has been cultivated from time immemorial. The ancient Greeks and Romans were well acquainted with it. It is said to have been brought to Sicily by the Arabians. It was introduced into Virginia in 1647—probably the upland rice. The true water rice (*oryza sativa*) is said to have been introduced into South Carolina accidentally, by a vessel from Madagascar, which put into Charleston bay in 1694. Four years after that, in 1698, sixty tons were shipped thence to England. In Louisiana, its cultivation began in 1718. It is now cultivated in thirteen Southern States, both swamp and upland varieties. Upland varieties have been cultivated as far north as Missouri, Illinois, and even Iowa. In 1850, seven hundred pounds were raised in the latter State. It is not to be supposed that upland rice can be raised with profit in the North. The cultivation has always been experimental, like that of cotton, and will always remain so, unless some varieties of the wild rice of the

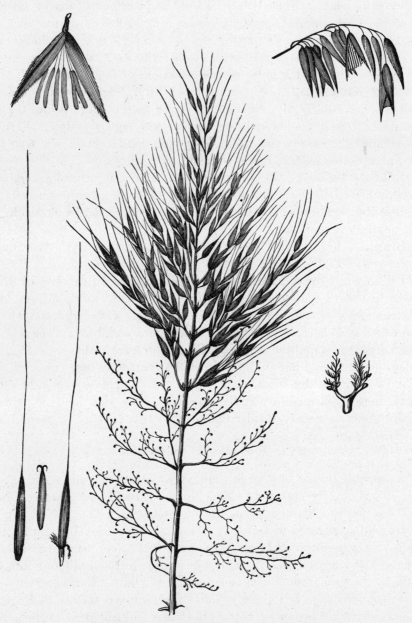

WILD RICE OF THE NORTHWEST. (*Zizania Aquatica.*)

North (*Zizania Aquatica*) may perhaps be improved. Of this, three species are common, in still-running water and ponds, as far north as Minnesota. These species are *Z. Aquatica*, *Z. Miliacia* and *Z. Fluitams*. It is a true rice, as the engraving of *Z. Aquatica* will show, which includes the inflorescence and grain.

The following description of this plant is from the reports of the Department of Agriculture, at Washington: "The Sioux call it pshu, and the Chippewas man-om-in. It is a constant article of food with the northern Indians of the lakes and rivers between the Mississippi and Lake Superior. This plant delights in mud and water five to twenty feet deep. When ripe the slightest wind shakes off the grains. After being gathered it is laid on scaffolds about four feet high, eight wide, and twenty to fifty long, covered with reeds and grass, and a slow fire is maintained beneath for thirty-six hours, so as to parch slightly the husk, that it may be removed easily. Its beard is tougher than that of rye. To separate it from the chaff or husk, a hole is made in the ground a foot wide and one deep, and lined with skins. About a peck of rice is put in at a time; an Indian steps in, with a half jump, on one foot, then on the other, until the husk is removed. After being cleaned the grain is stored in bags. It is darker than the Carolina rice. The hull adheres tightly, and is left on the grain, and gives the bread a dark color when cooked. The husk is easily removed, after being exposed to heat. In Dakota the men gather this grain, but all other grain the women collect. An acre of rice is nearly or quite equal to an acre of wheat in nutriment. It is very palatable, when roasted, and eaten dry."

Upland Rice.—The upland rice is cultivated precisely like oats, being usually sown broadcast. It is more than probable, however, that a system of drilling and gang cultivation, such as is practiced with the sugar beet, would render this industry highly remunerative in the South, since it would not add more than five dollars per acre to the cultivation, while it would raise the product, per acre, from ten to fifteen bushels to thirty or forty and more bushels per acre.

XVII. True Water Rice or Commercial Rice.

The Carolinas, and other suitable rice districts south, are the best rice-producing region of the world. The rice of Mississippi and other Gulf States has, however, never equaled Carolina rice, raised on the alluvial lands on the brackish water system of the State, when proper measures are taken to keep out salt water, which is fatal to the crop. The cultivation is fully described in communications to the Department in Washington, from which we condense. The plan on the Savannah river is described in the following article:

XVIII. Cultivation of Rice in Carolina.

Main canals having sluices at their mouths are dug from the river to the interior about twenty feet in width; and, as they very frequently extend across the whole breadth of the swamp, they are more than three miles in length. The rice plantations are subdivided into fields of about twenty acres each. The fields have embankments raised around them, with sluices communicating with the main canal, that they may be laid dry or under water separately, according as it may be required. Open ditches are dug over the grounds for the purpose of allowing the water to be more easily put on or drawn off.

In all cases the water is admitted to the fields as soon as the seed is sown, and when the young shoot appears above ground, the water is drawn off. In the course of a week the crop usually receives another watering, which lasts from ten to thirty days, according to the progress the vegetation makes. This watering is chiefly useful in killing the land weeds that make their appearance as soon as the ground becomes dry. But, on the other hand, when the field is under water, aquatic weeds, in their turn, grow up rapidly, and to check their growth the field is once more laid dry, and the crop is then twice hand-hoed.

By the first of July the rice is well advanced, and water is again admitted and allowed to remain on the fields until the crop is ripe. This usually takes place from the first to the tenth of September, and the water is drawn off the day before the crop is to be reaped, or long enough to dry the land for this operation.

Large Capital Necessary.—Large capital is necessary in the cultivation of rice, as well as good judgment, especially on tidal lands. The banks must be kept in order, the drains and canals must be kept clear, and the sluices and valves must always be in repair. The rice-swamps also are unhealthy, and it is difficult to get labor when other work can be had; hence, higher prices must be paid. Nevertheless, properly managed, the crop is remunerative, and rice plantations used to be the highest-priced lands in the South,—the best lands lying between twenty-nine and thirty-five degrees north latitude. The best variety is that known as Golden or Carolina rice.

XIX. Management of Rice Fields.

The authority previously quoted describes the management as follows. It is the most concise and comprehensive we have seen: Rice plantations are located above the junction of salt and fresh water, from the fact that rice, being an aquatic plant, requires a vast amount of fresh water during its growth; salt water being fatal to it at all stages. These swamps are usually reclaimed by means of banks or levees, which are made high and strong enough to bar out the river. Smaller embankments, called check banks, subdivide that portion of the plantation lying between the main river embankment and the high land, into squares or fields, generally from fifteen to twenty acres in area. These squares are all subdivided again into beds or lands, of twenty-five or thirty feet width, by a system of main ditches and quarter-drains. Canals from twelve to thirty feet wide and four or five feet deep, are sometimes cut from the river embankment, through the center of the plantation, to the high land, for the purpose of introducing or draining off the water to or from those fields situated far back from the river.

Flood-gates.—Flood-gates or trunks having doors at both ends are buried in the embankments on the river, as well as in the canal embankments and the check banks, those at the outlet of canals being so constructed as to permit the flat-boats to pass into the river. By means of these flood-gates or trunks the whole system of irrigation is carried on under the complete control of the planter, and the lands are flooded or drained at will. The canals and ditches being all carefully cleaned out, down to the hard bottom, the banks neatly trimmed and free of leaks—the floodgates and trunks all water-tight, either to hold out or hold in water—the planter commences his operations, as early in the winter as possible, by plowing.

These lands, being yearly enriched by alluvial deposits from the river, do not require deep plowing, four or five inches being generally sufficient to furnish a good seed-bed, and on account of the numerous ditches subdividing the fields, a single mule plow is always preferable. When lands are plowed early in the winter and nicely shingled, it is of very great advantage to put in a shallow flow of water, and suddenly draw it off, in severe weather, for the benefit of freezing the furrow slices. But it is not a good practice to flood deep, as the weight of water packs the land, which becomes run together by the action of the waves, and renders good harrowing afterward an impossibility.

XX. Cultivating the Crop.

HARROWING is usually begun only a few days previous to planting, in order that the seed-bed may be as fresh as possible, to encourage germination, and by its pliancy, permit the young roots to expand rapidly and take good hold on the soil, in order that the plant may resist the birds and a tendency to float. The operation of harrowing is one of the most important to the crop, and no consideration should induce the planter to slight it, as this is the opportunity afforded him for killing his potent and pernicious enemy—grass—the great obstacle all the summer-time. By breaking up every clod now, and exposing its roots and seeds to the action of the sun, half the battle is won. Immediately after the harrow comes the crusher, which implement is not abandoned until the field is reduced to garden tilth.

Seed and Seeding.—About the tenth or fifteenth of March, up to the tenth or fifteenth of May, the process of drilling is carried on, seeding from two and a half to three bushels of clean seed per acre. At this juncture two antagonistic systems are encountered, one known as covered rice and the other as open-trench rice. Both have their advocates. The first system, or covered rice, is where the grain is covered up in the soil two or three inches deep, as fast as it is drilled in, which thus protects it from birds, floating away, etc. The other, open-trench, consists in leaving the rice entirely uncovered in the drill, and taking the risks alluded to, in order to save time and labor, the grain being soaked in thick clay water before seeding, to hold it to the ground.

The seed being deposited, the flood-gates are immediately opened, and, if it be covered rice, and the ground pretty moist, the water is taken in as rapidly as the capacity of the gates may afford; and when it has attained a depth of twelve or eighteen inches, or deeper, if the check banks can bear it, the water from the river is then shut off, and the inside gate is closed, to hold in what water is on the field. The trash now rapidly rises and floats toward the banks, and it must be immediately hauled up with rakes, before it settles down on the rice. In the course of a few days the seed is carefully examined, and as soon as the germ or pip appears the water is drawn off the field to the bottom of the ditches, and kept out until the rice has two leaves.

XXI. Flooding the Crop.

If the grain is planted by the open-trench system, as soon as the seeding is done, the water is led into the field gradually, until the land sobs and the rice sticks, then it is flooded slowly until the previously mentioned depth is attained; the water is then held until the rice has good roots, or begins to float, and is then drawn off carefully. Here all difference in the culture ceases. The rice having two leaves — or earlier, if the field is inclined to be grassy — the water is again let in to the same depth as before, completely submerging the plants, and is held to this gauge from seven to ten days, the planter being governed by the weather. If warm, seven; if cool, ten days. Then a lead is put in the gate, and the water let off gradually, until a general verdure is seen floating all over the field. At this point the water is stopped, and a mark set upon the gate as a gauge-mark. To this gauge the water is rigidly held for sixty or sixty-five days from the day it first came on the field. This flow, when properly managed, effectually destroys all tendency to grass, and promotes a vigorous growth of rice.

It sometimes happens that, during this flow, the crop takes a check and stops growing. In this event to take off the water is fatal, as it will produce foxed rice; it must be held firmly to the gauge, and in a few days the plant will throw out new roots and go on growing. If the maggot attack it in this flow the water is drawn off for a day or two and replaced. And where water is abundant and easily handled, the maggot can generally be avoided by beginning, about the thirtieth day, to change the water once a week. To do so skillfully, both gates must be simultaneously opened at the young flood. The stale water will thus rush out and fresh water come immediately back with the rising tide to float the rice leaves and prevent them sticking to the ground in their fall.

The Rice Maggot.—If the maggot gets serious the field has to be dried immediately and thoroughly. The maggot is a tiny white worm, which is generated by stale water, and attacks the roots of the plant, causing serious injury to the crop. The presence of the maggot may always be suspected by the stiff and unthrifty appearance of the field. If the land is fertile at the end of the sixty-day flow, it will be found, on drawing off the water, that the rice has attained a vigorous growth of about three feet, and is well stocked with tillers, while also, if the field is level, and the harrowing and pulverizing was thoroughly attended to before planting, no grass will be seen; nothing but rice and the clean soil beneath.

The field is kept dry now for about fifteen or twenty days, or until the land dries off nicely and the rice takes on its second growth. And if there be no grass it ought not to be disturbed with the hoe, as the hands, at this stage, often do more harm than good. This, however, does not apply to cat-tails and volunteers, which should, of course, be carefully pulled up by the roots, and sheafed and carried to the banks, to be disposed of by the hot sun. At the end of fifteen or twenty days, as above mentioned, the water is returned to the field as deep as the rice and banks can bear, never, however, topping the fork of the former. This water, where circumstances permit, is changed every week or two, by letting it off on one tide and taking it back on the next, and increasing the gauge with the growth of the rice.

XXII. Harvesting and Threshing.

When the heads of the rice are well filled and the last few grains at the bottom are in the dough state, it is fit to cut, and as little delay is permitted as possible, as the rice now over-ripens very rapidly, and shatters in proportion during the harvest. The water may be drawn off the field from three to five days before cutting the grain, and the land will be in better condition for harvesting. The rice is cut from twelve to eighteen inches from the ground, depending on its growth, usually from four to six feet high, and the gavels laid evenly and thinly upon the stubble, for the purpose of curing and permitting the air to circulate beneath it. Twenty-four hours in good weather is usually required to cure the straw, and the binding does not commence before this period, and never while the dew is on the straw. It is safer always to cut from sunrise to twelve o'clock, and bind the previous day's cutting from that hour to sunset. As soon as bound the rice is shocked up, and at the end of a week taken to the yard and stacked in ricks, thirty feet long, eight feet wide, and ten feet high. A stake, four feet long, is put into the rick at each end for daily examination, and as long as the stake does not become too hot at its point to be held by the hand, when suddenly drawn out, the rick is not to be interfered with, otherwise it is to be pulled down, aired, and re-stacked.

Threshing.—So soon as the temporary heat is over the grain is fit for the thresher. As soon as the rice is taken from the field attention is immediately given to sprouting volunteer and shattered rice, providing the crop has not been allowed to remain in the field for an indefinite period beyond the week alluded to above. This is best accomplished by instantly flooding the field quite shallow, so as to promote fermentation, and drying it again every twelve or fifteen days, for a day or two at a time. This process is continued until freezing weather sets in, and if the season has not been remarkably cool it will be found that most of this grain is destroyed. Threshing is performed by steam power. The main building is commonly built on a brick foundation, about sixty feet long by forty feet wide, having two stories and an attic; the first story being fourteen and the second twelve feet high, with what is called by workmen a square roof. At the side of this building is the engine-house and boiler-room; and in front of the main building, a little distance off, is the feeding-room, which is connected with the second story of the same by a covered way which protects the feeding cloth. In the second story is placed the thresher. The rice is brought in sheaves from the ricks to the feed-room, where hands are stationed for the purpose of placing it on the feed-cloth in close succession. The revolutions of the cloth thus keep a continuous stream of grain flowing into the cylinder, which in turn is relieved by the rakes seizing the straw, and after tossing out the grain they throw it out of a window in the rear into wagons below, kept ready to receive and carry it away. A good engine, with machinery, will thresh and clean, ready for market, one thousand bushels of rice per day.

XXIII. Hulling for Market.

The rice as it comes from the machine is called rough rice. Twenty-five bushels of good, well-cleaned, rough rice, weighing forty-five pounds per bushel, will make a tierce of six hundred pounds of clean rice fit for the market. The cleaning is usually done at city mills that are fitted for the purpose. The rough rice is first carried between very heavy stones, running at a high speed, which partially removes the rough integument, or hull chaff. This chaff is passed out of the building by spouts, and the grain by similar means conveyed into the mortars, where it is beat or pounded for a certain length of time by the alternate rising and falling of very heavy pestles, shod with iron. These are operated by a revolving cylinder, armed with powerful levers, which, passing into a long opening in the pestle, about fifteen feet in length, raise it and let it fall suddenly into the mortars below. From the mortars, elevators take the rice to the fans, which separate the grain from the hulls. Thence it goes through other fans that divide it into three qualities, known as whole rice, middling rice, and small rice. The grain is finally passed through a polishing screen, lined with gauze wire and sheepskins, which revolve vertically at great velocity, giving it the pearly whiteness in which it appears in commerce. From the screen it falls immediately below into a tierce, which is kept slowly rotating, and struck on two sides with heavy hammers, all the time it is being filled, for the purpose of obtaining its greatest capacity. The tierce, as soon as full, is removed and coopered ready for market.

Good strong land, well managed, will average from forty to fifty bushels of rough rice—ten to twelve hundred pounds of clean-hulled rice—per acre. The rough rice averages one dollar per bushel of forty-five pounds.

XXIV. Rice in the Mississippi Delta.

In the delta lands of the Mississippi the cultivation of rice is somewhat different from that of the tide lands of the Atlantic coast, and much less elaborate. The Mississippi usually begins to swell in the delta region about the end of February, and continues to rise until the first of June, from which time it again gradually subsides. It is thus in flood during the hot season. A ditch, having a sluice at its mouth, is dug from the river toward the swamp. The land immediately behind the levee being the highest, is cropped with Indian corn and potatoes; but at a little distance from the river, where the land is lower and can be flooded, it is laid out in narrow rice-fields, parallel to the river, inclining off from the river's edge. The narrow strips are banked all around, so that they can be laid under water after the rice is sown. The land is plowed in March, and shortly afterward it is sowed and harrowed. As soon as the young plants appear above ground, the water is admitted for the purpose of keeping the weeds in check. The crop grows rapidly, and the depth of the water is gradually increased, so as to keep the tops of the plants just above the surface. There is a constant current of water flowing from the river into the fields and over the swamp, so that there is no stagnation, and the fields are not laid dry until the crop is ready to cut. The only labor that is bestowed in the cultivation of the crop is to pull up by hand the weeds, which are mostly grasses; and this operation is effected by men going to the fields knee-deep in water. The produce varies from thirty to sixty bushels per acre.

XXV. Some Special Crops.

FLAX, millet, Hungarian grass and canary-grass are largely cultivated in some sections for the seed. The only difference in the preparation of the soil from that for the cereals, is that the greatest care must be taken to bring the land into the highest possible tilth for sowing. Flax is sown at the rate of three pecks to one and a half bushels per acre, just before corn-planting time. Millet, Hungarian and canary-seed must not be sown until corn is well up and the nights warm, since it is easily checked by cold. Millet and Hungarian seed are sown at the rate of one-half bushel to three pecks, and canary-seed at the rate of three pecks to one bushel per acre. When fully ripe it may be cut and stacked, loose, for threshing, or harvested by binding and shocking. The cultivation of the more important special crops will be treated of in succeeding chapters.

CHAPTER III.

MEADOW AND PASTURE GRASSES.

I. THE VALUE OF GRASS.—II. WHAT IS GRASS?—III. HOW TO KNOW GRASS.—IV. TESTING THE VALUE OF SPECIES.—V. THE VALUE OF ACCURATE KNOWLEDGE.—VI. WELL-KNOWN CULTIVATED GRASSES.—VII. GRASSES FOR HAY AND PASTURE.—VIII. A LIST OF GOOD GRASSES.—IX. VALUABLE NATIVE WESTERN GRASSES.—X. DISAPPEARANCE OF NATIVE GRASSES.—XI. VALUABLE INTRODUCED GRASSES, SOUTH.—XII. BERMUDA GRASS.—XIII. GUINEA GRASS.—XIV. BROME OR RESCUE GRASS.—XV. SEEDING MEADOWS.—XVI. THE ALPHABET OF AGRICULTURE.—XVII. SOWING FOR HAY AND FOR PASTURE.—XVIII. THE CELEBRATED WOBURN EXPERIMENTS.—XIX. A SUMMARY OF MEADOW GRASSES.—XX. ABOUT PASTURES.—XXI. GENERA, SPECIES AND VARIETIES.—XXII. FAVORITE PASTURE GRASSES.—XXIII. BENT GRASSES.—XXIV. ORCHARD GRASS.—XXV. GRASSES FOR VARIOUS REGIONS.—XXVI. CLOVER IN ITS RELATION TO HUSBANDRY.—XXVII. THE SEED CROP.—XXVIII. VALUABLE VARIETIES OF CLOVER.—XXIX. DUTCH, OR WHITE CLOVER.—XXX. ALSIKE, OR SWEDISH CLOVER.—XXXI. CLOVERS FOR THE SOUTH—ALFALFA.—XXXII. JAPAN CLOVER.—XXXIII. MEXICAN CLOVER.—XXXIV. IMPORTANCE OF THE PULSE FAMILY.—XXXV. INTERCHANGE OF GRASSES BETWEEN NATIONS.

I. The Value of Grass.

THE grass crop of the United States has a greater real value than any other one crop raised. The corn crop represents a greater apparent money value, and so does the wheat crop, the corn crop for 1881 having a money value of $759,482,170, and the wheat crop $456,880,427, while the hay crop was estimated at only $415,131,366. But the hay harvest is comparatively a small portion of the grass crop. In the average, hay is fed to stock in this country scarcely four months in the year, even allowing for horses and mules in the cities eating hay all the year round. On the other hand we see immense grazing areas, and millions of live-stock which subsist and grow fat throughout the whole year on grass, which they gather for themselves. Besides, hay does not form more than half the food of farm stock during the winter months. On the whole, therefore, it is safe to say that the pastures and meadows of the country undoubtedly represent four times the value of the hay crop. This would make $1,660,525,464 yearly, a sum greater than the combined values of the corn, wheat, rye, oats, barley and buckwheat crops by $189,568,261, the crops of the cereals named being computed for 1881 at $1,470,957,200. The cry used to be, "Cotton is king." Later it was, "Corn is king." Let us not forget that grass is king and always will be.

II. What is Grass?

IT is found wherever other vegetation exists, from the polar regions to the equator. It is found growing in the crevices of inaccessible mountain cliffs, and in the most arid plains and watery marshes. But its true home is wherever life is most congenial to man. There it covers the soil with a cool mantle of green, giving of its wealth to the husbandman with the least outlay of expense. A good grass country is a fertile one.

In its ordinary meaning, as used in the foregoing article, grass is simply the herbage eaten by stock. By a broader scientific classification, the grass family includes all the great food grains—all the genera of plants, whose seeds not only furnish the most important portion of the direct sustenance of the human family, but whose herbage feeds all the animals used as food for man. This vast order of plants contains, according to the late Dr. Darlington, some two hundred and thirty genera, and not less than three thousand species. Of these the *Poa* sub-family, the *Phalaris* sub-tribe, and the *Panicum* sub-tribe are most important in agriculture. The first contains wheat (*Triticum*), rye (*Secale*), barley (*Hordeum*), oats (*Avena*), rice (*Oryza*), and the largest number of the meadow and pasture grasses; the *Phalaris* sub-tribe contains that grass of doubtful value, sweet-scented vernal grass (*Anthoxanthum*), and canary-grass (*Phalaris*); the *Panicum* sub-tribe gives us Indian Corn (*Zea*), sugar-cane (*Saccharum*), sorghum, and many others, among them the famous Gama grass (*Tripsacum*), the fox-tail grass (*Alopecurus*), and the panic grasses (*Panicum*).

American Varieties.—Few countries are richer in natural grasses than the United States, yet we have comparatively few varieties in general cultivation. England has two hundred varieties of grass in cultivation; the United States less than twenty generally disseminated. Yet we have about six hundred species native to our soil. Why have we so few cultivated varieties? We have depended upon England for the trial grasses. England is a moist, cool country; the United States comparatively a dry, hot one, in its summer temperature. Hence the mistake in adopting English varieties that flourish under the continual dripping of rain and a comparatively cool atmosphere. Yet two of the most valuable of so-called English grasses, one a most valuable hay grass (timothy), the other one of the most valuable for hay and pasture (orchard grass), are distinctively and truly native American grasses.

III. How to Know Grass.

The following description of grass, by Professor W. J. Beal, of the Michigan Agricultural College, a most accomplished botanist, is concise and exact. As a first lesson in botany Prof. Beal says: "Take in your hand a straight stalk of Indian corn —for this is a true grass. The leaves are on the alternate sides of the stem, one at each solid joint, making two ranks or rows from top to bottom. As you look at a straight stem of grass, the leaves may appear to you right and left, or they may appear on the side of the stem next to you and away from you. The leaves have no teeth or notches along their edges. They can be stripped into many fibrous threads, *i. e.*, the veins are nearly parallel with each other. Observe further that the lower part of each leaf forms a sheath which surrounds the stem. This sheath may overlap, but never grows fast to the stem, except at the joint, nor does it close opposite the main part of the leaf so as to form a tight tube. It is naturally split down to the joint, and may be unrolled like a scroll without tearing away part of it. In botany, it does not generally do to put very much stress upon the structure of the leaves, but in this case it is of much importance.

Bearing Flowers.—"All grasses in their healthy condition bear flowers at some time of their life. The plan of the flowers is well illustrated by a head of chess, or rye, or wheat, or June-grass. Usually two or more flowers (sometimes one) are included within two short leaf-like bodies called glumes. Each perfect flower has from one to six (usually three) stamens, with anthers attached by a point (versatile). Outside the stamens are usually two small, green, leaf-like bodies called palets, one of which has two ribs, with a thin strip uniting them. In the center of the flower is never more than one pistil, making a fruit which is called a grain. The embryo, or germ of the plant, is just at one end, on the side, and is the part first eaten out by mice and squirrels in Indian corn. This is by no means all that applies to the botany of grasses, but enough for our present purpose.

The Corn Flower.—"In Indian corn the flowers are of two kinds, each incomplete in itself. On the tip of the stalk the branching tassel produces an abundance of fine dust called pollen. On the side of the stalk are one or more short branches, covered with a cluster of leaves, from the top of which extend a large number of slender threads. The branches become the ears, the leaves about them the husks; the threads are often called silk. One of these long delicate threads runs down to each one of the embryo kernels of corn. That each may become a kernel, it is first necessary that a grain of pollen from the tassel should fall upon the silk. Hence it grows or thrusts out (the pollen does) a very delicate prolongation all the way down to the young kernel."

No person, after mastering these three paragraphs, can be deceived in the examination of any plant to determine whether it is a true grass or not. The final study of each species is of less consequence to the practical farmer. Few botanists understand them fully. The study is difficult from the minuteness of the organs, and the species and varieties are interminable and difficult to master.

Grasses Illustrated.—This chapter is fully illustrated with exact representations of the more valuable grasses, including root, stalk, leaf and blossom or seed-head. For readers who may be interested in botany we also give full-page illustrations, enlarged, of the inflorescence of many grasses. The scientific names alone are given. Those treated of especially will also have the common name appended when known.

GRASSES AND CLOVER.

Varieties are produced by the pollen of one falling upon another. If it were not for the prepotency of certain species in fertilization, inextricable confusion would ensue. Hence, in artificial fertilization the utmost care is used to ensure cross-fertilization. (See Artificial Cross-Fertilization, Chapter IV., Section XIX., Part I.)

Flowering Illustrated.—Referring to the full-page plates showing sections of the seed-heads and single flowers of a large number of grasses, any person can identify a grass by dissecting the flower and comparing it with the illustrations until it is found. Then, if you wish to find its common English name, look in the lists of grasses given on pages 156 and 157. It must be remembered, however, that many of our native grasses have no English name. This is especially true of the varieties now lately discovered. When the signs ♀ and ♂ are affixed it shows that the plant has the male organs distinct from the female organs. As a rule, however, in the grasses, both organs are contained in the same blossom. Buffalo grass (*Buchloa*) shows the departure from this rule. Hemp and spinach have separate and distinct sexes. That is, there are distinctively male and female plants, each sex having its appropriate organs on its distinctive plants. [See pages 136—141.]

IV. Testing the Value of Species.

EVERY progressive farmer should have an experimental plat of ground on which to test the qualities of different kinds of seed, and their adaptability to the soil and climate of his farm. Some of the agricultural colleges are adopting this plan, thus greatly simplifying the task of the farmers in this respect. The farmer's experiments should be conducted as follows:

Experimental Patches.—Sow a patch with the experimental seed, here and there, in the field when it is being seeded down, being careful to stake and register the places so sown. Be careful that strictly pasture grasses be not sown with meadow grasses, and *vice versa*. Timothy, for instance, is a meadow or hay grass, and one of the best we have; as a pasture grass, it is one of the poorest, if for no other reason than that it forms a bulb at the surface of the earth. Hence it will neither stand close feeding, close cutting, nor tramping. Blue-grass is the most valuable pasture grass we have where the soil is adapted to it. It forms a firm, smooth, compact sod, starts early in the spring, and grows late in autumn, furnishing the most valuable late autumn and early winter pasture, and kindly bears close feeding. It is not, however, a meadow or hay grass. Clover, on the other hand, is a good pasture plant, and a good meadow plant; sown alone it is one of the most valuable known as pasture for swine. In all other cases, however, it should form a mixture with the true grasses.

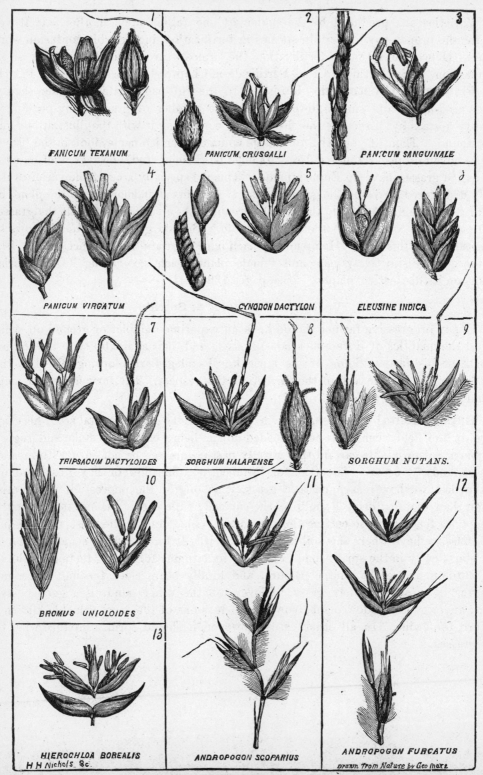

FLOWERING ILLUSTRATED. See page 135.

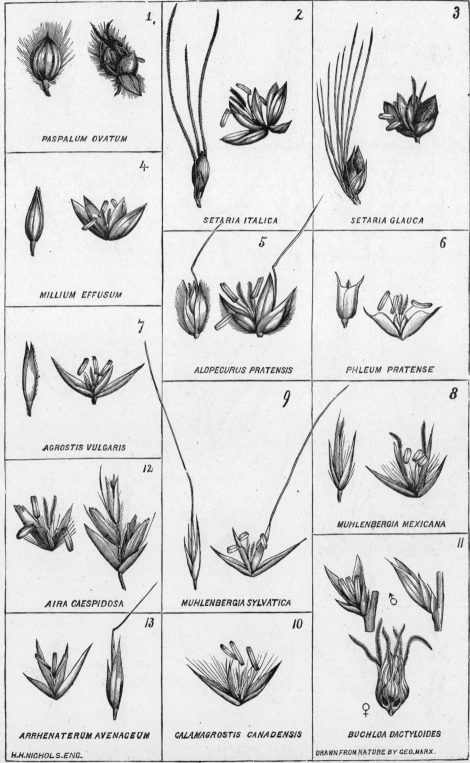

FLOWERING ILLUSTRATED. See page 135.

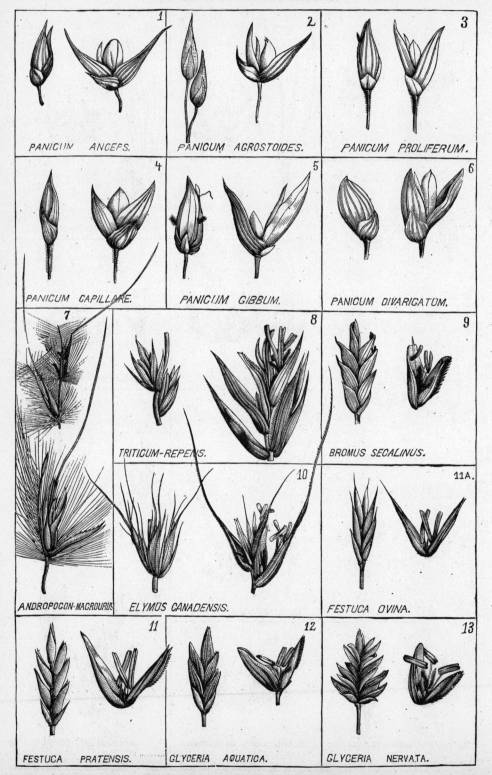

FLOWERING ILLUSTRATED. See page 135.

MEADOW AND PASTURE GRASSES.

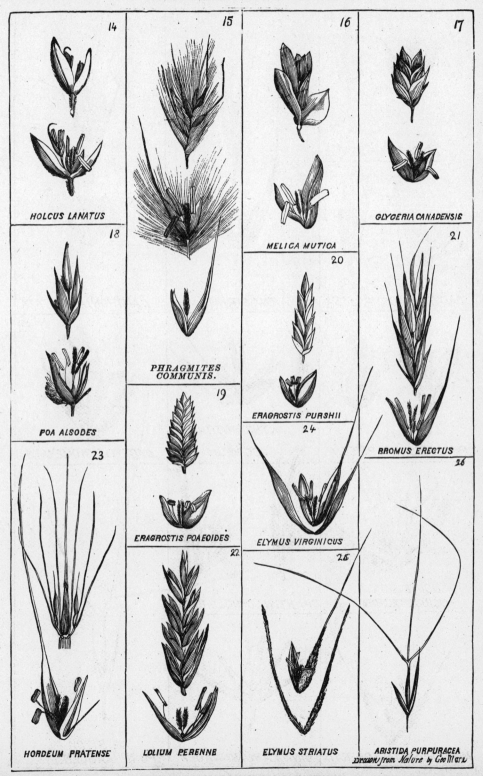

FLOWERING ILLUSTRATED. See page 135.

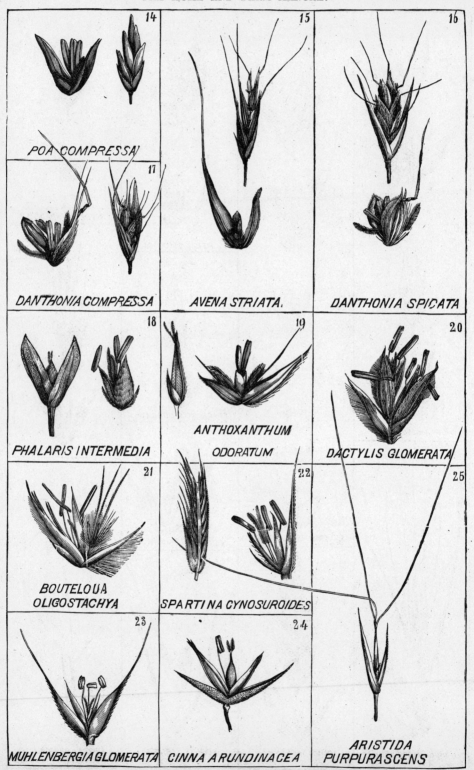

FLOWERING ILLUSTRATED. See page 135.

MEADOW AND PASTURE GRASSES. 141

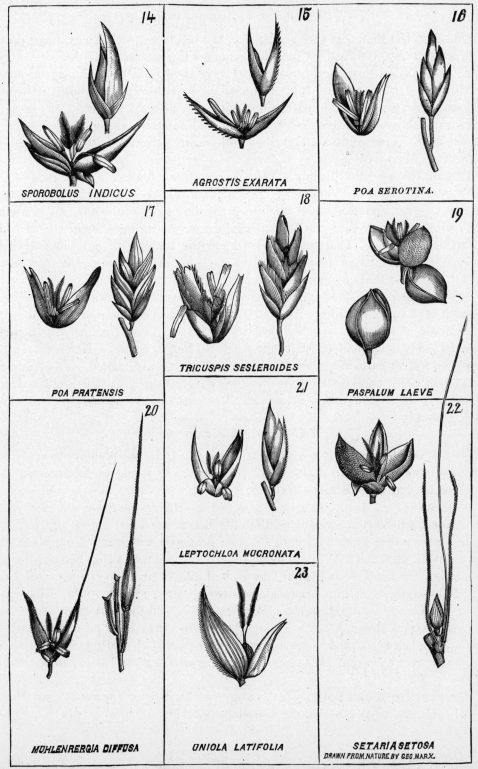

FLOWERING ILLUSTRATED. See page 135.

V. The Value of Accurate Knowledge.

A BETTER test than that given in Section IV, would be to set apart a plat divided into squares of say, ten feet. Upon these squares the several grasses to be experimented with may be sown, and a record of the relative value of each may be arrived at. Where a certain kind of soil, as moist or dry, strong clay, loam or sand is designated for a given variety, such soil may then be selected for trial, but, as a rule, those varieties that are recommended as doing well on a diversity of soils will be the most valuable. Lists of these, and also well-known varieties adapted to particular soils, will be given farther on.

Perhaps the reader will reply, What is the value of all this? Anybody knows grass. Ah! Do they? Very few, even of farmers, do. Nine-tenths of farmers would call clover and the other trefoils, grass, but they would not call the cereal grains, sorghum, sugar-cane and Indian corn, grass. A technical error like this does not cost them money. That they do not know the number of our valuable native grasses, and their relative adaptability to soils, and their relative value for pasture and hay, does cost them money—lots of it!

It is a great mistake to suppose a grass to be economically valuable for pasture because it is so for hay, and *vice versa*. The local names of the grasses are also often badly mixed. Timothy is called herds-grass in the New England States and in Michigan, and red-top is called herds-grass in Pennsylvania. In Great Britain, Timothy is called cat's-tail grass, a specific and descriptive name. The same confusion of names is true of many grasses distinctively valuable in the South. Hence, we shall give the specific name—known to all seedsmen—in parenthesis, in treating of the grasses.

VI. Well-known Cultivated Grasses.

THE economical value of a species consists of, 1, its adaptability to a climate; 2, its adaptation to the soil; 3, its feeding qualities, and 4, its productiveness.

Grasses for Dry Soils.—Many grasses of the first value in the cool and moist climate of England, are greatly lessened in value in the dryer summer climate of the New England States, and this disability becomes greater as we proceed west. For the reason that Timothy grass (*Phleum pratense*) will stand drought and heat comparatively well, and also the extreme cold of our winters, it has become the great hay grass of the northern portion of the United States, and especially of the Northwest. It also stands well on nearly all soils except dry, gravelly ones. The same is true of the Blue-grass of Kentucky (*Poa pratensis*), and the Blue-grass of the North (*P. compressa*). Rough meadow grass (*P. trivialis*), and Fowl meadow grass (*Poa serotina*) are better adapted to moist meadows; and also all the poa family, as well as Orchard grass (*Dactylis glomerata*), are not averse to partial shade in hot summer climates.

Grasses for Moist Soils.—The bent grasses or Red-top (*agrostis*), and Meadow Foxtail (*Alopecurus pratensis*), are especially valuable on moist, rather, open soils, but it is incorrect to suppose that any valuable grass is natural to a wet soil. No valuable grass ever grew naturally on a permanently wet soil. This is another of the things that the farmer must disabuse himself of. He must distinguish between wetness and moisture. The first means saturation with water; the other, saturation with the vapor of moisture—two very different things. A well-drained soil may be moist; it should never be wet.

VII. Grasses for Hay and Pasture.

In the North, Timothy and Red clover must form a large proportion of all the meadow (mowing) grasses raised. Timothy is in no sense a pasture grass, since tramping and close grazing destroy its bulb, which is situated just at the surface of the soil. So does very close early mowing. In the United States pasturing usually follows mowing. That is, after a meadow has been mowed for one or two or three years, it is then often grazed until again ready to be broken up. Herein lies the value of a variety of grasses in seeding. Not only will a greater harvest of hay be given, but in pasturing, as the Timothy is killed and the clover dies, the other grasses sown will take their place. Hence short descriptions of some of the best should be given, with soils adapted to their growth. Of Timothy—a better name would be Cat's-tail—nothing more need be said. It is known and universally cultivated all over Canada and the United States, except in the South, where it is not adapted to the climate.

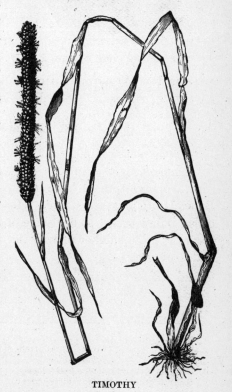

TIMOTHY

VIII. A List of Good Grasses.

The following list should be good in all average soils, and, except Timothy, are all good pasture as well as meadow grasses:

Bent grasses—Red-top (*Agrostis vulgaris*); upright bent grass (*A. stricta*).

Fescue grasses—Meadow fescue (*F. pratensis*); hard fescue (*F. duriuscula*); sheep fescue (*F. ovina*); purple fescue (*F. rubra*); spiked fescue (*F. loliacea*).

Foxtail—Meadow Foxtail (*Alopecurus pratensis*).

Rye grass—Perennial Rye grass (*Lolium perenne*).

Cocks-foot grass—Orchard grass (*Dactylis glomerata*).

Timothy or Cat's-tail grass—(*Phleum pratense*).

Poa grasses—Rough meadow grass (*Poa trivialis*); Fowl meadow grass—(*P. serotina*).

THE PROPER FORM OF STACK.

Their Qualities.—The list might be largely increased, but the idea of this work is always to keep within the bounds of practical experience. Of the grasses named, the Bent grasses make good grazing on arable soils, stiffening the sod. They are, in fact, better grazing than hay-grasses. Some of the fescues, notably Sheep's fescue, is native to the Rocky Mountain region. Meadow Foxtail is a native grass, and an excellent grass. Rye-grass is not always hardy until well established, but should be a good grass within 38° to 41° north latitude. Orchard-grass is a native, and very valuable on good loams or sandy soils. The two poas named are also natives; the latter being a common wild-grass in the West, on bottom lands, and even forms a considerable percentage of the native grasses in the Rocky Mountain valleys.

IX. Valuable Native Western Grasses.

An immense number of species of native grasses have been catalogued for the Government by the various military and scientific expeditions sent out from time to time. Many of these have not yet received English names, as some Brown, Gama, Bunch, Buffalo grasses, etc. Here, again, confusion comes in. Sheep's fescue is called Bunch grass, so are some of the Gama grasses. The following table, gathered from the large number of varieties natural to the far West, will explain itself. [See plates of grasses.]

SPECIES.		MISSOURI RIVER REGION.	ROCKY MOUNTAIN REGION.
1. Broom grass—(*Andropogon furcatus*),	Per cent.	40	16
2. Broom sedge—(*Andropogon scoparius*),	" "	20	10
3. Indian grass—(*Sorghum nutans*),	" "	20	12
4. Drop seed—(*Sporobolus heterolepis*),	" "	12	1
5. Buffalo grass—(*Buchloe dactyloides*),	" "	5	5
6. Mesquit grass—(*Bouteloua oligostachya*),	" "	0	10
7. Cord grass—(*Spartina cynosuroides*),	" "	2	2
8. Fescue—(*Festuca ovina*),	" "	0	20
9. Fescue—(*Festuca macrostachya*),	" "	0	5
10. Kalm's, Brome—(*Bromus Kalmii*),	" "	0	8
11. Fowl meadow—(*Poa serotina*),	" "	0	8
12. Feather grass—(*Stipa viridula*),	" "	0	6

The reader, by reference to the plates of dissections of grasses, will find 1, 2, 3, 5 and 11 under their scientific names, as given in the table.

Where Found.—In the above list it will be seen that 1, 2, 3 and 4 comprise the bulk of the grasses of the Missouri River region; and 1, 2, 3, 4, 6, 8, 10, 11 and 12 of the Rocky Mountain region. The first three, and the last, are native to the whole prairie, or the bottom-land region of the West; the first three comprising the bulk of the wild-hay cut. None of them, however, seed abundantly, except in moist seasons. No. 9 is an annual species and one of the bunch grasses of the mountain region.

Buffalo Grass.—No. 5 is the famous or true Buffalo grass, and may be recognized at once by its low, dense, tufted growth; also by its stolons, from which it spreads rapidly. It seldom if ever attains to the height of over two or three inches, except with its male flower stalks, which sometimes reach two or three inches above the leaf growth. There are, however, several plain and mountain grasses called Buffalo grass. The true Buffalo grass (*Buchloe*) grows most abundantly in the central region of the plains, and affords nutritious grazing for domestic animals; yet its value as a winter forage plant is not to be overlooked, as its stolons remain green during the winter months, and, combined with the dead leaves, afford to close-grazing animals a reasonably good living. In southern Kansas, the plant reaches its eastern limits, about one hundred miles west of Fort Scott.

Gama Grass.—No. 6 of the table. The Muskit, Mesquit, Gama, or Gramma, grass, contains a number of species, the one named being the principal. The name by no means applies to the species or to the several species of the genus exclusively, but is given by the mountain men to

PRAIRIE BLUE-JOINT, OR BROOM GRASS.
(*Andropogon Furcatus.*)

several other species of different genera. It is a most valuable species for grazing purposes, but grows too thinly and too short to be cut for hay. It abounds chiefly in the mountain regions and the adjacent plain districts, and may be readily distinguished from species of other genera by its peculiar spikelets of flowers all arranged on one side of the rachis, and pointing in one direction. It supports on its stalks from one to three or four, and sometimes five of these spikes, which are purplish, or of an indigo-blue tinge. Its general height is about twelve inches, but in sterile locations much less. The leaves and stems are smooth, having no hairs. It is perennial. Much of the beef of the Southwest is claimed to be the product of this grass.

No. 10 is a slender, tall grass, with a handsome head of drooping or nodding spikes of flowers. Where it grows plentifully it gives excellent pasturage. Fowl Meadow Grass, No. 11, is well known, and No. 12 has a plentiful supply of leaves, and affords much mountain grazing.

X. Disappearance of Native Grasses.

In the settlement of a new country the natural grasses soon disappear under close grazing and the constant tramping of herds. Hence the early necessity of turning attention to the preservation of the more valuable varieties by cultivation and saving their seeds; else recourse must be had to varieties already under cultivation. Experiment should be made in both directions, and especially with the grasses of the prairie and plains region of the West and Southwest. A natural grass region must necessarily contain species valuable for perpetuation. The hot, dry character of our summers, intensified as we go west, and our cold winters, often with bare ground, is not conducive to "the struggle for life" of moisture-loving plants against rank weeds. Hence the far West must be content with comparatively few species. What is wanted on the great plains region is, plants that will do there, as far as may be, what Blue-grass has done for the central Western States, or those east of the Mississippi. Grasses with creeping root stalks would naturally be suggested, if strong and growing perennial.

BUFFALO GRASS—(*Buchloe*).

The Lyme Grasses.—Some of the Lyme grasses, or wild rye (*Elymus*), have been recommended, and among them the following: Virginia Lyme Grass (*E. Virginicus*). A hardy species of early growth, producing an abundance of large, succulent leaves when young; a widely distributed species in America. Cultivation greatly accelerates its growth. It is a promising species. Siberian Lyme (*E. Sibericus*). Native of Europe and America, in the colder latitudes; would probably succeed well in the northern districts, where it is native. Canada Lyme (*E. mollis*), found on the shores of Lake Superior, and north. Grows early and spreads by its running root-stalks; foliage, when young, tender and juicy. It thrives rapidly well in a variety of situations different from its habitats (sandy shores); probably not adapted to a dry soil, but well worthy of trial; leaves broad, rather short, with a glaucous hue and strong wheat grass flavor. Some of the couch grasses (*Triticum*) will undoubtedly prove valuable.

XI. Valuable Introduced Grasses South.

Broom grass (*Andropogon scoparius*) and other varieties of broom sedge, are said to contribute largely to the pastures of the dry, pine-woods regions of the South while yet young and tender.

Drop seed (*Muhlenbergia diffusa*) forms the bulk of the woods pasture, after the rains set in; not especially valuable, but abundant, and imparting an agreeable flavor to butter.

Wild fescue, or oats grass (*Uniola latifolia*) is thought to be valuable; it is early, with rich foliage, vigorous in the South, and even found as far north as Pennsylvania, delighting in damp, sandy loams, where it forms large tufts, affording abundant and early feed. When cultivated, it gives good crops of hay.

Gama Grass of the South.—This is well known as a rank, strong, growing grass. If cut before the seed-stems shoot up, it makes good hay, and may be cut several times in a season; the fodder is said to be equal to corn fodder when cured. The seeds do not vegetate readily, and it is generally propagated from the roots, by setting out strips two feet apart each way, when if on good soil they will soon meet. It is a tall perennial grass, with solid culms, broad and flat leaves, and with flower-spikes from four to eight inches long, produced from the side joints or from the top, either singly or two or three together. The upper portion of these spikes is staminate or male, and the lower portion pistillate, and producing the seeds. It grows from three to six feet high, with large broad leaves, resembling those of Indian corn.

XII. Bermuda Grass.

(*Cynodon dactylon*) is undoubtedly one of the most valuable of the grasses of the South, but long detested by cotton growers, as were all of the persistent grasses when the South was engaged in producing only special crops. Now that diversified farming is more generally carried on, it is acknowledged to be one of the most valuable of Southern forage plants. It rarely or never produces seed in the United States, and is

GAMA GRASS OF THE SOUTH.
(*Tripsacum dactyloides.*)

propagated by chopping the roots into pieces, sowing them, and plowing them lightly under. It is a common pasture grass of the West Indies and other warm winter climates.

Mr. Charles Mohr, of Alabama, in a communication to the Department of Agriculture, says it thrives in the arid, barren drift-sands of the sea-shore, covering them by its long, creeping stems, whose deeply penetrating roots impart firmness to a soil which else would remain devoid of vegetation. It is esteemed one of the most valuable of our grasses, either in the pasture or cured as hay. Col. T. C. Howard, of Georgia, says, while we have grasses and forage plants that do well when nursed, we have few that live and thrive here as in their native habitat. The Bermuda and

Crab grasses are at home in the South. They are not only live, but live in spite of neglect; and when petted and encouraged, they make such grateful returns as astonish the benefactor. In relation to killing this grass, Colonel Lane says: "Upon any ordinary upland I have found no difficulty by close cultivation in cotton for two years. It requires a few extra plowings to get this sod thoroughly broken to pieces."

Dr. J. B. Killebrew, late Commissioner of Agriculture of Tennessee, and well known for his careful investigations of grasses valuable in the South, writes that in Louisiana, Texas, and in the South generally, it is, and has been, the chief reliance

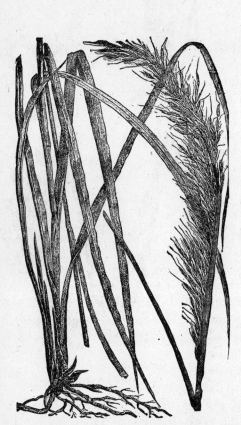

INDIAN GRASS—(SORGHUM NUTANS.)

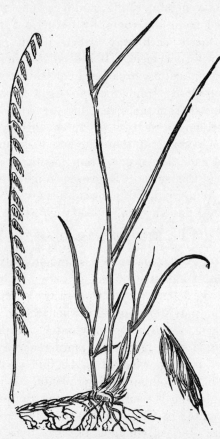

MESQUIT GRASS.

for pasture for a long time, and the immense herds of cattle on the southern prairies subsist principally on this food. It revels on sandy soils, and has been grown extensively on the sandy hills of Virginia and North and South Carolina. It is used extensively on the southern rivers to hold the levees and the embankments of the roads. It will throw its runners over a rock six feet across, and soon hide it from view, or it will run down the deepest gully and stop its washing. Hogs thrive upon its succulent roots, and horses and cattle upon its foliage. It has the capacity to withstand any amount of heat and drought, and months that are so dry as to check the growth of Blue-grass will only make the Bermuda greener and more thrifty, as has often been seen.

When grazed it is said no other plant will smother it out. Tramping does not injure it, but if ungrazed, Col. T. C. Howard says, "Broom Sedge will certainly kill it out in three or four years." When it is required to return the field to cultivation, the land should be plowed very shallow, in the autumn, so as to set the furrows on edge as much as possible, to allow the action of frost and air. This is said to kill it as certainly as it will sugar-cane.

XIII. Guinea Grass.

This grass (*Panicum jumentorum*) is sometimes confounded with Johnson grass (*Sorghum halapense*), incorrectly called Guinea grass. The true Guinea grass is perennial, strong and vigorous, extending quickly by its creeping roots, by which it is always propagated. Where there is danger of freezing, roots sufficient for the succeeding year's crop must be protected by gathering and covering in a trench with earth, secure from frost, like the ratoons of sugar-cane. Set out in the spring, they will quickly fill the soil.

In Alabama, if planted in April, the first cutting of the forage may begin late in May, and thereafter it may be cut every five or six weeks, until killed by frost, each succeeding crop being better and better. If cut when eighteen or twenty inches high the forage is sweet, tender, and said to be easily made into hay.

Panicum Varieties.—There are a number of grasses in the South of the Panicum sub-tribe, as for instance Cocks-foot or Barn-yard grass, which makes fair fodder if cut early. Slender crab-grass, a native species of Southern crab-grass, is not valued. The true crab-grass (*P. sanguinale*), Professor Killebrew thinks a fine pasture grass, although it has but few base leaves and forms no sward, yet it sends out numerous stems, branching freely at the base, and serves a most useful purpose in stock husbandry. He says: "Northern farmers would congratulate themselves very much if they had it to turn their cattle on while the clover fields and meadows are parched up with summer heat. It fills all their cornfields, and many persons pull it out, which is a tedious process. It makes a sweet hay; horses are fond of it, leaving the best hay to eat it."

We may say that northern farmers can do very well without it, in the cornfields at least. However valuable in the South, in the North new corn and other summer forage plants well supply its place during the droughts of summer.

There are a number of panicum grasses in the South that are undoubtedly valuable, Texas millet (*P. Texanum*), would seem to be one of these. It is thus described: "An annual grass two to four feet high, sparingly branched, at first erect, becoming decumbent and widely spreading, very leafy, sheaths and leaves finely soft—hairy, margin of the leaves, rough; leaf blades six to eight inches long and one-half to one inch wide, upper leaves reaching to the base of the panicle, or nearly so; panicle six to eight inches long, strict; the branches alternate, erect, simple, three to four inches long, with somewhat scattered sessile spikelets. A grass of vigorous, rapid growth. It is very leafy, the leaves broad, rather thin, sprinkled with short, soft hairs. It grows two to three feet high, but the spreading stalks are often four feet or more in length, growing very close and thick at the base, and yielding a large amount of food."

XIV. Brome or Rescue Grass.

ANOTHER grass that is gaining reputation in the South is a member of the chess family, a grass of many names, among them Brome grass, Schrader's grass, Rescue grass, etc. Its botanical name is *Bromus uniloides*, a so-called winter grass. In the Gulf States it seems to be much esteemed. Mr. Charles Mohr, of Mobile, says of late years it has been found spreading in different parts of Alabama, making its appearance in February. It grows in tufts, its numerous leafy stems ranging from two to three feet high; it ripens seed in May, and affords in the earlier months of spring a much-relished nutritious food, as well as a good hay. It will thus be seen that the South is well provided with valuable grasses, if the farmers will properly make use of what they have. In the hill region of the South and in the two Virginias, and the upland country of Tennessee, Kentucky, Missouri, and in Northern Arkansas, the clovers and northern varieties of grasses generally do well.

SWEET-SCENTED VERNAL GRASS.

XV. Seeding Meadows.

THE quantity of grass-seed sown per acre by the best farmers, and the number of varieties used in seeding meadows, and especially those that are to be pastured, seems to many men to be a great waste. The waste, however, comes from seeding on ill-prepared ground. No grass-seeds are large, and most of them are very minute. If left on top of the soil they often become so dry that they do not germinate. If sown too deep the germs never reach the surface, or only do so to die. The aim of every man should be to get the best return for his outlay. It never was yet gotten either by stinting seed or by slovenly cultivation. Rich soil, a fine tilth, and plenty of seed, will give heavy windrows of hay and deep pastures.

As a rule, from ten to twelve pounds of seed are enough if the crop is intended strictly as a seed crop. For mowing, ordinary thick sowing would be about twenty pounds, mixing according to the varieties sown, say timothy twelve pounds, clover eight pounds, or orchard grass seven pounds, timothy seven pounds, and clover six pounds. For mowing alone, timothy, red-top, orchard grass, meadow foxtail, fowl meadow grass, and red clover will be the basis. For mowing and pasture add bluegrass and white (Dutch) clover for grazing, except for cattle, leave out timothy, but for pasture the more varieties the better.

XVI. The Alphabet of Agriculture.

A BIG herd makes a bare meadow.
Bad grass, bad farming; bad farming, bad crops.
Cultivate grass, and win wealth.
Dank meadows give dreary dreams.
Excellent herbage is an excellent heritage.
Fat pastures make fat pockets.
Grass is the governor, clover the crown, of agriculture.
Heavy meadows make happy farmers.
In the lea lies a lever of wealth.
June-grass is a jolly good joke, say the kine.
Kindly cattle come of good grazing.
Lean kine are lean milkers
Mean grass shows mean farming.
"Nodding grass" is wealth to the owner.
Old pastures, say the sheep, if you please.
Pastures prudently managed get better with age.
Quick grass, quick profits.
Rather than stint your meadow, stint your grain.
Sweet pastures make sound butter; soft hay makes stout wool.
Tall grass, thickly set, fills big barns.w
Up to my ears in sweet grass, says the steer.
Vain are the hopes of the farmer if the grass does not win.
Wealth leaves when the fodder fails.
Xanthium, the clot bur, never helped the grass.
Yellow hay never comes to him who is zealous.
Zeal in the meadow means weal in the wear.

The adage (and it is older than the Christian era), No grass, no cattle; no cattle, no manure; no manure, no crops, covers the whole ground. It is as true to-day as when first spoken, and will continue to be so as long as agriculture lasts.

XVII. Sowing for Hay and for Pasture.

INFORMATION under this head is most clearly presented in tabular form. The first table on next page shows the weight of seed and depth of germination. Those following on same page give the quantity of seed to be sown of each variety for hay, and for hay and pasture as adapted to various soils, with the total number of pounds to be sown per acre.

NAME OF PLANT.	Pounds per Bushel.	Seeds per Ounce.	Depth (inches) at which greatest No. of seeds will germinate.	NAME OF PLANT.	Pounds per Bushel.	Seed per Ounce.	Depth (inches) at which greatest No. of seeds will germinate.
Timothy, clean,	56	74,000	0-¼	Wood meadow grass,	15	173,000	0-¼
Orchard grass,	12	40,000	0-¼	Kentucky blue-grass,	14	243,000	0-¼
Red-top,	12	425,000	0-¼	English rye grass,	20-28	115,000	¼-½
Meadow foxtail,	5	76,000	0-½	Italian rye grass,	15-18	27,000	¼-½
Tall oat grass,	7	211,000	½-¾	Rough-stalked meadow,	15	217,000	0-¼
Sweet-scented vernal,	6	71,000	0-½	Red clover,	60	16,000	¼-½
Crested dog's-tail,	26	28,000	¼-½	White clover,	60	32,000	0-¼
Hard fescue,	10	39,000	0-¼	Lucerne,	60	12,000	¼-½
Sheep's fescue,	14	64,000	0-¼	Millet,	48	5,000	½-¾
Tall fescue,	14	20,500	0-¼	Hungarian,	50	6,000	½-¾

GOOD MEADOW SOILS	Seed for hay, pounds.	Hay and pasture, pounds.	GOOD MEADOW SOILS.	Seed for hay, pounds.	Hay and pasture, pounds.
1—Timothy,	12	8	8—Red-top,	3	3
2—Red clover (biennial),	8	4	9—Rye grass,	4	2
3—Red clover (perennial),	0	4	10—Fowl meadow,	2	4
4—Orchard grass,	6	8	11—White clover,	0	4
5—Meadow fescue,	2	3			
6—Meadow foxtail,	0	3			
7—Blue-grass,	0	5	Total,	27	48

This gives seven varieties for hay, and eleven for hay and pasture.

FOR LANDS SUBJECT TO OCCASIONAL OVERFLOW.	Seed for hay, pounds.	Hay and pasture, pounds.	SANDY LOAMS AND OTHER DRY SOILS.	Seed for hay, pounds.	Hay and pasture, pounds.
1—Fowl meadow,	8	4	Blue-grass,	0	5
2—Alsike,	6	6	Red clover,	6	6
3—Tall fescue,	5	5	Sheep's fescue,	0	4
4—Rough-stalked meadow,	4	4	Purple fescue,	4	4
5—Blue-grass,	0	3	Orchard grass,	6	6
6—Red-top,	4	4	Tall oat grass,	6	4
7—Timothy,	5	2	Hard fescue,	4	4
8—Fiorin,	0	2	Rough-stalked meadow grass,	4	4
9—Meadow soft grass,	4	4	Crested dog's-tail,	2	2
10—Perennial clover,	0	3	Red-top,	0	4
11—White clover,	0	3			
Total,	36	40	Total,	32	45

Sweet-scented vernal grass is often recommended for pasture, and much has been said about its imparting a fine flavor to butter. There is little truth in this, and the grass is not liked by cattle. It starts early in the spring, but its place may well be supplied with some more useful variety.

Other mixtures could be given, adapted to various soils, and the list of varieties might be largely increased. Here, however, every one should experiment for himself, with a view to obtaining varieties adapted to his soil and climate. To assist toward this we give a list of varieties suited to different soils, which may be added to the

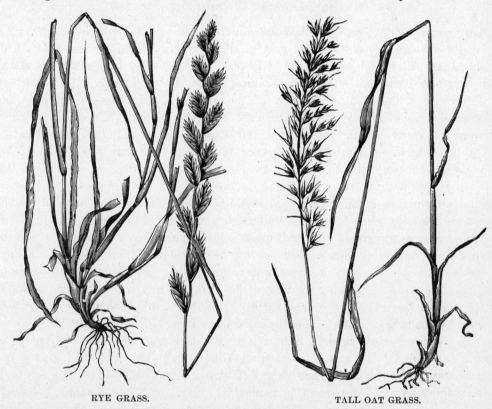

RYE GRASS.   TALL OAT GRASS.

lists given above, sowing them separately, as heretofore recommended, in portions of the pasture or meadow, marking them so that you may observe their hardiness and value for stock. For, as a rule, that which animals will eat green they will eat dry. The varieties are additional to those given in the tables.

Grasses adapted to soils sometimes wet, or undrained: Sweet-scented soft grass, Spiked fescue grass, Red meadow grass, Fiorin, Narrow-leaved creeping bent, Alsike and White clover.

Grasses which are adapted to good arable loams are: The Fescue grasses, nearly all the poa, or Blue-grass kind, the most of the Bent grasses (*Agrostis*), and Rye grasses (*Lolium*).

Timothy, Red-top, Orchard grass, Tall oat grass, Smooth-stalked meadow grass, Hard fescue, Sheep's fescue, and Red and Purple fescue; and of clovers, Red, Crimson, Perennial, White clover and Alfalfa, all do well on loams. They also do well on what are known as dry soils.

Grasses that do fairly well in boggy and swampy (not marshy) soils are: Upright bent, White bent (fiorin), Tall oat, Meadow fescue, Floating fescue, Meadow foxtail, and Smooth-stalked meadow grass.

XVIII. The Celebrated Woburn Experiments.

The results of the celebrated Woburn experiments in the cultivation of grasses are given for several reasons. They present all the better-known grasses now in cultivation in America, besides many of little or no value. The tables will be found useful because: 1, they give the true botanical name; 2, the English name; 3, where first described; 4, the natural duration of the grass; 5, height in a wild state when in flower; 6, time of flowering at Woburn; 7, time of ripening; 8, the soil in which they were cultivated; 9, the natural soil and situation where found; 10, the kind of roots; 11, produce per acre when in flower; 12, produce per acre when in seed; 13, loss or gain by cutting when in flower; 14, loss or gain by cutting when in seed; 15, produce of aftermath or rowen; 16, the general character of the grass. To impart the same information in any other way would be tedious, both to author and reader. The natural situation when found will indicate the proper soil in experimenting. The difference between the green and cured grass will be about the same here as where the experiments were made. Since, although our hay is higher dried than English hay, yet, grass in England contains more water than ours. To determine the time of flowering at any place, knowing the time of flowering of any species at your locality, find the time of flowering of the same variety at Woburn, and then add or subtract as indicated by the tables. Thus, Timothy ripens in Chicago, Illinois, July 15; at what time will Meadow foxtail ripen? Subtract fifteen days from the time of ripening as found in the table; for, Timothy ripened at Woburn July 30. This will give a fair approximation. In England, what we call Timothy is called Meadow Cat's-tail, its specific English name. The scientific names, however, are the true names the world over. The English botanist who classified the grasses may be excused for crediting so many of the valuable species to Great Britain; for instance, Timothy and Orchard grass: for at that time even the great naturalist Buffon supposed everything was so inferior in America, that even all the animals had shorter tails here than in Europe. [See pages 156, 157.]

XIX. A Summary of Meadow Grasses.

In the United States, by the term meadow, is generally understood a field devoted both to hay and pasture, though the term strictly means a field where hay crops are grown. In the West, especially, there are few permanent pastures, and these only on hilly and inaccessible places. Our climate is not so well adapted to permanent pastures as that of some other countries, and the white grub, which lives in the grub (larval) state three years in the soil, is often utterly destructive to grass. The only remedy is rooting out by hogs, and subsequent plowing up. Our natural system, therefore, is to keep the land in meadow and afterwards in pasturing until it is again ready for the plow in the regular rotation. It is probably the most economical system we can adopt. With thick seeding, Timothy, Clover, Red-top and Orchard grass being the basis, we may add any pasture grasses we please finally to take the places of the short-lived varieties, or, as the meadow fails from cutting and grazing. Timothy will not stand close cutting nor grazing; clover dies out at the end of the second or third year, except the perennial clover, white clover and alsike clover—neither of which is so valuable for hay or pasture as the biennial variety. For horses, Timothy, Blue-grass and Orchard grass make the best hay, since they are free from dust. For cattle, these and Fowl-meadow, Smooth-stalked meadow, Tall fescue and Meadow foxtail are good, and they all do well on dry, arable land. On moist or rather wet lands, Alsike clover may be substituted for Red clover. These grasses should do well in the South—north of the Gulf States—and in the mountain regions of the Gulf States. In the Gulf States, generally, the main dependence must be upon Bermuda, Crab, Crow-foot, Gama grass, Japan clover (*Lespedeza*), and the other forage plants heretofore mentioned.

XX. About Pastures.

MANY farmers suppose that any land too poor or too swampy for anything else, is good enough for a pasture. Poor land will indeed bear grass, and if there is not manure enough to enrich the whole farm, the best use the unfertilized land can be put to is to make it bear what grass it will. There is this about grass on thin soil, and it will apply especially to rocky and hilly land too rough for plowing: the grass grown thereon will be sweet and nutritious. On the other hand, the grass grown on swamp, boggy or marsh lands will be coarse and unnutritious. If drained, such land may became valuable, not only for pasture and meadow, but for the cereal grains. In any event, the field must be fully covered with soil to make a pasture, and thin, poor land always requires thicker seeding than stronger soil, because, on poor soils, plants spread less easily than on rich ones. Again, the thicker the sward, the less rank, and the sweeter the grasses are.

XXI. Genera, Species and Varieties.

THE word *genus* means the sub-family to which certain species belong. Species again are divided into varieties. Varieties have all the characteristics of the species, but vary slightly one from another. Species have many points of resemblance, though differing in minor characters, and are grouped into genera, these again into families or orders, until the whole vegetable kingdom is at length separated into two great series, comprising, one, flowering, and the other, flowerless plants. Thus the great order of plants, *Gramineæ*, comprise some two hundred and thirty genera, and more than three thousand species, with varieties innumerable.

TABLE OF THE GRASSES EXPERIMENTED ON AT WOBURN

	Systematic Name and Authority.	English Name and Native Country.	Where figured or described.	Natural Duration.	Inches high in a wild state	Time of flowering at Woburn	Time of ripening the Seed at Woburn	Soil at Woburn.	Natural Soil and Situation, as in Smith's Flora Brit.
1	Anthoxanthum odoratum L.	Sweet-scent. vernal gr..Brit.	E.B. 647	Peren.	12	Apr. 29	June 21	Brown sandy loam	Meadows
2	Holcus odoratus..Host G.A.	Sweet-scented soft gr...Ger.	Host, N.A.	Peren.	14	Apr. 29	June 25	Rich sandy loam	Woods, moist mead.
3	Cynosurus cæruleus...E. B.	Blue moor grass......Brit.	E.B. 1613	....	20	Apr. 30	June 20	Light sandy soil	Pastures
4	Alopecurus alpina...E. B.	Alpine foxtail grass..Scot.	E.B. 1126	Peren.	6	May 20	June 24	Sandy loam	Scotch mountains
	Poa alpina......E. B.	Alpine meadow grass..Scot.	E.B. 1003	Peren.	6	May 30	June 30	Light sandy loam	Scotch Alps
5	Alopecurus pratensis..E. B.	Meadow foxtail grass..Brit.	E.B. 848	Peren.	24	May 30	June 24	Clayey loam / Sandy loam	Meadows
6	Poa pratensis........E. B.	Smooth-stalked meadow grass..........Brit.	E.B. 1073	Peren.	18	May 30	July 14	Bog earth and clay	Mea. and pastures
7	Poa cærulea var. pratensis......E. B.	Short bluish meadow grass..........Brit.	E.B. 1004	Peren.	14	May 30	July 14	Bog earth and clay	Meadows
8	Avena pubescens...E. B.	Downy oat grass......Brit.	....	Peren.	18	June 13	July 8	Rich sandy soil	Chalky pastures
9	Festuca hordeiformis, or Poa hordeiformis..H. C.	Barley-like fescue grass ....Hungary.	....	Peren.	18	June 13	July 10	Manured sandy soil	Corn-fields
10	Poa trivialis........E. B.	Roughish meadow gr..Brit.	E.B. 1072	......	20	June 13	July 10	Man. light br. loam	Meadows
11	Festuca glauca......Curtis.	Glaucous fescue grass..Brit.	....	Peren.	12	June 13	July 10	Brown loam	Chalky pastures
12	Festuca glabra......Wither.	Smooth fescue grass..Scot.	....	Peren.	9	June 16	July 10	Clayey loam	Mountains
13	Festuca rubra.......Wither.	Purple fescue grass...Brit.	....	Peren.	12	June 20	July 10	Light sandy soil	Mea. and pastures
14	Festuca ovina......E. B.	Sheep's fescue grass..Brit.	E.B. 585	Peren.	6	June 24	July 10	....	Dry pastures
15	Briza media.........E. B.	Common quaking gr...Brit.	E.B. 340	Peren.	16	June 24	July 10	Rich brown loam	Pastures
16	Dactylis glomerata..E. B.	Rough-head cock's-foot grass..........Brit.	E.B. 335	Peren.	24	June 24	July 14	Rich sandy loam	Soft moist soils
17	Bromus tectorum ....Host G. A.	Nodding pencilled brome grass..........Eur.	....	Annual	12	June 24	July 16	Light sandy soil	Hedges
18	Festuca cambrica......Huds.	Cambridge fescue gr...Brit.	....	Annual	14	June 28	July 16	Light sandy soil	Dry pastures
19	Bromus diandrus....E. B.	Upright brome grass..Brit.	E.B. 1006	Annual	18	June 28	July 16	Rich brown loam	Corn-fields
20	Poa angustifolia......With.	Narrow-leaved mea. gr Brit.	....	Peren.	24	June 28	July 16	Brown loam	Meadows
21	Avena elatior......Curtis. Holcus avenaceus..Wil.en.	Tall oat grass or Knot grass.....Brit.	E.B. 813	Peren.	50	June 28	July 16	....	Arable lands
22	Poa elatior.........Curtis. Avena elatior, var.	Tall meadow grass...Scot.	....	Peren.	30	June 28	July 16	Rich clay loam	Meadows
23	Festuca duriuscula...E. B.	Hard fescue grass....Brit.	E.B. 470	Peren.	12	July 1	July 20	Light sandy loam	Pastures
24	Bromus erectus......E. B.	Upright peren. br. gr..Brit.	E.B. 471	Peren.	36	....	....	Rich sandy soil	Chalky pastures
25	Milium effusum......E. B.	Common millet grass..Brit.	E.B. 1106	Peren.	40	July 1	July 20	Light sandy soil	Woods
26	Festuca pratensis...E. B.	Meadow fescue grass..Brit.	E.B. 1592	Peren.	30	July 1	July 20	Bog soil and coal ashes	Meadows
27	Lolium perenne.....E. B.	Perennial rye grass...Brit.	E.B. 315	Peren.	24	July 1	July 20	Rich brown loam	Loamy pastures
28	Poa maritima........E. B.	Sea meadow grass....Brit.	E.B. 1140	Peren.	12	....	....	Light brown loam	Salt marshes
29	Festuca loliacea....E. B.	Spiked fescue grass..Brit.	E.B. 1821	Peren.	36	July 1	July 28	Rich brown loam	Moist pastures
30	Aira cristata.........E. B.	Crested hair grass....Brit.	E.B. 648	Peren.	9	July 4	July 28	Sandy loam	Sandy pastures
31	Cynosurus cristatus..E. B.	Crested dog's-tail gr...Brit.	E.B. 316	Peren.	24	July 6	July 20	Manured br. loam	Pastures
32	Avena pratensis....E. B.	Meadow oat grass....Brit.	E.B. 1204	Peren.	24	July 6	July 20	Rich sandy loam	Pastures
33	Bromus multiflorus..E. B.	Many fl. gr. brome gr..Brit.	E.B. 1884	Annual	24	July 6	July 20	Clayey loam	Poor past., hedges
34	Festuca Myurus.....E. B.	Wall fescue grass.....Brit.	E.B. 1412	Annual	9	July 6	July 28	Light sandy soil	Walls
35	Aira flexuosa.......E. B.	Waved moun. hair gr..Brit.	E.B. 1519	Peren.	9	July 6	July 28	Heath soil	Dry soils & heaths
36	Hordeum bulbosum Hort. K.	Bulbous barley grass..Italy.	....	Peren.	24	July 10	July 28	Man. clayey loam	Loamy pastures
37	Festuca calamaria....E. B.	Reed-like fescue grass.Brit.	E.B. 1005	Peren.	40	July 10	July 28	Clayey loam	Hedges
38	Bromus littoreus..Host G. A.	Seaside brome grass..Ger.	....	Peren.	20	July 12	Aug. 6	Clayey loam	Sea-shores
39	Festuca elatior......E. B.	Tall fescue grass......Brit.	E.B. 1593	Peren.	36	July 12	Aug. 6	Black rich loam	Meadows
40	Festuca fluitans.....E. B.	Floating fescue grass..Brit.	E.B. 1520	Peren.	18	July 14	Aug. 12	Str. tenacious clay	Ponds
41	Holcus lanatus.........W.	Meadow soft grass....Brit.	E.B. 1169	Peren.	24	July 14	July 26	Strong clayey loam	Moist meadows
42	Festuca dumetorum....W.	Pubescent fescue gr...Brit.	H.D. 700	Peren.	12	July 14	July 28	Black sandy loam	Woods
43	Poa fertilis......Host G. A.	Fertile meadow grass..Ger.	....	Peren.	24	July 16	July 28	Clayey loam	Meadows
44	Arundo colorata...Hort. K.	Striped-leaved reed gr. Brit.	E.B. 402	Peren.	40	July 16	July 28	Black sandy loam	Moist loams
45	Phleum nodosum......With.	Bulbous-stalked cat's-tail grass..........Brit.	....	Peren.	18	July 16	July 20	Clayey loam	Dry pastures
46	Phleum pratense.....With.	Meadow cat's-tail gr...Brit.	....	Peren.	24	July 16	July 20	Clayey loam	Mea. and pastures
47	Hordeum pratense....E. B.	Meadow barley grass..Brit.	E.B. 409	Peren.	24	July 20	Aug. 8	Man. brown loam	Meadows
48	Poa compressa......E. B.	Flat-stalked mead. gr..Brit.	E.B. 365	Annual	12	July 20	Aug. 8	Man. gravelly soil	Walls
49	Poa aquatica........E. B.	Reed meadow grass...Brit.	E.B. 1315	Peren.	72	July 20	Aug. 8	Str. tenacious clay	Ditches
50	Aira aquatica........E. B.	Water hair grass......Brit.	E.B. 1557	Peren.	9	....	....	Water	....
51	Aira cæspitosa......E. B.	Turfy hair grass......Brit.	E.B. 1453	Peren.	24	July 24	Aug. 10	Str. tenacious clay	Clayey pastures
52	Avena flavescens...E. B.	Yellow oat grass......Brit.	E.B. 952	Peren.	18	July 24	Aug. 15	Clayey loam	Pastures
53	Bromus sterilis......E. B.	Barren brome grass...Brit.	E.B. 1030	Annual	24	July 24	Aug. 20	Sandy soil	Rubbish
54	Holcus mollis........Curtis.	Creeping soft grass...Brit.	....	Peren.	30	July 24	Aug. 20	Sandy soil	Sandy pastures
55	Poa fertilis var. B. Host G. A.	Fertile meadow grass..Ger.	....	Peren.	24	....	....	Brown sandy loam	Meadows
56	Agrostis vulgaris......E. B.	Fine bent grass.......Brit.	E.B. 1671	Peren.	18	July 24	Aug. 20	Sandy soil	Mea. and pastures
57	Agrostis palustris....E. B.	Marsh bent grass.....Brit.	E.B. 1189	Peren.	20	July 28	Aug. 20	Bog earth	Marshy places
58	Panicum dactylon....E. B.	Creeping panic grass..Brit.	E.B. 850	Peren.	24	July 28	Aug. 28	Man. sandy loam	Arable lands
59	Agrostis stolonifera..E. B.	Fiorin of Dr. Richardson..Brit.	E.B. 1532	Peren.	24	July 28	Aug. 28	Bog soil	Moist places
60	Agrostis stolonifera var. angustifolia.	Narrow-leaved creeping bent..........Brit.	....	Peren.	24	July 28	Aug. 28	Bog soil	Moist places
61	Festuca pennata.....	Spiked fescue.........Brit.	E.B. 730	....	24	July 28	Aug. 30	Man. light san. soil	Meadows
62	Agrostis canina......E. B.	Brown bent...........Brit.	E.B. 1856	Peren.	9	July 28	Aug. 30	Brown sandy loam	Clayey pastures
63	Agrostis stricta......Curtis.	Upright bent grass....Brit.	....	....	9	July 28	Aug. 30	Bog soil	Clayey pastures
64	Agrostis nivea........E. B.	Snowy bent grass.....Brit.	....	....	9	Aug. 10	Aug. 30	Sandy soil	Clayey pastures
65	Agrostis fascicularis var. canina....Curtis.	Tufted-leaved bent gr. Brit.	....	....	9	Aug. 10	Aug. 30	Light sandy soil	Clayey pastures
66	Panicum viride......Curtis.	Green panic grass....Brit.	E.B. 875	Annual	36	Aug. 2	Aug. 15	Light sandy soil	Sandy
67	Agrostis lobata......Curtis.	Lobed bent grass.....Brit.	....	Annual	20	Aug. 6	Aug. 20	Sandy soil	Sandy pastures
68	Agrostis repens......With.	Black or creeping-rooted bent, bl. couch..Brit.	....	Peren.	26	Aug. 8	Aug. 25	Clayey loam	Arable lands
69	Triticum repens.....E. B.	Creeping-rooted wheat gr. or couch gr..Brit.	E.B. 848	Peren.	30	Aug. 10	Aug. 30	Light clayey loam	Arable lands
70	Alopecurus agrostis.....	Slender foxtail grass..Brit.	E.B. 1172	Annual	8	Aug. 10	Sep. 1	Light sandy loam	Road-sides
71	Bromus asper........E. B.	Hairy stalked br. gr...Brit.	E.B. 1310	Annual	48	Aug. 15	Sep. 10	Light sandy soil	Moist sand. places
72	A. mexicana.......Hort. K.	Mexican bent grass...S. Amer.	E.B. 1356	Peren.	24	Aug. 15	Sep. 25	Black sandy soil	Rich pastures
73	Stipa pennata........E. B.	Long awned fea. gr...Brit.	E.B. 909	Peren.	20	Aug. 15	Sep. 25	Heath soil	Peat bogs
74	Melia cærulea......Curtis.	Purple melic grass....Brit.	....	Peren.	20	Aug. 29	Sep. 30	Light sandy soil	Sandy pastures
75	Phalaris canariensis...E. B.	Common canary gr....Brit.	E.B. 750	Annual	26	Aug. 30	Sep. 30	Clayey loam	Cultivated fields
76	Dactylis cynosuroides..Lin.	Amer. cock's foot gr..N. A.	....	....	24	Aug. 30	Oct. 20	Clayey loam	Loamy pastures

MEADOW AND PASTURE GRASSES.

ARRANGED IN THE ORDER OF THEIR FLOWERING.

	Kind of Roots.	Produce, at the Time of Flowering, per Acre, in lbs.				Produce, when the Seed is Ripe, per Acre, in lbs.				Loss or Gain by Cutting when in Flowering, Nutritive Matter, in lbs.		Loss or Gain by Cutting when in Seed, in Nutritive Matter, in lbs.		Produce of the Latter-math, per Acre, in lbs.		General Character.
		Grass.	Hay.	Loss in drying	Nutritive matter	Grass.	Hay.	Loss in drying	Nutritive matter	Loss.	Gain.	Loss.	Gain.	Grass.	Nutritive matter	
1	Fibrous	7827	2103	5723	122	6125	1837	4287	311	188			188	6806	3828	An early pasture grass.
2		9528	2441	7087	610	27225	9528	17696	2233	1600			1600	17015	1129	The most nutritive of early flower-
3	Fibrous					6806			398							Not deserving culture. [ing grasses.
4	Fibrous	5445	1452	3993	85											Not worth culture.
	Fibrous	5445			127											A good grass for lawns.
5	Creeping	20418	6125	14293	478	12931	5819	7111	461					8167	255	One of the best meadow grasses.
		8507	2552		132											
6	Creeping	10209	2871	7337	279	8507	3403	5104	199		79	79		4083	111	Good early hay grass.
7	Creeping	7486	2246	5240	233											
8		15654	5870	9783	366	6806	1361	5445	212		154	154		6806	212	A good pasture grass on a rich soil.
9	Fibrous	13672	4083	9528	478											
10	Fibrous	7486	2246	5240	233	7827	3522	4304	336	102			102	4764	223	A most valuable grass in moist rich
11	Fibrous	9528	4811	5717	446	9528	3811	5717	223							A good hay grass. [soils.
12	Fibrous	14293	5717	8576	446	9528	3811	5717	186	260			260	6125	47	A tolerably good pasture grass.
13	Fibrous	10209	3557	6651	239	10890	4900	5989	340	101			101	3403	79	Good lawn grass.
14	Fibrous					5445			127					3403	66	Good lawn grass.
15	Fibrous	9528	3096	6431	409	9528	3335	6193	483	74			74	8167	255	
16	Fibrous	27905	11859	16045	1089	26544	13272	13272	1451	362			362	11910	281	A most productive grass, but coarse.
17	Fibrous	7486	3930	3556	350											Of little value.
18	Fibrous	6806	2892	3913	239											A good lawn grass.
19	Fibrous	20418	8677	11740	957											
20	Fibrous	10376	7810	10566	1430	9528	3811	5717	701		649	649				Excellent hay grass.
21	Creeping and Knot					16335	5717	10617	255					13612	265	A vile weed in arable lands.
22	Creeping and Knot	12251	4287	3617	669											A vile weed in arable lands.
23	Fibrous	18376	8269	10106	1004	19075	8575	10481	446		558	558		10209	199	A good grass for hay or pasture.
24	Fibrous	12931	5819	7112	555											Not worth culture.
25	Fibrous	12251	4747	7504	334											Of little value.
26	Creeping	13612	6405	7146	957	19057	7623	11434	446		510	510				Excellent early hay grass.
27	Fibrous	7827	3322	4494	305	14973	4492	10481	643	337			337	3403	53	A well known and esteemed grass.
28		12251	4900	7350	861									12251	191	
29		16335	7146	9188	765	10890	4492	6397	553		212	212		5403	66	One of the most valuable grasses for
30	Fibrous	10890	4900	5989	340											A good lawn grass. [hay and past.
31	Fibrous	6125	1837	4287	406	12251	4900	7350	478	71			71			A good lawn grass.
32		6806	1871	4934	239	9523	2858	6670	148							
33	Fibrous	22460	12353	10107	17154											Unfit for culture.
34	Fibrous	9528	2858	6670	223											A very inferior grass.
35	Fibrous	8167	3164	5002	191											Fit for lawns.
36		23821	9826	13994	1302											Of little use.
37		54450	19057	35392	3828	51046	12123	38293	2392		1435	1435				Early and prolific.
38		41518	21278	20540	973	38115	15246	22869	2084		1111	1111				Early, prolific and coarse.
39		51046	17866	33180	3988	51046	17866	33180	2392		1595	1595		15654	978	An excellent meadow grass.
40	Creeping	13612	4083	9528	372											An aquatic or amphibious gr. of good
41		19057	6661	12395	1191	19057	3811	15246	818		372	372				Early and productive. [quality.
42		10890	5445	5445	170											
43		14975	7861	7111	1052											An early grass.
44	Creeping	27225	12251		1701											Productive.
45	Creeping	12251	5819	6431	478											
46	Creeping	40837	17355	23481	1595	40837	19397	21439	3668		2073	2073		9528	297	An excellent hay grass.
47		8167	3267	4900	478											Early and nutritive.
48		3403	1446	1956	265											
49	Creeping	126596	75957	50638	4945											Most prolific, but coarse.
50		10890	3267	7623	382											
51		10209	3318	6891	319											An excellent lawn grass.
52		8167	2858	5308	478	12251	4900	7350	430		47	47		4083	79	A valuable grass.
53	Fibrous	29947	16845	13102	2339											Of little value.
54	Creeping	34031	13612	20418	2392	21099	8439	12659	1153		1238	1238				A valuable grass.
55		15654	6653	9000	733	14973	8235	6738	1169	436			436	4764	111	A valuable grass.
56		9528	4764	4764	251											An early grass.
57	Creeping	10209	4594	5615	438	13612	5445	8167	584	146			146			Useful.
58		31308	14088	17219	9783											
59	Creeping	17696	7742	9732	967	19057	8575	10481	1042	74						Useful on bogs.
60	Creeping	16335	7350	8984	765											
61		20418	8167	12251	398											
62		6125	2688	3437	239											
63		7486	2713	4772	175											
64						4764	1310	3454	148							
65		2722	680	2041	85											
66		5445	2178	3267	127											Of no value.
67		6806	3403	3043	319											
68	Creeping	6125	2679	3445	287											A vile weed on poor arable lands.
69	Creeping	12251	4900	7350	382											A vile weed in arable lands.
70		8167	3164	5003	223											
71	Fibrous	13612	4083	9528	425											Unfit for culture.
72		19057	6670	12387	595											Deserves trial.
73	Fibrous	9528	3454	6074	409											Not worth culture.
74		7486	2807	4679	172											A good lawn grass.
75	Fibrous	54450	17697	36752	1876											Grown for its seeds.
76		69423	41664	27769	1898											

The Grass Family.—Taking the arrangement of Gray, the *Gramineæ*, or Grass Family, has for its tribe, 1, *Poaceæ*. This contains fifty-two sub-tribes, or genera, and includes not only all our true cereal grains, but also Indian corn, sugar-cane, and the most valuable grasses. A genus then is an assemblage, or number of species, agreeing structurally and physiologically, in flowering, fruitage and perpetuation, having also a general resemblance in habit. A species comprises plants precisely alike in every character, capable of uniform, invariable and continuous perpetuation by natural propagation. A variety is the variation produced by accidental change in a species, and is not capable of uniform, invariable and permanent continuance by natural propagation. A hybrid is a plant produced by the mixture of two species. As a rule it is infertile—that is, not capable of propagation by seed. Hence hybrids and varieties are propagated by cuttings, grafting or budding. We have been thus precise in defining the meaning of terms, since, except among botanists, much confusion exists in relation to a subject which should be known to every farmer.

Let us trace Blue-grass, for instance, because it is easily recognized. It belongs to the poa sub-family, or *Poaceæ*. Its generic name is poa, the ancient Greek name for grass, while its specific name is pratensis. Thus we have what may be called its surname, Poa, and its given name Pratensis, which we translate " green meadow grass," the name Kentucky Blue-grass being merely a local name, but generally adopted. It is indigenous all over the North, on suitable lands, from the New England coast westward, and, in fact, the seed is said to have been carried into Kentucky from Indiana, by the soldiers in General Harrison's famous campaign, in which the power of the great western Indian tribes was broken.

XXII. Favorite Pasture Grasses.

BLUE-grass, where it is at home, or on good calcareous loams, is the best pasture grass known, giving early and late feed, but failing under the heats of summer, especially when dry. The Spear or June grass (*poa compressa*, from its flattish stalk) is of fully as much value on soils adapted to it; these are dry knolls, sandy loams, and dry, compact sandy soils. It is found from Northern Wisconsin to Tennessee. Rough-stalked Meadow grass (*P. Trivialis*) is an excellent pasture grass, except that it does not like hot suns. In cool, partially shaded soils, it stands the tramping of stock, and makes excellent pasture, and if not cut for hay stands the sun fairly well.

BLUE-GRASS

XXIII. Bent Grasses.

This is the name given to what we call Red-top. There is a large family, many of them excellent grasses, and especially noted for their tenacity of life. The common Red-top (*A. vulgaris*) is called Herds-grass in Pennsylvania, Burdin's grass in some sections of New England, and Red-bent and Summer-dew grass in other localities. The bent grasses thrive in soils deficient in lime, and hence in such soils supply the place of Blue-grass, to which it is next in point of value as pasturage. The whole family, however, are somewhat deficient in aftermath, but start early in the spring.

White Bent or Fiorin (*A. Alba*) has had a variable experience in the United States. It likes a moist, fertile soil, and gives more late feed than Red-top. Creeping Red-top (*A. stolonifera*) is said to be a variety of this grass, and was called Fiorin by Richardson. It thrives on dry land and ought to be valuable. The Red-top grasses must not be confounded with the dreaded quitch or couch grass (*Triticum repens*), which belongs to the wheat tribe. In some localities it is well thought of, but it should never be permitted on arable lands.

RED-TOP.

The bent family, on the other hand, though persistent, are not difficult to kill.

XXIV. Orchard Grass.

There are two varieties of this most valuable grass, called in England Cock's-foot. One, *Dactylis glomerata*, is the variety best known in the United States. For dry, somewhat open rich soils it is one of the most valuable of grasses. Stiff retentive clays are not adapted to it. It is a moisture-loving plant, but not the moisture of saturation. It also thrives well in the shade, and hence its name of Orchard grass. It is apt to form into tussocks, if sown thinly, but not when sown thickly, except on soil not adapted to it. It is indigenous to the whole of Europe, Northwestern Africa, Asia Minor and the United States, and is second, perhaps, to no other pasture grass in the country, since it starts early in the spring. It springs up quickly after being grazed or mown, and gives abundant pasturage in the autumn. Orchard grass flowers with Red clover, resists drought well, and in flesh-forming material is superior to Timothy, ranking with it as ten to seven in the scale of value.

XXV. Grasses for Various Regions.

ALL the United States north of, and including, Virginia, Tennessee, Kentucky, Missouri and Kansas, and also Canada, are natural grass countries. All the meadow and pasture grasses will thrive there, and the less known are worthy of trial. The following is a summary of grasses recommended for the following named States, south and on the Pacific coast, from returns sent to the Department of Agriculture, from practical cultivators in the States named:

ORCHARD GRASS.

Alabama—Orchard grass, Kentucky blue-grass, timothy, herds-grass (red-top), Johnson grass, alfalfa and California clover.

California—Timothy, large red clover, the millets, orchard grass, Italian rye grass, white clover, Guinea grass (*Panicum jumentorum*), Bermuda and alfalfa.

Florida—Bermuda, alfalfa, Guinea grass (*Panicum jumentorum*), orchard grass, Johnson grass and clover.

Georgia—Kentucky blue-grass, orchard grass, herds-grass (called red-top in New England), timothy, the clovers and alfalfa, in the order named.

Idaho and Montana—All the grasses for bottom lands, and alfalfa for "bench lands."

Louisiana—Kentucky blue-grass, orchard grass, Bermuda, timothy, herds-grass (red-top), the clovers and alfalfa.

Mississippi—Orchard grass, herds-grass (red-top), the clovers, Kentucky blue-grass and the millets.

Texas—Alfalfa, Bermuda, timothy, the clovers, orchard grass, Johnson grass and the millets, in the order named.

Washington Territory and Oregon—Italian rye grass, orchard grass, the clovers, tall meadow oat grass, Kentucky blue-grass, Texas mesquit and Bermuda.

XXVI. Clover in Its Relation to Husbandry.

CLOVER has a threefold relation to husbandry: as a seed crop, as a forage crop, and also for its wonderful power of renovating the soil. Its proper soil is a thoroughly drained loam or loamy clay. Soils that in drying out crack badly, or those subject to heaving, are not adapted to clover. Argillaceous, granitic, drained calcareous loams, red and other well-drained clays are all congenial to its growth, and, in fact, nearly all soils, except those quite sandy or wet. The first winter it is apt to suffer if seeded in the fall, and hence should be sown in the spring. When sown alone it usually blossoms the first season; if seeded with wheat, generally not until the next season. If the seed is to be taken, it should be sown alone, and for plowing under the same rule will apply. As soon as it is well up it should have two bushels of land-plaster (gypsum) per acre, especially on granitic soils. When sown for hog-pasture, for cutting green, or for turning under, not less than sixteen pounds of seed should be given per acre. If intended for seed, twelve to sixteen pounds will be sufficient. For plowing under, the Mammoth Red Clover (*T. Pratense var.*) is the best. It grows four to six feet high, and produces enormously in root and top. Cattle do not like it much, but hogs do.

XXVII. The Seed Crop.

MAMMOTH RED CLOVER.

Sow as early in the spring as the soil will admit, on land prepared and leveled in the best manner, covering one-half inch deep. The first flowering is apt to blast, hence this is cut for fodder, and the later or summer growth is taken for seed. It is generally mown with a machine, allowed to dry in the swath, raked into windrows, and, when thoroughly dry, either threshed directly with a clover huller or else stacked and threshed later. The yield is all the way from four to eight, or even ten bushels an acre. Six bushels is a good yield, and from the high price it bears, it is a good-paying crop in places remote from the great markets. The best seed is raised in the West, for there it is not infested with the seeds of Canada thistle, and other pestilent weeds.

XXVIII. Valuable Varieties of Clover.

CLOVER in England is called trefoil from its botanical name *Trifolium*, three leaved or lobed. Botanists number about 160 species. At least eight of these belong to the Northern States, and there are a number of species in California and the Southern States. The varieties most generally cultivated are Red clover (*T. pratense*), and White, or Dutch clover (*T. repens*).

Red Clovers.—Of the red clovers the more valuable are: biennial clover (*T. pratense*), perennial clover (*T. pratense, perenne*), meadow clover, medium or cow clover, as it is indifferently called (*T. medium*), and buffalo clover (*T. reflexum*). There are two so-called buffalo clovers: *T. stolonifera*, creeping buffalo clover, having a white blossom, and not valuable. The first mentioned has a large red blossom and hairy stem, and is worthy of extended trial. It is a southern rather than a northern variety.

The clover known as common red clover has a large, spindle-shaped root. The stems of the plant are somewhat hairy; the leaves oval or obovate, often notched at the end, and having on the upper side of the leaf a pale spot. The flower heads are egg-shaped and set directly on the end of the stalk, rather than on the branches. The perennial variety, sometimes called cow grass, is reported to be indigenous to rich English pastures. It is denied by some that there is any difference between this and the common biennial variety.

Medium Clover.—Medium clover (sometimes called cow grass) is specifically known as zigzag clover, the stems are zigzag, rather smooth, the leaflets oblong, not notched at the end, and without spot. The flowers are a deeper purple and larger than the first-named species. The root is more fibrous or creeping, and the flowers are later. Otherwise it is the same, except it is perennial in its habit.

Another variety of clover thought well of in Europe, is the crimson clover (*T. incarnatum*). It is an annual of great beauty, a native of Southern and Central Europe, and is sown there in autumn as a forage crop for the succeeding summer. It may be valuable south.

XXIX. Dutch, or White Clover.

THIS IS a perennial plant growing in a great variety of soils, but preferring a moist loam. If not indigenous to the United States, it is thoroughly acclimatized all over the North and West, and in Canada. It is one of the most important of our pasture plants in moist seasons; it shows itself but little during dry ones. In moist seasons every field and lawn will be gay with its blossoms, and in some portions of Iowa it has such complete possession of the soil that large amounts of seed are saved.

Mr. Charles Fox, in his text-book of agriculture, says there are two varieties of White clover essentially different in habit, the English variety being a strong growing plant, well adapted to meadows, frequently standing eighteen inches high when supported; and that the American variety, grown side by side, was short, adhering to the ground, wholly unfit for the meadow, which we all know it to be. He believes one a variety of the other, and considers them, economically, very distinct. Some English catalogues also claim that the American seed gives clover inferior in growth to the original variety. It is unnecesary to say this is simply an advertising trick.

WHITE CLOVER.

XXX. Alsike or Swedish Clover.

ON soils too moist for Red clover, Alsike clover (*Trifolium hybridum*) should be sown. There is no doubt of the value of this clover, or of its hardiness. On rich land it is apt to lie partly flat at mowing time, and hence the land should be made smooth and level for the seed. Ten to fifteen pounds is sufficient to seed an acre fully. In its general appearance and habit of growth, it is intermediate between the common red and the white clover. It is valuable for pasturage, growing rapidly after being eaten down, and furnishing a large amount of food during the season. The stems remain soft and succulent, never becoming so woody and hard as the red clover. It partakes of the creeping, spreading habit of the white clover, and yields well when cut for hay, being thick and close, although not so tall in growth as the red clover. It may be mown for hay when the blossoms are mature. Its aftermath is dense and heavy, and both the hay and green plants are relished by cattle. It is a fibrous-rooted plant, like Dutch clover, and might be described as a giant white clover, if it were not for its blossoms. These, when they first appear, are only faintly tinged, but as they become fully opened they assume a pale-red tinge. This clover requires three years to attain its full vigor of growth. It is then essentially richer than red clover, containing, according to the analysis of Wolff and Knopf two per cent more of flesh-formers than red clover.

ALSIKE CLOVER.

XXXI. Clovers for the South—Alfalfa.

THE best name for this plant is Lucerne. Its botanical name is *Medicago Sativa*. In warm climates it is undoubtedly one of the most valuable of forage plants, especially where it may be irrigated in hot weather. There seems, then, almost no limit to its production. In the North, it is not valuable, although it survives the winter up to forty degrees, but it cannot there compete with the true Red clover. In California it is the great forage plant, often growing wild. It is also widely naturalized in South America. In fact, in warm climates, on porous soils, it has the most extreme vitality, standing the severest drouths, but producing largely only under the influence of moisture.

For Hay.—Of its cultivation and value South the Rev. C. S. Howard has written fully. He says that no grass or forage plant in cultivation at the North will yield nearly so much hay as lucerne at the South. In good seasons, and on land sufficiently rich, it can be cut four or five times during the year. An acre of good lucerne will afford hay and cut green food for five horses the whole year. Ten acres will supply fifty head of plantation horses.

Soil and Cultivation.—It is useless to attempt the cultivation of lucerne on poor land. It will live, but it will not be profitable. There are certain indispensable requisites in the cultivation of lucerne. The ground must be good upland; it must be made very rich; it cannot be made too rich. It must be very clean. When the lucerne is young it is delicate, and may be smothered with the natural weeds and grasses of a foul soil. Land which has been in cotton, worked very late, if made sufficiently rich, is in a good state of preparation for lucerne. The manure you put upon it must be free from the seeds of weeds. Great depth of cultivation is necessary in preparation of the soil for lucerne. Ten pounds of seed are required for an acre, sown broadcast.

Either early in autumn or early in February are good seasons for sowing lucerne. The seed should be lightly harrowed in, and then the surface should be rolled. Lucerne lasts a great number of years, the roots ultimately becoming as large as a small carrot. It

ALFALFA OR LUCERNE.

should be top-dressed every third year with some manure free from the seeds of weeds. Ashes are very suitable for it. The lucerne field should be as near as possible to the stables, as work-horses, during the spring and summer, should be fed with it in a green or wilted state. As lucerne is much earlier than red clover, it will be found a useful adjunct in hog-raising. Hogs are very fond of it, and will thrive on it in the spring, when it is cut green and thrown to them.

XXXII. Japan Clover.

This is another leguminous plant of the South that, of late years, has attracted more and more attention. Cattle eat it readily, and sheep greedily. It has been called bush clover, from its habit. There are a number of plants of this genus *Lespedeza*. This one, *L. striata*, stands the winters as far north as Tennessee and

JAPAN CLOVER.

Kentucky. It is a low growing perennial plant and spreading. The leaves are small, three-lobed like clover, and numerous, growing on the poorest soils and standing the extremest drouths. It has now extended pretty well over the South, below thirty-six degrees, from the Atlantic to the Mississippi. It is said to be a native of Eastern Asia and introduced from there.

XXXIII. Mexican Clover.

This is another southern plant recently introduced. It is not a true clover, but belongs to the same family of plants as coffee and ipecacuanha. Its botanical name is *Richardsonia scabra*. It has become extensively naturalized in some parts of the South. Under favorable circumstances it grows rapidly, with succulent, spreading, leafy stems, which bear the small flowers in heads or clusters at the ends of the branches, and in the axils of the leaves. The flowers are funnel-formed, white, about half an inch long, with four to six narrow lobes and an equal number of stamens inserted on the inside of the corolla tube. The stem is somewhat hairy, the leaves opposite, and, like other plants of this order, connected at the base by stipules or sheaths. The leaves are oblong or elliptical and one or two inches long.

There is some conflict of opinion as to its value. It has been known in Florida for fifty years, and was regarded as a great pest among cultivated crops, which it certainly is. Since more diversified agriculture has prevailed in the South, adverse opinion is changing, and it is now regarded as of great value for feeding stock.

Mr. Matt. Coleman, of Florida, says attention was first called to it there from the cavalry horses feeding on it greedily, and adds: "Hearing of this, I procured some of the seed and have been planting or cultivating it in my orange-grove from that time to the present as a forage plant and vegetable fertilizer. I find it ample and sufficient. It grows on thin pine land, from four to six feet, branches and spreads in every direction, forming a thick matting and shade to the earth, and affords all the mulching my trees require. One hand can mow as much in one day as a horse will eat in a year. Two days' sun will cure it ready for housing or stacking, and it makes a sweet, pleasant-flavored hay. Horses and cattle both relish it. The bloom is white, always open in the morning and closed in the evening. Bees and all kinds of butterflies seek the bloom."

XXXIV. Importance of the Pulse Family.

The pulse family, to which clover belongs, contains a large number of plants valuable to man both as food for himself and as forage for stock. Among those necessary to man as food are peas and beans, in their variety. For various purposes in the arts, for medicine, and other purposes, many might be named. The pulse family is a vast one, comprising more than four hundred genera. Among them are logwood, sandal-wood, the locust, indigo and liquorice. The tamarind, senna, the peanut, Gum Arabic, all belong to the pulse family—leguminous plants—besides the soy bean (a species of *Dolichos*), from which soy is made. Besides the clovers proper, we have in this family of plants, peas, including the chick pea; beans, including the celebrated (so-called) cow-pea of the South—which, however, is not a pea, but is a bean (*Dolichos*)—there being many varieties, white, yellow, greenish, grey, red, purplish, black and spotted. The white varieties are eaten as food; but as food for cattle, pasture and hay, and as a crop in the rotation for plowing under, they have fully as much value in the South as has clover in the North.

XXXV. Interchange of Grasses Between Nations.

BEFORE leaving this subject finally we wish to say a word on the value of the interchange of seeds and plants between different countries. The fact that a plant is indigenous to a country does not prove that it is useful there. Some of the more valuable forage plants of the South, such as Alfalfa, Bermuda grass, Guinea grass, Japan clover, etc., are introduced species. The same is true of the North. It is more than probable that some of our western indigenous plains species may prove of value in Australia, since that climate, like our far western one, is dry and hot in summer. Australia has given to California the valuable Eucalypti. Our western grasses stand hot sun, and many of them extreme drought, as the Gama and other so-called Bunch grasses. The climate of Australia is mild in winter, which our western plains are not, but there, and in the hotter southwestern regions, may be found grasses that may yet prove of great value there, as many foreign varieties have been found valuable here.

CHAPTER IV.

SOILING, FODDER AND ROOT CROPS.

I. SOILING COMPARED WITH PASTURING.—II. SOILING INDISPENSABLE IN DAIRY DISTRICTS.—III. SOILING AS AGAINST FENCING.—IV. HOW TO RAISE A SOILING CROP.—V. CORN AND SORGHUM FOR SOILING.—VI. THE CLOVERS AS SOILING CROPS.—VII. MILLET AND HUNGARIAN GRASS.—VIII. PRICKLY COMFREY.—IX. THE ADVANTAGES OF SOILING.—X. RESULTS OF SOILING IN SCOTLAND.—XI. ROOT CROPS FOR FORAGE.—XII THINGS TO REMEMBER IN ROOT CULTURE.—XIII. PREPARING FOR THE ROOT CROP.—XIV. SOWING AND CULTIVATING.—XV. HARVESTING ROOT CROPS.—XVI. PITTING AND CELLARING THE ROOTS.—XVII. THE ARTICHOKE.

I. Soiling Compared with Pasturing.

SOILING is the system of cultivating, cutting and feeding forage green, as distinguished from pasturing in the field. It is only practiced in older-settled districts where land is comparatively scarce and dear, and manure plentiful, labor cheap, and the stock kept principally for use on the farm. It prevails in some portions of Europe, notably in Holland and Belgium, and in Great Britian. In the United States it can hardly be said to prevail, to the exclusion of pasturing in any of the farming districts. Its advantages are that no food is wasted, all the manure is saved, and all the land of the farm is thus enabled to produce its maximum of crops. Soiling, however, is coming to be regarded as of more and more importance year by year, in all those sections of the United States that are subject to summer drought, to carry the stock over those seasons when pasturage is scant. Another advantage is that it gives working cattle a daily portion of green food, so essential to their health, without the labor of gathering it for themselves. In this view there can be no doubt of its economy, since the daily cutting and hauling is comparatively light, and the animals will do more than enough additional labor to pay the cost.

II. Soiling Indispensable in Dairy Districts.

IN all the great dairy districts soiling is coming to be regarded as indispensable, during July and August, in order to keep cows up to their full flow of milk, and also to enable the dairyman to protect them from torturing flies and mosquitoes; thus they may be pastured in the early morning and in the evening, giving needed exercise, and kept under shelter during the heat of the day and at night. The question of profit and loss must be decided by every one for himself.

When pastures are flush it would be folly to cut and cart fodder, but instances are rare where the same quantity of stock can be kept full-fed during the heat of summer as in the spring and autumn. This can only be done where irrigation is practiced. So far as fattening stock is concerned, corn is the cheapest feed, undoubtedly, in the corn zone of the United States. Corn, or better, meal may form a portion of the daily feed of milch cows, but they must have succulent feed and an abundance of it, in order to keep them to their full flow of milk. Hence, some system of partial soiling should be adopted by every farmer who keeps milch cows, an important part of his regular farm economy.

III. Soiling as Against Fencing.

WHERE only enough cows are kept to furnish the family with milk, as in many districts where the "no-fence law" is in operation, and where stock is herded in summer, there is no doubt of the economy of soiling. The cost of fencing the farm into fields, in order that the pastures may enter into the regular rotation, is saved; in fact, the interest on this outlay would many times pay the cost of cutting and carrying the fodder for the few animals fed.

On a farm in Central Illinois, requiring the labor of sixty-five horses to work it, and where five cows were kept for milk and butter, all were fed green food in addition to their daily rations of grain during the summer. Two men and a team cleaned the stables, hauled away the manure, cut the grass and fed it, and took the entire care of the cows, besides doing various chores. The horses got about forty pounds of grass a day, what hay they would eat (very little), and the cows were fully fed on grass. The labor of not more than one man was required simply to cut and haul the fodder. This was sown: rye, clover, common meadow-grass, and later, sown corn-fodder. The cutting began when the crop was twelve to fourteen inches high, except the corn, which was allowed to grow two feet high, and all was cut with a mowing machine, and raked and loaded by hand.

IV. How to Raise a Soiling Crop.

ANY land for a soiling crop to be cut green should be as rich as possible. The more luxuriant the growth, the better the swath, thus making a great saving in the labor of cutting and gathering. It will not do to depend on one variety for stock, for they soon tire of a single diet. A patch of Red clover—in the South alfalfa—one of Orchard-grass, one of Rye-grass, and one of rye may be provided. These will make your first cuttings, and if the ground is heavily manured, and there is plenty of moisture, these may be cut over once in four or five weeks and give a good swath.

Millet, corn and sorghum should follow, to eke out these, and you will have a variety that stock will never tire of. It is better for swine to be fed in this way than to allow them to run in a pasture, since then you are not obliged to ring them, exposing them to the risk of becoming impregnated with contagious blood, and other infections. The grasses may be top-dressed to keep up the fertility. This should always be done with compost manure, not less than four loads of forty bushels each per acre per year, and if two bushels of plaster and one hundred pounds of superphosphate be added, it will pay. It will also pay to have the soiling crop field as near the barn as possible. If you doubt the soundness of this advice, take a meadow of mixed grasses, top-dress it with twenty loads of compost manure, plaster, and superphosphate, and watch the results. It will be a swath you cannot put a "scythe into clear up to the heel."

No Wilted Fodder.—Whatever the soiling crop, be it corn, sorghum, millet, clover, meadow grass, or cow peas, the soil must be rich, else you will fail. You will also fail unless you cut it when quite green, or in its most succulent state. If you think it does not pay to haul the crop when it is heavy with moisture, you may wilt it. But good milk is not made from wilted plants. They are distasteful to stock, and are eaten only under compulsion. Above all, do not let the cuttings lie on the wagon until they heat and begin to turn yellow. Distribute it as soon as hauled. Green grass and other fodder is in just the right state for heating. It will begin to get warm in half an hour if left in a pile. And certainly no humane man will force his stock to eat disgusting or distasteful food when it can so easily be avoided. We should almost as soon think one would take pleasure in having his family eat stale food. The taste of animals is fully as delicate as that of man. At least we may infer as much from watching them graze when not pressed by hunger.

V. Corn and Sorghum for Soiling.

SORGHUM.

Green corn and sorghum are among the most prolific plants that can be grown for soiling. Once the roots get established they feed greedily, and soon force the plants into dense growth if the soil is fully manured. Corn and sorghum should form a large portion of the feed, during August. These two plants with millet and Hungarian grass will carry the stock, with a fair variety of food, until the fall pastures become flush.

The preparation of the soil must be looked to. If the ground has been plowed early in the season, after heavy manuring, give it a light coat of compost, or at least half-rotted manure. Plow this under not more than four inches deep, bring your soil as level and smooth as possible and into the most perfect state of tilth. If you are certain your soil will be clean from weeds, sow three bushels of corn to the acre, broadcast, cover it in with the cultivator and roll the surface. If sorghum is to be sown, use a bushel and a half of seed to the acre, cover with the cultivator and then roll.

If you are afraid of weeds interfering drill the corn, one and a half bushels to the acre, or sorghum three-fourths of a bushel, by going over the ground twice with the corn planter set to drill—the last time so the rows will come between the previous ones. This will leave the rows twenty-two inches apart, wide enough to work between with a narrow harrow or five-toothed cultivator.

In fact, by this system about all the cultivation may be done with the harrow, as directed in the chapter on Indian Corn. We have never failed to get a smothering crop by drilling two bushels per acre with a grain drill set to feed properly and deep, and then harrowing with a light harrow, once each way, just as the weeds appear. Sorghum may be sown in the same way, from the grain drill and harrowed, or planted as directed for corn from the corn planter, if it have a drill attachment, which all good machines should have.

VI. The Clovers as Soiling Crops.

THERE are only two of the trefoils that give general satisfaction as soiling plants. These are Biennial clover (*Trifolium pratense*), and Alfalfa (*Medicago sativa*). The first thrives in the North, the latter in the South and in California; and in other mild rainless summer climates where the crop may be irrigated. The Yellow alfalfa or Burr clover of California (*M. maculata*), is recommended as most valuable in the South, or wherever it is hardy, and can thus have time to become established. It is not valuable in the West north of the latitude of Kentucky.

The Cow-Pea for Forage.—The pulse family produces another plant that is of the greatest value in the South, viz, the Cow-pea—which, by the way, is a misnomer; it is really a bean (*dolichos*), of which there are many varieties, some being used for the table. The well-known asparagus bean is a dolichos. The Cow-pea, while of less value for soiling, is of great value to the South as a forage crop. In fact, it is one of the most useful of plants there as fodder, and for its value as a fertilizer when turned under green. All farm stock eat it. It will even thrive on poor soil, leaves the land in the best possible condition, and when sown in the corn-field makes excellent pasture for stock after the corn is gathered. But care must be taken that greedy or hungry stock are not allowed to graze it heavily at first, else they will fill themselves so full that they will suffer from hoven or bloat, from the gas generated, and perhaps die if not promptly relieved.

HUNGARIAN GRASS.

VII. Millet and Hungarian Grass.

THERE seems to be a general feeling against Millet and Hungarian grass for cutting and feeding green. There can be no possible reason for this except that these plants cannot be sown successfully until both the days and nights are warm. Hence it is not available for feeding until late, when there are other forage plants in plenty.

Pearl Millet.—This plant (*Penicillaria Spicata*), also called Egyptian, or East Indian Millet, produces enormously and is coming more and more into favor in the South every year. It contains a large leaf surface. It does not mature its seeds in the North, but up to the latitude of about 40° attains full size for cutting. In the South, if cut when young, it will produce a second crop.

Brown Dhoura.—Thist is a sorghum, also called Indian millet, Chocolate corn, Guinea corn, and Pampas rice. It is a valuable crop South, for its grain as well as for its fodder. Both these plants, and, in fact, all the sorghums and Indian corn, should be drilled two feet apart, which will allow cultivation with a five-tooth cultivator or scarifier, by drawing the implement closely together.

Common and German Millet.—Common millet and German or golden millet are both excellent fodder-plants, and both of them are rapid-growing crops; the first has a close head, the latter more open but with plenty of leaves. These may be sown as late, in the North, as the first of July, and make a crop of hay. All the fodder crops are, however, often somewhat difficult to cure, late in

PEARL MILLET

the season, and both these and Hungarian grass are better sown from the first to the tenth of June. The usual seeding is three pecks to the acre, but for hay or fodder one bushel per acre is better. If you wish to make seed, drill one-half bushel of seed per acre, in drills two feet apart, covering the seed not more than half an inch deep.

VIII. Prickly Comfrey.

This plant, of which much has been said, for and against, we do not think of much value where there are other fodder plants to choose from. It has thick, broad, succulent leaves, produces enormously on rich land—up to forty or more tons of green fodder in a season,—endures the severest droughts, and in seasons of average moisture may be cut four or five times in a season. It is propagated by divisions of the roots, which may be set three feet apart, requiring nearly five thousand plants to the acre. Stock do not take kindly to it at first, but must be taught to eat it. It is strictly a plant for dry soil, and should never be planted on low, wet ground. It is, indeed, a moisture-loving plant, but it must be the moisture of a fairly drained soil, for its roots range deep. Hence its power of withstanding drought. Those intending to experiment with this plant should remember that it is not the common comfrey of the United States, but an allied species (*Symphytum asperrinum*) from the Caucasus. It is propagated and sold by nurserymen, and is also kept by some seedsmen.

PRICKLY COMFREY.

IX. The Advantages of Soiling.

THE advantages of exclusive soiling, that is, cutting and feeding green crops in summer, will never be acknowledged in the United States except near cities where land is valuable, and on restricted areas. The late Josiah Quincy is the father of soiling in America; his claims, true enough, were:

1, It saves land; 2, it saves fecinng; 3, it economizes food; 4, it keeps cattle in better condition and greater comfort; 5, it produces more milk; 6, it increases the quantity and quality of manure; 7, there is better docility and discipline of animals where it is used; 8, there is less breaking of fences; 9, there is increased order in all business of the farm.

Mr. Quincy's testimony in relation to soiling and the crops grown in Massachusetts, the State where his operations were carried on, is, that one acre soiled from will produce at least as much as three acres pastured in the usual way, and that "there is no proposition in Nature more true than that any good farmer may maintain upon thirty acres of good arable land, twenty head of cattle the year round, in better condition, and greater comfort to the animals, with more profit, less labor, less trouble, and less cash advance for himself than he by the present mode expends upon a hundred acres." He further says, "My own experience has always been less than this, never having exceeded seventeen acres for twenty head.

"To produce a sufficient quantity and succession of succulent food—about one and a half or two square rods of ground to each cow to be soiled—sow as follows:

"As early in April as the state of the land will permit, which is usually between the fifth and tenth, on properly prepared land, oats at the rate of four bushels to the acre.

"About the twentieth of the same month sow, either oats or barley, at the same rate per acre, in like quantity and proportions.

"Early in May sow, in like manner, either of the above grains.

"Between the tenth and twentieth of May sow Indian corn (southern dent being best), in drills, three bushels to the acre, in like quantity and proportions.

"About the twenty-fifth of May sow corn, in like manner and proportions.

"About the fifth of June repeat the sowing of corn, as above.

"After the last-mentioned sowing, barley should be sown in the above-mentioned quantity and proportions, in following successions, on the fifteenth and twenty-fifth of June, and in the first week in July, barley being the best qualified to resist the early frosts."

X. Results of Soiling in Scotland.

Mr. Brown, of Mankle, Scotland, a farmer of extensive operations, made the following experiment in order to ascertain the comparative merits of soiling and pasturing cattle. In the spring he took forty-eight Aberdeenshire bullocks which had been wintered in his farm-yard, and separated them into two equal lots, one of which he put to grass, while the other was soiled. The latter were fed on Swedish turnips until the clover was ready for cutting, and then the clover was given sparingly for a week, in order to avoid danger from over-eating, after which a full supply was allowed.

The animals thrived exceedingly well until the grass got hard and withered. About the last of July, the clover having ripened, vetches were substituted, which were continued until the second crop of clover was ready for cutting. Ten of the soiled lot were sold in August, and the remainder of the two lots in September. The results are thus stated: The forty-eight cattle cost in purchase and wintering, £503 2s. The best ten of the soiled lot sold at £17 5s. each; the remainder of the two lots sold at £14 5s. each; the soiled lot thus bringing £377, and the grazed lot £342, a difference of £35 in favor of the soiled cattle. It required one and three-quarters acre of Swedish turnips, eight acres of clover, and three acres of vetches to furnish the food consumed by the twenty-four soiled cattle. The result of soiling exhibited decidedly the larger profit.

XI. Root Crops for Forage.

Turnip culture revolutionized the agriculture of England. The cultivation of Indian corn in connection with western grass, making cheap beef, bids fair to revolutionize agriculture there again. The climate of the United States, with the exception of a small portion of the extreme north, is unsuited to the cultivation of the white or round turnip on account of heat and summer drought. Even rutabagas are generally hot, tough and stringy.

Another great objection to root crops is their cost, since they require much hand labor, and the roots cannot be grazed in the fields, as they can in England. And yet the need of some succulent vegetable food in winter has been so widely felt, that among our best farmers more or less attention has been paid to carrots, parsnips and to the sugar and mangel beets. We have found that mangel wurtzel beets fully met the requirements for both cattle and sheep. We are as fully satisfied that ensilage, to be treated of in the next chapter, will fairly perform all that is claimed for roots, except in some special cases. Hence it will be necessary to treat only of the general requirements for the cultivation of root crops.

XII. Things to Remember in Root Culture.

IMPORTANT things to be remembered in the cultivation of roots are:

1. They cannot be successfully raised on land recently treated with green manure. Why? It inevitably causes the roots to grow forked, reducing their value, and largely increasing the cost of gathering and cleaning. Hence the land, unless compost manure is used, should have been manured heavily one or two previous seasons.

2. A root crop should never be raised except on land made as rich as possible with manure. Why, again? Because it costs as much to cultivate a poor acre as a rich one. The cultivation of roots involves an outlay of thirty dollars or more per acre for labor alone.

3. The cultivation of roots should never be undertaken on lumpy, trashy land, or on land otherwise difficult to work. It adds too largely to the manual labor of making the crop. These points carefully remembered, it will not be difficult to attain the best results for the least outlay.

XIII. Preparing for the Root Crop.

THE chief expense in the cultivation of root crops is hand-weeding the rows, and thinning—singling as it is called—the plants. Hence the necessity of perfectly clean land, and of having the seed sown in absolutely straight equidistant rows, on soil entirely free from lumps or trash, and thoroughly friable. The orifice of the drill that delivers the seed should also deposit it in knife rows, that is, one single narrow line. In this way the hand cultivator may be run within an eighth of an inch of the rows, and in large fields gang implements may be used, by which two or more rows may be cultivated at a time.

These latter, however, are never used except in the most extensive market gardens, where forty or more acres of roots are grown, or where beets are grown by the hundreds of acres for making sugar. The writer has raised them thus, putting beets in the pits at a cost of three dollars and forty cents per ton, on an average yield of eleven tons per acre. It must be remembered that beets for sugar making are never to be much over one pound each in weight. On highly manured land, fifty tons of beets per acre, and of carrots and parsnips thirty or more tons per acre, may be raised.

To bring the soil into the best condition, it should be deeply fall-plowed. In the spring, when the soil will work thoroughly friable, it may be lightly replowed, harrowed, leveled with the leveler, and ground fine with the plank machine described in a previous chapter. The accompanying cuts show Fig 1, field carrot; Fig. 2, long mangel-wurzel beet. The shaded edges show the ground line.

XIV. Sowing and Cultivating.

FIELD beets may be sown in drills thirty inches apart, and, in field culture, carrots and parsnips in rows two feet apart. This will allow the horse cultivator to run between the rows, and after the plants have gained considerable size no handwork need be done. Six pounds of beet-seed will be required to the acre, to ensure a stand against all contingencies.

Each capsule of the seed plant contains from two to four seeds, and hence whether the land be weedy or not the plants must be singled. In all root crops the first thinning may be done with a narrow hoe or other implement; the subsequent thinning by hand. A wheel hoe (hand cultivator) will pay for itself every year in the cultivation of a single acre; and with such an implement one hand will keep from five to six acres free of weeds, going twice in each row, at every cultivation. Carrots and parsnips will require about four pounds of seed per acre, to be sure of a stand, and allow for what the insects may destroy.

Singling.—When the plants are up about three inches they must be singled, by hand, the beets to stand from nine to ten inches apart, and carrots and parsnips five to six inches apart. In thinning, steady boys may be employed. They must go down on hands and knees, astride of the rows, the spaces having been previously marked for them with the point of a hoe or a gang implement, cutting narrow lines. In extensive cultivation we have done this with a horse machine going across the rows.

Weeding.—The weeders, whether boys or girls, must be properly instructed. Being on the hands and knees, the weeds, if any, having been pulled, one hand secures the bunch of plants to be operated on, while the other removes superfluous plants. If the plants are strongly rooted, it may be necessary to guard the plant to be left by holding the finger before it, close to the ground. The weeding is difficult to describe, but not difficult to learn. The overseer should practice and experiment himself, so he may be able to properly instruct those under his charge. The subsequent cultivation is simply to keep down weeds.

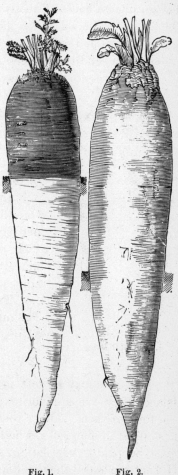

Fig. 1. BELGIAN CARROT.

Fig. 2. MANGEL WURZEL.

XV. Harvesting Root Crops.

One of the most expensive operations, next to weeding, is harvesting. Beets may be easily pulled by plowing a deep furrow away from them with a one-horse plow. They should be laid in regular rows, two rows with the tops pointing together.

Thus they are easily and quickly topped, by a man with a spade, ground sharp, topping one row going one way and another the other. The tops are more easily gathered up. The beets are then to be placed in long piles and covered with the leaves, or else hauled directly to the pit or cellar. Roots of any kind should never be allowed to be wilted by the sun.

Parsnips and carrots are dug by hand or plowed out. In plowing, begin on one outside row, and turn a deep furrow away from the row, running about seven inches from the row. Pass around the field and plow another furrow as on the first side. Returning to the first furrow, plow another furrow as deep as possible, and as close to the row as you can work. If you are a first-rate plowman, you can hit it fairly; if not, you will here find it out. The roots are then to be pulled, or lifted with the spade, topped, and carried to the pits.

PARSNIPS.

In Europe there are various machines for digging roots. One that we made, and that would loosen five acres of beets per day, was simply two very heavy, properly curved coulters, each of them running under a row of beets. The digger was attached to the beam of a gang-plow, and drawn by four horses.

Two varieties of parsnips are shown in the illustration. The one on the right is the hollow-crowned parsnip; the one on the left is an intermediate variety between short and long.

XVI. Pitting and Cellaring the Roots.

PARSNIPS may be left in the ground all winter without injury. In the Channel Islands they are a favorite crop for feeding milch cows, and all stock are fond of them. Other roots must be housed or covered. Parsnips are best piled in long ricks, whether above ground or in trenches. Attention must be paid to ventilation, so that the roots shall not sweat and heat. If kept too warm they will sprout.

All roots are ruined by freezing, except parsnips, salsify, onions and rutabagas. These when frozen must be thawed out naturally in the pits, kept dark, before being opened, in order to escape injury. The pits may be about three and one-half feet wide and three feet deep below ground, running to a sharp apex above, with small bundles of straw reaching from the bottom to the top at proper intervals. The whole should then be covered with straw, six inches thick; with a covering of earth at least six inches in depth or sufficient to carry off rain.

CARROTS.

Let the straw ventilators extend above ground. At the approach of hard weather give another covering of six inches of straw and ten inches of earth over this, and the pits will be safe from any ordinary winter weather; but when a good crust has frozen, if very hard weather is feared, cover all with green manure litter.

The cuts of carrots represent, the one at the right, the Danvers, the one at the left, the half long or intermediate varieties. For field cultivation the large red, and the white (Belgian) varieties, are mostly used. They grow with a considerable portion of their length out of the ground, thus making their harvesting easier. The last cut shows the red Altringham carrot, one of the long-rooted varieties.

XVII. The Artichoke.

A CHAPTER on soiling, fodder and root crops would not be complete without mention of the artichoke. It was introduced into Great Britain from Brazil before the potato, but never met with much favor as an esculent. Until its value as food for swine was discovered in the West, it was used principally for pickling.

It is very hardy, remaining in the ground all winter uninjured by our severest cold weather, and springing up the next season without farther cultivation than that given by hogs in rooting out the large tubers. They are, however, better, if the sections of the tubers are dropped in furrows four feet apart, and plowed once or twice, the yield in such cases sometimes reaching four hundred bushels to the acre.

RED ALTRINGHAM.

When partly grown, the tubers are round, but as they attain full size, they become irregularly elongated. The color of the skin and flesh is white, the stalks produce few branches, attain the height of six, and even eight feet, and bear

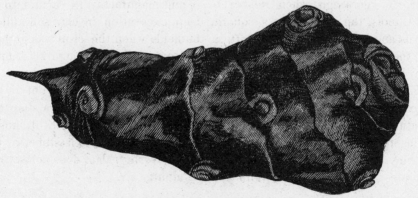

JERUSALEM ARTICHOKE.

yellow flowers, similar to the garden sunflower, but much smaller. The cut represents a tuber of the Brazilian variety, reduced in size. A native variety found growing in rich sandy bottoms of the West—the natural soil of the artichoke—is brown, smooth and long like the sweet potato; they are eagerly sought by swine.

CHAPTER V.

SILOS AND ENSILAGE.

I. WHAT IS ENSILAGE?—II. SILOS AND ENSILAGE LONG KNOWN IN EUROPE.—III. TWO METHODS ILLUSTRATED.—IV. THE FATHER OF ENSILAGE.—V. FERMENTATION SHOULD BE AVOIDED.—VI. WHAT ENSILAGE MAY DO.—VII. THE HISTORY OF ENSILAGE.—VIII. FEEDING VALUE OF ENSILAGE.—IX. ENSILAGE IN THE UNITED STATES.—X. EFFECTS OF FERMENTATION IN THE SILO.—XI. SIZE OF SILOS FOR CERTAIN NUMBERS OF STOCK.—XII. HOW TO BUILD A SILO.—XIII. PRACTICAL EXPERIENCE AND RESULTS.—XIV. PERFECT FOOD AND RATIONS.—XV. SOME STATEMENTS OF THE QUANTITY FED.—XVI. COST OF ENSILAGE IN MASSACHUSETTS.—XVII. BUILDING A MODEL SILO.—XVIII. PRACTICAL CONCLUSIONS FROM CAREFUL EXPERIMENT.

I. What is Ensilage?

ENSILAGE is a French word signifying the art of compressing into silos—pits, trenches, etc.—green crops, or other succulent vegetation; the word literally meaning the forage so preserved. Silo is the French name of the pit, trench or chamber in which the ensilage is stored. A silo, then, is simply a vat, cistern, or underground trench, water-tight at the bottom and sides, in which any vegetable substance liable to ferment may be kept fresh by exclusion of the air. The structure may be either entirely above or below ground, or partly above and partly below the surface. It is not even necessary that the silo be made water-tight in dry soil, nor is it necessary that it be bricked or stoned up in firm soil.

Thirteen years ago, when raising beets and manufacturing them into sugar, in Central Illinois, the writer dug a square, deep excavation in clay soil, filled it with the crowns and leaves of the beets left in the fields when the crop was gathered, and covered them with two feet of earth. The leaves were thoroughly tramped while filling in, and the earth occasionally was pounded down as the whole settled. Twelve inches more of earth was added afterwards, and the ensilage came out in good condition, except a crust a few inches thick at top and sides, but in a state of vinous fermentation—what the German laborers called " wine sour." Still, unsupported earth silos are not to be commended. Properly supported with stone, or brick and cement, they are cheaper in the end. The earth covering, also, does not settle evenly, gas generates, and the ensilage puffs up and admits air.

II. Silos and Ensilage Long Known in Europe.

THE art of preserving succulent food in tight cisterns has been known for many years. Brewers' grains have been so preserved. It is claimed that silos were known to the ancient Romans, but there is no good authority to show that green fodder was kept by them in this way. It is also asserted that the Mexicans so preserved their grain, both green and matured, by this method. The Mexican Indians and the Indians of North America did often keep their corn in the husk in underground caves, but not under pressure enough to exclude the air. A dry situation was chosen, and the grain was kept in much the same way that is now in general use for preserving roots—nothing more.

TALL CORN, CLOSE SHOCKED.

DWARF CORN, IN TWO TIERS.

DWARF CORN, IN THREE TIERS.

The practical application of air-tight silos for preserving green forage fresh and sweet, is a modern idea. The honor of the discovery belongs to the French. It is only within the last few years that the process has received the careful attention of experimenters in the United States.

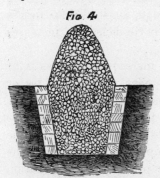
BEFORE COVERING.

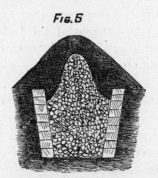
AFTER COVERING.

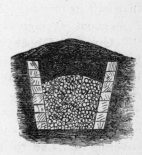
FINAL COMPRESSION.

III. Two Methods Illustrated.

The series of cuts numbered from one to six, exhibit two methods of forage mowed green. The first three show the proper plan of bundling Indian corn or sorghum, and of stacking together, when the fodder is to be cured in the field dry for winter use. Forage thus stacked will shed rain perfectly; this plan is in fact nothing but compact shocking. But this final binding must not be done until the bundles which form the shock are thoroughly air-dried. Fig. 1 shows simple close shocking of tall corn. Figs. 2 and 3 show the dwarf varieties; 2 in two tiers, and 3 in three tiers of bundles.

Earth Silos.—The French plan of ensilage is exhibited by the cuts numbered 4, 5 and 6. Fig. 4 shows the ensilage when first piled and tramped, and before covering. It also shows the proper form of root pits. [See chapter on Raising and Pitting Roots.] Fig. 5 shows the ensilage as compressed both vertically and laterally by the weight of the earth piled on top, and Fig. 6 shows the ensilage as finally reduced by the earth pressure.

Crevat's Experiments.— M. Crevat, after several years of experiment, recommends pits of the following dimensions: Depth, 2.30 meters (7.55 feet); length, 8 meters (26.25 feet), at the surface of the ground, sloping down to 7.40 meters (24.28 feet) on the bottom; breadth, 2.60 meters (8.53 feet) at the top, and 2 meters (6.56 feet) at the bottom. Each pit has a capacity for about 40 cubic meters (about $1,412 \frac{2}{3}$ cubic feet) of fodder. M. Crevat has found reason to deepen the trenches and to contract their width, in order to lessen the expense of covering them with earth. The sides and ends are sloped, in order to allow an oblique, as well as a vertical, pressure from the superincumbent earth, and to make the upper surface of the fodder convex. In each of these pits about $10\frac{1}{2}$ tons of green fodder may be packed. Two or three days' drying in the hot sun will reduce it about a third in weight. Many farmers prefer to dry the material in order to render it more easy of transportation. The trench is filled and the fodder piled up above the ground to a height equal to its depth under the surface. The earth is then thrown upon the mass before fermentation commences. Two feet depth of soil will depress the pile at least a yard by simple pressure. After some days of fermentation it shrinks to less than half its original volume. The weight of the material, by condensation, increases from about 800 pounds per cubic yard to over 2,000 pounds.

General Observations.—In some cases, the silos are mere pits, with walls of bare earth. In other cases, they are lined with brick or cement, either on sides or bottom or on both. Where the soil is excessively damp, the walls are built entirely or partially above the surface, and embankments are made for their support. It is found necessary to exercise special care in covering the pits to entirely exclude the air. The dislocations in the fermenting fodder will often open fissures through the covering soil, and the air thus admitted will transform the process of fermentation into one of putrefaction. Sometimes decidedly alcoholic fumes have been given off through the crevices in the covering. One case is noted in which the ensilage entirely failed, on account of using sand instead of earth as a covering. Different opinions prevail in regard to the propriety of cutting or chopping the maize into small fragments before packing in the trenches. In case the maize has become over-ripe, it is urged that cutting facilitates fermentation, which will render the harder portions as easy of mastication and digestion by farm animals as the softer portions.

IV. The Father of Ensilage.

To Mr. Auguste Goffart, a member of the Central Society of Agriculture of France, belongs the credit of a system of experiments by which green fodder, cut small, was kept in water and air-tight excavations in almost as good condition as when cut. In fact, the slight fermentation and breaking down of fiber, and the desiccations which the forage undergoes when thus stored, undoubtedly renders the coarser portions of the provender more digestible, the change being analagous, in a sense, to cooking.

In 1852, Mr. Goffart built six underground silos of cemented masonry. They were small, having a capacity only of two cubic meters each. [The French meter is three feet three and one-third inches.] Maize, Jerusalem artichoke, beets, sorghum, turnips potatoes and straw were experimented with; but not until 1873 did he have real success, and then only by a fortunate discovery. This was exclusion of air by strong pressure. Hence to this gentleman is undoubtedly due the perfection of this very valuable method of saving fodder for the winter months.

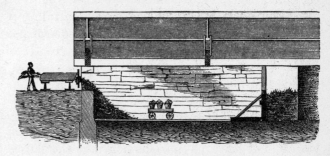

ONE OF THE EARLIER SILOS.

The Best Results.—His testimony, and no one is better qualified to speak authoritatively, is that a silo built upon the ground gives the best results during the season from December to March, but that underground silos are better for spring and summer feeding.

He recommends silos sunk two meters (about six feet six inches) below ground, and raised the same distance above. He feeds the upper portion during winter and the lower portion later. In the United States, however, ensilage will not be used to the exclusion of pasturage, and hence it will be economical for us to cut from top to bottom, section by section.

V. Fermentation should be Avoided.

In an address at Blois, France, Mr. Goffart held that: "It is important to avoid all kinds of fermentation during and after ensilage. Fermentation can be produced whenever desired, and a few hours suffice to give all its useful effects. Take each evening from your silo the maize required for the next day's feeding, and in fifteen or sixteen hours after, however cold and free from fermentation when taken out, it will be quite warm, in full fermentation, and the animals will eat it greedily. Eight hours later it will have passed the proper limit and it will spoil rapidly."

With the cheap French labor, the cost of the ensilage, exclusive of that of raising the crop is about twenty cents per ton prepared and placed in the silo.

VI. What Ensilage may Do.

It will give us succulent food in winter and also enable any farmer to tide over the droughts of summer by saving the material in underground silos, and this at a minimum cost. By this means many waste products of the farm, such as clean, bright straw, may be added to the green forage to assist in taking up the superabundant moisture, thus reducing the whole to one homogeneous mass.

Corn fodder is not a perfect food in itself; it may be made so by the addition of other matter, mill-stuff, bran, etc. To the dairyman ensilage is of great value, since it will keep the cows up to a full flow of milk continuously. To the shepherd it will allow the use of succulent food, so needful to sheep in winter. It will assist the breeder of young stock in keeping them in full growth and vigor in winter. It will be equally valuable to the breeder and feeder of fine stock, who, notably, spend large sums in artificial feeding stuffs.

VII. The History of Ensilage.

A COMMUNICATION to the Department of Agriculture at Washington, some ten years ago, shows that the preservation of fodder in silos has been practiced in Austro-Hungary for nearly eighty years, and in Germany previous to its employment in France, but since its introduction into the latter country it has been nowhere so elaborately developed as there. As we have already shown, the perfect results in France have only been reached through the most careful and laborious experiments, carried on through a long series of years.

VIII. Feeding Value of Ensilage.

M. PASQUAY has deduced some valuable facts in relation to the feeding value of ensilage. It was found that maize fodder (green) has a feeding-value equal to 22 per cent of that of hay; rye fodder, 38 per cent; grass (green), 34 per cent; bright wheat-straw, 48 per cent. In a good forage ration for a milch cow, the ratio of nitrogenous to non-nitrogenous matter should be as 1 to 5, or even as 1 to 4.5; for young animals, weighing between 250 and 300 pounds, as 1 to 3.3; for animals of 450 pounds, as 1 to 4; for oxen in absolute repose, as 1 to 8. Maize-forage cut green does not meet this requirement, as it shows a proportion of 1 to 9.24. The maize preserved with a mixture of straw, as at Cerçay, approximates the standard, showing proportion of 1 to 4.81. Its increased per cent of fatty matter represents also a great advantage, being six times greater than in the green maize.

Other Facts.—M. Goffart finds that his preserved fodder is sufficient without any other food to keep his animals in fine condition. M. Houette, of the department of Yonne, has found by experience that the maize should be cut for preservation in silos as near as possible to its maturity, when it is more nutritive, the ears more developed, the stalks more firm, and the watery element less predominant. Being finely chopped before pitting, its fermentation in the silo will soften it and render it as palatable to animals as the freshly cut maize. He has been able to keep stock upon it to the last of May, and once as late as July, the fodder being in a condition but imperceptibly changed from that of its primary fermentation in the silo. Some question has been raised as to the propriety of feeding fodder spoiled in the pits, but while no indications of injury from feeding it have been developed, it is justly considered that it is more available as a plant-food than animal food; hence it is thrown upon the manure-pile.

Maximum Yields.—The comparative maximum yields of various fodder plants in France, by M. Leconteaux, is summarized, as showing extreme results, but those obtained, of the root crops noted, have often been largely exceeded. The results are given in the table:

NAMES OF PLANTS.	GRASS YIELD PER ACRE.	EQUIVALENT IN HAY.
	Tons.	Tons.
Caragua maize (a tall species of Indian corn),	66.96	16.73
Sugar beets,	35.68	11.63
Rye-grass with liquid manure,	35.68	8.97
Marcite meadows of Italy,	28.85	7.21
Rutabagas,	21.41	5.35
Potatoes,	9.81	4.90
Cabbages,	17.84	3.56

IX. Ensilage in the United States.

Mr. Francis Morris, of Oakland Manor, Maryland, had his attention called to the subject through a French newspaper early in 1876. On the first of August, that year, he sowed five acres of corn in drills, at the rate of one bushel of seed to the acre. Three silos were bricked up inside a stone barn, each being ten feet deep, four feet wide, and twenty-four feet long (a single silo 12 x 12 x 24 would have been better). Early in October, the corn being in tassel, it was cut with a mowing-machine, drawn to the silos, cut into inch pieces, and mixed with about one-fifth its bulk of cut straw. The whole was placed in the silos, and well packed by tramping as it was put in. It was covered with boards heavily weighted with stone, and when thoroughly pressed the weights were taken off, the whole surface covered with straw, and this with clay, well rammed down, to exclude air. On Christmas day a silo was opened, and the ensilage given to the milch cows of the farm. Two of them refused to eat it the first day; the others took kindly to it, and the second day all ate. After that, he says, horses, mules, oxen, cows, sheep and pigs all ate it from choice.

Had Mr. Morris known at that time the superior methods now used to exclude air, his success would have been still better. It was, however, the first fairly carried out practical experiment in curing ensilage in the United States. Within the last few years dairymen in the West have eagerly seized upon the idea, and each year sees more and better silos built in all the great dairy districts. The system may profitably be extended among those interested in other branches of agriculture.

X. Effects of Fermentation in the Silo.

M. GRANDEAU, a French experimenter, analyzed various specimens, as shown in the table below—the specimen from Bertin being taken from the silo of Monsieur Goffart, with the following results:

COMPONENT PARTS.	GREEN MAIZE.	MAIZE FREE FROM STRAW, PRESERVED IN SILOS, CHATEAU BERTIN.	MAIZE MIXED WITH STRAW, PRESERVED IN SILOS AT CERÇAY.	STRAW AND CHAFF FROM CERÇAY.
Water,	81.28	81.28	60.71	14.50
Sugar,	0.58	0.15	1.89	..
Azotized matters,	1.22	1.24	3.74	4.88
Non-azotized matters,	10.40	9.58	14.59	34.52
Fatty matters,	0.25	0.36	1.50	1.50
Crude cellulose,	4.98	4.91	8.70	35.50
Ashes,	1.29	2.25	8.43	9.10
Acid,	....	0.23	0.44	....
	100.00	100.00	100.00	100.00

M. Grandeau gives proportion of azotized to non-azotized matters as follows: In green maize, 1 to 9.24; in maize preserved free from straw, 1 to 8.14; in maize preserved mingled with straw, 1 to 4.81; in straw, 1 to 7.38.

The fodder preserved with straw at Cerçay shows a remarkable reduction in its percentage of water in straw. It also shows a saccharine element three times greater than that of green maize, while that preserved at Bertin free from straw retains but a fourth of its original quantity. The Cerçay fodder also tripled the amount of azotized matter in the green maize, finding a large supply in the associated straw, while in the Bertin specimen it was but slightly increased. Again, the Bertin fodder decreased its proportion of non-azotized matter, while that of Cerçay borrowed largely from the straw. Both kinds of preserved fodder enlarged their proportion of fatty matter; that of Bertin less than fifty per cent, and that of Cerçay sixfold. Of crude cellulose, the Bertin shows a slight decrease, while the Cerçay about doubled its percentage. The proportion of ash increased twofold in the Bertin, and nearly sevenfold in the Cerçay. Both kinds showed a perceptible development of acid, acetic and lactic. In the Cerçay fodder, the maize was mixed with half its quantity of straw. As the result of his investigations, M. Grandeau came to the conclusion that the combination of straw with maize added very considerably to the nutritive value of the fermented fodder.

In a subsequent statement, M. Grandeau explains that the specimen called green maize in his analysis was partly desiccated by contact with air and sun-heat. This would more particularly affect its percentage of water. From a specimen freshly cut, he obtained, by analysis, the following percentages: Water, 86.20; sugar, 0.43; azotized matter, 0.90; non-azotized matter, 7.67; fatty matter, 0.18; crude cellulose, 3.67; ashes, 0.95.

M. Grandeau's conclusion in favor of the mingling of straw with the maize called forth considerable criticism, but after a careful reconsideration of the question he adhered to his opinion. The transformation of the buried fodder embraces two important elements of advantage: 1, the transformation of a part of the starch and cellulose into sugar; and 2, the enlargement of the azotized matter by the destruction of a portion of the fecula of the cellulose.

XI. Size of Silos for Certain Numbers of Stock.

WHILE ensilage may be kept more or less perfect in simple pits, eventually the cost is reduced by the best constructed silos. The silos need not be expensive, but they must be built thoroughly, because thorough building is cheap building in the end. The cut shows sections of a double underground silo.

Rations for a Cow.—A cow will consume as a full ration from fifty to sixty pounds of ensilage a day. A cubic foot of ensilage weighs forty to fifty pounds, according to the material and pressure employed. One and a third cubic foot daily will keep a cow; one cubic foot will feed a sheep a week, and fully one and a half cubic feet will be required daily for an ox. To feed a cow six months will require about two hundred and fifty cubic feet of ensilage.

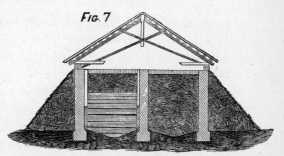

SECTION OF DOUBLE SILO.

If you have two cows, a silo 10 x 10 feet, and ten feet deep, will hold about twenty-five tons, and be ample for six months' feeding. A silo 10 x 10 x 30 feet would keep six cows, on nearly full rations, or double that number when other food is used to supplement the ensilage—a practice we should advocate in the West, when other food is cheap. Hence the dairyman feeding half rations, that is, half ensilage and half other food, could feed thirty cows for six months from a double silo, each compartment being 12 x 12 feet and thirty feet long. An extra silo would tide double this number of cows over the usual six weeks of summer droughts.

XII. How to Build a Silo.

BUILD it so it may be entered directly from the feeding stables, and one-half below ground. Eight feet below and four feet above, the upper portion banked up with the excavated earth, would be better. There must be a double door at least five feet high and three feet wide for ease in emptying the silos, after the first section is removed to a level with the bottom of the door. The ensilage nearest to the door should be supported by planks fitting into an inset in the wall, next to the door, to be removed one by one as the ensilage is taken out.

The Foundation.—The drainage under the foundation should be made perfect by means of a layer of coarse gravel and tile leading to a point below the bottom of the silo. The bottom and sides should be built of stone, laid in the best hydraulic cement. If stone cannot be had, hard-burned brick is the next cheapest good material. The bricks should be laid in cement and the whole inside and bottom thoroughly plastered with cement.

The Superstructure.—This may be of lumber and the roof of shingles; and if it is carried up one story above the silo, it will afford a useful workshop or room for storage of any kind. It will pay to finish the building in this way, as the extra expense will prove to be true economy in the end. The excavation may be mostly done with the plow and scraper, and the incline afterwards filled in by the scraper as the wall is built up. The cuts given will serve as a guide, so that any bricklayer and carpenter can do the work. Directions for raising the crops to fill the silo will be found in the chapter on Soiling and Fodder Crops.

XIII. Practical Experience and Results.

In 1882 the Commissioner of Agriculture, Hon. George B. Loring, sent out twenty-six questions in relation to silos and ensilage, to which nearly one hundred answers were returned, from fifteen States, east, south and as far west as Nebraska; also from Canada. The information given in these answers was summarized under twenty-six heads, making one of the most complete and practical papers ever issued by the department. Here it is:

1. LOCATION OF SILO.—A few have been built at a distance from the stables, but generally the silos are located with reference to convenience in feeding, in, under, or adjacent to the feeding-rooms. Local considerations will determine whether the silo should be below the surface, or above, or partly below and partly above. This is not essential. Where the stables are in the basement of a bank barn, the bottom of the silo may be on the same level, or a few feet below, and the top even with the upper floor. This arrangement combines the greatest facilities for filling, weighting and feeding.

2. FORM OF SILO.—With rare exceptions the silos described show a rectangular horizontal section; a few have the "corners cut off," and one is octagonal. (The cylindrical form, of which there is no instance in the accompanying statements, seems to have obvious advantages. If under ground, a cylindrical wall is self-supporting against outside pressure, and may be much lighter than would be safe in any other form. If of wood and above ground, the walls may be stayed with iron bands. In any case, for a given capacity, the cylindrical form requires the least possible amount of wall.)

A given weight of ensilage in a deep silo requires less extraneous pressure, and exposes less surface to the air, than it would in a shallow silo. For these reasons depth is important. If too deep, there is danger of expressing juice from the ensilage at the bottom.

Where the ensilage is cut down in a vertical section for feeding, a narrow silo has the advantage of exposing little surface to the air.

3. Capacity of Silo.—The silos reported vary in capacity from 364 to 19,200 cubic feet. If entirely full of compressed ensilage the smallest would hold 9.1 and the largest 480 tons, estimating 50 pounds to the cubic foot. Practically, the capacity of a silo is less to the extent that the ensilage settles under pressure. This should not exceed one-fourth, though in shallow silos, or those filled rapidly and with little treading, it is likely to be much more. A temporary curb is sometimes added to the silo proper, so that the latter may be full when the settling ceases.

4. Walls of Silo.—For walls underground, stone, brick and concrete are used. The choice in any case may safely depend on the cost. In firm soils that do not become saturated with water, walls are not essential to the preservation of ensilage. Above ground, two thicknesses of inch boards, with sheathing paper between (the latter said, by some, to be unnecessary), seem to be sufficient, if supported against lateral pressure from the ensilage.

5. Covering.—A layer of straw or hay will serve in some measure to exclude air, but is not necessary. Generally boards or planks are placed directly on the ensilage. The cover is sometimes made in sections two feet or more wide; oftener each plank is separate. The cover is generally put on transversely, having in view the uncovering of a part of the silo while the weight remains on the rest. Rough boards, with no attempt at matching, have been used successfully. A little space should be allowed between the walls and cover, that there may be no interference as the settling progresses.

6. Weight.—Any heavy material may be used. The amount required depends on various conditions. It will be noticed that practices and opinions differ widely. The object is always to make the ensilage compact, and thereby leave little room for air, on which depends fermentation and decay. In a deep silo the greater part is sufficiently compressed by a few feet of ensilage at the top, so that there is small percentage of waste, even when no weight is applied above the ensilage. Screws are used by some instead of weights. The objection to them is that they are not self-acting, like gravity.

7. Cost.—The cost of silos, per ton of capacity, varies from four or five dollars, for walls of heavy masonry and superstructures of elaborate finish, and fifty cents or less for the simplest wooden silos. Earth silos, without wall, can be excavated with plow and scraper, when other work is not pressing, at a trifling cost.

8. Crops for Ensilage.—Corn takes the lead of ensilage crops. Rye is grown by many in connection with corn—the same ground producing a crop of each in a season. Oats, sorghum, Hungarian grass, field peas, clover—in fact, almost every crop valuable for soiling has been stored in silos and taken out in good condition. There are indications that some materials have their value enhanced by the fermentation of the silo, while in others there is loss. The relative values for ensilage, of the different soiling crops, can only be determined through careful tests, often repeated, by practical men.

All thoughtful farmers would be glad to get more value from the bulky "fodder" of their corn crops than is found in any of the common methods. There are accounts of plucking the ears when the kernels were well glazed, and putting the fodder into the silo. The value of such ensilage, and the loss, if any, to the grain are not sufficiently ascertained to warrant positive statements.

9. PLANTING AND CULTIVATION.—Thorough preparation before planting is essential. Corn, sorghum and similar crops should be planted in rows. The quantity of seed-corn varies from eight quarts to a bushel and a half for an acre. A smoothing-harrow does the work of cultivating perfectly, and with little expense, while the corn is small.

10. WHEN CROPS ARE AT THEIR BEST FOR ENSILAGE.—The common practice is to put crops into the silo when their full growth has been reached, and before ripening begins. Manifestly one rule will not answer all purposes. The stock to be fed and the object in feeding must be considered in determining when the crop should be cut. On this point must depend much of the value of ensilage.

11. YIELD OF ENSILAGE CROPS.—Corn produces more fodder per acre than any other crop mentioned. The average for corn is not far from twenty tons—which speaks well for land and culture. The largest yield from a single acre was fifty-eight tons; the average of a large area on the same farm was only twelve and a half tons.

12. KIND OF CORN BEST FOR ENSILAGE.—The largest is generally preferred; hence seed grown in a warmer climate is in demand.

13. SWEET CORN FOR ENSILAGE.—It is conceded by many that the fodder of sweet corn is worth more, pound for pound, than that of larger kinds, for soiling. Some hold that the same superiority is retained in the ensilage, while others think that the advantage, after fermentation is on the other side. The sweet varieties generally do not yield large crops.

14. PREPARING FODDER FOR THE SILO.—The mowing-machine is sometimes used for cutting corn in the field—oftener the work is done by hand. Various cutters, having carriers attached for elevated silos, are in use, and are generally driven by horse, steam or water power. Fine cutting—a half-inch, or less—is in favor. It packs closer, and for this reason is likely to keep better than coarse ensilage. Fodder of any kind may be put in whole, and, if as closely compressed as cut fodder, will keep as well, if not better; but it requires much greater pressure.

15. FILLING THE SILO.—During the process of filling, the ensilage should be kept level, and well trodden. A horse may be used very effectively for the latter. Some attach much importance to rapid filling, while others make it more a matter of convenience. With the packing equally thorough, rapid filling is probably best.

16. COST OF FILLING THE SILO.—The cost, from field to silo, is variously reported, from thirty-five cents—and in a single instance ten or twelve cents—for labor alone, to two dollars and upwards per ton; though the higher amounts include the entire cost of the crop, not the harvesting alone. There is a general expectation that experience will bring a considerable reduction in the cost of filling.

It is probable that, with a more general adoption of ensilage, the best machinery will be provided by men who will make a business of filling silos. This could hardly fail to lessen the cost and bring the benefits of the system within the reach of many who otherwise would not begin.

17. TIME FROM FILLING TO OPENING SILO.—The ensilage should remain under pressure at least until cool, and be uncovered after that when wanted.

18. CONDITION OF ENSILAGE WHEN OPENED.—In nearly all cases the loss by decay was very slight, and confined to the top and sides where there was more or less exposure to air.

19. DETERIORATION AFTER OPENING.—Generally the ensilage has kept perfectly for several months, showing no deterioration while any remained in the silo, excepting where exposed for a considerable time. It is better to uncover a whole silo, or compartment of a silo, at once, and thus expose a new surface each day, than to cut down sections.

20. VALUE OF ENSILAGE FOR MILCH COWS.—Ensilage has been fed to milch cows more generally than to any other class of stock, and no unfavorable results are reported. There can be little doubt that its greatest value will always be found in this connection. Several feeders consider it equal in value to one-third of its weight of the best hay, and some rate it higher

21. EFFECTS ON DAIRY PRODUCTS.—There is a marked increase in quantity and improvement in quality of milk and butter after changing from dry feed to ensilage, corresponding with the effects of a similar change to fresh pasture. A few seeming exceptions are noted, which will probably find explanation in defects easily remedied, rather than in such as are inherent.

22. VALUE FOR OTHER STOCK.—Ensilage has been fed to all classes of farm stock, including swine and poultry, with results almost uniformly favorable. Exceptions are noted in the statements of Messrs. Coe Bros. and Hon. C. B. Henderson, where it appears that horses were injuriously affected. It should be borne in mind in this connection that ensilage is simply forage preserved in a silo, and may vary as much in quality as hay. The ensilage that is best for a milch cow may be injurious to a horse, and that on which a horse would thrive might render a poor return in the milk-pail.

23. DAILY RATION OF ENSILAGE.—Cows giving milk are commonly fed fifty to sixty pounds, with some dry fodder and grain.

24. METHOD OF FEEDING.—Experiments have been made in feeding ensilage exclusively, and results have varied with the quality of ensilage and the stock fed. It is certain that ensilage of corn cut while in blossom, or earlier, is not alone sufficient for milch cows. It is best to feed hay once a day, and some grain or other rich food, unless the latter is supplied in the ensilage, as it is when corn has reached or passed the roasting-ear stage before cutting. Ensilage, as it is commonly understood, is a substitute for hay and coarse fodder generally, and does not take the place of grain.

25. THE CONDITION OF STOCK, fed on ensilage, both as to health and gain in weight, has been uniformly favorable.

26. Profitableness of Ensilage.—There is hardly a doubt expressed on this point—certainly not a dissenting opinion.

XIV. Perfect Food and Rations.

Mr. Samuel Adams, of Massachusetts, in relation to the method of feeding, says: "For milch cows I should feed fifty pounds ensilage, ten pounds hay; if shorts were not too high, would feed two quarts per day, and if the dairy product was in demand would give a little meal of some kind."

In the West more grain and less ensilage would naturally be fed. When corn is used as ensilage we should recommend bran, mill-feed and oil-cake in preference to the corn in the grain. A mixture of one-quarter prickly comfrey (strong in nitrogen), one-half corn fodder, and one-quarter rowen hay, with fifty pounds of bran to the ton of green fodder, will make a perfect food for milch cows. Then, fifty pounds of ensilage and five pounds of corn-meal would make a rich ration for each cow.

XV. Some Statements of the Quantity Fed.

Mr. Jason Allen, Massachusetts, says: From the 10th of November to the 10th of January, I fed forty-five pounds of ensilage and five pounds of hay; from the 10th of January to the 10th of March, sixty pounds of ensilage and four pounds of middlings to dry stock, and two pounds more to milch cows; from the 10th of March until May 1, sixty pounds of ensilage, one foddering of oat straw, and six pounds of grain to cows in milk. He regards sixty pounds of ensilage, with six pounds of grain for an average-sized cow per day as a full ration.

Col. LeGrand B. Cannon, Vermont, fed an average of eighty-five pounds per head for three-year-old steers, daily, for five and a half months, with three pounds of grain daily. Cattle fed as stated made a greater gain and were in better health and condition than others fed on twenty pounds of chopped hay and three pounds of grain.

Mr. Wm. B. Eager, of Nebraska, fed forty pounds per day per head. The midday feed was of cut dry corn-fodder, or cut millet-hay, with ground feed. Occasionally for trial we fed meal upon the ensilage, but abandoned it and fed ensilage alone, and meal upon dry food, or cattle would not eat it. The herd of over three hundred milch cows came out in better flesh than when taken from pasture.

Professor S. C. Armstrong, principal of the Hampton Normal School, in Virginia, only experimented with thirty tons, which was fed in one month's trial. The quantity fed was sixty pounds per day, with other food, and about three quarts of wheat bran. There was no perceptible change in the condition of the stock. He says ensilage works well with other food; does well as a substitute for roots.

Prof. J. McBryde, of Knoxville, Tennessee, says: All our milch cows receiving ensilage showed a notable improvement in milk. Butter made from milk of cows fed on ensilage of excellent flavor. Three yearling steers fed exclusively on long forage; one weighing four hundred and twenty-eight pounds received a daily ration of twenty pounds of hay; gained twenty-two pounds in twenty-eight days. Another, weighing four hundred and fifty-seven pounds, received ten pounds hay and twenty pounds ensilage; gained twenty-eight pounds. A third, weighing four hundred and forty-two pounds, received forty pounds ensilage and gained thirty-eight pounds. Two pounds ensilage gave better results than one pound hay. It is plain that animals should be fed on mixed rations of ensilage and matter rich in albuminoids.

Mr. L. W. Weeks, of Wisconsin, says: I feed milch cows from forty to fifty pounds at two feedings, morning and night, mixed with two pounds of corn-meal each feed, and same weight of some nitrogenous food, as oat-meal, barley meal, or mill feed. At noon I feed hay, oats in straw chaffed, or barley straw chaffed; and stock always gaining in condition, coming out in spring in high flesh, and healthy. Since feeding ensilage I have had no trouble with garget or other unhealthy condition of udder. In my experience ensilage has proved a gain in profit of certainly forty per cent over any method of dry feeding that I know, besides enabling one to carry three times the amount of stock possible on the same amount of land with dry feeding.

Mr. John D. Whitman, of Iowa, testifies that ensilage is fully equal to half its weight in hay; the effects very similar to that of green grass, and extra good for calves.

If on full feed, seventy pounds per day is fed; a less amount with some grain and hay is better.

Mr. George A. Pierce, of Canada, says that ensilage is very valuable for dairy stock, entirely taking the place of roots and largely that of hay.

Immediately on feeding the ensilage the butter gained in quality and quantity. No fault was found by the purchaser.

Fed a few steers on ensilage and they did well; calves and young stock did remarkably well. He fed sixty to seventy pounds per day, mixed with meal. The stock began to gain as soon as they were fed ensilage; formerly had hay, roots and some grain. He found ensilage a great advantage over the system of feeding hay and roots in winter.

Dr. John Q. Sutherland, of St. Louis County, Missouri, was one of the first practical farmers in the West to adopt the ensilage system, after Mr. Morris's experiments had demonstrated its value. He kept thirty-two head of milch cattle, which he fed on ensilage both in the winter and in the season of scant pasturage at midsummer, making the ration fifty-five pounds per day of corn ensilage to each cow. This he supplemented with a sufficient quantity of dry food—chopped hay and bran—in winter, to keep the animals in good condition, but in midsummer little besides the ensilage was needed. His cows were always up to a full flow of milk, and the butter excellently flavored.

Thus the testimony of practical farmers from widely separated locations is conclusive as to the value of ensilage.

XVI. The Cost of Ensilage in Massachusetts.

The following statement of Mr. Jason Allen gives a fair exhibit of the cost of ensilage, and may be taken as a fair approximation where much hand labor is employed. In the West and Southwest the cost ought to be considerably less:

Plowing three acres,	$ 9 00
Harrowing,	3 00
Commercial fertilizers,	20 00
Fifteen loads of manue, one-half charged to the corn,	15 00
Planting,	4 50
Seed-corn,	5 00
Replanting,	3 00
Harrowing twice,	3 00
Hoeing by hand,	6 00
Cultivating by hand,	3 00
Use of land,	8 00
Raising crop,	$79 50
Cutting and storing in silo,	47 75
Whole cost,	$127 25
Cost per ton,	$2 70

In the West, corn fodder can be raised ready for cutting at a cost of one and a quarter days for man and team per acre.

XVII. Building a Model Silo.

Mr. Bisbee, of New York State, thus describes his plan of building: The top of the silo is even with a plateau, the bank descending fifty feet to the stable, and very steep. The ensilage is taken out by a hoisting apparatus over the top of the stone wall, and carried in a car on a gently descending grade into a small house, built on the roof of the stable, where the bottom falls out, and the ensilage drops to the floor over the stable. The silos were built double; seven and eight feet wide, respectively, by 24 x 15 feet deep. The material was stone. Outside walls dry, thirty inches thick at the bottom, and twenty at the top. Division wall twenty inches, laid in cement, and all walls plastered with cement. The walls were built by masons, in accordance with their notions of fitness, with the result of an extravagant cost. Above the silo walls is a curb of matched boards, six feet high, for settling-room—of course a roof covers the whole. The silos were covered with hemlock planks, and weighted with stone, fourteen inches thick, and earth banked at ends of plank. The cost was between $700 and $800.

Mr. N. Gridley, New York, has a silo 32 x 12 x 10 feet deep, built of concrete, 14 to 16 inches thick. Posts set in the ground and lined with two-inch plank, put in as wall is built. Any kind of stone used. One part cement to five parts sand, mixed while dry, then wet so that it will pour from the pails in which it is carried. After the posts and planks are taken down the walls are plastered with cement, made with less sand. Bottom covered about two inches with gravel. The cost was $200, including light frame building over it.

XVIII. Practical Conclusions from Careful Experiments.

Col. LeGrand B. Cannon, of Vermont, previously quoted, gives some well digested observations and experiments, which we append. He considers ensilage profitable, and believes it is entirely healthy, taking the place of roots. It is easily digested, as is shown by the uniform temperature of the animals and the condition of the skin and hair.

Observations.—The claims made by many writers in regard to ensilage are extravagant; that it has certain advantages cannot be denied.

First. Not more than fifteen to twenty-five tons can be depended upon per acre.

Second. It is more certain as a crop than hay.

Third. Twice as many animals can be kept on the same acreage.

Fourth. It is largely a substitute for roots.

Fifth. The labor of feeding ensilage is much less than hay.

Sixth. The space required to store ensilage is not one-quarter that required for hay.

Experiment.—I fed ninety three-year-old steers, divided in three lots; cattle and feed weighed monthly.

First lot. Fed twenty pounds hay with three pounds grain daily; run in yard with shelter.

Second lot. Kept in warm stable and stanchions; fed seventeen and a half pounds of hay, one peck mangolds, and three pounds grain.

Third lot. Fed eighty-five pounds ensilage with three pounds grain; this lot gained one-quarter pound a day more than No. 2, and one-half pound more than lot No. 1. The cost five per cent in favor of ensilage.

CHAPTER VI.

TEXTILE CROPS AND FIBERS.

I. COTTON: ITS HISTORY AND CULTIVATION.—II. THE FAMILY TO WHICH COTTON BELONGS.—III. THE SOILS FOR COTTON.—IV. HISTORY OF COTTON CULTIVATION IN THE UNITED STATES.—V. INCREASING IMPORTANCE OF COTTON.—VI. COTTON BY STATES.—VII. THE CLIMATE FOR COTTON. VIII. THE BEST COTTON STATES.—IX. THE CULTIVATION OF COTTON.—X. PREPARATION OF THE SOIL.—XI. TENDING THE GROWING CROPS.—XII. FLAX AND ITS CULTIVATION.—XIII. PROPER SOIL FOR FLAX—SEEDING.—XIV. HARVESTING FLAX.—XV. HEMP AND ITS CULTIVATION.—XVI. RAISING A CROP OF HEMP-SEED.—XVII. RAISING HEMP FOR LINT.—XVIII. THE TIME TO HARVEST HEMP.—XIX. ROTTING AND BREAKING FOR MARKET.—XX. CONCLUSIONS ON FLAX AND HEMP.—XXI. JUTE AND ITS CULTIVATION.—XXII. GROWTH AND HARVESTING OF JUTE—XXIII. PREPARING JUTE FIBER.—XXIV. THE RAMIE PLANT IN THE UNITED STATES.—XXV. SOIL AND PLANTING.—XXVI. RAMIE IS A PERENNIAL PLANT.

I. Cotton: Its History and Cultivation.

THIS wonderful plant, which has revolutionized the clothing manufacture of the world, is a native of the tropical and sub-tropical regions of Asia, Africa and America. Until the invention of the cotton gin, it could not be economically utilized. By hand labor, only a few pounds of the lint could be separated from the seed in a day. By the use of the cotton gin, three thousand pounds a day may be prepared for baling and market.

Though manufactured into cloth more than three thousand years ago, described by Herodotus, who lived four hundred and forty years before Christ, and mentioned by Strabo in the first century after Christ, as being manufactured into printed cloths, flowered, and of brilliant hues, it was not until nearly the middle of the present century that the production and manufacture of cotton had become so cheapened that it could be generally used. Pliny called it *Gossypium*, the scientific name by which it is now known.

II. The Family to Which Cotton Belongs.

THE mallow family contains some of the most important vegetable fibers of the world, and cotton is the most important of them all. The Hibiscus or rose mallow is its nearest relation. Linnæus recognized five species: 1, *Gossypium herbaceum;* 2, *G. arboreum;* 3, *G. hirsutum;* 4, *G. religiosum,* and 5, *G. Barbadense.* De Candole describes thirteen species and mentions six others. Dr. Rogle refers all the varieties to four primary species. The divisions generally recognized, however, are the three first of Linnæus, which are known by the English names, herbaceous, shrub and tree cotton, the first named being the most important. Of these there are many hybrids and varieties.

III. The Soils for Cotton.

Long-Staple.—The delicate, long-stapled, sea-island cotton, is grown in a very narrow belt, lying along the coast of South Carolina, Georgia and part of Florida. The soil is dark gray, sandy with a powdering of peat, shells, wood, twigs and leaves.

Uplands.—The upland or green seed cotton is raised from a variety of soils, but the rich alluvial soils of the bottom lands of the Mississippi Valley are the best. The region of Georgia, Alabama and Mississippi, underlaid by rocks of the cretaceous (chalk) system—soft argillaceous limestone and the sandy soils underlaid by metamorphic rocks, sandstones and chert limestones, also make first-rate cotton soils. But whatever the surface soil, it must have good and deep drainage, and not a large amount of vegetable matter to produce the best staple.

Avoid Humus Soils.—Soils that are light (fluffy), that dry out easily on the surface, that are composed mainly of decayed vegetation, and deficient in drainage, should never be selected for cotton. Rich, deep humus soils, however great a crop of stalks they may raise, will never give good results in fiber. So again, soils that are cold or wet in the subsoil, or those subject to flooding, will not grow cotton. Scab, rot, insects and other contingencies will destroy the crop.

Mineral Soils the Best.—Soils that are silicious and aluminous, rich in potash and other mineral matter, are always to be sought for cotton. A dark-colored, warm, finely comminuted upland, or second bottom, is always to be preferred, if not too rich in vegetable matter.

IV. History of Cotton Cultivation in the United States.

The aborigines of Mexico and the South American Pacific slope cultivated cotton and wove it into cloth. The savage tribes of the United States knew nothing of its use. The plant is supposed to have been introduced, about 1664, from Barbadoes. A South Carolina planter clothed his negroes in 1778 with cotton prepared entirely by hand, though the spinning-jenny of Arkwright was invented in 1769, and that of Hargreaves in the next year. Little cotton was raised in the United States up to the year 1793, when the invention of the cotton gin by Whitney started a revolution in cotton industries. By this originally crude instrument, instead of one pound of lint per hand, 350 pounds could be cleaned in a day.

The first cotton shipped from North America was one bag of the staple sent abroad in 1740, and no more was shipped for fifty years. During the Revolutionary war cotton was cultivated in small patches and woven at home. In 1793 cotton was first planted as a marketable crop.

In 1795 1,000,000 pounds was exported. In 1804 the first long-stapled cotton was raised in South Carolina. In 1826 the pioneer cultivator of this improved lint sold only sixty bags, but he got $1.25 a pound for it. In 1785 the seed of the short-stapled, or upland, cotton was introduced into Georgia from the West Indies. From that time on, improvements of the steam engine and new inventions in spinning and weaving machinery have caused a steady increase in the annual production for export and home use.

V. Increasing Importance of Cotton.

IN 1792 the export of cotton was 138,328 pounds; in 1840, 744,000,000 pounds; in 1860, 1,765,115,735 pounds, or 4,412,789 bales of 400 pounds each, but the quantity produced in 1860 was 2,079,230,800 pounds, or 5,198,077 bales. This production had fallen off somewhat in 1870, when the quantity produced was reported as 3,011,996 bales, or 1,204,798,400 pounds. During the war the production dropped to almost nothing. For the first eleven years after the war the average crop was about 3,300,000 bales, which is almost exactly the average for the eleven years immediately preceding the war. The largest crop made in America previous to 1860 was 4,669,770 bales in 1859, which fell far short of the crop of 1880–81. Of late years the increasing demand for cotton and the better prices obtained, have caused a rapid increase in the quantity raised, a large proportion of the crop being raised on small farms and by white labor.

The crop of 1880–81 was 6,589,329 bales, the largest ever produced up to that time, though it may soon be looked upon as no more than a fair average crop. Of this, 4,596,279 bales were exported, and the home consumption was 1,891,804 bales. The crop of 1882–83 was 5,435,845 bales. A comparison of seventeen-year periods, one ending with 1860–61, and the other with 1881–82, shows:

	CROP IN BALES.	EXPORTATION, BALES.	HOME CONSUMPTION, BALES.
First period of seventeen years,	51,330,790	39,913,005	11,422,799
Second period of seventeen years,	63,377,375	46,892,528	21,494,210

VI. Cotton by States.

STATES.	ACRES.	YIELD PER ACRE.	POUNDS OF LINT.
Virginia,	61,985	1̄8	11,033,330
North Carolina,	1,050,543	180	189,097,740
South Carolina,	1,587,244	183	290,465,652
Georgia,	2,844,305	152	432,334,360
Florida,	260,402	117	30,467,034
Alabama,	2,534,388	150	380,158,200
Mississippi,	2,233,844	190	424,430,360
Louisiana,	887,524	235	208,568,140
Texas,	2,810,113	240	674,427,120
Arkansas,	1,110,790	233	258,814,070
Tennessee,	815,760	170	138,679,200
Missouri, Indian Territory, etc.,	79,793	180	14,362,740
	16,276,691	187	3,052,837,946

THE production of the twelve cotton States, with the acreage of each State, average yield of lint per acre, and the pounds of lint produced, with totals for the year 1882, are shown by preceding table.

Suppose careful cultivation should raise the average to that of Texas, it would increase the yearly total to nearly one billion of pounds.

VII. The Climate for Cotton.

Moist and Sunny.—Cotton has been called a child of the sun. It requires a strong heat and plenty of sunshine, but the heat must be moist. There must be abundant moisture with continuous sunny weather. Between latitude 30° and 33° in the United States these conditions exist, where the soil contains plenty of sand, or is so well drained as quickly to percolate superabundant moisture.

Its Northern Boundary.—North Carolina is the northern boundary of its profitable cultivation on the Atlantic coast, while the bottom lands of Tennessee are practically its northern limit on the east banks of the Mississippi. High prices, at times, have tempted its cultivation in Kentucky and Missouri, somewhat largely in Kansas, on the eastern shore of Maryland, in southern Delaware, southern Indiana and Illinois, and even in southern Iowa. It may be ripened up to 40 degrees north latitude, but north of Tennessee it will not pay.

West of the Mississippi.—Arkansas contains much fine cotton land; so does the Indian Territory, and in Texas, as is well known, the soil and climate, in wide districts, is admirable.

The great Cotton States, in the relative order of production are, Georgia, Texas, Alabama and Mississippi. The three other States, in each of which, in 1882, over 1,000,000 acres were cultivated, were South Carolina, Arkansas and North Carolina. Louisiana and Tennessee produced less than 1,000,000 bales.

VIII. The Best Cotton States.

So far as the climate is concerned, the best regions for cotton cultivation in the United States are in the lower parts of Georgia, Alabama, Mississippi, Louisiana and Texas. In these regions, there is comparatively little frost, and the winter is always mild, with considerable heat in the summer; but this is tempered, to a great extent, by the pleasant and salutary effects of the sea breeze, which sets in from the Gulf or the Atlantic for a great part of the day. There are heavy dews at night, and frequent showers occur, in the spring as well as in the summer. In the interior and more northern portions of these States (which are in some parts elevated from five hundred to one thousand feet above the level of the sea), frost is expected in October, and often continues until April; sometimes it occurs even in May, so as to injure, but does not then usually destroy, the plant. The heat of summer, though frequently high, still is tempered by the influence of the ocean or the Gulf of Mexico, and of the numerous great rivers, as well as by the dews and occasional showers.

The cultivation of cotton is generally commenced about the beginning of April, when the land is still saturated with the winter rains, and difficulty is sometimes experienced in getting the land sufficiently dry; otherwise, a good shower is essential when cotton is first sown, and it is desirable also to have occasional showers during the planting, plowing and hoeing seasons. The bolls begin to open about the middle of July, and continue to do so until the appearance of frost, from the middle to the end of October.

The whole region west of the Alleghany mountains is adapted to cotton much further north than is the region east of the Alleghanies. The direction of the valleys favors the drawing-in of the warm air of the Gulf of Mexico, and the summer climate is thus modified even up to Wisconsin and Minnesota, so that Indian corn, tobacco, melons and other fast-growing semi-tropical products are freely produced.

IX. The Cultivation of Cotton.

Of late years the cultivation of cotton has been much simplified by improved implements for ridging, sowing the seed and cleaning the crop. The principal points to be remembered have been stated by an experienced cultivator in the middle cotton region as follows:

Lands should be deeply and thoroughly plowed long enough before planting to allow the spring rains to settle the soil. If not plowed previously, particular pains should be taken to secure uniform and deep pulverization. If rough and full of clods, the harrow should follow the plow.

The usual practice among successful cultivators is to form beds with the turning plow, as foundations for the ridges, turning furrows both ways toward the centers.

Ridge planting is almost universally practiced; yet the custom of planting in hills, as with corn, has obtained, and may be preferable in otherwise suitable lands that are inclined to be too moist and cold, giving a better exposure of the fibrous side roots to the action of the sun. An increased elevation given to the ridge has essentially the same effect.

If land has been fallow, or in sod, it should first be thoroughly broken up with a heavy plow, and then bedded with a smaller one, harrowing after the first plowing. This not only pulverizes thoroughly, but leaves grass and weeds far beneath the surface. It will not do to slight the work at this stage; the success of the crop depends upon its character. If done well, half the battle of the season is over.

When the ridge is ready to open for seeding, great care should be taken to get a perfectly straight furrow, to facilitate "scraping out" superfluous cotton and grass. A very light and narrow plow should be used, making a furrow not exceeding an inch in depth. Unless the soil is very light and dry, the seed should not be covered half an inch. A wooden instrument for making the seed-bed is frequently used to advantage instead of a plow.

The distance between ridges and between the plants must depend upon the probable size of the plants, which varies from eighteen inches to half as many feet in height. The largest yield is secured by so graduating the distance that the plants will cover the ground and slightly interlock their branches. In good soils the ridges should be four feet apart, and the plants fifteen inches; in lighter, three and a half, and twelve inches; in very rich lands the ridges might be four and a half feet, and the plants fifteen to eighteen inches. This direction is for good cotton soils. If a stinted growth only is expected, plants may be set nearer; some of our amateur planters think six inches will do, but counsels so extreme should not be heeded.

Improved Implements.—This was the general plan followed up to within twenty years ago, and in small fields where much of the work must be done by hand it would be good practice now; at all events, the land must be cleared of trash. The large planter will of course avail himself of all improved implements possible to facilitate and cheapen labor. This is written principally for that class who cultivate comparatively small areas. It would be better for all these to hire the improved implements if they cannot buy.

One thing must be remembered: cotton will not make a crop unless the land is in good heart; strong land will grow heavy grass and weeds; unless these are kept under, the crop is a failure, for the roots of cotton must have the soil and the tops the sun. The distance of the rows from each other, and the intervals of the plants in the row, should vary with the fertility of the soil. The rule for uplands is four feet between the rows and twenty inches between the plants. In lands that are strong enough to yield a bale to the acre the spaces should be five feet one way by three the other; while in the valleys of the Southwest, such as the Red river and Brazos bottoms, the luxuriance of the growth is such as to require an interval of eight feet between the rows and of five or six feet between the plants, and even then the branches interlock so that it is difficult to walk between the rows.

X. Preparation of the Soil.

If the soil be poor and warm, manures must be applied, but large quantities of barn-yard manure are not advisable, especially in soils that already have much humus. Phosphatic manures, lime and potash would here be indicated. The question of manure must be settled, each man for himself, according to the cost. The preparation of the soil recommended by a writer having wide experience in Louisiana is as follows:

Burying the Stalks.—Where the field is foul with weeds or the stalks of last year's crop, it is best to bury them under the middle of the cotton ridge, and it is recommended to run a double furrow at intervals of four feet in hill lands and at intervals of five or six feet in bottom lands. This can be done early in January, whenever the ground is not too wet. Behind each plow are two hoe hands, to break up and pull down into these ditches all the dead growth of the surface, one from the right side and the other from the left. Let it be well pulled down into the trench and covered with some earth to keep it in place. Then, about the middle of February, according to situation, latitude and drainage of the land, let the double plows set in to break up the entire surface. If the rows for burying the trash are uniformly laid off, they can be used as the foundation of the bed or ridge. Throw a deep furrow from each side into the trench, filling it and covering all the trash, and continue to plow out the intervening spaces or middles, as they are called, until the whole surface is turned under. This plowing should be deep and thorough.

Plowing and Fertilizing.—If compost or barn-yard manure is applied to the field it should be done just before plowing, and scattered broadcast. If concentrated fertilizers, such as ashes, guano, gypsum or superphosphates are to be used, they are best applied in the drill. In respect to fertilizers for the cotton field, it may be here remarked that cotton is a moderate consumer of the salts that nourish plants, especially if the seed is returned to the soil, as it should be, in the form of manure. Potash and phosphate of lime should abound in any manure applied to cotton. The effect of an abundance of potash in the soil will be to secure thrifty plants. The effect of phosphoric acid is to produce plenty of seed and a strong fine staple, and this is what brings the money.

When to Plant.—One rule, and a good one, is to plant cotton as soon as the oak leaves are the size of a squirrel's ear. Another good rule is to plant when the hickory buds have opened to show the leaves. This indicates that the days and nights are so warm that there is little danger of frost.

The amount of seed will depend upon the danger of destruction by cut-worms, etc. It is always best to plant seed liberally, but not, as some do, to manure the soil with them. The plants inevitably spindle before being thinned, and once spindled, you have lost your profit.

XI. Tending the Growing Crop.

The First Cultivation.—The first cultivation of the crop should begin about fifteen days after the planting, or as soon as the weeds begin to start. If the instructions given above, with regard to evenness of rows and of the intervals between the plants are followed, the first cultivation may be easy and rapid. A light plow should be run close to the line of plants, cutting away the weeds and grass and stirring the earth to a moderate depth. The hoes follow, smoothing the inequalities produced by the plow, and clearing the intervals between the clumps of young plants. In the ordinary mode of planting, when the seed is scattered thickly through the drills, this first cultivation is called "chopping out."

The Second Cultivation.—Two weeks after, the hands should go over the crop again, thinning out the young plants to a stand. This is sometimes done at the first cultivation, especially in strong soils. This second cultivation should be the most thorough of any, the thriftiest plants only being spared, and the rest being pulled up with care so as not to displace the roots of those allowed to remain. A little fresh earth is thrown around the roots of the young plants, and the entire ridge, as well as the intervals between, should be made perfectly clean. On a good soil, with favorable seasons, the growth will now be rapid, and the subsequent cultivation can be effected mainly by horse implements; but very deep cultivation, except in bedding up the land in the spring, is never beneficial. It breaks the lateral roots of the plants, and this retards the development of the pod and curtails the picking season; hence, the best implement for cultivating cotton is one which, instead of turning the soil, scrapes and pulverizes the surface of the earth.

The Scooter Plow.—The implement in common use is very well adapted to this purpose, and consists of a common scooter plow or bull tongue with wings attached three or four inches above the tip, and set in such a way as to pass just beneath the surface and throw a little ridge of fresh earth close to the stems of the plants. They often, when skilfully used, clean the surface so thoroughly that the hoes can pass over the crop very rapidly. Sometimes early in the month of July, on a good soil, the plants will be so far advanced that the branches will touch and perhaps lock across the middles. Many planters think that little is gained by running the plows after the crop attains this growth, but the more the ground is stirred, if lightly, the more readily will the heat of the sun penetrate the soil and fall upon the roots of the young plant, and this is what is required to hasten their development; but the cultivation must vary with the season and condition of the soil. As a rule the cultivation further north should be discontinued sooner than south, since, if the plants get a check they will blossom earlier and the crop thus be saved from frost.

Destroying Insects.—Insect depredators are a great drawback to the successful cultivation of cotton. They must be carefully watched for and destroyed. For those which feed on the foliage, spraying by hose, with a mixture of Paris green, arsenic or London purple. The last is probably best, as it more easily mixes with water. The illustration (p. 202) shows how to apply by means of a fountain pump, the mixture to be kept in motion by a dasher passing through a hole in the barrel. On small fields the poison may be applied by hand, mixed with damaged flour, by dusting from a suitable can. One pound of Paris green, or rather more of London purple, to a barrel (forty gallons) of water is sufficient; or, one pound of London purple to thirty pounds of flour. The poison must be pure and of known strength.

XII. Flax and its Cultivation.

A CONSIDERATION of the cultivation of flax naturally follows that of cotton, not because it is next in importance as a textile crop, for in the United States hemp holds that place; but because in seed and fiber it is germane thereto. The fiber has been used only since the introduction of machinery for making coarse tow and twine, the seed being the principal object. So much hand labor is required that the fine lint will be little used until the population of the country becomes more dense.

The Seed.—In the West, however, it is a favorite crop on new lands remote from market, since the price of the seed renders transportation over great distances practicable. In 1850 the production of flaxseed was 562,000 bushels; in 1860, 611,000; in 1870, 2,500,000, and the straw was estimated as equal to the production of 75,000,000 pounds of fiber.

SPRAYING WITH POISONED WATER TO DESTROY THE COTTON WORM.

The Fiber.—This in coarse bagging would cover, as baling, the whole cotton crop of the country. The textile strength of flax is rated as double that of East India jute, and yet, not one-fifth of the flax crop is utilized, although there are many mills in the West for working the straw, the reason being that other lines of manufacture are more profitable. It is, however, an important product for the seed alone, the crop of 1881 being computed at 8,000,000 bushels, from 1,127,300 acres, an average yield of a fraction over seven bushels of seed per acre. Over 800,000 acres of this being in the States of Iowa, Indiana, Kansas and Illinois.

XIII. Proper Soil for Flax Seeding.

It has been said that good barley land is good flax land. This means that flax likes a deep, open, warm, moist loam. In the West, new prairie and old turf lands are much used. Recent timber clearings are desirable if suitably drained, or any good corn land, or rich silicious soil in good tilth. Flax will grow well in any moist, deep, strong loam, upon upland. A light, sandy soil should be avoided, as well as very low lands or river bottoms, upon which flax is very liable to mildew. Flax should be put in after some hoed crop, to be free from weeds. A weedy soil, in any location, should not be thought of in connection with flax, even when raised for seed alone. If fiber is also an object, the time and labor will be wasted on such land.

Preparing the Soil.—On old land it is better that a pretty deep plowing be given in the autumn, and the area lightly replowed just before sowing. For the fiber, deep plowing is essential. To strengthen the fiber, three or four bushels each of superphosphate, of lime, plaster, ashes and salt should be applied per acre. The soil must be brought into the best possible tilth, the seed sown evenly, and covered not more than half an inch deep. The usual quantity to be sown per acre, when seed alone is the object, is from one-half bushel to three pecks. One bushel per acre has given us the best results, since the ground is quickly covered and the crop ripens more evenly.

Selecting the Seed.—The quality of the seed must be looked to. It should be clean, bright and heavy. The best time to sow flax is just prior to that of corn planting, or when the trees are beginning to green. A change of seed is necessary, since, in the West, the oily qualities rapidly deteriorate. East Indian seed is said to be the richest in oil, and next, that from Riga (Russian), and Rotterdam (Holland) is recommended.

Quantity of Seed to Sow.—If lint and seed both are the object, one and one-half to two bushels should be used. In Europe, where fine lint is raised, three and even four bushels are sometimes sown. In this country, however, where the principal object is seed, one and one-half bushels to the acre is the maximum. The yield will always vary with the season and the quality of the land. The maximum may be stated at twenty-five bushels, but half this quantity per acre is an average even on rich soil.

XIV. Harvesting Flax.

The time to harvest is when the lower portions of the stalk turn yellow, or when the seed-bolls show signs of shedding. Cut with a reaper that will rake off in gavels, since it is not necessary to bind it, the idea being to cure it as quickly as possible. Set the gavels up, one leaning against another, in regular rows sufficiently far apart for a wagon to pass between.

Thresh with a machine having beaters instead of teeth, since the straw is apt to tangle in the latter.

If the seed is not sold immediately, it should be spread and turned occasionally until fully dry, or it will heat.

If the straw is to be sold for tow, it should be spread, in October, the product of about two to three acres upon one of grass land (unless very heavy), and then left until ready for the mill, say a month or longer. The water-rotting of flax for fine fiber requires much labor, pools of soft (river or pond) water, and much manipulation. There is, however, so little likelihood that the preparation of water-rotted flax will soon become an industry of importance in the United States, that the directions are not worth the space necessary for description.

XV. Hemp and its Cultivation.

Like that of flax, the hemp industry may be said to be declining. When great navies of sailing vessels traversed the ocean, vast quantities of linen were used for sails, and of hemp for cordage. Steam has decreased the number of sails, and most of them now in use are more cheaply made from cotton, and the fibers of other plants, and iron cordage has largely taken the place of hemp. In some portions of the country, however, it is still an industry of some importance, but water-rotting of hemp is not practiced in the United States, on account of the labor required in the process. When intended for lint it is simply dew-rotted.

The Soil for Hemp.—The soil for hemp must be rich, deep, warm, loamy, and well drained—such land as will produce with good cultivation, fifty or more bushels of corn per acre. The seed will ripen perfectly up to forty degrees of latitude, and usually up to forty-three degrees in the Mississippi valley. The cultivation of the crop for seed is practiced in some sections, and with profit.

XVI. Raising a Crop of Hemp Seed.

Land intended for seed must be in good tilth and well prepared by careful plowing. It should be laid off in straight rows, four feet apart each way, and planted in hills seven or eight seeds to the hill; the same rules observed for cultivating corn will apply in the after-culture of hemp seed; when the plants reach the height of six or eight inches, they should be thinned to from three to four plants.

Male and Female Plants.—Hemp plants are divided into male and female, the former producing the pollen or impregnating powder, the latter bearing the seed.

A very little observation will enable the grower to distinguish between them. As soon as the distinction can be made, the male should be drawn up by the root, when cheap labor can be had, leaving, however, here and there, one that the female plant may be properly impregnated; the female is to be retained until its seeds are perfected, when it is to be harvested by cutting at the ground and removal to cover; when cured, the seed may be threshed with a flail, cleaned, winnowed, and put up in barrels or sacks, perfectly dry, and out of the way of rats and mice until sold.

XVII. Raising Hemp for Lint.

If lint is the object, it is necessary that the seed be raised as directed in the preceding section, for the lint crop is cut before the seed is formed. The soil must be prepared by deep and careful plowing, as directed for flax, and as carefully brought into a state of perfect tilth.

The ground must be free from weeds, or once carefully weeded by hand after the crop is up. Sow from fifty to seventy pounds of seed per acre, preferably from a broadcast seeder, or from a centrifugal seeder, to insure even distribution.

The seed ahould not be covered more than half an inch, and it is better, after sowing, to roll the land with a light roller. The sowing should take place at or immediately before corn-planting time. The plant, after it is up, is not affected by light frosts, but the seed itself is liable to rot in cold ground. In good weather the plants will show in a few days.

XVIII. The Time to Harvest Hemp.

When lint is the object, as is always the case in thick sowings, the time for cutting is indicated in two ways: 1, the crop changes from a deep green to a paler hue, and, 2, the leaves die and drop, beginning at the bottom. Hemp is of two sexes as before stated. The male plants bear the pollen and the female plants the seed, as in the case of spinach. That is diœcious, having staminate and pistillate flowers on distinct plants. The male plants ripen two weeks before the female.

In the United States, the pulling of the male plants before the female ripen costs more than will pay the benefit. A good indication of the time for cutting is given by the pollen of male plants rising in clouds from the field.

Cutting.—The cutting is usually done by a heavy hook made for the purpose, but large level fields may be cut by a reaping machine, made especially for this use. J. L. Bradford, of Kentucky, a noted hemp raiser, thus describes the process of harvesting, rotting and breaking: If the crop is to be cut with the hook, the operator is required to cut at once through a width corresponding to the length of the hemp, and as close to the ground as possible, spreading his hemp in his rear in an even and smooth swath, where it remains exposed to the sun's rays until the stalk is properly cured, and the leaves sufficiently dry to detach easily.

The hemp can be shocked with more compactness without the leaves than with them, and any operation having an influence upon the future security of the staple from dampness or atmospheric influence is certainly important; the perfect detachment of all the leaves should, then, in nowise be omitted. No time should be lost, after the stalk is cured, in getting the crop up and into neat shocks; every additional day's exposure to sun, wind, rain or dew, is deteriorating its quality and subtracting from its quantity. The brighter the stock can be secured, the better.

XIX. Rotting and Breaking for Market.

THE same rule will apply to hemp that obtain in securing good hay. The operator, in taking up the hemp, uses a crook, often a rude stick cut from the branches of the nearest tree, about the length and weight of a heavy hickory walking-cane, having at the end of the stick a small branch making a hook. With this primitive but very effective tool he can rapidly draw the stalks into bunches of the proper size for sheaves. In operating, he throws his rude hook forward to its full length and suddenly draws it toward him, each motion making a bunch. This he raises quickly from the ground and with his hook, by a few well-directed strokes, divests the plant of its leaves. He then binds his sheaf with its own stalks, and passes on to repeat the operation.

Shocking.—Other laborers follow and place the hemp into neat, close shocks of convenient size, securing the top by a neat band made of the hemp stalks themselves, after the manner of shocking corn. Here it is suffered to remain until the whole crop is thus secured as soon as possible, selecting clear, dry weather for the operation. The whole crop is to be secured by ricking or stacking. The same rules are to be observed in stacking as with grain, the object being to keep the crop secure and dry until the proper time for rotting arrives. In the latitude of Kentucky about the middle of October is the proper time. The crop must be retained in the rick or stack until the summer heats and rain have passed, and frost appears instead of dew.

Rotting.—The whole crop is then removed from the rick, and hauled back to the same ground on which it grew, there to be spread in thin swaths for rotting, where it remains without turning until properly rotted. This is indicated by the fiber freely parting from the stalk, and the dissolution by the action of the elements of the peculiar substance that causes it to adhere thereto. This stage is only to be learned to perfection by practical experience; yet the novice must have some information to enable him to begin, and it is easily acquired by a little observation.

Bunching.—When the operator finds his hemp sufficiently rotted, the wooden hook is again brought into requisition for once more drawing the swaths into convenient bunches. The hemp will have lost much of its weight, and can be bunched and shocked with less labor than at first; besides, at this last shocking, the binding is to be omitted entirely, the hemp is to be carefully and neatly handled, all tangling to be avoided, and placed again in shocks, and firmly bound at the top.

Breaking and Dressing.—Then comes the last and crowning operation—breaking and dressing the fiber or lint for the market. The peculiar break to be used, like the knife or hook for cutting, needs no description, being manufactured in hemp regions, at a cost of about five dollars each, and from long experience has been found perfectly adapted to the uses required. The beginner would save time and money by ordering a sample break, from which any carpenter can manufacture as desired.

The crop is broken in Kentucky and Missouri, directly from the shock in the open field by the removal of the break from shock to shock as fast as broken. In the North, owing to the severity of the climate, it would probably be necessary to remove the rotted hemp to the barn, where the labor of breaking could be more certainly performed. The coldest and clearest weather is the best for this operation; in fact excess of dampness in the atmosphere suspends this labor altogether. The breaking process is laborious, yet more depends on the skill than on the strength of the laborer.

XX. Conclusions on Flax and Hemp.

THE rich lands of the corn zone of the West are far better adapted to the cultivation of hemp for the fiber, than more southern latitudes. If water-rotting were practiced, the fiber would undoubtedly be the best in the world. Ponds and streams are plentiful, and the high price of ordinary unskilled labor is the only thing against this process. Other crops, however, are as yet more profitable in this region, and will continue so to be until the population becomes dense enough to cheapen labor. Therefore we do not advise the raising of either flax or hemp for fine fiber. Both flax and hemp raising for the seed are profitable, to a limited extent, up to, and even above, the 40th degree of latitude.

XXI. Jute and Its Cultivation.

JUTE is a plant known in the South for years, under experimental cultivation. It belongs to the mallow family. The fiber is in many respects superior to that of hemp. When American skill and ingenuity shall have found means to prepare the fiber cheaply for the loom, the cultivation of this plant will be a source of great wealth in the South. The time may soon come when the fibers can be cheaply separated.

Jute (*Coochrrus*) is an annual, the two species cultivated in the United States being *C. capsularis* and *C. olitorius*. The first named being the better. The plants grow from five to seven feet high, and the quantity of seed sown, broadcast, is, on rich land, prepared as for flax or hemp, from fifteen to twenty-two pounds per acre. The plants are cut about three inches above ground, one month before the seed ripen.

The best fiber is raised on deeply drained, moist, rich land. The seed should be sown as directed for hemp: that is, evenly distributed. The produce is all the way from 2,000 to 4,000 pounds per acre. The Gulf States contain the region best adapted to the cultivation of jute, which is thus described by a planter of Louisiana, one of the pioneers in its cultivation:

XXII. Growth and Harvesting of Jute.

THE ground being well tilled and the seed properly sown, on wet days if possible, the jute is left alone like wheat. No other care than that of drainage is necessary until maturity. The cost of that first operation cannot exceed four dollars per acre, if the material is adequate and the management judicious. That expense, of course, does not include the value of the seed, because, after the first outlay, planters will provide themselves with it from the low lands, or from the weak spots of the plantation. In the bottoms, when we plant in drills for seed, a subsequent plowing or two will be necessary in the intervals to neutralize the encroachments of grass. In Louisiana that labor is a necessity principally for the purpose of combating the tall weed called wild indigo, which occupies the low grounds. That weed, also fibrous, is the only plant that keeps pace in growth with jute; all other plants are distanced and smothered by the shade of the jute. In the field, planted broadcast, no parasite can resist the vigorous and absorbing influence of jute. Even the hardy and noxious plant, commonly called coco in Louisiana, is destroyed after two seasons of broadcast cultivation.

Harvesting.—The best period for cutting crops of jute is during the stage that precedes the blossoming, or, at least, the seeding. The fiber is then fine, white and strong. The monthly sowing graduates the maturing of the successive crops, which facilitates labor. April planting can be harvested in July, May planting in August, and June planting in September. Any late growth can be harvested in October, and even after, if no frost interferes. The plant stands green until frost dries it up; but even then it can furnish a good material for paper. The cutting operation is done with a mower or a reaper. The albumen of the plant makes it easier to cut than dry wheat. The reaper gathering the stems, bundles are made and carried as fast as possible to the mill, where the textile is rapidly separated.

XXIII. Preparing Jute Fiber.

As fast as the fiber is turned out by the decorticating machine it is plunged into large vats filled with pure water, and left exposed to the heat of the atmosphere. Kept under at least one foot of water, the filament is disintegrated by the dissolution of the gums or resins which united it in a sort of ribbon. That process of fermentation or rotting takes about a week in summer. With care and attention to the proper degree of rotting the fiber comes out almost white, lustrous, and fine like flax. The disintegration is known to be complete when the fiber assumes a pasty character.

Then the rotted hanks are withdrawn, carefully washed in clear water, and hung up to dry in the shade. Care must be taken that the filament be well covered with water during the fermenting period, because atmospheric agencies tend to communicate to it a brownish color. After a few days of good weather it is ready to be shaken and twisted for baling like other textiles. That new process of rotting the separated filament, instead of whole stalks, combines different profitable results—the advantages of economy in labor, in value, and also in the integrity of product.

Experiments in South Carolina, Florida and the Gulf States have shown that, wherever in the Southern States there is a hot, damp climate, and a moist soil of sandy clay or alluvial mold, jute can be profitably raised. April plantings were cut in July, and the June plantings in September. Some of the stalks reached the height of fifteen feet, and the yield was in several cases at the rate of 3,500 pounds to the acre, yet this is probably an exception.

XXIV. The Ramie Plant in the United States.

RAMIE, sometimes, but incorrectly, called china grass, is known botanically as *Boehmeria tenacissima*, and was introduced first into the United States in 1855. Since that time it has attracted more or less attention, but has never been cultivated with profit, on account of the cost of separating the fiber. Several machines have been invented for this, some mechanically, some with the aid of chemicals, but so far perfect success has not been attained.

The Ramie Fiber.—The fiber is of great luster and fineness, between flax and silk. It is used with silk and worsted in the preparation of fine fabrics. In China and India the cost of separating the fiber by hand is about $150 per ton; and the price in England, for the best prepared ramie, $350. A successful machine for reducing this fiber would bring the inventor a great fortune. The Department of Agriculture (Report for 1879) gives the following, which will be of value to intending cultivators:

XXV. Soil and Planting.

1. Whether for nursery purposes or for cultivation, the land must be sufficiently elevated to receive the benefit of natural drainage, because the roots will not live long in a watery bottom.

2. The soil must be deep, rich, light and moist as the sandy alluvia of Louisiana. Manure supplies the defects in some lands in these respects.

3. The fields must be thoroughly cleared of weeds, plowed twice to the depth of eight to ten inches, if possible, harrowed as much as a thorough pulverizing requires, and carefully drained by discriminate lines of ditches. Water must not be allowed to stand in the rows of the plant.

The land being thus prepared, planting becomes easy and promising. December, January and February are the best months in which to plant. Roots, ratoons and rooted layers are the only available seed. They are generally four or five inches long, carefully cut, not torn, from the mother plant. The dusty seed produced by the ramie stalks in the fall can be sown, but it is so delicate and requires so much care during the period of germination and growth that it seldom succeeds in open land.

Furrows five or six inches deep and five feet apart are opened with the plow. The roots are laid lengthwise in the middle, close in succession if a thick stand of crop is desired, but placed at intervals if nursery propagation is the object in view.

The first mode will absorb 3,000 roots per acre, but will save the labor of often filling the stand by propagation. The second mode will spare three-fourths of that amount of roots, but will impose the obligation of multiplying by layers. Being placed in the furrow closely or at intervals, the roots are carefully covered with the hoe. Pulverized earth and manure spread over the roots insure an early and luxuriant growth in the spring. When the shoots have attained a foot in height they are hilled up like potatoes, corn and all other plants that require good footing and protection from the fermenting effect of stagnant water. The intervals between the rows being deepened by the hilling have also a draining influence, which can be rendered still more effective by ditches dug across from distance to distance, say fifteen feet.

XXVI. Ramie is a Perennial Plant.

Good crops are obtained by thickening the stands. The stems are then abundant, fine, straight, and rich in fiber. Close planting is then necessary, inasmuch as is prevents the objectionable branching of the stalks. The period at which the plant is ripe for cutting is indicated by a brownish tinge at the foot of the stems. The first cutting may be unprofitable on account of the irregularity and sparseness of the growth; but if the stand is razeed and manured over the stubbles the ensuing cuttings will be productive. For that purpose the field must be kept clear of grass until the growth is sufficiently dense to expel the parasites by its shade. That necessary density is obtained by means of the important laying process. This consists in bending down, right and left along the growing stand, the highest switches, and in covering them with earth up to the under tip, which must not be smothered. One of the causes of the perennity and of the vigor of the plant is the nourishment it draws from the agencies of the atmosphere. Consequently the leaves of the layers should never be buried under ground. When properly performed, laying is very profitable; it creates an abundance of new roots, and fills up rapidly the voids of the stand.

RAMIE—ROOTS AND STEM, THREE MONTHS' GROWTH.

After two years the plants may be so thick as to spread out in the rows. Then the plow or the stubble-cutter has to chop in a line, on one side, the projecting ratoons. If well executed, this operation leads to notable advantages. It extracts roots or fractional plants suitable for the extension of the cultivation elsewhere; it maintains, as a pruning, a vigorous life, and develops a luxuriant growth in the stand. If always applied on the same side of the row, this sort of stubble-cutting has the remarkable advantage of removing gradually the growth toward the unoccupied land in the intervals, and of pushing it into a new position without disturbance.

That slow rotation preserves the soil from rapid exhaustion, and the ramie from decay, through the accumulation of roots under ground. Of course this lateral plowing will not prevent the opposite row from receiving the benefit of hoeing after each crop. Experiments made in Louisiana have demonstrated the efficiency of that method, to which are due the preservation and propagation of the plant in that State, while it has been destroyed in other sections for want of similar care.

CHAPTER VII.

SILK AND SILK-WORMS.

I. SILK CULTURE IN AMERICA.—II. SILK-PRODUCING INSECTS.—III. FROM THE EGG TO THE MOULT.—IV. VARIETIES OF THE SILK-WORM.—V. KEEPING AND HATCHING THE EGGS.—VI PREPARING TO FEED THE WORMS.—VII. FEEDING AND CARE OF SILK-WORMS.—VIII. MOULTING OR CASTING THE SKIN.—IX. WINDING FRAMES ON WHICH THE WORMS SPIN.—X. KILLING THE WORMS.—XI. REELING THE SILK.—XII. MARKETING COCOONS AND EGGS.—XIII. FOOD OF THE SILK WORM.—XIV. RAISING MULBERRY TREES.

I. Silk Culture in America.

THE art of rearing and feeding the larvæ of the silk-worm and reeling the fiber of the cocoons in which the egg has been previously killed, by baking, is called Sericulture. When the cocoons are raised in a small way by the children of a family, the baked cocoons are usually sold to the manufacturers. The following condensed history will be interesting:

In Virginia.—The industry has been followed in America spasmodically since the cultivation of the mulberry was first encouraged in Virginia, by James I. The coronation robe of Charles II. was spun from cocoons raised in Virginia. A state robe was made for Queen Caroline, in 1735, from silk grown in Georgia, and in 1749 the export of cocoons from the American Colonies, South, reached 1,000 pounds. The industry flourished steadily under a royal bounty until 1766, when the export was 20,000 pounds. The bounty being withdrawn the industry declined.

In the North.—In the latter part of the century the North became interested. In 1770 Dr. Franklin sent seeds, cuttings of mulberry trees and silk-worm eggs to Pennsylvania for distribution, and the next year a silk manufactory was established in Philadelphia. In New Jersey, New York, Connecticut and Massachusetts, mulberry groves were more or less extensively planted, but the Revolutionary war put an end to the industry. It was revived in the beginning of the present century, so that over 10,000 pounds of cocoons were, in 1819, produced in Mansfield, Connecticut, alone. In 1840 the total domestic silk production in the United States was estimated at 60,000 pounds, worth over $4 per pound. In 1844 the make was 400,000 pounds, worth $150,000. From this time the industry again languished, and in 1850 only 14,673 pounds were produced.

The writer in 1832-36, then a boy, fed silk-worms in New Jersey from the street and pasture mulberry trees, the remains of the pre-revolutionary days, and in 1839 in northern Illinois from leaves gathered from the native mulberry, occasionally found growing on the sandy bottoms of the Calumet river. With the decline of the

Multicaulis mania, which began in 1840 and raged some years, silk culture declined, and until within the last ten years but little attention was paid to it.

Where Most Profitable.—The fact is, feeding the mulberry silk-worm cannot be profitable north of the 40th degree. Kansas, Mississippi and California are now the three principal nuclei of this growing industry. This last State, would, from its mild climate, seem eminently adapted to the industry. At the late French Exposition, California cocoons were among the finest shown. The leaf of the mulberry there attains the highest excellence in texture, and there is as a natural result excellence in the fiber of the cocoons. During the feeding season, from June to October, there is exemption from rain, and no electric or other meteorological changes check the growth or kill the silk-worms. Loss there is said yet to be almost inappreciable, while in all other silk-producing countries it is rated at from twenty to thirty-three per cent.

Care Necessary.—The rearing of silk-worms entails much care and labor, and hence all should be warned not to rush into this industry without due investigation, and a careful study of its possibilities.

II. Silk-Producing Insects.

ALL insects which weave cocoons for the protection of their eggs produce a species of silk. Spiders produce silk exceedingly fine and strong, but they are solitary, kill and eat each other when colonized, and are not gregarious. The common basket worm, which feeds indiscriminately upon deciduous trees, forms silk, but it must be carded, thus destroying its principal value. The ailanthus silk-worm (*Attacus cynthia*) is a good silk-producing insect; is acclimated in this country, but experiments with it so far have not been conclusively satisfactory. Our native polyphema or cecropia yields good silk.

Various insects of the genus Bombyx from Senegal, China and Bengal, and one from Japan, which latter feeds on the oak, have been received by the Department of Agriculture in Washington, but are considered of no greater value than our cecropia. The probability is that the true mulberry silk-worm (*sericaria mori*) will continue to be in the future, as it has been for ages past, in some of its many cultivated varieties, the most profitable. The perfect insect is a moth—scaly-winged insect—and belongs to the family called spinners. These have been broken by cultivation into numerous varieties, the qualities sought to be intensified being subjection to confinement, and quality of the silk product.

III. From the Egg to the Moult.

THE intending cultivator of silk must study the nature and habits of the insect, the proper food, the manner of feeding, curing the cocoons and reeling the silk. This we propose to present in the most condensed form possible, since to give all these things in full detail would require a volume larger than this whole book. The practical points are all that is necessary for the beginner. The habits of the insects and their proper management have been carefully studied by Dr. Riley, United States Entomologist. From his reports we take the following facts:

Stages of the Worm.—The silk-worm exists in four states—egg, larva, chrysalis, and adult or imago. The egg of the silk-worm moth is called by silk-raisers, the seed. Its color when first deposited is yellow, and this color it retains if unimpregnated. If impregnated, it soon acquires a grey, slate, lilac, violet, or even a dark-green hue, according to variety or breed. It also becomes indented. When diseased it assumes a still darker and dull tint. As the hatching point approaches, the egg becomes lighter in color.

SILK-WORM LARVA FULL GROWN.

Just before hatching, the worm within becoming more active, a slight clicking sound is frequently heard, which sound is, however, common to the eggs of many other insects. After the worm has made its exit by gnawing a hole through one side of the shell, this last becomes quite white. Each female produces on an average from three to four hundred eggs, and one ounce of eggs contains about 40,000 individuals. It has been noticed that the color of the albuminous fluid of the egg corresponds to that of the cocoon, so that when the fluid is white the cocoon produced is also white, and when yellow the cocoon again corresponds.

SILK-WORM MOTH.

The Moults. The worm goes through from three to four moults, the latter being the normal number. The periods between these moults are called ages, there being five of these ages, including the first from the hatching and the last from the fourth moult to the spinning period. The time between each of these moults is usually divided as follows: The first period occupies from five to six days, the second but four or five, the third about five, the fourth from five to six, and the fifth from eight to ten. These periods are not exact, but simply proportionate.

COCOON

The time from the hatching to the spinning of the cocoons may, and does, vary all the way from thirty to forty days, depending upon the race of the worm, the quality of the food, mode of feeding, temperature, etc.; but the same relative proportion of time between moults usually holds true. The preparation for each moult requires from two to three days of fasting and rest, during which time the worm attaches itself firmly by the abdominal prolegs. In front of the first joint a dark triangular spot is at this time noticeable, indicating the growth of the new head; 'and when the term of sickness is over, the worm casts its old integument, rests a short time to recover strength, and then, freshened, supple and hungry, goes to work feeding voraciously, to make up for lost time.

IV. Varieties of the Silk-Worm.

DOMESTICATION has had the effect of producing numerous varieties of the silk-worm, every different climate into which it has been carried having produced either some changes in the quality of the silk, or the shape or color of the cocoons; or else altered the habits of the worm. Some varieties produce but one brood in a year, no matter how the eggs are manipulated; such are known as Annuals. Others, known as Bivoltins, hatch twice in the course of a year; the first time, as with the Annuals, in April or May, and the second, eight or ten days after the eggs are laid by the first brood. The eggs of the second brood only are kept for the next year's crop, as those of the first brood always either hatch or die soon after being laid.

The Trevoltins produce three annual generations. There are also Quadrivoltins, and in Bengal, a variety known as Dacey which is said to produce eight generations in the course of a year. Experiments, taking into consideration the size of the cocoon, quality of silk, time occupied, hardiness, quantity of leaves required, etc., have proved the Annuals to be more profitable than any of the polyvoltins.

Varieties are also known by the color of the cocoons they produce, as Greens, or Whites, or Yellows, and also by the country in which they flourish. The white silk is most valuable in commerce, but the races producing yellow, cream-colored or flesh-colored cocoons are generally considered to be the most vigorous.

Japanese Eggs the Best.—Owing to the fearful prevalence of *pebrine* among the French and Italian races for fifteen or twenty years back, the Japanese Annuals have come into favor. The eggs are bought at Yokohama in September, and shipped during the winter. There are two principal varieties in use, the one producing white and the other greenish cocoons, and known respectively as the White Japanese and the Green Japanese Annuals. These cocoons are by no means large, but the pods are solid and firm, and yield an abundance of silk. They are about of a size, and both varieties are almost always constricted in the middle. Another valuable race is the White Chinese Annual, which much resembles the White Japanese, but is not as generally constricted.

V. Keeping and Hatching the Eggs.

THE eggs should be kept in a cool, dry room in tin boxes to prevent the ravages of rats and mice. They are most safely stored in a dry cellar, where the temperature rarely sinks below the freezing point, and they should be occasionally looked at to make sure that they are not affected by mold. If, at any time, mold be perceived upon them it should be at once rubbed or brushed off, and the atmosphere made drier. If the tin boxes be perforated on two sides and the perforations covered with fine wire gauze, the chances of injury will be reduced to a minimum. The eggs may also, whether on cards or loose, be tied up in small bags and hung to the ceiling of the cold room. The string of the bag should be passed through a bottle neck, or piece of tin, to prevent injury from rats or mice. The temperature should never be allowed to rise above 40° Fahr., but may be allowed to sink below freezing point without injury.

Hatching.—They should be kept at a low temperature until the mulberry leaves are well started in the spring, and great care must be taken as the weather grows warmer to prevent hatching before their food is ready for them, since both the Mulberry and Osage Orange are rather late in leafing out. One great object should be, in fact, to have them all kept back, as the tendency in our climate is to premature hatching. Another object should be to have them hatch uniformly, and this is best attained by keeping together those laid at one and the same time, and by wintering them as already recommended, in cellars that are cool enough to prevent any embryonic development. They should then, as soon as the leaves of the food plant have commenced to put forth, be placed in trays and brought into a well-aired room where the temperature averages about 75° Fahr.

Heat and Moisture.—The heat of the room may be increased about two degrees each day, and if the eggs have been well kept back during the winter, they will begin to hatch under such treatment on the fifth or sixth day. By no means must the eggs be exposed to the sun's rays, which would kill them in a very short time. As the time of hatching approaches, the eggs grow lighter in color, and then the atmosphere must be kept moist artificially by sprinkling the floor, or otherwise, in order to enable the worms to eat through the egg-shell more easily. They also appear fresher and more vigorous with due amount of moisture.

Ventilation.—The building in which rearing is to be done should be so arranged that it can be thoroughly and easily ventilated, and warmed if desirable. A northeast exposure is the best, and buildings erected for the express purpose should, of course, combine these requisites.

VI. Preparing to Feed the Worms.

WHEN the business of rearing silk-worms is carried on extensively, proper buildings and appliances must be provided. The directions in this volume are intended simply for the use of families who incline to the industry, and have an airy spare room where the rearing may be carried on. Here only simple means will be employed. The author used to feed them in a well-ventilated attic, and had no trouble with disease, which has been so troublesome to silk-growers of late. Each day's hatching should be kept separate, that the moultings may be equal and regular in each colony.

Shed Feeding.—Open or shed feeding has been employed with some success for family establishments. This, however, confines the whole business, particularly in the Northern States, to one or two crops in the season. In the South more can be successfully fed.

These sheds may be cheaply made, by setting posts in the ground, from six to eight feet high, with a roof of shingles or boards. The roof should project two feet over the sides. There should be some protection to the ends and sides of the shed, as strong cotton cloth sewed together, with small rods across the bottom which will answer as weights, and also as rollers, which by the aid of a pulley may be rolled up or let down at pleasure.

The width of the sheds must be governed by the size of the hurdles or feeding trays used. The length according to the extent of the feeding contemplated.

Feeding-Shelves.—In fitting up the hurdles or feeding-shelves for a building twenty feet wide, it will require a double range of posts two and a half or three inches square, on each side of the center of the room, running lengthwise, and the length of the shelves apart, in the ranges, and each two corresponding posts, crosswise of the ranges, about the width of the two shelves apart. On each double range across the posts are nailed strips, one inch or more in width and about fifteen inches apart, on which the trays or hurdles rest, which may be drawn out or slid in as may be found necessary in feeding. The aisles or passages of a building of the above width will be four feet each, allowing two feet for the width of each single hurdle.

The hurdles are of twine net-work. A frame is first made five feet long and two feet wide of boards seven-eighths of an inch thick, and one and a half inches wide. There should be two braces across the frame at equal distances of five-eighths by seven-eighths of an inch square. On a line about half an inch from the inner edge of the frame are driven tacks nearly down to their heads, at such distances as will make the meshes of the net about three-quarters of an inch square. Good hemp or flax twine is passed around these tacks, forming a net by passing the filling double over and under the warp, or that part of the twine that runs lengthwise. This twine should be somewhat smaller than that running lengthwise. On a damp day the twine becomes tight; then give the netting two good coats of shellac varnish. This cements the whole together and renders it firm and durable. The varnish is made by dissolving a quantity of gum shellac in alcohol in a tin-covered vessel, and placed near the fire. It should be reduced, when used, to the consistence of paint.

Another set of frames are made in the same way and of the same size, and covered with strong cotton cloth; this is secured with small tacks. Upon these the net frames rest, which serve to catch the litter that falls through from the worms. Hurdles made and supported in this manner admit of a free circulation of air, and the litter is then less liable to mould or ferment, and can be removed and cleaned at pleasure.

VII. Feeding and Care of Silk-Worms.

The eggs upon the papers or cloths will hatch in an atmosphere raised gradually from seventy to seventy-five degrees, in nine or ten days. But few worms will make their appearance on the first day, but on the second and third the most will come out; should there be a few remaining unhatched on the fourth day they may be thrown away, as they do not always produce strong and healthy worms. When the worms begin to make their appearance, young mulberry leaves cut into narrow strips should be laid over them, to which they will readily attach themselves; these should be carefully removed and placed compactly upon a cloth screen or tray, prepared for them, and other leaves placed upon the eggs, for the worms that still remain, which should be passed off as before. A singular fact will be observed, that all the worms will hatch between sunrise and before noon of each day. Care should be taken to keep the worms of each day's hatching by themselves, as it is of the greatest importance to have the moultings and changes of all the worms as simultaneous as possible.

Young and tender leaves should be selected to feed the worms with; these should be cut with a sharp knife into pieces not exceeding a quarter of an inch square, and evenly sifted over them. They should be fed in this way six or eight times in twenty-four hours, as near as possible at regular and stated periods, and it should be unnecessary to say that all access by rats, mice and birds must be shut off from the feeding-place.

Move and spread the worms every day, except when they are moulting. Feed often with fresh leaves, give all the air you can, so that they do not blow away. After the first moulting, feed with short, tender twigs. They are easily moved and spread with the twigs in the morning when they are hungry. If they are neglected while young it is useless to feed them when they are old. After they pass the second moulting, if fed with care, they will eat the leaves so clean that they will need to be moved but once between each moulting, and that should be done just before they moult; but should their bed become foul, move them by all means.

VIII. Moulting or Casting the Skin.

If the worms are well fed, not too thick on the papers, and the weather warm, they will moult nearly at the same time; that is, each day's hatching, and when they are kept separate and the papers marked first, second and third day, etc., you can feed them as they ought to be fed, and when they commence winding you can put up the bushes for them to wind in, as each lot commences. They will not all need them at once, as they would if all ages were mixed. When all the frames commence winding at once they cannot be attended to in time, and many worms will be lost if there is no place provided for them. They will crawl over the frames and waste their silk; even if they make a cocoon it will be of but little value. After the third moulting, feed with branches as long as they will lay on the frames. Keep the bed as even as possible. Let no leaves hang over the frame, lest some of the worms crawl out on them, others will cut them off, and leaves and worms will fall together to the ground.

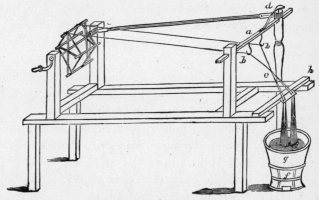

PIEDMONTESE SILK REEL.

When the worms get too large to lift with the branches, and they want moving, place five strips, three-eighths or one-half inch square, across the frames (the frames are three by four feet), the strips are three feet four inches long; so as to extend two inches over the frame on each side. Sift lime lightly over the whole bed till it is all white, worms and all; then lay branches lengthwise of the frame across the five strips. After feeding a few times the worms will all be on a new bed; they will not stay among the lime in the old bed. They are then ready to move.

Have a few duplicate frames ready; lay two sheets of heavy brown paper that will cover the frame: if you could get one large enough to cover the frame it would be better. Give the worms a good feed, and as they come upon the upper bed, place two strips, four feet long, under the ends of the five cross strips. Two persons can then raise the worms up, while the third person slips the frame and the old bed out and puts one of the duplicates in its place. The worms can then be let down, and they will keep eating as if nothing had happened. If it becomes necessary to move them suddenly, have four or five sharp-pointed sticks, slip them through the bed of branches just below the worms; then proceed as before. Pick off what few worms remain on the frame; throw off the litter, and the frame is ready for the next move.

IX. Winding Frames, on which the Worms Spin.

A LIGHT frame, the length of the hurdles, two feet and four inches wide, and made of boards one and a half inches wide, is filled crosswise with thin laths about one inch apart in the clear, and answers the twofold purpose of winding frame and mounting ladder.

When the worms are about to spin, they present something of a yellowish appearance; they refuse to eat and wander about in pursuit of a hiding-place, and throw out fibers of silk upon the leaves. The hurdles should now be thoroughly cleaned for the last time. The lath frames are used by resting the back edge of the frame upon the hurdle, where the two meet in the double range, and raising the front edge up to the under side of the hurdle above, which is held in its place by two small wire hooks attached to the edge of the hurdle, showing an end view thus:

LATH FRAME.

A covering of paper or cloth should be applied to the lath frames. In using the hurdles and screens, remove the screen from under the hurdle, turning the other side up, and letting it down directly upon the winding frame. During the spinning the temperature of the room should now be kept at about eighty degrees, as the silk does not flow so freely in a cool atmosphere. The frame resting upon the back side of each hurdle renders this side more dark, which places the worms instinctively seek, when they meet with the ends of the laths and immediately ascend to convenient places for the formation of their cocoons. From these frames the cocoons are easily gathered, free from litter and dirt, and when they are required they are put up with great expedition.

Next to lath frames, small bunches of straw afford the best simple accommodation for this purpose. Take a small bunch of clean rye straw about the size of the little finger, and with some strong twine tie it firmly about half an inch from the butt of the screw; cut the bunch off about half an inch longer than the distance between the hurdles. They are thus placed upright with their butt ends downwards, with their tops spreading out, interlacing each other, and pressing against the hurdles above. They should be thickly set in double rows about sixteen inches apart, across the hurdles.

After the most of the worms have arisen, the few remaining may be removed to hurdles by themselves. In four to six days the cocoons may be gathered. While gathering, those designed for eggs should be selected. Those of firm and fine texture with round hard ends are the best. The smaller cocoons most generally produce the male, and those larger and more full at the ends, the female insect. Each healthy

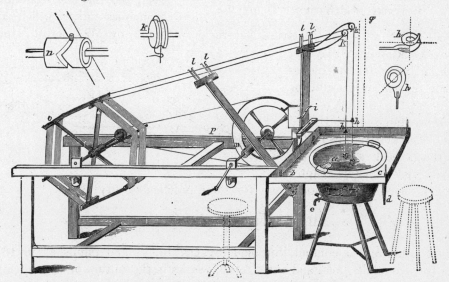

FRENCH SILK-REELING MACHINE.

female moth will lay from four hundred to six hundred eggs. But it is not always safe to calculate on one-half of the cocoons to produce female moths. Therefore it is well to save an extra number to insure a supply of eggs.

X. Killing the Worms.

This is done by subjecting the cocoons to heat. If baked, the cocoons are put in shallow baskets, placed in iron pans, and subjected to a temperature of two hundred degrees (not more), and kept there until the humming noise within, entirely ceases. Or the cocoons may be put in air-tight boxes, and taken to a steam mill, and the steam turned on for about twenty minutes. They may also be killed by subjecting them to a very hot sun for several days, carrying them under cover at night. But they must afterwards be thoroughly air-dried to prevent putrefaction.

XI. Reeling the Silk.

In family sericulture it will not pay to reel. The cocoons had better be sold direct to the manufacturer, although carefully reeled silk is worth, per pound, enough more to pay fairly for the labor. One pound of reeled silk is made from three and two-thirds pounds of dry cocoons, in which the worms have been choked. It will hardly be necessary to describe the reeling of the silk, since those who go into silk-worm raising extensively enough to permit reeling, can readily make themselves acquainted with the whole matter by instructions from the sellers of the machine. The illustration shows a French reel, and the one combining the principles upon which all are founded.

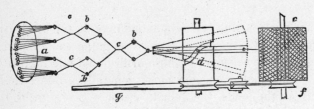

PLANE VIEW OLD FRENCH REEL.

The automatic electrical silk-reeler invented by Mr. Serrell, formerly of New York, now of Lyons, France, may do for the industry what the cotton gin did for cotton, and the reeling of silk become profitable in the United States. The illustrations of the old French reel and of the Piedmontese reel in this chapter are given more as curiosities than for their practical utility.

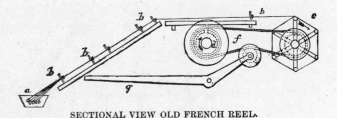

SECTIONAL VIEW OLD FRENCH REEL.

XII. Marketing Cocoons and Eggs.

The following is a computation made by Dr. Riley, and published in 1878, of results for two adults, man and wife, for six weeks in the culture of silk-worms:

Average number of eggs per ounce, 40,000.
Average number of fresh cocoons per pound, 300.
Average reduction in weight for choked cocoons, 66 per cent.
Maximum amount of fresh cocoons from one ounce of eggs, 130 to 140 pounds.
Allowing for deaths in rearing—26 per cent, being a large estimate—we thus get, as the product of an ounce of eggs, 100 pounds of fresh or 33 pounds of choked cocoons.
Two adults can take charge of the issue of from 3 to 5, say 4, ounces of eggs, which will produce 400 pounds of fresh or 133 pounds of choked cocoons.
Price per pound of fresh cocoons (1878), 50 cents.
Four hundred pounds of fresh cocoons, at 50 cents, $200.
Price per pound of fresh cocoons (1876), 70 cents.
Four hundred pounds of fresh cocoons, at 70 cents, $280.

Actual sales in Marseilles, December, 1878, of choked cocoons, 15 francs per kilogram, or $1.66 per pound, which for 133 pounds choked cocoons would be $220.78.

Price per pound of choked cocoons (1876), $2.25; 133 pounds of choked cocoons at $2.25, $299.25.

Freight, packing, commissions, and other incidental expenses, say $25, making as the return for the labor of two persons for six weeks, at the present low prices, $195.78.

Calculating on the basis of $1.50 per pound of choked cocoons, which as shown in the following estimates, a reeling establishment in this country could afford to pay, we get approximately the same amount, viz, $199.50.

The same gentleman also gives estimates upon raising eggs as follows:

Average number of eggs in an ounce, 40,000.

Maximum number of cocoons for one ounce of eggs, 40,000.

One-half of these, or 20,000, are females.

Number of eggs laid by each female, say 300.

Quantity of eggs from one ounce, 6,000,000, or 150 ounces.

Deducting as probable loss from all causes combined, one-half, we have 75 ounces.

Price of eggs in Europe, $2 to $5; say, $3 per ounce.

Amount realized on one ounce, $225.

On the basis of the first estimates two adults could take charge of the issue from four ounces of eggs. These would yield the sum of $900, and, even after allowing for the first cost of eggs, trays, commission, freight (which is light), extra time and labor (say another month), and incidental expenses, it leaves a very excellent return.

XIII. Food of the Silk-Worm.

BESIDES their regular food, silk-worms will eat young, tender lettuce and Osage orange. It is a waste of time, however, to attempt to rear worms on lettuce. If they hatch too early, they may be fed on lettuce for a few days, until other leaves appear. Very young leaves from the tree must not be fed to the worms, except when they are themselves young. Wet leaves must in no case be fed. Hence, in rainy weather a supply of leaves must be on hand, or else they must be artificially dried before being fed. So, in feeding Osage orange, after the worms become partly mature, the soft and terminal leaves must not be fed. Dr. Riley says, neither of our indigenous mulberries is suitable for food. He is probably correct. The experiment made by the writer, in Illinois, in 1839, with the red mulberry, was not successful.

The white mulberry (*Morus Alba*), the variety known as multicaulis, and the black mulberry (*M. Nigra*) are valuable, and in the order named. The Moretti, a dwarf variety of the white mulberry, is said to be valuable from its abundance of large leaves. The Russian mulberry has lately been extolled, but not enough experiments have yet been made to determine its value.

XIV. Raising Mulberry Trees.

To raise silk-worms one must first have the leaves. The mulberry, especially the white and multicaulis, grows readily from layers or cuttings. Set these in rows four feet apart, for ease in horse cultivation, and six inches apart in the row. When large enough to transplant, or when one or two years old, take out the plants, leaving the trees in the original plantation in squares four feet by four. As the plants again crowd, take out every other row and transplant. Thus in a short time you may have plenty of trees for foliage. In transplanting it is better to cut down the trees to within one foot of the ground. Transplant ten or fifteen feet apart, and keep them dwarfs by annual cutting back.

CHAPTER VIII.

SPECIAL CROPS—HOPS, TOBACCO, PEANUTS AND SWEET POTATOES.

I. HOP GROWING IN AMERICA.—II. COST OF RAISING.—III. ESTABLISHING A HOP YARD.—IV. THE PROPER SITUATION AND SOIL.—V. PREPARING FOR THE CROP.—VI. TRENCHING THE SOIL.—VII. SETTING THE PLANTS.—VIII. CARE OF THE HOP YARD.—IX. CULTIVATION IN CROP YEARS.—X. PICKING THE HOPS.—XI. DRYING THE HOPS.—XII. MANAGEMENT IN THE KILN.—XIII. THE CULTIVATION OF TOBACCO.—XIV. SOILS AND SITUATION FOR TOBACCO.—XV. THE TRUE TOBACCO BELT.—XVI. RAISING THE CROP SOUTH.—XVII. TRANSPLANTING, CULTIVATING AND WORMING.—XVIII. THE SEED BED.—XIX. RAISING PLANTS NORTH.—XX. PREPARING THE LAND.—XXI. PLANTING AT THE NORTH.—XXII. PROPER WAY TO TRANSPLANT TOBACCO.—XXIII. CULTIVATION.—XXIV. CUTTING AND CURING TOBACCO.—XXV. THE TOBACCO HOUSE.—XXVI. TWELVE RULES FOR TOBACCO GROWERS.—XXVII. PEANUTS, OR GOUBERS.—XXVIII. THE CULTIVATION FOR PEANUTS.—XXIX. GATHERING THE NUTS.—XXX. AFTER-MANAGEMENT AND CARE OF SEED.—XXXI. SWEET POTATOES.—XXXII. FIELD CULTURE OF SWEET POTATOES.—XXXIII. KEEPING SWEET POTATOES IN WINTER.—XXXIV. GARDEN CULTIVATION.

I. Hop Growing in America.

THE hop prevents fermentation, and adds an agreeable bitter to beer, ale and porter. To these qualities it owes its commercial value, England, Germany, Austria and the United States being the principal sources of supply. In 1840, the quantity produced in the United States was 1,238,502 pounds. In 1850, this had increased to 3,496,850 pounds, of which the State of New York alone gave 2,536,299 pounds. In 1860, 11,010,012 pounds were grown, and two years later the crop exceeded 16,000,000 pounds.

About this time, western farmers awoke to the profits of hop culture, and in 1867 the crop had, in some parts of this region, reached enormous proportions. Wisconsin, which has much valley land favorable to the cultivation, produced during the year named, 7,000,000 pounds, of which Sauk county alone gave 4,000,000, worth $2,500,000.

In 1869, the hop crop of the United States was no less than 25,456,669 pounds, and from that time the acreage steadily increased. In 1876, it was over 60,000 acres, almost equaling that of England, then the greatest hop-producing country of the globe. In 1877, the year of largest production, the yield was 110,000 bales, of which 95,000 bales were exported. The acreage of the great hop-growing countries is as follows: United States about 70,000 acres; England, 68,000 to 70,000 acres; continental Europe, 76,000 acres. The year 1878 was a disastrous one to hop raisers, insects and meteorological conditions combining to ruin much of the crop.

II. Cost of Raising.

IN the West the product has often been enormous; 1,000 pounds per acre not being unusual, and the cost to the farmer as low as six cents a pound. In the

Eastern States, the cost varies from twelve to fifteen cents, while in Kent, the famous hop district of England, the average cost of producing one hundred pounds is estimated at $24.30, or twenty-four and one-third cents per pound. The following figures given by a prominent hop-grower in Sauk county, Wisconsin, in the flush time of 1876, will explain the epidemic in the West. The hop yard contained four acres, the capital invested, including land, fixtures, poles, kilns, presses, etc., was $2,000. For 1877, his second year, his statement is as follows: interest on capital, ten per cent, $200; cultivation, setting poles, etc., $100; harvesting, curing, etc., $943; total expenses, $1,243. Receipts for 11,520 pounds of hops at sixty cents per pound, $6,912; net receipts for hop roots, $3,040; total receipts, $9,952; net receipts, $8,709, or 435 per cent on original investment.

III. Establishing a Hop Yard.

THE principal reasons for failures of the hop crop, in the West, especially, are inexperience in the preparation of the land, careless cultivation, and the neglect of prompt measures to prevent the depredations of insects. A crop of hops pays a large sum of money per acre, and much labor must be spent to get it. The hop plant is always propagated by sets, or sections of the roots, and never from the seeds unless the operator wishes some new variety.

IV. The Proper Situation and Soil.

NEVER put the hop yard in a situation where there is not a free circulation of air, and at the same time exemption from violent winds. Avoid all cold, tenacious, poor or wet soils. Any of these will cause failure. In England the best hops are produced in the Farnham district, upon the outcrop of the upper green-sand, and on a deep diluvial loam lying in the valleys beneath; in East Kent, upon a rich, deep loam, resting upon the upper chalk and plastic clay; in Mid Kent, upon the ragstone rock of the lower green-sand; in West Kent, chiefly upon an outcrop of the upper green-sand and gault, and in the Hill Grounds, upon the upper chalk; in the Weald of Kent and Sussex, upon Hastings sand of the Wealden formation; and in the Worcester district, upon the marls of the new red sandstone.

In the United States a deep, rich, sandy loam, tolerably firm, thoroughly well drained, rich in lime, the phosphates, potash and humus is the best—soil that will produce large crops of wheat, and one that will not heave from freezing and thawing. If you have these conditions, or can make those you do not naturally have, including protection, "go ahead." If not, "go slow."

V. Preparing for the Crop.

THE best English authorities have established the following rules in preparing the soil for hops, setting and cultivating, which we have adapted to American practice: Having chosen the site for a new plantation, the ground is trenched, or subsoil-plowed, and the holes dug, early in October. The plants are raised by cutting off the layers, or shoots, of the preceding year. These should have been bedded out in the preceding March or April, in ground previously trenched and well manured, which, by autumn, will have become what are termed "nursery plants," or bedded sets; or the cuttings themselves are planted out the same year; but this plan is not recommended, although less expensive, since, in a dry spring, there is great risk of their dying.

If the nursery plants be used, it is desirable to set them early. When cuttings are used, they are planted in squares, or triangles, at equal distances, generally from six to seven feet apart. The triangular planting possesses an advantage over the square, as, when three poles to a hill are employed, it allows the hop cultivator more completely to move all the ground on the outside of the poles, which is a matter of some importance. With regard to distances, as a general rule, six feet is preferred for square planting, and six and one-half feet for triangular. For very fertile grounds, the distances are further increased, sometimes to nine feet in square planting, having poles from twenty to thirty feet in length. In all these matters, however, the exercise of judgment is required.

VI. Trenching the Soil.

TRENCHING is considered to be, in the first instance, the preferable mode of preparing the ground, especially when meadow or pasture land is to be broken up, where, indeed, it is almost indispensable. In this, as in every other case where trenching is adopted, care is taken not to bury the surface-soil too deeply, but leaving it within reach of the spade, when the ground is dug over the following year. Very deep trenching for hops, even when the top-soil is not buried deeply, is not advisable, and unless the soil is rich, plenty of well-rotted manure should be trenched under. Very deep plowing answers well, if the land be taken from arable cultivation, provided it be in a clean condition.

When the ground has been trenched or subsoiled, if it be in "good heart," no manure is required at the time of planting; but if the ground be poor, it is desirable to dig small holes, about a foot square and fifteen inches deep, and put into the bottom of each hole a spit of good dung compost, or a few rags, hair, or any kind of animal refuse, but on no account to use guano or the salts of ammonia at this period. When large holes are dug as a substitute for trenching, it is almost always advisable to put in some manure, which should be mixed up with the soil, instead of being placed at the bottom of the holes.

VII. Setting the Plants.

IF nursery or bedded sets are employed, one, two or three plants may be used to form a hill, according to the strength of the plants. One is sufficient, if it be a large, strong, healthy plant, and if great pains and attention be bestowed upon the subsequent management. When cuttings are used, it is safest to plant five to each hill, which should be dibbled in around one as a center. Each cutting should have an inch of earth between it and its fellow. In the planting of new grounds, attention should be paid to the introduction of a sufficient number of the *male plants*. One hill in two hundred, or about six on an acre, are considered ample. They ought to be planted at regular and known intervals, in order that, in subsequent years, the cuttings saved from these grounds may not become indiscriminately mixed. The introduction of these male plants is a matter of extreme importance, and ought on no account to be neglected; for it is an established and indisputable fact, that the grounds which possess them are more prolific, and bring the hops to maturity earlier than those plantations which are deficient in them, and, in addition to these advantages, the hops are of a better quality.

The subsequent cultivation of a new plantation requires constant attention. The ground must always be kept quite clear of weeds, and should have a good depth of pulverized soil. In the latter part of the spring, a light pole about six or seven feet high above ground, should be placed to each hill, if planted with "nurseries," and about four feet high if planted with cuttings; to these, the young vines, as they shoot out during the summer, must be tied up. At the end of May, or the beginning of June, unless the ground is new and rich, a dressing of guano and superphosphate of lime should be applied, at the rate of 300 pounds of the former and 100 pounds of the latter per acre. This should be placed in equal quantities around each hill and hoed in, taking care not to allow any of the mixture to come in contact with the plant. Another and similar manuring should be applied in July, and after this, the hills should be earthed about six inches. The above quantities of fertilizers may appear extravagant, but it must be borne in mind that young hops cannot be too strong; for, unless they be very strong, they will not come into full bearing the next year. This recommendation is the result of a long and extensive experience. The cost, too, is often repaid in the same year, by the growth of 200 or 300 pounds of hops per acre. When the hops from these nursery grounds are picked, the vines must not be cut, but the hops must be gathered from the sticks, as they stand, into small baskets. The vines and poles of this young plantation should not be removed until late in autumn, or when the plants have entirely ceased growing. Whatever the age, nothing should be done except when the soil will work in a perfectly friable condition. It is especially dangerous to the crop to work the soil when wet.

In the West, where land is not so valuable as in England or the Eastern States, we advise wide planting. It gives greater ventilation and ease of horse cultivation. In England, where the climate is moist, planting is done in raised hills. In the West, if the land is well drained, level cultivation is best. Dwarf varieties should be selected, since they are richer in the constituents which make hops valuable.

In setting the plants, manure should not be put in the hill, especially new, unfermented barn-yard manure, but a richer soil may be added on thoroughly worked old compost if the ground is not rich. The roots of the sets should be spread out carefully, fine mold put around them, the soil pressed firmly and the earth heaped over them. Each hill should have two poles. In England the number is determined by the kind of hop. The Farnham, Canterbury White, and the Goldings are strong growers, and require large poles, from fourteen to twenty feet long. The Grape varieties are smaller and need poles not exceeding ten to fourteen feet in length.

VIII. Care of the Hop Yard.

LAY out the ground in regular rows seven or eight feet apart by plowing or checking perfectly straight furrows each way. Manure should not be used in the hill when setting the hops, but, if necessary, very rich earth may be added. It is usual to place five cuttings in a hill. Three plants may be allowed for the distances here given, though two plants to stand are enough. The first year the yard may be planted with corn, potatoes, or any similar crop, between the vines, the hops being tied temporarily to short poles as previously directed, and the cultivation may be hill or flat according to the drainage or other features of the field. In well-drained soils not too retentive, flat culture is the best. In the autumn two good shovels full of well-rotted compost manure over the crowns will serve to protect the plants during winter, besides enriching the soil and giving the plants a vigorous start in the spring.

IX. Cultivation in Crop Years.

AFTER the first season, the hops should occupy the whole soil. Two poles are allowed to each hill; these should be sharpened true and set deep enough with the bar to prevent danger of being blown down when weighted with hops. They are better if inclined apart at the tops.

When the hops appear above ground two of the best should be selected for each pole, and, when they reach a height of two feet, be tied thereto with stocking-yarn, bast, prepared rushes, or other suitable material. All other vines should be cut just beneath the surface of the ground. The cultivation is simply to keep the surface of the soil clean and mellow, to destroy all weeds and supernumerary vines that may appear, to tie the vines to the pole until they twine and support themselves, and to watch for and destroy all insects that may appear.

X. Picking the Hops.

THE English rule is that the hops are ripe when the seed has changed from a bright straw color to a pale brown, and emits its peculiar fragrance. Another rule is to pick when the hop becomes hard and crisp to the touch; when the extreme petal projects prominently at the tip of the hop; when the color is changed from a light silvery green to a deep primrose or yellow; and when, on opening the flower, the cuticle of the seeds is of a purple color, and the kernel, or seed itself, hard, like a nut. Even after the hop has attained a lightish-brown color no real injury to its quality will have accrued, and, for many purposes, such hops are most esteemed in the market; but after the hops generally attain a dark-brown hue there will be a great loss, both in quality and weight. When in a proper stage of ripeness, four pounds of undried hops will make one of dry, and five pounds, scarcely ripe, are required to make one when dried. Before picking time the hop-grower should secure all necessary aid; and that aid, when promised, under no circumstances should fail, as it so often does, in the harvesting of other crops. The hops are commonly picked in large boxes, containing from twenty-four to forty bushels. These boxes are divided lengthwise by a thin partition, and then subdivided into quarters. They are raised a little from the ground, and have handles at the ends.

One man and four girls are allowed to each box. Each hand deposits the hops in his or her own division of the box, and a good hand can pick twenty bushels in a day without difficulty. They are generally paid by the quantity, at so much for the box-full. It is the business of the man to supply the boxes with poles, which he raises from the ground as needed, cutting the vines about a foot high; to see that the picking is properly done, to remove the empty poles, clear them of the vines, and stack them in a systematic manner. In picking, the hops should be kept free from stems and leaves, and all blasted or immature ones should be rejected. The boxes should be emptied at least once a day; at all events, no hops should be left in the boxes over night.

HOP KILN OR DRY-HOUSE.

The picking finished, the poles are stacked wigwam fashion and bound at the tops, or else stacked so that nothing but the lower ends will appear; they must be kept from the ground.

XI. Drying the Hops.

In California and other dry, sunny climates, hops are sometimes dried in the sun, but in the end, it is everywhere cheapest to build a kiln or dry-house. This may be a simple affair, the lower room containing a stove, with as much radiating pipe as possible, and a room above with a slotted floor, upon which the hops are dried on cloths. A regular kiln, such as is used for curing malt, is better, when charcoal, coke or anthracite coal can be used for fuel. Hops being from three-fourths to four-fifths water, soon spoil if kept in bulk in a green state. Hence the kiln is worked day and night, and the hops pressed into bales, of two hundred pounds each, as soon as dry.

The Drying Kiln.—The best form of kiln for drying hops is undoubtedly one square and tight to prevent the escape of the heated air except at the ventilator in the roof. Paper orifices, regulated by sliding doors, are left near the ground to admit cool air to be warmed. The heat for ordinary farm use may be stoves, with plenty of pipe running around the heating-room. The illustration shows a dry-house twenty-two by thirty-two feet, with a kiln sixteen by sixteen feet. The stove-room is twelve by twenty-two and two and one-half feet lower than the level of the kiln. The drying-floor should be ten feet from the ground so that there may be no danger of scorching the hops in drying. This floor is formed of slats about one and a half inch each in width, and the same distance from each other. They are covered with a strong, coarse cloth, of open texture, so as to admit of a free transmission of the heated air from the kiln below. The drying-room should be of comfortable height for a person to work in it, and the sides should be lathed and plastered so that there may be no irregularity of the heat in the different portions of the room in high winds.

The cloth for the drying-floor should be well stretched over the slats and firmly nailed. On this floor the hops are spread to the depth of six or eight inches. The proper thickness will depend somewhat on the condition of the hops; if they are very full of moisture, they should be laid on quite thin; but if gathered when fully ripe, and in fine weather, a depth of ten inches will be allowed.

XII. Management in the Kiln.

The hops being spread as evenly as possible, the fires are immediately kindled in the kiln, and the temperature regulated to one uniform degree of heat. This, however, may be quite high at first, as there will be at that time but little danger of scorching the hops if the floor is sufficiently high. If the hops are rusty, or discolored from any other cause, it is usual to burn a little sulphur under them, which will bring them to a uniform appearance. This is done as soon as the hops are well warmed through, and feel somewhat moist. Great prejudice formerly existed against the use of sulphur in drying hops, but no objection is now made by the brewers, and it is generally thought that the use of it improves the appearance of all hops, and that it also facilitates the drying. During the drying process the fires should be kept up, and there should be a free supply of fresh air below, sufficient to keep up a regular succession of heated air from the kiln, passing through the hops and out at the ventilator, carrying with it the vapor expelled from the drying hops.

Dried by Hot Air.—Mr. Morton, the well-known English authority, states the principal points in drying hops. The great object with the hop-drier, he says, is to get rid of the condensed vapor from the green hops as quickly as possible, and the dry-houses should be so constructed as to effect this object perfectly. It must be borne in mind that hops should be dried by currents of heated air passing rapidly through them, and not by radiation of heat. This is a distinction of the utmost importance, since success is entirely dependent upon a strict adherence to the former principle. In order to accomplish this effect, the space above the hops must be kept hot, and all the lower parts of the kiln cold, whereby the greater density of the cold air will force the rarefied air above, carrying with it the vapor from the hops, through the aperture or cowls upon the summit of the building. To aid this ascent of the heated air passing through the hops, a stream of heated air is sometimes thrown above the hops through a tube, thus adding greatly to the heat of the current passing through the hops, and giving it a greater ascending power.

After Drying.—When sufficiently dried the hops should be allowed to cool off a little, if time can be afforded, otherwise there will be great danger that they will break in moving, or a portion of them shell off and waste. Ten or twelve hours are required to dry a kiln of hops. Two kilns may be dried in twenty-four hours by keeping the heat up through the night. A twenty-foot kiln will thus dry four hundred bushels in a day, as they come from the vines, making about seven hundred and fifty pounds of hops when dry. Do not let the heat slacken, but rather increase it, until the hops are nearly dried, lest the moisture and sweat which the fire has raised fall back and discolor the hops. For these reasons chiefly it is that no cool air should be suffered to come into the kiln while the hops are drying. After the hops have lain about seven, eight or nine hours, having left off sweating, and leap up when beaten with a stick, then turn them with a malt shovel or scoop made for that purpose; let them remain in this situation for two or three hours more, until every hop is equally dried. They must not be turned while they sweat, for that will scorch and cause them to lose their color; the fire may be diminished a little before they are turned, and renewed again afterwards; the heat should be kept as equal as possible. It may be of service to use a thermometer, by marking upon it the degree of heat proper for drying hops, as soon as that degree is ascertained by experiment.

The Cooling-Room.—Mistakes are often exceedingly detrimental to the hops, and great attention is required by the drier, night and day, until finished. When they are thoroughly dry, which is known by the brittleness of the inner stalk (if rubbed and it breaks short), the fire should be put out and the hops taken from the kiln into the cooling-room. Here they should be spread out, not exceeding twelve inches in depth, and in a day or two will be ready to bale. Care should be taken to exclude a drying air from the cooling-room. The hops being dried, the next process is to bale them. This should not be done immediately after they are taken from the kiln, but they should be allowed to lie a few days in the store-room, till they become a little softened, otherwise their extreme brittleness will cause them to be much broken in baling, and the sample be thereby greatly injured.

We have been particular in describing all the minutiæ of cultivating, curing and baling hops, for they are important. No one should undertake this industry unless he is prepared to carry out the directions to the minutest detail.

XIII. The Cultivation of Tobacco.

The tobacco crop of the United States is every year increasing in importance, and it is the belief of the writer, that in the valleys of some of the hill regions of the South will yet be found soils that will produce leaf equal to Havana tobacco. The product is sought the world over. It is one of the great money crops of the United States, and its area of production is constantly spreading wherever suitable soils are found, from Massachusetts to California, and from Wisconsin to the Gulf States. To show the value of the tobacco crop it may be mentioned that in 1869 the crop of the United States amounted to about 324,000,000 pounds against, in round numbers, 412,000,000 pounds in 1870; 410,000,000 in 1871; 505,000,000 in 1872; 502,000,000 in 1873; 358,000,000 in 1874; 520,000,000 in 1875; 482,000,000 in 1876; and 581,500,000 in 1877. Afterwards this enormous production fell off, and in 1880 it was 446,296,889 pounds, worth $36,414,615.

It is not safe for farmers to rush into the business unless they have a soil and climate suited to the crop, and have also informed themselves thoroughly upon the best modes of cultivation and management. Proper houses for curing and packing the tobacco must also be provided. The plant will ripen wherever the Concord grape will, but it does not therefore follow that any soil that will produce the Concord grape will produce good tobacco. There is no plant that is more susceptible to influences of soil and situation than tobacco.

XIV. Soils and Situations for Tobacco.

Tobacco requires a deep, rich, thoroughly drained, friable soil, strong in potash and niter. A rich humus loam is usually rich in these constituents, if it be produced from a granite soil. Sandy loams are preferable, but whatever the soil, the situation must be protected from sudden changes of temperature, and especially from blowing winds, which would bruise the delicate leaves by whipping them about. Hence, protected valleys are always sought.

If the soil is not naturally rich in potash, nitrogen and the phosphates, it must be made so. Valley lands, protected from high winds, are excellent, and if manure can be had cheap, rather, light sandy lands, if not too dry, will make good crops. However good the land, manure will help it, since it costs little if any more to take care of an acre of good land, producing up to 2,000 pounds per acre, than one producing 800 to 1,000 pounds, and the large, choice leaves of uniform quality will sell for much more per pound, than the light, thin leaves. In fact, the measure of success in tobacco culture lies in the difference between six or seven cents per pound, and twenty-five or thirty cents per pound. The first will lose money; the latter will make money fast. No acre should produce less than 1,500 pounds, if the crop is going to pay. Not even then will the farmer make money if he raises five and six cent tobacco.

XV. The True Tobacco Belt.

The belt of country in which the best tobacco is grown in the United States lies between thirty-six and forty degrees, though much of the best cigar tobacco is grown in the West, well north, in Wisconsin, about the latitude of forty-three degrees. The best manufacturing tobacco, plug and chewing, is raised in Kentucky and Missouri. Virginia and North Carolina raise fine tobacco for smoking in pipes, and some of the Florida soils are celebrated for a cigar tobacco, second only to that of Cuba.

The principal producing States are Virginia, Kentucky, Tennessee, Missouri and Ohio. Kentucky produces by far the largest quantity. In the Centennial Exposition twenty-one States were represented, which besides the hung-leaf exhibited ninety specimens of pressed leaf, the best sample being from Virginia. In cigar tobaccos, Connecticut produces the best, and Wisconsin the next best tobacco, if we except Florida, which produces comparatively a small quantity, but of a high grade for cigar wrappers and fillers.

What the country west of Arkansas and Missouri may do in the cultivation of tobacco (excepting California, which produces an excellent article), is yet to be learned. It is thought that Arizona contains lands that will produce leaf of the best quality.

XVI. Raising the Crop South.

The principles of tobacco raising are, of course, the same everywhere. Proper attention to the condition of the soil, judicious selection of plants, careful setting, thorough cultivation, effective precautions against the cut-worm in the spring, careful worming during the season of the tobacco worm, topping in season, removing suckers and pruning (removing the leaves next the ground) are all necessary. How this is done is well told by a southern tobacco planter of large experience. He says:

Plenty of Plants.—Select good land for the crop; plow and subsoil if in autumn to get the multiplied benefits of winter's freezes. This cannot be too strongly urged. Have early and vigorous plants and plenty of them. It were better to have one hundred thousand too many than ten thousand too few. To make sure of them give personal attention to the selection and preparation of the plant bed, and to the care of the young plants in the means necessary to hasten their growth, and to protect them from the dreaded fly.

Manure Liberally.—Collect manure in season and out of season, and from every available source—from the fence corners, the ditch-bank, the urinal, the ash pile. Distribute it liberally. Plow it under (both the home-made and the commercial) in February, about four inches deep, that it may become thoroughly incorporated in the soil, and be ready to answer to the first and every call of the growing plant. Often (we believe generally) the greatest part of manure applied to tobacco—and this is true of the bought fertilizer as well as that made on the farm—is lost to that crop from being applied too late. Don't wait to apply your dearly-purchased guano in the hill or the drill from fear that, if applied sooner, it will vanish into thin air before the plant needs it. This is an exploded fallacy. Experience, our best teacher, has demonstrated that stable and commercial manure are most effective when used in conjunction. In no other way can they be so intimately intermixed as by plowing them under—the one broadcasted on the other—at an early period of the preparation of the tobacco lot.

Spring Cultivation.—Early in May (in the main tobacco belt between the thirty-fifth and fortieth parallels of north latitude), re-plow the land to about the depth of the February plowing, and drag and cross-drag, and, if need be, drag it again, until the soil is brought to the finest possible tilth. Thus you augment many fold the probabilities of a stand on the first planting, and lessen materially the subsequent labor of cultivation. Plant on lists (narrow beds made by throwing four furrows together with the mold-board plow) rather than in hills, if for no other reason than that having now, if never before, to pay wages in some shape to labor, whenever and wherever possible horse-power should be substituted for man-power—the plow for the hoe. Plant as early as possible after a continuance of pleasant spring weather is assured. Seek to have a forward crop, as the benefits claimed for a late one from the fall dews do not compensate for the many advantages resulting from early maturity. Make it an inflexible rule to plant no tobacco after the tenth of July, in the tobacco belt we have named. Where one good crop is made from later planting ninety-nine prove utter failures.

TOBACCO PLANT IN BLOSSOM.

XVII. Transplanting, Cultivating and Worming.

Take pains in transplanting, that little or no re-planting shall be necessary. The cut-worm being a prime cause of most of the trouble in securing a stand, hunt it assiduously, and particularly in the early morning when it can most readily be found. Keep the grass and weeds down, and the soil loose and mellow by frequent stirring, avoiding as much as possible cutting and tearing the roots of the plant in all stages of its growth, and more especially after topping.

There are few cultivated plants more beautiful when in blossom than the tobacco plant, as the illustration will show. The white line shows the proper place for topping, to be varied according to the circumstances, and to be explained hereafter. When at all practicable—and, with the great improvement in cultivators, sweeps and other farm implements, it is oftener practicable than generally supposed—substitute for hand-work in cultivation that of the horse. The difference in cost will tell in the balance-sheet at the close of the operation.

Worming.—Attend closely to worming, for on it hinges in no little degree the quality and quantity of tobacco you will have for sale. A worm-eaten crop brings little money. So important is this operation that it may properly claim more than a passing notice. Not only is it the most tedious, the most unremitting and the most expensive operation connected with the production of tobacco, but the necessity for it determines more than all other causes the limit of the crop which in general it has been found possible for a single hand to manage. Therefore bring to your aid every possible adjunct in diminishing the number of worms.

TOBACCO WORM MOTH.

Killing the Moth.—Use poison for killing the moth in the manner so frequently described in treatises on tobacco, to wit: by injecting a solution of cobalt or other deadly drug into the flower of the Jamestown or jimson weed (*Datura stramonium*), if necessary, planting seeds of the weed for the purpose. Employ at night the flames of lamps, of torches, or of huge bonfires, in which the moth may find a quick and certain death. In worming, spare those worms found covered with a white film or net-like substance, this being the cocoon producing the ichneumon fly, an enemy to the worm likely to prove a valuable ally to the planter in his war of extermination. Turn your flock of turkeys into the tobacco field, that they, too, may prey upon the pest, and themselves grow fat in so doing. If these remedies should fail, sprinkle diluted spirits of turpentine over the plant through the rose of a watering pot, a herculean task truly in a large crop, but

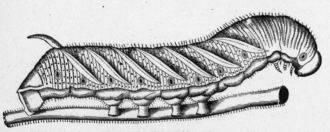

TOBACCO WORM, LARVA.

mere child's play to the hand-picking process, for the one sprinkling suffices to keep off the worms for all time, whereas hand-picking is a continual round of expensive labor from the appearance of the first worm until the last plant has been harvested.

Turkeys.—The writer's experience in raising tobacco in the North is that hand-picking is the only sure means of killing the worms. It costs money and time, but the difference in the leaf and the crop, in price, is what ensures profit. Turkeys are indefatigable hunters of the tobacco worm, and they will kill them after their hunger is satisfied.

Topping.—The topping of the plants must be attended to in season, just at the time the buds appear. From eight to twelve leaves should be left to each plant, according to the richness of the soil. This will give strong growth, but the grower must exercise his judgment here. It is better to have few strong leaves than more weak ones.

XVIII. The Seed-Bed.

GROWING the plants is one of the most important things in tobacco culture. Without good, healthy plants, failure is pretty certain. In the South a warm, sheltered situation of well-drained land is selected. This should be carefully dug over in the autumn or winter when dry, and so covered with brush that the soil may be burned deep enough to kill the weeds lying near the surface. The beds may be burned over in February, March or April, according to the locality, and immediately sown, since the seed will not sprout until the earth has the proper temperature.

Sowing.—After burning, hoe and rake the surface thoroughly, to a depth of two or three inches, and leave the surface fine and smooth. Mix the seed with dry ashes, at the rate of a large tablespoonful for each eighty square yards, and sow evenly, rake lightly, or better, tamp the whole carefully over with the rake, the handle being held upright. Cover carefully with brush, but not so as to exclude the sun. As soon as the plants require weeding, remove the brush carefully, at the same time thinning the plants where they stand too thick. In this way you will get fine, well-rooted plants for setting.

PUPA OF TOBACCO WORM.

XIX. Raising Plants North.

IN the North, raising plants is more difficult. The season is so short that the crop is often late in ripening. If the plants are placed in too warm a border they are apt to become chilled or killed by frost, and are seldom large enough for setting by the first to the tenth of June. We have always had the best success by raising the plants in a cold frame—a compartment of boards sixteen inches high at the back, sloping to ten inches in front, covered with sashes, and containing four or five inches of fine compost soil.

Caring for the Plants.—The seed may be sown in this bed about the first of April, and are easily cared for, readily protected against the fly, by dusting with soot or fumigating with smoke. They are also thoroughly protected against frost at night, or too much wet, and are easily watered when necessary. If given plenty of ventilation to keep them growing slowly and healthily, and if exposed to the full influence of the air during the day, for two or three weeks before setting, they make stronger and better rooted plants than any grown out of door

The writer's plan was to start in a small hot-bed, and when the plants had leaves as large as a mouse's ear, to prick them out, two inches apart, in a cold frame. Thus we always had plants of uniform size for the first setting, and a bed of pricked-out plants, in a warm, open border, furnished plants for succession and filling in, when the stand had been destroyed by worms or otherwise. Every square foot will contain thirty-six plants, and the space required for an acre is not as large as at first would seem necessary.

Transplanting.—Transplanting in the North should not be undertaken until cucumbers and melons will germinate and grow promptly, or until both days and nights are warm. In untoward seasons plants are set until about the first of July. From the first to the fifteenth of June is the proper time in Wisconsin, and a little earlier in Northern Illinois; about the first of June is the best time in the New England States. Of seed-leaf tobacco from five thousand to six thousand plants are set per acre, and of Havana six thousand to seven thousand, according to the size of plants your seed will produce.

XX. Preparing the Land.

In the North the ground for tobacco should always be deeply fall-plowed, turning under a liberal quantity of barn-yard manure. If the land has not been previously manured, twenty loads of fine manure should also be carted and spread in winter to ensure richness near the surface. About ten days before planting-time this should be turned under about four inches deep and the surface brought to an uniformly fine tilth.

Marking the Land.—Mark the land in straight lines three feet apart, and with a single horse-hoe or double mold-board plow run through these marks, thus bedding up the land. Run a harrow over the ridges lengthwise and then a plank to bring all fine and smooth. You will then have a succession of flat, slightly raised beds upon which to plant.

If you mark these beds crosswise, three feet apart, you will get 4,840 plants per acre. This is space enough for the largest Connecticut or Maryland tobacco. If you mark your squares two feet you will have 7,200 plants per acre, and this is close enough for the smallest Havana plants. Thus you may graduate your distances to accommodate plants of any size that one season's cultivation will show, according to the richness of the soil or variety of tobacco cultivated. It should always be remembered that the closer you grow your plants, according to soil, the better the crop as a rule.

Smaller Squares.—Or if you choose to make your beds three and a half feet apart, then by marking across the beds thirty inches apart you will get 4,976 plants per acre. Marked two feet apart there will be 6,223 plants; if twenty inches apart in the row 7,467 plants per acre.

XXI. Planting at the North.

DIRECTIONS for planting, general cultivation, care, harvesting, drying, stripping and packing tobacco will apply to all parts of the country, allowances being made for differences in latitude and other conditions. Many tobacco growers pull the plants when the leaves are the size of one's thumb-nail, and simply press them into the ground when it is wet. This is never done, North or South, by the best cultivators. We give the best plan, and in the end it is the cheapest, because it is the best. Never wait for rain in transplanting anything, provided the ground is in a fairly moist condition; that is, friably moist, or not really dry. Water the plants thoroughly in the bed, and as soon thereafter as the soil can be worked, take out enough plants for the day's planting, beginning at about three o'clock in the afternoon, or if the weather is cloudy, work all day. Plant as late at night as you can see. Havana seed plants have long roots like cabbage, and are not so easy to set as Connecticut seed-leaf, which has fibrous roots. Hence it is better to prick them out.

XXII. Proper Way to Transplant Tobacco.

WHEN the plants have leaves about the size of a silver dollar, let a careful hand take them up and bring them to the field as wanted, arranging them in baskets on wet moss, so they can be easily handled and covered with a fold of damp cloth. The ground being properly marked, the planter takes a plant, makes a hole for it with one or two fingers, inserts the roots, pressing the earth firmly around it, but leaving a depression to hold a little water. This an assistant supplies from a water-can with a spout. If the earth is in good condition a gill is enough for each plant; if the soil is pretty dry more water must be given, and always in the depression which is left about the roots.

Thus the hands, as many as are necessary, go on working with a deftness that is learned only by practice, and covering a large area of ground in a day. When the water has dried down, other hands, girls preferably, smooth the soil nicely to the plants, covering the watered surface with fine dry mold.

The plants should be set as shown in the annexed cut; they will seldom suffer for want of moisture, and in ordinary weather will grow right along. It is the cheapest way in the end, and by no means slow, for an active hand will set 5,000 plants in a day. And when the work is thus done, in the best possible manner, there will be no baking or drying of the soil about the roots. This is true of plants of every kind and should be remembered. In setting on these raised beds the tobacco plants should be rather below the level of the surface, for as soon as the crop is fairly growing a little earth should be dressed up to them.

TOBACCO PLANT PROPERLY SET.

XXIII. The Cultivation.

ALL other cultivation should be the same as that for corn or other hoed crops, thorough and frequent. No weeds dare be allowed at any time. In an average season the plant will mature sufficiently by the early part of August to dispense with further cultivation, but until the ground is fairly shaded, the cultivation must be thorough. When the plants have from fourteen to sixteen leaves, or when they begin to throw up the blossom shoots, pinch off the tops, and from time to time, as the suckers appear, pinch them out before they attain a length of three inches. If the suckers are allowed to remain, they will reduce the growth of the true leaves.

The necessity of watching for, and killing cut-worms, after the plants are set, and destroying the horn (tobacco) worms as fast as they appear has already been pointed out. These must be attended to or the crops will be seriously damaged and may be ruined. No man should undertake to cultivate tobacco or any other special crop unless he is prepared to spend the time and money necessary to do everything in the best manner, since, upon this depends the ultimate profits. In the North the worm, larva of the tobacco sphinx, and also that of the tomato worm, an allied species, appear about July first, and feed on the leaf until the crop is secured. In fact, they frequently, if not picked off clean, cling to the leaves after the stalk is hung up. Usually, from three to four weeks from the time of topping, the plant will mature and be ready to cut.

Uniform size of leaves, and a stiffness of the leaf, making it liable to break by bending and handling, are the surest signs of maturity. The lower leaves change color, and in some varieties the leaves present a spotted or mottled appearance. This must be carefully studied, and the beginner would do well to employ a man competent to judge, and who is also familiar with handling, hanging, drying, stripping, bulking and packing the crop. Otherwise, the beginner should experiment in a small way until he learns. Nowadays, however, the crop is usually sold in bulk, in the North, the buyer attending to the casing and shipping himself.

XXIV. Cutting and Curing Tobacco.

THE time to cut must be determined by the condition of the crop. Sometimes it ripens unevenly. In this case, the portion that is ripe must be selected first. If the crop ripens up handsomely, it is better to cut altogether, since the inferior plants left are apt to be whipped and injured by the wind. The stalk is severed with a heavy knife (similar to a corn stalk knife), just above the ground, and at a single blow. Each stalk is laid on the ground to wilt, but it must not be long exposed to the sun, especially if it is hot; nor must it be cut with the dew on.

Cut after the dew is off, but not during the middle of the day, when the sun is bright, as you must guard against burning while it is undergoing the wilting process, preparatory to spearing and handling in the removal to the shed. When wilted, so the plants may be handled without breaking the leaves, they are speared, spiked, or strung by the butts upon laths four feet long. Four or five plants are strung to each lath, and hung on proper frames, on a wagon or sled, for removal to the house. Some persons hang in temporary sheds in the field or near the house, for partial curing, but it is not a good plan. The house should be large enough for the whole crop. It should have ventilators at the top to pass out the foul air, and ventilators at the bottom to admit fresh air in windy weather.

XXV. The Tobacco House.

Ventilation and Drying.—The tobacco house may be arranged for four or five tiers of stalks. It should rarely or never be higher. The illustration shows the general arrangement—beams for hanging the tobacco, lath doors or shutters for ventilation, etc. The ventilation is important. In damp weather the house must be closed. In windy weather the leaves must not be blown about. If the dry heat of charcoal, coke, etc., is used for drying, it should be conveyed in pipes running at proper intervals through the house and not within eight feet of the leaves. Care must also be used in curing, according to whether the demand is for light or dark tobacco.

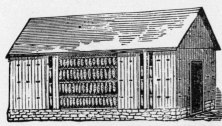
TOBACCO HOUSE.

Stripping.—In December, and from that time on, when the weather is moist enough so the tobacco will be pliable, or "in case" for handling, stripping may commence. The tobacco should be assorted into three qualities, first, second and third, corresponding to best, second-best and inferior, and all leaves in a "hand" should be of uniform length. This assorting must be by competent hands. One man may take the best, passing the stalk to another, he selects the seconds, and another the inferior. These tied in bundles of twelve to sixteen leaves, and bound at the butts by a single leaf, constitute a "hand" of tobacco, as shown in the cut. Twelve leaves make a "hand" of the best wrappers; from fourteen to sixteen are often put into "hands" of seconds and thirds.

Bulking.—This is a nice job, and requires the utmost care and watchfulness to bring the tobacco into the proper condition. If too damp, it will get hot, if too dry it will not warm up sufficiently to bring out the fragrance and color of prime condition. It will pay to hire a competent man until the owner is familiar with the process. The object is to have the tobacco dry out slowly, and to remain in proper condition until ready for packing in cases, in the North, or in hogsheads in the South. These boxes, or cases, contain four hundred pounds, while the

"HAND" OF TOBACCO.

hogsheads contain one thousand pounds each. The bulking is done between the sides of a frame as high as the bulk is to be carried, and wide enough to allow the leaf-tips to lap, one on the other, with the butts at the sides. A bulk three and a half feet high, and twenty feet long, should hold four thousand pounds. Bulk each sort by itself. In bulking, take the "hands" one at a time, laying them straight, over-lapping the tops one-third on the other, keeping the whole even, and pressing with the knees as you proceed, until the task is finished. Then remove the side-pieces, and cover all with blankets, weighting them down with boards if necessary.

The bulks may remain until sold to the packer, watching it carefully to see that it does not heat unduly. If the fermentation is too strong the tobacco will blacken;

MOTH OF TOMATO WORM.—DESTROYS TOBACCO IN THE NORTH.

if too little the flavor will not develop. To get the right effect requires the greatest care and judgment.

XXVI. Twelve Rules for Tobacco Growers.

1. THE land must be rich and in good condition generally; potash and nitrogen are essential to the crop, as well as friability and permeability of the soil.

2. The seeding in the seed-bed must be thick and even; to be afterwards thinned, to enable the plants to grow stocky and strong for transplanting. It is well to allow an ounce of seed for every two acres of tobacco, to allow for destruction by the fly and other insect enemies.

3. Do not transplant until the weather is permanently warm—the nights as well as the days. If once the plants are chilled the crop is injured.

4. After planting out watch carefully for cut worms, at daylight in the morning, and wherever a leaf is attacked find every worm before you quit.

5. Transplant promptly from a reserve bed of extra plants whenever you find a plant missing.

6. Cultivate thoroughly, always being careful not to break or injure the leaves. Careless workers will destroy more than the value of extra wages paid to careful men.

7. Watch for the appearance of the tobacco moth, which lays the eggs. It is well to have a plat of Jimson weed near, or of tobacco plants in flowers, to attract the moths. The flowers may be poisoned with a solution of cobalt, such as is used for killing flies.

8. When the worms—larvæ of the tobacco moth, hatched from the eggs laid on the tobacco leaves—appear, go over the field twice a day, carefully, to kill them. Also hunt for the patches of eggs on the leaves and destroy them.

9. Top the field to twelve or fifteen leaves, as soon as the buttons—flower-buds—have generally appeared, and pinch out suckers before they grow three inches in length. Take off all lower leaves that sweep the ground.

10. Cut the crop when ripe, preferably with a sharp saw, and never allow plants to wilt when the sun is hot. Handle very carefully, to prevent injury in carrying to the house and in hanging.

11. Watch the ventilation in the house. The leaves must not hang near enough for one stalk of leaves to touch another. The wind must not blow them about and the vapors must be promptly carried away through ventilators at the top of the house.

12. In stripping, keep each grade by itself. Bulk carefully, and watch daily to see that it does not overheat.

XXVII. Peanuts or Goubers.

The peanuts, as known in commerce, are called ground peas, goubers, or pindars, locally, according to the part of the country where they are raised. The botanical name of the plant is *Arachis hypogœa*. Until comparatively a few years ago the supply came principally from the East Indian Islands and along the African coast. In these regions quantities are still raised for the oil, which is excellent for lubricating and burning, and equal, for culinary and table purposes, to olive oil. The nuts are also largely used for adulterating chocolate, and especially chocolate condiments.

The gouber crop in the South is yearly increasing, especially in Tennessee. Virginia, North Carolina, and the other Atlantic States, South, produce the best nuts, a sandy, or at least fairly arenaceous soil, containing plenty of lime, being necessary to produce full-meated nuts. Under good cultivation, the yield is sixty or seventy bushels, and from that to eighty bushels, per acre. When raised for forage, the vines make excellent hay; the product is about half a ton per acre, cured. With proper care they may be raised as far north as forty degrees, but they are essentially a southern plant, being killed by the least frost. Except as a curiosity, they are hardly ever raised north of Virginia.

XXVIII. The Cultivation of Peanuts.

The cultivation is simple, and yet peculiar. The blossoms, when fertilized, hang down, grow into the ground and pierce it until the firm soil is reached, where the pods form and ripen. Hence, the necessity of shallow cultivation. The soil should be plowed in autumn, and in the spring only surface-plowed, not more than three or four inches, to kill weeds.

When all danger of frost is over, the soil is bedded up and prepared, as for tobacco, leaving only a slight furrow-mark between the rows. In the center of each of these beds, in a straight line, plant two seeds, at distances of eighteen inches; also have reserve plants, to fill the places of those that may be destroyed by cut-worms, etc.

The cultivation is simply to keep down the weeds, preserving the shape of the beds until near the time of blossoming.

PEANUTS—PLANT AND TUBERS.

A narrow cultivator is then run through the rows, followed by a horse team to earth up the plants. The earth is afterwards leveled to present a flat hill, in which the nuts are to form. If weeds or grass thereafter appear they must be pulled by hand. The illustration shows the vine, the root, and the nuts formed under the surface.

XXIX. Gathering the Nuts.

THE crop is not harvested until the vines are touched by frost, for the longer the vines grow the greater the number of sound pods, except in the extreme South, where the vines ripen fully. Hands follow the rows and loosen the nuts with pronged hoes or flat-tined forks. They are followed by others, who pull the vines, shake the earth from them and leave them turned to the sun to dry. In dry weather they will thus be sufficiently cured for shocking. The shocking is done somewhat after the manner employed for beans; or they may be finally cured as beans sometimes are on scaffolds under sheds.

Shocking.—The Tennessee plan is to provide stakes seven feet long, made sharp at both ends; then lay two fence rails on the ground as a foundation, but with supports underneath to afford free access to the air. The stakes are stuck in the ground at convenient intervals between the rails, the stacks built up around them, and finished off by a cap of straw to shed the rain. The diameter of the stack is made to conform to the spread of a single vine. After remaining about two weeks in the stack the picking should begin, taking off none but the matured pods. These are to be carried to the barn, and prepared for market by finishing the drying process, and then fanning and cleaning. The most tedious part of the work is picking. An expert discriminates at a glance between the mature and immature pods, but cannot pick more than two and a half or three bushels per day.

XXX. After-Management and Care of Seed.

UNLESS the management in the barn is carefully conducted there is great danger, where there is much of a bulk, that the peas will become heated and moldy. The condition in which the early deliveries are often made on the market renders this caution necessary. In fact, there is as much slovenliness in the handling of this crop as there is in regard to any other, perhaps more; for the reason that so many inexperienced persons engage in the culture every year. Until the pods are thoroughly seasoned the bulk should be frequently stirred and turned over. A certain classification, in respect to quality, obtains in peanuts as in every other article of agricultural produce. The descriptive terms in general use are inferior, ordinary, prime and fancy: but these are not so definite as to admit of no intermediate grades.

Seed Peanuts.—A matter of primary importance is to provide seeds of good quality for planting; and in order to be assured of their excellence the planter should either raise them himself or buy them of a person on whose fidelity he can rely. If, after the vines are dug and they are lying in the field, they should be exposed to frosty weather, the germinating principle would be destroyed or impaired. As a merchantable article, however, their value is not affected. Neither should the nuts become the least heated or moldy; nor should they be picked off the vines while wet, or before they are thoroughly cured. It is obvious, therefore, that the most careful attention is requisite in this matter. Previous to planting, the pods should be carefully shelled, and every faulty bean thrown out; not even the membrane inclosing the seed should be ruptured. It takes about two bushels of peanuts in the pod to plant an acre.

XXXI. Sweet Potatoes.

THE sweet potato is another special crop that year by year becomes more important, especially since cheap railway facilities admit transportation for long distances, and improved methods enable the farmer to preserve them in good condition through the winter, and until late in the spring. There is now only about two months in the year when they may not be readily bought in the northern markets.

Sandy soil, or a rather firm, sandy loam is the best for this crop. In soft land, especially if plowed deep, the tubers grow long and stringy. The potatoes are never planted directly in the hill, but are grown from "slips"—sprouts three to five inches long—obtained by bedding the potatoes in a hot-bed, covered with boards, to shed rain, and protect them from being chilled at night. As the sweet potato is killed by the slightest frost, the plants should never be set out in the field until the days and nights are warm. Planted from the first to the middle of June, good crops are raised up to, and even north of, forty-two degrees in favorable situations.

A central Ohio farmer gives, in a nut-shell, all that is necessary for field cultivation. For the garden it is cheaper to buy the plants than to raise them, two hundred plants being enough for a family of moderate size. Our authority, a thorough practical farmer, says:

XXXII. Field Culture of Sweet Potatoes.

"My plan is to place logs on a sloping piece of ground, say ten or twelve feet apart. I then drive small stakes, or pegs, in rows three feet apart, and eight inches high. The object is to have not more than seven or eight inches depth of manure, which should be fresh horse-dung, a mixture of hay, straw, corn-fodder, etc., trampled down level with the tops of the pegs. I then put a coat of loam, three inches deep, upon the top of the manure, which answers for the dressing the subsequent year. I then place my tubers on, cover them from two to three inches deep, and then lay on boards, so as to keep them effectually covered from rain or cold until the plants are up.

Drawing the Plants.—During the day, I let them have the sun, until I am sure they cannot be injured by frost. I sometimes water them, but not before the heat has somewhat subsided in the bed, which I ascertain by putting my forefinger through the covering. A very little warmth from beneath is sufficient; there is more to be apprehended from too much heat than too little. Some place a covering of saw-dust on top of the bed; but this is entirely unnecessary. In this latitude, the beds should be made as early as the tenth or twentieth of April. The plants will be ready for drawing, from the eighth to the twentieth of May.

Preparing the Ground.—I select ground, for growing the tubers, that will produce good corn. To manure just before planting will cause the plants to run to vines. Good loam, with or without sand, such as we call "second year's land," lying to the sun, yields best. It need not necessarily be sandy, to produce the greatest yield; on the contrary, good loamy land produces tubers of the best flavor. I plow the ground well, when dry, and harrow thoroughly. It would even be better to cross-plow it. Then I throw two "moles" together, about four feet apart, and see that the ground is well pulverized, in order that the list may be clear from clods, sods and trash, and that the land is in the best order to receive the plants. The time for transplanting is when the ground is what we call "dry."

Planting.—The mode of planting is to make a hole with the hand, or otherwise, of the proper depth to receive the young plant; and, when it is placed in the hole, I pour in half a gill of water, so that the earth may settle around the fibrous roots; then, I draw the dry earth around the plant, and compress it a little with a hoe. In less than twenty-four hours the plant will be as vigorous as though it had never been removed. On good land, the distance of the plants apart should be from eighteen to twenty inches; for thin land, fifteen inches will be sufficient. The yield, in this section, is from 100 to 150 bushels to the acre. I should state that the plants require to be hoed about as much as corn. The vines should be thrown on the ridges, out of the way, while dressing. In digging, I use a large, long, flat, three-tined fork, to throw the tubers out of the ground. When dug, I spread them to dry and wilt somewhat, preparatory to putting them up for winter, which requires much care.

XXXIII. Keeping Sweet Potatoes in Winter.

Sweet potatoes are easily kept through the winter in a room where the temperature is about fifty degrees. A temperature materially lower than this will make them "frost-bitten," and if the room is much warmer than fifty degrees, it will sprout them. Sweet potato houses are built secure from frost, heated to the proper temperature, and the potatoes are kept in bins one over another, each containing about a barrel.

Any room of the temperature stated will keep them, if the potatoes have been handled without bruising. They may be packed either in barrels or boxes, and kiln-dried or thoroughly sun-dried sand poured over them to fill the interstices, or boxes of uniform size, separated one from the other by an inch space may be piled one above the other. In this way the potatoes will keep sound until spring.

XXXIV. Garden Cultivation of Sweet Potatoes.

Where a few plants are raised for family use in the autumn and early winter, the earth may be thrown up either into pretty high ridges or hills, and the plants set as directed, at any time after the season becomes permanently warm. Watch for cut-worms, keep the soil clean, prevent the vines rooting from the joints, by occasionally lifting with the handle of a rake, and in the autumn you will have fresh potatoes that will come in well for family use. Every farmer should plant from 200 to 500 vines.

CHAPTER IX.

CROPS FOR SUGAR-MAKING.

I. SUGAR AND ITS MANUFACTURE.—II. CANE AND OTHER SUGARS COMPARED.—III. HISTORY OF BEET SUGAR IN THE UNITED STATES.—IV. OUR TWO GREAT SUGAR PLANTS.—V. THE VARIOUS SACCHARINE PRODUCTS.—VI. THE THREE SUGARS COMPARED.—VII. CULTIVATION OF SUGAR-CANE.—VIII. CULTIVATION OF SORGHUM.—IX. WHEN TO CUT SORGHUM CANE.—X. CUTTING AND HANDLING THE CANE.—XI. SPECIFIC GRAVITY AS A BASIS OF VALUE.—XII. SPECIFIC GRAVITY AND COMPOSITION OF JUICES.—XIII. TABLE OF JUICES.—XIV. VALUE OF SORGHUM DURING WORKING PERIOD.—XV. FOUR IMPORTANT POINTS.—XVI. VALUABLE CANES SOUTH. —XVII. THE REAL TEST OF VALUE.—XVIII. TABLE OF COMPARATIVE VALUES DURING WORKING PERIOD.—XIX. THE MANUFACTURE OF SORGHUM.—XX. MAKING SUGAR ON THE FARM.—XXI. GENERAL CONCLUSIONS.—XXII. MAPLE SUGAR.—XXIII. TAPPING THE TREES.—XXIV. BOILING AND SUGARING.—XXV. SUGARING OFF.—XXVI. TO TELL WHEN SUGAR IS DONE.

I. Sugar and its Manufacture.

SINCE sugar has come to be thought one of the necessities of life, various plants containing saccharine sap have been utilized for the manufacture of syrup, or sugar, or both. Sugar-cane and the sugar beet have been the most important of these; the maple-tree standing next, until within the last few years, during which time, improved processes of separating tree sugar from the glucose of sorghum have come into use.

From the author's earlier experiments, forty years ago, in concentrating the juices of the corn-stalk, and of water-melons, we were convinced that these plants would never afford merchantable sweets. Not so after experimenting with sorghum, in 1856. The saccharine material was there; the question remained, how to separate it cheaply. This has now been so answered by the chemist as to make it seem probable that within a very few years the West will be able to produce sugar from sorghum as satisfactorily as Europe has done from the sugar beet. How important this is will be understood when we mention, that, notwithstanding the gradual increase of the sugar production in the very narrow Gulf belt of the United States, which itself is only partially adapted to the production of cane sugar, this country grows only one-seventh of the sugar it uses.

II. Cane and Other Sugars Compared.

IN 1875, we had occasion to investigate the sugar production of the world. The Island of Cuba alone produced 700,000 tons, yearly, and our Southern States only 75,000 tons. The annual production of cane sugar for the world was 2,186,000 tons, yearly. Of beet sugar Europe produced 1,317,626 tons, of which the little country of

Belgium gave 79,796. This European sugar thus amounted to more than half the cane-sugar production of the globe, or more than one-third of the world's total sugar crop.

The manufacture of true sugar from sorghum has not yet reached large figures, but since, in 1882, a single factory produced 319,000 pounds, while the aggregate of the whole country was only 500,000 pounds, it is simply a question of time when this industry must become one of the most important in the West.

The manufacture of beet sugar has disappeared from the United States, only one factory, in California, being, we believe, now in operation. The Chatsworth, Illinois, enterprise failed long ago, and, lately, the one in Maine stopped. The causes of failure lie, principally, in the cost of manual labor, the strong nitrogenous quality of the land, and the difficulty in procuring competent persons for the direction of the intricate processes of manufacture. The author knows of what he writes, for he had charge of the initial enterprise in the United States, that at Chatsworth, Illinois, during the last three years of its existence.

III. History of Beet Sugar in the United States.

There have been, in all, seven large beet-sugar factories started in the United States during the last twenty years. Two in Illinois, two in Wisconsin, one in Maine and two in California. True success has been reached by none, and all but one have suspended. The manufacture of beet sugar requires an abundance of living water, intricate machinery, large capital, cheap labor in the production and working of the crop, and men of exact and scientific knowledge in the management of the factory. The same enterprise and money employed in the production of sorghum, will produce double the results. The time may come when the production of beet sugar may be profitable in the United States, yet with the cheaper labor of Europe, the industry does not flourish there as in former years.

IV. Our Two Great Sugar Plants.

The sugar-cane (*Saccharum officinarum*) and the sorghum cane (*Sorghum vulgare*) in some of its varieties, are, therefore, destined to produce the future sugars of the United States. The cultivation of the first-named plant is limited to the Gulf coast—Florida, Mississippi, Louisiana and Texas.

Sorghum.—Sorghum culture in the United States is only limited by the finding of soil suited to the plant. The rich sandy soils of Minnesota have so far produced some of the best samples of sugar. Farther south, it is at home on any soil that will make good crops of Indian corn, though arable lands and well-drained sandy loams will always be found to give the best grades of sugar. This is true of all plants producing saccharine juices. Strong nitrogenous soils always act against the crystallizability of sugars, and in the case of beets, are a bar to its production. Hence, as sugar producers, we shall have to deal only with the two plants mentioned, and with the maple-tree.

V. The Various Saccharine Products.

Cane Sugar.—Cane sugar occurs in the ordinary sugar-cane, in the sap of the maple and the juice of the beet, without the admixture of any other kind of sugar. It crystallizes readily from a pure solution, in large oblique prisms. It rotates the plane of polarization to the right. It does not precipitate suboxide of copper from an alkaline solution of that metal (Fehling's test) at the boiling temperature. By being heated with acids, or by being boiled for a long time with water, it is converted into a mixture of grape and fruit sugar.

Grape Sugar.—Grape sugar, or glucose, constitutes the white powder seen upon the outside of old raisins; it also forms the sediment arising in old honey. It is found in connection with cane and fruit sugars in many fruits, and may be made artificially by the action of acids upon cane sugar, starch or wood. Made thus it is used in Europe for adding to wine musts which are weak in sugar. It crystallizes with difficulty, forming cauliflower-like masses which, under the microscope, appear like fine needles or blades, and in some conditions as six-sided tablets. It also polarizes to the right, but to a less degree than cane sugar. It is less sweet than cane sugar, one pound of the latter producing the same degree of sensation of sweetness as from two to two and a half pounds of grape sugar. At the boiling temperature it precipitates the copper of Fehling's test. While cane sugar has to pass into grape and fruit sugars before fermentation takes place, grape sugar ferments without further change.

Fruit Sugar.—Fruit sugar occurs, as its name partly implies, in acidulous fruits with grape and cane sugars. It occurs also in molasses, as before stated. It is not capable of crystallization, but exists as a syrup, or, when dried, as a transparent candy. It is as sweet as cane sugar. It rotates the plane of polarization to the left. At the boiling temperature it removes the copper from Fehling's test solution, like grape sugar. It ferments without passing into any other kind of sugar. These are the most prominent differences between the three sugars.

As "polarization to the right or left" cannot be sufficiently explained without many words, the unscientific reader is requested to accept that, "cane, grape and fruit sugars behave differently towards polarized light."

VI. The Three Sugars Compared.

A GREAT want of clearness rests in the public mind as to grape and fruit sugars, arising from the carelessness with which scientific men use the terms, employing the words "grape sugar" or "uncrystallizable sugar" either to pure grape sugar, to pure fruit sugar, or to a mixture of the two. The mixture of grape and fruit sugars arising from the action of acids, ferment, or water upon cane sugar is called "inverted" sugar, "grape" sugar and "uncrystallizable" sugar; being thus named differently by different persons. "Inverted sugar" is the proper name, which is derived from the change of action upon polarized light from right to left.

The practical results of our present chemical knowledge of the sugars may be briefly stated, as follows: Grape sugar is practically uncrystallizable in the manufacture of cane sugar, as it remains in the molasses; it is also much less sweet than cane sugar. Fruit sugar is as sweet as cane sugar, but does not crystallize. Cane sugar may be transformed into inverted sugar (which is a mixture of grape and fruit sugars) by means of acids, long boiling with water, and fermentation, etc.; but neither of these last sugars can be changed again into cane sugar by any process known in chemistry. For practical purposes the difference of composition of the three sugars, as shown by their organic analyses, need not be discussed here. It is, however, important to note that they form compounds with salts, and that these combinations with the salts naturally in the vegetable juices associated with the sugars do not crystallize. In the compound of cane sugar with lime the cane sugar is not destroyed or "inverted" by boiling, but grape or fruit sugar in combination with lime are rapidly destroyed by boiling.

VII. Cultivation of Sugar-Cane.

It is not likely that the cultivation of the true sugar-cane can ever become a great industry in the United States. The small yield, even in the best sections of Louisiana, which gives but from 1,200 to 1,800 pounds of sugar to the acre, as against 3,000 to 5,000 pounds in the Mauritius, and occasionally even up to 7,000 pounds per acre in Cuba, is against it. That would suffice, to say nothing of the malaria of the sugar plantations; the larger cost of cultivation; higher price of labor, compared with tropical climates, where peon or slave labor is used.

Of late years, attempts have been made to introduce the plan of delivering the cane direct to central factories for working. What this may accomplish in time remains to be seen. The system has worked well in the French West Indies, and large profits have been made.

VIII. Cultivation of Sorghum.

The cultivation of sorghum from the first preparation of the soil until the cane is ripe, is identical with that of Indian corn, with these exceptions: The soil should be reduced to a finer tilth than is generally made for corn; the cane being delicate in growth, more care must be used in cultivation when it first comes up; and the crop will be undoubtedly better for one thorough hoeing of the plants.

If planted in check-rows, three and a half feet by three would be about right, four or five plants to remain in the hill. If drilled, which is much the best plan, the plants may stand eight or ten inches apart. When the seed is to be planted by hand, the seed should always be soaked until the germ is ready to appear. A pocket-full of kiln-dried corn-meal, in which the hand may be dipped occasionally, will help to prevent the seed from clinging.

If the land is inclined to be wet after rains, the soil should be listed up for planting on; if well drained use level planting. The seed should not be covered more than an inch, and in the case of sprouted seed, half an inch is better. Never plant on trashy land. It should be as clean as a garden. When this is the case, you may drill the seed, and plenty of it, always pressing the soil pretty firmly to the seed. When the land is thoroughly dried, you may put a sharp-toothed harrow on the land, crossing the drills at right angles to thin the plants. Lumpy soil should never be planted with sorghum. Lumps, as a rule, result from plowing the land when too wet.

The cultivator must decide for himself distance to plant, whether in hills or drills. For sorghum, good barn-yard manure, some phosphate (never nitrogenous manure) and gypsum should be used.

IX. When to Cut Sorghum Cane.

It is fully established now that the cane must not be cut until the seed is about ripe, or fully developed and hard. According to a late report to the Commission of Agriculture, taken from the results of 2,739 analyses of sorghum, the percentage of juice extracted from the stripped stalks gradually increases up to a certain point of ripeness, and then gradually decreases to the close of the season.

The Process of Ripening.—The specific gravity of the juice, the percentage of sucrose (true sugar), the percentage of solids not sugar, and the exponent regularly increase, with one or two exceptions, until the close of the season; the percentage of glucose (syrup product) in the juice as steadily decreases from the first.

Hence, the cane should be allowed, as heretofore stated, to ripen. The want of knowledge on this point has done more to prevent the cultivation of sorghum than all other things combined. Farther on will be given, with other matter, a table fully explaining this important point.

X. Cutting and Handling the Cane.

The cane should be cut near the root with a suitable knife, laid in piles, separated into convenient handfuls, the cuts leveled and presented on a table or suitable form, for cutting off the heads and that portion of the stalk not useful for crushing. The leaves can then be stripped before grinding, or not, as preferred. We favor stripping, but it is a question of the cost of labor. The cane should be kept from wet, and worked as soon as possible after cutting. Until worked it must not be piled so close as to heat.

XI. Specific Gravity as a Basis of Value.

The sugar-maker must take nothing for granted, unless the results, as shown by careful and accurate experiment, warrant it. The individual cannot make these for himself. The Department of Agriculture has done much useful work in this direction.

The table on page 253 is of practical value to those engaged in sugar-making from sorghum. By reference to it the sugar-boiler can determine the composition of any juice of which he knows the specific gravity. These figures are averages drawn from all the analyses recorded, and although the different canes differ somewhat in the composition of the juice for the same specific gravity, still these differences are not so great as to be of much practical importance.

In examining these tables it should be remembered that the results are valuable in proportion to the number of analyses from which each figure has been derived. If only those figures are examined which are based on ten or more analyses, it will be seen that the recorded results are very seldom exceptional.

Among other points shown by this table, the following are important:

1. The amount of juice obtained seldom falls below sixty per cent. of the weight of the stripped stalks; this percentage does not vary greatly throughout the season.

2. The amount of crystallizable sugar (sucrose) in the juice is at first little over one per cent, but it regularly increases with the increase of specific gravity. No one relationship is more evident than this close correspondence between the increase of specific gravity and percentage of sucrose in the juice; the average increase of sucrose for an increase of .001 in specific gravity (between 1.030 and 1.086) is 0.233 per cent. The following table shows the average increase of cane sugar corresponding with an increase of .001 in specific gravity of the juice:

    Between 1.030—1.039 = .164 per cent sucrose.
    Between 1.040—1.049 = .167 per cent sucrose.
    Between 1.050—1.059 = .229 per cent sucrose.
    Between 1.060—1.069 = .250 per cent sucrose.
    Between 1.070—1.079 = .142 per cent sucrose.
    Between 1.080—1.086 = .164 per cent sucrose.

3. It is a noticeable fact that the "solids not sugar" increase regularly and with almost the same rapidity that the glucose diminishes. One point, however, seems to be strongly suggested, namely, that the decrease in glucose bears a much closer relationship to the increase of organic solids not sugar than to the increase of crystallizable sugar. In other words, it seems at least possible that the commonly accepted idea that cane sugar is formed in plants only through the intervention of glucose may be a mistaken idea. This point is a very interesting one and worthy of careful study in the future.

4. The percentage of total solids regularly increases, with a few exceptions, with the increase of specific gravity; the average increase for each gain of .001 in specific gravity is 0.17 per cent of solids.

5. Experience has shown that the percentage of crystallizable sugar in the total solids of the juice should exceed 70 in order that good results may be had.

XII. Specific Gravity and Composition of Juices.

An inspection of the table indicates that these juices attained that percentage (see column headed "Exponent") when the specific gravity 1.066 was reached, and this exponent was maintained, and even exceeded, until the specific gravity 1.086 was passed. After this the exponents are somewhat variable, because specific gravities above 1.086 were not attained until quite late in the season, when the plants had nearly or quite ceased growing; also, the number of experiments for these higher specific gravities was smaller than for the lower figures. It is safe to say that the profitable working period for sorghum canes begins when the juice attains the specific gravity 1.066, and continues until the specific gravity 1.086 is reached, and frequently even longer. During this period the canes here examined furnished on an average 61.9 per cent of juice from the stripped stalks. A good mill should furnish not less than sixty per cent on the large scale. Several manufacturers are willing to contract for mills to furnish sixty-five per cent.

On the supposition that a good mill, yielding at least 60 per cent of juice from the stripped stalks, is used, the amount of sugar which should be obtained from 100 pounds of stalks is found by referring to the figures in the last column corresponding with the specific gravity of the juice obtained. For example, each one hundred pounds of stripped stalks, the juice of which has the specific gravity of 1.073, should actually furnish 5.51 pounds of cane sugar. Even better results than these have actually been obtained in several instances. In the same manner the yield of sugar can be calculated from the weight of the juice by reference to the figures under the heading of "Available percentage of sucrose in juice."

The study of the table will be interesting to all readers, and of great value to those who raise the canes, and especially so to the sugar maker. There are so many integers of value in the conversion of sacharine juices into sugar, and so many contingencies to be met in all sacharine juices containing glucose combined or in connection with true sugar, heretofore very imperfectly understood, and not yet perfectly known, even by the best chemists, that great difficulty has been experienced in particular cases in working even the true sugar-cane of the South.

We are year by year coming to understand that money cannot be made in the working of sorghum in the hitherto crude methods of the farm. The abandonment of those methods and the better system now adopted have given the most gratifying results. Hence the value of the tabular information here given—dry reading except to those especially interested, but presenting at a glance just the information that is useful to those studying sugar making with a view to becoming experts.

Take the item of density of juice, and its relation to the sugar product. Once the quality of the cane is known from actual working, and the average density of the juice, a fair estimate of the outcome can be made from other canes that have had similar conditions as to soil, culture, age, etc.

So of all the tables given in this volume; they are intended to show in the most condensed forms the actual practical deductions on the absolute practical facts, obtained from many experiments in a constant direction.

CROPS FOR SUGAR-MAKING. 253

XIII. Table of Juices.

Specific gravity.	Per cent of juice.	Per cent of glucose.	Per cent of sucrose.	Per cent of solids not sugar.	Total solids in juice.	Exponent.	Available per cent sucrose in juice.	Available per cent sucrose in stripped stalks at 60 per cent juice.	Number of analyses.
1.019	61.32	.67	2.20	3.12	5.99	36.73	.81	.48	1
1.021	58.30	3.91	.54	.68	5.13	10.53	.06	.04	2
1.022	69.04	3.06	1.46	1.11	5.63	25.93	.38	.23	1
1.023	47.36	3.27	1.15	1.29	5.71	20.14	.23	.14	3
1.024	60.49	3.85	1.02	1.73	6.60	15.45	.16	.10	1
1.026	62.78	4.04	.98	.91	5.93	16.53	.16	.10	1
1.027	57.08	3.41	2.09	1.61	7.11	29.40	.61	.37	3
1.028	46.61	3.98	1.79	2.34	8.11	22.07	.40	.24	8
1.029	57.72	4.34	1.55	1.53	7.42	20.89	.33	.20	6
1.030	45.44	3.98	2.36	1.82	8.16	28.92	.58	.35	11
1.031	56.01	3.83	2.16	2.05	8.06	33.00	.88	.53	12
1.032	60.97	3.95	2.66	1.95	8.16	26.47	.57	.34	17
1.033	60.13	4.52	2.50	1.93	8.56	25.40	.60	.36	28
1.034	66.96	4.24	2.26	1.98	8.67	28.84	.72	.43	13
1.035	60.22	4.11	3.29	1.59	9.38	35.08	1.15	.69	23
1.036	64.28	4.56	3.12	1.75	9.27	33.66	1.05	.63	23
1.037	60.12	4.42	3.56	1.88	9.73	36.59	1.30	.78	25
1.038	61.37	4.43	3.43	1.85	9.74	35.22	1.21	.73	21
1.039	61.30	4.14	4.00	1.77	9.99	40.00	1.60	.96	25
1.040	62.78	3.94	4.41	1.92	10.17	43.36	1.91	1.15	18
1.041	62.41	4.21	4.30	1.91	10.43	41.23	1.77	1.06	26
1.042	59.40	4.13	4.69	1.92	10.73	43.71	2.05	1.23	23
1.043	64.72	4.26	4.95	1.92	11.13	44.48	2.20	1.32	22
1.044	63.98	3.79	5.23	2.17	11.13	46.74	2.42	1.45	17
1.045	64.54	3.87	5.51	2.19	11.47	48.04	2.65	1.59	24
1.046	64.34	3.76	5.72	2.10	11.58	49.34	2.82	1.69	30
1.047	65.03	3.99	6.28	2.15	11.86	52.95	3.33	2.00	31
1.048	65.18	3.43	6.28	2.03	12.10	50.25	3.06	1.84	36
1.049	62.88	3.62	6.34	2.23	12.19	52.01	3.30	1.98	37
1.050	66.17	3.32	6.99	2.29	12.60	55.48	3.88	2.33	48
1.051	62.81	3.12	7.18	2.26	12.56	57.17	4.10	2.46	43
1.052	64.36	3.12	7.64	2.46	13.28	57.61	4.32	2.59	42
1.053	63.95	3.18	7.58	2.31	13.31	56.95	4.32	2.59	43
1.054	63.33	3.42	7.74	2.27	13.13	58.95	4.57	2.74	49
1.055	65.66	3.38	8.12	2.24	13.74	59.09	4.80	2.88	55
1.056	63.66	2.96	8.61	2.40	13.97	61.63	4.92	2.95	52
1.057	62.74	2.99	8.90	2.34	14.23	62.54	5.57	3.34	56
1.058	64.10	2.78	9.18	2.53	14.49	63.35	5.82	3.49	76
1.059	63.93	3.05	9.80	2.44	14.77	62.90	5.84	3.50	53
1.060	63.15	2.65	9.80	2.67	15.12	64.81	6.35	3.81	100
1.061	64.86	2.73	9.88	2.75	15.36	64.32	6.36	3.82	76
1.062	63.35	2.51	10.24	2.77	15.52	65.98	6.76	4.06	73
1.063	64.74	2.65	10.16	2.95	15.76	64.47	6.55	3.93	84
1.064	63.48	2.43	10.64	2.95	16.02	66.42	7.07	4.24	64
1.065	61.08	2.07	11.19	2.83	16.11	69.46	7.77	4.66	81
1.066	63.58	2.08	11.46	2.72	16.26	70.48	8.08	4.85	74
1.067	60.98	1.99	11.80	2.87	16.66	70.83	8.36	5.02	69
1.068	63.25	1.97	11.84	3.00	16.81	70.43	8.34	5.00	56
1.069	61.15	1.85	12.30	2.98	17.16	71.68	8.82	5.29	75
1.070	63.45	1.84	12.59	3.00	17.43	72.03	9.09	5.45	82
1.071	62.37	1.81	12.54	3.26	17.61	71.21	8.93	5.36	89
1.072	61.81	1.68	12.94	3.21	17.83	72.58	9.39	5.63	82
1.073	62.46	1.68	12.83	3.20	17.88	72.56	9.19	5.51	75
1.074	61.44	1.69	13.22	3.37	18.28	71.76	9.56	5.74	75
1.075	61.78	1.71	13.47	3.37	18.55	72.32	9.78	5.87	67
1.076	61.49	1.47	13.66	3.54	18.67	72.62	9.99	5.99	68
1.077	60.41	1.62	13.75	3.58	18.95	73.16	9.98	5.99	45
1.078	61.18	1.50	13.88	4.04	19.42	72.56	9.92	5.95	52
1.079	60.80	1.51	14.01	3.67	19.19	73.01	10.23	6.14	46
1.080	60.58	1.57	14.01	3.74	19.32	72.52	10.16	6.08	41
1.081	60.47	1.43	14.21	4.10	19.77	73.03	10.26	6.16	25
1.082	59.71	1.14	15.06	4.05	20.25	74.37	11.20	6.72	29
1.083	59.27	1.50	14.71	4.23	20.44	71.97	10.59	6.35	17
1.084	60.07	1.48	14.84	4.13	20.45	72.37	10.77	6.46	25
1.085	58.74	1.22	15.65	4.56	21.46	72.92	11.41	6.85	12
1.086	53.68	2.35	13.83	4.38	20.56	67.26	10.96	6.58	14
1.087	59.08	1.38	15.32	4.69	21.40	71.59	10.87	6.52	3
1.088	59.08	1.22	15.65	4.38	20.92	72.37	11.41	6.85	9
1.089	57.72	.80	16.25	6.32	23.37	69.53	11.30	6.78	1
1.090	55.57	1.19	15.87	4.78	21.84	72.66	11.53	6.92	3
1.092	54.55	2.75	14.76	4.70	22.21	66.45	9.81	5.89	1

XIV. Value of Sorghum During Working Period.

The chemist of the Department of Agriculture in Washington has made an exhaustive comparison between all the principal varieties of sorghum, and many varieties of corn, including rice or Egyptian corn, doura, sweet corn, dent and horse-tooth corn. From the table given on page 256 it will be possible to judge accurately as to the comparative values of the different canes for sugar. These values are applicable more especially to the latitude of Washington, and it will be seen later that certain canes which do not stand high in the list, when grown in this section, are very likely to prove valuable where the growing season is longer.

Again, those which mature quickest and also have a long working period are the ones especially recommended for culture in more northern latitudes.

In this table the canes are arranged in the order of their comparative value, as shown from the large number of analyses recorded. It must not be inferred, however, that it is possible to state positively that this order may not be somewhat modified by future experience; it certainly would be somewhat changed were any one characteristic of the juice used as the basis of comparison to the exclusion of all others. It has been attempted to give due weight to all the factors which tend to show the good or bad qualities of the canes.

XV. Four Important Points.

Among the points which have the most direct bearing on the determination as to the value of any cane for any locality are the following:

1. Other things being equal, that cane is best adapted to any locality which most quickly reaches the working stage, and longest continues workable. It will be noticed that, judged by this rule, the first eight varieties in the table on page 256 are superior to those that follow. It appears, also, that these varieties matured in from seventy-seven to eighty-nine days, and continued workable from eighty-seven to one hundred and seven days, or, on an average, over three months. It is very important to have sufficient time in which to work up the crop.

2. The average purity of the juice is another very important consideration. This is shown by the column headed "average exponent;" by this term is meant the percentage of pure crystallizable sugar in the total solids of the juice. As has already been stated in the discussion of the table of specific gravities, the exponent should not fall below seventy for the best results.

3. The average available sugar in the juice has very much to do with its value. The figures in this column were calculated by multiplying the figures in the column showing "average per cent sucrose in juice" by the corresponding figures for "average exponent."

4. The pounds of juice per acre have much to do with the amount of sugar that can be obtained.

XVI. Valuable Canes South.

As will be seen, the various canes do not differ very materially in the percentage of juice they can furnish; hence, the pounds of juice per acre depend more directly upon the number and weight of canes which can be raised. By reference to the tables for each variety, it will be seen that several of the varieties standing low in this list (Honduras, Honey Top, etc.) furnish canes much heavier than those standing near the first of the list; hence, if an equal number of such heavy canes could be grown on an acre, the amount of juice must be correspondingly greater.

If, then, the quality of the juice from heavy canes is as good as that from the light, and the season for working is greater, the heavy canes would be preferable, because they would furnish the larger amount of sugar per acre. Unfortunately, this is not the case in this latitude. The first two columns in the table show that the heavier canes do not attain their full growth and maturity, in time to be worked up into sugar.

It is fully believed that these heavy canes are well adapted to the more southern parts of the United States, and that in those regions they will reach full maturity in time to leave an ample working period. In fact, several examinations of canes sent from South Carolina a year ago confirm these statements.

If it be supposed, for sake of comparison, that an equal number of canes of each variety can be grown on an acre of land, the results given in the last three columns will show what amounts of stripped stalks, juice and available sugar can be obtained on an acre from each variety of corn and sorghum. The number of stalks per acre has been placed at 24,000, which is believed to be a fair estimate.

In comparing these figures with those in the three columns just preceding them, which represent actual results of analyses, it will be seen that the figures do not differ greatly.

XVII. The Real Test of Value.

AFTER all, the real test of value for any cane is the amount of crystalized sugar that can be actually separated from the juice obtained from the stalks grown on an acre. This amount will depend very greatly on the quantity and quality of the canes, and upon the promptness and care with which they are worked up after cutting. The figures here given in explanation of the various points which have been discussed have been derived from very carefully conducted work, and they are offered as fair statements of what can and should be attained by careful workers.

Among the essential points worthy of repetition are the following:

1. Select a cane that matures quickly, and has as long a working period as possible.

Do not work the cane too early; the seed should be well matured and quite hard, and the juice should have a specific gravity of 1.066 or higher.

3. After cutting the canes, work them up without great delay. It is best to draw directly from the field to the mill as may be needed.

XVIII. Table of Comparative Values During the Working Period.

	NAME. VARIETIES OF SORGHUM.	SOURCE OF SEED.	Number of days to maturity.	Number of days for working.	Number of analyses.	Average per cent sucrose in juice.	Average per cent glucose in juice.	Average per cent other solids in juice.	Average exponent.	Average per cent available sugar.	Average per cent juice.	ACTUALLY OBTAINED. Stripped stalks per acre.	ACTUALLY OBTAINED. Juice per acre.	ACTUALLY OBTAINED. Available sugar per acre.	COMPUTED AT 24,000 STALKS PER ACRE. Stripped stalks per acre.	COMPUTED AT 24,000 STALKS PER ACRE. Juice per acre.	COMPUTED AT 24,000 STALKS PER ACRE. Available sugar per acre.
												LBS.	LBS.	LBS.	LBS.	LBS.	LBS.
1	Early Amber,	D. Smith,	77	99	80	12.42	1.55	2.98	73.15	9.11	60.02	27,073	16,249	1,480	25,520	15,317	1,395
2	Early Amber,	Plant Seed Co.,	80	99	70	12.00	1.51	3.18	71.72	8.67	61.33	29,808	18,281	1,585	24,480	15,023	1,302
3	Early Golden,	A. B. Swain,	80	104	76	11.47	1.76	3.09	70.24	8.12	60.03	24,611	14,774	1,200	24,480	14,695	1,352
4	Golden Syrup,	W. H. Lytle,	87	82	67	12.48	1.42	2.99	73.65	9.24	61.36	15,822	9,708	897	24,480	14,023	1,388
5	White Liberian,	D. Smith,	88	101	39	13.43	1.31	3.17	74.98	10.08	63.82	32,165	20,528	2,069	31,920	20,371	2,053
6	Early Amber,	S. E. Evans,	89	96	24	13.21	1.54	3.28	73.23	9.69	59.02	27,962	16,503	1,599	23,760	14,023	1,359
7	Black Top,	D. W. Aiken,	87	87	35	12.69	1.21	3.07	74.75	9.51	61.35	21,907	13,440	1,278	22,800	13,977	1,329
8	African,	W. E. Parks,	87	107	83	11.50	1.46	3.14	74.38	8.13	62.92	21,716	13,664	1,111	27,840	17,517	1,424
9	White Mammoth,	Amos Carpenter,	102	83	32	13.51	1.18	3.45	74.50	9.13	62.31	29,341	18,282	1,851	31,680	19,740	1,990
10	Oomseeana,	Blymyer & Co.,	115	77	54	12.16	1.49	3.07	72.43	8.81	64.15	19,522	12,523	1,103	27,840	17,859	1,573
11	Regular Sorgho,	Blymyer & Co.,	101	93	71	11.80	1.49	3.03	72.27	8.70	60.77	26,611	16,172	1,407	30,720	18,669	1,624
12	Hybrid,	E. Link,	101	84	33	14.24	.93	3.43	76.08	10.84	63.53	34,477	21,903	2,374	42,240	26,835	2,909
13	Sugar Cane,	J. W. Barger,	108	77	28	13.82	1.49	3.13	74.18	10.27	62.32	21,117	13,150	1,350	21,600	13,461	1,382
14	Oomseeana,	D. W. Aiken,	104	88	35	12.84	1.12	3.31	74.21	9.57	62.04	22,825	14,160	1,355	28,080	17,420	1,667
15	Neazana,	W. H. Lytle,	136	58	43	13.16	1.93	3.18	72.13	9.48	61.58	23,467	14,451	1,360	26,400	16,257	1,441
16	Goose Neck,	P. P. Ramsey,	111	72	44	12.26	1.46	2.99	72.58	7.58	62.12	27,362	16,997	1,288	30,480	18,934	1,435
17	Early Orange,	——— Hedges,	117	79	53	13.18	1.58	3.39	72.45	9.56	61.67	48,758	30,069	2,875	35,520	21,903	2,094
18	Neazana,	Blymyer & Co.,	129	65	46	13.45	1.95	3.11	72.77	9.78	60.52	20,156	12,198	1,193	25,200	15,241	1,491
19	New Variety,	E. Link,	108	84	31	12.84	1.19	3.35	73.93	9.50	65.22	30,731	20,042	1,904	28,320	18,470	1,755
20	Chinese,	D. Smith,	137	57	31	13.18	1.81	3.58	70.66	9.22	60.43	30,956	18,707	1,725	32,720	19,773	1,823
21	Wolf Tail,	E. Link,	118	56	21	11.72	1.23	2.98	71.87	8.65	62.09	31,493	19,554	1,691	30,960	19,223	1,663
22	Gray Top,	H. C. Sealey,	135	59	33	13.03	1.47	3.54	72.19	9.42	63.00	29,587	18,809	1,772	28,800	18,144	1,709
23	Liberian,	Blymyer & Co.,	131	38	25	13.18	2.05	3.22	71.23	9.39	62.56	45,580	28,269	2,654	45,120	27,983	2,628
24	Liberian,	W. H. Lytle,	134	48	36	12.92	2.09	3.37	70.31	9.08	62.56	44,913	28,088	2,550	44,400	27,777	2,522
25	Oomseeana,	W. I. Mayes & Co.	127	67	36	13.62	1.74	3.40	72.50	9.88	61.89	35,414	21,918	2,165	42,480	26,291	2,588
26	Sumac,	W. Pope,	*152	31	14	14.24	1.67	4.18	70.82	10.09	60.15	39,919	24,011	2,423	39,360	23,675	2,389
27	Mastodon,	D. W. Aiken,	128	60	23	11.24	1.68	3.03	69.93	7.95	64.27	20,413	13,119	1,043	47,760	30,595	2,440
28	Imphee,	D. W. Aiken,	155	87	9	14.21	1.76	3.61	72.56	10.31	61.67	37,031	22,837	2,354	37,920	23,385	2,411
29	New Variety,	J. W. H. Salle,	172	7	5	13.99	2.02	3.73	70.88	9.92	58.57	26,090	15,287	1,516	25,920	15,181	1,506
30	Sumac,	J. H. Wighton,	168	20	6	14.40	1.80	3.40	73.53	10.58	60.84	39,815	24,223	2,563	38,960	22,486	2,388
31	*Honduras,	Arsenal,	148	29	14	10.32	2.26	3.09	63.76	6.81	57.09	25,335	14,464	985	29,760	16,990	1,157
32	†Honey Cane,	J. H. Clark,	133	43	21	10.80	2.56	2.51	67.76	7.37	65.08	30,301	30,301	2,233	53,760	34,987	2,579
33	*Sprangle Top,	W. Pope,	153	38	20	11.21	2.61	2.94	66.79	7.51	65.91	46,634	30,736	2,308	44,880	29,580	2,221
34	Honduras,	E. Link,	157	10	4	12.83	1.80	2.95	72.98	10.06	65.06	45,695	29,729	2,991	50,740	33,011	3,321
35	*Honeytop, or Texas cane	Brussels, Mo.,	163	20	7	12.98	2.11	3.92	66.27	8.86	64.68	47,246	30,559	2,708	51,220	33,129	2,939
36	*Honduras,	L. Brande,	164	22	6	11.67	2.03	3.22	69.00	8.06	66.59	46,421	27,912	2,250	51,840	34,510	2,782
37	*Sugar Cane,	C. E. Miller,															
38	Hybrid,	J. C. Moore,	99	8	6	8.84	2.37	2.32	65.39	5.79	64.60	13,839	8,940	518	17,280	11,163	649

* The juices of these five canes did not reach the exponent 70. (See remarks later.)
† The juice of this cane in some cases reached an exponent above 70, but did not average it. (See remarks later.)

A knowledge of the varieties of the sorghum cane is so important that illustrations of some of those most valuable described in the preceding table are given, and are well worth careful study; they are all accurate representations.

HEAD OF LIBERIAN CANE.

HEAD OF NEEAZANA CANE.

HEAD OF WOLF-TAIL CANE.   HEAD OF BLACK-TOP CANE.

HEAD OF RICE OR EGYPTIAN CORN. [259] HEAD OF HYBRID CANE.

HEAD OF WHITE MAMMOTH CANE.

HEAD OF EARLY AMBER CANE. [261] HEAD OF GRAY-TOP CANE.

HEAD OF OOMSEEANA CANE.   HEAD OF GOOSE-NECK CANE.

HEAD OF HONDURAS CANE.

XIX. The Manufacture of Sorghum.

Sorghum of a superior grade will never be manufactured on the farm, or upon a small scale. The true cane sugar cannot be so made at all. The machinery and fixtures not only cost too much money, but require an accurate knowledge, only attained by long study and practice. It will cost from $20,000 to $50,000 for the buildings and machinery alone that are necessary for making the best sugar. A single vacuum pan (and there should be two) costs from $3,000 to $5,000.

How it is Done.—The juice is crushed from the cane, roughly filtered, heated in a tank to a hundred and twenty degrees Fahrenheit, treated with milk of lime thoroughly stirred in, until litmus paper dipped in shows a purple or bluish-purple color; the quantity of milk of lime, of a given strength, being determined when you know the acidity of the juice, etc. The heat is then quickly raised until the boiling point is nearly reached, or not much over two hundred degrees Fahr.

After Boiling.—The boiled juice is then skimmed, and the clear liquor syphoned off, treated with carbonic acid gas, to take out the excess of lime, if any; passed through a slam press—that is, forced through thick layers of linen cloths—to remove the fecal matter; condensed over a steam coil to what is known as thin juice; filtered through bone-black, and then reduced to thick syrup—forming mush sugar—in a vacuum pan, that, under the air pump, will work at one hundred and forty or one hundred and fifty degrees Fahr.

Granulation.—When granulation takes place the mush is broken, and, if too hard, moistened with a little syrup, ground through the sugar mill, and the pure sugar thrown out by means of centrifugal action. This is, essentially, the process of making crystal sugar, either from cane, beets or any other sugar-producing plant.

XX. Making Sugar on the Farm.

There are a number of patent methods nowadays, all more or less valuable in large works. Fair brown sugar, and especially syrup, may be made on the farm, on any of the better class of evaporators. After the cane is gathered, stripped and crushed as heretofore directed, proceed as below, which is the course recommended by Dr. Jackson, of Boston, Mass:

Filtering and Liming.—In the first place, it is necessary to filter the juice, as it comes from the mill, in order to remove the cellulose and fibrous matters and the starch, all of which are present in it when expressed. A bag filter, or one made of a blanket placed in a basket, will answer this purpose. Next, we have to add a sufficiency of milk of lime (that is, lime slaked and mixed with water) to the juice, to render it slightly alkaline, as shown by its changing turmeric paper to a brown color, or reddened litmus paper to a blue. A small excess of lime is not injurious.

Boiling.—After this, the juice should be boiled, say for fifteen minutes. A thick, greenish scum rapidly collects on the surface, which is to be removed by a skimmer, and then the liquid should again be filtered. It will be of a pale straw color, and ready for evaporation. It may now be boiled down quite rapidly to about half its original bulk, after which the fire must be kept low, the evaporation to be carried on with great caution, and the syrup constantly stirred to prevent it from burning at the bottom of the kettle or evaporating pan. Portions of the syrup are to be taken out, from time to time, and allowed to cool, to see if it is dense enough to crystalize. It should be about as dense as sugar-house molasses, or tar.

Crystallizing and Draining.—When it has reached this condition, it may be withdrawn from the evaporating vessel, and be placed in tubs or casks to granulate. Crystals of sugar will begin to form generally in three or four days, and sometimes nearly the whole mass will granulate, leaving but little molasses to be drained. After it has solidified, it may be scooped out into conical bags, made of coarse, open cloth, or of canvas, which are to be hung over the receivers of molasses; and the drainage being much aided by warmth, it will be useful to keep the temperature of the room at 80° or 90° Fahr. After some days the sugar may be removed from the bags, and will be found to be a good brown sugar.

Refining.—It may now be refined by dissolving it in hot water, adding to the solution some whites of eggs (say one egg for 100 pounds of sugar), mixed with cold water, after which the temperature is to be raised to boiling, and the syrup should be allowed to remain at that heat for an hour. Then skim and filter, to remove the coagulated albumen, and the impurities it has extracted from the sugar.

Decoloring.—By means of bone-black, such as is prepared for sugar refiners, the sugar may be decolored by adding an ounce to each gallon of the saccharine solution, and boiling the whole together. Then filter, and you will obtain a nearly colorless syrup. Evaporate this, as before directed, briskly, to half its bulk, and then slowly until dense enough to crystallize, leaving the syrup, as before, in tubs or pans to granulate.

Whitening.—This sugar will be of a very light-brown color, and may now be clayed, or whitened, by the usual method—that is, by putting it into cones and pouring a saturated solution of white sugar upon it, so as to displace the molasses, which will drop from the apex of the inverted cone. The sugar is now refined as loaf sugar.

The methods here described are the common and cheap ones, such as any farmer can employ. It may be advantageous, when operations of considerable extent are contemplated, to arrange a regular system of shallow evaporating pans for the concentration of the syrup, similar to those now used in Vermont for making maple sugar.

Vacuum Pans.—It is evident that no ordinary methods can compete with those of a regular sugar refinery, where vacuum pans are employed, and evaporation is consequently carried on at a very low temperature. If the planter should raise sufficiently large crops to warrant the expense of such an apparatus on his farm, he would not fail to manufacture larger quantities of sugar, and to operate with perfect success in sugar making; but this can be done only in the Southern, Middle, or Western States, where extensive farming is common. Those who wish to have their brown sugar clarified can send it to some of the large refineries, where the operations may be completed and the sugar put up in the usual form of white loaves.

Syrup.—A very large proportion of our agricultural people will doubtless be satisfied with the production of a good syrup from this plant. They may obtain it by following the methods described in the first part of this paper, or they may omit the lime and make an agreeable but slightly acidulous syrup, that will be of a lighter color than that which has been limed.

This syrup is not liable to crystallize, owing to the presence of acid matter. The unripe canes can be employed for making molasses and alcohol, but, as before stated, will not yield true cane sugar.

XXI. General Conclusions.

LET no person suppose that syrup, much less sugar, can be made without serious study of the art. Nothing must be left to chance. The eye of the master must be untiring. But, with a little care, both sugar and syrup of fair quality may be made at comparatively little expense.

Where it is practicable to raise the cane and cart it to sugar works, which are not more than five miles distant, to be worked on shares, that will, undoubtedly, be the better way.

When works are erected and the cane can be sold for cash, after being divested of the seed, which is as valuable for feeding, pound for pound, as corn, this will be profitable.

We have not written to induce any person to undertake the manufacture of sugar in a small way, for we do not believe it can be made profitable.

Still, we repeat that when capital and skilled labor undertake the manufacture of sorghum in a business way, the rich sandy, and sandy-loam region north and south, east, and especially west of the Mississippi, will supply sufficient sugar, with the cane sugar south, to enable the United States, instead of being obliged to import nearly all the sugar consumed within her territory, to become an exporter of sugar to other countries.

XXII. Maple Sugar.

IN making maple sugar, the syrup is concentrated by any of the modern condensing pans in use for sorghum. The sap is obtained by tapping maple-trees, the sugar-maple producing sugar, and the red (soft) maple and ash-leaved maple yielding syrup liberally.

Boiling the Sap.—The quantity of sap that can be boiled in a given time depends on many circumstances. Sap will boil much faster on a clear day than on a cloudy or stormy one, and weak sap will boil away faster than that which is stronger.

Sap Buckets.—The buckets used to catch the sap are made both of wood and tin, the wooden ones being generally used. These are made of pine lumber, hooped with iron, and painted with oil paint on both sides; at the top of the bucket, on the outside, is an ear made of sheet iron, through which is a hole large enough for the spike to pass on which it is hung. The spouts used for carrying the sap from the tree to the bucket are principally made of wood, although metallic ones are better.

The spikes for hanging the bucket on the tree are made of wrought iron, and are about two inches in length, with the head on one side of the nail to prevent the bucket from slipping off. All these may be bought ready-made, or rough substitutes, including troughs for holding the sap, may be made on the farm, and the sap boiled in a common potash kettle.

XXIII. Tapping the Trees.

A COMMON half-inch bit is used for tapping the tree, though many use one seven-sixteenths of an inch for that purpose, and a one-half inch bit for boring the second time. In all sugar lots where the surface of the land will admit of a team being used, the sap is drawn from the different parts of the lot to the sugar-house, on sleds.

Gathering Tubs.—For this purpose, a gathering tub, holding three or four barrels, is used. This tub is made with a head in both ends, the diameter of the bottom being much larger than the top, to prevent it from tipping when filled. In the top of the tub a hole is cut large enough to turn in the sap; a lid is made to fit this hole, so that when the tub is full it can be closed tight, to prevent the sap from being wasted in going to the house. The tub is fastened on the sled with stakes or chains.

Storing Tubs.—The tubs in the house, for storing, are usually about the size of the gathering tubs; they have but one head, and the tops of these are the largest. Both the gathering and storing tubs are made of spruce or pine planks, hooped with iron, and usually painted on the outside. The storing tubs should be painted on the inside like the buckets, to prevent them from becoming sour and discolored with mildew. Whenever storing tubs or buckets become sour, they should be immediately washed clean, before putting more sap in them.

XXIV. Boiling and Sugaring.

THE reduction of the sap is carried on precisely as recommended for sorghum. No clarifying is needed, except filtering, since, the weather being near the freezing point when maple-sugar is made, it does not change readily. If it should do so, a little milk of lime may be used to neutralize the acid; but this portion should be kept by itself.

The sooner the sap is converted into sugar after it leaves the tree, the better; and especially is this the case when the weather grows warm; for the sap is liable to sour in the buckets, and also in the store tubs. When the weather is quite warm—as it sometimes is, for a day or two—sap will sour in twenty-four hours. At such times, the boiling should be forced to the utmost extent, night and day, if necessary. At no time should much sap be allowed to accumulate on hand, if it can possibly be avoided. After the sap has been gathered, if there is dirt in it without ice, it may be strained as it runs into the pans. After the boiling has commenced, it should be kept up without cessation until it is reduced to syrup, or as thick as it can be strained through a flannel or cloth when taken from the fire.

XXV. Sugaring off.

SUGARING is best done in a deep kettle when the syrup has settled after the first boiling. Pour off that part which is clear into the pan or kettle to be used in boiling it, leaving the sediment in the tub. By turning some hot sap into this, it can be settled again, and either boiled down by itself or with the next lot of syrup. After the syrup is placed on the fire, it should be kept boiling with a steady fire until it is done.

Running Over.—Sometimes, while boiling, it is inclined to run over. To prevent this, put a piece of butter the size of a marble into it, and sometimes it will be necessary to put in a second or third piece before it will settle. A very good way is to take a stick long enough to reach across the vessel; lay this stick across the top of it, and from the stick suspend a piece of fat pork; when the syrup rises against the pork, it has the same effect as the butter. If neither of these methods will prevent the syrup from running over, the heat of the fire must be reduced until it boils steadily.

How Long to Boil.—The degree of hardness to which the sugar needs to be boiled depends on the subsequent treatment. If it is to be put into tubs and drained, it should be boiled only enough to have it granulate readily; if it is to be put into cakes, it should be boiled so much that it will not drain at all; it is necessary to boil it as long as it can and not burn.

XXVI. To Tell When Sugar is Done.

THERE are various ways of telling when the sugar is boiled enough. A convenient and good way is, when snow can be obtained, to have a dish of snow, and when some of the hot sugar is put on the snow, if it does not run into the snow, but cools in the form of wax on the surface of the snow, it is done enough to put into tubs to drain. But when it is to be caked or stirred, it should be boiled until, when it is cooled on the snow, it will break like ice or glass.

When snow cannot be obtained, stir some of the sugar in a dish, and as soon as it will granulate, it is done enough to drain; when it will form bubbles, feathers or ribbons, on being blown, it is done enough to cake or stir. To try it in this way, take a small wire or stick and form one end into a loop; dip this loop into the sugar and blow through it to produce the forms described. When the sugar is done, it should be taken from the fire immediately, and cooled. It is then ready to be put up in any way that may be wanted.

In case the sap is taken from any of the trees named, except the sugar-maple, it is boiled at once into syrup, that when cold will be of good consistency.

Part III.

ARRANGEMENT OF FARMS.

HOW TO SECURE COMFORT AND PROFIT
IN THE HOMESTEAD.

FENCING AND DRAINAGE ART.

FARM IMPROVEMENTS ILLUSTRATED AND EXPLAINED.

ARRANGEMENT OF FARMS.

CHAPTER I.

COMFORT AND PROFIT IN THE HOMESTEAD.

I. PIONEER FARMING.—II. IMPROVING THE FARM.—III. TRUE SUCCESS IN FARMING.—IV. LOOK TO THE DETAILS.—V. THRIFT AND UNTHRIFT ILLUSTRATED.—VI. HOW TO SELECT A CLAIM OF LAND.—VII. COMMENCING THE FARM.—VIII. THE CROPS TO RAISE.—IX. THE SECOND YEAR'S CROPS.—X. THE THIRD YEAR'S WORK.—XI. WIND-BREAKS AND GROVES.—XII. STARTING THE ORCHARD.—XIII. HOW TO CLEAR A TIMBERED FARM.—XIV MAKING A CLEARING AND BUILDING THE HOUSE.—XV. CARRYING UP THE SIDES.—XVI. PUTTING ON THE ROOF.—XVII. BUILDING THE FIREPLACE.—XVIII. CHINKING THE HOUSE.—XIX. DEADENING TIMBER.—XX. THE WORK OF IMPROVEMENT.

I. Pioneer Farming.

BEFORE the great prairie region of the West was available to settlement, opening a farm was a far more serious business than since that time. Forty or fifty years ago, when the pioneers began to settle in the prairie region, the obstacles to be encountered were also far greater than in this day of cheap transportation, when railway lines pierce a wilderness of grass in advance, almost, of civilization, and are ready to carry off the products of the settlers.

MODERN PRAIRIE BREAKER.

The pioneers of to-day know little of the discomforts and privations of those who opened the timbered farms of the country North and South, or of the difficulties encountered in the settlement of the prairie region of Indiana, Illinois, Wisconsin,

Iowa, Eastern Kansas, and the earlier settled portions of Texas. In that day implements were clumsy and means of transportation primitive. Only think of the dreary teaming, sometimes for over a hundred miles, to obtain a few necessaries, clothing and medicines; camping at night on the road, and returning worn out with a journey of two or three weeks!

Nowadays the pioneer farmer generally markets his crops and receives his supplies by rail, and he can do five times the work with the improved implements that he could forty years ago. In place of the clumsy plow, drawn by a long string of oxen, and doing poor work, the prairie is turned by a plow, drawn by two horses, the sharp share and rolling coulter slicing even furrows of the tough sod as thin as may be desired, so that in the autumn the whole is easily re-plowed for that universal crop of the pioneer—wheat.

II. Improving the Farm.

THE second year finds him with a mellowing soil, upon which any of the cereal grains may be sown, and upon which corn may be raised at the rate of sixty acres to the hand and team; so he goes on breaking more land year by year. Other settlers gather around him, cave or sod houses are exchanged for more comfortable dwellings, small though they may be.

Orchards grow green, young groves are planted, ornamental trees, shrubs and vines find a place in the door-yard, or cluster about the porches and windows of the dwelling.

Meadows are laid down, improved stock crop the sweet grass, under-drains laid along the valleys form cool rills upon the once arid prairie, or the wind-mill pumps the water from the greater depths.

A settlement becomes a hamlet, then a village—a city—and the busy hum of machinery tells how, in a short ten years, perhaps, this wonderful transformation has been accomplished. How? By the indomitable will and industry of a people, the division of labor, and the intelligent application of machinery to the various industries of life.

The Pioneer's Cottage.—Few persons know how easily and cheaply simple structures are built. A house of the kind represented in the illustration need not cost over fifty dollars, and the ground plan will show how a lean-to roof may cover bed-room, pantry and wood-shed, or two bed-rooms and pantry. This building is 12 x 14 feet, and seven feet high to the eaves. It will take material as follows; we give frontier prices for lumber:

800 feet inch boards, at $30.00 per M.,	$24 00
4 sills, 6 x 6; 4 beams, 2 x 6; 4 rafters, 2 x 6, and 12 2 x 4 joists,	6 00
2 panel doors at $2.50,	5 00
3 windows at $1.50,	4 50
Nails and roofing paper,	10 00
	$49 50

COMFORT AND PROFIT IN THE HOMESTEAD.

If sixteen-feet boards are used, cut in the middle for the vertical siding; the building will be seven feet in the clear. The sills may be mortised for the beams; then nail two boards to the sills, at each corner, perpendicularly, and stayed at the bottom with inch blocks. The beams for the second floor may rest on stancheons cut seven feet long; the rafters sawed and nailed; the sides boarded up; the floors laid; if more than the lower floor be used, the elevation may be twelve feet, for attic bed-room, and the roof made of tarred paper, which is manufactured expressly for this purpose. A veranda may be added at any time, and again, a lean-to, as shown in the ground plan. This shows the original building k-v. B is a bed-room, p, pantry, and w-s, wood-shed, comprising the lean-to. Any person, with a saw, a square, a hammer and a short ladder, can do this work, with one assistant. It would be better to shingle the roof, and this should be done at the first opportunity.

A PIONEER'S COTTAGE.

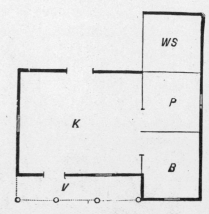

GROUND PLAN WITH LEAN-TO.

III. True Success in Farming.

Success in farming nowadays depends more upon correct methods than upon grinding hard work. Good farmers do not go out in the morning and begin the day's work in a haphazard way. If plowing is to be done, no time is to be lost in scouring the plows, while teams and hands are waiting. The plows have been thoroughly cleaned, rubbed dry and the metal parts slightly, but evenly, oiled. If the plow has been out of use for some time, the metal has been thinly painted with lamp-black and kerosene oil, and put away where this coating would not be rubbed off. Thus, the first furrow turned is as good as the last.

Every tool should thus be kept in condition for service and duplicates of bolts ready to meet any small loss. The farmer should also be able himself to do riveting and minor repairs, and bad weather utilized for grinding or filing the cutting surfaces. Work should be systematized; done at the hours for work, and there should be other hours for rest and amusement. There should be a place for everything, and everything in its place. A time for labor, and a time for play.

274 THE HOME AND FARM MANUAL.

IV. Look to the Details.

It is attention to details that makes the whole system of labor perfect. Water furrows should be drawn at the proper time in the fields; lands laid out correctly for plowing; the furrows straight and equal in depth and width, according to the soil and the requirements of the crop. On a well conducted farm there is no slighting of work at the corners, or in the final plowing of headlands, and the hands are required to use constant care that every hill of a row is properly cultivated.

If a field of grass or grain is to be cut, the first swath will be straight and the second will be perfect.

"SHALL I MOVE THE BARN OR THE MANURE PILE?"

There will be no shirking or weaving by the team; they will have strength for their work, from proper care and feeding, and will have been taught by kind, but decisive training, just what is expected of them. They will be driven straight out at the end to the proper place to stop. They will be brought about so the machine will enter correctly and cut its full width at the first movement of the knives.

The track clearer will be adjusted exactly right on the grass, will not interfere with the working of the machine at the next round, and yet will be evenly spread to the sun. The sheaves of grain will be bound in equal bundles and of proper size, according to the ripeness and stoutness of the grain. The grass will be raked into straight windrows; the hay-cocks even and of uniform size; the shocks of grain in straight lines through the field, firm, and carefully capped. There is profit here; there has been no preventable loss, and all things have been done in the cheapest manner—cheapest, because most economically consistent with good work. So with every labor of the farm.

V. Thrift and Unthrift Illustrated.

SHALL we give the other side of this picture? It may be seen in every neighborhood. There are men whose work is never done in season, nor well done at any time. Their implements are always " lying about loose," but too often the owner may be found " tight " enough at the village grocery. They are of that class who insist that " farming don't pay." Their farms are mortgaged, gradually run down, and are absorbed by their more enterprising neighbors. They " don't believe in book larnin," yet they have faith enough in their calling to think they may succeed in a new country.

THE SUCCESSFUL FARMER'S MODEL BARN.

The out-door indications are generally an index to the inner life. The surroundings of the man who " never has time" will not be unlike the opposite picture. His implements will lie around; his animals will rest where they can. He saves manure carefully—just where it is thrown out from the barn-stable. At last, the accumulations, which have been trodden under foot, increase, until a mountain rises, accessible only by strong-winged fowls. Something must be done. The indolent farmer says: "Yes, Johnny, I calc'late we must stratin out that manure. We can't git the barn-door open any more."

Sensible son.—" Why don't ye move the barn, dad? It'll be a heap easier."

Will the barn be moved? No. Will the manure pile be carted to the fields? No, there is no time. It will be "stratined out," and the mortgage, constantly accumulating, will, at length, straighten out the indolent farmer.

The Careful Farmer's Barn.—Let us look at another picture. There is neither waste nor extravagance here. Careful management and business tact have kept Farmer Skillful steadily on the road to success. First, a small barn was built. At the end of a few years it was shored up, a stone foundation put under it, and it was filled with stock. All manure made was hauled to the fields, and the yard kept perfectly clean. The central figure in the illustration shows the first barn. Additions were gradually made, until about the time Farmer Indolent, in the same neighborhood, was "calc'lating" to "stratin out that manure," Farmer Skillful's barn and yard presented the appearance shown in the companion illustration.

There is nothing extravagant about this; nothing for show, but everything is solid and substantial. It fronts east, the main building is 35x45, the south wing, the first addition made, is 24x45, the north wing 30x50, and both lap on the main building ten feet. The basement walls are eight feet high by two feet thick. There is a central shed under each wing for manure, which is regularly carted out. This gives complete shelter for the store stock.

VI. How to Select a Claim of Land.

WHEN settling in a new country there are many things that require careful thought. The intending settler should know something about soils, texture and composition; drainage, water supply, above and below ground; summer and winter climate, and the general adaptability of the land to present and future crops to be raised. Much of this may be learned from books, but, so much is written that the beginner is befogged. A few salient points, however, may well be borne in mind:

Rough Land.—Do not choose rocky, unevenly broken land, rough steep hills, nor strongly rolling land, unless the principal feature is to stock, and not then unless the price is sufficiently low to offset the expense of bringing it into subjection.

Unhealthful Soils.—Avoid land abounding in wet holes and marshes. These will be unhealthful. They may, indeed be drained, but this the new settler can rarely do.

Valuable Soils.—Seek land gently rolling, if possible, or, if level, see that the soil is naturally dry; that is, that the impervious clay or rock lies far enough below the surface to ensure drainage.

Vegetation as an Index to the Soil.—Observe the character of the vegetation on the surface. The prairie dock, or compass plant, shows a rich, moist soil, adapted to Indian corn or other soft-land crops. Hazel brush, the woody-rooted red root, amaranth, indigo weed, and short grasses indicate a good wheat soil—firm, fertile and dry. Horse weed, wild artichoke, and others of the sunflower tribe, show a rich, deep, warm soil, such as is usually found on arable river bottoms. All the sedges indicate cold, wet land. Thistles are found on rich, dry bottoms. Wild redtop, and the taller and more slender of the wild grasses, indicate good meadow land.

Hence, by carefully observing the natural vegetation of a country—for each soil grows its characteristic plants—a pretty good idea may be formed of the value of the land. For further information on this subject see chapters on grasses and on drainage.

How to Test the Soil.—Provide yourself with a small ground auger. With this you may judge of the nature and value both of the surface and sub-soil. If it is black and sandy, or loamy and friable, a chocolate, or even light brown, it is usually first rate.

The color of the soil, is, however, not always an indication of its value. A black soil indicates a humus soil if very light in gravity, or if heavy in weight it may contain charcoal and humus combined. Many light-colored soils are excellent and lasting, especially for the cereals and grass. If the subsoil is a stiff, tenaceous, pasty clay, reflect before selecting it, unless it lies at a considerable depth below the surface. If hard-pan, reject it.

Selecting the Situation.—Do not select the highest points for permanent residence. They are less subject to malaria, but usually less fertile than the lower levels. He who can get the upland for the home and the low-lying land for crops, is fortunate. If there is an open grove, and living water, and if the vegetation is strong and varied in character, go no farther. The best possible home has been found, for arable lands contain the greatest variety of plants, while particular plants are confined to soils that have some peculiar texture or qualification. We repeat—the chapter on drainage, and that on grasses, will largely assist the intelligent reader in determining many points of value here.

VII. Commencing the Farm.

THE first thing after securing the claim, is to make some kind of shelter. No matter how rough, let it be comfortable, and above all do not shut out the light. Let the situation be as elevated as possible, though it be only temporary. The turning and working of the sod, and the decay of rank vegetation always produce more or less miasma.

Enclosures and a stable, of some kind, for stock, are next in importance. The stables must be proof against prairie storms, and the cold of winter.

As soon as the grass is up three or four inches, set the breaking team to work, and keep on breaking—two inches or less is the right depth—until the first of July, or until the sod becomes too dry. It does no harm to break prairie when wet. It works easier, and this shallow breaking effectually kills all vegetation.

VIII. What Crops to Raise.

A SMALL piece of tender sod may be broken five inches deep for a garden spot, and perhaps to sow wheat, oats, rye, barley and flax sufficient for the next year's seed. A greater area of deep breaking than this is a waste of labor. A piece of tender sod may also be broken, three inches deep, for potatoes, which should be planted in every third furrow, and about eighteen inches in the row. By turning over the sod in the autumn, a nice crop is often thus obtained, and one that requires no labor in cultivation.

Plant as much corn as possible on the shallow breaking, by chopping seams in and through the sod, three feet apart, on every third furrow. Drop three grains in a place, step on it firmly to press the seam to the corn. It will give you feed and fodder, and sometimes pretty sound corn.

As soon as the sod is thoroughly dead, you may begin to "back-set" the sod not planted to corn. This means turning the sod back, and as much deeper as the team can comfortably work, always remembering to keep the plowshare and coulter as sharp as possible, both in breaking and back-setting—the share by filing and the coulter by grinding on a good sharp-gritted stone. The time so spent is well spent.

The ground thus re-plowed, may be sown to winter wheat and rye, and the corn land, re-plowed as soon as the corn is gathered, may be reserved for spring wheat, barley, flax and oats. Raise no more oats than is necessary for feeding, and no barley unless you can save it bright. No. 1 barley is worth about as much as wheat; discolored barley little more than oats. The chief dependence for many will be wheat and flax, since these two grains bear longer transportation than any other crop.

Three Horse Teams.—In breaking prairie, and in fact, in all difficult plowing, use three strong horses abreast if you have them. The third horse is a good investment. Such a team walks right along, even and true, without difficulty, and will do half as much again plowing per day, with the same driver, as two horses will. Plowing day after day, although not exhausting to the driver, is so to the team, and very few pairs of horses can be found that will stand a six weeks' or two months' campaign of breaking without failing. The third horse will make a difference in favor of twenty to thirty acres in the amount of breaking during the season, according to the quality of the team.

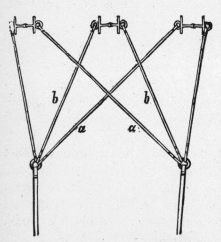

REINS FOR THREE HORSES ABREAST.

Driving Three Abreast.—How to drive three horses abreast is shown in the figure. When the team becomes perfectly manageable, the lines a, a, and b, b, may be dispensed with, and the middle horse simp.y tethered to the inside bitts of the outside horses.

Hitching Three Abreast.—How to hitch three horses abreast, in a simple way, is shown by the next cut. The evener is the same as that used for two horses; the whiffletrees have each a long and a short arm, the long arm twenty-four inches in length and the short one, made of bar iron or steel ½ by ¾ inch. Fasten by a bolt, so it will play; hitch as shown in the cut, being careful that each trace draws evenly, and it will work at plowing or with any other similar draft.

Turning Flat Furrows.—To turn flat furrows, the mold-board of the plow must be of the proper shape. This is shown on page 271. The furrow must have the proper twist in passing over the mold-board. This is shown in the accompanying cuts, and the greatest possible thickness of furrow, in relation to its breadth, to do good work, is also shown. Observe well what has been said in relation to depth of furrow, in prairie-breaking. In turning meadow sward, deeper breaking is advisable. A vertical view is also shown of the plow and furrow, with the common three-horse whiffletrees in ordinary use.

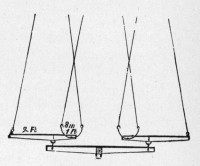
THREE-HORSE DRAFT.

IX. The Second Year's Crop.

The second year, sow all the breaking of the previous year to wheat, principally. If flaxseed can be sold, also sow liberally of this, but the seed soon deteriorates and should be often changed. (See article on flax). Sow the wheat as early in the spring as the soil can be worked. Later, sow what oats you need; then barley, if the price is fair, and later, just before corn-planting time, sow the flax.

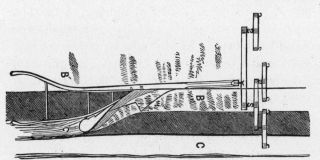
VERTICAL VIEW OF BREAKING PLOW.

Diversity of Crops.—The advantage of a diversity of crops is, if one fails you can have another to depend on. You also prevent your harvest coming on all together, which is an important point. A small area—the best you have—should be reserved for the garden, for potatoes, and for a field of corn sufficient, with the oats, to feed the team and make meal. Upon this land haul all the manure you have made, and plow it well in.

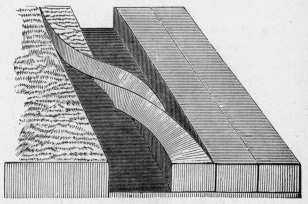
FLAT FURROWS—BREAKING.

From this time on, break all the new land possible, and chop in corn. It will make excellent feed for milch cows, calves, steers, working oxen and hogs. Backset this at the proper time for the next year's work.

X. The Third Year's Work.

THE third year's work will be a repetition of that of the first year, except that the whole of the first year's breaking should be sown to wheat, and with it, timothy, orchard grass and red clover; each variety by itself. The object being, first the seed, because these seeds bring good prices and will bear longer transportation than any other product; they also bring cash.

The First Pastures.—The timothy will, for two years, afford a good seed crop, as well as fall pasture for cattle, which alone must be thus fed, and they not permitted to eat too close. The clover will afford one crop of hay cut early, the second crop being taken for seed. The orchard grass will make valuable late pasture. This should be lightly pastured in the spring, and only when the land is dry. The same rules will apply to timothy—that is, when it is up a good height, it may be pastured down once if you have stock enough to do so quickly. It must, however, be eaten even to give good seed. Clover must not be pastured in the spring, but may be cut for feeding to hogs and calves, or to milch cows, at night, if necessary.

Permanent Pastures.—From this time, the crops may be more and more diversified, until all the land is broken and subdued. As the area of wild feed and hay diminish, meadows of mixed grasses, and a permanent pasture of blue-grass for spring, fall and winter feeding should be laid down. This should be done as soon as possible, because prairie hay is not economical, since all prairie grasses come late, and die with the first frost.

XI. Wind-Breaks and Groves.

IN prairie countries, the question of timber is an important one. However cheap other fuel may be, trees are needed for poles, fence-posts, wagon-racks, levers, foundations for stacks and, more important than all, shelter for cattle and for the fields; this last not the least in importance.

The Timber Plantation.—The timber plantation should be placed where it will be easy of access, and where, at the same time, it will afford shelter for the farm buildings and stock. Planted timber has these advantages: you have the desired varieties just where you want them. Ash will give you timber where strength is required; catalpa is valuable for posts, stakes, etc.; pine, larch and spruce for beams and light poles; chestnut, hickory, butternut and black walnut for their nuts, and all of the latter for their timber. In forming these, their uses must be borne in mind. Ten to twelve years bring the nut trees into bearing. The same length of time forms the most impervious wind-breaks of evergreens, and will give split posts from the catalpa. Willows and cottonwood are valuable at five or six years old, and all yearly increase in value. (See Part VI., Chapter VII.)

How to Start a Grove.—Plant your grove as you would a corn field—in rows four feet apart, but thickly in the rows. These may be gradually thinned to form wind-breaks, until the trees stand four feet by four. As they begin to crowd, take out each alternate row, one way, and then, again, the other. They will now stand eight feet apart. Still another thinning, at two operations, will leave them sixteen feet apart. When finally thinned to thirty-two feet each way you have a noble grove, that has paid its cost many times, and is still worth more than any equal area on the farm. Your wind-breaks have grown into noble timber, beneath which stock may find shade in summer and shelter in winter. The increased crops from your fields have many times repaid their cost, and the farm itself has become of far greater value than the bare acres would be.

XII. Starting the Orchard.

The wise man, beginning a farm far from nurseries, will provide himself, not later than the second spring, with material for an orchard, and will have prepared sufficient land for his permanent garden of small fruits, or at least for the plants.

ORCHARD AND WIND-BREAKS.

What are called maiden trees—trees one year old from the graft—may be ordered, or budded trees of the previous year.

The Trees to Plant.—Root grafts of apples, pears and cherries; budded peaches; cuttings of grape, currant and gooseberry; young roots of raspberry and blackberry; eyes of rhubard, for the kitchen garden; and cuttings of cottonwood, white willow and mulberry; seeds of catalpa and the nut trees mentioned may be obtained. Later come seeds of ash-leaved, white-leaved and sugar maples, and of ash. The apples, pears, peaches, cherries, and the cuttings of the cottonwoods and willows may be planted in well-pulverized soil, in rows four feet apart by twelve inches in the rows; the cuttings, except those mentioned above, in two-feet rows by three inches in the row.

Put nuts in rows eighteen inches apart by six inches in the row, and plant the other tree seeds thicker.

Transplanting.—A year or two years later, remove to the position where they are permanently to stand. The first trees may remain in the rows three years. The other plants and the rhubarb should be taken to the garden the succeeding year.

The Result.—In the end you will see the economy of all this, when you find yourself three or four years in advance of your more tardy neighbors. Do not, however, hide your knowledge. Perhaps some will join you, and thus save expense in buying and transportation. Information as to the proper care of all these will be found in other chapters.

XIII. How to Clear a Timbered Farm.

THE clearing of a timbered farm is a very different affair from opening up a prairie, and yet, aside from the hard labor of chopping and logging, not an unpleasant task. A man may not accomplish results so fast, but some comforts can soon be attained. When the timber is valuable, money may be earned at once by chopping and delivering the logs at the mill, either by hauling direct or to the nearest stream to be rafted.

Saving Valuable Timber.—If not valuable for timber the trees may be cut, logged together in the usual way, and burned, the ashes sold for making potash, or leached and boiled on the farm. If there be no present sale for them, the valuable logs, especially walnut and pine, should be rolled into triangular heaps, well raised from the ground, with skids between each layer, covered with a crotch and pole roof, and this again with bark, to shed the rain. They will thus remain for many years, with a little looking after, until increasing population demands the erection of mills for sawing. Of the oak, hickory, maple and other valuable hard woods, the first may be converted into posts, the second into firewood and the others into rails.

XIV. Making a Clearing and Building the House.

THE first thing to be done is, of course, to build a shelter until a clearing sufficiently large for the house may be had.

Trees for Shade.—When making this clearing always preserve the largest and finest of the forest trees for shade and beauty around the house. This is often neglected, and the result is, when the want is felt, small trees must be planted and a generation pass before these reach a size which could so easily have been attained from the first.

The Other Trees.—These are chopped, logged together and burned, reserving such logs as will run from ten to twelve inches through for building the house, which should be about fourteen by sixteen feet. When necessary, another building of the same size can be added, with a gallery and porch between them.

Laying Foundations.—The logs having been cut and hauled, the four ground logs are laid and leveled; saddles are chopped on the front and rear logs; the two end logs are laid on these and notches are cut to correspond to the saddles of the longer logs. Thus, if straight, and none others should be taken, the logs will come nicely together so as to be easily chinked. For a log house 14 x 16 inside, the longer logs should be eighteen feet long, and the shorter ones sixteen feet, or the width of the logs on each side, and almost six inches more at each end for projection.

The four bottom logs being laid, smaller logs, faced on one side, are placed along the ground. Upon these the floor is to be laid, and they should be so arranged that, when the lower log is cut one-third away and faced to form the door-sill, this floor will come within an inch of that face—all this is, however, sometimes left until the last work in finishing.

XV. Carrying Up the Sides.

The first layer having been finished, continue to roll up the logs, spotting them true at the corners so they will lie closely together, until a height of six and a half feet, or more, is reached above the floor.

Places for the door, windows and fireplace are now to be sawed out, and the ends of the logs held, by temporary slabs, nailed or pinned on, until permanent ones can be placed.

A Building Bee.—It is better still that all these pieces should have been hewed, the door and hinges, and the window frames gotten ready to be hung, for then those who have come, perhaps from a distance, to the "raising bee," may without delay assist in doing much valuable labor. The writer remembers helping at one raising when the settlements were so thin that two of the men came forty miles, swimming their horses over two rivers. The fireplace may well be four, or even five feet wide, and four feet high; fuel will be cheap, and much splitting of wood will thus be saved.

The chimney and fireplace are, of course, to be built outside of the house. If the measurements have been correctly made, all this sawing-out may be avoided, but unless every log is most securely stayed, accidents are apt to happen.

XVI. Putting on the Roof.

When the house has reached a height of not less than seven feet above the ground floor, and eight is better, timbers are laid across and spotted, or notched and saddled, to the upper logs, so that when the floor is laid it will be level. These logs will sag less if they are faced down on two sides, like scantling. A ridge-pole is then raised and stayed, and upon this the rafters are laid.

The roof may be one-quarter pitch, or even less. A good way, however, to prepare the foundation for the roof, is to carry it up with the gables, by placing the roofing logs, as the gables are raised, until the center is reached. Upon these shingles may be directly laid. The best possible form of roof, is to halve logs of the proper length and form them into troughs. These are securely spotted and pinned to the log forming the ridge and to those forming the sides of the house. Other troughs are inverted over these to break joints. No water will ever enter such a roof, if the logs are sound.

The other way of forming a roof is with "shakes"—rough shingles—split out with a froe, as thin as possible, one shingle being lapped over the other, and each successive layer extending some distance down along the other one. When regular rafters are used, the ends of the house are often formed of these shakes or clapboards. The lower floor is usually laid with logs hewed thin and squared at the edges, and the upper floor, with clapboards jointed on the beams. In fact, so comfortable are log-houses—

warm in winter and cool in summer—that some wealthy people are beginning to lavish money on them, after the Norwegian style.

XVII. Building the Fireplace.

This should be of stone, if it can be procured. The chimney may be of sticks thickly plastered with mud. These sticks should be split out of hard wood, be almost two inches square, and are laid up cob-house fashion, in two parallel lines, one within the other. The space between is filled with thoroughly tempered clay, and the out and in sides thickly plastered with the same material.

If carefully built, not carried to within six feet of the hearth, the clay well settled between the lines of sticks, and once or twice plastered, as it cracks, in drying, there is little or no danger of fire. In fact, the whole may be built thus, from the ground up, if a good backing of stone is given to the fireplace. It should, however, be not less than five feet wide by three feet deep, inside measure, and gradually drawn in, to a height of five feet, until the inside measurement of the chimney will be not more than two feet by eighteen inches.

ADZE-EYED MATTOCK.

XVIII. Chinking the House.

The doors, windows and fireplace being complete, and the floors laid, the spaces between the logs should be carefully chinked with pieces split running to an edge. These should be pinned to hold them in place, and the whole thoroughly plastered with well-tempered clay, thin enough to fill all cracks.

Sometimes chopped spagnum or other moss is mixed with the mortar, and it is not a bad plan, since it serves, like hair, to bind the mortar. A ladder is arranged to give passage to the loft, and the house is then ready for the family, and welcome it will be to those who have been camping out for a month or two. A frame is made in one corner of the cabin for the bed, unless a veritable bedstead makes a part of the furniture. The fireplace may be omitted if the emigrant has a stove, though even if so, the fireplace is most healthful, and will be cheerful in winter. In summer the stove may occupy a shed, built of clapboards, and covered with bark. Nails may be used in place of pins, and shingles may be shaved. The log-house may be built of squared timber, halved together at the ends.

A frame house would certainly be cheaper if the material can be bought or traded for. We are describing the manner of life where mills are not, and nails hardly to be thought of; and, so far as real comfort is concerned, in an humble way, there are many buildings of greater show less comfortable than a carefully built log-house.

XIX. Deadening Timber.

The first crop is often raised under deadened timber, where the larger trees have been girdled by cutting out a narrow circle around the trunk down through the sap-wood, before the buds have started. The smaller timber is chopped and burned with the dry leaves and trash; all saplings, say those under four inches through, and all bushes are grubbed with a mattock, to add to the fire. The best form of mattock is shown in the cut. Then the crop is sown or planted without plowing, and harrowed or hoed in, so far as small grain is concerned, and the corn and potatoes dropped where a place offers, and cultivated entirely by hand.

This is tedious and slow, but all the heavily timbered farms of the country were originally opened in this way. The only revenue while the crop was growing was in the potash made from the ashes of the burned timber. There are, however, now but very few localities in the United States where the timber will not pay handsomely for the labor. Oak and hemlock bark is sought far and wide by tanners. The logs are sawn into timber and lumber by portable saw-mills, and the cord-wood finds a ready sale.

XX. The Work of Improvement.

The work of clearing and preparing the timber goes steadily on from year to year. Field is added to field, each being seeded to grass as soon as possible, until the smaller stumps can be drawn out. Up to the time when grass can be produced the stock subsist on what they can find in the summer, on mast in the autumn, and on the tender twigs and buds of the trees chopped in the winter. Hogs, except in a very inclement climate, will manage to live the year round, since the ground seldom freezes deeply in the dense forest, and nuts and roots furnish them with food.

As field after field of grass is added, the calves grow up into cows, and butter and cheese are made. The idea in clearing timber farms being that all the stock possible must be carried; the only care necessary being not to keep too many animals until grass and hay can be made. It takes a great deal of browsing to support a cow, and it is a make-shift at best.

Feeding grain raised in the laborious manner named must not be thought of. Only the necessities of the family should be looked after. When grass is produced add to your live stock by every possible means. It is indeed hard labor to "hew a farm from the forest," yet it has many comforts not to be enjoyed by those who open and improve a prairie. Nevertheless we advise no one to take the timber farm, from choice, if the prairie may be had.

Still, if the prairie farm is not available, do not refuse the timber because you are not a chopper. Two months' practice will enable you to swing the "woodman's axe" deftly, and in three months you can carry the broadaxe "true to the line."

CHAPTER II.

FARMS AND THEIR IMPROVEMENT.—LEASING.

I. SOILS INDICATING VARIETY OF CROPS.—II ADAPTATION OF SOILS TO CROPS.—III. ADAPTATION OF CROPS TO LOCALITIES —IV. STARTING A DAIRY.—V. WHEN TO SELL THE CROP.—VI. STUDY THE PROBABILITIES.—VII. WHEN TO HOLD THE CROP.—VIII. HOW TO SELECT A FARM.—IX. IMPORTANT THINGS TO CONSIDER.—X. SITUATION OF THE FARM.—XI. SOME THINGS TO BE REMEMBERED.—XII. LEASING A FARM.—XIII. FORMS OF LEASE AND CERTIFICATE.—XIV. PLAN FOR LAYING OUT A FARM.—XV. A GARDEN FARM.

I. Soils Indicating Variety of Crops.

THE crop best suited to the soil may generally be told by the natural or wild vegetation found upon it. Hazel brush and red root (the hard, woody-rooted species) are indications of a good wheat soil. Why? They tell of a rich, and at the same time, firm soil. All the cereal crops will do well on such land. As a rule, our upland prairie soils are rich in the phosphates and potash. Heavy-timbered lands usually have what may be called hard or firm soils, well adapted to wheat, rye, sorghum, sweet potatoes, onions, and, when there is moisture enough, to flat turnips and the pasture grasses.

The lower lands, covered with timber, often resemble what are known as soft soils; that is, they contain much humus, and are adapted to Indian corn, the common potato, garden crops, and the meadow grasses.

II. Adaptation of Soils to Crops.

UPON the prairie, both firm and soft soils are found. When stiff or gravelly clay comes near the surface, they are hard lands, adapted to winter wheat, barley, rye and oats, to sorghum, sweet potatoes and onions. When well fertilized, with barn-yard manure, they are also good for garden crops and the cultivated grasses.

This hard soil will be prolific in a variety of rather short grasses, and of the low-growing flowering plants. The lower lands, where the grass is taller, and where sunflowers, various species of helianthus, iron-weed, wild artichoke and all that flower-bearing class thrive, or where are found the compass-plant or rosin-weed, are humus, and, generally, potash soils. Such lands are adapted to Indian corn, the common potato, to oats, and in a lesser degree, and when not subject to wet in the spring, to barley and spring wheat. Sorghum also does fairly well on some of these low-lying soils, especially when they have been drained. Drainage also fits them well for the cultivated grasses. As a rule, the greater the variety of wild plants natural to the soil the better is the soil for general agriculture, for such variety shows that the soil is not only well drained, but that it contains a variety of the elements of plant-food in a soluble condition.

III. Adaptation of Crops to Localities.

WHEN far from a market it would be unwise to cultivate much corn or oats, unless the farmer has stock to feed it to. If his stock is cattle and hogs he would still be wrong in raising large fields of oats to fatten them. His crops, as before stated, must be of wheat, barley, linseed, grass and clover seed. Live stock being his main object, hogs, being the most easily multiplied, come first, then horned cattle, and lastly, as the country becomes more settled and free from predatory animals, sheep. All these may be driven long distances on foot to reach a market.

As stock increases, Indian corn is more and more largely raised and pastures are steadily increased. Later a regular rotation is established; then railroads come to lessen the cost of transportation, and crops become more and more diversified from the increasing demand as towns and villages spring up.

IV. Starting a Dairy.

THE wise man is he who foresees what this increasing population naturally requires, and meets the demand. He, for example, obtains a herd of really good milking cows in advance of his neighbors. If there is not a good demand for milk, he has the best procurable machinery and implements put in for converting the milk into butter or cheese, and establishes his market. Others see that there is profit in this, and may solicit him to manufacture their milk also, and this often grows to a large and profitable business, the farmer almost unwittingly lapsing into the tradesman and manufacturer. A careful study of the various crops will, however, indicate many other lines of possible profit that may be built up by a careful study of soil, climate, locality, and the crops adapted to the increasing wants of a growing community.

The Dairy-House.—A dairy-house, even though only an adjunct to the farm, is almost a necessity upon a large, well-managed country-place, and we herewith give a plan for the building required.

The cut, however, contains several errors. The corner of the barn-stable is drawn too near, for, although the dairy must always be fairly near the milking-yard, it should be removed from all animal or other foul odors; cleanliness being a necessary element of success. The eaves should also project more than is shown, or the dairy must be protected from the southern sun by an awning. The roof of the gallery or porch protects it there, and the porch itself serves for spreading tins and other utensils to dry and sweeten in the air.

The Building Described.—The building occupies 24 x 30 feet of ground, and eight feet between floors. This will be sufficient for the product of forty cows, or with improved fixtures, of more than that number. The lower floor should be divided into two rooms; one for butter, the other for cheese. The attic is the curing-room for cheese, and has, besides the ventilator in the roof, open spaces in the sides which may be closed by wickets, and which serve to admit the air.

DAIRY-HOUSE ELEVATION.

An abundance of fresh, cold water, or of ice, must be at hand, as well as the necessary fixtures, including presses for cheese-making, tanks for setting cream, etc., all of which should be of the best patterns. If the soil will allow, a cellar should be dug, and, in all cases, the drain-pipes and drainage must be of the most thorough character.

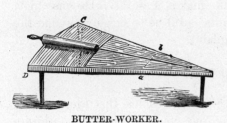

BUTTER-WORKER.

Dairy Fixtures.—May now be bought complete, from certain dealers, who will supply everything required in a factory or creamery, from a steam engine to the simplest implement. The cut shows one of the many forms of butter-workers used. It is simply a slab, preferably of slate or marble, and has no sharp corners or cracks to be cleaned.

V. When to Sell the Crop.

The knowledge of when and how to sell the crops of a farm, is among the most important of the many elements that go to make success in farming. The man who blindly accepts the prices bid by the grain buyer in his local market is apt to come out loser. The farmer can tell the value of his crop on the farm as well as anybody else, if he keeps himself informed as to freights, prices in central markets, insurance and storage.

If he does this he can always get full value for his products, even if there is no competition among buyers, or if there is only one buyer in the market.

The farmer can always ship direct to some reputable house in a central market, or he can combine with neighbors and do so. This will soon bring the local buyer to terms, since he can make some profit, as between the transportation rates that he gets and those which the farmer gets, who ships in smaller quatities.

The farmer can often contract his whole crop for less money per bushel than he will sell one bushel for, and still make money. If he has kept himself informed on prices, and the probable crop of a given commodity, he may often contract to deliver at a stipulated time, and get far more money per bushel than his neighbor, who sells his crop in a kind of "hit or miss" way, a few bushels at a time, just as he feels in the humor for "going to town."

VI. Study the Probabilities.

The ideal time to sell is when the market is at the highest, or as near this point as possible. You cannot find this out by asking the village buyer. You could no more expect him to tell you that which might take money out of his pocket, than your neighbor could expect you to tell him of a trade that would take money out of your pocket.

The man who waits day by day for the market to go higher, and then refuses to sell when it has fallen somewhat from the highest point, is very apt to sell at the lowest price. In like manner, the farmer who ships on his own account and holds until the highest market price is past, usually finds his produce reaching tide-water about the time the lowest market is reached.

The fluctuations of the market are caused by so many varying influences that the wisest often are deceived. But the farmer who carefully figures profit and loss, who carefully studies the probable markets, who makes up his mind what should be a fair price, and sells when that price is reached, seldom makes failures. He may not, indeed, make "a hit" every time, but the "good hits" will be so largely beyond the bad ones that he will have no serious cause for complaint.

VII. When to Hold the Crops.

In a country so large as the United States there is little chance that there will be a failure of a given crop all over the country in any one year. The transportation facilities and the increasing railway extension year by year, preclude famine prices in any locality. The man who holds a crop of grain, wool, or other perishable commodity, hoping to get famine prices, always gets beaten.

The time to hold a crop is from seasons of general plenty and low prices to a season of scarcity and high prices. In this, storage, interest on the capital employed, insurance against fire, insects and shrinkage in weight must always be taken into account. If the present price be so low that it will pay to hold in spite of all these items of expense, do so, but as a rule the best price in any one year is the best price for all time.

VIII. How to Select a Farm.

The selection of a farm more or less improved requires not only taste and judgment, but the ability to estimate correctly the cost of making all those minor improvements which render a place desirable, as well as estimate the natural advantages of the farm. Few of the farmers who have made these improvements are able to tell what they have cost, and not many who buy farms really appreciate their true value. A piece of land with partly worn fences and inferior buildings is really worth less than the raw prairie adjoining it.

Never buy a farm with indifferent fences and buildings, unless the natural advantages are such as to compensate for these deficiencies. A good orchard, wind-breaks in suitable positions, a grove, careful drainage, a man can afford to pay liberally for. A stream of water on natural situations, for ponds; natural protection by hills or timber from storms are always to be taken into account. The proximity of large bodies of water is often of great value in tempering a climate. A soil may be unproductive from bad management, and yet the real fertility may be little impaired.

If the proposing buyer has made soils and their management a study, he may profit thereby in buying such a farm. There are plenty of these for sale in the best neighborhoods. It is for this reason that eventually the best farms are gradually concentrated in the hands of a few keen persons. Their previous owners were poor farmers, and the soil is rather inert from bad management, than worn out by severe cropping.

IX. Important Things to Consider.

The first consideration in seeking a new home, is soil; the next is situation; the third is water, and the fourth, drainage. We put drainage last, for the reason that very few farms really lack the slope necessary to natural drainage, though the difficulty arises sometimes that the outfall of the drainage lies on land beyond the limits of the farm.

Testing the Soil.—The fertility of the soil can be judged pretty accurately by the vegetation growing thereon; the texture by various means; the depth by boring. If the surface soil, upon being wetted and kneaded, becomes pasty and tough, resembling putty, it indicates a stiff clay. If the subsoil contains some gravel, however little, it will indicate sufficient sand, also, for easy percolation of water. A black deep soil indicates humus, suitable for corn, oats, potatoes and grass. If it may readily be compressed under foot, spring wheat, barley and rye will do well on it. If the soil is a chocolate color, or lighter brown, it indicates considerable sand. If of a light or ashen color (silicious clay), a wheat and fruit soil will be indicated. In this way a careful examination will enable any thinking man to decide pretty accurately as to the composition and capability of the farm.

Plow and Pasture Land.—The relative proportion of plow and pasture land must also be taken into consideration, though as a rule a good grain farm will be a good grass farm. Remember always that good grass is the foundation of successful farming.

X. Situation of the Farm.

The situation of a farm is not considered by the average farmer as of great importance, yet a good building site is often worth half the price of the farm. It should be sufficiently elevated to be above danger of miasma, and yet easy of access, and even on the largest farms not more than a quarter of a mile from a main road. The rise to the buildings should be easy, and, if possible, the whole farm should lie in view of the top of the house or barn.

If a stream meanders through the farm or is capable of being turned to supply a pond, so much the better. Or if pipes may be laid to reach the house and barn it will be a valuable improvement.

Oak, maple, hickory, black walnut, wild cherry and ash are the most valuable timbers. If the fences are in good order, and if the house—however small and rough—is well cared for as to its surroundings, you may be pretty sure the land has not suffered seriously.

But, when you find a farm thoroughly in order, with buildings and fixtures, the full value will have to be paid. The owner will pretty well understand what it has cost, and the man who has once put his farm into perfect home condition is seldom in a hurry to leave it. Yet good, natural features unimproved are not rare to those who have the eyes to see them. Excellent places are plentiful in timbered districts where homes can be made, but the farms in timbered districts will average smaller than in prairie regions, and where there is an original growth of hickory, burr oak, black walnut, sugar maple, white wood (tulip tree), the soil and subsoil will generally be excellent.

XI. Some Things to be Remembered.

In Malarious Districts.—Do not select a farm in a malarious district, unless the situation is high. Even then it is better that you satisfy yourself as to the probable health of your family before buying. A very low-priced farm in such a district is to be very carefully investigated before buying.

Costly Improvements.—Never buy a farm with costly improvements in buildings and planting, if they are not suitable to your wants, unless the price is so low that you can afford to alter and reconstruct. A run-down farm, if the soil is there, is the place to improve to your liking; but be sure that the farm is adapted to the crops you intend to cultivate. A high and dry farm is not adapted to grass, neither is moist, cold land, subject to every sweep of the wind, adapted to fruit. But, if protected by wind-breaks, the wet, cold land, after drainage, will be excellent for grass; and, if not too tough, will make good grain land. If your system of farming requires large amounts of manure, a location near some city will greatly cheapen the cost of getting manure. The other remedy, and a good one always, is to keep plenty of stock.

Rocky Farms.—Do not buy a rocky, hilly or stumpy farm, unless pasturage is your object, and then the price should be low. For stock, except sheep, the moderately level land is always the best. If the soil is stiff and wet, under-draining will cure it, and such soils when under-drained, are generally the most productive; but it will cost from fifty to eighty dollars for every acre you thoroughly under-drain. All this must be figured in.

Too Much Land.—Do not buy too much land. The necessary repairs must be made, implements bought, the farm must be stocked, and a proper sum reserved for working capital. Fifty acres to each hand to be employed is fully as much as a good manager should undertake to work, even in the West, where the obstacles to thorough cultivation are less than in most other countries, unless stock-feeding is to be the principal object.

Foresight Necessary.—Before you decide finally, remember that farming requires fully as much thought as any other business; but all the requirements may soon be mastered by application. Farming is no longer the drudgery it was fifty years ago. The comforts and the elegancies are by no means to be overlooked. The man with five thousand dollars, or more, of capital, especially if he have a growing family, needing schools, may do far better to invest the money in an improved or partly improved farm, with schools, churches and society, rather than isolate them by going to the far West, and buying himself "land poor."

XII. Leasing a Farm.

WE do not advise any man who has money enough to stock a farm, however moderately, to lease. If, however, it is desirable to lease a farm, on account of the advantages offered by society, settlements and markets, nothing should be left to chance. Everything must be in black and white, and so plainly stated in the lease that there can be no room for dispute. Every permanent improvement made by the tenant should be paid for by the landlord, and every improvement made by the landlord, at the request of the tenant, will become an additional consideration.

Some leases are so carefully drawn, that the number of loads of manure to be made yearly is a condition, and to what particular crops these are to be applied; even the rotation of crops is often stipulated. The object, both of the tenant and landlord, is to get as good terms as possible for himself. In making a contract, of whatever kind, avoid all unnecessary words, and be sure that the meaning is clear.

XIII. Forms of Lease and Certificates.

WE append here forms of a lease and certificates of landlord and tenant. Forms are usually printed, to be filled in, and it is necessary that this printed wording be carefully noted, as heretofore suggested. Here are the forms:

INDENTURE OF LEASE.

THIS INDENTURE, made the——day of———, in the year one thousand eight hundred and—, between ———, of——, in the county of——, and State of——, [*State business of owner or agent*] of the first part, and———, of——, in the said county, farmer, of the second part, WITNESSETH: that the said party of the first part, for and in consideration of the rents, covenants, and agreements hereinafter mentioned, reserved and contained, on the part and behalf of the party of the second part, his executors, administrators and assigns, to be paid, kept and performed, has leased, demised and to farm let, and by these presents does lease, demise, and to farm let, unto the said party of the second part, his executors, administrators and assigns, all [*insert full description of premises*]: TO HAVE AND TO HOLD the said above mentioned and described premises, with the appurtenances, unto the said party of the second part, his executors, administrators and assigns, from the——day of———, one thousand eight hundred and—, for and during, and until the full end and term of——years thence next ensuing, and fully to be complete and ended, yielding and paying therefor, unto the said party of the first part, his heirs or assigns, yearly and every year, during the said term hereby granted, the yearly rent or sum of———dollars, lawful money of the United States of America, in equal quarter [*or*, half] yearly payments—to wit: on the—— day of [*name the months intended*], in each and every year during the said term: Provided, always, nevertheless, that if the yearly rent above reserved, or any part thereof, shall be behind or unpaid, on any day of payment whereon the same ought to be paid, as aforesaid; or if default shall be made in any of the covenants herein contained, on the part and behalf of the said party of the second part, his executors, administrators and assigns, to be paid, kept and performed, then and from thenceforth it shall and may be lawful for the said party of the first part, his heirs or assigns, into and upon the said demised premises, and every part thereof, wholly to re-enter and the same to have again, repossess and enjoy, as in his or their first and former estate, anything hereinbefore contained to the contrary thereof in any wise notwithstanding.

AND THE SAID PARTY OF THE SECOND PART, for himself and his heirs, executors and administrators, doth covenant and agree, to and with the said party of the first part, his heirs and assigns, by these presents, that the said party of the second part, his executors, administrators or assigns, shall and will, yearly, and every year, during the term hereby granted, well and truly pay, or cause to be paid, unto the said party of the first part, his heirs or assigns, the said yearly rent above reserved, on the days, and in the manner limited and prescribed, as aforesaid, for the payment thereof, without any deduction, fraud or delay, according to the true intent and meaning of these presents: [*if so agreed, add:* and that the said party of the second part, his executors, administrators, or assigns, shall and will, at their own proper costs and charges, bear, pay, and discharge all such taxes, duties, and assessments whatsoever, as shall or may, during the said term hereby granted, be charged, assessed or imposed upon the said described premises;] and that on the last day of the said term, or other sooner determination of the estate hereby granted, the said party of the second part, his executors, administrators, or assigns, shall and will peaceably and quietly leave, surrender, and yield up, unto the said party of the first part, his heirs or assigns, all and singular, the said demised premises. AND THE SAID PARTY OF THE FIRST PART, for himself, his heirs, and assigns, doth covenant and agree, by these presents, that the said party of the second part, his executors, administrators, or assigns, paying the said yearly rent above reserved, and performing the covenants and agreements aforesaid, on his and their part, the said party of the second part, his executors, administrators, and assigns, shall and may at all times during the said term hereby granted, peaceably and quietly have, hold and enjoy the said demised premises, without any let, suit, trouble, or hindrance, of or from the said party of the first part, his heirs or assigns, or any other person or persons whomsoever.

IN WITNESS WHEREOF, the parties hereto have hereunto interchangeably set their hands and seals this———day of———one thousand eight hundred and———

Signed, sealed, and delivered }
   in the presence of   [Signatures and seals.]
[Signature of witness.]

LANDLORD'S CERTIFICATE OF LETTING FARM WITH STOCK AND TOOLS.

THIS IS TO CERTIFY, that I have, this———day of———, 18—, let and rented unto———, of———, in the county of———, a certain farm situate in the town and county aforesaid, and bounded as follows [description]: with the appurtenances, and also with the use, profits, and behoof of the following named stock and farming utensils, cattle, horses, and stock now being or to be on the said premises within described, on and from the———day of———, 18—. during the time below stated—viz, [here describe all machinery, farm stock and farming utensils], on the said farm now remaining and being, and the sole and uninterrupted use and occupation thereof, for the term of———from the———day of———, 18—, at the yearly rent of———dollars, payable on the first day of January, with the refusal of the same for [state time of extension] years more at the same rent, upon the said———giving me notice in writing of his intention to renew the lease on or before the ———day of———, 18—.

[Date.]   [Signature.]

TENANT'S CERTIFICATE.

THIS IS TO CERTIFY, that I have, this———day of———, 18—, rented of———, of———, his farm, and have agreed to the following covenants, viz, [here describe every individual thing that is to be performed.] I also promise to use the horses, oxen, wagons, sleighs and other tools carefully, and to return them in as good condition as they are now, the necessary wear excepted, together with possession of the farm and buildings, on the———day of———, 18—.

[Date.]   [Signature.]

Chattel Security.—It is sometimes required that tenants pledge chattels as security for payment of rent, etc., or that they give security therefor. The first annexed form shows a tenant's pledge; the second, security for rent, etc. Append to the tenant's certificate of letting the farm, or the lease, the following:

TENANT'S CERTIFICATE PLEDGING HIS CHATTELS ON THE PREMISES AS SECURITY.

And I do hereby pledge and mortgage to the said———all my personal property of what kind soever which is or may be on the premises aforesaid, for the faithful performance of the covenants herein, hereby authorizing the said———, in case of a failure on my part to perform all or any of said covenants, to take said property so pledged, and sell the same, and out of the proceeds of such sale to pay and discharge all rent, damages and expenses, which may at such time be due, and to pay over to me or my assigns the surplus moneys arising from such sale.   [Signature of tenant.]

SECURITY FOR RENT AND PERFORMANCE OF COVENANTS.

In consideration of the letting of the premises above described, and for the sum of one dollar, I do hereby become surety for the punctual payment of the rent, and performance of the covenants in the above written agreement mentioned, to be paid and performed by——as therein specified; and if any default shall at any time be made therein, I do hereby promise and agree to pay unto the landlord in said agreement named the said rent, or any arrears thereof that may be due, and fully satisfy the conditions of the said agreement, and all damages that may accrue by reason of the non-fulfillment thereof without requiring notice or proof of demand being made.

Given under my hand and seal, the——day of——, 18—. [*Signature and seal.*]

ANOTHER FORM.

In consideration of the letting of the premises above described [*or*, for value received], I guarantee the punctual payment of the rent [and performance of the covenants] in the above agreement mentioned to be paid and performed by said lessee, without requiring any notice of non-payment or non-performance, or proof of notice or demand being made, whereby to charge me therefor.

[*Date.*] [*Signature.*]

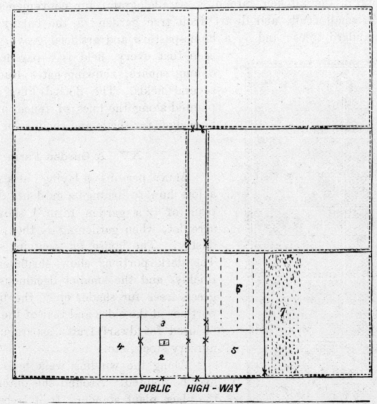

DIAGRAM OF FARM.

XIV. Plan for Laying Out a Farm.

The careless, haphazard method so often practiced in laying out a farm and in dividing it into fields is well worthy of improvement. A careful study of the peculiarities of the farm is necessary in selecting the building site, laying out the farm roads, conveying water to the fields, and in deciding as to the proper sites for permanent pastures and wood-lot, if such is to be planted.

As a rule it is better that the farm-house and other buildings should occupy a central position. Yet circumstances may prevent this. The configuration of the land may even render it necessary that it be on one corner of the farm. A ridge running diagonally through the farm may render this necessary. However it may be, the fields should be arranged both as respects rotation, and also so that each one may be reached by the farm road.

Map of the Farm.—When no serious obstacles intervene, the diagram will show the arrangement of the house lot, garden, orchard, grove and home pasture, etc., in a location central upon one side of the farm, with private road, showing how every field may be reached: 1, is the house, back from the public road; 2, the lawn and house lot; 3, the kitchen garden, a parallelogram, for convenience in horse-cultivation; 4, small fruit, and dwarf fruit tree garden; 5, the barn-yard; 6, the orchard of standard trees; and 7, a home-pasture and artificial grove. It will be seen that every field is a parallelogram, or oblong square, showing gates leading to the several fields. The dotted lines show trees planted along the lines of fence, or they may be hedges for shade and shelter for stock.

XV. A Garden Farm.

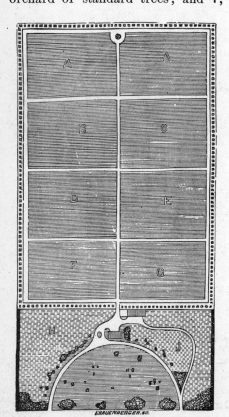

A GARDEN FARM.

Many persons in laying out a place are at a loss how to arrange a garden. The ground plan of "a garden farm," shows a forty acre lot, when gardening is the principal industry. The house lot contains eight acres. The dark portions show shrubbery, planted thickly, and the smaller deciduous and evergreen trees for shade, etc.; the light shaded portions at the sides and rear of the house may be used for dwarf fruits, asparagus, rhubarb, nursery stock, etc.

Along the winding walk beds of flowers may be planted. Around the outer border of the place plant standard fruit trees, for shade and shelter. A road runs through the farm with a circle at the end for turning about, and there is a road around every field to cart manure in, and to cart out the produce; this road space is also necessary for turning when working the land.

These roadways may all be in grass, so that they may be utilized for hay-making or pasturing. A A, may be in orchard trees and small fruits; and B, C, D, E, F, G, will give ample area for gardening and other crops, and at the same time allow for rotation.

If this plan is to form simply the farm-house place and lawn, and kitchen garden, etc., H and J may contain the garden and small fruits, F and G the home orchard and pasture, and the rest of the plat used, as desired, as part of the farm.

CHAPTER III.

FENCES, HEDGES AND GATES.

I. RELATIVE COST OF FENCES AND BUILDINGS.—II. COST OF FARM FENCES IN THE UNITED STATES.—III. THE COST OF FENCE PER ROD.—IV. WORM, OR VIRGINIA FENCES.—V. STAKING AND RIDERING THE FENCE.—VI. POST AND RAIL FENCE.—VII. PREPARING THE TIMBER.—VIII. MORTISING THE POSTS AND SHARPENING THE RAILS.—IX. SETTING THE POSTS.—X. FASTENING THE RAILS AND FINISHING.—XI. HOW TO BUILD A BOARD FENCE.—XII. STRINGING A WIRE FENCE.—XIII. SOD-AND-DITCH FENCE.—XIV. COMPOUND FENCES.—XV. PORTABLE FENCES.—XVI. FENCING STEEP HILLSIDES.—XVII BARS AND GATES.—XVIII. THE SLIDE AND SWINGING GATE.—XIX. SWING GATES AND SLIDE GATES EXPLAINED.—XX. SELF-CLOSING SLIDE GATES—XXI. SOUTHERN STRAP-HINGE FARM GATE.—XXII. DOUBLE-BRACED GATE.—XXIII. ADJUSTABLE SWING GATE.—XXIV. HOW TO PREVENT POSTS FROM SAGGING.—XXV. ORNAMENTAL GATES.—XXVI. FLOOD AND WATER GATES.—XXVII. STREAM GATE AND FOOTWAY.

I. Relative Cost of Fences and Buildings.

THOSE States of the American Union having cast-iron laws regulating the kind of fence and the space between boards or rails, would do well to repeal them. It is a generally accepted fact that the fences of the country cost more than the buildings. They must be renewed, on an average, about once in twelve years. The Secretary of the Wisconsin State Agricultural Society, a few years ago, estimated the cost of the perishable fences of the State to be $40,000,000, reckoning one rod of fence at 85 cents. More recently, a careful and unprejudiced observer, Mr. David Williams, of Walworth county, Wis., says:

"I have, with the assistance of a number of well-informed farmers of this county, made a careful computation of the first cost, annual deterioration, per cent, and cost of annual repair. There are sixteen townships, or seven hundred and fifty-six square miles, in the county. Estimating one-sixteenth as lake, ponds, or abandoned lands, gives five hundred and forty square miles, or 345,600 acres of improved or inclosed land. This, if fenced into 40-acre lots, will require five rods of fence to the acre (a careful estimate gives 25 acres as the average size of fields), or 1,728,000 rods of fence, exclusive of ornamental and village fences. Estimating one-eighth of this as division fence, and therefore duplicated in the foregoing estimate, and to include also temporary and comparatively worthless fence, will give in even numbers 1,500,000 rods of farm fence for the county, 100,000 rods for each township (one-sixteenth of the total area having been thrown out of the estimate as lakes, ponds or abandoned lands) of improved or inclosed lands. From carefully prepared data, I find about two-fifths to be highway fence, making 600,000 rods of highway fence for the county, and 40,000 rods for each township.

"Estimating the cost of this fence at $1 per rod gives $1,500,000 for the county, and $100,000 for each township. Two-fifths of this for highway fences gives $600,000 for the county and $40,000 for each township, or a total cost of all farm fence of $4.34, nearly, per acre, and a cost of $1.73 per acre of highway fence. Estimating 10 per cent on first cost for annual deterioration and repairs, and 7 per cent interest on first cost, gives $275,000 as the aggregate *annual* cost of farm fence for the county, and $18,333.33 for each township. Fully two-fifths of this are for highway fence. If to this sum be added the cost of village fences—mainly made necessary by the pernicious habit of using the highway as a public pasture—the total cost of fence for the county will be swelled to the considerable sum of $1,750,000, and the annual cost to $297,500."

II. Cost of Farm Fences in the United States.

THE cost of the farm fences in the United States has been estimated at $1,350,000,000, and their annual maintenance at $250,000,000. Thirty years ago, the annual cost of repairing fences in Pennsylvania was about $10,000,000. The annual cost of fencing in New York State was placed at $13,500,000. In Illinois and some other Western States, fencing is not compulsory. The people of a county or township can decide by vote, whether they will have fences at all, or what shall constitute a lawful fence. This should be the rule everywhere.

No-Fence Laws.—When tried in Livingston county, Illinois, some years ago, the no-fence law worked excellently. We have had cattle herded within a few rods of standing corn, and they would take the road, morning and evening, quietly enough under the care of the herdsmen. It is cheaper to fence cattle in than to fence them out, especially in all neighborhoods where pasturage is not the principal industry.

III. The Cost of Fence per Rod.

THE statute laws of the United States do not require fences. It is simply the law of custom, and many of the circumstances which originally gave countenance to the local laws have become obsolete. Let us take a section of land, which is six hundred and forty acres. To enclose this, will require 1,280 rods of fence, which at a dollar per rod, would cost $1,280. To divide this into 160-acre lots, would require 640 rods more of fence, or $640. To divide these again into forty-acre fields would require 1,280 rods more, or, in all, 3,100 rods of fence. The annual repairs on this will be at least ten per cent of the cost, or $310.

The smaller the farm the more the fences cost, since the fields are smaller. But the cost of the fence is not all; they diminish the cultivatable area. If enclosed by a hedge, at least half a rod in width of land will be lost. In fact little can be grown within eight feet of a hedge on either side. If a board or wire fence is made, the loss will be at least four feet, to say nothing of the harbor for weeds. In a section of land the loss amounts to nearly ten acres.

IV. Worm, or Virginia Fence.

THE old-fashioned worm (Virginia) fences are seldom found outside timbered districts. They are unsightly, take up too much land, and are great harborers of weeds. They should be replaced at the first opportunity with post-and-rail, or other straight fence. Their only advantage lies in the ease of getting out the material, laying it up and transferring it when no longer wanted.

How to Build.—The width—allowing eleven feet as the length of the rails—should be four and a half feet, certainly never under four feet, from outside to out-

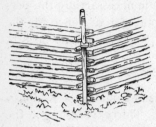

PANELS OF VIRGINIA FENCE.

side of "the worm," as the zigzag line of the fence is called. Set two lines of stakes this distance apart. Then, the rails having been delivered as near the outside of one of these lines as possible, lay a rail on the ground, so that each of its ends projects beyond one of the stakes six inches. Then lay another rail, so that its butt shall overlap by six inches one end of that first laid, and, being placed at an angle of, say, 130° to 140° thereto, so the other end of the second rail shall lie in line with the other row of stakes. Reverse the angle of opening with each succeeding rail until you have one laid along the whole line, much in the shape of a number of capital W's, but more open, so laid as to touch, or, rather, to slightly overlap.

Blocks eighteen inches long and six inches thick should be provided for every panel of fence. Place one of these blocks under each end, and bring every corner true to the line of stakes as you go back. Then, lay successive rails, from the point where you first began, to the far end, and walking back to the place of beginning, so proceed until you get six or seven rails in height, according to the intended height of the fence. As the fence is laid keep the corners vertical (in a direct line up and down), and of equal height one with another, by laying the big end of the rail one way or the other, as occasion may require.

V. Staking and Ridering the Virginia Fence.

THE fence is now ready for the stakes and top rails, or riders. The stakes are better if sharpened at one end and with the other end square. Make the holes for

LOCKED VIRGINIA FENCE.

the stakes by thrusting the spade in the ground from, and diagonally with the corners, to receive the stakes, and at a proper distance so the stakes will not only lie kindly but at the same time receive and lock the rail. A little practice and observation will soon enable any one to do this deftly. When two stakes are set lay one end of a rail in the fork of the stakes, and the other on top of the next rail. Set two more stakes, lay on the rail as before, and so proceed—having reserved two of the heaviest rails for riders in laying the fence—until the whole line is staked. Then walk back and lay the remaining rail on top of all. The stakes are sometimes set in the middle of the panel instead of at the corners, but whatever

FENCES, HEDGES AND GATES.

may have been said as to the advantage of this plan, the fence is neither so strong against animals nor the wind, as when staked at the corners.

The cut entitled "Panels of Virginia Fence" illustrates perpendicular staking and capping, which make a neat and firm fence. Here see an illustration of four sections of Virginia fence with the caps laid parallel with the line of the fence. By this method one man can set the stakes, since both are on the same side of the fence.

PERPENDICULAR STAKING AND CAPPING.

Straight Rail Fence.—Another simple and good way of making fence of rough rails is to prepare stout stakes, and also cross-pieces fourteen inches long. These are set in the ground, a cross-piece nailed on near the ground, the rails laid in, another cross-piece nailed, two more rails laid, the next two stakes set, and so, proceeding precisely as directed for laying up the rails for Virginia fence. In this case no caps are needed, but it is better to have the tops of the stakes square so that a cap-piece can be nailed on top to hold the last rails.

STRAIGHT FENCE WITH STAKES.

Seeding to Grass.—If the strip of ground on which a Virginia fence is laid is seeded down to blue-grass, red-top, orchard grass or some other persistent grass, say ten feet wide, it will prevent the growth of weeds and furnish valuable pasturage late in the autumn, and in the spring when the fields are not occupied with crops. The same rule will apply to all fences, whether wood, stone, hedge, ditch or wire.

VI. Post-and-Rail Fence.

THE next most common fence, when timber is plentiful, is of heavy posts properly mortised, and with rails properly sharpened to lap together in the mortises.

The Best Posts.—The best posts are first, red cedar, and then catalpa, osage orange, locust, yellow or white cedar, bur oak and swamp oak, lasting in the order mentioned. The first two will last indefinitely; the next two from twenty to thirty years, and the last two from ten to fifteen years, according to size and nature of soil. Here it should be remembered that a post should be not less than six inches square at the surface of the ground, and eight inches is better. From the place where the bottom rails are inserted the posts should be hewn flat down to three and one-half or four inches thick, preserving the full width of the post. The cut explains our meaning, and shows proper width of rails and spaces in inches.

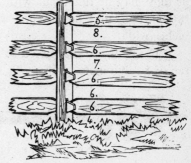
POST-AND-RAIL FENCE.

VII. Preparing the Timber.

The posts should be split out of large logs, if possible, by first quartering and then taking out the heart pieces, leaving the quarters eight inches, if possible, from the bark to the center; the pieces are then divided into posts not less than six inches through at the bark and seven and a half feet long, to allow them to be set three feet in the ground. At three feet from the butt end score in and hew down the sides of that portion that is to be above ground, to a thickness of about four inches, or perhaps three and a half. Stack them up until you have a sufficient number.

Then select the straightest-grained logs possible, each eleven feet long, quarter them, take out the heart, leaving the quarters five or six inches deep from outside to inside, according to the timber. Split these into flat rails not less than three inches thick at the bark. Stack them up until you have enough, counting five rails to each panel. If the top of the fence is to be a pole pinned on, four rails to each panel will be sufficient.

VIII. Mortising the Posts and Sharpening the Rails.

The bottom of the first mortise in the post should be four inches from the ground line, and each mortise should be four inches deep by two and a quarter inches wide. The bottom of the second mortise may be sixteen inches from the ground, the third, twenty-five inches, the fourth, thirty-six inches, and the fifth forty-eight inches from the ground line, allowing that the rails will average five inches in width. This will form a fence proof against even small pigs, and four and a half feet high. The cut shows a four-rail fence, giving nearly the same space, but only four feet high. To increase the height, nail or pin on a pole at the top.

Gauge for Mortising.—To get all the posts alike, carefully mortise a piece of inch board of the same length as the post for a gauge. Set it against the broad face of each post, and mark off the places for the mortises, having the top of the upper mortise at least four inches from the top of the post. The mortises are made by boring a hole for the top and bottom of each mortise with a two and a quarter inch auger (two inches will do), and knocking out the centers with a mallet and chisel, or they may be taken out with a thin, narrow-bitted axe made for the purpose; or the mortises may be cut out with a post-hole axe—a very thin axe with a bit two inches wide.

The rails are sharpened to a taper edge, the same on both sides, the scarf being ten or twelve inches long. The posts are hewed, by laying them across two pieces of timber two feet from the ground. A bed is cut in one log to receive the post, bark side down, so it may be fastened with a wedge. For sharpening the rails, a bed is cut in both logs, and the rail laid in and wedged so the ends of the rail will just project beyond the logs for hewing. There are various frames in use for boring the

POST-HOLE AUGER.

mortises in the posts. The best is that used by carpenters, with a crank on each side. The cut shows what is known as the Kentucky post auger, to be worked by a boring machine, or in usual way with a cross-handle; only part of the shank is shown.

IX. Setting the Posts.

WE have given seven and a half feet as the length of the posts. Eight feet would be better, for then the posts may be sharpened at the lower end and driven down six inches or more. This will prevent all heaving, for it is well known that a pointed post is not lifted by the frost like a square one. Another advantage is, that the posts, being of equal length, may be driven so that the tops will come exactly to the line; but, in the case of a post-and-rail fence a block must be used to drive on, so hollowed out on the under side that the force of the blow comes at the edges of the post, and *never* over the mortise, else the post will be split. And no driving must be allowed except to fix the post in line.

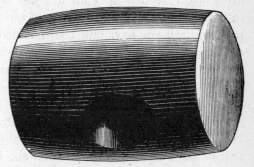

CAST POST MAUL.

Aligning the Fence.—The line of the fence being actually determined, set a line of strong pegs exactly where the outside of the posts is to come—the presumption being that every mortise is bored in the middle line of the post. The center of the fence will then be three inches inside this line. Six inches outside this line of pegs drive another line of strong pegs, so they are two or three inches above ground. Dig or bore the post-hole so it cuts two inches beyond the pegs, which are of no further account, once the hole is made plumb; then the outside pegs are your guide.

X. Fastening the Rails and Finishing.

DRIVE the posts in the bottom of the hole, so the bark side will measure six inches to the outside peg left standing. Tamp in the earth solid to the top, keeping the post plumb, and each post will come exactly in line. One post being set, enter the rails, set another post, entering the rails in this, driving them properly together. Fasten the post, as before directed, lap in another panel of rails, and so proceed until the whole string is finished.

If the whole fence is not exactly plumb when finished, straighten it by tamping the earth here and there as necessary, but if the work has been well done, the whole will be in a perfectly straight line. This plan is the simplest and best for any fence where posts form a part, but for boards, poles, slabs or wire, one side of the posts must be hewed fair, for nailing upon.

XI. How to Build a Board Fence.

BEGIN precisely as directed in making the post-and-rail fence, having the posts sharpened with a good taper. If the ground is soft, as is generally the case in prairie soil in spring, the posts may be driven entirely from the surface or from twelve to eighteen inches below, as the case may be. [See cut of post maul on page 303.] If the soil is dry dig or bore as far as may be necessary and drive the rest of the way, remembering that a driven post always stands firmest and is least liable to heave.

Cut gains in a piece of board so they will correspond with the spaces required for the fence. With this mark where each upper line of each board is to come. Measure a board, so that the ends will just lap to the center of the first and third posts, saw off square and nail on with two ten-penny fence nails at each end and center. So proceed until you have one board nailed along the string of fence. Starting again at the end, work back breaking joints, that is, make the center posts for the first line the lap posts for the second line of boards, and so proceed until the whole fence is of the required height. Set your posts so that the tops will be four inches above the upper boards of the fence.

If the fence is to be capped and battened, saw off the posts so the saw gap will be just even with the outer line of the top board, and three inches higher at the rear of the post. Saw the battens six inches longer than the extreme height of the fence, point one end and drive them down until the top of the batten is on a line with the slope required for the cap-board. Nail on the battens with twelve-penny nails, and the cap-boards with the same, using eight-penny nails for towing into the upper board of the fence. This you will find the best fence ever made; the wind will not shake it, and if it is four and a half feet high, it is secure against every animal except jumping horses. For such rogues it must be five feet high, and sometimes more.

The same rule will apply to slabs and poles scarfed at the ends. The pieces must be made to meet nicely at the middle of the lap posts, and be securely pinned or spiked on. They are generally a nuisance, and are better adapted to being laid up between double stakes set vertically, with blocks between each pole, and the tops of the stakes, gripped together and fastened with a cap piece or strip of board nailed securely.

The accompanying cut will answer a twofold purpose. Without the triangular key pieces it shows a five-board fence, capped and battened. The cut entire shows a portable board fence, made in panels, and fastened with double battens. The top of the triangle rests upon a projecting piece, as shown, and the bottom on its own feet. The bottom board rests in a slit of the triangle, and the projecting piece, between the fourth and fifth rail, rests in a notch of the triangle, and projects beyond the next panel, holding all firm. It is easily taken down and put up, and may serve as a temporary fence whenever wanted.

PORTABLE BOARD FENCE.

XII. Stringing a Wire Fence.

SET the posts as directed for a board fence. Mark the places where the wires are to be strung on the posts. Never use a single wire. The contraction and expansion will bother you more than it is worth, whether it be barbed or not. Use two wires twisted together, or flat strips twisted. The objection to barbed wire is more theoretical than practical. Few animals are ever hurt with the barbs, and seldom is any but a breachy animal hurt at all, unless it may be horses or colts at play, or Texas or other semi-wild cattle, which do not see the wire. For such, a pole on top will serve as a warning.

END-POST AND BRACES FOR WIRE FENCE OR TRELLIS.

To put up the fence take the reel of wire, placed on a wagon so it will revolve. Fasten one end to the end-post, which must have been well braced to stand the strain, the posts being placed nine, ten or twelve feet apart, according to the strength of the fence required. Drive the wagon forward until the end of the fence is reached, strain to the required tension upon a roller turned by hand-spikes or other suitable device. Drive the staples properly into the posts, and so proceed until the fence is of the required height.

The manufacturers of each special wire usually furnish all necessary fixtures and full directions for putting up. In all prairie regions a good strip or double wire fence, whether barbed or not, is the cheapest and best stock fence made, if put on solid posts not too far apart. The cut shows the manner of bracing the end-post for a wire fence, and also the end-post of a long grape trellis and trained vine.

XIII. Sod-and-Ditch Fence.

SOD-AND-DITCH fence, if surmounted with a willow, osage-orange or three-thorned locust hedge, are among the best fences on all low grounds requiring drainage or subject to overflow. The ditches need not be deep, simply sufficient to carry off the surface water, and should lie on each side of the bank.

How to Build.—The bank should not be less than six feet wide at the bottom, measuring from the line of each ditch. Stretch a line from the inside of the ditch, and with a sharp spade mark to the line. Chip this so as to show a narrow three-cornered strip, four inches wide at the top, as a guide. Cut the other side in like manner. With the spade take out narrow spits of sod, which should be laid regularly in a correct line, grass side out and about four inches from the side of the ditch to form a berme, which eventually will disappear.

In digging any ditch, the workman must be careful that the bottom of the spade does not cut the slope of the ditch. The outside of the spit must be two or more inches from the outside lines, and when one spit is finished, the sides of the ditch must be nicely pared to correspond to the required slope.

Banking the Sod.—The first spit of sod being taken out to the full depth of the spade, throw out a corresponding spit from the other ditch, laying one spit of turf on another as you proceed. This done, the balance of the earth in the ditch may be thrown into the center below the lines of sod, and the slope of the ditches evenly pared and the loose dirt thrown out, thus finishing the ditch and bank, which, if it is to be planted, should be slightly rounded on top.

The Hedge.—If this is to be surmounted by a barrier of willow-trees, get branches of uniform size and two to three feet long. Set them in a line, all slanting one way, eighteen or twenty inches apart on the top of the bank, ramming the earth firm about the lower ends. If preferred, cuttings a foot long may be planted. If osage orange or honey locust is to form the barrier, one-year-old plants should be set, and the clay thrown up from the bottom of the ditch should be enriched with about four inches of good soil mixed with compost manure.

XIV. Compound Fences.

COMPOUND fences are those composed of two or more materials as a ditch-and-bank fence, surmounted with a barrier of plants, as already described, or of wood, stone, brush, or other material. Of course, stone walls will never be built except as the loose stones on the surface of the field are required to be gotten rid of. Brush-and-stake fence should never be used. They are perishable and harborers of weeds.

The best barrier, except a hedge, or one of trees, is barbed wire strung upon the top of a bank. This forms a perfect fence, and cannot injure stock unless they are decidedly breachy. Such are well out of the way, though they are often not so much to blame as their masters. They are generally made breachy by bad fences or starvation.

XV. Portable Fences.

PORTABLE fences are seldom used in the United States, though they are sometimes useful for dividing off flush pastures where fattening stock are fed, thus making them eat off the best of the grass regularly, to be followed by store cattle to finish up. They are also useful for penning off or for pasturing sheep. In the article Board Fences is shown one form of portable fence. Portable fences are generally patented, and as permanent fences are worthless.

The best unpatented fences we have ever used are strings of wire fence eight or ten rods long, with a small, round, smooth, carefully sharpened hard-wood post at each end, and strengthened at intervals of half a rod with upright slats of hard wood, an inch and a half square, to which the wire is stapled. The sections may be dragged to any desired spot while lying flat; they are easily set up, the post driven, and if two guys of wire, fastened to slanting stakes, are used occasionally on the side from which the pressure may come, they are perfectly secure. It is not a bad plan to have enough of this fence to reach across a field. When not used for feeding off meadows, they are excellent for small pastures for calves, lambs or hogs, and also for confining hogs upon parts of a meadow or other land infested with the white grub. These swine will thoroughly destroy, by rooting out and eating, every grub in the infested land.

XVI. Fencing Steep Hillsides.

STEEP hillsides are fenced either with post-and-rail, board or wire fences, but the posts should be set fully six inches deeper than upon level ground, and in no case must the posts incline with the slope. They must be placed perpendicular. A fence for hillsides that will stand, and one that is also useful for marshy places, may be made with poles and stakes, as shown in the accompanying cut, or with boards or slabs, or any other rough material. This fence will need no explanation, except to say that the stakes must be well sharpened so that they may be driven easily. Hold the stake with one hand, the end in a proper position on a block, and sharpen with a hand-axe, kept quite keen.

FENCE FOR A HILLSIDE.

XVII. Bars and Gates.

ONE of the tests of careful farming is the farm-gates. Bars are simply a makeshift for something better. They are never safe unless pinned or otherwise fastened so they cannot be rubbed down by stock. They are dangerous to stock in passing through unless entirely removed; annoying if a team is to be driven through them, and cost more in time, in a single year, than would pay for a good gate. They also cost about as much to make and put up as a single gate, that any farmer can make with a square, a saw, an auger, an axe and a handful of nails.

We shall elaborate somewhat on the subject of gates, and illustrate with a number of the more simple and practical, as well as those complex. The subject of gates is an important one to farmers, and hence a careful study of the various plans will be desirable.

XVIII. The Slide and Swing Gate.

FARMERS have been swindled out of large amounts of money, first and last, by the holders of a patent that never should have been issued. The principle of the Slide and Swing Gate is older than the oldest inhabitant, and not patentable. The primitive gate is made as follows: Set two stakes diagonally together and wide enough apart to admit the gate with plenty of play, but so that when the gate is closed the end will press against each stake. At the other end the gate rests on strong pins, so hooked that when movable pins are inserted the whole will be held firm. The gate is made by nailing inch boards upon two inch posts at each end. The braces extend from the bottom board to the board next the top. In the middle a strong batten extends from top to bottom and is securely nailed with

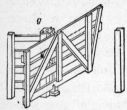

SLIDE AND SWING GATE.

clinch nails. The gate is slid back on a pin in the rear post under the top board. When slid back until the central batten is reached, swing it around so that the team can pass. Twelve feet is wide enough for a load of hay, and a person easily passes through by swinging it slightly. The cut shows a more strongly built gate of this description.

XIX. Swing Gates and Slide Gates Explained.

THE forms of gates and the various attachments are so numerous that a full explanation of the principal ones would fill a volume. The following cuts very nearly

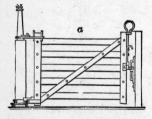
GATE SWINGING ON RINGS.

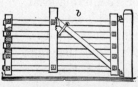
GATE WITH STRUT.

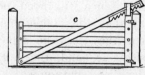
GATE WITH TIE LATCH.

explain themselves, and each one has its practical value. The annexed cuts show: at *a*, gate with adjustable hinges, operating on rings on the post, so the gate may be keyed up if it sags. The fastening consists of a movable spring latch. The gate is made of boards, let into the end posts, or griped between two hard-wood pieces at the end, firmly bolted, and with a single brace extending diagonally from top to bottom.

At *b* and *c* are shown two feasible modes of raising gates when the posts are set in ground where it is difficult to prevent sagging. At *b* the gate is raised by the movable diagonal strut, the gate resting on the common hook-and-eye hinge. At *c* is another form, the gate being raised at the outer end by means of the tie slot, one set of notches resting in others reversed, thus firmly held except when lifted out.

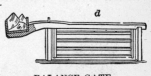

BALANCE GATE.

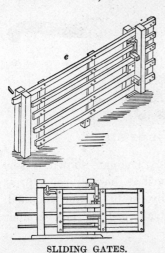

SLIDING GATES.

At *d* is shown the old-fashioned and strong gate swinging on a pin, and so loaded at one end as to balance. It is not liable to sag, since the weight is perpendicular on the post, and is therefore not difficult to swing. It is usually latched by the top bar fitting into a gain in the post.

At *e* and *f* are shown two sliding gates. The one at *e* rests on rollers supporting slats which traverse longitudinally. One end slides between two vertical posts, upon which the guide rests to keep it in position when closed, a bar working in a slot at the rear keeping it closed. At *f* is shown a gate the rollers of which keeps it level whether open or shut. Any farmer, if he is handy with tools, can make any of these gates, the cuts explaining themselves.

The next illustration shows two views of a swing gate, that in opening, rises upon an incline on the lower hinge, so that the front part, when open, shows as in the lower view. It is held open by the roller, after passing up an incline, passing into a slight depression, thus holding it open until again closed by hand. When released the gate is surely closed by its own gravity. The upper cut shows the gate closed. The various forms here given may be built according to any of the cuts, the slats being either horizontal or vertical. For small gates we show a form of upper hinge, that will shut the gate by its own gravity. There are now many forms of these kept in stock by hardware merchants. A latch is also shown, not easily opened by stock.

RISING GATES.

XX. Self-closing Slide Gates.

Any slide gate may be made self-closing by having the rollers pass up a slight incline, and may be held open by the rear roller passing into a slight depression. If the gate is only partially opened, it will, of course, close itself by its own gravity. The illustration of a self-closing slide gate is intended to show two things: the manner

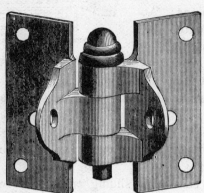

SELF-SHUTTING UPPER HINGE.

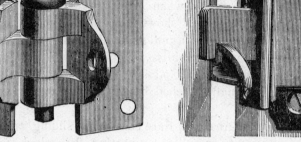

GATE LATCH—HALF SIZE.

of building a square-picket fence, and a gate to correspond. The gate shown is for a footway. For a carriage gate, there must be another post at the rear, or a projection from the fence to carry the rear roller, as shown by the brace and inclined plane beyond the fence. This gate has the following advantages for village and

SELF-CLOSING SLIDE GATE.

suburban residents: it will not sag, and, therefore, large posts are not necessary; it clears all obstacles in its path, and closes surely, when relieved, by its own weight. This feature will recommend it to all who have suffered from wandering stock, and a self-fastener may be attached so no animal can open it.

A Simple Slide Gate.—A simple slide gate is made thus: Two posts are united by a cross-piece below. One of the posts has a slit in its top to receive the cross-bar. This bar passes at its other end into a mortise in the other post, and is fixed by a pin upon which it moves. The other end is made long enough to be shaped into a projecting handle. A perpendicular piece is attached to the cross-bar, connected by a pivot having at the lower end a ring that runs on the cross-bar.

Rollers.—While upon the subject of slide gates, it should be remarked that, except for the very lightest gates, or all those that are required to stand considerable pressure, rollers must be employed, upon which the gate runs. These are now sold of various styles and weights, from those moving the heaviest barn-doors with ease to those adapted to light hand gates. There are so many patterns for gates and doors of various kinds, that it will not be difficult for the purchaser to be suited. The cut shows one of the more simple forms, applicable to gates and barn-doors, giving front, rear and side views.

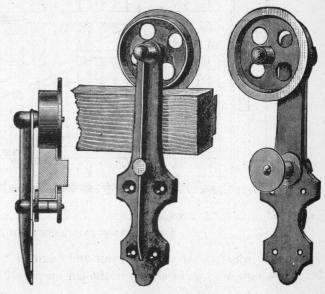

ROLLER HANGERS.

XXI. Southern Strap-Hinged Farm Gate.

The old-fashioned strap hinge assists largely in strengthening the swing gate, and in preventing it from swagging. For heavy gates of this kind the posts must be large and deeply set. In the South the heart-wood of the pitch-pine unites strength with stiffness, and is generally used; the form of gate shown in the illustration

HEAVY STRAP HINGE.

annexed, is a favorite plantation gate, because it is simple, light, strong and durable. In the North, oak or beech for the ends, with yellow pine bars and brace, and hemlock pickets may be used. The cut explains itself. By the scale below, each portion of the gate may be accurately determined. The manner of bracing—the diagonal brace dove-tailing into the upper, helps to prevent sagging. The latch guard and holder are of wood, and its simplicity of construction will commend itself

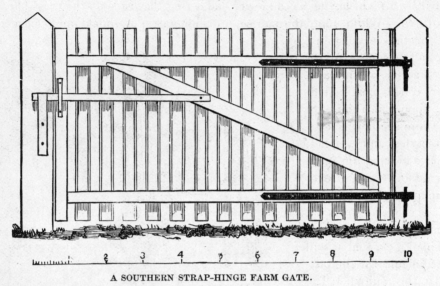

A SOUTHERN STRAP-HINGE FARM GATE.

to any man who can use a square, saw and hammer. The usual form of strap hinges for gates, now used, is as shown in the cut on page 311.

XXII. Double-Braced Gate.

The cut here shows the manner of double-bracing a gate, dividing it into four equal triangles, the form here exhibited being a hand gate with spring fastening. This bracing gives the greatest strength and solidity. The dimensions for a gate nine or ten feet long are as follows: Bottom board three inches wide, lower space three inches, the rest of the boards are six inches wide, except the top board, which is seven inches. The spaces as shown, are six inches, except that between the two upper boards, which is four inches; but these may be varied according to circumstances. The front rail is 3x3 inches, and the rear rail 3x4 mortised nearly through, leaving only sufficient wood in the front and rear rail to protect the boards and hold them firm. The boards must fit the mortises well, be driven home and held with pins. The braces are four inches wide, one on each side, firmly nailed with clinch nails. Any farmer can make this gate, and no animal can break it.

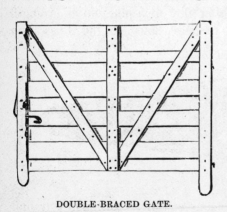

DOUBLE-BRACED GATE.

XXIII. Adjustable Swing Gate.

When deep snow accumulates in winter, some plan must be used to allow passage through gateways without shoveling great masses of snow, to be again accumulated, perhaps, in five minutes. The accompanying illustration shows an easily constructed and simple gate, and one easily raised to a considerable height above the ground, in the case of deep snows. The upright *A* is six or seven feet long, and three and a half inches square, round at the top and bottom. The dark points indicate places for iron pins upon which the hinges (clasps passing around the post and upright *A*) play. The hinges are shown at *B*. The second slat is cut short, as shown, to allow raising the gate. It may be necessary to keep cattle and horses from passing through the gate, and yet allow the passage of sheep and swine.

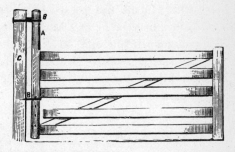

ADJUSTABLE SWING GATE.

This may be done by raising the gate to the required height. The gate thus made swings freely both ways, and may be fastened by a spring latch falling into a slot in the post. The eyes of the hinges must be large enough to allow the upright to pass freely up and down. The lower eye may connect with a shank passing through the post, *A*, and drawn tight by a nut.

XXIV. How to Prevent Posts from Sagging.

Gate-posts sag because they are often too small or made of soft wood, and are not properly braced under ground. The post for a heavy gate should be of hard wood not less than eight inches square, set four feet deep, perfectly plumb and braced under ground. For a heavy gate dig a trench five feet long for the heel-post of the gate. Frame the bottom of the post into a two or three inch hardwood plank, so the plank will project under the gate three feet from the post. Frame also a strong brace for the bed-piece to the post, tamp the earth solid about the post and it will never sag. For very heavy posts it is sometimes framed and braced at the front, and at the right and left side, when the gate is to swing both ways. But for heavy gates the slide arrangement is better.

XXV. Ornamental Gates.

Ornamental gates are now made in many designs, of a great variety of material, and from patterns always ready to be shown by manufacturers' agents, so that it is not necessary to elaborate them.

ORNAMENTAL GATES AND FENCE.

The illustration shows one pattern of ornamental gate in a square picket fence, including cased and capped posts. Any carpenter will understand from this design the manner of putting up any ornamental fence, the material being furnished, since it is simply a question of casing and fitting each piece in its appropriate place.

XXVI. Flood and Water Gates.

WHEN farms are crossed by streams, water gates and fences are necessary. The approaches may be of any kind, preferably such as may be removed easily from the low grounds, in time of flood. In the current, the gates must be self-acting, so as to give the least resistance to the water, and arranged to free themselves readily from trash and debris brought down by the floods. Two forms of flood gate are shown in the cuts. The first is apt to catch and hold all trash, though swinging freely; the

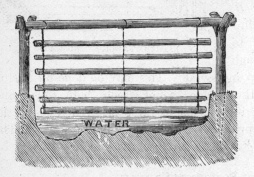

INCORRECT FORM OF WATER GATE.

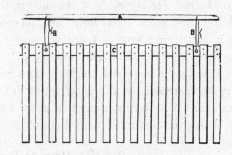

CORRECT FORM OF WATER GATE.

other will allow obstructions to be freely disengaged and pass away. The first is simple, effective as a barrier to stock, easily removed when necessary, even to driving the sharpened crotched posts, but incorrect in principle, simply because the slats are put on the wrong way. They are constantly catching trash, and consequently, often choked.

The second plan is correct, since it freely clears itself. A is the supporting pole, B B, the flexible hinges, wires or chains supporting the frame, C, the cross-piece upon which the slats are firmly bolted or nailed with clinch nails. In this form, when aquatic birds are to be prevented from passing, the slats may dip into the current; if not, they should be just above the ordinary stage of water.

XXVII. Stream Gate and Footway.

IT is often desirable to combine a stream gate and footway, and at the same time to arrange to raise the gate up in time of floods to allow free passage to the water, especially where there is a fall, or a swift current liable to sudden rise and fall, and perhaps carrying heavy trash. To do this the posts must be firmly set. The main figure shows the gate attached to upright posts with lever for raising the gate; it can be used across a stream fifty feet wide. Iron rods a quarter of an inch thick pass through the long and short pieces as shown; $c, c, c,$ are sections of chain ending in solid shanks passing through the revolving beam and fastened by nuts. When the gate is to be raised, as shown in the end figure, the lever revolves the beam, the gate

STREAM GATE AND FOOTWAY.

slides back thereon, and when the whole comes to a horizontal position it is secured. The end view shows the gate suspended on crotched posts, valuable when the banks are subject to washing.

CHAPTER IV.

FARM AND ORNAMENTAL HEDGES.

I. THE POETRY OF HEDGES.—II. ADVANTAGES AND DISADVANTAGES OF HEDGES.—III. HOW TO PREPARE THE HEDGE-ROW.—IV. SETTING THE HEDGE.—V. FINISHING THE PLANTING—CULTIVATION—VI. TRIMMING THE HEDGE.—VII. ORNAMENTAL HEDGES.—VIII. ORNAMENTAL PLANTS FOR HEDGES.—IX. HOW TO PLANT THE HEDGE.—X. CARE OF DECIDUOUS HEDGES.—XI. TREES FOR BARRIERS AND PROTECTION.

I. The Poetry of Hedges.

IN the equable and moist climate of England, the hawthorn, holly, privet and other shrubs or small trees, are well adapted to ornamental hedges. They bear close cutting and training, and are perfectly hardy; the last two holding their foliage all winter. In our extremes of heat and cold, the rigorous winters of the northern States are apt to prove, if not fatal, at least most injurious, while western droughts and southern summer suns are almost as destructive. Around the hawthorn, from its ebon bud until its flowers scent the gale; from when its shed petals whiten all around till winter sends, in berries, second bloom and decks its thorny boughs with gleaming scarlet, the poets, from Father Chaucer to the Idyls of the King, have hung their garlands. But, for us English of the West, the hawthorn tree has lived only in the poet's verse; and now, alas! in this prosaic age, English hedges are becoming a thing of the past. Like the poet and the painter they produce nothing but beauty; they cumber land which can grow gold in corn; they harbor weeds hard to uproot; they require time and labor to keep them in repair, for, left untrimmed and untended, they lose, not only their beauty, but their use; unsightly gaps appear, and now the practical English farmer, under the close competition of America, finds he must sacrifice poetry to pelf. Thus America, hedgeless by climate, in revenge kills the holly and the hawthorn of English fields.

II. Advantages and Disadvantages of Hedges.

THERE is a practical use for the hedge in the protection it gives, especially in the timberless districts, to the fields and stock, and to this we may add the pleasure a well-kept hedge affords the eye. The disadvantages are, they are costly to keep in order; they harbor weeds; they take up much valuable land; they prevent evaporation from roads, keeping them wet and muddy, and, if not carefully trimmed, they are unsightly. The question of fencing is one of the most important the farmer in any district has to meet, and this becomes more and more serious as we advance beyond the Mississippi, upon the vast plains, that were once considered a desert, but are now found to be among the most productive lands of the West. But ingenuity has solved the problem of enclosing regions far distant from timber, through improvements in wire fencing.

The Osage orange has played an important, in many districts an indispensable, part in the settlement of many prairie regions of the West. It may do so still in some remote regions, but neither this plant nor the three-thorned locust (*Gleditschia*), the only two hedge plants really adapted to the West, will be able to hold favor with those who regard space and cleanliness in fencing. But, as in any other operation, every farmer must judge for himself as to the economy of hedging. We believe that, simply as protection to fields, and as shelter to stock from sweeping winds—lines, or, better, clumps of trees along boundaries and principal fields, will prove more useful than hedges.

III. How to Prepare the Hedge-Row.

WHETHER hedging ever again regains its hold upon public taste or no, it will be used on many farms and, eventually, in an ornamental way on every farm. The osage orange will only thrive on dry soil; wet land is certain death to it. Hence, in preparing for a hedge of this plant, it is necessary to raise a slight ridge, even upon high ground; over low places this must be of some height, and have a waterway beneath, where the accumulation of water is to be carried away. In fact, all hedge plants do better on a slight ridge in prairie land, that in spring is always partially or fully saturated with water.

This ridge may be entirely made with the plow, harrow and leveler. Eight feet in width is none too much. Plow first as deeply as possible by throwing out the soil, leaving the dead furrow where the hedge is to stand. This should be done in the autumn. In the spring, as soon as the soil is in good condition, plow the furrows back, and again, deep. Three plowings should form the ridge, except in low places, where earth may be added with the scraper. Harrow and level until the tilth is perfectly fine and smooth, and leave the ridge to settle until wanted for planting.

IV. Setting the Hedge.

THE hedge plants having been bought and sorted into best, second-best and culls, the hedge-row is economically prepared as follows: Draw a straight line along the center of the ridge. With a steady horse throw out the earth with a bull-tongue plow or other implement that will move the earth to either side. Pass back and forth in this line, correcting it until it is perfectly straight and true. Upon a strong garden line, not less than two hundred feet long, sew strips of red flannel, at such distances as you wish your plants, say ten to twelve inches. One man, with a bright, sharp spade, walking backwards along the line, thrusts in the spade, obliquely at every mark, presses the handle from him, and an assistant inserts the root. The spade is withdrawn, the earth is stepped on to compress it firmly about the bottom of the roots, leaving the plant fixed, and slanting somewhat, in the direction the workmen are going, the spademan working backwards. Care must be taken that, when the earth is finally filled in around the plants, they are covered about an inch above the yellow portion of the root. To enable this to be done accurately, the line should be supported at proper intervals, at the desired height. The object in opening the trench is, to save labor in planting, and by this means. it may be accurately and speedily accomplished.

V. Finishing the Planting—Cultivation.

WHEN thirty or forty rods of hedge has been set the bull-tongue may be used to carefully cover back the earth to the plants, after which they may be brought into line and the earth firmly packed around them, the sides of the ridge being left rough the better to kill the first weeds that start. So proceed until you have all your plants set, first the best, then the second best, throwing away all inferior plants. If you have raised the plants set all culls in nursery rooms for future use, and if you buy them stipulate for No. 1 plants and accept nothing else. These may be divided into firsts and seconds. The first season's cultivation may be wholly with the straddle-row cultivator, and the plants must be kept earthed so that the yellow root does not show. The second year's cultivation may be done with any implement that will throw shallow furrows to or from the plants. No trimming is necessary the first two years. The object is to get a strong root, especially in the North, where the plant is liable to damage in winter, until it is three or four years old. After this time it is nearly as hardy as the oak, on dry soils.

VI. Trimming the Hedge.

IT has been found not a good plan to attempt to keep Osage orange in shape by trimming and shearing, as practiced for ornamental hedges. The most that can be done is to keep the upward growth within bounds by cutting back in the spring, and perhaps again in midsummer, to a height of five feet. This may be done with the

CORN KNIFE.

common corn or cane knife. When the stems of the plants have reached a diameter of about two inches it may be laid down. This is done by trimming up the sides so a man can work. The stems are then sawed two-thirds through with a rather fine saw, or cut and bent over in line by means of a very heavy pole, worked by a man on each side of the hedge, so the stems will lie at an angle of about 25 degrees. If they tend to rise they may be weighted down when necessary. This will reduce the height of the hedge to three feet or less, but the new growth will soon present an impenetrable barrier, and the following year the hedge will have attained its full height. All that will be required thereafter will be to prevent the branches rising above five feet by cutting the hedge to this height in the spring before the leaves start, and again at midsummer. This trimming may be done with a sharp corn knife, or with a similar tool made for the purpose, as heretofore shown. The strong limbs may be hooked off with a bill-hook, as shown in this cut, or with a hedge-clipper to be shown presently. For keeping out stock a well-kept

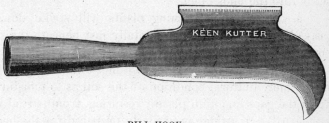

BILL-HOOK.

Osage-orange hedge is impenetrable, and for this purpose it is useful if not ornamental.

VII. Ornamental Hedges.

In preparing for ornamental hedges the directions given for Osage orange will apply. Every hedge should be set on raised land. The height of the ridge should

NORWAY SPRUCE AND ARBOR-VITÆ HEDGE.

be such as may be made by two plowings. In moist land three plowings will not be too much. Some of the advantages of setting the hedge on ridges are:

1. The hedge will be more likely to escape winter killing, the exemption being due to the fact that the roots are above the level of saturation.

2. Operations can be commenced and completed from ten days to two weeks

earlier, in all localities where the natural drainage is insufficient, and the plants can be set before the buds open.

3. The roots of young plants will strike down obliquely in ridged ground, instead of extending out horizontally just beneath the surface soil, and attain a growth corresponding with the increase of available soil.

4. The young plants make a more uniform growth when ridged, in consequence of the more uniform condition of the soil as to moisture, and will generally be exempt from the gaps and thin places, resulting from partial winter killing.

5. When a ridge is properly prepared for the hedge, the roots of the hedge-row will form a more fibrous growth, which will be made chiefly in central parts of the ridge soil, instead of the roots growing long and straggling. If, in the course of years, however, straggling roots should be found to require pruning at a distance of eight or ten feet from the hedge-row, they will present less obstruction on a ridge than when grown upon level ground.

WHITE OR EVERGREEN THORN.

6. When a hedge becomes strong enough to turn stock, it is desirable to check its growth, which can be done by cutting off the ends of the roots on the sides of the ridge with a pruning plow, or with a revolving colter, and this without endangering the life of the hedge, the large amount of root growth in the deeper, central parts of the ridge being sufficient for the plant.

7. A ridge eighteen to twenty-four inches above the level will add thirty to forty per cent to the effective height of the hedge; and, in combination with the latter, will form a barrier that will turn stock, thus constituting an effective fence from one to two years sooner than when planted on low, level ground; and, at the same time, equally contributing toward the effectiveness of the hedge in its incidental capacity as a wind-break.

VIII. Ornamental Plants for Hedges.

AMONG the plants for ornamental hedges in the West, none are more common than the Arbor Vitæ. A better plant would be the hemlock, if it did not refuse to grow when transplanted, except here and there, in a favored position. Privet (*Ligustrum vulgare*) is hardy up to about forty-two degrees, and once established, it is beautiful, being evergreen. It stands the shears admirably, forming a low hedge. Within the grounds for a border to walks, or as a low division hedge, nothing is prettier. In the South the Cherokee rose makes an admirable hedge, especially when covered with its profuse blossoms. The evergreen thorn, or pyracanthus, also is handsome, but it must have a deep soil, and will not stand a temperature much below zero. The illustration shows a section of white evergreen thorn (*cretagus pyracanthus alba*). In mild latitudes it is evergreen, and nearly so north up to forty-two degrees, in protected situations. The illustration will also show the manner of forming any hedge, by upright and lateral shearing, and is applicable to all that class of plants with small flexible shoots.

HEDGE CLIPPER.

The Japan quince is hardy in the North when well established, and fully so up to forty-two degrees. It is handsome always, and lovely in spring, with its deep-red blossoms. The purple-leaved barberry makes a good ornamental hedge, and is quite hardy when established, even in Wisconsin and Minnesota. But no person should undertake ornamental hedging unless prepared to give it proper care in trimming and shearing. In this respect, no hedges are more easily kept than Arbor Vitæ.

IX. How to Plant the Hedge.

ORNAMENTAL hedges are always started from rooted plants. The soil having been prepared as heretofore recommended, stretch a line marked with red flannel at the proper distance for the plants, twelve inches for evergreens, and eight inches for privet and barberry. Select plants twelve inches high, of evergreens, and two-year-old plants of deciduous species. Cut the latter down to within about eight inches of the ground. Set them exactly in line, about as deep as they hitherto stood, spreading the roots properly and pressing the earth firmly about the roots with the hands. This is essential.

If the ground is dry, give the whole a good watering, and when the water has settled about the roots, draw the dry earth over all and rake smooth, preserving the regular crown of the surface of the bed. If they can be thickly mulched the first season, so much the better. If not, the soil must be raked over often to keep up the tilth and kill weeds. In any event a good covering of mulch should be given the first winter. The first season, and perhaps the second, no trimming will be required. With evergreens none will be required until a height of about three feet has been reached.

Then it may be clipped at the top to prevent too strong upward growth, allowing the hedge to increase in height only a few inches yearly, until the ultimate height of four or five feet has been reached. Probably no ornamental hedge should ever be allowed to outgrow a height of four or four and a half feet. The arbor vitæ requires little or no trimming until it reaches the desired height, and then only to keep it in fair shape. In fact, except in the most perfectly kept places, a little irregularity will take away that air of primness that most people dislike. But be careful not to go into the other extreme, for shabbiness in an ornamental hedge is always annoying. The question of shabbiness is just here. If the hedge has not been properly taken care of while young, the bottom will kill as to leaves, and in the end the hedge will present the appearance of an inverted pyramid. The true hedge must be broad at the base, and narrow or square at the top. With this explanation the whole matter of hedging should be sufficiently clear, at least so far as evergreen hedges are concerned.

X. Care of Deciduous Hedges.

THE common mistake with persons unused to ornamental hedges is that they are in too much of a hurry to get their hedge. They want height, when breadth is really the key to success. It takes years to form a perfect hedge, but, once formed, it remains a beautiful object so long as it is cared for. The hedge should not be allowed to increase in height more than a few inches yearly, the object being lateral growth. Hence, the spring cutting must be ruthlessly carried to within four or five inches of that of the previous season. Here the good sense of the operator must be on the alert to discover weak places that require thickening, and perhaps places which, the growth becoming tangled, require thinning; in fact, the working of deciduous hedges must be something like that of evergreen hedges, except that the cutting must be more severe. Shearing may be practiced with ornamental hedges, but for the ordinary farm hedge, the corn-knife and bill-hook will be the principal implements required. The cuts given in this chapter will show how both evergreen and deciduous hedges should appear when fully formed, and the general principles stated must suffice, since to particularize as to each individual species would fill a volume, and be interesting to few.

XI. Trees for Barriers and Protection.

AMONG the trees that bear cutting well, and are hardy everywhere, none is better than the Norway spruce. It may be stopped in height anywhere from five feet upward, and is excellent for shutting out unsightly objects. The purple-leaved beech is another deciduous tree that bears cutting well, and is handsome, the leaves often remaining all winter. The golden, and especially the white willow, may be topped and made pollards of at any height, and are especially available where quick growth is desirable.

OSAGE ORANGE AS A TREE.

The hardy catalpa (*C. Speciosa*) is desirable for road-side planting, where the trunks are eventually to be used as posts for stringing wire. So is the three-horned locust, and the Osage orange especially so, if you plant it at regular distances, and can afford to wait for it. It is one of the most beautiful of trees, with its glossy leaves and magnificent fruit, and as valuable as it is beautiful. How beautiful, let the accompanying illustration show. It is true it is rather a slow-growing tree, but always elegant in shape, as it is glossy in foliage and magnificent in its fruit. When grown, no timber is more valuable. It is dense, strong and lasting, and has the remarkable property of not shrinking and swelling when exposed to the weather. In fact, in any district of our country, any observant person has only to look about him to discover a variety of trees of valuable characteristics when planted on the farm. This subject, and also plants for ornamental hedging, will be further treated of in Part V., Chapters V. and VI.

CHAPTER V.

DRAINAGE AND THE DRAINER'S ART.

I. THE IMPORTANCE OF DRAINING.—II. THE ANTIQUITY OF DRAINAGE.—III. ANCIENT WRITERS ON DRAINAGE.—IV. DRAINAGE AMONG THE GREEKS.—V. DRAINAGE DEFINED.—VI. DRAINAGE AMONG THE ROMANS.—VII. DRAINAGE BY FRENCH MONKS.—VIII. SOME FATHERS OF MODERN DRAINAGE.—IX. THE ORIGIN OF TILE.—X. PRACTICAL MEN ON TILE DRAINAGE.—XI. A DRY SURFACE MAY NEED DRAINAGE.—XII. WHAT AN OHIO FARMER SAYS.—XIII. DRAINING IN INDIANA.—XIV DRAINING IN MICHIGAN.—XV. ILLINOIS EXPERIENCE.—XVI. A RIGHT AND A WRONG WAY FOR OPEN DRAINS.—XVII. STOCK WATER FROM DRAINS.—XVIII. HOW TO EXCAVATE THE POND.—XIX. DRAINAGE AND FENCES.—XX. THE FORMATION OF UNDERDRAINS.—XXI. VARIOUS MEANS OF DRAINAGE.—XXII. STONE-LAID DRAINS.

I. The Importance of Draining.

THE railways of the West early saw the advantage of tile drainage, and the Chicago, Alton & St. Louis, and the Illinois Central set the example of transporting tiles, for the farmers, at nominal rates, well knowing that the increase in the productivity of the soil would soon amply repay them in increased freights. The farmers of Indiana, Illinois and Iowa are now eagerly working at subsoil drainage, and thousands of acres are there yearly added to those already so improved. The "Prairie Farmer," the oldest agricultural journal in the West, was the pioneer in thus calling attention to the importance of this subject, as was the "Country Gentleman," the oldest agricultural paper east of the Alleghanies. Nor do we remember a single instance, where a paper of this class, east, west or south, has failed to labor in this direction. The "Breeder's Gazette" was, we think, the first paper, especially devoted to stock, which gave this subject of drainage attention. In it, the author of this book furnished a series of articles from which, as he finds them, on review, to be just, he will largely quote. This subject will be profusely illustrated, for a picture often shows, at a glance, what it would take pages to describe in words.

II. The Antiquity of Drainage.

MANY think drainage to be a modern art. It was ancient in the palmy days of Greece and Rome. In fact the necessity of draining is one of the first things that presented itself to man whenever, in the settlement of a country, the high and least fertile lands have been occupied, or where advancing population required the lower and more fertile lands for cultivation. Hence, in ancient civilization, when from the necessity of providing for mutual defence by congregation on restricted areas, and where from lack of inventive talent cultivation was restricted principally to manual art, such lands as could be most easily worked soon became the most valuable.

Carrying off the surplus water in countries where the rainfall was ample, and the provision of irrigating ditches in arid sections would naturally and easily be suggested. From this the transition to covered or underground drains is easy.

III. Ancient Writers on Drainage.

Among the ancients whose writing on drainage have come down to us, Cato, Varro and Virgil mention only open ditches or drains. But in the days of Tiberius, Columella, who lived then, wrote both of open and covered drains. It does not necessarily follow, however, that the art of builing underdrains was then first discovered. On the contrary, they then first began to come into general use, or at least to be well known. The drains mentioned by Columella, besides the ordinary ditches or open drains, are stated to have been three feet deep and filled to half their depth with stones or pebbles, and also with fascines (brush tied in small and regular bundles) properly laid to allow free percolation of water, and covered with earth.

IV. Drainage Among the Greeks.

While no authentic records of a system of thorough drainage have come down to us from the Greeks, the vast subterranean canals built by them to carry off enormous accumulations of surface waters to prevent floods; the outlet of the Lake Stymphalide which, when obstructed, covered a surface of about thirty miles; the Alpheus, that in its course several times disappeared, and which, according to tradition (incorrect of course), found its way at length to Sicily. The Plain of Orchamenes, when its underground outlet became filled, formed a marsh. (The Plain of Phenea was drained, according to various authorities, either by an earthquake at a remote period, or by a prince, was accomplished by two abysses which draining the whole rendered the country healthy.)

The Valley of Artemisium became marshy whenever its subterranean water-way was obstructed at the gulf which was its outlet. This water-way extended (according to Pausanias, the authority for all these statements, except the first named, which is by Strabo) to Genethlium, a city at the head of the Lake Dine. It will be noticed that these works are mostly accorded to a traditional age, or else to providential interposition. Were they not the remains of a previous civilization of which the Greeks knew nothing? If really built by this people, or even the remains of previous civilization, is it credible to suppose that the builders were unacquainted with the more simple and easily accomplished thorough drainage of the soils which they could only have imperfectly drained without thorough drainage? We may thus suppose the Greeks were not the builders:

These vast underground outlets, so far as they are authentically stated to have been built by the Greeks, were undertaken as a hygienic measure and at the public expense. This, however, is only another proof that the vast areas drained, to have met the object intended, must have been supplemented by a more or less thorough system of minor drains. Is it to be supposed that the civilization that could compass the drainage of vast areas would stop short of the more important one of rendering these drained soils fit for cultivation? Certainly not. Until this was accomplished, the original idea—health—would not have been accomplished. Again, then, the probability is that the Greeks were not the builders, but that previous people, the heroes of which were worshipped as gods.

V. Drainage Defined.

WE will here speak only of the drainage of the soil for agricultural purposes; that is, the drainage of the soil of fields to enhance their fertility in the production of crops. The subject of drainage in its entirety is one that has occupied the attention of the most eminent men in every civilized age. It embraces many branches, among the more important of which are house drainage [see chapters relating to architecture], sewage drainage, the drainage of marshes, having for the primary object the prevention of malarious diseases, and farm drainage, which renders a soil originally wet and unhealthy to agricultural crops arable, fertile and healthy through the constant removal of excessive moisture.

Thorough drainage not only does this, but in times of drouth it actually causes the accumulation of moisture in the drain pipes and in the pores of the soil, by condensation of the moisture of the air, through a well-known meteorological law. In isolated cases this condensation has been so great as to cause the dropping of water from the pipes at times when the moisture could have been produced by no other cause.

Drainage, in a general sense, then, may be defined as being a system by which surface water may be quickly carried away. This applies more particularly to open ditches or to covered water-ways having a surface inlet and an outlet from whence the water may pass freely off.

Underdraining.—Underdraining is that system having subsoil water-ways formed of tiles, stones, gravel, brush, poles, slabs, or even, in stiff clays, an earth channel, protected artificially, by which the surplus water of the soil is quickly percolated and carried away.

Thorough draining is that system of underdrainage whereby a tract of land otherwise unsuited to cultivation may be rendered uniformly arable and fertile. Drainage is as old as civilization. Our system is built upon an improved form of the old Roman system. Modern—thorough—drainage has only been generally possible since books, and especially journals devoted to agriculture, have been common, through the general education and consequent general intelligence of the masses.

It may help us to a better understanding to show something of what ancient civilization knew of agricultural underdraining, and the progress of the art up to the introduction of thorough drainage by means of continuous pipes having inlets for moisture at the joints and through their pores.

VI. Drainage of the Romans.

The Romans in Columella's time understood not only the proper slope and correct bottoming of open ditches, but, besides underdrains of stones and brush, they were acquainted with the use of clay pipes for carrying water underground. They are probably the real inventors of the system; and ancient fields so drained have been found in lower Austria, Saxony, and in other countries having communication with the Roman empire.

Hidden Drains.—Columella taught that when a soil was wet ditches were to be dug to allow the water to run off. He also taught the use of both wide-open drains and hidden underdrains. Of these latter we quote from the text as given by the late J. H. Klippart: "One will dig out trenches of three feet in depth, which shall be half filled with small pebbles or pure gravel, and then the whole will be covered with the earth which was taken out of the trench. Should there be neither stones nor gravel, then fascines, formed of branches tied together, of the same shape and capacity of the trench, may be placed into it so as to fill up the cavity. When the fascines have been sunk into the bottom of the canal, they must be covered with leaves of cypress, pine, or of any other tree. Then shall be superadded the earth extracted from the trench, and the whole will be strongly compressed. At both ends must be placed, in the form of a buttress (as it is done for small bridges), two large stones, surmounted with a third one, in order to consolidate the sides of the ditch, and favor the fall and exit of the water."

Open Trenches.—A drain as carefully made now would leave little to be desired, as it is the most perfect form of drain other than that of tile. Palladius, who wrote long after his eminent predecessor, and who evidently had access to his writings, says: "When the lands are wet they will be dried up by digging trenches everywhere. Every one knows how to make open trenches, but here is the way to make hidden trenches: One must cut out across the field ditches of three feet in depth, which are half filled with small stones or gravel, after which they are filled up with earth from the digging and leveled. But the ends of those causeways must lead in declivity into an open ditch, whither the water will run without carrying away the earth of the field. Should there be no stones, one will lay at the bottom of the ditch fascines, straw or briers of any kind whatever."

From these extracts we find that in the time of Columella, who lived about the year 42 of the Christian era, that not only the making of open drains, but also underdraining was well understood; and that when Palladius (Rutillius Taurus Æmillianus) lived (about the fourth century) that open drains were so common that all were supposed to know how to construct them. Even in the time of Columella the proper depth, economically (three feet), of these underdrains was also known.

VII. Drainage by French Monks.

During the dark ages that succeeded the fall of Rome, agriculture, with literature and art, languished. The monks held what was known of higher agriculture in their own hands. That they did not neglect drainage is evidenced from the fact that excavations on an estate undergoing repairs, in the latter part of the last century, at Maubeuge, France, showed a system of thorough drainage at a depth of four feet, of hard-glazed earthenware pipes, ten inches long and four inches in diameter, very hard and vitrified, evidently made by hand and lathe. It is not known when they were constructed. The grounds were originally occupied by a convent of monks, the chapel, which remained, being of the pure gothic type. The convent fell before the fury of the republic of 1793, but the garden, renowned from time immemorial, escaped.

Indestructible Drainage.—According to a memoir of G. Hamoir, member of the Agricultural Society, there were under this garden "two complete and regular pipe drains, extended throughout the whole garden, at the depth of four feet. One of the drains had all its pipes radiating to a sinking well situated in a central position; the other was made of pipes, all parallel, ending at a collecting pipe which discharges into a cellar." There is no record of the construction of this system of drainage, but there were tombs placed over the drains in 1620. This system, admirable in its work and correct in its engineering, must have been made anterior. When? It may never be known. But it teaches us, to-day, that if those drains worked perfectly and without repairs for hundreds upon hundreds of years, it settles the matter, once and for all, that perfect drainage is substantially indestructible. Hence, drainage is, and should be considered in the light of a permanent investment.

Thus, in estimating the value of drainage, we have to estimate simply the annual interest upon the first cost or capital originally invested. Hence, again, it will pay to do this work in the most complete manner.

VIII. Some Fathers of Modern Drainage.—Springs.

As in the days of Rome, we have Columella, who was contemporary with Christ, and Palladius, who wrote in the fourth century, so in modern times, Oliver de Serres, in France, wrote, in 1600, his "Theory of Agriculture," including a complete description of underground drainage, and strongly recommended its utility. He advised a depth of four feet, to cut off the source of springs; thus giving the idea that in that day the art was carried out more to cut off the subterranean water of springs than to remove the surplus or superfluous water of rains. This author describes minutely the making and the filling of the drains, using stone, brush and even straw. Speaking of this material he says: "Straw, thus employed, will last a long time, for it is admitted that, being enclosed within the earth, and without the effects of air, straw remains sound an hundred years. I am a witness that some sound straw was found entire in the midst of an old, ruined house, and the wall appeared to be the work of former ages;" and adds: "Therefore, use it without scruple, with the understanding that if it should rot at the end of a hundred years those who will come then may change it if they have a mind to."

IX. The Origin of Tile.

This French drainage, however, was not thorough drainage with tile. The invention of underground drainage has been claimed as English. They cannot even claim priority in the use of tile. It dates back to the days of the monks in France, of which no history of the time when it was laid is given, unless it may yet be found in monkish records. To the English, however, is due the elaboration of a system ending in the almost universal adoption of porous pipes, and within the last fifty years brought to a high state of perfection, in engineering, mechanical appliances, exactness of the work, and great excellence in the manufacture of tile. In no country, perhaps, has so strong an impetus been given as within the last few years in the West. And now that the economical perfection of shape has been reached in round tile, both as to its outer surface and its caliber, there is no reason why, through the multiplication of manufactories of tile, any person having a soil not naturally well drained, when the improvement of the crops raised will more than pay the annual interest on the capitalized cost of drainage, should hesitate to use them. We repeat, none such should hesitate to undertake drainage in a more or less thorough manner, upon a proper survey of the particular requirements of the soil for the intended crops.

X. Practical Men on Tile Drainage.

Tile drainage is an ancient art, lost during the dark ages, and slowly revived in modern days. Even in England, where the great value of farming land, the abundance of capital and the low rate of interest all favored its extension, the spread of its use has been slow. Let us give a short review of its history and the testimony of practical men in different parts of our country as to the benefits derived from it.

Mr. John Johnston's Testimony.—No person in the United States has probably exerted a wider influence in the early introduction of draining and persistent effort in doing the work thoroughly than Mr. John Johnston, who settled near Geneva, N. Y., and who, at a time when tile works were almost unknown in the United States, was obliged to import tile of the old-fashioned horseshoe pattern from England. He not only paid for his farm through the enhanced products per acre (and this in the face of the sneers of those about him, that he was " burying his money by putting crockery in the ground "), but he kept on buying and making tile pottery until he had 210,000 tile in the soil, paying for his original purchase of land and adding to his farm, until he had over 300 acres in such a state of cultivation and productiveness that his would-be sympathizers might well hang their heads in shame that they could not have seen when the first laid tile began to draw that he was " sowing money to reap one hundred-fold; " in other words, getting one hundred per cent yearly profit, and this when tile cost him $24 per thousand to make, for for this was forty years ago.

Mr. Johnston says, among other things, that tiling paid for itself in ten years. One field of twenty acres that hardly produced ten bushels of corn per acre before draining, produced after draining an average of eighty bushels per acre; and it was also found that half the manure sufficed on drained land to that required on undrained land—that is, on land that previously was at times, during the growing season, sodden with water.

XI. A Dry Surface May Need Drainage.

Nor is land apparently dry on the surface exempt from the necessity of drainage. Mr. Johnston had a field of thirteen acres on the shore of a lake, with a bluff bank from thirty to forty feet high descending nearly perpendicularly to the lake. The soil seemed dry, yet was not satisfactory in its crops. He engaged men to open a ditch, with the understanding that if water entered within eight hours he would have the whole field drained. The top soil was hard and dry, so much so that a pick had to be used. At the depth of a foot it was so wet and soft that it was easily spaded. As the ditches were opened, water flowed in and ran away from the outlet. It was thoroughly drained, and then commenced regularly to produce sixty to seventy bushels of corn per acre, and proportional crops of other grain. He testifies that he never saw a farm of one hundred acres but some portion of it would pay for drainage. Every man in the West who has ever done any draining knows this will apply to all soils where the top and subsoil is not sand or gravel.

XII. What an Ohio Farmer Says.

In the Ohio reports for 1878, Mr. S. J. Woolley testifies that drainage with him has increased the richness of his meadows and pasture, and it has not only improved the wealth of the people but alleviated the condition of animals. This has too often been stated to be disproved. He says that a forty-acre field which, before draining, produced not more than eight bushels per acre, produced, after being drained, from sixty to eighty bushels per acre, and with much less labor of man and team. A thirty-acre field on the same farm was sown to wheat; it winter-killed badly and was injured by rust. Since being drained it has produced large crops of superior wheat, and the crops have not been affected by rust.

So, also, on drained land, potatoes were large and of fine quality, when before they were inferior and suffered from rot. The subsoil of this farm was a tough, sticky blue clay, difficult to plow and work. Since it was drained, the clay has become friable, loose, easy to work and has changed color, so that now it is a fine, black loam, and works easily the whole season. But the writer adds, that if swampy timbered lands are suddenly drained, most of the timber will die—the oaks and hickories first, the change being first noticed in the tops. The young timber, however, accommodating itself quickly to the change, suffers but little, afterwards growing more thriftily than ever. In this connection we add: Land intended for planting timber should not be tile-drained. The roots will inevitably, sooner or later, choke the drains. Here dependence must be had on surface or open drains, or else the drains may be filled with brush.

Samuel Israel, Esq., near Mount Vernon, Ohio, in relation to a farm of beech land, the subsoil a tenacious clay, abounding in low places holding surface water during the spring, fall and winter, testifies to the value of draining, and specifies that over $40 per acre had been realized from sixteen and one-half acres drained and planted in potatoes, and this without manure.

How many farmers are there who might give similar testimony of lands which in dry seasons produce large crops, but which, one year with another, are wet and sodden in the spring, and often, as last season, remain too wet to work until well into the summer? Had it not been for the wonderfully genial and pleasant autumn weather extending into November, of 1882–83, the corn crop of the State of Illinois would have been almost totally ruined in these years.

XIII. Draining in Indiana.

THE wonderful results from drainage have been most thoroughly shown in Indiana. Commencing about ten years ago, the demand for tile has continued greater than the supply from year to year. At length the tile-makers of the State organized an association in which is discussed the best form of tile, the quality of clay for working, and the gathering of statistics relative to drainage, and drainage processes have been undertaken, which have been of great value generally, and have done much to spread correct information, in connection with the efforts of the State Board of Agriculture in the same direction. Of the thousands upon thousands of acres there drained, the testimony is constant as to the money value of drained over undrained soils. Each recurring wet season goes more and more to intensify the belief that drainage is the one thing most needed to produce the best results in tillage upon our prairie soils, when they are flat and where they are underlaid with clay. Illinois, later, fell into line, and has now an active tile-makers' association.

If such soils were tile-drained to a depth of three feet, the tiles laid in lines corresponding to the natural slope of the soil, the whole of the land would be drained. If the underlying water were pretty uniform through the under surface, the drains would, even if laid three feet deep, require to be, perhaps, thirty feet apart, in the case of strong, tenacious clay, and from this to forty or fifty feet in soils less tenacious.

Again, in draining many soils, especially where stock-raising is the principal object, the most that will be necessary will be to run a main drain down the gently sloping valleys to carry off the superabundant water, to prevent long saturation and consequent slow evaporation at the surface, keeping the soil cold and sodden below the water line. This single drain will, as a rule, relieve all such lands, unless the valley is very wide and flat. When such is the case, lateral drains must be laid to connect with the main drain, striking it not at right angles, but somewhat obliquely.

XIV. Draining in Michigan.

As long ago as 1867 a committee of the Michigan Agricultural Society, reporting on the subject of drainage, gave an account of the profits therefrom, the land in question comprising twenty-five acres of swale land, producing the coarsest vegetation—bog grass, flags, rushes and other worthless plants. About 2,400 feet of tile was laid, at an aggregate expense of $480. The grass product of the field the next season, after draining, gave $1,570; expense of crop, $541.25; drainage, $480; $1,021.25, or a net profit of $548.75. The second year the land produced crops of the value of $1,425; expense of crop, $550, or a net profit of $875, equal to a rent of $35 per acre on land originally worthless.

The increase in value on arable lands, requiring only partial drainage to bring the whole into a homogeneous state, will be fully as great, or greater, according to the outlay, since a comparatively small portion only, will need drainage, often not more than one acre in ten of the whole farm.

XV. Illinois Experience.

In 1875 Messrs. Spalding & Co., Riverton, Illinois, made a report to the Department of Agriculture at Springfield, Illinois, of the thorough draining of eighty acres of strong loam, the subsoil being strong clay and reddish clay. The land drained naturally to the south, and was intersected by three low ridges. Most of the land was considered to be well adapted to ordinary farm crops, but the low lands were too wet for cultivation, and it was decided to drain the whole, wet and dry alike. Mains of five-inch tile were laid between the ridges, at a depth of three and one-half and four feet; laterals of two, three and four inch tile were laid to connect. The next season the whole was planted in nursery trees and plants of the varieties usually grown, including ornamental stock. As we have seen this nursery several times since, there was no seeking to find a good spot for this or that stock; the whole surface was alike dry and friable, and remains so, perfectly, to-day. The proprietors estimate that the value of the land has increased from one hundred to two hundred per cent, and Mr. Spalding assures us that wet seasons have fully proven the importance of drainage, since their success in producing stock of a high character is fully apparent. There is no weak, indifferent stock from being grown on water-soaked land.

In 1866 the writer had a field of twenty acres that had been thoroughly mole-drained some years previous, but which had ceased to be dry enough after heavy rains for garden crops. It was decided to tile-drain it. The land was so nearly level that one line of tile, in particular, had only one inch fall in three hundred feet. The tile could only be laid two feet eight inches deep, on account of the depth of the ditch at the outfall. The work was done in the best manner, and the next season the soil was in good condition, although it was a rainy year, and no difficulty was experienced in raising any of the crops usually grown in a market garden. The diagram of drained field, Chap. VI, page 351, will illustrate the drainage of a portion of this field, the drains thirty feet apart, except there being no seipage, the top drains A B, and lower drains, C E, were not used, the outlets being directly into a ditch on the road. F G, also, is a covered drain.

Mr. Patrick, of Du Page County, Illinois, is known as one of the best farmers in his section. He believes in drainage. Upon his farm there are many circular sloughs, or wet places, which retain water for a long time. He has adopted a plan of his own which has worked well with him, and which will work well in all similar cases. The cut will illustrate his system of draining. It will also serve as a lesson to show the necessity of careful thought by the owner of a farm in deciding upon the plan which will prove most economical when only partial drainage is needed. He has to protect against seipage, and also to provide for draining the low places and carry away the water. Hence, he lays a line of three-inch tile (two-inch tile, as shown in the cut, will do, except in rare cases) entirely around the slough, with laterals between. In this case, the only question is, whether less tile will suffice to take the seipage according to his plan than to run the laterals far enough into the bank to take it in the usual way. He says the first crop after the tile was laid fully reimbursed the expense of

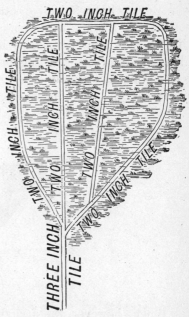

DRAINAGE OF SLOUGHS.

draining, besides leaving him in possession of these drained acres, as being the most productive on his farm.

XVI. A Right and a Wrong Way for Open Drains.

In digging drains there is always a right and a wrong way. An open ditch must have slope sufficient to its sides so they will not founder down from the washing of water or the action of frost. When they are intended simply for carrying away super-abundant water from beyond, they may be made entirely by the plow and scraper, and this also allows vehicles to be easily driven over them, and this surface need not be lost. With a little care they may be seeded to grass, and this again will prevent washing. Our plan for making these carriers is to plow a strip of land from twelve to sixteen feet wide, according to the depth required, turning the furrows to the outside, and so continue until nothing further can be gained in moving the earth with the plow. Then, with a scraper, carry the earth out, spreading it equally on each side over the outer surface.

Then plow again and again, and scrape until the required depth is gained at the center. When considerable depth is attained the chain to the doubletree must be lengthened. If the slopes of this carrier are harrowed to a good tilth, and a little fine manure spread on the surface, it may be seeded with red-top or other suitable grass.

When the land along which this carrier is to be made is sinuous, or not in a straight line, the windings must be followed measurably, but every advantage should be taken to cut off the turns as much as possible, to bring it straighter. Here a little work with a spirit or other level may be advantageous to discover how much may be cut off at the turns. These higher points may be deepened afterwards with the plow and scraper, to bring the fall as equal as possible. If a main underdrain is ever to be laid, this preliminary work will not be lost. You will be enabled to place the main deeply—an important matter—and the original carrier will serve to carry off the water of spring, or of heavy flooding rains quickly and easily.

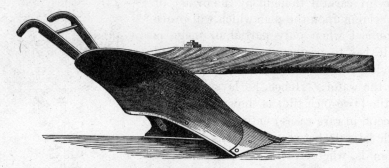

DEEP-TILLER PLOW FOR WORKING DITCHES.

When, however, the declivity at certain points is considerable, and these carriers are deemed necessary, such places should contain stones, pebbles or brush to prevent washing. In this way the carriers may sometimes be available as a surface out-fall for lateral underdrains; but, as a rule, the covered mains of tile are better and cheaper in the end.

XVII. Stock Water from Drains.

One advantage of underdraining is that the mains will often furnish permanent stock water, where otherwise the supply would not be available except from wells. When the mains are laid for considerable distances in long sloughy valleys, or in valleys to connect one system of wet-land drains with another, all that is necessary when a drinking place is wanted is to excavate about the required place, the sides sloping equally to the required depth, and put in a water-box. The cut shows the water entering from the pipe at in-flow, which should be higher than the pipe in the other side at the out-flow.

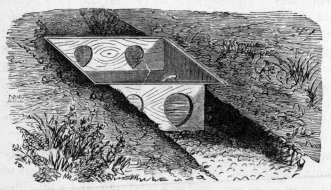

WATERING-BOX FROM UNDERDRAIN.

The ovals show the water line, and are for smaller animals to drink from. The standing places and sides must, of course, be protected from poaching. Of course the water will stand in the box on a level with the outflow pipe. This box should also be protected so that stock may not get into it.

Another plan answering the same purpose is to excavate a passage down to a level with the top of the tile, and, say sixteen to twenty or more feet wide. A trough is bedded down in place of the tile, and covered so that only a space along the top is left sufficiently wide for stock to drink through. The water enters at one end of the trough out of the tile, filling it, and passes out at the other end by means of a waste pipe connecting with the line of tile below. These drinking places may thus be multiplied indefinitely, according to the necessities of the stock, and the longer the line of tile the larger will be the water supply.

Another plan entirely feasible is to allow the tile to empty into a pond, excavated at some suitable place, and of greater or lesser extent, according to what is required. If in a valley, select a spot where there are good banks on each side. But the point where the tile empties into the pond at the upper part must never be covered by the water of the pond. Hence, the dam at the lower end must be far enough away so that whatever the height of water at the dam at the lower end, the level will never reach up to within two inches of the incoming tile, else there may be stoppage from the collection of silt.

XVIII. How to Excavate the Pond.

When the pond is small, the earth to form the dam may be taken from the upper side of the pond and scraped to the lower side, to form the embankment of the dam. Thus you may get a solid body of earth, wide enough so the water will not affect it, and the pond will be of uniform depth throughout — an important matter, since the evaporation from a deep body of water is no greater, for a given surface, than from a shallow surface. A pond should never contain less than four feet of water, and if deeper still better. The dam must be protected from the burrowing of animals, as muskrats, etc. [See Part V., Chapter VI.— Fish and Fish Ponds on the Farm.]

A dam for not more than four or five feet of water may be so protected by driving down plank through the bank, side by side, across the bank from one side to the other, after the embankment is formed to the required water line. Thus one may always have a reservoir of water for supply during droughts, and the water may, if necessary, be carried in pipes to any distance, if it be not higher than the water of the pond.

Field Drainage Illustrated.—The accompanying diagram shows the drainage of a field of Mr. S. E. Tod, New York. In the main drain may be placed watering boxes (see page 334), or a pond may be excavated between H and D to take the place of the main drain. These points the owner of the land must decide. In the case illustrated there would be water enough to amply supply the evaporation of a large pond where fish might be kept. Mr. Tod says the soil was a heavy clay loam, and the subsoil a retentive calcareous clay. The ditches were made about forty feet apart over the entire field. During a portion of the time a small stream of water that would all pass through a four-inch tile, flowed over the surface in the valley from B to L, where a main ditch was sunk to a minimum depth of three feet, in which a course of four-inch tiles was laid. As there was a valley at B O and at D D, sub-mains of three-inch tiles were laid as represented to connect with the main drain. The parallel ditches were then made up and down the slopes as nearly as practicable. From A the water would run most readily to B B. At G H short

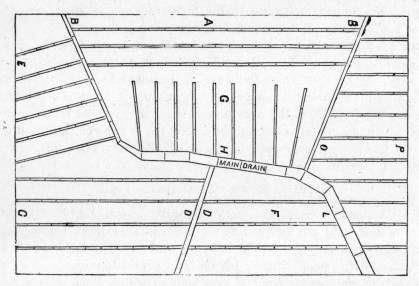

DRAINAGE MAP—POND MAY LIE IN A PORTION OF G H.

branches were made up the slope, in which one and a half inch tiles were laid. At P O the ditches all ran directly up the slope. From F the water ran either toward the main L or the sub-main D. From C the descent was more uniform toward D D. Hence, parallel ditches were made as represented. The object in laying out the ditches in so many directions was to have them extend, as nearly as practicable, directly up and down the slopes.

So far as farmers are concerned, in the drainage of land, the whole matter is more simple than is practiced by experts in thorough drainage. It is generally done by piece-meal, as occasion requires. Open ditches are first used. These subsequently form the main drains. Then the laterals are run into these, and the whole filled up as directed elsewhere. The plat shows this natural system of drainage, and we have used it to illustrate how a pond may be made as described. With the directions given, and the simple instruments designated, no person need err in accomplishing any ordinary piece of drainage.

XIX. Drainage and Fences.

In many localities where fencing material is scarce, a ditch and embankment is a most economical way of making a barrier against cattle. A light wire fence on top will afford the most perfect protection. When protection is only sought from the outside, the ditch must be outside, but when it is required that cattle or other stock be kept from encroaching from either side, there should be a ditch on each side.

This ditch should be not less than two and a half feet inside, and twenty inches is deep enough for all practical purposes. The slope should be considerable, not less than forty-five degrees, or even more, according to the nature of the soil; a firm, hard soil, or one containing a good deal of firm clay, requiring the least inclination. The best tools we have ever used for open ditches, are the common digging spade, to be hereafter shown, and the long-handled round-pointed socket shovel, here shown, for throwing out loose earth.

We have never had stock break over a ditch and sod bank, when the ditch on each side was two and a half feet across, and the bank on top protected with a fence two feet in height. If white or yellow willow, or Osage orange be planted on top, the barrier will be impassable to any stock. Division fences can also be made in this way, and thus in many instances subserve the double purpose of drainage and fencing, and at the same time, if the bank be planted, it will afford admirable protection to stock against sleet, driving rain and wind.

These ditch banks, and the sides of the ditches as well, should be made to bear grass, and to the end that they be as quickly and completely covered as possible, the banks should be carefully graded, the sods taken from the surface of the ditches forming as much of the bank as possible, and used next the ditch, preferably. Then, if some very fine manure be given to the top of the bank, grass-seed—blue-grass, orchard grass, red-top, etc.—should catch nicely. This may be assisted, however, by grafting (bedding in pieces of sod here and there), and this is the preferable manner of grassing the sides of the ditches, the object of grassing the banks being to prevent the growth of weeds; and if the ditches are properly made, stock will not injure the ditches materially in grazing the banks.

ROUND-POINTED SHOVEL.

XX. The Formation of Underdrains.

Tile should always be used if they can be had at not too great a cost. Where the cost is excessive various other drainage material may be employed, but, in all underdraining, the shape of the ditches is important. They should be as narrow as will allow a man to work in them, since it is a saving of labor in casting out and throwing back the earth. In digging, except in such soils as will not hold themselves up, the sides may be carried down nearly straight until towards the bottom, which in no case should be much, if any, wider than the tile—not more than three inches for a tile of two-inch caliber—and still less proportionally for larger tile.

When a subsoil plow is to be used for loosening the earth in the drain, it should have but one handle, and be small enough so a light mule can work it. By passing back and forth several times, the soil may be loosened to a depth easily worked with a shovel. A small mule will work in a space of sixteen inches wide, with a little practice, and when this plan is adopted a passage must be left at suitable intervals, so the animal can reach the surface to turn around. This may be governed by the length that can be worked in a day, the mule plowing one portion while the hands are shoveling out another.

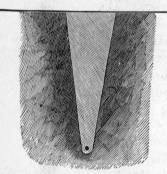

CROSS SECTION OF UNDERDRAIN.

When the work is to be done entirely by hand, or, at least, except the first eighteen inches, the cut here given will show the least width proportionally that may be used; but this only by the most expert drainers, with narrow tools for bottoming. However wide the top may be started, and carried down, the last spit, at the bottom, must conform to the width of the drain-tile or other drainage material to be used.

XXI. Various Means of Drainage.

NOTHING but tile should be used, where they may be had, unless in the case where stone lies on the surface of the soil, and which must be gotten rid of. But it may happen that drainage may be necessary where tile must be carried long distances, and thus cost too much. Again, in the draining of orchards and woodlands the tile are apt to become filled with the roots of the trees. Thus, stone, poles, slabs, and in very tough soils shouldered earth-drains may be used. Hence we give representations of several forms to meet every case.

For orchard drainage and woodland drainage where stone is plenty, the ditch may be filled with stone to within twenty inches of the surface, if necessary. In woodland, nearly to the surface. Where brush is plenty, it may be tied in small bundles and laid regularly in the drains, the larger ends pointing down, to form a waterway. That is, the inclination of each succeeding series of bundles should be laid one on another, that the larger ends will point downward to the mouth of the drain, each successive layer being covered by the succeeding one. Here again the draining may be filled to within twenty inches of the top, with brush, well tramped down, which the intelligence of the operator should soon enable him to accomplish deftly. The forms of these drains may be illustrated in the series of cuts showing underdraining other than with tile. This class of drains being made wider than for tile. For ease in referring these drains will be numbered.

XXII. Stone-Laid Drains.

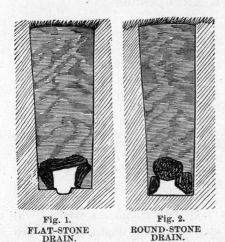

Fig. 1.
FLAT-STONE DRAIN.

Fig. 2.
ROUND-STONE DRAIN.

Fig. 1 shows the first form of drain, with irregular flat stone, and a channel for water, where the flow is light. These may also be made with round rough stone, as shown at Fig. 2. If any of the forms we give could have the bottom covered with clean gravel pounded into the clay, and be left dishing, it would save washing. The manner of filling all drains is alike. When the joints of stone or slab drains are not earth-tight, fine brush, marsh reeds, or slough hay should be used as a covering to keep out earth. Wet, pasty clay should never be used next the drainage material. It will leave holes. The finest of the earth should be used next the drainage material, and when sufficient has been put over to ensure no displacement, it should be well tramped or pounded, to thoroughly settle it.

CHAPTER VI.

DRAINAGE AND THE DRAINER'S ART—Continued.

I. SLAB AND POLE DRAINS.—II. TILE DRAINS.—III. LAYING OUT THE WORK.—IV. DRAINING TOOLS.—V. GRADING THE DITCH.—VI. LEVELING THE BOTTOM.—VII. CHALLONER'S LEVEL.—VIII. LEVELING FROM THE SURFACE.—IX. ALTERING THE GRADE—SILT WELLS—X. THE WATER CARRIED BY TILE.—XI. CAPACITY OF SOILS FOR WATER.—XII. VELOCITY OF WATER IN TILES.—XIII. CONNECTING LATERALS WITH MAINS.—XIV. DRAINING A FIELD.—XV. WHEN IT PAYS TO DRAIN A FARM.—XVI. SINKS AND WALLOWS.—XVII. SPRINGS, SOAKS AND SLOUGHS.—XVIII. DRAINING LARGE AREAS.—XIX. LANDS REQUIRING DRAINAGE.—XX. WET WEATHER PLANTS.—XXI. HOW TO KNOW LANDS REQUIRING DRAINAGE.—XXII. IMPORTANCE OF DRAINAGE TO STOCKMEN.

I. Slab and Pole Drains.

THESE should be laid four feet apart. No drainage with rough material should be less to be permanent, nor should it be less than this depth for orchard drainage. Thus the drain will last indefinitely, especially when there is a constant supply of moisture to keep the brush, poles, etc., moist. Figs. 3 and 4 show the manner of forming these, the first with poles laid end to end, and covered with slabs, laid in the same way, the other of three poles to form the waterway. Still other forms are shown, adapted to strong clay subsoils, where neither tile, stone, poles nor brush can be obtained.

Fig. 3.

Fig. 4.
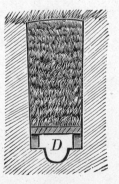
Fig. 5.
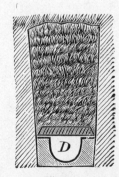
Fig. 6.

SLAB AND POLE DRAINS.

Again the channel may be cut in the clay with a shoulder, as shown in Figs. 5 and 6, and covered with sections of plank (hemlock preferable) cut in lengths so they will fit, and pressed tightly against the sides of the ditch. Hence the necessity of carrying the width exactly. If this is not done they should be wedged tight with slivers, or the earth tamped carefully at the ends to preserve them from moving. *D D* shows two different forms of waterways.

Another form is to use two large tile, one on each side, covered with flat stone or slabs, to allow the passage of water in the middle, and also through the tile when a large quantity of water is flowing. See Fig. 7. We see no advantage in it. It would be better to use one large tile at once, and also cheaper. We give the same advice in this case as, when consulted by a shiftless client, did a celebrated lawyer. It was as to the advisability of getting married. The answer was, "don't." The same answer will apply to any form of draining except tile, where these can be bought.

II. Tile Drains.

THE form, therefore, of the drain should be as perfect as possible as to the sides and bottom. There is nothing lost in paring all smooth, and the bottom of the drain must be absolutely correct; no low places, no high places. The gradient (fall) must be as equal as possible. Hence the time spent in the engineering is time or money well invested. It is not necessary here to enter into the minutiæ of this. If the fall is so light that particular exactitude is required, the leveling stakes may have to be set by a professional engineer. This, however, will hardly be required, except in the case where a large field is to be elaborately drained. With an ordinary fall, a spirit level will answer, or where large tile are used the water line may be corrected by admitting water into the ditch, and excavating the high places where the water dams, and correcting others where the flow is not correct. But, before the tile is finally laid, it is better that the bottom dry sufficiently so that the tile may be properly laid.

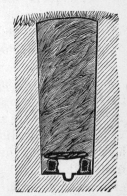

NO. 7.—TILE DRAIN.

Different Kinds of Tile.—There have been various forms of tile recommended. The old horseshoe was the one thought best up to about forty years ago. Experience has shown the fallacy of this. Then the sole tile took its place. The flat sole was supposed to give bearing to prevent its being thrown out of place. Then experiment showed that the displacement of tile was owing to other causes, just as the filling-up of horseshoe tile was shown to be from the squeezing-up of the bottom of the ditch into the tile. The perfection of shape in the orifice of tile is undoubtedly the egg-shaped caliber; but this involves laying the tile always one side down. The round tile is as near perfection as may be. Any side may be laid down, and thus tolerably perfect joints are secured. When absolutely perfect joints are necessary, as in the case of quicksand, collars should be used.

III. Laying Out the Work.

WE have already stated that, in nearly all, except that known as "thorough" drainage (where the tiles are laid at regular intervals all over the field, so that every part has equal drainage), very little engineering, if any, will be required, but what may be accomplished without the services of a professional. Where, however, a drainage engineer may be secured without too much expense, his labor will not only facilitate but really cheapen the work.

When ditchers go in companies, contracting, as they sometimes do, at a fixed price for work, the foreman of the gang is sometimes competent to undertake the planning and laying-out of the drains. But no sensible person will take chances here, unless he knows the man; for in drainage, the excellence and value of the work is determined by the weakest part, or by the poorest-laid tile. One bad, all is bad. This, however, applies especially to cases where the nature of the soil is difficult, or the inclination is not well defined, or intricate. All such must come under the supervision of a competent drainage engineer.

Practical Study.—All other drainage may be supervised by any intelligent man who will inform himself by a little study of some practical work on drainage. In this, however, the farmer wants to "skip" all that part that deals in contour lines. This and digging the ditches, and laying the pipe without putting foot in the bottom of the ditch, is finer art than will come in ordinary farm drainage. All the wonderful Birmingham tools may also be given the go-by. A sharp pick for hardpan, a common short-handled spade, a short-handled shovel, a long-handled, round-pointed shovel, a long-bitted spade, running to not more than three inches wide at the lower edge, and a rounded finishing (bottoming) scoop (all of them kept bright and sharp) will dig and finish the narrowest ditch.

Fixing Gradients.—In draining, the eye is no guide whatever in determining levels or in fixing gradients. If you happen to be conceited in this respect, a few trials with a spirit level will easily disabuse one of having a correct eye in establishing gradients. Even a common square, fixed movable on a standard, and with plumb bob attached, will satisfy one of this.

IV. Draining Tools.

SOME sort of level, therefore, is necessary: a carpenter's level will, generally, answer every purpose; although we will, hereafter, describe a better instrument. The tool shown in the cut is, from A to A (Fig. 1), eight feet long, and is five inches wide at the middle; B B are legs four feet long; and C C, braces. The spirit level may be fastened to the middle of A A with screws, and, the level being true, sighting along it will show the rise and fall of the land; that is, the level being adjusted true, when the sight touches the ground the rise will be equal to the height of the top of the level. Or, if there is no wind, a pretty correct level may be obtained by using the long arm of a true steel square, fixed on a stand-

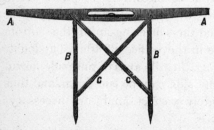

DRAINER'S LEVEL.—Fig. 1.

ard, to sight over. This must be adjusted by a plumb-bob along the shorter end of the square.

V. Grading the Ditch.

A SPIRIT level set on a standard so it may be moved in any direction, with a rifle sight at the eye end, and a circular sight, crossed at right angles by fine hairs at the other, will be better; but, when great accuracy is required, a drainer's level is the best of all. Looking through this when the spirit bulb shows it to be level, an assistant holding a rod graduated to half-inches some distance beyond, the target of the rod will show the difference in the level between the height of the eye piece of the level and the target. This must be noted on stakes driven in the ground at the points operated on, and also entered in a book of reference. Thus proceeding the whole length may be taken and the average fall obtained from end to end.

The Right Slope.—Fig. 2 will illustrate our meaning. A B represents the natural level of the land. If the drain is dug by guess it may take the course indicated by C D, but probably will not be so nearly correct. O G represents a horizontal line through the hill. What is wanted is the line E F. To get the average, drive a stake at H and another at K, so they will be exactly three feet above the soil. So place your level that you can sight along it from H to K. Drive stakes, say at six feet intervals, until they come in exact line.

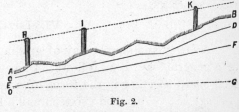

Fig. 2.
FINDING THE GRADE.

Then, if your ditch is to be four feet deep, seven feet from the top of each stake will give you the true gradient, E F.

Another Way.—If the ditch is to have an average depth of three feet, then set a stake to reach four feet above the surface at one end and the same at the other. Stretch a small, strong cord from end to end, supported at proper intervals by other stakes, to keep it from sagging. Mark, on short stakes driven every two rods, the

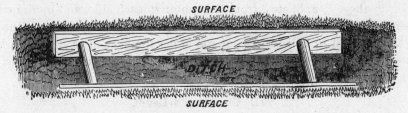

Fig. 3.
PROTECTING THE BANKS.

required depth of the drain at such points. If the line, for instance, at a given point is only three feet above the surface, the ditch should be four feet deep there. If five feet above the surface, the ditch must be only two feet at that point, and so for inches and fractions of an inch. Thus the workman may easily ascertain the approximate depth of excavation, graduating the width of the ditch so at the bottom it may correspond to the size of the tile to be laid.

The banks of a ditch are often soft and will cave. Then the sides must be protected as shown in Fig. 3.

Tools Illustrated.—Few tools are necessary in digging a drain. Figs. 4 to 8 will show a full equipment, including a scoop for bottoming the ditch. Fig. 9 shows a flat bottoming tool, and Figs. 10 and 11 two forms, one for cleaning the bottom from the top of the ditch, the other to clean the bottom by drawing the earth toward you. They will not be required in ordinary draining. All these tools we have grouped together for comparison. Fig. 4 is a very long-bitted spade for deep, narrow spits. Fig. 5 is the ordinary digging spade. Fig. 6 is a half-round tool for working in deep ditches for the bottom spit. Fig. 7 is a wider half-round spade. Fig. 8 is the scoop for bottoming the ditch true.

The cut of the German ditching spade, which we give, shows the best implement we have ever seen for fast work in place of, or supplementary to, the common spade. Figs. 4 and 5.

VI. Leveling the Bottom.

An easy and simple way of grading the bottom of the ditch, and accurate enough in a general way, is as follows: To get the required regular fall, fasten two strips of inch-board together in the form of a triangle, so that the feet may measure exactly one rod (sixteen and a half feet) across. Stay these strips by a cross piece two-thirds of the way from top to bottom. Mark this to inches and half-inches. To do this set the instrument on a perfectly level floor, and a plumb-bob, suspended from the top, will mark the center. Then set an inch block under one end and draw a line where the plumb line shows the difference as indicated on the cross-piece. Now set a two-inch block under the end and mark that. So proceed one inch at a time until one-half of the cross-piece is marked. Reverse the implement and proceed the same with the other side. Number these 1, 2, 3, etc., from the center to each side, and by drawing a half-line between each mark the half-inches will be obtained. Thus we have the power of obtaining the gradient to half-inches per rod along the bottom of the ditch, by setting the instrument with one foot on the true grade and excavating at the other end until the plumb-bob shows the proper gradient.

GERMAN SPADE.

VII. Challoner's Level.

Colonel Challoner, in a communication to the Royal Agricultural Society of England, describes a simple means of grading the bottom of the ditch by means of a mason's level, made on an extended scale. It is as follows:

He first ascertains what amount of fall can be obtained from the head of the drain to the outfall. Suppose the length of the drain to be ninety-six yards with a fall of two feet. This would be a fall of one-fourth inch to each yard. He takes a common bricklayer's level twelve feet long, attaching to the bottom of it, with screws,

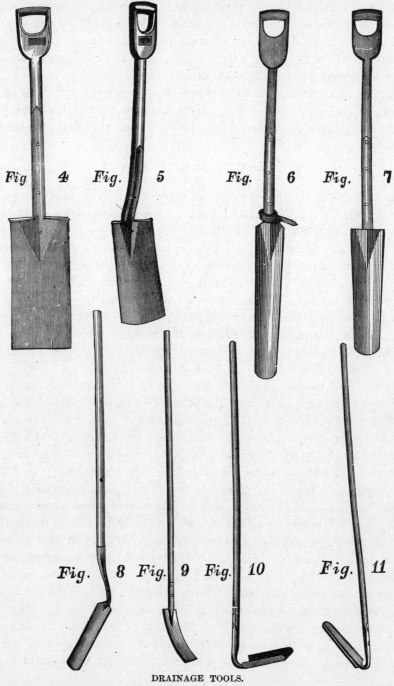

DRAINAGE TOOLS.

a strip of wood one inch wider at one end than at the other, thus throwing the level one inch out of a true horizontal line. The drain being at the proper depth at the outfall, apply the broadest end of the level to this plane, the other extending up the ditch. When the plumb-bob indicates the level to be correct, by hanging vertically, the one-inch fall in the four yards has been gained. Thus the drain may continue to be tested quite up to the head, and an unbroken, even and continuous fall of two feet in the whole ninety-six yards will be obtained.

An improvement on this level would be an arrangement by which this gradient could be changed by one-fourth inches. This might be accomplished by fastening a movable strip along the side of the bottom piece, to be held by thumb-screws at any desired gradient.

VIII. Leveling from the Surface.

The rise and fall at the top having been determined by a spirit level, or better an engineer's level, the gradient of the ditch may be obtained by Challoner's level. [See Article IV, V, pp. 342, 3.] The line of the ditch having been marked out and the surface thrown out on one side, prepare a lot of stakes dressed smooth on one side, twelve or more inches long, one inch by one and a half inches square. Drive a stake at the mouth of the ditch, down so the top will be six inches above the surface, and six inches from the side of the ditch. Drive another twelve feet from this, and at the same distance from the edge of the ditch, so that six inches is above the earth. Having obtained the rise in the whole length of the ditch, reduce to inches. Divide this by one hundred and forty-four, or by the length in inches of whatever leveling instrument you use. This will give the rise in inches.

Arrange the level to this gradient. Set one end of the level on the first peg, and the other on the next peg. If the plumb-bob hangs vertical the gradient is correct. If the stake is too low, raise the end of the level until the plumb-bob is right, and mark upon the stake the difference in inches and fractions of an inch, adding the mark —, when the level has to be raised. If the stake is too high, lower the leveler alongside until the plumb-bob is right, and mark the difference on the stake in inches and fractions, using the mark +. Now suppose a stake showed two and a half inches —. The ditch must be just so much less at this stake than at the preceding one. If two and a half inches +, it must be just so much deeper from the top of this stake (or six inches above the surface) than at the preceding one; and so for any other quantity. Or, the correct depth may be figured at once and marked. Suppose the ditch is to be three feet deep. If the stakes, being six inches above the ground line, show one inch variation, that must be added to three feet six inches, or subtracted from, as the case may be, and marked on the stake; then the workman, by measuring this distance from the top of the stake, knows when he is deep enough. The operation of leveling may also be performed as well with the other level described, but if this be one rod long, the computation must be in proportion.

Thus, any person who can sight a rifle fairly well may take his levels with a common spirit level, if he have not a drainer's level. And he may, if he can multiply and subtract, accurately determine the gradients, and figure the work so any farm hand can dig the ditches correctly.

IX. Altering the Grade—Silt Wells.

AN important principle in draining is that when the grade of the tile is altered so the gradient (fall) is less than that above, the caliber of the tile must be increased so the same amount of water may be carried. In this case it is also safe to place a silt well at the point where the grade changes, to catch silt, which always accumulates more or less at such places. The cut shows one made of sections of vitrified pipe, the incoming water and the outflow, lower. Below the water line is shown silt, which must be cleaned out from time to time with a scoop something like that shown at Fig. 10, drainage tools, on page 345.

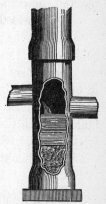

SILT WELL.

Silt Wells Explained.—It is always safe to put in these wells, unless the utmost care is taken in leveling; for the less swift the water the greater the tendency to deposit silt. Hence the axiom that the running of clear water at the end of the drain is evidence of perfect work. If the water be not clear it is certain that muddy water is entering the drains. If so, silt must be deposited. Hence the necessity of true grades, for if there are depressions the silt will collect here and obstruct the flow. The steeper the grade the less the danger from this cause, but in every case it is constant and serious.

Of Rock or Stone.—In mains where there is a heavy flow of water, the silt wells should be correspondingly large, and built of brick or stone, and should extend almost three feet below the level of the water therein, to give room for the precipitation of silt, if any, which is to be removed whenever it accumulates unduly. When the mains are smaller, sections of vitrified pipe, with cover, may be placed one on another, as shown in the cut, so the incoming main shall be higher than the outlet. For the large wells, strong casks may be sunk one on another, but these, of course, will only be temporary.

If objection be made to these silt wells at every change of grade, please remember that the additional cost is in no comparison to the permanent value. It is better to have the silt deposited where it may be easily removed than to have a drain stopped, the ground saturated from overflow, and the impossibility of repairing it perfectly without considerable expense.

X. The Water Carried by Tile.

THE amount of water that a given tile will carry is governed by the inclination or fall of the tile. Now the fall of a drain is not a matter of choice. It is governed by the conformation of the land. It is laid down as a rule that a three-inch tile, with a fall of three feet in one hundred feet, will discharge more water per hour than a tile of six-inch caliber with a fall of one inch to the one hundred feet.

Grading.—The fall necessary to carry water in tile may be very low, but the less the fall the more careful must be the grading and alignment. A fall of one inch in a hundred feet will allow perfect drainage. A fall of four inches in one hundred feet will keep the tile channels clear if there be not much silt and the grade is uniform. An inch to the rod gives ample fall, and for this reason, in the description of implements, the rod measure was used. Preference, however, is to be given to numbers of ten and their integers.

A Single Tile the Best.—To carry a given amount of water, economy is always in favor of a single tile of proper caliber rather than the use of several smaller tiles to make up the caliber required. The capacities of tiles compare as the squares of their diameter. Hence it is said that four tiles, each of three-inch caliber, are equal to one tile of six-inch caliber, as $3 \times 3 \times 4 = 36$, the capacity of four tiles of three-inch or $6 \times 6 = 36$, the capacity of one six-inch tile. This, however, is fallacious in practice. One five-inch tile will really carry more water than four three-inch tile; there is less friction and greater swiftness. Hence always use a single tile of whatever caliber is needed, rather than several smaller ones to make up the required capacity.

XI. Capacity of Soils for Water.

It is well known that the faster water runs the greater its scouring force. Thus the quicker water is filtered from the soil the more surely will it carry away with it silt and fertilizing matter. Hence light sands are called hungry (infertile). The point of saturation of soils is important in this connection as showing the great difference in their capacity for holding water. Soils are saturated in about the following proportions; that is: Sandy soil will be saturated when it holds from 24 to 30 per cent of water, calcareous sands holding the most. Sandy loams hold 38 to 40 per cent of water; clay loams 45 to 48 per cent of water, and peat up to 80 per cent of water. Thus soils will be saturated with varying quantities of water from the lesser to the greater amount given, according to their composition. It is easy, therefore, to see that sandy soils, resting upon an impervious clay near the surface, may be those most needing drainage; but on the other hand, the sandy soils will give up superabundant moisture by evaporation quicker than clay soils, and they will also filter them quicker; yet no agricultural (fertile) soil will filter water too fast, naturally.

Insect Borers.—But soil, when not saturated, is always filled with insect life. During drouths they always bore down until moisture is reached. Hence a soil, well drained, is always honeycombed with these insect chambers, all leading ultimately to the tile as the source of constant moisture. Thus we see the simple reason why a drained soil filters quickly. Now, if the surface tilth is such as to obstruct these channels, then direct filtration ensues; the valuable properties of the water are given up to the surface soil, where it is needed, and thus we have the reason why a drained soil is continuously fertile. It is always receiving fertility from the original and great source of fertility, the atmosphere.

XII. The Velocity of Water in Tiles.

The velocity of water in a pipe is determined by the caliber of the pipe giving greater or less friction, the rate of descent (inclination of the pipe), the smoothness of the inside, and also, the length of the drain; that is, a very short drain will not run so freely as a longer one. Another important thing must here be observed. Rough tile here and there in the line will obstruct the flow; hence the necessity of carefully sorting the tile. If those rough inside, soft, crooked, or otherwise unsound, must be used, let it be upon some portion not entering the mains of general drainage; it is better that they be discarded. If the contract be for perfect tile, no respectable maker will ever put in these imperfect ones.

Tile to the Acre.—It has been estimated that an acre of land when wet may contain 1,000 hogsheads of surplus water. A line of tile eighty rods long, laid three feet deep, will drain one rod on each side, of pretty stiff clay, or one acre in forty-eight hours. This is soon enough to fully carry away the surplus water of a rain. A pipe of two-inch caliber, laid at an inclination of one foot in fifty feet, it is said, will discharge more than 1,100 hogsheads in forty-eight hours.

Rate of Discharge.—The following table, from an English source, will be interesting, as showing the discharge from pipes of smooth caliber:

DIAMETER OF ORIFICE.	RATE OF DESCENT.	VELOCITY OF WATER PER SECOND.	HOGSHEADS DISCHARGED IN 24 HOURS.
2 inches,	1 foot in 100	32 inches	400
2 inches,	1 foot in 50	32 inches	560
2 inches,	1 foot in 20	51 inches	900
2 inches,	1 foot in 10	73 inches	1,290
3 inches,	1 foot in 100	27 inches	1,170
3 inches,	1 foot in 50	38 inches	1,640
3 inches,	1 foot in 20	67 inches	3,100
3 inches,	1 foot in 10	84 inches	3,600
4 inches,	1 foot in 100	32 inches	2,500
4 inches,	1 foot in 50	45 inches	3,500
4 inches,	1 foot in 20	72 inches	5,600
4 inches,	1 foot in 10	100 inches	7,800

This is given more to show ratio than, even, the approximate quantities of water carried. A hogshead is an indefinite quantity. It may mean 52½ imperial gallons, or, in this country, 110 to 120 wine gallons. Taking it in its English sense, probably correct, the two-inch pipe would carry, at its lowest inclination, 400 hogsheads in twenty-four hours, or 100 tons, which is an inch of rainfall to the acre.

XIII. Connecting Laterals with Mains.

The better class of tileries manufacture pipe of various large caliber, with a hole or branch for lateral pipes. When these can be bought, it is altogether better and cheaper than to chip the holes. When they cannot, the holes must be made by hand. Care is needed here not to break the tile. None but the soundest pieces should be used for this purpose. A very narrow-bitted tomahawk, with a sharp spike on the head, like an ice pick, but straight, is used. Begin at middle of the tile, and with a succession of light blows mark the circle to correspond to the size of the pipe to be entered. Follow this up, deepening the circular channel, chipping out from the inside of the circle, until the channel is nearly through the tile. It may then be broken in. This hole should be made so the lateral pipe may enter diagonally the way the stream is to run. If it flow in square it will cause back water, and, of course, a stoppage of the flow. The stream should enter the main near the top.

ICE HATCHET.

It is rather a nice job making these inlets, and, therefore, the work should be entrusted to a careful hand. In place of a regular tile pick an ice hatchet may be used, as shown in the cut. It may be done during rainy weather, or when work cannot be done out of doors. Whenever such connections are made they must be carefully guarded, by placing over the connection, first, broken crocks of tile, and above that gravel, to guard against the introduction of silt or muddy water. In this way soft or imperfect tile may be utilized, and also for covering the joints; but collars are always to be preferred, when the extra expense may not be considered too great. When it can be done, these connecting tile should be bought ready-made. The cut shows the connection and best form—the cylinder tile.

CONNECTION OF LATERAL WITH MAIN.

XIV. Draining a Field.

The diagram of a Drained Field shows the drainage where the fall was all in one direction. It has been alluded to in Section XV. of the preceding chapter. It is described as follows: A field lay on a slope, as in the diagram, which, on the tableland, was a large tract of swampy ground, chiefly woodland. This slope descended about six inches per lineal rod; and there seemed to be no reason why the soil should be so wet when such land ought to be dry. The entire slope, over an area of many acres, was rendered very wet by the water that came to the surface from the swampy land above.

The first step toward draining that field thoroughly was to sink a three-foot ditch, with a stoned throat, across the upper end from A to B, from B to C, and from A to E, letting the water discharge into a deep gutter at one side of the highway.

As there was a low place at F, a ditch was sunk from F to G. The deep "catch-water drain" across the upper end, from A to B, cuts off a large proportion of the water. Yet six or eight yards down the slope the water would soak out from the catch-water drain, rendering the soil on the lower side of the slope as wet as ever. It is probable, also, that the veins which conducted the water from the swamp to the lower part of the field were not yet reached by the catch-water drain A B. Hence other drains were made about forty feet apart, as shown by the dotted lines up and down the slope, ending four or five rods below the catch-water drain. These latter drains collected the surplus water in the most thorough manner. For the catch-water drain two-inch tiles would have been preferable to

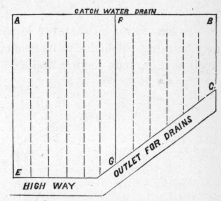

DIAGRAM OF RAINED FIELD.

stones, and one-inch to one and one-half inch tiles would have been sufficiently large for the parallels, as represented by the dotted lines. At E was the lowest point. Hence, if there had been no deep highway gutter along E, G, C, it would have been necessary to make a drain in that place to receive the water from the parallels As the slope below, which was wet also, belonged to another person, it would be necessary for him to sink parallels directly up and down the slope, ending two or three rods below the highway, if he would succeed in carrying out the system of thorough drainage that was commenced above the highway, and which is fully represented by the above diagram.

XV. When it Pays to Drain a Farm.

In draining a farm, the first thing to be done is to carefully go over the land and estimate the amount of drainage necessary to bring all the fields to one uniform state, so the low or wet lands may be rendered as dry and pervious to water as are the naturally drained portions. A rule, and a good one, is that any soil in which water will stand in a hole two feet deep twenty-four hours after a heavy rain, needs draining. While this is true, if the land is intended for wheat and all that class of plants requiring quick percolation of water, it will not apply in every case. Another integer comes in. The value of the land and the value of the crops to be raised must be taken into consideration. In one situation, where tillage land is worth forty, fifty or sixty dollars an acre, for the reason that a class of crops may be raised that will pay interest on these sums, lands too wet for the finer crops will pay for drainage; but in other situations, more remote from market, where the best lands are not worth more than five, ten or twenty dollars per acre, drainage would not pay. Hence, every man must be his own judge as to the advisability of draining, after all. Yet this does not falsify the statement, in the abstract, that all soils require draining where the water does not readily settle away from holes dug to the depth as stated. Hence the careful survey of the farm in order to estimate rightly the amount of drainage necessary.

XVI. Sinks and Wallows.

EVERY farm, in whatever location it may be, requires some draining. A field may be fairly tillable as to its general character, and yet certain portions, often a very small portion, may require drainage, to bring the whole into a homogeneous state. A hole, for instance, not more than an acre in extent, may render a considerable area around it wet and unfit for cultivation. This not only causes waste from the fact that this area is unproductive, but it increases the cost of tillage. The plowing of the land must end at this wet portion; the planting and cultivation must cease here. There must be turn rows here for the team to come about. In gathering the crops, the team must stop here, or else go around the place. This will amount in any case to more than the plowing and cultivation, and hence, wherever these holes occur, and however low the value of the land, these places will pay for drainage. These are called sinks, cat-holes or wallows.

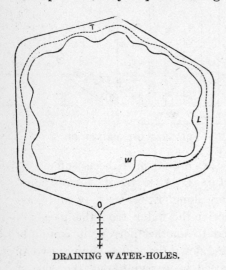

DRAINING WATER-HOLES.

Often a single line of tile, or two or three lines, ending in a small main, carried along to where the water may have free flow, will be all that will be needed. No engineering will be needed. Carry a narrow ditch, three feet deep, directly up to and through the low place, grading the bottom so the water will flow equally along the bottom, lay the lateral drains as they may be needed, communicating with the main. Cover all carefully with the excavated earth, the clay below, being careful that the tiles are not displaced in so doing, tramping or pounding the whole solid, so there will be no settling; and this wet place will be as good or better than the best of the field. In Section XV. of the preceding chapter we have shown the drainage of a slough, or, as such are often called, a "cat-hole." The diagram shows a very usual form in the West. W, shows the water-line; L, the water-saturated soil; and T, the edge of the dry land. Sometimes a line of two-inch tile laid around the edge next the dry bank and meeting in a main at O, will dry these sloughs. When not, tile must be laid straight through, as shown in Section XV., Chapter V.

XVII. Springs, Soaks and Sloughs.

SOMETIMES there is another class of work of equal importance with that spoken of in the last chapter. Water entangled between impervious strata will flow along, underground, until the strata, being broken, the water will rise to the surface, forming a "soak," or blind spring, often saturating the surface for a considerable distance. These must be drained by laying parallel lines of tile along the line of greatest descent and communicating with a main, to carry away the water to a natural outlet.

Sometimes these soaks occur at intervals along a declivity. In this case, drain the lower soak thoroughly. It will sometimes free the water from above, and the whole will become dry. If not, drain the others in succession until you have accomplished your object, always providing a main sufficient to carry off the accumulated water.

In the West, however, this class of drainage water is rare. As a rule, the spouting is near the bottom of the declivity, or in the valley itself. The reason is, that the valley has been gradually filled with the wash of the hills, through which the water rises easily.

Draining a Slough.—Valleys and intervales are sometimes called sloughs, but incorrectly so. A slough is a miry, low place, where water collects, as at the bottom of what might have once been a lake. A slough must have a regular system of lateral drainage, more or less extensive, according to the area inclosed by higher lands.

These valleys, that hold the water of percolation, may often be drained by laying a single deep line of tile along the center to a natural outflow, with lateral drains only here and there, where the valley broadens, or is more than usually wet. Here, again, no scientific leveling will usually be required. The leveling may be done with the simple instruments heretofore described, or by the flow of water, being careful, however, to put in a silt well wherever a change of inclination must be made. These have been heretofore described, their principal object being to allow the deposit of silt at these points to be cleaned out whenever occasion requires. Another advantage is, that the flow of water may at any time be determined by examination. For this reason, peep-holes (smaller wells) are put in occasionally, where the silt wells are not necessary.

XVIII. Draining Large Areas.

BESIDES this system of partial drainage, and which may be applied to lands of low value, there is no doubt that where land is worth $50 an acre and upward, it will pay to drain the whole farm, except such high portions or sandy soils as are never wet. It will undoubtedly pay to so drain all stiff clays, or that class of soils through which water does not percolate readily. In this case, it will be altogether cheaper to employ a drainage engineer to make the levels and supervise the work, especially when the fall may be difficult; and on the same principle that we employ an architect in the erection of a superior building; it will be found a good investment in the end. Yet, the drainage of the average farm may be undertaken with but very little expense for supervision. To illustrate this we give the outline (exaggerated), but prepared from actual experience, of the drainage of a quarter section (160 acres), as shown in the illustration, Draining 160 Acres. The dark lines, next the mains, show the lands requiring drainage on each side of the mains; the unshaded portions, the naturally dry land. The ponds to the right and left were originally water ponds, always wet, except through the severest drouths. The central portion shows a wet slough, extending beyond the boundary of the farm. The size of tile used is designated in the cut, except in the lower shaded portions, where they were of two-inch tile, running diagonally to the mains; not always on the line of greatest descent. The mains were

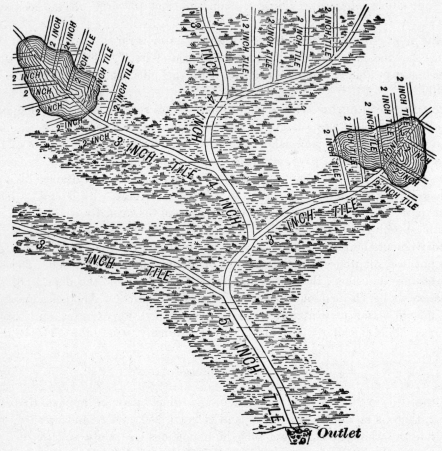

DRAINING 160 ACRES.

laid three and a half and four feet deep, and the two-inch tile three feet deep. In this case, the lines of greatest descent, were, as a rule, at right angles, or nearly so, to the mains, and are shown in the cut by the hair lines, and when this was the case, the laterals were curved where they entered the mains, to give the water an impetus towards the outlet, to prevent back water.

Greatest Descent.—It will be seen, by reference to the cut, that in the drainage of the upper section showing the lateral drains, the lines of greatest descent (the hair lines) run partially across the lines of tile; that is, the tile enter the mains not entirely according to the lines of steepest grade. This was because the fall was ample, and the idea was to intercept oozing water.

This, according to some authorities, is incorrect; and, as a general rule, it is so; but often it is economical in the draining of isolated, low places or ponds, the idea being that tile enough are used to catch all the water, and what escapes the higher level will be caught on the lower level.

XIX. Lands Requiring Drainage.

THE expense of thoroughly draining land three feet deep will be from sixty to one hundred dollars per acre, varying according to cost of labor, tile and the obstacles to be encountered in prosecuting the work. Since, then, this represents a permanent work, it is capital employed. Thus, any person may figure what lands will or will not pay for drainage, according to the interest required on the investment. If the increased crops will pay the interest on the outlay, draining will pay. With corn at forty cents per bushel, fifteen additional bushels per acre in yield will pay the interest on one hundred dollars. If wheat is worth one dollar per bushel, an additional yield of six bushels per annum will pay the interest. There is no wet farm where it will not pay more than this. If but a small portion of the whole land requires drainage, the economy of draining will become more and more apparent; for, while a farmer might not be able to drain an entire farm, any farmer can, from year to year, drain field after field, until soon the drainage will render the wet portions of the soil as good, or better, than the naturally dry portions.

Lands requiring draining:

1. On general principles, as heretofore stated, any land should be drained where the water stands in holes two feet deep twenty-four hours after a soaking rain.

2. Any soil that in winter or spring becomes water-soaked, so that plowing may not be carried forward twenty-four hours after a heavy rain, or where the soil remains wet during forty-eight hours of dry weather after the frost has completely left the ground.

3. Any meadow or pasture that becomes packed from the treading of stock forty-eight hours after a heavy rain, or that remains soft after the frost is out of the land; for it must be remembered that none of the superior grasses flourish on wet soil. Moist soils they like, and a drained soil is always moist. Again, grasses do not root below the standing-water line; they do, however, root deeply where the land is either naturally or artificially drained. Hence, draining will allow this deep rooting. The low-land vegetation will disappear, and the superior grasses will take their place.

4. Drainage prevents surface washing. Water falling on a soil, if it be sufficiently porous, sinks directly down until it finds the point of continual saturation, or of absorption by the soil. However steep the hillside, this is constant, so long as the rainfall is not greater than the power of absorption. For the reason that drainage renders the soil friable and porous, it absorbs the water, and hence one reason why in time of drouth it retains moisture. Every drop of rain or dew assists the deep

reservoir of friable earth beneath, and the hygroscopic water of the atmosphere is constantly being separated in the minute cavities of the soil.

5. Drainage assists all low places or swamps; in fact all soils where the plants named in the next article flourish.

XX. Wet Weather Plants.

SOME of the many of these are spearmint, cursed and bristly crowfoot, marsh marigold, marsh cress, cuckoo flower, pale and spotted touch-me-not, poison sumach, swamp rose, loosestrife, water purslane, cow parsnip, cow bane, water hemlock, poison hemlock, button bush, golden rod, tick-seed sunflower, cardinal flower, great lobelia, brooklime, knotweed, swamp and green dock, arrow head, great purple orchis, lady slipper (*cypripedium*), blue flag, blue-eyed grass, white hyacinth, wild yellow lily, spike rush, club rush, wool grass, cotton grass, white grass, water foxtail, blue-joint grass, water cord-grass, manna grass, vanilla grass, reed canary grass, millet grass (*millium effusum*) and Indian grass. Many other plants might be named, but this is sufficient to show the wet-land plants of a variety of situations. So far as timber land is concerned, the soils indicating need of draining are those where hemlock, swamp and water oak, beech, sometimes maple, ash, elm, and all that class of trees and shrubs natural to such soils, will show the need of drainage.

Other Soils to Drain.—Springy places. These should be drained at least so as to cut off the supply of water from saturating lower portions of a field.

Sandy or other porous soils resting on stiff clay near the surface. These are among the most valuable of soils when not sodden with water. Undrained they are cold, weedy and unserviceable; once carefully drained, they often become the most fertile, as they are the most easily worked of soils.

All clayey and impervious soils. The object here is to quickly liberate the water of saturation, either that from rain, or that slowly passing up from below—for the word impervious is used comparatively—since no soil is absolutely impervious, and stiff clays, under the action of thorough drainage, gradually become placable, friable and of the most lasting fertility.

XXI. How to Know Lands Requiring Drainage.

M. BARREL, author of a great French work on drainage, has put this so tersely and graphically that we give a translation. The author says: "Whenever, after a rain, water remains in the furrows; wherever stiff and plastic earth adheres to the shoes; wherever the foot of man or horse makes cavities that retain water like so many little cisterns; wherever cattle are unable to penetrate without sinking into a kind of mud; wherever the rays of the sun form on the earth a hard crust, slightly cracked, and compressing the roots of plants as in a vise; wherever, three or four days after a rain, slight depressions in the ground show more moisture than other parts; wherever a stick, forced into the ground eighteen inches deep, forms a well-like hole having standing water at the bottom; wherever tradition consecrates, as advantageous, the cultivation of lands by means of convex, high, large ridges." In all these the author affirms that drainage will be advantageous. How many farms are there where, on large portions, cultivation cannot be had at all, and yet the owner perhaps might scout at the idea of drainage; in fact, would insist that it would ruin his land? It would be, in fact, the only salvation of the farm gradually being ruined by having to be plowed when out of condition.

Twelve Propositions.—The late John H. Klippart, one who united practical knowledge to an active, observing mind, in a chapter on soils and their properties, discusses the advantages of underdraining, so far as theory (not hypothesis) in its proper sense is susceptible of demonstration, and asserts the following twelve propositions:

1. That drainage removes stagnant waters from the surface.
2. It removes surplus water from under the surface.
3. It lengthens the seasons.
4. It deepens the soil.
5. It warms the soil.
6. It equalizes the temperature of the soil during the season of growth.
7. It carries down soluble substances to the roots of plants.
8. It prevents "heaving out" or "freezing out."
9. It prevents injury from drought.
10. It improves the quality and quantity of the crop.
11. It increases the effects of manures.
12. It prevents rust in wheat and rot in potatoes.

A Good Test.—While this is all correct as a rule, as applying to soil, it is none the less true that it applies only to soils requiring drainage, and a good and safe rule as to whether a soil really does require drainage, is, that it does not come into condition for working soon after heavy rains. If a soil does not free itself perfectly from the plow-share, and fall friably therefrom in plowing within forty-eight hours after a saturating rain, it requires draining to reach the most economical results in tillage. The only question then to be decided, is, whether the crops to be cultivated will pay the interest on the sum invested. There is really but little land in the West, comparatively speaking, but would be better for draining.

XXII. Importance of Drainage to Stockmen.

THOSE interested in breeding and rearing stock, might, from a superficial view of the subject, suppose that to them the drainage of the soil was of little importance. Not so. It is of fully as much importance to this class as to any other class of farmers. The mere lengthening of the season for a week or ten days in the spring, and the same length of time in the fall, is a most important matter, since it shortens the winter, and consequently, the foddering of stock for the same length of time. The more important question, however, lies here; that while wet soils, and even the worst marshes, will bear plants, and even grass, yet, it is not such as will be eaten by stock, unless they are starving. Again, upon wet land, when the superior plants, as bluegrass, orchard grass, red clover, etc., will exist only on the higher portions, and then perhaps not in the highest perfection, or where they may be found growing generally over the field but sparsely, the loss to the value of the pasture is immense. Unfortunately, we have in this country but few examples, that are well authenticated, to show, from careful experiments, a fair exhibit in such cases. Therefore, let a single one, from high foreign authority, suffice:

The Versailles Experiment.—A certain piece of land connected with the Agricultural Institute of Versailles, France, was drained. Before draining, it produced (except in so small proportion as to be practically worthless for any purpose) only noxious plants, or those not eaten by cattle. We reproduce the statement, giving the Latin names as presented, with a translation showing our common names, where known, which are not found in the original, and also the proportional number of plants each, as accurately determined:

COMMON NAMES.	SCIENTIFIC NAMES.	PROPORTIONAL NUMBER
Common rush	Juncus communis	100
Plantain	Plantago lanceolata	83
Colchicum	Colchicum autumnale	67
Scouring rush	Equisetum arvense	50
Crowfoot	Ranunculus (two varieties)	50
Sedge	Carex riparia	50
St. John's wort family	Hypericum tetrapterum	50
Mint family	Ajuga genevensis	33
Thistle	Cirsium palustre	33
Cuckoo flower	Cardamine pratensis	33
Common agrimony	Agrimonia eupatoria	33
Valerian family	Valeriana dioica	17
Marsh marigold	Caltha palustris	17
Sorrel-dock	Rumex acetosa and crispus	17
Red and white clover	Trifolium pratense and repens	1.2
Orchid	Orchis latifolia	0.8
Vernal grass	Anthoxanthum odoratum	0.4

The interesting point in this is the exceedingly small number of plants eaten by stock—the clover and vernal grass—and also their proportional number, and most important, the number of poisonous plants. *Colchicum* is one of these. Cattle will not eat it unless mixed with hay. A very small portion will kill them. It is in this respect that the value of drainage comes in to the stockman. A piece of land, for instance, is worthless without draining. If only one-half, or even one-quarter, were plants not eaten, would it not pay to drain, in order to have them superseded with valuable plants? Yes! for such is surely the effect of drainage. The question really answers itself, and thus we leave the subject of drainage, for really in this respect, as in any other relating to any economy, the good sense of the proprietor must be exercised; for after all, it is the pocket-book that must be interviewed, before any decision can be made, when improvements of any kind are projected.

PART IV.

RURAL ARCHITECTURE.

ILLUSTRATED PLANS AND DIRECTIONS FOR VILLAGE AND COUNTRY HOUSES.

BUILDING MATERIAL AND THE BUILDER'S ART.

INCLUDING EVERY GRADE OF RESIDENCE, OUT-HOUSES, GARDEN AND ORNAMENTAL STRUCTURES.

MECHANICS AS APPLIED TO THE FARM.

RURAL ARCHITECTURE.

CHAPTER I.

PLANS AND DIRECTIONS FOR COUNTRY HOUSES.

I. BUILDING ACCORDING TO MEANS.—II. IMPROVING THE OLD HOMESTEAD.—III. AN ELEGANT COUNTRY HOME.—IV. FARM AND SUBURBAN COTTAGE.—V. WHEN TO BUILD.—VI. THE PROVIDENT FARMER'S MARRIAGE SETTLEMENTS.—VII. HOW TO BUILD.—VIII. WHAT TO BUILD.—IX. TASTE AND JUDGMENT IN THE DETAILS.—X. WHERE TO BUILD.—XI. A HILLSIDE COTTAGE.—XII. ICE-HOUSE AND PRESERVATORY.—XIII. THE WATER SUPPLY.—XIV. HOUSE DRAINAGE.—XV. VENTILATION.

I. Building According to Means.

HOWEVER small and rude the beginning of a country home may be, the house and barn should be so planned that additions may be advantageously made, or else they must be so placed as not to interfere with the erection of better structures when increasing wealth shall allow. Our aim is to give directions for the erection of simple buildings, of a cost not exceeding $1,000, that any intelligent person can understand, and any carpenter build, without the aid of an architect. For a house costing over the sum named the fees of an architect will be money well spent, and in the end an economy, since he should be able to save the builder more than he will ask for the working plan.

We will, however, present a few designs for more elaborate dwellings, and would give the following as general directions: Farm-houses should always be of solid and substantial appearance, avoiding florid and useless ornament. Land being cheaper than in towns, it is well to use it freely; do not build tall and narrow structures, which are always unsightly, and give unnecessary stairs to climb. The farm-house should be roomy, with high ceilings, solid and cheerful. When the means of the owner will allow, let it be imposing.

II. Improving the Old Homestead.

IN Part III., Chapter I., we have shown the cottage of the prairie pioneer. As years have passed he will, from the profits of the farm, have added other simple structures—a wood-shed and store-room in the rear; then a porch in front; trees and shrubs have been planted, and, as children have increased, a larger, plain building has been added at the end of the original cottage, as shown in the illustration of a "Farm-house Built from Increasing Profits." There is comfort here, though of the homeliest kind. How now may this comfort be retained or enhanced, and more of beauty added? In our climate of hot summer suns and dripping snows we may secure both by so extending the roof that it shall overhang the building at the eaves

ASHLAND, THE HOME OF HENRY CLAY.

and gables by some twenty inches, with dormer windows and neat cornices on the gables, as shown in the illustration of "The Old House Remodeled." These architectural effects will be the more easily obtained if the gables are comparatively steep to begin with; or the pitch of the roof may be increased in remodelling.

The Ground Plan.—The ground plan, besides kitchen, living-room, etc., may contain a parlor, 12 x 14 feet, as in Fig. 2, instead of the two bed-rooms, 7 x 12, shown in Fig. 1. Then the laundry and wash-room should be removed to the rear of the kitchen, as shown in Fig. 2, and the room, 6 x 8 feet, adjoining the parlor turned

FARM-HOUSE BUILT FROM INCREASING PROFITS.

into a bed-room. Then the wood-shed will separate the parlor and bed-room from the laundry, with its odors of soap-suds, soiled linen, etc., on wash-days, besides being convenient to all three of those rooms when fuel is wanted. Fig. 2 shows only one bed-room in the ground plan, but, however the lower floor may be divided, the second story will contain ample space for sleeping-rooms. In both plans the pantry occupies the same position, convenient to the kitchen and living-room, with the second-story stairs on two sides of it.

III. An Elegant Country Home.

The house may be still further improved by setting it back from the road, but we happen to know that it is in contemplation soon to build the new house near the

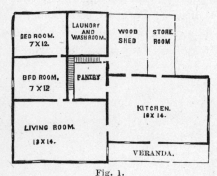
Fig. 1.
GROUND PLAN OF FARM-HOUSE.

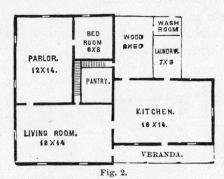
Fig. 2.
GROUND PLAN OF FARM-HOUSE.

sheltering artificial grove on the higher land with the beautiful meadow lands in front. Ample barns now stand there hidden by the house. Then the modest house will be moved back far enough to give a pretty lawn in front, the improvements suggested will be made and if there is a married son or daughter the whole given to the young

THE OLD HOUSE REMODELED.

couple. When the alterations are completed, and the new roof with its overhanging eaves and cornices is finished, the upper story will contain three good sleeping-rooms.

The Ground Plan.—If the original farm-house is large enough it may be turned round and remodeled upon the ground plan shown in Fig. 3. The parlor, 13x16 feet, opens upon a veranda in front. The library or study has places for books at B, B, at one of which a door might, if desirable, connect with the dining or living room. The two closets in the dining-room are so placed as to give a kind of bay-window effect, both pleasant and convenient. The kitchen, a bed-room for the family and a child's bed-room, are all placed in connection. The door between the latter and the back entry glazed to admit light to that part of the entry behind C. The second floor has five good bed-rooms with a closet to each.

The Cost.—To thus alter the old house and give the high pitched roof, pierced with windows, should not cost over two thousand dollars. Thus from the shelter of the pioneer has grown the pretty residence shown in the illustration of The Old Homestead Remodeled.

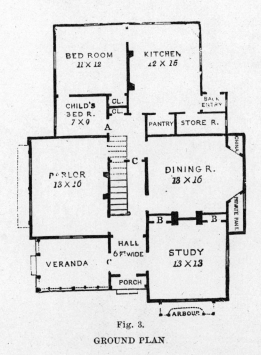

Fig. 3.
GROUND PLAN.

The rustic veranda and trellis over the windows are intended not merely as a support for vines, but rather as giving an air of rural refinement and poetry to the house, with small expense. They are added from time to time by the owner, aided by some farm hand, expert with the saw and hammer. They should be constructed of cedar poles with the bark on, which may be had in many places for a trifle, and which, if neatly put together, will be more becoming to such a cottage as this than more elaborate carpentry work. As for the lawn, remember that the grass should be unbroken. Trees improve, shrubs ruin it. Flower beds should be massed in colors and placed so that the lawn may be a clear expanse of sod, not patched with disturbing tints.

IV. Farm or Suburban Cottage.

A GOOD form of house for the successful farmer to build after getting pretty well ahead, is shown in the illustration of "Farm or Suburban Cottage," with the accompanying plans of the first and second floors. It is plain, substantial, well-arranged, and would also make a good suburban home for a city man. It is as complete inside as it is tasteful without, and under ordinary circumstances its cost would be about $2,500.

Fig. 4 shows the plan of the ground floor: A, front veranda, 10x16; B, hall, 7x20; C, parlor, 12x18, with bay-window, 4x9; D, dining-room, 15x20; E, library, 12x15, with square bay-window, 4x8; F, kitchen, 11x12; G, pantry, 8x8; H, store-room, 10x12; I, coal-room, 7½x8; K, wash-room, 7½x8; L, veranda, 8x16; M,

FARM OR SUBURBAN COTTAGE.

veranda, 4x30; N, cistern, 9 feet diameter; O, well; *c, c,* closets; *s, s,* shelves; *b,* bath; *f,* back stairs; *t,* sink; *p,* pump

Fig. 5 gives a map of the upper floor of the cottage, with a hall, seven feet wide at the stairway. C, C, C, C, are closets; D, linen closet; E, attic stairs; F, servants'

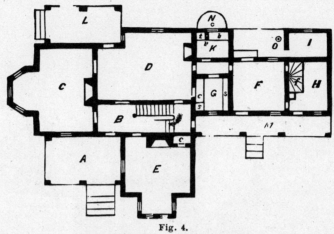

Fig. 4.
FIRST FLOOR OF FARM COTTAGE.

bed-room, 11x20; G, garret; B, bed-room, 15x15; H, bed-room, 12x15; K, bed-room, 12x18; R, R, R, R, roofs to verandas.

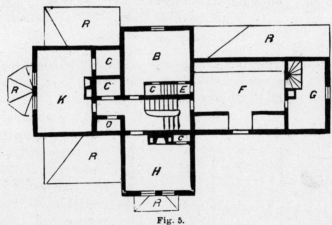

Fig. 5.
UPPER FLOOR OF FARM COTTAGE.

The Grounds.—The grounds about the house are ample and carefully arranged. The water supply comes from the hills above, allowing irrigation, with plenty of head for fountains, and for baths, closets, etc., in the house, and for watering and cleaning in the stable and barns. Although the house stands far back from the public road, it need not be approached by a lane, but better through an open park, set in blue-grass, and utilized as a pasture. The drive-way, sixteen feet wide, and foot-paths, nine feet wide, may all be made by the farm hands under proper direction. A portion, including a fountain and foot-path, is shown in the annexed cut. This park, like the fields, may be planted with deciduous and evergreen trees natural to the climate. For information on which head, see chapter on Trees further on.

PARK OF THE FARM COTTAGE.

Barn and Carriage-House.—The barn and carriage-house shown in the plan is a building 41 x 81 feet, with enclosed sheds of twenty feet, and built with overhanging eaves, and plain but heavy cornice to the gables. In the ground plan, $c\ c\ c\ c$ are four box stalls for the carriage and riding horses; d, four, and h, eight double stalls for farm horses; k, the carriage-house, with stalls for vehicles; a, room for farm-wagon and implements; b, an open shed for grooming horses; l, a closet for farm harness; $j\ j$, carriage-harness rooms; $e\ e$, feeding passages; n, feeding-tube for box stalls. There are also four closets for robes, saddles, etc. The entrance to the carriage-room is at the end, and the other doors and windows are marked by the open spaces. The second story is high enough to give ample space for hay, feed, etc., with appropriate chutes for delivering below. The

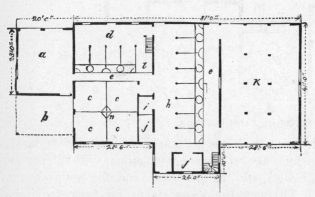

FARM STABLE AND CARRIAGE-HOUSE.

illustrations we have given in this connection will, if carefully studied, tell more than a hundred pages of descriptive print. One thing always remember: Never build until you are sure your circumstances will warrant it.

V. When to Build.

WE repeat, never undertake a building until you have more than enough money to finish and to furnish it. It is far better to occupy the old house a year or two longer, or to add a cheap shed or two, for the time being, than to build a new house or elaborate barn on borrowed money. It may be necessary to borrow for the building of the first house; never for the second, unless storm or fire has destroyed your home.

When to build, then, is when you have the money to do so, or can get it by the sale of surplus crops or stock. If you have more land than you can work to advantage, it may be wise policy to sell a part, to make your home comfortable.

VI. The Provident Farmer's Marriage Settlements.

THE man who adds farm to farm, and lives in a hovel, is neither a good citizen, a good father nor a good farmer. Let us speak unto you a parable: A certain man had a farm, which he thought he worked well. So thought his neighbors. To him were born three children. When the first of these married, one-quarter of the farm went as a marriage portion. By industry, the father raised as much on the three-quarters as he previously had on the whole, and was well satisfied.

Another child married and received a like portion to the first. Improved implements of cultivation, he found, to his surprise, still enabled him to live as well as ever.

The last child, a daughter, was to be married. "Ah, wife," said the father, "the child must have the same portion as the others. How shall we, in our declining age, live on the produce of the quarter of the farm?"

"We are but two," replied the wife, "and shall want but little. The new implements, the enriching of the soil, and the draining, gave us as much from the half as we once had from the whole. Increase the dressing, drain more thoroughly, plow and subsoil still deeper, cultivate yet more faithfully, exchange the poorest of the stock for pure breeds, and see what that will give."

This was done, and the third year thereafter surprised the farmer with more money as the result of his labor than he had ever received from the whole of the original acres of the farm. The application will serve him who intends to build a new house, and has not money enough, but does possess salable surplus land. A half-tilled farm and fine buildings do not go well together.

VII. How to Build.

WHATEVER the structure, careful calculation will show its cost in the amount of excavation, stone, brick, cement, sand, lime, timber, scantling, joists, lumber, shingles, nails, sash, glass and labor. The farmer, perhaps, cannot calculate all these; the architect can, and any master carpenter should be able to do it, if the size of the structure, and the money to be expended are stated. Whatever you build, build well; that is, substantially and of good material. Sound knotty lumber is as good as clear, where it is covered up. Unsound, shaky lumber should be used nowhere.

Special care must be taken that the bricks are firm and well burned. If the clay is of poor quality, or not well burned, the bricks soon crumble and shell. They are not only unsightly, but are costly in the end, and often dangerous. The cost of walls too thin for their purpose is always money wasted. As in draining, the value of the whole work is estimated by the quality of the tile, and the poorest tile laid; so in brick or stone work, the value of the wall is in the quality of the material, the integrity of the mortar, and the honesty of the work.

There is a right and a wrong way to lay stone walls. Hence, it is not economy to employ a contractor simply because he is cheap. His work may be as bad as his price is low. Any man, by intelligent observation and study of the plans and specifications furnished, may be able to judge how the work is going on.

How to build, then, may be summed up thus: Build according to your means. Whatever the structure, use good material; it is cheapest in the end. Have the work carefully planned, and then see the whole carried out properly.

VIII. What to Build.

WHATEVER you build make it comfortable. Remember that the house is woman's kingdom. Therefore, consult her. You sleep and eat there; she lives her life almost wholly within doors. None understand the arrangement of the rooms, pantries, closets, dressers and drawers so well as she.

Every farm-house should have a dairy-room, distinct and by itself; a bath-room, for cleanliness is health as well as comfort. When water cannot be brought direct from the windmill tank, it is easily supplied by a small force-pump and pipe leading to the cistern. Remember the woman who is obliged to cleanse the hard water of the well; and save her useless labor by providing the necessary cisterns for soft water. If nothing better can be had, two or three oil hogsheads, first thoroughly cleansed, may suffice, but never forget the permanent cistern when building the new house.

IX. Taste and Judgment in the Details.

IT is folly to build a house larger than you require, or to have the rooms unduly large, unless your means will permit the hiring of necessary servants to keep all in order. Many a woman on whom the unaided work has fallen has been made prematurely old through such slavery, and often is herself to blame. Do not hesitate to spend money in making both the outside and the inside of your house pleasant to the eye.

Do the doors touch on top or bottom after the new house has been built and furnished? Do the walls crack? In either case you have been neglectful or have been deceived; the foundations are unsound. Does the wood-work shrink? You have used green lumber instead of dry. Do the windows rattle and let in drafts of air? It is from the same cause. Does the water enter to discolor walls and ceiling? The siding or the roofing is to blame.

If you have a carefully written contract, and the builder is responsible, you can, perhaps, recover after a tedious lawsuit. It is better not only to have a carefully written contract, but also to watch how the details are carried out. If you have deliberately cheated yourself in refusing to pay for honest material you are not to be pitied. The pity should all be given to the wife and children of the family, who really are the chief sufferers.

HILLSIDE COTTAGE.

X. Where to Build.

So build as most economically to serve the various uses of the farm, and at the same time obtain, if possible, a view of the surrounding country. The farmer has the whole farm from which to select a building site. If he dumps his house and other buildings down next the road, only because it is a road, he shows bad judgment. If he builds upon a pinnacle simply because it is high, or in a hollow alone, because the wind cannot reach him there, he makes a great mistake. One should take into account, in carefully estimating the value of a building site, central situation on the farm, freedom from exposure to the full effects of the wind, air and elevation in relation to healthfulness, and the advantages to be obtained from perfect drainage; the adaptability of the site to the proposed farm buildings, and the economy of the water supply. All these must be duly weighed, and they are often difficult of solution. If the doubt is of drainage or of roads, the surveyor should be able to advise; if in relation to the house, the architect should be consulted. Take counsel, also, of a landscape gardener—we mean a real landscape gardener, one who can look far beyond the simple details of planting and decoration, by taking the natural beauties into account, and the proper means for heightening them.

XI. A Hillside Cottage.

A FARMER of moderate means who had originally built a very modest house and a comfortable barn in a valley with higher lands on each side wished to build a house where the family could carry on the dairy under the same roof. The new house was built on a hillside with a view of the pastures from the back porch. The barn lay to the west of this with the cultivated fields in front. The building was modest enough; a substantial bracketed cottage fronting the south, as shown in the illustration. And here let us say that a modification of this plan would well suit the hill country of the South. The roof might contain dormer windows, and be raised at least two feet to give height and ventilation, and a gallery extend the length of the south front in place of the porch. In this plan the windows opening on the veranda at the north, and on the small balcony at the end, are long and hung on hinges. The basement contains dairy, dining-room and kitchen, and beyond, but connected with the kitchen, is a vegetable cellar and fuel-room. The plans of the ground floor and attic are given in Figs. 6 and 7. The attic rooms are ten feet in the highest part, but only two feet nine inches at the side.

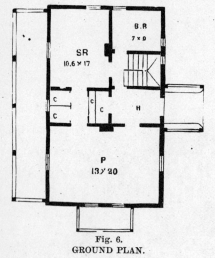

Fig. 6.
GROUND PLAN.

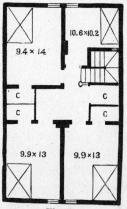

Fig. 7.
PLAN OF ATTIC.

The outside of the building is built of upright boarding and battens; a finish that we like only for barns, but it was built at a time when there was a craze for this style.

The Barn.—The barn is 50 x 36 feet, with a well-lighted basement, containing stalls for cattle. The plan of the main floor explains itself. The stalls are provided with stancheons and manure drops, and the animals fed through the chutes. A plan of the basement of a barn, Chapter IV., Sec. III., will explain the arrangement of stancheons and drops for manure. The horse barn is a building by itself, comprising stables, 18 x 34 feet, shown at *D*, in diagram, and wagon and carriage-yard, *A*, 20 x 30 feet; *B*, 5 x 10 feet, is the harness-room; *c*, stairs to loft; *P*, a projection around two sides of stable, so horses may be fed from the passage without going into the stalls. The heads of the horses are thus kept from the sides of the building, ensuring freedom from drafts. In the projection from the end of the stable, are the feed-bins, *b b*, and a trough, *F*, for mixing feed.

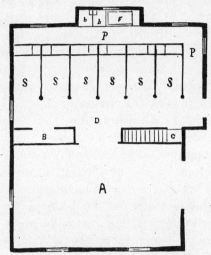

GROUND FLOOR OF HORSE BARN.

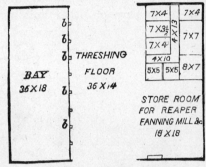

MAIN FLOOR OF FARM BARN.

XII. Ice-House and Preservatory.

ONE form of ice-house, 12 x 12 feet, with preservatory beneath, is shown in the accompanying illustration. In the preservatory were kept the cans of milk for raising the cream, the churned and packed butter, and other articles like canned fruits, etc. A leg of mutton is shown hanging in the preservatory. There is no special objection, if the room is very cold, to keeping fresh meat where milk is. It is better, however, that it be kept in a separate ice-box. The illustration nearly explains itself. The walls are of stone. Then comes an air-space, eight inches wide. Then boarding, with sawdust between, as shown. The floor, of galvanized iron, slopes to the center, and the drainage is carried away by a pipe passing through the dairy-room, where it helps to cool the tanks for milk. The ventilation, as well as the manner of conducting cold air to the preservatory, is shown by the arrows.

XIII. The Water Supply.

IF a stream runs near the site of your house, have the necessary levels taken to find

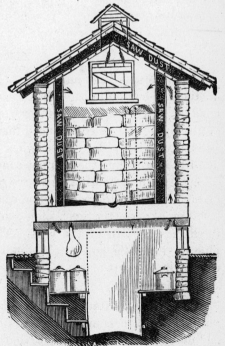

ICE-HOUSE AND PRESERVATORY.

if water can be, by gravity, conducted in pipes to the homestead. If so, the advantage will be great. Besides the comfort, there will be water for the stock, and beauty be obtained in the future, by a pond and fountain. If this natural supply cannot be obtained, you must depend upon a wind-mill and wells for stock water, which same mill may be so arranged as to force soft water to the house from an ample cistern at the barn. This cistern should be in two parts, with a filter between. It is only a question of the first cost, and the necessary pipes and faucets. The reservoir-cistern may be in the barn or other building that will furnish the necessary head. Water will rise, in a pipe, to the same height as the "head," as the elevation of the source of supply is called, but many persons suppose that it may be thrown, in an unconfined stream, to the same height. This is a great mistake, for the jet of a fountain never springs as high into the air as the reservoir which supplies it with water. The cut shows a form of reservoir and fountain that looks better on paper than it does in practice. The hogshead B, may be filled by a wind-mill, and the water running through to the pipe, C, will throw a jet into the air at the basin, but it will be a puny one. A wind-mill will furnish stock water, and water for the house, but a hogshead reservoir for supplying a fountain is a failure.

POOR RESERVOIR FOR A FOUNTAIN.

XIV. House Drainage.

DRAINAGE is most important. It must be perfect. The pipes and mains conducting the house drainage should not only be of ample size to prevent choking, but have the best possible traps to prevent foul gases from rising back into the building. These pipes must be so arranged that they can be "flushed out" with water when necessary. The cellar drains, when these are necessary, and the out-going main may be of tile, providing they do not connect with the other house drainage; if they do, a strong trap must be used where the drain-water of the cellar empties into the house main. All other drains should be water-tight at the joints, and as a further precaution, a soil-pipe, perfectly tight, should lead from the bottom to at least four feet above the highest portion of the roof. This pipe connects at or near the bottom with the pipe service of the house. It is indispensable to carry off the effluvia which always collects in confined places.

XV. Ventilation.

No less important is ventilation. It is true, in the country, where the air is pure, ventilation is not so serious a matter as in the city. Windows may be raised and doors left open in summer; but this is at no time the best way for the general ventilation of buildings, and in winter is not to be thought of. In the winter proper ventilation is difficult. An architect who does not fully understand this question and the best means to be employed, has not wholly learned his business. Do not hamper him here with objections as to cost. However elaborate and expensive the structure, money stinted in ventilation and drainage always proves a costly error. A leak, however slight, in the soil-pipe will, perhaps, let out death to the family. If the earth within, or near, the foundations of the house becomes saturated with the drainage, the germs of low malarial fevers, and all that class of diseases, are scattered through the house. Do not believe, because your nostrils are not offended, that there is no danger. The most deadly miasma often gives no indication to the sense of smell. So with ventilation. There should be some system by which the air vitiated by breathing, and the other emanations of the body, may be freely carried off. Do not employ an ignorant man who professes to be an architect, because he works cheaper than a master of his profession.

CHAPTER II.

BUILDING MATERIAL AND THE BUILDER'S ART.

I. BUILDING MATERIAL.—II. HOW TO MAKE UNBURNED BRICK.—III. SPECIFICATIONS OF FARM AND OTHER BUILDINGS.—IV. OUTLINE OF SPECIFICATIONS FOR HOUSE OF WOOD WITH STONE OR BRICK FOUNDATIONS.—V. MASONRY AND MASON'S WORK.—VI. CARPENTRY AND CARPENTER'S WORK.—VII. PAINTER'S WORK.—VIII. TINNER'S AND PLUMBER'S WORK.—IX. CONTRACT FOR PERFORMANCE OF OBLIGATIONS.—X. HOW TO CONSULT AN ARCHITECT.—XI. GLOSSARY OF SCIENTIFIC TERMS USED IN ARCHITECTURE.

I. Building Material.

A MAN who is going to build, must consult, not only cost of materials and labor, but be governed by many circumstances. Good burned brick are the most lasting of all things used, and where abundant, are often the cheapest in the end. In some regions of the far West, "adobes," or unburned brick, are used, and if the roof projects sufficiently to prevent rain reaching the walls, and if between the ground and the adobe walls, stones, hard-burned bricks, or even planks saturated with bitumen, tar or rosin are laid, these sun-dried bricks last well. As in many places they are the only building material easily obtainable, we give directions for making them. Adobes, if protected as we have indicated, answer for any required structures, in regions where fuel for burning brick is difficult to obtain, and they are especially useful for temporary buildings, to be used until better material can be had. The adobe may be made from clay containing limestone or other small gravel, which would render it unfit for burning, but the more tenacious the clay, the better will be the wall. In making any brick, it is better that the clay, when dug from the bed, be well "weathered," to break down and disintegrate it before using, but less so for the adobe brick than that clay which is to be burned.

II. How to Make Unburned Brick.

THE clay is put into a pit and brought to the proper consistency for treading. Two bundles of straw, cut into six-inch lengths, are added for each one hundred brick. The mass is then thoroughly trodden by cattle, after which it is formed in molds of plank, whose bottoms are not air-tight. These molds are twelve inches long, six wide and four deep, inside measure. They must be well sanded as emptied before being again filled, which filling is prepared by hand, and the surplus clay struck off by an iron "straight-edge." When taken from the molds the adobes are set upon edge on the drying ground, and the second day turned over. In three days they should be dry enough to pile under cover, and are then left for two weeks or more to "cure."

Building the Wall.—The walls of the building are laid in alternate courses of "headers" and "stretchers." "Headers" being brick laid endwise to the weather. "Stretchers" are brick laid with the side to the weather, that is, to the outside of the wall. The bricks, being one foot long, determine the thickness of the wall, and the first course should run through. In carrying up the wall, joints must be broken, not only as regards layers of "stretchers," but with the "headers" also.

III. Specifications of Farm and Other Buildings.

It will not be necessary to enter into any detailed statement as to the construction of buildings, farther than what has been made heretofore. The illustrations and definitions, except so far as may be necessary to show how changes might be made for the better, will be all that will be necessary. In the illustrations we begin with the simpler and proceed to the more elaborate, for the reason that the majority want simple structures, and here will be the proper place to give the general specifications: First, for an ordinary modern building; and second, for those for a first-class building of brick with stone foundation. Many items specified are not needed on the farm, but are necessary in suburban and other houses. In these we have followed, in a general way, the comprehensive and elaborate directions as given in a valuable, complete and costly practical architectural work, known as "Palliser's American Country Homes."

IV. Outline of Specifications for House of Wood, with Stone or Brick Foundations.

Dimensions.—Drawings and details must be accurately followed according to the scale given, and preference must always be given to the scale rather than to the dimensions. The building must be in exact proportion and in size as shown on the architect's plans, and as figured in the drawings—as, for instance, height of cellar, first floor, kitchen, second floor over main part of house, height of second floor over kitchen, third floor, etc., always in the clear, and built in exact accordance with the plans and specifications.

Note.—In the specifications, in some instances, we shall give size, quality and dimensions as being appropriate for buildings of two stories. Where blanks are left they are to be filled with figures and names as required. The whole being intended for what is known as a balloon-frame, to be still further treated of hereafter.

V. Masonry and Mason's Work.

Excavating.—Do all necessary excavating required for cellar, area and foundations until firm and solid ground is reached, and always be entirely certain to go deep enough to be beyond the reach of frost.

Stone Work.—Build the foundation walls sixteen inches thick, of good flat stone, of firm bed, well bonded through the wall, laid up in clean, sharp sand, lime and cement. Mortar made in the proportion of one part of cement to two of lime. Lay by, and full to, a line on the inner face, the joints of which must be flushed and pointed at completion. Put like foundation under all piers, chimneys and exterior steps.

Drains.—First quality cement drain-pipe, as per plan, are to be connected, if in a town, with the sewer; or in the country, a drain built for the purpose, and these pipes must be properly graded and trapped and the joints cemented tight.

Underpinning.—From the top of the stone wall, at grade level, extend up two feet in height with eight-inch wall of best hard-burned brick, and clean, sharp sand and lime mortar; face walls with selected brick of even color, laid in red mortar, close joints, jointed, properly cleaned down at completion, and finished with black joints, window-sills of (state the kind of stone).

Piers.—Build piers in cellar, as shown in the plan, of best hard-burned brick, laid in clean, sharp sand and lime mortar, and cap with flat stone the size of piers.

Chimneys.—Build chimneys as shown, plastered on inside and out, furnished with proper stove collars, and with ventilating covers where required; turn arch to fireplace, and turn trimmer arch under hearth; hearth to be (state the material of hearth) bedded in cement. Top out the chimneys above the roof, as shown, with selected brick in like manner to underpinning.

Lathing.—All stud partitions, ceilings and work that is furred off, on first and second floors, to be lathed with (state kind of lathing), and joints to be broken (state how the joints are to be broken, as for instance, every tenth lath).

Plastering.—All walls, partitions and ceilings, throughout first and second floors, to be plastered, one coat of brown, well-haired mortar, and finish. State whether white, hard, or other finish is required. All walls to be finished straight and plumb. All angles to be maintained sharp and regular in form, and the plastering in all cases to extend to the floor, ceiled surface, or base board, as the case may be.

VI. Carpentry and Carpenter Work.

Timber.—All timber must be put together in the most substantial and workmanlike manner known to the trade. State of what kind of wood all timber is to be, when not otherwise specified. This is important, since in every locality there is much inferior material.

Framing.—State the kind of framing. If a hollow frame, as follows: The frame to be what is known as a hollow frame, well nailed together; second floor girts to be notched into and well spiked to studs. Do all necessary framing around stairways and chimneys, properly mortised and tenoned together.

Frame Timber.—The size must be carefully stated, and also the kind of material. The following may represent that for an ordinary sized two-story house: Girders, 4x6 inches; sills, 3x7 inches; posts, 4x5 inches; girts, $1\frac{1}{4}$x4 inches; plates, 2x4 inches, doubled and well spiked into ends of studding. Frst-floor timbers, 2x8 inches; second-floor timbers, 2x6 inches, all to be 16 inches apart, from center to center of timbers; header and trimmer beams, 3 inches thick; roof-rafters 2x5 inches, by 24 inches to centers, apart; door and window studs, 3x4 inches; intermediate studding, 2x4 inches, and 16 inches each to centers; studdings in partitions, 2x3 inches, and 16 inches each from centers to centers. Veranda sills and cross sills, 3x6 inches; floor timbers, 2x6, and 20 inches from centers; plates, 4x5 inches.

Bridging.—Bridge floor timbers with 1x2 inch cross-bridging properly cut in between timbers, and nailed at each end with two ten-penny nails.

Furring.—Fur overhead on rafters for rooms on second floor, and also any other furring required.

Sheathing.—Cover all sides of the frame with tongued and grooved boards, not to exceed six inches in width, and nail through each edge to every stud with ten-penny nails.

Lumber.—All lumber must be of white pine (unless otherwise specified), free from knots, shakes and other imperfections impairing its strength and durability. Water table seven-eighths inch thick, furred off one inch and capped with a leveled and rabbeted cap for clapboards to lap. The corner boards, casings and bands to be one and one-quarter by six inches, bands to be rabbeted top and bottom for clapboards and beveled on top.

Clapboarding.—All sides to be covered with clear pine clapboards four and one-half inches wide, nailed with eight-penny box nails, and to have not less than one and one-quarter inch lap, and underlaid with water-proof sheathing felt, which also place under all casings, water table, etc., so as to lap and make a tight job.

Cornices.—These are to be formed on three by five inch rafter feet, spiked on to rafter at plate; gutter formed on same and lined with tin, so as to shed water to points as are indicated in plan. Plancier to be formed by laying narrow pine matched boards, face down on rafter feet, large boards two inches thick and as shown, and all as in the detail drawing.

Window-Frames.—These are to be made as shown in the drawings. Cellar-frames of two-inch plank rabbeted for sash; sash hinged (or not as the case may be) and to have suitable fastenings to hold open or shut; all other sashes to be double hung with best sash-cords and cast-iron weights, and to be glazed with best sheet glass (or other as the case may be); also specifying the number of sash and size of lights. Also state thickness of sash. All sash to be made of best clear, thoroughly seasoned pine. Window sills to be two inches thick. (For ordinary windows one and three-eighths inch is the usual thickness.)

Blinds.—State whether blinds are to be outside or inside, and of how many folds. Outside blinds should never be used where they can be avoided. All wood work should be primed with best white lead and linseed oil, as soon as exposed to the weather.

Door-Frames.—Outside door-frames of plank, rabbeted, and furnished with 2-inch oak, or other hard-wood sills.

Porches.—These vary much in character. They should be constructed in accordance with the detail drawings, including columns, rails, newels, panels, etc. steps should be $1\frac{1}{8}$ inch thick, with $\frac{7}{8}$ inch risers, to have cove under nosings; floors laid with 1⅛x4 inch flooring, blind-nailed to beams; the joints served with white lead; ceiling to be ceiled with narrow-beaded battens of even width and molded in angles.

Roofs.—Cover all roofs with best sawed pine shingles, laid preferably on roofing boards, solidly nailed to the rafters. If laid on strips, these may be 1x2 inches, nailed to the rafters with ten-penny nails. The shingles to be laid to break joints, nailed with white metal nails, two to each shingle, the nails to be well covered by the succeeding lap, to make a perfectly weather-tight roof. It pays to paint shingles with mineral paint before laying.

Floors.—Kitchen floors are better laid with three or four inch wide ash strips seven-eighths inch thick. Sound yellow and Georgia pine are next best. The principal floor of the house should be laid with best pine flooring, 1x6 inches. The second floors of seven-eighth inch pine, 6 inches wide. All floors to have joints broken, to be well driven home, and securely blind-nailed. Kitchen and other floors requiring mopping and scrubbing, are better if the flooring is payed with white lead, to make it water-tight.

Partitions.—All partitions should foot on girders, and have 3x3 inch plates to carry second floor; all angles formed solid, and all partitions bridged at least once in their height. The *grounds* to screed plaster to should be seven-eighths inch thick, and left on.

Wainscoting.—Wainscot walls for kitchen, when used, may be three feet high, if with beaded battens; if not, of ⅞-inch flooring, well driven home, and blind-nailed. They should be furnished with beveled and molded cap.

Casings.—These must be described in detail; the following will serve as an example: Casings in front hall and living-rooms—to be cut and stop-chamfered—1¼x6 inches; all doors and windows elsewhere to be cased before plastering with ⅞-inch casings, and finished with a ⅞x1¾ inch band mold. Put down 7-inch beveled base in front hall and bed-rooms after plastering; door jambs to be ⅞ inch thick, rabbeted for doors and headed on edges; windows to be finished with neat stool and apron finish.

Doors.—State whether the doors, and which of them, are to be panel, sliding and sash doors. Sash doors are used for entre-sal, and sometimes for outside doors. Six-panel, ogee, solid molded doors are usual for inside single doors. Saddles of doors should be, preferably, of hard pine.

Stairs.—Cellar stairs should be of plank, without risers; second-floor stairs, 1¼-inch tread, ⅞-inch risers, properly put together and supported.

Sinks.—Ceil up under sink with narrow beaded work to match wainscoting; hang doors for closets underneath; place appropriate hooks; ceil up splash board, 16 inches high, and place drip board.

Pantries and Closets.—Pantries should have a counter shelf and at least four shelves above, with appropriate pot hooks beneath counter shelf. China closets with counter shelf, drawers underneath, and appropriate shelves above. Wardrobe closets, to be fitted up with shelves, double wardrobe hooks, on molded strips.

Door Furniture, etc.—The door furniture must correspond to the specifications and drawings. Locks: mortise locks, brass fronts and keys, with stop-locks and shove-bolts for all outside doors. The stops should be of hard wood with rubber tips; hinges, of loose joint, but of size and strength appropriate to the doors.

Mantels.—These may be of marble, slate, pottery, or of hard wood, according to the nature of the building.

Cellar.—The partitions may be of brick or of wood (brick is best), divided into rooms. with suitable doors, and furnished with shelves, bins and other fixtures according to the necessities of the case.

VII. Painter's Work.

All wood-work outside and inside, should receive two coats of the best white lead and raw linseed oil. Accept nothing adulterated. Paint that remains sticky, and is affected by atmospheric changes, is a nuisance. Clapboards should be painted in some light neutral colors, with darker trimmings. Grain the wood-work in kitchen in oak or maple, inside blinds, doors, etc., preferably in imitation of some light-colored handsome wood; bed-rooms are best painted in one color; chamfers and cut work should be picked out in appropriate colors. Paint the roof a dark-slate color; tin work, Indian red or other rather dark color. The whole to be in accordance with the design.

VIII. Tinner's and Plumber's Work.

Tinning.—All tinner's work should be of the best material, soldered in rosin; gutters lined with tin, tin leaders to convey water from gutters to grade level, to be firmly secured to the building, and to be graded in size to correspond to the amount of water to be carried.

Sinks.—These should be of cast iron, to be supplied from five-eighths inch tin-lined lead pipe, with five-eighths inch brass cocks; waste-pipes two inch, of cast or wrought iron, gas-tight, properly caulked at joints if of cast iron; if of wrought iron, the pipes must be screwed tight, the joints first payed with red lead, trapped and closely connected with the drain; the waste-pipe to extend through the building and above the roof for vent. All water-closet (if any) and other drain fixtures to be according to the best scientific skill.

In conclusion, we repeat, that all work of whatever kind must be in accordance with the specifications and design, in all the departments. Hence the importance of a specific contract and an honest architect.

IX. Contract for Performance of Obligations.

The party of the first part (the contractor) agrees for the consideration named, and according to the plans and specifications, to build complete, in a thoroughly finished manner (furnishing all good and sufficient material if so to be stipulated), within a given time, and in accordance with the direction of the architect or master builder, stipulations being plainly stated as to the number of payments, and time, including the final payment, reserving all payments until the lapse of time necessary to provide against mechanics' liens, or liens for material furnished, certificates being furnished from the proper law officers that such do not exist. The work previously quoted, and from which we have generalized the matter relating to plans and specifications for building, gives a form of contract which leaves no loop-holes. This contract reads as follows, and may be altered to suit any given circumstances:

ARTICLES OF AGREEMENT made and entered into this———day of———in the year One Thousand Eight Hundred and———by and between———of the———, of———, County of——— and State of ———, as the part——of the first part, and———of the———of———County of———, and State of ———, as the part——of the second part, witnesseth:

First.—The said part——of the first part, do——hereby, for——— heirs, executors, administrators, or assigns, covenant, promise and agree to and with the said part——of the second part, ———heirs, executors, administrators or assigns, that———, the said part——of the first part, ———heirs, executors, administrators or assigns, shall and will for the consideration hereinafter mentioned, on or before the ———day of———, in the year One Thousand, Eight Hundred and———, well and sufficiently erect, finish and deliver, in a true, perfect and thoroughly workmanlike manner, the———for the part——of the second part, on ground situated———, in the———of———, County of———, and State of———, agreeably to the plans, drawings and specifications prepared for the said works by———, architect, to the satisfaction and under the direction and personal supervision of , architect, and will find and provide such good, proper and sufficient materials, of all kinds whatsoever as shall be proper and sufficient for the completing and finishing all the———and other works of the said building mentioned in the———specifications, and signed by the said parties, within the time aforesaid, for the sum of ———Dollars.

Second.—The said part——of the second part do——hereby for——— heirs, executors, administrators or assigns, covenant, promise and agree to and with the said part——of the first part———heirs, executors, administrators or assigns, that———, the said part——of the second part———heirs, executors, administrators or assigns, will and shall, in consideration of the covenants and agreements being strictly executed, kept and performed by the said part——of the first part, as specified, well and truly pay or cause to be paid, unto the part——of the first part, or unto———heirs, executors, administrators or assigns, the sum of———Dollars, lawful money of the United States of America, in manner following:

First payment of $———. Second payment of $———. Third payment of $———. Fourth payment of $———. Fifth payment of $———.

When the building is all complete, and after the expiration of———days, being the number of days allowed by law to lien a building for work done and materials furnished, and when all the drawings and specifications have been returned to———, architect.

Provided, That in each case of the said payments, a certificate shall be obtained from and signed by———architect, to the effect that the work is done in strict accordance with drawings and specifications, and that he considers the payment properly due; said certificate, however, in no way lessening the total and final responsibility of the part——of the first part; and, *Provided further* that in each case a certificate shall be obtained by the part——of the first part, from the clerk of the office where liens are recorded, and signed and sealed by said clerk, that he has carefully examined the records and finds no liens or claims recorded against said works or on account of the said part——of the first part. AND IT IS HEREBY FURTHER AGREED BY AND BETWEEN THE SAID PARTIES:

Third.—That the specifications and the drawings are intended to co-operate, so that any works exhibited in the drawings and not mentioned in the specifications, or *vice versa*, are to be executed the same as if they were mentioned in the specifications and set forth in the drawings, to the true intent and meaning of the said drawings and specification, without extra charge.

Fourth.—The contractor, at his own proper costs and charges, is to provide all manner of labor, materials, apparatus, scaffolding, utensils and cartage of every description needful for the due performance of the several works; and render all due and sufficient facilities to the architect for the inspection of the work and materials.

Fifth.—Should the owner, at any time during the progress of the said works, require any alterations of, deviations from, additions to, or omissions from the said contract, he shall have the right and power to make such change or changes, and the same shall in no way injuriously affect or make void the contract; but the difference shall be added to or deducted from the amount of the contract, as the case may be, by a fair and reasonable valuation.

Sixth.—Should the contractor, at any time during the progress of the said works, refuse or neglect to supply a sufficiency of material or of workmen, or cause any unreasonable neglect or suspension of work, or fail or refuse to comply with any of the articles of agreement, the owner or his agent shall have the right and power to enter upon and take possession of the premises and provide materials and workmen sufficient to finish the said works, after giving forty-eight hours notice in writing, directed and delivered personally to the part——of the first part, and the expense of the notice and the finishing of the various works will be deducted from the amount of contract.

Seventh.—Should any dispute arise respecting the true construction or meaning of the drawings or specification, the same shall be decided by——, architect, and his decision shall be final and conclusive; but should any dispute arise respecting the true value of any extra work, or of works omitted by the contractor, the same shall be valued by two competent persons—one employed by the owner and the other by the contractor—and these two shall have the power to name an umpire, whose decision shall be binding on all parties.

Eighth.—No work shall be considered as extra, unless a separate estimate in writing for the same shall have been submitted by the contractor to the architect and the owner, and their signatures obtained thereto.

Ninth.—The owner will not, in any manner, be answerable or accountable for any loss or damage that shall or may happen to the said works, or any part or parts thereof, respectively, or for any of the materials or other things used and employed in finishing and completing the said works.

Tenth.—The contractor will insure the building before each payment, for the amount of the payment to be made; and the policy will not expire until after the building is completed and accepted by the architect and owner. The contractor will also assign the policy to the owner before the payment will be made.

Eleventh.—Each artisan and laborer will receipt the architect's certificate, that he has been paid in full, and the contractor will make oath according to the architect's certificate, that all bills have been paid, and that there are no unpaid accounts against the works.

Twelfth—Should the contractor fail to finish the work at or before the time agreed upon, ——shall pay to the part——of the second part, the sum of——Dollars per diem, for each and every day thereafter the said works shall remain unfinished, as and for liquidated damages.

IN WITNESS WHEREOF, the said parties to these presents have hereunto set their hands and seals, the day and year above written.

Witnesses, { ——   Part——of the First Part { —— [SEAL.] [SEAL.]
Witnesses, { ——   Part——of the Second Part { —— [SEAL.] [SEAL.]

X. How to Consult an Architect.

It is an architect's business to give the greatest amount of room, with the best architectural effect, for the money spent. In all superior buildings the architect's charges are well earned. The architect may charge — for full working plans, all details of drawings for exterior and interior work and fittings, specifications and forms of contract—two and a half per cent on the cost of erecting and completing the building. That is, on a building to cost $2,000 the charge would be, say, fifty dollars. If they prepare complete bills of quantities of materials, $\frac{3}{4}$ of one per cent additional may be charged. If the architect is required to superintend and supervise the building personally and the contractors' bills, the charge will be in accordance to the distance traveled or various local necessities. If you decide to consult an architect, state:

The outside amount of money to be expended on the building complete.

The nature of the ground, size and shape of lot, which way the building is to front. A rough draft is essential if it can be made. What material will be used in the construction, and the price of material and labor.

What number and size of rooms on each floor, also any special disposition to be made of certain rooms in relation to use, scenery, views, or otherwise, and what the building is to be used for. State precisely location and character of the grounds and surroundings, and anything necessary to be considered (that may occur) relating to design, location and arrangement of rooms.

What are the means of drainage, the improvements required, in heating, water (hot and cold), bath, water closets, and artificial lighting.

About the outside finish, as porches, verandas, bay-windows, towers, etc.; also provisions for water service, cisterns, etc.

Fences and out-buildings required; and also name any work you wish to do, or material you wish to furnish.

Give post-office, county and State, with name. The whole to be legibly written.

Thus the architect can work understandingly, and it may save the cost of a personal visit. But if possible, see the architect personally, and be prepared to answer intelligently such questions as he may propound, in addition to those stated. This you may do easily if you have studied the matter in this work relating to buildings, and also landscape effect.

XI. Glossary of Scientific Names Used in Architecture.

ABACUS.—The upper member of the capital of a column, on which the architrave is laid.

ABUTMENT.—Masonry, earth or timber, at the end of a bridge, or the solid part of a pier supporting an arch.

ARCADE.—A covered walk along the side of or within a building, with columns on the outer edge, supporting arches.

ARCH.—A curved, self-sustaining structure, supported by the key-stone and abutments; the beginning of the arch is called the spring of the arch; the middle, the crown; the distance across, the span; and vertically, the height.

ARCHITRAVE.—The lower of the three members of the entablature, resting immediately on the columns.

ASTRAGAL.—A small moulding, with semi-circular profile, as an ornament on the top or bottom of a column.

ATTIC.—The upper story or garret of a building. An attic base is the base of a column, with double mouldings.

BALCONY.—A projection from the exterior wall of a building, inclosed with a railing, usually placed before a window or glass door in the second story.

BALOON FRAME.—A strong frame made with few mortises and tenons, spikes and nails holding all firmly together.

BALUSTER.—One of the upright portions of a railing, miscalled Banister.

BALUSTRADE.—A range of balusters, connected by a rail on the top, and commonly called a railing.

BANISTER.—See Baluster.

BARGE-BOARD.—The projecting board placed at the gable, so as to hide the horizontal timbers of the roof, more properly called verge-board.

BATTEN.—A narrow strip of board, for covering the exterior joints of vertically boarded buildings. A batten-door is made of boards, with battens nailed on across them as stiffeners.

BATTLEMENT.—A wall on the top of a building.

BAY.—The space between posts or buttresses; in barns a low space for storing hay.

BAY WINDOW.—A window, curved or angular, set in an exterior projection from the walls of the house, and having its base on the ground.

BEAD.—A moulding whose vertical section is semi-circular; a moulding ornamented like beads.

BEARING.—The span of a beam or rafter, or that part which is without support.

BOND.—Mode of laying bricks or stones, to break the joints. When the stretchers and headers, as they are called, are in alternate and separate courses, it is termed English bond; when alternately in the same course, Flemish bond.

BOND-TIMBER.—Timber laid in a wall horizontally, for tying it together.

BOUDOIR.—Private ladies' room, for calls, dressing-room, etc.

BOX-SHUTTERS.—Shutters folding into cases.

BRACKET.—A support for shelves, stairs, balconies, projecting roofs, etc.

BREAST OF A CHIMNEY.—The contracting part of the back, opposite the throat.

BRICK-TRIMMER.—A brick arch, abutting on the wooden trimmer, under the slab of a fireplace, to prevent the communication of fire.

BRIDGE-BOARD.—The notched board on which the steps of wooden stairs are fastened.

BUTTRESS.—A prop or support of masonry against the sides of a building, to resist pressure and stiffen walls.

CAMBER.—Convexity or arch on the upper side of a beam.

CAMPANILE.—A tower on a building, serving as a belfry.

CAPITAL.—The upper, projecting, and ornamental part of a column.

CASEMENT.—Applied to windows divided into two parts by the mullion, and hung on hinges.

CAVETTO.—A concave molding, whose profile is the quarter of a circle.

CESS-POOL.—A well or cistern under the mouth of a drain, to receive the sediment.

CLAPBOARD.—See Siding.

CLUSTERED COLUMN.—One made of several united.

COBBLE-STONE.—A round stone, often used for walls of buildings by imbedding in regular courses in mortar or cement.

COLONNADE —A range of columns.
COLUMN.—A pillar consisting of base, shaft or body, and capital.
COMPOSITE ORDER.—A compound of the Ionic and Corinthian orders.
CONSOLE.—A bracket.
COPING.—The capping stone or brick covering of a wall, wider than the wall itself, to throw off the water.
CORBEL.—A projecting piece of wood or stone from a building.
CORINTHIAN ORDER.—An order of Grecian architecture.
CORNICE.—The upper projecting division of an entablature; any molded projection which crowns or finishes the part to which it is attached.
CORRIDOR.—A gallery or passage.
COTTAGE ORNEE.—An ornamental cottage, where expression or appearance is the chief object.
COURSE.—A continuous horizontal range of stones or brick in a wall.
COVE.—The concavity of an arch or ceiling.
CROSS-BRIDGED.—The cross-bracing placed between a series of timbers or joists.
CUPOLA.—A spheroidal roof or dome; a small structure on the top of a dome.
CURB-ROOF.—Gambrel roof, a roof with the lower half inclined at a steeper angle.
CYMA.—A wave-form member or part of a cornice; also termed ogee.
DEAFENING.—A floor covered with mortar placed beneath a floor, to exclude sound, and prevent the passage of flames.
DETAILS.—Applied to the drawings of the separate parts of a building; working drawings.
DORIC.—An order of Grecian architecture; intermediate between the Tuscan and Ionic, combining simplicity, strength and chasteness.
DORMER WINDOW.—A window standing vertically on a sloping roof.
DOVE-TAIL.—A joint made for connecting wood, the parts cut in the form of a dove's tail expanded, with a corresponding hollow.
DOWEL.—A pin used in connecting two pieces of wood.
DRESSINGS.—Parts to decorate plainer work, as the mouldings of a window.
DRIP-STONE.—A projecting window-cap, usually hollowed beneath that the rain may drop from it.
DUMB-WAITER.—A cupboard or platform running on pulleys, to convey dishes, food, etc., from one story to another.
ELEVATION.—A drawing of the face or principal side of a building, every part seen exactly in front; differing from a perspective view, which is seen from one point.
ENTABLATURE.—The whole of the parts of an Order, above the column, including the architrave, frieze and cornice.
FACADE.—The front of a building.
FASCIA.—One of the parallel bands used to break the monotony of an architrave.
FILLET.—A narrow, flat band, used for the separation of one molding from another.
FINIAL.—In Gothic architecture, the top or finishing of a pinnacle or gable.

FLASHING.—Lead or other metal let into the joints of a wall, so as to lap over gutters and prevent the rain from injuring the interior works.

FLOAT.—A long straight-edged board used to render a plastered wall perfectly straight.

FOILS.—A term applied to rounded or leaf-like forms seen in Gothic windows, niches, and the like.

FOOTING.—Spreading courses at the base of a wall.

FRIEZE.—The middle part of an entablature, between the architrave and cornice.

FUNNEL.—The stack or upper part of a chimney; the shaft.

FURRING.—Slips of wood nailed to joists and rafters, to bring them to an even surface for lathing.

GABLE.—The triangular end of a house above the eaves.

GAIN.—The beveling shoulder of a joist or other lumber.

GALLERY.—A common passage to several rooms in an upper story; a long apartment for paintings, etc.

GAMBREL ROOF.—See curb roof.

GINGERBREAD-WORK.—A profusion of fanciful ornamental carvings; this is always in very bad taste.

GIRDER.—The principal beam or timber in a floor.

GIRTH.—Horizontal connecting timber in an upright frame.

GOTHIC ARCHITECTURE.—The style of architecture denoted by the pointed arch. It admits of great variation in all its parts; the roof may be castellated or pointed, or with broad projecting eaves. A still greater variety exists in the windows, among which are the arched, triple lancet, rose, square-headed, oriel, triangular and other forms.

GRAINED.—Painted in imitation of the grain or texture of wood.

GROIN.—A line made by the intersection of two arches, crossing each other at any angle.

GROUND-SILL—GROUND-PLATE.—The lower and outer timber, supporting the posts.

HALL.—A large public room; the first large room within a building; the narrow entrance of a dwelling house, designated as the entrance hall.

HAMMER-BEAM.—A horizontal timber, in place of a tie-beam, just above the foot of a rafter; used in pairs to strengthen Gothic frames.

HARMONY.—In large buildings, where variety prevails, it is that which brings all the varied parts into an agreeable relation to each other.

HEADERS.—Bricks laid crosswise in a wall, in contradistinction from stretchers, laid lengthwise. See Bond.

HIP.—The sloping angle of a hipped roof.

HIP-KNOB.—A finial, pinnacle or other ornament on the point of a gable, or on the hips of a roof.

HIPPED-ROOF.—A roof with sloping ends.

HOOD.—A projecting covering over a window or door, for shade and to throw off water.

Hood-molding.—The molding over a Gothic window, called also label-molding.

Hydraulic Cement.—Mortar made of water lime, which hardens like stone under water; used for cisterns, cellar bottoms, etc.

Intertie.—A horizontal piece of timber between two posts, to keep them together.

Inverted Arch.—Arch curving downwards, to give a firm foundation to piers.

Ionic Order.—A Grecian Order of Architecture.

Italian Architecture.—An irregular and beautiful style of modern architecture. Has projecting eaves, arcades, balconies, ornamental chimney tops, campaniles, etc.

Jack Timbers.—Those shorter than the rest in the same row or line, by being intercepted by something else.

Joggles.—Pieces of hard stone introduced to stiffen the joints of masonry.

Joint.—The place where two pieces of timber come together.

Joist.—The smaller timber of a floor.

Key.—A piece of wood let into another, across the grain, to prevent warping.

King-post.—The middle post of a framed roof, reaching from the center of the tie-beam to the ridge; called crown-post.

Label.—The outer moulding over a window or doorway, descending a short distance on each side.

Lancet-window.—A window in Gothic architecture, acutely pointed at the top.

Landing.—The floor at the head of a flight of stairs, or portion of a flight.

Lintel.—The head-piece of a door or window frame.

Lodge.—A small house or tenement connected with a larger. A gate lodge or porter's lodge is one placed near an entrance gate to an estate.

Louver-window.—A window open to the sound of bells within, but with blinds to exclude rain.

Mansard-roof.—A French roof, inclining back slightly from the perpendicular, with a roof of low pitch above.

Miter.—The junction of two boards, at an angle, by a diagonal fitting.

Modillion.—A carved horizontal bracket.

Mortise.—A hole cut in a timber to receive a tenon, or corresponding piece of another timber.

Mouldings.—The ornamental contour given to angles of cornices, window-jambs, etc., or to ornamental lines or borders generally.

Mow.—The loft of a barn.

Mullion.—The upright post or bar, dividing the two or more parts of a window.

Newel.—The column about which the steps of a spiral staircase wind.

Notch-board.—The board which receives the ends of the steps of a flight of stairs.

Ogee.—See Cyma.

Oriel-window.—A projecting window, supported on a corbel or other projection; a bay-window; or has a foundation resting on the ground.

Ovolo.—A convex molding, whose profile forms about a quarter of a circle on its lower inclined side.

PANEL.—A sunken space, most commonly applied to the portion of a door between the upright pieces (styles) and the horizontal pieces (rails).

PARLOR.—The sitting-room or living-room of a family; more commonly restricted to a room for visitors.

PAVILION.—A word variously applied in rural architecture; a broad, highly finished veranda on the better class of dwellings.

PEDESTAL.—The lower part or base of a column, consisting of the die or square trunk, the cornice or head, and the base or foot; also, the support of a vase, statue, etc.

PEDIMENT.—The triangular or circular part of a portico, between the roof and top of the entablature.

PENDANT.—An ornament hanging from the vault of a roof, in Gothic architecture; more commonly from the peak of a gable—the lower part of the ornament being the pendant, and the portion above the roof the hip-knob or finial.

PIAZZA.—A covered walk on one or more sides of a building, supported on one side by pillars. It is used nearly synonymously with veranda; the latter implies more shade and seclusion, often having lattice-work in front.

PIER.—Usually the pillar-like masses of masonry from which arches spring.

PILLAR.—A general name for a permanent prop or support; a column is an ornamental pillar, usually round, and belonging to one of the Orders of Architecture.

PINNACLE.—The summit or apex; usually a square or polygonal pillar, at the angles of Gothic buildings, terminating at a point, and embellished with ornament.

PISE.—A wall constructed of stiff earth or clay, rammed in between moulds as the work is carried up. In countries where frequent rains prevail, it cannot be very durable, unless covered, and is similar in character to walls made of unburnt brick.

PITCH OF A ROOF.—The proportion between the height and the span. If the rafters exceed in length the width of the building, the roof has a "knife-edge pitch:" if equal to the width, it is Gothic; if two-thirds, it is termed a Roman pitch; flatter it is a Grecian pitch. Generally the pitch is designated by number; if the height of the ridge is one-fourth of the span of the roof, it is termed "quarter pitch;" if one-third the span, "third pitch," etc.

PLAN.—A drawing of the horizontal section of a building, showing the distribution, form and size of the parts.

PLATE.—See Roof, as showing much in little space.

PLINTH.—A projecting, vertical-faced member, forming the lowest part of the base of a column or wall.

POINTING.—Trimming with mortar the joints of a wall of masonry.

PORCH.—An appendage to a building, forming a covered approach to a door or entrance.

PORTE COCHERE.—A carriage porch, or covered entrance to a large dwelling, under which a carriage may drive; literally, a covered carriage-way.

PORTICO.—A covered space or projection, supported by columns, at the entrance of a building.

PURLINS.—Horizontal pieces of timber to support rafters

PUTLOG.—A horizontal timber to support a scaffold.

QUARTERS.—Upright posts in partitions, to which lath are nailed.

RABBET (REBATE).—A cut made on the side or edge of a board, to receive the edge of another cut in the same manner.

RAIL.—This is a horizontal piece of timber, as between the panels of a door, or over balusters, etc.

REEDING.—A small convex molding.

RIBBING.—The timber work sustaining a vaulted ceiling.

RIDGE-POLE, OR RIDGE-PLATE.—The horizontal timber or board sustaining the upper ends of the rafters.

ROMANESQUE.—A style of architecture, adopted during the later period of the Roman Empire. It is prominenly marked by arches and columns, and also by irregular forms.

ROOF.—The upper covering of a building, consisting mainly of two parts, viz: the framing or trussing, and the covering of shingles, or other material. The different forms are a curved or French roof, a roof with an ogee curve, a gable, hip, and gambrel or curb roof.

ROOM.—Interior division of a dwelling, entered by a door. The first room (in houses containing all these different apartments) is the vestibule, or lobby, or ante-room, when used as a reception-room. The second, the hall, or first large room within the building. There are the library, study, or office, or a room with these variously combined; the parlor or family room, sometimes used as an every-day living-room, in other instances as a breakfast-room, or a room for company only; the drawing-room, or room specially for the reception of company, or into which the company retire from the dining-room. In the smaller houses the parlor and drawing-room are one. The dining-room and kitchen are distinct; and appended to the kitchen may be the laundry or wash-room, the store-room or pantry, for provisions; the iron closet, for the coarser utensils; the scullery or sink-room, where utensils and dishes are cleaned and kept; the bath-room; the nursery; the boudoir, or ladies' private dressing-room, or for the private reception of company; and bed-rooms, the larger of which may have dressing-rooms attached, and closets. In the largest and most expensive dwellings, all these rooms are found separately; but as dwellings become smaller, the purposes of two or more are combined in one.

ROUGH-CAST.—Rough mortar or cement for the exterior walls of buildings, mixed with pebbles, small shells, etc.

RUBBLE.—Small rough stones, used for walls or filling between walls.

RUSTIC-WORK.—Building with the faces of stone left rough, and the joining sides wrought smooth; ornamental wood structures, with the bark on.

SAFETY-ARCH.—An arch built solid in the substance of a wall, to sustain any unusual weight on that part; a discharging arch..

SALOON.—A lofty, spacious apartment; state-room; reception-room.

SCARF-JOINT.—A joint made by cutting away corresponding portions of timbers.

SHAFT.—The principal or central part of a column; the chimney above the roof.

SHOE.—The projecting part of a water-pipe at bottom, to throw the water from the building.

SIDING.—The exterior side covering of boards to a building.

SILL.—The lower, horizontal timber of a frame, door or window.

SPECIFICATION.—An exact written description of the different parts of a building to be erected.

SPRINGER.—The base of an arch; the rib of a groined roof.

STACK.—A number of chimney shafts combined in one.

STILE.—The vertical piece in framing or paneling.

STRETCHERS.—Bricks laid lengthwise in a wall.

STRUT.—An oblique timber in a frame, serving as a brace. The term brace is usually applied to smaller and shorter pieces.

STUCCO.—Fine plaster for covering walls, and for interior decorations. The best is made of two parts of sharp and perfectly pure sand, and one part of purest lime, the latter slacked with water to a fine powder, sifted and mixed with the sand. Outer walls, stuccoed, should have broad projecting eaves to throw off water.

STUD.—A piece of timber inserted in a sill to support a beam—a term usually applied to the upright scantling of a frame.

SURBASE.—A cornice or series of mouldings above the pedestal; also applied to the board which passes horizontally around the walls of a room, to protect them from injury.

TERRA COTTA.—Architectural decorations, vases, chimney tops, etc., made of a mixture of pure clay and broken flints, crushed pottery and other materials, and burned to the hardness of stone.

TIE.—Timber serving to bind walls or other parts together.

TRACERY.—In Gothic architecture, the ornamental, feathery or foliated upper parts of an arched window, formed by the branching of the mullions: the intersecting rib-work on a vaulted ceiling, etc.

TRAP.—A small water reservoir in a drain-pipe, to intercept bad odors, and retain sediment.

TRIGLYPH.—An ornament repeated at equal intervals in a Doric frieze.

TRUNCATED GABLE.—A gable with a portion of its roof drooping in front.

TRUSS.—A horizontal timber supported by bracings above, so as to form a long span without posts below.

TURRET.—A small tower, usually attached to and forming part of another tower.

TUSCAN.—The simplest order of architecture, formed in Italy in the fifteenth century.

VALLEY.—The receding angle formed by the meeting of two inclined sides of a roof.

VENETIAN BLIND.—A window blind made of slats of wood strung together so as to be raised or lowered by a string.

VENETIAN DOOR.—A door having panes of glass on each side for lighting the entrance hall.

VENETIAN WINDOW.—One formed of three apertures separated by slender piers, the center one being the largest.

VERANDA.—A covered walk on the side of a building, of an awning-like character, with slender pillars, and frequently partly enclosed with lattice-work. It is usually understood to be more secluded than a piazza. Arbor veranda is a frame covered with foliage.

VERGE-BOARD.—The gable ornament of wood-work—often called barge-board.

VESTIBULE.—See Room.

VILLA—A country house of superior character.

VOLUTE.—A scroll or spiral ornament, which forms the principal distinction of the Ionic capital, and is also found in the Corinthian and Composite. See Ionic Order.

WALL-PLATE.—See Roof.

WATER-CLOSET.—A privy, supplied with a stream of water, or water-pipe, to keep it clean.

WATER-LIME.—A species of lime that when made into mortar (see Stucco), will become hard under water.

WEATHER-BOARD.—A board on the gable from the ridge to the eaves; the outer boards of a building nailed so as to overlap and throw off rain.

WEATHER-MOLDING.—A molding or drip-stone, over a door or window, to throw off the rain.

WELL-HOLE.—The space enclosed by the walls of a circular stair-case.

WORKING-DRAWINGS.—Drawings of different parts of a building, according to accurate measurement, including plans, elevations, profiles and sections, by which the builders are to be guided.

WAINSCOT.—The wooden lining on the interior surface of a wall.

CHAPTER III.

RURAL BUILDINGS, OUT-HOUSES AND GARDEN STRUCTURES.

I. FARM HOUSES AND COTTAGES.—II. COTTAGE FOR FARM HAND.—III. SQUARE COTTAGE.—IV. SUBURBAN OR FARM COTTAGE.—V. A PRETTY RURAL HOME.—VI. A CONVENIENT COTTAGE. —VII. FARM-HOUSE IN THE ITALIAN STYLE.—VIII. ENGLISH GOTHIC COTTAGE.—IX. PLAN OF RURAL GROUNDS.—X. SCHOOL-HOUSE AND CHURCH ARCHITECTURE.—XI. CHILDREN'S WIGWAM.—XII. RUSTIC SEATS AND SUMMER HOUSES.—XIII. SOME RURAL OUT-BUILDINGS.—XIV. POULTRY HOUSES AND CHICKEN COOPS.—XV. GLASS STRUCTURES.—XVI. SMOKE-HOUSES.—XVII. THE FARM ICE-HOUSE.—XVIII. PRIVIES AND THEIR ARRANGEMENT.

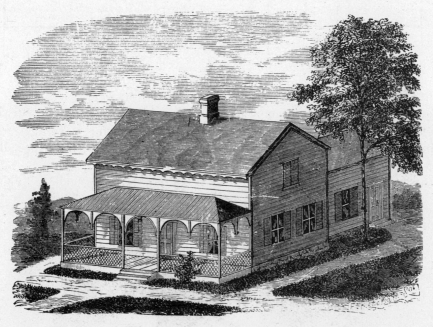

A PLAIN FARM-HOUSE.

I. Farm Houses and Cottages.

WE have attempted to give a comprehensive description of what was absolutely necessary in the erection of farm and suburban dwellings, barns, carriage houses and stables, as well as some important considerations relating to the production of landscape effects. To carry out the matter fully, it will only be necessary to present illustrations accompanied with diagrams to enable any one, in connection with those heretofore given, to select a plan within his means. Wealthy men who desire to build, can afford to pay for elaborate drawings. The plans we give can

394   THE HOME AND FARM MANUAL.

be carried out by any master workman. Some of the more simple of these, such as minor out-buildings, summer houses, seats and fixtures, can be constructed, and even

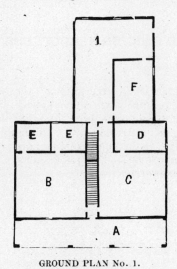

GROUND PLAN No. 1.

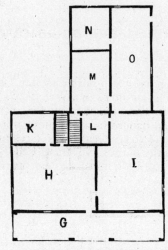

GROUND PLAN No. 2.

elaborated, by any person able to use ordinary tools. Our first illustration shows a plain farm-house, with little attempt at ornament.

In this house the rooms are of fair size, and suitable for a working family of seven persons. The veranda is tasteful. The eaves should project farther all round,

COTTAGE FOR FARM HAND.

and the gables be furnished with a handsome cornice. Then it would be no less comfortable, but far more attractive. One hundred dollars added to the original cost would accomplish all this. The diagrams on page 394 show two plans of dividing the first floor. They are: In plan No. 1, A, veranda; B, living room, 13x12 feet; C, kitchen, 13x12; D, pantry, 8x11½; E E, bed-rooms, 6x7½, too small, but doors connecting with kitchen and living room may be left open; F, laundry, 9x12 feet; 1, wood-shed.

Plan No. 2. G, veranda; H, living room, 17x12 feet; I, kitchen, 18x11; K, bed room, 8x10; L, pantry; M, laundry; N, store-room; O, wood-shed. The upper floor may be divided by a hall through the middle, and if the elevation is made higher, to admit attic windows, front and rear, or half-dormer windows, will make four good bed-rooms.

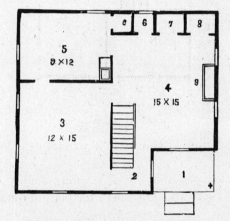

FIRST-FLOOR, SQUARE COTTAGE.

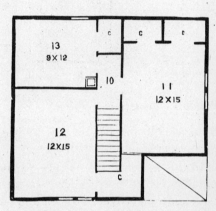

SECOND-FLOOR, SQUARE COTTAGE.

II. Cottage for Farm Hand.

THE design of a cottage for farm hand is made with a view to economy of space. It would be appropriate for a farmer of small means, or for the married farm hand of a well-to-do farmer. The enclosed porch, 7½x7½ feet, forms an entry or vestibule to the parlor, 13½ feet square. In this case, the kitchen, 13½x16½, serves also as the living room. The bed-room is 13½x9, with closet; pantry 6½x8½. The passage is two feet eight, and the stairs two feet four inches wide. These last are in a projection not shown in the cut.

III. Square Cottage.

THE plan of cottage for farm hand may be modified, and gain room, in comparison with the cost. The first and second floors are shown in the diagrams. The square form of building is better than any other in relation to the economy of space, heating, and relative cost of construction. Hence, square houses are favorites, where strict economy must come in. We have illustrated a number of square forms, or oblong square, and for the reason that this book is intended for the masses, whose buildings are frequently constructed without the direct assistance of the professional architect.

This cottage would make a capital farm-house, if carried to an attic above the

SUBURBAN OR FARM COTTAGE.

second floor, and covered with a hipped roof, that is, one sloping equally to each of the four sides. The attic being converted into bed-rooms. To add still more to the appearance of a dwelling of this height, the eaves should over-hang, and the center of roof support an observatory. The lower floor would contain parlor, dining-room and kitchen, with necessary closets; the dining-room having a handsome bay window. The second floor contains two parlor bed-rooms, and another of nice size, with ample clothes-presses and closets, and the attic might be divided into double bed-rooms for farm hands.

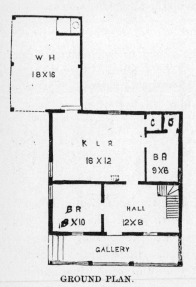

GROUND PLAN.

IV. Suburban or Farm Cottage.

This is a tasteful, economical and cosy cottage, adapted, in point of architecture, to a rolling or hill country. The hall is to be used as a sitting-room or parlor, and the front bed-room may be converted into a library. The kitchen and living room is 18x12, and the rear building combines a wood-house, laundry and water-closet. The rooms are nine feet high in the clear, and whether built of wood, brick or stone, the house is handsome. The upper story

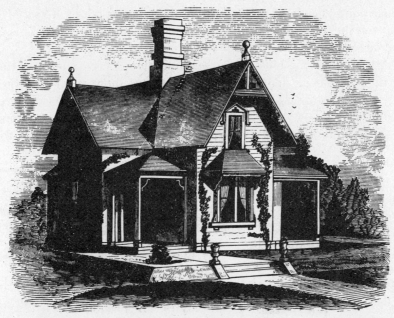

A PRETTY RURAL HOME.

has two feet of perpendicular wall, which, with sharp roof, gives plenty of air, and may contain two large sleeping-rooms of unequal size, each lighted by a handsome side window, and one of them by a dormer.

V. A Pretty Rural Home.

This house is adapted to a family of moderate means, doing business in a city and living in the country, or for a well-to-do business man or retired farmer, with small family, in a suburban town.

The elevation and ground plan here given fully explain it. The upper story consists of four bed-rooms and a bath-room. Ground plan: 1, porch; 2, lobby; 3, parlor; 4, library or boudoir; 5, outside porch; 6, dining-room; 7, kitchen; 8, scullery. It will be seen that the porch, 5, might easily be arranged as a conservatory.

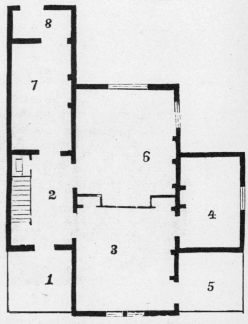

GROUND PLAN OF RURAL HOME.

VI. A Convenient Cottage.

This house combines convenience with utility and economy of space. It may be cheaply built, for the reason that there is no costly ornamentation. This, however, may be added outside and in, for it is the finish of the average house that costs money. It will be seen that while the halls are large enough to be convenient, all that can be spared from them has been added to make the rooms more spacious. The opening, usually filled with folding doors, is eight feet square, making the parlor and dining-room a large saloon, thus greatly adding to the hospitable look of the house, and giving large space.

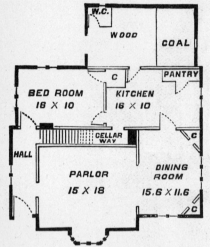

GROUND PLAN OF CONVENIENT COTTAGE.

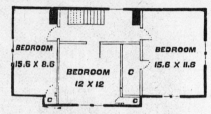

SECOND FLOOR OF CONVENIENT COTTAGE.

The stairs, enclosed between two walls, are more cheaply built. Each room has a closet, and with one exception, has straight edges. The house is ten feet between joists on the

FARM-HOUSE IN THE ITALIAN STYLE.

first story and nine feet above; the plan also provides for a cellar and stone cistern.

VII. Farm-house in the Italian Style.

Low pitched roofs projecting over the walls mark the Italian style of architecture. It is adapted to mild sunny climates, not subject to violent winds, heavy rains, or deep snow. The elevation gives walls of ample height, both above and below. The tower adds dignity to the building, gives the noble porch below, an office or library in the second story and an observatory above, making a nice summer sleeping-room, and giving also quick access to the roof in case of fire. From the porch one door

A CONVENIENT COTTAGE.

opens to a hall, and thence to the living-room and to the parlor and its bay-window. There is also an ample kitchen, with pantry and china closet, laundry and wash-room. The second story is divided into sleeping apartments.

Buildings of this class are favorites with suburban residents, of limited means, and especially as summer residences. They are cheap, may be made attractive at small cost, but if erected on a farm, or for a permanent residence, should be more substantially built, than if only used as a summer home. Particular care must be taken to secure warmth in winter by protecting the sheathing boards with the best building paper, especially on the prairies, where the wind searches every crevice.

400   THE HOME AND FARM MANUAL.

VIII. English Gothic Cottage.

It has not been the aim of this work to deal in elaborate architecture, and hence

ENGLISH GOTHIC COTTAGE.

in presenting the plan of an English gothic cottage, we have been guided by the elegant solidity represented in the stone cottage illustrated, as adapted to the retired business man or farmer. There are thousands of far more costly and elaborate farm structures in the United States, and the taste for such is constantly increasing with increasing wealth. The builder of costly structures, as we have heretofore said, should consult a good architect. It will save much more than the cost of his com missions. But the master carpenter will be sufficient for the more simple homes. When elaborate ornamentation of the grounds is intended, it will pay to consult a competent landscape gardener, while again the more simple ornamentation may be done by the farm hands under the direction of the intelligent farmer who has studied this work. In carrying out the details, the ground plan fully explains itself. The vestibule opening into the parlor and library by its wide sliding doors, will afford magnificent space for special occasions. The hall (dining-room), also in connection with the vestibule, will enable the proprietor to dispense large hospitality. Not the least attractive features of the whole are the projecting eaves,

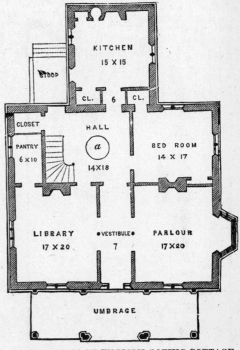

GROUND PLAN OF ENGLISH GOTHIC COTTAGE.

and the elegantly grouped chimneys, while the latticed casements give an added charm. The second story should form four spacious chambers, arranged as spare rooms. It would be a pity to cut them differently for the sake of additional apartments. When there are many guests, they may generally be so quartered together that additional beds in each room will accommodate many.

While preserving the same general form, it will readily be seen that the plan given above, is adapted to extensive modification. We have given a somewhat unusual arrangement of the ground floor, but especially adapted to the retired farmer, living at his ease, principally on the revenue of his farm. The bed-room may be made a second parlor or family room, and connect by folding doors, thus almost throwing the whole lower floor into one grand reception room for special occasions.

IX. Plan of Rural Grounds.

One more illustration and our pictures of rural domestic comfort will be complete. This is a plan of rural grounds showing water and a broken slope from the house to the public road. The front is an ornamental hedge of hemlock, unfortunately a plant that does not generally thrive. When it does, it makes the most magnificent of evergreen hedges.

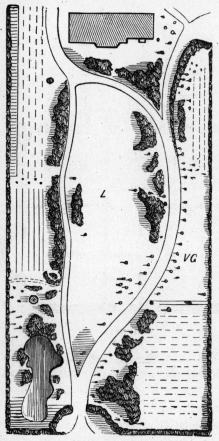

PLAN OF RURAL GROUNDS.

The illustrations in an earlier portion of this chapter will give a variety to select from. In this view we show grounds adapted to a suburban residence, the sides planted with a hedge of various ornamental plants suitable, but by taking away these so the lawn may gradually give way to pasture and then to fields of grain, it may be adapted to the farm. The public road is at the north, and as you enter at the left hand is the pond, as shown in outline; at the south end is rock-work planted with shrubs, vines, etc., and so more or less of rocks, vines, shrubs, etc., dot the banks of the pond, while trees of elegant shapes cast their shadow over the water. Continuing south beyond the pond is, eighty feet from it, a rustic summer-house with vines entwined upon it. The straight lines mark rows of grapes, while bordering the foot-path is a belt of shrubs. Then you reach the indication of trees, marked by dots; a mass of flowering shrubs is planted against the foot-path, backed up with dwarf apple and pears until within about twenty feet of the line boundary, which space is devoted to strawberries. Going back now to the entrance, on the right of the carriage road, we have beds and masses of rock-work, evergreens and flowering shrubs, with elm, weeping birch, etc., while bordering the carriage-way, most of the way to the house is an orchard of cherries, quince and pear, and the vegetable garden. On the lawn the flower-beds are shown, cut out of and surrounded by grass. The rear portion of the place is blocked and planted in line with fruit trees, and in the rear of the house are planted evergreen trees for screens, shelter and ornament.

X. School-house and Church Architecture.

On the principle that every building should be adapted to the use for which it is intended, there would seem to be room for improvement in our public buildings in all our smaller cities and villages. The great mistake is in the failure to provide ventilation, correct acoustic facilities, perfect heating arrangements, comfortable seats, and ready egress in case of panic from fire or other accidental causes. In country districts the school-house is always the place of holding caucuses, society meetings, clubs, singing societies, public amusements, and often it is used on Sundays as a church. Hence it should be the best building in the neighborhood; not only built in the best manner, but pleasant in its surroundings. The lot should not be less than a full acre in extent, thoroughly and substantially fenced, carefully planted with trees and shrubbery, and, except the play-ground, laid out with walks and flower-beds and ornamented with flowers. The situation should be commanding, on high or well-drained land, near a public highway, and as near the center of the district as possible.

SCHOOL AND MEETING HOUSE COMBINED.

Then, if the teacher have taste and practicality, and the trustees business discretion and firmness, the place will become one of the most attractive in the neighborhood, and all will work together in keeping it so.

The illustration of school and meeting house combined will serve to convey our meaning, and may be used for a variety of purposes, as a school-house and union church, or as a church edifice solely.

The next illustration is that of a neighborhood or district school-house, such as should be found in every neighborhood. The diagram shows the inside arrangement.

In furnishing either a country church or school-house, attention should especially be paid to the comfort of the scholars, communicants and attendants generally. Avoid upholstery, but not necessarily carpets in the aisles and other passage-ways. These

NEIGHBORHOOD PRIMARY SCHOOL-HOUSE.

should always be covered with carpets or matting to prevent noise. The inside arrangement of a primary school is shown in the subjoined plan. The school-house standing back sixty feet from the road. The building is 40x33 feet with 12 feet posts. A, lobby leading to entrance; *a*, stove, the coal or wood bin being in the lobby at *c*, and the wood-box at *d*. The closet, B, contains a sink and washing conveniences, and as many hooks as possible for hanging wraps, etc. C is the teacher's platform, 20x5 feet and seven inches high above the general floor of the interior. The windows, eleven in number, are hung as to both sashes with weights. The room is provided with sixty desks, of three sizes, to accommodate scholars of larger and smaller size. The center aisle is three feet wide and the side aisles two feet wide. The aisles and passage in front of the desk

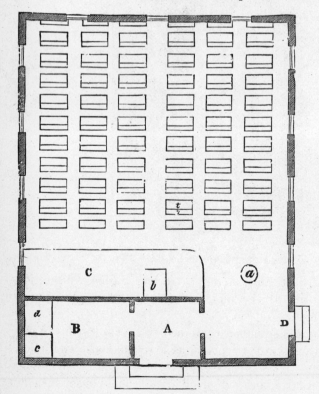

INTERIOR OF PRIMARY SCHOOL-HOUSE.

are covered with cocoa matting to deaden the sound. The teacher's platform is carpeted, and the most of the scholars furnish small pieces of carpet or rugs for their feet. Hence there is no clatter of feet on the floor.

XI. Children's Wigwam.

In reference to the decoration of school-lots, one feature is worthy of notice in connection with the school building we have just portrayed. It is in a district where a

SCHOLARS' WIGWAM.

good many boys are raised. The children, instigated by the teacher—a lady as refined and delicate in her culture as she is exacting in her kindly discipline—are enthusiastic cultivators of the flowers in the school-yard, as they are industrious drivers of the lawn mower when the grass gets above regulation height; the team of which usually consists of a girl and a boy at the handle. The children conceived the idea that a summer-house would be nice, and so the teacher evolved the above house, from hop-poles constructed by the pupils, with some help from a clever carpenter, in fitting and raising.

A RUSTIC SEAT.

How do you like it? The children enjoy it immensely! Especially so, since they *believe* they did a large part of the labors of thinking out the plan and erecting it. And this brings us to a cognate subject.

XII. Rustic Seats and Summer-Houses.

The garden and lawn are incomplete, if not supplied with some kind of seat, and when these have to be bought, there is such an infinity of designs to select from that all may be satisfied according to the contents of the purse. The charm, however, of all these ornamentations, whether of shade or comfort, is their rustic character. Here is one rustic enough, and at the same time, comfortably arranged for a tete-a-tete.

One of the prettiest effects we have ever seen was an elegant summer-house of woven wire, appropriately situated on a fine lawn, shaded here and there with large trees, and planted with shrubbery, a cut of which we give on opposite page.

What do you suppose, dear reader, was contained within this summer-house? As surprising and as pleasant as thing as a school-yard, with trees, walks, flowers

SUMMER-HOUSE OF BARK.

SQUARE SUMMER-HOUSE.

and a rustic summer-house. It was neither more or less than the veritable stump we have illustrated, with a rustic seat running all around it. The ladies voted it positively delightful—in fact, "too cute for anything." But every person has not the bank account of our friend of the elegant summer-house, who first had to buy the stump and pay railroad transportation on it. It is not necessary, as the preceding elegant designs will show. Anybody who can peel bark in June, lay it under pressure to dry flat, and cut and fit the pieces, can build either of these two elegant designs that we give herewith out of many we have seen. Why not try? There is nothing of either of them but bark and poles, not even the furniture.

It is not necessary to describe how to do the work. In fact, the illustrations are the best description that can be given. It is simply a matter of taste, ingenuity and

AN ELEGANT SUMMER-HOUSE.

judgment. We think the circular house especially fine, particularly in its light and graceful appearance, added to by its roof of bark cut in scallops, and by its center-table and seat, covered with bark. It is surprising how many fine combinations of color may be gotten entirely out of bark.

XIII. Some Rural Out-Buildings.

The out-buildings of every farm must correspond to the branches of agriculture followed. The gardener will want structures of glass, and every farmer should have one structure besides the hot-bed or cold-frame containing a good deal of glass. That is a poultry-house. We give an illustration. The main building will serve the ordinary farmer, and the extension may be added on either side to suit the growing numbers. The care of fowls has been fully treated in a previous work, "Pictorial Cyclopedia of Live Stock and Complete Stock Doctor," issued by the publishers of this volume. The poultry-house should always be provided with a bath of dust, feeding boxes and fresh drinking water. A drinking fountain that will be continually supplied, so long as the water in the barrel lasts, is shown. A modification of this will also supply grain for food. A small tube extends from the barrel above nearly to the bottom of the reservoir below. The barrel being air-tight the water is only given as needed, or when the water gets so low as to allow the air to enter through the tube. A larger tube, to admit grain, will also give satisfaction, since the grain piling up around the outside of the tube will prevent the giving down more than is wanted.

DRINKING FOUNTAIN.

XIV. Poultry-Houses and Chicken-Coops.

The poultry-house, page 409, will suffice for the wants of any farmer. Besides this, coops and enclosures for young chickens and ducks are necessary.

We give two illustrations of the more simple forms, which explain themselves, and which any farmer can make. They may be covered with bark, thatched with prairie-grass, or, as in lower view, page 409, where grouped, be separated by wattles or reeds as shown, or, indeed, the whole built of reeds.

There is one economy in the keeping and rearing of young chickens, and especially ducks, not generally estimated; that is their value as indefatigable hunters of insects. The younglings, therefore, should have full liberty to range as soon as large enough to run freely, the mother being confined, as shown in cut.

A WICKET COOP.

The old-fashioned barrel-coop is familiar to many housewives who have been compelled to improvise something in haste. The improved form, as shown, may perhaps tempt husband, hired man, or the boys to prepare a supply in the workshop at odd times.

BARREL-COOP.

XV. Glass Structures.

The gardener and seedsman, besides hot-beds and cold frame, may need a greenhouse or propagating pit. A good lean-to form, against a wall, is shown on page 410.

The general nursery-man, and especially the seeds-man, often require extensive ranges of glass for propagating plants that will not reach full maturity without some early forcing. The market gardener also requires these ranges of glass for forcing

POULTRY HOUSE.

early vegetables and other plants for removal, later in the season, into the open air. The illustration will show a propagating house in connection with the dry-house for curing seeds, and ranges of glass for forcing plants. These ranges allow the passage of a wagon between them for carrying in heating manure and removing the spent. The dry-house is provided with proper heaters and ventilators for the perfect curing of seeds, and to keep them from dampness during extended foggy or rainy weather.

CHICKEN AND DUCK ENCLOSURE.

XVI. Smoke-Houses.

Every farm homestead should be provided with a fire and thief proof smokehouse, or one at least secure from common depredators. We give companion pictures, one of brick, the other of wood. Our advice is to build of brick, with an iron door, and without light save through the ventilators, which latter would be better if placed at the peak than in the gable, as is shown in the frame building, or under the eaves as in the brick building. However ventilated, these should be protected by wire screens to keep out insects. Air should be admitted at the bottom by means of pipes protected from rats and mice by screens. The house should be ten

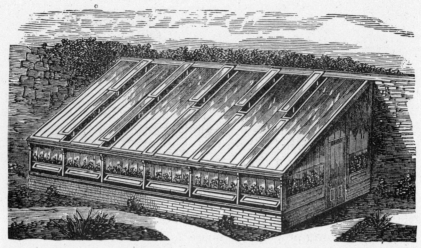

LEAN-TO PROPAGATING PIT.

or twelve feet high, to give plenty of space between the fire and meat, and if the smoke is introduced from the outside through a pipe so much the better. [See cuts p. 413.]

PROPAGATING AND DRY-HOUSE.

XVII. The Farm Ice-House.

EVERY farm homestead should have an ice-house as one of the out-buildings. The construction is exceedingly simple, merely two walls fourteen inches apart filled

RURAL BUILDINGS, OUT-HOUSES AND GARDEN STRUCTURES.   411

FARM ICE-HOUSE.

with sawdust, drainage underneath, and a floor of poles filled in and covered with sawdust for the ice to lie on. There must be a double door on one side, for putting in ice and for ventilation, as shown, under the eaves, but no ventilation—all must be tight—at the bottom and sides. A cube of ice of eight feet, that is, eight feet on every side, will keep perfectly and supply a moderate family for a year. Lay the ice in square blocks to the eaves, the height of the house to be determined by the width, and cover the top with a foot of sawdust, or eighteen inches of hay, allowing a free circulation of air above the ice as shown in the cut. Thus ice will keep as well on the farm as in the more elaborate building of the city man. A cube of ice twelve feet square will supply an ordinary farm dairy and one of sixteen feet a large creamery.

XVIII. Privies and their Arrangement.

THE practice of building privies with vaults dug deep in the earth, holding the accumulation of generations, with no means of cleaning except those most repulsive, added to the danger of contamination of the water of wells, and infection of the air therefrom, has long engaged the attention of sanitary engineers. These old-fashioned and repulsive buildings of the farm-house have now been abandoned in many instances for something better. Twenty-five years ago we had built, within fifty feet of the house and well, a building entirely concealed by a screen of shrubbery, and entered from a flight of three steps, and with all the lower apparatus above the ground and easy of access. From this no odor was apparent either inside or out. It was cleaned by drawing the privy-box, running on rollers, from a door in the rear, and emptied several times a week into a wheel-barrow.

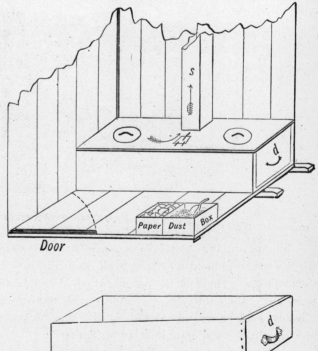

OUTLINE OF FARM EARTH-CLOSET.

The following is an improved and modified plan. The sketch shows the interior of the privy, elevated high enough from the ground to permit the box being drawn out directly on to a barrow and wheeled away. D, is the dumping-box, S, the air-shaft running up through the roof. The paper and dust box, with scoop, is shown on the floor, the opening of the door being at one side. The best dust is, of course, finely pulverized and air-dried clay, but dried road-dust is the most easily obtained and good enough. It may be re-dried under a shed, and used over and over again, until thoroughly saturated, and then used as manure. There is no disagreeable odor whatever from it.

RESERVOIR EARTH-CLOSET.

Whenever the seat is used, one or more scoopfulls of dry earth are to be thrown over the deposit. That is all, and certainly cleanliness and health should cause its adoption. There is no patent attached to it, and the odor from the old-fashioned nuisances calls loudly for a change. The use of dry earth in preventing odor from offensive substances is as old as the wanderings of the Israelites in the desert, and therefore is venerable from antiquity.

Another form is the French earth closet, in which the dry earth is held in a reservoir above, and a portion set free by raising the seat-cover. The earth is deposited in the pan shown beneath, by raising the rod, M, by which the earth-slide is pushed out from under the reservoir and drops its contents in the pail. After trying both, we prefer our own plan, since it is just as efficient and never gets out of

BRICK SMOKE-HOUSE.

FRAMED SMOKE-HOUSE.

repair. Sifted coal ashes, especially that from anthracite coal, answers well as a deodorizer, but more is required than of dust from a clay road.

CHAPTER IV.

BARNS, STABLES AND CORN-CRIBS.

I. GROUPING FARM BUILDINGS.—II. A COMPLETE CATTLE-FEEDING BARN.—III. HORSE AND COW BARN WITH SHED.—IV. SUBURBAN CARRIAGE-HOUSE AND STABLE.—V. SHEEP BARNS AND THEIR ARRANGEMENT.—VI. HOG BARNS.—VII. GRANARIES, CORN-HOUSES AND CORN-CRIBS.—VIII. RAT-PROOF GRANARY AND CORN-CRIB.—IX. CORN-CRIBS WITH DRIVEWAY.—X. SECTION OF WESTERN CORN-CRIB.

SUBURBAN CARRIAGE-HOUSE AND STABLE.

I. Grouping Farm Buildings.

THE barn will often contain the granary, sometimes the horse-stable, and perhaps, even the cow-stable. By this arrangement, there is saving in the original cost of the barn-yard buildings, with the further advantage, that the feed is always at hand. On larger farms, the same plan, is in a measure, followed. There may be separate buildings for each kind of domestic animal, for horses, for cattle, for sheep and for swine, but each of these should contain a supply of the necessary food. Some of the most complete barns have the stables for cattle in the basement, the horse-stables on the main floor, mills for grinding feed, cutters for hay and straw, pulping machines for roots, and the silos connected with the barn by a

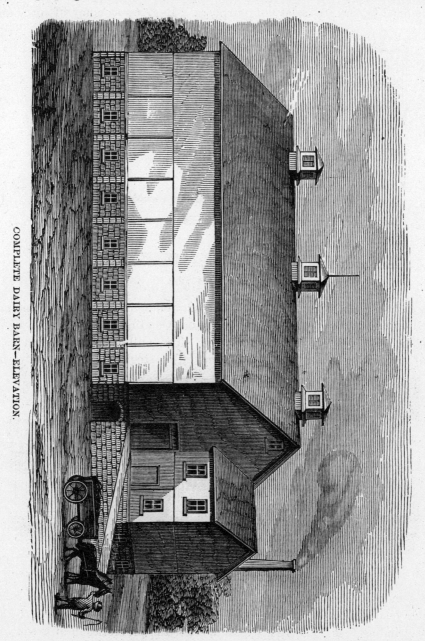

COMPLETE DAIRY BARN—ELEVATION.

covered passage. There should also be a steam engine, for driving the machinery and for pumping water, when this cannot be brought in pipes from higher ground. This is true economy, however many structures may be needed for surplus produce.

II. A Complete Cattle-Feeding Barn.

This is especially adapted to farmers keeping a large herd of milch cows, but it may be modified to serve a variety of purposes. It gives economy of space and ease

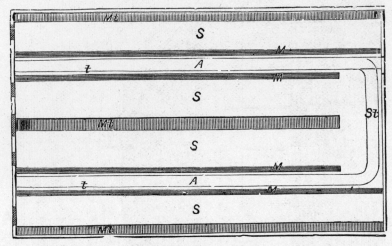

THE STABLE FLOOR.

in feeding, since all the operations, including pumping of water and grinding feed, are carried on under one roof. The engine-house and mill are shown in the illustration on page 415, connected with the barn. The mill would be better, especially in case of fire, if separated from the barn. The stock is kept in the warm, well-

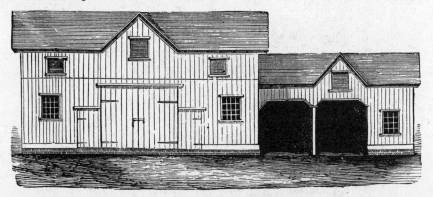

HORSE AND COW BARN.

lighted basement. The distinguishing feature of the plan is the free use of tramways for moving feed, chutes for conveying forage from the lofts above, and the arrangements for grinding feed. There are also silos for the preservation of green forage.

The barn is 96x56 feet, and the mill 24x20 feet. The barn will stable 120 cattle, the upper portion being used for storing fodder, that is delivered below, cut or uncut, through appropriate chutes.

The Stable Floor.—In this plan S S S S are the stalls for cattle; M M M M,

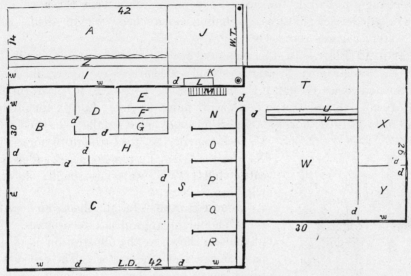

GROUND PLAN OF BARN AND SHEDS.

mangers; Mt Mt Mt, manure spaces; A A, tramways, and St switch track. If it is desired to stable the cattle on the floor above, the cellar may be used for manure, which may be dropped down directly from the floor through trap-doors. In this case the cellar may be used for any number of swine.

A Modified Plan.—A modification of the foregoing plan, where fewer cows are to be fed, is shown where only the center is used, having shed-room outside. A is the alley; B, stall floors; C, ditch or drop for manure; D, walk; E, stanchions; F, outside sheds; G, pieces of stone 2x4 feet; H, column under crosssills of barn; ✕, doors; XIX, windows.

III. Horse and Cow Barn, with Shed.

This barn has features that may commend it to a class of suburban folks, partly farmers and partly business men, especially those who like to keep a small herd of Jersey or other milk-

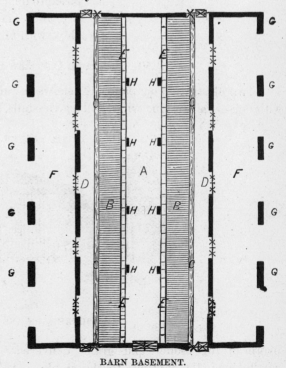

BARN BASEMENT.

ing cows. It provides for a sleeping-room for the man who has charge of the stable. With some changes it will do for horse-keepers, or for a barn for a country hotel. It contains all necessary fixtures, including open shed, wagon shed, tool-room, feed-bins, water, harness room, etc.

The Ground Plan.—In the plan, page 417, A is the stable, 8x28 feet, for nine cows; B, man's-room; C, carriage-room; D, harness-room; E, bin for bran or shorts; F, shelled corn; G, oats; H, passage-way; I, passage to platform floor; J, open shed, 10x14; K, platform floor, with pump at end; L, box for mixing feed; M, stairs to loft; N to R, horse-stalls; S, passage behind stalls; T, shed for cattle; U, feed-trough; V, feed-trough to horse-shed; W, X, wagon-shed; Y, tool-room; Z, feed-troughs in cattle-stalls; W T, watering-trough; d, doors; w, windows.

FEED-BOX.

Feed-Boxes.—In all sheds and cattle-yards there should be appropriate feed-boxes. The plan of the one shown in the illustration is old, but we have never seen one better. The especial value of this form of box is that four animals can feed from it at once, and being on different sides, are not disposed to quarrel. The posts are 3x3 inch hard-wood scantling, and the boards are nailed solidly, preferably, with ten-penny fence nails.

IV. Suburban Carriage-House and Stable.

The illustration on page 414, shows a carriage-house in center of the building, with horse-stalls in one wing, and cattle-stalls in the other, and contains ample room

Fig. 1.

SHEEP BARN AND SHEDS.

overhead for fodder and grain, delivered below through chutes and tubes. The small door is the entrance to the stables. If the room is not desired in the other wing for cows, it may be converted into a tool-house and work-shop. The interior arrangement will readily suggest itself from an examination of the other plans of stables given.

Wagon Jack.—Every farm-barn or stable should have some means of lifting an axle-tree without hunting for a rail and support. They may be readily bought, but Fig. 2 shows a simple and effective one

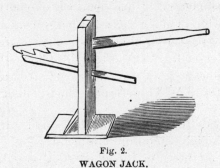

Fig. 2.
WAGON JACK.

Fig. 1.
IMPROVED WAGON JACK.

that any farmer can make for himself. If, however, you prefer a better one, purchase one such as is illustrated in Fig. 1. It will lift a heavy axle.

V. Sheep Barns and their Arrangement.

It is well known that sheep will not bear close and constant confinement like cattle. They must have not only exercise, but plenty of air. Their natural

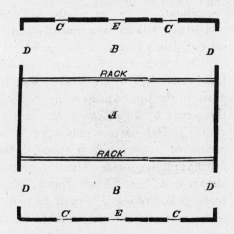

PLAN OF SHEEP BARN AND YARDS. Fig. 1.

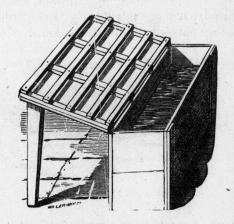

SHEEP-DIPPING BOX

habitat is in mountain regions, where the air is bracing. In the care of sheep this must always be kept in mind. Hence, sheep-barns must have an abundance of ventilation, with large yards attached for exercise when the weather is favorable. In fact, their fleeces amply protect them from extreme cold when not exposed to storms. In all mild climates they thrive better under open, protected sheds. The mutton breeds, and especially the long-wooled breeds, and more especially the New Leicester, require better protection than the hardy American Merino.

In the North, and always in the Northwest, sheep barns are essential. Yet sheep barns are simple structures that any carpenter should be able to build from the plans we present. The plan, Fig. 2, is the most complete; and plan, Fig. 1, more simple. Protection by timber belts is a feature of the first, and should be carried out in all structures for animals.

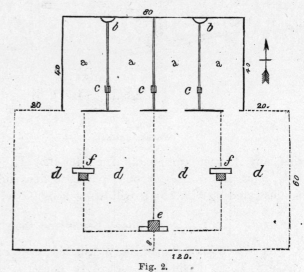

Fig. 2.
GROUND PLAN OF SHEEP BARN AND YARDS.

Ground Plans.—The plan, Fig. 2, shows at a, a, a, a, stables 20x40 feet each; b, b, watering tubs; c, c, c, doors in partitions; d, d, d, d, sheep yards. The two central yards are inside the barn, and amply ventilated by wickets; they are 30x50 feet each. The outer yards are 30x60 feet each. All the yards should contain racks such as are shown in the illustration. The water is carried into the barn at e, and distributed thence to f, f, and b, b, by pipes, each trough or tub supplying two yards. A yard 40x40 will accommodate 150 sheep if they have plenty of ventilation, but racks must be placed all round the outside.

SHEEP RACK FOR OPEN YARD.

The plan, Fig. 1, comprises a central barn: A, for hay, with bins for grain in the upper part, delivered into the yards by tubes, as is the hay by chutes. The barn stands upon abutments of stone, giving space underneath for a sheep-run. The barn proper, shown at A, is a balloon frame, 24x60 feet; B B, are sheds, 18x60 feet, with racks and feed boxes; D D D D, are doors, ten feet wide, to admit a team and wagon; C C C C, are windows, hung on hinges, for ventilation; E E, are small slide doors, to open into yards outside the sheds. The outer posts of the sheds are 4x4 inches and 8 feet long, the roof extending from the roof to meet them.

Sheep-Dipping Box.—No sheep barn is complete without some means of dipping sheep, and especially lambs, in tobacco water or other medicated mixture, to free them from ticks, etc. The illustration shows a simple, easily-made apparatus, with dipping slats, that may be constructed by any farmer who can make a water-tight box. When the box becomes leaky, paint it thickly on the inside with water lime.

VI. Hog Barns.

When few hogs are kept, and especially in the great cattle-feeding districts, where hogs follow the stock as gleaners, the protection for swine is usually only such as may afford shelter while the animals are at rest. Where swine are kept in per-

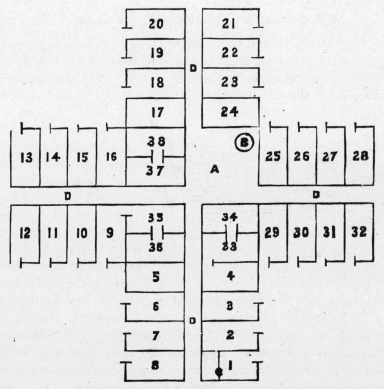

SQUARE HOG BARN, WITH EXTENDED WINGS.

manent stables, and especially where they are fed on ground food, steaming or other cooking apparatus must be provided, with suitable conveniences for feeding. However small the number kept, there must be feeding pens and sleeping apartments with yards attached, and the pens must be regularly cleaned. Where not too many hogs are kept the ground floor of an extensive hog-feeding barn may be used.

The floor plan shows a square building of two stories, with pens in a lean-to, as represented in the square A, and pens 34–33, 36–35, 37 and 38. The boiler is at B. There is an upper story for corn, meal and other feed, delivered below by tubes.

Where a large number of hogs are kept the space for pens and yards may be extended indefinitely. In this case, the pens 37 and 38 may also form part of the floor of the main building, which comprises also pens 33, 34, 35 and 36, these being used as boar pens, or for hospital service. The main building must be ventilated by a shaft running through the second-floor to the apex of the building at the center, which has a hipped roof. The pens in extensions, covered by span-roofs, each have a passage-way leading to sleeping apartment outside, and thence into yards. D D D D represent alleys five feet wide for the feed wagon, which is placed upon low iron wheels, the forward ones of which turn completely under, so the vehicle may be turned round in its own length.

VII. Granaries, Corn-Houses and Corn-Cribs.

The typical corn-crib of new countries is simply a pen of rails carried up ten feet. It may be either square or flared toward the top; it generally has only a rail floor and is often left entirely open to the rain on top. This certainly is a wretched way to keep corn, when the first intelligent thought would suggest a covering of rails and hay. A step in advance would be something like the illustration, which, only carried up seven feet, may be extended nine or ten feet.

CORN-CRIB OF POLES.

Such a crib six feet wide will keep corn as well as the best, and will be secure from rats and mice if an inverted pan, or flat stone is placed on top of the posts, next the crib, the posts being two feet high. If the corn is not dry when put in it may mould in a six feet crib in open moist winters. This may be prevented even in cribs nine feet wide—a not unusual width for the great store cribs of the West—by placing a $\wedge$-shaped ventilator four feet wide at the bottom and half the height of the crib, running to an apex at the top. This is made of five-inch fence boards, with spaces of five or six inches; and very little corn will drop through. It gives a free passage of air from end to end, and circulating through the corn above. We once saved a crop of many thousands of bushels, in Central Illinois, in cribs ten feet wide in this way, and during a winter when large quantities of corn were lost in cribs eight and nine feet wide, the open moist weather extending into March.

VIII. Rat-Proof Granary and Corn-Crib.

Every farm should have a rat-proof granary and corn-crib combined, for keeping grain, and there should be separate bins for seed-corn and seed-grains, with hooks for hanging bags of the smaller seeds. We give a plan for ventilated granary, with explanations, and with description of a modification which we used, when engaged in active farming.

The building is 24x18 feet long, with 14-feet posts, which extend two feet below the first floor, resting on stone piers, as shown, and protected from rats by galvanized iron plates between the posts and floor. The studs are two feet apart, set $1\frac{3}{4}$ inches inside of the face of the sills and posts, with strips of plank 2x3, with notches cut in them, upon which siding is nailed horizontally, and pitching down to shed rain. The first story has a height of seven feet between floors. The granary is in the center of this story and is made of matched flooring six feet wide by eighteen feet long, and extending through the upper floor three feet into the second floor; it is divided into appropriate bins for wheat, oats and barley, with doors on top for emptying in grain,

VENTILATED GRANARY.

and traps below for delivering it. Above these main bins are separate bins, four feet wide, extending nearly to the peak, and properly divided for seed-grains, with chutes for delivery. There is a passage-way three feet wide all around the grain-bins, leaving space for corn-crib three feet wide around the outside, flared to four and a half feet when it meets the platform of the second story, and thence carried up square. This narrow crib is divided for containing selected seed-corn according to varieties, and the wall space is used for any purpose required, as for hanging bags of smaller seeds.

Another plan, Fig. 1, 20x14 feet is entirely occupied as a granary, except the enclosed stairway and door. Two bins are shown, 10x14 feet each, which may be divided in two. The grain is elevated into the second story, and taken out of traps on the outside of the building.

The next plan, Fig. 2, shows combined corn-crib and granary, the inside walls sloping inward to meet at the second floor ten feet high, the loft being used wholly for corn except a passage-way around the granary, C, 5½x15 feet. C, 4½x20, is corn-crib; H is hall, and 3 is a large window for light and ventilation, with door at opposite end.

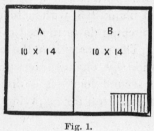

Fig. 1.
GROUND PLAN OF GRANARY.

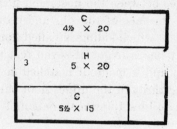

Fig. 2.
CORN-CRIB AND GRANARY.

IX. Corn-Cribs with Drive-Ways.

Many farmers who raise comparatively little corn build enclosed corn-cribs, into which wagons may be driven. The figure, Skeleton of Crib, will explain itself and

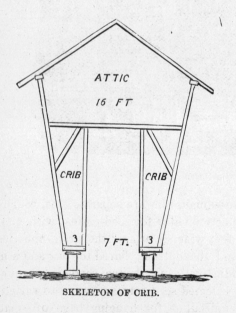

SKELETON OF CRIB.

CRIB, EXTENDED INWARD.

also show the manner of framing. It would be more roomy and in no wise interfere with driving space if the inside also were flared, as shown in the next figure, which also explains itself.

We do not recommend either of these. They are harborers of rats and mice, which enter from the wagons left standing under them. They are given more to show incorrect ideas of economy, and as such are good object lessons.

X. Section of Western Corn-crib.

The last illustration in this chapter represents the form of crib generally used in the great corn-growing region of the West. They may be seen of this form, the correct one, but oftener with straight sides, all over the West, and of varying length up to one hundred feet. Sometimes row after row of them are seen with passages between for wagons. There is usually no attempt made to keep out rats and mice, which freely swarm about and within them, affording sport for boys with their ratting dogs. When a permanent building of any kind is decided on it should be built on correct principles. A sheet of galvanized iron between the posts, thirty inches above ground and the floor of the crib, will be ample protection from rats. Hence we have simply shown the proper form which gives the greatest capacity and ventilation with the best protection from rain, admissible in cribs for corn.

WESTERN CORN-CRIB.

CHAPTER V.

MECHANICS AS APPLIED TO THE FARM.

I. THE FARM WORKSHOP.—II. MECHANICS' TOOLS ON THE FARM.—III. ARRANGEMENT AND CARE OF TOOLS.—IV. HOW TO KEEP FARM IMPLEMENTS.—V. SHARPENING TOOLS.—VI. PROPER WAY TO FILE AN IMPLEMENT.—VII REPAIRING COMMON IMPLEMENTS.—VIII. THE FARM PAINT SHOP.—IX. PUTTING UP ROUGH BUILDINGS.—X. SHINGLING A ROOF.—XI. MAKING A HAY-RACK —XII. STONE FENCES.—XIII. MOVING HEAVY STONES.—XIV. FOR AND AGAINST STONE WALLS.—XV. HOW TO BUILD THE WALL.—XVI. THE BALLOON FRAME IN BUILDING.—XVII. HOW TO BUILD THE FRAME.

I. The Farm Workshop.

EACH farmer must decide for himself how much purely mechanical work it will repay him to perform or have done on the farm. Where population is dense, the division of labor must necessarily be more minute than where it is scattered. Hence, in thickly settled districts, the farmer may find it cheaper to buy everything he does not grow on the farm rather than make it himself. On large estates there are generally carpenters, a blacksmith, and other artizans hired by the year; often a book-keeper, engineer and miller are required, until at last these employes, together with the farm laborers proper, and their families, form the nucleus of a village. We have seen all this happen in Illinois, and once on a farm of less than 3,000 acres. In the South, on some of the large estates, especially on sugar plantations, where the crop must be manufactured, and, in the North, where-ever sorghum is produced in large quantities, it will repay the planter to do much of the repairing at home.

In thinly settled districts the farmer should himself know how to do simple repairing. Making rails and posts and fitting them for use, is strictly a mechanical art, yet on timbered farms this is also a part of the necessary farm labor. On every farm some fencing is always to be done; there are gates to be made and hung, and rough sheds to be put up. The repair of the ordinary tools used is a natural application of mechanics to agriculture. The tightening, and even fitting, of horse-shoes, is often important. This only requires dexterity and observation to render its performance easy; and the same may be said of simple repairs to iron-work.

If skilled labor is near it will be cheaper, as a rule, to hire mechanics for all important repairs; yet every farmer should have some simple tools and a workshop. Many needed mendings and changes may be done at times unfitted for out-door labor. If the workshop be kept supplied with the necessary materials for such work there is little time that may not be profitably employed by those necessary to work the farm.

II. Mechanics' Tools on the Farm.

The necessary tools are chopping-axes, hatchets, hammers, a broad-axe, grindstone, an oil-stone, augers from one inch to two inches, a brace and bits, a set of chisels, rip-saw, one or more cross-cut saws, a tenon saw, square, a spirit-level, two-

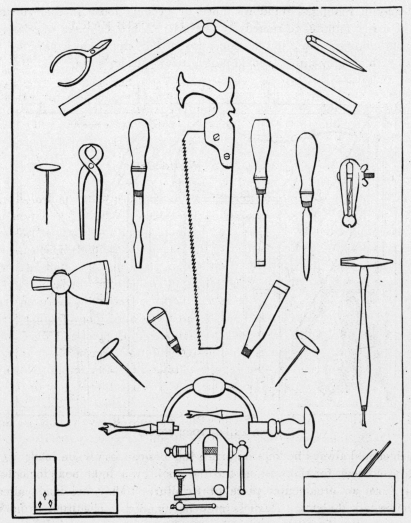

A FAMILY SET OF TOOLS.

foot rule, tape-line, dividers, jack-plane, jointer and smoothing-plane, screw-driver, awls, a drawing-knife, a vise to hold boards and one for holding iron implements, a set of files for wood and one for iron work, an iron "claw," sand-paper, wire, and an assortment of nails, a few of which should be of wrought-iron to be used in clamping, screws and lumber. With these all simple repairs may be made at home.

With perseverance the necessary skill will soon come, and there are many things to be done, that take less time in the doing, than would the sending for a skilled workman. Thus, certain mechanical work is as necessary for the farmer to know how to do as to plow or reap. A shed or lean-to is to be built. It may be done at times when the land is unfit for working. A door is out of level; in ten minutes it may be rehung. Windows pinch or become loose in their fittings; it is the work of a few minutes to remedy the defect. A broken pane of glass, either in house or outbuilding, may, if one depends entirely upon mechanics, be a serious matter, yet with a putty-knife, a little putty, a few glazier's tins, and the necessary

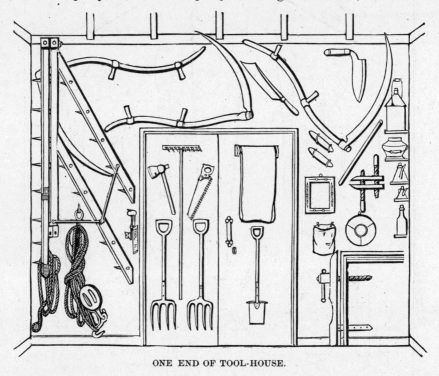

ONE END OF TOOL-HOUSE.

glass, which should always be kept on hand, the loss can easily be made right. The wife requires a bench for the washing-tub; a stool, or a light box, for covering and stuffing to form an ornamental piece of furniture. They are made almost while they are being talked about. Harness may be mended. The irons from a broken whipple-tree or other implement, may be fitted to a new wood. Rustic structures may be made, and valuable work done, from time to time—even an important building, under direction of a regular builder. To accomplish all this successfully, tools must be kept in perfect order, and not be lent, except to those who know how to use them, and such persons generally have their own. A neighbor may, perhaps, think it hard to be denied; he may prefer to use your bright, sharp tools in place of his rusty and dull ones. Why should he not take care of his own tools?

MECHANICS AS APPLIED TO THE FARM.   429

III. Arrangement and Care of Tools.

THERE should be a place for every tool and every tool should be in its place. Such tools as will not easily rust may be arranged on the wall over and around the work-bench, but all tools with bright surfaces, as saws, chisels, etc., unless inside a case, should be kept in a chest, in their appropriate niches, and if not to be used for some time, lightly oiled when put away. Thus kept, the implements are always bright, only requiring to be wiped for use, when wanted.

Our cuts illustrative of tool-keeping are: first, an inside view of closet for the simple tools necessary for the farmer of few acres, as shown in the cut entitled, "A Family Set of Tools," and also the four walls of a complete tool-house attached

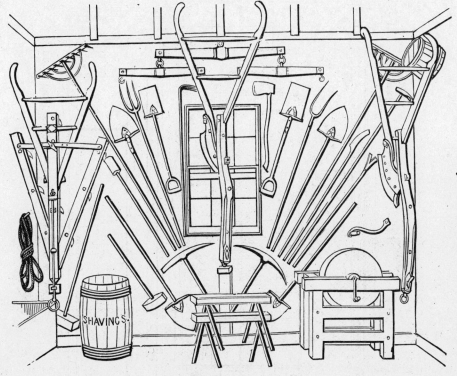

SECOND END OF TOOL-HOUSE.

to a work-shop, 30 x 14 feet, and which is now in use upon a farm of 2,400 acres in Illinois. This contains all the minor hand tools and implements required by the farmer. See pages 427—431.

IV. How to Keep Farm Implements.

ALL farm implements should be kept under cover and cared for when not in use. The mowing machine, reaper, plows and all other implements having bright surfaces, should have these covered with a mixture of kerosene and lampblack, when put away.

It is easily rubbed off when they are again wanted, and the surfaces thus retain their

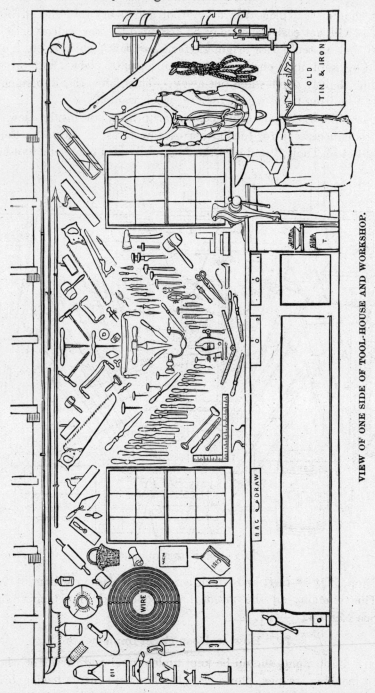

VIEW OF ONE SIDE OF TOOL-HOUSE AND WORKSHOP.

polish. When left in the field over night they should be rubbed with an oiled cloth.

MECHANICS AS APPLIED TO THE FARM. 431

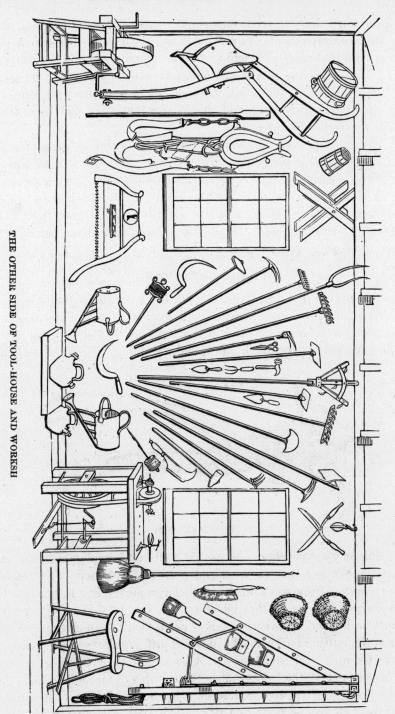

THE OTHER SIDE OF TOOL-HOUSE AND WORKSHOP

Only pure oil, unsalted, should be used. A pint will last long and save many dollars.

You will be surprised, on trial, how small a space is really required to store all tools and small farm implements from the weather. An open shed will do for wagons, sleds, harrows, and that class of machinery; but a closed room is necessary for plows and other implements having bright surfaces. If they are exposed under an unenclosed roof the moisture of the atmosphere is apt to rust them in damp weather, to say nothing of injury from dust and the danger that they will be stolen by night prowlers while the farmer is asleep.

In this day of improved implements successful farming cannot be carried on without perfect tools and implements. They cost much money; with care they wear a long time; without care their life is short. The abuse of implements costs ten times their wear. A wise man looks to economy. Study the object-lessons presented, and learn to economize by care. When first a tool is properly hung on the wall, as represented, outline its form with paint; or, better, paste up its name clearly written. It will save time in properly replacing them.

V. Sharpening Tools.

THERE is no excuse for dull tools. A file will keep the plowshares and cultivators sharp. A grindstone and whetstone will keep the mower and reaper sickles in order. Steel teeth wear much longer in a harrow than iron ones, and are as easily repointed; the first cost is not much more, and a dull harrow means lost time. Carry

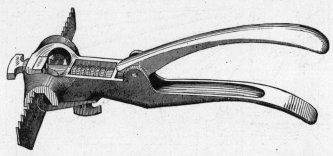

IMPROVED SAW-SET.

out this principle in your purchase and use of tools and implements, and it will save you fully twenty-five per cent in wear and tear of implements and teams, besides bettering the quality of the work done. This large percentage is clear profit.

How to sharpen tools is another question. It can be told only in general terms, the detail must be learned by practice, but is not difficult. In grinding a surface, as that of an axe, the cutting part must be beveled off regularly and equally, and the edge then whetted on a stone until keen. A broad axe or chisel is ground from one side only, thus preserving the bevels; it is then simply " faced " on the side containing the steel. In fine, every tool should be ground according to its structure. Formerly, scythes were all ground upon one side. The best are now made to grind on both sides alike, and when so, it is stated on the tool.

VI. Proper Way to File an Implement.

In filing always do the cutting by thrusting the file from you. In the reverse motion it should not press the tool, because this cuts the edge of the file. In filing a saw preserve the form of the teeth. A cross-cut (hand-saw for cutting across the grain of the wood) is filed diagonally; a rip-saw, more nearly square across. The form of the teeth, it will also be observed, is quite different in the two. There is no mystery in filing. It is simply a question of accuracy. In saws every alternate tooth is to be filed in one direction, and every other tooth in another; observation will easily show this. A spade or shovel is edged from the front; a plowshare is filed from the upper side, and, as a rule, the shares of cultivators from the bottom. The wear will show

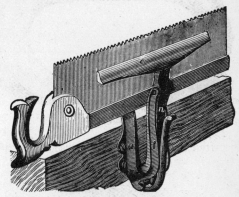

A SAW CLAMP.

when this is not correct. A hoe is filed from the bottom, and tools for edging and paring garden-walks filed or ground on both sides alike. These general rules will enable any farm-hand to acquire the necessary skill, and this will be worth many dollars yearly in wages. It is skilled labor, and skilled labor always commands an extra price. On the farm the manner of holding a tool of any kind for filing must be arranged according to the conveniences. In filing saws, they must be held from springing, else they cannot be filed correctly. They must also be set true. Hence we illustrate a clamp for holding, and a simple and perfect implement for setting the teeth true.

VII. Repairing Common Implements.

We have said that the farmer should have a supply of lumber, which must always be kept perfectly dry. Handles for tools, wagon tongues, and various fixtures can be bought, either ready-made or sawed in the rough. They should be kept on hand, then the work of fitting is often less than that of going to the shop. You have your work-bench fitted with vise and claw, for holding the wood to be worked. A taper bit, or a larger and a smaller bit will form the hole to receive the hasp of a rake, fork, etc. The drawing knife, a bit of glass and sand-paper will fit the end for the ferule (see Singletree). It is the work of perhaps ten minutes. A wagon tongue is worked to proper shape, and the irons of the broken one fitted, and so with the addition of

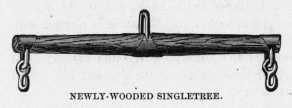

NEWLY-WOODED SINGLETREE.

a little paint you have a wagon tongue as good as new.

Let us illustrate here in a single direction. An ox-chain is broken. You are in the woods, far from the shop. You put in a wooden toggle to last home. You have open links, that you bought of your hardware merchant. The chain is mended quicker than it has taken to tell.

Exercising Ingenuity.— The exercise of a little ingenuity will enable the farmer's boy, after he gets used to handling tools, to alter and construct many articles of comfort and use. The cut shows the seat and rails of a common wood-bottomed arm chair, grafted on a stand with a drawer in the bottom to hold blacking brushes or any object

OPEN LINK.

THE ARM-CHAIR TURNED BACK.

THE ARM-CHAIR CLOSED.

of that kind. It may hold small implements for mending many things about the farm, as riveting hammers, awls, etc. Here, again, we illustrate a very useful device for trapping ground moles that may be easily constructed. The sharp spikes fall right upon each side of the narrow portion of the trap. The trigger is also represented by itself, showing the handle passing into the run-way of the mole, and the pressure of the animal in burrowing underneath forces the trigger up, which allows the top board, hinged at the back, and pretty heavily weighted, to fall, and the mole is pierced with the spikes. It is about the only sure trap for moles, which work under ground entirely, and besides being useful as an exterminator of these pests of the farmer, will afford employment for the boys during leisure hours.

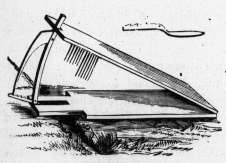

MOLE TRAP.

VIII. The Farm Paint-Shop.

TIME and money can be saved by having paints and brushes. They are now sold by dealers, in quantities to suit, ready mixed and colored, requiring only a little oil or turpentine for use. Buy only pure lead and oil, except for fences and rough buildings. There the better class of the mineral paints may serve. Adulteration is nowadays most shameless, and in none more flagrant than in paints. If you buy your paint or your brushes of a respectable manufacturer or dealer under the guarantee that it is pure, it should be good. It will cost you more than an inferior article, but the best is always the cheapest in the end. If a package of paint is not all used up it may be saved intact by covering it with linseed oil and closing tight. Brushes may be kept from day to day in cold water. When the job is done clean them thoroughly in turpentine, dry, and hang in a dry place so the bristles may point downward. For the farm a brush two inches in diameter, one an inch in diameter, and a sash brush will be sufficient. Brownish red is the best general color for implements requiring paint.

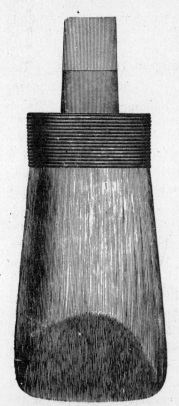

PAINT-BRUSH—BEST

SASH BRUSH.

IX. Putting up Rough Buildings.

In making rough buildings very little framing is necessary. Nails and proper bracing will hold the building together. Studding set squarely on sills and toed fast by driving nails diagonally through the studding and into the sills, with blocking nailed around them, is stronger than mortising, for rough buildings. Each one must

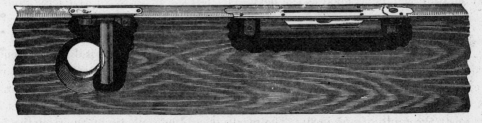

SECTION OF ADJUSTABLE PLUMB AND LEVEL.

be in exact line, set vertical (with a plumb and level—see cut), made fast and held so. The floor joists are made level by means of a long straight strip (straight-edge) and a carpenter's level. If an upper floor is necessary, they are laid upon stringers firmly nailed to the studding. The flooring must be firmly driven together, by placing a block of wood against the edge of each successive board, and driving against it. When driven firm, nail. Sheathing should be nailed to the studding before the clapboards are put on, unless the boards are to be placed vertically. In this case, strips must extend across the studding, upon which to nail the vertical boarding. Then cut out the windows and doors, unless these have been previously cased in. The roof, as a rule, should be one-quarter pitch, and sheathed over the rafters to receive the shingles. (See chapters relating to Building.)

X. Shingling a Roof.

MORE persons fail in shingling a roof than in other rough building work, yet it is really very simple. If you begin at the *top* of the roof to shingle, you will not be the first man who has done so. BUT DON'T! Always begin at the bottom. Break the joints by laying the center of a shingle over the crack of two others, or a wide shingle to cover the cracks of narrow shingles. The rafters should be laid level; the shingles laid with not more than one-quarter of their length exposed to the weather, and nailed above the lap. Very wide shingles have three nails, the average, two, and very narrow shingles one nail each.

Each line of shingles must be laid true to the line, one with the others, the lower course being laid about two inches over the edge of the lower sheathing board. The details of shingling are as follows: Stretch a line at the proper distance beyond the lower roof-board, lay the butts of the first course of shingles to this line, narrow and wide, just as they come, discarding such as are shaky, wormy or rotten. This course laid, stretch the re-chalked line along the row of shingles the proper number of inches above the lower edge, draw it tight, snap it, and you have the mark for the next course. Nail on this course, always having a shingle cover a crack by at least one inch. So proceed, course by course, moving your foot-rest up the roof when you can no longer nail from the scaffold on the side. When you have reached the peak, saw the last shingles square with the slope of the other roof. Shingle the other side, saw these off fair, cover the peak with two strips, nicely jointed together, and the roof will be as good as the best.

XI. Making a Hay-Rack.

LET a carpenter make one with iron bolts, if you can afford it. If not, one as strong as the best, if not so handsome, is easily made by laying two 2 x 8 inch joists, twelve feet six inches long, on the bed-pieces of the wagon; across these lay three 2 x 4 inch scantling; mark the bottom pieces so these three scantlings may be let into the joists the depth of one inch. The marks should be, one six inches from the front end, one in the middle, and one at six inches from the rear. The scantlings which should be seven feet long, are to be then securely pinned or bolted to the bed-pieces; along the outside of the scantling securely nail a board six inches wide, one inch thick, and inside of where the hind wheels come, nail another four inch wide board.

Over the hind wheels form an arch and cover it with slats; nail a cross-piece front and rear, put a "ladder" in front six feet high, and with three rungs, playing on a roller through the bed-piece, so it may be turned down. This rack or ladder—as a hay or grain rack is sometimes called—will hold all that two horses can draw, and will be strong; how handsome it is will depend upon the skill of the builder.

XII. Stone Fences.

FENCING with stone walls is not to be advised in any case, except when it is absolutely necessary to remove the stones from the land. In some hill regions of the United States, the quantity of loose stone in the soil is a most serious obstacle to cultivation. The stones must be gotten rid of; they are a nuisance piled in the field, and are too heavy to haul long distances. In such cases it may be economy to form them into stone walls. The stones of fields are generally those called boulders. That is, stones that have been more or less worn by abrasion in being moved about by the forces of nature. They are of all sizes, from a man's fist to those weighing tons. The larger ones must be reduced by blasting or other cleavage, or buried in pits dug so deep as to take the stone below the possible reach of the plow.

XIII. Moving Heavy Stones.

ANYTHING from the size of a man's head or somewhat less to those two men can lift may be laid into a wall, and the larger ones that can be moved by oxen and a stone boat, may form the foundation. The stones may be rolled onto the boat and also into their places in the foundation wall, by means of a rolling or sliding hitch of a chain. Anything that one or two pair of oxen can move, may be accomplished by passing a chain around the stone and over the hook, so the hook comes next the ground, or better, partly under the stone. The rolling hitch is made by passing the chain once or more around the stone and then over the hook—in this case the hook being next the ground on the side farthest from the team. Thus the chain will form a purchase, identical with the same hitch in rolling logs. The reason why oxen are better than horses is they move slowly and steadily, and will generally continue a pulling strain longer than horses or mules, unless the horses have been specially trained for the purpose. Oxen also come about more readily, and there is not the hamper of whipple-trees and harness.

SIMPLEST FORM OF STONE BOAT.

XIV. For and Against Stone Walls.

ABOUT the only argument that can be made in favor of stone walls for fencing is, that if well laid they last forever. The next is that if stones cumber the soil to the serious detriment of cultivation they can thus be made useful when removed. On the other hand, they require skill in laying. They occupy more land than even a Virginia worm-fence. They harbor the roots of weeds and noxious plants, difficult to exterminate, and if not laid in the best manner they are constantly falling down. Hence the argument narrowed down to this, is: If you have stony fields, and the distance prevents their hauling to some ravine or other waste place where they may be dumped, they may be laid into a wall. In the New England and other rocky Eastern States, stone walls, many of them built a century ago, are being torn down and carted away. The land is too valuable now to permit their continuance. There is but little land between the Alleghenies and the Rocky mountains where it is necessary to build walls to get rid of the stone. In the mountain regions of the Far West the value of the land will eventually cause the walls that are already built to be torn down. Hence, again, the warning we give to fully discuss the economy before building. The stone will some time be valuable in forming roads, concrete and other foundations for buildings, and possibly even for the walls of rough buildings themselves.

XV. How to Build the Wall.

1. The foundation must be on solid earth to prevent heaving. On gravel or firm sand it may be quite near the surface; otherwise it must rest on the subsoil.

2. The foundation stones, or at least those lying contiguous to each other, should be of nearly the same size, and should extend the full width of the wall. If the boulders are rough and of uneven size, lay the roughest side down, so that the flattest side may come up, on which to lay the first course.

3. If the stones are so small that more than one course is required to form the wall, they must be tied together at short intervals with stone reaching through the wall. If not, pieces of hard, lasting wood, not less than one by two inches in thickness, may be used for this purpose.

4. The wall must be laid in any event so as to break joints. If this is not carefully observed the wall will certainly fall.

5. Small stones must never be chinked into the face of the wall, and this face must be carried up fair and square, or with a slight but perfectly equal gradient.

6. If the wall is let to be made by contract, see that the contractor understands his work and does it properly. A man may not know how to lay a wall, and yet by studying the rules we have laid down may be perfectly competent to decide whether it is properly laid.

7. In making the contract be certain that the stipulations embrace all the points you wish carried out, and then do not alter it. It is extras that enable all contractors to make big profits in any work, since for this they can charge arbitrary prices.

XVI. The Balloon Frame in Building.

The balloon frame is essentially an American institution, and has been an important factor in developing the prairie regions of the West, the Pacific Slope, and all the country lying between. These frames are strong, cannot be torn apart by the wind, and are cheap.

The late Solon Robinson, from whom we received many early ideas on agriculture, who, as a resident of northern Indiana, witnessed the early development of that portion of the West, and who, in the youth of the writer of this, used jocularly to call him the "boy farmer," has truly observed: "If it had not been for the knowledge of balloon frames, Chicago and San Francisco could never have risen, as they did, from little villages to great cities in a single year."

Later, Mr. Geo. E. Woodward, a celebrated architect and civil engineer, of New York, said: "The balloon frame belongs to no one person, nobody claims it as an invention, and yet, in the art of construction it is one of the most sensible improvements that has ever been made." It is one of those things gradually suggested by the lack of heavy timbers, and has been found altogether superior to them, not only for wooden dwellings, but also for barns, where great weight is not required to be carried on the beams. The light pieces are not weakened by cutting. The bearings are short, forming a continuous support for each piece from foundation to rafter. It is braced in every direction naturally, and cut nails have proved not only cheaper but stronger than mortise and tenon in braces, beams and other supporting parts of ordinary wooden buildings.

XVII. How to Build the Frame.

Mr. Woodward sums up the advantages of balloon, or basket frames, as follows:

1. The whole labor of framing is dispensed with.
2. It is a far cheaper frame to raise.
3. It is stronger and more durable than any other frame.
4. It is adapted to any style of building, and better adapted to all irregular forms.
5. It is forty per cent cheaper than any other known style of frame.
6. It embraces strength, security, comfort and economy, and can be put up without the aid of a mechanic. The two last items are of especial value. The latter particularly so, where skilled labor is difficult to get.

Mr. Woodward gives the following plain directions:

We hear and read much about the policy of cutting mortises, tenons, gains, etc., in the various pieces which go to make up the balloon frame. It is our opinion, based upon a long and thoroughly practical experience,

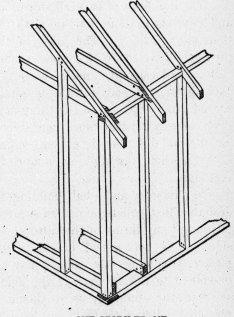

ONE-STORY FRAME.

that he who does much of this will have some misspent time to account for hereafter, besides weakening his building and hastening the decay of his frame.

1. A line must be cut in the studding for the side girt, unless the dwelling be lined. Gains are sometimes cut in floor joists for the purpose of locking them over partitions that run through the height of the building. Rafters projecting over the sides should be notched, to give them foothold on the plate. These causes would, as a general thing, constitute all the cutting necessary.

2. In building houses one and a half story high, never cut a gain for the side girt, on which to rest the upper story floor joists, unless the thrust of the roof be well guarded against by secure collar beams. We prefer, when we cut this gain, to use studding one inch wider for the sides. When the building is lined, the side girt rests on top of the lining, and no cutting is necessary.

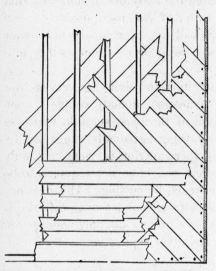

DIAGONAL LINING, OUTSIDE AND IN.

3. Unplastered buildings of a moderate size are strong enough if the girt be nailed directly to the studding, without cutting the gain in recess.

4. Buildings of two full stories are abundantly strong with two by four studding and gains cut in them for side girt; the third floor ties the top of the studding, so there is no yielding. The joists of the third floor should be placed upon the plate, the ends beveled to the same pitch as the rafters, and each joist nailed at both ends to each rafter. He advises the building of the second story full for a dwelling-house. It gives more strength, more convenient room, and the real difference in expense is practically nothing, where the studding is more than five feet high.

5. In story and a half buildings it is very desirable that collars be put on securely, so as to prevent any thrust of the rafters; when the side girt is not gained in, as in small unplastered buildings, the collars may be nailed or spiked to the rafter. If the side girt is set into the studding, as it should be in a plastered building not lined inside, it makes a weak point in the studding, practically reducing them from two by four to two by three, and the collars should be put on in such a manner as to guard against any thrust whatever.

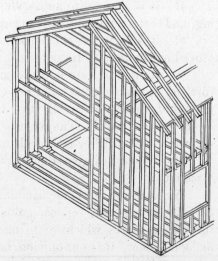

ISOTHERMAL VIEW OF BALLOON FRAME, AFTER WOODWARD.

The size of the building and the judgment of the constructor will indicate the best course to pursue. Buildings of one, two, or more full stories have no collars; the joists of the upper floors tie the top of the building and take the thrust of the rafters. In the usual mode of inside lining, one strip laps the stud. The ends of the lining of the adjoining side are nailed to a strip fastened to the stud to receive them.

6. In the construction of a barn twenty-four by thirty, alternate studs on the sides two by four and two by five are recommended, the side girt to be nailed to the narrow stud and let one inch into the wide stud. When the studding is more than five feet above the second floor of a barn, two or three tie-strips across the foot of the rafter will make all snug. There should be tie or collar beams on all rafters.

PART V.

HORTICULTURE.

VARIETIES AND CULTIVATION OF FRUITS, FLOWERS AND SHRUBS.

THE PRACTICAL ART OF GRAFTING AND BUDDING.

LANDSCAPE GARDENING AND LANDSCAPE TREES.

COMMON SENSE TIMBER PLANTING.

INCLUDING

FISH AND FISH CULTURE.

HORTICULTURE.

CHAPTER I.

ORCHARD, VINEYARD AND SMALL-FRUIT GARDEN.

I. THE FARM ORCHARD AND GARDEN.—II. ARRANGEMENT OF THE HOME ORCHARD.—III. HOW TO PREPARE FOR AN ORCHARD.—IV. LAYING OUT THE ORCHARD AND PLANTING.—V. WHEN TO BUY TREES AND WHEN TO PLANT THEM.— VI. WHAT VARIETIES TO PLANT. — VII. APPLES, THEIR CULTIVATION AND VARIETIES.—VIII. PEARS, THEIR VARIETIES AND CULTIVATION.—IX. THE FORMS OF FRUIT EXPLAINED.—X. PEACHES.—XI. NECTARINES.—XII. THE CHERRY.—XIII. PICKING AND PACKING ORCHARD FRUITS. — XIV. THE SMALL FRUITS —XV. THE VINEYARD.— XVI. THE GRAPES FOR FARMERS.—XVII. CULTIVATION OF THE CRANBERRY.

I. The Farm Orchard and Garden.

IT is a well-known fact that farmers as a class, especially in the West, are more poorly supplied with fruit than the average townspeople. One reason is, an impression prevails that the cultivation of fruit requires great care and attention, and that the proper soil for fruit can be found only in certain districts. The same may be said of the garden for vegetables. But if the farmer would give his orchard and his garden the same attention that intelligent farmers give their stock and corn fields, an abundance of fruits and vegetables might be had the year round at less than half what the average citizen has to pay for them. The mistake made by farmers in planting a home orchard, and especially in the arrangement of the vegetable garden, is that they follow the directions of writers of fifty and a hundred years ago. They should employ the same methods that they do in their corn fields—long rows and horse cultivation—for all but the minor plants of the garden; and for these improved implements of hand-cultivation should be used. Clean cultivation is necessary in the home orchard, for the orchard for home use must be separate and distinct, and its management different from that of commercial orchardists.

II. Arrangement of the Home Orchard.

THE small fruits and the vineyard may come in the same plot of ground and yet give ample room for all. A plat of land about thirteen rods by twenty-five is just five rods over two acres, and the shortest way across will enable most of the work to be done by horse-power. By beginning at the farther end the larger trees, as apple, pear, cherry, peach, plum, quince, etc., according as the climate and situation will allow, may come in successive rows, to be followed by grapes, blackberries, raspberries, gooseberries, currants and strawberries. To make the farther part of the orchard easily accessible, a pathway ten feet wide should remain unplanted through the

CANADA REINETTE APPLE.

middle, which will not interfere with the cultivation, for no grass should be allowed in the farm orchard. It must receive the same clean cultivation as the corn-field.

Next the house may come the vegetable garden, divided by the same broad path,

TETOFSKY APPLE.

so that the cart, the wagon or the wheelbarrow may freely pass along from one side to the other; or a space sufficient for a "turn-row" to be left on each side would be better; and, in that case, the turn-row may be permanently seeded down to grass or clover, to be cut for soiling. Hence, there is no waste space left whatever, and there are no weeds to seed, in any portion of the garden. In the space next the house, or in the kitchen-garden proper, allot the most sheltered spot for a hot-bed, or a cold frame, and also as a border for the early cultivation of some special crops, as cress, radish, lettuce, plants of cabbage, cauliflower, etc., to be followed by egg-plant, lima beans, okra and other heat-loving plants. Then the first spaces, next the small fruits, may be devoted to pie-plant (rhubarb), asparagus, sage, tansy, mint and other perennial plants, and the balance, commencing with crops requiring poling or staking, sugar corn, early potatoes, etc.; the smaller annual crops may succeed each other. Thus you may have what will not only make a pleasing feature of the homestead, but also a plat of ground that may be cultivated at a minimum cost, and which will turn out a maximum crop, if made rich enough with manure, and the soil is properly cultivated.

It may be objected by some, that a row clear across such a patch of some varieties of plants will not be needed. Suppose not, piece it out with some variety, requiring the same space of row, always remembering to cramp nothing. Thus, if you want half a row of raspberries, and the same of blackberries, let the width be that for blackberries. Currants and gooseberries may be pieced out in the same manner. So may the asparagus and rhubarb; carrots and parsnip; cabbage and cauliflower; radish and lettuce; dwarf beans and dwarf peas; muskmelon and cucumber; bush, or patty pan squash, and many other things that might be named, and which will naturally suggest themselves to the observing man in the first season's cultivation.

III. How to Prepare for an Orchard.

As a rule, in the West, the soil, if undrained, is at some seasons saturated with water for weeks. Many persons make the mistake of digging deep holes in such soils in which to set the trees. Nothing could be more fatal. With the plow and subsoiler make the orchard "one great hole." That is, deepen the soil, and cast it into high beds corresponding with the width of the rows of orchard trees.

There is no better time than immediately after harvest for preparing the soil. If not naturally drained it must be artificially drained, as a prerequisite to the best success. Upon prairie soils, plow the land in one of the directions in which the trees are to be set, as deep as the soil will admit, following in the furrow with a subsoil plow, and loosening the earth below to as great a depth as possible, leaving the surface rough. If the soil is plowed both ways, to form squares, so much the better.

Just before cold weather, but always when the land is dry and friable, proceed along the tops of the ridges where the trees are to be set, and cast two deep furrows apart so as to leave a " land side " in the middle. The ground should be left now until spring. Then, when the earth is in a good and friable state for working, set the trees so that the necks will be from one-half to one inch deeper than they stood in the nursery—in heavy soils the same depth they stood in the nursery—being sure they row both ways.

IV. Laying Out the Orchard and Planting.

BEGIN at one side of the field, and extend the row as far as you wish, setting stakes exactly in line, and so the last stake will be some distance beyond the last row of trees desired. Then from the place of beginning, run a line at a right angle from the first. This may be done by adjusting two ten feet rods—by means of a square—at right angles to each other, and bracing them. Then set a stake at the corner, and another one at the end, and standing some distance behind the first stake, extend the line by means of other stakes, and as far as desired. So proceed from each corner until you intersect the first line run, correcting any errors that may occur. Then measure and stake accurately, the distance required for each tree, with a chain or tape line, entirely around the piece, and then the intervening spaces across the orchard plat. Thus, if you have done your work correctly, the stakes thus set will line both ways.

The Distance Apart.—The distance at which apple trees should be set is entirely a matter of taste with the planter. At thirty feet apart with good cultivation, many varieties will meet and interlash their branches, at the end of eighteen or twenty years, if the cultivation has been good, and the trees remain healthy. Our own experience is, that twenty feet is a proper distance for apple trees, but we should set every alternate row with sorts that bear early, and by liberal cultivation and root pruning, force them into bearing, and wear them out at the end of twelve years, and then grub them out, leaving the alternate rows to occupy the entire land, plowing the earth from the centres, towards the remaining trees. Thus they will eventually be placed upon beds well elevated, and sloping gradually to the centres, until near the dead furrows. If the remaining trees interfere too much in the rows every alternate one may be taken out, and at last you will have your trees forty feet apart each way, the proper distance when they get age.

Planting.—To set your trees, provide yourself with a fence board, say nine feet long, notched in the middle, and containing an inch and a half hole at accurate distances from each end—say six inches. Place the notch against the stake where the tree is to set; thrust a short stake through the hole at each end, and remove the board, allowing the outside stakes to remain. Dig the hole and so proceed until you have the whole completed. Or, having two guage boards, exactly alike, one hand can be digging while another is setting. In planting the tree all that is necessary is to slip the holes in the board over the pegs, and the notch in the middle will mark the exact place where the tree is to stand.

In digging the holes, be sure you have them large enough to accommodate the roots without crowding, leaving a good, broad mound in the centre, upon which to set the tree. This is easily accomplished by drawing the earth to the centre, after the hole is dug, tramping it solid, and then smoothing the mound to your satisfaction.

On Drained Soils.—The remarks here made apply to our ordinary undrained prairie soils. If the soil is artificially or naturally drained do not raise the beds, unless the blue clay comes very near the surface. If so, the raised beds will be of advantage to give a deeper soil for the trees while yet young. If the roots run deep and unchecked, this will operate against early bearing. Hence the excellent success

EARLY FAT.

of the late Dr. Hull, of Alton, by means of root pruning. Mr. B. F. Johnson, of Champaign, a graphic writer and discriminating observer, lately assured the author that the most regular bearing farmers' orchards that had come under his eye, were on firm soils, underlaid with strong clay, and which in the spring were even wet. The reason is a natural one. If the soil is not permanently wet, the trees receive a check during summer droughts that throw them into bearing.

V. When to Buy Trees and When to Plant Them.

We prefer to order and receive the trees in the autumn for obvious reasons. They should be shipped as soon after the fall of the leaf as possible. Having received them, cut all ends of lacerated roots as clean as possible, and the trees being

HIGBY SWEET APPLE.

pruned into shape, heel them in, in some place where the winter sun will not fall on them in the middle of the day. To do this dig a trench on some well-drained spot, large enough to contain all the roots, and about a foot deep, throwing the earth to the south. Lay the roots into this trench, the trees as closely together as possible, and at an angle of about 45 degrees. Cover the roots with mellow earth, dug from in front of where the roots lie, and cover the stems, also, well up to the branches. They are then safe for the winter.

Much has been said first and last about the proper time to plant orchard trees. If you are ready to plant and your soil has been properly prepared, there is no objection to fall planting, if it is properly done. The great difficulty with fall-planted trees is, first, they are not protected from being swayed about by the wind. If fall planted, this must be attended to by carefully staking and tying. Then raise a sharp mound of earth about the tree. This will assist in holding it firm. The second, and principal objection to fall-planted trees is the loss by winter evaporation, and especially by our cold, drying winds. Hence, we should guard against this by protecting all such trees from the wind and sun as much as possible the first winter. One of the means to

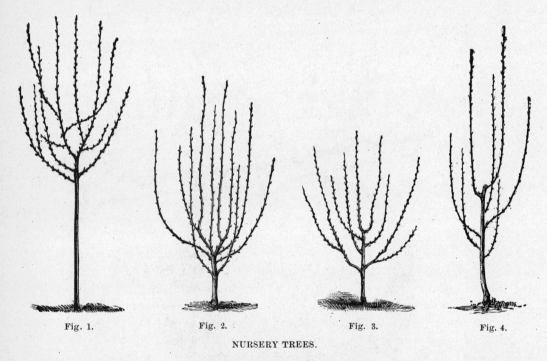

Fig. 1.   Fig. 2.   Fig. 3.   Fig. 4.

NURSERY TREES.

meet this end is a more severe pruning (cutting back) than is usual with spring-planted trees. Other means will be readily suggested by the planter's own observation. On the whole, we prefer spring planting in the West. As to the form of trees as received from the best nurseries, the illustration, Fig. 1, shows the regular standard form of orchard trees, the trunk three to four feet clear of branches; Fig. 2, dwarf apple or doucin stock; Fig. 3, dwarf apple or paradise stock; Fig. 4 is a pear tree, pruned, to produce a pyramidal form, being two or three years from the bud, and showing one season's growth after cutting back. These will also show the several forms of trees, standard, low heads, etc.

VI. What Varieties to Plant.

THERE is this to be remembered in planting a home orchard. While the cultivator must select measurably of those varieties that are hardiest and most prolific, he may, nevertheless, give himself larger latitude in selection than the purely market orchardist, whose selection runs to few varieties and those which will give the largest crops of fruit. The farmer raises fruit for himself, and may be content with a smaller crop and better fruit. The illustrations of fruits given in this chapter are not intended to convey the idea that they are the best for general cultivation. They are superior fruits, in repute in particular localities, and are given here to show forms and characteristics.

VII. Apples, their Cultivation and Varieties.

WHATEVER the cultivation of the general orchard, the care of the home or garden orchard should be as good as that of the garden. The cultivation, however, must be superficial—only enough to keep the surface soil in tilth. About the tenth of July it is not a bad plan to sow buckwheat, and just before it comes into bloom plow it lightly under. It keeps down weeds and when turned under helps to enrich the soil.

Pruning, Etc.—The pruning is important. Whenever you see a twig that is liable to give trouble by crossing another, take it out, whatever the season of the year. But do not prune too much—in the West, especially. The thicker the head the better, provided it does not get so dense as to exclude proper light and air from the leaves. The form of the tree must also be studied, and its natural habit be complied with. Insects must be watched, and the proper means taken to destroy them. These will be indicated in the chapter relating to insects. The bark of trees seldom gets mossy and bark-bound where the soil is cultivated. If it does, it should be scraped, as to the rough bark, and washed with soft soap, or with a solution of potash and water. If leaf blight attacks the trees, cut it away at the first indication and burn the twigs. If trees die from the effects of a hard winter, take them up and plant again. The first ten bearing years of any apple or other long-lived tree is better than all that comes after. The profit is in young, thrifty trees, not in old ones. In the West, the average productive life of an apple orchard is less than twenty-five years. Trees five years old of early bearing sorts will produce fruit, the later bearing varieties will range longer, even up to ten years for such varieties as Northern Spy.

We append brief descriptions, some of them after Elliot, of the varieties illustrated. These, as before stated, being given not only as excellent varieties in certain localities, but to illustrate also forms and peculiarities of the fruits.

The following descriptions apply to the fruits illustrated, and are given as descriptive of form, color and other characteristics, and as an index to the study of fruits.

Canada Reinette.—Synonyms: Canada Pippin, Portugal, Canadian Reinette, Janaurea, Pomme de Caen, De Bretagne, Reinette de Grosse du Canada, German Green, Wahr Reinette, Reinette du Canada Blanche, Grosse Reinette d'Angleterre,

Reinette du Canada à Cortez, White Pippin (erroneously), Yellow Newton Pippin (erroneously)

FRUIT.—Size, large to extra large; form, varying, generally roundish flattened, slightly oblique, angular, much ribbed, especially toward the crown or calyx; sometimes the form is almost oblong and quite smooth; color, light greenish yellow, with frequently a faint blush of red on the sun-exposed side; many small dark green specks, surrounded with light green suffused beneath the skin; stem, short, set a little inclined on one side in a deep, open cavity; slightly russeted; calyx, with short

SUMMER ROSE APPLE.

divided half-open segments; basin, with prominent ribs; flesh, yellowish-white, jucy, crisp, tender, sharp, sub-acid, sprightly, aromatic; core, small, compact; seeds very dark brown, almost black, season, December to May.

Tetofsky.—FRUIT: size, medium; form, nearly round, slightly oblate conic, smooth; ground color, a pale yellow, beautifully striped with red and overspread with a fiber-like whitish bloom; flesh, white, crisp, juicy, slightly acid, and with an agreeable fragrance, and early ripening; tree, short, vigorous, upright, with a broad, distinct foliage that marks it at once to the beholder. It is very hardy, an early and regular bearer, and forms a roundish, conical, open head, with branches abounding in fruit-spurs. This is an old apple, and it is strange that it has not been more appreciated, especially by those who live in climates trying to the vitality of fruit trees. It has been grown in Maine, in most of the northern localities in Canada, and nearly fifty years since was fruiting at Salem, Mass.

Early Joe.—The fruit in size is below medium; form, roundish flattened, very regular; color, pale yellowish green, overspread with broken stripes and splashes of

GRIMES' GOLDEN.

pale and pale red; stem, of medium length, rather slender, set in a deep open cavity, somewhat russeted; calyx, small, nearly closed; basin, shallow; flesh, yellowish, white, tender, crisp, with a delicate rich pear flavor; core, medium, with an open center; seeds, abundant, short, pyriform; season, July and August.

TREE.—In the nursery this is a slow, stocky grower, but after becoming established in good soil in the orchard it made moderately vigorous and healthy shoots, and forms an open spreading, rather irregular tree of only medium size. It produces very abundantly and may be noted for its dark colored foliage, as well as by its quite dark reddish brown annual shoots. Its origin is claimed for Ontario County, New York.

REMARKS.—This is one of the most delicious of all the summer apples. It is comparatively little known or grown at the West, probably from the trees growing so slowly in the nursery as to make their cultivation unprofitable. Although of small size, the trees are such good bearers, and the fruit so firm for carriage, that were it once grown and offered, its superior quality would undoubtedly make it always command a ready sale. Like all our early summer ripening varieties when grown at the South its size is much increased and the tree becomes larger and stronger.

Higby Sweet.—Synonyms: Lady's Blush, Trumbull Sweet, Fenton Sweet.

FRUIT.—Size, medium or above; form, roundish conical, flattened at ends, often one side enlarged or slightly oblique; color, clear pale yellow, with a faint tinge of red in the sun, and a few obscure, suffused, reddish dots; stem, medium, often short and rather slender, usually set a little on one side of an open rather deep cavity; calyx, small to medium, generally closed; basin, deep, abrupt, slightly furrowed; flesh, white, very tender, juicy, delicate, rich, sweet; core, small, compact; seeds, plump, roundish ovate, sharply pointed; season, October to December.

TREE.—A vigorous, healthy, upright grower while young, with moderate sized shoots. In the orchard it makes a round, regular, open head, and forms a rather large tree, producing almost annually and abundantly a fair even-sized fruit. Originated in Trumbull County, Ohio.

REMARKS.—This is, comparatively, a new variety, but a very hardy tree, productive, and one of the most delicate and pleasant of all the sweet apples. It is especially suited for table use, or for cooking, or for other uses about the homestead, but is too tender for shipment.

Summer Rose.—Synonyms: Woolmans, Harvest, Lippincott.

FRUIT.—Size, below medium; form, roundish, flattened regular; color, glossy, pale yellow, blotched, splashed and streaked with two shades of rich red; a few minute dots; stem, varying from stout to slender; cavity, narrow, pretty deep; calyx, with recurved segments, partially or quite closed; basin, broad, open, pretty deep, slightly furrowed; flesh, fine grained, white, tender, crisp, juicy, sprightly, agreeably subacid; core, medium to large; seeds, abundant, short, plump, full, round, ovate; season, July and August.

TREE.—A vigorous, healthy grower, with short jointed, stout shoots, forming a small or medium sized tree, with an irregular spreading head; very productive. Origin, New Jersey.

REMARKS.—The Summer Rose usually proves one of the most desirable of early summer fruits for family use. It is an early bearer, continues a long time in ripening, and although not so rich as Early Joe, or Garden Royal, it nevertheless has a sprightliness that makes it always admired, and fits it well for the dessert or cooking. It is valuable as a market sort, where quality is ranked before size.

Grimes' Golden.—This apple, originally from Virginia, was introduced thence to Ohio, and later West. It is a deservedly popular and fine dessert, and also a good cooking fruit.

TREE.—A vigorous growing, healthy, spreading, productive and early bearing sort.

QUALITY.—The late Dr. Warder classed it as being too good for aught else than the dessert. We endorse it without hesitation. The core being small and closed; flesh, firm, yellow, very fine grained and juicy; subacid, aromatic, spicy and refreshing. It is a winter apple, in season from January to March.

VIII. Pears.—Their Variety and Cultivation.

It has been said that pears cost the amateur ten times as much to cultivate as to buy them. It is certain that pears are only successfully cultivated for market in widely isolated localities. Still we have seen pears growing in farmers' gardens and bearing regularly, in many apparently unfavorable localities, judging from the lack of orchards near. The pear is long-lived, and resists severe winters. Its great drawback is blight, except on soils peculiar to the tree. In Michigan, for instance, some sections of Indiana and Illinois, and elsewhere in the West, are trees planted by the French missionaries of the last, and even preceding, century, "hale and hearty yet." Still, it is not to be denied that "pear culture" *is* "treacherous." It is best for the amateur to experiment with dwarf pears; we have had good success with them in garden culture.

BONNE DU PUITS ANSAULT PEAR.

Pears on Quince Stock.—Here is a list of good sorts of pears for cultivation on quince stocks: Summer.—Andre Desportes, Bartlett, Beurre Giffard, Brandywine, Tyson, Petite Marguerite, Clapp's Favorite. Autumn.—Beurre Hardy, Beurre Clairgeau, Belle Lucrative, Urbaniste, Duchesse d'Angouleme, Doyenne Boussock, White Doyenne, Beurre Superfin, Flemish Beauty, Louise Bonne of Jersey, Seckel, Howell. Winter.—Beurre d'Anjou, Easter Beurre, Doyenne d'Alencon, Lawrence, Josephine de Malines.

GATHERING PEARS.—One of the most important points in the management of pears, is to gather them at the proper time. Summer pears should be gathered at least ten days before they are ripe, and autumn pears at least a fortnight. Winter varieties if they will hang so long, may be left until the leaves begin to fall. When pear trees are heavily laden with fruit, they should be thinned when about one-third grown; else the fruit will be poor and the trees injured.

The pears illustrated are, two of them, of recent introduction; the others, pears that have received favorable mention from many growers.

Frederic Clapp.—Of this pear, Hon. Marshal P. Wilder, the life-long President of the American Pomological Society, gives the following description: "Form generally obovate, but somewhat variable; size above medium, skin thin, smooth and fair, clear lemon yellow; flesh fine grained, very juicy and melting, flavor slightly acidulous, rich and aromatic; season October 15th to November 1st, remaining sound at core to the last; quality very good to best, and will be highly esteemed by those who like acidulous pears. Of this pear the committee of the Massachusetts Horticultural Society have reported favorably for years. Of its quality they state in 1873: 'It was pronounced decidedly superior to Beurre Superfin, and is regarded by all who have seen it as the highest bred and most refined of all the many seedlings shown by Messrs. Clapp.' It is probably a cross between Beurre Superfin and Urbaniste, the tree resembling in habit the latter variety, and may safely be commended as worthy of trial by all cultivators of the pear." Tree a vigorous or free grower and somewhat spiny.

FREDERIC CLAPP PEAR.

Bonne du Puits Ansault.—Another recent pear well spoken of is Bonne du Puits Ansault, a pear of 1865, of Mons. Leroy, France. Of this pear, Messrs. Ellwanger & Barry say: "Medium size; melting, juicy and very fine grained; one of the finest in quality of all pears, superior to Seckel. Tree a poor grower, which necessitates top grafting to obtain good standard trees. Bears when quite young."

Howell.—Size, medium to large; form, obovate, pyramidal, very regular; color, greenish, becoming pale, lemon-yellow or straw color at maturity, many small russet dots, and on the sunny side a faint blush, sometimes deepening into a clear red cheek; stem, about one and one-quarter inch long, curved, moderately stout and inverted without depression; calyx, open in a shallow, smooth, regular basin; core, small; seeds, round, oval, plump; flesh, white, fine grained, juicy, melting, sweet and pleasantly perfumed; season, September.

Tree.—An upright, vigorous grower, with roundish, broad, oval foliage, an early bearer on the pear stock, and succeeding among the best when worked on quince.

HOWELL PEAR.

This fruit originated in 1829, in the garden of Thomas Howell, New Haven, Conn., from seed of a hard winter pear, which had, growing on one side of it, a summer Bon Chrêtien, and on the other a White Doyenne.

Dix.—Fruit:—Size, large; form, oblong pyriform; color, pale yellow, becoming deep yellow when well matured, with many distinct irregular-sized russet dots and patches, and considerably russeted around the stem; stem, rather short, stout, thickest at each end, set obliquely or with a raised lip on one side, with little or no depression;

DIX PEAR.

calyx, small for the size of the fruit; basin, shallow; flesh, yellowish-white, moderately fine-grained, juicy, melting, rich, sweet, slightly perfumed; core, marked with a dark, gritty circle, and the same extending toward the stem; season, October and November.

Tree.—A vigorous, upright grower, with pale yellow, slender shoots, sometimes thorny, quite hardy, unproductive while young, but an abundant bearer when the tree becomes of mature age, say ten to fifteen years from planting; originated in Boston, Massachusetts, in the garden of Madam Dix, and fruited for the first time in 1826.

PARADISE D'AUTOMNE PEAR.

Remarks.—Although the Dix is comparatively a long time before coming into bearing, so far as I can learn, it proves an earlier bearer and a better fruit South than in its own locality; and such is its vigor and hardihood, that it is yet one of the most valuable sorts for extensive orchard planting; for, when it once commences bearing it produces abundantly, of a regular, even, large fruit, desirable for table or market.

Paradise d'Automne.—Synonyms: Autumn Paradise, Calebasse Bosc, Maria Nouvelle, Princess Marianne.

FRUIT.—Size, large; form, obovate, obtuse, pyriform, with an irregular, uneven surface; color, dull yellow, mostly overspread with a bright cinnamon russet, deepening on the sunny-side; stem, rather long and slender, largest at ends, and obliquely attached to the fruit by fleshy wrinkles, without depression; calyx, rather large, open, with reflexed segments; basin, abrupt, furrowed; flesh, yellowish-white, slightly granulous, juicy, buttery, melting with a delicious, rich, vinous, aromatic flavor; core, small; seeds, full, long, pointed; season, September and October.

TREE.—A vigorous, strong grower, with long, reddish brown shoots, dotted with many large, whitish gray specks; at first the tree is quite upright, but it soon becomes half pendulous, spreading, open, and rather straggling; an early and very abundant bearer; of foreign origin.

REMARKS.—As a standard orchard fruit, the Paradise d'Automne is by many eminent pomologists regarded as even superior to the Beurre Bosc, which it somewhat resembles both in tree and fruit. It is a variety that as a standard comes early into bearing, and produces abundantly a fruit that in quality has few to surpass it.

Every farmer who wants this fruit should try pears, for in any locality some varieties are pretty sure to give satisfaction. The cultivator once he finds the sorts most healthy on his grounds will retain them, of course; but, in experimenting, do not be afraid of new sorts on account of blight. They are not necessarily more liable to blight than the older varieties. But do not be deceived by representations that certain pears are blight-proof. There is no blight-proof pear.

Little Marguerite.—Medium size, skin greenish yellow, with browish red cheek, and covered with greenish dots. Flesh fine, melting, juicy, vinous, and of first quality. Tree, vigorous, upright, and an early and abundant bearer. Succeeds as a standard or dwarf. The finest pear of its season, and worthy of special attention. Ripens latter part of August.

LITTLE MARGUERITE PEAR.

IX. The Forms of Fruits Explained.

THE forms of fruits are an interesting study. Many of the more common forms are illustrated in this work. To define the various forms of fruits, as designated by specific terms we quote from "American Pomology," by the late lamented Dr. Warder. These apply to the various orchard fruits:

The form may be *round* or *globular* when it is nearly spherical; the two diameters, the axial and transverse, being nearly equal. *Globose* is another term of about the same meaning. *Conic*, or *conical*, indicates a decided contraction toward the blossom end; *Ob-conic* implies that the cone is very short or flattened. *Oblong* means that the axial diameter is the longer, or that it appears so, for an oblong apple may have equal diameters. *Oblong-conic*, that the outline also tapers rapidly toward

GEORGE IV. PEACH.

the eye. *Oblong-ovate*, that it is fullest in the middle; and like *Ovate*, which means egg-shaped, that it tapers to both ends. *Oblate*, or flattened, when the axial diameter is decidedly the shorter. *Obtuse* is applied to any of these figures that is not very decided. *Cylindrical* and *truncate* are dependent upon one another, thus, a globular, or still more remarkably, an oblong fruit, which is abruptly truncated or flattened at the ends, appears cylindrical in its form. *Depressed* is an unusually flattened oblate form. *Turbinate* or top-shaped, and *pyriform* or pear-shaped, are especially applicable to pears, and seldom to apples.

When these forms are described evenly about a vertical axis, as shown by a section of the fruit made transversely, or across the axis, the specimen may be called *regular* or *uniform*; if otherwise, it is *irregular, unequal, oblique* or *lop-sided*, in

NOBLESSE PEACH

which last cases the axis is inclined to one side. If the development at the surface is irregular, as in the Duchesse d'Angouleme and Bartlett pears, the fruit is termed *uneven*.

When a transverse section of the fruit, made at right angles to the axis, gives the figure of a circle, the fruit is *regular;* if otherwise, it may be *compressed* or flattened at the sides; *angular, quadrangular,* sulcate or *furrowed,* when marked by sulcations; or *ribbed*, when the intervening ridges are abrupt. *Heart-shaped* is a form that applies more especially to the cherry than to any other kind of fruit.

Size is a character of but second-rate importance, since it is depending upon the varying conditions of soil, climate, overbearing, etc. It has its value, however, when it is considered as comparative or relative. The expressions employed in this work

NECTARINE OR SMOOTH PEACH—ELRUGE.

to indicate size, are: very large, large, medium, very small, small, making five grades.

The characters of the skin and surface are generally very reliable, though the smoothness of the skin as well as the coloring depend upon both soil and climate. We find, however, that a striped apple which has been shaded, though pale, will always betray itself by a splash or stripe, be it ever so small or rare, nor will any exposure so deepen and exaggerate its stripes as to make it a self-colored fruit; and no circumstances will introduce a true stripe upon a self-colored variety.

X. Peaches.

They are as easily cultivated as corn, south of 40 degrees, and pretty much all over Michigan up to latitude 43 degrees. The only serious drawback is the disease called yellows, and this generally exists in the more sandy districts. "Curl" in the leaf is another disability, but not so fatal as the yellows, as deadly to the peach as is glanders to the horse. The only remedy is to grub the trees whenever found, and in pruning always clean your knife-blade with a solution of carbolic acid after pruning one tree and before commencing on another.

When you plant a new orchard, always be sure you get your budded trees from a nursery not affected with the yellows. A peach orchard should never be allowed to grow up in grass if you wish good fruit. The curculio and damp, hot weather often cause rot, but no person, on account of any of these causes, should refuse to set peach trees wherever the climate is favorable to carrying the trees through the winter. Don't raise seedlings because they come up and grow themselves. Buy budded trees or bud them yourself. This you may easily do from instructions given in the chapter on Budding and Grafting.

George the Fourth.—Fruit: Size, medium to large; form, roundish, divided by a broad deep suture, making one half appear larger than the other; skin, yellowish-white, dotted with dark bright red, and shading into a rich dark red cheek where fully exposed to the sun; flesh, whitish, pale red next to the stone, melting, juicy, with a rich, luscious flavor; stone, small, separating freely from the flesh; season, last of August. [See page 463.]

Tree.—A moderately vigorous grower and a regular, uniform, moderate bearer, producing its fruit evenly distributed and all of unqualified excellence; the flowers are small and the leaves have obscure globose glands; originated in New York City.

Remarks.—Although the peach, like the strawberry, may be termed an evanescent fruit, yet there are a few old varieties whose excellent qualities surpass all those of more recent origin, for instance, the one here described. It is not a profuse bearer, and hence its buds are so generally well perfected that it often sustains uninjured a greater degree of cold than many other varieties, and when it fruits all the specimens are nearly equally good. The large Early York, Haines' Early, Walters' Early, and one or two more popular market sorts, undoubtedly spring from this; and while possessing some superior qualities for the market orchard, have none of them the richness and delicacy of this sort for table use.

Noblesse.—Synonyms: Lord Montague's Noblesse, Millistre's Favorite, Vanguard, Noblest, Double Montague.

Fruit.—Size, above medium to large; form, roundish, sometimes with a hollow at the apex and with a small point. Sometimes it is roundish oblong, and the point at the apex quite prominent. Skin, pale greenish-white, marbled and streaked with two shades of dull red in the sun, occasional faint blotches of red on the shaded side; flesh, greenish-white, very juicy, melting with a rich, delicious flavor; stone, large, obovate, pointed, separates freely from the flesh and without any stain of red; season, early in September. [See page 464.]

TREE.—A moderately slow grower at the North, and somewhat liable to mildew when not in good ground. At the South it grows more vigorously, and does not mildew. The flowers are large, and the leaves serrated without glands. Originated in France.

REMARKS.—The Noblesse is one of the old varieties, whose good qualities have as yet been unsurpassed by any of recent origin. It is of the richest and highest flavor, and being entirely white at the stone, is quite desirable for canning or preserving.

XI. Nectarines.

NECTARINES belong to the peach tribe, but have smooth skins. They are seldom raised, on account of the delicacy of the trees, and the ravages of the curculio on the fruit, the depredations of this insect being fully as fatal as with the finer varieties of plums. The apricot, on the other hand, is allied to the plum, and like the nectarine, but little cultivated. In the South they are grown to a limited extent, but so far, California seems to be the home of these two delicious fruits. Their cultivation is identical with that of the peach. We give an illustration of the nectarine, Elruge. with description.

Elruge.—Synonyms: Common Elruge, Anderson's, Oatland's, Claremont, Temple, Spring Grove, Peterborough, (incorrectly.)

FRUIT.—Size, medium to large; form, roundish, inclining to oval; suture deepest toward the apex; skin, smooth, of a pale greenish ground, becoming, when well ripened in the sun, nearly covered with a deep violet or blood red, distinctly dotted with minute brownish specks; flesh, greenish-white, slightly stained with pale red next the stone, from which it separates freely, very juicy, melting, rich, and high flavored; stone, medium size, oval, slightly pointed, quite rough and of a pale color; season, early in September.

TREE.—The tree is a vigorous, hardy and healthy grower, with crenated leaves, having uniform glands; flowers, small, and of a pale, dull red; of English origin.

REMARKS.—The nectarine is one of the choicest of our stone fruits, and the trees are as easily grown and more hardy than the peach, while to insure the crop of fruit no more care is requisite than to insure that of the plum—the curculio being the only obstacle to success. The variety figured and described here is one of the very best and hardiest. [See page 465.]

XII. The Cherry.

THERE is no fruit more easily cultivated than the cherry, and none more liable to disaster than the sweet varieties. Its great enemies in the West are black knot and the curculio. Do not attempt to raise any cherries but Early Richmond (Early May) and Late Morello, except in those districts where the sweet varieties are healthy. Michigan and some portions of Ohio are the only States in the West where sweet cherries are generally a success. Every farm should have a hundred trees of Early

Richmond and Late Morello. The latter for the reason that they often give a crop when the first fails, and they are about two weeks later. The illustration shows the Late Morello at the bottom, and Early Richmond at the top.

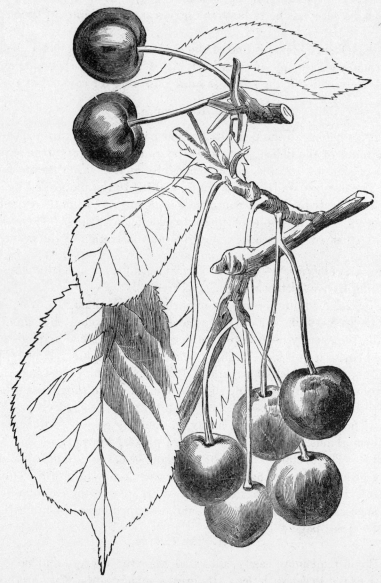

EARLY RICHMOND AND LATE MORELLO CHERRIES.

As we go East and especially South the sweet varieties may be more freely cultivated. East of the Alleghenies their cultivation is general. The inference then is that the chief difficulties in the West are too hot and dry summers, and too cold and dry winters. This is borne out by the fact that Michigan is congenial to the sweet cherry; its climate is moist, comparatively cool in summer and mild in winter. When the cherry is raised for family use, we should bud on Mahaleb stocks, since there are no suckers and the fruit is larger, But on Morello the bearing is more profuse and the tree comes into bearing earlier. These reputable Northern varieties of sweet cherries are illustrated and described below.

Black Eagle.—FRUIT: Size, above medium to large; form, obtuse, heart-shaped; surface, smooth, even, regular; color, reddish purple, becoming nearly black

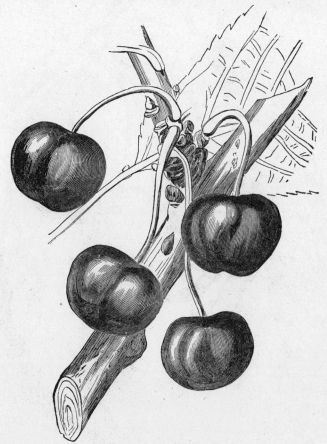

BLACK EAGLE CHERRY.

at maturity; stone, medium length, rather slender, inserted in a round regular basin; flesh, deep purple, almost or quite tender, with a rich high flavored juice, superior to any other black cherry, except Black Hawk; season, early in July.

TREE.—A short jointed, stout, strong grower, with large leaves, producing only moderately while young, but abundantly when the trees have acquired some age. The fruit is borne in pairs and threes. It is an English variety, originated by the daughter of Mr. Knight, in 1806, from seed of the Bigarreau, fertilized by the May Duke.

REMARKS.—This is one of the richest in quality of all the sweet cherries, and also one of the most hardy trees. Its unproductiveness while young has almost

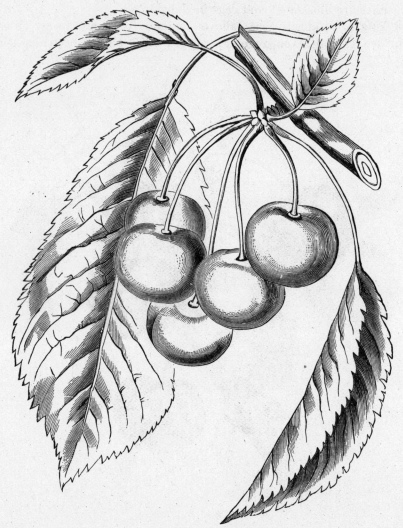

GOVERNOR WOOD CHERRY.

thrown it out of cultivation, but it is a variety that should be retained, and one or more trees planted in every orchard for family use, where sweet cherries will thrive.

Governor Wood.—This cherry is large, roundish, heart-shaped; color, rich light yellow, mottled or marbled with a beautiful carmine, that when fully ripe and exposed to the sun becomes a clear rich red; it has a suture half round, followed on the opposite side by a dark line; flesh, light pale yellow with radiating lines, half

KNIGHT'S EARLY BLACK CHERRY.

tender, juicy, sweet, with a rich, high flavor; pit, roundish, ovate, considerably ribbed; stem, varying in length and size; season, middle of June.

TREE.—A vigorous, healthy, strong grower, forming a round regular head, very productive; flowers, large; foliage, abundant.

This cherry, originated by the late Professor Kirtland, of Ohio, has been very generally distributed and fruited all over the United States, and also by some of the best pomologists in France and England, and everywhere the testimony is that it ranks among the very highest in every particular.

Knight's Early Black.—FRUIT: Size, medium or rather above; form, obtuse heart-shape; surface, a little uneven; suture, broad, open, half round, with a knobby projection opposite; color, purplish red, becoming nearly quite black when fully ripe; stem, stout to medium, inserted in a deep round basin; flesh separates freely from the pit, is tender, juicy, rich and sweet; pit, medium; season, last of June.

TREE.—A stocky, strong grower, with short jointed wood, oblong leaves, and flowers of middle size. Originated by Mr. Knight, of England, in 1810. For amateur garden culture it is one of the finest, making a tree of only moderate size. [See page 471.]

XIII. Picking and Packing Orchard Fruits.

BEFORE leaving the subject of orchard fruits, something should be said about picking. Never shake the harder fruits from the tree, unless they are intended simply for their juice. Pick by hand, in smooth baskets, and handle without bruising until they are in the packages properly closed for market. Then they will remain intact. If barreled, press in the head so it will squeeze down hard upon the first layer of apples or pears. Although this may indent, it will not rot the fruit. It is shaking about in the package that destroys fruit. In picking, provide yourself with a proper ladder. The form shown in the illustration is the proper one, and it also makes a good step-ladder for a variety of purposes.

FRUIT-PICKING LADDER.

XIV. The Small Fruits.

Blackberries.—These should be planted six feet apart between the rows, by three and a half feet in the row, and cut off when the canes are four feet high.

Raspberries.—These are planted the same distance apart between rows as blackberries, by three feet in the row. They are cut off at a height of three feet. Currants and gooseberries are planted four feet apart between rows, by about three feet in the row. The cultivation should be clean.

Strawberries.—The strawberry is universally cultivated. No farm-garden should be without them. Any land rich enough to bring forty to fifty bushels of corn per acre, under good cultivation, will do. The ground should be plowed deeply and thoroughly well pulverized. Mark the land, if for field culture, the distance as for corn. If for garden cultivation, the rows may be three feet apart. For field culture, the land may be marked both ways, and one good plant placed at each intersection, spreading the roots naturally, placing the plants so the crowns will not be above the surface, giving a little water to the roots if the soil be not fairly moist, and after the water has settled away, drawing the dry earth over all. For garden culture, one plant to three feet of space will be sufficient, unless the plants are to be raised in stools, and the runners kept cut out, when a plant to each two feet will be about right, if you want extra large berries.

The cultivation is simple. The spaces between the rows, about two feet wide, may be kept clean with the cultivator. In the rows the weeds may be kept, early in the season, clean with the cultivator; later, when the runners have encroached on the rows, the weeds must be pulled out, if necessary, but on fairly clean soil the cultivation will not be difficult. Beds of the previous year, and which should be in full fruit this season, may be kept clean between the rows with the cultivator. The weeds will not trouble much until the crop is gathered.

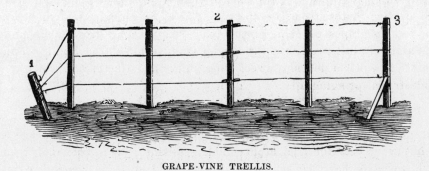

GRAPE-VINE TRELLIS.

XV. The Vineyard.

THERE is no reason why any farmer in the West, up to the line of forty-four degrees, if he have good corn land, should be without plenty of Concord grapes in their season. Farther South, other and sweeter varieties may be grown. We should rather say, the more delicate varieties, for the Concord grape contains a full average of sugar mixed with its acid. There are four principal species of American grapes in cultivation, which, with their crosses, constitute the large number of varieties for out-door cultivation. In California, the European varieties are also grown.

The American species are, 1, *Vitis labrusca*, of which the Concord is a type; 2, *V. æstivalis*, of which the varieties Devereaux and Elsinburgh are examples; 3, *V. vulpina*, of which the Scuppernong of the South may be accepted as a type, and 4, *Vitis cordifolia*, of which Clinton is a well-known example. The Labrusca has the greatest number of varieties in general cultivation North; Æstivalis is in repute in the Atlantic States for wine, Catawba being a well-known variety; Vulpina contains a number of reputable varieties at the South; Cordifolia represents the most healthy class of northern grapes, the leaves not being subject to mildew, or the fruit to rot. It is sour, but makes an excellent acid wine. Vulpina is not hardy north of Virginia, but is especially adapted to a low, warm country. It is deficient in sugar, but when galized, by the addition of sugar to the juice, gives a wine rich and of unequaled perfume. *Vitis vinifera* is the European species, but is little cultivated outside of glass structures, except in California.

If it is decided to train on trellises, the cut will show a simple form. 1, is a firmly fixed slanting post, to sustain the tension; 3, is the end-post braced. The wires are stapled on, except at 2, where they pass through holes in the post. They are tightened by any usual means of tension. See cut on page 473.

Within the past twelve or fifteen years, the grape has taken its place among the fruits regularly cultivated by farmers and gardeners in this country, and bought for daily use by all classes in our principal markets. The hot-house standard has ceased to be the test as to what constitutes a palatable grape, and nearly every section produces, in abundance, the varieties of this wholesome and delicious fruit. The firm-pulped, thin-skinned grapes of California, now found upon the fruit stands of every city in the United States, from Boston to New Orleans, already rival in delicacy and flavor the famous grapes of Malaga; the juicy scuppernong grows abundantly, and with comparatively little care, upon the light sandy soils of the South Atlantic and Gulf coast. It yields a delicious sweet wine, in the manufacture of which a considerable industry is springing up.

XVI. The Grapes for Farmers.

THE grapes for the farmer's garden are not the new and untried varieties brought out every year for trial at five dollars a vine. It is true that from these successive varieties have come the well-established sorts in general repute. If experiments are interesting, make them by all means, but stick to well-established varieties until you have found a better. Our choice would be that Concord should have a place everywhere. It is a good grape, north, south, east and west. Then decide as to the other varieties to fill up the complement of the vineyard, earlier and later. Many of Rodgers' hybrids are worthy of trial. If you are at a loss, ask the advice of some practical cultivator near you. In the North, especially north of forty degrees, we should plant principally of Concord, with Delaware and Clinton to fill up the vineyard. In the cultivation, avoid close summer pruning. As to soil, land that will produce forty bushels of corn per acre, will give good crops of grapes if the subsoil is not wet.

Cultivation of the Vine.—In the cultivation of the grape avoid close pruning in summer, pinch the side growth of the current year to about two buds on each spur, and the vine being in fruit, prune none at all after the middle of July, except to clip off superabundant growth. This may be done with a corn-knife. Avoid, also, all fancy training. Close pruning and fancy training, advocated by so many theoretical writers, has done more to suppress the cultivation of the vine than the want of superior varieties. For ourselves we gave up, many years ago, the trellis for simple stakes, either bowing the vine or twisting and tying it around the stake. Our plan with young vines is to set one year old plants 8x8 feet for the stronger growing varieties, and 6x6 feet for the weaker, as, for instance, Delaware.

The first season we simply tie the vine to a slender stake, cutting back in the fall to three eyes. The next spring we rub out two of these eyes, reserving the strongest shoot. This is tied to the permanent stake, which may be three or four inches in diameter and six feet high, although five feet is enough.

When the vine has reached a height of six feet it is pinched off at that height, the laterals as they put out are pinched off, beyond the first bud. When this bud makes growth to the extent of one bud, it is stopped again beyond that bud. It is sometimes (generally) pinched back once more. This leaves a succession of three buds, for fruiting the next year, on every spur.

Covering the Vines.—We believe in laying down the vines and covering them with a little earth each winter, in the North. It gives better fruit and saves occasional winter-killing. Vines six or seven feet long and studded with fruit, three bunches for each spur, and planted 8x8 will give tons of fruit per acre in good seasons. The aim of the cultivator is to keep the vine going by encouraging new fruit spurs each year that the bearing may be continued for years.

VINE OF TWO CANES TRAINED TO STAKE.

The time will come when the vine must be cut down at such a point as to induce the formation of a bud near the ground, and upon which to form a new bearing vine. This may be so arranged that about one-quarter, or up to a half, may be cut back each year; or two shoots may be reared from each vine, one for fruit, to be cut away in the fall, and the other to succeed it the next year. Our own plan, however, is to allow only one strong shoot, and renew the whole vine when necessary. Thus we have always got more and better fruit from the same area.

Many persons, however, are wedded to the trellis system. It is not to be denied that it has some advantages. Hence we give a cut of a section of trellis and vine. D D, are the canes of the first year's growth; b b b, are fruited canes of last year cut away; a a a, are shoots of last year pinched back to three buds to each spur (as we have recommended), and form the fruiting canes of the current year.

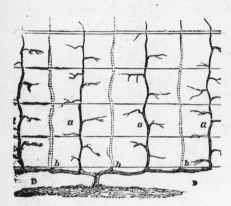

TRELLIS AND VINE, SHOWING ALTERNATE OR RENEWAL SYSTEM.

Terminology of the Vine.—For the reason that there is confusion relating to nomenclature in the several parts of the vine, Mr. J. E. Starr, before the Illinois Horticultural Society, stated the terminology of the vine as follows:

1. The ROOT—that part growing under the ground.
2. The STEM—that part between the root and the first departure.
3. The CANE—wood of last season's growth—prepared for fruiting.
4. SHOOTS—wood of the current season's growth, growing from the stem to the spur.
5. LATERALS—wood growing out from the canes, sometimes grown from shoots.
6. SUB-LATERALS—wood growing from the laterals.
7. SPURS—wood cut back for the production of new wood.
8. ARMS—canes laid down for fruiting.
9. PERMANENT ARMS—canes trained along the trellis permanently, for producing either fruit or wood.
10. LEAVES.
11. CLUSTERS.
12. TENDRILS.
13. JOINTS—spaces between the buds.

The above is the generally accepted definition. The late Dr. Hull, one of the most scientific cultivators in the West, used the following terminology. Either expresses the sense perfectly:

1. A YOUNG CANE—a shoot of the current year's growth.
2. A CANE—the ripened wood of the young cane.
3. BRANCH OR STEM—the cane which produces fruit.
4. FRUIT SHOOTS—wood from the branches that produces the fruit.
5. SUB-SHOOTS—wood (shoots) from the fruit shoots. But new canes and lateral canes may be called CANES.
6. SPURS—canes cut back for producing fruit or new wood.
7. DORMANT BUDS—those which do not grow till the second year.

In other particulars there is no essential difference between Dr. Hull's and Mr. Starr's nomenclature.

XVII. Cultivation of the Cranberry.

In many of the sandy marsh districts bordering the Atlantic and the great lakes of the West, cranberry culture has assumed vast proportions, and the berries are now largely exported to Europe. Any farmer who has a piece of land that can be flooded or made dry at will, may cultivate this fruit. The water supply is absolutely necessary for flooding, to preserve the marshes in winter, but more especially for killing insect pests, which otherwise will surely destroy the crop. Many might utilize natural sandy marshes with large profit.

The best soil is muck with a coating of sand on top. Clay and loam soils will not answer. Hence never plant cranberries on a drift formation, and the sand should be sharp (a silicious sand).

When a situation has been selected for a cranberry bog, the first thing to be done is to level it. A levelling instrument is not necessary. All that is required is a strip or plank ten or fifteen feet in length, the edges jointed and made exactly parallel; with this and a common carpenter's level the work may be quickly done. Stakes of a foot or more in length, cut off square at the top, should be provided. Begin by driving one of the stakes so that the top will correspond with the supposed surface of the bog when completed. With this as the standing point, run several lines of stakes through and across the bog. If more convenient, the tops of the stakes may be elevated six or more inches above the proposed level. This operation is important because stakes show where material is to be removed, and where filling is to be required; and by making a little calculation the earth to be removed may be made to exactly correspond with the amount required for filling. But this is not the principal advantage; it requires much less water to flow a bog that has a level surface than one that is uneven.

If the bog is extensive, and cannot, without too much expense, be reduced to one common level, there is no objection to having different grades with low dykes between them. It is said that in building railroads nothing is ever lost by spending much time in engineering. This remark has force and truth in it when applied to cranberry bogs. The money and time spent in laying out the work to be done is always economically expended. All that is to be done, and how it is to be done, should be known before work is commenced. In many bogs it would be economical to employ an experienced engineer, and have marked stakes put up, profiles and working plans drawn. With such marked stakes and drawings, the workman knows when he has filled his barrow, where he is to tip the contents. There will be no mistakes, no alterations to be made, and in the end money will be saved.

The depth of sand required to be spread on the surface depends upon the depth of the peat. If the latter is only a foot or two in thickness, five inches of sand is considered sufficient; if it is several feet, at least a foot of sand is required to make a good bog. The more sand there is used, the longer it requires to bring the vines into a bearing state; but when brought into that state they continue to bear for many years.

Planting.—The planting is generally done in the spring, by covering pieces of the vine, say three inches long, in the soil, about two inches deep, eighteen inches apart, three pieces in a place. A better way on prepared soil, would be to open

SHORT VINE, CULTIVATED CRANBERRY.

narrow furrows, two feet apart, and strew the vines, cut into sections in the cutting-box, rather thickly therein and covering lightly. If in planting in this manner, care is taken to leave out one end of the vine, the best means will have been employed.

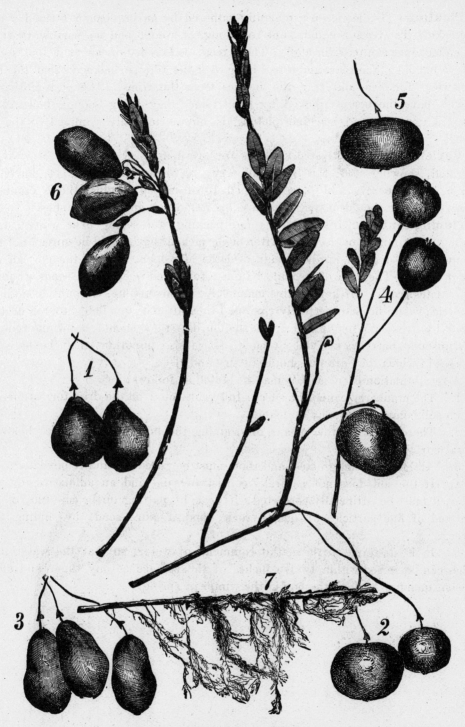

VARIETIES OF THE CRANBERRY

Picking.—The berries are generally gathered by an implement termed a cranberry rake. The teeth are about one foot long, of wood, and so narrow that only the slender vines can pass through. These teeth end in a box about a foot square, having a handle at the rear, and a bail hung over the top, so balanced that the teeth may sweep forward and tear the berries from the vines. The best cultivators, however, have the berries picked by hand, since they bring enough more to pay the extra expense, and machine picking is now principally confined to the wild marshes.

Varieties.—The cultivated varieties are now many. Some of the best we have illustrated, showing at 1, the Bell cranberry; 2, Cherry; 3, Bugle; 4, Early Red Bell; 5, Cheeseberry, and 6, a modified form of the original Bugle cranberry. 7, shows the vine rooted in spagnum moss, as found in many wild marshes.

Curing and Packing.—Curing and packing, for saving over winter, or for shipping, must not be neglected. After being picked they should be spread not more than five inches deep in hurdles made of laths left open, so that the air can draw through them. These hurdles should be piled for three weeks in a room where the air circulates freely. If kept in this manner for a longer time, it would do no harm; the berries would become perfectly ripe, and thereafter be less liable to rot or to be injured by frost. When taken from the hurdles they should be winnowed, and every unsound berry picked out by hand. It is also important that the barrels or packages in which they are put should be dry and clean.

A recapitulation of essentials may be stated as follows:

1. The cranberry cannot be successfully cultivated on the drift formation—that is, on a soil composed of clay or loam.

2. There must be the means of draining the bog eighteen inches below the general surface.

3. All bushes, wild grasses, and roots must be pared off and removed.

4. If the soil does not naturally consist of sand and an admixture of peaty matter, it must be artificially corrected; if peat, by putting on beach-sand, or sand composed of fine particles of quartz rock; and if pure sand, by adding peaty matter.

5. It is desirable to have the command of water, so that the water in the ditches can be raised within twelve inches of the surface at any time, and also in sufficient quantity to flood the bog in the winter or spring.

CHAPTER II.

GRAFTING AND BUDDING.

I. GRAFTS, CUTTINGS AND SEEDLINGS.—II. THE GRAFTER'S ART.—III. HOW TO GRAFT.—IV. TOOLS FOR GRAFTING.—V. GRAFTING BY APPROACH.—VI. GRAFTING OLD ORCHARDS.—VII. CUTTING AND SAVING SCIONS.—VIII. GRAFTING WAX.—IX. BUDDING.—X. WHEN TO BUD.—XI. HOW TO PREPARE THE BUDS.—XII. HOW TO BUD.—XIII. SPRING BUDDING.—XIV. TIME TO CUT SCIONS.—XV. GRAFTING THE GRAPE.

I. Grafts, Cuttings and Seedlings.

THE object of budding and grafting may be briefly stated. If the seed of a fruit be planted, the tree or shrub growing from that seed will not bear fruit like that from which the seed was taken. If, then, you wish to grow a certain choice apple, peach, pear or other fruit, it is useless to keep the seed or "stone" of that particular fruit and plant it. If it be an apple seed that you plant, the tree growing from it would, certainly, produce apples; but they would, almost certainly, be of quite another variety, and, perhaps, of a very inferior quality. It is, in fact, by thus planting seeds and growing what are termed seedling-trees, or "seedlings," that new varieties are produced. Most are valueless—one in several thousand may, by some chance, produce a new and, perhaps, splendid variety of the fruit. It is very difficult to make fruit-tree cuttings (that is, branches cut off and put in the ground) grow. If you cut a branch from a willow and put the cut end into the earth it will at once form roots and become a tree; a branch from a fruit-tree so treated dies.

If, then, you have a tree bearing a certain choice variety of fruit, you may cause the reproduction of the same choice variety thus: cut from the tree bearing the good fruit "grafts" or "buds;" then take certain branches, or the main stem, of a valueless tree of the same species and, having destroyed its natural shoots, "bud" or "graft," as hereafter described, with your cuttings. By some law of nature the sap which comes up the stem of that tree will, on entering the new wood made by the portion grafted, produce fruit exactly like the tree from which the cutting has been taken. Thus, with a stem of the same species, but of a different variety, you may produce a tree all the top and branches of which will yearly give you a fruit unlike what the original tree grew, but like to that grown upon the one from which your shoots came. Note—the tree grafted and the one from which the graft is taken must be of the same species. You cannot graft an oak with an apple, an apple tree with a peach, a plum tree with a pear; but any variety of the same species may be grafted with another.

II. The Grafter's Art.

Any boy or girl on a farm may easily learn to graft and bud. Except in the more unusual kinds of grafting, that art is exceedingly simple, and budding requires only nicety and care. Grafting is uniting a portion of a shoot (scion), containing one or more buds, upon a "stock" or a root with a view to their union, and subsequent growth. If varieties came true from seed, grafting and budding would be less important than they now are. If fruits could be readily propagated from cuttings, there would be little use for grafting or budding, but they do not. Hence, grafting and budding will always be necessary.

Grafting on the Farm.—On the farm, it will yearly be desirable, there being but few orchards that will not require some change of varieties. If a tree, bush or vine proves barren or long in coming into bearing, it may be made to fruit, by grafting on it some earlier bearing variety. An unprofitable grape vine may be root-grafted with a better sort. Stone fruits may be budded to varieties better adapted to the climate and situation. Young seedlings are to be raised and grafted or budded as required; grafting being usually employed for the apple and cherry, and budding for the other orchard fruits. The peach, pear and plum should be budded, though all the fruits may be grown by grafting, and the plum is, perhaps, as often thus propagated as by budding.

III. How to Graft.

The usual modes of procedure are by "cleft grafting" and by "saddle grafting"—the latter being little used except where the "stock" and "graft" are of nearly the same size. The cut will explain the manner of fitting this latter graft; No. 1 showing the two parts prepared, and No. 2 the parts accurately fitted. The whole art is to so fit the parts that the liber or inner barks come naturally together, for this insures the passage from one to the other of the "cambium," that is to say the gummy fluid between the sap-wood and bark, from which both wood and bark are formed. To ensure this the graft, when inserted as hereafter shown, is generally slightly crossed with the stock by pointing the top of the graft somewhat inwards, by which the union of the two is at some point made certain. The whole is then covered with "grafting wax" to exclude moisture and air. If the grafting be done in the spring, between the rising of the sap and the putting out of the leaves, a proper union will soon take place.

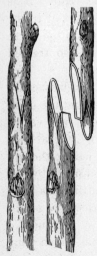

No. 2.   No. 1.
SADDLE GRAFTING.

Root Grafting.—This is performed in precisely the same manner as stock grafting; it being simply the proper union of the "scion" upon a piece of root, say six inches long, and preferably, that portion of the root of the year-old seedling next the crown of the plant. Root grafting is usually performed late in winter. It is not necessary to describe the performance here, since it does not pertain to the farm, but is a part of nursery work now generally done with machines. Hence the root grafts may be bought of nurserymen cheaper than the work can be done on the farm by hand.

IV. Tools for Grafting.

The tools necessary for grafting on the farm, are a sharp panel saw, a keen pocket knife for paring the stocks and sharpening the grafts, a grafting chisel or a butcher knife, and a mallet for splitting the stocks, a wedge for holding the split stock apart while placing the graft, and grafting wax for spreading over the cut surfaces. A grafting chisel and wedge combined is shown on page 484, which any blacksmith can make, but in our own grafting and budding we always use a knife like that shown in the illustration of a Budding, Pruning and Grafting Knife, the hooked, open blade of which makes a clean split without tearing the bark; an important thing to be remembered. The half open blade shows the proper form for use in budding, and is also the best shape for use in spaying and gelding animals. If no better means are at hand, the point of an old scythe, cut off and ground sharp, will serve to split the stock, and is of the right form to prevent tearing the bark.

Processes Illustrated.—In grafting, the illustrations we give will show the whole art better than words. Fig. 1 shows a portion of the scion properly sharpened; Fig. 2, the stock split and the grafts, of three buds each, properly placed; Fig, 3 shows the end section of the stock and the manner of placing the scion, so the inner bark, or liber, of each may accurately meet. The manner of pointing the scion in at the top is shown by the projection at the bottom of the scion in Fig. 2. [See next page.]

Waxing.—Nothing remains to be described except waxing. The wax must be soft enough easily to cover all the parts, and care must be taken that the scions are not moved from their places. Not one graft in ten need miss, but to ensure one, and also hold the cleft secure, two are usually inserted, as shown. If both grow, one should be cut away.

V. Grafting by Approach.

Sometimes it is necessary to graft "by approach." For instance, a tree may

BUDDING, PRUNING AND GRAFTING KNIFE.

have been so gnawed by mice or rabbits in the winter as to prevent the rise of sap. In this case, when the tree cannot be immediately mounded up above the wounded part, select in the spring branches, pared and shown in the cut, Fig. 1, and fit in well at top and bottom so there will be pressure, nail fast with slender nails, cover the whole with cow manure and clay, tempered together, and secure with strong muslin firmly wrapped around the whole and then fastened. The cut, Fig. 2, shows another plans of grafting by approach that may be used in any tree, and is often used in working the grape. A shows the branch to be grafted on; B a section of the stock to be grafted; C the manner of paring the scion; D the bark of the stock opened; E F the parts united and bound, and their relative positions.

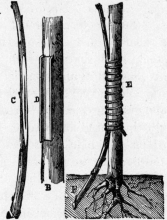

Fig. 1.
GRAFTING BY APPROACH.

Fig. 2.
GRAFTING BY APPROACH.

VI. Grafting Old Orchards.

An old orchard of unprofitable varieties may be grafted with superior fruits. Only one-third of any such tree should, however, be grafted in one season. Thus it will take three seasons to complete the operation. Cutting off and grafting the top

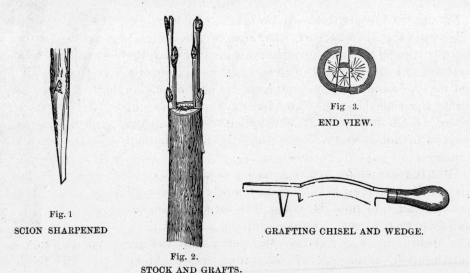

Fig. 1
SCION SHARPENED

Fig. 2.
STOCK AND GRAFTS.

Fig 3.
END VIEW.

GRAFTING CHISEL AND WEDGE.

first gives the grafts there the best possible chance, while the necessary reduction of the top throws the sap into the remaining side branches, which fits them well for grafting the following year; and, the third year, the lowest branches, being made ready in the same way, may be grafted successfully. By this mode it will be seen that, when the grafts are put in on the side branches, they are not shaded by the heavy shoots above them, and they have an unusual supply of nourishment to carry them forward. Those who have attempted to graft the whole head of a large tree at once are best aware of the great difficulty in the common mode of getting the grafts to take on the side limbs.

VII. Cutting and Saving Scions.

SCIONS may be taken at any time in winter when the trees are not frozen. Select healthy twigs, of the current year's growth, with a terminal bud to each; let the sticks of your scions be of nearly equal lengths; tie with three bands into small bundles, not more than three inches through. Correctly label each bundle according to the variety. This should be done with a tag, wired on, and also by sticking a sharpened slip of wood in the top of each bundle, with the name plainly marked. You may thus easily know the variety contained in each when the bundles are packed away. Set these bundles in moist sand, moist sawdust, or in moist (not wet) moss. Place in a cellar that will not freeze, and one that is secure from mice. The cuttings will then keep in good order until wanted. Scions of the peach or plum should be cut in autumn, since these trees are liable to be injured by severe weather.

VIII. Grafting-Wax.

ALMOST every professional grafter has his own formula for making grafting-wax. Many use, instead, a mixture of blue clay and fresh cow-dung, kneaded and beaten until it will work like putty. This really forms one of the best applications for grafting, in a small way, upon nursery stocks; a ball of the mixture being formed all around the mutilated stalk and graft. Grafting-wax is simply a compound of rosin, tallow and beeswax, in such proportions as to admit of being easily applied when softened by warmth, but not liable to melt and run in the sun's rays. A good grafting-wax is made of three parts of rosin, three of beeswax, and two of tallow. A cheaper composition, but liable to adhere to the hands, is made of four parts of rosin, two of tallow, and one of beeswax. One of the best and cheapest consists of one pint of linseed oil, six pounds of rosin, and one pound of beeswax. These ingredients, after being melted and mixed together, may be applied, when just warm enough to run, by means of a brush; or may be spread thickly with a brush over sheets of muslin, or thin, tough paper (manilla tissue paper), which are afterwards, during a cold day, cut up into plasters of convenient size for applying; or, the wax, after cold, may be worked up with wet hands, and drawn out into thin strips or ribbons of wax, and wrapped closely around the inserted graft. This is the better way on the farm, and in all cases the wax should be closely pressed so as to fit closely to every part, and leave no interstices, since it is indispensable that every portion of the wound on the stock and graft be excluded from the air.

IX. Budding.

In relation to budding, the late Dr. Warder, in "American Pomology," wrote: "It has been claimed in behalf of the process of budding, that trees, which have been worked in this method, are more hardy and better able to resist the severity of winter than others of the same varieties, which have been grafted in the root or collar, and also that budded trees come sooner into bearing. Their general hardiness will probably not be at all affected by their manner of propagation; except, perhaps, where there may happen to be a marked difference in the habit of the stock, such, for instance, as maturity early in the season, which would have a tendency to check the late growth of the scion placed upon it—the supplies of sap being diminished, instead of continuing to flow into the graft, as it would do from the roots of the cutting or root-graft of a variety which was inclined to make a late autumnal growth. Practically, however, this does not have much weight, nor can we know, in a lot of seedling stocks, which will be the late feeders, and which will go into an early summer rest."

X. When to Bud.

The time for budding is before the tree has perfected its terminal bud, or during that season when the bark may be separated from the wood. The late F. R. Elliott, in an essay before the Ohio State Agricultural Society, sums up the whole matter concisely. He says:

"The time for insertion of buds into the stock for the purpose of changing the kind of fruit, varies with the habit and character of both the tree to be propagated and the stock on which it is to be worked. All buds, in order to be successful, must be well ripened—that is, the tree on which they have formed, must have made its terminal bud, or, in other words, the growth of the shoots must present a continuation of perfect formed leaves to its point. This ripening of buds occurs earlier in some varieties than in others; usually early summer fruits ripen their buds earlier than winter sorts. Next, the stock in which the bud is to be inserted must be in a vigorous, healthy condition, but apparently about to close its season's growth. Here again comes the necessity before alluded to, of selecting in the seed-bed the different habits of the young plants relative to early or late maturity. Through our northern middle States, the usual time to commence budding the apple and pear is about the 10th to 15th of August. Further south they are in condition in June; and so on, all the intermediate time, according to latitude and season; some seasons being earlier than others by six to ten days.

"Such stocks as grow late in the season, should be budded late, because as new layers of wood are constantly forming with every bud of extension in growth, it follows, if the bud is inserted too soon, it must be covered and destroyed. On the other hand, if the bud be inserted too late in the season, the cambium has acquired consistency, the ripened flow of sap is checked, and the bud having no powers in unison, dies. The quince, therefore, from its habit of growing very late in the season, should be the last to bud.

XI. How to Prepare the Buds.

"If it is necessary, in order to have the buds ready to meet the growth of the stock, that the scion or branch from which buds are to be taken should be made to hasten its maturing of the buds, then pinch off the end of the shoot one or two buds. In from eight to twelve days the remaining buds will have ripened and fitted themselves for forming either new plants or branches. If this pinching is done early in the season, and the branch left to remain on the tree, the result is, the buds, after ripening, send out new branches, and make a sort of second growth. If scions have to be brought from a distance, or if it is desired to keep them several days, wrap them first in damp moss, or failing that, a damp cloth; then inclose that in paper, and the whole in oiled silk. Other material will answer, provided moisture and a cool temperature be kept around and next them.

"When cutting the scion, select healthy branches of medium size in growth, with full and perfect buds. Cut off the leaves, leaving the foot stalks as shown in Fig. 2, next page, as soon as you have taken the branch from the tree, as the leaves rapidly exhaust the matter in the bud when its connection is separated from the root and thus impair vitality.

Materials Necessary for Budding.—"Bands or strings for tying the bud in its place after setting, are requisite; these may be of bass matting, such as is used in wrapping sheet iron or furniture, and which may be also procured by getting the bark of our common basswood in the spring or early summer, and laying it awhile to soak in water, when the outer rind readily peels off, and the inner bark peels into thin, strong, flat strips, that tie easily when wet; woolen-yarn, cotton-wicking and many other materials are used. Anything thin, soft and strong will answer. Bass bark is the best, and should always be wet just before using. A knife (Fig. 1) with a thin blade, sometimes rounded at the point, and at the opposite end a wedge-shaped piece of smooth ivory or bone is used.

XII. How to Bud.

"Holding the stock in your left hand, and with your knife in the other, make first a perpendicular slit, two or three inches from the ground, on the north or east side of the tree, and about one inch long; then at the top of that slit, make a cross cut (Fig. 3); then with the ivory end of the knife raise the bark a little at each corner below the cross cut, being careful not to injure the cambium or new layer of soft matter just underneath. Next take the scion in hand, set in the

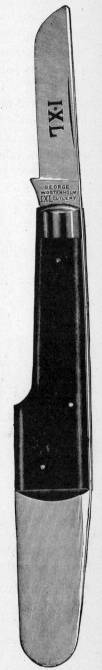

Fig. 1.
BUDDING KNIFE.

knife half an inch above a bud (Fig. 2), resting the thumb of the knife-hand on the scion just below the bud, and make a drawing cut just deep enough into the scion to cut out the bud and a very thin piece of wood with it (Fig. 4), bringing out the knife again half to three-quarters of an inch below the bud. This is the usual length, although the bark may be shorter or longer without injury or benefit. Now put the lower end of the bud into the top of the opening made in the stock, and slide it down, until the bud is a little below the cross cut, and then cut off the end at the cross cut. Next take your strip of bark or tieing, which should be (for ordinary sized stocks) about half a yard long, and commencing at the bottom, wind around with a steady, even, tight, but not hard strain, until you have covered the whole of the cut, and left nothing to view but the bud (Fig. 5); then tie, by holding the last turn around the stock or tree in one hand, winding again with the other, tucking the end under as it comes round and drawing all tight.

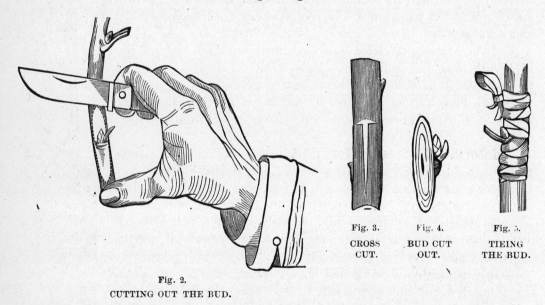

Fig. 3. CROSS CUT.　　Fig. 4. BUD CUT OUT.　　Fig. 5. TIEING THE BUD.

Fig. 2. CUTTING OUT THE BUD.

Terminal Budding.—"A variety of budding, called terminal budding, is sometimes practiced, in order to use the last bud on a shoot. It is performed in the same manner as above, only the tying has to be performed all below the bud.

XIII. Spring Budding.

"Although the summer is the time for most of budding operations, yet it sometimes happens that a new sort is obtained late in the spring, and, being valuable, it is desirable to make every bud become a tree. The scions are, therefore, kept until the trees have made leaves, when the buds are inserted in the usual manner, and as soon as they are united the top of the stock is bent over or cut off, and the sap being forced into the new set bud, it makes a good growth the same season.

"The distance from the ground at which a bud should be inserted is a matter of some disagreement among tree growers. In all dwarf trees, however, it is essential that it should be near the ground, on account of the necessity of burying all the stock in the soil when it is transplanted. Orchard trees are sometimes benefitted by being worked high up on the stem; as some of the best varieties appear to suffer injury when worked near the earth, and to do well when worked high on a hardy seedling stock. Budding may also be used to change the limbs or branches of a bearing tree, as any branch of not over half to three-quarters inch diameter will answer to bud if it is in a thrifty growing condition.

Care after Budding.—"In about a week or ten days after budding the buds should be looked over, and if any have failed they may yet, if still growing, be budded over.

Loosing Strings.—"In about two or three weeks the strings will require to be loosened. Some will have to be retied, as if not so done the rapid growth will break the strings before the bud becomes well united. Generally the strings may be all cut loose in from three to four weeks after setting the bud. This is done by passing a knife perpendicularly at the back of the bud or opposite side of the tree.

Protection of Buds in Winter.—"In some locations it is found advisable for the protection of the bud to earth up around and over them during the first winter. This is done in large nurseries by a careful plowman; in small grounds with the hoe or spade. This is not necessary except in very cold, bleak situations, or in soils that upheave badly in winter. During the winter the trees should be looked over, and record made of the number and kinds of buds that have failed.

XIV. Time to Cut Scions.

"At any time during the winter, or from the fall of the leaf until the swelling of buds in spring. Whenever the weather is mild scions may be cut for the purpose of engrafting. The best time is immediately after the fall of the leaf.

Selection and How to Keep.—"Select well-ripened, thrifty shoots, of medium sized wood, of the previous season's growth, from healthy trees. In cutting the scion leave one or two buds of the last year's growth on the tree. It injures the tree less. Scions cut at any time during the winter may be kept either by packing them in layers with saw-dust, moss or clean sand in any cool place, such as a cellar, or they may be buried in dry sand out-doors, and shielded from sun and rain by boards. They should always be kept moist, but never wet. It is better to have them too dry than too wet—as, if too dry, they can be buried in the earth and recovered; while, if too wet, there is often a premature swelling of the bud that destroys them, or they are injured by saturation."

XV. Grafting the Grape.

Grafting the grape does not differ, essentially, from other grafting. The ordinary cleft-grafting upon the growing root is usually performed. Wax is not used since, the graft inserted, the earth is then drawn around the graft, which protects it from air. The vine, however, must be grafted either in the winter, the early spring before the sap has started, or so late that, the leaves being out, will prevent flooding of the graft and its drowning from the bleeding of the root. Any one who has pruned the vine in spring will easily understand this.

Vines may be laid down and grafted as late as mid-summer. Our advice is: keep your grafts in a cool place until the vine is in full leaf; insert your scions in the root, but leave at least a portion of the original vine, or stock, growing; wait until your grafts have become securely to united the stock and have put forth leaves of their own; then cut away the old vine. The leaves having formed on the new part the sap will be, at least partly, absorbed by them, and the " bleeding " be less.

CHAPTER III.

VEGETABLE GARDENING.

I. ECONOMY OF THE GARDEN.—II. HOW ONE MAN BECAME A GARDENER.—III. STARTING A MARKET GARDEN.—IV. TROUGHS FOR FORCING PLANTS.—V. THE NUMBER OF PLANTS TO RAISE.—VI. THE HOT-BED.—VII. LAYING UP THE HOT-BED.—VIII. MARKET AND KITCHEN GARDENING.—IX. WATER AND VENTILATION.—X. HOW TO HAVE EARLY RHUBARB.—XI. "TAKE TIME BY THE FORELOCK."—XII. WHAT TO RAISE FOR MARKET.—XIII. ECONOMY IN CULTIVATION.—XIV. PREPARING VEGETABLES FOR MARKET.—XV. HOW TO RAISE POTATOES.—XVI. "PLANTING IN THE MOON."—XVII. POTATOES ILLUSTRATED.

I. Economy of the Garden.

STRANGE to say, the family table of nearly every other man of equal means is better supplied with vegetables than that of the farmer. So few, indeed, have good gardens, that the class may almost be said to do without fresh vegetable food. Why is this so? The majority with whom we have talked, have freely admitted their short-comings in this respect, but excused themselves by saying they

THE WEALTH OF THE GARDEN.

could not afford the time to "potter" in the garden. Here lies the principal difficulty. It is pottering work according to the old fashioned way of cultivating everything in narrow rows and small beds. Read over again, Chapter I, Section I, of this Part, then apply the same common sense here that you do in the cultivation of field crops, and you may raise half the food of the family on a single acre, at an average outlay of about forty dollars. The same labor and money applied to the corn field, will raise, say, six acres of corn. It will produce $200 worth of garden stuff, none too much for the average farmer's family.

Garden Cultivation.—Potatoes, early corn, okra, cabbage, early peas, summe squash, etc., may be grown in three-feet rows; late peas and tomatoes, in five-feet rows; muskmelons, cucumbers, etc., in six-feet rows; watermelons, in eight-feet rows, and squashes and pumpkins, in twelve-feet rows. Asparagus and pie plant should have four feet between the rows; beets, cauliflowers, early cabbage, carrots, parsnips and onions, two feet; and the smaller plants, such as the radish, lettuce, spinach, and all the so-called bedded plants, eighteen inches between the rows. All these garden vegetables, except the bedded plants, may be cultivated almost entirely with the horse and cultivator, the thinning being about the only work that need be done by hand. The whole cultivation of bedded plants, and the close cultivation, when young, of all others, except the gross growers, may be managed with the hand cultivator.

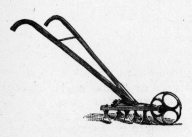

HAND CULTIVATOR.

II. How One Man Became a Gardener.

The following account of how one woman, to whom the writer gave advice, succeeded in having a good garden, will show that any one may do the same if the husband consents.

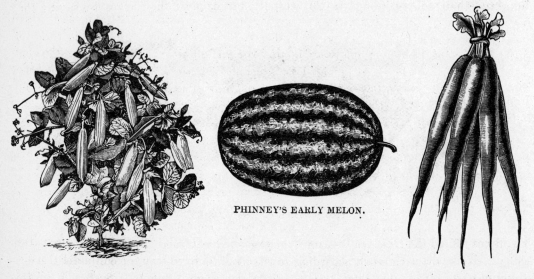

EARLY DWARF PEAS. PHINNEY'S EARLY MELON. LONG SCARLET RADISH

Mrs. Speedwell, the wife of a well-to-do farmer, near the largest city in the Mississippi valley, was lamenting to a neighbor, Mrs. Goodman, that her great trouble was the want of fresh fruits and vegetables, when the following colloquy ensued:

"Why do you not ask your husband to lay out a garden and have it taken care of?"

"Oh, it is of no use. My husband says it is cheaper to buy our vegetables. He cannot spare time nor have the bother; the consequence is the vegetables are never fresh, for I have to order them from the city. I really would not mind working in the garden myself if I only knew how; but, indeed, it is about all I can do, with the help of Jane, to take care of the flowers and my house."

"Oh," replied her friend, "it is easy enough. Do it by proxy. Buy a good practical book on fruit and gardening, and get your husband interested. Get him to have the garden plowed, harrowed, and nicely raked, and hire the hard work done under your own supervision. Any intelligent man can take care of a garden if he is properly directed."

"Yes, the men know enough—at least to keep out of the garden, but they dearly like good fruits and vegetables. I have often envied you your nice garden. Your husband really has a talent that way."

"Has he? Yes, that is true, but it is an acquired one."

"Indeed! How did it come about?"

"Well, I was once in the same strait that you are. At length I took the initiative. After coaxing and coaxing I induced my husband to have me a piece of ground manured, plowed and harrowed. By hook and by crook I had it cultivated. I planted strawberries and some other small fruits. We had plenty of vegetables, and the second year, strawberries, some raspberries and blackberries and a few currants. My husband never went near the garden, but I could see the satisfaction with which he enjoyed its products, and I noticed, also, that we never had any trouble with the hired men. In fact I get many an hour's work from them, and they an occasional nice dish in pay. The third spring I took my husband into the garden and lamented that it had really grown beyond my powers, and how sorry I was that the good living we had enjoyed must now cease."

"That would be a pity," he said, "but, really, it is of no use to set a hired man to work here, and I really have not the time to do it. Nor is it right that you should have the care and supervision of it. I suppose we must go back to the old way of getting stale things from the city."

"An apparently bright thought struck me. I say apparently, for, of course, I had the plan all laid out. If, said I, a good man could be spared me three days a week, during April, May and June, I think I could manage it."

"Oh," my husband replied, "hire a man constantly if you like," and, after going into a brown study for a few moments, he said, "Jenny, I think I'll learn gardening myself, and after this year, take the whole thing off your hands. That garden is worth more than any ten acres on the farm. And it really seems not to cost much beyond a little clever care."

"But then, every man is not like your husband nor mine. It is really surprising how so many farmers are content to live year after year without the comfort of a good garden, with its health-giving fruits, and wealth of vegetables, which should constitute more than half the living of the family."

MARTYNIA.

"That is true enough," replied Mrs. Speedwell, "though, perhaps, a great deal of the fault really lies with the wives, in not knowing how to manage their husbands, as you do. A man, of course, must have a human heart to be managed by kindness. I can, at all events, now see my way clear to have a good garden, and I think I see how I may get my beloved flowers taken care of also. By the way, I wish to speak for some of your Martynia again this year, for pickles. Next season I hope to have a garden of my own."

III. Starting a Market Garden.

The general principles relating to gardening will apply to either market or kitchen gardening, as raising of vegetables for family use is usually called. The young man who expects to succeed in raising vegetables for the market, whether exclusively in the open air or in connection with the use of hot beds, must be guided by business tact quite as much as in any other business. He must carefully look over the ground to inform himself as to what will pay best and how much of each variety will be needed. If facilities for shipping to large markets are good, the glass, (hot beds) may be increased in proportion, since not only will glass-grown lettuce and radish be in demand, but this increased area of glass will be needed for starting tomato, egg plant, cucumber, melons and summer squash. Crops raised entirely under glass had better be confined pretty much to lettuce, as there is always a demand for this salad plant, while that for radish, unless most carefully grown, is slow. The product is always inferior, while lettuce properly grown under glass sells better than that grown in the open air, and the improved means of transportation now allow lettuce to reach, in good condition, a destination 200 or 300 miles north of where it is grown.

The Local Market.—If, on the other hand, the market be local, the gardener must fully understand just how much of a certain article he can sell, and this must be found by inquiry. The hotel, or hotels of a village or city and the well-to-do inhabitants will furnish the customers for forced lettuce, radish, mint and rhubarb. These two latter must be provided for in advance. The second season will furnish mint roots for forcing, while at least three years will be required to furnish rhubarb for the same purpose. Local markets, however, will furnish customers for but very little mint, and not much glass-forced rhubarb, but it may consume a very considerable quantity forced in a warm and partially lighted cellar. Thus while but little forcing exclusively under glass will be required, enough must be provided to at least start such plants as will be required to be transplanted into hot beds, covered with muslin.

IV. Troughs for Forcing Plants.

Many things, in fact, as squash, melons, cucumbers, okra, Lima beans, sweet

SWEET POTATO AND VINE.

potatoes, etc., are better started over a gentle heat of fermenting manure, protected at night by a covering of boards, than in any other way. All except the last-named are planted in the frame, in troughs of boards filled with earth not earlier than ten days before they should be outside. They may thus be kept until the weather is permanently warm before transplanting, the vines, for instance, until they begin to show signs of running, and then may be transplanted safely, when all possible danger of frost or cold nights is passed, or say, when the same vegetation, planted outside, is making its first rough leaves. Thus you may have your plants fully hardened by exposure to the open air before transplanting, and at the same time be fully two weeks in advance of those grown entirely in the open air. These open air hot-beds are so important to all, and especially to farmers, who thus may force all their vegetables, that a full description of the process will be important.

Making the Trough.—The troughs for the plants are made by sawing ordinary cull fencing and siding into lengths of three feet each; nail a piece of siding against the edge of a piece of fencing and you have the half of a square trough. So proceed until you have enough of these half troughs to cover the surface desired. The front of one trough forms the back of another, and thus you have a series of sections running the whole length of the bed that will be each six inches wide. Each one of these three feet troughs will form six squares of six inches each, or thirty-six square inches in all; enough space for the largest transplantable plant. They may be loaded into a wagon as they were in the bed, and be safely driven long distances to the field, and, if carefully broken apart, even squashes and Lima beans may be thus transplanted and continue their growth without feeling another change than the chill incident to removal, and if they have been properly hardened (exposed to air in the bed), this will be slight.

The Frame.—The frame to protect the bed should be made of rear boards, say eighteen inches high by one foot for those at the south side, and with strips across from front to rear to properly stiffen them. Thus the bed may be made of any length by using 2x6 scantling as uprights where one board is joined to another, 2x4 scantling being sufficient for the end pieces, upon which to nail the end boards to connect the front and rear. Three-inch strips of inch boards nailed across the top from front to rear at intervals of six feet will fully strengthen the whole. Thus a bed six feet wide and sixteen feet long will contain sixty of the troughs we have mentioned, and each trough six plants, or 360 plants in all. Every farmer should prepare at least this number.

We illustrate some of the more important vegetables on next page.

V. The Number of Plants to Raise.

THE proportion may be as follows, but should be varied to suit the particular wants of the planter: Tomatoes, 150; egg plant 15; cabbage, 25; cauliflower, 25. All these should be transplanted into the troughs, from hot-bed green plants, and

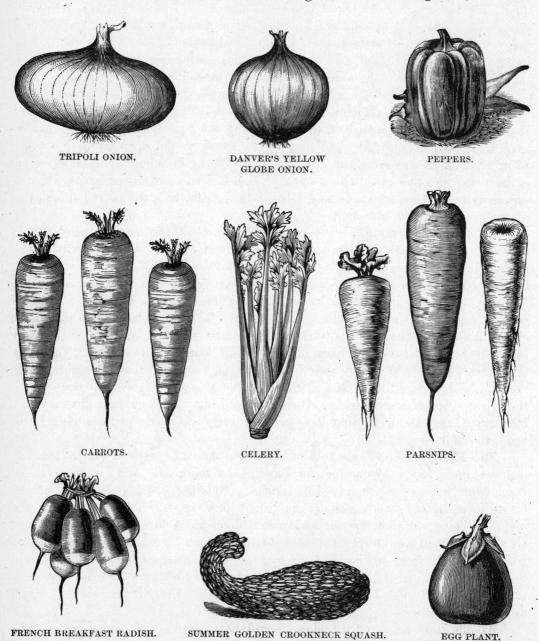

TRIPOLI ONION.   DANVER'S YELLOW GLOBE ONION.   PEPPERS.

CARROTS.   CELERY.   PARSNIPS.

FRENCH BREAKFAST RADISH.   SUMMER GOLDEN CROOKNECK SQUASH.   EGG PLANT.

form one compartment. The seeds of the other plants, summer squash, winter squash, water-melon, musk-melon, cucumber, of each 15; Lima beans, 50, and okra,

20; being planted in regular rotation. The reason for this is that the transplanted plants will need extra protection, even cabbage and cauliflower being better for it, and the two latter will be great, lusty plants when transferred to the open ground the middle of May, in latitude of, say 42 degrees. If these boxes are placed back in the frame, celery may be pricked in three inches apart and will furnish a nice lot of plants by the first of July.

VI. The Hot-bed.

The heating material—fermenting horse-manure—should be laid in the bottom of the bed, about fourteen inches thick, and an inch of earth put over all to form a level surface for the troughs, which must be filled with the best compost of one-half friable loam—sandy loam preferable—and one-half thoroughly rotted manure intimately mixed. Old hot-bed manure and earth laid in a heap the previous season and turned two or three times is the best. Now, if the farmer has, in addition to this, a hot-bed frame of four sashes, three feet four inches wide by six feet long, he may not only raise lettuce for his family, but also have space enough for tomato, egg-plant, cabbage, cauliflowers, kohl rabi, lettuce, celery, etc., for transplanting, either into the open frame we have mentioned, or directly into the open air, or both, for the frame will accommodate extra

THE FARM HOT-BED.

early samples, while the general crop of hardy plants may go safely into the open air as soon as danger of severe white frosts is over. But remember always, never transplant from the hot-bed into the open air without at least ten days hardening to the weather. Hence, for the farm garden, the necessity of the open frame, for, without it, the air required for cabbage, for instance, might be fatal to the tender plants.

Spring Forcing.—For spring forcing, a special hot-bed may be made up about the first of January, and the plants slowly grown until large enough to prick out, say about the middle of February. The plants should be pricked out, dibbled or fingered in two and a half inches each way, and with not less than two feet of bottom heat, green horse-manure, which has been turned often enough before putting into the bed not to heat violently. This must be taken from a heap sufficiently large not to have become chilled. If there is any suspicion that it will be slow, if dry it may be assisted by using hot water to moisten it. The management of the manure can only be learned by experience. Until this experience is gained, it is better not to put on the compost earth in which the plants are to grow, until satisfied that the heat is rising kindly. If a heat of sixty degrees is indicated inside the bed in the morning, it is just right. When one becomes expert the earth may be put on immediately after laying up the manure, and the bed be planted the second day, if still, and the outside temperature is right—that is, above the freezing point.

VII. Laying up the Hot-Bed.

THE best manure is that made by horses fully fed on grain, and bedded with hay or straw. Manure made with sawdust bedding is apt to heat violently and burn itself out quickly. If, however, it be tempered with tan bark, it makes an excellent heating material. The next best tempering material is the chaff of straw. If neither of these can be obtained, the manure must be tempered by turning often enough (every three days) to get the rank heat out of it, and cause it to heat slowly and equally. This may be assisted by pretty hard and evenly tramping the material when laying it up in the bed. Once the proper management of sawdust or shaving manure is learned, it makes an excellent material. In laying up the manure, choose a still day, as wind is pretty sure to destroy the heat. In preparing the manure for the bed, it is often necessary to add water, to bring it to a proper state of moisture, and one object in turning it several times in the heap is to promote a uniform state of moisture in the pile. If, however, water is added, it should be warm—unless the manure be very hot—and given from a fine rose water pot.

The Pressure.—All that is required in laying up is that the hard lumps, if any, be thrown out, and that the whole be laid equally, so it will settle evenly. The less firm the manure is laid, the faster and stronger it will heat. If there be too much pressure, it will heat too slowly. Hence, the amount of tramping is important. Our plan is to tramp the whole lightly, especially next the sides, three or four times in laying up a bed two feet thick, in addition to that employed with the fork in discovering if it is laid true. The manure all in, the surface is made perfectly smooth and even, and the finer manure that may not have been taken up on the fork, is then spread evenly over the surface, and six inches of the prepared earth is thrown over all. If the earth is frozen, such portions should lie around the edges of the frame, that they may thaw gradually and perfectly, before the whole is finally raked smooth.

The Soil for Hot-Beds.—The best soil for the bed is a rich, light, sandy loam, to which has been added one-third of its bulk of fully decomposed horse manure. That formed of rotted sods is the best earth material. A clay soil should never be used. The soil must be one that will, under no circumstances, become pasty. Once you have formed your soil, preserve it carefully from year to year in a conical heap when removed from the beds in summer, adding, from time to time, such fresh soil and hot-bed manure as may be needed. Six inches of soil is sufficient for forcing general crops, and eight inches for asparagus. If rhubarb is forced, the entire roots are to be bedded upon three inches of earth, as closely together as possible, afterwards filling in so the crowns are just covered, packing the earth well about the roots. If the earth settles away from the crowns, in three or four days, add more. Neither of these, however, generally pay for forcing nowadays by the market gardener; on the farm they may. They may be transported long distances from the South. When forced, a half lighted cellar, or the space under the staging of a green-house is utilized. The cellar is kept at about fifty degrees, or somewhat less. More than this average heat should not be given in the hot-bed.

Winter Forcing.—Double walls of the hot-bed for winter forcing should extend to the top upon which the glass is placed, and the tan bark filling between the walls should have a width of at least six inches. The glass should be carefully bedded in thin putty, carefully trimmed, and well glazed. All the points and fittings should be fairly tight. The lower end of the frame should be about eight inches above the ground, the rear twelve inches. From ten to twelve inches between the soil and glass is sufficient for any of the plants generally forced, since the nearer they are to the glass the less liable will they be to "draw," as spindling up is called.

VIII. Market and Kitchen Gardening.

MARKET gardening differs from kitchen gardening only in degree. In the kitchen garden only such articles are raised as may suit the taste of the family. The market gardener must not only cater to the taste of all, but larger quantities must be raised, and here again the forcing of plants is entered into more largely, according to the size and wealth of the city or village, while the farmer is generally content with forcing plants for transplanting rather than the raising of a crop entirely under glass.

For the village market gardener, the 1st of January or the 1st of February will usually be time enough for starting the hot-bed. The farmers will not usually do so until the 1st of March in latitude forty degrees and north. If only a cold frame, or better, a slight hot-bed is required, the 1st of April will be quite soon enough for starting the bed. In this may be sown lettuce, mint, parsley, cabbage, cauliflower, kohl rabi, celery and tomato. Egg-plant, for the family garden, is better sown in a box, to be kept in a warm, sunny window until large enough to prick out in the bed, and the same is true of tomatoes when not more than 100 or 200 plants are wanted.

KOHL RABI.

Pricking Out.—Pricking out plants in the frame is simple enough. Having reduced the soil to the finest possible tilth, thinning out all lumps that can be gathered with a fine rake, place a wide board in the bed to stand on, and lay off your spaces with a form, made by inserting three-eighths inch short pegs, three inches apart, in a strip of wood, the rows to be three inches apart. With the fingers deepen the hole, drop in the roots of the plants and press the earth rather firmly to it, and so proceed until the required space is filled. Tomato, cabbage, and all that class of plants may be set so a good part of the stalk is covered, and when two inches high will be fully large enough. Lettuce and that class of plants may be pricked in when of a size to handle, but must not be set below the first leaves. A little practice will soon make one adept at planting. When the plants begin to fairly cover the ground, take out every other row, and transplant six inches apart in another part of the bed, and again every other plant in the remaining rows. At this distance they may remain until ready to be finally set out of doors, but cabbage, cauliflowers, kohl-rabi, lettuce, parsley, and all pot herbs, may be transplanted to the open air when they cover the ground at three inches apart. Many good gardeners do not allow these more than two inches and a half each way.

IX. Water and Ventilation.

WHEN a bed is first set it may need some shade the first few days, and if so the glass may be kept pretty close. When the plants begin to grow, do not water too often; water when needed, just so the moisture will not drip much into the heating manure below, and always with water slightly warm, if possible, and preferably in the afternoon, about an hour before covering the bed for the night.

Ventilation is an important matter. The more air is given, the better and more healthy the plants will be. Therefore, give air whenever the outside temperature will admit, and always tilt the sash so the wind will not blow directly on the plants. When the wind is considerable, the ordinary frame will usually get sufficient air through the cracks, except the sun is very warm. This is another of those points where no definite rule is applicable. The proper amount of ventilation must be gained by experience. The beginner is apt to give too little. Do not be afraid of air, but avoid strong drafts. The slower a plant is grown in the hot-bed, the stronger it will be when finally transplanted into the open air.

GROWING CUCUMBERS SUSPENDED FROM GREEN-HOUSE TRELLIS.

Many gardeners, who do a large business, and are near to large cities, have great green-houses for forcing vegetables, principally lettuce. The cut shows cucumbers, of the long English fancy sorts, grown on trellises next the roof. We have grown them over twenty inches long, but they are in no wise superior to our short varieties.

X. How to Have Early Rhubarb.

THE market gardener has his soil as rich as manure will make it. He sets out one-third of the space to be occupied by rhubarb every spring—giving him a three years' rotation. All the large roots not required for eyes for the new bed, and that are dug from the older portion of the oldest plantation are reserved for forcing. The farmer should do the same. In lieu of this a bed should certainly be set in the most sheltered position possible. Cover this in the autumn with fine manure, to be dug in the spring; set old barrels without heads or bottoms over the crowns of the plants, in the spring, and it will be found profitable. If you get one cent a pound for the surplus it will pay. Try it.

It is to be hoped every farmer has carefully read the directions for planting the home orchard and garden, so as to economize labor, and has already made preparations to begin in time. Do not allow other work to prevent so doing. Better hire a little extra help if necessary. Cheap help, properly directed, can handle manure and do all the rough work, and any of the better class of seedsmen's catalogues will direct as to sowing and general care; but, make rows clear across the garden, and give space enough between the rows to save hand labor. This is important—one year will give you strawberries; two years will give you rhubarb; three years asparagus, currants, gooseberries, raspberries and blackberries; three to five years will give you cherries and peaches and some apples and pears, and from that time on an abundance.

XI. "Take Time by the Forelock."

Do not be in too great a hurry to begin in the spring, but when the soil is in good condition to work—that is when it is quite friable—lose no time. Plant peas, beets, carrots, parsnips, onions, salsify, radish, lettuce, etc., as early as the soil may be worked; corn, bush-beans, and other half-hardy plants as soon as danger of white frosts is over, and at the same time transplant from the hot-bed or cold frame, lettuce, cabbage, cauliflower, kohl-rabi; also, parsley, sage, mint and other pot herbs. Squash, cucumbers, melons, pole beans, okra, tomato and egg-plant should not go out until the nights as well as the days are warm, and all these should be transplanted from the troughs from cold frames we have named, rather than be planted by seed. Once the planting is done, it is simply to keep the soil clean and gather the produce. If you provide yourself with the same class of modern garden implements that you do with farm implements, this cultivation is neither onerous nor difficult.

OKRA.

A plank, a leveler and a smoothing harrow will fit the garden nicely. A modern seed sower will sow any seed correctly and evenly, and a wheel hoe will accomplish the cultivation next the rows better than it can be done with a hand hoe, and after the plants get up a little, any good, one-horse cultivator will work in rows not less than two feet wide.

XII. What to Raise for Market.

In gardening for market, the grower must be guided by the circumstances governing the case. He must raise that which will suit the taste of his customers. This he must find out and cater for. It is what makes success. The reverse makes failure.

There are, however, certain lines of products that sell always and everywhere—potatoes, green corn, tomatoes, lettuce, radish, onions, beets, carrots, parsnips, salsify, string and shell beans, peas, asparagus, rhubarb, cabbage, cauliflower, kohlrabi, are all staples, but here again the quantity desired must always be governed by the demand of the locality.

Yet, the gardener, if he have tact and business energy, may easily get rid of additional supplies of even these staples, by introducing them among families who are in the habit of buying few, if any, vegetables. In other words, the demand in any community should be an increasing one, and will be, if only the gardener understands his business.

Take lettuce, for instance. Many families will not buy hot-bed lettuce, because they erroneously suppose it must necessarily cost many times more than that raised in the open air. This every gardener knows is not so when the plants are started under glass and transplanted. Even if such things were sold at first at little or no profit, customers would be secured who would afterwards look for this class of products before they could be raised entirely in the open air, and again one sale would lead the way to others and different classes of vegetables.

The same will apply to all vegetables that the gardener, with his superior facilities, can bring forward several days sooner than the kitchen gardener, and here must his full exertion be used to induce sales of all the vegetables usually raised by the inhabitants of the smaller towns and villages.

Coming now to those vegetables not generally raised in village or farm gardens, and to the common vegetables raised for a continuous supply, green corn will always be found to sell very early and very late. So will string beans, peas, cabbage and cauliflower. Lima beans should always sell, and for the reason that the average cultivators of kitchen gardens will not give the necessary care to ensure success, the great fault being too early planting. In this direction lie also sweet potatoes, okra, early cucumber, melons, squash, egg plant, etc. With okra and egg plant tact must be used. The healthfulness of the first must be asserted and the delicacy of the second must be stated, and not content with this, samples may be given, with careful directions for cooking. Every person who has a garden should raise all these; where there is a competent gardener, some of them can probably be bought fully as cheaply as they can be raised.

Mint, sage and other pot herbs should always form a part of the stock of the village market gardener. If he have them the surplus can always be shipped, after drying in the shade, and sooner or later customers will increase at home.

Small fruits should not be neglected. Strawberries, blackberries, raspberries, currants and gooseberries should play an important part in the profits of the garden. What are not sold may be canned, and if not to be sent beyond the village market, the common sealing-wax, now sold by all grocers, for self-sealing, will keep any fruit perfectly. Celery should be persistently kept before customers during the season. Once a customer gets used to buying it you may always be sure of selling to him thereafter. Above all have a good stock for winter and spring. It is so easily raised as a secondary crop, and so easily saved, that it is a wonder that all who cannot buy it of their gardener should neglect to raise it at home.

XIII. Economy in Cultivation.

ONE great advantage that the market gardener has over those who raise their vegetables in confined village gardens, is that a far greater economy of labor may be had in the first case than in the latter. The market gardener can plant in long lines, and do the most of the cultivation by horse power. So can the farmer, for land is plenty. The village gardener must do most of the work by hand. The market gardener and the farmer can plow, harrow, level and sow by machinery. A wheel hoe will do all the weeding except thinning out. The one-horse cultivator and harrow and horse hoe will do all the cultivating of any crops planted twenty inches or more apart. We should give fully this space to onions, and two feet to all other root or bulb crops. Too close planting is, in fact, the great mistake made by the average country market gardener. We have always given three feet between rows to asparagus, and made money by so doing in cultivation and size of the stems.

Another mistake made is in stinting manure. If well prepared (composted) you cannot have too much of it. After the first year it will begin to tell, and once the soil is full of manure, your profit is certain. It costs no more to cultivate a well manured acre than a hungry one, in fact, not so much, since the plants grow faster, and sooner cover the ground. Therefore, do not be afraid of manure.

XIV. Preparing Vegetables for Market.

PREPARING and bunching the vegetables for market is of the first importance. Wash all roots clean, bunch in convenient hands all that are sold with their tops attached and tie securely at the necks. Lettuce should be cut with a portion of the root attached, the head divested of all dying leaves, washed and neatly packed, so it will show its full size and length when taken out. Pot herbs should be washed by rinsing, and neatly tied in small bunches. Tomatoes should not be washed, but rubbed off, if necessary, when they are dry.

Egg plant and all other smooth fruits should be rubbed with a soft cloth to polish them. Peas and beans should be rinsed free of dirt and grit. Cucumbers and fruits that are covered with a bloom should be cleaned by sprinkling, so as to preserve this as far as possible intact. Cabbage and cauliflower should be divested of all supernumerary leaves, but the most of the leaves should be left on kohl-rabi.

Green corn may or may not have a part of the husk removed, but the butt should always be cut or broken fairly close to the ear. You cannot deceive any one with long shanks, even if you wished to, which of course you do not. It is attention to minor details in gardening, as in any other business, that wins success.

It should be unnecessary to say that the garden should always be free from weeds, and the drives cleaned of trash every day. This, besides being economical, is also a good advertisement, for especially in country places, a well-kept garden will be more or less a place of resort, and if you have a taste for flowers it will gradually become more and more so, and these gradually will also come to add to the profits, for bouquets and plants will be bought.

XV. How to Raise Potatoes.

POTATOES on the farm are too often raised in some corner, where nothing else can be grown, and hence most farmers have neither enough in quantity nor the best in quality. There is no crop that pays better for extra care than the potato. If those farmers who think they can better afford to buy than to plant them will try the

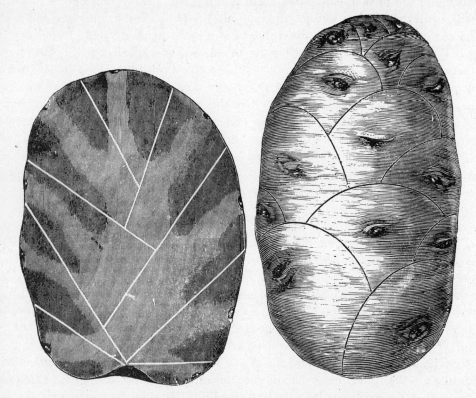

DIAGRAMS SHOWING HOW TO CUT POTATOES TO SAVE SEED.

following plan, they will be convinced of their mistake: Select a piece of land the second year from the sod, or else a piece of stubble land enriched with not less than forty loads of half-rotten manure per acre, all which has in the fall been plowed under to the depth of, say, eight inches. In the winter draw on ten more loads of fine manure to the acre; plow this under about four inches deep, and, as early as the ground will work friably, harrow the land.

Furrow out three feet apart and drop a piece (cut as shown in the diagrams, herewith) every ten inches in the furrow. If cut in the ordinary way drop a piece every fifteen inches, stepping on every piece. Cover by throwing two light furrows forming a ridge. When the weeds appear, harrow the ground lengthwise of the rows, with a two-horse harrow turned upside down. If weeds again appear before the potatoes are up, again throw up the two furrows to cover, and again harrow down just as the potatoes appear above ground. Do not be afraid of killing the potatoes. We have harrowed them when they were three inches high. The after cultivation consists simply in slightly earthing up the plants from time to time with a shovel plow, until the buds appear. If weeds appear after that, the large ones must be pulled out by hand. In a good season the crop will look like the cut, "Potatoes as they should grow."

POTATOES AS THEY SHOULD GROW.

To harvest them, if you have a large field, get a potato digger; if not, plow two furrows away from each row, and dig the balks with a potato fork. That is about all there is to potato culture, except storing.

When they are dug, put in compact piles in the field, cover liberally with potato tops, or slough hay, and six inches of earth well smoothed down to turn rain. Just before cold weather take them to the cellar, or place in pits for the winter, as directed in Part II., Chapter VIII., Section XXXIII.

In planting potatoes, do not delay too long. Plant about the time you sow oats, for unless the potatoes get their growth before drought and heat sets in you reduce the crop. Potatoes will not stand drought or heat. They are a moisture-loving and cool weather plant. Good crops however, are, sometimes raised if planted about the twentieth of June. Then they have the advantage of the late rains and cool nights of autumn to mature. Above all do not "plant in the moon," hoping to get a superior crop.

XVI. Planting in the Moon.

WE generally plant early potatoes just as soon as the frost is out of the ground and the soil settled, and for the general crop as soon thereafter as convenient—always before corn-planting time. Why? Because, if you do not get a good growth to your vines before hot and dry weather sets in, the crop is pinched. With earlier varieties it insures the crop ripe in August, and with late varieties, like peachblows, it gives them what they require, the whole season to grow and the cool autumn to ripen in.

There is just this much in the moon theory and no more: plants do grow faster in light nights, probably, than in dark ones; and thus such plants as come up quickly, planted when the moon is new, may, with favorable weather, grow faster; and potatoes, which are slow in germinating, planted after the full of the moon, and coming up when the nights are light, may, under favorable circumstances, seem to grow faster; but that there is any difference in the outcome, no careful experiments made have ever shown.

XVII. Potatoes Illustrated.

A word now in relation to the illustration of potatoes on this page. They were not selected as representing the best kinds. Early Ohio and Chicago Market

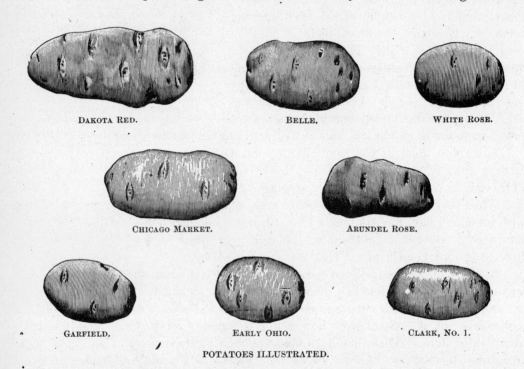

POTATOES ILLUSTRATED.

are old, well-known and excellent early varieties; the others are newer. We have chosen them as showing the principal forms assumed by the potato from nearly round to irregularly round, and also as indicating relative sizes. As a rule, medium sized varieties are best; the very large ones inferior.

CHAPTER IV.

GARDEN FLOWERS AND SHRUBS.

I. THE FLOWER GARDEN.—II. HOW TO CULTIVATE FLOWERS.—III. SELECT LIST OF FLOWERS FOR GENERAL CULTIVATION.—IV. BIENNIAL AND PERENNIAL FLOWERS.—V. SUMMER FLOWERING BULBS.—VI. FLOWERING PLANTS AND VINES—ROSES.—VII. FLOWERING SHRUBS.—VIII. CLIMBING AND TRAILING SHRUBS —IX. FLOWERING TREES.—X. EVERLASTING FLOWERS AND ORNAMENTAL GRASSES.—XI. WATER PLANTS.—XII. TRELLISES.

I. The Flower Garden.

THE vegetable garden is properly the province of the master of the farm; the flower garden is preëminently the home of feminine art and taste. However

CRESTED MOSS ROSES—HALF SIZE.

small the village or city lot, flowers may smile up from it to the sun, and call down blessings from the admiring passer-by upon the fair and skillful hands that have tended them. In no place do they more improve and refine the surroundings, and show the beauty of feminine taste and culture than upon the farm. An ample lawn studded with ornamental shrubs and trees, a bed cut out here and there, or a smiling parterre where these lovely gifts of nature have been taught by care and skill to bloom, fill the soul with an harmonious joy. All long for the beautiful. All love flowers. But many, ignorantly, suppose that flowers cost so much in time and money that only the wealthy may enjoy them. They are within the easy reach of all. The common flowers are, as a rule, as beautiful as the rarer ones. Perennial flowering plants, or their seeds, once planted, remain year after year, increasing in beauty with each successive season until they arrive at their full perfection.

II. How to Cultivate Flowers.

THE principal mistake made in the cultivation of flowers, is permitting them to be smothered while young, by weeds. Many of the common varieties, which sow themselves by their seeds, remaining in the ground during the winter, manage to make a pretty successful struggle with their enemies, the weeds, but the plants are

PANSIES

MOSS PINK. (PHLOX SUBULATA.)

so crowded as to detract much from their beauty. This is why, in the country, so much attention is paid to perennials, that is to say, the plants or flowers that live from year to year. If the directions here given are followed, the cultivation of annuals will cost less labor, they will come much earlier to perfection, and be, in every respect, better than if sown in the open ground. All that is necessary is to prepare, about the time field plowing begins, a small hot-bed, as described in the chapter on Vegetable Gardening; or a cold frame (a bed covered with sashes), in which latter case, the seeds should not be sown therein, until just after the time for planting spring wheat.

Having prepared the bed and put in six inches of clean, fine mold, sow the flower seeds in lines four inches apart between rows, putting down a peg marked with the name of each variety sown. When large enough to transplant, pick the plants out into another frame, place them two inches apart, and here let them stand until they are ready to go out of doors. Take up with earth about the roots, lay on trays, and they are easily and safely carried to where they are to grow.

There are but few annuals that, treated in this way, will not transplant kindly. Sweet peas, candytuft, etc., should, however, be sown where they are to stand. The larkspurs, poppies, mignonette, heliotrope and cypress vine, which are somewhat difficult to transplant, may be pricked out in troughs, similar to those described in the chapter on Vegetable Gardening, each trough having a pasteboard, or other division, thrust down along the middle, to separate it into two parts. Plants difficult of removal may be grown in these simple troughs, and be quickly and easily transplanted. This may be done at any time in the evening, unless the soil is too wet to work. If so, wait until it is dry enough. To transplant in dry weather, give the plants in the bed a good soaking the morning before transplanting, which, as stated, had best be done at night.

PERENNIAL DAISY.

CALADIUM.

Leave a little depression, water the roots, and, when the water has disappeared, draw the dry earth over all. They will hardly shrink. You will have forwarded the season of flowering fully three weeks, and produced your flowers, even of the more hardy sorts, far cheaper than if you had sown the seed outside. You will also be able to grow many things usually bought as plants from the florists, such as verbena, pinks, daisy, pansy, etc. You may, early in July, have in full bloom China and other garden pinks, and nearly all the class that, sown outside, do not usually bloom until the second year. You may have balsams, candytuft, alyssum, mignonette, nasturtium, phlox, zinnias, morning glory, in splendid condition before your neighbors, who have sown in the open air, can see theirs among the weeds.

A second crop of mignonette, candytuft, annual phlox, alyssum, balsams, etc., should, later, be sown inside to produce autumn bloom. In your hot-bed start gladiolus, dahlia, tigridias and other bulbs and tubers, and also such roots of perennials, including Bengal and other tender roses, as you have kept through the winter in boxes placed in a light cellar.

III. Select List of Flowers for General Cultivation.

The running notes below will give all the information necessary to the grower in addition to that already stated. Any respectable seed catalogue will give information as to special varieties.

AGERATUM.—Cuttings may be started under glass. If seed is sown cover lightly; set plants six inches apart; nice for winter flowering in the house.

AMARANTHUS.—Ornamental foliage plants; fine in masses, and in mixed shrubbery, borders and centers; sow in hot-bed and transplant.

ASPERULA.—Dwarf, desirable for shady situations and moist soil; fine for bouquets; plant six inches apart.

ASTER.—Showy for borders; flowers in autumn; sow in cold frame; transplant tall varieties sixteen inches apart, dwarf varieties seven inches apart, in good deep soil.

BALSAM.—Showy and desirable; easily cultivated; prune by pinching out the terminal buds; sow in hot-bed, cold-frame or window box; transplant into a deep rich soil, twelve inches apart; set dwarf sorts separate from tall varieties.

CACALIA. (TASSEL FLOWER).—Tassel-shaped flowers in clusters on slender stalks; nice for bouquets; sow in cold-frame; transplant to ten inches apart.

CALANDRINIA.—Sow seeds in slight hot-bed and transplant to light soil; it flowers freely, and is perennial if protected in winter.

CALENDULA. (POT MARIGOLD).—Very pretty; flowers toward sunset and does not open on cloudy days; hence one of its names, rainy marigold.

CANDYTUFT.—Fine for edging of beds and bouquets; for early flowering sow seed in fall and protect during winter with mulch; thin plants to four inches apart in the spring; it is difficult to transplant.

CELOSIA. (COCKSCOMB).—Start early in hot-beds or window-boxes, and transplant into small pots, to remain until the flowers begin to appear, then set out in warm garden soil fifteen inches apart.

CLARKIA.—Sow in March, under glass, and again later in the open air; they flourish in any soil free from wet; thin to a foot apart.

CONVOLVULUS. (MORNING GLORY).—There are dwarf and also running species, all of them handsome; they may be sown in the open air at corn-planting time, or earlier in a cold frame and transplanted.

DIANTHUS. (PINKS).—Among the most elegant of plants are carnation, clove pink, China pink and sweet William; sow in hot-bed; transplant dwarf varieties six inches apart; tall, twelve inches apart; if not kept too warm they are good house plants.

DELPHINUM. (LARKSPUR).—These have finely cut leaves and beautiful flowers of scarlet, pink, purple, blue and white; double white is fine for bouquets; sow in the autumn or in the spring.

ESCHSCHOLTZIA.—Showy flowers of yellow and cream white; will not bear transplanting; thin out to eight inches apart.

Gaillardia.—If sown early under glass the bloom can be kept up the whole summer; the seed germinates slowly; do not transplant until all danger of frost is past.

Gilia.—Low growing, profuse in bloom; the best effect is produced by them in masses, or in borders on rock-work; the flowers are nice for bouquets; sow in fall and cover lightly during winter; thin to six inches apart.

Lobelia.—Very pretty for baskets or vases; sow seed in hot-bed or frame; dwarf varieties are useful for borders or pots; transplant six inches apart.

Lupin.—Hardy and easy to grow; sow the seed in the open ground where wanted to bloom; they can not be transplanted.

Marigold.—The varieties all are showy, and produce fine effects in masses; hardy, and continue in bloom the whole season; sow seed in frame or hot-bed; transplant two feet apart; dwarf varieties, twelve inches apart.

Mesembryanthemum.—Pretty plants of dwarf habit, fine foliage, suitable for basket or pot culture on the border; sow seed under grass; transplant eight inches apart.

Mignonette.—Delightful for its fragrance; sow under glass and transplant in the open air eight inches apart; sow in the open ground in May for succession or late bloom.

Mirabilis. (Marvel of Peru).—Foliage and flowers are beautiful. For early flowering, sow in hot-bed or box, or may be sown where wanted to bloom; thin out two feet apart.

Nasturtium.—Dwarf and running varieties; the latter used for hanging-baskets in winter. Dwarfs, pretty, low-growing, profuse flowering plants. The green seeds, like martynia pods, are valuable for pickling. Sow in hot-bed and transplant in open air, eight inches apart.

Nemophila. (Baby's Eyes).—Loveliest of blue-eyed flowers. They are low, hardy annuals. Sow in frames, transplant six inches apart; thrive best in cool, shady places. Seed sown in the fall will succeed well.

Pansy. (Viola Tri-color).—Nothing prettier; bloom the first season, in June, if sown early in hot-bed and transplanted. Requires protection during the winter if in open-air beds.

Petunia.—Indispensable, and elegant in masses; fine in the window garden. The seed may be sown in hot-bed or cold frame; transplant eighteen inches apart; the plants do not always come true from seed; they are of every shade of color, and bloom from early spring until frost.

Phlox, Drummond's. (Annual).—Among the most beautiful of garden flowers, and of infinite variety of colors. The seeds for early flowering should be sown in the hot-bed or the cold-frame, and transplanted one foot apart, as too close planting produces mildew. Or plant out doors where wished to grow. The pretty moss pink is one of the perennial phloxes.

Portulaca.—One of the most brilliant of sun-loving flowers; low-growing, creeping plants, flowering abundantly. Sow in hot-bed and transplant, six inches apart in the open air.

RICINUS. (CASTOR-OIL BEAN).—Plants of green and purple foliage, of tropical and striking effect. A centre-piece of ricinus, with plants of canna next, and a row of caladium plants outside, will form a bed truly tropical in effect.

SCABIOSA. (MOURNING BRIDE).—Bright-colored, annuals, adapted for beds and for bouquets. The German dwarf varieties are double; sow in frame or in open border. Set the tall varieties fifteen inches apart, dwarf, a foot apart.

HYBRID TEA ROSE—LA FRANCE.

STOCK FLOWERS.—Stock or gilliflower will never go out of favor, being abundantly flowering, with colors running through all the shades of crimson, lilac, rose, white, etc. Rich soil is requisite to keep Stocks double; they are planted in May or sown earlier in the hot-bed, and set out, twelve inches apart. The annual Virginia Stock is fine for edgings, but it does not transplant easily.

ZINNIA. (YOUTH AND OLD AGE).—The varieties of this Mexican plant are magnificent in color. The flower is nearly as double as the dahlia, and lasts a long time. Sow the seed under glass in early spring, and transplant to the open ground, when danger of frost is over.

In this list we have included some plants that are biennial, but treated as annual plants as blooming the first year. The double daisy is also not an annual, but will bloom the first season if sown very early and twice transplanted before being set out. They are difficult to winter out of doors, although in Europe they are hardy as far north as Sweden. When there is plenty of snow, many tender plants and flowers may be safely wintered.

CHARLES LEFEBVRE ROSE.

IV. Biennial and Perennial Flowers.

ACANTHUS; aconitum (monk's hood); adonis; alyssum (distinct from sweet alyssum, an annual); aquilegia (columbine); hollyhock, of which the double varieties are elegant and in great variety of color; iberis (candytuft); linum, or ornamental flox; lobelia; lychnis (flame flower); myosotis (forget-me-not); œnothera (evening primrose); perennial poppy; perennial pea; penstemon; perennial phlox; pinks, including picotees and sweet William; primula (primrose); pyrethrum; rudbeckia; salvia; veronica (speedwell); and violet (sweet violet). These varieties may be named among the useful ornamental plants which either flower the second season or are perennial.

V. Summer Flowering Bulbs.

Of bulbs, tubers and roots, which include many that must be kept over winter in dry sand to prevent freezing, we may note: amaryllis (magnificent lily-like plants); gladiolus (sword-lily), and all the hardy lilies proper; dahlias, caladium, dicentra (bleeding heart), Maderia vine, Japan spirea, tritoma, tuberose (must be started very early, planted out when the nights are warm and freely watered), tulips and hyacinths.

COUNTESS OF SERENE ROSE.

VI. Flowering Plants and Vines—Roses.

Among the ornamental vines are adlumia (Alleghany vine); ampelopsis quinquefolia (Virginia creeper); bignonia radicans (trumpet-vine); celastrus scandens (climbing bitter-sweet); clematis flammula (European sweet, white); clematis vitalba (virgin's bower, white). And, among shrubby plants, the hardy roses, and all the flowering shrubs.

Moss Rose.—Among the most beautiful of plants are roses, and none are more so than those that are not strictly hardy. An especially charming rose, charming even by comparison with its lovely sisters, is the moss rose. We illustrate, on page

507, one variety, the Crested Moss, at half the natural size. Its special merit is its long fringe of moss, giving the buds a most unique and beautiful effect.

Tea Roses.—The hybrid tea roses do well bedded out in summer, blooming profusely and having most exquisite odor. The cut, page 512, shows one of the finest, "La France," at half its natural size. In color a delicate silvery rose, changing to silvery pink; very large, full, of fine, globular form; a most constant bloomer; it is among the most useful of roses and unsurpassed in the delicacy of its coloring.

LOUIS VAN HOUTTE.

Perpetual Roses.—The most valuable of any of the rose species are, without doubt, the hybrid remontant, or hybrid perpetual roses. Many of these are hardy enough to grow well far into the North with but slight protection, and south of forty degrees most of them need none even in winter except in very exposed locations. They are constant, charming in bloom and color, exquisite in fragrance, and should be cultivated everywhere. One of the best, although it has the one failing of fading quickly, is Charles Lefebvre, shown on page 513, one-half its natural size. The special merits are its fine color, finished shape and the beautiful wavy form of its petals. The color is reddish crimson, very velvety and rich; it is large, full and beautifully formed, is a free bloomer, but as we have said, fades quickly. Its parentage is on one side the "Gen. Jacquiminot," and on the other the "Victor Verdier" rose.

Autumn Roses.—One of the finest of autumn roses, and also one of the best for forcing is the Countess Serene, shown two-thirds its natural size. It should be in every collection, however small. It is in color a silvery pink, has great beauty of form, delicate mottling, and other merits which should command attention. Page 514.

FLOWERS OF WHITE FLOWERING DOGWOOD. NATURAL SIZE

The "Louis Van Houtte" is a rather tender variety, but prolific, and is decidedly the peer of any crimson rose known. We illustrate of full size on page 515. In color it is a deep crimson maroon; it is of medium size, full, semi-globular form; has large foliage, fewer thorns than the other dark roses, and is highly perfumed. Its special merits are fine durable color, great sweetness, free blooming, and excellence in form.

But magnificent as roses are, and admired as they are by all, they are subject to many drawbacks, of which insect pests are not the least. Lice and slugs must be watched for, and hence we favor the perpetuals, and the still more tender tea roses.

They may be wintered in the cellar, and will give ample satisfaction bedded out. Yet the hardy June roses must by no means be neglected. These should find a place in every garden.

VII. Flowering Shrubs.

ALL may have shrubs and clinging vines. From the better flowering sorts, of which we illustrate many, as to their flowers and leaves, there is large room for selection. In this connection we shall have little to say of trees. They will naturally come in when treating of Landscape Gardening. The shrubs figured will be taken alphabetically, the English name following the proper scientific term.

FLOWERS OF JAPAN QUINCE. NATURAL SIZE.

CORNUS. (DOGWOOD).—The variegated cornelian cherry is illustrated, as being somewhat rare, and as belonging to the genus. The members of this species are valuable shrubs when planted singly, in groups or in masses. Some are distinguished by their elegantly variegated foliage, others by their bright colored bark. The variety in question is a small tree, native to Europe, which produces clusters of bright yellow flowers early in spring, before the leaves make their appearance. The foliage, beautifully variegated with white, makes this decidedly the prettiest variegated shrub in cultivation.

WHITE FLOWERING DOGWOOD.—This is an American species, of spreading irregular form, growing from sixteen to twenty-five feet high. The flowers produced in spring before the leaves appear, are from three to three and a half inches in diameter, white and very showy. They begin to appear just as the magnolia flowers are fading, and are invaluable for maintaining a succession of bloom in the garden border or on the lawn. They also are very durable, lasting in favorable weather more than two weeks. Besides being a tree of fine form, its foliage is of a grayish green color, glossy and handsome, and in the autumn turns to a deep red, rendering the tree one of the most showy and beautiful objects at that season. It may be regarded, all things considered, as one of the most valuable trees for ornamental planting, ranking next to the magnolia among flowering trees, and only second to the scarlet oak (which it almost equals) in brilliant foliage in autumn.

ROSE-COLORED WEIGELIA.

VARIEGATED CORNELIAN CHERRY.

CYDONIA. (QUINCE). JAPAN QUINCE.—The flowering varieties of the Japan quince rank among our choicest shrubs. Although of straggling growth, they bear the knife well, and, with proper pruning, may be grown in any form. As single shrubs on the lawn, they are very attractive, and for the edges of borders or groups of trees, they are specially adapted. Their large, brilliant flowers are among the first blossoms in spring, and they appear in great profusion, covering every branch, branchlet and twig, before the leaves are developed. Their foliage is bright green and glossy, and retains its color the entire summer, which renders the plants very ornamental, especially for ornamental hedges. It is sufficiently thorny to form a defense, and at the same time makes one of the most beautiful flowering hedges. There are a number of varieties: scarlet, blush, rosy red, and the double flowering. The cut on page 517 shows the flowers of full size.

DENTZIA.—We are indebted to Japan for this valuable genus of plants. For hardiness, fine form and luxuriance of foliage, and in attractive bloom, they are among the most popular of plants. The flowering season in the North, is the latter part of June, the beautiful racemes being from four to six inches long. The illustration shows Pride of Rochester, an American variety, producing large double white flowers; the back of the petals being slightly tinged with rose. It excels all of the older sorts in size of flower, length of panicle, profuseness of bloom and vigorous habit; blooms nearly a week earlier than its parents, the double dentzia crenata.

DIERVILLA. (WEIGELIA).—This is another valuable genus from Japan. Shrubs of erect habit while young, but generally spreading and drooping as they acquire age. They produce in June and July superb, large, trumpet-shaped flowers, of all shades and colors, from pure white to red. In borders and groups of trees

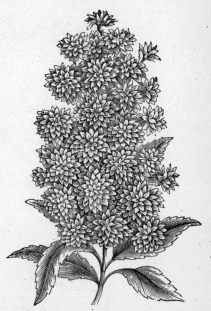

PRIDE OF ROCHESTER DENTZIA BLOSSOMS, ONE-THIRD SIZE.

they are very effective, and for margins the variegated leaved varieties are admirably suited, their gay colored foliage contrasting finely with the green of other shrubs.

There are many varieties, all beautiful. The figure on page 518, the rose-colored variety, an elegant shrub, with fine rose-colored flowers, introduced from China by Mr. Fortune, and considered one of the finest plants he has discovered; of erect, compact growth; blossoms in June.

One of the most valuable properties of the weigelias is that they come into blossom after the lilac, and when there is some scarcity of flowering shrubs.

FORSYTHIA. (GOLDEN BELL).—These are pretty shrubs, of medium size. All natives of China and Japan. The flowers are drooping, yellow, and appear very early in spring, before the leaves. The best very early flowering shrubs. The one we illustrate is in its growth, upright; foliage, deep green; flowers, bright yellow, and shows the natural size of the flowers. (See next page).

HALESIA. (SILVER BELL).—This is the common snow drop, a pretty companion to the golden bell. A beautiful large shrub, with pretty, white, bell-shaped flowers

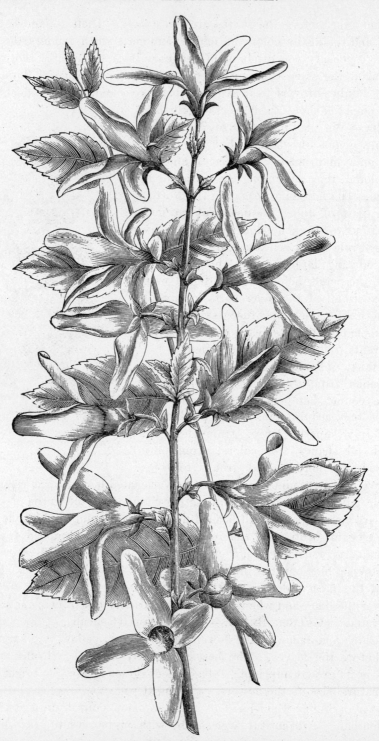

FORTUNE'S FORSYTHIA.—NATURAL SIZE.

DOUBLE-FLOWERING PLUM.

in May. It is distinguished by its four-winged fruit, which is from one to two inches long. One of the most desirable shrubs.

HYDRANGEA.—The native species are handsome shrubs of medium size, with fine large leaves, generally of a light green color, and perfectly hardy. The introductions from Japan and China are particularly interesting and valuable. *H. paniculata grandiflora* is remarkable in foliage and flower, and being perfectly hardy, is of great value. The other Japanese varieties, like the *H. hortensia*, require protection in winter. They should be grown in pots or boxes and wintered in the cellar, and in summer placed along walks under the shade of trees. *H. otaksa* is specially adapted for this purpose. The foliage is a beautiful deep green color; the plant produces immense trusses of rose-colored flowers in profusion in July; free bloomer.

HYDRANGEA OTAKSA.

The hydrangeas are worthy of cultivation wherever they can be given shade, good drainage and plenty of water.

PHILADELPHUS. (SYRINGA, OR MOCK ORANGE).—The syringa is an invaluable shrub. Of vigorous habit, very hardy, with large handsome foliage, and beautiful white flowers, produced in the greatest profusion at the blossoming season; it merits a prominent place in all collections of shrubbery. Most of the varieties, except those of dwarf habit, form large sized shrubs, twelve to fifteen feet high. They can, of course, be kept smaller by pruning. The dwarf sorts do not yield many flowers, but are such pretty, compact plants as to be very useful where small shrubs are desired. All of the varieties flower in June, after the wiegelia. By planting the late flowering sorts,

SYRINGA, OR MOCK ORANGE—NATURAL SIZE.

the season may be considerably extended. The syringa illustrated has very large, white, fragrant flowers, upright habit, is very free flowering and a valuable sort.

PRUNUS. (PLUM).—Among the plums are embraced some of the most charming early spring flowering shrubs. *Prunus triloba* or the double-flowered plum, as

SILVER BELL.—NATURAL SIZE.   MEADOW SWEET.—SPIREA EXIMIA.

it is commonly called, and the double-flowered almonds produce in remarkable profusion, perfectly double, finely formed flowers of most attractive colors. At the blossoming season each little tree appears like one mass of bloom, forming a most beautiful and interesting object, whether planted singly upon the lawn or in groups.

As the almond and plum flower at the same time, they can be massed very effectively. Both are hardy and of fine habit. The double flowering variety illustrated is a highly interesting and desirable addition to hardy shrubs; flowers double, of a delicate pink, upwards of an inch in diameter, thickly set on the long slender branches; flowers in May. (Page 521).

LANCE-LEAVED SPIREA.

SPIRÆA. (MEADOW SWEET).—The spireas are all elegant, low shrubs, universally admired and grown in some of their numerous varieties. They are of the easiest culture, and their blooming extends over a period of three months. We illustrate two fine varieties—eximia, page 523, a well-known sort, and the double-flowered, lance-leaved variety. We also give a list of ten sorts, which flower in the order named, beginning in the middle of May and ending, in the North, about the middle of August: (1.) Prunifolia fl. pl., Thunbergii. (2.) Niconderti. (3.) Chamædrifolia. (4.) Cratægifolia, lanceolata, lanceolata fl. pl., lanceolata robusta. (5.) Ulmifolia. (6.) Opulifolia aurea crenata. (7.) Fontenaysii, salicifolia, sorbifolia. (8.) Billardi. (9.) Ariæfoiia. (10.) Callosa and callosa alba, callosa superba. The marks fl. pl. mean double flowering.

Spirea Japonica.—This is a new and very handsome species from Japan. A dwarf growing, but very vigorous, narrow-leaved variety, with rose-colored flowers, appearing in great profusion in mid-summer and autumn. When small shrubs are desired, this will be an acquisition. The cut gives a faithful representation of the foliage and bloom.

Rhodotypus. Kerrioides.— This is another fine plant from Japan, a country that has given us many elegant hardy and half-hardy shrubs and flowering plants. It is a very ornamental shrub, of medium

JAPANESE SPIREA.

size, with handsome foliage and large single white flowers in the latter part of May, succeeded by numerous small fruits.

Viburnum. (Snowball).—This tribe is well known. The cut shows, at one-third its size, V. opulus, or bush cranberry, variety, sterilis, the common guelder rose of the garden. The plicate viburnum is better in every respect. Of moderate growth; handsome, plicated leaves, globular heads of pure white neutral flowers, early in June, it surpasses the common variety in several respects. Its habit is better, foliage much handsomer, flowers whiter and more delicate. One of the most valuable flowering shrubs.

GUELDER ROSE (VIBURNUM).

VIII. Climbing and Trailing Shrubs.

Of climbing and trailing shrubs there is a great variety. Ampelopsis (Virginia creeper); Dutchman's pipe (aristolachia); and clematis, or virgin's bower, being well known.

CLEMATIS JACKMANNI—HALF-SIZE.

CLEMATIS.—These should be in every collection, and will stand the severest winters if protected over the roots with mulch. They are elegant, slender branched

HALL'S JAPAN HONEYSUCKLE.

shrubs, of rapid growth, handsome foliage and beautiful large flowers of all colors. Either in the open ground as pillar plants, bedding plants, single plants, in masses, or about rock-work, or cultivated in pots or tubs, the clematis cannot be excelled.

The cut shows the blossoms of clematis Jackmanni, half-size. There are six distinct types of these beautiful plants, all excellent and flowering at various seasons.

HEDERA. (IVY).—The ivies are not hardy in America, either North or South, except in peculiar situations. They suffer from the sun in winter. Hence, ampelopsis takes its place.

LONICERA. (HONEYSUCKLE OR WOODBINE).—This is hardy, and there are many beautiful varieties. Among them none is better than Hall's Japan honeysuckle. A strong, vigorous, almost evergreen sort, with pure white flowers, changing to yellow. Very fragrant, and covered with flowers from July to December; holds its leaves even until January. The illustration on page 526 shows a spray and blossoms of this beautiful plant.

CHINESE WISTARIA.—QUARTER SIZE.

MENISPERMUM. (MOON SEED).—This is a pretty, native, twining, slender-branched shrub, with small yellow flowers and black berries.

SILK VINE. (PERIPLOCA).—This is another beautiful, rapid growing vine, climbing to a height of thirty to forty feet. The foliage is glossy and the clusters of flowers a purple brown.

WISTARIA.—These are beautiful climbers, and there are many varieties, with various colored flowers. The Chinese wistaria is one of the best. In fact, is one of the most elegant and rapid growing of all climbing plants; attains an immense size, growing at the rate of fifteen or twenty feet in a season. Has long pendulous clusters of pale blue flowers in May and June, and in autumn. The illustration shows the leaves and flowers quarter size.

IX. Flowering Trees.

THESE are of many kinds, and should not be neglected. They are especially useful on lawns of large extent, such as should always belong to the better class of farms. The magnolia should find a place everywhere, from the magnificent varieties adapted to the far South to the smaller shrub-like varieties of the North

MAGNOLIA SPECIOSA.—The illustration shows, in half size of nature, the flowers of magnolia speciosa, the showy-flowered magnolia. This tree resembles the *M. Soulangeana* in growth and foliage, but the flowers are a little smaller, and of a lighter color, fully a week later, and remain in perfect condition upon the tree longer than those of any other Chinese variety. These qualities, combined with its hardiness, render it, in our esteem, one of the most valuable sorts. Other magnificent varieties obtained by hybridization are:

FLOWERS OF MAGNOLIA SPECIOSA.—HALF SIZE.

MAGNOLIA CONSPICUA (CHINESE WHITE MAGNOLIA). (CHANDLER, OR YULAN MAGNOLIA).—A Chinese species of great beauty. The tree is of medium size, shrub-like in growth while young, but attains the size of a tree in time. The flowers are large, pure white, very numerous, and appear before the leaves.

MAGNOLIA NORBERTIANA. (NORBERT'S MAGNOLIA).—A hybrid between *M. conspicua* and *M. obovata*. Tree, vigorous and of regular outline; foliage, showy; flowers, white and dark purple. One of the best.

MAGNOLIA SOULANGEANA. (SOULANGE'S MAGNOLIA).—A hybrid, in habit closely resembling *M. conspicua*. Shrubby and branching while young, but becoming a fair sized tree. Flowers, white and purple, cup shaped, and three to five inches in diameter. Foliage, large glossy and massive. It forms a handsome tree worked upon the *M. acuminata*. One of the hardiest and finest of the foreign magnolias. Blooms later than *conspicua*.

There is one thing, however, always to be remembered: In ordering any tree or plant of your nurseryman describe your situation and locality. State some plants that are hardy and half-hardy with you, and then the advice given in relation to varieties may be worth money to you and save disappointment.

ÆSCULUS. (HORSE CHESTNUT).—Among the horse chestnuts is a double flowered variety that is magnificent, as any person will admit who has seen this tree as we have, fifty feet high in its glory of blossoms. Not the least value of this tree is that the blossoms are sterile, and hence the saving of the usual "horse chestnut litter" on the lawn.

DOUBLE-FLOWERING CHERRY. (CERASUS).—There are few trees which combine beauty with usefulness as do the cherries. We illustrate the flowers of the double-

RACEMES DOUBLE FLOWERED HORSE CHESTNUT—ONE-FIFTH SIZE.

flowering cherry one-quarter size. They flower in May in the North, and are especially beautiful. The flowers are so numerous as to conceal the branches, and present to the eye nothing but a mass of bloom, each flower resembling a miniature rose. The drooping varieties of the ornamental cherries are especially adapted to beautifying small grounds. As single specimens on the lawn they are unique and handsome, and require only to be better known in order to be extensively planted. See next page.

THE DOUBLE-FLOWERING THORN. (CRATÆGUS).—There are no trees or shrubs more attractive than the flowering thorns. They are small trees, attractive in shape and foliage, where they have room, and by judicious pruning can be brought into small limits. They flower in May and June. The foliage is varied and attractive, flowers very showy and often highly perfumed. The fruit is ornamental. There are numerous varieties, all of which are hardy and will thrive in any dry soil.

DOUBLE FLOWERING CHERRY—QUARTER SIZE.

A HANDSOME CRAB-APPLE. (PYRUS).—Few know the beauty of the ornamental crabs. These trees will bear investigation. We illustrate one of the best, with its double, rose-colored and fragrant blossoms, of which the size is shown in the cut. It is the best ornamental crab known. See page 532.

THE CATALPA.—The catalpa has a value as a timber tree. The wood is one of the most lasting known, but it is, in all its varieties, an ornamental tree also, with its magnificently large leaves; and it is, probably, as valuable as the sun-flower in arresting miasma. One species, catalpa speciosa, a western forest tree, hardy up to forty-two degrees North, standing the hard winters there, and to our mind a much finer

FLOWERS OF DOUBLE FLOWERING THORN—NATURAL SIZE.

tree, in every respect, than the southern species, catalpa bignonioides; the northern,

FLOWERS OF CATALPA SPECIOSA—QUARTER-SIZE.

the speciosa, being an upright tree and the other of a straggling form. Its blossoms also open from two to three weeks earlier than bignonioides. There are a number of ornamental varieties, among them golden catalpa, a medium-sized tree with heart-shaped leaves, golden in spring but turning green later; catalpa Bungei, from China, a dwarf with large glossy foliage; Japan catalpa, of medium-size, deep green glossy foliage. Flowers fragrant, cream colored, speckled with purple and yellow; seed pods long and very narrow; flowers about four weeks later than catalpa speciosa.

CHINESE DOUBLE-FLOWERING CRAB—NATURAL SIZE.

THE PERSIMMON.—(DIOSPYROS).—The persimmon is not without beauty, as a small shade tree on the lawn, and the children will not forget it, when its ripe fruits, mellowed by the frost, are lying in the grass. It was a great favorite with the late venerable Arthur Bryant, who loved all that belonged to the forest and grove, as well as his brother, the poet. When horticultural friends visited him, as many did, the handsome persimmon trees on his lawn were admired by these "children of older growth," especially in the later autumn. The tree has a wide range from the great lakes to the gulf, and some varieties in every situation bear excellent fruit. The Japanese persimmon bears superior fruit, but is not hardy North.

X. Everlasting Flowers and Ornamental Grasses.

The ornamental grasses and everlasting flowers (so-called), that is, flowers that retain their color and shape in drying, are considered indispensable in all good collections. Among these should not be omitted panicles of oats and heads of other grains carefully dried, while green, in the shade, and then bleached, if desired, with the fumes of burning sulphur. There are many varieties of some of the species of everlastings mentioned. We give some of the better ones. These are: Acroclinium, white and red; ammobium, white; gomphrena, (globe amaranth), white, flesh-colored, pink and white, and orange; helichrysum, rose, red, white, yellow and crimson; helipterum, white and yellow; rodanthe, white and yellow, purple and violet, rosy purple, etc.; statice, yellow, blue and rose; waitzia, yellow; and xeranthemum, purple, light blue and white.

Besides these, that admirable and truly magnificent plant, Statice Latifolia, with its large trusses of lilac flowers, is most desirable where it is hardy. The cut gives its characteristics of foliage and blossomheads. There are a number of varieties besides the one shown; as S. alba, grandifolia, maritima (sea pink), and S. undulata.

Ornamental Grasses.—Among the giants in this class, are pampas grass, (gynerium) not hardy in the North, and erianthus ravennæ, hardy with slight covering, and fully as fine. The smaller ornamental grasses which we recommend are: agrostis nebulosa, elegant, fine and feathery; arundo donax, perennial, yellow striped leaves; avena sterilis, (animated oat); Briza maxima, one of the best of

STATICE LATIFOLIA.

the ornamental grasses, also geniculata; brizopyrum siculum, pretty; bromus brizæformis, perennial; chrysurus cynosuroides (lamarckia aurea), yellowish, feathery spikes; coix lachryma (Job's tears); hordeum jubatum (squirrel tail grass), fine; lagurus ovatus (hare's tail grass), dwarf, showy heads; pennisetum longistylum, very graceful; stipa pennata (feather grass), magnificent; tricholæna rosea, beautiful, rose tinted. The illustration shows how pretty these are in a simple basket.

ORNAMENTAL GRASSES.

XI. Water Plants.

Where there is water, various plants will make pretty additions to the scenery.

Wild rice and other aquatic plants found in the streams may be used, though the wholesale nurseries will supply anything wanted, or, your nearest nurseryman will order them for you. Nothing, however, is prettier than the water lilies so abundant in the West. The double, fragrant water-lily (nymphæa odorata) should come first. The heart-shaped lily is, to our mind, no less pretty, though not so well known nor so much admired. The nelumbo, or American water-lily, bears an edible bean, and is the American representative of the sacred lotus of the Nile. The water-lilies are easily cultivated by sinking the root, tied to a large stone, into the mud of a pond, or, if the bottom is hard, by tying the root to the top of a stone and covering with muck.

WATER LILIES.

XII. Trellises.

Fig. 1. FAN TRELLIS.

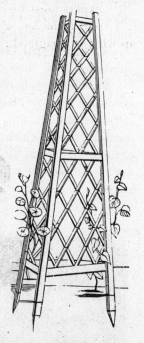

Fig. 3. OBELISK TRELLIS.

Trellises are of various forms and easily made by any one who has a little mechanical skill. Those partly of wire and partly of wood are, many of them, of elegant forms, and are sold by all horticulturalists. When climbing plants or shrubs are grown at some distance from any building, some support must be given. Simple, strong stakes, the rugged stump, or even top of a tree may, with great effect, be covered by ampelopsis, or any of that class of runners. Other climbers must have trellises to conform to their habit, and the height of these trellises must be governed by the plants employed. The three forms we give will illustrate our meaning and fully explain themselves. The fan-shaped trellis is quickly made, of the required size, by slitting the board or siding partly through, and then spreading and fastening. It is shown at Fig. 1. Fig. 2 shows a trellis of uniform width, and Fig. 3, a square trellis, which is contracted at the top and useful for a variety of twining plants.

Too many suppose that floral beauty must be confined to the parterre. Not so. You will find, in the list we have here shown, climbing plants, shrubs and trees which are among the loveliest of nature's gifts. Again, we repeat: be sure of your climate, and that your shrubs, trees, vines and flowers will flourish in it.

Fig. 2. STRAIGHT TRELLIS.

CHAPTER V.

LANDSCAPE GARDENING AND LANDSCAPE TREES.

I. THE LANDSCAPE GARDENER'S ART.—II. STUDYING EFFECTS.—III. DESIGN FOR A VILLAGE LOT.—IV. DESIGN FOR SECLUDED GROUNDS.—V. TREES AND TERRACES—TREE PROTECTORS.—VI. LAYING OUT CURVES OF WALKS AND DRIVES —VII. LAYING OUT AND PLANTING FLOWER BEDS.—VIII. LANDSCAPE EFFECTS.—IX. TREES FOR LANDSCAPE PLANTING.—X. TROPICAL PLANTS.

I. The Landscape Gardener's Art.

THE landscape gardener's art, in all its details, is too abstruse and wide a subject for a book of this general character. When extensive work is to be done, an expert must be employed, but the owner will, of course, have ideas to be carried out, as he has when building a house. Some landscape gardeners are so puffed up with an imagined importance of their calling, and have such a contempt for ideas

ENGLISH OAK.

not emanating from themselves, that they will endeavor to argue away those of their employer, and indeed, often willfully ignore them altogether. In nine cases out of ten, such men are ignorant pretenders. In any event it is well to avoid such, and employ one who can see the value of a suggestion or clearly show its impracticability, and suggest improvements thereon. Avoid quacks in employing men of this or any other profession.

Let us first consider the wants of that large class who live in villages of greater or lesser extent. The kitchen garden, the lawn with its few shrubs and trees, and the parterre of the owners constitute the farm. The mistake of the greater number of this class is in trying to do too much. We have seen more correct taste displayed, and better effects produced, in some village lot, where the principal ground-work was grass, and a few trees interspersed with some graceful shrubbery, and simple winding paths, than in great country parks. The flower-beds cut from the green turf, and the little vegetable garden, hidden by vine-covered trellises, is oftentimes more pleasing to the eye searching for rural beauty than many places of greater pretence, where much money has been wastefully expended. One mistake common to all classes is, they plant too many trees, and trees, too, of a kind that eventually grow to very great size, hiding an otherwise pleasant prospect. In this respect, please remember the adage, if you "plant thick, thin quick." Not many trees of the size of the English oak, shown on page 535, or of the mulberry herewith shown, could find room on the largest lawn, yet the germs of these giants, were, in the one case, once contained in the acorn, and in the other, in a minute seed.

MULBERRY TREE.

II. Studying Effects.

Let us take, for example, a lot 100 by 200 feet, and study what effects can be produced in its adornment. Suppose, in the first place, that the house is set about fifty feet back. If the road-way in front is adorned with maples and elms, so much the better. A winding path, starting from near the corner of the lot, should approach the front of the house, and be carried around to the rear, other paths being cut as necessity or taste may dictate—some, perhaps, to an arbor, others to rustic seats. Deciduous trees should be principally used in front, and evergreens at the side and rear, because the former do not impede a view. Among the paths, small side-beds may be cut for flowers. Remember that a few well-kept flower-beds are better than many unsightly ones. Here and there, in graceful confusion, spread over the lawn, may be placed circular beds, containing peonies, clumps of lilies, and other hardy perennial flowering plants, or, perhaps, such tall-growing tender ones as the dahlia or gladiolus. Half-climbing roses, tied to narrow wire trellises or stakes, produce a very pretty effect.

On a lot the size of the one we are supposed to be adorning, one or two large shade trees are better than twenty. As a shade tree, the linden tree—basswood—is one of the most beautiful. But if these cannot be procured, then the walnuts, the horse-chestnuts, and, of course, the elms and maples are always desirable. The whitewood, or tulip-tree, is a more upright grower, and when aged, is magnificent, especially when viewed from a distance. Among evergreens, the white-pine, hemlock (where hardy), Norway spruce and firs, are easily obtained and desirable. The larch is also a beautiful tree. It is sometimes classed among evergreens, but is deciduous.

If the lot is smaller, and the house necessarily placed near the street, of course, none but shrubs and small-growing trees are admissible—such as spireas, and upright honeysuckles, among the deciduous, and Siberian arbor-vitæ, red-cedar, or even the white-cedar, if kept sheared, making pretty evergreens. The shade trees upon a small lot should not be so large as upon a greater one. Among the smaller-growing sorts is the white birch, with its silvery bark, its toothed and pointed leaves, and gracefully-drooping branches. The ash-leaved maple is a medium-sized and pretty tree, but is not very long-lived. The moosewood, or striped maple, is not large enough for shade, but it is very pleasing to the eye, on small lots. The sassafras tree is also of the middling sized, and not without beauty. It loves a sandy soil. The magnolias, of course, are magnificent when hardy. The common dogwood also is very attractive, especially when covered with its beautiful blossoms and fruit.

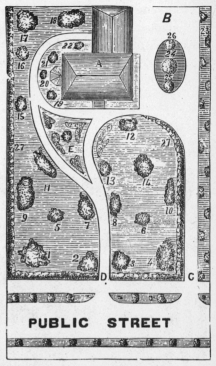

DESIGN FOR A VILLAGE LOT.

III. Design for a Village Lot.

The design we give is for a village lot of only half an acre, yet so planned that the house is more retired from public view and dust than is many a mansion upon a plot five acres in extent. The public street is planted with elms and hard maples, alternately. The lot is surrounded with an arbor-vitæ hedge, except along the carriage road, C, where the hedge is of privet. D shows the main walk leading to the house, A. B is the house-yard; E, the flower garden, the beds for which are cut in the grass; all else is lawn. The strip to the right of the carriage road is planted with evergreens, as shown, and this planting must be of just sufficient density to break the view.

The Planting.—The planting may be as figured on page 537, but it may be changed as desired: 1, American mountain ash; 2, Japan quince; 3, deutzia; 4, European mountain ash; 5, Kilmarnock willow; 6, European weeping ash; 7, Siberian arbor vitæ; 8, red cedar; 9, red bud or Judas tree; 10, Tartarian honeysuckle; 11, Weigelia; 12, Mahonia; 13, rosemary-leaved willow; 14 and 15, deutzias; 16, spirea; 17, purple-leaved berberry; 18, sumac; 19 to 22, next the house, are hybrid perpetual roses; 23, evergreens along the border; 24 is a Norway spruce; 25 and 26 are junipers; 27, hedges. These are all common plants to be had at any nursery.

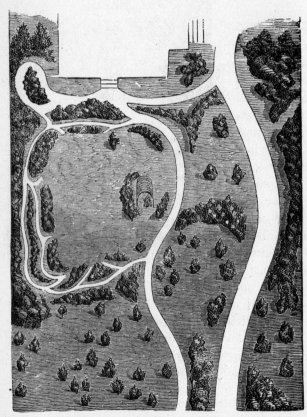

PLAN FOR SECLUDED GROUNDS

Some of the handsome shrubs described in the preceding chapter may very properly take the place of those we have mentioned, and 5, 6, 14 and 17 be replaced by some of the trees yet to be described.

IV. Design for Secluded Grounds.

The plan for secluded grounds is better suited to small farms near villages, or to the better sort of village homes. The grounds may extend in every direction, according to the taste or means of the owner. The house is indicated by the blank space, with a terrace and steps leading down therefrom. The planting is massed to give privacy to the family, and yet not shut out the views. The house has a public entrance on the front, and also one at the side, thus avoiding the necessity of carrying the main drive across the front, which is sometimes undesirable. The grade is presumably nearly level for some two hundred feet from the house, and thence descends to the public road over rolling ground. The lawn is bordered with beds of flowers, and flowering shrubs are massed here and there as shown. Masses of shrubs are also grouped along the line where the land breaks from the level land.

V. Trees and Terraces—Tree Protectors.

In planting trees dig carefully, and carry them with a ball of earth attached.

We do not advise transplanting very large trees. They are seldom satisfactory; especially constructed wagons are necessary and the loading and transporting costly. In large cities the means of doing this work may always be hired, and men found who have all the lifting machinery; but these facilities are lacking in the country. The cut shows a simple stone boat and means for loading any tree that a horse can haul with its ball of earth. Place the cross-piece, M, of the standards, R R, against the trunk, fasten as shown, and it will be easy, by means of a horse attached to the ropes, B, to swing the tree on the boat. (See page 540.)

MASSING FOR EFFECT IN HEIGHT.

The work of digging around trees of considerable size is best done in the winter. When planting let the tree have an ample hole, and fill around it with the best possible soil, leaving a slight depression so it may be watered when necessary.

Massing Trees.—The cut illustrates the manner of massing trees to give the appearance of a hillside. The close, low-growing shrubs in the foreground gradually increase to small, and from that to larger trees. The diagram, Sodding Terraces, shows correct and incorrect forms of slopes for terraces. Fig. 1, the proper slope, both to prevent washing and to preserve the line of beauty. Fig. 2 shows a usual but most wretched form; Fig. 3, the common straight line, which should, however, never be employed except for very short slopes.

Protecting Trees.—Protect all trees liable to be approached by animals. The illustration of a tree protector shows one of the most simple and effective means of doing this, and this protector serves, also, to keep the tree from being blown about. To accomplish the latter object stay the tree to the protector by bands, either of leather or of twisted hay.

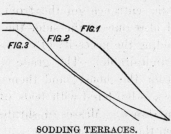

SODDING TERRACES.

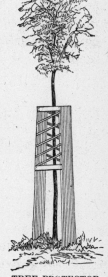

TREE PROTECTOR.

VI. Laying Out Curves of Walks and Drives.

In laying out walks and drives on extensive places, it is better that the guide

540   THE HOME AND FARM MANUAL.

pegs for curves, and those for grades, be made by an engineer. On small areas any person who can handle a spade and shovel and measure correctly, can do the work. Make a proper slip of board, six feet long and two inches wide. Put a hook in one end and a slight notch in the middle; on the other end nail a strip twelve inches long, bored with holes an inch apart. Lay the rod down in the direction the walk is to

TRANSPLANTING TREES.

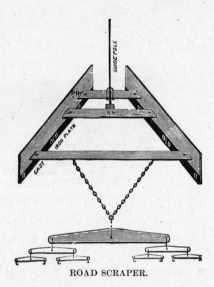
ROAD SCRAPER.

take, the notch at the first peg. If the walk is to deflect one inch in three feet place a peg on one side or the other in the first hole according to the deflection. Then move the hook to the next peg, and the notch to the last peg set, and put a peg again at the hole one inch from the center as before. The deflection will be constant as long as you continue in this way, but if you wish to increase or reverse it change the pegs out, or else upon the other side as the case may be; or notches may be cut at regular distances at the end of the rod and the pegs set by these.

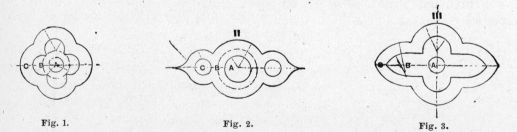

Fig. 1.   Fig. 2.   Fig. 3.

Keeping the Roads in Order.—All roads however made—and they should be of the best material, covered at least six inches deep with good gravel that will cement together—should be crowning in the center, and then kept up to grade. The cut of road scraper shows a large instrument for rounding up public earth-roads. A smaller one, to be worked by two horses, will be useful in any country place for keeping walks and drives in shape.

VII. Laying Out and Planting Flower Beds.

THE formation of the more simple beds will easily suggest themselves by the aid of the diagrams.

Fig. 1 is a simple bed and may be planted as follows: A, coleus la nigra; B, geranium, scarlet; C, verbena venosa; the border of yellow coleus.

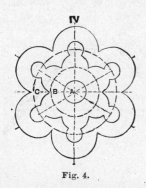

Fig. 4.

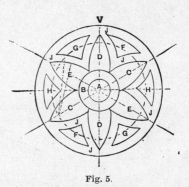

Fig. 5.

Fig. 2 shows a form of flower bed which may be advantageously planted as follows: A, scarlet geranium; B, white verbenas; C, blue verbena; border, with alternanthera.

Fig. 3, A, coleus beacon; B, coleus wonderful; C, coleus harlequin; border, coleus facination.

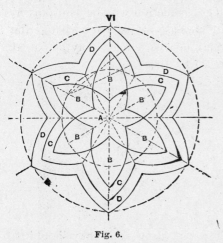

Fig. 6.

Fig. 4, A, coleus vershaffeltii; B, verbena, mixed; C, sweet allyssum.

Fig. 5, A, canna, discolor; B, caladium esculentum; C, scarlet geranium; D, white geranium; E, yellow coleus; F, lobelia, blue; G, sweet allyssum; H, portulucca, mixed; I, achyranthus lindenii, and the border of cineraria maratima.

Fig 6, A, coleus, la nigra; B, coleus, south park gem; C, verbena venosa; D, achyranthus metallica, and the border of coleus charter oak.

These will give good effects as examples of carpet bedding, to be varied at the taste of the operator. For irregular or other beds, flowers of various kinds may be introduced in masses or ribbons as desired.

The full page illustration carefully drawn by the engineer of Lincoln Park expressly for this work will be interesting as showing what is acknowledged to be the

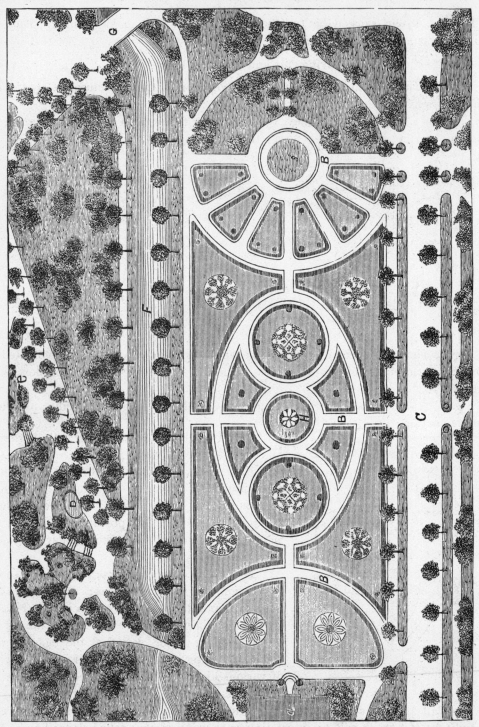

THE GREAT FLOWER GARDEN IN LINCOLN PARK, CHICAGO.

finest example in the United States of floral work on a large scale. The beds will

LANDSCAPE GARDENING AND LANDSCAPE TREES. 543

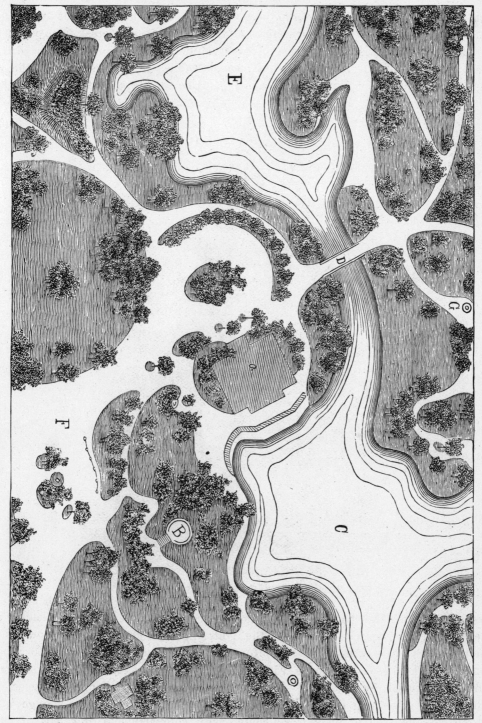

A LANDSCAPE EFFECT IN LINCOLN PARK, CHICAGO.

also show how effects may be produced on a smaller scale. Those here shown being one inch to one hundred feet, will serve to give an idea of the magnitude of the work at Lincoln Park, the whole flower-bed representing an area of eight acres. (See page 542.)

The explanation is: a, entrance to conservatory; B, walks; C, main drive at the west; D, bear pits; e, wolf dens; F, canal leading to lake, shown in the view following, page 543; G, bridge over canal; g, place for proposed fountain; H, large

HEAVY-WOODED PINE (PINUS PONDEROSA).

vase and carpet bed. The other round figures are carpet beds; the dark shading next the walks are the borders of flowers; the lighter portions grass. Let us now describe some of these carpet beds for the information of those who may wish to produce something like them. See also Figs. 4, 5 and 6, page 541. The principal colors will be produced by means of masses of alternantheras, achemenas, escheverias, othona, with something like oxalis for a border. Or, when it is large enough, a vase of upright and trailing plants for a center, with compartments of coleus of appropriate colors and habit of growth. Or there might be a center plant surrounded by coleus, achyranthus and lobelia, with an inside border of verbena venosa, and an outside border of gnaphalium lanatum.

For ribbon beds take any bright-colored geranium for a center, with some dwarf, high-colored zonale geranium for another line, and let the outside be some low-growing coleus. Nothing is prettier than dwarf sweet alyssum for an outside border

LAWSON'S CYPRESS.

if the inside lines of plants are not too large, for massing; it is a brilliant white with plenty of perfume. A pretty bed may also be made of tea roses pegged down, with mignonette interspersed, and for perfume nothing is better than heliotrope. We might extend indefinitely but the indications we have given will suffice.

VIII. Landscape Effects.

THE landscape effects that may be produced where there is room are numberless. We have given exhibits of what may be done on small areas. The full page engraving of A Landscape Effect in Lincoln Park, page 543, represents eight acres, the scale one inch to the hundred feet, showing portions of lakes, drives, hill and valley. A, is the boat-house; B, a hill with look-out from the top, over Lake Michigan; C, part of lower lake; D, bridge; E, part of upper lake; F, part of the main drive, with picketing ground for carriages to the right; G, summer house. Persons of moderate means may produce effects on a smaller scale, while those whose purses are long—and it takes long purses for elaborate landscape effects—may "act accordingly."

IX. Trees for Landscape Planting.

In planting grounds for landscape effect the owner must be guided by circumstances. Hence it will only be necessary to figure some of the better trees, with running notes as to their size and peculiarities.

Magnolias.—The magnolia is always beautiful in foliage and in flower. The glaucous magnolia is almost an evergreen,

MAGNOLIA GLAUCA.

growing in the East, in protected situations, up to forty-two degrees north latitude, and up to forty degrees in the West. It is a shrub-like tree, and hence, valuable for small places.

Arbor Vitæ.—Among the cone-bearers or true evergreens, the Siberian arbor vitæ is an ornament to the smallest lawns. It is not a Siberian tree, as its name would indicate, but a

SIBERIAN ARBOR VITÆ.

sort of the common arbor vitæ, and is the best of all the genus for this country;

exceedingly hardy, keeping its color well in winter; growth compact and pyramidal, makes an elegant lawn tree; of great value for ornament, screens and hedges.

Heavy-wooded Pine.—This is a noble tree, growing to the height of one hundred feet, and hardy in the West. It is a native of the mountains of Oregon and California, a rapid grower, with leaves eight to ten inches in length and of a silvery green color. (See cut on page 544).

Lawson's Cypress. (*Cupressa Lawsoniana*).—This is one of the most magnificent of Californian trees, but not hardy North. In its native habitat it is very large, and with its elegant drooping branches and slender, feathery branchlets, is beautiful; one of the very finest of the cypress tribe. The leaves are of a dark, glossy green, tinged with a glaucous hue. East, this tree is considered tolerably hardy, but is apt to lose its tips in the winter. There are fine specimens on Long Island, and we believe also in Rochester, N. Y. In hilly regions South it would be valuable. The timber is good, clear, easily worked and of strong odor. See cut, page 545.

White Spruce.—The spruces and hemlocks are both useful and ornamental in high degree. Unfortunately, the hemlock does not generally do well when standing alone. The Norway spruce is everywhere as hardy as the oak, and is generally known and planted. The white spruce is not so well known, but is very valuable, hardy and beautiful. It is a native tree of medium size, varying in height from twenty-five to fifty feet, of fine pyramidal form. Foliage silvery gray, and bark light-colored.

WHITE SPRUCE.

X. Deciduous Trees.

There are so many deciduous trees to select from that no one need go astray. Among them none are more beautiful than the scarlet maple, with its varied autumn shades and tints. The maples are also noted for their endless variety of foliage and coloring. Some of these we illustrate, as to their leaves, believing they should be better known.

Maples.—The crisp-leaved maple is an elegant tree of medium size and compact growth, foliage deeply cut, crimped and quite distinct from any of its class. The tripartite-leaved maple is another nursery tree, and certainly handsome; a vigorous, upright grower, with deeply lobed foliage, the leaves being cut nearly to the midrib and three-parted. The young growth is conspicuously marked with white spots, and particularly noticeable in winter. (See cuts on page 555).

Wier's cut-leaved maple originated in Illinois with the gentleman whose name it bears. It is a variety of the silver-leaved maple, and one of the most remarkable and beautiful trees, with cut or dissected foliage. Its growth is rapid, shoots slender and drooping, giving it a habit almost as graceful as the cut-leaved birch.

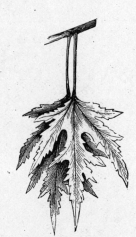

LEAVES OF WIER'S CUT-LEAVED MAPLE, REDUCED.

LEAVES OF CRISP-LEAVED MAPLE—NATURAL-SIZE.

The foliage is abundant, silvery underneath, and on the young wood especially deeply and delicately cut. The leaf stalks are long and tinted, with red on the upper surface. It ranks among the most interesting and attractive lawn trees, and may be easily adapted to small places by an occasional cutting back, which it will bear to any degree necessary, as well as a willow.

Sycamore.—There are no nobler trees than the sycamores. Among the more curious of this species is one with tricolored foliage, a charming variety of the

European sycamore, and said to be identical with the variegated-leaved sycamore. Its leaves are distinctly marked with white, red and green, retaining their variegation all the summer. One of the finest variegated-leaved trees.

Ash.—The ash is valuable for timber, but all kinds do not deserve a place on the lawn. Those most valuable for

LEAVES OF ACUBA-LEAVED ASH. QUARTER SIZE.

LEAF OF A TRICOLORED-LEAVED SYCAMORE, QUARTER SIZE.

timber are the white, the green, the blue and the black ash. An American variety is beautiful as an ornamental tree, and valuable for grouping with purple-leaved trees, the variation being permanent. We refer to the acuba-leaved ash, having

LEAF OF FERN-LEAVED BEECH.—NATURAL SIZE.

gold-blotched leaves like the Japan acuba. Another variety, *punctata*, resembles it closely.

Fern-Leaved Beech.—The beeches are noted for their rich, glossy foliage and elegant habit. The purple-leaved, fern-leaved and weeping beeches are three remarkable trees, beautiful even while very young, but magnificent when they acquire age. As single specimens upon the lawn, they exhibit an array of valuable and attractive features not to be found in other trees. The fern-leaved beech is a tree of elegant habit, and delicately cut fern-like foliage. During the growing season its young shoots are like tendrils, giving a graceful, wavy aspect to the tree. It is one of the finest lawn trees. On page 549 we figure the leaf, showing its great beauty of form.

FERN-LEAVED BEECH.

Weeping Beech.—This magnificent tree should be in every collection, on account of its curiosity both as a summer and winter tree. It is of Belgian origin, remarkably vigorous, picturesque, and of large size. Its mode of growth is curious. The trunk or stem is generally straight, with the branches tortuous and spreading; often ungainly in appearance, divested of leaves, but when covered with rich, luxuriant foliage, of wonderful grace and beauty.

The Birch.—The birches are all both handsome and useful. One variety has the well-known fragrant bark, and the bark of others is used by the Indians in making canoes. The wood is valuable for fuel, and no tree is prettier on the lawn

WEEPING BEECH.

than the cut-leaved weeping birch. Beyond question it is one of the most popular of all weeping or pendulous trees. Its tall, slender, yet vigorous growth, **graceful**

CUT-LEAVED WEEPING BIRCH.

drooping branches, silvery white bark, and delicately cut foliage, present a **combination** of attractive characteristics rarely met with in a single tree.

The Linden, or Lime.—Every woodman knows this tree as bass wood. It is valuable for lumber when it can be kept dry; its bark makes the bass matting of commerce. The lindens are all beautiful, and their flowers fragrant and much sought

WHITE-LEAVED WEEPING LINDEN

after by bees. The ornamental varieties are numerous. Among them are the gold-barked, fern-leaved, and the white-leaved weeping linden, the latter of which is illustrated herewith. Its large foliage and slender, drooping shoots are unsurpassed. Another variety, the white-leaved European linden—a Hungarian species—

is a vigorous growing tree, of medium size and pyramidal form, with cordate acuminate leaves, downy beneath and smooth above. It is particularly noticeable among trees by its white appearance. Its handsome form, growth and foliage render it worthy to be classed among the finest of our ornamental trees.

YELLOW WOOD—VIRGILIA LUTEA.

Yellow Wood—(*Virgilia Lutea*).—This is one of the most beautiful of American trees; tree and foliage fine and blossoms sweet. It is a tree of only moderate growth, broadly rounded head, foliage compound like that of the locust tribe, and of a light green color turning to a warm yellow in autumn; flowers pea-shaped, white, sweet scented, appearing in June in great profusion, in long drooping racemes covering the tree.

SCARLET MAPLE.

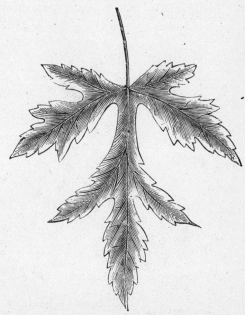

LEAF OF MAPLE, VARIETY TRIPARTITUM.
HALF NATURAL SIZE.

The Elms.—What is more magnificent than an elm, with its graceful outlines and varied forms! The English think their elm the finest of their native lawn trees. But it cannot compare with our drooping forest elm. Yet the English elm has a rugged beauty of its own.

X. Tropical Plants.

The Ivory-nut Plant.—This plant is not hardy in the United States outside a small portion of Florida and California, and is only met with in conservatories. It is a plant of curious habit, and its nuts have some value in the manufacture of small articles in imitation of ivory. The tree is a native of the northern regions of South America, extending just across the Isthmus of Panama, large groves of it having been discovered not long since in the province of that name. It banishes all other vegetation from the soil it has taken possession of, has the appearance of a stemless palm, and consists of a graceful crown of leaves twenty feet long, of a delicate pale green color, and

ENGLISH ELM.

divided, like the plume of a feather, into from thirty to fifty pairs of long narrow

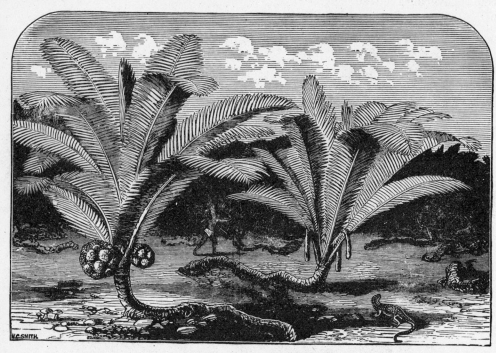

THE IVORY-NUT PLANT.

leaflets. It is not, however, really stemless, but the weight of the foliage and the fruit gives it this appearance sometimes. Where it is seen it is like a large root, stretching along the ground for nearly twenty feet. The long leaves are employed by the Indians to cover the roofs of their cottages. The fragrance of the flowers is most powerful, and the tree produces a large, roundish fruit, from eight to twelve inches in diameter, and weighing, when ripe, about twenty-five pounds. The seeds of this fruit constitute the vegetable ivory of the commercial world.

DECIDIOUS CYPRESS OF THE SOUTH.

PERSIMMON TREE.

Above are shown, also, the decidious cypress of the South, and the persimmon tree—the latter described on page 532.

CHAPTER VI.

FISH AND FISH PONDS.

I. FISH ON THE FARM.—II. FISHES FOR CULTIVATION.—III. RIVER AND POND FISH AND THEIR TIME FOR SPAWNING. —IV. THE FAMILIES OF RIVER AND POND FISH.—V. RULES FOR THE TRANSPORTATION OF FISH.—VI. ARTIFICIAL FISH BREEDING.—VII. HATCHING THE FISH.—VIII. FISH-HATCHING BOXES.—IX. BREEDING FISH IN PONDS.—X. CARP BREEDING.—XI. HOW TO FORM THE POND.

I. Fish on the Farm.

THE breeding and care of food fishes has, within the last few years, been given so much attention, that a volume of the character of THE HOME AND FARM MANUAL would be incomplete without some practical information on the subject. It is appropriately placed in the part relating to horticulture, because artificial ponds and lakes are intimately connected with landscape gardening.

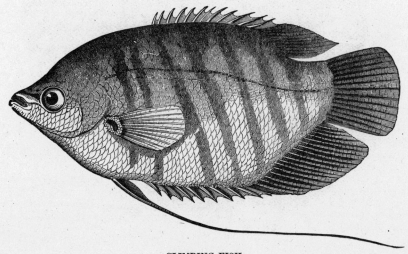

CLIMBING FISH.

The Federal Government has spent large sums in experiments upon fish breeding and their general management, and many of the States now have regularly organized fish commissions, for stocking the local waters. Shad have been introduced into the Ohio River, where they were heretofore unknown. These fish must annually seek salt water, and they naturally return to their homes for spawning. This they have successfully done through the turbid waters of the Mississippi, and returned to the places where they were originally introduced. This shows the success of this enterprise. A few years ago eels were unknown in the great lakes above the Falls of

Niagara. During the present season they were found in great numbers about a wreck in Lake Michigan.

Some Usual Mistakes.—If the introduction of other fish, notably of brook trout, has failed in western rivers and ponds, the failure is what any intelligent person might have predicted. They are essentially a clear and cold water fish, and thrive in no other kind of water. In waters, ponds or streams, fed by cold springs, they have been most successfully bred. Five years ago, we, with a party of horticulturists, ate

A FISH NURSERY.

large, handsome speckled trout at the home of Dr. Pratt, near Elgin, Ill. These had been bred by him, and were taken from his hill-side ponds. There, at any time, they might be seen, by dozens, sporting in the clear waters of his cedar-shaded miniature lakes.

A fatal mistake in fish culture has been made by many in the introduction of the pickerel of the West. By every possible means these predatory tyrants should be excluded. They relentlessly and ferociously pursue all fish smaller than themselves, and will soon decimate the finny inhabitants of an artificial pond. Whoever cultivates pond fish, must use the same discretion as the breeder of other farm stock. That is, he must breed to a purpose, and for a purpose.

II. Fishes for Cultivation.

TROUT may be bred in any clear pond fed either by springs or by a brook, and in which the temperature of the water seldom rises above sixty-five degrees; or in any clear water abounding in deep, shaded holes. Those who have large lakes may

POND AND FISH-WAY.

breed various fishes, since such waters have an inlet and outlet, and are adapted to a variety of species. The owner of small ponds must content himself with perch, cat-fish, roach, dace and carp. In the natural spawning of fish comparatively few eggs are hatched—less than one in ten, and many of them are destroyed before they reach an eatable size. In the economy of nature enough are hatched to keep up the supply, but when the seine, the gill-net, the traps of the fisherman, and the endless fishing tackle of the anglers are constantly at work, not only are stream and lake fish destroyed, but even the salmonidæ, which frequent our coasts and estuaries, have already been so decimated that the enhanced price has, at length, brought about legislation for their protection and preservation.

Of game fish we have little to say. The pond fish to be cultivated we have already indicated, and these, with a list of the principal fish of interior lakes, ponds

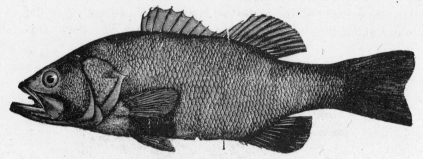

BLACK BASS OF THE WEST.

and streams must suffice. Those fishes which are of a strongly predatory character will be marked thus *; the game fishes admissible marked with a †.

STRIPED OR BRASSY BASS OF THE MISSISSIPPI.

III. River and Pond Fish and Their Time of Spawning.

We give herewith a list of lake and pond fishes of the West, with their weight and the approximate number of eggs each will spawn:

Yellow perch, weighing 3½ ounces, gave of eggs,					9,943
White fish,	"	2 pounds,	"	"	25,076
Herring,	"	5¾ ounces,	"	"	265,650
Roach,	"	12 "	"	"	480,480
Brook trout,	"	8 "	"	"	600
To this list we add the shad, weighing ——, gave of eggs,					25,000

We give below the period of spawning of some of our principal river, lake and pond fish. It is calculated for the New England States, but will be useful everywhere, since a change of the calculation for differences of latitude will give an approximation:

† Perch pike (lucioperca Americana), last of April and first of May.
* Pickerel (Esox reticulatus), last of April and first of May.
Yellow perch (perca flavescens), April and May.
White perch (merone Americana), June.
Roach (pomotis appendix), May.
Sunfish (pomotis vulgaris), May.
Sucker (catostomus), May.
Rock bass (centrarchus æneus), May.
† Bottom pike (lucioperca *var.*), May.
Mullet (catostomus), June.

BROOK TROUT.

Black bass (grystes fasciatus), June.
Hornpout catfish (Pimelodus), September.
Trout in brooks (salmo fontinalis), October and November.
Trout in artificial ponds (salmo fontinalis), February and March.
White fish (coregonus albus), October and November. Deep water lake fish.

IV. The Families of Pond and River Fish.

Some of the principal families of fish contain a number of species. We again mark the pike family * as not to be countenanced, and the salmon family † as adapted only to deep or very cold, pure waters. Only the common names (as known in the West) are given.

Perch family.—Yellow perch, pickerel of the lake, sunfish, rock bass, grass bass, black bass of the lake and black bass of the Ohio River—two distinct fish, although bearing the same name; dwarf bass.

HOG FISH FAMILY.—White perch of the Ohio River, sheeps-head of the lake, hog fish, blenny-like hog fish, spotted hog fish, variegated hog fish.

CARP FAMILY.—Carp of the Ohio, mullet of the lake, Missouri sucker, white sucker, red-horse sucker, buffalo sucker, brook sucker, spotted sucker, mud sucker, white sucker, black sucker, rough-nosed dace, stone-roller, silver shiner, large shiner, red-bellied shiner, red-bellied shiner of the lake, white and yellow-winged shiner, horned chub, red-sided chub, gold shiner, flat shiner, chub-nosed shiner, flat-headed chub, mud minnow.

* PIKE FAMILY.—Masquallonge pike, black pike.

CATFISH FAMILY.—Blue catfish, yellow catfish, channel catfish, mud catfish, bullhead, yellow backtail.

† SALMON FAMILY.—Mackinaw trout, speckled or brook trout, shad of the lake, white fish.

SHAD FAMILY.—Gold shad, hickory shad, larger herring, lesser herring, moon-eyed herring, (dog fish, duck-bill gar, alligator gar, common gar). The last named, in parentheses, are worthless and savagely predatory.

V. Rules for the Transportation of Fish.

ONE of the best authorities on the subject, Mr. Coste, early called attention to the care necessary in the removal of fishes from one part of a country to another. These rules have since been generally adopted. By these means fish may be successfully brought over the ocean, or transported by railway from the Atlantic to the Pacific. The rules to be observed are:

1. Very young fishes should be selected.
2. These fishes should be distributed among several receptacles.
3. Care should be taken not to crowd too many together in one receptacle.
4. The water should be renewed partially or entirely whenever it becomes necessary.
5. It should also be aerated from time to time.
6. The fishes should be fed whenever they shall seem to require it.
7. The remains from the food which has been given to the fishes should be carefully taken up from the bottom of the receptacle, and removed within eight hours after feeding; the dejections and other impurities which would injure the water should also be removed.
8. Finally, the several receptacles should be kept in different places, and under various conditions.

VI. Artificial Fish Breeding.

THE breeding of fish—hatching and rearing artificially for transportation and the stocking of ponds—requires accurate knowledge, and constant care and attention. The water must be pure. It is not admissible on the farm, because the young fry are now so safely transported, that it is cheaper, in those States where there are no fish commissioners to supply them, to purchase what are needed. They may also be obtained, gratuitously, upon application to the United States Fish Commissioners at Washington; nevertheless, we give cuts showing an indoor and an outdoor apparatus, for such amateurs as may wish to undertake the labor as a pastime.

Spawning.—The fish, ready for spawning, must of course first have been bred in proper ponds, whence they may be taken in hand-nets, or else procured from natural waters. It is a fact well known to naturalists that the eggs of fishes are fertilized or impregnated *after* they are deposited by the female parent. The fact that it was possible to obtain the eggs unimpregnated, and yet in a condition to be susceptible of impregnation, and the further fact that the *milt*, or spermatic fluid, could be obtained in such a condition as to be preserved for many days without loss of its virtue, suggested the idea of artificial impregnation, and successfully practiced first in France, Germany, England and Scotland, and later in America. The process of this artificial impregnation is as follows: Having placed on a table or other convenient place a perfectly dry porcelain or other non-corroding dish, then immerse the hands in water, and hold the female fish with say the left hand, the pressure of the hand being immediately behind the gills; hold the fish upright; some ova may escape by the action of gravity and muscular contraction. Dip the right hand in water, and clasp the body of the fish in such a manner that the thumb may be gently pressed along the abdomen. If no eggs are extruded by a gentle pressure, replace the fish in its element, because either it is not sufficiently ripe or else is diseased. Never handle a breeding fish with *dry* hands, for the reason that the glutinous covering of the fish adheres to the dry hands, to the very great and absolute injury to the fish. But when the fish is fully ripe the eggs extrude with a very gentle pressure. In no event must the abdomen be *squeezed*, because squeezing might rupture the air-bladder or injure other of the viscera.

ARTIFICIAL EXTRUSION OF EGGS.

When the abdomen has been emptied of all the eggs, then seize the male and treat him in the same manner that the female was treated. A few drops of milt or spermatic fluid will be the result of this process. The milt should be dropped from the body of the fish on the eggs or ova directly, and as soon as the milt is dropped pour sufficient water to cover the eggs, and stir them with a quill, glass rod, or tail of the male fish. There is no objection to the dish which receives the eggs and milt containing a very little water, but it is not now used by the best cultivators.

While the stirring of the mass is going on, the eggs are undergoing great changes; prior to the introduction of the milt or zöosperms, they were in a manner agglutinated and in a flaccid condition; now they have become enlarged, and now translucent; each egg, no longer coherent, is an individuality, and by one of those mysterious processes by which Nature works, they are become hard to the touch, so that they will roll about like shot on a smooth surface. Here, now, we have the vivified germ, the embryo fish. In this state they are taken, cleansed in one or two waters, and carefully placed upon a bed of gravel or upon wire-cloth trays, and with a feather evenly distributed over the surface, the object of such spreading being to allow the clear, living water to come continually in contact with all the eggs—well-oxygenized water being as essential to a normal, healthy development of the embryo, as it is material to the life and growth of the fish in its subsequent stages. Now, with pure and perpetually-running water, filtered, if necessary, by one or more flannel screens, with clean tools, clean surroundings, and with clean hands, we enter upon the work of incubation, a labor lasting five, ten, twenty, forty, eighty, one hundred and twenty days, or even longer, depending upon species, and upon quality and temperature of water.

Dead eggs, easily distinguished, whenever discovered, are to be at once removed, as they produce a byssus that sends out its clammy, fibrous arms to destroy every living egg within their reach, and all sediment and substances of every sort foreign to the before-named conditions of their health and growth are to be sedulously guarded against.

VII. Hatching the Fish.

THE eyes first appear, then a faint embryonic structure, and soon after a dim outline of the coming fish may be seen, growing more and more visible each day, until some morning you see the wreck of a habitation floating down the current, and a tiny creation, most unmistakably alive, settled down amid the interstices of the gravelly bed, or meshes of the wire tray, a third, or a half, or perhaps, three-fourths of an inch in length. About the most perceivable thing of this new birth, is a bag or sac attached to the belly of the fish. This sac, with the salmo quinnat, is of a rich pinkish color, resembling one or two drops of blood incased in a semi-transparent membranous bag. At birth, it is larger than the fish itself, rendering all movements of the new comer exceedingly awkward and clumsy. This is the umbilical vesicle, or yolk sac—Nature's store-house for the supply and sustenance of the fish during its tender infancy. Until this sac is absorbed, the fish will eat nothing, seems to desire nothing but to be let alone, content with the pabulum stored in its little knapsack, from which it daily, hourly draws that nourishment, the provision and pottage of birthright. Day by day the sac becomes smaller, till it can scarcely be perceived with the naked eye; then the fish begins to move about, as if in quest of something to satisfy its hunger. This yolk sac, with the salmon and trout and some other species, lasts from thirty to forty days; with other varieties, not so long, During the existence of the umbilical vesicle the fish are known as alevins; afterwards, up to certain periods of growth, minnows or fry. The sac being absorbed, the fry should be fed two or three times a day, or oftener, in limited quantities, will do no hurt. Various kinds of food are given—bonny-clabber, yolk of an egg, boiled calf's or beef's heart, boiled hard and grated; liver of all kinds (except hog's liver), chopped or grated so fine as to become the consistence of thick blood, mixed with a little sweet cream, are used as food, while the fry is very young. Under proper care and feeding, the fish will come on rapidly, so that in a few days or weeks they will do to be removed from their hatching-troughs and planted in the lakes and rivers, there to grow and to bear testimony that fish culture is neither a myth nor a phantasm, but an ocular, tangible and gustible reality.

VIII. Fish Hatching Boxes.

On this page is a cut of one of a series of out-door hatching boxes used by Mr. Francis, of England. He says: A spring, from which a rill flowed, was first obtained; there was a considerable fall in the run of the water, which was very advantageous; nevertheless, the plan here adopted can be applied more or less to any stream. We first bricked up the little rill so as to form a reservoir (1) and raise the water to a higher level; we covered the reservoir in with a large stone to keep out dirt and vermin, and placed at the lower end of it a zinc shoot, (2) over which the stream flowed. Immediately under this we placed our first box, a fac-simile of which is given.

OUT-DOOR HATCHING BOX.

It was made of elm, four feet long, and fifteen inches wide in the clear, and ten inches deep. At the upper end of the box a projecting zinc trough (3) was fixed to catch the water, this trough being about three-quarters of the width of the box itself. At each end of every box a piece was cut out six or seven inches in width, and through these the water flowed into each box. (These openings were not carried all across the boxes, as the shoulders left made an eddy very favorable, as quiet resting places, to the young fry when first hatched.

If the stream be at all strong, artificial eddies should be created by sticking small pieces of perforated zinc upright in the gravel at intervals along the sides and across the stream; behind these the helpless fry can be in safety. The top cut, which first received the water, being secured from foes without by being covered with perforated zinc through which the water flowed, and the further end one having a zinc shoot to deliver the water, and also a perforated zinc face, not only to keep foes out, but the fish in. Fastened over the cut, in the lower end of the first box, was a short zinc shoot (5) to convey the water into the next box over the corresponding cut, so that no water should run to waste between the boxes. Thus, when No. 1 was fairly placed on a brick foundation so as to receive the water in the zinc trough, all that was required was to insert the shoot at the other end of the box into the corresponding cut of No. 2 box, and slide No. 2 safely and closely up into its place, and so on with Nos. 3, 4 and 5, etc.

These boxes were then partially filled with coarse gravel of the size of gooseberries, and some larger, even to the size of plums, for the more irregular their shape the better, as there will be more interstices between them, in which the ova can be hidden, and the little fish, when hatched, can creep for safety. The gravel was at a level of about an inch below the cut which admitted the water, an inch depth of water being quite sufficient to cover them. Each box was furnished with a lid, comprising a wooden frame-work, and a perforated-zinc center. This lid was made to fit closely by means of list being nailed on all around. It was padlocked down, to keep out inquisitive eyes and fingers. Boxes in exposed places should always be covered in, if not with coarsely-perforated zinc, yet, with fine wire netting, or water mice will get in, and various birds, as moor-hens and dab-chicks, will pick out the spawn, while a king-fisher, should he discover them, will carry off the fry by wholesale. The stream was then turned on, and flowed steadily from box to box throughout the boxes, and finally discharged itself by the end shoot into the bed of the rill.

IN-DOOR HATCHING BOX.

It need not be imagined that a full stream is necessary, for a small amount of water is sufficient. Indeed, a flow of water, say through a half-inch pipe, would be enough, perhaps, though it is advisable, while the ova are unhatched, to have more, so that there shall be more stream and movement in the water, and consequently, less time for deposit to settle; so that we had on, perhaps, as much as a stream three-quarters of an inch in diameter. When the fish are hatched, half that quantity would be preferable, as they are not well able to struggle against a stream, and would be carried down, perhaps, to the end box, and so, against the perforated-zinc face, where they would stop up the holes, and finally be smothered.

We give this, mainly to show the care and difficulties of fish breeding. No person should attempt it who has not abundance of leisure, patience and a real love for the work. For such amateurs as have these requirements, we show a simple in-door hatching box, that may be supplied by the house hydrant. In the cut showing In-door

Hatching Box, 1, is a frame-work of glass rods; 2, tank with eggs resting on gravel; 3, catcher; 4, hand net.

IX. Breeding Fish in Ponds.

The ponds once stocked, it will be only necessary to provide some simple, artificial means for spawning that the fish will seek for themselves. An example is illustrated on next page of an Artificial Spawning Bed. These will be necessary only where natural spawning beds are not present. To make the artificial bed, a framework of poles and laths interspersed with twigs, boughs and acquatic plants is laid in

SUCCESSION OF HATCHING BOXES.

the pond, so as to form an irregular structure. This is weighted down with stones, and may be of any required form. When not in use, it can easily be drawn out upon the bank. (See page 568.)

X. Carp Breeding.

We have given information enough to enable our readers, who have ponds sufficiently deep, to stock them with bass, perch, sun-fish, etc. If they have ponds not more than three or four feet deep they may cultivate that valuable pond fish the carp, if they can form a hole therein six feet deep where the fish may in winter lie secure from freezing. These can live only in water so shallow that aquatic plants, which are their only food, may grow. Where much underdraining has been done the accumulated water may serve to feed a small pond. If you have a spring on the farm carry that to the pond if possible. There is one thing worthy of being remembered; Water that supports both animal and vegetable life is never stagnant and hence is fit for stock-water. A stagnant pond is not.

XI. How to Form the Pond.

THE German carp is a pond fish, requiring warm water; in cold weather it hyberates in the deep part of the pond. This fish has long been artificially grown in Germany, having been brought originally from China, where it has been cultivated from time immemorial. Its growth is exceedingly rapid when the condition of the water is proper for producing vegetation. If the pond is fed by springs, or from deep-laid tiles, the cold water will naturally settle and flow along the bottom. There should be an outlet near the bottom through which the cool water may be drawn off for use in watering stock, etc. The pond should be, say four feet deep, with one or two pools six feet deep or more. A waste-way should be provided for the surplus water, and a

ARTIFICIAL SPAWNING BED.

sluice for drawing off the water of the pond, when necessary, for cleaning, etc. In the mud of the deep pools, just described, the fish pass the winter, in a torpid state. In this deep portion they will congregate when the water is drawn off, the sluice-way being protected by a grating. This waste-way may be made at any point along the dam, and is simply a continuous box, say six inches or more in diameter. The passage-way for the water being about twelve inches below the top of the dam. When the water reaches the proper level it will pass up from the bottom, over the top and through the sluice, either directly to the trough, or may be conducted any distance to it, or to a series of troughs at any point below.

It is well-known that water is a poor conductor of heat. That is to say, the heated water at the surface extends down only a short distance. The carp like warm water. This is also conducive to aquatic vegetation, the principal food of carp. Hence, the necessity of drawing off the surplus water from below. If game fish are required, take the water directly from the top, since the requirement would be that the water be kept as cool as possible. Where drainage water is used to supply a carp pond, the point where the water enters from the mains must be higher than the highest water-mark, to prevent the deposit of silt, and it would be better if the water of drainage could be conducted directly, in pipes, along the bottom of the pond, to the deep pools. The water out-flowing would be cooler for stock; the warm water of the shallow portion of the pond would retain its heat, and an abundance of food be provided for the carp.

CHAPTER VII.

COMMON SENSE TIMBER PLANTING.

I. THE ECONOMY OF TIMBER.—II. WHAT TIMBER REALLY DOES FOR A COUNTRY.—III. WHAT TIMBER TO PLANT.—IV. OUR EXPERIENCE IN TREE PLANTING.—V. THE POETRY OF THE FOREST.

I. The Economy of Timber.

IN the mountain country strict legislation should prevent the destruction of the forests. There the rainfall should be soaked among the roots, and the leaves and twigs retained at the foot of trees, whose shade prevents the sun's rays reaching the ground in full force. Then the water flows gradually away, giving to the country perennial rivulets. Remove the trees and this same water forms torrents which sweep before them the soil, leaving the hillsides bare, and, the rainfall over, the sun bakes the land into a desert. The rivers rise in a night from wretched streams, twining along through enormous beds of sand and silt to resistless currents bearing destruction to everything in their course. In a level country all the conditions are different, and in this, an agricultural work, the fair, arable land is all of which we consider the needs. So far as regards the great acreage of the lands of the United States the necessity for dense belts of timber is not as great as it has frequently been represented to be. The work of our horticultural and forestry societies has kept the people alive to the importance of the subject, although some of them have published assertions not borne out by later and more careful investigations. The writer once held somewhat extravagant notions which a more careful study of the subject did not confirm. One was upon the amount of forest per square mile of a country necessary to promote the best results in the cultivation of field crops. He finally settled upon the French standard of one-quarter of the area of a country. This we have modified from time to time until now we assert, without hesitation, that one-tenth of the area of a country, equally distributed over farms as they may be in the prairie and plains region of the United States, will be amply sufficient, not only for the uses for which timber is designed, but also to insure the best results in the cultivation of crops.

For Cutting into Lumber.—In the production of valuable timber the above proportion will be fully equal to one quarter of the area of a country in wild forest, especially where such timber is grouped into large and broken areas of country. In cultivated forests valuable varieties are planted, and upon soil congenial to them. They are planted at proper distances, and the result is the product, per acre, for each year bears the same ratio of product that any cultivated crop does over a wild one. So far as the protection to crops is concerned, that afforded by carefully planted trees, of the proper kinds, placed scientifically where the greatest benefit may be derived from their growth, greatly exceeds that received from larger but widely scattered forest areas. And since the cultivation of timber is now generally conceded to be fairly remunerative, there is no reason why all, who have no timber, should longer delay its planting.

II. What Timber Really Does for a Country.

TIMBER, generally distributed, modifies the climate in that the rains are more frequent and moderate, and the climate is, as a rule, more humid. The sun does not create so great a heat during the day, the radiation is less at night, and hence, the temperature is more equable. The average rainfall of a country may be the same with or without timber. With timber it will be more equally distributed. A country may have the same average yearly pressure of wind with or without timber, but the probability is that a timbered country will be less subject to tornadoes, tearing every work of man from their pathway. It is not laborious this planting of timber—not more so than the intelligent cultivation of corn.

Do we hesitate to wait so long as we must for the returns of our labor? We have to wait three and four years for our steers and colts to grow into usefulness, and from five to ten years on our orchards to pay for the labor bestowed on them. Five to ten years will give us poles, posts and fire wood from our planted trees, and twenty years will give us timber.

III. What Timber to Plant.

To RAISE timber for profit we must plant, first, those varieties which enter most largely into the economy of civilized life—conifers, white pine, Norway spruce, Scotch pine, arbor vitæ, and European larch; of decidious trees, white ash, black walnut, sugar, silver and soft maple, and the hardy Western catalpa (*C. speciosa*). Second: In planting a forest, in a prairie country, set apart a portion of the land for trees of rapid growth, to use before the slower growing species come forward. Among these rapid growers the cottonwoods, white and golden willow, and silver maple are the best.

We have already stated, in the appropriate chapters, that trees should be planted to ornament the landscape and to shelter stock and buildings. We have no very correct data in this country on the growth of forests. This we know—that they will reach a given size in two-thirds the time it requires in Europe where forestry has been systematically carried on for many years.

Time of Growth.—The late Dr. Warder, one of the fathers of forest planting in America, stated that the coppice growths in European forestry are often utilized in periods of ten or fifteen years; in our own country, too, we have many trees of short rotation, and some of the most useful and most profitable trees are of this character.

The black locust may be harvested after it has grown from twenty to thirty years.

The catalpa speciosa in the same period will make good cross-ties and fence posts.

The ailanthus very soon attains a useful size, and for certain purposes has been very highly commended, both in this country and in Europe. Prof. C. S. Sargent is advising its extensive plantation, and some years ago it was spoken of as the most promising tree for the arid plains of the Southwest.

The forests of Scotch pine in Germany are allowed sixty years to reach their useful size for fuel and timber.

The birch there reaches its maturity in about half a century.

The willow, used for charcoal needed in the manufacture of gunpowder, may be cut after growing twenty years or even less.

Chestnut, in its second growth, is most profitably cut every twenty or twenty-five years.

The wood of the wild cherry soon reaches a profitable size for many purposes, though for saw-logs and lumber the trees should be larger.

Many individual trees, planted by the pioneers upon the broad plains of Nebraska, within the few years the white men have occupied the so-called "American Desert," have already attained to useful size and will yield each a cord of fire-wood.

Protection and Fuel.—The protection from cold winds afforded by groves of wood is also an economic consideration. A well-sheltered dwelling-house requires less fuel to warm it in cold weather than an exposed one. Animals, if well sheltered, need less grain and forage. Crops, too, are benefited by shelters. I have repeatedly, in cold backward seasons, noticed the difference between corn protected by woodland and that on the open prairie.

From ten to twenty years, varying with the rapidity of growth of the trees planted, is, with proper care, a sufficient time to raise a grove of timber large enough to constitute an efficient protection, and to supply wood for most of the purposes for which it is needed on a farm. As has been often said, the trees will be growing while he who planted them is sleeping. No man who had grown a shelter of that sort would have it removed for ten times the cost of raising it.

For Other Uses.—The trees most suitable to plant for economic uses are to be noticed. Where it is desirable to have wood for shelter and use as speedily as possible the white willow and silver maple are perhaps the best, as they are easily raised and grow rapidly. The white willow thickly planted, produces long straight poles, which are very serviceable in making fence and will last a long time. It may also be made useful as a screen for plantations of other trees which can not so well endure the buffets of unchecked winds on the naked prairie.

The trees should be planted four feet apart each way, and thinned as occasion requires. The poles cut in thinning will be found useful. At eight feet apart the larch will, in twelve or fifteen years, grow large enough for posts and railroad ties. Grown thus thickly, the tree shoots up perfectly straight, with small side branches, so that the entire trunk is available for use. I have trees eighteen years planted, more than fifty feet high and straight as arrows. Some have supposed that as the American larch is commonly found in swamps the European species has the same habitat. This is a mistake—it should be planted on dry ground; it will thrive in rocky, barren soils; poor, broken land suits it better than rich, flat prairies.

IV. Our Experience in Tree Planting.

Our individual experience with tree planting may be stated as follows: Cottonwoods planted sixteen years, along the street, will measure from sixteen to twenty-two inches in diameter near the ground, and will average nearly a cord per tree if felled. Had they been planted sixteen feet apart and kept free of weeds for three years, their average would have been about sixteen inches each in diameter, and they would have made over one-half cord each of wood; certainly eighty cords per acre. Wild cherry planted at the same time are twelve inches through near the ground.

Balsam Firs.—A row of balsam firs, planted at the same time, eight feet apart for a wind-break, are from nine to fourteen inches through, about forty feet high, and so thickly clothed from the ground up, that a man cannot separate the branches to break through them. A Norway spruce, standing single, is thirty feet high, and fourteen inches through next the ground, the branches regular from the roots up, and of a circumference of sixty feet.

Black Walnut.—A black walnut of the same age is thirteen inches through near the ground, and a white walnut—butternut—is of the same diameter; both have borne regular crops—a half-bushel each per year—for the last six years. Golden willows are twenty inches in diameter of trunk, and linden, twelve years planted, are eleven inches in diameter.

Hard and Soft Timber.—Thus we see that hard wooded timber will in this time be large enough for any of the ordinary uses of the farm, and that soft and fast growing timber will, in from twelve to sixteen years, become large trees, and will be worth for fuel, in any region where fuel is scarce, fully $350 per acre for cord wood.

White willow will grow nearly as fast as cottonwood, and is really an excellent substitute for hard wood for fuel and rails, until better can be grown.

When a Nuisance.—A good deal has been said, first and last, about the nuisance of planting lines of trees along the roadside as wind-breaks; that thus they cause snow to lodge in the road, often rendering it impassable in winter, and keeping it wet and miry for a long time in the spring. There is a good deal of force in this argument, and we have always held that it should not be done, unless the planting were so open as to allow fair passage to the wind.

Another objection, and in many cases a serious one, is, that crops suffer for three or four rods next such rows of trees in the fields adjoining. The remedy, however, is simple. Do not plant trees unless the crop next them may be pasture or meadow. Grass will do fairly next trees where other crops will not. Why? Simply because there is generally moisture enough in the spring and fall for the grass, but the roots of trees in the summer absorb a large share of the moisture, to the detriment of crops of grain, and especially corn.

The Remedy.—How, then, shall we obviate snowdrifts from roadside plantings?

Either by planting at such distance, and of such trees, as create but little impediment to the wind, or, else, by planting lines of sufficient width to catch the snow within their own area.

Trees planted at a distance of from thirty to sixty feet, with the limbs sufficiently high to allow the passage of teams under them, do not collect drifts.

Lines of trees for wind-breaks should never be less than sixty to one hundred feet in breadth; but if of this width they will catch the snow-fall and drift within their own shade.

The great mistake made by railway companies in planting along the lines of their roads is, the lines of trees have not been of sufficient width to catch the snow, but have in many instances allowed it to sift through and fill the very cuts they were intended to protect.

Now protection from drifts would have been certainly accomplished if either deciduous trees, or better, evergreens, had been planted in strips of sixty to one hundred and fifty feet wide, according to the annual snow-fall, and experience; the latter width being sufficient for any climate except some mountainous districts, such as are found, for instance, upon the line of the Union Pacific. So, a wind-break sixty feet wide, when tolerably grown, will protect any of our railways in the West, and the road-bed, if well graded, will not suffer seriously from mud in the spring.

V. The Poetry of the Forest.

As a Western man, and one who, as man and boy, has seen Illinois, (now a great and populous State) redeemed from the grip of the savage Indian, we have learned to appreciate the value of forests, and love their varying beauty. Who, among the dwellers upon the great prairies of the West, has not looked back sometimes, with longing, to his childhood home, with its forest-clad hills and purling streams, upon whose moss-clad banks many a childhood hour has been dreamed away? Who does not, in imagination, revisit the mountain stream, where the trout flashes his silvery sides; or the rocky cliff, within whose niches delicate mosses, ferns, blue-bells and violets tremble in the breeze?

> "The groves were God's first temples,"

and here, before man learned

> "To hew the shaft and lay the architrave,
> Amidst the cool and silence he knelt down
> And offered to the Mightiest, solemn thanks.'

Has it ever occurred to the farmer of the West, how much more lovely this beautiful land would be if they could be permitted to return, and as upon wings, view the landscape, after the lapse of three or four centuries had enabled each farm to have its own little forest, and each home its sheltering grove—to see the climate modified and softened;—the gardens blooming with what to us are exotics, and the orchards dropping ripe and delicate fruits that now we cannot hope for;—to see the grateful herds, at midsummer, slaking their thirst from rills born within cool forest-crowned slopes whose mighty monarchs

> ' Wave their giant arms athwart the sky!—''

or in the valley, where sycamore, and maple, and beech, and linden, and the tulip tree, rear their leafy heads; or beside the streamlet, where the willow and the aspen lave their roots; or against the bank, where the wild cherry and dogwood, the thorn and crab-apple fill the air with fragrance;—where

> " All meek things,
> All that need home and covert, love your shade.
> Birds, of shy song, and low-voiced quiet springs,
> And nun-like violets, by the wind betrayed;—"

or, wandering, love the foliage as we pass, where

> " Honeyed lime,
> Showers cool, green light o'er banks where flowers weave
> Thick tapestry; and woodbine tendrils climb
> Up the brown oak, and buds of moss and thyme—
> And the white poplar, from its foliage hoar,
> Scatter forth gleams like moonlight, with each gale
> That sweeps the boughs."

Then the landscape would gain beauty from the tulip tree—

> " That tuneth its harp-leaves to the wind gust;—"

or from the light, quivering aspen—

> " That bowed not its head when the Redeemer passed,
> And so shivers and trembles until He returns."

The forest has its charms too, in winter, when

> " The wood's soft echoes mock the baying hound;"

and—

> " All day long
> The woodman plies his sharp and sudden axe
> Under the crashing branches."

Do we really estimate how pleasant a land we have—this prairie country—stretching away and away, like a vast, undulating ocean concreted by the hand of its Maker into firm land; verdant with emerald slopes; gorgeous with flowers, lacking only one thing to make it perfect—Trees.

Do those who swelter under our torrid summers, or lament our arctic winters, realize that with one acre in ten or twenty planted in timber, how perfect this land would become in all that makes a fertile country? The heat of summer tempered; our winters shorn of their terrors. How many know that, with each farm containing its little forest, each home its sheltering grove, the climate would be modified and softened; the gardens bloom with what to us now are exotics; orchards drop ripe and delicate fruits that now we cannot hope for; or fields bear a wealth of grain that would not be laid low with devastating storm, torn and tangled by tornadoes, or swept away by devouring floods. All these terrors we now sometimes experience.

Give us, then, trees about each homestead. Is it not pleasant, the pictures which our best-loved poet, Bryant, pen-paints of the forest, where

> "The century-living crow,
> Whose birth was in their tops, grew old and died
> Among their branches, till at last they stood
> As now they stand, massy, and tall, and dark,—
> Fit shrine for humble worshiper to hold
> Communion with his Maker."

These shall be green with gladness in the spring-time; glowing under the summer sun, they will shelter grateful heads and happy homes; fling their banners of purple, and crimson, and gold, in the autumn breeze.

If we plant forests, poets of generations who succeed us may sing, as the poet Hempstead has sung of those we are now transplanting from seeds of the Pacific Slope. Of the great California redwoods, he says:

> "They were green when in the rushes lay and moaned the Hebrew child;
> They were growing when the granite of the pyramids was piled;
> Green when Punic hosts at Cannæ bound the victor's gory sheaves,
> And the grim and mangled Romans lay around like autumn leaves.
> From their tops the crows were calling when the streets of Rome were grass,
> And the brave Three Hundred with their bodies blocked the rocky pass;
> In their boughs the owl was hooting when upon the hill of Mars
> Paul rang out the coming judgment, pointing upward to the stars.
> Here, with loving hand transplanted, in the noonday breeze they wave,
> And by night, in silent seas of silver-arrowed moonbeams lave."

To enjoy the shade of trees in the West, we must plant them. If we would seek the shelter of the woods at noon-day, we must make it. Would we leave the noblest heritage to our children that the Western farmer can,—a growing grove of timber,—all that is necessary is, each spring, *plant trees!* PLANT TREES!

PART VI.

INSECTS AND BIRDS
IN
THEIR RELATION TO THE FARM.

INSECTS INJURIOUS AND BENEFICIAL.

ILLUSTRATED CLASSIFICATION OF INSECTS.

REMEDIES AND PREVENTIVES AGAINST DAMAGE.

BIRDS TO BE FOSTERED OR DESTROYED.

INSECTS AND BIRDS.

CHAPTER I.

ENTOMOLOGY ON THE FARM.

I. PRACTICAL VALUE OF ENTOMOLOGY.—II. DESTROYING INSECTS ON NURSERY TREES.—III. ORCHARD CULTURE IN RELATION TO INSECTS.—IV. CARE OF TREES IN RELATION TO INSECTS.—V. PREDATORY BIRDS AND INSECTS.—VI. THE STUDY OF INSECTS.—VII. CLASSIFICATION AND ANATOMY OF INSECTS.—VIII. DIVISIONS OF INSECTS ACCORDING TO THEIR FOOD.—IX. NOXIOUS AND INJURIOUS INSECTS.

I. Practical Value of Entomology.

AS a science, entomology interests few; but it has done as much for the farm and garden as any other of the many departments of learning which are, more or less, connected with agriculture. To the tireless labors of students of insect life, the farmer, the gardener and the pomologist are deeply indebted for careful investigations into the habits and characteristics of injurious insects and their foes. They have taught us remedies for the ravages of the insect pests, and the means of extending the benefits derived from their predatory and carnivorous enemies. Farmers have, as a rule, given little attention to the study of insect life, probably because few of these insects feed upon grain. They have, however, been active in the destruction of birds, which the more skillful of the horticulturists have always protected. This thinning out of bird life has increased the insect world to such a degree, that the army worm, chinch bug, weevils, wire worms, corn worms and other noxious forms have become a scourge.

Why have farmers persistently destroyed the warblers, a class of birds that live largely on insects? Simply because of their ignorance, for if, at certain seasons, these eat a little of the green corn, at all other times of the year they are a blessing to the farmer. The same is true of the mole, of mice, shrew mice, bats, skunks and owls, of some hawks, and even of our little striped prairie squirrel. He, indeed, has a bad name among farmers, yet he hunts persistently for the May-beetle (white grub), both in the larval and pupal state.

II. Destroying Insects on Nursery Trees.

A COMMISSION of horticulturists was, a few years ago, appointed in Illinois, to take into consideration the best means of destroying injurious insects; to solve, if possible, the question of what birds were beneficial, and also to investigate fungus growths and their influences. With this commission, were Dr. Cyrus Thomas, State

Entomologist; Prof. T. J. Burrell, of the Illinois Industrial University; Prof. A. S. Forbes, of the State Normal University, at Normal, Ill., now State Entomologist; and also some of the prominent horticulturists of the State. They went carefully over the whole ground chosen, but this was limited to the nursery and orchard, because more injury had been there felt than in any other one direction. The rules they gave will, however, be applicable to the farm and garden so far as the soil is concerned, and with the special directions given later, will enable any one to combat, successfully, the growing danger for insect pests.

Preparation of the Soil.—Ground to be used for seeds, grafts or cuttings should be cleaned of rubbish, and cultivated without a crop of any kind, (fallow plowed), and plowed late in the fall at least one season before planting. This will, in a great measure, free the soil from noxious insects, notably the white grub, larvæ of the May beetle, the wire worms, larvæ of the elators, (these wire worms would not be entirely destroyed, but generally), all the noxious worms known as cut worms, particularly those that are destructive early in the spring, the striped plant bug—though this being an insect that flies strongly, the remedy would only be partial—nearly all the caterpillars that pass the winter in the larvæ or pupa state, and would also be a great help in freeing the land from leaf-destroying insects of various kinds.

By clearing off rubbish is meant, the cleaning of the ground of all such matter as cannot readily be plowed under and would not readily decay, so as to make plant food; and in localities where the disease known as "rotten root" prevails, all decaying wood should be carefully removed.

Grafts, Seeds, Seedlings and Cuttings.—All these should be carefully examined for the eggs of noxious insects, such as bark-lice, leaf-lice, root-lice, etc. Apple seedlings should be carefully examined, and if any indications of the woolly-root aphis or woolly apple-root plant-lice are seen on them, the whole lot of seedlings should be carefully washed in strong soap-suds, and packed away for a few days in saw-dust before grafting them. Scions for grafting should not be used if infested with eggs of leaf-lice, aphides (very small, shiny black globules stuck on the surface), and bark-lice—small scale-like things, the worst of which is shaped like the one-half of an oyster shell, known as the oyster shell bark-louse, and of the color of the bark, others more flattened and whitish—unless they can be entirely removed. It is only safe to entirely reject such scions; if only a few out of thousands are used that are infested with these bark-lice, they will infest the whole nursery; a few scions of a special variety might be cleared from them.

Cultivation the First Year.—Thorough and clean cultivation from planting until the first of August for grafts and seedlings, and other things to be taken up in the fall, until they are taken up. This clean culture not only keeps noxious insects from accumulating, but gives the plants vigor to withstand their attacks, and also of mildew (fungi). The plants should not be too crowded in the rows, as this would induce a weak growth, rendering them liable to attacks of mildew and insects. Apple grafts should not be cultivated later than the middle of July to the first of August, and at the last working should have corn or oats sown thickly among them, or the seed for a crop of fall turnips; this to check and ripen up for winter. No sod-forming grasses should be allowed to grow near the nursery, as it affords breeding places and hiding (wintering) places for many noxious insects, such as white grubs (young of the May beetle), wire worms (larvæ of elators), cut-worms (various larvæ of agrotis), army worms (larvæ of lucania), and of numerous species of leaf-hoppers.

Lime and Sulphur.—Insects and mildews (fungi) injurious to the leaves of seedlings and root grafts, can be kept in subjection or destroyed by a free use of the following combination of lime and sulphur—it may be called the bi-sulphate of lime: Take of quick or unslaked lime four parts, and of common flowers of sulphur one part (four pounds of sulphur to one peck of lime); break up the lime in small bits, then, mixing the sulphur with it in a tight vessel (iron best), pour on them enough boiling water to slake the lime to a powder, cover in the vessel close as soon as the water is poured on. This makes also a most excellent whitewash for orchard trees, and is very useful as a preventive to blight on pear trees, to cover the wounds in the form of a paste when cutting away diseased parts; also for coating the trees in April. It may be considered as the one specific for many noxious insects and mildew in the orchard and nursery. Its materials should always be ready at hand; it should be used quite fresh, as it would in time become sulphate of lime, and so lose its potency. Whenever dusting with lime is spoken of, this should be used. This preparation should be sprinkled over the young plants soon as, or before, any trouble from leaf-lice (*Aphides*), mildew, mould (*fungi*), thrips or leaf-hoppers appear, early in the morning, while the dew is on. This lime and sulphur combination is destructive to these things in this way: Firstly, by giving off sulphuric acid gas, which is deadly poison to minute life, both animal and fungoid; and the lime destroys, by contact, the same things. Besides, its presence is noxious to them. Neither is injurious to common vegetable life, except in excess, except the lime to the foliage of evergreens.

Cultivation the Second and Subsequent Years.—Cultivation of the soil similar to that of the first year. The sulphur and lime to be used the same as the year before. The first thing to be looked after in the winter or early spring is the larvæ of the leaf-crumpler. This is readily discovered; as a little bunch of dried leaves strung to the twigs by a silken thread, the larvæ will be found in this, encased in a little horn-shaped case. These should be carefully picked off and destroyed. It is best not to burn or crush these, as many of them are parasitized; if not destroyed, the parasite, being mature, will develop, and go on with its good work; while if the Physita larvæ are merely "dumped" on the ground a little way from any trees, it at this time not being mature, will certainly starve. Hand picking is the only remedy for these crumplers, as well as for other small, noxious, and sometimes very destructive caterpillars, to wit: the lesser apple-leaf folder, known by its usually folding together the apex of a leaf, and feeding between the folds; the apple-leaf folder and the apple-leaf skeletonizer. Of these, the first one is protected from destructive applications by the folds of the leaf on which it feeds; and the last one by a web which it spins, and under which it feeds. The last may be seen by a brown patch near the base of the leaf; these are the larvæ of quite small moths, and as these are generally two or more brooded, great care should be taken to destroy the first brood in the larval state, by searching for and crushing them between the thumb and fingers, as it appears there is no other way of reaching them. If the first brood are all destroyed there can be no second brood. They should be carefully sought for from about the middle of June to the first of July, and destroyed. The lime and sulphur will be of great help in holding them in check. Borers are sometimes injurious in the nursery, but will be found under the next head.

III. Orchard Culture in Relation to Insects.

TREES for the orchard should be sound and free from insects, and the eggs of insects. If infested with bark-lice they should be rejected; if with root-lice, their roots should be thoroughly washed in strong soap-suds. Eggs of the white tussock moth should be destroyed. The orchard should have a high, dry, well-drained situation, soil neither very rich nor poor; the trees thus make a stronger matured growth, therefore better able to withstand climatic changes and the attacks of insects and fungi. If such a proper selection is made, it needs no other preparation than such as would be sufficient for planting a corn crop. The cleaning of the ground should be the same as recommended for the nursery

Trees for Planting.—Trees for planting in the orchard should have branches low enough to partially shade their trunks, and when planted should lean considerably to the southwest. This leaning should be done to prevent the damage to the southwest side of the trunks, commonly known as "sun scald," which invites attacks of the flat-headed apple tree borer.

Cultivation of the orchard before coming into bearing, or the first five or eight years may be as follows: It should be planted with corn, if the land is rich; if thin, in potatoes or other suitable hoed crop. When the trees are old and large enough to bear, the orchard should be sown in red clover, but in no case should small grain or any of the perennial grasses be sown in it. If the growth is too vigorous, check it, and throw the trees into bearing; clover harbors and feeds less noxious insects than any grass-like plant that makes a close sod, and should always be used for this purpose. Plowing should always be done, if possible, before the starting of growth in the spring; the summer culture with the harrow and cultivator alone, so as not to destroy the roots in the growing season. Late fall plowing, though not recommended as a rule, will be found of great use occasionally, in destroying and keeping in check the canker-worm, grub of the May beetle, the climbing and other cut worms, tarnished plant bug, and the bugs and chrysalids of other noxious insects that pass the winter in the grub or larval state. Such plowing should in all cases be shallow, and if the canker-worm or grub is present, great care should be taken to turn and break up all the soil; where it cannot be reached under the trees with the plow, the spade should be used.

Orchardists should bear in mind that continued healthy vigor of the trees throughout their lives, not too rampant, is essential, to enable them to withstand the attacks of noxious insects, and to enable them to recuperate when attacked; therefore, if the trees at any time present a sickly or unthrifty condition, the soil should be manured or cultivated, or both.

Mulching.—Whenever mulching is applied, the portions immediately about the collar of the tree should be mixed with ashes or lime, to prevent noxious insects from working and harboring there.

IV. Care of the Trees in Relation to Insects.

Twigs during the winter season, upon which the eggs of injurious insects are seen, should be cut off and burned, if the eggs cannot be readily and completely destroyed in some other way—such as hand-picking, crushing, etc. These include the eggs of the tent caterpillars, which are found 300 to 400 cemented in a bunch, surrounding a twig; eggs of the tree hoppers, which are found in little slits immediately beneath the outer bark; eggs of leaf-lice, minute, shining, black globules scattered upon twigs of the last year's growth.

Cleaning the Trees—A general cleaning of the trees in winter should be carefully attended to. Pick off all cocoons, leaf-crumplers, basket or drop worms, eggs of the Tussock moth found on sides of deserted cocoons, etc. Old trees, neglected heretofore in orchards, should have their trunks carefully scraped of all rough bark; but if the directions below given are followed, there will be no scales of bark to scrape off. The trunks and larger branches of the trees should be washed at least once each year with soft soap (thinned somewhat), or a strong alkaline wash, between the middle of May and middle of June, according to latitude, and if washed off soon by heavy rains, should be renewed; this renders them free of rough scales of bark, which harbors many noxious insects and parasitic plants, and is an effectual remedy against attacks of the flat-headed apple-tree borer, which is often very destructive. In the North, the round-headed apple-tree borer preys upon the tree near the surface of the ground; therefore, in April or May a small portion of the soil should be removed from the collar of the tree, and this should be filled in and mounded up a little with lime or ashes. This mounding up prevents the beetle laying her eggs so low down on the trunk as that access to the larvæ would be difficult; besides, the lime, ashes and the soap, as alkaline washes, are all obnoxious to her. This beetle deposits her eggs during the months of June, July and August, but mostly in June and July, on the trunk of the trees near the ground—rarely in the branches or under the surface. The trees should be carefully examined for the borers during the last half of September, and if any are found their burrows should be opened and the larvæ (young borers) killed. If they have penetrated deeply, they may be punched to death with a flexible twig or wire. Their burrows may be readily found by the reddish-brown castings thrust from them. This is the most insidious foe of the apple orchardist, in the Northwest, and too much pains cannot be taken in destroying them.

Flat-Head Borer.—The larvæ of the flat-headed apple-tree borer may be found in the trunk and at the base of the larger branches and in or near any patches of sun scald, wounds or abrasions, and near the base of dead limbs, by a little discoloration of the bark; and by taking off a thin shaving of the discolored bark, the larvæ may be seen and destroyed. This work should be done in the months of August and September. If the larva has penetrated deep into the wood, it will often be found best not to cut it out, as the wound made in doing so would be worse than the injury done by the grub if left alone. A place attacked by this grub on a tree is most liable to be so again the next year, therefore it should be carefully watched.

Bark-Lice.—Trees infested by bark-lice may be partially cleared of them by washing with an alkaline wash (and thoroughly dusting with the lime and sulphur compound), at the time the young are hatched out, which is from May to June 15, according to season and latitude. These young lice are very small and delicate; when hatched from the eggs under the scales they crawl around for a few days and then fix themselves to the bark, and never move again, and soon become covered with a hard shell, which acts as a protection, and a most perfect one, too, during the rest of their existence; hence they are hard to destroy, except when in the young (moving) state. These lice, though small and insignificant looking, are exceedingly hurtful to trees infested with them; their immense number, sucking the vital juices of the trees during the entire season, fearfully weakens them, and too great pains can not be taken to prevent their gaining a foothold in the orchard, and to destroy them if they should.

Fine lime dust, sprinkled over the trees at this time, is not only useful for these young bark-lice, but also for leaf-lice and the codling-moth.

Nests of the tent caterpillar, fall web-worm, and other insects that feed and nest together, should be carefully removed from the trees and destroyed. Be careful that the caterpillars are in their nests when taking them off.

For the destruction of the canker-worm, the rope bands, with encircling tin bands, are recommended. The late fall plowing, already spoken of, destroys a large portion of the chrysalids of this insect, and when done for this purpose, should be as late as September and the soil all carefully turned over and broken up, especially under the trees near the trunk, and left as rough as possible.

Protection of the Fruit from Insects.—The codling-moth, or core-worm of the apple is best combated by thoroughly dusting the trees from above downwards, so that it may fall into the calyx of the young apples, with the lime and sulphur preparation finely powdered. This should be done immediately after the petals of the flowers fall, and the young fruit begins to form. If delayed too long, the worm will have entered the apple, and be out of reach of the lime. The lime dust is also very distasteful to the moth. The egg from which this apple-worm is hatched is laid by a small, nocturnal moth, in the calyx or blossom end of the apple; it soon hatches and the young caterpillar eats its way into the core of the apple, where it feeds about thirty-four days, when it eats its way out and seeks a secure hiding-place in which to spin its cocoon, and undergo its transformation. In its cocoon it soon changes to a pupa, and in about twelve days emerges as a moth, ready in a few days to lay its eggs in the apples. The use of the bands spoken of below is to entrap the insect while in its cocoon. The best means for entrapping the larvæ of this insect are the cloth, paper or woolen bands. These may be old cloths of any kind, carpets, woollen or other cloth garments of any kind, torn in strips three to four inches wide, coarse, cheap straw or felt paper, untarred building paper, veneering cut as for berry boxes, etc., in strips three to six inches wide, and wrapped around the tree two or more times, and tacked or tied fast. If the trees are clear of rough bark, the larvæ will seek these bands, as being the best available place in which to spin their cocoons, and the bands can be examined and the insects destroyed. These bands should be placed around the trees as early as the tenth of June for latitude 40 degrees; a week earlier for the South, and must be examined as often as once in twelve days, and the insects destroyed during the balance of the season. The last brood of worms, on leaving the apples, spin their cocoons, but remain as worms during the winter, and do not change to the pupa state until the following spring; hence all barrels and bins in which apples have been stored will be full of these larvæ in their cracks and crevices. They should be hunted out and destroyed. Many moths may be destroyed on the windows or cellars where apples have been stored.

V. Predatory Birds and Insects.

It may be laid down as a principle that all insects and all birds that feed entirely on vegetation are injurious, and that all insects are beneficial that feed on their kind. Birds that feed on their kind are, however, often most injurious; for the rapacious ones (hawks and others of predatory habit) feed on the smaller, insect-eating birds.

We have shown that the horticulturist finds many birds inimical that are beneficial in agriculture. Hence, a wise discretion must be exercised; and hence, also, the necessity that each should know his friends from his foes. This necessitates some study both of entomology and of birds.

The common birds and animals are known because we readily see them; not so with insects. Many of these work under ground during the destructive stages of their lives, and, all being small, and most of them hidden while in the egg state and during the feeding period, they are difficult to distinguish. We will illustrate some of the more destructive, and describe others, so the farmer and horticulturist may easily distinguish friend from foe.

VI. The Study of Insects.

The classification of insects has been carefully done by the late Dr. Le Baron, while entomologist of Illinois, and we shall follow this author in our general and condensed description.

All insects are not bugs, nor are they worms. All bugs and worms are not insects; yet, originally, spiders, worms, and even crustacea (snails and other animals covered with shells) were so termed. But the term insect is now restricted to the hexapods (six-footed species), beetles, bees, bugs, grasshoppers, locusts, fleas, etc., produced from eggs, whatever the secondary mode of propagation may be.

Since, then, many of these insects are destructive to the farmer's crops, it is important he should not only recognise them, but know where they are to be found

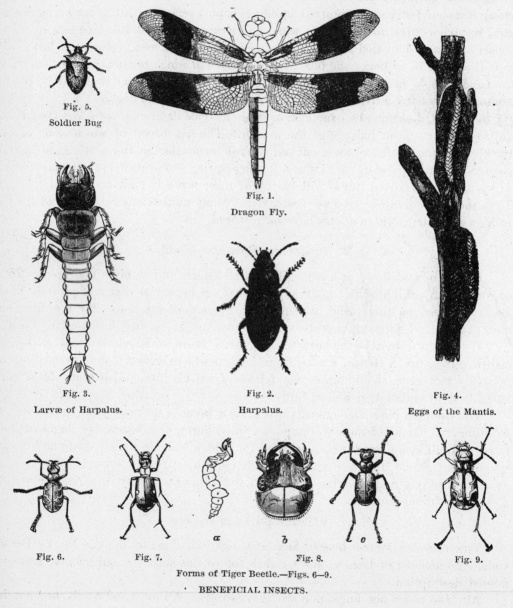

Fig. 5. Soldier Bug.
Fig. 1. Dragon Fly.
Fig. 3. Larvæ of Harpalus.
Fig. 2. Harpalus.
Fig. 4. Eggs of the Mantis.
Fig. 6. Fig. 7. Fig. 8. Fig. 9.
Forms of Tiger Beetle.—Figs. 6—9.
BENEFICIAL INSECTS.

in the egg state, and how they may, most easily, be destroyed, what the mature insect is, moth, butterfly or beetle, and the best means for their destruction, to prevent,

ENTOMOLOGY ON THE FARM. 587

so far as possible, the laying of the eggs. He should, also, be able to easily distinguish the predatory, or carnivorous, species, such as lady-birds, tiger-beetles, soldier-beetles, ichneumon-flies, lace-wings and others which are insect eaters and should be carefully preserved. Hence, the value of an outline of entomology.

Insects Illustrated.—The illustrations on page 586, showing several beneficial insects, will be worth a careful study. The lady-birds are insect eaters, in both the larval and perfect state. In the perfect state they eat the eggs of insects, being thus

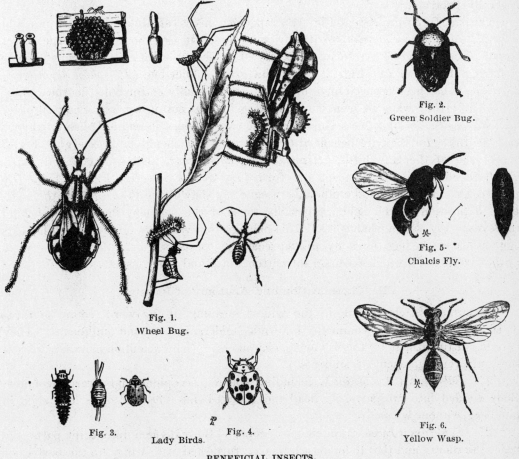

Fig. 1.
Wheel Bug.

Fig. 2.
Green Soldier Bug.

Fig. 5.
Chalcis Fly.

Fig. 3.   Fig. 4.
Lady Birds.

Fig. 6.
Yellow Wasp.

BENEFICIAL INSECTS.

doubly valuable. The tiger beetles are also insect eaters, in the larval and perfect state. Fig. 1 shows one of the many species of dragon flies (*Libellula*). The larvæ are aquatic and predaceous, as are also, the perfect insects (flies). Their food consists of various insects, which they catch "on the wing," and from their habits, they may well be called the swallows of the insect tribe. Fig. 2 is a species of harpalus, an indefatigable destroyer of insects, and Fig. 3, its larva. Fig. 4, shows the eggs of the mantis, a great destroyer of insects. This, in the Southern States of the American Union, is popularly called the devil's rear horse, and in the North, the praying mantis. It is a long, slender insect, and gets its more northern name from an odd-looking habit it has of folding together its fore legs. At Fig. 5 is shown the spined soldier bug (*arma spinosa*)—untiring hunters and destroyers of larvæ, which they impale and suck dry. Figs. 6, 7, 8 and 9, are several forms of tiger beetles (*cicindeldæ*), which both as larvæ and as perfect beetles prey on insects. Fig. 8, *a*, shows the larvæ, and *b*, an enlargement of jaws. All these forms of beetles should be carefully protected.

On page 587 we show, at Fig. 1, a rapacious insect called devil's horse, or wheel bug (*Reduvius novenarius*), in its various stages. It eats both insects and insect eggs indiscriminately. At the top of Fig. 1 is shown a very little (young) fellow attacking an egg mass. Fig. 2 shows the green soldier bug (*Rhaphigaster hilaris*, Fitch). It has been accused of injuring plants, especially cotton bolls, in the South. This is an error; when so seen it was seeking for the cotton worm, its natural prey.

All the classes of insects shown on the two pages, 586 and 587, are worthy of careful study, for they are beneficial to the farmer and should be preserved. Fig. 3 shows one of the lady birds (*Hippodamia convergens*), also its larva and pupa. They are all carnivorous except the form (*Epilachna borealis*) Fig. 4. This is a plant eater. Do not mistake others, subsequently shown, for this. Notice the number and position of the spots, especially those of the segment immediately behind the eyes. Fig. 5 is a Chalcis fly. This belongs to a valuable family, which lays its egg either in the egg or body of depredatory insects. Fig. 6 is a yellow wasp (*polistes bellicosa*), which feeds on the cotton worm and other insects.

VII. Classification and Anatomy of Insects.

INSECTS, as a class, and, in the widest meaning of the word, comprise three divisions, or sub-classes, commonly known as spiders, insects and millipedes. They may be distinguished by the following characters: (The scientific names in parenthesis are for the benefit of students.)

1. Sub-class (*Arachnida*), including spiders, scorpions and Acari, or mites. Body divided into two parts, the head and thorax being united in one: legs eight in number; without wings.

2. Sub-class (*Insecta*), or insects proper. Body divided into three parts, the head, the thorax and the abdomen; legs six; furnished with wings, in the perfect or *imago* state

3. Sub-class (*Myriapoda*), commonly called millipedes or centipedes. Body divided into many parts or segments, varying from ten to two hundred; legs numerous; usually either one or two pairs of legs to each segment of the body; never have wings.

Exceptions.—The exceptions to these characters are very few. In the Arachnida (spiders, etc.) some of the most minute (*Acari*) have but six legs. Insects proper are always six-legged in their last or perfect state; and they also have, generally, six true legs in their larval state; but some larvæ have no legs, and the larvæ of the Lepidoptera, commonly called caterpillars, have, in addition to their six true legs, several pairs of false legs, or pro-legs, which assist in locomotion. There are a few exceptional cases in which insects are destitute of wings. The fleas (*Pulices*), the lice (*Pediculi*), and the little family of insects known as spring-tails (*Thysanoura*), never have wings. In some rare instances the females are wingless, whilst the males have wings. This is the case with some species of the lightning-beetles (*Lampyridæ*), and with the canker-worm moth and the tussock-moth, and a few other species amongst the Lepidoptera. Similarly exceptional cases are also found in other orders of insects.

The Nervous System.—The nervous system of insects consists of a double cord extending the length of the body, and lying upon the inferior or ventral side of the internal cavity. The two threads which compose this cord do not lie side by side; but one above the other. The lower thread swells at intervals into little knots of nervous matter, called ganglia. In insects of an elongated form, such as some of the Neuroptera, and the larvæ of the Lepidoptera, there is a ganglion at each segment of the body, making thirteen in all; but in most mature insects the ganglia become more or less consolidated. In the butterfly (*Papilio*), there are ten ganglia, counting the brain as one; in the bee (*Apis*), there are eight; in the May-beetle (*Melolontha*), there are five, and in the Cicada there are but two. The upper of the two nervous threads runs nearly in contact with the lower, but is destitute of ganglia. These two threads seem to represent the double and more compact cord which constitutes the spinal marrow of the higher or vertebrated animals. The upper simple thread is supposed to furnish the nerves of motion, and the lower and ganglionic thread, the nerves of sensation. The fibers which compose these cords separate at the anterior extremity of the body, so as to embrace the œsophagus or gullet, above which they again unite.

The Muscles.—The muscles of insects possess a wonderful contractile power in proportion to their size. A flea can leap two hundred times its own length, and some beetles can raise more than three hundred times their own weight. This remarkable strength may probably be attributed to the abundant supply of oxygen by means of the myriad ramifications of the air tubes. Insects are evidently endowed with the ordinary senses which other animals possess, but no special organs of sense, except those of sight, which have been discovered with certainty.

The Eyes.—The eyes of insects are of two kinds, simple and compound. The simple or single eyes are called *ocelli*, and may be compared in appearance to minute glass beads. They are usually black, but sometimes red, and are generally three in number, and situated in a triangle on the top of the head. In insects with a complete metamorphosis, these are the only kind of eyes possessed by them in their larval state, and in these they are usually arranged in a curved line, five or six in number, on each side of the head. In some insects which undergo only a partial metamorphosis, as for example the common squash-bug (*Coreus tristis*), the ocelli are wanting in the larval and pupal states, but become developed in the last or perfect stage.

The compound eyes of insects present one of the most complex and beautiful mechanisms in the organic world. They are two in number, but proportionately, very large, occupying in many insects nearly the whole of the sides of the head and, in the dipterous order especially, often present across their disks, bands of the richest tints of green, brown and purple. These eyes are found to be composed of a great number of lesser eyes or eyelets, in the form of elongated cones so closely compacted as to form apparently a single organ. The larger ends of these cones point outwards, and by their union form the visible eye. Their smaller extremities point inwards, towards the brain, to which they are connected by means of a large optic nerve. When one of these little eyes is examined through a strong magnifying glass, it is seen to be composed of a very great number of little facets, sometimes square, but usually six-sided, each one of which represents the outer and larger extremity of one of the component parts. These facets vary greatly in number in the eyes of different kinds of insects. In the ants there are about fifty in each eye; in the sphinx moths about 1,300; in the house fly, 4,000; in the butterfly, upwards of 17,000; and in some of the small beetles of the genus Mordella, it is said that more than 25,000 facets have been enumerated in one compound eye; so that if we suppose that each of these component parts possesses the power of separate vision, one of these insects must have 50,000 eyes. How vision is effected, or how a unity of impression can be produced by so complex an organ, we are unable to conceive.

Insects are evidently affected by loud noises, and moreover, as many insects have the power of producing voluntary sounds, it is reasonable to suppose that they possess the sense of hearing. No organ, however, which has been generally admitted to be an organ of hearing, has been discovered.

Insect Transformations.—Nothing in the history of insects is more remarkable than the striking changes of form which many of them undergo, in the course of their development. Whilst other animals progress from infancy to maturity, simply by a process of growth, and by such gradual and imperceptible changes only as their growth necessitates, many insects assume totally different forms in the course of their development, so that they could never be recognized as the same individuals, if this development had not been actually traced from one stage to another. These changes are called the metamorphoses or transformations of insects. All insects, in their growth, pass through four stages, designated as the egg state; the larva, or caterpillar state; the pupa, or chrysalis state; and the imago, or perfect and winged state. The metamorphoses of insects are of two principal kinds, complete and incomplete. In the complete metamorphosis the larva bears no resemblance to the imago, and the insect, in the intermediate or pupa state, is motionless and takes no food.

Nomenclature.—The Head.—It often becomes necessary to refer to different parts of an insect's head, and they are therefore designated by particular names indicative of their situation. These are—the hind-head, (*Occiput*); the crown, (*Vertex*); the fore-head, (*Frons*); the face, (*Facies*); the cheeks, (*Gena*). The appendages of the head are the horns, (*Antennæ*); the eyes, (*Oculi*); and the parts of the mouth, (*Trophi*, or oral organs). All insects have two more or less elongated and usually many jointed antennæ situated one on each side of the head, and varying greatly, in different kinds of insects, in length and in the form of their component joints. Insects have very short antennæ in their larval state, and in some perfect insects, such as the water-beetles (*Gyrini* and *Hydrophili*), the antennæ are not longer than the head, whilst in others, such as some of the longicorn beetles, they are more than twice as long as the whole body, and in some of the small moths of the genus *Adela*, they are five or six times as long. The uses of the antennæ are not known; they are supposed to be instrumental in the sense of hearing.

Variations.—The most common variations in the forms of the antennæ are expressed by the following terms: *Filiform*, or thread-like; long and slender, and of the same or nearly the same width throughout. *Setiform* or *setaceous;* bristly or bristle-like; long and slender, but tapering toward the tip. *Moniliform*, or bead-like, when the joints are about the same size; and *Serrate*, or saw-toothed, when each joint is somewhat triangular, and a little prominent and pointed on the inner side. *Pectinate*, or comb-toothed, when the inner angles of the joints are considerably prolonged. *Bi-pectinate*, or double comb-toothed; pectinate on both sides. *Clavate*, or club-shaped; gradually enlarging towards the tip. *Capitate*, or knobbed, when a few of the terminal joints are abruptly enlarged. *Lamellate*, when the joints which compose the knob are prolonged on their inner side, in the form of plates.

The Thorax.—The thorax is the second, or middle, division of the bodies of insects. Though apparently single, it is really composed of three pieces which seem as though soldered together. These pieces are more distinct in some insects than in others, but they can always be distinguished by impressed lines upon the surface called *sutures*. The three pieces of the thorax are distinguished as the fore thorax, the middle thorax, and the hind thorax; or, in scientific language, the *pro-thorax*, the *meso-thorax*, and the *meta-thorax*. In the Coleoptera the pro-thorax is very large, and forms the large upper part or shield, to which we usually give the general name of thorax. In this order of insects, the meta-thorax is invisible above, and the only part of the meso-thorax seen from above is the triangular piece between the bases of the elytra, called the *scutellum*. In many insects (*Hymenoptera* and *Lepidoptera*) the pro-thorax is much reduced in size, and forms only a narrow rim, which is usually called the collar. The under side of the thorax is called the *sternum* or breast plate. Each of the three divisions of the thorax has its sternum, designated respectively as the *pro-*, *meso-* and *meta-sternum*. In many insects, and especially the Coleoptera, each section of the sternum is divided by sutures into a middle piece, *sternum* proper, and a side piece, *episternum*.

The Wings.—The appendages of the thorax are the organs of motion, namely, the wings and the legs. The great majority of insects have four wings. The anterior pair are attached to the upper part of the meso-thorax, and the posterior pair to the meta-thorax. The wings are thin, membranous, transparent organs, in some cases folded when at rest, and supported by ribs or veins running across them. These veins are found to correspond in their number and complexity to the rank of the insect in the scale, and from the ease with which they can be seen, they furnish admirable characters for the purpose of classification. In some insects, such as the grasshoppers, the fore-wings are thicker and less transparent than the hinder pair, and have nearly the consistency of parchment; and in one large order of insects, the Coleoptera, or beetles, the fore-wings become converted into the hard opaque pieces known as the *elytra* or wing-cases. The elytra take no part in the flight, but serve only to cover and protect the hinder or true wings, which are folded under them when at rest. In one large order, the insects have but two wings, and are named from this character *Diptera*, or two-winged insects.

The Legs.—Insects have six legs, attached in pairs to the under side of each of the three segments of the thorax. The leg consists of four principal parts, the hip (*coxa*), a short piece by which the leg is attached to the body; then an elongated piece called the thigh (*femur*, plural *femora*); then another elongated piece called the shank (*tibia*); and lastly the foot (or *tarsus*); which is composed of a number of smaller pieces or joints; of which five is the largest and most common number. The feet of insects terminate almost invariably, in a pair of sharp, horny claws (*ungues*); and between these, at their base, is often one or two little pads (*plantulæ*) by means of which flies and many other insects adhere to glass, or any other surface which is too smooth and hard for the claws to catch upon. The Lepidoptera have but one plantula, and the Diptera have two. Besides the parts of the legs here enumerated, there is a small piece attached to the hind part of the hip, called the *trochanter*. This is usually small and inconspicuous, but in the hind legs of the ground-beetles (*Carabidæ*) it forms a large egg-shaped appendage, one of the most characteristic features of this family.

The Abdomen.—The abdomen is the hindermost of the three divisions of an insect's body. It is sometimes attached to the thorax by the whole width of its base, in which case it is called *sessile*. But it is often attached by a slender petiole or foot-stalk, when it is said to be *petiolated*. The abdomen is composed of a number of rings, one behind another, each ring usually lapping a little upon the one following it. The normal number of rings or segments of the abdomen is considered to be nine, and this number is actually present in the earwig (*Forficula*) and a few other insects; but in the great majority of insects, several of the terminal segments are abortive, and only from five to seven can usually be counted. In the females of many kinds of insects the abdomen terminates in a tubular, tail-like process, through which the eggs are conducted to their place of deposit, and which is therefore called the *ovipositor*. In some insects the ovipositor is simple, short, straight and stiff, as in some of the Capricorn beetles; but in others, as the Ichneumon flies, it is long, slender and flexible, and composed of three thread-like pieces, which when not in use, are separated from each other, giving these insects the appearance of being three-tailed.

VIII. Division of Insects According to their Food.

In the classification of insects according to the nature of their food, all may be divided into two classes—the carnivorous insects, or those which eat animal food (*Sarcophaga*); and the herbivorous insects, or those which subsist upon vegetable substances (*Phytophaga*). Each of these classes is again divisible accordingly as the insects which compose it take their food in a fresh and living state, or in a state of decay. The former are called predaceous insects (*Adephaga*) when they live upon animal prey; and the latter are designated by the name of scavengers (*Rypophaga*). Those insects which eat living animal food are still further divisible into predaceous insects proper, which seize and devour their prey, and parasitic insects, which live within the bodies of their victims and feed upon their substance. Those insects which feed upon decaying animal matter present three divisions: First, general scavengers, which devour particles of putrescent matter wherever they may be found; second, those which live exclusively in or upon the bodies of dead animals (*Necrophaga*); and thirdly, those which are found exclusively in animal excrement (*Corpophaga*).

Herbivorous Insects.—The herbivorous insects may be divided in a similar manner into those which eat fresh vegetable food (*Thalerophaga*) and those which subsist upon vegetable matters in a state of decay (*Saprophaga*). They can also be usefully classified according to the particular parts of the plant which they devour, into lignivorous or wood-eating insects (*Xylophaga*); the folivorous, or leaf-eating insects (*Phyllophaga*), and the fructivorous, or fruit-eating insect (*Carpophaga*).

The above Greek terms in parenthesis have been used chiefly in connection with insects of the Coleopterous order, in which these diversities of food habits exist to a much greater extent than in any of the other orders, but the terms themselves are of general signification, and being very concise and comprehensive, they might, not improperly, be used in speaking of insects in all the orders, so far as they are applicable.

Changing with Age.—In attempting to classify insects according to the nature of their food we meet with a peculiar difficulty, owing to the remarkable change which some species undergo in this respect in passing from the larva to the perfect state. Most caterpillars, for example, feed upon leaves, whilst the butterflies and moths which they produce subsist upon the honey of flowers, or other liquid substances. Some two-winged flies (*Asilidæ*) feed upon the roots of plants in their larval state, but become eminently predaceous in their winged state. Another remarkable example is furnished by certain coleopterous insects (*Meloidæ*), which are parasitic in their larval state, but subsist upon foliage after they have assumed the beetle form.

The question, therefore, arises, to which stage of the insect's existence shall the precedence be given in this respect? At first view it would seem that the perfect state ought to govern, but when we take into account that insects are comparatively short-lived in this state; that having arrived at maturity they require but little food; and that some insects take no food whatever at this stage of their lives; whereas all the growth of an insect takes place whilst it is in the larval state, and consequently it is in this state that they feed so voraciously; when we consider this, it seems more reasonable that in classifying insects upon this basis, the food habits of the larva should take the precedence.

IX. Noxious and Injurious Insects.

The terms noxious and injurious are often used indiscriminately, but strictly speaking, noxious insects are those which are endowed with some poisonous or otherwise hurtful quality; and these are divisible into two classes according as they are hurtful to mankind directly, such as the mosquito, flea, and bed-bug; or are hurtful to the domesticated animal, as the horse-fly, the bot-fly, and the various kinds of animal lice. The insects which attack man directly are annoying rather than seriously hurtful, and this is usually the case with those which molest the domesticated animals; but these sometimes multiply so as to seriously impoverish the animals which they infest.

The term injurious, as distinguished from noxious, is properly applied to all those insects which damage mankind indirectly, but often to a most serious extent, by depredating upon those crops, cultivated, upon which we depend for subsistence and profit. It is worthy of remark that by far the greater proportion of the damage caused by injurious insects is effected by species of very small size, whilst the large species are generally harmless. The two most serious fruit insects, the codling-moth and the plum-curculio, are both below the medium size, and the apple bark-louse, the apple-aphis, the Hessian-fly and the wheat-midge, are so minute that they would not be noticeable were it not for the wide destruction which they cause to some of our most valuable crops, in consequence of their excessive multiplication.

CHAPTER II.

INSECTS, INJURIOUS AND BENEFICIAL

I. PLANT-LICE.—II. SCALE INSECTS.—III. PLANT-BUGS.—IV. GENERAL MEANS FOR DESTROYING BUGS.—V. REMEDIES FOR CHINCH-BUGS.

I. Plant-Lice.

PLANT-LICE are among the most injurious of any species, since they are found on every variety of plants. The cut, Fig. 1, shows the plant-louse of the pear tree magnified. In relation to the insects illustrated we, as heretofore, follow the best authorities, condensing simply to get the gist of the matter, and such only as will be of practical benefit to the farmer and horticulturist.

The *Aphides*, or plant-lice, are exceedingly injurious, inserting their beaks into the tender shoots and leaves of plants and then sucking out their sap. These insects are generally of very small size, having antennæ of five to seven joints, and a long three-jointed beak, or proboscis, for puncturing plants, and then sucking out the sap. Their bodies are soft, rounded or flask-shaped, and apparently only consist of a skin filled with a liquid; their legs are long and very slender, and many of them have two upright processes or tubercles on the hinder part of the abdomen, from which a sweet gummy substance is occasionally ejected, which is eagerly sought for by ants and other small insects. The wings are generally transparent, and the upper pair are much larger than the lower, and are furnished with strong nerves or veins, which pass outward from the costal or outer marginal vein; these wings are very much deflexed at the side of the body when the insect is at rest. Early in the spring the eggs are hatched, and the young plant-lice puncture the plant, suck the sap, and increase in size, the whole brood consisting of individuals capable of reproducing their species without any connection with a male by a species of gemmation or budding forth. These summer broods are wingless. The second generation and several others pursue the same course, being sexless, or at least without the trace of a male among them, and so on indefinitely until the autumn, when winged individuals are produced, which lay eggs for the spring brood of sexual individuals. Bonnet obtained nine generations and Duval seven by this process of gemmation in one season, and Packard states that *Aphis dianthi*, the plant-louse of the pink, continued to propagate by gemmation without any males for four years, in a constantly heated room. It has been supposed that the final autumnal set of plant-lice were males and females alone, but Dr. Burnet states that on examining the internal organs of the winged individuals many of them were not females

Fig. 1.
Pear Tree Louse.

proper, but simply the ordinary gemmiferous or summer form. As there are peculiar plant-lice infesting different plants, the number of species must necessarily be very great.

Apple Plant Louse.—(*Aphis mali*).—The females deposit their eggs on twigs and bark in the autumn; the insect is hatched out the next spring, and feeds upon the sap of the tree. The first broods are all females, which in a short time, give birth to living young by the process of gemmation. These also produce other young ones, which are all females as long as the summer lasts, and it is only in the autumn that males are produced, which uniting with the females, become the parents of the eggs for the following spring brood, thus bearing living young all the summer, and laying eggs in autumn while the parent insects die. These insects, as larva, pupa, and perfect insect, are found generally on terminal shoots and on the under side of leaves. The male is winged, has a blackish thorax, and is 0.05 to 0.08 inch in length to the end of abdomen. The female (see Fig. 2, magnified) is green, with a row of

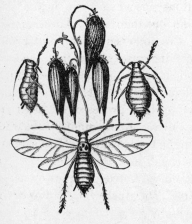

Fig. 3.—Magnified.
Grain Louse.

Fig. 2.—Magnified.
Apple Plant Louse.

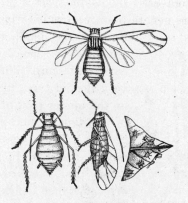

Fig. 4.—Magnified.
Cotton Plant Louse.

black marks down each side, and has no wings, and is rather larger than the male. These insects eject honey-dew from two projecting tubercles on each side of the hinder part of the abdomen, which is greedily eaten by ants and other small insects.

Grain or Oat Plant-Louse.—(*Aphis avenæ*).—This does much injury to grain, and especially to oats, but is also found on wheat, rye, and other cereals. Their habits are much the same as other plant-lice, excepting that it is said that although their honey-tubes are well developed, these insects emit no honey, and in consequence, are not followed by ants. The feet and knees are generally of a darker or nearly black color; length 0.05 inch. (See Fig. 3 magnified).

Cotton Plant-Louse. (*Aphis gossypii*).—This is a great nuisance to the planter, especially when the plants are very small. Sucking out the sap they distort the stems, and frequently kill the plants before they have attained sufficient maturity and strength to withstand their repeated attacks. Their habits are much the same as the rest of the aphides, and their colors vary from green to a decided yellow, striped with black on the upper side of the thorax.

Woolly Apple-Tree Blight.—(*Eriosoma lanigera*).—These destructive insects are shown in Fig. 5, enlarged, and also the natural size on the portion of wood shown. The eggs are deposited in crotches or cracks of the branches or bark, often at or near the surface of the ground, or on new shoots springing from the parent tree. As larva, pupa, or perfect insect they are equally injurious, sucking the sap, and, when numerous, do much injury to the trees. These insects are 0.10 to 0.12 inch in length, and are gregarious, feeding in societies, which, when seen from a short distance, resemble small bunches of cotton adhering to the trunk or branches of the tree. The young are produced alive all summer, but in the fall the females lay eggs which withstand the winter and hatch into young lice the following spring.

Root-Lice.—There are various forms infesting the roots of plants and trees south of forty degrees. Fig. 6 shows the natural size, and magnified, of plant root-lice (*Rhozobius*), infesting verbenas. Those who grow flowers in-doors, would do well to make an examination of drooping plants. We illustrate the general form of root-lice, and their manner of work on the roots of many plants and trees.

Remedies.—The remedies for all the plant-lice is scrubbing or washing with a strong solution of soap, dilute tobacco water, or when on the bark of trees, with a moderate solu-

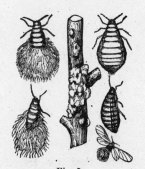

Fig. 5.
Woolly Louse.

Fig. 6.
Root Louse.

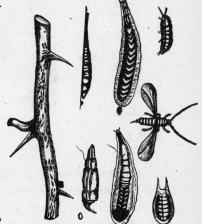

Fig. 7.
Orange Scale Insect.

tion of potash water. The remedy for root-lice, shown at Fig. 6, is to wash the roots of the trees in a solution of potash and water, with scrubbing, if the plants are infested with them, when received from the nursery. They do not breed north of forty degrees.

II. Scale Insects.

THERE are many scale insects infesting the bark of trees, encased in a shell or cover. Various plants, as the apple, orange, etc., have their peculiar species. The apple-tree in the North is often so affected with these bark-lice (scale insects), as sometimes to kill the trees. They are well known to most people who have badly-tended orchards. They are only destroyed, when under their scales, by scraping off. In the North, they transform in June, and move forward, when they may be killed by spraying with emulsions of kerosene, as described presently.

Orange Scale Insects.—(*Mussel Scale*).—This insect infests the bark and sometimes the leaves of the orange trees, and is especially destructive to the orange groves of Florida. Fig. 7 shows the insect, male and female, and the scale magnified, and also in natural size at the bottom of the upper scale; also natural size of scales as shown on the twig.

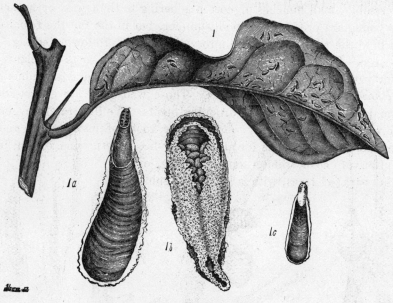

Fig. 8.
MYTILASPIS CITRICOLA.

The subject of orange scales has attracted so much attention that we have reproduced elaborate delineations from a report to the government. The following are the explanations of the plate: Fig. 8 shows *Mytilaspis Citricola* (Packard); 1, scales on orange, natural size; 1 *a*, scale of female, dorsal view; 1 *b*, scale of female, with ventral scale and eggs; 1 *c*, scale of male, all enlarged. Fig. 9 shows another species, *Mytilaspis Gloveri* (Packard); 2, scales on orange twig and leaf, natural size; 2 *a*, scale of female, dorsal view; 2 *b*, scale of male; 2 *c*, scale of female with ventral scale and eggs, all enlarged.

Remedies.—These we give as stated by Dr. Riley in his report to the department. Mr. H. G. Hubbard, the special agent conducting the experiments, recommends an emulsion of kerosene (as the results of numerous experiments), consisting of refined kerosene, 2 parts; fresh—or, preferably, sour—cow's milk, 1 part; or a percentage of oil, $66^2/_3$. When cow's milk cannot be obtained, and it is often hard to get in some parts of Florida, the following is recommended:

Kerosene,	8	pints	= 64 per cent.
Condensed milk,	1½	"	} = 36 per cent.
Water,	3	"	

It is prepared as follows by Mr. Hubbard: "Mix thoroughly the condensed milk and water before adding the oil; churn with the Aquapult pump until the whole solidifies and forms an ivory-white, glistening butter as thick as ordinary butter at a temperature of 75° F. If the temperature of the air falls below 70°, warm the diluted milk to blood heat before adding the oil.

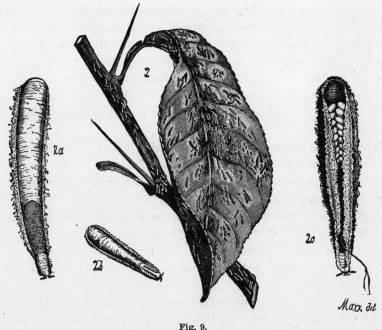

Fig. 9.
MYTILASPIS GLOVERI.

"In applications for scale insects the kerosene butter should be diluted with water from 12 to 16 times, or 1 pint of butter to 1½ gallons (for chaff scale); 1 pint of butter to 2 gallons (for long scale). The diluted wash resembles fresh milk, and if allowed to stand, in two or three hours the emulsion rises, as a cream, to the surface. The butter should therefore be diluted only as needed for immediate use, and the mixture should be stirred from time to time.

"A wash prepared in accordance with the above directions will kill with certainty all the coccids and their eggs under scales with which it can be brought into direct contact. No preparation known to me will, however, remove the scales themselves from the tree, or in any way reveal to the unassisted eye the condition of the insects within. This can be ascertained only by microscopic examination of detached scales. Time alone, and the condition of the tree itself, will indicate the result of an application. Kerosene, it is true, loosens the scales from the bark, so that for a time they are readily brushed off, but they afterwards become more firmly adherent, and are very gradually removed by the action of the weather.

"Upon trees thickly infested a large proportion of the scales are so completely covered up by the overlapping of other scales, or the webbing together of leaves by spiders and other insects, that the wash cannot be brought into direct contact with them, and they are only reached, if at all, by the penetrating action of the oil. This takes place gradually, and the number of bark-lice killed increases for some time after an application, reaching the maximum in the case of kerosene about the fifth day. In long scale the oil penetrates the outer end, killing first the eggs at the broad and thin outer end, but its action is gradually exhausted and several pairs of eggs in the middle of the scale are often left alive. It is, therefore, impossible, in a single application, to destroy every scale upon an orange tree. This can, however, be accomplished by making two or three applications at intervals of four or five weeks. The mother insects being nearly or quite all killed by the first treatment, and the surviving eggs having in the interval all hatched, a second application, if thorough, will clear the tree.

"The great difficulty experienced in reaching every part of the tree renders it absolutely necessary that any liquid used should be applied in fine spray and with considerable force. An ordinary garden syringe does not accomplish this and can never be used satisfactorily against scale insects.

"Although I have thought it advisable to recommend several applications, a single very thorough spraying with a good force pump will, in most instances, prove entirely effectual in clearing the tree, since, if only an occasional egg or coccid escapes, the great army of parasites and enemies will be almost sure to complete the work.

"As has been already said, diluted kerosene does no injury to young growth or to the bark of the orange trees. It, however, causes the older leaves to drop, and where the tree is badly infested with scale, or otherwise out of condition, the defoliation is sometimes complete, especially if the wash is applied in the sun. The death of moribund branches and twigs is also hastened. Beyond this the injury, if such it be considered, is imperceptible, and dormant trees are invariably stimulated to push out new growth in two or three weeks after treatment.

"Even in midwinter, if the weather is mild, sprouts will show themselves, and this is perhaps the only objection to its use at this season, for it is clearly not desirable to start the buds at a time when there is danger of frost. During the past winter (1881–'82) I have experimented with many young trees, using emulsions containing from forty to eighty per cent of kerosene, and in no case has any real injury resulted, although some trees in very bad condition have lost a portion of their twigs and smaller branches that had been long infested with scale and were in a dying condition. In the spring, when the trees are in full growth and covered with tender sprouts, they may be sprayed with the diluted emulsion recommended above, without danger of checking their growth."

Other Preparations.—Mr. Mathew Cooke, Chief Executive Horticultural Officer of California, under date Sacramento, June 1st, 1882, gives the following, in relation to scale insects:

In regard to remedies, I have tried, and recommended others to try, various experiments, and have been successful beyond doubt.

1. Nursery trees dipped in a solution of one pound of American concentrated lye to each one and a half gallons of water will be perfectly cleaned of *A. perniciosus*, *A. rapax*, or any other of the Aspidiotus except *A. conchiformis* (*Mytilaspis pomorum*).

2. Nursery trees dipped in a solution as above, but one pound to each gallon of water, will be perfectly cleared of *A. conchiformis* (*M. pomorum*).

3. The roots of nursery trees should be dipped in soap and sulphur (soap two parts, sulphur one part), one pound to each gallon of water.

4. Fruit trees, apple, pear, quince, cherry, plum, etc., washed before the sap begins to run or buds begin to swell with one pound of American concentrated lye to each gallon of water, will be effectually cleaned of Aspidiotus and Lucanium Scale insects.

5. Cherry and plum trees covered with Aspidiotus Scales and red spider were washed, as an experiment, with two pounds of American concentrated lye to one gallon of water. Insect life all destroyed, and trees bearing a large crop of fruit at present.

III. Plant Bugs.

PLANT bugs are, next to lice, the most destructive to vegetation of all the insect tribes, being provided with a beak for sucking the juices of plants. They are mostly active in all stages of their existence, from the young hatched from the egg to the full-grown insect. Some of the bugs, as the bed-bug, for instance, never have wings; none do in the pupa or nymph state. In the perfect state they acquire wings and fly to scatter their species abroad. The following descriptions are after Mr. Townend Glover, and here, we follow his classification, which is partially that of Amyot and Serville, not a bad one for the unscientific reader, since it is formed wholly on certain marked peculiarities in the structure of the insect as visible to the naked eye. Those we present are of two primary divisions, bugs frequenting the land (*Geocorisæ*), and next those frequenting or living in water (*Hydrorisæ*).

With Large Shield.—The first family of the land-bugs is distinguished by the great size of their scutel, or shield. These insects are generally of moderate or large size, and have a long four-jointed beak, or piercer, with elongated five-jointed antennæ. Among these we frequently find several plant-bugs, which present the appearance of small beetles, as the scutel covers most of their back, and the wings are almost entirely concealed by this covering as with a coat of mail.

Fig. 1.
Corymelæna.

A good example of this class is *Corymelæna*, which is a small, almost round, black bug, Fig. 1, is abundant on strawberries, raspberries, cherries, and almost all other soft fruits. When they are numerous they cause the stems of young fruit trees to wither up and perish from their punctures. They are also said to injure grape vines.

The genus *Tetyra* is also distinguished by its very large scutellum, which covers the whole of its abdomen, leaving only the side of the wing covers exposed.

TETYRA BIPUNCTATA.—Fig. 2 is a medium-sized, or rather large bug, of a brownish-gray color when dried, and is figured merely to show the size of the scutellum.

Fig. 2. Tetyra Bipunctata.　　Fig. 3. Cabbage-Bug.　　Fig. 4. Gray Tree-Bug.　　Fig. 5. Green Tree-Bug.　　Fig. 6. Brownish-Gray.

The following plant-bugs, with large scutella, may be classed as some of those most destructive to the foliage and shoots of various plants and trees.

HARLEQUIN CABBAGE BUG.—*Strachia (Murgantia) histrionicha*, Fig. 3, commonly known as the harlequin cabbage-bug, from its mottled, bright, and harlequin-like colors of black, striped and variegated with bright red or orange, in all their stages, from the egg up to the adult insect, are very destructive to the cabbage, turnip, mustard and other cruciferous plants. The eggs we have are oblong and very beautiful, being banded with dark-colored rings. These eggs are generally deposited in bunches of ten or twelve on the under side of the leaves. Twelve to twenty-four days after the deposition of the eggs, the perfect insect is developed, and there are two broods or more annually in the extreme Southern States. They pass the winter as perfect insects, under stones, moss, or bark. Nauseous washes, such as whale-oil soap, even if they did drive away the insects for a time, would render the cabbage unedible for mankind; and poisons such as Paris green, if taken by the insects, would certainly be most dangerous to the consumers, even if washed off with half a dozen waters. Insect powder will kill them. These insects are destroyed by *Leptoglossus phyllopus*, figured at No. 12.

GRAY TREE-BUG.—A large speckled gray tree-bug, resembling in color the bark of a tree, *Brochymena arborea*, Fig. 4, is not uncommon on trees. It feeds on the sap of trees, and hibernates under bark and logs. A Southern species.

GREEN TREE-BUG.—*Nezara hilaris* (*Rhaphigaster pennsylvanicus* of Fitch), Fig. 5, is a large green tree-bug which feeds on the sap of trees. This insect is of a somewhat flattened form, of a grass-green color, edged all round with a yellow line, interrupted at each joint with a small black spot. Besides feeding on the sap of forest-trees, it punctures the leaves of the grape-vine and hickory-trees.

ACANTHOSOMA NEBULOSA.—Fig. 6 is a medium-sized brownish-gray plant-bug, feeding on the sap of trees and plants.

EUSCHISTUS (PENTATOMA) PUNCTIPES.—Fig. 7, is a middle-sized plant-bug, of a brownish-gray color, common on thistles, mulleins, and other weeds, and lives on the sap of plants. Many species of *Pentatoma* are insects of medium or large size, found on shrubs or trees, and live generally on the sap; but they are also somewhat beneficial by transfixing caterpillars with their beaks to extract their juices, and eventually killing them. Their eggs are usually of an oval form, and attached by a glutinous substance at one end to leaves or branches, the other end being furnished with a cap or cover, which the young larvæ burst off when they hatch out.

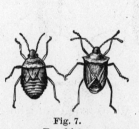

Fig. 7. Euschistus.   Fig. 8. Podisus Cynicus.   Fig. 9. Podisus Spinosus.   Fig. 10. Stiretrus Dian

PODISUS CYNICUS.—Fig. 8, (*Arma grandis* of Dallas), or the large tree-bug of Fitch, is of a dull pale-yellowish or brownish color. It feeds on the sap of the apple, oak, and other trees. The insect is somewhat the shape of a pumpkin-seed, and has a conspicuous sharp spine projecting outward on each side of the thorax.

PODISUS SPINOSUS.—Another smaller species, *Podisus* (*Arma*) *spinosus*, Fig. 9, or the spined tree-bug, a brownish or grayish plant-bug, nearly the color of tree-bark, injures leaves of apple and other trees by sucking out the sap; but it is very beneficial to the farmer or gardener by destroying the Colorado potato-beetle. The spined tree-bug is said also to destroy the American gooseberry saw-fly (*Printiphora grossulariæ* of Walker) and other insects.

STIRETRUS DIANA, (*anchorago*) Fab., Fig. 10, a beautifully marked plant-bug of a purple black color, with red or orange ornamental marks on the thorax and scutel, was found in Maryland busily employed in killing and sucking out the juices of the larva of the squash-beetle (*Epilachna borealis*), and no doubt it destroys also any other soft-bodied larva it can overcome, and should be protected.

STIRETRUS FIMBRIATUS.—A near relative of the last insect, the ground colors of which are orange or yellow, with black ornamentations, Fig. 11, is very rapacious and carnivorous, as it feeds almost entirely on other insects, including the Colorado potato beetle. It destroys caterpillars of the black asterias, swallow-tail butterfly, which are so injurious to parsley, parsnips, celery, etc., in our gardens, and probably also the social caterpillars in the web nets, which disfigure our shade and fruit trees.

Fig. 11. Stiretrus.

Geocorisæ (land bugs).—The second family of the land bugs is that of the *Supericornes*, so called because the antennæ are inserted upon the upper side of the head, above an ideal line drawn from the eyes to the origin of the labrum.

Some of the insects of this family are beneficial to the farmer by destroying other injurious insects, among which may be classed *Leptoglossus phyllopus*, (*Anisoscelis albicinctus*), Fig. 12, a reddish-brown or blackish bug, with a distinct dirty-white or yellowish band across its wing covers. It may easily be recognized by the singularly broad, flattened, leaf-like projections on its hind shanks. When young, the insects are of a bright red color.

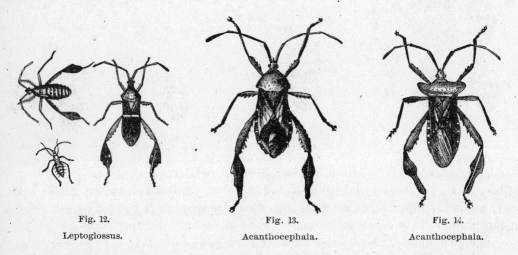

Fig. 12. Leptoglossus.   Fig. 13. Acanthocephala.   Fig. 14. Acanthocephala.

ACANTHOCEPHALA (RHINUCHUS AND METAPODIUS, SYN).—This genus is the largest and most powerfully-developed of the *Heteroptera* in this country, and is generally found in the Southern States. The insects frequent cotton fields, but have never been detected in the act of piercing cotton bolls or of destroying other insects, so far as we know.

ACANTHOCEPHALA (METAPODIUS) FEMORATA.—Fig. 13, so called from its swollen spiny thighs, is a large reddish-brown or blackish insect, quite abundant in the Southern cotton fields. It is very slow in its motions, and appears to be fond of basking in the sun. The thighs are strongly developed and spiny, especially on the under side, while the shanks have broad, thin, plate or leaf-like projections on their sides, which give these insects a very peculiar appearance. The eggs are smooth, short, oval, and have been found arranged in beads like a necklace, on the leaf of white pine. The full-grown insect is stated to injure cherries in the Western States by puncturing them with its beak and sucking out the juices; thus proving it at least in one instance to be a feeder on vegetable substances.

ACANTHOCEPHALA DECLIVIS.—Fig. 14 resembles the previously named insect in general size and form, but differs materially in the shape of the thorax, which is much broader, and projecting outward and forward. It also has strong, spiny hind thighs and the peculiar flattened plate-like shanks.

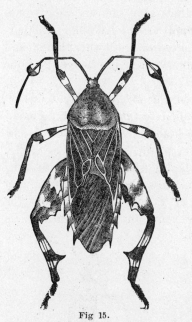

Fig 15.
Pachilis Gigas.

Fig. 16.
Alydus.

Fig. 17.
Alydus.

Fig. 18.
Squash Bug.

The natural history and habits of these insects has been very little studied; but, of all the specimens taken in the Southern States, our authority never yet took one in the act of killing other insects.

PACHYLIS GIGAS.—Fig. 15 is one of the largest and most gaudily-colored heteropterous insects found in this country, and as yet appears to be essentially southern and rather scarce. Its markings are of a bright red orange on a black ground; the contrast between the two colors being very marked and distinct, rendering the insect plainly visible at a great distance.

ALYDUS EURINUS.—Fig. 16, a slender bug, with several sections of the upper part of the abdomen of a bright red color when the wings are opened, occurs in late summer and autumn, sometimes in great numbers, on golden rod and other herbaceous plants, growing near the edges of woods, also on the *Rhus glabra* or smooth sumach. Fig. 17 is the female.

SQUASH BUG.—One of the most destructive plant bugs in this family is the squash bug, *Anasa tristis*, Fig. 18, (*Coreus* and *Gonocerus tristis* of some authors). The eggs are deposited in little patches, fastened with a gummy substance to the under side of the leaves of squashes and other *Cucurbitaceæ*, in June and July, etc., until late autumn. These eggs are not deposited all at one time on the plants, but in successive broods during the whole season. The larvæ, pupæ and perfect insects, all being active, indiscriminately attack the leaves, and cause them to wither up by sucking out the sap and apparently poisoning the foliage. They moult their skins several times before attaining the winged or perfect state, and become more oval in form as they grow older; and as there are successive broods during the whole summer, they do much injury to the squash and pumpkin vines. These insects sometimes collect in masses around the stem near the earth, and injure the plant itself by extracting the sap with their piercers. When handled or disturbed, they give out an odor somewhat similar to an over-ripe pear, but which is too powerful to be agreeable. The perfect insects, late in the autumn or when cold weather begins, leave the plants, and hibernate, or pass the winter under bark of trees, in moss, or in crevices in stone walls, and in old fences.

RHOPALUS.—A small plant-bug, *Rhopalus lateralis*, Fig. 19, probably feeds on the sap of plants, as Mr. Walsh states that an insect allied to this is one of the commonest bugs near Rock Island, Illinois, and ruins the buds of the pear-tree. The antennæ are clubbed at the end.

NEIDES (BERYTUS) SPINOSUS.—Fig. 20 is a remarkably slender bug, with very long, slim, hair-like legs and antennæ, and is figured to show the singular form and structure of the insect. Another species, *N. elegans* of Europe, is taken about the roots and young stems of the rest-harrow (*Ononis arvensis*), and with regard to its

Fig. 19. Rhopalus.   Fig 20. Neides.   Fig. 21. Lygæus.   Fig. 22. Lygæus.

habits Wedwood states that as the larvæ and pupæ were discovered in company with the imago, it appears evident this was its food-plant.

Infericornes.—The third family are distinguished by the antennæ being inserted below an ideal line drawn from the eyes to the origin of the labrum, or below the middle of the side of the head. The third joint of the beak is longer than the fourth.

In the *Lygæides*, the antennæ are four-jointed; the terminal joint not being thinner or forming a terminal club. They are generally rather small or of moderate size, and several species are beautifully marked; being black, variegated with bright crimson, red, orange or yellow. They are mostly found on plants.

LYGÆUS TURCICUS.—Fig. 21 is common in Maryland, and is of a black color, ornamented with bright red, and has been observed once or twice preying on the small caterpillars feeding on the *Asclepias*, or milk-weed.

Another species, *Lygæus fasciatus*, Fig 22, of an orange and black color, has also been found in great abundance in Maryland on flowers of the *Asclepias*, in company with caterpillars of *Euchetes egle*, a medium-sized moth, or miller, and it probably feeds also upon them.

Lygæus bicrucis, Fig. 23, a plant-bug of a bright-red and black color, with white edges on the elytra and thorax, was taken under bark in winter, showing that this class of insects hibernates in the perfect state in sheltered situations.

These three examples will suffice to show the general form of the genus *Lygæus* in this country.

Fig. 23. Lygæus.

OPHTHALMICUS. — Fig. 24, is figured merely to show the singularly broad head and projecting eyes of one genus of the *Infericornes*, and so different from the rest. Most probably it is a plant-feeder.

Fig. 24. Ophthalmicus.

Fig. 25. Nysius.

Fig. 26. Chinch-bug.

NYSIUS RAPHANUS.—Fig. 25 is a small plant-bug of a brownish color when dried, injurious to radishes, mustard, grape, cabbage, potatoes, and cruciferous plants. There are two or three broods annually in some of the States. The insect has a very disagreeable smell, and sucks the sap of plants, causing them to wilt. The leaves attacked show little rusty circular specks, where the beak has been inserted, which form little irregular holes that look more as if caused by a coleopterous insect, the common flea-beetle.

Chinch-Bug.—The chinch-bug, *Micropus* (*Rhyparochromus devastator*,) (*Micropus*) (*Blissus*) *leucopterus*, Fig. 26 (enlarged and natural-size), is one of our most destructive insects to wheat, corn, etc., in some of the Western States, and has done considerable damage to the crops. The eggs, to the number of about 500, are laid in the ground about June, on or among the roots of plants; and the young larvæ, which are of a bright-red color, are said to remain underground some time after they are hatched, sucking the sap from the roots, and have been found in great abundance at the depth of an inch or more. The full-grown insects measure about one-twelfth of an inch in length, and are of a black color, with white wings, and may be known by the white fore or upper wings, contrasting with a black spot in the middle of the edge of the wing.

Remedy.—The most feasible method for the destruction of the chinch-bug is burning over the stubble and grass fields in the autumn and winter, and especially to rake up the corn standing in the field, after husking, and burn them. An insect, the false chinch-bug, much resembling the true, is said to kill it. Quails eat them; two or three lace-wing flies are also said to destroy them.

Cecigenæ.—The fourth family are destitute of oceli, hence their name. They frequent plants and shrubs.

Largus succinctus, Fig. 27, of a rusty-black color, with the borders of the upper wings edged with dull orange or yellow. We have found this insect hibernating under moss, stones or bark, in mid-winter in Maryland, but have never yet caught it in the act of injuring plants, although it probably is a vegetable-feeder.

Another plant-bug of this family, *Dysdercus (Pyrrhocoris) suturellus*, Fig. 28, male and female, natural size and enlarged, with eggs, is the too well-known red-bug, or cotton-stainer of Florida, which in some seasons does so much injury to the cotton fiber in the bolls of the plant, when in the field, by sucking out the sap from the boll and seed, and voiding an excrementitious matter over the opened bolls, which produces an indelible stain on the fiber and renders it totally unfit for the market.

Fig. 27.
Largys.

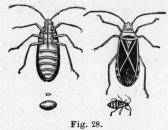

Fig. 28.
Red-Bug.

Remedy.—These insects being in the habit of collecting together where there were splinters or fragments of sugar-cane on the ground, advantage was taken of this fact to draw them together by means of small chips of sugar-cane laid upon the earth near the plants, where they were at once destroyed by means of boiling water. They also collect around heaps of cotton-seed, where they may readily be destroyed at the commencement of cold weather. Small heaps of refuse trash, dried corn-stalks, or especially of crushed sugar-cane, may be made in various parts of the plantation in the vicinity of the plants; under these the insects take shelter from the cold, and when a sufficient quantity of the bugs are thus drawn together, the various heaps may be fired, and the insects destroyed with the trash. A very cold morning, however, should be selected, and the fire made before the insects have been thawed into life and vigor by the heat of the sun; and especially all dead trees, rotten stumps, and weeds in the vicinity of the field should be burned or otherwise destroyed, as they afford a comfortable shelter for all sorts of noxious insects, in which they can pass the winter in a semi-dormant condition.

Bicelluli.—The fifth family contains plant-bugs having two basal cells in the membrane of the wing. The last joint of the antennæ is very fine and setiform.

Fig. 29.
Campyloneura.

The group *Capsides* contains insects of active habits. The females have ovipositors nearly half the length of their bodies, somewhat saber-shaped, and received in a slit on the under side of the abdomen. These small plant-bugs are very active, running and flying with agility. They frequent plants, trees and fruits, upon the juices of which they appear almost exclusively to subsist. Some of the species are especially fond of fruit, such as raspberries, which they suck with their rostrum, and impart a very nauseous taste to the fruit.

An exception to their general plant-feeding habits, however, is shown in one species, *Campyloneura (Capsus) vitripennis*, (Fig. 29) or the glassy-winged soldier-bug of Riley, which is said to be beneficial by destroying the leaf-hoppers of the vine-leaf, *Erythroneura vitis*, (incorrectly called the thrips). The insect is of a pale greenish-yellow, the head and thorax are tinged with pink, and the upper wings are transparent, with a rose-colored cross. It lives also on the wild chicken-grape, and attains its full growth in August, and destroys small caterpillars by sucking their juices, according to Professor Uhler. Most probably many other species of the *Capsides*, hitherto considered as plant-feeders, also occasionally vary their diet by sucking out the juices of other insects.

Fig. 30. Resthenia

As the *Capsides* in general are very injurious to vegetables, as well as numerous, we will give a few figures of them in order to give the student some general idea of their size and form.

RESTHENIA (CAPSUS) CONFRATERNA.—Fig. 30, is of a black color, with red thorax, and is somewhat common on weeds and low herbage. It is very active, either running swiftly away and hiding, or flying away when disturbed.

CALOCORIS (CAPSUS) BIMACULATUS, Fig. 31, is also a common insect of a green and brown color, and is very common in Maryland on weeds.

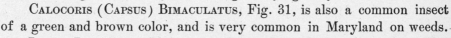
Fig. 31. Calocoris.

LYGUS LINEATUS (CAPSUS AND PHYTOCORIS QUADRIVITTATUS), Fig. 32, or the four-striped plant-bug, is a very common insect south, and is of a green or yellowish color, with four black lines on its wing-covers. Larvæ, pupæ, and perfect insects puncture leaves, abstract the sap, and produce a blighted appearance of the foliage of currants, parsnips, potatoes, mint, weigelia, dentzia, &c.; sometimes causing them to wither up entirely.

One of the most common small plant-bugs south is *Lygus lineolaris* (*Capsus oblineatus*, Say), Fig. 33, or little lined plant-bug of Harris. This insect is of a black and brownish yellow color, and is very common on almost all kinds of plants. It appears in April, but is more abundant during the summer, when it injures plants by sucking their sap. The punctures made by them appear to be poisonous to vegetation. This insect injures pear-twigs, and the stalks of grape-vines, potatoes, strawberries, fruit trees, such as quinces, &c., and is very fond of congregating on the flowers of cabbage. It is stated to have injured the crops in Illinois very considerably. Dr. Le Baron says that it destroys the Colorado potato-beetle, and the "American Entomologist" reports it as destroying the eggs of other insects as an offset to the great amount of damage it does to the crops. It has been found in the perfect state in winter.

Fig. 32. Lygus Lineatus.

Ductirostri.—The sixth family contains plant-bugs, which, when at rest, have their beaks, or piercers, in a groove, or duct, under the body.

The first group contains a singular, small, greenish insect, marked with brown, *Phymata* (*Syrtis*) *erosa*, Fig. 34, having raptorial, crooked, sickle-shaped fore feet, with which it catches and holds its prey while it leisurely sucks out the juices. This insect stings severely: it lies in wait in flowers or among leaves, where hidden from observation by the similiarity of its color to the places it frequents, it seizes any unfortunate insects that may happen to alight near its hiding-place. One of these insects was taken in the very act of sucking out the juices of a small blue butterfly;

the bug itself being completely concealed among the petals of a rose, the butterfly only appearing in sight, which was seized as a specimen and drawn out, with the bug still clinging to it. Many other bugs of the same species were afterwards observed lying in wait in various flowers for any roving insects that might be attracted to them. It is said to prey on small bees and wasps, and also is beneficial by destroying plant-lice, or *Aphides*.

The group *Tingides* are small, flattened, singularly-formed insects, living on various plants and trees.

A good example of this group is *Tingis juglandis*, an insect found abundantly on the butternut, birch and willow, where it pierces the leaves and sucks the sap. This insect resembles a flake of white froth; its whole upper surface being composed of a net-work of small cells, with an inflated egg-shaped protuberance like a small bladder on the top of the head and thorax. The wing-covers are square, with rounded corners.

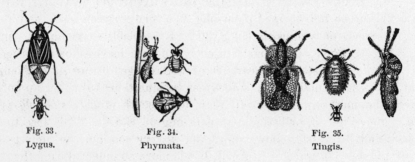

Fig. 33.
Lygus.

Fig. 34.
Phymata.

Fig. 35.
Tingis.

TINGIS ARCUATUS.—Fig. 35, is distinguished by the arcuated edge of the hemelytra, or wing-covers, with brown bands. They live on the sap of plants and trees, and one species closely related to it was found on the quince-bushes in Mississippi and Florida, where the bushes were literally swarming with them, in all stages of larvæ, pupæ, and perfect insects, and some of the trees were very much injured, if not totally destroyed by them. They were also very troublesome to mankind by their stinging propensities.

ARADUS AMERICANUS.—Fig. 36, a small, flat, brown or blackish bug, is very common under bark of trees.

THE BED-BUG.—*Acanthia lectularia* (*Cimex lectularius*), Fig. 37. The eggs are white, oval, slightly narrowed at one end, and terminated by a cap, which breaks off when the young escape. The young ones at first are very small, white and transparent. It takes eleven weeks before they attain their full growth, and they are said to cast their skins several times before attaining maturity. It is probable, however, that the temperature and food have much influence in accelerating or delaying their final change into the full-grown imago, or perfect insect. The insects are gregarious in habits, and herd together in cracks and chinks, in

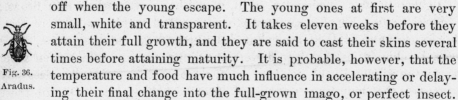

Fig. 36.
Aradus.

Fig. 37.
Bed-Bug.

corners of bedsteads, etc. Professor Verrill states they return constantly to the same hiding-places morning after morning, like birds returning to their roosts. These insects are very tenacious of life, and have been kept in hermetically-sealed glass bottles for more than a year without food, and were yet lively, and had a good appetite.

The Seventh Family.—(*Nudirostri*), contains bugs having the beak or piercer naked or free, entirely disengaged, and not in any duct, as in the last family. The habits of most of them are raptorial, preying upon other insects, and as such they are generally beneficial to the farmer.

PIRATES BIGUTTATUS.—Fig. 38, sometimes called the spotted corsair, is a large, slowly-moving bug of a blackish color, with legs, antennæ, and markings on wing-covers of a dull orange color, with two spots on the wing-covers, and is said to be carnivorous, destroying other insects, and probably destroys bed-bugs also, as one was found between the mattresses of a bug-infested bed, and the insect itself is closely allied to *Reduvius personatus*, mentioned below, which is known to feed upon bed-bugs.

REDUVIUS PERSONATUS.—Fig. 39, is a brownish bug, not rare in Europe in houses, where it is generally found dead and hanging in spiders' webs. Burmeister says that the spiders do not seize it, as its puncture is very poisonous, but let it encumber their webs until it dies of hunger.

Fig. 39.
Reduvius.

Fig. 38.
Spotted Corsair.

The insect is stated to exhale a disagreeable odor, something like that of mice. It hibernates without taking any food, when its body becomes meager and flat; but on the return of fine weather, it recovers from its lethargy, and commences to hunt for such insects as form its prey. The larvæ and pupæ cover themselves with a mask or coating of dust and dirt even to the legs and antennæ, and so disguise themselves as scarcely to be distinguished from the places they frequent, and prey upon the common bed-bugs.

MELANOLESTES (PIRATES) PICIPES.—Fig. 40, a medium-sized black bug, is said by Walsh to be found underground, where no doubt it feeds on subterranean insects. In Maryland it is found under stones, moss, logs of wood, etc., and is capable of inflicting a severe wound with its rostrum, or piercer. It feeds on other insects, and is slow and deliberate in its motions. *M. abdominalis* is distinguished by its red abdomen, which generally shows on each side of the wing-covers.

APIOMERUS (REDUVIUS) SPISSIPES.—Fig. 41, a carnivorous plant-bug of a brown color, with light-yellowish markings, is known as a destroyer of insects, and has also been reported to the Department as killing honey-bees. These insects when in their search for prey are very slow and cautious in their movements, as if they were aware that any rapid or sudden motion would frighten their victim away.

MILYAS (HARPACTOR) CINCTUS.—Fig. 42, is a medium-sized raptorial bug, with a spine on each side of its thorax, and is of a yellowish-brown color, with mottled or banded legs. It feeds upon all insects it can overcome, and is therefore very useful as an insect-destroyer. It has been reported as destroying the Colorado potato beetle, and also the small caterpillars of the apple-worm, or *Tortrix*.

Fig. 40.
Melanolestes.

Fig. 41.
Apiomerus.

Fig. 42.
Milyas.

Fig. 43.
Sinea.

Fig. 44.
Ectrichodia.

SINEA MULTISPINOSA.—Fig. 43, so named from its prickly or spiny appearance, somewhat resembles the last-named insect in general appearance and habits. It is of a brownish color, and wanders about on plants and shrubs, seeking what insects it can overcome, and has also been reported as killing the larvæ of the above-mentioned Colorado potato-beetle. It also is very useful by destroying the *Aphides*, or plant-lice, and other insects.

Fig. 45.
Hammatocerus.

ECTRICHODIA CRUCIATA.—Fig. 44, a carnivorous plant-bug, of a black and scarlet or orange color, has the same habits and propensities as the *Reduvius*. It kills all the insects it meets in its wanderings, and sucks out their juices.

HAMMATOCERUS (NABIS) FURCIS.—Fig. 45, is a very large and powerful predatory plant-bug, of a black and orange or yellowish-red, with the upper part of the wing-covers of a yellow color. It has much the same habits as the nine-pronged wheel-bug, *Prionotus cristatus*.

BLOOD-SUCKING CONE-NOSE.—*Conorhinus (Sanguisuga) variegatus*, Fig. 46. This insect insinuates itself into beds in the Middle and Southern States, and sucks the blood of mankind, causing great pain and inflammation. It hibernates in the pupa and perfect state, and is stated to feed not only on human blood, but also the insect that causes the blood to flow, namely, the common bed-bug (*Acanthia lectularia*). From its blood-sucking propensities, there is very little doubt but what it also destroys insects.

EVAGORUS RUBIDUS.—A much slighter-formed raptorial plant-bug, with longer and slimmer legs, and also of a black and red or orange color, called *Evagorus rubidus*, Fig. 47, preys upon other insects, and was found to be very useful in destroying the myriads of plant-lice upon the orange trees near Palatka, Florida.

INSECTS, INJURIOUS AND BENEFICIAL.   613

DIPLODUS LURIDUS (EVAGORUS VIRIDIS).—Fig. 48 is a slender insect, somewhat related to the last-mentioned species, the larva of which is very common on fruit trees. It is said to be wingless, and covered with a glutinous substance, to which little pieces of dust and dirt are commonly seen to adhere. The perfect insect is winged, and said to destroy the plum-curculio (*Conotrachelus nenuphar*).

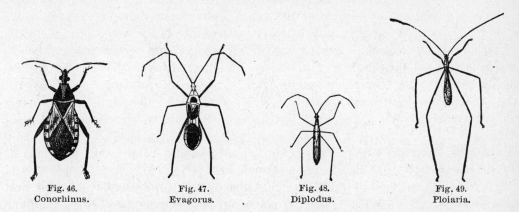

Fig. 46. Fig. 47. Fig. 48. Fig. 49.
Conorhinus. Evagorus. Diplodus. Ploiaria.

PLOIARIA VAGABUNDA.—Fig. 49 is a very slender plant-bug. It has very short anterior legs, or rather arms, while the two posterior pairs are very long. When walking, it moves very slowly, with its fore-legs (which are perhaps useful in climbing or to seize its prey) applied to its body, while the antennæ being bent at the extremity, which is rather thick, are made to rest upon the surface on which the insect moves, and to supply the place of fore-legs. The insect is found on trees; it vacillates or trembles, and balances itself continually like a *Tipula*, or long-legged crane-fly. Dr. Geer says it is found in houses, and walks slowly but flies easily and quickly. Burmeister says that the larva covers itself with dust and lives on prey. In England the insect lives in thatch.

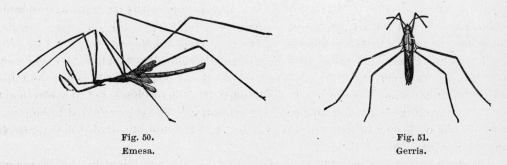

Fig. 50. Fig. 51.
Emesa. Gerris.

EMESA LONGIPES (BREVIPENNIS of Say).—Fig. 50 is an exceedingly thin and slender carnivorous plant-bug. These insects feed on other insects, and resemble the thinnest bits of sticks fastened together. The antennæ are long and delicate. The fore-legs are raptorial, with long, thin coxæ, admirably adapted for seizing and holding their prey, which consists of other insects. The body is long and thin; the wings are either wanting (in some species) or reach only to near the middle of the abdomen.

Heteroptera.—This contains the eighth family of the insects that row on the surface of the water, hence the name *Ploteres*, or rowers; their four hind feet being formed for gliding on the surface of the water, and are sometimes erroneously called in Maryland water-boatmen (see *Notonecta*). These insects are very active, and skim the surface of the water with great velocity. When gliding over streams and ponds, their hind feet act conjointly as a rudder, and the longer middle feet are used somewhat as oars, not dipped into, but merely brushing over, the surface.

The insect of *Gerris conformis*, Fig. 51, was taken in Maryland on the surface of slowly-running water in the act of devouring a dead fly, which was floating on the surface.

GERRIS LACUSTRIS.—Fig. 52 is a smaller species, also common in Maryland on the surface of water, and also feeds on other insects.

The second section of the sub-order *Heteroptera*, *Aydrocorisæ*, contains only three families, viz: Family 1, (or 9,) *Bigemmi*, bugs having two ocelli; family 2, (or 10), *Pedirapti*, water-bugs, having raptorial fore-legs for seizing and holding their prey, and family 3, (or 11), *Dediremi*, water-bugs having their posterior tarsi generally like oars, and formed for swimming and diving; the anterior feet are not raptorial.

GALGULUS OCULATUS.—Fig. 53 is a representative of the group *Galgulides*. These insects have broad heads, with peduncled eyes; their antennæ are four-jointed, but concealed beneath the eyes; the ocelli are present; the body is short, broad and flattened, and the legs are formed for running. These insects, at the first glance, resemble

Fig. 52. Gerris.

Fig. 53. Galgulus.

Fig. 54. Naucoris.

miniature toads. They are probably predatory in habit, preying on other insects, and appear to form a link between the aquatic and terrestrial species.

NAUCORIS POEYI.—Fig. 54 is a rather small, yellowish-brown water-bug, with two raptorial fore-feet and four hind-feet, which the insect uses for walking in the water and running, although they are not ciliated. These insects frequently leave the water during the night to scour round the country. The eggs are said to be glued to the blades of leaves or water-plants in April, and the bugs feed on all the insects they can capture when in the water.

NEPA APICULATA.—The water scorpion, Fig. 55, is a good example of the group. It feeds upon other insects, and also most probably on small fishes. Kirby and Spence state that a *Nepa*, put into a basin of water with several young tadpoles, killed them all without attempting to eat them. It is therefore very evident that they will destroy young fish, and should be extirpated in or near any fish-breeding establishment.

Fig. 55.
Water Scorpion.

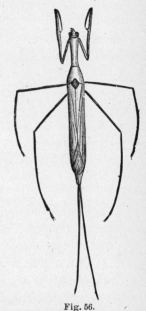

Fig. 56.
Ranata Quadridenticulata.

RANATRA QUADRIDENTICULATA.—A singularly-formed, large, brownish-gray water bug of the family *Pedirapti*, is *Ranatra quadridenticulata*, Fig. 56. The body is of an elongated form, with a double tube at the end for respiration; the eyes are prominent; the two fore legs are raptorial; the four other legs are long and slender, and the prothorax is greatly elongated. These insects, living in the water, are compelled to come to the surface for air, which they obtain with the assistance of the before-mentioned two appendages placed at the end of the anus. They are very voracious, feeding on other insects, aquatic larvæ and small fish. They fly from pond to pond in the evening or at night, especially when the waters begin to dry up. These insects are mostly found at the bottom of stagnant water, as they swim badly. Westwood mentions a European species which is said to carry, attached to their feet, very small grains of a lively red color, which are surmised to be the eggs of an aquatic mite.

IV. General Means for Destroying Bugs.

In the foregoing we have figured principally the rapacious or insect-killing plant-bugs, and this as object lessons so they may not be destroyed. It is of really more consequence that we preserve predatory insects, as we should insect-eating birds, than that we kill injurious ones. Certainly insects should not be destroyed indiscriminately. For the plant-eating bugs there has been found only one successful remedy, except hand picking for the larger species—that is, a solution of Paris green or London purple in water. This, however must not be used on vegetables that are to be eaten. For the smaller species, and the minute fleas and beetles that prey on vegetation, soot, sulphur, powdered charcoal, etc., are recommended. They are, however, the most difficult to exterminate of almost any of the destructive vegetable-eating insects. So far as all that class of mites and scale insects that infest green-houses are concerned, scrubbing, and the use of emulsions will be indicated for scale. For lice and other mites, fumigation with tobacco, burning sulphur, etc., will be proper. For window plants the kitchen sink will be available for washing, and the little net figured in Part VIII, Chapter II, relating to the parlor and library, Section XIII, will be found useful. Out of doors in the field these pests are not so easy to manage, so far as plant-bugs are concerned.

V. Remedies for Chinch-Bugs.

ROLLING the ground, although it will kill many, is not a perfect remedy, for the reason that the inequalities of the surface prevents their being all crushed. The only practical remedy yet found is burning the stubble and dead grass late in autumn and early winter, and especially fence corners and rubbish piles, where they congregate to winter; also all dry corn stubble, as heretofore indicated. When the insects are migrating a trench plowed round the field and deepened with a winged shovel plow, leaving the sides dusty, will prevent their migrations. A log drawn back and forth in the trench, or better, a smooth implement, heavy, like that shown in the cut, will crush and kill them. It has been recommended to make the implement of sheet-iron, with chimney and fireplace. This might be useful to keep the trench dry after rains, but would hardly keep enough heat to kill the chinch-bugs. Resort must be had to pressure. The same rules which apply to chinch-bugs will apply with still greater force to the destruction of the army-worm. Sowing Hungarian-grass is said to have protected from the ravages of the chinch-bug, the insects preferring the tender grass to the older grain.

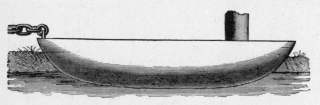

DEVICE FOR KILLING CHINCH-BUGS, DESTROYING ARMY-WORMS, ETC.

CHAPTER III.

INSECTS DESTROYING GRASS AND GRAIN.

I. INSECTS THAT PREY UPON GRASS.—II. INSECTS INJURING CLOVER.—III. CLOVER-LEAF BEETLE.—IV. THE ARMY WORM.—V. VAGABOND CRUMBUS.—VI. INSECTS INJURING GRAIN.—VII. THE SORGHUM WEB-WORM.—VIII. SUGAR CANE BEETLE—IX. THE SMALLER CORN-STALK BORER.—X. THE RICE-STALK BORER.—XI. GRASS-WORM OF THE SOUTH.—XII. CORN BILL BUG.—XIII. THE CORN OR COTTON-BOLL WORM.—XIV. REMEDY FOR THE COTTON-WORM, SOUTH.—XV. POISONS FOR WORMS—XVI. THE HATEFUL GRASSHOPPER OR LOCUST.—XVII. REMEDIES AGAINST THE GRASSHOPPER.

I. Insects that Prey upon Grass.

THE insects most injurious to grass are those which destroy the root. Of these the white grub, which is the larva of the May beetle, is the worst. Next in destructiveness come those boring into the stalk, of which the clover-stem borer will illustrate the type, and those which, like the army-worm, feed on rye. The chinch-bug, already figured and noticed, and whose depredations are principally on grain, is, also, exceedingly destructive in some seasons. The Hessian fly is another insect which eats the young plant down into the roots. The means of destroying these last is by burning the stubble where they hibernate. None of this class of insects like wet ground, hence the dryest parts of a field are attacked first. The Hessian fly is a minute, two-winged insect, and two-brooded; it is well known to every farmer. Insects attacking grass, feed also, as a rule, on grain. We figure some of the newer pests first, and then proceed to those attacking grain.

II. Insects Injuring Clover.

Clover-Stem Borer (LANGURIA MOZARDI).—This insect has been known only during the last four years. Where abundant, the stalks of red clover show small discolored spots. If these are cut into, a grub one-sixteenth of an inch, or, if full-grown, nearly one-third of an inch will be found. The stems attacked are gradually weakened, and often fall; thus seriously injuring the crop

THE REMEDY.—The apparent remedy for this pest is to cut the clover early and again late. Waste clover remaining in the field will afford protection. Hence, when the borer is found, destroy all clover which cannot be cut or kept close pastured. This insect has two parasites: A small black chalcid, and an ichneumon fly. The cut of Clover-Stem Borer shows, at the top, section of stalk with young larvæ; to the right, the grub or larva; next, the pupa; then the eggs, and to the left, the beetle; all enlarged. The hair-lines show natural length, and the small ovals the real size of the eggs.

Clover-Root Borer. (HYLESINUS TRIFOLII).—This insect seems first to have been noticed in New York, about 1878. It hibernates in all of the three stages, though, in the autumn, the beetles greatly predominate. In the spring, these beetles issue from the ground and pair. The female deposits her eggs, from four to six in number, in a cavity at the crown of the roots. The young hatch in about a week, feed their way into the plant, and

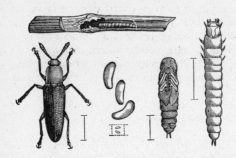

CLOVER-STEM BORER.

CLOVER-ROOT BORER.

channel both root and stalk, as shown in the cut of Clover-Root Borer. It will be seen that if this insect should become common (we do not know that it has yet appeared in the West) it must prove destructive.

The cut explains itself; *a*, the work of the insect; *b*, larva; *c*, pupa, and *d*, perfect beetle enlarged.

THE REMEDY.—No perfect means have yet been found. One which naturally suggests itself would be to plow infested clover under after the eggs are laid. It is to be hoped that parasites will appear to keep this insect in check, since it will be difficult to manage it either by picking or poisoning. A telephorid larva has been found preying on it.

Clover Leaf Midge.—(*Cecidomia trifolii*)—This is a minute insect lately found attacking the leaves of white clover. A similar, if not identical insect has long been known in Europe. The size of the insect (fly) is shown by the minute figure under the larvæ. Its length is only .059 of an inch. The young fold the leaflets together, and fasten them on the midrib, by delicate threads of silk; here galls are formed. There is no well defined remedy for this insect.

If found and examined under a powerful lens its several stages will be seen as shown in the accompanying cut.

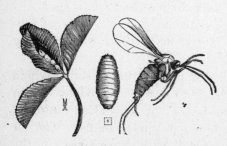

CLOVER LEAF MIDGE.

III. Clover Leaf Beetle.

The technical name is *Phytonomus punctatus*. This insect feeds, both in the larval and beetle state, on clover and alfalfa. There are many of this genus preying on a variety of plants. The pest we describe feeds on clover generally, including the white. Near Barrington, N. Y., last year, it was so destructive, that it was reported scarcely a plant could be found not infested with it. It has not yet made its way West. It is to be hoped it will not. The cut of Clover Leaf Beetle shows the insect in its various stages. These include back and side views of the beetle, the hair-lines showing the natural size. [See cut page 621.]

Remedies.—We quote, in relation to remedies, from the last report of Dr. Riley, Entomologist to the U. S. Department of Agriculture. He says: "Our experience and observations during the winter show that this Phytonomus hibernates principally in the young larval state, and that any mode of winter warfare that would crush or burn these larvæ hibernating in the old stalks would materially reduce the depredations of the species the ensuing summer. Clover stubble is, however, not so easily burned in winter, and whether rolling could be advantageously employed will depend very much on the smoothness of the field and other conditions.

The extreme timidity of the larva as well as of the beetle, and the protected position of the insect in all stages, render the application of pyrethrum, or any other remedy acting upon contact, entirely useless. To poison the clover with London purple or Paris green would no doubt be effective, but can be safely applied only wherever the clover is not used for fodder.

Should the Phytonomus be very bad in a field, it would be well to plow the clover under rather than to allow such field to become a source of contagion. This should be done in the month of May, when the insect is mostly in the larval state, and when all eggs from the beetles that hibernated have been hatched. To plow the field when the Phytonomus is in the imago state would have no other effect than to disperse the beetles over other fields.

Natural Enemies.—Of the various species of Ichneumon flies known in Europe to prey upon the larvæ of Phytonomus, none have been observed so far in this country, and to this immunity from the most efficient natural checks the undue multiplication of the species is no doubt to be attributed. Of other insect enemies only one has been actually observed so far, viz., the larva of a small beetle, *Collops quadrimaculatus*, which was found feeding upon the eggs sent from Barrington in January. Mr. Schwarz found three dead larvæ on the plants, and from the manner in which they were killed he thinks that they were sucked out by soldier-bugs, several species of which were seen in the fields, but none in the act of sucking Phytonomous larvæ. Several ground-beetles (*Harpalus pleuriticus*, *H. pennsylvanicus*), a *Pterostichus* larva, and numerous specimens of a large red mite (genus *Trombidium*) are found under the infested plants, and these probably prey upon the Phytonomus in its earlier stages, but no proof thereof can be given at present. Ants do not seem to trouble the larvæ, as on several occasions specimens of the latter were found in the middle of the ants, which build their colonies under small stones and sticks in the field. This species is in all probability extensively fed upon by tiger-beetles (*Cicindelidæ*), which, both in the larval and beetle states, doubtless attack and devour the Phytonomus larvæ, whether when they feed or crawl over the ground, or in the ground to pupate; for we found, during August, on Mr. Snook's farm, that the ground in the infested clover-fields was in many places literally riddled with holes of larvæ of *Cicindela repanda*, most of them apparently nearly full-grown, and many just having changed to the perfect beetle.

IV. The Army Worm.

BESIDES the insects already mentioned, there are few committing serious depredations. The army-worm deserves particular notice. It is a universal feeder, and a species quite similar is as destructive to cotton in the South as is the army-worm to grain in the North. There are many so-called army worms. In fact, some persons call every smooth caterpillar found in grain or grass, the army-worm. The true species which multiplies in such amazing numbers as sometimes to utterly devastate a country, is the *Leucania Unipuncta*, of Haworth. The illustration will enable it to be easily recognized. The favorite place for depositing the eggs is along the inner base of the terminal blades of grass or grain, and preferably, in the rankest tufts of the grass or grain. The moth remains concealed by day and flies at night, depositing her eggs in the early part of the evening; as many as seven hundred or more being laid by a single female.

Army Worm, Moth, Pupa and Eggs.

Army-Worm Larva.

The worm life may be stated at from twenty days in the South to about four weeks in the far North. The now well established fact that the insect hibernates in the larval or worm state, renders their destruction comparatively easy by burning, in the autumn or early winter, all infested stubble. Wet winters are also unfavorable to their well being. Army-worm years always follow dry seasons, though the season that finds them in great numbers may not be a dry one.

The marching of the worms is in search of food, and not a normal habit. Their instinct of concealment is so strong that a person may pass daily over an infested field and suspect their presence only by a greater or less number of bare patches.

Remedies.—We have already stated that the burning of stubble and trash where the insects hibernate is the most effective means of destroying the army-worm. The worms may be prevented, as a general thing, from passing from one field to another by judicious ditching. It is important, however, that the ditch should be made so that the side toward the field to be protected be dug under. About every three or four rods a deep hole in the ditch should be made, in which the worms will collect, so that they can be killed by covering them with earth and pressing it down. They may also be destroyed by burning straw over them—the fire not only killing the worms but rendering the ditch friable and more efficient in preventing their ascent. Coal-oil has been used to good advantage, and the worms have a great antipathy to pass a streak of it.

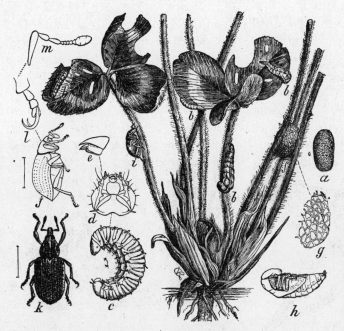

CLOVER LEAF BEETLE.

Many correspondents successfully headed them off by a plowed furrow six or eight inches deep, and kept friable by dragging brush in it. Along the ditch or furrow on the side of the field to be protected, a space of from three to five feet might be thoroughly dusted (when the dew is on) with a mixture of Paris green and plaster, or flour, so that every worm which succeeds in crossing the ditch will be killed by feeding upon plants so treated. This mixture should be in the proportion of one part of pure Paris green to twenty-five or thirty parts of the other materials named. If used in liquid form, one tablespoonful of Paris green to a bucket of water, kept well stirred, will answer the same purpose, as also will London purple, which has the merit of being cheaper. These substances should, of course, be only used where there is no danger of poisoning stock, poultry, or other animals. Logs or fences over running streams or irrigation ditches, should be removed, otherwise the worms will cross on them.

Lumber may be used to advantage as a substitute for the ditch or trench by being secured on edge and then smeared with kerosene or coal tar (the latter being more particularly useful) along the upper edge. By means of laths and a few nails the boards may be so secured that they will slightly slope away from the field to be protected. Such a barrier will prove effectual where the worms are not too persistent or numerous. When they are excessively abundant they will need to be watched and occasionally dosed with kerosene to prevent their piling up even with the top of the board and thus bridging the barrier.

Where the crop of a field has been completely destroyed by the worms, the plan of killing them by heavy rollers has been tried. This, however, is an expensive remedy, and is not as satisfactory as might be supposed. Experiments have proved

LAMP FOR KILLING NIGHT-FLYING MOTHS.

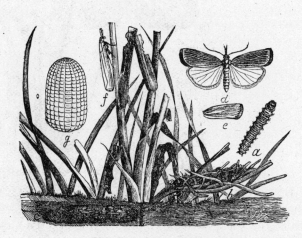

THE VAGABOND CRAMBUS.

that even where the ground was level, the rollers soon became irregularly covered with mud composed of earth and of the juices of the crushed worms, so that the effect was much the same as if the ground had been uneven, and many worms escaped in consequence.

A means for killing all night-flying moths, which we, years ago, practiced most successfully, when largely engaged in gardening, is shown in the cut of Lamp for Killing Night-Flying Moths. It is simply a light placed over a large pan of water, upon which floats a little kerosene oil. The insects, attracted by the light, fly through it, drop into the oil, and die. Fires kept burning in fields up to ten o'clock at night would serve the same purpose, and each female destroyed means the destruction of hundreds of eggs.

V. Vagabond Crambus.

This is an insect (*Crambus vulgivagellus*) that has but lately attracted attention in the State of New York. It was first found by Prof. Lintner, in 1881, in Jefferson county. The people supposed it to be the army worm. We figure it so that it may easily be known if it appears in the West. The cut shows all stages of the insect, with eggs enlarged at *g*. In the State of New York, some pasture-lots were almost entirely ruined, a dozen or more worms being found in a space as large as a man's hand. Indeed, the moths would rise up before the feet in a cloud, and a field be laid waste and turn brown in two weeks.

The Remedy.—They hibernate in the larval state, and, as full grown larvæ, do much damage the next spring. Burn over the fields infested in late autumn and winter.

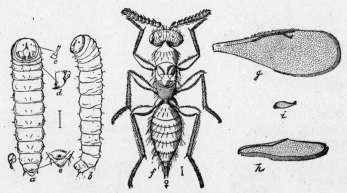

WHEAT ISOSOMA. (ISOSOMA TRITICA.)

VI. Insects Injuring Grain.

Leaf-Hoppers.—This class of minute insects, of the order Homoptera, injure plants by punctures. We figure one which is destructive to grain by puncturing the bases of the leaves of winter wheat, causing the plants to turn yellow and die. Sometime they appear in immense numbers. They are active, jumping, of brownish color. The whole order is destructive to plants of various kinds, many species having their relative insects.

Remedies.—The general remedy is dusting with sulphur, soot, dust and similar articles. The cut shows the destructive Wheat Leaf-hopper (*Cicadula exitiosa*). Bonfires attract this insect at night. See lantern shown in Section IV.

Wheat Leaf-hopper.

The Wheat Isosoma.—This minute insect does immense damage because it is generally overlooked. It is comparatively new, and is allied to the joint-worm. There are a number of species depredating on grain and grass. The species we figure is new to science, having been first described in March, 1882. So far as discovered there is but one annual brood; this hibernates in the larval and pupal state in stubble and straw.

THE REMEDY.—The readiest remedy is to mow the weeds in the autumn, on wheat stubble infested, and burn with the stubble before winter. This plan is proving destructive in some of the Eastern American States.

VII. Sorghum Web-Worm.

A NEW species of web-worm, infesting the heads of sorghum, and probably its congeners, broom-corn, etc., has been discovered. It is a lepidopterous insect of the family Bombycidæ, named *Nola sorghiella*. It was first found on sorghum vulgare, (rice corn, pampas rice, etc.,) in Kansas, and in immense numbers. We have reproduced figures of the insect in its several forms, together with enlarged drawings of the several parts. These insects confine

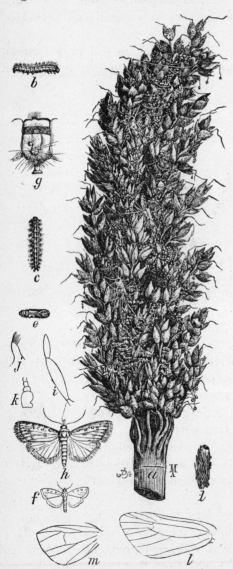

SORGHUM WEB-WORM.

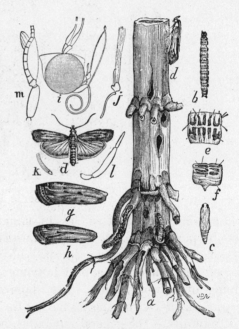

THE SMALLER CORN-STALK BORER.

their depredations entirely to the heads of the plants, sometimes entirely and, again, only partially destroying the grain. The moths issue, late in June and in July, a week or more after spinning the cocoons. *a*, shows head of sorghum; *b*, larva, side view; *c*, back view; *d*, cocoon; *e*, chrysalis; *f*, moth, natural size. The other letters refer to the several parts enlarged.

VIII. Sugar-Cane Beetle.

The technical name is *Ligyrus rugiceps*. Five years ago a black beetle made its appearance which proved most destructive to sugar-cane, especially so to native cane in Louisiana. Its habits are not, as yet, well known, but the presumption is that its depredations will be, like those of the May-beetle, not constant. So far as we know, the ground beetles, the larvæ of some click beetles, an ichneumon fly, and a beetle belonging to the same family with the Ligyrus, have been found destroying them.

Sugar-Cane Beetle.

IX. The Smaller Corn-Stalk Borer.

This is another newly-discovered insect, which most seriously depredates on corn, by boring. It would seem, so far, to be confined to the South, destroying many stalks and thus necessitating re-planting. Besides this, it works through the entire season as late as October, doing much damage. The insect is known as *Pempelia lignasella*. The explanation of the illustration is as follows: *a*, is the stalk showing the work of the grub or larva; *b*, larva; *c*, pupa; *d d*, moth, natural size, at rest and with wings expanded. The other letters refer to the several parts enlarged, and are of no consequence except to scientific persons.

Dr. Riley describes the insect at length. From his description we condense: The moth issues in about ten days after the larva has transformed to pupa. It has the singular habit of feigning death, and is not readily disturbed. The corn or other object upon which it may be resting can be handled quite roughly, and it even allows itself to be touched, when it will either remain in position, or will only move for a short distance, and will rarely attempt to fly. If, however, the corn on which it rests be shaken too suddenly, it will drop to the ground, draw the legs and antennæ close to the body, and will remain in this position motionless for a considerable length of time, even if quite roughly moved about. It rests in an upright position with the wings close to the body with their tips on the corn; the antennæ are laid backward on the dorsum and are not readily seen. Its flight is quite swift but of short duration.

This insect appears to be at least two-brooded in the Southern States.

Preventives.—It will be impossible to find a perfect preventive for the damage done by this insect, since it hibernates, as we have just stated, in all three states of larva, pupa, and adult. It seems extremely probable, however, that plowing up and burning the stubble will greatly reduce the numbers of the worms. The earlier this is done the more effectual will it prove.

X. The Rice-Stalk Borer.

This is another new insect. It is found boring into rice stalks in the Southern States. Its generic name is *chilo aryzellus*. The explanation of the cut, next page, is: *a*, larva in split stalk, side view; *b*, larva, back view; *c*, pupa; *d*, female moth, natural size. It is allied to the species feeding on the sugar-cane. Every burrow examined contained either larvæ, pupæ or fresh pupa skins at the time that harvest had already commenced. In the volunteer rice, however, another brood is probably developed.

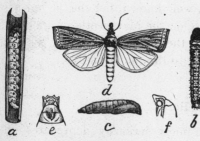

RICE-STALK BORER.

The duration of the pupa state varied from seven to twelve days, and the moths issued from August 20 to September 5. The moth is a very pale yellowish or straw-yellow color, with golden cilia to the front wings, a few golden scales scattered over the disk, and a series of seven black dots on the hind margin. It has an average expanse of a trifle more than an inch.

Enemies.—Dipterous larvæ were found destroying a pupa inside the stalk, and in a single instance, there has been bred from them *Phoraaletiæ* Comstock, a fly whose larvæ were supposed to be parasitic but which seem to be more scavengers than parasites.

Preventive Measures.—The borer, in the fields Mr. Howard examined, occurred in about one-fifth of the blasted stalks. It was sufficiently abundant, in fact, to make its destruction a matter of some importance. The later brood, if there is one, must take to the volunteer rice around the edges of the fields, or to the large grasses growing upon the embankments, though none were found in such. It is the custom, some time during the winter, to burn the stubble over the entire plantation. Great care is however taken not to allow the fire to reach the trash near or upon the embankments, as the soil of which these are made is of such a character as to burn readily, and their bulk would be greatly reduced by such a burning. Instead, then, of burning the weeds and volunteer rice along these banks, they are simply cut. It is probably here that the insect hibernates, either as larva or pupa, and it will be necessary to cut most carefully the wild rice and grass close to the ground and carry it to some safe place where it can be thoroughly burned.

XI. Grass-worm of the South.

Another insect depredating on rice is the common grass-worm of the South (*Laphygma frugiperda*). The moths, when abundant, lay their eggs on the growing rice-stalks. Flooding will, of course, kill the larvæ. The insect is said also to be hurtful, not only to grass, but to cabbage, strawberry plants and beans. A solution of Paris green, or London purple,

Fig. 1.
Grass-worm of the South.

Fig. 2.
Moth of Grass-worm.

will kill them when it can be applied. Cabbage and other plants to which poison cannot be applied, may be treated with pyrethrum, which will also be indicated for the cabbage butterfly in the North. Young ducks, young chickens and turkeys should also be liberally employed in the extermination of all the insects pests of the garden. Of those mentioned, young ducks are the most agile and, for their size, the most voracious fowls we have ever employed. We give two cuts of grass-worms. In Fig. 1, *a* is the larva, natural size; *b*, head from front; *c*, a middle joint from above; *d*, the same joint, side, enlarged. In Fig. 2, at *a*, is shown moth of grass-worm (L. frugiperda), natural size; *b*, wings of the variety fusca; and *c*, wings of the variety obscura.

XII. Corn Bill-Bug.

A CURCULIO-LIKE beetle has of late years done much damage to corn. This is especially true of the South. This insect is *sphenophorus robustus*. It punctures the

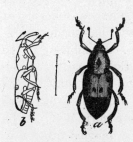

Corn Bill-Bug.

stalk and sucks the sap near the ground. It has been known for years in various States East, West and South. The egg is laid in the stalk, just at the surface of the ground, sometimes below, and the pith is eaten at these points; this of course greatly weakens the stalk. Dr. Riley says that wherever the larva had reached its full size, the pith of the stalk was found completely eaten out for at least five inches. Below ground, even the hard, external portions of the stalk were eaten through, and in one instance everything except the rootlets had disappeared, and the stalk had fallen to the ground.

In a great majority of instances but a single larva was found in a stalk, but a few cases were found where two larvæ were at work. In no case had an ear filled on a stalk bored by this larva. The stalk was often stunted and twisted, and the lower leaves were invariably brown and withered.

The cut of Corn Bill-Bug shows a closely allied species, *Rhodobœnus 13-punctatus* (Illiger), which has a more slender form than the corn bill-bug described above. *a*, gives back view, with markings; *b*, the side form of the insect.

Preventives.—With regard to preventives, a most perfect one will be found, as already indicated, in pulling up and burning the stubble during the winter, or, preferably, as early as possible after harvest. With reference to this remedy Glover says a very perceptible decrease of the bill-bug has been observed where the practice of burning the roots has been followed, and, if persevered in, might nearly eradicate them in the course of a few years.

XIII The Corn or Cotton-Boll Worm.

THE description of the army-worm will do fairly for that of the cotton-worm of the Southern States. Although the species are distinct, their work is the same. They devastate fields by eating the leaves of plants. The boll-worm, corn-worm and tomato worm (*Heliothis armigera*) are identical. This also feeds upon various other plants, such as the red-pepper, Jamestown-weed (*Datura*), the ground-cherry (*phrysalis*), and, in Europe, upon tobacco. Here the young ears of Indian corn are its favorite food, and in the South, where the insect is three-brooded, it attacks the cotton boll, probably because the corn is then too hard for the third brood. The cut shows the boll-worm and its work on the tomato. In the North the insect does comparatively little injury—principally to late sweet corn and tomatoes. In the South, however, it has become a great pest. For this reason we append the suggestions of Judge Laurence Johnson, of Holly Springs, (who has made a careful investigation of the subject) to the Department of Agriculture. This contains much that is of practical value.

CORN OR COTTON-BOLL WORM.

Preventives.—*Heliothids*, as known, pass the winter in the pupa state in the earth, in cotton and corn fields, where the full-grown worm drops. As often as possible, then, change the cropping, and never plant cotton after corn if it can be avoided; nor should it be planted near corn if the crop can be pitched otherwise. When a cotton-field becomes much polluted sow it down in wheat or oats, or plant in corn, to be followed by one of these. Green corn is the great nursery of this plague, and next to the corn is a great crop of Southern cow-peas.

The worst infested field I observed this year was a small one in which there had been a bad stand of cotton in the spring, and to mend it corn was planted in the missing places. By unskilled working more damage was done to the stand, and to mend this again cow-peas were dropped in the gaps. No arrangement could have suited *Heliothis* better. The peas supplied the moth shelter during the day, and their favorite repast at fall of evening.

Some old and formerly large and successful planters tell me that their practice to top cotton, about the 10th of August, and burn the young shoots was a check to the boll-worm. By this practice no doubt many eggs and young larvæ were destroyed.

Natural Enemies.—Their natural enemies afford some degree of protection. Birds might be fostered, by putting up martin boxes about in the fields. The blue-birds are fine hunters of the worms, but I have never seen them catch the moth. They will take to any kind of a box if the martins do not. These are great fly-catchers, as is well known, and fly late—the very time for crop-destroying moths of all kinds. But of all birds, the most effectual are domestic turkeys and chickens.

Turkeys range through a cotton-field, looking up into the leaves, and well hid must be the worm they do not find. Their value has long been known in tobacco-fields. Chickens, on the other hand, not so good after worms, are exceedingly active in pursuit of the moths. When two small fields, near me, and daily visited this summer, became naturally infested with aletia, the last of August and first of September, the neighboring chickens and turkeys were there from morning until evening. They never allowed aletia to get more than half grown. Even when, the 20th of September, I brought hundreds of aletia larvæ into one of the fields for experiments with pyrethrum, the turkeys hunted them out, and, with superior interest and eyesight, in a few hours none were left except two, which were old enough to web up before they were found out.

How they should find the boll-worm so often I do not know, but as a fact it was vain for me to mark stalks with young *Heliothis* upon them with a view to future observations. The turkeys were there from morning until night, and no *Heliothis* dared to show his head, as they often do at close of day, without danger from these vigilant guards. Practically, I was compelled to cage all I proposed to watch. To the great planting interest these facts can be of little value. It would require flocks of immense numbers, and to be herded about over the fields, to accomplish anything proportionate to what is above related of small patches near habitations. Jays, blackbirds, woodpeckers, and crows destroy vast numbers of *Heliothis* in corn about the time the grain begins to toughen, but these allies levy toll also on the crop.

XIV. Remedies for the Cotton-Worm, South.

THE observations upon the army-worm will apply everywhere to the cotton-worm. In the South, eggs of the cotton-worm, *Aletia argillacea*, are deposited in the cotton fields. Unlike the army-worm of the North (*Leucania unipuncta*), the cotton-worm seems rarely a traveler, though, if the necessity came so to do, it might develop ability in this line. According to Dr. Riley, the cotton-worm makes its first appearance in the southern portion of the Cotton Belt between the middle of April and the middle of May. Near the Gulf of Mexico, seven or more successive broods have been produced before being cut off by frost. It is thus easy to see that, the more that is done to destroy the earlier broods, the less the injury from the later. In the article on cotton cultivation we have figured a machine for spraying any hoed crops from above. The illustration, Spraying Cotton from Below, shows a machine for a different method. This is much more effective for cotton, since the worms feed naturally on the under side of the leaf. (See next page.)

In relation to poisons Dr. Riley says: By the ordinary method of sprinkling poison from water-pots, or in broadcast sprays from barrel pumps, about forty gallons of water containing one pound of Paris green or two-fifths of a pound of London purple, kept well mixed by stirring or shaking, may be applied, to the acre. When a bellows atomizer is used to diffuse it more finely and more thoroughly, which is much preferable, less than half that quantity of poison and water to the acre will give

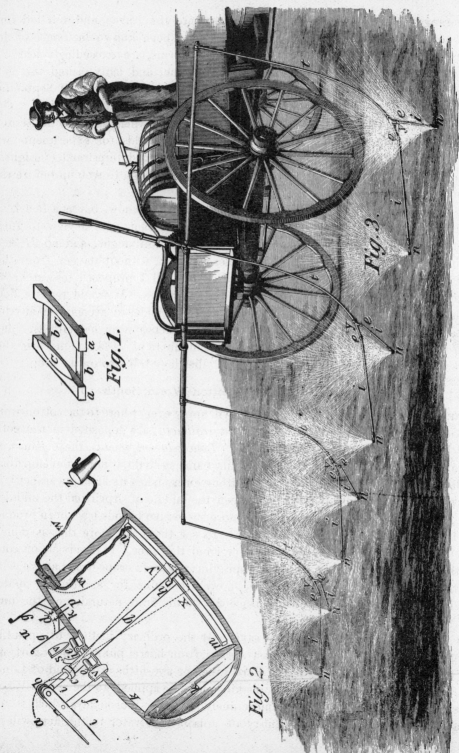

SPRAYING COTTON FROM BELOW.

equally good results. In sifting on dry poison by such sifters as are usually employed, one pound of the Paris green to thirty-five pounds of such mixture of flour and ashes, or one pound of London purple to forty-five pounds of such mixture, are proper proportions to use. The flour is adhesive, holding the poison fast to the leaves and coating the particles of poison so that they come less in contact with the surface of the plant, and hence it helps to prevent their caustic action or burning of the leaves. The ashes have a still greater ameliorative effect in preventing the caustic action, and on this account it is well to use as much as one-third ashes to two-thirds flour to form the mixture. With this preparation the poison cannot be too thoroughly mixed. Better devices for mingling these homogeneously with each other are still to be sought. The best now easily prepared by the planter consists of a barrel with a number of rods put through it endwise, and a great number of large spikes driven through its sides to project far into the cavity.

XV. Poisons for Worms.

These are put into the barrel through a large hole, which is then closed, while the barrel is hung upon an axis and rotated until thoroughly mixed.

It should be added that in case the poisons recommended are in any instance not obtainable, the pure arsenic or arseniate of soda may be resorted to, since these have been used to advantage, though not always with the best satisfaction. Although these substances are cheap, their caustic effect on the plant is greater. The mixture now most used consists of twenty grains of arseniate of soda and 200 grains of dextrine, dissolved in one gallon of cold water. Four ounces of this mixture to forty gallons of water can be sprinkled on each acre. The common arsenic water, which every druggist knows how to make, will answer well. To make it from the white arsenic (arsenious acid) and common baking (carbonate of) soda is cheaper than to buy the arseniate, although the arseniate method of preparation involves less time and labor. One-fifth of a pound of sal soda to a pound of arsenic should be boiled in a gallon of water until dissolved. The solution is permanent, no stirring or shaking being necessary to keep the poison mixed. One quart of the solution to forty gallons of water is used on each acre.

In applying poison with blowers, a much smaller quantity of the poison and its dilutents will be sufficient, and when the poison is blown onto the under surfaces the adhesive element is no longer needed.

Both Paris green and London purple, when not adulterated, and where properly applied, have always given satisfactory results. The latter seems to act a little slower than the Paris green; perhaps, because the worms do not eat it so quickly, for they refuse to eat poisons until they become very hungry, but it is much the cheapest, and being a finer powder, is susceptible of a much thinner distribution than it usually gets. If very thinly and evenly applied, it will be eaten sooner, and when used in due time will prove equally as effective as the Paris green. And it is likewise commendable to administer any poison whatever that is to be used, so early as to destroy the worms

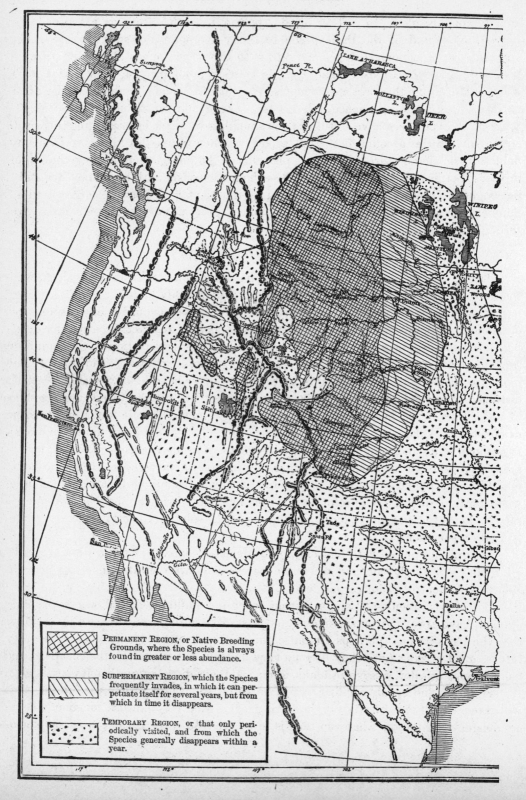

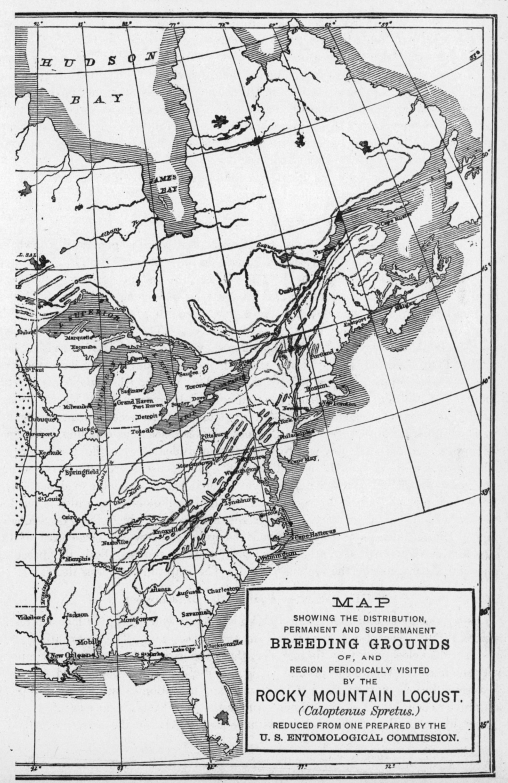

before they reach destructive size, and before they appear on the upper surfaces of the leaves. Planters must be urged to watch carefully the under surfaces of the foliage throughout the cultivating season. The very young worms are less easily seen than the small spots of light color made by their gnawing off little patches from the lower surfaces of the leaves. As soon and whenever the young ones have started, apply the poison immediately beneath the foliage. The plowman or "weed-chopper" should be taught how to see the young worms and be carefully trained to find them. At the same time he should have hanging from his shoulder or plow, a light bellows atomizer, charged with poison, ready for use.

The cut of machine explains itself. A fountain or other force-pump must be used, and the proper fixtures provided.

XVI. The Hateful Grasshopper or Locust.

ALL grasshoppers are not locusts, but locusts *are* what is generally known as grasshoppers. The insects usually called locusts, in the eastern portion of the United States, are not locusts at all, but cicadas. These are an entirely different insect, appearing regularly, and at stated periods of fourteen or seventeen years, on trees. Those periods represent the time the larva lives, according to the species, in the ground, before transforming into the perfect insect. They do far more damage when feeding on under-ground roots than when they make their periodical visits.

The True Locust.—This is far more formidable, devastating vast regions. The hateful grasshopper, well named by the late Dr. Walsh, (the *caloptenus spretus* of entomology), differs so little from the common red-legged locust of Harris (*caloptenus Femur rubrum*), a common so-called grasshopper, that, except to entomologists, there is no apparent difference. Hence, those who have seen our common red-legged grasshopper, and who has not, will form a very good idea of what the dreaded locust of the region west of the Mississippi is like. The locust is preyed upon by every insect-eating thing, and that it is a good food for man we have reliable evidence. The oriental locust is not more destructive and is not unlike our own, which is but too well-known to all in the districts west of the Mississippi, which they occasionally ravage. Hence, nothing need be said about it here. To show the large area of the United States visited by these pests, we insert a map, showing both the breeding grounds of the insects and the regions visited. The map fully explains itself.

XVII. Remedies Against the Grasshopper.

As previously stated, their habits are known to all who live in the locust-infested districts. Full descriptions are, therefore, not necessary. The eggs are laid in masses, and so closely do they lie together, that bushels may be found on a comparatively small area. Where the ground is light and porous, prolonged and excessive moisture will cause most of the eggs to perish, and irrigation in autumn or in spring may prove beneficial. Yet, experiments prove that it is by no means as effectual as is generally believed, and as most writers have assumed to be the case. In pastures or in fields where hogs, cattle, or horses can be confined when the ground is not frozen, many, if not most of the locust eggs will be destroyed by the rooting and tramping. The eggs are frequently placed where none of the above means of destroying them can be employed. In such cases they should be collected and destroyed by the inhabitants, and the State should offer some inducement in the way of bounty for such collection and destruction. Every bushel of eggs destroyed is equivalent to a hundred acres of corn saved.

Destroying the Eggs.—One of the most rapid ways of collecting the eggs, especially where they are numerous and in light soils, is to slice off about an inch of the soil by a trowel or spade, and then cart the egg-laden earth to some sheltered place where it may be allowed to dry, when it may be sieved so as to separate the eggs and egg masses from the dirt. The eggs thus collected may easily be destroyed by burying them in deep pits, providing the ground be packed hard on the surface.

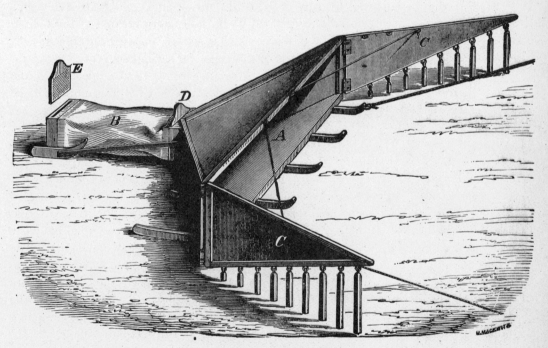

THE RILEY LOCUST-GATHERER.

Killing the Young.—The young grasshopper after hatching will congregate on timothy in preference to other grasses or grain, and a strip of timothy around a corn or wheat field, poisoned with a solution of Paris green, would kill the insects. In fact, any of the means given for the destruction of the army-worm would be more or less effective.

Trapping.—This can easily be accomplished, especially when the locusts are making their way from roads and hedges. Thus, the use of nets or seines, or long strips of muslin, calico, or similar materials, converging after the manner of quail nets, have proved very satisfactory. By digging pits or holes three or four feet deep, and then staking the two wings so that they converge toward them, large numbers may be secured in this way after the dew is off the ground, or they may be headed off when marching in a given direction. Much good can be accomplished by changing the position of the trap while the locusts are yet small and congregate in isolated or particular patches. Many machines have been made for trapping locusts. Of these each district has its favorite. The cut of the Riley Locust Gatherer shows the one invented by Dr. Riley, United States Entomologist. It is simple and effective. D is a slide-door, to be closed when the receptacle, B, is full of locusts; E is a slide-door, to be raised when the locusts are to be passed into a bag for killing; A shows the contraction towards D, whence the insects naturally run as the machine is drawn forwards. [See page 635.]

CHAPTER IV.

OTHER DESTRUCTIVE INSECTS AND THEIR ENEMIES.

I. INSECTS INJURIOUS TO TREES.—II. INSECTS INJURIOUS TO CONIFEROUS TREES.—III. INSECTS INJURING THE GRAPE.—IV. INSECTS INJURING FRUIT TREES.—V. LEAF ROLLERS.—VI. APPLE TREE CASE BEARER.—VII. THE ORANGE LEAF-NOTCHER.—VIII. FULLER'S ROSE BEETLE.—IX. INSECTS INJURING PLANTS.—X. SNOUT BEETLES.—XI. THE WHITE GRUB OR MAY BEETLE.—XII. THE SPANISH FLY OR BLISTER BEETLE.—XIII. BENEFICIAL INSECTS—LADY BIRDS.—XIV. SOLDIER BEETLES.—XV. TIGER BEETLES.—XVI. OTHER BEETLES AND PARASITES.—XVII. CONCLUSIONS.

I. Insect Injurious to Trees.

TO describe all the insects infesting our crops would require volumes. Our aim has been simply to show by accurate engravings some of the most destructive, and some of the more recently discovered, with brief descriptions; preventives against their ravages, and the remedies when known. The name of insects infesting our orchard and forest trees is legion. The more ruinous ones are generally known, but there are some which are destructive and yet not widely disseminated. Some of these we illustrate, besides giving their names and habits and the best means for getting rid of them.

Catalpa Sphinx.—Of late years the catalpa has been much planted in the West. This tree is exempt from attacks of insects as a rule, but its peculiar moth, the catalpa sphinx, deserves notice. In the illustration, page 638, *a* shows the egg mass; *b*, the newly hatched larvæ, feeding; *c*, larvæ one-third grown; *e*, *f* and *h*, differently marked larvæ; *j*, the pupa taken from its case; *k*, the moth, natural size. At *l*, is an egg enlarged to show its form. The illustration shows the general characteristics of all the species, of which there are many.

Remedies.—The catalpa sphinx has several parasites, and they are also devoured by many kinds of birds. In fact its natural enemies keep it pretty well suppressed. The egg masses should be destroyed wherever found. All large moths, of whatever species, should be trapped and killed, for the larvæ are voracious. The worms being gregarious may be hunted and killed in numbers, especially when molting. Spraying the infested trees with lime-water will kill the larvæ, but a solution of Paris green or London purple is more effective. In fact the latter is the remedy for all insects infesting the leaves of plants and trees, when the fruit may not thus be rendered dangerous to man.

Osage Orange Sphinx.—(*Sphinx Hageni*).—This insect is here figured, since it is yet rare, and also because it is one of the most beautiful of sphinxes. Dr.

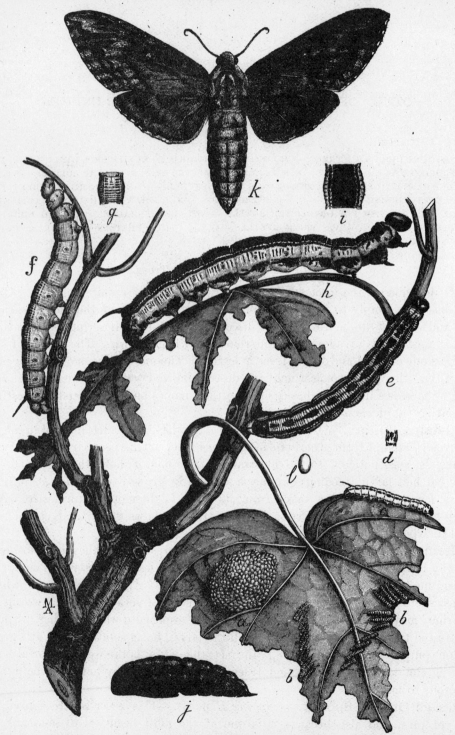

THE CATALPA SPHINX.

OTHER DESTRUCTIVE INSECTS AND THEIR ENEMIES.

Riley says the general color is light brown, with olivaceous shades, and marked with black and white, as indicated in the figure. There is a small white spot, surrounded by black, near the middle of the front wings, and a large white patch immediately outside of this, as well as another at the tip of the wing, the latter bounded behind by an oblique, wavy, black line. The wing is crossed by four transverse black lines outside of the central spot, one of which runs into that spot, and two or three nearer the base. The outer margin is strongly shaded with white, and the fringes alternately of the ground color and white. The hind wings are smoky brown, lighter toward the base, crossed by an indistinct darker band. The under side of the

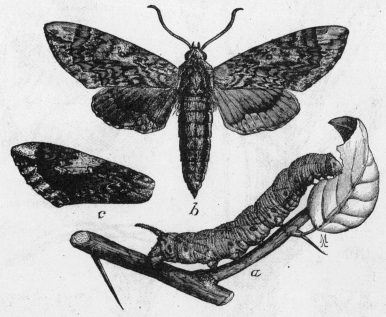

OSAGE ORANGE SPHINX.

wings is cinereous, crossed by darker lines. The middle of the thorax is of the color of the fore-wings, the edges whitish, with a black line running through the white portion. Abdomen brownish cinereous, with dorsal, sub-dorsal, and traces of lateral black lines, as shown in our figure. The variation is great, some specimens being very light, others almost black.

II. Insects Injurious to Coniferous Trees.

Of late years great injury has been done to evergreens by insects that have increased wonderfully, probably through the large production of such trees, and also, without doubt, from the slaughter of birds, and general inattention to the ravages of the insects.

Pine-Tree Borer. Comstock's Retinia.—This insect bores into the twigs and young growths of pitch pine. The cut of Pine-Tree Borer, Fig. 1, shows butterfly, larva and pupa, the hair-lines giving the natural size. On the branch is shown the gall and exuded gum, and also a section of the same below. The remedy is to cut off all infected parts and burn them. For when trees are kept on the lawn it is necessary to protect them from the depredations of insects.

Another species, the frustrating retinia (*R. frustrana*), which infests pitch and other pines of the South, is shown in Fig. 2. This insect ruins both the delicate twigs and the base of the leaves.

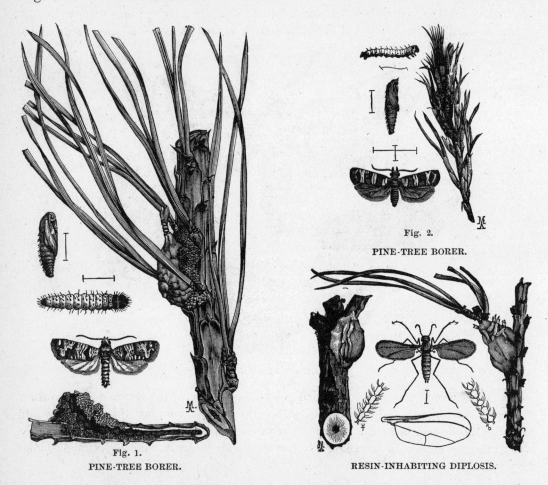

Fig. 1.
PINE-TREE BORER.

Fig. 2.
PINE-TREE BORER.

RESIN-INHABITING DIPLOSIS.

Resin-Inhabiting Diplosis. (*D. resinicola*).—This insect damages pitch pine, North and South, and any one having handsome lawn evergreens cannot be too watchful of them. The illustrations we give must suffice for all evergreen trees. The figure of Resin-Inhabiting Diplosis shows the work of the insect. Its home, when transforming, is in the resinous lumps shown, but when feeding, the insect burrows in the soft part of the bark. It will be seen by the hair-lines showing natural length, that the insect is little more than a midge; but it is destructive.

The Pine-Leaf Miner.—(*Gelechia pinifoliella*). This is a new species, and another of "the destructive little things." The cut shows all that is necessary to enable one to recognize it. The hair-lines mark the size of the insect. It mines the leaves of many kinds of pines, principally the yellow and pitch pines. There has been no remedy indicated. In fact, insect-

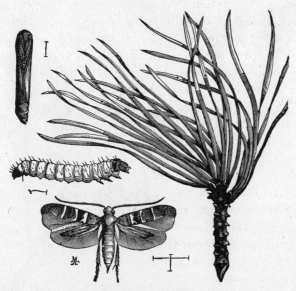

PINE-LEAF MINER.

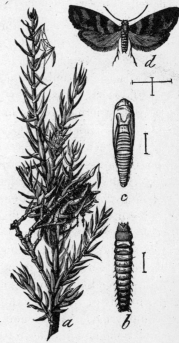

JUNIPER WEB-WORM.

eating birds and parasites are the best means of destroying this class of insects. A minute chalcid is said to puncture and deposit its eggs in the body of the pine-leaf miner.

The Juniper Web-Worm.—(*Dapsilia rutilana*). Various web-worms prey upon evergreens and especially upon the junipers. Since this worm lives entirely within its web, except when eating, it is difficult to extirpate. It is one of the worms that eat the foliage and should, therefore, be treated by spraying with a solution of Paris green.

III. Insects Injuring the Grape.

The phyloxera are the most deadly. The many insects of the caterpillar tribe depredating upon the grape must be gotten rid of by such methods as may be suggested. Showering the vines with a solution of Paris green of the strength recommended for the cotton-worm would be indicated when the vines are not in fruit. Borers must be extracted, galls taken off, aphides and other lice cleaned away, and eternal vigilance generally practiced, unless the cultivator is content with few varieties and such as are most exempt from the ravages of insects.

The Flea-Beetle.—Few enemies of the grape are more destructive in a general way than the flea-beetle (*Graptodara chalybea*). The remedy for their destruction is to remove and destroy, in the autumn, all rubbish about the vines, and strew unslaked lime or unleached ashes liberally about. In the spring, shower the canes and young foliage with Paris green, and in the morning, when the air is chilly, the

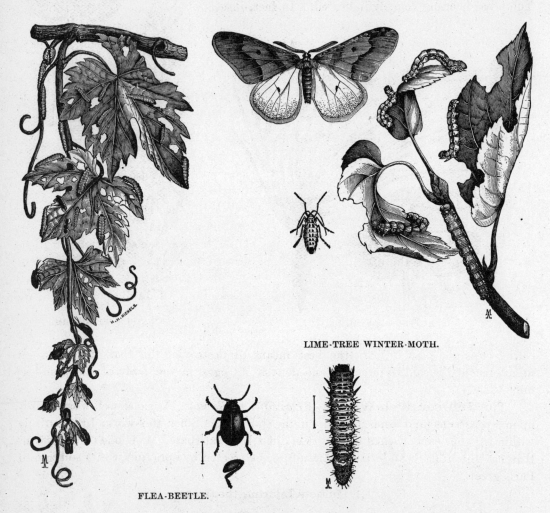

LIME-TREE WINTER-MOTH.

FLEA-BEETLE.

beetles may be shaken onto sheets and killed. The cuts show the beetle and larvæ enlarged, also the thigh which gives it leaping powers. The hair lines show the natural size of the beetles, and the larvæ are at work on the leaves.

IV. Insects Injuring Fruit Trees.

Lime-Tree Winter-Moth, (*Hybernia tiliaria*).—This insect is little known, but it is a fair specimen of the large family called span-worms, destructive to so many kinds of trees. Like the canker-worm of the apple tree, the female is wingless; it is shown to the left of cut of Lime-Tree Winter-Moth, the male (moth) being shown on top and the larvæ upon the leaves.

Remedies.—The females of the canker-worms being without wings, they cannot lay their eggs, if they can be prevented from climbing up the trees. Hence, bands covered with tar and oil, of the right consistency to entangle the feet of the insect in attempting to pass, are efficient. Probably printers' ink is the best for smearing the bands, since it retains its viscidity for a long time. If the worms get a lodgement in the top of the tree, they may be jarred, and they will spin down upon a silken thread. By passing a pole rapidly under the tree, they may be entangled and destroyed. Ants are great destroyers of canker-worms, A friend of ours keeps his large orchard free of them by cultivating the common prarie ant.

V. Leaf Rollers.

Two insects deserving notice here are the apple-leaf sewer (*Phoxopteris nebeculana*), which we illustrate, and the strawberry leaf-roller (*P. fragraria*). The cut of the Apple-leaf Sewer shows the insect as moth and larva, the hair-lines showing the natural size. The work of the insect in sewing the leaf together is also shown. It passes the winter rolled up in the dried leaves.

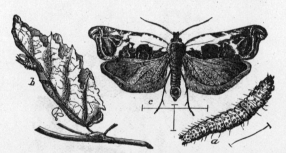

APPLE-LEAF SEWER.

The Remedy.—Pick off all dried leaves on the trees in the winter and burn them. On strawberry plants the leaves may be sprinkled with hellebore-powder and water.

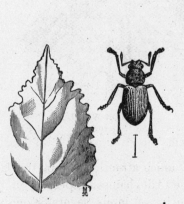

ORANGE LEAF-NOTCHER.

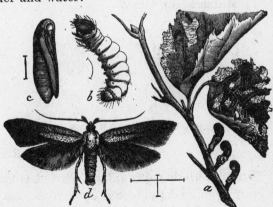

APPLE-TREE CASE-BEARER.

VI. Apple-Tree Case-Bearer.

This insect, known also as apple-tree caleophora (*C. malivorella*), destroys the buds, and later in the season it sometimes reduces the leaves to skeletons. It gets its name from the curious cases shown at *a*, in the cut, these being the shields under which the insect moves.

Remedy.—The insect hibernates in the cases. The remedy is to scrub the twigs with alkaline wash, or perhaps the kerosene emulsion, recommended for the orange scale, would be preferable. The cut shows the insect in all its stages, the hair lines showing natural size.

VII. The Orange Leaf-Notcher.

THIS insect injures the foliage of orange trees by notching the leaves as shown in the cut, on page 643. Its scientific name is *Artipus floridanus*. It is pale greenish-blue or copper color, densely covered with white scales.

Remedy.—In Florida they are destroyed by jarring the trees and collecting the beetles in sheets, as is done in all plum and peach raising districts to destroy the curculio. The orange scale insects have been already noticed.

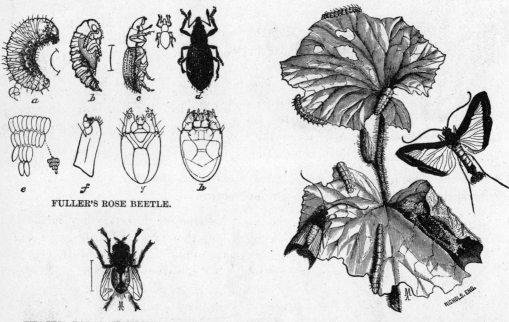

FULLER'S ROSE BEETLE.

THACINA PARASITE ON MELON WORMS.

MELON WORM.

VIII. Fuller's Rose Beetle.

THE rose beetle (*Aramigus Fulleri*) has attracted much attention within the last ten years from its depredations upon roses east of the Alleghanies, especially upon tea roses propagated under glass. The cut shows the insect in its several stages of development, the small beetle and the hair-lines showing the natural size. The figures in lower row are enlargements of the several parts. It is a snout beetle, and like all the curculio tribe, shy. The damage is done by the larvæ, which eat the leaves

The Remedy.—Pick or shower them off. Also, one precaution is to thrust folds of oil paper in slits of small pegs and stick the pegs into the ground about the plants. The eggs will be laid in the folds, and may be destroyed.

IX. Insects Injuring Plants.

The Melon Worm.—This insect (*phakellura hyalinatalis*) is one of the most destructive to melons of the caterpillar tribe at the South. The cut shows the insect in its various stages. They not only eat the leaves of muskmelons, but also bore into the fruit. They are preyed upon by a tachina fly, shown in connection with the worm, and also by an ichneumon fly.

The Remedy.—The only remedy is hand-picking and the use of pyrethrum powder.

Asparagus Beetle.—This beetle (*crioceris asparagi*) is most destructive in gardens where it gets a foothold. The beetle and larva are shown enlarged, with hair-lines showing natural size. The worm lives on the tender bark of the asparagus, and sticks pertinaciously, no matter how hard the stalks may be shaken before bunching. Hand-picking in the spring is the usual remedy, but if all seedlings in the patch were hoed up, and no wild asparagus allowed to grow, the beetles would only have the young shoots on which to lay their eggs. Hence, since these are cut for market or use every other day, the eggs would not hatch.

ASPARAGUS BEETLE.

SNOUT BEETLE.

X. Snout Beetles.

The snout beetles are among the most destructive insects; they include the weevil. In their winged state they are hard-shelled, varying in size from the minute to the imbricate snout beetle (*epicœrus imbricatus*), as represented in cut of Snout Beetle. This beetle is somewhat of a universal feeder, attacking the twigs of various fruits, also cabbage, radish and other cruciferous plants, beans, corn, beets, and plants of the watermelon and cucumber tribe.

Remedies.—On trees it may be taken by jarring onto sheets; on vegetables, by hand-picking.

Sweet Potato Root-Borer.—(*Cylus formicarius*).—This southern species seems to be widely distributed all over the tropical and sub-tropical regions of the world, and is found in the Gulf States. It bores into the roots.

SWEET POTATO BORER.

DISTENDED MAY BEETLE.

646   THE HOME AND FARM MANUAL.

The Remedy.—As soon as the potatoes are found to be infested with this

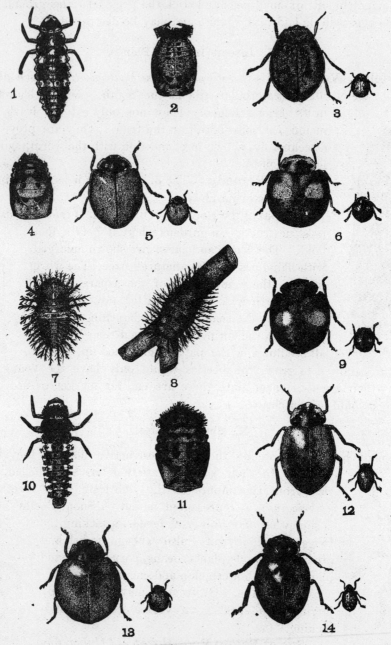

LADY BIRDS OF CALIFORNIA.

borer, the most effectual remedy is to dig the crop without delay, and feed the infested roots to stock.

XI. The White Grub or May Beetle.

These are among the most destructive of insects. The larvæ live a variable length of time in the ground, eating the roots of grass and plants. We have seen a meadow infested with the common May beetle of the North so destroyed that the sod might be rolled up like a carpet. The distended May beetle (*Lachnosterna fareta*) is a southern species. It injures garden crops, especially beans. Like all night-flying insects, it is attracted by a light, and this seems the readiest method of destroying it. For cut see page 645.

XII. The Spanish Fly or Blister-Beetle.

These insects, of which there are a large number of species, are most destructive to plants of various kinds, seeming to be almost universal feeders. All of them seem to have the blistering qualities of the so-called Spanish fly (*cantharis*). The figures will give a good idea of the species North and South. The illustration shows Nuttall's blister-beetle (*Cantharis Nuttalli*). This will serve to give their general character, although the most of the genus are smaller and some of them handsome.

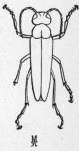
Nuttall's Blister-Beetle.

The Remedy.—They may be driven from fields like grasshoppers, on to straw and stubble, and there killed by setting fire around the edges. They have many insect enemies and are said to prey upon other insects themselves.

XIII. Beneficial Insects—Lady-birds.

The lady-birds, belonging to the family Coccinellidæ, are among the most useful of insects, their only food being other insects, and especially the eggs of other insects. They should be most carefully protected, and care should be taken not to mistake them for the single vegetable-eater figured page 587, Fig. 4. This vegetable-eater somewhat resembles Fig. 1, shown on this page (*Hippodamia Maculata*), and also the twelve-spotted diabrotica, Fig. 2. But a comparison will show the plant-eater to be

Fig. 1.   Fig. 2.   Fig. 3.   Fig. 4.   Fig. 5.   Fig. 6.
LADY-BIRDS.

rounder, larger and critically different. The cuts show some of the principal varieties Fig. 3 is *Coccinella Munda;* Fig. 4, *C. Venusta;* Fig. 5, the true marked coccinella; Fig. 6, the pupa of the latter species suspended from a support.

California Varieties.—The preservation of the lady-birds is so important that we illustrate some of the less known on page 646, magnified, the small figures at the side showing the natural size of the beetles. Seen under a good pocket lens they should look like the large figures. They are well worth careful study.

These are species found in California where they do good work destroying aphis, scale and other insect pests. Fig. 1 is the larva of the ashy-gray lady-bird (*Cycloneda, abdominalis*); Fig. 2, its pupa; and Fig. 3, the beetle. It has seven black spots on the thorax and eight on each wing-cover. Fig. 4 is the pupa of the blood-red lady-bird (*C. Sanguinia*), and Fig. 5, the beetle, varying in color from brick-red to blood-red. This species is common all over the country. Fig. 6 is C. Oculata, the history of which at present seems little known; but please remember, again, that all lady-birds are to be carefully cultivated. Fig. 7 is the pupa; Fig. 8, the larva; and Fig. 9, the beetle of the cactus lady-bird (*Chilocarus cacti*). The larva destroys many insects, particularly bark and leaf (scale) lice. Figs. 10, 11 and 12, show the larva, pupa and beetle of the ambiguous hippodamia (H. Ambigua). It is probably one of the most beneficial of the California species, because one of the most abundant. Fig. 13 shows the five-spotted coccinella of California (*coccinella 5-notata*). Fig. 14 is *Hippodumia Convergens*, a species common all over the United States.

XIV. Soldier Beetles.

HERE is another class that prey on other insects, both in the larval and perfect state. The strong jaws as represented in the enlarged head as seen at *b*, are well

Fig. 1.
Soldier Beetle.

Fig. 2.
Soldier Beetle.

adapted to killing, as their agility is to catching their prey. Fig. 1 shows the variety known as (*Chauliognathus Pennsylvanicus*) found plentifully in the South. The larva, *a*, is also shown with mandibles and claws enlarged at *b*. It is the larvæ that are carnivorous. The beetle shown at *i*, Fig. 1, like all the fire-fly family, are vegetable feeders. The one in question seems to live on the pollen or nectar of flowers. The yellow margined soldier beetle (*C. marginatus*), Fig. 2, is a distinctly southern species.

XV. Tiger-Beetles.

THE tiger-beetles are indefatigable hunters and are well named. No insect that they can master seems to come amiss to them. Fig. 1, next page, is the Carolina tiger-beetle (*Tetracha Carolina*). Fig. 2, the Virginian (*T. Virginica*). These two are said to be the only representatives of the genus tetrarcha in North America.

XVI. Other Beetles and Parasites.

THE insect represented by Fig. 1 is the rapacious Soldier-bug (*Sinea multispinosa*). It is well named, for from plant-lice, which it eats when young, to canker-worms and Colorado potato-beetles, which it attacks when older, it eats indiscriminately all it can devour.

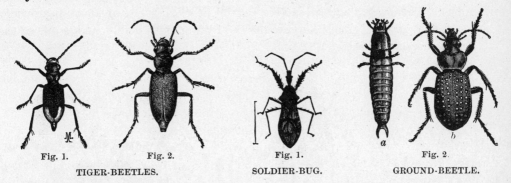

Fig. 1.　　　Fig. 2.
TIGER-BEETLES.

Fig. 1.
SOLDIER-BUG.

Fig. 2.
GROUND-BEETLE.

Take a good look, also, at Fig. 2. It is a Ground-beetle (*Calosoma calidum*), with larva. Then again we show, Fig. 3, *Calosoma scrutator*, a different looking insect, but no less beneficial.

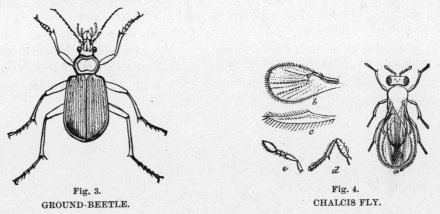

Fig. 3.
GROUND-BEETLE.

Fig. 4.
CHALCIS FLY.

The Chalcis and Tachina flies next demand attention. Fig. 4, is *Trichogramma minuta*, enlarged which lays its eggs in the eggs of butterflies and moths, notably of the cotton-worm. Fig. 5 is another chalcis-fly, (*Exiophilus mati*), which destroys the woolly apple-louse.

Next, Fig. 6, is a curious fly, *Epax apicaulis*. They are among the most beneficial of the Southern insects, and may be called insect swallows from their rapidity and rapacity in devouring the cotton-moth.

Fig. 7 shows Lebia grandis, one of the predatory beetles, and Fig. 8 the same enlarged. Fig. 9 is a Tachina fly, parasite on the army-worm. [See next page.]

Fig. 5.
CHALCIS FLY.

Fig. 6.
EPAX APICAULIS.

XVII. Conclusions.

WE have given delineations of many new insects, and many older but not well-known ones. The lesbias are all active, small beetles, some of them beautifully marked. Lebia grandis feeds on the larva of the potato beetle, and is one of the most valuable common insects.

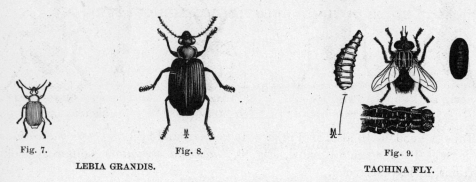

Fig. 7.   Fig. 8.   Fig. 9.
LEBIA GRANDIS.   **TACHINA FLY.**

The study of entomology is regarded as among the little things. Not so! Insects are among the most interesting and beautiful of created things. They occupy an important place in the economy of nature, among other uses being the fertilization of flowers by them. The destruction of their natural enemies, the birds, causes the undue propagation of mischievous insects, so that artificial means become necessary for their destruction.

CHAPTER V.

BIRDS IN THEIR RELATION TO AGRICULTURE.

I. BIRDS IN THE ECONOMY OF NATURE—II. WHAT BIRDS SHALL WE KILL?—III. FOOD OF SOME COMMON BIRDS.— IV. BIRDS CLASSIFIED BY THEIR FOOD.—V. BIRDS THE NATURAL ENEMY OF INSECTS—VI. BIRDS TO BE CAREFULLY FOSTERED.—VII. BIRDS OF DOUBTFUL UTILITY.—VIII. BIRDS TO BE EXTERMINATED.—IX. DESTROYING INSECTS.

I. Birds in the Economy of Nature.

NATURE preserves the economy of her system, by pitting the several races of animated nature one against another. All that breathes, including man, is animal, and all that vegetates is plant. With all it is a perpetual struggle for existence. Animals breathe the air, consume oxygen, giving off carbonic acid. Plants exhale oxygen and consume carbonic acid. Here is one great and beautiful economy in nature, to sustain equilibrium. The vegetable-eating animals are eaten by man and the carnivorous animals, while some animals—man, bears, swine, etc., are both plant and flesh-eating.

The smaller birds feed on insects, berries and seeds; some almost exclusively on the first and last, and some on both. Even fishes are divided into vegetable and flesh eaters. If a soil becomes barren, plants cease to live; if food becomes scarce, animated nature migrates to where it is more plenty. But man steps in with his dense population in advancing civilization, and changes the economy of nature. Hence, man must provide the remedy. Birds are killed or driven away, and insect life increases. Forests are destroyed, and tornadoes, cyclones and waterspouts devastate the land.

Shall not the birds be killed, or forests cut down? Unquestionably, but the wise man will do so with a wise discrimination. If all men acted wisely, we should have less of trouble in many directions in agriculture. The practical answer to the question, so far as birds are concerned, is to kill those birds that are injurious at the time of their depredations, usually after the nesting season, and preserve them at all other times.

There are two views from which birds are looked at generally. The scientific view is that they are one of the forces of nature to balance other forces of nature. The æsthetic view is that they are beautiful creatures, given by Providence to beautify and help glorify the earth.

II. What Birds Shall We Kill?

The thoughtful man might well ask in all seriousness, Is there any bird which we may safely pursue to extermination? Upon careful investigation, he would say, No! But assist Nature in checking that which may become an injury in the new order of things which artificial man creates. The philosopher could never argue the indiscriminate destruction of any birds, since he finds them great helps in the destruction of insects, so difficult to man. When birds increase to such an extent as to become seriously destructive to crops, he kills, but at all other times he fosters them.

Certain birds are not at all injurious to the farmer, as the cat-bird, robin, cherry-bird, etc. On the other hand, there are birds particularly destructive to the farmer at certain seasons, as the red-winged blackbird. The first birds mentioned are especially obnoxious to the fruit-grower, while the blackbird is not. So the quail, prairie-fowl, and all others of the gallinaceous family, eat grain largely at such seasons as they cannot find insects; and yet there are none that should be more carefully fostered than not only these, but all that class of land-birds usually denominated game-birds.

Know Your Friends.—We do not sufficiently study to know which are our friends and which our enemies; or when they may be doing us exclusive good, and when, perhaps, serious injury.

Let us take robins, for instance. They love fruit, and take what they want during its season, if not carefully watched. What else? A family during the nesting season, or when the young are hatched, will consume 200 larvæ or worms each day. This insect-food for three months, if allowed to multiply, would, in three or four years, produce insects enough to consume every green thing on a square mile of land.

The stomach of the robin in March contained worms, grubs of the terrestrial species of insects, and seeds; in April, insects, worms and grubs; in May, the same; and so in June, with the addition of cockchafers; in July and August, many sorts of worms and fruits; in September, larvæ and seeds. In October, these birds migrate South. Why? They can no longer procure their natural food—insects.

Moreover, the wise man will not urge blind and indiscriminate war against birds, any more than upon insects; for there are many insects that are cannibals, eating other insects, and, therefore, worthy of preservation. The wise man will kill such birds, after the hatching season, as depredate to any serious extent upon his crops; and it is surprising what a very little amount of killing will suffice to drive them from crops, if done before they become fairly habituated to the diet. They love life, and will not stay when their lives are in danger. Thus, if persistent shooting be kept up in a field of grain or corn for two days, so that blackbirds cannot eat, they will abandon the place for that of some person who thinks his crop not worth protecting.

If we give you a list of some of the more common varieties of birds, and the food they eat, you may be able to form a pretty clear idea of their general value to the cultivator, and will easily see that the indiscriminate slaughter of birds is anything but wise. In fact, the total extermination of any one species would probably bring with it evils of another nature, extremely difficult to be overcome.

Birds are found where shelter and food are most plentiful. In prairie countries, birds are confined to but few specimens—grouse, quail, swallows, martins, meadow-larks, and the rice-bunting, making up a large portion of them. Plant trees, and their numbers increase indefinitely; for thus they have shelter. Protect them, and a little killing during the ripening of crops does not prevent their return with the returning spring.

It is related that a shoemaker of Basle, Switzerland, once put a collar on a swallow in the autumn, containing this inscription:

"Pretty swallow, tell me whither goest thou in winter?" The next spring the same courier, by the same means, brought this answer: "To Anthony, at Athens. Why dost thou inquire?"

Poetry is pleasant. Theory, speculation and sentiment are well enough in their way. Practical knowledge, through careful observation, is what the practical man wants when bread and butter are concerned.

III. Food of Some Common Birds.

WHAT some of the more common birds eat will enable the practical man to answer the question asked at the head of this article in the most practical manner possible.

FRUIT-EATERS.—Robins, cat-birds, thrushes, blue-jays, cherry-birds, orioles, and others of that class, are well known as fruit-eaters, as well as great destroyers of insects. They are undoubtedly worthy of destruction by the horticulturist when fruit is ripening—never so by the farmer.

THE EAGLE.—Of the rapacious birds, or those of prey, the eagle, notwithstanding the airs he puts on as the "Bird o' Freedom," is a thief and robber, and not especially brave.

OWLS.—Of owls, the mottled owl (*Scops Asio*) and the saw-whet owl (*Nictale Acadica*): The first lives about barns in winter, if allowed, and both are inveterate mousers. In the summer, they also eat large moths and beetles.

HAWKS.—The hen-hawk will sometimes take a chicken; but their principal food is mice, frogs, grasshoppers, and various vermin. The broad-winged hawk eats mice, grasshoppers and other insects. The rough-legged hawk eats principally field-mice. The sparrow-hawk, early in the season, lives on birds and mice; later, principally on grasshoppers. The swallow-tailed hawk lives on snakes, lizards, frogs, beetles, grasshoppers and other insects.

The red-tailed buzzard often makes bold attempts on barn-yard fowl; but his principal food is squirrels, rabbits, rats and mice.

THE CROW.—The American crow, so universally and wrongfully detested by farmers, eats mice, frogs and various insects; in fact, is an almost universal feeder; but does not eat so much corn as he is charged with.

BLACKBIRDS.—The red-winged blackbird has been universally detested by farmers. It pilfers the ripening grain when pressed by hunger. In the spring, its food is cut-worms, wire-worms, caterpillars, and the larvæ of noxious butterflies. It is surprising what a little amount of shooting will drive them from the fields where they flock after breeding time.

The purple grackle, or crow-blackbird, is less destructive. It pulls some corn, it is true, but its principal food is injurious larvæ, much of which it gets by following the plowman if allowed.

CLIMBERS.—The climbers include woodpeckers and cuckoos. They are nearly all beneficial. The yellow-billed cuckoo especially—quite common in Northern Illinois—are persistent devourers of that pest of orchards, the American tent-caterpillar.

THE ORIOLE.—The Baltimore oriole lives largely on beetles, especially the plum and pea curculio, and the long-snouted nut-weevil. It is very properly detested by fruit-growers, from its mischievous habits among fruits. On the other hand, the orchard oriole is a most industrious bird, and well worthy of protection—ridding our orchards of hosts of noxious insects.

WRENS AND BLUEBIRDS.—The house-wren lives exclusively on insects. The white-breasted nuthatch and the American creeper live on tree-insects alone. The bluebird eats the larvæ of insects exclusively, especially those of the canker-worm and the codling-moth.

FINCH, LARK AND PLOVER.—The finch family, comprising about seventy species, sometimes spread in flocks over large tracts of country, destroying immense numbers of insects, larvæ, and the seeds of weeds. So the warblers, as a rule, are of immense benefit, through the destruction of multitudes of noxious insects. The meadow-lark subsists principally upon larvæ which are found underground. The plover, which nests on the prairie and in corn-fields, eats grasshoppers, beetles, snails, etc.

LAND AND WATER BIRDS.—The land game-birds, although classed generally as graniverous, are of undoubted benefit, since their food consists more largely of insects than anything else. The water-birds—waders—as a class, are neither especially beneficial nor injurious to the farmer. The swimming birds—ducks, geese, etc.,—are decidedly destructive, since, when they make excursions upon the land, it is generally for the purpose of raiding upon fields of grain.

IV. Birds Classified by Their Food.

BIRDS may be classified very properly, into omnivorous—those living on a variety of food; insectivorous—those living entirely on insects; and granivorous, or those which make a large proportion of their food to consist of grain and seeds.

OMNIVOROUS.—Of the first class, crows, blackbirds, jays, orioles, starlings, cedar-birds and titmice are among the principal families. These, as a rule, feed indiscriminately on berries, seeds, grain and insects.

INSECTIVOROUS.—Of insectivorous birds, those feeding almost exclusively on insects—woodpeckers, swallows, night-hawks, shrikes, bluebirds, fly-catchers, creepers, the warblers (Sylviæ), and wrens are especially noteworthy. Few of these are hurtful in any sense, and all of them are extremely beneficial, both to the farmer and horticulturist. A few of them take a ripe berry now and then, but never to such a degree as to cause them to be missed.

GRANIVOROUS.—These comprise the gallinaceous tribes, including our poultry-birds, grouse, partridges, quail, etc., grosbeaks, tanagers, finches and buntings. Many of them depredate severely upon the husbandman at times, and then may very properly be killed. It must, however, always be borne in mind, that many of those classified as being pre-eminently grain-eaters, as grouse, quail, blackbirds, finches, etc., are also largely insect-eaters, and feed their young almost exclusively on insectivorous larvæ.

V. Birds the Natural Enemies of Insects.

It must be remembered, also, that birds are the natural enemies of insects; that, were it not for birds, insect-life would soon increase to such an extent as to devour every green thing from the face of the earth; that, even with their help, insects do so increase in certain successions of prolific seasons as to overrun and devastate large areas of country — locusts, chinch-bugs, Hessian-flies and army-worms, notably among farmers; and codling-moths, canker-worms and curculio, among fruit-growers; that, in a state of nature, birds do hold insect-life in suitable check; that, as a country settles up, birds are destroyed to a considerable extent, while through the systematic cultivation of the earth, a large supply of insect-food causes a corresponding increase of insects, and, for cogent reasons, without a corresponding increase of birds—one reason being that no species of birds are the special foes of insects prevalent in particular seasons; another being, that insect-propagation is so extremely prolific when the conditions are suited thereto. Therefore, let us use a wise discretion in the protection of birds, and as careful a discretion in killing them.

Birds destructive to one class of cultivators may be beneficial to another, and *vice versa*. When they are taking more than their share, one class may be killing them, while the other is protecting; and here, again, the natural law of self-preservation becomes apparent. Robins, cat-birds, orioles, thrushes, etc., are not generally regarded as destructive, but rather as being beneficial; and for the reasons that the culture of fruit is not the predominant industry, and orchards, at the time of ripening their fruits, are, in many instances, watched. In the fruit regions these birds are justly regarded as being the most inimical of the feathered creation.

We hold that the extermination of any of our birds would be a serious evil; that if the cultivator would take the same pains to prevent injury by birds as he must in certain seasons to prevent the ravages of insects, and often failing even with the help of birds, he would soon come to know, and to discriminate between those most truly his friends, and those more destructive to his crops.

We have carefully stated the foregoing facts in order that the farmer need not kill birds valuable in husbandry because they are reported obnoxious to the fruit-grower. Each will kill the class that is injurious to his crops. The following test, the results of the extended and most careful study of Prof. S. A. Forbes, as Curator of the Museum, and Professor of Natural History in the State Normal University of Illinois, and now State Entomologist, will show: 1, birds of the greatest value as destroyers of injurious insects; 2, birds of doubtful utility in horticulture; and 3, birds that should be exterminated by fruit growers.

VI. Birds to Be Carefully Fostered.

Blue birds, tit-mice or chicadees; warblers (small summer birds, with pleasant notes, seen in trees and gardens); martins; swallows; vireos (small birds called green-necks); all birds known as woodpeckers except the sap-sucker. The sap-sucker is *entirely injurious,* as it is not insectivorous, but feeds on the inner bark cambium and the elaborated sap of many species of trees, and may be known from other woodpeckers by its belly being yellowish, a large black patch on its breast, and the top of its head of a dark, bright red; the male has also a patch of the same on their throats, and with the inner margins of the two central tail feathers white. This bird should not be mistaken for two others, most valuable birds which it nearly resembles, to wit: The hairy woodpecker and the downy woodpecker. These two species have the outer tail feathers white, or white barred with black, and have only a small patch of red on the back of the heads of the males only. The yellow-hammer, or flicker, is somewhat colored with *yellow*, and should not be mistaken for the sap-sucker; it is a much larger bird. The red-headed woodpecker sometimes pecks into apples, and devours cherries, and should be placed in the next division. The wrens, ground robin, known as chewink, meadow lark, all the fly-catchers, the king bird, or bee-catcher, whip-poor-will, night hawk, or goat-sucker, nut-hatcher, pewee, or pewit. All the blackbirds, bobolink, American cuckoos, plovers, upland snipe, grosbeaks and other finches, quails, song sparrow, scarlet tanager, black, white and brown creepers, Maryland warbler, indigo bird, chirping sparrows, black-throated bunting, thrushes, except those named in the next class, and all domestic fowls except geese.

VII. Birds of Doubtful Utility.

This class includes those which have beneficial qualities, but which have also noxious or destructive qualities, in the way of destroying fruits, other birds and their nests, and whose habits are not fully determined. Thus the robin, brown thrush and cat-bird are very valuable as cut-worm eaters, but also very obnoxious to the small fruit-grower. The blue jay is not only destructive to grains and fruits, but very noxious in the way of destroying the nests, eggs and young of smaller and better birds. I think that, notwithstanding his great beauty and sauciness, that he should be placed in the list of birds to be exterminated. Robin, brown thrush and cat-bird, shrike, or butcher bird, red-headed woodpecker, jay-bird or blue-jay, crow and the small owls, pigeons and mocking birds. There are many other birds that should be placed in one or other of the foregoing lists; but it will be safe, as a rule, to preserve all birds not named in VII. and VIII.

VIII. Birds to be Exterminated.

SAP-SUCKER, or yellow-bellied woodpecker, Baltimore oriole, or hanging bird, cedar bird, or wax-wings, hawks and the larger owls.

The farmer should foster all the birds named in VI., and also those in VII., except the blue-jay, and the oriole in VIII. As a fruit-grower we should accept the list as written. The list of birds beneficial in husbandry might be widely extended.

Frederick the Great, of Prussia, because sparrows ate his royal cherries, ordered their extermination throughout his Kingdom. This was faithfully carried out. Insects so increased, especially caterpillars, the eggs of which sparrows eat, that the small fruits could scarcely be raised at all. Then his royal wrath evaporated, and he went to the cost of importing these birds for the benefit of his well-loved subjects.

The blue jay is a nuisance to farmers; it eats corn and is quarrelsome with other birds, driving them away, yet for the fruit-grower it eats the eggs of the tent-caterpillar of our apple orchards.

IX. Destroying Insects.

The investigations of M. Prevost, acting for the French government, demonstrated that those birds generally regarded as being insect-eaters, are not as a rule the most beneficial to the farmer, but that for the most part, the birds which render the greatest service are those against which the popular prejudices are strongest. Thus the sparrows, the starling and crows are the great destroyers of the cockchafers, and so our crows and black-birds are of the May-beetles, and we are finding that many birds we have deemed to be our enemies are really our best friends. There is another fact to be remembered also. That is: nearly all birds, during the period of reproduction, whatever may be their natural food at other times, are almost entirely insect-eaters, and that they feed their young almost exclusively with insect food. Then the amount of insect food a young bird will consume is enormous. Dr. Wyman took from the crop of a young pigeon a mass of canker-worms that was more than twice the weight of the bird itself.

Part VII.

FARM LAW AND ITS PRINCIPLES.

LEGAL FORMS AND OBLIGATIONS.

AGRICULTURAL LAW.

STOCK, GAME AND FENCE LAW.

SECURING A HOMESTEAD.

HIRING HELP, ETC.

FARM LAW.

CHAPTER I.

PRINCIPLES IN RURAL LAW.

I. LAW GOVERNING FARMERS' ANIMALS.—II. LIABILITY FOR INJURY BY DOGS.—III. TRESPASSING UPON PROPERTY.—IV. DIVISION FENCES.—V. RAILWAY FENCES AND TRESPASS.—VI. RAILWAYS RUNNING THROUGH FARMS.—VII. PUBLIC ROADWAYS.—VIII. THE RIGHTS OF THE PUBLIC IN THE ROAD.—IX. AVOIDING OBSTRUCTIONS IN THE ROAD.—X. RIGHT OF WAY OVER LANDS OF OTHERS.—XI. LIABILITY OF THE FARMER FOR HIS SERVANTS.—XII. RIGHTS RELATIVE TO WATER AND DRAINAGE.—XIII. LIABILITY OF DEALERS.—XIV. HIRING HELP—SPECIFIC WAGES. XV. WHAT IS A FARM?—XVI. GETTING A FREE FARM.—XVII. THE PUBLIC LAND SYSTEM.—XVIII. PRE-EMPTION, HOMESTEAD AND TIMBER-CULTURE ACTS.—XIX. LAND TAKEN UNDER THE THREE ACTS.—XX. THE DESERT LAND ACT.—XXI. LAND YET OPEN TO SETTLERS.

I. Law Governing Farmers' Animals.

THE principles governing the law of the various States in relation to animals may be stated as follows: The farmer is responsible for the good behavior of his animals when beyond his care, and also for any injury or hurt done to persons by animals belonging to him, and known to be vicious, even although the person injured was a trespasser on the premises at the time. If you keep a vicious dog, and he tears a person hunting in your fields, you are liable in damages. If a vicious bull, ox or cow injures a person on your premises, you must pay the damage, even although the sufferer may have been a boy stealing apples from your orchard—but the person trespassing is also liable for his trespass. In fact, a person may be liable to more than money damage if he put a vicious animal in an enclosure to injure any special person who may have annoyed him by his trespasses across the field. A case is reported in which the farmer had to pay $500, for two broken ribs received by another in this way.

If an animal injures a person in the open highway, either by assault or by frightening a horse so he shall run away, to the damage of the driver or occupants of a carriage, the owner of such animal is liable. The owner of a horse, was, in one case, convicted of manslaughter, because having turned his horse out on the road, and some children switching him, the animal kicked one of them, causing death.

II. Liability for Injury by Dogs.

WE have already said that the owner of a vicious dog is responsible for any injury the animal may do a person, even on his owner's grounds. If a dog, even in play, runs out into the street and frightens a horse or other animal so that he does injury, the owner of the dog is liable. However good-natured a dog may be, if he turns and bites children, who tease him, his owner is liable. The old English law held that a man could have no ownership in a dog. This became the rule in some of our States. The more enlightened rule now is that a dog is property, and to steal a dog a crime. Hence, in some States, although a person may not be the owner of a dog, merely harboring the animal renders the harborer liable for any damage the dog may do.

In States where dogs are licensed, a dog not licensed may be killed "whenever or wherever found." It is the law in some States, and should be everywhere, that a dog licensed or not, if he attack a person, outside his owner's premises, may be killed, but your neighbor's dog has the right to pass peaceably over your premises, and you may not kill him simply for so doing. But if a dog is found beyond the enclosure of his owner, worrying or killing farm animals, he may be killed. Again, a dog may be killed if he haunt your premises, by day or night, to the disturbance of the family, either by his noise or otherwise. It is unnecessary to give the authorities from which these law-facts originally came. But by way of illustration, take the case of the dog, snapping and biting a boy who struck at him, in 124 Massachusetts, 57; and 38 Wisconsin, 300. The jury gave damages because they thought the boy acted naturally, but in 65 Illinois, 235, if it had been a mature man instead of a boy, damages would not have been given.

III. Trespassing Upon Property.

Hunters.—One of the most common mistakes of gunners, trappers, etc., is that that they have the right to hunt and fish where they please. All persons have the right to fish in navigable streams and lakes, subject to the laws on fishing. In small streams and ponds, the right belongs solely to the owner of the adjacent land, for he owns the water also, until it has passed from his premises.

Trespassing Animals.—The owner of the land has the right to kill trespassing animals, wild or tame, invading his premises to destroy his stock, and this, notwithstanding a law against killing wild animals, as mink, otter, etc., except at certain seasons. In fact, a man has a right to kill wild animals, especially destructive ones, upon his own land at any time, and all wild animals are destructive in some way— even rabbits.

Berrying, Etc.—Persons may not trespass upon the land of another to pick berries or other wild fruit. Custom does not make law. Permission must first be had in order to avoid the payment of damage. It would, however, be considered mean to molest children berrying in the woods or enjoying a pic-nic. Still it is best always to ask permission.

Posting Notice.—But because you are annoyed by trespassers you have no right to set traps, spring-guns, etc., to do injury, without first giving notice. It is also necessary, if you have a vicious animal on your premises, to post notice thereof.

Some persons post the notice—"Man-traps and spring-guns;" "Beware the dog," etc., on general principles, though the traps are imaginary and the dog innocent of harm.

Neighbor's Fowls.—You may not shoot, nor lay poison for your neighbor's fowls, which trespass and scratch up your garden, even though you return the fowls back upon your neighbor's land. If they have been poisoned by you, and the neighbor should eat them, it might go hard with you in court. So if you should injure a boy caught stealing fruit, you would be liable for malicious mischief.

Injury to Stock.—You would be liable for damage to your neighbor's stock if you threw dangerous substances, as glass, bits of wire, or other trash into his field, or even deposited it in your own fence corners, and the fence being broken, the animal was injured. For instance, a cow swallowed bits of a wire fence that the owner had allowed to fall to pieces and get rusty. These pieces had fallen in a neighbor's field. A jury decided that the owner of the wire must pay for the cow. The same rule would follow from injury from trimmings of osage orange, or other substances thrown in the road.

IV. Division Fences.

THE primary object of fences is not to protect against the invasion of other animals, but to keep in one's own. Yet, it is not lawful for you to put up a fence except it be entirely on your own land. Hence, when a division fence is to be built, if on the line, both persons join in building it, and it belongs jointly to them, and cannot be removed except by the consent of both. There must be an agreement as to which portion each person shall keep up.

Then, if the adjoining fence-owner does not do his duty, and your cattle walk over his imperfect fence, he has no redress; but if they stray beyond your neighbor's, upon the land of a third party, you are liable for the damage to that party, and even if this third party had not a proper fence. If you turn your cattle into the highway and they stray upon the property of another, even though he has no fence, you are liable. You are equally responsible if persons straying through your fields let down the bars or leave gates open by which your cattle do injury to the property of another. But if one is carefully driving cattle along the road, and without the driver's fault they break away and trespass, there is no liability if driven back as quickly as possible; for you have the right to drive your cattle along the road though you may not lawfully turn them into it to shift for themselves.

V. Railway Fences and Trespass.

WHILE it is the fact that by the common and general law, every man is bound to restrain his stock from doing injury to another, the manner of restraint by a fence is due to the statute. Hence, unless the statute clearly requires it, a man need not build a fence if he has no stock to confine. In relation to fencing railways by the corporations, the character of the fence is defined by law. The general law requires the company to maintain a suitable fence along the whole line. It does not mean that the fence shall be the same along the whole line. It may be necessary that some portions be better and stronger than others. It must be suitable.

If Stock is Killed.—Now, if the stock of a land owner adjacent gain access to the track, by reason of the unsuitable character of the fence, and are injured or killed, the company is responsible; but if the animals stray from premises beyond, and find their way to the track over this contiguous land, the railway company is not liable for damages. Nor would the company be liable under the common law for cattle straying along the public highway, without the care of their owner, if killed by a passing train on a crossing.

VI. Railways Running Through Farms.

The conveyor of lands for a highway does not lose ownership in the lands, except as to its use for the purposes of a highway. Railway companies generally acquire simply the right of way, and not the absolute ownership of the land. The reason of this is that in the case of the abandonment of the highway, or the franchise of a railway company, the land may revert to the one who sold it, or the person owning the original tract at the time of the abandonment, unless it is expressly stipulated to the contrary. But if a person gives an absolute deed, it, of course, would be binding. This should never be done, under any circumstances, because the abandonment and sale of a strip, might cause great damage to a farm, if resold to other parties, and the conveyance for road purposes is sufficient for the railway or road commissioners.

The exclusive rights in the trees, and vegetation on the surface and the minerals below belong to the conveyor, and not to the company who only hold the right of way. And if a stranger take anything belonging to the soil proper therefrom, he is liable, not to the railway company, but to the owner of the soil; but the owner of the land does not therefore have the right, as against the company, to enter upon and remove the turf, soil, or anything growing thereon, or to disturb the same without the permission of the company, and the company have the right to cut down and remove whatever may be within their line, if it may at any time interfere with their use and operation of the road. No person has the right to take any property from the line of a railway, nor make use of the same without the permission of the company, and the person so doing, if it be personal property, is liable to the company and not to the seller of the land.

VII. Public Roadways.

Roadways are made for the use of the public. Their right is simply the right of way or passage over the roads, by themselves, teams, vehicles, stock, etc. The road officers may use the soil, gravel, etc., of a road for repairs, or for transportation to some other portion of the road, but not for their own private use. The owner of the land bordering on a road, owns the soil, trees, grass, or any valuable thing on or under the surface. No man has any more right to remove anything therefrom, except the owner of the soil, or the properly constituted officer, than from any other private property. Neither has any person a right to deposit and leave vehicles, wood, timber or other property on the public road. If he does, the owner of the adjacent land may remove them, and if injured or lost, the owner has no redress. Stock may not be stopped to feed on the public highway. No person may hitch a team in the highway, to the detriment of trees or other property. If he does, the animal or animals may be removed. No one has the right to stand in the road to abuse another, to throw missiles at your animals, without liability for trespass; and if obscene language is used, the person using it may be driven off even by force. Fruit trees standing even in the road, belong to the owner of the adjacent land and the fruit as well; and a well standing partly in a field or yard, and partly in the road, belongs exclusively to the owner of the land.

VIII. The Rights of the Public in the Road.

The road is for the convenience of the public at large. The owner of the adjacent land has no right to obstruct it. He can not use the road to deposit trash; can not place any structure, even a pig-pen, thereon; nor leave any vehicle standing thereon. If he does, even if not in the traveled path, and a person or animal runs into them at night, and is injured, the owner is not only liable for the damage but for obstructing the highway. The owner of adjacent land must place his fence entirely outside the road, and not half over the line. Neither can he place terminal posts half on his own and half on his neighbor's land unless he has liberty so to do. But if the road is discontinued the land reverts to him, and he can enclose it again as a part of his premises.

IX. Avoiding Obstructions in the Road.

It is generally supposed that no person has the right to leave the road for any purpose and pass over the adjacent land. There are, however, occasions when he may do so. If deep snows have fallen, or the road is drifted so full that it is impassable; if there is a washout that can not be passed, or if from any cause the road is absolutely impassable, a traveler may have the right even to remove a fence and pass over fields, to a point beyond the obstruction, and he is not liable for trespass. But in doing this, he must be careful not to do unnecessary injury. For instance: A fence may not be broken down so as to injure it seriously; it must be carefully taken down. Hence it would not be safe to tear down a permanent wall or destroy a living hedge.

X. Right of Way Over Lands of Others.

If a person be shut out from the public highway by the intervening lands of another, he has the right to a private roadway, by the most practicable route, over such intervening land to the public road. This right must be acquired by one of three means:

1. The right by continued use. To acquire this right, the roadway must have been peaceably used for a period of fifteen or twenty years; that is, continuously or regularly, and under a claim to this right of use; but this right extends only to a definite road and for the definite purpose of passing directly to and from the public road. No person should attempt to acquire this right by use simply to avoid paying a just compensation. To gain this right it is not necessary that one person should have traveled, but successive owners, if there have been such within the prescibed period. It is not peaceably used if done under protest, however long the use be continued; and the right, once acquired, if for a specific purpose, as hauling wood or other commodity from another lot, the right ceases when this specific use no longer continues.

2. If you sell all your land fronting the highway, retaining that lying away therefrom, you reserve the right to pass to and from your new home to the public road. So, if you sell to another that portion away from the road, the law gives him the right to cross your land to the road, if he be otherwise cut off. It makes no difference whether the right of way is stipulated in the deed or not. Yet, it is proper in every case to have the whole clearly stipulated.

3. The other way, and the proper one, is to buy your right of way, and no sensible man would object to granting it in this way. The difficulty in this case generally arises in dissension as to the price. In such a case common sense and humanity would dictate that no advantage be taken on either side—both should make all reasonable concessions.

However the right to a private road may be acquired, you have no rights outside the line of the road. The seller, in the absence of any stipulation to the contrary, has the right to put in suitable gates, or bars, at the entrance and exit. And if the bars are left down by the person who acquires the road, or by his family, servants or visitors, he is liable for all damage resulting therefrom, either by himself, family, servants, or visitors.

XI. Liability of the Farmer for His Servants.

THE liability of the farmer for the acts of his workmen, is more onerous than is generally supposed. It might not be thought that if a horse be driven by one's hired man, and the horse cast a shoe, which flies and breaks a window, the owner of the team would be liable; but it is true that he is liable. So you would have to pay for the damage done by your team running away, if damage is done. It was a large bill that a gentleman had to pay whose coachman allowed his team to run away and crush through a plate glass window, and into a jewelry store filled with costly articles.

If you lend a servant a team, and with it give him an order to execute, and he gets drunk, or from other preventable cause allows the team to do injury, you are liable. But if he borrows a team simply for his own pleasure and commits injury, you are not liable. If your hired man injures himself through your negligence, you are responsible to him.

If you, or your hired man, by your orders, set fire to brush or trash on your own field, and the fire damages another, you are liable. So you are if you set him to chopping, and he accidentally chops down or into trees beyond your land line. In short you are liable for any act of injury by your hired man in the performance of something you have set him to do. Therefore never hire a man who gets drunk, and to set him a good example, never get drunk yourself. Under this head, however, the line is so closely drawn as to liability or not, that instead of depending upon general rules, in most cases it is better to consult a lawyer.

XII. Rights Relating to Water and Drainage.

There are certain cases where one person may flow another person's land, as in the case of the mill owner, etc. It is a franchise that has been paid for. You may do what you please with your water, so long as it does not flow back upon the land of another. If it does, and the privilege has not been paid for, he may take down so much of the dam as will relieve his land from the overflow. This flooding is often done innocently, from ignorance of the extent to which the back-water will rise.

If a stream has become obstructed by drift-wood, etc., below, so that the water backs upon your land, you may remove this natural obstruction, leaving the material on the bank. This, however, does not apply to surface water accumulated by heavy rains. But water is no longer surface-water after it has been gathered into a natural or artificial channel. Surface-water originating on one's own land, the owner may detain it if he can.

Against an overflow by floods, a person has the right to embank, even to the detriment of his neighbor. His neighbor is at liberty to do the same, but it is not permissible to place impediments in the bed of the stream, to the detriment of another. Spring-water and underground water belongs to the soil. Your neighbor may have a well fed by springs on your land. You may cut the source of these springs and convey the water where you please. So long as it remains on your own land he has no redress.

You have no right to turn your drainage water onto your neighbor's land; but the same rule works here as in the acquirement of private roadways.

XIII. Liability of Dealers.

The adulteration of every article of use or sale, and frauds in contracts, etc., are growing evils. So far as contracts are concerned, unless the farmer is able to understand them, to be assured that they are correct, or cannot be separated into parts for his discomfiture, he should have nothing to do with them without first consulting his lawyer, to know that they are all right. Trust no stranger in any event. Beware of lightning-rod peddlers, jockeys, and confidence men generally, who want to make you rich in a trade. Sign no contracts for anything for future delivery, unless the men are known to your bankers as solvent, and especially sign no paper that may have a double meaning.

The liability of a dealer will compel him to make good any deficiency. If you buy grain or other seeds, of a given name, they must be true to name. If expressly warranted pure and fresh, the dealer is not only liable for the purity of the seed, but if they fail to grow, or turn out something else, the dealer is not only liable for the value of the seeds, but for whatever loss the farmer or gardener may have suffered from these causes. Seedsmen intend to be honest as a rule. They sometimes make mistakes. If so, the innocent purchaser does not have to stand the loss, but the seedsman or dealer.

XIV. Hiring Help.—Specific Wages.

IF a man or woman is hired without specific agreement as to wages, he or she is entitled to the current price of that particular labor. If a laborer hires for a specific time, and quits before the expiration of that time, even though he may have been hired at so much per month for that time, he cannot recover any of the money from the master. But if the farmer has paid him money on account from time to time, or has given him a note or notes in lieu of money, the farmer cannot recover this. But if the farmer voluntarily discharge the laborer, his wages must be paid.

If a laborer hire for a specific time at a fixed price and works, say until that season when wages may be much higher per day or month than the price agreed on per day or month for the whole time, and then quits, the master can recover this extra amount for the remainder of the time, and the laborer can not set off the value of the work already done. So if a man or firm be hired to do a specific work at a specific price, as digging a well, building a wall or a house, and leaves it unfinished, without good excuse, he is entitled to nothing.

If a laborer has good cause for refusing longer to work he may do so, and the master must pay for the work actually done. Sickness of the laborer, a dangerous epidemic in the family or neighborhood, improper treatment, bad food, etc., are valid causes for quitting. If the laborer is arrested and imprisoned for crime, it is no bar to his receiving pay for the work already done. As to what constitutes cause for quitting, outside of those mentioned, and in fact when litigation settles the matter, it must be determined by the jury.

If another person entice your workman away before his legal time has expired, you have recourse upon the person enticing the man. The law would hold that he who interferes with another man's business must pay all the damage accruing from the inconvenience, and if done maliciously this might add special damage; but one person may offer inducements for a man to leave an employer where the person was only working from day to day, or when his time had expired.

XV. What is a Farm?

A FARM is any considerable piece of land, described, by metes and bounds, by monuments, blazed and distinctly marked trees, government or other surveyor's stakes, properly recorded. The extent of a farm is determined by the length of the boundary lines, by visible objects or those that may be found, and those visible monuments, trees, rocks, stakes or stones, naturally or artificially placed. These control all other agents. When described by metes and bounds, the number of acres wrongly stated in the deed, would give no cause for redress, even though they were far less than stated in the deed, or even though the seller fraudulently overstated the number of acres.

In buying a farm, if the seller overstate, even fraudulently, how much grass it will carry, how much stock it will pasture, or how much wood it will go to the acre, the buyer has no redress. But if, with a view to selling, he should fraudulently state that the farm *had* produced a specific quantity of any article in a year, knowing it to be false, he would be liable to an action at law, so very close is the line drawn between mere talk and *actionable* talk.

A man may have a farm to-day and none to-morrow. If a stream carry away a part of or all of his farm, the loss lies with himself, although he may know where his farm is deposited. It thereafter belongs to the man fortunate enough to acquire the accretion, so long as it remains.

When a farm is bounded by a stream, the owner's right goes to the middle of the current, not always to the middle of the water. This should be remembered in determining what islands in a stream belong to one or to his opposite neighbor. If the land is bounded by a large lake, navigable river, bay or gulf, his rights extend only to low water mark. Farther, his rights are merely those common to all. But in tide waters, there may be flats; in this case it will depend upon his deed in its accuracy whether he owns to high or low water mark.

If a boundary line runs to a specific object, as a tree, rock, fence, etc., it runs to the middle of the object, unless specifically stated otherwise. If so, examine the record, to know that the next man actually owns up to your line; in fact, in buying any piece of property in which a deed, or other contract passes, it is well for your lawyer to pass on its merits. As a rule, the fee for such a service will be well invested.

XVI. Getting a Free Farm.

ALL public lands are virtually open to free settlement, the fees under the Homestead, Timber Culture and Desert Land Acts being light. At public land sales the price ranges from one dollar and twenty-five cents to one dollar and fifty cents per acre, but the best of these lands are always gobbled up by railways and syndicates, and they often wrongfully dispossess the poor settler. But this need not be if the man knows his rights and asserts them. Formerly pre-emption was the only means of acquiring title by actual settlement. Then the settler had to prove title and also pay one dollar and twenty-five cents per acre when the land was placed in the market. It is still a favorite means by which speculators secure water privileges, and valuable tracts of timber land, by means of fraudulent affidavits. They also pre-empt water fronts under the Desert act, and often by armed mob-force drive off actual settlers. Colonies of actual settlers may command respect from these pirates by organized force.

XVII. The Public Land System.

A COMPLETE and condensed compilation of the principal means by which public lands may be secured, brought down to the year 1883, is given below. To get a clear understanding of this it must be remembered that the public lands are surveyed into a series of lines of townships running north and south, each township consisting of thirty-six sections of 640 acres, or one square mile, each. The area of a township is, therefore, 23,040 acres. Each line of townships is called a range, the ranges being numbered from east to west, and the townships north and south. Each section is divided into quarters of 160 acres each, and these again divided into quarters of forty acres each. The public lands are divided into two great classes, the minimum price of one class being one dollar and twenty-five cents per acre, and of the other two dollars and fifty cents per acre. The latter class consists mainly of alternate sections reserved by the Government in land-grants to railroads. Public lands are not now placed in the market subject to purchase for cash, the general policy of the Government being to hold the lands for actual settlement only. The principal laws under which titles can be perfected are the "Homestead," "Pre-emption," and "Timber-culture" acts.

XVIII. Pre-emption, Homestead and Timber-Culture Acts.

UNDER the Pre-emption law the settler must pay the Government price for the land. The maximum amount of each grant is 160 acres. To secure this the claimant must first become a resident of the land—by claim-shanty or otherwise—and within three months after settlement must file a declaratory statement of the facts at the nearest land office. For this filing he pays two dollars. He must reside on the land for at least six months, and within thirty-three months from the date of settlement he must submit final proof of actual residence and improvement, and pay for the land—two dollars and fifty cents or one dollar and twenty-five cents per acre, according to class. Any time before the thirty-three months expire the settler may convert his claim into a homestead by payment of the homestead fees.

The Homestead Law.—Under the homestead law any citizen or intending citizen of either sex, over the age of twenty-one, single or the head of a family, may obtain 160 acres of public land free by five years' actual settlement and residence thereon. The only payments the settler is required to make are the land office, patent, and commission fees, amounting altogether to about twenty-six dollars, of which sum eighteen dollars is paid at the time of entry and eight dollars at the end of five years when the title is perfected. Until last year the settler was allowed only eighty acres of the two-and-a-half dollars land, or land within the railroad limit, but the law has been so amended that 160 acres can now be secured. Soldiers and sailors who served during the war are allowed to deduct the time of such service—not exceeding four years—from the five years' residence required before completion of title—

a privilege which extends to widows or minor orphan children of all those who if alive could claim this allowance. After six months' actual residence and cultivation, the settler has power to prove up and purchase the land at the Government price, instead of residing thereon the remaining four and one-half years required to complete his title. This is what is known as commuting an entry. Under this act, therefore, any man, however poor, may become the owner of a farm of 160 acres for twenty-six dollars— a farm which at the end of five years should be worth at least $1,000. Special provision has been made for people who have been unfortunate in business, or burdened with debt, who wish to start anew, this act expressly providing that "no lands obtained under the provisions of this chapter shall in any event become liable to the satisfaction of any debts contracted prior to the issuing of the patent therefor." According to recent rulings under this law a man and woman, each having a homestead entry, may pool their rights by marrying without invalidating either claim. Lands entered under this law are exempt from taxation until title has been completed. One person cannot relinquish a claim to another; relinquished lands revert to the Government. A single woman's rights are unaffected by marriage—so far as this act is concerned—provided the requirements of the law are complied with. A married woman making an entry, who has been deserted by her husband, will, upon final proof, receive the patent in her own name, notwithstanding the husband's return.

The Timber-Culture Act.—Under the timber-culture act actual residence is not required, and the same amount—160 acres—can be secured. The party making an entry is required to break or plow at least five acres the first year and five more the second year. The first five acres are to be cultivated during the second year, and planted with timber seeds or cuttings during the third year. The second five acres are to be cultivated the third year, and similarly planted the fourth year. Not less than 2,700 trees, seeds or cuttings must be planted on each acre, and at the time of final proof there shall be growing not less than 675 living and thrifty trees to each of the ten acres. A tree crop, if destroyed one year, must be replanted the next. At the expiration of eight years from entry, final proof can be made and patent obtained. The fees are fourteen dollars at time of entry and four dollars at final proof. This land is exempt from taxation or execution for eight years.

XIX. Land Taken Under the Three Acts.

UNDER these three acts, any qualified applicant may obtain 480 acres of land at a nominal cost. A person cannot file under the homestead and pre-emption laws at the same time, actual residence being necessary in each case, but is at liberty to enter a pre-emption and tree-culture claim together, and after proving up on the pre-emption by six months residence or longer, may take a homestead, and thereby get possession of the 480 acres within a year of his first settlement. Every son and daughter over twenty-one can do the same. A pre-emption settler may mortgage his land to pay the government price for it. A pre-emption claim cannot follow a homestead and tree-culture claim, as persons already holding 320 acres of land are barred from the privileges of the pre-emption act. Where the government alone is concerned, the laws will be liberally construed; where adverse rights are involved, a strict construction of the statutes is necessary.

XX. The Desert Land Act.

THERE is, however, another land act, under which large areas have been sometimes taken, and actual settlers ousted through terrorism. It was intended to be beneficent, and might be so under proper restrictions. This act is applicable to all lands, exclusive of timber and mineral lands, which will not produce an agricultural crop without irrigation. This act is taken advantage of principally in the far West Territories, where there are large areas of arid land requiring irrigation to make them productive. Under this act, a person may obtain one section—640 acres—of desert land at $1.25 per acre by three years' irrigation—twenty-five cents per acre to be paid at time of entry, and the remainder of one dollar per acre on final proof at the end of three years. Actual residence is not required.

Under these laws, an immense number of fraudulent land entries have been, and are being made, the Land-Office Commissioner being comparatively powerless to enforce the law or investigate complaints. The work of the General Land-Office being limited by the size of the Congressional appropriations, it has been found impossible to inquire into a tithe of the alleged frauds, and the groundwork for a great structure of future litigation is now being laid out. Great quantities of valuable coal and iron lands, forests of timber, and the available agricultural lands in whole regions of grazing country have been monopolized by persons who have caused fraudulent pre-emption and commuted homestead entries to be made by their agents and employes, and the commissioner, in his latest report, states his inability to stop this, owing to the limited facilities of the land-office. He strongly recommends the repeal of the pre-emption law, on the ground that it is being largely made a shield for fraudulent entries, and that the passage of the homestead law leaves it unnecessary. Formerly, the pre-emption system afforded the only means by which settlers could acquire title to homes on the public domain, but, with the passage of the homestead act and the recent supplemental legislation, which placed homesteaders on an equal footing with pre-emptors the special utility of the pre-emption law for bona-fide settlers has wholly ceased.

XXI. Lands Yet Open to Settlement.

THE bulk of the land yet open to settlement, is either mountain land, desert land, or the vast areas in the far Northwest, including Alaska, much of it inhospitable for cultivation. Alaska alone comprises about 370,000,000 acres. The tide of immigration is now setting into Dakota, Minnesota and the farther valley lands of the Northwest. Dakota alone contains 150,000 square miles (a square mile contains 640 acres), two-thirds of which is unsurveyed, and with much desert land and mountain land at the West. The word desert lands with us means land requiring irrigation, but often naturally producing grass. The Northwest mountain valley lands are also of vast extent. Government surveys are progressing at the rate of about 50,000,000 acres per annum, or nearly 1,000,000 acres per week. The total area of public lands surveyed in the several States and Territories from the commencement of surveying operations by the Government until the end of the fiscal year 1882 was 831,725,863 acres. The estimated area unsurveyed is about 983,000,000 acres, figures too vast to be appreciable except by comparison.

CHAPTER II.

LAWS RELATING TO AGRICULTURE.

I. NEEDED REFORMS IN FARM LAWS.—II. LAWS THAT EVERY FARMER SHOULD KNOW.—III. FISH AND GAME LAWS.—IV. GAME LAWS IN OLD AND NEW STATES.—V. LAWS RELATING TO DOGS.—VI. STOCK AND ESTRAY LAWS.—VII. STOCK LAWS OF THE NEW ENGLAND STATES.—VIII. STOCK LAWS OF THE MIDDLE STATES.—IX. STOCK LAWS OF THE SOUTHERN STATES.—X. STOCK LAWS OF THE WESTERN STATES.—XI. STATE LAWS RELATING TO FENCES.—XII. FENCE LAWS IN GENERAL.—XIII. FENCE LAWS IN NEW ENGLAND.—XIV. FENCE LAWS IN THE MIDDLE STATES.—XV. FENCE LAWS IN THE SOUTH.—XVI. FENCE LAWS IN THE WESTERN STATES.—XVII. FENCE LAWS OF THE PACIFIC SLOPE.

I. Needed Reforms in Farm Law.

THE laws relating to agriculture, taking agriculture in its broad sense, form no small part of the general statutes of a nation. The day is probably not far distant when the people will demand a simplification of our laws generally, by which unnecessary verbiage may be expunged: 1, That they may be simplified to conform to fundamental principles, so that any man of average comprehension may understand the nature of any particular law. 2, To do away with the practice of brow-beating, intimidation, and badgering of witnesses, by which they are made to say what they do not mean; and to simplify pleadings, by which facts only shall be kept in view. It is true that such is the general purpose of law, and under the rulings of the judge, much that we have mentioned as desirable may be accomplished; but cannot always be done even if the judge desires. So many abuses have crept into our courts that the covering up of facts, and special pleas of counsel on either side, often so befog a jury that, notwithstanding the charge of the judge, they often find it impossible to eliminate from their minds that which has speciously been instilled by the pleaders. The more carefully trained mind could not follow and retain speech after speech, each one of a week's duration, presenting the most diverse arguments for and against, and sift the true from the false.

II. Laws that Every Farmer Should Know.

IN relation to law, in its connection with some departments of rural affairs, we can only be expected to generalize. The laws relating to birds, game, stock, dogs, fences and roads will receive special attention. Much of this matter has been made easy to us through the labor of Hon. J. R. Dodge, for many years, and now, the Statistician of the Department of Agriculture at Washington. To bring the whole matter clearly together, we present them, by States, so far as we have been able to collect them, and under their separate and distinctive heads. It must be understood that these laws, like all others, are in many cases being changed from year to year. The general character of the law must suffice. What we give will enable the reader to get the general information required. Specific information must be sought from a competent attorney.

III. Fish and Game Laws.

Maine.—The penalty is one dollar for taking larks, robins, partridges, woodpeckers, or sparrows, between March first and July first; and ten dollars to the owner of lands, with the liquidation of all damage suffered for any trespass committed, between March first and September first, in hunting or killing the birds named.

New Hampshire.—The law prescribes a fine of one dollar for killing, taking, or having in possession, at any season of the year, any robin, thrush, lark, bluebird, oriole, sparrow, swallow, martin, woodpecker, bob-o-link, yellowbird, linnet, flycatcher; or warbler, or rail, yellowleg, or sandpiper, between March first and August first. The fine is three dollars for each snipe, woodcock, or plover, between March first and August first; or for each partridge, or grouse, or quail, between March first and September first. One dollar additional is assessed for each bird, if taken in defiance of a published notice by the owner of the land—one-half for the use of the complainant, and the other half to the town or city. The action of the law may be suspended for one year, at any time, by vote of a town or city, so far as relates to such town or city.

Vermont.—The law makes the fine one dollar in each case for taking, wounding, or killing, or for the destruction of the nest or eggs of the robin, blue-bird, yellowbird, cherry or cedar bird, catbird, kingbird, sparrow, lark, bob-o-link, thrush, chickadee, pewee, wren, warbler, woodpecker, martin, swallow, night-hawk, whippoorwill, groundbird, linnet, plover, phœbe, bunting, hummingbird, tattler, and creeper.

Massachusetts.—In this State the penalty is two dollars each for killing, at any time, robins, thrushes, linnets, sparrows, bluebirds, bob-o-links, yellowbirds, woodpeckers, or warblers; the same for killing birds on salt marshes, the owner excepted; five dollars for killing partridges or quail, between March first and September first; woodcock, between March first and July fourth; five dollars for trapping or snaring any birds at any time, save partridges; twenty dollars for killing grouse or heath hen at any time, and ten dollars to the owner of the grounds, and a search warrant authorized for any one suspected of the offense; and twenty dollars for hunting deer with hounds or dogs in Plymouth or Barnstable counties. There is a fine of one dollar for killing between sunset and one hour before sunrising, any plover, curlew, dough-bird or chicken bird. Any city or town may vote to suspend, within its limits, any of the provisions of this law.

Rhode Island.—In this State there is a penalty of two dollars in each case for killing, destroying, selling, buying, or having in possession any lark, robin, wood duck, gray duck, or black duck, between February first and September first, or quail, partridge, or woodcock between January first and September twentieth; snipe, between May first and September twentieth; grass plover, between February first and August first; grouse, or heath hen, between January first and November first, and swallow, or box martin, between May first and October first; twenty dollars in each case for killing woodcock between January first and July first. In addition, five dollars may be imposed, to be paid to the owner of the land, for the first offense, and ten dollars for the second offense, besides a liability to damage for trespass. Action must be brought within three months.

Connecticut.—The law in Connecticut provides a fine of three dollars for killing, selling, or possessing, or destroying a nest of eggs of woodcocks between the first day of February and the first day of July; pheasants, partridges, or ruffed grouse, between the first day of February and the first day of September; quails of any species, between the first day of February and the first day of October; wood duck, widgeon, black, gray, broad-bill, canvas-back or teal duck.

The fine is one dollar for killing, or trapping, a nightingale, bluebird, Baltimore oriole, finch, thrush, lark, sparrow, catbird, wren, martin, swallow, or woodpecker, at any time, or a robin or bob-o-link, between the first of February and the first of September.

The taking of brook or lake trout between September first and January first, is fined one dollar. It is also forbidden under a penalty of ten dollars, to take pheasants, partridges, or quails, on the land of any other person.

New York.—The laws relating to game have been frequently modified, and now are probably among the best, in a general way. Insectivorous and other birds are protected between February first and October. The fine is placed at five dollars for each wood-cock, between January first and July fourth; ruffed grouse, between January first and September first; quail, between January first and October twentieth; wood, black, gray and teal duck, between February first and August first (excepting upon the shores of Long Island). It is forbidden to catch quail or ruffed grouse with a snare at any time; and it is unlawful to take prairie fowl within ten years, under penalty of ten dollars for each one killed or taken.

Five dollars each is the penalty for taking trout between September first and March first. A penalty is incurred of one hundred dollars and damages for putting lime or drugs in any lake, pond or stream, by which fish may be injured. Owners of dams, if two feet or more in height, on the tributaries of Lake Ontario, Champlain, or the river St. Lawrence, are required to provide a sluice at an inclination of not more than thirty degrees, suitably constructed and protected, as a passage-way for fish.

Deer are prohibited game from February fifteenth to August first, in all counties except Clinton, Franklin, St. Lawrence, Jefferson, Lewis, Herkimer, Hamilton, Essex, Warren, Fulton and Saratoga, (where the prohibition is taken off only in October,) and in Kings, Queens, and Suffolk, where November is the only month for their pursuit.

For fishing, except with hook or line, in certain interior lakes, the fine is twenty-five dollars. A similar penalty attaches to trespass in fishing, after public notice has been given.

Pennsylvania.—In this State it is forbidden under penalty of two dollars, to trap, kill, or shoot any blue-bird, swallow, martin, or other insectivorous bird, at any season of the year, and the same penalty attaches to the destruction of eggs or nest of any of the birds mentioned in law. A fine of five dollars is laid for killing rail or reed birds between June first and September first; pheasant, between February first and August first; woodcock, between February first and July fourth; partridge or rabbit, between February first and October first, and a similar penalty is incurred by buying these birds out of season to sell out of the State.

New Jersey.—A fine of five dollars each is imposed for killing any partridge, water-fowl, grouse, quail, or rabbit, between January first and November first, or woodcock between January first and July fifth; to be recovered with cost of suit, and in default of payment imprisonment for sixty days may be adjudged.

A penalty of fifteen dollars is laid for placing decoys for geese, ducks or brant, at a distance of more than three rods from ice, marsh, meadow bank or sand-bar, or for hunting them with a light at night; and it is made unlawful to kill geese, ducks or brant, between April fifteenth and October fifteenth, in or about the waters of Barnegat bay or Manasquon river.

The fine is five dollars each for killing geese, ducks or brant, between April first and December first, at Cape May. A trespass, after having been once forbidden to enter lands, renders one liable to a fine of three dollars.

Delaware.—It is unlawful for non-residents to catch or kill any wild goose, duck, or other wild fowl, under a penalty of not less than fifty and not more than one hundred dollars. Citizens do not rest under this prohibition. The plan of procedure in prosecution is set forth so that any boat, gun, or decoy, used in violation of this law, may be seized and confiscated, and the penalty for resisting an officer is fixed at one hundred dollars.

The law does not prohibit persons from killing game on their own premises, but it is unlawful for others to kill a partridge, pheasant, robin, or rabbit, between February first and October fifteenth, (in Newcastle County, between January first and October fifteenth); woodcock, between February first and July first. The penalty is one dollar for each bird killed. A person not a citizen of the State, gunning upon land not his own, without permission of the owner, is liable to a fine of five dollars for each bird or other game. The penalty of hunting or killing deer is two dollars.

Ohio.—In this State the penalty is from two to ten dollars, for killing insectivorous birds, or disturbing their nests. The same penalty is incurred for killing, between February first and fifteenth, any dove, wild rabbit, or hare, yellow-hammer, or flicker. From five and to fifteen dollars may be imposed for killing wild turkey, quail, ruffed grouse, prairie chickens, or wild deer, between April fifteenth and September first; woodcock, between February first and July fourth, and wood duck, teal, or other wild duck, between May first and September fifteenth. Exposing for sale or having in possession, incurs the same penalties, and the costs of prosecution are in all cases to be paid by the offender.

Michigan.—The penalty for killing small birds in Michigan is fixed at five dollars each, and for wild turkey, partridge, or ruffed grouse, between February first and September first; for woodcock, between March first and July first; prairie chicken or wild duck, goose or swan, between February first and August fifteenth; for quail, from January first to October first. It is made unlawful to destroy nest or eggs. The fines go to the school library fund. Indians and the inhabitants of the upper peninsula are exempt from the effect of these provisions. Illinois has no general bird law. In a portion of the counties it is made unlawful to hunt or kill deer, turkey, grouse, prairie hen, or quail, between January fifteenth and August fifteenth.

Wisconsin.—Five dollars is the penalty for killing grouse or prairie chicken between December first and August twelfth; or partridge, ruffed grouse, or quail, between December first and the first Tuesday of September. It is unlawful to kill or take woodcock in Iowa between the first of January and first of July; prairie hen or chicken, between first of January and first of August; or quail, ruffed grouse, pheasant, wild turkey, or deer, between the first of January and first of September.

Minnesota.—In this State the penalty is five dollars each for killing at any time a whip-poor-will, nighthawk, bluebird, finch, thrush, lark, linnet, sparrow, wren, martin, swallow, bob-o-link, robin, turtle dove, catbird, or other birds; five dollars for each woodcock, from January first to July fourth; partridge or ruffed grouse, between January first and September first; ten dollars for trespass in sporting; twenty-five dollars for killing each deer, elk, or fawn, or having the skin of one in possession between January first and August first. A fine of five dollars is also imposed for each speckled trout taken, except in Lake Superior, Mississippi, Minnesota, St. Croix and Root rivers.

IV. Game Laws in Old and New States.

It will be seen that in the newer States the laws do not cover so wide a scope as in the older ones. In other words laws are not made until the necessity arises, first—for protecting insectivorous birds, then game birds, and lastly, fish and four-footed game. On the other hand, in comparatively unsettled regions, bounties are given for the destruction of wild beasts. Sooner or later, however, it becomes necessary to enact and enforce laws for the preservation of birds beneficial to the farmer, and a careful comparison of the laws of several States will enable farmers to suggest to legislators what animals should be protected and what destroyed.

California's Experience.—In California it is only within the last few years that anything has been done for the preservation of birds. The State began to be overrun with insects injurious to vegetation. Then the people began to move in relation to the protection of insectivorous birds; but some birds, at certain seasons of the year, when insects are scarce, will eat fruit.

Clamor Against Useful Birds.—The thrush family, including the mocking-bird, robin, etc., will eat fruit. Some ignorant fruit-growers clamor against them. Some ignorant farmers clamor against the crows and all that class of birds, because they pull corn; against owls and hawks, because, when they have exterminated mice and other vermin, they pick up a chicken once in a while; against the skunk for the same reason; against the prairie chicken and quail because they eat grain; against birds in general, forgetting that they pay ten-fold for all they eat, in the destruction of insects. Then there is a clamor for the legislature to do something, for entomologists, and for means for destroying insects, after themselves have caused the destruction of birds. Even the English sparrow, which is a general scavenger, and like the bob-o-link, likes green grain, in common with the sparrow tribe in general. Yet the English sparrow and others of his ilk, largely live in winter on the eggs of insects, when they can find them, thus nipping the insect evil " in the bud." The bob-o-link has been known to save a cotton field from destruction by insects, but he likes young rice.

V. Laws Relating to Dogs.

Most men like a dog. It is natural, for however worthless a dog may be in general, his master is his god, and if there is anything the average man likes, it is to be worshipped. Then again, dogs are clever to children, and therefore, children like them. Yet ninety-nine out of every hundred dogs are practically worthless—at least for the purpose for which dogs are generally kept on a farm—to watch property: First, for want of proper training, and second, from mixing up all known breeds together. If dogs, like other animals, were kept for a specific purpose, all difficulty might be avoided, and the sheep owner would not suffer from the depredations of worthless curs; the poultry-yard would not be decimated by bastard " fice," nor cattle worried and harried by the low-down yellow dog, with perhaps just enough bull-mastiff in him to intensify the savage propensities in these brutes, added to the sneaking disposition to run away in time to save their hides from the shot-gun; and this brings us to dog laws.

Massachusetts.—The law relating to dogs in Massachusetts is, in the main, one of the best in the country. There, dogs are taxed from two to five dollars each; owners are made responsible, under heavy penalty, for their registry and taxation; assessors must make accurate lists, and evasions of the listing are heavily fined; refusal or neglect of officers to execute the law incurs a penalty of one hundred dollars; and untaxed dogs are killed without mercy, and district attorneys are required to prosecute officers who neglect to destroy them.

Many of the older States, of late years, have so amended their dog laws that a very little agitation would cause them to be made perfect

Maine.—In this State the law is a good one, if carried out; but a saving clause, by which a township may nullify the law within the township, makes it worse than useless. Without the *saving clause* it would provide as follows: Dogs inflicting damage subject their owners to fines of double the amount of the damage done, to be recovered by an action of trespass. Any person may lawfully kill a dog that assaults himself or other person while walking or riding peaceably, or is found worrying, wounding, or killing any domestic animal. Any person finding a dog strolling out of the inclosure of his owner may, within forty-eight hours, make oath before a magistrate that he suspects such dog to be dangerous or mischievous, and notify the owner by giving him a copy of the oath; and if the dog shall be found again at large, he may be lawfully killed; and if he shall thereafter wound a person or kill a domestic animal, the owner shall be liable to treble damages and costs.

New Hampshire.—In New Hampshire the amended law of 1863 provides that double the amount of damage by dogs shall be recoverable from the owner by an action of debt; or a complaint may be made to the selectmen of towns, who are required, upon proof made within thirty days, to draw an order upon the treasury, which is registered and made payable, in whole or in part, from the fund accruing from the dog tax, on the second Tuesday of March annually.

Vermont.—The law of Vermont is good enough for those who own dogs, and since the owner is liable for damage done in any State under the organic law, it is good enough for sheep owners. The law is as follows:

The listers in several towns in this State shall in each year set all dogs in their respective towns in the grand lists to the owner or keeper of the same at the sum of one dollar each; and no person shall be entitled to have the amount so assessed deducted from their lists in consequence of any debts owing.

Every owner or keeper of a dog shall, when called upon by the listers for their lists, notify them of the dogs by him owned or kept; and every owner or keeper of a dog who shall neglect or refuse to notify the listers as aforesaid, shall forfeit and pay to the town in which he resides the sum of two dollars, to be recovered in an action on the case in the name of the treasurer of such town, before any court competent to try the same, with full costs.

It is hereby made the duty of the owner or keeper of a dog, whether set in the lists or not, to cause a collar, with the name of the owner or keeper plainly written thereon, to be worn on the neck of each dog by him owned or kept; and it shall be lawful for any person to kill any dog running at large off the premises of the owner or keeper not having on such collar; and the owner or keeper of such dog shall recover no damage for such killing.

Rhode Island.—By the law of 1860, a dog might be killed with impunity if found without a collar bearing his owner's initials, or worrying or wounding sheep or other stock out of the inclosure of his owner. Any person might make oath to any case of injury, or to the special illfame of any particular whelp, and if the allegation was sustained, the dog must be confined, or the life of the animal was forfeited. A late amendment requires dogs to be collared, registered, numbered, described and licensed, with the payment of one dollar and fifteen cents for each male, and five dollars and fifteen cents for each female dog, before the last day of April, and one dollar additional for each dog after that date, and previous to the first of June. Any person keeping a dog contrary to the provisions of the law is liable to a fine of ten dollars.

New York.—The laws, as amended by that of 1862, impose a tax of fifty cents for the first dog, two dollars for each additional; three dollars for the first female dog, and five dollars for each additional. The assessors are required to annex to the assessment roll the names of persons liable, and supervisors must return them, when, if failure in paying the tax occurs, it becomes the duty of the collector, and the privilege of any other man, to kill the dog. The collector has a commission of ten per cent on fines, and one dollar for each dog killed. The previous enactment provided that the owner of dogs killing sheep should be liable for injuries perpetrated; and in case the owner should not be found, the loss should be paid out of the fund arising from the dog tax.

Delaware.—This State has had carefully-considered dog laws since 1811. The old laws are repealed, but the better features are contained in the new law. The owner of a dog which shall kill, wound, or worry a sheep or lamb, shall be liable to pay the owner of such sheep or lamb, the full value thereof, and it shall be lawful for any person to kill such dog. It shall be lawful for any person to kill any dog running at large in Newcastle county, beyond the owner's premises, without a collar upon his neck with the owner's name upon it. The law of 1862 requires an assessment list, of persons owning dogs, to be returned to the levy court. The tax is placed at fifty cents for each male, and one dollar for each additional dog, and two dollars for each female dog, which shall procure the fund from which damages shall be paid, not to exceed three dollars for each lamb, and five dollars for each sheep injured or killed, the remainder, if any, to go into the school fund. A dog not on the assessment list, which may be wandering, or caught worrying sheep, may be killed. Persons paying taxes upon dogs are deemed to have property therein, and may recover damages for theft of or injury to such dogs.

Pennsylvania.—The laws of this State seem to have had for their object, to hold the dogs to good behavior. Dogs may kill sheep, but not the second time; but the owner is liable, if he knows the dog has killed sheep. It is to be hoped that the State has either expunged the law altogether or sensibly amended it.

Ohio.—The solons of this State, unless the law has been amended, simply contented themselves with politely informing dogs that they should not run at large at night. The owners were held to no proper responsibility for their restraint. It is to be hoped that this great agricultural State will move in the proper direction. We, however, have not seen anything indicating that a general law has been enacted for proper protection against dogs. Yet, we suppose the farmers are still left their natural protection against marauding dogs—the shot-gun.

Indiana.—A license is required, at fifty cents for the first male dog, one dollar for each additional dog, and one dollar in every case for a female dog. All unlicensed dogs are declared nuisances, and may be lawfully killed. Accruing funds are set apart for the payment of damages suffered from injuries to sheep in the several townships.

The sufferer has his option of the following remedies: Within ten days after having knowledge of such depredations, he may substantiate it to the satisfaction of the township trustee, and draw the amount at the end of the current year, or a *pro rata* proportion if the fund is deficient; or he may recover, by suit, full damages from the owner of the dog. A fine of from five to fifty dollars and liability to damages, recoverable by the owner, are the penalties for killing licensed dogs that maintain a fair canine character.

Michigan.—In this State, since 1850, the law authorizes the destruction of dogs attacking any kind of domestic animals, except on the premises of the owner of the dog, and such owner is liable for double the amount of damage done by the dog. When notified of such damage, neglect of the owner to kill the dog is punishable by a fine of three dollars, and one dollar and fifty cents additional for every forty-eight hours thereafter, until such dog shall be killed. Supervisors, upon complaint of a citizen, verified by his oath, are required to prosecute and recover the fines imposed by this act. An act was passed March twentieth, 1863, requiring township assessors to ascertain the number of dogs liable to be taxed, and the names of their owners; and if such owners refuse for ten days after demand to pay the taxes assessed, it becomes lawful to kill the dogs so taxed.

Wisconsin.—By the law of 1860, dogs are required to be numbered, collared, registered and licensed on payment of one dollar for males and three dollars for females; and police officers, constables and marshals are required to kill and bury all unregistered dogs, and to receive twenty-five cents for such service. A person may be fined fifty dollars for removing a collar. Persons suffering loss from dogs are paid full damages at the first of April, if the tax fund is sufficient; if not, pro rata; and the owner of the dog is liable to the town for the full amount. The fine for keeping unregistered dogs is five dollars. Officers neglecting or refusing to obey the law, are fined twenty dollars for every twenty-four hours of such neglect. Towns may increase the license not more than one dollar, and the penalty not more than ten.

Minnesota.—This State early recognizing the fact that dogs should be held as amenable to the law as their masters, or rather that the masters should be so for them, enacted a law, in substance as follows: Every owner or keeper of a dog shall cause such dog to be registered, numbered, described, and licensed, paying one dollar for each male and two dollars for each female. The township or city clerk shall conspicuously post a list of all licensed dogs, and furnish one to constables and chief of police. Failure to license shall make one liable to a penalty of ten dollars. Stealing or poisoning a dog is punishable by a fine not exceeding fifty dollars, and killing subjects to a liability for damages double the value of the dog. Constables and police officers shall and any person may, kill any unlicensed dog; any one may also kill a dog assaulting him, or worrying sheep out of the inclosure of his owner. Within thirty days after suffering injury or loss of sheep by dogs, proof of damages may be presented to the county auditor, who may draw an order upon the treasurer, payable from the fund accruing from taxes of dogs, when the city or town may sue and recover full damages from the owner of the dog. It is made the duty of the mayor and aldermen of cities, and the supervisors of towns, to require the destruction of unlicensed dogs, and officers refusing or neglecting to perform these duties are liable to a fine of twenty-five dollars for the benefit of schools. All of these penalties may be recovered, on complaint by any householder, before any justice of the peace of the county. Money remaining after the yearly payments from the tax fund is turned over to the school fund.

VI. Stock and Estray Laws.

ALMOST every State has stock laws of some kind, differing principally in relation to stock running at large. In some States, the law leaves it optional with counties or districts, to decide whether fences shall or shall not be maintained. If not, stock of all kinds must be kept on the owner's premises.

Estrays.—In some cases, there is only fencing against cattle and horses; or, sheep and swine must be kept close. So far as laws in relation to estrays are concerned, they do not differ in essential respects. If an animal is found running at large, in violation of law, it may be taken up and impounded, where public pounds have been provided; or it may be held by the person taking up, on his own premises. If the owner is known, notice must be given him at once; if unknown, the animal must be advertised for a specified time; and no owner making claim, it must be sold to the highest bidder. The person taking up an estray is entitled to a reasonable compensation for maintaining the beast. In some States, after a certain time, the estray becomes the property of the person taking it up, the prescribed legal notice having been given. When an animal is found doing damage on the land of another, the fences being constructed according to law, it may be held as security for damages. In all cases where the owner is known, he must be notified of the facts, and a reasonable time allowed him to reclaim and to inspect damages.

VII. Stock Laws of the New England States.

Maine and New Hampshire.—In Maine and New Hampshire, towns may make by-laws concerning the running of animals at large. The laws of Maine provide that persons injured by beasts may sue for damages, and distrain the animal. New Hampshire allows the owner of stock impounded for doing damage, four days to respond to notice of the fact; and if he fails to answer, the animals may be sold and the amount of the damage be deducted from the proceeds.

Vermont.—In Vermont, twenty days are allowed for redemption. Ungelded animals are not allowed to run at large. Rams must be restrained from August first to December first, and be marked with the initials of the owner's name; and if found at large, a forfeit of five dollars is due for each one taken up, to the person so taking up. The owner of such animals is responsible for all damage done by them. Sheep infected with foot-rot or scab, must be diligently restrained, and for all damages resulting from neglect of this provision, the owner is responsible, and is also subject to a fine of ten dollars. Any person finding such diseased animals at large, may take them as forfeit, and no action at law, or in equity will lie for their recovery. Any person who shall drive, or in any manner bring into the State, any neat cattle, knowing them, or any of them, to have the pleuro-pneumonia, or to have been exposed to that disease, is liable to a forfeit of a sum not over five hundred dollars, or to imprisonment in a county jail for not more than twelve months, nor less than one month. Towns may establish regulations, appoint officers or agents, and raise and appropriate money for the purpose of preventing and arresting the spread of pleuro-pneumonia.

Massachusetts.—In this State it is provided that when a person is injured in his crops or other property by sheep, swine, horses, mules, or neat cattle, he may recover damages in an action of tort, against the owner of the beasts, or by distraining the beasts doing the damage; but if it be found that the beasts were lawfully on the adjoining lands, and escaped therefrom in consequence of the neglect of the person who suffered the damage to maintain his part of the division fence, the owner of the beasts shall not be liable for such damages.

The laws of this state in relation to pleuro-pneumonia are very strict, and are made more and more so from time to time. The selectmen of towns and the mayor and aldermen of cities, in case of the existence of pleuro-pneumonia or any other contagious disease among cattle, shall cause the infected animals, or those exposed to infection, to be secured in some suitable place or places, and kept isolated, the expense of keeping to be paid, one-fifth by city or town, and four-fifths by the State. They may prohibit the departure of cattle from any enclosure, or exclude them therefrom; may make rules in writing to regulate or prohibit the passage of any neat cattle to, or through their respective cities or towns, or from place to place, and arrest and detain them at the cost of the owners. They are authorized to brand infected animals, or those exposed to infection, with the letter P on the rump. For selling an animal so branded, there is liability to fine not exceeding $500, or imprisonment not exceeding one year. Notice of any suspicion of the existence of contagious disease must be given, with a penalty for neglect or refusal. A board of commissioners is appointed for the State, with authority to use any measure to control the introduction of diseased cattle into the State, or the spread of the disease. The rules and regulations made by this board supersede those of the selectmen of towns, and mayor and aldermen of cities. The moving of cattle into other States without permission is prohibited.

The law of 1867 provides that no cattle diseased, or suspected of being diseased, shall be killed, except by order of the governor. The owners of cattle ordered to be killed are indemnified.

Rhode Island.—In Rhode Island, animals trespassing on lands are held a year and a day; and if a horse, must have a withe kept about his neck during that time. Each town is required to erect and maintain at its own charge one or more public pounds, and it is lawful for any freeholder or qualified elector or field driver, and it is made the duty of every surveyor of highways, to take up and impound any horse, neat cattle, sheep, or hogs, found at large on any highway or common. Provisions of the act extend also to goats and geese. In 1860, in view of the dangerous disease which had become prevalent in other States, the general assembly enacted that neat cattle might only be brought into the State from places west of the Connecticut river, upon thoroughfares leading into the western and southern portions of the State, under regulations established by a board of commissioners, until they should prohibit importations from any of said places. For a violation of the provisions of the act, penalty was provided, not exceeding three hundred dollars for each offense, and liability to indictment, and, on conviction, imprisonment not exceeding one year. In case of the introduction of a number of diseased cattle at the same time, the introduction of each animal is to be deemed a separate and distinct offense.

Town councils are empowered to take all necessary measures to prevent the breaking out or spreading of any infectious diseases among the neat cattle in their respective towns, and to prescribe penalties in money, not exceeding five hundred dollars. A board of commissioners is provided for, to be appointed by the governor, consisting of one person from each county, to see that the law is faithfully executed. It is made the especial duty of the board to endeavor to obtain full information in relation to the diseases known as pleuro-pneumonia, and to publish and circulate the same, at their discretion; and in case the disease should break out, or there should be a reasonable suspicion of its existence in any town, they are required to examine the several cases and publish the result of their examination, in order that the public may have correct information. If satisfied of its existence in any town, they must give public notice of the fact in printed handbills, posted up; and, thereafter, any incorporated company or person who may drive, carry, or transport any neat cattle out of the town into any other town in the State, is liable to the penalties above stated. Any person who sells or offers to sell any cattle known to be infected with pleuro-pneumonia, or with any disease dangerous to public health, is liable to indictment, and, on conviction, to punishment by fine not exceeding one thousand dollars, or imprisonment not exceeding two years. The act of March twenty-sixth, 1864, provides that any person knowingly bringing into the State any neat cattle or other animal suffering from any infectious disease, or who knowingly exposes such cattle or other animal to other cattle and animals not infected with such disease, shall upon conviction pay a fine of not less than one hundred dollars, and not exceeding five hundred dollars.

Connecticut.—In this State, the law allows owners of sheep to keep flocks in common, and to make their own rules and regulations concerning their care and safety. No horses, asses, mules, neat cattle, sheep, swine or geese are allowed to go at large in any highway or common, or to roam at large for the purpose of being kept or pastured on the highway or commons, either with or without a keeper. Any person may seize and take into his custody and possession any animal which may be trespassing upon his premises, provided the animal enter from the highway, or through a fence belonging to the owner of the animal, or through a lawful fence belonging to any other person. He must give immediate notice to the owner, if known, and may demand for every horse, mule, ass, ox, cow or calf, twenty-five cents; and for every sheep, goat, goose or swine, ten cents; together with just damages for injuries occasioned by such animals, if applied for within twenty-four hours after such notice shall have been given. If the owner is not known, the animal shall be sold by the town clerk, after due public notice.

VIII. Stock Laws of the Middle States.

New York.—The laws of all the States east of the Alleghanies have, of late years, been very strict and carefully drawn in relation to animals liable to infection with contagious diseases, and are framed generally after those of Massachusetts, where the first case of pleura-pneumonia occurred. In relation to the general laws for stock, the laws of New York allow any person to seize and take into his custody any animal which may be in any public highway, and opposite to land owned or occupied by him, or which may be trespassing upon his premises. Notice must be given to a justice of the peace, or a commissioner of highways of the town in which the seizure has been made, who shall post up notices in six public places, that the animal will be sold in not less than fifteen nor more than thirty days. The surplus money, after payment of all charges, is subject to the order of the owner for one year. The owner, before sale, may pay all charges and take the animal. If the animal has been trespassing by the willful act of another than the owner, to effect that object, the owner is entitled to the animal upon making demand, after paying the compensation fixed by the justice or commissioner, but no other costs; and the person committing such willful act will be held liable to a penalty of twenty dollars.

New Jersey.—Town committees, upon notice of the existence of any disease supposed to be contagious, are required personally to examine the cause, and if the symptoms which characterize contagious diseases are exhibited, shall cause such animals to be removed and kept separate and apart from other cattle and stock, five hundred feet distant from any highway, and the same distance from any and all neighbors. If any die of the disease, or are killed, they must be buried immediately, five hundred feet distant, etc., as above. No cattle that have been sick, and have recovered from any supposed contagious or infectious disease, shall mix with other cattle, or be removed, unless permission has been given by the town committee. Any person knowingly storing a hide, or any other portion of a diseased animal, is subject to a fine. The town committee are authorized to prohibit the importation or passage of cattle from other places into or through their respective towns. After notice of prohibition, owners are liable to a fine of one hundred dollars for every animal driven into a township. A fine of one hundred dollars is imposed for every animal sold and known to be diseased. The act of 1866 authorizes the Agricultural Society of the State to take measures for preventing the introduction or increase of rinderpest, and any other disease among cattle, at their discretion; animals affected with glanders are authorized to be killed. Cattle must not be marked by cropping both ears; nor must either ear be cropped more than one inch.

Pennsylvania.—The running of cattle at large is controlled in Pennsylvania by towns and counties, through special legislation. The sale of cattle or sheep affected with pleuro-pneumonia, or any other contagious or infectious disease, is punished by a fine not exceeding $500, or imprisonment not exceeding six months. Animals must not be sold alive from, or slaughtered on, premises where disease is known to exist, nor for a period of two months after disease shall have disappeared from the premises. Cattle and sheep are not allowed to run at large where any contagious disease prevails. Constables of townships are required to take up and confine any animals so found, until all costs are paid.

Delaware.—In Delaware cattle are forbidden to run at large in certain districts. Stallions over eighteen months old are not permitted to be at large.

IX. Stock Laws of the Southern States.

Maryland.—In this State it is provided that any person aggrieved by trespass upon his premises of any cattle, hogs or sheep in the possession or care of a non-resident, may impound them, and have the damages sustained by the trespass, valued on oath by two disinterested citizens of his county, and the animals may be sold for the damages and costs.

Virginia.—In this State, if any horses, cattle, hogs, sheep or goats enter into any grounds inclosed by a lawful fence, the owner or manager shall be liable to the owner of the ground for all damages; and for every succeeding trespass by such animals the owner shall be liable for double damages; and, after having given at least five days' notice to the owner of the animals of the fact of two previous trespasses, the aggrieved party shall be entitled to the animals if again found trespassing on the same lands. Horses diseased and unaltered, are not allowed to be at large. Every person shall so restrain his distempered cattle, or such as are under his care, that they may not go at large off the land to which they belong; and no person shall drive any distempered cattle into or through the State, or from one part of it to another, unless it be to remove them from one piece of ground to another of the same owner; and when any such cattle die, the owner thereof, or person having them in charge, shall cause them to be buried (with their hides on) four feet deep. Any justice, upon proof before him that any cattle are going at large, or are driven in or through his county or corporation, in violation of law, may direct the owner to impound them; and if he fail to do so, or suffer them to escape before obtaining a certificate that they may be removed with safety, they shall, by order of the justice, be killed and buried four feet deep, with their hides on, but so cut that no one may be tempted to dig them up. For the protection of sheep, special laws have been passed, taxing dogs in certain counties, and for their restraint in those counties.

The Carolinas.—In North Carolina, if cattle are driven from one part of the State to another, they must be certified to be healthy, sound, and free from any infectious distemper; the granting of such certificate by any justice, without affidavit, is a misdemeanor in office. Stallions and mules over two years old are not allowed to go at large, under a penalty of twenty dollars. Damages for injury done by trespassing animals are recoverable as in other States.

In South Carolina, horses, cattle, hogs, sheep or goats breaking into any field having a crop of any kind growing or ungathered, with a lawful fence, may be seized and kept confined until notice is given to the owner, within twenty-four hours of the seizure, who shall be bound to pay the owner of such field fifty cents a head for each horse or mule, and twenty-five cents for every head of cattle, hogs, etc., before he is entitled to have the animal delivered up to him. For the second breaking, within one month after the first, the owner is liable to the person injured for all damages sustained, in addition to the fine. Full satisfaction lies for injuring any animal found in any field where the fence is not a lawful one.

Georgia.—If any trespass or damage is committed by stock in the State of Georgia on any lands not protected by lawful fences, the owner of the animal is not liable to answer for trespass; and if the owner of the premises should kill or injure the animal in any manner, he is liable in three times the damages. When fences are made pursuant to law, and any animal breaks in, the owner of the inclosure shall not kill or injure him for the first breaking, and not until after notice is given to the agent or owner, if possible, but the owner shall be liable for double the damage done by his stock.

Florida.—In many Southern States what constitutes a lawful fence is stated with the utmost minuteness, as to height, spaces, etc. In Florida there can be no trespass or damage if the fence is not a lawful one; nor in such case can stakes, canes, or other devices to maim or kill cattle, sheep, swine, etc., be used, under a penalty of ten dollars for each offense, and full damages. Marks upon stock are required.

Alabama.—Any person is allowed to take up any horse, mare, jack, neat cattle, hog or sheep found running at large, if the owner is unknown. If any stallion or jackass over two years of age, is found at large it must be taken before a justice, who shall cause it to be advertised. The taker-up is entitled to five dollars from the owner, and reasonable compensation for keeping. If such stallion or jackass is not claimed within three months it may be gelded.

Mississippi.—In this State it is provided that every owner of cattle, horses, mules, hogs, sheep or goats shall be liable for all injuries and trespasses committed by breaking into grounds inclosed by legal fence. If any person, whose fence is not a lawful one, shall hurt, wound, lame or kill, by shooting or hunting with dogs, or otherwise, any cattle, etc., that may have broken into his inclosure, he shall pay the owner double damages. A ranger is elected in each county to attend specially to estrays, of which he is required to keep a record. When any person finds horses, mules, jacks, cattle, sheep or hogs straying upon his land, he may take them up and forthwith send them to the owner, if known; if unknown, he must give notice to the ranger, or some justice of the peace. The owner of all estrays appraised at ten dollars and not exceeding twenty dollars, is allowed six months, and if less than ten dollars, three months, from the date of certificate of appraisement, to claim and prove his property. It is not lawful for any drover or other person to drive any horses, mules, cattle, hogs or sheep of another from the range to which they belong; but it is made his duty if any such stock join his, to halt immediately at the nearest pen, or some other convenient place, and separate such stock as does not belong to him, or to the person by whom he may be employed. For neglect, a forfeit of twenty dollars for every offense is provided, and liability to all damages. Any person may confine and geld any stallions that are above the age of two years, found running at large, at the risk of the owner, but this will not apply to stallions usually kept up, or to those which accidentally escape. Any animal addicted to fence-breaking may be taken up by owner of land, who may recover seventy-five cents a day for keeping, provided owner has been notified, if known; but condition of fence may be shown in mitigation of damages. Double damages may be recovered for injury to animals where fence is not a lawful one. Defacing or altering marks of animals subjects to a penalty of imprisonment in the penitentiary for not more than three years, or fine of not more than $500, and imprisonment in county jail for not more than one year, or both.

Texas.—No neat cattle belonging to non-residents are allowed to be taken into Texas for grazing or herding purposes, under pain of forfeiture to the county into which they shall have been so taken. Severe penalties for altering the brands of animals are provided in that State.

Arkansas.—If any horse, cattle, or other stock break into any inclosure, the fence being of the required height and sufficiency, the owner of the animal shall, for the first offense make reparation for true damages; for the second offense, double damages; and for the third the party injured may kill the trespassing beasts, without being answerable. If any stallion or jack over two years old is found running at large, the owner may be fined two dollars for the first offense, and ten dollars for each subsequent offense, and is liable for all damages that may be sustained. Any person may take up such animal, and, if not claimed within two days, may castrate, and recover three dollars for doing so; but the life of the animal must not be endangered. If any such animal cannot be taken up, he may be killed, if notice be first put up at the court-house, and at three other of the most public places in the county for ten days, accurately describing the animals. In Tennessee, stallions and asses over fifteen months old are not allowed to run at large under penalty to the owner of not less than five dollars, or more than twenty-five dollars. The animal may be taken before the nearest justice of the peace, who shall give public notice. If not claimed within three months the animal may be gelded at the risk and expense of the owner. The party taking him up is entitled to five dollars and reasonable expenses for keeping.

Kentucky.—Breachy and mischievous bulls may be taken up and altered; a jack or stallion may be gelded if found at large, allowing the owner, if known, at the rate of twenty-five miles a day to reach the place where the animal is held, and recover the animal; when the owner is not known, the animal is dealt with as an estray, and may be ordered by a justice to be gelded. If the owner of any distempered cattle permits them to run at large, or drives them through any part of the State, he is liable to a fine of ten dollars for each head; and if any die the owner must cause them to be buried, subject to a penalty of five dollars for neglect in each case.

X. Stock Laws of the Western States.

Of late years more or less has been done in the several States west to prevent the introduction of animals with infectious, and especially contagious diseases, since the outbreak of pleuro-pneumonia in the Eastern States. Illinois has a State veterinarian whose duty it is to prevent any diseased cattle entering the State, and the State veterinarian has now a general jurisdiction over diseased stock since the outbreak of glanders, and may condemn and cause to be killed horses found with this horrible disease. Congress has passed a general quarantine law for imported live stock, and the Cattle Commission of the United States have the supervision of them.

West Virginia.—There is no law in force in West Virginia to prevent cattle from running at large; but if they break into an inclosure and destroy any grain or crops, the owner is liable; provided the fence is a lawful one. A law exists to prevent diseased sheep from traveling on the highway.

Ohio.—It is unlawful in the State of Ohio for any one to sell, barter or dispose of, or permit to run at large, any horse, cattle, sheep, or other domestic animal, knowing them to be infected with contagious or infectious disease, or to have them indirectly exposed thereto, unless he first duly informs the party to whom he may sell as to the facts. The fine for so doing is not less than twenty dollars nor more than two hundred dollars, with costs, or confinement in the county jail not more than thirty days. For allowing infected animals to come in contact with animals belonging to another, a fine is provided of not less than fifty dollars nor more than five hundred dollars, with costs of prosecution, or confinement in the county jail not less than ten nor more than fifty days. If any horse, mule, ass, or any neat cattle, hogs, sheep, or goats, running at large, break into or enter an inclosure other than inclosures of railroads, the owner is liable for all damages, and the animal so breaking into or entering an inclosure is not exempted from execution issued on any judgment or decree rendered by any court. For allowing any such animal to run at large in any public highway, or upon any uninclosed land, or for herding any of them for the purpose of grazing on premises other than those owned or occupied by the owner or keeper of the animals, the party offending is liable, for every violation, to a fine of not less than one dollar nor more than five dollars. But a general permission may be granted by the commissioners of any county for certain animals to run at large, and in counties where there is no such general permission, township trustees may grant special permits, such general and special permits terminating on the first Monday of March of each year; and special permits are revokable at the discretion of the trustees, upon three days' notice in writing to the owner of animals. Special permits must be directed to individuals, and for particular animals described therein. The owner of trespassing animals is liable for all damages upon premises of another without reference to the fence which may inclose the premises. Any person may take up and confine an animal found at large contrary to law, and the owner may reclaim the same within ten days. The fees are as follows: For taking up and advertising each horse or mule, one dollar; neat cattle, seventy-five cents each; swine, fifty cents each; sheep or geese, twenty-five cents each; and reasonable pay for keeping the same. It is unlawful for the owner or keepers of any animals knowingly to permit them to enter the inclosure of any railroad, or having entered, to remain therein; or to lead or drive any such animals within the inclosure, or along or upon the track of any railroad, at any other place than a regular street, road or farm crossing or way.

Illinois.—In Illinois, as in a number of other Western States, counties or townships may define by vote whether they will have fences or not, and how much. The owner of animals breaking through a legal fence is liable to full damages for the first trespass, and to double damages for any subsequent trespass. Where the fence is insufficient, and the land owner injures or destroys animals, he is answerable in damages. Stallions over one year old are not permitted to run at large; but if so found may be gelded, if the owner does not reclaim them, one day for every fifteen miles' distance of the animal from home being allowed, after notice. Diseased horses, mules and asses must be kept within the owner's inclosure, under penalty of twenty dollars damages. Estray hogs must be sold from November first to March first. To convey any Texas or Cherokee cattle into the State between the first day of October and the first day of March renders the party so doing liable to a fine not exceeding two thousand dollars nor less than five hundred dollars, and imprisonment at the discretion of the court. Any and all fines are paid into the county treasury, subject to the order of the board of supervisors, or county court, for the purpose of being divided pro rata among persons who may have suffered damage or loss on account of any such Texas or Cherokee cattle. All persons or corporations are liable to injured parties for any damage arising from the introduction, by any of them, of any diseased cattle. It is made the duty of any circuit or county judge, or justice of the peace, upon oath of any householder, setting forth that Texas or Cherokee cattle are spreading disease among the native cattle, to forthwith issue a warrant to any sheriff or constable of the county, commanding him to arrest and impound such cattle, and keep them by themselves until the first day of October following. Texas and Cherokee cattle are defined to mean a class or kind of cattle, without reference to the place from which they may have come. In Indiana, laws regulating the running at large of stock are local in their application, county boards designating what animals may or may not run at large. However, when any animal is found at large contrary to local law, and has been taken up, the owner may reclaim it within ten days, after which time the animal may be sold. The laws of the State are in effect prohibitory against bringing in diseased cattle, and it is the duty of the State veterinarian to see that the law is enforced.

Missouri.—The State of Missouri has created a board of cattle inspectors to prevent the spread of the Texas or Spanish fever. The county court of each county is authorized to appoint three competent and discreet persons to act as a board for the inspection of cattle supposed to be distempered or affected with the disease known as the Texas or Spanish fever. They may stop any drove of cattle. If they adjudge cattle to be diseased or distempered, and in a condition to communicate any contagious or infectious disease, they are required to order the cattle to be removed from the county without delay, upon the same route upon which they came in, if practicable. If the owners comply with the order, they will not be further liable; but if they or the persons having the cattle in charge, willfully delay or neglect to do so, the president of the board will direct the sheriff to drive the cattle out by the route they came in, or to kill them, if the board think it necessary in order to prevent the spread of the disease. The parties owning, or in charge of the cattle ordered to be removed or killed, are liable for all the costs that may accrue in case of examination, removal or killing. The act to prevent the introduction of diseased cattle into the State provides that no Texas, Mexican or Indian cattle shall be driven or otherwise conveyed into any county in the State between the first day of March and the first day of December in each year, but this does not apply to any cattle which have been kept the entire previous winter in the State. Cattle may be carried through the State by railroad or steamboat, provided they are not unloaded, but the railroad companies or owners of the steamboat are responsible for all damages which may result from the Spanish or Texas fever, should the same occur along the line of transportation; and the existence of such disease along the route shall be prima facie evidence that the disease has been communicated by such transportation. For every head of cattle brought into the State contrary to law, a fine of twenty dollars may be recovered, or the party may be imprisoned in the county jail not less than three nor more than twelve months, or may be subjected to both fine and imprisonment. It is lawful for any three or more householders to stop any cattle which they may have good reason to believe are passing through any county in violation of the act.

Michigan.—In Michigan it is not lawful for any cattle, horses, sheep or swine to run at large on the highway, except in those counties or parts of counties where it shall be otherwise determined by the board of supervisors in such county. Where the law is in force, any person may seize and hold in his possesion any animal found running at large, and give notice to a justice of the peace or a commissioner of highways, who is required to post up notices describing the animal. The animal must be sold at public outcry in not less than thirty nor more than sixty days after date of notice; but the owner may redeem the animal by paying costs and compensation for keeping—redemption to be made within one year. An animal found trespassing by the willful act of another, may be taken by the owner on demand, after paying reasonable compensation, but the person committing the act is liable to a fine of twenty dollars. Any person taking up a beast going at large contrary to law, or contrary to any by-law of a township, is entitled to fifty cents per head for all horses, mules, asses and neat cattle, and ten cents per head for all sheep, goats and swine. When any person is injured in his land by animals, he may recover damages in an action for trespass against the owner of the beasts, or by distraining the beasts doing damage, unless the animals shall have been lawfully on adjoining lands, and shall have escaped therefrom in consequence of the neglect of the person who has suffered the damage to maintain his part of the division fence.

Wisconsin.—The laws of Wisconsin permit towns to make regulations concerning the running of animals at large. The owner or occupant of lands may distrain all beasts doing damage within his inclosure, and when any distress shall be made, the person distraining is required to keep such beasts in some place other than the public pound until his damages are appraised; and within twenty-four hours he shall apply to a justice of the peace, who shall appoint three disinterested free-holders to appraise the damage sustained. If within twenty-four hours after appraisement the damages are not paid, the animals may be placed in the public pound, to be there maintained until the amount of damages and costs is recovered by due process of law. If the owner of any sheep infected with contagious disease, permits any of them to go at large out of his own inclosure at any season of the year, he shall forfeit the sum of five dollars for each and every such sheep, to the person who may enter complaint, for each time they are so found running at large. If the owner neglects to restrain such sheep, any person is authorized to take them up and put them in some safe place other than the public pound. Rams are not permitted to go at large between July fifteenth and December first, and the owner forfeits ten dollars to the person taking up the animal for each time so found abroad.

Minnesota.—The electors of each town in the State of Minnesota have power at their annual meetings to determine the number of pound-masters, and the location of pounds, and regulations for impounding animals, and to fix the time and manner in which cattle, mules, asses and sheep may be permitted to go at large, provided that no cattle, horses, mules nor asses be allowed to go at large between the fifteenth day of October and the first day of April. The owner or occupant of lands may distrain all beasts doing damage upon his lands during the night-time, from eight o'clock in the evening until sunrise; and when any such distress is made the distrainer shall keep such beasts in some secure place other than the public pound, until his damages are appraised, unless the same is made on Sunday, in which case, before the next Tuesday morning thereafter he shall apply to a justice of the peace of the town, who shall appoint three disinterested persons to appraise damages. No damage can be recovered by the owner of any lands for damage committed by any beasts during the day-time, until it is first proved that the lands were inclosed by a lawful fence. Distress may be made at any time before the beasts doing damage escape from the lands, and without regard to the sufficiency of fences. The owner of any horse or other animal, having the disease known as the glanders, who knowingly permits such animal to run at large, or be driven upon any of the highways of the State, or any hotel-keeper, or keeper of any public barn, who permits any such animal having such disease to be stabled, such person shall be deemed guilty of a misdemeanor, and upon conviction before any justice of the peace, shall be punished by a fine of not more than one hundred nor less than twenty-five dollars.

Iowa.—In Iowa, no stallion, jack, bull, boar or buck is permitted to run at large. Persons aggrieved are allowed to distrain any such animals and compel the owner to pay damages. If the animal is not redeemed within seven days, seven days' notice of its sale at public auction must be given, the proceeds to apply on damages after deducting costs. If any domestic animal, lawfully on adjoining land, escapes therefrom in consequence of the neglect of the person suffering the damage to maintain his part of the division fence, the owner of the animal is not liable for any damages. If beasts are not lawfully upon the adjoining land, and came upon it, or if they escaped therefrom into the injured inclosure, in consequence of the neglect of the adjoining owner to maintain a partition fence or any part of one, which it was his duty to maintain, then the owner of the adjoining land shall be liable as well as the owner of beasts. Fence-viewers appraise all damages. An act of April eighth, 1868, forbids any one to bring into the State, or to have in possession any Texas, Cherokee or Indian cattle. Transportation on railroads through the State is not forbidden, nor the driving through any part of the State of such Texas or southern cattle as have been wintered at least one winter north of the southern boundary of the State of Missouri or Kansas. The penalty of violation is a fine not exceeding $1,000, or imprisonment in county jail at the discretion of the court, not to exceed six months, together with all damages that may accrue by reason of such violation of the law. Any one driving or importing diseased sheep into the State, knowing the disease to be contagious, is deemed guilty of misdemeanor, and is punishable by fine of not less than fifty dollars nor more than one hundred dollars. The same fine is imposed upon any person who may turn out of his inclosure, or sell sheep, knowing them to be diseased.

Kansas.—In Kansas, when a majority of the electors in any township petition county commissioners for orders to confine animals during the night-time, such orders shall be made and notice thereof given. The owner is liable for depredations of animals during the continuance of such orders, without regard to condition of fences. Persons damaged in their property have a lien upon the stock. If any stallion or jack over the age of two years is found at large, the owner, if known, must be notified of the fact; and if he fails or refuses to confine the animal, he is liable to a fine of five dollars for the first offense, and ten dollars for each subsequent offense, and all damages. Stallions and jacks, not used for breeding purposes, may be castrated by the person taking them up, if the owner fails, after three days' notice, to reclaim the same, and pay damages; or such animals may be killed after six days' notice. Any bull, boar or stag found running at large may be taken up at any time or place. Electors of townships may decide whether swine may run at large or not, at least ten voters having petitioned for the submission of the question. No horse, mule nor ass diseased with glanders is allowed to be at large, under a penalty of not less than five dollars nor more than $100. Knowingly to import or drive into the State sheep affected with contagious disease is a misdemeanor, with a fine not to exceed $200. The same penalty is provided for any owner allowing such sheep to run at large, together with responsibility for damages to other owners. Rams must be restrained between June fifteenth and December fifteenth, under penalty of five dollars for each day allowed at large. Electors of townships determine whether or not sheep shall run at large. In February, 1867, a sanitary measure was passed for the protection of cattle from the ravages of the Spanish fever. Stock from Texas and the Indian Territory brought into the State between the first day of March and the first day of December in any year, are not to be driven through the State, except in the remoter parts on the plains, and then not within five miles of any highway or ranch, except by consent of the owner of the latter. Violation of the law is treated as a misdemeanor, and the first offense is punishable by fine of $100 to $1,000, and imprisonment from thirty days to six months; for subsequent offenses the penalties are doubled.

Nebraska and the Territories.—In Nebraska, cattle and other stock are restrained in particular counties. The legislation concerning cattle, etc., is of a local character in all the far-west States and Territories and also in the State of California.

Oregon.—The laws of Oregon interdict the running at large of any stallion, jack or mule, over eighteen months old, within the months of April, May, June, July, September and October. If not kept for breeding purposes, the animal may be gelded. If kept for breeding purposes, the distrainer may return him to the owner, and recover two dollars. The owner of such an animal is liable for damages. Animals affected with contagious diseases must not be brought into the State under a penalty of not less than fifty dollars nor more than five hundred dollars for the introduction of each animal so diseased.

XI. State Laws Relating to Fences.

Laws relating to fences are constantly undergoing changes by amendment by the several State legislatures, especially in the Western States. The general idea is to simplify the laws as to what may constitute a lawful fence, and as a rule, to give to localities within the State the power to vote upon the matter, even to doing away with fences altogether. This seemed necessary in much of the vast prairie region; but cheap transportation by railways, and the low price (in contrast with that of lumber) of wire fencing, is operating again to induce the employment of fencing material more than formerly.

Hedges and fences of living trees are largely employed in the settlement of a new country. These again give way, as the country becomes thickly settled and the land valuable, for the reason that living fences require much room for the extension of roots. Yet in all prairie countries living fences will in time pay their cost if made of valuable timber, since for all ornamental work a tree standing alone, and open to the action of storms, becomes more valuable in its grain than those standing closely together, and straighter grained. The following synopsis of the fence laws, originally compiled for the Department of Agriculture at Washington, as heretofore stated, will serve to show the general scope of these laws in the various States.

XII. Fence Laws in General.

In the older States the laws regulating fences are substantially alike. As to height, a legal fence is generally four and a half feet, if constructed of rails or timber. Ditches, brooks, ponds, creeks, rivers, etc., sufficient to turn stock, are deemed equivalents for a fence. In case a stream or other body of water is considered inadequate to the turning of stock, the facts are investigated by officers known as fence-viewers, who will designate the side of the water upon which a fence shall be erected, if the fence be deemed necessary, the cost to be equally borne by the parties whose lands are divided. Occupants of adjoining lands which are being improved are required to maintain partition fences in equal shares. Neglect to build or to keep in repair such fences subjects the negligent party to damages, as well as double, and in some States treble, the cost of building or repairing, to the aggrieved party. A person ceasing to improve land, can not remove his fence unless others interested refuse to purchase within a reasonable time.

A provision in the laws of several of these States, which is well calculated to serve the interests of neighbors, saving the expense of fence building, is one permitting persons owning adjoining lots or lands to fence them in one common field, and for the greater advantage of all, allowing them to form an association, and to adopt binding rules and regulations for the management of their common concerns, and such equitable modes of improvements as are required by their common interest; but in all other respects, each proprietor may, at his own expense, inclose, manage and improve his own land as he thinks best, maintaining his own proportion of the general inclosure.

XIII. Fence Laws in the New England States.

The laws regulating fences in the New England States differ only in a few particulars. The required height of a fence in Maine, Massachusetts and New Hampshire is four feet; in Vermont, four and a half feet; in Rhode Island a hedge with a ditch is required to be three feet high upon the bank of the ditch, well staked, at the distance of two and a half feet, bound together at the top, and sufficiently filled to prevent small stock from creeping through, and the bank of the ditch not to be less than one foot above the surface of the ground. A hedge without ditch to be four feet high, staked, bound and filled; post-and-rail fence on the bank of a ditch to be four rails high, each well set in post, and not less than four and a half feet high. A stone-wall fence is required to be four feet high, with a flat stone over the top, or surmounted by a good rail or pole; a stone wall without such flat stone, rail or post on top to be four and a half feet high.

In each of the New England States there are plain provisions in regard to keeping up division fences on equal shares, and penalties for refusal to build them, and when built for neglect to keep them in repair. Fence-viewers in the respective towns settle all disputes as to division fences. Owners of adjoining fields are allowed to make their own rules and regulations concerning their management as commons. No one not choosing to inclose uncultivated land can be compelled to bear any of the expense of a division fence, but afterwards electing to cultivate, he must pay for one-half the fence erected on his line.

XIV. Fence Laws in the Middle States.

New York.—In New York the provisions for the maintenance of division fences are similar to those of New England; but, whenever a division fence has been injured by flood or other casualty, each party interested is required to replace or repair his proportion within ten days after notification. When electors in any town have made rules or regulations prescribing what shall be deemed a sufficient fence, persons neglecting to comply are precluded from recovering compensation for damages done by stock lawfully going at large on the highways, that may enter on their lands. The sufficiency of a fence is presumed until the contrary is established; assessors and commissioners of highways perform the duties of fence-viewers.

Pennsylvania.—In Pennsylvania, towns and counties obtain special legislation as to the running of stock or other cattle at large.

New Jersey.—Fences in New Jersey are required to be four feet two inches in height, if of posts and rails, timber, boards, brick or stone; other fences must be four and a half feet, and close and strong enough to prevent horses and neat cattle from going through or under. Partition fences must be proof against sheep. Ditches and drains made in or through salt marshes and meadows for fencing and draining the same, being five feet wide and three feet deep, and all ditches or drains made in or through other meadows being nine feet wide at the surface, and four and a half feet wide at the bottom, three feet deep, and lying on mud or miry bottom, are considered lawful fences. Division fences must be equally maintained. If one party ceases improving, he cannot take away his fence without first having given twelve months' notice. Hedge-growing is encouraged by law.

Delaware.—In Delaware, a good structure of wood or stone, or well-set thorn, four and a half feet high, or four feet with a ditch within two feet, is a lawful fence; in Sussex County four feet is the height required. Fence-viewers are appointed by the Court of General Sessions in each "hundred." Partition fences are provided for as in other States.

XV. Fence Laws in the South.

Maryland.—There is no general law in Maryland regulating fences, the law being local and applicable to particular counties.

Virginia.—In Virginia a lawful fence is five feet in height, including the mound to the bottom of the ditch, if the fence is built on a mound. Certain water-courses are specified as equivalent to fences. Four feet is the height of a legal fence in West Virginia, and five feet in North Carolina. In the latter State persons neglecting to keep their fences in order during the season of crops are deemed guilty of misdemeanor, and are also liable to damages. Certain rivers are declared sufficient fences.

South Carolina.—Fences are required to be six feet high around provisions. All fences strongly and closely made of rails, boards, or post and rails, or of an embankment of earth capped with rails, or timber of any sort, or live hedges five feet in height, measured from the level or surface of the earth, are deemed lawful; and every planter is bound to keep such lawful fence around his cultivated grounds, except where a navigable stream or deep water-course may be a boundary. No stakes or canes that might injure horses or cattle are allowed in an inclosure.

Georgia.—The laws of Georgia provide that all fences, or inclosures commonly called worm-fences, shall be five feet high, and from the ground to the height of three feet the rails must not be more than four inches apart. All paling fences are required to be five feet from the ground, and the pales not more than two inches apart. Any inclosure made by means of a ditch or trench must be three feet wide and two feet deep, and if made of both fence and ditch, the latter must be four feet wide and the fence five feet high from the bottom of the ditch. All water-courses that are or have been navigable are deemed legal fences, as far up the stream as navigation has ever extended, whenever, by reason of freshets or otherwise, fences cannot be kept; and the streams are subject to the rules applicable to other fences.

Florida.—The fences in Florida are required to be five feet in height, but where there is a ditch four feet wide the five feet may be measured from the bottom of the ditch. If the fence is not strictly according to law, no action for trespass or damages by stock will lie.

Alabama.—In Alabama all inclosures and fences must be at least five feet high, and, if made of rails, be well staked and ridered, or otherwise sufficiently locked; and from the ground to the height of three feet the rails must be not more than four inches apart: if made of palings, the pales must not be more than three inches apart; or if made with a ditch, four feet wide at the top; the fence, of whatever material composed, must be five feet high from the bottom of the ditch, and three feet from the top of the bank, and close enough to prevent stock of any kind from getting through. No suit for damages can be maintained if the fence is not a legal one. For placing in an inclosure any stakes, poles, poison or anything which may kill or injure stock, a penalty of fifty dollars is provided. Partition fences must be equally maintained.

Mississippi.—Fences in Mississippi are required to be five feet high, substantially and closely built with plank, pickets, hedges or other substantial materials, or by raising the ground into a ridge two and a half feet high, and erecting thereon a fence of common rails or other material two and a half feet in height. Owners of adjoining lands, or lessees thereof for more than two years, are required to contribute equally to the erection of fences, if the lands are in cultivation or used for pasturing. No owner is bound to contribute to the erection of a dividing fence when preparing to build a fence of his own, and to leave a lane on his own land between himself and the adjoining owner; but the failure to erect such fence for sixty days is deemed an abandonment of intention to do so, and determination to adopt the fence already built.

Texas.—In Texas, every gardener, farmer or planter is required to maintain a fence around his cultivated lands at least five feet high and sufficiently close to prevent hogs from passing through it, not leaving a space of more than six inches in any one place within three feet of the ground.

Arkansas.—Fences in Arkansas must be five feet high. In all disputed cases, the sufficiency of a fence is to be determined by three disinterested householders, appointed by a justice of the peace. Division fences are provided for as in the majority of the other States.

Tennessee.—In Tennessee, every planter is required to make a fence around his cultivated land at least five feet high. When any trespass occurs, a justice of the peace will appoint two freeholders to view the fence as to its sufficiency, and to ascertain damages. If a person whose fence is insufficient, should injure any animal which may come upon his lands, he is responsible in damages. In case of dispute between parties as to a division fence, a justice of the peace will appoint three disinterested freeholders to determine the portion to be maintained by each. No owner, whose fence is exclusively on his own land, can be compelled to allow his neighbor to join it.

Kentucky.—In Kentucky, all sound and strong fences of rails, plank or iron, five feet high, and so close that cattle or other stock cannot creep through, or made of stone or brick four and a half feet high, are deemed legal fences. Division fences cannot be removed without consent of the party on adjoining land, except between November first and March first, in any year, six months' notice having been given.

XVI. Fence Laws in the Western States.

Ohio.—The laws of Ohio provide that whenever a fence is erected by any person on the line of his land, and the person owning the land adjoining shall make an inclosure on the opposite side, the latter shall pay one-half the value of the fence as far as it answers the purpose of a division fence, to be adjudged by the township trustees.

Indiana.—The laws are simple and founded on common sense in this State. Any structure or hedge, or ditch, in the nature of a fence, used for purposes of inclosure, which shall, on the testimony of skillful men, appear to be sufficient, is a lawful fence.

Michigan.—Fences in Michigan must be four and a half feet high, and in good repair; consisting of rails, timber, boards or stone walls, or any combination of these materials. Rivers, brooks, ponds, ditches, hedges, etc., deemed by fence-viewers equivalent to a fence, are held to be legal inclosures. No damages for trespass are recoverable if the fence is not of the required height. Partition fences must be equally maintained as long as parties improve their lands. When lands owned in severalty have been occupied in common, any occupants may have lands divided. Fences extending into the water must be made in equal shares, unless otherwise agreed by parties interested. If any person determines not to improve any portion of his lands adjoining the partition fence, he must give six months' notice to all the adjoining occupants, after which he will not be required to keep up any part of the fence. Overseers of highways act as fence-viewers.

Illinois.—According to the statute laws of Illinois, unless decided otherwise by counties or districts by a popular vote, fences must be five feet high. The laws regulating division fences are similar to those of the New England States. In cases of dispute three disinterested householders decide as to the sufficiency of any fence. Proprietors of commons may make their own regulations. Line fences are protected on public highways.

Missouri.—In Missouri all fields must be inclosed by hedge or fence. Hedges must be five feet high; fences of posts and rails, posts and palings, posts and plank, or palisades, four and a half feet; turf, four feet, with trenches on either side three feet wide at top and three feet deep; worm (Virginia) fence at least five feet high to top of rider; or, if not ridered, five feet to top rail, and corner locked with strong rails, poles or stakes. Double damage may be recovered from any person maiming or killing animals within his inclosure if adjudged insufficient.

Wisconsin.—A legal fence in Wisconsin is four and a half feet high, if of rails, timber, boards or stone walls or their combinations, or other things which shall be deemed equivalent thereto in the judgment of the fence-viewers. While adjoining parties cultivate lands they must keep up fences in equal shares; double value of building or repairing may be recovered from delinquents. The law regulating division fences is similar in most particulars to those of the New England States and Illinois. Overseers of highways perform the duties of fence-viewers.

Minnesota.—In Minnesota four and a half feet is the legal height. Partition fences are to be kept in good repair in equal shares. In case of neglect, complaint may be made by the aggrieved party to the town supervisors, who will proceed to examine the matter, and if they determine that the fence is insufficient, notice will be given to the delinquent occupant of land; and if he fails to build or repair within a reasonable time, the complainant may build or repair, and may recover double the expense, with interest at the rate of one per cent per month, in a civil action. No part of a division fence can be removed if the owner or occupant of adjoining land will, within two months, pay the appraised value. When any uninclosed grounds are afterward inclosed, the owner or occupant is required to pay for one-half of each partition fence; the value thereof to be determined by a majority of the town supervisors. If a party to a division fence discontinues the improvement of his land, and gives six months' notice thereof to the occupants of adjoining lands, he is not required to keep up any part of such fence during the time his lands are unimproved, and he may remove his portion if the adjoining owner or occupant will not pay therefor. County commissioners are the authorized fence-viewers in those counties that are not divided into towns.

Iowa.—A legal fence in Iowa is four and a half feet high, constructed of strong materials, put up in a good, substantial manner. In all counties where, by a vote of the legal voters, or by an act of the general assembly, it is determined that hogs and sheep shall not run at large, a fence made of three rails of good, substantial material, or three boards not less than six inches wide and three-fourths of an inch thick, such rails or boards to be fastened in or to good substantial posts, not more than ten feet apart where rails are used; or any other fence which, in the opinion of the fence-viewers, shall be equivalent thereto, is deemed a lawful fence, provided that the lowest or bottom rail shall not be more than twenty nor less than sixteen inches from the ground, and that the fence shall be fifty-four inches in height. The respective owners of inclosed lands must keep up fences equally as long as they improve. In case of neglect to repair or rebuild, the adjoining owner may do so, and the work being adjudged sufficient by the fence-viewers, and the value determined, the complainant may recover the amount, with interest at the rate of one per cent per month. If an owner desires to throw his field open, he shall give the adjoining parties six months' notice, or such shorter notice as may be directed by the fence-viewers.

Kansas.—In Kansas, fences may be of posts and rails, posts and palings, or posts and planks, at least four and a half feet high; of turf, four feet, and staked and ridered, with a ditch on either side at least three feet wide at the top and three feet deep; a worm fence must be at least four feet and a half high to top of rider, or if not ridered, four and a half feet high to top rail, the corners to be locked with strong rails, posts or stakes. The bottom rail, board or plank in any fence must not be more than two feet from the ground in any township, and in those townships where hogs are not prohibited from running at large it must not be more than six inches from the ground. All such fences must be substantially built and sufficiently close to prevent stock from going through. Stone fences are required to be four feet high, eighteen inches wide at the bottom, and twelve at the top. All hedges must be of sufficient height and thickness to protect the field or inclosure. A wire fence must consist of posts of ordinary size for fencing purposes, set in the ground at least two feet deep and not more than twelve feet apart, with holes through posts, or staples on the side, not more than fifteen inches apart, and four separate lines of fence wire, not smaller than number nine, to be provided with rollers and levers at suitable distances, to strain and hold the wires straight and firm. Owners of adjoining lands must maintain fences equally. In case of neglect of one party to build or repair, another party may do so and recover the amount expended, with interest at the rate of one per cent per month. A person not improving his land is not required to keep up any portion of a division fence. The trustee, clerk and treasurer in each township act as fence-viewers, to adjust all disputes concerning fences.

Nebraska.—A legal fence in Nebraska is any structure, or hedge, or ditch in the nature of a fence, used for the purpose of enclosure, which is such as good husbandmen generally keep. Division fences must be equally maintained. A party may remove his portion of division fence by giving sixty days' notice. If removed without such notice the party so doing is liable for full damages. Where a fence is injured or destroyed by fire or flood, it must be repaired within ten days after notice by interested persons. Justices of the peace are ex officio fence-viewers.

XVII. Fence Laws of the Pacific Slope.

California.—The legal fences in California are described with great minuteness. Wire fences must consist of posts not less than twelve inches in circumference, set in the ground not less than eighteen inches, and not less than eight feet apart, with not less than three horizontal wires, each one-fourth of an inch in diameter, the first to be eighteen inches from the ground, the other two above at intervals of one foot, all well stretched and securely fastened from post to post, with one rail, slat, pole or plank of suitable size and strength, securely fastened to the post, not less than four and a half feet from the ground.

Post-and-rail fence must be made with posts of the same size and at the same distances apart and the same depth in the ground as above required, with three rails, slats or planks of suitable size and strength, the top one to be four and a half feet from the ground, the other two at equal distances between the first and the ground, all securely fastened to the post. Picket fences must be of the same height as above, made of pickets not less than six inches in circumference, placed not more than six inches apart, driven in the ground not less than ten inches, all well secured at the top by slats or caps.

Ditch and pole fence—the ditch must not be less than four feet wide on the top and three feet deep, with embankment thrown up inside of ditch, with substantial posts set in the embankment not more than eight feet apart, and a plank, pole, rail or slat securely fastened to posts at least five feet high from the bottom of the ditch. Pole fence must be four and a half feet high, with stakes not less than three inches in diameter, set in the ground not less than eighteen inches, and when the stakes are placed seven feet apart, there must not be less than six horizontal poles well secured to the stakes; if the stakes are six feet apart, five poles; if three or four feet, four poles; if two feet apart three poles, and the stakes need not be more than two inches in diameter; if one foot apart, one pole, and the stakes need not be more than two inches in diameter. The above is a lawful fence so long as the stakes and poles are securely fastened, and in a fair state of preservation.

Hedge fence is considered lawful, when by reliable evidence it shall be proved equal in strength, and as well suited to the protection of inclosed lands as the other fences described.

Brush fence must be four and a half feet high and at least twelve inches wide, with stakes not less than two inches in diameter, set in the ground not less than eighteen inches, and on each side, every eight feet, tied together at the top, with horizontal pole tied to the outside stake five feet from the ground. In the case of partition fences, if one party refuse or neglect to build or maintain his share, the other may do so and recover the value. Three days' notice to repair is sufficient. The sufficiency of a fence is to be determined by three disinterested householders.

CHAPTER III.

LAW FORMS RELATING TO BUSINESS TRANSACTIONS.

I. GUARDING AGAINST SWINDLERS.—II. RULES OF GUIDANCE IN BUSINESS.—III. RULES IN RELATION TO BANKING.—IV. INDORSEMENTS.—V. FORMS OF NOTES.—VI. JUDGMENT NOTES.—VII. DUE-BILLS, RECEIPTS, ORDERS, ETC.—VIII. SOME DEFENSES WHICH MAY DEFEAT PAYMENT OF NEGOTIABLE PAPER.—IX. REMARKS CONCERNING NOTES.—X. DRAFTS EXPLAINED.—XI. REMITTANCES.—XII. OBLIGATION FOR MARRIED WOMEN.—XIII. DRAWING UP IMPORTANT PAPERS.—XIV. SHORT FORM OF LEASE FOR FARM AND BUILDINGS.—XV. AGREEMENTS BETWEEN LANDLORD AND TENANT.—XVI. WILLS.—XVII. POWER OF ATTORNEY —XVIII. MORTGAGE—SHORT FORM.—XIX. WARRANTY DEEDS.—XX. BILLS OF SALE.—XXI. BONDS.—XXII. ARBITRATION.—XXIII. AWARD OF ARBITRATORS.—XXIV. COUNTERFEIT MONEY.—XXV. GOOD BUSINESS MAXIMS.—XXVI. SOME POINTS ON BUSINESS LAW.—XXVII. DEFINITIONS OF MERCANTILE TERMS.—XXVIII. BUSINESS CHARACTERS.

I. Guarding Against Swindlers.

THERE is a class of adroit swindlers whose business in life is the obtaining from farmers their signatures to notes, or other obligations to pay money, for which no consideration has been given. A common plan is to excite the cupidity of the intended victim by making him agent for the sale, in his State or county, of some seemingly desirable patented article. A note, or order for goods, is then offered for his signature. This reads fairly enough, for it calls for payment only after the newly appointed agent has received and sold a certain number, or amount, of the goods or article: examined more closely it will be seen, however, that the obligation may be cut in such a manner as to make it a negotiable note in which nothing is said in regard to conditions; or, if an order for goods, it becomes without reservation. This note, or order, is then sold, and, passing into the hands of an innocent party, or one whose collusion with the swindler cannot be proven, however much it may be suspected, it becomes good in law against the maker. By this time the swindler has disappeared; no goods have ever come to the farmer, and he finds that the "patent rights" either are imaginary or the property of another. The note is, however, in the hands of "innocent" parties and must be paid to the uttermost farthing.

Never, then, sign an agreement to pay money until the goods are received, or the person to whom you give the agreement is vouched for, under bond, by some responsible person, well-known in your community. A business man, who has anything of value to sell and who intends to deal honestly, will find no difficulty in satisfying business men of his soundness. If he cannot do this, and give you a sound and respectable local endorser, have nothing to do with him. Those who act on this principle only, and have a knowledge of proper forms of business contracts, will save themselves from all possibility of difficulty and loss from this kind of fraud.

Beware of any paper not drawn in the ordinary, established and simple manner; and, that our readers may be conversant with this, we here give some simple forms that will cover all ordinary cases of contract. If at a loss, or the money to be paid is a considerable sum, a consultation fee paid to a respectable lawyer, whom you know to be such, will always be money well invested. See Chapt. II, Part X.

II. Rules of Guidance in Business.

1. Do not enter into any sort of business unless you are thoroughly conversant with it.

2. Trade only for cash; or at least, never give a note to pay money merely because some one confidentially tells you immense profits must arise from handling a certain article, and offers to entrust you with its sale. A farmer may indeed run into debt for an implement or machine when he is sure the saving in the crop will repay the outlay. Not otherwise. If you have not the money for a really needed article, buy it from a well-known, responsible dealer, not from some unknown adventurer, who may prove a smooth-tongued swindler. Any respectable farmer can obtain proper credit from the local dealer or from the bank at reasonable rates, and know what he is buying.

3. The man who pays cash can always obtain a discount more than sufficient to cover all interest and collecting expenses. No man gives another credit without the expectation of increased gain.

4. Keep your business to yourself, and never employ another to do that which you can as well do yourself.

5. If you make a business engagement, or even a social engagement, let nothing but some unsurmountable obstacle prevent you from keeping it.

6. Trust nothing of importance to memory alone; a memorandum book and a pencil should always be carried in the pocket.

7. It is safe to keep copies of all letters relating to business, or others matters of importance. Endorse them with their titles, and place them away properly classified. If you have many business letters a copying book is economical.

8. However generous you may be from motives of humanity, never concede more than is just in matters of business, and especially in becoming security for another.

9. A man's promise should be inviolable. Hence, say "No!" with decision, when it is to resist temptation. Say "Yes" with caution if it implies a promise to perform; that is to say, be frank, self-reliant and punctual.

10. Never draw a note except to the order of the person to whom it is given. Never endorse for a friend. This may seem hard, but if you can spare the money, or its loss would be a trifle to you, get the amount and lend him the cash. If you cannot afford this you cannot afford the chance of being obliged to pay the note. Many of the best people in the world have sanguine temperaments, feel sure of success, but fail, and their friends are ruined.

III. Rules in Relation to Banking.

1. In opening an account with a bank, get some one known to its officers to introduce you.

2. In all your dealings with bank officials, as in those with your lawyer, doctor or clergyman, be candid and state facts exactly as they are. No business man ever gained anything in the end by double dealing. If, for instance, you have paper for discount, state its nature exactly; whether it be accommodation, a renewal of a note, or other kind.

3. Never wrangle or contend unduly, nor consider bank officials arbitrary, because your note or request is declined. They know their business as well as you do your own.

4. A man's promise should be inviolable. Hence, if you say yes, mean it; if no, let it be founded on the right to say no.

5. It is always better to do business directly with a bank than with a third person who has an account there. As a rule, it is safer to keep your money in deposit at a bank than at your residence.

6. Never draw a check for a greater amount than you have to your credit in the bank.

7. Never send a check to a distant place intending to deposit funds to meet it before it is returned. A telegraphic inquiry to your bank may cause its dishonor.

8. Never exchange checks with any one, and never give your check under the stipulation that it is not to be used until a given time. It may cause you dishonor.

9. Make all checks payable to the order of the party with whom you transact business.

10. Never take a distant check from any one to pass it, as an accommodation, through your bank, giving your check in exchange.

11. Never give a check to a stranger; it may be tampered with and passed, thus entailing loss on the bank.

12. In sending a check to a distance, make it payable to order of the person named, and give residence of payee, that is, the person who is to receive the money.

13. In indorsing to the order of another never indorse in blank, but make payable to the order of the person named.

14. Every person doing business at a bank should have checks in blank of the form used. A check is simply the order of the depositor to the bank to pay money.

Form of Check.—Checks, like all other papers in regard to business transactions, should be carefully and plainly written in those portions not printed. The proper form is as follows:

$100.00. ST. LOUIS, MO., September 15th, 1883.

National Bank of St. Louis, pay to John Doe, or order, One Hundred (100) Dollars.
No. 79. RICHARD ROE.

NOTE.—The number, as for instance 79, is needed because this must correspond with that number on the stub of your check-book, showing amount and purpose of that check.

IV. Indorsements.

In all matters relating to business we use the old law names of fiction, John Doe and Richard Roe; also, fictitious names of institutions, etc., as though they were real, wherever the proper names of persons should be introduced. The person who agrees to pay, or orders the payment of money, is the maker; the person to whom the obligation is given is the payee. If a note, check, or other paper is made payable to the payee, "or bearer," it is due and to be paid, without question, to the person who presents it. If made payable to order, the person holding it cannot transfer it, without writing his name across the back of the paper. Then he is the indorser, and the person to whom it is transferred is the indorsee or owner, and entitled to receive payment of the amount named on the face.

Indorsement in Blank.—There are five different ways of indorsement. The most common way is the indorsement in blank, which is simply writing the name across the back. Thus indorsed the note or draft is transferred by delivery from hand to hand, like a bank bill, and so long as it so continues it is payable to bearer, his indorsement being last.

Indorsement in Full.—When a person who has received a note or check, payable to order, wishes to send it away to another person, it should be indorsed in full, thus:

Pay to John Doe or order.
Richard Roe.

Then none but the person to whom it is ordered paid can demand payment. It is important, as ensuring safety in transmitting funds.

Qualified Indorsement.—This releases the indorser from responsibility. It may become necessary in some cases, since an indorsement, unless qualified, renders the indorser liable for the amount. A qualified indorsement would be as follows:

Without recourse.
John Doe.

A Restrictive Indorsement.—It may be necessary to restrict the negotiability of an instrument to a particular person, or for a particular person. The indorsement would then be:

Pay to John Doe, only.
Richard Roe.

Conditional Indorsement.—A conditional indorsement is made when something is to be performed before the person is paid, as the presentation of vouchers, the receipt of a degree or honorary title, the attainment of a certain age, etc. This must then be expressed, and the payee, before receiving the payment, must show that the obligation has been fulfilled or complied with. The following will show one instance when such indorsement might be needed:

Pay the within named sum to John Doe, when he shows certificate that he has passed the necessary examination to qualify him to enter the Galen Medical College of St. Louis, Mo.
Richard Roe.

V. Forms of Notes.

A NOTE, if made payable to bearer, is negotiable without indorsement. If made payable to order, it is negotiable only by indorsement. A note is not negotiable if made payable only to an individual. And time is the essence of the note.

A note secured, must be signed by the maker as principal, and by the surety as surety.

A joint note must be signed by all the parties drawing it.

A note written jointly and severally renders any one of the persons signing it liable for the whole sum.

The note should state the rate of interest to be paid, and the place where it is to be paid. With these explanations the forms of notes given below may be readily placed as to their character.

NOTE PAYABLE TO BEARER.

100\frac{00}{100}$.   ST. LOUIS, MO., Sept. 15, 1883.

Three months after date I promise to pay John Doe, or bearer, *One Hundred (100) Dollars*, value received, interest at six per cent per annum.

RICHARD ROE.

This is all there is to any note. This is negotiable without indorsement. If payable to order it is negotiable only by indorsement. If made payable simply to the individual, it is not negotiable, and if made payable on demand or without time, it is payable when presented. If payable at a bank, the name of the bank must follow the words or order, and if with interest that, and the rate must be stated. A note, principal and surety, is written as follows, and in this one shows the form of inserting cents.

NOTE WITH SURETY.

100\frac{25}{100}$.   ST. LOUIS, MO., Sept. 15, 1883.

Three months after date I promise to pay John Doe, or order, One Hundred $\frac{25}{100}$ (100$\frac{25}{100}$) Dollars, with interest, value received.

RICHARD ROE, Principal.
RICHARD ROE, Jr., Surety.

Joint Notes.—A joint note is written precisely like the foregoing, except the words "we jointly promise," or "jointly and severally promise," are used instead of "I promise." The two names, or more, of several persons making a joint note, are signed at the bottom in the same manner as any single maker would sign.

Special Forms of Notes.—Some of the States have peculiar forms. For instance, in Indiana, after the word with interest, it is usual to state " without any relief whatsoever from value or appraisement." In some of the Eastern States, the words " without defalcation " are used, and in Missouri, the words " negotiable and payable without defalcation or discount," are added.

VI. Judgment Notes.

SOME States have particular forms of judgment notes, and these are printed ready for filling in. A judgment note is a confession of judgment in law, and execution may be had thereon immediately, if not paid at maturity. A short form of judgment note is as follows:

SHORT FORM OF JUDGMENT NOTE.

$100.00.   ST. LOUIS, MO., Sept. 15, 1883.

One year after date I promise to pay John Doe, or order, One Hundred (100) Dollars, with interest at the rate of six per cent per annum, from maturity until paid, without defalcation. And I do hereby confess judgment for the above sum, with interest and costs of suit, a release of all errors, and waiver of all rights to inquisition and appeal, and to the benefit of all laws exempting real or personal property from levy and sale.

RICHARD ROE.

VII. Due-Bills, Receipts, Orders, Etc.

DUE-BILLS are simply acknowledgments of debt, payable at any time. An I. O. U. is an acknowledgment of debt, but does not amount to a promissory note, and is as follows:

FORMS OF DUE-BILLS.

ST. LOUIS, MO., Sept. 15, 1883.

MR. JOHN DOE:

    I. O. U. One Hundred $\frac{50}{100}$ Dollars.

RICHARD ROE.

ANOTHER FORM OF DUE-BILL.

$175.   ST. LOUIS, MO., Sept. 15, 1883.

Due John Doe, or order, One Hundred and Seventy-five (175) Dollars.

RICHARD ROE.

Forms of Receipts.—A receipt for any specific thing should state the fact, as for rent, the value of a horse, rent of a farm, etc. Here is one form:

RECEIPT FOR RENT.

$50.00.   ST. LOUIS, MO., Sept. 15, 1883.

Received, from John Doe, Fifty Dollars for rent of house, No. 1220 St. Louis Street, for month ending August 31, 1883.

RICHARD ROE.

If the rent paid is for a year, or quarter, this should be stated; if for a farm, or field, or horse, this also should be stated. The words "in full" are often inserted, and there is no objection to so doing.

RECEIPT ON ACCOUNT.

A receipt on account should read:

$100.   ST. LOUIS, MO., Sept. 15, 1883.

Received of John Doe, One Hundred Dollars on account.

RICHARD ROE.

If in full, after the sum named, add, *in full of all demands to date.*

Forms of Orders.—Farmers and others hiring men, especially married men, often have occasion to give orders for merchandise. The form is as follows:

$25.00.   St. Louis, Mo., Sept. 15, 1883.
Mr. John Doe:
   Please pay to the bearer Twenty-five Dollars, in merchandise, and charge same to account of yours.
                                                                    Richard Roe.

If the request for merchandise is in full of all accounts, the order should state the fact.

VIII. Some Defenses which may Defeat Payment of Negotiable Paper.

If any negotiable paper has been sold for value by the original holder, it must be paid. No defense will defeat payment, even if no value has ever been given, originally, for the paper. Hence, every person should be careful in giving an obligation for money, to know that it is correct. The legal defense against payment may be:

1. Want of consideration, either total or partial.
2. If obtained through compulsion, as threats of imprisonment, which would be unlawful; the fear of injury to person, to reputation, to property, etc.
3. If it was obtained by fraud.
4. If it was found, or obtained by having been stolen.
5. If the obligation has been misapplied, still remaining in the hands of the original holders.
6. If the consideration has been illegal or if the obligation is illegal on its face.
7. If not collected when due without unnecessary lapse of time, or delay. For then, the presumption is against its validity.

But, again, remember, if the person to whom the obligation was given parts with it, for value, to innocent holders, and the transfer has been legitimately made, no defense will operate to prevent the liability of the maker of the paper. It must be paid.

IX. Remarks Concerning Notes.

The prejudice against giving notes, which some merchants as well as farmers seem to have, is not well founded. A note is nothing more than a written acknowledgment that a debt is due and payable at a stated time. It adds nothing—morally or legally—to the obligation of paying the account according to terms. On the other hand, there are several good reasons why wholesale merchants like to have notes, prominently these: The accounts can be closed on the books; in case of death or insolvency, the trouble of proving up the account is saved; notes can be used as collateral in getting loans. Wholesale merchants rarely let notes get beyond their control, and in case a customer wants an extension of time it is usually granted as readily on a note as on an open account.

Making Notes.—It may not be out of place to note here the difference between a piece of paper drawn so many days after date and one stating the time in months. For instance, a note drawn March 15, ninety days after date, would be due on June 13–16, while one drawn March 15, three months after date, would be due on June 15–18, two days later. In the first case the actual number of days are counted, and as two of the months —March and May—have thirty-one days, the note falls due two days sooner than if it was drawn at three months. Law and custom allow three days' grace on all time paper, unless it reads expressly "without grace." The last day of grace is the day on which it can be protested.

Canceling Notes.—Some persons cancel notes by tearing off the signature, or by destroying them altogether. It is not advisable to do so. If the person to whom the note was payable should deny that it had been paid, and make oath that he lost or misplaced the note, it might cost a good deal of trouble to prove that it really was paid. The proper way to cancel a note is to write across the face or back of it how and when it was paid.

Take a Receipt.—When paying an account with a draft, persons frequently make a remark like this: "Never mind a receipt, the draft is a receipt," meaning thereby that the draft, with the indorsement of the person to whom it was given, proves that the money was paid. That is a mistake. The draft proves that so much value passed between the parties, but it does not prove that it was given for a certain purpose, unless it so states on its face. The party who received it might claim that he cashed it, or that it was given for a different purpose altogether. A receipt should always be taken, regardless of how the payment is made.

Receipts "In Full."—Parties are sometimes very anxious to have a receipt in full of account. Such a receipt is no better than one on account. If the payment is in fact in full, the account will show it. If a payment is made, which at the time is supposed to be in full, and afterwards errors are discovered which show a further just claim, the receipt in full will not prevent its collection.

X. Drafts Explained.

At Sight, etc.—Drafts may be drawn at sight, so many days after date, or so many days after sight. A draft dated January 10, ten days after date, is due January 23; while a draft dated the same day, and drawn at ten days' sight, does not become due until thirteen days after it is accepted, or in case acceptance is refused, thirteen days after demand has been made

Drafts should never be held longer than necessary. In case the party drawn on should fail while the draft is in your possession, you would be responsible for the amount if it could be shown that you did not use due diligence in forwarding the draft to its destination for collection. This is a point of considerable importance, not generally understood.

Protest.—If a note, draft or acceptance is not paid on the day it is due, or if the acceptance of a time draft is refused, it is liable to be protested. This means that a legally authorized person—a notary—certifies to the fact that the demand has been made, and that payment or acceptance has been refused. When there are indorsers on a piece of paper, protest is necessary, and they must be notified of such protest with the least possible delay, otherwise they are released from responsibility.

When there are no indorsers except the owner of the paper, protest is not necessary, but even then it is sometimes advisable. Debts have been saved by instructing protest on a piece of paper. The person drawn on may be in a failing condition. If the demand for payment is made by a business house, he may put them off on some pretext or other, but if the notary presents the paper he will strain a point, and pay it because he is not quite ready to quit.

XI. Remittances.

By Check or Draft.—Remittances can be made by check or draft, by post-office money order, or by sending money by express or registered letter. There are certain precautions to be observed in remitting by these different methods, which will be briefly pointed out. When checks or drafts are sent, the party sending them should see that they are properly indorsed by those who held them before him; then he should indorse them, not by merely putting his name across the back, but in this manner.

Pay to the order of
JOHN DOE & CO. [The firm to whom sent.]
RICHARD ROE. [Sender.]

If this is done, the check or draft cannot be collected by the wrong party, except by forging the name of John Doe & Co.; whereas, if it is indorsed in blank, anybody who might happen to get hold of it might collect it, and the last indorser would be the loser. All checks, even those payable to bearer, should be indorsed by the persons handling them, so that in case—for any reason—they are not paid, they can be traced back.

"His Mark."—When the person who should indorse the draft cannot write, he indorses by X mark, and such indorsement must be witnessed, the witness signing his name and stating his residence, in this manner:

his
PETER X WALTON.
mark.

Witness:
THOMAS JOHNSON,
Jamestown, Mo.

By Express.—Money should never be advanced on drafts, except to parties whose responsibility is above question. If parties want drafts cashed, they should wait until sufficient time has elapsed for the drafts to reach their destination and be heard from, before receiving the money. Remittance by money-order is quite safe, and it is only necessary to exercise care in having them made payable to the right party. To send money by express is also quite safe. The express company is responsible for the amount receipted for. To send money by registered letter is comparatively safe, but as the government does not hold itself responsible for any loss that may occur, no very large amount should be risked in one letter.

Charges on Remittances.—Remittance by any of these modes should be made in such a manner that the full amount due the person or firm to whom it is sent will come into their possession. If drafts on country banks are sent, they should be drawn *with exchange*, or include a reasonable sum, say one-fourth on one-half per cent on their face. Express charges on money should be prepaid. Merchants in the city expect to receive the amount due them without reduction, and while some may accept the amount less the cost of collection, without complaint, it always causes dissatisfaction, and tends to disturb the kindly feeling that should exist between them and their customers. Bills are supposed to be payable in the city where the seller does business, and in funds current there.

XII. Obligation for Married Women.

THERE are forms which married women use in giving an obligation which differ somewhat in different States. A usual form, after the regular writing of the obligation, is to add: " And I hereby charge my individual property and estate with the payment of this note "—or other obligation, as the case may be, which must precede the signature.

XIII. Drawing Up Important Papers.

ALL important legal papers should be drawn by a lawyer, or other person familiar with the proper forms. Chattel mortgages, for instance, vary in every State. As a rule, in every important transaction it is better to pay a consultation fee, or the regular legal fee for drawing up articles of agreement, deeds, wills, mortgages, powers of attorney, etc. Besides, in this way you get legal advice as to what to do with the paper. More simple contracts may be drawn without consultation with a lawyer. We give some of these forms in blank, to be filled, which will show the general character.

XIV. Short Form of Lease for Farm and Buildings.

THIS INDENTURE, Made this first day of March, one thousand eight hundred and eighty-three, between———, of the township of———, county of———. and State of———, of the first part, and——— of the said township and county, of the second part.

WITNESSETH, That the said———, for, and in consideration of the yearly rents and covenants hereinafter mentioned, and reserved on the part and behalf of the said———, his heirs, executors and administrators, to be paid, kept, and performed, hath demised, set, and to farm let, and by these presents doth demise, set, and to farm let, unto the said———, his heirs and assigns, all that certain piece, parcel or tract of land situate, lying and being in the township of———, aforesaid, known as [*here describe land*], now in the possession of———, containing one hundred acres, together with all and singular the buildings and improvements, to have and to hold the same unto the said———, his heirs, executors and assigns, from the———day of———next, for, and during the term of———years, thence next ensuing, and fully to be complete, and ended, yielding and paying for the same, unto the said———, his heirs and assigns, the yearly rent, or sum of———dollars, on the first day of———in each and every year, during the term aforesaid, and at the expiration of said term, or sooner if determined upon, he, the said———, his heirs or assigns, shall and will quietly and peaceably surrender and yield up the said demised premises, with the appurtenances, unto the said———, his heirs and assigns, in as good order and repair as the same now are, reasonable wear, tear, and casualties, which may happen by fire, or otherwise, only excepted.

IN WITNESS WHEREOF, we have hereto set our hands and seals.

Signed, sealed and delivered in presence of } (Signatures of parties to contract.) [L. S.]
(Signature of witness.) [L. S.]

XV. Agreements Between Landlord and Tenant.

LANDLORD'S AGREEMENT.

This certifies that I have let and rented, this——day of——18—, unto John Doe [here describe premises and where situated], and all appurtenances; he to have the free and uninterrupted occupation thereof for one year from this date, at the yearly rental of——dollars, to be paid monthly in advance; rent to cease if destroyed by fire, or otherwise made untenantable.

RICHARD ROE.

THE TENANT'S AGREEMENT.

This certifies that I have hired and taken from Richard Roe, his [describe the premises], with appurtenances thereto belonging, for one year, to commence this day, at a yearly rental of [state amount], to be paid monthly in advance, unless said house becomes untenantable from fire or other causes, in which case rent ceases; and I further agree to give and yield said premises one year from this—day of——18—, in as good condition as now, ordinary wear and damage by the elements excepted.

Given under my hand this day. JOHN DOE.

NOTICE TO QUIT.

TO RICHARD ROE:

Sir:—Please observe that the term of——, for which the house and land, situated at [describe the premises], and now occupied by you, were rented to you, expired on the——day of——18—. As I desire to repossess said premises, you are hereby requested and required to vacate the same.

Yours truly, JOHN DOE.

ST. LOUIS, MO., Jan. 1, 1880.

(Three months' notice is usually given.)

TENANT'S NOTICE OF QUITTING.

Dear Sir:—The premises I now occupy as your tenant, at [state locality], I shall vacate on the——day of——18—. You will please take notice accordingly.

Dated this——day of——18—. RICHARD ROE.

TO JOHN DOE.

If payment of rent is refused, the landlord gives notice for surrender, and states the reason.

XVI. Wills.

WILLS may be drawn up by any person if properly witnessed. When the property is to be equitably divided between the wife and children no will is necessary. Intricate wills should always be drawn by a competent lawyer. The following is a short form of will, to be varied according to circumstances:

FORM OF WILL.

KNOW ALL MEN BY THESE PRESENTS, That I,——, of the town of——, in the county of——, and State of——, being of sound mind and memory, do make and publish this my last will and testament.

I give and bequeath to my sons [give their names], [state the bequest], each, if they have attained the age of [state the age], before my decease; but if they shall be under the age of——at my decease then I give to them——each, the last-mentioned sum to be in place of the first-mentioned.

I give and bequeath to my daughters [state the names and add the bequests.]

I give and bequeath to my beloved wife——, all my household furniture, and all the rest of my personal property, after paying from the same the several legacies already named, to be hers forever; but if there should not be at my decease sufficient personal property to pay the aforesaid legacies, then so much of my real estate shall be sold as will raise sufficient money to pay the same.

I also give, devise and bequeath to my beloved wife———, all the rest and residue of my real estate, as long as she shall remain unmarried, and my widow; but on her decease or marriage, the remainder thereof I give and devise to my said children and their heirs, respectively, to be divided in equal shares among them.

I do nominate and appoint my beloved wife———, to be the sole executrix [or name the executor] of this my last will and testament, hereby revoking all my former wills.

IN TESTIMONY WHEREOF, I hereunto set my hand and seal, and publish and decree this to be my last will and testament, in presence of the witnesses named below, this——— day of———, in the year of our Lord one thousand eight hundred and———.

[Name of maker of will.]   [L. S.]

Signed, sealed, declared and published by the said——— as and for his last will and testament, in presence of us, who, at his request and in his presence, and in presence of each other, have subscribed our names as witnesses hereto.

———, residing at———, in——— county.
———, residing at———, in——— county.

Codicils.—A codicil to a will may be made later. The following would be the form:

Whereas I,———, of the——— of——— in the county of——— and State of———, have made my last will and testament in writing, bearing date the——— day of———, in the year of our Lord one thousand eight hundred and———, in and by which I have given and bequeathed to [here state the bequests, and also the changes to be made, signing and having the codicil witnessed as in the will.]

XVII. Power of Attorney.

A POWER of attorney is always best—as in the case of other important documents—drawn by a good lawyer; but we give the forms of various documents that the reader may be conversant therewith. A short form is as follows:

KNOW ALL MEN BY THESE PRESENTS, That I, ———, of———, in the county of———, and State of———, have made, constituted and appointed, and by these presents do make, constitute and appoint ———, of———, my true and lawful attorney, for me, and in my name, place and stead, and to my use, [here insert exact subject-matter of the power], and to do and perform all necessary acts in the execution and prosecution of the aforesaid business in as full and ample a manner as I might do if I were personally present.

IN WITNESS WHEREOF, I have hereunto set my hand and seal, the——— day of——— in the year one thousand eight hundred and———

[Signature of the maker.]

Signed, sealed and delivered }
   in presence of   }
[Signature of witness.]

If the attorney-in-fact is to have power of substitution and revocation, then, at the end, after the words personally present, add the words:

"With full power of substitution and revocation, hereby ratifying and confirming all that my said attorney, or his substitute, shall lawfully do, or cause to be done, by virtue hereof."

Then sign and witness as above

What Powers of Attorney are for.— 1. To collect rents, debts, etc., receive dividends, legacies, etc.

2. Transfer or sell anything.

3. To mortgage land, renew leases, and to act generally in the place of the person granting the power.

XVIII. Mortgage—Short Form.

THIS INDENTURE, made the———day of———in the year one thousand eight hundred and——— between———, of———, in the county of———, and State of———, manufacturer, of the first part, and———, of———, in the said county, farmer, of the second part, WITNESSETH, that the said party of the first part, for and in consideration of the sum of———dollars, grants, bargains, sells and confirms unto the said party of the second part, and to his heirs and assigns, all [here insert full description of the property]; together with all and singular, the hereditaments and appurtenances thereunto belonging, or in any wise appertaining. THIS CONVEYANCE is intended as a mortgage, to secure the payment of the sum of———dollars, in [here state terms of payment], according to the condition of a certain bond, dated this day, and executed by the said party of the first part to the said party of the second part; and these presents shall be void if such payments be made. But in case default shall be made in the payment of the principal or interest, as above provided, then the party of the second part, his executors, administrators and assigns, are hereby empowered to sell the premises above described, with all and every of the appurtenances, or any part thereof, in the manner prescribed by law; and out of the money arising from such sale, to retain the said principal and interest, together with the costs and charges of making such sale; and the overplus, if any there be, shall be paid by the party making such sale, on demand, to the party of the first part, his heirs or assigns.

IN WITNESS WHEREOF, the said party [or parties] of the first part has [or have] hereunto set his hand and seal [or their hands and seals], the day and year first above written.

Signed, sealed and delivered }
   in the presence of   }   [Signature and seal.]
[Signature of witness.]

A mortgage must be not only witnessed, but acknowledged like any other deed, and recorded.

Assignment.—A mortgage may be assigned, as follows:

I hereby assign the above [or within] mortgage to ———
WITNESS MY HAND AND SEAL, this———of———

[Signature.]   [Seal.]

Release.—It may be released, thus:

I hereby release the above [or within] mortgage.
WITNESS MY HAND AND SEAL, this———day of———

[Signature.]   [Seal]

Foreclosure.—The proceedings by default, and sale by mortgage, must be in accordance with the laws of the State in which it is executed.

Mortgages are given for various purposes; as security for payments, for debts, notes, the fulfillment of leases, etc. They should state explicitly what they are for.

XIX. Warranty Deeds.

A WARRANTY Deed is an important document, and should, as a rule, be drawn by a lawyer. Printed forms are used for various conditions specified.

SHORT FORM OF WARRANTY DEED.

KNOW ALL MEN BY THESE PRESENTS, That I, ———, of———, in the county of——— and State of ———, in consideration of———dollars, to me paid by———, of———, in the county of———, and State of———, the receipt whereof is acknowledged, do grant, bargain, sell and confirm unto the said———, his heirs and assigns, forever, all [here insert description], with the appurtenances. And I do, for myself and my heirs, executors and administrators, covenant with the said———, his heirs and assigns, that at the time of making this conveyance I am well seized of the premises, as of a good and indefeasable estate, in fee-simple, and have good right to bargain and sell the same, as aforesaid, and that the same are free from all encumbrance whatsoever; and the above-granted premises, in the quiet and peaceable possession of the said——, and his heirs and assigns, I will warrant and forever defend.

IN WITNESS WHEREOF, I have hereunto set my hand and seal, the——day of——, in the year one thousand eight hundred and——

Signed, sealed and delivered in presence of }   [*Signature and seal.*]
[*Signature of witness.*]

Erasures, etc.—If erasures or interlineations are made, these must be explained in a document attached to the deed, and then as follows:

IN WITNESS WHEREOF, the said party of the first part has hereunto set his hand and seal, the day and year first above written.   [*Signature and seal.*]

Sealed and delivered in the presence of ——(the word——, on the——page, was erased, the words——written over an erasure; on the——page, the words——interlined in——places, and the word——canceled on the ——page, before——.)   [*Signature of witness.*]

For instance, a specified number of acres may be changed to more or less, or added after acres, etc.

XX. Bills of Sale.

THESE are given for almost every description of property. A short form is as follows:

I, John Doe, of——, in consideration of——dollars paid by Richard Roe, of——, hereby sell and convey to said Richard Roe the following personal property, [schedule of articles], warranted against adverse claims.

Witness my hand this——day of——A. D. 18—

                                                         [Signature.]   [Seal.]

Executed and delivered in presence of——

XXI. Bonds.

THESE are given for various purposes, as a condition to convey land or other property. They may be given with sureties; that is, certain persons named in the bond become sureties with the principal or principals. Or a bond may be given for the payment of a lost note or other obligation. If so, the conditions must be explicitly stated; and the sum named should be double that named in the condition at the bottom before the signature. A general form of bond may be as follows:

FORM OF BOND.

KNOW ALL MEN BY THESE PRESENTS, that I, John Doe, of——, in the county of——, am held and firmly bound to Richard Roe, of——, in the county of——, in the sum of——dollars, to be paid to the said Richard Roe, to the payment whereof I bind myself and my heirs firmly by these presents. Sealed with my seal.

Dated the——day of——A. D. 18—

The condition of this obligation is such, that if I, the said John Doe, shall pay to said Richard Roe, the sum of——dollars and interest, on or before the——day of——, 18—, then this obligation shall be void.

                                                              JOHN DOE.

Executed and delivered in presence of——

XXII. Arbitration.

This is a means of settling disputes with or without legal advice, rather than litigate in the courts. It is specially to be commended as being peaceable and involving comparatively little expense. The usual way is to submit the question or claim, if simple, to a single person, chosen by the parties; or, if more complicated, each of the parties chooses an arbitrator, and they a third person or umpire. The decision may be oral, but is better written, and made in important matters returnable to some court.

The arbitrators constitute a tribunal and exercise final jurisdiction between the parties, and are not bound by legal rules in the admission or exclusion of testimony, unless so stated in the agreement of submission. They are guided by their judgment as to what evidence will assist them in arriving at a just conclusion, not only as to the awards, but also as to costs to be recovered, from one or the other party, or to be divided between them. The statutes usually authorize this and the agreement of submission should do the same. The form of submission may be as follows:

FORM OF SUBMISSION.

Know all men, that we——, of——, and——, of——, do hereby promise and agree, to and with each other, to submit, and do hereby submit, all questions and claims between us [if there is a specific question or claim describe it], to the arbitrament and determination of [here insert the name of the arbitrators], whose decision and award shall be final, binding, and conclusive on us; [if there are more arbitrators than one, and it is intended that they may choose an umpire, add]: and, in case of disagreement between the said arbitrators, they may choose an umpire, whose award shall be final and conclusive [or if there be three arbitrators chosen to act together add:] and, in case of disagreement, the decision and award of a majority of said arbitrators shall be final and conclusive.

In witness whereof, ——have hereunto subscribed these presents, this——day of——, one thousand eight hundred and——.

In presence of—— [Signatures.]

Arbitrators' Oath.—It is proper that an oath be administered to the arbitrators to faithfully and justly decide the question or questions at issue. This may be administered by a justice of the peace or by a judge of a court of record.

XXIII. Award of Arbitrators.

The following is the form of award. When only one person is chosen, only his name should appear, as, for instance, I, John Doe:

To all to whom these presents shall come: We——, to whom were submitted, as arbitrators, the matters in controversy existing between—— as by the condition of their respective bonds of submission, executed by the said parties respectively, each unto the other, and bearing date the——day of ——, one thousand eight hundred and——more fully appears.

Now, therefore, know ye, that we——, the arbitrators mentioned in the said bonds, having been first duly sworn according to law, and having heard the proofs and allegations of the parties, and examined the matter in controversy by them submitted, do make this award in writing; that is to say, the said——shall on or before the——day of——next ensuing the date hereof [here insert the award made or whatever is to be performed.]

And also, the said arbitrators do hereby further award, that all actions depending between the said ——and—— for any matters arising or happening before their entering into said bonds of arbitration, shall from henceforth cease and determine, and be no further prosecuted or proceeded in by them or either of them, and that neither party recover costs against the other.

Finally, said arbitrators do further award, that the said——and——shall, within the space of—— days next after the date of this award, execute, each to the other, mutual releases of all actions and causes of action, suits, debts, damages, accounts and demands whatsoever.

IN WITNESS WHEREOF, ——have hereunto subscribed these presents, this——day of——, one thousand eight hundred and——

In presence of—— [Signatures.]

XXIV. Counterfeit Money.

THE art of detecting counterfeit money is not easily acquired, and it is especially difficult to impart it to others by means of writing. The first rule to observe is to become well acquainted with all the characteristics of good money.

Greenback Paper.—The paper on which the government prints its bills is of superior quality. Its texture is firm and strong; it feels smooth and solid to the touch. Counterfeiters make every effort to produce paper that resembles the genuine, but never succeed, so that an expert can almost invariably tell a good bill from a bad one by the touch alone. The U. S. Treasury notes of 1862 and 1863, and the old issues of National Bank notes having the red pointed seal, are printed on plain paper. All U. S. Treasury notes of 1869 to 1879 and National Bank notes having the scalloped seal are printed on paper which has silk or jute fiber interwoven all through it. Notes issued since 1880 are printed on fiber paper having a red and blue silk thread running lengthwise of the bill near the top and bottom. The fiber paper has not been successfully imitated. Attempts have been made to imitate it by pasting fiber on the surface or by marking lines resembling it with pen and ink, but these methods do not deceive a close observer.

The Ink.—The next important point to study is the ink used in printing the red and blue numbers and the seals. That on genuine bills is of superior quality, and remains bright and clear so long as the bill lasts. Counterfeiters have not been able to produce inks like those used by the government, and the numbers and seals on every counterfeit are inferior in color and printing. The whole appearance of good bills of every denomination should be carefully studied.

The Face of a Friend.—One should know the face of a bill as he knows the face of a friend. It would be no easy matter for a person resembling another to make you believe he actually was that other person. Just so with counterfeit bills. If you once know the genuine, the counterfeit cannot deceive you. The engraving of pictures and letters on the bill should be closely studied. On genuine bills it is perfect; the lines are even and straight.

The Engraving.—On counterfeits it is usually scratchy and irregular, though not always. Some counterfeit plates are engraved by expert workmen, who have no superior. The lathe work which surrounds the large numbers, and is found on the back of bills, should also be carefully examined. This is made from a die engraved by a machine that never produces the same design more than once. The government itself could not produce a duplicate. The counterfeiter has to imitate the design by hand, which, owing to its intricate pattern, is next to impossible. There are numerous special points by which certain bills are detected, but it does not come within the scope of this work to enumerate them. Those who want full information on the subject should subscribe to a good counterfeit detector.

Spurious Coin.—As there is a large amount of spurious coin in circulation, it is proper to refer briefly to some of the points that distinguish good from bad. Gold and silver possess a clear ring which distinguish them from the inferior metals. A person with a good ear for distinguishing sounds can usually detect a counterfeit by its ring alone, or rather by its lack of ring. The metals used in counterfeit coin are relatively lighter than gold and silver, and any one who has a good sense of touch can detect a counterfeit by its light weight and by a certain soapy feeling entirely different from the genuine. The outlines of devices on the coin are not usually so distinct and sharp in the counterfeit as in the genuine. Close observation is the best teacher. Compare a counterfeit piece of money, whether paper or coin, with a good piece of the same denomination, and the points of difference will be readily detected.

XXV. Good Business Maxims.

EVERY man is the architect of his own fortune.

Caution is the father of security.

Never boast of your success.

Speak well of honorable competitors; of dishonorable ones say nothing.

Systematize your business, and keep an eye on expenses. Small leaks sink large ships.

Never fail to take a receipt for money paid, and insist on giving a receipt for money received.

Keep copies of all important letters.

Be prompt in all things.

Avoid going to law.

Apply the golden rule to your business transactions, and you can't go far astray.

Josh Billings remarks: "I like to see a fellow practice his religion when he measures corn, as well as when he hollers Glory Hallelujah."

XXVI. Some Points on Business Law.

IGNORANCE of the law excuses no one.

It is a fraud to conceal a fraud.

The law compels no one to do impossibilities.

An agreement without consideration is void.

Signatures made with lead pencil are good in law.

A receipt for money paid is not legally conclusive.

The acts of one partner bind all others.

Each partner individually is responsible for the whole amount of the debts of the firm.

Contracts made on Sunday cannot be enforced.

Contracts for advertisements in Sunday papers are invalid.

A note drawn on Sunday is void.

A contract made with a minor is void.

A contract made with a lunatic is void.

Principals are responsible for the acts of their agents.

Agents are responsible to their principals.

A note obtained by fraud, or from a person in a state of intoxication, cannot be collected.

If a note be lost or stolen, it does not release the maker; he must pay it.

An endorser of a note is exempt from liability, if not served with notice of protest with the least possible delay.

There are, of course, exceptions to the above-mentioned general rules, which special cases may develop.

XXVII. Definitions of Mercantile Terms.

Acceptance.—The written agreement to pay a draft according to its terms.

Account.—The systematic arrangement of debits and credits under the name of a person, species of property or cause.

Assets.—Resources; available means.

Balance.—A term used to note the difference between the two sides of an account, or the sum necessary to make the account balance.

Bill.—The general name for a statement in writing, used in a variety of ways

Bills Receivable.—Written obligations or promises to pay money due the concern.

Bills Payable.—The concern's written promises to pay.

Capital.—Investment in business.

Days of Grace.—The time allowed by law and custom between the written date of maturity of a note or draft and the date upon which it must be paid.

Discount.—Consideration allowed for the payment of a debt before due.

Draft.—An order for the payment of money.

Drawee.—The person on whom a draft is drawn.

Drawer.—The person who draws the draft.

Exchange.—The fundamental principle of trade. Paper by which debts are paid without the transmission of money. Premium and discount arising from the purchase and sale of funds.

Favor.—The polite term for a letter received. A note or draft is in *favor* of the person to whom it is to be paid.

Honor.—To accept or pay when due.

Indorse.—To subscribe to a thing; to write one's name across the back of a note or draft.

Interest.—Compensation for the use of money.

Invoice.—A bill of goods bought or sold.

Ledger.—The chief book of accounts.
Liability.—A debt or claim against a person.
Maturity.—The date on which a note or draft falls due.
Maximum.—The highest price or rate.
Minimum.—The lowest price or rate.
Net.—That which remains after a certain reduction.
Net Proceeds.—The amount due a consignor after deducting charges attending sales.
Note.—An incidental remark made for the purpose of explanation. A written obligation to pay money.
Par.—Equal in value.
Principal.—An employer. The head of a commercial house. The amount loaned on which interest accrues.
Protest.—The formal notice that a note or draft was not paid when due, or that the acceptance of a draft was refused.
Stock.—Capital in trade. The title given to the property of a business.
Surety.—Indemnity against loss. A person bound for the performance of a contract by another.
Tender.—An offer for acceptance. A legal tender is an offer of such money as the law prescribes.
Usury.—Illegal interest.
Voucher.—A written evidence of an act performed.

XXVIII. Business Characters.

@	at.	#	number.
a/c	account.	+	sign of addition.
%	per cent.	—	sign of subtraction.
1^1	one and ¼	×	sign of multiplication.
1^2	one and ½	÷	sign of division.
1^3	one and ¾	=	sign of equality.

Part VIII.

HOUSEHOLD ART AND TASTE.

BEAUTIFYING THE HOME.

DRESS AND TOILET ART.

THE NURSERY AND SICK-ROOM.

RULES FOR THE PRESERVATION OF HEALTH.

REMEDIES AND PREVENTIVES OF DISEASE.

COOKING FOR THE SICK, ETC., ETC.

HOUSEHOLD ART AND TASTE.

CHAPTER I.

HOUSEHOLD ART AND TASTE.

I. BEAUTIFYING THE HOME.—II. FURNISHING THE HOUSE.—III. THE PARLOR FURNITURE.—IV. THE DINING-ROOM.—V. THE KITCHEN.—VI. THE BED-ROOMS.—VII. THE CELLAR.—VIII. THE WATER SUPPLY.—IX. SOFT-WATER CISTERNS.—X. LAYING DOWN CARPETS.—XI. PAINTING AND KALSOMINING.—XII. ARRANGEMENT OF FURNITURE.—XIII. HOUSE CLEANING.—XIV. SWEEPING AND DUSTING—RENOVATING CARPETS.

I. Beautifying the Home.

WHEN a man builds a house his first duty, after the family is comfortably settled, should be to make the surroundings pleasant. In the smallest village lot there is room for decoration. The walks should be graded and made firm and dry; the garden laid out and planted, vines shrubs and the necessary shade trees planted. There may not be room for shade within the inclosure, but trees should always be set next the street as soon as the house is built, unless finished too late in the spring. In that case the tree planting should be done the next autumn or in the following spring. Do not forget to plant a few flowers: Pæonia, bleeding heart, bellflower, larkspur, French honeysuckle and phlox are hardy, herbaceous perennials. The lilies, hyacinths, tulips, crocuses, narcissus, etc., are hardy bulbs. Train the common honeysuckle, the woodbine, any of the hardy climbing roses, or the ampelopsis over the porch or along the veranda. The hardy shrubs for the lawn are without number. In a previous chapter we have given a list of valuable fruit-trees and shrubs, as well as annual and perennial plants for the farm, orchard and garden. There is nothing that will so endear their home to the children, and make them love it, as the light labor of assisting to keep it trim and fair.

II. Furnishing the House.

The furniture of the house should correspond with the condition of the owner. Tawdriness must always be avoided. Do not try to ape some one wealthier than yourself, by buying cheap, flashy garniture. Plain, substantial furniture for those in moderate circumstances will look better and command more respect than cheap display. Study harmony. Never furnish a house by buying inferior, second-hand furniture or hangings, if it can be avoided. Especially let all bed-room furniture and bedding be new. Second-hand pictures, if good, are admissible. The first wear of everything else is generally the best.

Carpets and Bedding.—Never buy a flimsy carpet at any price. Do without until good ones can be purchased. If you can buy a good Brussels, with the pile dense and close, it will last a generation with proper care. In bedding, start with new, clean, honest material. Never let any young person sleep on a feather bed; it will cause undue heat, weaken the action of the skin, and cause those who lie in it to become susceptible to cold, besides other, more serious, evils of over-heating. If the bed or bedding be narrow, the occupant will not rest well, because proper movement cannot be made.

Use the same careful discrimination in the selection of all furniture that you would in any other matter. Have less, if necessary, but have that good, rather than crowd the rooms with inferior material. It is easier to add to a small number of good articles than suffer the annoyance of mistakes in over-furnishing with cheap stuff.

Hygiene of Bedding.—The system, so prevalent in America, of sleeping on feathers, and of placing two or more in a bed together, cannot be too strongly condemned. Healthy children, and all others not invalids, should sleep on hard mattresses, of which the best are made from curled horse-hair. These are, however, expensive, and many good substitutes may be bought; one of the best of these is the clean wood-fiber, called excelsior. Have springs, or woven wire, under beds if you choose: never feathers, except for very elderly people, who have grown too used to them to change. Never let a grown-up person, and, above all, never let an old person or an invalid, sleep with a child; it will destroy the child's vitality. So far as possible, give to each member of the family a bed. Not only is this better for the general health, but often, in case of illness, prevents contagion. Avoid stoves, especially coal stoves, in sleeping-rooms. If a light is needed, never use a turned-down kerosene lamp, for its fumes are injurious. Use hard beds, ventilate by open fires or otherwise, and cover well with coarse woolen bed-clothes, and half the illness in the family will disappear.

III. The Parlor Furniture.

THE parlor, like every other room in the house, should be furnished for wear. No sensible person furnishes a parlor to be shut up and remain unused, except upon great occasions. It is the place for the family to gather in when leisure allows, and not a place to be opened only when "Mrs. Grundy" calls. Hence, the furniture must be bought with an eye to use.

GLASS CASE FOR HOUSE-PLANTS.

The Pictures.—The pictures, whether oil, water-color, good chromos, prints or photographs, should correspond to the condition of the owner. A few really good engravings or paintings are better than any number of cheap ones. If you already have these, paper or kalsomine the walls to correspond. If there are engravings, composition frames, or those of walnut, rosewood or bird's-eye maple with gilt moldings, will be appropriate. There is no better place to study effect than in a well-arranged picture gallery, yet how few persons visit one of these for this purpose.

You may also there get some good lessons in hanging, with reference to light, etc. If the room is low, hang on nails behind the pictures, so the wire or cord is not seen.

The Curtains.—Curtains are pleasant things to have in every window of the house. They temper the light, keep out cold drafts, prevent the direct rays of the sun from entering when not wanted, and should be of material to correspond with the other furniture of the room.

The Parlor Carpet.—The parlor should, of course, have the best carpet in the house. In rooms of ordinary size avoid large figures. They cause a carpet to cut to waste and make the room look small. Also, avoid glaring colors. That so many such carpets are made shows that taste in the masses needs cultivating. The manufacturers are not to blame. They simply cater to the demands of the public.

House-Plants, etc.—Flowers and plants are in order everywhere, inside the house and out. The parlor, however, unless it be the living-room also, is not the place for their cultivation. We do not believe in dark parlors, yet in these rooms

AQUARIUM.

there is hardly sun enough admitted for the best growth of plants. Place such as may be easily moved in nice vases, and use them when in their best condition to ornament the room; those kept in wardian and other glass covered cases will also do well. The illustration of Glass Case for House-plants on page 727, shows a pretty design that may be kept in the living-room, and is easily moved from one room to another as may be desired. Ferns are admirable; none are prettier, in a collection, than the walking fern, shown on opposite page. Ferns will not bear the sun nor live in a dry atmosphere. For house cultivation they are usually kept in glass cases. An aquarium is pretty anywhere; especially so in the dining-room. All these we have mentioned may, with proper care, be freely rolled along a carpet from one room to another.

IV. The Dining-Room.

The dining-room requires little furniture, but that should be good and as handsome as you can afford. Stuffed furniture is out of place here, even if the dining-room is also used as a living-room. A sideboard, with proper conveniences, should be had if possible. The carpet ought to be bright rather than dark, and correspond to the other furniture, and the pictures in harmony with the surroundings. Here again the skill of the housewife may be used to have the proper closets for china and table ware convenient. In these days of inexpensive and excellent plated ware, a very little money, comparatively speaking, will add largely to the comfort and economy of the table service. Do not overdo the matter, though here and in the kitchen one may

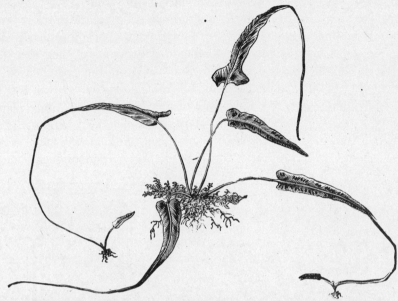

WALKING FERNS.

be pardoned any reasonable expense that will add to the real comfort of the family.

PINEAPPLE AND FRUIT.   CASE OF FERNS.

What Taste may do.—The dining-room is an excellent place to display taste, and this is especially the case if it be also used as the living-room. Let some of the pictures be suggestive of good living, game-birds, fish, fruit pieces, etc. An aquarium, plants, etc., as described in the section relating to the parlor, will also be appropriate for the dining-room. In the South, fruits not quite hardy enough for growing out of doors may be used. Among a collection we once saw was a growing pineapple as a dining-room decoration, removed from the greenhouse.

If one has a greenhouse many beautiful things may be grown for temporary removal to the dining-room; if not, some of the fruit-bearing house-plants may be used; such, for example, as Solanum or Jerusalem cherries. Among table decorations bouquets of flowers, or at the least some green thing, always suggest refinement. The fern case shown is appropriate for this purpose. The dining-room should be

WINDOW-PLANTS IN DINING-ROOM.

well lighted and cheerful, especially so when also used as the living-room. Break the glare of the sun, when necessary, with curtains. The illustration of Window-plants in Dining-room shows a pretty effect. To produce this costs little besides the necessary care of the plants, which may be made a labor of love as well as an educator to the children.

V. The Kitchen.

This is the most important room in every house of moderate expense, if not in all houses. The furniture should be ample and of the best manufacture consistent with the means of the owner. All kettles, stew and sauce pans should be of good tinned ware, or of stone or other silicate finish. Granite and other enamel coatings are now made so cheaply that they soon pay their cost in the ease of cleaning. The sink should be ample; the stove provided with a hot-water apparatus, the pantry and other closets easy of access. Let the floor be of hard wood, and covered with a good oil-cloth, if you can afford it, or, if not, well painted; rugs may be used in places where the work usually stands, for a woman's feet should not be in constant pressure either on oil-cloth or upon the bare floor. Here again the good sense of the mistress of the house may be shown in furnishing, both with a view to comfort and for the economy of work.

If servants are employed, they must be instructed in the proper care of the kitchen utensils, or there will be much waste from breakage or misuse. Hence, the necessity that the mistress fully understand how things should be done. If she must do the work herself, it will be a pleasure to be able to do it deftly and neatly; for light-handed neatness is the crowning glory of housekeeping.

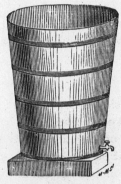
TUB FILTER.

TOWEL RACK.

Every kitchen should be provided with a filter for water, especially where rain-water is used for drinking, as is the case in many districts. The cut of Tub Filter shows a home-made affair, but as good as the best where ice forms a part of the filter over the straining cloth and under the dust cloth or cover. A towel rack is also indispensable. The one shown in the cut needs no explanation; any man can make it.

VI. The Bed-Rooms.

NOWHERE can better taste be displayed than in the sleeping apartments. As to carpets, every housekeeper must decide for herself. We should prefer a painted floor, and rugs so laid that the occupant would have no occasion to step on the wood after the shoes are removed. The rugs can be taken out in the sun, shaken and aired while the floor is mopped clean. Carpets in bed-rooms are unknown on the Continent of Europe in the grandest of private houses, and in such rooms, are traps for dust, germs of disease and death.

The furniture, however rich, should always be simple, and of solid material, to avoid dust and dirt, and be readily cleaned. The old-fashioned carpets and bed-curtains should be avoided. If possible, have in the bed-room only the bed, a rug and a few chairs; dress and undress in an adjoining room, which may be made as pretty as possible and kept much warmer in winter than the one which is used for sleeping in. In this dressing-room place the wardrobe, chest of drawers, pretty odds and ends, as well as the wash-hand stand and other such conveniences. This, too, is improved by a bright and pretty carpet, pictures and other wall decorations; the bedroom must have none of these. They only hold germs of disease, and dirt.

VII. The Cellar.

EVEN the smallest cottage should contain this important adjunct, if the nature of the soil will, or can be made to admit of proper drainage therefrom. The arrangement of the cellar is no less important. It should be fairly lighted, and be divided off into proper rooms according to the size of the house. In the smallest cottage the vegetable cellar should be separated from the rest, and proper ventilation should be looked to, else the odor will certainly reach every part of the house above. In large houses the laundry often occupies a portion of the cellar. If so, it should be provided with conveniences for hot and cold water, perfect ventilation, stationary tubs, with means for draining off wash-water, a sink and other fixtures.

VIII. The Water Supply.

EVERY farm-house—where there is a windmill for raising water—should have the necessary tanks for soft water for the house. These tanks may be in the barn or on other suitable elevations, from which the water may be conducted in pipes provided with faucets. In laying the pipes the greatest care should be taken that they are nowhere within reach of frost. If they are they become a source of constant annoyance in winter, and often of considerable damage to the building. Architects more often fail in providing against damage from frost than in any other respect. Plumbers are never mindful in this matter. They simply do their work according to the plan given. See, therefore, that no water pipes run next the outer wall of the building; that they are always where they can easily be gotten at, and, as an extra precaution, that they are always encased in some non-conducting material when there is any danger from frost; and, also, that in very cold weather a small flow may escape from the discharge-pipe connecting with all, during the night, so a constant current may be kept up. The supply-pipe should be brought, underground, well beneath the frost line, to the center of the cellar beneath the house. A wooden box, of boards not less than one foot wide, should receive the pipe at the depth of about two feet under the cellar floor and conduct it to the story above. This box must have the pipe in its center and the space between be packed in sawdust. In northern climates the pipe must extend to no room not kept warm in winter. It is better to do without it than have it freeze and break.

IX. Soft-Water Cisterns.

CISTERNS for rain-water should always be placed where they will not freeze. This is especially necessary where they are built of brick or stone and cement, for we suppose no one nowadays will consent to have a cistern plastered up directly on the earth or clay. It is cheap, and it is as worthless as it is cheap. In some localities, drinking-water is so difficult to obtain, that cisterns which collect the rain-water furnish the only supply. In this case, the cistern should be in two parts with a filter between. When it is built in this way, and the water comes from clean roofs, the water, though insipid, is pure and healthful. The living water of wells is, however, better in every respect, but no well-water is entirely safe, unless means have been taken to keep out surface drainage, and they are liable to be contaminated by the seepage, from sewers, the out-houses or the barnyard.

The danger from this latter cause is much greater than many people suppose. The earth is always honey-combed with the borings of insects and small animals, which always carry their burrows to the nearest water, generally the well. Seepage once entering these cavities inevitably finds its way to the well. Unfortunately, the most deadly germs are often not to be detected by the taste or smell. For this reason, many persons prefer cistern water to that from the well. When danger is suspected from wells, the water should always be boiled. No filtering will take out the deadly germs. Indeed, half the disease of the world would be avoided if all the water drunk was first boiled.

X. Laying Down Carpets.

Laying down the carpets is a task always dreaded by women. In fact, no heavy carpet should ever be undertaken by them. That it is a man's work always, may be easily discovered by anybody who has laid one. For this reason, in cities, the merchant who sells, undertakes to cut, fit and lay the carpet. In the country this is not always possible. With expensive carpets it will always be better to employ some one who thoroughly understands the art. An implement for stretching the carpet is always convenient and often indispensable.

CARPET STRETCHER.

First of all, the carpet must be cut into suitable lengths for the room, allowing for the proper matching of the breadths. It is then sewn together, breadth upon breadth, until the proper width is obtained. Then tack it down upon two sides, one way with the length and the other across the breadth, being careful to stretch all equally. It must be cut and fitted to inequalities, when necessary, and if a bay window is to be carpeted, this must be allowed for in cutting. The other two sides are then tacked down, the stretching always being carefully attended to, so that when finished, it will lie perfectly flat and without wrinkles.

XI. Painting and Kalsomining.

It is always better, if you can afford it, to hire both painting and kalsomining done by competent mechanics. If you do this, a perfect understanding must be had that the material shall be of the best, and that it shall not be dropped about the floors and over the furniture. There is no reason whatever why a person who understands his business should mess up a house with either paint or kalsomine. No person who takes little enough on the brush at a time of properly mixed material, need make a slop. If a dirty wall is to be kalsomined, never allow the size or first coat to be put on until the wall has been washed. This is done with a large sponge dipped in warm water, and pressed until nearly dry. This will wash off the dirt without dripping. With good brushes any one can soon learn to kalsomine, and to do common painting.

A mistake, too often made, is not mixing at first enough material of the required color to do the whole work. If you keep mixing a little at a time you never have

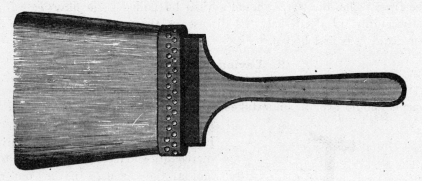

KALSOMINE BRUSH.

your walls of a uniform tint. So far as paints are concerned, they may be bought ready mixed and of any color. Graining and ornamental painting should never be attempted by an inexpert. In this case the very best workmen are always the cheapest.

Whitewashing.—Whitewash of lime is now seldom used for covering inside walls. There is no reason why it should not be. If properly made, it covers a surface almost as smoothly as the chalk of which kalsomine is composed, and is devoid of the disagreeable smell thereof. For covering rough buildings, fences, and other structures, where paint would be too costly, it is excellent, and, if properly made, is fairly water-proof, and may be applied by any one of ordinary intelligence and care. In order, however, that it may saturate the surface and hold, it should be put on hot, for which purpose the vessel containing it may be kept over a good-sized kerosene lamp or a low fire of charcoal.

Recipe for Whitewash.—One of the best washes we have ever used is made thus: To so much water as will fill a barrel to the depth of two inches add one-half bushel of pure, white quick-lime; then put in one peck of salt, previously dissolved in hot water; cover tightly to keep in the steam; when cold, strain through a fine sieve; heat it again and then add, hot, a thin starch paste made from three pounds of rice flour; stir; add one pound, hot, of strong glue; add half-pound of whiting, previously dissolved in hot water; dilute with hot water to consistency of cream; apply hot.

The glue is first soaked, then gradually dissolved in water, by placing the vessel holding it in another containing boiling water; used as directed, this is the most permanent wash we know of. About a pint of the mixture will cover a square yard of surface.

Colored Washes.—To make the above a cream-color, add yellow ochre until the desired shade is reached. For fawn-color, add four pounds of umber to one pound of lampblack. For gray or stone-color, four pounds of raw umber to two pounds of lampblack. Add to the whitewash until the desired shade is reached. To determine the color it must be seen dry, and not damp. Hence, when trying the color, let it dry to observe the tint.

XII. Arrangement of Furniture.

THE arrangement of furniture may make pleasant or mar the appearance of a room. Primness and precision should be avoided. If the chairs are set carefully against the wall at equal distances, if the sofa looks as though it had never been sat upon, if the center-table has a touch-me-not appearance, the general effect of the room will chill the visitor. The appearance of the room should be that of one used daily. The drapery about the windows should not be such as to shut out the light, but simply to tone down the glare. If the carpet is good, a fair amount of light will not hurt it, and a room that is always closed and dark, except when "company" comes, is sure to be musty, uncomfortable and unhealthful. If you cannot study out effects yourself, call in the aid of some one who has an eye for effect, and can produce like ones without copying. Observe effects in other houses, or take the advice of your upholsterer, always reserving to yourself the casting vote, as to how much you can afford to spend upon any particular room or object.

XIII. House-Cleaning.

IN house-cleaning you will save yourself and family much inconvenience, by not undertaking too much at once. Clean one room or one set of rooms at a time, and observe order in so doing. If your house is to be kalsomined or papered, this should be done first. House-cleaning is a time of severe labor, and any arrangement that will lighten the labor should be observed. Thus, the carpets may be taken up to be cleaned by the men, who may also kalsomine the walls. This will materially lighten the labor of the women. Many housewives prefer to hire extra labor, and this is decidedly the better way.

XIV. Sweeping and Dusting.—Renovating Carpets.

CARPETS should be brushed over at least once a day, and thoroughly swept once a week, when every movable piece of furniture should be moved. These should be thoroughly wiped or dusted every time the broom is used. If the carpet becomes dingy, it should be wiped with a damp sponge, and dried with clean flannel cloths. If there are grease-spots, they may be taken out by thoroughly pounding and mixing together equal parts of magnesia and fuller's earth. Make this into a paste with boiling water, lay it over the grease-spot, hot, and by the following day it will have absorbed the grease; it may then be scraped and brushed off. If, unfortunately, grease or oil has been spilled on the carpet, it should be taken out, if possible, before that is again swept. Ordinary stains may usually be removed with lemon-juice or dilute oxalic acid.

FLOOR BRUSH.

Not every person knows how to sweep clean without raising a great dust. If in sweeping you carry the broom, in its stroke, beyond you, it will inevitably make dust from the spring of the fibers of the broom. The spent tea-leaves should always be saved moist and scattered over the carpet for the regular sweepings, or salt should be strewn over the carpet. The strokes of the broom should be short, firm, and each should end when the broom has been drawn nearly up to a line with the person. Corners and the sides of the room should receive especial attention. As a preventive of dust a good carpet sweeper is valuable, but the movable furniture must be taken out to do good work, and the corners and edges cleaned with the broom, brush and dust-pan.

CHAPTER II.

THE PARLOR AND LIBRARY.

I. THE ROOMS FOR COMPANY.—II. GUESTS OF THE HOUSE.—III. ETIQUETTE OF THE PARLOR.—IV. ENTERTAINING VISITORS AND GUESTS.—V. DAILY DUTIES NOT INTERRUPTED BY GUESTS.—VI. GOING TO BED.—VII. SERVANTS AND PARLOR SERVICE.—VIII. DUTY TO CHILDREN.—IX. WHAT CONSTITUTES VULGARITY.—X. PARLOR DECORATION.—XI. DECORATION NOT NECESSARILY COSTLY.—XII. A ROCKING CHAIR.—XIII. A PRACTICAL FAMILY.—XIV. INGENIOUS AND USEFUL.

I. The Rooms for Company.

THE Parlor.—The apartment where guests are received may be one of the parlors, the reception-room, or the library, supposing the house to be large enough to contain all these. In England a distinction is made between the parlor and the drawing-room. In city houses, the parlor (from *parloir*, a place to speak in) is on the ground floor, and used as a reception-room and place to transact business, while the drawing-room (or withdrawing-room, as it was formerly called) is on the first floor, up one flight of stairs, and used more ceremoniously. In the United States we use the two words with the same meaning, as our houses have no such division. In large houses there are often two or more parlors, and the mistress of the establishment has, on the bed-room floor, what is called her boudoir, a private parlor for the reception of intimate, and, usually, female friends; as the study is, for men who have no office, the private room of the master of the house.

The Library.—The library should be solidly furnished, and contain, besides the bookcases, writing table and desk, easy chairs, lounges, sofas, etc. The books may be kept either in movable cases or those built permanently into the walls.

In smaller houses the parlor may serve also as a library, and often the "living-room" has to do duty as parlor, library and sitting-room.

II. Guests of the House.

It is the duty of the host and hostess to receive guests cordially and make them feel "at home." The tact of the individual must teach how to do this properly. It comes of the usage that can only be learned by contact with polite people. Rules cannot be laid down. They must be learned by observation. It is in perfect accordance with good taste among people of small means that the master, mistress or children of the household perform all the offices necessary to the comfort of guests, including those of the table. If there are servants, well and good. If not, such service is only that of a friend to friends.

III. Etiquette of the Parlor.

ETIQUETTE has been said to be the code of unwritten laws that governs the manners of people living in polite society. All society is "polite," whatever the station in life, provided good breeding is observed. Good breeding is the exhibition of gentleness, deference, suavity of manner, thoughtfulness, generosity, modesty and self-respect. Ease and cordiality, without freedom of manner, mark the gentleman or lady; freedom without ease, the vulgarian. If you receive a letter of introduction by a postman, acknowledge it immediately or call upon the stranger. If the person introduced brings it in person, receive the gentleman or lady courteously, and if a continuation of the acquaintance is desirable, give an invitation for another day, upon leave-taking.

All must exercise their own discretion as to introductions. In small parties, the guests are, as a rule, introduced to each person separately. In large gatherings not.

IV. Entertaining Visitors and Guests.

VISITORS are entertained by the ordinary gossip of the day, matters of local interest, society news, fashions, music, art, articles of taste, paintings, prints, poetry, and the general literature of the day. To entertain well, both parties should be well conversant with some of these. Gentlemen are interested in horses, fine stock, hunting, fishing, literature, art, science, and the out-door sports. The particular tastes of the visitor or guest being discovered, the drift into these channels is easy enough. In this, the visitor or guest should also come to understand the taste of the host and hostess, and then all comes easy enough.

V. Daily Duties not Interrupted by Guests.

ON the farm there are always routine duties that must occupy the attention of the host. Guests should be careful never to intrude upon these, and the host and hostess as careful, while attending to all necessary duties of the farm and household, to give as much time to guests as possible. No sensible person will ever make a long visit at a time of pressure of business. There are, however, many farmers whose leisure is ample at all times, and who keep servants enough to attend to all household duties, supervision only being necessary. With the majority, however, there is at most seasons, an absolute necessity for daily labor. Yet there are few who cannot entertain visitors during some part of the day, or find time to receive guests, and yet neglect no necessary labor. The guest who cannot at such times entertain him or her self, and even assist, had far better stay at home. The guest who has the happy faculty of keeping out of the way at proper times, and of doing service at others, we have never yet seen unwelcome in the farm household.

VI. Going to Bed.

NOT every person knows when to go to bed, nor when to get up. In the country, hours are necessarily early. The great charm is the early summer morning. To enjoy this, we must see the sun rise, and to be up early enough to do this, one should be in bed by nine o'clock at night. The routine duties of the farm make these hours imperative upon the host and hostess, if working farmers, and country life should require their observance by all in summer.

Once a guest has been shown to his chamber, courtesy would require that the service should not be daily performed. Yet, many like to continue chatting after leaving the parlor. Ladies, especially, find pleasant amusement with each other in a short bed-room talk over the events of the day and plans for the morrow.

VII. Servants and Parlor Service.

WHEN there are servants, the routine work must be performed by them. It is their duty to see that everything is done at the proper time. The parlors, library and dining-room are to be aired, swept and dusted before the family appear in the morning, and guests should scrupulously avoid being present at such times. Among the services to be performed is carrying hot or cold water to the guest chambers, attending to their occupants, and lighting these to bed at night, or, at least, bringing in the necessary candles to the parlor or bed-rooms and lighting them. In the winter, care should be taken that the bed-chambers are properly heated. In a country where fuel is so cheap as in the United States, there is at least no possible reason why a person should be obliged to undress or dress in a frozen room. There is neither economy nor wisdom in it.

VIII. Duty to Children.

CHILDREN should be early instructed in the ordinary amenities of life. At their time of life impressions are easily received and become a part of their character as men and women. Politeness and decorum in the parlor, dining-room or library costs nothing, and are as necessary when the parlor, dining-room and kitchen are one and the same room, as in more extensive houses. In this country we have no caste, and the child of the poorest parents may be called to the highest positions. If they are taught habits of cleanliness, decorum, and gentleness of manner, when young, it will cling to them through life, and go far to keep them out of bad company when they grow up. It gives habits of self-respect, and deters them alike from seeking bad company and from foolish or criminal expenditures. Children should by no means be curtailed of enjoyment. On the contrary, they should be given full liberty to indulge in innocent pleasures, guided by habits of self-restraint and self-reliance. They should feel they may always freely go to their parents for advice and sympathy. Thus they will come to regard home as the most pleasant place. They will naturally avoid rough and vulgar companions, and seldom, if ever, care for vicious or vulgar pastimes.

IX. What Constitutes Vulgarity.

IN general terms that is vulgar which is not in accordance with the usage of refined society. Loud laughter or loud conversation in public places is vulgar. Not to pay the proper deference to each other is vulgar. Not to assist a woman in any difficulty is not only vulgar but positively brutal. Not to apologize to any person whom you may accidentally jostle in a thoroughfare, especially if roughly, is vulgar and causes the action to be brutal. To refuse to accept an apology under like circumstances is vulgar. It is vulgar not to show self-respect. Arrogance is equally vulgar.

In fact, a dictionary of vulgar actions would fill a volume, and yet few persons are so lost to self-respect that they do not have a prick of conscience at a vulgar action, at least until their consciences are seared, or unless their education has been neglected at home. For conscience, although inherent in human nature, may be as much improved by education as any other faculty.

X. Parlor Decorations.

The furniture and decoration of the parlor should be as rich as you can afford. In any event the room should have an air of light and cheerfulness. Both it and the library should be comfortable, home-like rooms. Avoid glaring colors or cheap finery. Never buy stiff "sets" of furniture and never crowd your rooms with trash. The handsomest effects are often produced where each piece of furniture has been bought when wanted or it was convenient to do so. Let there be comfortable easy chairs and sofas, and always some chairs light enough to be easily lifted and moved. The wall-paper should be rich and generally light-colored, but of no pronounced pattern. Wooden mantels are handsome and often costly. If of marble these may be white or clouded. The wood-work should be light, unless rich dark woods are used, and the door-plates to match. Where the heating is by a stove, it may be steel, bronze or ormolu, and the fixtures for lighting the apartment at night should match. A chandelier makes a pretty center-piece for the ceiling, or if the room is long and large, two or more.

XI. Decoration Not Necessarily Costly.

The parlor must be, to be pretty, the room lived in. Never have a room too fine for yourself and your children to pass the evening in. It is part of the education of the latter to be in daily, familiar contact with the pretty household decorations that are always found in such rooms, when occupied by persons of taste, and an ordinary faculty of feminine ingenuity. We have seen better-kept and more healthy plants, in such a room where the family did the labor, and a prettier arrangement of them, than we ever saw in households where the care was left entirely to servants. One case in particular we recall. A large oriel window like that shown in the illustration made

AN ORIEL WINDOW.

one side of the room. The carpet was a good ingrain, in pretty figured squares.

There was not a costly piece of furniture in the room, and yet it had an air of comfort and of refinement that is within the reach of all.

Another room we remember where there was no bay window, was made fully as handsome, by the arrangement shown in the illustration of Living-room Window. The projection is inside, not out, and yet the effect is as pleasing, as unique; serving, not only to hold and protect the plants, but adding to and ornamenting the room. Except the ordinary furniture and the prints, everything,

DRAGON-LIKE ORNAMENT.

THE LIVING-ROOM WINDOW.

including the frames for the pictures, had been made by the members of the household. The chair covers and the back of the lounge were made of patchwork, resembling embroidery, which is named applique.

Some of the individual figures were grotesque enough, as for instance, a dragon with bat-like wings, came near the present fashionable rage for Japanese ornamentation. The backs of the chairs were covered in designs something like that in the illustration of Chair Cover, or rather, ornamented by pieces like those fastened to the back, for the cut really represents what the ladies call a tidy.

XII. A Rocking-Chair.

The rocking-chair deserves more than a passing notice, for it is "mother's chair," and made for her especial use. The boys, let us suppose, having reseated and cushioned the old rocker, the girls covered it. The seat and back were cushioned with hair, covered with soft grayish material. Then Turkish toweling was cut somewhat larger, to allow for fold and nailing. The figures were cut from dark cloth, and appliqued to this foundation with zephyr worsted, the fringe and tassels made of ravelings of the toweling colored red. The cross-stitch at the edges was worked with the same color. The dragon rampant was considered especially appropriate, since the grandfather had once been a sailor to the "China Seas."

XIII. A Practical Family.

The old grandfather had reason to be proud of his descendants. There were three boys and three girls, and each "wonderful for something," as the old gentle-

DESIGN FOR CHAIR COVER.

THE DRAGON CHAIR.

man expressed it. Dolly and Tom engineered the chair and other work of like kind, Aleck was good in carpentry work, and Sarah in designing. The Work-box and Seat was made by Aleck, Dolly and Sarah; the designs and fittings being by Sarah, and the embroidery by Dolly, of course, while Aleck made the frame. The

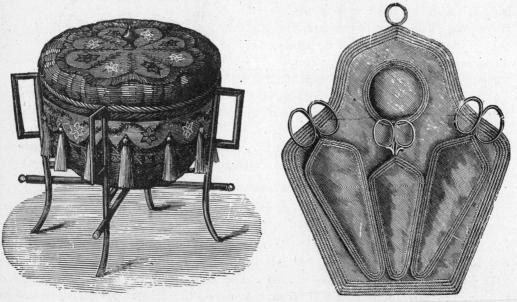

WORK-BOX AND SEAT.   SCISSORS CASE.

appliques were of velvet, caught down with a button-hole stitch of silk, and the embroidery in a large loose stitch, but delicately rendered.

Sarah was the "author," as Tom called it, of all the "cute fixtures." The Scissors Case was her work, of course. Tom was a genius in his way—a maker of

A PLANT CASE.

rustic work and a florist helper to Anna, the artist. The Plant Case was the joint work of Tom and Anna.

A PLANT FUMIGATOR.

PLANT-CASE BOTTOM.

It is well known that many plants, as ferns, orchids, etc., do best in a close, moist atmosphere; at least they will not thrive in a dry or changeable one, and so

744   THE HOME AND FARM MANUAL.

Tom promised, when he went away from home, that some day they should have a better case than the one originally improvised. In time it came, and with it the plants to fill it. The stand was of mahogany, lined with zinc and strengthened with brass. The top was of brass and French glass; the panels were painted by Anna.

Now, plants cannot be kept healthy at all times in a room without occasional fumigation to destroy insects. Perhaps you would like to see their fumigator. Simple enough, is it not, as shown in the illustration? A muslin cover draped over a wire frame, a little tin box, if you like, for the burning tobacco, a tube leading under the cloth and another tube for blowing to keep up combustion and drive out the smoke.

XIV. Ingenious and Useful.

While on the subject of simple things, here is another of Sarah's ingenious contrivances, a water-cooler.

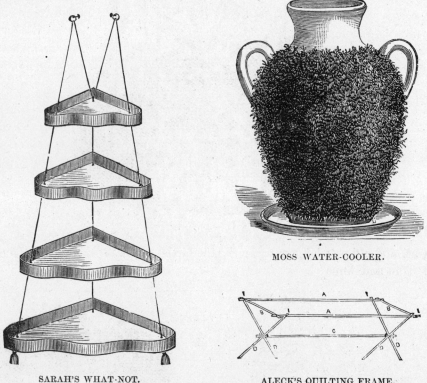

SARAH'S WHAT-NOT.   MOSS WATER-COOLER.   ALECK'S QUILTING FRAME.

A porous (unglazed) vase was bought and moss was fastened about it as shown. The water, percolating through the pores, helped to keep the moss damp, cooled the water, and, when much evaporation and consequent coolness were wanted, the vase was set in a draft of air. The family have ice now, and the old water-cooler is now a living vase. It is kept full of water, and various small seeds are sown among the moss, which is held in place with fine silvered wire. Garden cress and various ornamental grasses are pretty growing in this way.

A modification of the Wardian case may be had by getting a stand of terra cotta, putting a rim inside, within which a glass shade is set. Inside the shade are mosses, lycopodiums and ferns; and between the two rims a living fringe of foliage may be had with a little care; for, the rim being kept filled with water, cut flowers may be maintained there as shown in the illustration of Plant-case Bottom.

Sarah's What-not was a light affair, but heavy enough for the light articles it was to hold. It was made of card-board, small figured gilt wall-paper, velvet bordering, picture-frame cord and tassels.

Aleck's Quilting Frame must also be described: The legs, $d\ d\ d\ d$, were of hard wood, one and one-quarter inch thick, three inches wide and three and one-half feet long; the bar, c, was two and one-half inches "eight square;" the rollers, $a\ a$, to which the quilt is attached, two and one-half inches "eight square,' with a strip of cloth on one side on which to fasten the quilt, the rollers passing through the legs at 1 1 1 1; the cross-pieces, $b\ b$, were twenty-seven inches'long, of inch stuff mortised at each end, which held the quilt stretched. These slipped on and off, and when not in use the frame could be folded up and put away. It has on more than one occasion been made to serve as a cot.

CHAPTER III.

THE DINING-ROOM AND ITS SERVICE.

I. DINING-ROOM FURNITURE AND DECORATION.—II. TABLE ETIQUETTE.—III. CARVING AT TABLE.—IV. CARVING FOUR-FOOTED GAME.—V. CARVING BIRDS AND FOWLS.—VI. CARVING FISH.—VII. THE SERVICE OF THE TABLE.—VIII. SOME DISHES FOR EPICURES.—IX. QUEER FACTS ABOUT VEGETABLES.—X. THE USE OF NAPKINS.

I. Dining-Room Furniture and Decoration.

IN all we have said of the ornamentation of farms, of the home, of household art and taste, and in all we shall say, we wish again to have borne in mind that it is intended for those whose yearly increasing means allow them to gratify their tastes. Those struggling to pay debts, or those still employed in bringing their farms into condition, should spend little in display; all is needed for mere comfort. While we have endeavored to show how comfort may be secured at light cost, we wish at the same time to educate taste, and show how, as wealth increases, the money may be spent to the best advantage.

The dining-room furniture has already been spoken of. The paper on the walls should not be of gaudy pattern, but may be rich and warm in tone. Massive moldings and cornices should be used in the better class of houses. The arrangements for heating should be perfect, for no one can enjoy a meal in a room that is insufficiently heated or over-heated. The dining-room table may be as massive and handsome as possible, even to solid mahogany if the purse will allow. The chairs ought to be strong and at the same time as light as is consistent with strength. Either mahogany or oak will be handsome enough, and our preference would be for the latter wood without reference to its lesser cost; there is no limit to the ornamentation that may be put upon wood, in carving, etc. Whatever it be, all must harmonize. Let it be all oak, all cherry, all mahogany, to correspond with the graining or the solid woodwork of the room. If the chairs are upholstered, leather is the best. A sideboard of some kind for glassware, china, etc., and which is to be placed near the head of the table, may be considered indispensable in a large dining-room. In all farmhouses of the better sort these will be found economical in the end.

Decoration of the Table.—The illustration of Completely Arranged Dinner-table is given more for the hints and suggestions it contains than for close imitation. With handsome silver and cut glass the wealthy can, of course, make up a magnificent dinner-table. The floral work shown in the illustration is thus done: In the dish at the bottom of the center-piece place the flowers of the scarlet cereus, and about it cluster stephanotus, with spikes of cyperus, alternating with delicate fern-leaves above. In the compartment half-way up the stand are various flowers, including pale small geraniums, lily of the valley, maiden-hair, some spikes of ornamental grasses, with lycopodium trailing over the edge. In the funnel-shaped top is a bouquet of various flowers, with vines drooping therefrom. The end-pieces are low pots of ferns covered with foliage. There is a small bouquet at every plate.

Of course, excessive decoration will not be indulged in by persons of good taste, however wealthy. The every-day decoration is in accordance with the every-day expenditure, but within this limit some attempt should be made by all, every-day, to beautify the table with flowers and in other ways that refined taste may suggest.

COMPLETELY ARRANGED DINNER-TABLE.

II. Table Etiquette.

IF you are not conversant with table etiquette, observe the actions, in a quiet way, of those whom you suppose to be most conversant therewith. Full directions for table etiquette and all other matters of personal deportment will be found in Part IX. of this work. No person is expected to take wine at a dinner party unless it is usual for them so to do. If you do, always use the proper glasses for the wine served. If fish knives and forks are provided, use them; if not, use only your fork and a small crust of bread. Never put the uneatable portions of fish, flesh, fowls, or other debris off your plate, and especially do not lay any such thing on the cloth. Avoid remarks upon the quality or value of the food or service. If asked to take wine with any of the company, and you do not wish to be excused on the plea that you do not drink wine, have your glass replenished. It is not usual now to take wine in this way, and no person is expected to take wine as a matter of course. Never take a bone in your fingers; the meat can always be separated with the knife. If, unfortunately, you get a bone or other substance in your mouth that may not be swallowed, remove it deftly in your napkin; but such should never be the case, if you are careful and do not hurry in your eating. If you *must* cough or sneeze, do so in your handkerchief and as quietly as possible; and, above all, never drink hastily or with your mouth full.

III. Carving at Table.

EVERY person should know how to carve; practice until you are reasonably perfect. It is never done by main strength. A good knife, moderately long, pointed and

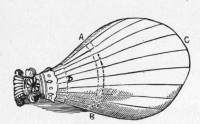

A DRESSED HAM.

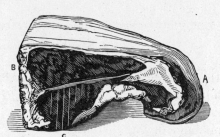

SIRLOIN OF BEEF.

keen must be used. All meat should be separated at the joints by the butcher. Birds are served whole. Fish are divided with the fish slice, and the flakes should not be broken in serving. A good rule in carving meats is to cut in rather thin slices, across the grain, serving some fat with the lean.

Carving a Ham.—In carving a boiled or baked ham begin nearly midway from the small end and carve in thin slices across the ham, following in succession to the larger part. Cut across from A to B, as shown in the illustration. For those who prefer the hock, carve at D. When a fair amount of meat has been carved, cut the remainder, in the direction of C, D, in thin slices.

FILLET OF VEAL.

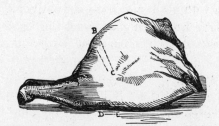

LEG OF MUTTON.

Roast Beef and Veal.—If a roast of beef or veal is made into a fillet by the butcher, that is, the bone removed and the meat rolled, the carving is easy. Cut a slice off the entire top that you may have a piece to serve with that from the inside. If there is stuffing skewered in, serve some of this also; serve fat if it is liked.

In a sirloin the carving is easy. The cuts should be as indicated by the white lines; with the tenderloin, also serve liberally of slices from the top.

Leg of Mutton.—The under, or the thickest part of a leg of mutton, should be placed uppermost and carved in slices moderately thin from B to C. If the knuckle is asked for, serve it. When cold, the leg should be carved from the upper side. The cramp-bone, considered by some a dainty, is removed by putting your knife in at D and passing it around to E.

In all carving use firm strokes of the knife, so as to make clean cuts. Ladies should make carving a study. Their carving knives and forks are smaller than those used by gentlemen. All broiled or fried meats are easily carved, but each piece should be shapely and, if possible, have some "tid-bit" or fat with it.

IV. Carving Four-Footed Game.

HARE, rabbit, squirrel, and that kind of game, should be unjointed by placing the knife properly in, turning it back, thus disclosing the joints, when it may be separated into proper pieces. Cut moderate pieces from the shoulder to the end of the loin, and divide the head last, by severing from the neck, removing the lower jaw, cutting through the division, from the nose to the top of the skull, and laying it open. Serve dressing and sauce, or gravy with each piece.

ROAST TURKEY.

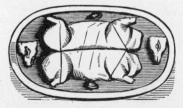

ROAST PIG.

Roast Pig.—To carve a roasted pig—which should be sent to table garnished as shown with head and ears—sever the pieces as shown in the diagram, divide the ribs and serve with plenty of sauce. The ear and jaw are favorite parts with some. If a joint is too much, separate it into smaller pieces.

V. Carving Birds and Fowls.

SMALL birds are either divided in halves and served in halves, or served whole, as are quail, larks, etc. Larger land game is carved as are fowls.

Carving Turkey.—A turkey is carved by first taking thin slices from the breast, as at A B on each side, until the whole breast is removed. Then take off the legs, dividing the thigh from the drum-stick, and if a disjointer is at hand, use it to separate the joint. Take off the wings and separate them at the joints. Serve with dressing from C, and gravy. It is not usual to separate the bones of the rack of a turkey. A boiled turkey is carved in the same manner as a roasted one, but the trussing being different—the legs are drawn in to the body—it is somewhat more difficult.

Chickens.—Barn-yard fowls are carved as is a turkey, so far as removing the joints is concerned. Then remove the merry-thought—wish-bone—by inserting the knife and passing it under the bones; raise it, and the separation is easily effected. The breast is served generally in slices, with other parts. To divide the fowl, cut through the ribs down to the vent, turn the back uppermost, put the knife in about the center, between the neck and rump, raise the lower part firmly, but not with haste, and the separation is made. Then turn the neck or rump from you, take off the side-bones, and the carving is complete. If the fowl is a capon, the breast is sliced as in a turkey. Young fowls are generally served without slicing the breast.

 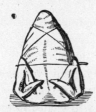 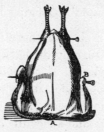 

BOILED FOWL—BREAST.   BOILED FOWL—BACK.   PHEASANT.   PARTRIDGE.

Grouse, Ducks, Pheasants.—Grouse are carved like chickens. So, also, are pheasants, but none of the breast is taken with the wings. The larger ducks are carved in the same way.

Partridges.—If the party is of gentlemen only, the partridges are carved by dividing the bird into halves, cutting down through the center, lengthwise. If the party is of ladies and gentlemen, separate the legs at the thigh, and divide the bird into three parts, leaving the leg and wing on each side together. The breast is then divided from the back, the breast either helped whole, or divided in two, and helped with any of the other parts. The cut shows the manner of trussing.

Pigeons, etc.—The breast of ducks is the choice part, and is served in slices. Teal, widgeon and other small ducks are sometimes divided into halves, and thus

PIGEON—BREAST.   PIGEON—BACK.   ROASTED GOOSE.

served. Pigeon, woodcock, and the larger snipe, are sometimes divided into half, but generally served whole. The cuts show the manner of trussing pigeons, both breast and back view.

THE DINING-ROOM AND ITS SERVICE.

Geese.—In carving geese, follow with your knife the marks shown in the cut, A to B; remove the wings, and the legs also if required. The dressing is taken from the apron beyond B.

VI. Carving Fish.

ALL large fish are divided into slices with the fish slice. It requires more tact than knowledge. Take thin slices from the back and serve with pieces of the belly.

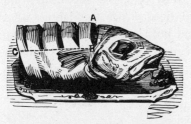

CODFISH—HEAD AND SHOULDERS.

Cod.—Cod's head and shoulders is served by taking slices across the back down to a line with the fin, serving some of the sound, which lines the back and is taken by passing the knife under the back-bone; serve also some of the liver.

Flat Fish.—Flat fish are served whole, if small. The larger flat fish are served in flat slices, serving a part of the fin with each piece. If too large to be served whole, serve the halves by taking out the back-bone. The same rule will apply to all other fish.

Serve all pan fish whole, if small enough, or cut in halves by dividing along the back-bone. Bony fish, when baked, should have the back and belly slit up, and each slice drawn gently downward, by which means less bones are served. If there is dressing, serve with the fish; also sauce.

PAN FISH.

A PIECE OF SALMON.

Salmon.—In carving a piece of salmon, take thin slices, as shown from A to B, and serve with each piece some of the belly taken in the direction from C to D. The best part is the upper or thick flesh.

VII. The Service of the Table.

BREAKFAST is and should be one of the pleasantest meals of the day. Here the family assemble, if not before, and courtesies and pleasant chat are interchanged over the fragrant coffee, chops and rolls. The linen should be of the cleanest, the silver bright, and the china dry and polished. The plates should be hot in winter; rolls should be hot, unless cold bread is preferred, and the dry toast just from the fire. Buttered toast should be served as soon as it is made. Let there be flowers to adorn the table at each meal, if possible. If you have an urn for hot water, place the coffee-pot and the tea-pot in front of it, the coffee cups and saucers at your right and the tea-cups at the left; the cream and hot milk at the right, the slop-basin and milk-pitcher at the left, and the sugar-bowl handy. There should be a fruit-plate and breakfast-plate for each person.

Tea is a more simple meal, but none the less pleasant, since it closes the labors of the day. Luncheon also is a pleasant meal, because simple and lively with chat.

Dinner is the stately meal of the day, unless simply a family meal. It is then, one of the most pleasant. In fact, why should not every meal be pleasant, and without ceremony? Pleasant conversation and plenty of time at meals make digestion easy. There is no reason why a meal should be bolted because of the labors of a hurrying season. It is bad for digestion, and that is bad both for the temper and the health. The farmer, of all the laboring classes, deserves that his table should be pleasant to look at, and adorned with flowers and green things. He has it all within himself to make it so.

VIII. Some Dishes for Epicures.

WE have it from the most ancient authorities that their meals were not dull. The household of Job, for instance, would seem to have been hospitable and merry, for we read: "His sons went and feasted in their houses, every one their day (birthday); and sent and called for their three sisters to eat and drink with them." The tables, however, probably were not provided with cooked peacock, feathers, tail and all. That was a whim of later and more degenerate days, even if of a higher civilization. The Romans—who were fond of the dish just spoken of and many others still more curious, as for instance a pig baked on one side and boiled on the other—fed the thrushes destined for their gourmand tables, with figs, wheat and aromatic grains. A French epicure has said that small song-birds should be eaten the last of November, for then the feeding on juniper berries gives their flesh the much-admired bitter flavor. So with us, the grouse of Pennsylvania—our "prairie hen"—is supposed to be much finer than anywhere else, from the mountain berries they feed upon.

All wild birds are supposed to be more nutritious and digestible than domesticated ones, probably because they contain more fibrine and less fat. The same may be said of venison. It produces "highly stimulating chyle," hence the digestion is easy and rapid.

IX. Queer Facts about Vegetables.

HERE again, it seems curious that the tables of farmers are not better supplied with what they may so easily and cheaply raise. Three great Roman names came from their vegetables. Fabius, the great general, we should call General Bean; the great orator, Cicero, was Vice-Chancellor Pea, and the house of Lentilus got the name from the lentil. Gray peas were said to have formed the principal refreshment at the circus and theater. This refreshment would rather surprise the average circus audience nowadays. Instead, we have the familiar circus lemonade, made without lemons, peanut and prize-package bawlers. With the fall of the Roman Empire vegetables went out of fashion, and with them went pretty much all civilization.

The Romans are said to have raised asparagus stems which weighed three pounds each. To match this the Jews raised radishes of a hundred pounds weight, and so hollow that a "fox and cubs might burrow therein."

The history of cabbage is curious. The Egyptians deified it, but then they had many curious gods. They took cabbage first at their feasts. The Greeks and Romans took it as a tonic after drunkenness. Cato thought it a panacea for all the ills of man. It was thought to be a specific for paralysis. Hippocrates prescribed it boiled, with salt, for the colic, and in Athens it was thought a most excellent thing for young nursing mothers. Yet it is comparatively a short time only since cabbage became common in England. It is hardly fashionable in the United States except raw, and with oysters.

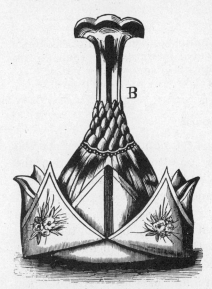

NAPKINS FOLDED ABOUT A DECANTER.

X. The Use of Napkins.

BECAUSE the great Duke of Wellington was obliged to envelop his whole chest in a napkin to prevent catching too much of the soup and other dishes in his waistcoat, is no reason why clean eaters should do so. The place for the napkin—except with gluttons, old men and infants—is on the lap. The first thing on sitting down to table is to unfold the napkin, spread it on the lap, and the last thing after dining is to lay it beside the plate or slip it in the ring if there be one. When placed in a ring they are simply folded square. There are many fancy shapes for folding, but simple styles are preferable. The cuts, B 3 and B 4, show fancy

 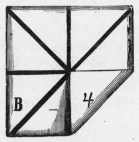 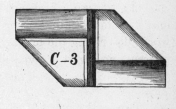

FOLDING NAPKINS.

ways of folding, the black lines marking the several folds. C 3 shows the miter fold. The decanter requires no explanation. The arrangement will be easily seen from the cut, and may be varied in many pleasing forms, to suit vases or other table ornaments.

CHAPTER IV.

DRESS, AND TOILET ART.

I. DRESS, ANCIENT AND MODERN.—II. THE REAL PURPOSES OF DRESS.—III. CLOTHE ACCORDING TO CIRCUMSTANCES.—IV. MENDING CLOTHES.—V. ALTERING CLOTHES.—VI. THE KIND OF CLOTHES TO WEAR.—VII. TASTE IN LADIES' DRESS.—VIII. SOMETHING ABOUT COLOR.—IX. TOILET-ROOM AND BATH.—X. GARMENTS NEXT THE SKIN.—XI. THE CARE OF CLOTHES.—XII. THE CARE OF BRUSHES AND COMBS.

I. Dress, Ancient and Modern.

IN classic Greece, and in fact among all polite ancient nations, the form of the wearer gave shape to the dress. In our days clothes too often make the figure of the person who wears them. Among the ancients, a simple piece of cloth was allowed to drape itself negligently over the form, now and then disclosing its proportions. Now it is the art of the tailor and dress-maker which makes the fashionable man or woman. The one padded and corseted, puffed out here, drawn in there; the other a mystery of cotton, whalebone, steel and bustle, beneath which the wearer " moves and has her being," sighs, languishes and breathes. In the country, indeed, fashion does not go so far. Simple taste may there find exercise. Happy is the city man or woman whose taste and wealth allow a few months of natural life in the country.

True Taste in Dress.—The highest art in dress is that which, while conforming in a measure to the prevailing fashion, exercises taste in the natural adornment of the body. It is a sad sight to see suddenly rich people in city and country in gaudy and vulgar finery; they deserve only the ridicule they meet in their attempts to ape the aristocrats abroad. The real gentleman or lady is never so dressed as to attract special attention. Hence the real art of dress, whatever the station, or however rich, may be summed up in two words: Unobtrusive simplicity. Almost equally few are the words describing the manners of the gentleman or lady. They are simply: Modesty, unassuming dignity, and self-possession.

II. The Real Purposes of Dress.

THE real purposes of dress, beyond that of satisfying the natural instinct of modesty and gratifying the universal passion for personal adornment, are to properly prevent the loss of animal heat in winter, to facilitate the escape of the animal heat in summer, and to protect the person from the extremes of the weather. The reason why it is necessary to comply measurably with the demands of fashion, is, that any one who departs too far from the general custom in dress, becomes so conspicuous as to attract that attention which every right-minded person seeks to avoid. That lady was perfectly dressed of whom this story is told: Having visited a friend, the next day the friend was asked by a visitor (not perfect in dress or manners) what she wore. "Indeed," was the reply, "I did not notice her dress; only the charm of her manner and conversation." A delicate hint to the other, that if asked *how* she was dressed, it would not have been hard to remember.

No Warmth in Clothes.—The notion that the body receives warmth from the dress is altogether wrong. There is no actual warmth in clothing. It is simply its power of conserving or preventing the escape of the natural heat of the body that makes it seem warm. With the thermometer at seventy or eighty degrees, but little clothing is required. With the thermometer at one hundred degrees, the real science of dress is to facilitate the escape of heat by every possible means, as, for instance, a free circulation of air. The heat of the body comes from the food we eat. Hence, in the summer, heating food should be avoided. Perspiration is the principal natural means of reducing excessive heat of the body; and thus the person who works in a great heat perspires continuously and violently. If this were not so, the system would quickly give way and death ensue.

III. Clothe According to Circumstances.

It is not our purpose to write a dissertation on fashionable dress. We only propose to show something of the philosophy and economy of clothing. The good sense of every lady and gentleman will easily suggest what kind of clothes they shall wear. The pocket as well as the taste must here be consulted. But yet, it is not so much the material, as the manner in which it is made up and worn, that marks the person of refinement and sensibility—the true gentleman and lady. A person may be soiled with work that brings one in contact with dirt and grime, and yet not be offensive. Yet it is a fact, dirt does not stick so easily to some people as to others—probably for the reason that the same impulse towards cleanliness, that prompts bathing and a change of dress, immediately the work is done, prevents also undue contact with that which is offensive.

IV. Mending Clothes.

Many persons have a horror of patched clothes. If patching becomes necessary, there should be no sensitiveness upon the subject. We must clothe according to our condition in life. The opinion of the butterflies and fops who largely make up certain phases of fashionable life, is not worth a thought. It is a false presumption, that because a person has wealth, he or she looks down on those who are poor. The gentleman and lady never do. It is only "cads" and "snobs" who fear contact with those who keep the world moving. The world moves by the force of the labor that is done. It is those who mix mortar, lay brick and stone, build and finish houses, make machinery and utensils, set type and run the printing press, raise the provisions that feed the multitude—in short, the thinkers and workers of this world—who are the moving power. The mechanic and the manual laborer, as well as the merchant, the manufacturer and the financier, are the machinery which keeps the world in motion. The merely wealthy, who spend their lives in idleness, are as moths flitting about a candle—necessary perhaps, because the money they spend helps to make a market for the products of the workers. But the soiled laborer, with patched clothes, if temperate, industrious and honest, performs a more important part in this workaday world of ours.

Some must wear mended clothes. In fact, there are few sensible persons, however rich, but do so in some degree. In making up a dress or suit, reserve some of the stuff for the mending that must surely come. If this cannot be done, then select the nearest match you can. It is no disgrace to wear a coat of as many colors as that of Joseph. It may have been the height of fashion in Jacob's time; it is not so now. But if the gentleman or lady of wealth desire not to attract attention to their dress, certainly the gentleman or lady less wealthy should not seek to attract attention by the singularity of their patching and darning.

"A stitch in time" should be the motto. How easy it is in sorting the clothes for the wash to 'make a memorandum of such as require mending, if you cannot remember—a button off a garment, a frayed edge or button-hole, a rent, a missing string or band, towel to be cut in half and the edges resewed. The clothes being rough dried, mend them before starching and ironing.

V. Altering Clothes.

In these days of cheap sewing machines the work of making, mending and altering clothing is much simplified. Children's clothing requires constant care. Tailoring requires considerable strength. Men's clothing is more cheaply bought ready-made than it can be made at home. The mending is not difficult.

Altering Children's Clothes.—Patches may be so neatly inserted as scarcely to show. The legs of a child's trousers may be turned by cutting off, and reversing—changing right and left—so as to scarcely show. Children's skirts may be altered to fit the constantly growing form, letting out the tucks and bands. The material for children's suits or dresses may be gotten from the partly worn clothes of the adult members of the family. Garments thus made are just as good for the little ones to play about in the fresh earth and grass, just as good to be torn by briers and brush as new ones. Children need pretty free scope at play. It is healthy; it is good for both brain and muscle. The child who never rolled in the grass, never wanted to make mud pies or haul sand on a shingle, who never soiled or made a rent in the dress, never made any mark in the world as man or woman.

VI. The Kind of Clothes to Wear.

Children are easily kept comfortable if proper care is taken in the materials and making of their clothing. That young children should be girded in tight garments is absurd. The young girl who is laced together to bring her "form into shape," with dresses of the shortest, and whose legs from the knee are covered only with thin stockings in winter, for fashion's sake, has as foolish a mother as has the boy a father who refuses him flannel in winter, and makes him leave off his shoes and stockings in early spring, to "toughen him." That kind of fashion and that kind of toughening send a legion of children yearly into premature graves.

Dress your children comfortably, whatever the material, with light colors and thin texture in summer, and bright, warm colors and stout, close goods for winter. Keep good shoes and stockings on the feet, and light gear on the head, for it is as true now as it was in Franklin's time, that "he who keeps the feet warm and the head cool may bid defiance to the doctors." Comfortable, flexible, rather loose hats and caps, and strong, well-fitting shoes for the boys; and no tight bandaging with belts, stays and garters for the girls. Let them race and run to their hearts' content. Nature intended that all young things should do so, to develop and round them out. Let those who think that children should act as prim as the maiden of forty, take a lesson from the colts, calves and lambs. They are always at play when not feeding or resting. It is their education.

VII. Taste in Ladies' Dress.

Every lady should know what colors, and shades of color, harmonize with her hair, eyes and complexion; and what patterns will be in harmony with her form. Ladies instinctively understand this; gentlemen, as a rule, depend upon their tailors. To those ladies who are in doubt, we should say, consult your dressmaker or your milliner. However costly the material, it should be simple rather than glaring in color. Avoid strong contrasts in the colors of the dress. However fashionable a color may be, abjure it unless it is becoming to you. A fashion is as often started to hide some deformity or peculiarity in a leader of fashion, as for any other reason. Adapt new purchases as much as possible to the articles you already have, and always let them be in harmony with your height, age, station in life and complexion.

VIII. Something About Color.

The strong or primitive colors are red, yellow and blue. Yellow and red produce orange by simple union; yellow and blue produce green; red and blue make purple; orange and green, again, produce olive; orange and purple produce brown; green and purple form a slate-color. The cold colors are blue, green and purple. The warm colors are yellow, red and orange; olive, brown and slate are neutral colors. These are modified by light or shade. For instance, grass which in the bright sunshine appears almost yellow, in the shade is a cool, refreshing green. Take the three *primitive* colors, yellow, red and blue. Upon a disk of paper paint the lower half blue, the upper right-hand quarter red, and the upper left-hand quarter yellow. Fasten this upon any swift-whirling object, as a humming or peg top, and they form white. A cold or warm effect is produced by a proper combination of these colors, with reference to light. Warm effects are produced with white, yellow, orange, red, purple, indigo and black, and their combinations. Cold effects, by white, pale yellow, yellow, green, blue, indigo and black, and their combinations.

Table of Colors.—The following table will give a definite idea of color and their various combinations. Gray is produced by a combination of white and black. The three primitive colors being yellow, red and blue, the first compounds are orange, purple and green, and the second compounds brown, slate and olive, as previously stated. A careful study of these will show that effects of color in dress and trimmings, corresponds to the exercise of the painter's art. The lady who studies the combinations most closely with relation to her own figure, complexion, and color of hair and eyes, will produce the most pleasing effect in dress.

1 is yellow,	} Primitive colors.	$2+5$ are red and purple
2 is red,		$2+8$ are red and slate.
3 is blue,		$3+6$ are blue and green.
4 is orange,	} First compound.	$3+9$ are blue and olive.
5 is purple,		$3+5$ are blue and purple.
6 is green,		$3+8$ are blue and slate.
7 is brown,	} Second compound.	$4+7$ are orange and brown.
8 is slate,		$6+9$ are green and olive.
9 is olive,		$4+8$ are orange and slate.
$1+4$ are yellow and orange.		$5+8$ are purple and slate.
$1+7$ are yellow and brown.		$5+9$ are purple and olive.
$1+6$ are yellow and green.		$6+7$ are green and brown.
$1+9$ are yellow and olive.		$7+8$ are brown and slate.
$2+7$ are red and brown.		$7+9$ are brown and olive.
$2+4$ are red and orange.		$8+9$ are slate and olive.

Here we have the three primitive colors, their six pure compounds, and twenty-one additional tints, or compounds, by means of the couplets. Take the three original colors, put them together, regularly, and study the effect. It will be an excellent lesson in color, besides showing what an infinity of tints may be produced.

IX. Toilet-Room and Bath.

A LADY writer insists, and very properly, that a room intended expressly for toilet purposes, is necessary to every farm-house, large or small. In cities and large towns, a room of this kind is sometimes found; but in farm-houses and country residences, its necessity is very often overlooked. And at the same time, there is no reasonable plea why it should be; for on a farm, more than anywhere else, is such an apartment an absolute necessity. When the farmer and his boys come in from the field, tired and stained with toil, would it not be refreshing to have a room where they could all repair for a comfortable wash, without waiting each for his turn at the one basin, standing outside of the door, on the bench?

Such a room need not be large. A moderate-sized room, fitted up with basins, sinks, combs, brushes, towels, hooks or racks for hats and coats, and a glass; these are sufficient for ordinary use, and will save many moments of waiting the meal, when it is ready. Such a room can be spared in a farm-house, as well as not, and should have a door outside. In this same room every working member of the family can tidy up, and do it, too, without much delay. If there is no vacant room which can be converted to such use, a portion of the wood-house or any adjacent out-building can be set aside to this purpose.

The Bath.—We have said that every house should have conveniences for providing a bath. It should be so arranged that hot or cold water may be used, even if the hot water must be carried to the bath-tub by hand. Bathing should be performed often enough so the skin may be quite free from the odor and effects of perspiration. After the bath, the body should be rubbed with a towel rough enough so the skin may glow. It is not any more necessary, however, that the human skin be harrowed up, than it is necessary that the groom tear a horse's skin to pieces by a brutal use of the curry-comb. Thus, if brushes or harsh towels are used, use them gently.

Care of the Hair.—The scalp should be thoroughly brushed every day with a brush stiff enough to reach the skin, so the dandruff may be removed. Have nothing to do with nostrums to force the hair to grow, or to remove the hair. Cleanliness is the best hair tonic; depilatories are dangerous. If your hair is thin, use false hair as little as possible. There is no more certain cause of baldness than constantly wearing masses of false hair. Dyeing the hair, also, to produce some fashionable shade, is injurious in the extreme. All hair-dyes are poisonous, and the constant growth of hair at the roots requires a constant renewal of the dye. As a rule, nature has given hair that harmonizes with the complexion and eyes. Add all the helps to nature you please, but do not attempt to interfere too decidedly with it. If a woman is unfortunate enough to have a beard, or rather, if she considers this a misfortune, depilatories will not remove it. Nothing but the patient use of the tweezers will eradicate a beard—often, even this will not succeed perfectly.

X. Garments Next the Skin.

As a rule, woolen should be worn next the skin, where the object is quickly to pass off the perspiration. It is the best summer shirt, and the coolest, for workingmen, but it need not be as thick as a board. Cotton increases the warmth and perspiration, and has the property of retaining discharged humors, and passing them back into the system. Wool promotes perspiration, but by its gentle friction keeps the skin healthy, without clogging the pores. Linen gives a sense of coolness, but a fictitious one; it soon becomes damp. It holds the perspired matter, and the air striking the moist surface, chills the body. The action of flannel is to excite perspiration, quite necessary to the person at work, but it passes it through the material to the outside, where it is dissipated freely. Thin soft flannel for summer, and thicker flannel for winter, should therefore be the rule.

XI. The Care of Clothes.

NEVER brush clothes when wet. They should first become perfectly dry. Then lay the material flat, and brush thoroughly the right way of the cloth. Before brushing, rub out spots of mud. Remove all hard grease with the nail or the point of a knife, then cover the place with some absorbent paper (a piece of a blotter), and run a hot iron over it; change the paper, and repeat the operation until all the grease is absorbed. If the grease has soaked in, soft soap or ox-gall, or both, should be employed to remove it. Cloth suits that have become somewhat threadbare, are restored by second-hand clothes dealers, in the following manner: The cloth is first soaked in cold water for an hour or more. It is then laid flat on a board and rubbed the way of the cloth with a teasel-brush or partly worn hatter's card. The clothes are then hung up to dry, the nap properly laid with a hard brush and dressed smooth with a hot iron.

XII. The Care of Brushes and Combs.

HAIR brushes and combs should be regularly cleaned. A solution of bicarbonate of potassia or carbonate of soda is good for cleaning brushes, and if they are rinsed in bay rum afterwards, so much the better. If not, use pure soft water. Combs may be cleaned by washing in soft water and soap. A weak solution of carbonic acid, or of sulphate of soda, is also useful for cleaning brushes; but always thoroughly dry and air them after washing. The horrors concealed in a damp, dirty hair-brush can only be revealed by the microscope, but these minute germs are a virus from which scalp diseases originate and are disseminated. Hence never use the brushes and combs common to the guests of hotels and boarding-houses. The best tonic for the hair is frequent brushing with a dry brush, and without dressing for the hair. If anything is to be used as a dressing, take a little dilute bay rum, let the hair be hand-rubbed dry, and afterwards dry brushed.

CHAPTER V.

THE NURSERY AND SICK-ROOM.

I. TO PRESERVE HEALTH AND SAVE DOCTORS' BILLS —II. THE CARE OF CHILDREN.—III. NURSERY BATHING.—IV. DURATION OF AND PROPER TIME FOR BATHING —V. EXERCISE OF CHILDREN. —VI. STUDY AND RELAXATION.—VII. THE SICK-ROOM.—VIII. COOKERY FOR INVALIDS.—IX. TABLE OF FOODS AND TIME OF DIGESTION.—X. SOME ANIMAL FOODS IN THEIR ORDER OF DIGESTIBILITY.—XI. THE TIME REQUIRED TO COOK VARIOUS ARTICLES.—XII. COOKING FOR CONVALESCENTS—RECIPES AND DIRECTIONS.—XIII. JELLY OF MEAT.—XIV. OTHER SIMPLE DISHES.— XV. GRUELS.—XVI. TEAS AND OTHER REFRESHING DRINKS.—XVII. REMEDIES FOR THE SICK. —XVIII. DOSES AND THEIR GRADUATION.—XIX. DISINFECTION.—XX. TESTS FOR IMPURITIES IN WATER.—XXI. SIMPLE POISONS AND THEIR ANTIDOTES.—XXII. VIRULENT POISONS AND THEIR ANTIDOTES.—XXIII. HEALTH-BOARD DISINFECTANTS.—XXIV. HOW TO USE DISINFECTANTS.

I. To Preserve Health and Save Doctors' Bills.

THE preservation of health is of far greater importance than to dose a man after he is sick. To gratify "a false hunger," or slake "a false thirst," are only provocatives to disease. It should be remembered that we live not by what we eat, but by what we digest. Neither Walpole, who thought that with diet and patience all diseases might be cured; nor Montesquieu, who held that health, purchased by vigorously watching over diet, was but a tedious disease, was far from the mark. But a wise discretion in eating is better than all.

Heed the Stomach.—"What is one man's meat is another man's poison," is an old saying, and a true one. If every person would study his own individual powers, and learn to respect his stomach, to remember that he has no more right to overload it with improper food than he has to drink to intoxication, he would save himself many an unnecessary ache and ailment. Chronic dyspepsia never came of regular occupation, abundant exercise, early hours, and generous, but not imprudent diet. If you wake in the morning with a headache and lassitude, you have probably not taken the advice of the "self-monitor," which has its home in the stomach.

Conform to Nature.—Dean Nowell, although he may have blazed the way for red noses, did not grow strong by drinking ale. The Rajpoots who slay infants from pride do not kill so many infants as do Christian mothers with too much stuffing. The Bolton ass did not become fleet-footed by chewing tobacco and taking snuff. The New Zealand warriors were not made stout nor brave because their mothers thrust stones into their stomachs, as infants. And Brantome's uncle, who took gold, steel and iron, in powders, from weaning-time until twelve years of age, did not thereby acquire the strength to stop "a wild bull in full course."

First Principles.—A certain Kentucky man—and Kentucky men live much in the open air—minded the silent monitor of the stomach. He was at a first-rate hotel table, where the bill of fare was in French. After reading carefully the whole, he remarked, "I will go back to first principles, and take roast beef." He was not far wrong. A good constitution, roast beef, vegetables, fruit, tea and coffee in moderation, with liberal exercise, will keep any one in health.

If you have not the good constitution, get one by proper exercise, and a moderate but generous diet. When you have got the constitution, keep it by avoiding excesses. Moderation is said to be the first principle of digestion.

II. The Care of Children.

Gentle Firmness.—All who undertake the care of children, or who have the care of the sick, should cultivate the virtue of patience, soft speech and gentle kindness. There must be no swerving from duty, however distasteful it may be to the child or patient; firmness and gentle perseverance in the thing to be done should be the rule. The nursery should be provided with every possible appliance for the comfort of the infant, so that when sleeping it may lie soft, and warm in the winter and cool in the summer. The ventilation must be perfect; children, like birds, require an abundance of fresh air. When awake they should have a soft pallet where they can exercise their limbs to their hearts' content. It will save much unnecessary tending. Don't be afraid to toss them about when they have acquired strength to stand it, but be certain of your own coolness and muscle; then they will come to enjoy it.

As they begin to notice objects, provide them with toys to amuse them, hard rubber or ivory rings, wicker rattles, a toy balloon or other object to catch the eye and educate the sight. There is no reason why children should be constantly sick or ailing; nine times out of ten their ailments come from want of care, or rather from too much care and dosing.

Feeding the Infant.—The mother should be sure that her own milk is not made unwholesome to the child by worry or over-work, and that the milk of the cow or goat, if used, is perfectly healthful. When an animal is found that is known to be healthy, continue feeding the child on the milk of that animal. This selection is not difficult in the country. In the city it is only possible in certain cases, but every mother can and should keep her own temper equable.

III. Nursery Bathing.

The child should be regularly bathed in water fully as warm as the body. A bath-tub, or other vessel—a wash-tub will answer—in which the child can sit, when old enough, should be provided. The child will soon come to like its baths, and look forward to them with pleasure. The infants may be washed with tepid water, in a room where there are no drafts, and dried gently with the softest of towels. They will like it. Let them play and kick about in the bath pretty much as they like. A good large square of oil-cloth will prevent wetting the carpet. It is taken for granted, of course, that the child is in good health. If it is ailing, from any cause, consult your physician, but be sure your physician looks into causes carefully. The physician who does not carefully study a case, has mistaken his calling If the child is delicate, the bath should always be warm. If in ordinary health, the water should be tepid. The cold bath should not be employed for children, except under the advice of a physician. When the child is old enough to take care of itself, and proper friction is employed immediately after, followed by a brisk run, cold bathing is not always objectionable. Salt bathing may be artificially had by using a little sea salt in the water. It is excellent.

IV. Duration of and Proper Time for Bathing.

ALL baths, even by those in health, should be taken in a warm room, unless the regular swimming-bath is indulged in. Here the temperature of the air is pleasant to the naked skin, for nobody with moderately good sense will go in swimming in cold weather. In the following table, after Dr. Wooton, it should be understood that winter baths, both warm and cold, should be taken in a warmed apartment.

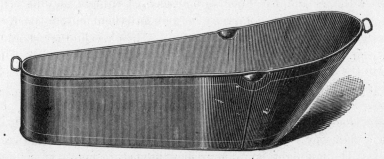

INFANT BATH-TUB.

With this understanding, the table may serve as a general guide in bathing for both adults and children.

TEPID BATHS IN SUMMER—FRESH AND SALT WATER.—Healthy people—time of bath: ten minutes; frequency: twice daily; period of day: before breakfast and retiring to rest. Weak people—time of bath: ten minutes; frequency: once daily; period of day: before breakfast.

COLD BATHS—FRESH WATER.—Healthy people—time of bath: ten minutes; frequency: twice daily; period of day: before breakfast and retiring to rest. Weak people—time of bath: five minutes; frequency: once daily; period of day: before breakfast.

COLD BATHS IN SUMMER—SALT WATER.—Healthy people—time of bath: ten minutes; frequency: once daily; period of day: before breakfast. Weak people—time of bath: five minutes; frequency: once daily; period of day: two hours after breakfast.

IN WINTER.—The same rules will apply for winter in a properly warmed room, except that weak people should take the cold bath, if at all, before breakfast, as directed for healthy persons, the duration five minutes, and once daily.

V. Exercise of Children.

THE more children are left free, always under careful supervision, the greater will be their enjoyment, and the more exercise will they take. Winter and summer they should have it. In the summer let them roll and frolic about the lawn, if the ground is dry. In damp weather, an old carpet may be laid down for them to play, and in damp or rainy weather they should have access to a garret or attic, or regular play-room, where they may romp and play. In the winter their exercise should not be omitted. Clothe them warmly, with mittens, ear-muffs, thick stockings and strong shoes, and let them run at will. The pure air will expand their lungs, and send a glow to their cheeks they can get in no other way.

VI. Study and Relaxation.

Do not drive your children to study too young. From six to eight years is early enough for regular school-going. Of course their education begins as soon as they begin to notice things and run about—in object-lessons and toy instructors, lettered blocks, etc.; but they should not really be put to school before the age mentioned, and not thus early if they languish under study. Until a child is ten years old regular study should not be permitted. It must be more play than study. Then, the wise teacher, up to the age of twelve, will mix plenty of play with study. From this time on the study may be more and more continued; but no labor, except a few light chores, should be included. Out of school hours let the child have play. Labor, however light, does not stand the child instead of play. "All work and no play makes Jack a dull boy." Hard study, with little exercise, fills the graveyard with young bones. The midnight lamp nourishes the mature man's mind; but dreamless sleep for the youth makes a healthy brain.

VII. The Sick-Room.

Cheerfulness and Quiet.—In the sick-room there should be no unnecessary noise, and, above all, no confusion. Neither should there be "solemn silence." Some cheerful conversation is often better than medicine. In any event, never allow a friend with long-drawn, solemn face, or a procession of them, to walk in, and, with a shake of the head, after gazing, to walk out again. Because a person is ill—even dangerously so—there is no reason why the nurse or visitors should carry on their faces the you-will-never-get-well look. It would dishearten any invalid that it did not exasperate, and neither disheartenment nor exasperation is good for the sick, even though it be said that when the sick are "strong enough to get mad," they are "strong enough to get well."

The Nurse.—The nurse should be soft-handed, deft in her work, of cheerful disposition, even tempered, and above all, intelligent. She should have delicate tact in cooking, for while every operation of cooking should be cleanly in the extreme, here the cooking, while simple, should be delicate. The beef tea must be pleasing to the eye as well as grateful to the palate. The steak or chops should be tender and cooked to a turn. The egg should be so boiled or poached as to be good to look at. Some tempting, simple, easily digested pudding, that comes as a surprise, is ten times more grateful than if the patient has been promised it, and then given the impatience of longingly waiting for it. A simple drink of water, if fresh from the well or spring, is always welcome; however pure it may be, it is nauseous if it has stood in the room until warm. It is all these little things, these attentions, that mark the careful from the careless nurse. In fact, no person should undertake the office of nurse unless loving kindness and self-abnegation are strong personal traits. With members of a family, these feelings are, of course, present. Happy is the patient who can always command such service.

In severe sickness, it is the physician, his medicines, and the soothing offices of the nurse, that bring the patient through. Here fully as much depends upon the nurse as upon the doctor. A time comes when no longer medicine but food is needed. With convalescence it is the cook who takes the place of the doctor, and here the nurse's best efforts are shown.

Weak Patients.—Very weak patients must be rallied; stimulants may be necessary. There may be a nervous difficulty in swallowing; the nurse should keep her wits about her. The physician may have ordered a fixed quantity, say a teacupful of some liquid food every three or four hours; the patient's stomach rejects it. Will the nurse follow the given rule? No. She will try a single tablespoonful, once an hour, or even a teaspoonful every fifteen minutes. Perhaps a stimulant is necessary. These are things—the knowledge of them—that every nurse should inform herself upon and be ready to act upon.

VIII. Cookery for Invalids.

General Rules for Cooking.—In addition to what has just been said it is only necessary to give these general rules for cooking:

1. There must be no smoke for broiling.
2. All soups should be made with the most gentle simmering.
3. All fruits and vegetables must be perfectly fresh.
4. An hour before cooking vegetables, put them in cold water to which a little salt has been added to free them from any possible insects. Wash clean, drain, and drop into water that is boiling fast. Take them from the water and drain the instant they are done.

These general directions relating to cooking will suffice as to the processes in invalid cookery. Some special recipes for dishes palatable and wholesome will be given presently. These will, of course, consist of the most simple dishes, not highly seasoned or spiced.

IX. Table of Foods and Time of Digestion.

The table given below is compiled to show the average time required for the digestion of different foods, but of course, it is only approximate, since in the real digestion of foods, no two systems will act precisely alike. The result will perhaps surprise many persons, who have been led into error in the supposed digestibility of certain foods. For instance, oysters are generally supposed to be among the most easily digested of foods. They are not even approximately so except when eaten raw. Roast goose is by many supposed to digest slowly, but this is a great mistake:

AVERAGE TIME FOR DIGESTION.

NAME OF ARTICLE.	HOURS.	MIN.	NAME OF ARTICLE.	HOURS.	MIN.
Apples, sweet,	1	30	Parsnips, boiled,	2	20
Apples, sour,	2	00	Mutton, roast,	3	15
Beans, green in pod, boiled,	2	30	Mutton, broiled,	3	00
Beef, fresh, roasted rare,	3	00	Mutton, boiled,	3	00
Beef, fresh, broiled,	3	00	Oysters, raw,	2	55
Beef, fresh, dried,	3	30	Oysters, roast,	3	15
Beef, fresh, fried,	4	00	Oysters, stewed,	3	30
Beets, boiled,	3	45	Pork, fresh fat and lean, roast,	5	15
Bread, wheat, fresh,	3	30	Pork, corned, boiled,	3	15
Bread, corn,	3	15	Pork, corned, raw,	3	00
Butter, melted,	3	30	Potatoes, boiled,	3	30
Cabbage, with vinegar, raw,	2	00	Potatoes, baked,	2	30
Cabbage, boiled,	4	30	Rice, boiled,	1	00
Cheese, strong old,	3	30	Sago, boiled,	1	45
Codfish,	2	00	Salmon, salted, boiled,	4	00
Custard, baked,	2	45	Soup, beef and vegetable,	4	00
Ducks, tame roasted,	4	00	Soup, chicken,	3	00
Ducks, wild,	4	30	Soup, oyster,	3	30
Eggs, boiled hard,	3	30	Tapioca, boiled,	2	00
Eggs, boiled soft,	3	30	Tripe, soused, boiled,	1	00
Eggs, fried,	3	30	Trout, fresh, broiled or fried,	1	40
Goose, roast,	2	00	Turkey, tame, roast,	2	00
Lamb, fresh, broiled,	2	30	Turkey, wild, roast,	2	15
Liver, beef, broiled,	2	00	Turnips, boiled,	3	30
Liver, beef, fried,	2	30	Veal, fresh, broiled,	4	00
Milk, boiled,	2	00	Veal, fresh, fried,	4	30
Milk, uncooked,	2	15	Venison, broiled,	1	35

X. Some Animal Foods in their Order of Digestibility.

These may be named about as follows. Not, however, in relation altogether to time of digestion:

1. Sweetbreads.
2. Venison.
3. Lightly boiled eggs.
4. New cheese, toasted.
5. Roast barn-yard fowl.
6. Oysters.
7. Lamb.
8. Wild duck and other water-fowl.
9. Boiled fish, not oily, as trout, perch, etc.
10. Roast beef, boiled beef and steak.
11. Roast veal.
12. Oily fish, as salmon, mackerel, etc., boiled.
13. Wild pigeon and hare.
14. Tame pigeon, duck and geese.
15. Fish, fried.
16. Roast or boiled pork.
17. Lobster, crab or clams.
18. Smoked, dried, salted or pickled fish.
19. Old strong cheese.

XI. The Time Required to Cook Various Articles.

Vegetables.—Carrots, parsnips, turnips, onions, salsify, rutabagas: Boil from forty minutes to one hour. Cabbage, beets, potatoes, string beans: Twenty minutes to one-half hour. Cauliflowers and squash: About twenty minutes. Green peas and asparagus: About fifteen minutes.

Fish.—The proper time in which any fish will cook properly, can only be learned by experience. Fish must never be under-done. When the bones separate easily from the flesh the fish is done. About seven or eight minutes may be given as the proper time to each pound, after the water boils. Cutlets of fish will require from five to ten minutes to fry, and somewhat longer to broil. Flat fish, the same. The cleaving from the bone may be observed.

Roasting and Boiling.—The time required for properly roasting and boiling meat is about fifteen minutes to the pound. Boiled meat will separate easily from the bone when done. When roasted meats are done the flesh will yield easily to the fingers. In fowls or game the flesh of the leg will yield and show it is ready to separate from the bone, and in roasting before the fire, jets of steam will come from the side next the fire, just before the joint of meat is done.

XII. Cookery for Convalescents—Recipes and Directions.

Extract of Beef.—This should be made the day before it is required for use, kept cold, skimmed of all fat and warmed up. Mince one pound of lean beef to each pint of extract required. Place in a jar with a closely fitting top (if luted, so much the better), or in a bottle tightly corked. Place the vessel in another of cold water and set on the stove where it will heat slowly. When the water boils, move to where it will simmer, adding boiling water as the water boils away, so as to keep the inner vessel pretty well submerged to the top. Let it cook for three or four hours; strain through a cloth, and when cold remove any fat that may appear. When warmed up for use, a teaspoonful of cream may be added to each teacupful of extract, or a very little corn-starch or arrowroot.

Beef Tea.—A weaker extract or broth may be made in a covered saucepan with a quart of water to each pound of chopped beef; simmer until the water has evaporated down to a pint.

Raw Beef Tea.—Made by allowing one ounce of fine chopped lean beef to each tablespoonful of cold water. Let them stand together fifteen minutes, strain and season to taste. For any of the above recipes the fiber of the meat may be scraped away with a knife.

Broths.—Broths are made and thickened in the usual way, but care must be taken to strain, and skim off all fat.

Breaded Chops and Cutlets.—The meat must be tender and lean. At least the fat should not be eaten. They are prepared by dipping the chop into melted butter, or better, the beaten yolk of egg, to which is added a very little melted butter. Sprinkle with fine crumbs of stale bread, and fry. If broiled, dip into melted butter instead of into the egg.

Broiled Fowl and Game.—The fowl must be young and tender. Divide it down the back, flatten it out, rub with a little butter (and pepper if allowed), and cook the inner side before the outer side. Salt when it is turned over to cook the skin side. Or, it may be partly roasted and then broiled or fried. Serve hot, and on a hot dish.

Roasting.—This should be done before a clear fire, if possible. It may be easily managed with a fire of anthracite coal in the stove, since the bird or meat to be served will presumably be small.

Boiling.—The rules heretofore given for boiling may be consulted, and some nice bit selected from that prepared for the family dinner. See directions for boiling vegetables. In boiling fish use as little water as will serve to cover it.

XIII. Jelly of Meat.

We do not think any meat jelly so good as that made from calves' feet. The jellies and meat extracts of commerce are never so good as those prepared at home; but they are good substitutes. The jelly made from the heels of older cattle come next to calf's-foot jelly.

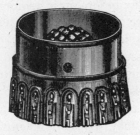

OVAL JELLY MOULD.

JELLY SIEVE.

Calf's-Foot Jelly.—It is made thus: Take two calf's feet, scald, and scrape the hair clean, take off the hoofs and remove the fat between the toes, and wash all thoroughly in warm water. Put the feet into cold water and let it gradually come to a boil, carefully skimming. Simmer six or seven hours, strain through a sieve and let it cool. When it gets firm, remove the fat from the top, the jelly from the sediment, and put the jelly into a saucepan. To each quart of jelly add six ounces of best granulated or loaf sugar and the shells and whites of five eggs, thoroughly beaten. Stir all together while cold, but not after it begins to heat. Let it simmer five minutes; throw in a teacup of cold water and let it boil five minutes more; cover the saucepan closely and let it remain where it will keep hot for half an hour. Have ready a flannel bag, wrung out of hot water, and pour in the jelly, keeping all quite near the fire to prevent the jelly setting before it is strained. If it is not clear the first time it runs through pass it through again.

The jelly bag is made larger at the top and narrows to a point, and is best of closely woven flannel, with the seams double sewed. Wine or lemon juice may be added before straining if desired.

Jelly of Cows' Heels.—This is prepared precisely as is calf's-foot jelly. When cooked, made cold, and the fat removed, a quart of ale or wine, the juice and rind of two lemons, a quarter of a pound of sugar and the whites of six eggs are added, and it is then finished like calf's-foot jelly.

Calf's-Foot Blanc Mange.—The stock of calf's-foot jelly is reduced, to bear the addition of milk, flavored with vanilla, lemon-peel or other flavor, and is improved by the addition of a little wine or brandy.

XIV. Other Simple Dishes.

Iceland Moss Jelly.—This is soothing in colds, coughs, catarrh, and pulmonary affections generally. Put four ounces of Iceland moss in a quart of water, stirring constantly while on the fire. When it has boiled about forty minutes or more, add two ounces of lump sugar and a wineglass full of white wine. Strain through a jelly-bag, and it will be fit for use when cold and firm.

A BLANC MANGE may be made by boiling in milk instead of water, omitting the wine, and flavoring with lemon, vanilla or other flavor.

Moss and Currant Jelly.—This is made by boiling slowly, in the same proportions as for the first recipe, omitting the wine, and straining it on a tablespoonful or more of currant jelly, mixing it well and putting it in a mould to cool.

Puddings.—These are made according to the recipes hereafter given. They should be of the simpler preparations, as batter, bread, rice, arrow-root, etc., and may be baked in a proper-sized shell or cup, and eaten with cream or wine sauce. A simple wine sauce is made with a little sherry and water, sweetened with soft sugar.

Baked Rice and Apples.—This is a wholesome invalids' dish. Pare, quarter, and core the apples, and stew them with a little cold water and sugar, in which there is also a little cinnamon and allspice, tied in a little bag for easy removal. Ten minutes should stew the apples. Turn them into a saucer, spread boiled rice over, and cover the whole with white of egg beaten to a froth. If the apples and rice are put together cold, they must be heated through in the oven before putting on the egg. Serve when well browned.

XV. Gruels.

GRUELS are made of corn-meal, oatmeal, rice or barley, generally of corn-meal or oatmeal. Rice gruel is used for relaxed bowels. Gruels are all made by mixing the meal with cold water, properly seasoning and turning into boiling water. When done, say in five minutes, strain, sweeten to taste, flavor and serve. Add wine or brandy if stimulus is necessary. Corn-meal and oatmeal, or other grits are better soaked for some time in cold water before cooking.

Gruel of Groats.—To a tablespoonful of groats mixed with cold water, add a pint of hot water. Boil ten minutes.

Oatmeal Gruel.—Stir two tablespoonfuls of oatmeal in a pint of cold water, and let it stand for some hours. Then, after stirring well, strain through a fine sieve, and cook the thin part, with constant stirring, until it has simmered from five to eight minutes. Season and flavor to taste.

Rice Water with Raisins.—Take six ounces of rice, two ounces of raisins, and two quarts of water. Simmer for half an hour. Strain and add to the liquid two tablespoonfuls of good brandy. This is good for dysentery and diarrhœa.

Apple and Other Fruit Waters.—Slice unpared apples, pears, etc., and cook with water, until the fruit is quite tender. Strain through clean muslin. To be taken cold.

MILK, PORRIDGE OR RICE BOILERS.

Orangeade or Lemonade.—Pare the rind thinly from four oranges, and put the rind in a pitcher. Take off and throw away the white slice; then remove the seeds, put with the thin peelings, add an ounce of sugar and a quart of boiling water. Let it stand until cold, setting it on ice if necessary; or bottle and hang down the well. Lemonade is made in the same way by substituting lemons for oranges, and adding more sugar. For ordinary use, either is made by squeezing out the juice, with a squeezer, and adding sugar and ice-water.

XVI. Teas and Other Refreshing Drinks.

Linseed Tea, for Gout, Gravel, etc.—As an accessory it is in good repute. Take one tablespoonful of flaxseed, one quart of water and a little orange-peel. Boil ten minutes in a clean porcelain kettle, sweeten with honey, add the juice of a lemon, to allay irritation of the chest. Omitting the lemon, it is good for irritation of the lungs, gout and gravel.

Chamomile Tea as a Strengthener.—Use one pint of boiling water to about thirty chamomile flowers. Steep, strain, sweeten with honey or sugar, and drink a cupful half an hour before breakfast, to promote digestion and restore the action of the liver. A teacupful of the tea, in which has been stirred a full dessert-spoonful of sugar and a very little ginger, is an excellent tonic and stimulant for an old person, taken two hours before dinner.

White-Wine Whey.—Let a pint of milk come to a boil; add half a gill of white-wine; allow the whole to come to a boil, and pour into a basin to cool. When the curd has settled, the whey is excellent for coughs and colds.

Hop Tea.—This is considered good as an appetizer and strengthener of the digestive organs. Take one-half ounce of hops, upon which is poured a quart of boiling water; let it stand fifteen minutes; strain, and give a small teacupful half an hour before breakfast.

Effervescent Drink.—Put the juice of a lemon, strained, in a tumbler of water, with sugar enough to sweeten it. Add half a teaspoonful of bicarbonate of soda, and drink while effervescing.

Sherbet.—Take one pound of best powdered sugar, two ounces of carbonate of soda and three ounces of tartaric acid. Mix all thoroughly and keep in a bottle corked tight. When wanted for use, put a teaspoonful of the powder in a tumbler, add a drop of essence of lemon, fill with ice-water, stir and drink.

XVII. Simple Remedies for the Sick.

Every family should know something of simple remedies, especially those who live far from physicians. Often some simple remedy given in time will cure, or, at least, carry the patient until permanent relief can be obtained. For this reason we give a variety of recipes collected from the best authorities, with appropriate doses, the doses given being for adults. For children's doses, see table of proportionate doses in the next section. The most of them are simple and easily procured. Castor oil is now much less used than formerly, but is too valuable in certain cases to omit.

Acid, Acetic.—Vinegar distilled from wood and purified, used as a lotion for its cooling properties, removing warts. It is not given internally, except in combination with other remedies.

Acid, Benzoic.—Used in chronic bronchitis. Dose: 5 grains to ½ drachm, twice a day.

Acid, Sulphuric.—(Diluted.) Sulphuric acid mixed with 11 times its bulk of water. Used in dyspepsia, also to check sweatings, salivation and diarrhœa; also as a gargle.

Acid, Tartaric.—Used in fevers with some soda of potassa, as an effervescing draught, instead of citric acid; the acid is dissolved in water as a substitute for lemon, juice, and added to soda. Dose: 15 to 25 grains.

Aloes, Barbadoes.—Used in dyspepsia and head affections; also as a common purgative. Dose ¼ grain to 5 grains, well powdered or dissolved in hot water.

Alum.—Used internally in hemorrhages and mucous discharges; externally as a wash in ophthalmia, or as a gargle in relaxed uvula. Dose: 10 to 20 grains.

Ammonia, Liquor of.—Ammonia condensed in water. Used, when largely diluted, in fainting, asphyxia, hysteria, spasms, acidities of the stomach; and externally as an irritant of the skin. Dose: 5 to 15 minims.

Assafœtida, Gum.—Used in hysteria, flatulence, colic, etc. Dose: 5 to 10 grains.

Borax, Biborate of Soda.—Used in intestinal irritation of infants. Externally applied to thrush, and to cutaneous diseases. Dose: 5 to 30 grains. Externally applied, dissolved in 8 times its weight of honey or mucilage.

CAMPHOR.—Used in hysteria, asthma, chorea, and generally in spasmodic diseases. Externally, in muscular pains, bruises, etc. Dose: 3 to 5 grains, in pills. When dissolved in water, as camphor mixture, the quantity is scarcely appreciable.

CAPSICUM.—Used in dyspepsia, flatulence, externally as an ingredient in gargles for relaxed sore throat. Dose: 3 to 5 grains, in pills; 2 drachms to 8 ounces form the strength for using as a gargle, diluted largely with water.

CASCARILLA BARK.—Stimulant, stomachic and tonic. Used in dyspepsia, flatulent colic, chronic dysentery and gangrene. Dose: 20 to 30 grains of this powder 3 or 4 times a day.

CASTOR OIL.—Mildly aperient. Used in colic and in those cases of constipation which will not bear drastic purgatives; also for mixing with gruel for the ordinary enema. Dose: A teaspoonful to 1 or 2 tablespoonfuls; an ounce is the proper quantity for mixing with gruel to make an enema.

SIMPLE CERATE.—Add 20 ounces of melted wax to a pint of olive oil, and mix while warm, stirring until cold. Used for covering blisters or other healing sores.

CHALK, PREPARED.—Used in acidities of the stomach and bowels, and to correct the irritation which is established in diarrhœa. Externally, as a mild application to sores and burns. Dose: 10 to 15 grains.

CHAMOMILE FLOWERS.—Tonic, stomachic and carminative. The warm infusion, when weak, is emetic. Externally, soothing. Used in dyspepsia, hysteria, flatulence, and also to work off emetics. Dose of the powder: 30 to 40 grains, twice a day.

CHARCOAL.—Vegetable. Used as an ingredient in tooth-powder; also to mix with other substances in forming a poultice for foul ulcers. Sometimes given internally. Dose 10 to 20 grains.

CINCHONA BARK.—(Yellow.) Astringent, tonic, antiseptic and febrifuge. Used in typhoid fevers, and in all low states of the system, being in such cases superior to quinine. Dose: 10 to 50 grains, in wine or wine and water.

CINNAMON BARK, OIL AND WATER.—Used as a warm and cordial spice to prevent the griping of purgatives, etc.

COD-LIVER OIL.—Prepared from the liver of the codfish. Nutritive, and acting also on the general system, from containing very small doses of iodine and bromine. Dose: 1 drachm carried up to 4 in any convenient vehicle, as infusion of cloves.

DECOCTION OF BARLEY.—(Barley water.) Wash 2½ ounces of pearl barley, then boil it in ½ pint of water for a short time. Throw this water away, and pour on the barley 4 pints of hot water; boil slowly down to 2 pints and strain. Soothing and nourishing. Used as a diluent drink in fevers and in inflammation of mucous surfaces, especially those of the urinary organs.

DECOCTION OF BARLEY (COMPOUND).—Boil 2 pints of barley water (see above) with 2½ ounces of sliced figs, 4 drachms of bruised fresh licorice, 2½ ounces of raisins, and 1 pint of water, down to 2 pints, and strain. Effect, the same as barley water, but, in addition, laxative.

DECOCTION OF BROOM (COMPOUND).—Take ½ ounce of broom, ½ ounce of juniper berries, and ½ ounce of bruised dandelion; boil in 1½ pints of water down to a pint, and strain. Diuretic, and slightly aperient. Used in dropsy. Dose: 1½ ounces to 2 ounces, twice or thrice a day.

DECOCTION OF CINCHONA.—Boil 10 drachms of bruised yellow cinchona in 1 pint of water for 10 minutes, in a closed vessel, then strain. Used in fevers, malignant sore throat and dyspepsia. Dose: 1½ ounces to 3 ounces, 3 times a day.

DECOCTION OF DANDELION.—Boil 4 ounces of bruised dandelion in 1½ pints of distilled water, to a pint, and strain. Used in torpid conditions of the liver, jaundice, habitual constipation, etc. Dose: 2 or 3 ounces, 2 or 3 times a day.

DECOCTION OF ICELAND MOSS.—Boil 5 drachms of Iceland moss in 1½ pints of water down to a pint, and strain. Used in consumption and dysentery. Dose: 1 to 2 ounces.

DECOCTION OF POPPYHEADS.—Boil 5 ounces of bruised poppyheads in 3 pints of water for ¼ hour, and strain. Used as a fomentation in painful swellings and inflammations.

DECOCTION OF QUINCE-SEED.—Boil 2 drachms of quince-seed in 1 pint of water, in a tightly covered vessel, for 10 minutes, and strain. Used in thrush and irritable conditions of the mucous membrane.

DECOCTION OF SARSAPARILLA (COMPOUND).—Mix 4 pints of boiling decoction of sarsaparilla, 10 drachms of sliced sassafras, 10 drachms of guaiacum-wood shavings, 10 drachms of bruised stick-licorice, and 3 drachms of mezeron-bark; boil ¼ hour, and strain. Used in cutaneous diseases, chronic rheumatism and scrofula. Dose: 2 ounces, 2 or 3 times a day.

EXTRACT OF HOP.—Physical properties. A dark-colored, bitter extract, without much smell. Tonic and sedative. Used in chronic dyspepsia and loss of sleep. Dose: 10 to 15 grains.

INFUSION OF CASCARILLA.—Macerate 1½ ounces of bruised cascarilla in 1 pint of boiling water for 2 hours, in a covered vessel, and strain. Stomachic and tonic. Used in dyspepsia, diarrhœa and general debility. Dose: 1 ounce to 2 ounces.

INFUSION OF GENTIAN (COMPOUND).—Macerate 2 drachms of sliced gentian, 2 drachms of dried orange-peel, 4 drachms of lemon-peel, in 1 pint of boiling water, for 1 hour, in a covered vessel, and strain. Stomachic and tonic. Used in dyspepsia and general debility. Dose: 1½ to 2 ounces, 2 or 3 times a day.

INFUSION OF HORSERADISH (COMPOUND).—Macerate 1 ounce of horseradish, sliced, and 1 ounce of bruised mustard-seed in 1 pint of boiling water 2 hours, in a covered vessel, and strain. Then add a fluid ounce of the compound spirit of horseradish. The same as the root. Dose: 1 to 3 ounces, 3 or 4 times a day.

INFUSION OF QUASSIA.—Macerate 10 drachms of quassia, sliced, in 1 pint of boiling water, 2 hours, in a covered vessel. Tonic and stomachic. Used in dyspepsia. Dose: 1½ to 2 ounces.

INFUSION OF ROSES (COMPOUND).—Put 3 drachms of the dried red-rose leaves into 1 pint of boiling water, then add 1½ fluid drachms of diluted sulphuric acid. Macerate for 2 hours, and strain the liquor; lastly, add 6 drachms of sugar. Therapeutical effects: Astringent, refrigerant, and antiseptic. Used as a drink in fevers; also a vehicle for sulphate of magnesia, quinine, etc. Dose: 1½ to 2 ounces.

LIQUOR OF ACETATE OF LEAD.—Used as a lotion to inflamed surfaces when largely diluted with water.

LIQUOR OF POTASS.—Used in acidity of the stomach and bowels; also in irritability of the stomach and of the bladder, and in cutaneous diseases. Dose: 10 to 30 drops, in beer or bitter infusion, or lemonade.

MAGNESIA, CARBONATE OF.—Used in dyspepsia with costiveness, in the constipation of children and of delicate grown persons. Dose: ½ to 1 or 2 drachms.

MERCURY, CHLORIDE OF CALOMEL.—Used in chronic diseases of the liver and general torpidity of the stomach and bowels; in dropsy, in combination with other medicines. *A most dangerous medicine when employed by those who are not aware of its powerful effects.* Dose: 1 grain twice a day as an alterative, 4 to 5 grains as an aperient, combined with or followed by some mild vegetable purgative.

MIXTURE OF IRON.—All mixtures of iron should be prepared by capable druggists.

POULTICE OF CHARCOAL.—Macerate, for a short time, before the fire, 2 ounces of bread in 2 fluid ounces of boiling water; then mix, and gradually stir in 10 drachms of linseed meal; with these mix 2 drachms of powdered charcoal, and sprinkle 1 drachm on the surface. Used in gangrene.

POULTICE OF YEAST.—Mix 5 ounces of yeast with an equal quantity of water, at 100°; with these stir 1 pound of flour, so as to make a poultice; place it by the fire till it swells, and use. Stimulant, emollient. Used in indolent abscesses and sores.

QUININE, SULPHATE OF.—Physical properties: Colorless, inodorous, lustrous, bitter efflorescent crystals, totally soluble in water previously acidulated with sulphuric acid. Stomachic, stimulant, febrifuge and tonic. Used in general debility, neuralgia, and after fever. Dose: 1 to 3 grains.

SODA, BICARBONATE OF.—Physical properties: A heavy, white powder, without smell, and tasting slightly soapy. Entirely soluble in water. Anti-acid. Used for acidities of the stomach. Dose: 5 to 30 grains.

SPIGELIA.—A very useful remedy for round worms. Dose: 10 to 20 grains of the powder, given fasting; or ½ to 3 ounces of the infusion made by pouring 1 pint of water on ½ ounce of the root.

SYRUP OF IODIDE OF IRON.—Is used because the iodide of iron is liable to injury from change. Alterative, and affording the effects of iron and iodine. Used in scrofulous diseases, and in cachectic states of the system. Dose: ½ drachm to 1 drachm.

WINE OF IRON.—Stomachic and tonic. Used the same as other iron medicines. Dose: 30 to 60 minims.

XVIII. Doses and Their Graduation.

ALL who have charge of sick-rooms where the physician is not in regular attendance, should understand the proportionate doses for various ages; but here, again, proper calculation must be made for development, constitutional differences, etc. The nurse should also know something of how certain medicines act on different systems. The following table will give a general idea of the proportionate dose to get ordinary action of medicine, allowing that a person of twenty-five to forty years of age requires a full dose.

TABLE OF PROPORTIONATE DOSES.

Age....80..65..50..25–40..20..16..12..8..5..2 Years.
Doses..⅝..¾..⅞.. 1 ..⅞..¾..⅝..½..⅜..¼
Age................ 12 .... 6 .... 2 to 1 Months.
Doses............ 1-5.....1-8.....1-15 1-24.

Adult women require about ¾ the full dose of men.

MEASURES FOR DOSES.—A tumbler is estimated to contain four or five fluid ounces; a wineglass one and one-half or two fluid ounces; a tablespoon about one-half fluid ounce; a teaspoonful one fluid drachm; a minim is one drop.

XIX. Disinfection.

With Clay or Loam.—Dry earths, strong loams and clay, reduced to powder, are cheap and perfect deodorizers of fetid substances, when the latter are covered with the earth. These are valuable in all cases when the substance does not nearly saturate the earth used.

Copperas.—For privy-vaults, cesspools, etc., especially those giving off the smell of sulphurated hydrogen (rotten-egg smell), use copperas in powder. It is cheap, and one or two pounds will destroy the smell of an ordinary privy-vault or cesspool. It is also the best cheap disinfectant for sinks, drains and all that class of fixtures giving off bad smells.

Carbolic acid or chloride of lime may be used in all cases when the smell of these agents does not reach the rooms of a building.

Earth Closets.—We give two simple forms of earth closet for sick-rooms. Fig. 1, a form with back; Fig. 2, showing the arrangement for depositing the earth on the "stool." Fig. 3 is a more simple form; Fig. 4, showing the seat opened. They are valuable in the country for invalids, who cannot at all times go out of doors. [See next page.]

To Disinfect Clothing.—Clothing may be disinfected by subjecting it to a dry heat just below that which will injure the cloth. Perfectly boiling water is usually sufficient to remove the contagion of diseases like small-pox, etc., but it is better after washing to subject the clothing to a heat of not less than 300° Fahrenheit. This may be done under pressure of steam.

Disinfecting the Sick-Room.—It is useful to know whether the air of a sick-room is pure or not. To discover this, dampen a piece of white linen with a solution of nitrate of lead. If impure, the cloth will be darkened. The following

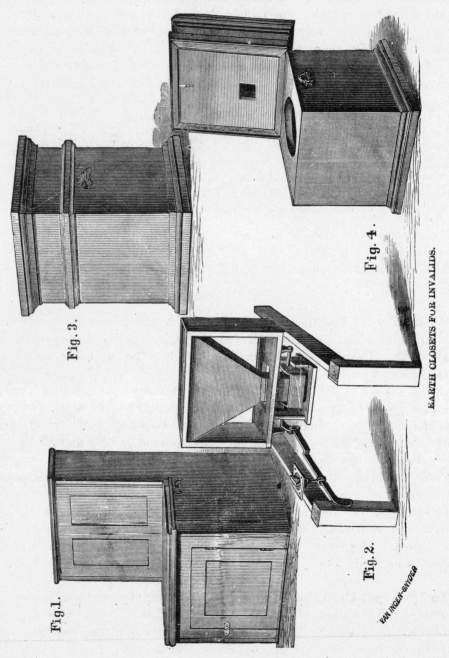

EARTH CLOSETS FOR INVALIDS.

table will show the relative value of some of the more common disinfectants, the first-named being taken at 100.

TABLE OF RELATIVE VALUE OF DISINFECTANTS.

Chloride of lime with sulphuric acid,	100
Chloride of lime with sulphate of iron,	99
Carbolic acid, disinfecting powder,	85.6
Slaked lime,	84.6
Alum,	80.4
Sulphate of iron,	76.7
Sulphate of magnesia,	57.1
Permanganate of potash, with sulphuric acid,	51.3

Hence, if nothing better is available, use air-slaked lime freely in powder whenever epidemic and contagious diseases are present. For cesspools use sulphate of iron in solution.

XX. Tests for Impurities in Water.

To tell if water be hard or soft, dissolve soap in alcohol, and drop a little in a glass of the water; it will become more or less milky, according to the hardness of the water.

Test for Iron.—A crystal of prussiate of potash, dissolved in water containing iron, will turn it blue.

Test for Copper.—If copper be present in water, a few drops of liquid ammonia will turn it blue.

Test for Lead.—If lead is suspected in water, add a little sulphuret of ammonia or potash. If there is lead in solution, water will assume a dark brown or blackish hue.

XXI. Simple Poisons and Their Antidotes.

EVERY person having the care of children, should be conversant with the best-known antidote for simple poisons. A specific for poisoning by poison oak (*Rhus. toxicodendron*), and other poisonous plants of that class, is to dissolve a handful of quicklime in water, let it stand half an hour, then paint the poisoned parts with it. Three or four applications will never fail to cure the most aggravated cases. Poison from bees, hornets, spider-bites, etc., is instantly arrested by the application of equal parts of common salt and bicarbonate of soda, well rubbed in on the place bitten or stung.

XXII. Virulent Poisons and Their Antidotes.

OIL OF VITRIOL, AQUA-FORTIS, SPIRIT OF SALT.—*Antidotes*—Magnesia, chalk, soap and water.

EMETIC TARTAR.—*Antidotes*—Oily drinks, solution of oak bark.

SALT OF LEMONS OR ACID OF SUGAR.—*Antidotes*—Chalk, whiting, lime or magnesia water. Sometimes an emetic draught.

PRUSSIC ACID.—*Antidotes*—Pump on back, smelling-salts to nose, artificial breathing, chloride of lime to nose. Strong prussic acid kills instantly.

PEARLASH, SOAP LEES, SMELLING-SALTS, NITER, HARTSHORN, SAL VOLATILE.—*Antidotes*—Lemon-juice and vinegar and water.

ARSENIC, FLY-POWDER OR WHITE ARSENIC, KING'S YELLOW OR YELLOW ARSENIC.—*Antidotes*—Emetics, limewater, soap and water, sugar and water, oily drinks.

MERCURY, CORROSIVE SUBLIMATE, CALOMEL.—*Antidotes*—White of eggs, soap and water.

OPIUM, LAUDANUM.—*Antidotes*—Emetic draught, vinegar and water, dashing cold water on chest and face, walking up and down for two or three hours.

LEAD, WHITE LEAD, SUGAR OF LEAD, GOULARD'S EXTRACT. —*Antidotes*—Epsom salts, castor oil and emetics.

COPPER, BLUE-STONE, VERDIGRIS.—*Antidotes*—Whites of eggs, sugar and water, castor oil, gruel.

ZINC.—*Antidotes*—Limewater, chalk and water, soap and water.

IRON.—*Antidotes*—Magnesia, warm water.

HENBANE, HEMLOCK, NIGHTSHADE, FOXGLOVE.—*Antidotes*—Emetic and castor oil, brandy and water if necessary.

POISONOUS FOOD.—*Antidotes*—Emetics and castor oil.

XXIII. Health-Board Disinfectants.

The instructions of the National Board of Health in relation to disinfectants and their use, with explanations as to disinfectants and deodorizers, are valuable. We have in XIX given simple means of deodorizing and disinfection. Deodorizers destroy smells; they do not necessarily disinfect. Disinfectants do not necessarily have odors, and some of the most virulent germs, as typhoid germs, may not, in water, be apparent to the sense of taste or smell. Disinfectants destroy the poisons of infectious and contagious diseases.

Some disinfecting agents recommended by the Board are:

"1. Roll-sulphur (brimstone) for fumigation.

"2. Sulphate of iron (copperas) dissolved in water, the proportion of one and a half pound to the gallon; for soil, sewers, etc.

"3. Sulphate of zinc and common salt, dissolved together in water, in the proportions of four ounces sulphate and two ounces salt to the gallon; for clothing, bed-linen, etc."

XXIV. How to Use Disinfectants.

"1. IN THE SICK-ROOM.—The most available disinfectants are fresh air and cleanliness.

"The clothing, towels, bed-linen, etc., should, on removal from the patient, and before they are taken from the room, be placed in a pail of the zinc solution, boiling hot if possible.

"All discharges should either be received in vessels containing copperas solution, or when this is impracticable, should be immediately covered with copperas solution. All vessels used about the patient should be cleansed with same solution.

"Unnecessary furniture, especially that which is stuffed, carpets and hangings, should, when possible, be removed from the room at the outset; otherwise they should remain for subsequent fumigation and treatment.

"2. FUMIGATION.—Sulphur is the only practicable agent for disinfecting the house. The rooms to be disinfected must be vacated. Heavy clothing, blankets, bedding, and articles which cannot be treated with zinc solution, should be opened and exposed during fumigation. Close the rooms as tightly as possible, place the sulphur in iron pans, supported upon bricks placed in wash-tubs containing a little water, set it on fire by hot coals, or with the aid of a spoonful of alcohol, and allow the rooms to remain closed for twenty-four hours. For a room ten feet square, at least two pounds of sulphur should be used; for larger rooms proportionally increased quantities.

"3. PREMISES.— Cellars, yards, stables, gutters, privies, cesspools, water-closets, drains, sewers, etc., should be frequently and liberally treated with copperas solution. It is easily prepared by hanging a basket containing about sixty pounds of copperas in a barrel of water.

" 4. BODY AND BED CLOTHING, ETC.—It is best to burn all articles which have been in contact with persons sick with contagious or infectious diseases. Articles too valuable to be destroyed should be treated as follows:

" Cotton, linen, flannels, blankets, etc., should be treated with the boiling-hot zinc solution; introduce piece by piece; secure thorough wetting, and boil at least half an hour.

" Heavy woolen clothing, silks, furs, stuffed bed-covers, beds and other articles which cannot be treated with the zinc solution, should be hung in the room during fumigation, their surfaces thoroughly exposed, and pockets turned inside out.

" Afterward they should be hung in the open air, beaten and shaken. Pillows, beds, stuffed mattresses, upholstered furniture, etc., should be cut open, the contents spread out and thoroughly fumigated. Carpets are best fumigated on the floor, but should afterward be removed to the open air and thoroughly beaten.

" 5. CORPSES.—These should be thoroughly washed with a zinc solution of double strength; should then be wrapped in a sheet, wet with the zinc solution, and buried at once. Metallic, metal-lined or air-tight coffins should be used when possible, certainly when the body is to be transported for any considerable distance."

CHAPTER VI.

CONTRIBUTIONS FROM FRIENDS ON HOUSEHOLD ECONOMY.

I. VALUE OF CONDENSED INFORMATION.—II. ORIGIN OF OUR HOUSEHOLD RECIPES.—III. ECONOMY IN THE KITCHEN—WASHING DISHES.—IV. THE DAMPER IN THE STOVE.—V. REGULATING COAL FIRES.—VI. THE USE OF WASTE PAPER.—VII. CLEANING SOILED MARBLE, ETC.—VIII. VERMINOUS INSECTS.—IX. CLOTH AND FUR MOTHS.—X. BOOK-DESTROYING INSECTS.—XI. KEROSENE. XII. THE LAUNDRY—SOME HELPS IN WASHING.—XIII. STARCHING AND IRONING.—XIV. BLEACHING LINENS, ETC.—XV. HOME-MADE SOAP AND CANDLES.—XVI. TO CLEAN SILVER.—XVII. SWEEPING.—XVIII. PAPERING, KALSOMINING AND PAINTING.—XIX. KALSOMINING.—XX. PAINTING.—XXI. SPRING HOUSE-CLEANING.—XXII. HOUSEHOLD HINTS—XXIII. TOILET RECIPES.—XXIV. HOME-MADE WINES.—XXV. HOME-MADE INKS.—XXVI. RECIPES FOR GLUE.—XXVII. THE DYER'S ART.—XXVIII. COLORING DRESS AND OTHER FABRICS.—XXIX. COLORING YELLOW, BLUE AND GREEN.—XXX. SCARLET AND PINK.—XXXI. COLORING BLACK, BROWN AND SLATE.—XXXII. WALNUT COLORING—BLACK WALNUT.—XXXIII. COLORING CARPET RAGS.

I. Value of Condensed Information.

RECIPES, to be of use, should be suited to the needs of those for whom they are intended. Elaborate preparations that can only be made by a chemist, or by the aid of scientific appliances beyond the reach of the masses, would be out of place in a book of the practical nature of this work.

Tables of useful facts are also of great value in every department of life, for the reason that they present at a glance necessary information that could not be otherwise given except by many pages of print. They are simple and valuable to have at hand when needed. Every person outside of cities and villages is interested, for instance, in knowing the number of plants that may be contained on a given piece of ground; the quantity of seeds required per acre or per rod; weights per bushel of various grains, and the number of seeds in an ounce or pound; how to judge of the quality of land by its vegetation, and scores of other things of like kind. These we have grouped together in this volume so as to be easily examined, and so classified that no time need be spent in hunting for them.

II. Origin of Our Household Recipes.

IN the course of the author's experience as agricultural editor and writer, many valuable, because simple, recipes relating to household art have come into his possession, partly through correspondence with the best housekeepers and partly through communications to the journals with which he has been connected. The best of this collection have been selected for reproduction in this work. The household departments have been prepared with the aid of a lady of long experience as a housekeeper, and well known for her patience and deft skill in nursing the sick.

III. Economy in the Kitchen—Washing Dishes.

An English lady says: There are so many modes of washing dishes, that some will take it as quite unnecessary that they should be told how to do it. The proper way is perfectly simple. Have a pan of hot water in which a little soap has been dissolved, and then use a mop made of an old linen towel, or candle wicking fastened to the end of a stick, and then transfer them to a pan of still hotter water, and drain a moment, and wipe dry. This gives them an elegant polish. They should be wiped as soon as they have been through the last water, else they have a streaked effect, which can be felt, if not seen. It is the custom in England to drain them in racks, but we think our own mode the best—at least, with the white ware so fashionable in this country, and which is so little known there.

The glassware should be washed first, then the silver, then the cups and saucers, etc., and the greasy dishes last. Never wash nor wipe more than one article at a time. When china is rough to the touch, it is simply because it is not cleansed. Hot water, and plenty of it, dry, clean towels and rapid wiping make the dishes shine like mirrors. You can wash glasses in quite hot water, by rolling them round in the water, filling them as soon as they touch it, thus making all portions of the glass equally hot. They will never crack if treated in this way. Dish-washing forms a large proportion of the daily life of the housekeeper, and anything which expedites it, and leaves time for other things, ought to be welcomed.

IV. The Damper in the Stove.

The following, on the use of the damper, by an editorial lady friend, although written for stoves in which wood is burned, contains information of equal value for those burning coal, whether hard or soft. The use of the instrument, must, however, be studied, since different fuels require different treatment. Concerning the damper, our contributor says: A damper in the stove is of great importance in a house—both as a matter of economy, and of comfort. It makes the hot air remain in the stove, and does not take in the outside heated air, which is done through the crevices and proper drafts. If the damper is shut, you instantly feel the heat on your face, showing that it is thus kept in the room. The circulation is thus stopped in the room, and a soft, pleasant atmosphere is the result.

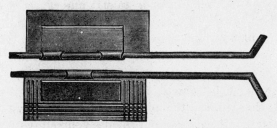

ADJUSTABLE STOVE DAMPER.

Economy in Fuel —The main item is, however, economy in fuel. Not more than half the quantity of wood is used, and yet an equal amount of heat is obtained. This is of some consequence to the purchaser, or to the person who has the wood to chop, and of course, has an extra amount to furnish when it goes roaring up the chimney. Then to the housekeeper, the fact that she gains more ashes by the use of a damper, is an inducement to use one, as they are not lost in the air.

Wood Fuel.—Dry hard wood is positively necessary, where a damper is used. Dry hickory makes the best coals. Maple and birch come next, though the flame is not so hot and bright. The beech blazes well, but is too much like soft wood. When the blaze is gone there is not much left of it. With a damper you can use soft maple. It is often the case that when there is not a damper, the fire is continually "going down," the heat is unequal, and the temperature of the room is being continually rendered cool—first dry, then damp, making it disagreeable and dangerous. This matter is of the utmost importance to the housekeeper, and should be attended to.

V. Regulating Coal Fires.

NEVER fill a stove more than half or two-thirds full of coal, even in the coldest weather. When the fire is low, never shake the grate or disturb the ashes, but add from ten to fifteen lumps of coal, and set the draft on. When these are heated through and somewhat ignited, add the amount necessary for a new fire, but do not disturb the ashes yet. Let the draft be open half an hour. Then shake out the ashes. The coal has thoroughly ignited, and will keep the stove at a high heat from six to twelve hours, according to the coldness of the weather. In very cold weather, after the fire is made, add coal every hour.

Use of Coal in Sick Rooms.—Mrs. M. G. L., of West Virginia, writes: You know what a racket is caused, even by the most careful hand, in supplying coal to a grate or stove, and how, when the performance is undertaken by Biddy, it becomes almost distracting. If you don't remember, take notice the first time you are ill, or have a dear patient in your care, or the baby is in a quiet slumber. Let some one bring in the coal scuttle or shovel, and revive your recollection. Well, the remedy we suggest is to put the coal in little paper bags, each holding about a shovelful. These can be laid quietly on the fire, and, as the paper ignites, the coals will softly settle in place.

You may fill a coal scuttle or box with such parcels, ready for use. For a sick-room, a nursery at night, or even for the library, the plan is admirable. Just try it. Besides, it is so cleanly. If you don't choose to provide yourself with paper bags, you can wrap the coals in pieces of newspapers at your leisure, and have them ready for use when occasion requires. Perhaps the "help" will kindly do it for you; or better still, the children, if the house is so sunshined, will attend to the wrapping, and think it fine fun.

Economy in Coal.—Mrs. N. M., of St. Charles, Missouri, says: In any fireplace not excessively small, a plate of iron set upon the grate will halve the consumption of coal, reduce the smoke and leave a cheerful, free-burning fire. Quite sufficient air enters through the bars, no poking is necessary and the fire never goes out until the coals are consumed. There is no ash and no dust, every particle being consumed.

Any householder can try this experiment and reduce his coal bill, say thirty per cent, at the cost of a shilling.

Care of Stoves.—Blackening and polishing stoves is hard work. Indeed, one of the best known lady writers on economy and household art, has said that a blackened stove may be a nuisance. It may be so in more ways than one. Few housekeepers, says our authority, have time to blacken their stoves every day, or even every week. Many wash them in either clean water or dish-water. This keeps them clean, but they look very brown. After a stove has been blackened, it can be kept looking very well for a long time by rubbing it with paper every morning. If I occasionally find a drop of gravy or fruit juice that the paper will not take off, I rub it with a wet cloth, but do not put on water enough to take off the blacking.

VI. The Uses of Waste Paper.

A CORRESPONDENT in Little Rock, Arkansas, truly says: Comparatively few housekeepers are aware of the many uses to which waste paper may be put. After a stove has been blackened, it can be kept looking very well for a long time by rubbing with paper every morning. Rubbing with paper is a much nicer way of keeping the outside of a tea-kettle, coffee-pot and tea-pot bright and clean, than the old way of washing them in suds. Rubbing with paper is also the best way of polishing knives and tinware, after scouring. This saves wetting the knife handles. If a little flour be held on the paper in rubbing tinware and spoons, they shine like new silver. For polishing mirrors, windows, lamp-chimneys, etc., paper is better than dry cloth. After it has been so used it is none the worse for kindling fires. Preserves and pickles keep much better, if brown paper, instead of cloth, is tied over the jar. Canned fruit is not so apt to mold if a piece of writing paper, cut to fit the can, is laid directly on the fruit. Paper is much better to put under a carpet than straw. It is warmer, thinner, and makes less noise when one walks over it. Two thicknesses of paper placed between other coverings on a bed, are as warm as a quilt. If it is necessary to step upon a chair, always lay a paper on it and thus save the paint or woodwork from damage. A fair carpet can be made for a room not in constant use, by pasting several thicknesses of newspaper on the floor, over them a coat of wall-paper, and giving them a coat of varnish.

VII. Cleaning Soiled Marble.

MUCH annoyance is frequently experienced from soiling marble table-tops, kitchen slabs or other marble objects. It is said that if slacked lime is mixed with a strong solution of soap into a pasty mass and spread over the spot, and allowed to remain for twenty-four or thirty hours, then carefully washed off with soap and water, and finally with pure water, the stain will be almost entirely removed, especially if the application be repeated once or twice.

Ox-Gall and Lye.—Another preparation consists in mixing an ox-gall with a quarter of a pound of soap-boiler's lye, and an eighth of a pound of oil of turpentine, and adding enough pipe-clay earth to form a paste, which is then to be placed upon the marble for a time, and afterward scraped off, the application to be repeated until the marble is perfectly clean. It is quite possible that with all our endeavors a faint trace of the stains may be left; but it is said that this will be almost inappreciable. Should the spots be produced by oil, these are to be first treated with petroleum for the purpose of softening the hardened oil, and the above-mentioned applications may be made subsequently.

Ink Spots on Marble.—Ink spots may be removed by first washing with pure water, and then with a weak solution of oxalic acid. Subsequent polishing, however, will be necessary, as the luster of the stone may become dimmed. This can be best produced by very finely powdered soft white marble, applied with a linen cloth first dipped in water and then into the powder. If the place be subsequently rubbed with a dry cloth the luster will be restored.

Grease Spots on Wood.—If one is so unfortunate as to get any sort of grease on floor or table, apply directly potter's clay, just wet with water so as to form a stiff paste. Spread it pretty thick upon the grease spot, and lay a thin paper over to keep it from being rubbed off. After twenty-four hours scrape it off and spread on fresh clay. It will gradually absorb the grease, and leave the floor or table clean; but it may need to be renewed several times. When the clay looks clean, wash off with soap and water. The clay is also good to take grease from clothing, applied in the same way.

To Clean Tin Covers.—Mix a little of the finest powdered whiting with the least drop of sweet oil, rub the covers well with it, and wipe them clean; then dust over them some dry whiting in a muslin bag, and rub bright with dry leather. This last is to prevent rust, which the cook must guard against by wiping them dry and putting them by the fire when they come from the dining-room, for if hung up once damp, the inside will rust.

VIII. Verminous Insects.

Cockroaches and Bed-Bugs.—Cockroaches are the plague of many housekeepers, and yet a little Paris green is death to them. Keep it in a common flour-dredging box, label it poison, and apply it weekly to their haunts. Bed-bugs or chinch-bugs can also be dispersed and utterly routed with this remedy; and both cockroaches and bed-bugs will flee from powdered borax. Travelers should always carry a paper of borax in their bags, and sprinkle it under and over their pillows, if they fear they shall become food for the last-named wretches.

Ants and Flies.—Sprigs of worm-wood will drive away large black ants; and none of them, whether black, brown or red, relish wintergreen, tansy, Paris green, cayenne or kerosene; so if they invade our pantries, we can, by a judicious application of some one of these articles, make the premises too unpleasant for them. Fly-paper should be kept around the house as early as the middle of May. Put it in every open window, and thus destroy every intruder. It must constantly be borne in mind, that the cobalt with which some fly-paper is saturated, and also Paris green, are two most deadly poisons. Keep them safely out of the way of children.

Expelling Flies From Rooms.—It is stated that if two and a half pounds of powdered laurel leaves are macerated or boiled in two gallons of water, until their poisonous quality is extracted, and with the solution a whitewash is made, by adding as much quicklime as can be slaked in it, and if a room be whitewashed with this preparation, flies will not settle on the walls for six months. If a paste, made by stirring together one pint of the powdered laurel leaves with a quarter pint of glycerine, be applied to windows and door casings, a room so prepared will soon be emptied of flies. Two applications of this paste are enough to keep even a kitchen clear of insects for a fortnight.

There is nothing disagreeable or deleterious to human beings in the odor of the wash or paste, though laurel leaves, or laurel water, taken into the human stomach, acts as a violent poison.

You may also drive flies out with a brush, but, unless something is done to render the place uninviting to them, they will return immediately. There are many weeds or plants emitting an empyreumatic odor which answer well for the purpose. None are more effectual than the wild chamomile (Mayweed). The odor of this plant is not at all disagreeable, and branches of the weed when in flower, or some of the dried flowers, scattered about a room, will soon rid it of all flies.

Another way is to throw some powdered black pepper on a hot shovel and carry it about the room. The generation of empyreumatic vapors in the same way from other spices will also, it is said, answer the purpose. A few drops of carbolic acid or creosote, on a cloth hung up in a sick-room or used in the dressings, would probably be effectual, but the odor is not usually so acceptable to one's olfactories.

The best thing of all is to keep them out. The author has never found any means of doing this so cheap, effectual and pleasant, as wire screens to all doors and windows. This will keep them out, with a little driving occasionally, if the doors all open outward. For mosquitoes and gnats, when it is impracticable to keep them out, as in the case of tents, used while camping out, a little brown sugar burned on coals we have found effective in some instances in driving them away, but of course a good mosquito-bar is the best.

IX. Cloth and Fur Moths.

The small moths so destructive to cabinets, tapestry, clothes, carpets, furs, grains, etc., are called tineans, and belong to the natural order lepidoptera. They have four membranous wings covered with imbricated scales, like fine powder—as the butterflies and moths. Among these are the carpet moth, the clothes moth, the fur moth, and the hair moth. These moths are nocturnal in their habits, flying in the evening. They do not lay their eggs in material in constant use, and therefore wardrobes, drawers, chests, etc., should be frequently examined, and the contents aired, and beaten to dislodge the eggs or larvæ.

In old houses subject to their depredations, the cracks in closets, and other exposed places, may be brushed over with turpentine or other odorous substances. Sheets of paper sprinkled with spirits of turpentine, powdered camphor, shavings of Russia leather or tobacco, are also preventives. Chests and boxes of camphor-wood, red cedar and Spanish cedar, are obnoxious to these insects, and are useful for preserving costly articles.

The cloth linings of carriages, etc., may be preserved from their depredations by being sponged on both sides with a solution of corrosive sublimate in alcohol, just strong enough so that it will not bleach a black feather white. The insects may also be killed by fumigating with tobacco smoke, or the fumes of sulphur. It is also said that if hemp, in flower, cut and dried, be placed in a cushion, it will prevent the ravages of moths for years.

Furs.—A good recipe for preserving furs from moths is: One ounce gum camphor and one of powdered shell of red pepper. Macerate in eight ounces of alcohol for several days, then strain. Sprinkle furs, etc., with this tincture, roll up in a clean cloth and lay away.

X. Book-Destroying Insects.

Books, large or small, made up of dry paper, are nesting-places for a variety of insects, hardly large enough to be recognized as living things. Besides making themselves homes between the leaves, they feast on the paste, binding, twine on the backs and the green mold that gathers on them if neglected. One species takes up residence in the binding, devouring as it goes. Another feeds upon the paste. Still another book pest that is sure to appear in a library, not overhauled and dusted occasionally, eats through a volume.

Bookcases should not be made light with glass doors. Wire-netting is far preferable, because the books are kept drier; fresh air is all-important. An upper story is superior to a basement, being less liable to gather mold, which is a forest of minute vines in which bookworms ramble for exercise. Twice in each summer the books should be exposed to a bright sunlight while dusting them, also exposing the open leaves to a fresh current of air.

XI. Kerosene.

Kerosene is volatile and its vapor is explosive. Only the best oils—from 150 degrees fire test up to 175 degrees—should be used. The oil does not explode. An oil may even extinguish a burning match when thrown into it, and yet be highly dangerous to be used as a burning fluid. It is the vapor of these oils mixed with air that is dangerous, as far as explosion is concerned. While a partly filled lamp has the portion above the oil filled with a mixture of vapor and air, it may explode. When a lamp is filled while lighted, the mixture of air and vapor in the can or filler explodes upon coming in contact with the flame; the oil itself does not explode, though it does serious injury when scattered by the explosion.

Test for Kerosene.—Dr. Nichols, the well-known chemist and writer on chemical science, advises the following test for kerosene: Fill a pint bowl two-thirds full of boiling water, and into it put a common metallic thermometer. The temperature will run up to over 200 degrees. By gradually adding cold water, bring down the temperature of the water to 100 degrees, and then pour into the bowl a spoonful of the kerosene, and apply a lighted match. If it takes fire, the article should be rejected as dangerous; if not, it may be used with a confident feeling of its safety.

XII. The Laundry—Some Helps in Washing.

The Germans, and especially the Belgians and Hollanders, are noted for their fine washing. Their method, which does away with the use of soda, is as follows: Dissolve two pounds of soap in about three gallons of water as hot as the hand can bear, and add to this one tablespoonful of turpentine and three of liquid ammonia; the mixture must then be well stirred, and the linen steeped in it for two or three hours, taking care to cover up the vessel containing them as nearly hermetically as possible. The clothes are afterwards washed out and rinsed in the usual way. The soap and water may be re-heated, and used a second time, but in that case half a tablespoonful of turpentine and a tablespoonful of ammonia must be added. The process will cause a great economy of time, labor and fuel. The linen scarcely suffers at all, as there is little necessity for rubbing, and its cleanliness and color are perfect. The ammonia and turpentine, although their detersive action is great, have no injurious effect upon the linen; and while the former evaporates immediately, the smell of the latter will disappear entirely in drying the clothes

Washing Summer Suits, etc.—Summer suits are nearly all made of white or buff linen, pique, cambric or muslin. Whatever the material, common washerwomen spoil everything with soda, and nothing is more frequent than to see the delicate tints of linens and percales turned into dark blotches and muddy streaks by the ignorance and vandalism of a laundress. It is worth while for ladies to pay attention to this, and insist upon having their summer dresses washed according to the directions which they should be prepared to give their laundresses themselves. In the first place the water should be tepid, the soap should not be allowed to touch the fabric; it should be washed and rinsed quickly, turned upon the wrong side, and hung in the shade to dry, and when starched (in thin boiled but not boiling starch) should be folded in sheets or towels, and ironed upon the wrong side as soon as possible. But linen should be washed in water in which hay or a quart bag of bran has been boiled. This last will be found to answer for starch as well, and is excellent for print dresses of all kinds; but a handful of salt is very useful to set the color of light cambrics and dotted lawns; and a little ox gall will not only set but brighten yellow and purple tints, and has a good effect upon green.—Adele.

To Cleanse Blankets.—Put two large tablespoons of borax and a pint bowl of soft soap in a tub of cold water. When dissolved, put in a pair of blankets and let them remain over night. Next day rub them out, rinse thoroughly in two waters, and hang them to dry. Do not wring them.—S. E. F.

To Wash Flannels.—I wonder if housekeepers know that flannel should never have soap smeared upon it, or be rubbed upon a board? A hot suds should be made, and the flannel should be squeezed through it, rubbing the dirtiest portions in the hands as lightly as possible. When the stains are softened, another warm water should be ready, into which dip the flannels, and squeeze them dry as possible out of it. Shake them well, and hang them out where the wind will not strike them hard; never hang them in the sun.—A. W.

Washing Fluid.—Three tablespoonfuls of soda, the same quantity of dissolved camphor (the same as kept for family use), to a quart of soft water; bottle it up, and shake well before using. For a large washing take four tablespoonfuls of fluid to a pint of soap, make warm suds and soak the clothes half an hour; then make another suds, using the same quantity of soap and fluid, and boil them just fifteen minutes, then rinse in two waters.—MAMIE, Lake County, Ind.

To Remove Acid Stains and Restore Color.—When color on a fabric has been destroyed by acid, ammonia is applied to neutralize the same, after which an application of chloroform will, in almost all cases, restore the original color. The application of ammonia is common; but that of chloroform is but little known. Chloroform will remove paint from a garment or elsewhere, when benzole or bisulphide of carbon fails.

To Preserve Clothes-Pins.—They should be boiled a few moments and quickly dried, once or twice a month, when they become more flexible and durable. Clothes lines will last longer and keep in better order for wash-day service, if occasionally treated in the same way.

To Remove Grease from Worsted.—Take one-quarter pound of Castile soap, one-quarter pound ammonia, very strong, one ounce sulphuric ether, one ounce spirits of wine, one ounce glycerine. To mix this cut the soap fine and dissolve in one quart of soft water, and then add four more quarts of water and all ingredients.

Two or three daily applications of benzine will also remove the grease spots. Apply with brush or woollen cloth. Do not make the application in a warm room, as the article is highly inflammable.—MAGGIE, Richland, Mich.

XIII. Starching and Ironing.

STARCH and iron shirt bosoms as usual, and when the articles are thoroughly dry, place one at a time on a narrow, hard and very smooth board, which has one thickness of cotton cloth over it, sewed tightly; have the polishing iron heated so that it will not scorch, and rub it quick and hard over the surface, up and down the bosom, using only the rounded part on the front of the iron. A still higher polish may be obtained by passing a damp cloth lightly over the smooth surface, and then rubbing hard and quickly with the hot iron. It needs a good deal of patient practice to do this admirably, but when once learned, it is as easy as other ironing. A polishing iron is small and highly polished, with a rounded part, which allows all the friction to come on a small part at one time, which develops the gloss that may be in both linen and starch. Collars and cuffs look nicely done in this way.

For Lawns.—Take two ounces of fine white gum Arabic powder, put it into a pitcher, and pour on a pint or more of water, and then, having covered it, let it stand all night. In the morning, pour it carefully from the dregs into a clean bottle, cork it and keep it for use. A teaspoonful of gum water stirred into a pint of starch made in the usual manner, will give to lawns, either white or printed, a look of newness, when nothing else can restore them, after they have been washed.

Gloss for Shirt Fronts, Collars and Cuffs.—To a pail of starch, a whole sperm candle is used. When the linen is dry, it is dipped in the cold starch and ironed in the ordinary way; then it is dampened with a wet cloth, and the polishing iron pressed over it. To this last manipulation the linen is indebted for the peculiar laundry gloss which all admire so much, but which many housekeepers have vainly striven to leave upon the wristbands and bosoms of their husbands' shirts.

XIV. Bleaching Linens, Etc.

The best method of bleaching or restoring whiteness to discolored linen is to let it lie on the grass, day and night, so long as it is necessary, exposed to the dews and winds. There may occur cases, however, when this will be difficult, and when a quicker process may be desirable. In these cases, the linen must be first steeped for twelve hours in a lye formed of one pound of soda to a gallon of soft boiling water; it must then be boiled for half an hour in the same liquid. A mixture must then be made of chloride of lime with eight times its quantity of water, which must be well shaken in a stone jar for three days, then allowed to settle; and being drawn off clear, the linen must be steeped in it for thirty-six hours, and then washed out in the ordinary manner. To expedite the whitening of linen in ordinary cases, a little of the same solution of chloride of lime may be put into the water in which the clothes are steeped; but in the employment of this powerful agent, great care must be exercised, otherwise the linen will be injured.—HOUSEKEEPER, Louisville, Ky.

Bleaching Cotton Goods.—A very good way, says Mrs. M. T. M., Auburn, Ill., to bleach cotton cloth is to soak it in buttermilk for a few days. Another way is to make a good suds, put from one to two tablespoonfuls of turpentine into it, before putting the clothes in. Wash as usual, wringing the clothes from the boil, and drying without rinsing. By using one tablespoonful of turpentine in the first suds on washing days, it will save half the labor of rubbing, and the clothes will never become yellow, but will remain a pure white. It is simple, and I never wash without it.

To Clean Merino.—Grate two or three large potatoes; add to them a pint of cold water; let them stand for a short time and pour off the liquor clear, when it will be fit for use. Lay the merino on a flat surface and apply the liquid with a clean sponge until the dirt is completely extracted. Dip each piece in a pailful of clean water and hang up to dry without wringing. Iron while damp on the wrong side. It will then appear almost equal to new.—NELLIE, Jefferson, Ill.

Removing Iron Rust.—Wash the stains in ripe tomatoes. Then hang in the sun to dry. After thoroughly drying, wash in clear water.

XV. Home-Made Soap and Candles.

Soft-Soap.—In making soft-soap, use a pine barrel, for a hard-wood barrel will warp and leak. A well-cleansed fish barrel is commonly taken for the purpose. Put in ten to twelve pounds of potash. and throw upon it two pailfuls of boiling water. Let it digest awhile and then put in two pounds of grease to each pound of potash. Have the grease hot. Let that digest awhile, then add a third pailful of hot water. Stir and digest awhile, then add another pailful of hot water. Keep doing this until the barrel is within six inches of being full. Stir occasionally until the whole is well mixed, It should stand three months before use. Stir occasionally during the first week. The longer it stands after making, the better the soap. We keep it a year before use.—W. Niagara Co., N. Y.

Another Way.—For one barrel of soap, take thirty pounds of grease, free from salt, rinds or bits of lean meat, and the lye from two barrels of good ashes. Put one quart of lime in the bottom of each barrel of ashes. Put boiling water on to leach with; have ready the soap barrel where it is to stand. When the lye begins to run, melt the grease in a little lye and pour it in the barrel. Heat the lye and fill it full, stirring frequently until cold. I always use the stove kettle, as that is free from rust, which makes white cloth yellow. Soap made in this way will be very light-colored and thick, and requires but little labor in making. If the lye is not strong enough to eat the grease, boil it awhile.—Mrs. A. G.

Cold-Made Soap.—Have lye strong enough to bear up an egg. Then stir in any soap grease until the lye is pretty well filled, and in a week, or ten days the soap will be fit for use. In the meantime, stir occasionally.—MRS. M. A. C., Labette Co., Kan.

Hard-Soap.—Five pounds soda ash, two and a half pounds white lime, one-half pound resin, ten pounds grease, eight gallons soft water. Boil five hours. Take the soda ash and lime, put them in your kettle, pour the water over, and boil one-half hour. Then let it settle, and turn off the lye. Lift out the lime and soda ash, turn over it more water, as it is yet quite strong, return the lye to the kettle, add the grease and resin, and boil five hours. This makes excellent soap.—MRS. E. A. H., North Benton, O.

Second Recipe for Hard Soap.—Pour four gallons of boiling water over six pounds of salsoda and three pounds of unslaked lime. Stir the mixture well and let it stand over night. Then drain it off. Put six pounds of tallow, or any kind of clear grease with it, and boil it two hours, stirring most of the time.—C. E. S., Carondelet, Mo.

Lard Candles.—Take twelve pounds of lard, one pound saltpeter, one pound alum. Pulverize and mix the saltpeter and alum; dissolve the compound in a gill of boiling water; pour the compound into the lard before it is quite melted. Stir the whole until it boils, and skim off what rises. Let it simmer until the water is all boiled out, or until it ceases to throw off steam. Pour off the lard as soon as it is done, and clean the boiler while it is hot. If the candles are to be run in a mould, you may commence at once, but if to be dipped, let the lard cool first and cake. Then treat as you would tallow.—NETTIE, Terre Haute, Ind.

Hardening Tallow.—Take the common prickly pear and boil or fry it in the tallow, without water, for half an hour, then strain and mould. I use about six average-sized leaves to the pint of tallow (by weight one pound of leaves to four of tallow), splitting them up fine. They make the tallow as hard as stearine, and do not injure its burning qualities in the least.—MRS. E. L. O., Waco, Tex.

XVI. To Clean Silver.

A LADY correspondent in Southern California sends the following: Silver is most susceptible of spotting and discoloration by sea air, the human perspiration, the presence of sulphureted hydrogen (as seen in an egg spoon left uncleaned), the excreta of cockroaches and other strong-smelling insects, and lastly, by the contact of mice; the latter cause has irretrievably injured new plated-ware, never used, but left on a sideboard accessible to these little vermin. It is the practice of the East-Indian jewelers never to touch silver and gold with any abrasive substance. The most delicate filigree work and wire constructions of silver are rendered snowy white by their simple manipulation. They cut some juicy lemons in slices; with these they rub any large silver or plated article briskly, and leave it hidden by the slices in a pan for a few hours. For delicate jewelry, they cut a large lime nearly in half and insert the ornament; then they close up the halves tightly and put it away for a few hours. The articles are then to be removed, rinsed in two or three waters, and consigned to a saucepan of nearly boiling soapsuds, well stirred about, taken out, again brushed, rinsed, and finally dried on a metal plate over hot water, finishing the process by a little rub of wash leather (if smooth work).

For very old, neglected or corroded silver, the article may be dipped, with a slow stirring motion, in rather a weak solution of cyanide of potassa, but this process requires care and practice, as it is by dissolving off the dirty silver you obtain the effect. Green tamarind pods or oxalate of potash are greater detergents of gold and silver articles than lemons, and are much more employed by the artisan for removal of oxides and fire-marks.

A strong solution of hyposulphite of soda, as used by photographers, is perhaps the safest wash, as it will in no way attack the metallic silver, but only the films of chloride, etc., on its surface.

XVII. Sweeping.

If brooms are wet in boiling suds once a week they will become very tough, will not cut the carpet, will last much longer, and always sweep as clean "as a new broom," if kept hanging up when not in use. A most admirable way of sweeping a dusty carpet is to have a pail of clean cold water stand by the door, into which the broom can be dipped, taking care to shake all the drops out of it, by knocking it hard against the side of the pail. Then sweep a couple of yards or so, wet the broom again, and sweep as before. When carefully done, and the drops are all shaken out, it will clean a very dirty carpet nicely, and you will be surprised at the amount of dirt removed. Sometimes you will need to change the water two or three times. In winter, snow can be sprinkled over the carpet and swept off, before it has time to dissolve. Some throw down tea-grounds, and sweep them off briskly. Fresh grass is an excellent cleanser of a carpet, strewn thickly about and swept hard. Moistened Indian meal has proved of good effect.—ELLA W., Lincoln, Neb.

XVIII. To Paper Walls.

Mrs. Annie R. White, for many years literary and household editor of the *Western Rural*, discourses as follows about the way to paper rooms: Don't try to paper with a carpet down. Make paste, cut bordering and the paper the day before. If the wall has been whitewashed, it must be washed in vinegar to neutralize the alkali in the lime. If papered before, and you wish the paper removed, sop with water and it will peel off. If convenient, provide a long board, wide as the paper, though a table or two will do. The paper must be measured, placed right side down on the board; then with a brush proceed to lay on the paste, not too thickly, but over every part, and be careful that the edges receive their share. When completed, double within three inches of the top, the paste sides being together; carry to the wall, mount your chair, and stick your three inches of pasted paper on the wall at the top. That holds it; now strip down the other end, and see that it fits just right; if not, peel down, make right, then press to the wall from the center right and left. Leave no air under, or when warm it will expand, bursting the paper.

Of course the paper must be matched; it will not do to measure by lines unless the walls are perfectly plumb. Small figures make less waste, and make a small room look larger. Stripes make a low room look higher, and if there are no figures between, or in the stripe to match, there is no waste, and no trouble in putting on. If a narrow border is the style, let it be bright, if the paper be neutral; but if that be bright, the border had better be dark and neutral. If the paste be made too thick, the paper will be apt to crack and peel off; if too thin, it will saturate the paper too quickly, and make it tender in putting it on. A counter-duster (Brussels brush) is nice to brush the paper to the wall. White clean cloths will do, but it will not do to rub the paper with this; being damp, the paint or color rubs off the paper. The tables must be dried each time after pasting, for the same reason. Paste under paper must not freeze, nor be dried too quickly. If whitewashing is done after papering, tack double strips of newspaper wider than the border all around the room, to prevent its soiling the paper.

Papering Whitewashed Walls.—If the walls are covered with thick, scaly whitewash, the result of years of additions, they must be scraped with a thin steel scraper—a hoe will do if carefully used. This will smooth them. Then wash them in weak lye and sweep off thoroughly when dry. Size the walls with glue water, one pound of glue to a pail of water, and the paper will stick and not peel off. The paste should be smooth rye flour paste, rather thin, but perfectly smooth. Starch paste is the next best.—PAINTER-TURNED-FARMER, Lincoln, Neb.

XIX. Kalsomining.

THERE are as many ways to kalsomine as there are to whitewash. The simplest mode we know of is to take ten pounds of Paris white, and soak it in cold water—just enough water to dissolve it well. Take one-eighth of a pound best white glue, soaked in cold water enough to cover. Let it soak three to four hours; or till well swelled. If there is much liquid by the time the glue is well swollen, take the glue out and put it in a saucepan over the fire, with a little water to keep it from burning. Mix the dissolved whitening thoroughly with the hand. Then add the melted glue, mixing well. This mixing needs to be done in a large vessel. Then pour into these ingredients a quarter of a pint of linseed oil, and on top of oil pour sufficient muriatic acid (perhaps ten cents' worth) to cut the oil, stirring it the while. After this is done, add cold water enough to the whole to thin it down to about a pailful of the liquid. Then mix a little ultramarine in a cup of cold water, and add to the whole, so as to remove the yellow tinge, and make it a bluish white. Apply with a clean whitewash brush, one or two coats. So says Mrs. O. A. N., who adds, that her husband does the kalsomining.

XX. Painting.

Best Time for Outside Work.—Paint houses late in the autumn or during the winter. Paint then will endure twice as long as when applied in early summer, or in hot weather. In the cold season it dries slowly and becomes hard, like a glazed surface, not easily affected afterward by the weather, or worn off by storms. But in very hot weather the oil in the paint soaks into the wood at once, as into a sponge, leaving the lead nearly dry, and ready to crumble off. This last difficulty might be guarded against, though at an increased expense, by first going over the surface with raw oil. By painting in cold weather, one annoyance might certainly be escaped—the collection of small flies in the fresh paint.

Recipe for Inside Paint.—A cheap inside paint, and by no means a bad one, especially where the smell of oil or turpentine would be objectionable, or in any case where lead paint is not desirable, may be made by taking eight ounces of freshly slaked lime, and mixing it in an earthen vessel, with three quarts of skimmed sweet milk.

In another vessel mix three and a half pounds of Paris white with three pints of skimmed milk. When these mixtures are well stirred up, put them together, and add six ounces of linseed oil. Mix these well, and it will be ready for use. This preparation is equal to oil paint, and is excellent for walls and ceilings. Any shade may be made by the addition of dry pigments.—PAINTER-TURNED-FARMER.

To Soften Putty.—To remove old putty from broken windows, dip a small brush in nitric or muriatic acid (obtainable at any druggist's), and with it paint over the dry putty that adheres to the broken glass and frames of your windows; after an hour's interval, the putty will become so soft that it can be removed easily.

XXI. Spring House-Cleaning.

Now is the time that tries women's souls, and no sound is heard o'er the house save the scrub-brush, the mop and the broom. The spring cleaning is at hand.

Blankets and Furs.—And first, there are all the woollens, blankets, etc., to be washed, and all that can be spared (for we dare not put them all out of sight, lest we provoke another snow-storm), are to be packed away in deep chests, and plenty of cedar boughs strewn over them, or else powdered camphor gum. The fortunate possessor of a cedar-wood trunk need have no apprehensions, but without that, the moth-millers will make sad havoc among your furs, woolens, etc., unless you guard them carefully.

The Carpets.—All carpets do not need to be taken up; those which do not, can be loosened at the edges, the dust-brush pushed under a piece, and a clean sweep of all the dust can be made. Then, wash the floor thus swept, with strong soap-suds, and spirits of turpentine after. Then, tack the carpet down. The odor is soon gone, if you open your windows, and you can feel safe for this summer, at least. Upholstered furniture can be treated to the same bath, if applied with a soft, clean cloth, and the colors will receive no injury. But before using it, brush the cushions with a stiff hand-brush and a damp cloth, so as to take away all the dust.

A good way to clean straw matting after it is laid, is to sprinkle corn-meal over it, or damp sand, and sweep it thoroughly out.

Windows Washed.—Windows are hard to wash, so as to leave them clear and polished. First, take a wooden knife, sharp-pointed and narrow-bladed, and pick out all the dirt that adheres to the sash; dry whiting makes the glass shine nicely. I have read somewhere, that weak black tea and alcohol is a splendid preparation for cleaning the window-glass, and an economical way to use it would be to save the tea-grounds for a few days, and then boil them over in two quarts of water and add a little alcohol when cold. Apply with a newspaper and rub well off with another paper, and the glass will look far nicer than when cloth is used.

The Beds.—When mattresses and feather-beds become soiled, make a paste of soft-soap and starch, and cover the spots. As soon as it dries, scrape off the paste and wash with a damp sponge. If the spots have not disappeared, try the paste again.—ANNIE R. W.

XXII. Household Hints.

Seventeen Facts.—A good housekeeper kindly sends the following maxims and recipes, "all warranted tried and approved:"

1. Simple salt and water cleans and preserves matting more effectually than any other method.
2. Tepid tea cleans grained wood.
3. Oil-cloth should be brightened, after washing with soap and water, with skimmed milk.
4. Salt and water washing preserves bedsteads from being infected by vermin; also, mattresses.
5. Kerosene oil is the best furniture oil; it cleanses, adds a polish, and preserves from the ravages of insects.
6. Green should be the prevailing color for bed hangings and window drapery.
7. Sal-soda will bleach; one spoonful is sufficient for a kettle of clothes.
8. Save your suds for the garden and plants, or to harden yards when sandy.
9. A hot shovel held over varnished furniture will take out spots.
10. A bit of glue dissolved in skimmed milk and water will restore old rusty crape.
11. Ribbons of any kind should be washed in cold suds and not rinsed.
12. If flat-irons are rough, rub them well with salt, and it will make them smooth.
13. If you are buying a carpet for durability, you must choose small figures.
14. A bit of soap rubbed on the hinges of doors will prevent them from creaking.
15. Scotch snuff, if put in the holes where crickets run out, will destroy them.
16. To get rid of moths and roaches from closets and bureau drawers, sprinkle powdered borax over and around the shelves, and cover with clean paper
17. To remove grease-spots apply a stiff paste to the wrong side of the material or garment; hang it up and leave it some time; the grease will have been entirely absorbed by the paste, which can then be rubbed off.

Furniture Doctored.—To take out bruises from furniture, wet the part with warm water; double a piece of brown paper five or six times, soak it, and lay it on the place; apply on that a hot iron till the moisture is evaporated; two or three applications will raise the dent or bruise level with the surface. If the bruise be small, merely soak it with warm water, and apply a red-hot iron very near the surface; keep it continually wet, and in a few minutes the bruise will disappear. To remove stains, wash the surface with stale beer or vinegar; the stains will be removed by rubbing them with a rag dipped in spirits of salt. Re-polish as you would new work. If the work be not stained, wash with clean spirits of turpentine and re-polish with furniture oil.

To Clean Looking-Glasses.—Wash a piece of soft sponge, remove all gritty particles from it; dip it lightly into water, squeeze it out again, and then dip it into spirits of wine; rub it over the glass, dust it with powdered blue or whiting sifted through muslin; remove it lightly and quickly with a clean cloth, and finish with a silk handkerchief. If the glass be a large one, clean one-half at a time, otherwise the spirits of wine will dry before it can be removed. If the frames are gilt, the greatest care must be taken to prevent the spirits of wine from touching them. To clean such frames, rub them well with a little dry cotton wool; this will remove all dust and dirt, without injury to the gilding. If the frames are varnished, they may be rubbed with the spirits of wine, which will take out all the spots and give the varnish a good polish.—MATTIE M., Cleveland, O.

Fastening Window Sashes.—A convenient way to prevent loose window sashes from rattling unpleasantly when the wind blows, is to make four one-sided buttons of wood, and screw them to the stops, which are nailed to the face-casings of the window, making each button of proper length to press the side of the sash outward when the end of the button is turned horizontally. The buttons operate like a cam. By having them of the correct length to crowd the sills of the sash outward against the outer stop of the window frame, the sash will not only be held so firmly that it cannot rattle, but the crack which admitted dust and a current of cold air will be closed so tightly that no window strips will be required. The buttons should be placed about half-way from the upper to the lower end of each stile of the sashes.

French Polish.—To one pint of spirits of wine add half an ounce of gum shellac, half an ounce of gum lac, and half an ounce of gum sandarac; place the whole over a gentle heat, frequently stirring till the gums are dissolved. Then make a roller of list, put a portion of the mixture upon it, and cover that with a soft linen rag, which must be slightly touched with cold-drawn linseed oil. Rub them into the wood in a circular direction, covering only a small space at a time, till the pores of the wood are filled up. Finish in the same manner with spirits of wine with a small portion of the polish added to it. If the article to be polished has been previously waxed, it must be cleaned off with the finest sand-paper.

Restoring Furniture.—An old cabinet-maker writes that the best preparation for cleaning picture-frames and restoring furniture, especially that somewhat marred or scratched, is a mixture of three parts of linseed oil and one part spirits of turpentine. It not only covers the disfigured surface, but restores the wood to its original color, and leaves a luster upon the surface. Put on with a flannel, and when dry, rub with a clean soft wooden cloth.

Rough on Grease.—The following will be found a most excellent preparation for taking grease-spots from carpets or other fabrics: Four ounces white Castile soap, four ounces alcohol, two ounces ether, three ounces ammonia, one ounce glycerine. Cut the soap fine; dissolve in one quart soft water over the fire; then add four quarts more soft water, after which add the spirits, and bottle. Cork tight. Apply with a stiff brush, and rinse. *

To Brighten Carpets.—Dissolve a handful of alum in a pail of water, dip your broom in, shaking it well, and sweep a small space. Then re-dip the broom, and sweep as before, until you have gone over the whole carpet. You cannot imagine how it will renew the colors in the carpet, especially green.—HOUSEKEEPER.

Laying Down Oil-Cloths.—Oil-cloths always come in rolls. The nearer we buy, says a correspondent, towards the last end of the piece the more they will shrink after laying them down. To prevent this, unroll them, place them smoothly on the floor wrong side up, and use them so for a week or even two. Then turn them, and tack them to the floor. This method prevents their pulling up and cracking, as we often see new oil-cloth do.

Cleaning Gold Chains.—Put the chain in a small glass bottle, with warm water, a little tooth-powder and some soap. Cork the bottle, and shake it for a minute violently. The friction against the glass polishes the gold, and the soap and chalk extract every particle of grease and dirt from the interstices of a chain of the most intricate pattern; rinse it in clear, cold water, and wipe with a towel.

To Whiten Ivory.—Boil alum in water; into this immerse your ivory, and let it remain one hour; then rub the ivory with a cloth, wipe it clean with a wet linen rag, and lay it in a moistened cloth to prevent its drying too quickly, which causes it to crack.

XXIII. Toilet Recipes.

To Remove Freckles.—Take one ounce Venice soap, one-half ounce lemon-juice, one-quarter ounce oil bitter almonds, one-quarter ounce deliquated oil of tartar, three drops oil of rhodium. Dissolve the soap in lemon-juice, and add the two oils. Place in the sun until it becomes an ointment. Then add the rhodium. Anoint at night with this ointment, then wash in the morning with pure water, or mixture of elder-blows and rosewater.—H. B., Zanesville, Wis.

Face Wash.—Take a small piece of gum benzoin, boil in spirits of wine until it is a rich tincture. Use fifteen drops in a glass of water, three or four times a day. Let it remain on to dry. It is very efficacious in removing spots, eruptions, etc.—MARY R., Cedar Falls, Iowa.

Curling False Hair.—Wind the hair on smooth round sticks about as large as a curling iron, fasten the ends firmly to the stick, then wind over the hair a strip of cloth, which must also be fastened at the ends, put in a dish of warm water sufficient to cover, and let it boil two hours. Remove from the water and place in a moderately heated oven to remain until nearly dry, when they should be placed in the sun or near the stove until they are perfectly dry, when they may be unwound from the sticks and brushed over the finger. If too dry or not sufficiently glossy, put a little oil on the brush. Care should be taken while the hair is in the oven that it does not become too warm.—"PERDU."

Stimulant for the Hair.—One of the best stimulants to promote the growth of the hair, when there is danger of baldness, and to hasten growth, is as follows: One pint alcohol, castor oil enough to take up the alcohol, two ounces spirits ammonia, one-quarter ounce oil origanum, one-quarter ounce tincture cantharides. Shake all well together before using. Apply about four times a week.

Cleansing the Hair.—Use a tablespoonful or two of common spirits of hartshorn, in a basin of water; then thoroughly wash the scalp and hair until they are clean; then wash with clean water, wipe dry, and apply a little light oil or pomade, if needed, to prevent taking cold.

Another good hair-wash is: Beat the whites of four eggs to a froth, rub well into the roots of the hair. Leave it to dry. Wash the head clean with equal parts rum and rosewater.

Dandruff can be removed by washing the head with buttermilk and thoroughly cleansing with pure soft water afterward.—FARMER'S GIRL.

Glycerine Ointment.—A glycerine ointment for chaps and excoriations is made as follows: One-half ounce spermaceti melted together with a drachm of white wax and two fluid ounces of oil of almonds by a moderate heat; the mixture is poured into a mortar, when a fluid ounce of glycerine is added to it and rubbed till the ingredients are thoroughly mixed and cold.

Court Plaster.—Soak isinglass in a little warm water for twenty-four hours; then evaporate nearly all the water by a gentle heat, dissolve the residue in a little proof spirits of wine, and strain the whole through a piece of open linen. The strained mass should be a stiff jelly when cool. Now, extend a piece of silk on a wooden frame and fix it tight with tacks and thread. Melt the jelly, and apply it to the silk thinly and evenly with a hairbrush. A second coating must be applied when the first has dried. When both are dry, cover the whole surface with two or three coatings of balsam of Peru, applied in the same way.

XXIV. Home-Made Wines.

TEMPERANCE writes from Benton Harbor, Michigan: "I think you will find these two recipes all right."

Unfermented Wine.—Take the pure juice of well-ripened grapes, put in a porcelain kettle with about one pound of best white sugar to each gallon of juice, and let it boil gently, skimming carefully. Let it simmer slowly till it is reduced about one-fifth. Then bottle or can while hot, and you have a rich, refreshing drink.

Elderberry Wine.—To every quart of the berries put a quart of water and boil for half an hour. Bruise from the skin and strain, and to every gallon of juice add three pounds of double-refined sugar and one-quarter ounce of cream of tartar, and boil for half an hour. Take a clean cask and put in it one pound of raisins to every three gallons of wine, and a slice of toasted bread covered with good yeast. When the wine has become quite cool, put it into a cask and place in a room of even temperature to ferment. When this has fully ceased, put the bung in tight. No brandy or alcohol should be added.

XXV. Home-Made Inks.

A GOOD black ink may be made as follows: One gallon of soft water, one-quarter of a pound extract of logwood, twenty grains bichromate potash, fifteen grains prussiate potash. Heat the logwood and water to a boiling point and skim well. Dissolve the potash in one-half a pint of hot water and put all together, stirring well. Boil three minutes; strain and it is fit for use. A few cloves put in each bottle will prevent it from molding.

Ink not Injured by Freezing.—Take about one handful of maple bark—the inside bark, the outside bark having been scraped off. Put it in three pints of water and boil until the strength is all out of the bark; then strain the bark out of the ooze. Put in the ooze half a tablespoonful of copperas, and boil five or ten minutes until the copperas is all dissolved. Keep stirring. This will make near one gill of good ink that will not be injured by freezing.—J. E. L., Cambridge, Ind.

Indelible Ink.—Four drachms nitrate of silver, four ounces rainwater, six drops solution of nut-galls, and one-half a drachm gum Arabic. This will make an ink which will not fade, and costs very little.

Indelible Inks for Brushes.—For using with a marking-brush, an ink may be made by diluting coal tar with benzine to a proper consistency, or equal parts of vermilion and copperas may be rubbed up with oil varnish. Either of these holds well on linen or cotton fabrics.

Ink for Zinc Labels.—An ink for zinc only, that will endure for years, cuts slightly into metal, has a black color, and is as legible after a dozen years as when newly written, is made as follows: One part verdigris, one part sal ammonia, half part lampblack, and ten parts of water; mix well and keep in a bottle with a glass stopper; shake the ink before using it. It will keep any length of time. Write it on the label with a steel pen, not too fine pointed. It dries in a minute or two.—NURSERYMAN.

XXVI. Recipes for Glue.

Isinglass and Spirits.—A strong and fine glue may be prepared with isinglass and spirits of wine, thus: Steep the isinglass for twenty-four hours in spirits of wine and common brandy; when opened and mollified, all must be gently boiled together and kept well stirred until they appear well mixed, and a drop thereof, suffered to to cool, presently turns to a strong jelly. Strain it while hot through a clean linen cloth, into a vessel, to be kept close stopped. A gentle heat suffices to dissolve the glue into an almost colorless fluid, but very strong, so that pieces of wood glued together with it will sooner separate elsewhere than in the parts joined.

A Strong Cement.—Mix a handful of quicklime with four ounces of linseed oil; boil them to a good thickness, then spread it on tin plates in the shade, and it will become exceedingly hard, but may be easily dissolved over a fire, as glue, and will join wood perfectly. This glue will resist fire and water.

Cheap Water-Proof Glue.—A glue that will resist water to a considerable degree is made by dissolving common glue in skimmed milk. Fine levigated chalk added to the common solution of glue in water makes an addition which strengthens it, and renders it suitable for sign-boards and things which must stand the weather.

Paste That Will Keep.—Dissolve a teaspoonful of alum in a quart of water. When cold, stir in as much flour as will give it the consistency of thick cream, being particular to beat up all the lumps; stir in as much powdered resin as will lie on a dime, and throw in a half dozen cloves to give it a pleasant odor. Have on the fire a teacupful of boiling water, pour the flour mixture into it, stirring well at the time. In a very few minutes it will be of the consistency of mush. Pour it into an earthen or china vessel; let it cool; lay a cover on, and put in a cool place. When needed for use, take out a portion and soften it with warm water. Paste thus made will last. It is better than gum, as it does not gloss the paper, and can be written on.—AMANDA D.I., Madison, Wis.

XXVII. The Dyer's Art.

THE time is long since past when spinning and weaving constitute an important part of rural economy. It will no longer pay even to dye old fabrics at home, except in those sections far removed from dyer's establishments. It will not pay at all, except for the most common fabrics. Rag carpets, however, have not gone out of fashion, and they never should. If tastefully made, they are pretty, and for kitchen and general family wear, certainly lasting. The recipes given by contributors fairly include all the regular colors, and will show that this department of rural art is still extensively practiced, for many still keep up the knowledge of the art as much because it amuses them as for any other reason.

XXVIII. Coloring Dress and Other Fabrics.

As to the stability of dyes imparted to silks, damasks and fabrics, used in furnishing, an eminent French chemist has found that the blue colors produced by indigo are stable; Prussian blue resists moderately the action of air and light, but not soap; scarlet and carmines, produced by cochineal and lac-dye, are last; the most stable colors on silk are produced by weld

Mordants.—In colorings it is sometimes necessary to employ mordants, or substances to "fix" color; they may even change a color; so, by mixing mordants, different shades are produced. But it will not be necessary to enter into this subject here. Where mordants are necessary, they will be given in the simple recipes. In relation to fixing colors generally, and this applies to washing, the following will be useful:

Take a large double handful of bran, put it in a saucepan and set it over the fire, allowing it to boil thoroughly in a quart of water. When thoroughly boiled, strain the bran, and throw the water into that in which you are about washing your lawn or chintz dress. Let the dress soak for an hour or so in it before washing. Instead of starch use a weak solution of glue-water, and iron on the wrong side.

XXIX. Coloring—Yellow, Blue and Green.

Yellow.—Dissolve one-half pound sugar of lead in hot water; dissolve one-fourth pound bichromate of potash in a vessel of wood, in cold water. Dip the goods first in the lead-water, then in the potash, then alternate until the color suits. This quantity answers for five pounds of goods.

Blue.—Dissolve one-fourth pound copperas in soft water, sufficient to color five pounds of goods; put in the goods and let them remain fifteen minutes; then take them out. Take clean soft water and dissolve two ounces prussiate of potash. Put in the goods when it is milk-warm. Let them remain in this fifteen minutes; then take out the goods, and add one ounce of oil of vitriol to the potash dye when it is only milk-warm; put in your goods again; boil for deep blue, and take out before boiling for lighter shades.

Green.—Take the yellow dyed by the above recipe, and dye by the recipe given for dyeing blue, and you will have a beautiful green.—N. B., Elm Grove, Mich.

Coloring Cotton.—To four pounds of rags take one and one-half ounces oxalic acid, two ounces of Prussian blue; let each soak over night in one quart of rain-water, then put together in as much warm rain-water as you want to color with. Put in the rags and let them be in twenty minutes. Wring out and dip in the following yellow dye.

Take six ounces of sugar of lead, four and a half ounces of bichromate of potash; dissolve in a pint of hot rain-water. Take as much hot rain-water as you want to color with. Dip first in the lead, then in the potash several times. Rinse in cold rain-water. Use tin or copper—no simmering is needed. The first makes a blue, the last a beautiful yellow, and both a durable green.—Mrs. Lizzie B., Rochester, Ia.

A Good Yellow.—Take bichromate of potash, one pound to a pailful of water; for blue, two boxes bluing. Color yellow first, then dip the goods, either cotton or woolen, into the blue dye, and you have a deep durable green. Scald thoroughly.—A. J. T., Algona, Ia.

Coloring Cotton Red.—Take two pounds of Nicaragua, or red wood, four ounces solution of tin. Boil the wood for an hour or more, turn off the dye into a tub or pail. Then add the tin, and put in your cotton. Let it stand five minutes, and you will have a nice red.—Mrs. H., Fort Atkinson, Wis.

Or this.—For four pounds of goods, take one pound of redwood. Steep in cold water over night, then let it come to a boil. Skim out the chips; wring out the the goods in the dye, then add sufficient muriate of tin to set the color; return the goods to the dye, let them remain until colored deep enough. Color in brass or tin.—Eliza, Atchison, Kan.

Coloring Cotton Green.—Dissolve six ounces of sugar of lead in hot water, four ounces bichromate potassa in warm water; dip the cloth in the sugar of lead, wring out, then dip in the potassa. Dip three or four times, till a bright yellow is obtained. When the cloth is dry, dissolve four ounces Prussian blue, four ounces oxalic acid. Dissolve separately in warm water; then turn together, and dip your cloth in the blue dye, and you will have a splendid green.

Prussian blue and oxalic acid make a beautiful blue for cotton. Dip three or four times for a deep shade. Rinse in salt-water.—Mrs. D. B., Northfield, Minn.

OR THIS.—For five pounds of goods dissolve nine ounces sugar of lead in four gallons rain-water. Dissolve in another vessel six ounces bichromate potash in four gallons of rain-water. First, dye your goods blue (if you wish a dark green, you must have a dark blue—if light green, a light blue). Dip the goods first in the lead-water, then in the potash-water, and then again into the lead; wring out dry, and afterward rinse in cold water.—MRS. E. M., Grand Mound, Iowa.

XXX. Scarlet and Pink.

Scarlet for Woollen Goods.—To each pound of goods take one ounce of pulverized cochineal, one-half ounce of cream-of-tartar, two ounces of muriate of tin. Use soft water. Color in tin or copper. Let the water get a little warm before putting the dye-stuff in. Stir well, so that all is dissolved, then put in the goods and let them come to a boiling heat and simmer until the right shade is obtained. A beautiful rose color can be made by taking out when at that shade. It will not fade by washing or wearing, but grow darker as all other scarlets do. This will not do for cotton or silk.—NETTIE, Paris, Kentucky.

Pink.—Take three parts of cream-of-tartar and one of cochineal, nicely rubbed together; tie a teaspoonful in a mustard bag. Put this with a quart of boiling water; dip in the articles to be colored, previously cleaned and dipped in alum water; if wished stiff, put in a little gum arabic.—C., Mansfield, Ohio.

XXXI. Coloring—Black, Brown and Slate.

Black.—Take one pound of extract of logwood. Put it in a kettle and fill it half-full of water. Dissolve it the day before it is wanted, and pour half of it in a kettle of water. Put in your yarn and boil half an hour. Have ready a quarter of a pound of copperas dissolved in another kettle and take out your yarn and pour in half your copperas water again. Put in your yarn. Let it remain five minutes. You will then have a nice black. When this is done, you can put in all of the rest of your dye and throw in all your old black and gray rags and color them over. You will thus have your rags in shape to take to the weaver's, and have a nice carpet.—MRS. J. N., Rockford, Illinois.

Brown.—For nine pounds of goods take one-half pound japonica, two ounces blue vitriol, one ounce bichromate of potash. Dissolve the japonica in enough soft water to cover the goods, and let them stay in all night. In the morning make a solution of the vitriol and potash; wring the goods out of the other dye and let them stand in this half an hour. The goods should simmer in both dyes. For light brown use a brass, and for a dark, a copper kettle, to make the dyes in. This is a good recipe for coloring dress goods, as well as carpet rags.—NELLIE B.

Brown With Catechu.—Take one pound of catechu extract and one-half ounce of vitriol; dissolve in rain water; the catechu put in water enough to wet your goods. Color in an iron kettle. Then put in your vitriol. Wet your goods in soap-suds before putting in the dye. This is a fast color.—BLUE GRASS BRUNETTE, Ky.

Slate Color.—Boil yellow oak bark in an iron kettle until the strength is extracted. Take out the bark, then add a very little copperas and you have a pretty color.—M. A. V., Nashville, Tenn.

XXXII. Walnut Coloring—Black Walnut.

With Walnut Barks.—Walnut bark will color any shade from a light tan to coal-black. Color the wool before carding as follows: Peel the bark from the body of the tree—the bark of the root is the best. Put it into a barrel, a layer of the bark and wool alternately, till you fill the barrel; then fill up the barrel with rainwater. Lay on the top heavy weights. Let it stand in the sun or some warm place till you get the shade required.

With Butternut Bark.—Another way to color yarn, cloth or carpet rags, is to boil a large iron kettleful of butternut-bark for four hours; take out the bark, put in a spoonful of copperas. If you wish a black put in more copperas or a little blue vitriol—too much vitriol rots the goods. Then while the dye is boiling, put in the goods and keep stirring and once every few minutes lift the goods with a stick into the air, then put them under. And so on keep watching and moving them till you get the shade required. If left folded or packed too tight they will spot.—SARAH A. B., Shellsburg, Iowa.

Nearly Black.—Put the bark in an iron kettle, and boil until the strength is all out; then skim out, and add about one teaspoonful of copperas to set the color, airing the goods while boiling. If you wish to color woollens, omit the copperas.—H. L. S., Bainbridge, Mich.

Butternut and Black-Walnut.—Peel the bark when the sap is up; put in a kettle, cover with water and let stand until it sours; then boil an hour, throw out the bark and put in the yarn, (woollen wet in soapsuds) cover it over with the bark and weight it down in the dye. Let stand for a day, then wring it and hang it out in the air for half a day. If it is not dark enough re-heat the dye, put back the yarn and let it stand as long again. It will be a nice brown that won't fade with washing. Black walnut colors the darkest. I believe it would color black by having the dye very strong and airing it often.—C. L., Adair, Mich.

Hickory-Bark Color.—Hickory bark will color a beautiful bright yellow, that will not fade by use. It will color cotton and wool. Have the bark shaved off or hewed off, and chopped in small pieces, and put in a brass kettle or tin boiler, with soft-water enough to cover the bark, and boil until the strength is out; then skim out the chips and put in alum. Have it pounded pretty fine. For a pailful of dye I should put in two good handfuls, and wet the goods in warm water so there will be no dry spots on them; wring them as dry as you can, shake them out and put them into the dye. Have a stick at hand to push them down and stir them immediately, so they can have a chance all over alike. If the color is not deep and bright enough, raise the goods out of the dye, lay them across a stick over the kettle, and put in another handful of alum. Stir it well and dip again. It will want to be kept in the dye and over the fire to a scalding heat about an hour, but keep stirring and airing, so they will not spot.

XXXIII. Coloring Carpet Rags.

Drab, Green, etc.—S. P., Lapeer City, Mich., who not only colors, but weaves her own carpets, gives the following: To color drab: Save your cold tea and put a little copperas in it. Boil it up and skim it, and then put in your goods and let them remain a short time.

To color cotton green: First color blue, and then put them in a yellow dye.

To color blue: For four pounds one ounce prussiate potash, one ounce copperas, one ounce of alum. Dissolve the alum and copperas in water enough to wet the goods. Then put them in the potash, and let them remain ten minutes; then put in the copperas and alum. Let the dye be hot.

To color yellow: Take eight ounces sugar of lead, four ounces bichromate potash. Dissolve the sugar of lead in hot water, in a jar. Dip the goods in the sugar of lead first, and then in the potash, alternately, till you have the color desired. This will color six pounds.

Yellow and Blue.—For each pound of cotton rags, take one ounce of sugar of lead, dissolve in warm water, put in a brass or copper kettle. Heat it to a scalding heat and put in the rags. Let them remain in half an hour; then dissolve one ounce of bichromate potash in warm water in a wooden dish. Take the rags out, dip in the potash, wring out and air. Repeat until you use the dye up. You will have a beautiful yellow. Be sure and use soft water.

To color blue: to five pounds of cotton rags take five ounces prussiate of potash, five ounces copperas and two ounces oil vitrol. Take the copperas and potash, put in a copper or brass vessel, heat it till well dissolved. Put in the rags, and scald from eleven to thirty minutes. Take out and cool. Add oil vitrol, then dip and take out. Hang in the shade. You can take more white rags and make a pale blue by dipping in after this. Put your yellow rags in this same blue dye, and you will have a nice green. Hold some in your hand and put in the dye in places, and it will be clouded yellow and green. This must be in soft water, also. Then hang in the shade, and when dry, rinse in warm water.—Mrs. J. N., Seward, Ill.

Green.—For five pounds of white cotton rags reeled in skeins, I take one pail of the inner bark of yellow oak, cut in fine chips, and boil it two hours in three pails of soft water, in either tin, brass or copper. Then skim out the bark and add one-fourth of an ounce of alum. While the dye is boiling, take three ounces of Prussian blue, tie it up in a strong cotton rag, and rub it in enough soft water to thoroughly wet the rags; squeeze and turn them in the blueing nearly half an hour. Wring them out and take as many as you can handle at one time and put them in the hot dye, stir them around a few seconds and take them out; then put in more until you have them all green. Do not leave them in the dye a minute for it will soak out the blue. The rags may need to be dipped more than once. This color will not fade. If light green is desired, use less blue.—Mrs. A. G., Ontario, Ind.

Green and Red.—To color carpet-rags green: to five pounds of cotton cloth, take one pound of fustic and four ounces of chip logwood; soak in a brass kettle over night; heat the dye, then add two ounces of blue vitriol; wet the cloth in suds. When the dye is boiling hot put in the cloth.

For coloring red, for five pounds of goods, take one pound of redwood, steep in cold water over night, then let it come to a boil, skim out the chips, wring out the goods in the dye, then add sufficient muriate of tin to set the color; return the goods to the dye, let them remain until nearly colored deep enough; color in brass or tin.—Ella T. B., Groveport, O.

Blue and Yellow.—I first color blue, then yellow. Take one ounce prussiate of potash, one tablespoon of copperas, one ounce oil of vitriol. Bring to a boil. Then put in the goods for twenty minutes, skimming often. This is sufficient for five pounds.

To color yellow, dissolve one and a half pounds sugar of lead in hot water, one and a half ounces bichromate of potash, dissolved in a vessel of wood in cold water. Dip first in lead water, then in the potash, and alternate until the color suits.—Mrs. H. A. B., Pompeii, Mich.

PART IX.

PRACTICAL, COMMON SENSE HOME COOKING.

KITCHEN ECONOMY AND KITCHEN ART.

OUR EVERY-DAY EATING AND DRINKING.

RECIPES FOR ALL STYLES OF COOKING.

EXCELLENT DISHES CHEAPLY MADE.

ECONOMY OF A VARIED DIET.

PRACTICAL, COMMON SENSE HOME COOKING.

CHAPTER I.

THE LARDER AND KITCHEN.

I. THE MEAT-ROOM.——II. HANGING, TESTING AND PRESERVING PORK, ETC.——III. MUTTON AND LAMB.——IV. CALVES AND THEIR EDIBLE PARTS ——V. BEEF ON THE FARM.——VI. THE KITCHEN. VII. THE FLOOR, WALLS AND FURNITURE.——VIII. CLEANLINESS INDISPENSABLE.——IX. KITCHEN UTENSILS.——X. CHEMISTRY OF THE KITCHEN.——XI. THE COMPONENT PARTS OF MEAT.——XII. A FAMOUS COOK ON BOILING.——XIII. BOILED AND STEWED DISHES.——XIV. HOW TO STEW.

I. The Meat-Room.

THE larder is the place where meat and other food are kept. On the farm it is especially necessary that a place be provided where fresh meat may remain sweet, to ensure a regular supply, and thus avoid a diet of salt meat in summer. There are pigs, sheep, lambs and calves available, and even the quarters of a fat heifer may be used on the larger farms. Or the meat, not required, may be economically distributed among neighbors not so well situated. It is only a question of ice and a suitable room, as shown in Part IV., Chapter III., Section XVII., "The Ice-House and Preservatory." The illustrations in this chapter will show, not only the dressed meat hung, but also, by figures, the manner of cutting it up. Another advantage of such a meat-room is that the meat placed in it, after killing, ripens slowly without tainting, and is both more tender and more nutritious.

II. Hanging, Testing and Preserving Pork, etc.

MEAT, after being killed and somewhat cooled, may be hung up whole, in quarters, or cut as shown in the diagram of Hog Figured for Cutting. In which of these ways it shall be hung will depend entirely upon the space in the cooling-house. Pork should be firm and white as to its fat; the lean flesh light in color and fine in grain; the skin fine and smooth. The fat must be without kernels, since these indicate that the pig may be "measly." If the flesh is clammy to the touch, it is bad.

Cooling.—Pork is at its best when it has become fully cold. All other animal meats of the farm require longer hanging to reduce the fiber, and this is especially true of game. Fowls require to be kept longer than pork, but not so long as mutton; veal and lamb coming next to pork in the shortness of time they should be hung before cooking. Next come fowls and next beef. Mutton and venison ripen, for cooking, more slowly than other meat.

"Ripe" Meat.—No meat should be allowed to taint in the remotest degree before being cooked. The term "ripe" is used to denote that stage when the meat acquires tenderness, and before any change toward taint has been acquired.

Cutting up a Hog.—The head should be taken off at the dotted line behind the ears as shown in the diagram. The curve, 1–2, is the line of cutting to get a shoulder of pork; 3, is the belly or bacon piece; 4, is the neck and long ribs or fore-loin; 5, the sirloin, called the hind-loin in pork; 6, is the ham. The other side

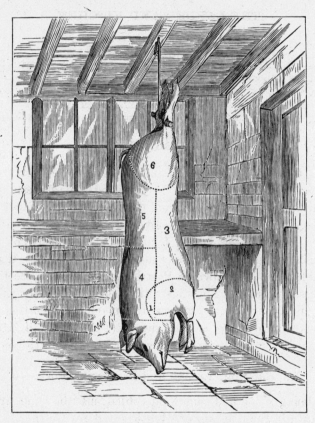

HOG FIGURED FOR CUTTING.

of the hog will give corresponding pieces. The roasting pieces of fresh pork are the spare-ribs, loin and leg. The other pieces are salted. The hind and fore-legs are made into hams and shoulders for smoking, and the side and flitches (belly) into bacon.

Good Bacon.—Good bacon has a thin rind, firm, pinkish fat when cured, and firm lean, adhering to the bone. Rusty bacon has yellowish streaks in it. If a clean thin blade or a skewer stuck into a ham or shoulder of cured smoked meat smells clean and without taint, when withdrawn, the meat is good, for the least taint will immediately be evident to the nostril.

III. Mutton and Lamb.

BOILED mutton and caper sauce (the garden nasturtium makes a good substitute for the caper), roast mutton and Worcestershire sauce, lamb and mint sauce, and lamb with green peas, are dishes good enough for anybody, and any farmer may have them.

When Mutton is Best.—A fat wether makes the best mutton, and the mutton is better if the animal is over three years of age when killed, than if younger. It

DRESSED CARCASS OF MUTTON.

should be dark and fat. A fat, grass-fed mutton of five years old may be had on the farm. It is seldom found at the butcher's. Those who do not know mutton ask for yearling mutton. This is neither mutton nor lamb.

Cutting up a Sheep.—The saddle of mutton is the best part; the haunch next. The saddle comprises the two loins undivided; the leg and loin, separated, are the next best pieces. Chops and cutlets are cut from the loin; the cutlets from the thick end; they are also taken from the best end of the neck and from the leg, though those from the leg really should be called steaks The leg is often salted like a ham of pork, and sometimes smoked; the breast is sometimes pickled and then boiled. The scrag is considered good stewed with rice. In the diagram of Dressed Carcass of Mutton, 1 is the leg, 2 the loin, 3 the ribs, 4 and 5 the neck, 6 the shoulder, 7 the breast. 1 and 2 together constitute the hind-quarter, and 3, 4, 5, 6 and 7, the fore-quarter.

Lamb.—A lamb should be young—six weeks to ten weeks old—the flesh of a pale red, and, of course, fat; a lean lamb is not worth killing. In selecting lambs, many will be found under-sized, but fat. They are the ones for the pot. All animals should be carefully bled in killing, and small animals hung up before their throats are cut. This is easy enough with lambs and sheep. All parts of the lamb may be roasted, but the thin and flank portions are best stewed.

DRESSED LAMB.

Cutting Up a Lamb.—The diagram of Dressed Lamb shows the several parts for cutting: 1, the leg; 2, the loin; 3, the shoulder; 4, the breast; 5, the ribs; 3, 4, 5, the fore-quarter; 1 and 2, hind-quarter. Lamb steaks are called chops and cutlets, and are taken from the same parts as in mutton.

IV. Calves and their Edible Parts.

VEAL should be young, say from six to eight weeks old; the flesh pale, dry and fine in grain. Veal makes the richest soup, and is much used for stock for that and gravies. All parts of the dressed animal may be used. The head is a delicacy. The feet make a firm jelly, and are good boiled or stewed. The loin, fillet and shoulder, are the usual roasting joints. The breast is also sometimes roasted, but is better stewed.

Cutting Up Veal.—The cutlets are taken from the loin and occasionally from the hind leg. In the diagram of Carcass of Veal: 1, is the loin; 2, rump end of

CARCASS OF VEAL

loin; 3, leg or round; 4, hock; 5, fore-leg; 6, chine; 7, neck; 8, shoulder; 9, ribs; 10, breast or brisket; 11, head.

V. Beef on the Farm.

Whether the farmer can afford to kill a heifer or a steer for summer meat will depend upon the size of the family, the number of hands employed, the facilities for preserving the meat, or those for selling or exchanging with neighbors. In the winter there is no reason why the family should not be liberally supplied, both on the score of taste and economy. Beef is the favorite meat, and it is economy to kill it at home rather than buy it cut ready for cooking, unless the butcher can serve the family every day. In the latter case it may be economy to sell the steer or heifer and buy back such meat as is wanted.

Cutting up an Ox.—In the diagram of Dressed Ox, 1 is the sirloin; 2, top, aitch or edge-bone as it is indifferently called; 3, rump; 4, round or buttock; 5, mouse or lower buttock; 6, veiny piece; 7, thick flank: 8, thin flank; 9, leg; 10, fore-rib (containing five ribs); 11, middle-rib (containing four ribs); 12, chuck-rib

DRESSED OX.

(containing three ribs); 13, shoulder, or leg of mutton piece; 14, brisket; 15, clod; 16, sticking piece or blood piece; 17, shin; 18, cheeks or head.

Choice Parts.—The ribs and sirloin are the best for roasting (the middle rib piece, 11, is the best of all). The best steaks come from the chump end of the sirloin, next to 2; it has a good tender-loin or fillet. The rump is the next best roasting piece, regarded by many epicures as the most choice. The soup pieces are the more bony parts, as 9, 16, 17, etc; 6, 7, 8, 13, 14 and 15 are corning pieces; 13 and 14, containing the brisket and the plates, are the best of these.

VI. The Kitchen.

The appointments should be as perfect as possible, and all that may save labor provided. The best stove or range and fixtures should be put in, and closets and pantries made with drawers for culinary articles; sinks with proper waste pipes and fixtures; towel racks, hooks and the many little things that go to lighten labor and make cooking a pleasure rather than a drudgery.

Why should the wife or daughter, who stands over the heated stove, be made to run perhaps two or three hundred feet for fuel when a very little time of the men in the morning and evening might supply the wood or coal box? Why should the cook ever be obliged to use green wood when proper forethought would supply fuel already seasoned?

Some ignorant people think that green wood makes a hotter fire than dry wood. It does nothing of the kind. It takes longer to burn, of course, and is more vexatious in every way.

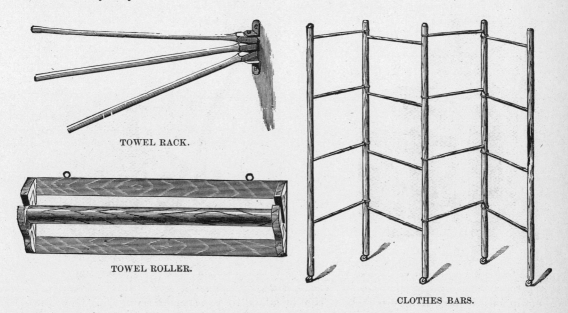

TOWEL RACK.

TOWEL ROLLER.

CLOTHES BARS.

VII. The Floor, Walls and Furniture.

The floor of the kitchen, unless covered with an oil-cloth, should be of ash, thoroughly seasoned, of full inch-thick stuff. The plank not more than four inches wide, being tongued, grooved, well driven together and "blind nailed."

The walls and ceiling, when not wainscotted, should be kalsomined, or painted and varnished so they can be easily washed. They may be papered, but if so, after the paper is thoroughly dry it should have two or three coats of varnish so that the walls may be washed as often as necessary.

The furniture should be solid and simple; the table, or tables, provided with drawers; the chairs wood-bottomed or cane-seated. The stove must be heavy to be lasting, with plenty of oven room, and with the addition of a warm oven to be most economical.

VIII. Cleanliness Indispensable.

In the kitchen absolute cleanliness is indispensable. It is virtue everywhere. Here it is a necessity. It is also economical, for however dirty a kitchen, cleaning time must come, and it is easier and more healthful to clean often than seldom. Every utensil should be thoroughly cleaned and dried each time it is used, and all bright surfaces carefully polished. We do not advise that the stove be blackened every day; a clean, unblacked stove in a clean kitchen is pleasant to look upon.

IX. Kitchen Utensils.

Avoid inventions that contain in one and the same implement everything from a cover-lifter to a meat-broiler. We give a very moderate list of utensils for a well-

REFRIGERATOR.

MEAT-CUTTER.

equipped kitchen; it may be taken from or added to, as occasion requires: One souppot; two vegetable-pots; one stew-kettle; one teakettle; one coffeepot; two enameled sauce-pans; two enameled stew-pans; one meat-broiler; one bread-toaster; two frying-pans; one Bain-marie-pan; one omelet-pan; two pudding-moulds; two jelly-moulds; one rolling-pin; one flour-dredge; one pepper-dredge; one salt-dredge; one meat-chopper; one colander; one fish and egg slice; one marble slab for pastry; one steamer for potatoes, etc.; one coffee-mill.

To these may be added, pans, ladles, knives, skewers, baking-pans and moulds; scales, meat-forks, wooden and iron spoons, fish-scalers, egg-beater, steak-beater, and in lieu of the marble slab, a smooth, hard-wood board for moulding bread. A "Bain-marie" may be improvised from any flat-bottomed pan that will hold one or a number of small sauce-pans, its principal use being when filled with boiling water and placed where it will keep hot, to simmer sauces, entrees, etc. The list might be

FAMILY MEAT-CLEAVER.

added to indefinitely. With those we have mentioned almost any dish may be prepared, except that of meat roasted before a fire.

X. Chemistry of the Kitchen.

COOKING is simply change produced chemically through heat. Condiments are for giving, or adding, zest to flavors. The chemistry of bread-making is to cause it to rise "light," through the action of carbonic acid gas, which is done by adding yeast, or the combination of an alkali and an acid. Soup-making consists in extracting the nutritious constituents of meat by long-continued and slow boiling.

SOUP DIGESTER.

MORTARS AND PESTLES.

Violent boiling should never be allowed with meat of any kind. For soups, stews and other dishes where the juices are to be extracted and form a component part of the soup, the meat should only be simmered. When the nutriment is to be retained *in* the meat, it should be put into boiling water and made to cook up quickly. This coagulates the albumen which surrounds the fiber of the meat and prevents the escape of the juices. A good mortar is often useful in the kitchen. The cuts show sizes ranging from 1 pint to 1 gallon.

XI. The Component Parts of Meat.

ANIMAL flesh, and, of course, this includes that of birds, is composed of the fiber, fat, albumen, gelatine and osmazone.

Fat and Fiber.—The fiber cannot be dissolved. The fat is nearly pure carbon, contained in cells covered with membrane. The application of a boiling heat bursts the cells, and the fat, which melts at a much less temperature than the boiling point of water, is set free and floats on the top of the boiling water.

The Albumen.—Albumen is a substance well known as composing the white of eggs; the albumen of flesh is similar. Its office is to keep the fibers from becoming hard. Under the influence of heat it coagulates, and prevents the fibers from becoming dry. It is more abundant in young than in old animals. The more albumen the flesh contains, the more tender it is. Hence, also, the flesh of young animals is whiter than that of old ones. If soup is to be made, the meat must be heated very slowly, in order that the albumen may not be coagulated too quickly; but if the meat is to be eaten, heat it to the boiling point quickly, to coagulate the albumen, and thus retain the gelatine and the osmazone, the latter of which gives flavor to the meat.

The Gelatine.—Gelatine is the glutinous substance of flesh. It is without flavor, but extremely nutritious; from it is made jelly. It has the power of dissolving bone. Powdered bones are completely dissolved in it. Bones contain much gelatine, two ounces having as much as a pound of meat. Hence, the economy of having much bone in the soup meat, or that for stews, etc.

The Osmazone.—Osmazone flavors meat. The flesh of young animals contains less than that of old ones; the flesh of young animals is more insipid. Roasting, baking, or other dry process of cooking intensifies the flavor of meat, by acting strongly on the osmazone.

Vegetables with Meats.—Many people are perplexed to know just what vegetables are most proper for different meats. Potatoes should be eaten with all meats. When fowls are eaten for dinner, the potatoes should be mashed. Carrots, parsnips, turnips, greens and cabbage are used with boiled meats. Mashed turnips and apple-sauce are indispensable to roast pork. Tomatoes are good with every kind of meat, and at every meal. Cranberry or currant sauce is nice with beef, fowls, veal and ham. Many like currant jelly with roast mutton. Pickles are suitable to be eaten with all roast meats, and capers or nasturtiums are nice with boiled lamb or mutton. Horseradish and lemons are excellent with veal, while no dinner-table is complete without a variety of relishes, such as Worcestershire sauce, chow-chow, mushroom or tomato catsup. Tobasco sauce is the best preparation of Chili pepper.

POTATO MASHER.

TINNED SKEWERS.

THE LARDER AND KITCHEN.

XII. A Famous Cook on Boiling.

Carême, a celebrated French cook, says of soup: The good housewife puts her meat into an earthen stock-pot, and pours on cold water in the proportion of two quarts of water to three pounds of beef. She sets it by the fire. The pot becomes gradually hot, and as the water heats, it dilates the muscular fibers of the flesh by dissolving the gelatinous matter which covers them, and allows the albumen to detach itself easily and rise to the surface in light foam or scum, while the osmazone, which is the savory juice of the meat, dissolving little by little, adds flavor to the broth. By this simple process of slow boiling or simmering, the housewife obtains a savory and nourishing broth and a *bouilli* (boiled meat), which latter is tender and of good flavor.

As to the reverse way of boiling, he says: If you place the *pot au feu* (or soup-pot), on too hot a fire, it boils too soon; the albumen coagulates and hardens; the water, not having the necessary time to penetrate the meat, hinders the osmazone from disengaging itself, and the sad result is, you have only a hard piece of boiled meat and a broth without flavor or goodness. A little fresh water poured into the pot at intervals, helps the scum to rise more abundantly.

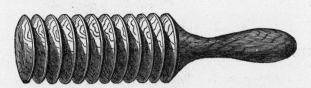
WOODEN STEAK MAUL.

MEAT BLOCK.

XIII. Boiled and Stewed Dishes.

Soup should be gently simmered at least four or five hours, but longer is better. The meat should be put in a thoroughly clean pot, with the amount of cold water heretofore directed; it should be frequently skimmed to remove the suet, and salted and peppered to taste; when vegetables are used they should be sliced; twenty minutes should cook them; rice, dumplings, etc., should be added the last of all to thicken it.

Joints of meat for boiling should be washed clean, skewered into shape, put into the saucepan, or into a kettle having a tight lid, then well covered with cold water, and set over a moderate fire. As the scum rises, skim; and keep the cover tight when not skimming. It must be skimmed at least once before, or just at the time the water begins to boil. If this is delayed, the scum will fall back upon the meat and disfigure it. Salted meat requires a longer time to boil than fresh meat, and salted meat should be freshened by soaking before boiling. Smoked and dried meats require a still longer time for boiling than those only salted.

There are two things to remember in boiling: 1, neither allow the water to boil violently, nor to cease simmering; 2, keep the meat covered, by adding boiling water, if necessary.

If the meat is required to be light in color, wrap it in a clean white linen cloth. The cloth intended for this purpose must be boiled in pure water after being taken from the meat, carefully dried, and not allowed to get damp, else it will be musty. The time for boiling is from fifteen to twenty minutes for each pound of meat, the boiling to be uniform throughout.

ROUND-BOTTOM POT.

SEAMLESS GRANITE STEW-KETTLE.

XIV. How to Stew.

STEWING is slowly simmering in a tight vessel. The liquid should never actually boil. The fire must be slow and the process continued until the meat is quite tender. If only the pure gravy is desired the meat is put into a close jar and this placed in a stewpan of water. If the meat is stewed in water this must be graduated, so that, when done, the gravy will be of the required thickness. If vegetables are used, twenty minutes will suffice to cook them, if they are properly sliced. Stewing is simply slow cooking or gentle simmering in a vessel closed as tight as possible, and with very little water. A digester is a form of stewpan, closed steam tight. This process will, if long continued, disintegrate the bones.

CHAPTER II.

SOME USEFUL RECIPES.

I. VEGETABLE SOUP.—II. CLEAR BEEF SOUP.—III. SOUPS OF VARIOUS MEATS.—IV. FISH SOUPS.—V. BOILED DISHES.—VI. STEWING.—VII. HOW TO MAKE STOCK.—VIII. TO CLARIFY STOCK OR SOUP.—IX. TO COLOR SOUPS.—X. ROASTED AND BAKED MEATS.—XI. BEEF A LA MODE.—XII. PREPARING THE ROAST.—XIII. ROAST SADDLE OF VENISON.—XIV. FOWL AND TURKEY.—XV. BAKED HAM.—XVI. BAKED BEANS.—XVII. BROILING AND FRYING.—XVIII. PREPARED DISHES BAKED.—XIX. PASTRY FOR MEAT PIES.—XX. INGREDIENTS FOR MEAT PIES.—XXI. DISHES OF EGGS.—XXII. STEAMED DISHES.

I. Vegetable Soup.

THIS is the soup most commonly made in the United States; it is excellent with the family dinner. Take a beef shank, crush the bone and put into cold water. Bring it to a boil and skim. Cook four hours, so that when done there shall be about two quarts of soup to each three or four pounds of meat. If turnips and carrots are used, put them in sliced as soon as the liquid is first skimmed. When

SOUP OR STEW POT AND LID.

SKILLET AND LID.

the soup is half done, add a pint of peeled tomatoes for each gallon of soup, and in an hour more half a pint of young okra sliced. Half an hour before it is served, add a quarter of a pound of sliced potatoes, and the same of green corn grated from the cob. The season of the year and the opportunities for getting vegetables will readily suggest substitutes. If okra or tomatoes cannot be had, thicken with a little flour and rice. If onions are liked, flavor with them, but lightly, and also with salt and pepper.

II. Clear Beef Soup.

ONLY the flesh of young animals should be used for soup. The flesh of very old animals will not make really good soup. Take four pounds of lean beef and a knuckle of veal; put into a suitable quantity of water; when it has been skimmed add two or three thin strips of pork and a tablespoonful of butter, two onions, stuck with six cloves each, and a blade of mace. Continue to skim as may be necessary, and then let it just simmer for at least five hours more. Drop in a small bunch of parsley half an hour before it is done. Before sending to the table strain through a clean cloth, and color suitably with burnt sugar.

III. Soups of Various Meats.

Soup is made of various meats, of chicken, and also of fish. Broths are thin soups, and the meat from which they are made is also to be eaten separately; yet most cooks are in the habit of calling them soups. Thus vermicelli, macaroni, noodle and okra soups are thickened with these ingredients, and are made with chicken, beef or veal.

Mutton Broth.—To six pounds of neck of mutton take three quarts of water, carrots, turnips and potatoes. Soak the mutton in cold water for an hour; cut off the scrags and all the fat, and put into the stew-pan with three quarts of water. It must be simmered three hours, skimming thoroughly. The carrots, turnips and onions, cut into suitable pieces, are added after the first skimming, and also four tablespoonfuls of pearl barley. Half an hour before taking from the fire add a little chopped parsley and serve hot. Serve the meat separately, divided into cutlets of two bones each.

Okra Soup.—Take two chickens, three strips of sweet bacon, or a quarter of a pound of ham, one quart of tomatoes, four pints of okra and two onions. Fry the chicken, bacon and onions in a skillet. When done, pour on water and rinse into the soup-kettle, with plenty of water; put in the tomatoes. Cook the okra in a sauce-pan. When the meat has cooked so it may be pulled from the bone, pour through a colander, add the bacon or ham, and the tomatoes from the colander; put back the soup again to boil, pull the chicken from the bone, add the okra and let it cook until thick enough. If the chicken is chopped fine before being put in, and the whole stewed down pretty thick and ladled upon rice, boiled just so the grains are separate and distinct, it makes an excellent gumbo, which is still better if a little sweet cream is added to the boiling rice, and the soup seasoned with cayenne pepper, black pepper and salt. But it is generally made by adding the rice and cream to the soup.

Cabbage Soup.—Take a large cabbage, three carrots, two onions, five slices lean bacon, salt and pepper to taste, two quarts of medium stock. Scald the cabbage, cut it up and drain it. Line the stewpan with the bacon, put in the cabbage, carrots and onions. Moisten with skimmings from the stock. Simmer very gently until the cabbage is tender. Add the stock, stew softly for half an hour, and carefully skim off every particle of fat. Season and serve. It takes one hour and a half to cook. This is a splendid soup for cool weather, and this quantity is sufficient for eight persons.

Corn Soup.—Take eight large ears, cut off the grains and scrape the cobs. Cover this with water, (not too much) and boil until perfectly done. Add two quarts of milk, let it come to a boil; stir in two tablespoonfuls of butter rolled in a tablespoonful of flour, let the whole boil ten minutes; have ready the yolks of three eggs well beaten, pour the soup on them, stirring hard all the time. Serve immediately after seasoning to taste.

Pea Soup.—This may be made with any meat. It is generally made with a fowl. Take half of a fried or broiled chicken and simmer for two and a half hours in a gallon of water and with a quart of clean shells of green peas. Strain through a colander and thicken with two tablespoonfuls of flour, one cup of cream; to be cooked half an hour before serving. A quart of green peas may be cooked with the soup, and when done, mashed and returned to the soup.

Bean Soup.—One teacupful of beans soaked over night is to be used with each quart of water and half a pound of meat. Simmer for four hours. Take out the beans, mash, and strain through a colander with the soup, leaving out the bits of meat and bone; return to the soup-pot and simmer a little longer, with stirring. Season to the taste. Pea soup may be made by using peas instead of beans.

PORCELAIN-LINED KETTLE—FOR FISH.

BRASS KETTLE.

IV. Fish Soups.

Soup may be made of any hard-fleshed fish. They should be carefully cleaned, skinned and cut into fillets Then cut out of the fillets, with a cutter, as many round pieces, an inch in size, as possible. Put the head, bones, and all the trimmings into a saucepan, with one quart of stock, a large handful of parsley, a piece of celery, one onion stuck with two cloves, a blade of mace, and pepper and salt to taste. Let this boil slowly from three to four hours, skim and strain the liquor, put it on the fire again, and when it boils, put in the cut pieces. When they are cooked, take them out, put them into the soup tureen with a little chopped parsley (blanched); then strain the soup into the tureen, and serve at once.

Eel Soup.—This may be made of two pounds of eels, one pound of other hard-fleshed fish, a bunch of celery, one onion, six cloves, a bunch of parsley and sweet herbs. Season with a blade of mace, and pepper and salt to the taste. The fish must be skinned as well as the eels, and well cleaned. Cover with a quart of water in the stew-pan, add the pepper and salt, the onion with the cloves stuck in it, the bunch of herbs, the celery cut up and the parsley minced. Let it simmer for an hour and a half, covered close. Then strain. This is stock, and may be thinned as desired. If brown soup is wanted, fry the fish in butter before boiling.

Rich Oyster Soup.—Take two quarts of the fish stock. Beat the yolks of two hard-boiled eggs, and the hard parts of two quarts of oysters in a mortar, and add this to the stock. Let it simmer for thirty to thirty-five minutes, add the rest of the oysters, and simmer five minutes. Then beat the yolks of six fresh eggs and add to the soup, stirring all one way until it is thick and smooth, keeping it hot, but not quite boiling, say at about 197 to 200 degrees. Then serve at once.

A Good Oyster Soup.—To every four dozen oysters, freshly opened, allow one quarter of a pound of butter, six ounces of flour, two quarts of veal or chicken soup, a quart of milk and seasoning, including a tablespoonful of anchovy sauce, more or less, to suit the taste. Put the butter in a stew-pan, and, when fully melted, add the flour and stir until smooth. Then add the liquor from the oysters that have been just blanched in their liquor, but not boiled, and pour in the soup. Season with a little cayenne and a blade of mace, with black pepper and salt. When all is well mixed, strain and boil ten minutes. Put the oysters that have been blanched and a gill of cream into a tureen, pour the boiling soup over them and serve immediately.

Clam Soup.—Wash four dozen clams, open them and let them lie on the half-shell until the water has run out. Chop them fine with celery, mace and pepper, and an onion if you like it. Put the liquor and all in a saucepan and thicken with two tablespoonfuls of butter rolled in flour. Simmer twenty minutes and then stir in the beaten yolks of five eggs. Serve in a tureen with slice of toasted bread

Plain Oyster Soup.—Take one quart of oysters to one quart of milk. Boil the milk and liquor from the oysters together. When it has fairly boiled, add a tablespoonful of butter and let it boil up again, using powdered crackers or flour to thicken while boiling. Put in the oysters and serve immediately. The butter is often, and, we think, preferably, added with the seasoning of pepper and salt, for butter when boiled loses its fresh and pleasant flavor.

Oyster Stew.—Oysters are stewed with milk, cream or water. When the liquor of the oysters is used, and this is stewed down considerably, it is called a dry stew. The ordinary stew is made as follows: Pick the oysters out of the juice with a fork, as dry as possible; stew the juice, thickening the milk or water of which the soup is to be made, and let it stand until thoroughly cooled; then drop the oysters in, and just as the cooled soup begins to show signs of simmering, empty out all together, and you will have a rich soup and plump oysters.

V. Boiled Dishes.

Fish Chowder.—The ingredients are: Cod, haddock, or any other firm-fleshed fish, and salt pork. Fry three or four slices of salt pork in a deep kettle. When crisp take it out and put into the kettle, first, a layer of sliced potatoes, then one of fish, and then one of onions, alternating with a layer of fish until all is used. Pepper it well, add boiling water enough to cover the whole, and boil half an hour. Put in half a pint of milk, and cook it five minutes longer, gently, to prevent burning. A brass kettle is often used when there is a large party.

Steamed Turkey.—Cleanse the fowl thoroughly; then rub pepper and salt well mixed into the inside of it. Fill up the body with oysters mixed with a small cupful of bread-crumbs. Sew up all the apertures; lay the turkey into a large steamer and place over a kettle of boiling water; cover closely, and steam thoroughly for two hours and a half. Now take it up; set the platter in a warm place, and turn whatever gravy there is in the steamer, straining it first into the oyster sauce which you have prepared, in the following manner: Take a pint of oysters, turn a pint of boiling water over them in a colander. Put the liquor on to boil, skim off whatever rises on the top. Thicken it with a tablespoonful of flour rubbed into two tablespoonfuls of butter; season well with pepper and salt. Add two or three tablespoonfuls of cream or milk to whiten it; and pour it over the turkey and platter; serve boiling hot. This sauce must be made while the turkey is still in the steamer, so that it can be poured over the turkey as soon as it is taken up.

Boiled Turkey and Fowl.—Select a fat, young fowl; prepare the dressing of cracker or bread crumbs, made fine; chop bits of raw salt pork very fine; sift in sage, savory, thyme, or any other sweet herbs you prefer; add to this pepper, salt, and considerable butter; mix with hot water. An egg is sometimes added. After the turkey is stuffed, wrap closely in cloth. Put in cold water to boil, having all parts covered. Boil slowly, removing the scum as it rises. A small turkey will boil in less than two hours. If you use oysters for the dressing, it is better to steam the turkey instead of boiling. When tender, take it up, strain the gravy found in the pan, thicken with flour; stew the oysters intended for the sauce, mix this liquor with the gravy, add butter, salt and pepper to suit the taste; a trifle of cream improves the color.

Boiled Corned Beef.—If the piece is very salt, let it soak over night. If young beef and properly corned, this is unnecessary. For boiling, put it in the pot, pour cold water over it after rinsing, letting the meat be well covered. The rule is twenty-five minutes to a pound for boiling meats, but corned beef should never be boiled; it should only simmer, by being placed where the simmering can be uninterrupted from four to six hours, according to the size of the piece. If it is to be served cold, let the meat remain in the liquor until cold. Tough beef can be made tender by letting it remain in the liquor until the next day, and then bringing it to the boiling point just before serving. For rump pieces this is a superior method. A brisket or plate piece may be simmered until the bones can be easily removed;

then fold over the brisket piece, forming a square or oblong piece; tie over it a piece of muslin, place sufficient weight on top to press the parts closely together, and set it where it will become cold. This gives a firm, solid piece for cutting from when cold.

A Boiled Dinner.—Select a good piece of fresh beef, not too fat, rub over it sufficient salt to "corn" it, but not to make it very salt; let it stand two or three days, judging of the time by the size of the meat; then wash thoroughly in cold water, put in the pot, cover with cold water and boil gently till quite tender. Add such vegetables as are desired, judging the quantity by the strength of flavor desired in the thick soup to be made from the water in which the whole is boiled; when done dish beef and vegetables, and serve hot.

Boiled Lamb, Mutton or Veal.—Wrap the joint, quarter or piece of meat in a wet cloth. Dust it with flour and let it remain so half an hour. Have the pot ready boiling; dip the joint in, first one end and then the other—then put it in the pot and cover closely. Let it boil gently but steadily, an hour and a half for lamb, and two hours for veal and mutton.

Sauce for Boiled Meats.—Drawn butter, with chopped parsley and sliced carrots, and pickled cucumbers. Boil carrots for a dish to eat with the lamb, etc. Slice into it some potatoes, parsley and onions, and with a little thickening, you have a good soup.

To Boil Rice.—Rice when done should have every grain perfect. It should not be a gluey mass. The way to do it is to drop the rice into plenty of boiling water, boil fast and with the lid off, and when just done drain into a colander before serving. This is the way to boil rice for serving with gumbo.

To Wash and Boil Rice.—Wash in several waters, rubbing gently between the fingers; drain, drop it into boiling water only sufficient to cook it by the time the water is boiled off, and so when done each grain will preserve its shape. This is the Chinese method.

VI. Stewing.

STEWING is the basis of all made dishes, and a most economical and savory manner of cooking. Its perfection depends upon the slowness with which it is done. A stew should never boil, nor even simmer. Two hundred degrees is the greatest heat admissible; 190 degrees is hot enough. Hence it is most safely performed by placing the stew-pan in another vessel of water—a Bain-marie. Stews should never be greasy nor very highly seasoned. The pot lid should be kept close, and an occasional shaking of the contents will save stirring.

Irish Stew.—Take a neck of mutton, trim off some of the fat, and cut into as many cutlets as you have bones; shape them, and sprinkle them with pepper. Peel six moderate-sized onions, and for every pound of meat take one pound of potatoes. Blanch the vegetables separately. Take a clean three-quart stew-pan, and add half a pint of water or stock. Arrange a layer of potatoes at the bottom of the stew-pan, then cutlets, then onions; then potatoes, then cutlets, then onions, and so proceed until you have the whole in. Stew at least two hours if you want it rich; or one hour if the meat is to be more solid.

Beef Stew.—Cut cold beef into small pieces, and put into cold water; add one tomato, a little onion chopped fine, pepper and salt, and cook slowly; thicken with butter and flour, and pour over toast. Or chop fine, cold steak or roast beef, and cook in a little water; add cream or milk, and thicken with flour; season to taste, and pour over thin slices of toast.

Onion and Meat Stew.—Slice some onions and fry brown. Pound the meat, fry it over a hot fire until it browns a little, turning each piece as soon as it has been a few seconds in the pan, to keep in the juice. Put it into a sauce-pan, pour water into the frying-pan, and put this brown liquor, with the fried onions, to the beef. Let it simmer slowly for an hour. Other seasoning may be added, according to taste, or to vary the dish, such as tomatoes, fresh or in catsup, sage and summer savory, or a grated carrot. Young green onions, such as must be thinned out, are good cut up in it; pepper and salt, and a teaspoonful of curry powder are a great improvement.

CONVEX STEW-PAN.

If onions are not liked they may be left out, and the stew made brown with other fried vegetables, or the meat itself may be first fried.

Beef Steak Stew with Jelly.—Take rump or round steak and pound it well, to make it soft, and lard it thoroughly. Put it in a stew-pan, in equal parts of white-wine and water, and add some slices from a leg of veal. Season it with spice, salt, garlic, thyme and parsley. Simmer all over a steady fire four or five hours. When sufficiently done, remove the meat, and strain the broth through a sieve; then pour it into another pan, and boil it down until it becomes a jelly. If it is wished that the jelly should be clear, the whites of two eggs may be beaten up in a tablespoonful of stock broth and added to it, and well mixed. It must then be boiled for seven or eight minutes. Some lemon is then to be added, and the contents of the stew-pan strained through a fine cotton strainer, taking care not to squeeze the cloth, or the dregs may be forced through the pores of the material. The filtered jelly is then put in a cold place to set. When it has become perfectly solid it is to be cut into nice pieces, which should be tastefully arranged on the dish, around the piece of meat. Sometimes the jelly is colored before being strained by the addition of a little cochineal powder.

Hotch-Potch.—The ingredients are: Neck or scrag of mutton, made into cutlets, cauliflower, carrots, green peas, onions, stock and turnips. This is something like the New England boiled dinner already described.

VII. How to Make Stock.

STOCK is the foundation of all meat-soups, sauces and purees. It is prepared as follows: To make three quarts of good beef stock, put into a saucepan or stock-pot one and a half pounds fresh shin of beef, half-pound of bones broken into pieces, with seven pints of clean, soft water. Let the contents come slowly to the boil, then remove all the scum by skimming. The addition of a little cold water at intervals will facilitate the rising of the scum by altering the specific gravity of the water; if the scum be not removed it will partially redissolve and spoil the clearness and flavor of the stock, and you will have the trouble of clarifying. After skimming well, add the following: one ounce of salt; one onion, with two or, at most, three cloves stuck in it; two leeks, say five ounces; half head of celery weighing half-ounce; turnip cut into quarters, weighing five ounces; carrot sliced, weighing five ounces; parsnip sliced, weighing one ounce; one teaspoonful of white pepper. The contents must now simmer at 180 to 200 degrees for four or five hours; then remove the fat by skimming, it can be used when cold for frying and other purposes.

Take out the meat, vegetables and bones, and strain the stock into a glazed earthenware vessel and keep it in a cool place free from dust; a piece of muslin gauze may be placed over it. Any remaining fat can be removed in a solid state when the liquor is cold. Stock, soup, broth, or stew should always be kept in earthenware vessels. The vegetables should not remain longer in the stock than is necessary to properly cook them, as they afterwards absorb the flavor. In spring and summer, when vegetables are young, they cook in less time, but a stock may be and often is prepared without vegetables.

A stock may also be prepared from previously cooked meat and bones, but the stock will not be so good or rich in flavor as when prepared from fresh meat and bones. The idea which must be ever present in preparing a stock or soup is absolute freedom from fat. Spare no pains in skimming, and a little kitchen-paper or blotting-paper laid on the surface will remove specks of fat which evade the spoon.

VIII. To Clarify Stock or Soup.

SOMETIMES stock will not clarify itself. To clarify stock or a soup, take the white and clean shell of an egg for every quart of soup; crush the shell in a mortar, and mix the shell and white of egg with a gill of cold water. Whisk the mixture well, and then add about as much of the boiling soup, still beating [up all together. Pour the mixture to the remainder of the stock in the saucepan, still stirring briskly till the whole comes to the boiling point. Remove from the fire, and let the stock remain ten minutes, or until the white of the egg or albumen separates; then strain carefully, and the broth is clarified. The albumen and egg-shells entangle the small solid particles floating in the soup. If care be taken in the preparation of a stock or soup it will not often require clarifying.

IX. To Color Soups.

It is sometimes desirable that stock should be of a bright golden color, although it is no better on that account. The point to remember in coloring, is not to alter the flavor of the stock or soup; burnt onions or carrots should never be used; they impart a disagreeable taste. The only proper coloring substance is caramel or burnt sugar, which may be prepared as follows:

Take a clean stew-pan or saucepan and put in half pound of pounded loaf sugar, and constantly stir it over the fire with a wooden spoon. When the sugar is thoroughly melted, let it come to the boiling point, and then boil slowly for fifteen minutes, with occasional stirring. When the sugar is of a dark-brown color add one quart of cold water, then boil for twenty minutes on the side of the fire. Let it cool; then strain it, and keep it in clean well-stoppered bottles, and it is ready for use. Caramel should be of a dark-brown color; if it boil too quickly it will become black, and will spoil the color and flavor of the broth. When you use caramel, put it into the soup tureen just before serving.

X. Roasted and Baked Meats.

In the United States, very little meat is roasted before the fire. This method is undoubtedly better than baking, but few families have facilities for roasting. The cook stove is now supreme, and no person will object either to a joint or a bird nicely roasted in an oven. But roasting before the open fire undoubtedly exalts the flavor of meat more than any other way of cooking. Only the best pieces can be used for roasting. The neck, tops of the ribs, shanks and tail make soup, all the odds and ends come in well for stewing, while the best roasting pieces are the ribs, the fillet, the sirloin and rump.

How to Roast.—To roast meat properly, the fire must be hot and steady. About two hours will be required for a roast of seven pounds of beef, and somewhat less for a leg of mutton. No time, however, can be given exactly, though fifteen minutes for each pound will be near the mark. Beef is usually liked rare, mutton often somewhat so, but pork and fowls should be thoroughly cooked.

Basting.—The meat should be basted from time to time, and if you wish the meat frothed, just after the last basting, dredge it very lightly with well-dried flour and give it time to crisp. The imperative rule for baking meats is to have a quick fire and baste frequently. Never parboil meat that is to be roasted. If it is frozen thaw it out in cold water before putting it in the oven, always wiping it dry after taking it from the water.

There is another thing that should be observed with all meats that are to be roasted, broiled or fried. They should be kept in a cool place after being killed, until ready for cooking. This breaks down the fiber and renders the flesh tender.

XI. Beef a la Mode.

This is a fillet or round of beef with rich stuffing, whether stewed or baked. A round of beef is prepared as follows: Cut out the bone and fill with rich stuffing of bread-crumbs seasoned with pepper, salt and onions; mix together a teaspoonful each of pepper, salt, cloves, mace and nutmeg. Make incisions in the beef and place thereon strips of salt fat pork rolled in the spices. Sprinkle the remainder over the beef, and cover the whole with strips of fat pork. Tie it all round with tape, and skewer it well, put in the oven in a dripping-pan containing plenty of water, say three quarts, and bake from four to six hours, according to the size of the piece, basting it well with butter or nice beef drippings mixed with a little flour. When nearly done skim the fat from the gravy and thicken to serve with the meat. The gravy may be seasoned with Worcester sauce, catsup or wine.

FLESH-FORK.

XII. Preparing the Roast.

Ribs of beef may have the bones cut out, the meat rolled compactly together and properly skewered. A loin should have the spine properly cut for convenience in carving. Fowls should be properly skewered, and the roasting should be done with only water enough to properly baste. The roast should be covered with a buttered paper to prevent burning until such a time as it may be ready to finish by browning. With these directions any cook should be able to do a plain roast.

XIII. Roast Saddle of Venison.

The side which is to come uppermost at table should be placed next the pan in baking. When half done, turn it over in the pan, and cut into it in several places on each side of the bone, nearly three inches deep, and fill with a stuffing of bread crumbs highly seasoned with pepper and salt. Pour over the meat half a teacupful of catsup, covering equally. Stir into half a teacupful of black molasses a tablespoonful of whole allspice, and a teaspoonful of brown sugar. Spread this equally over the meat. Then crumble stale bread over all, keeping the meat well basted all the while. Bake slowly until finished, for it burns easily. When taken from the oven, garnish with bits of jelly, and serve.

XIV. Fowl and Turkey.

Roast Fowl.—The dressing for roasted fowl should be of bread toasted crisp, spread with butter, and moistened with water; or if plain dressing, pound in a mortar. It should be rather highly seasoned. Add if you like, sage, thyme and parsley, and have the whole soft enough so it will fill the cavity compactly. The giblets, chopped fine, should always be served with the gravy.

Roast Turkey.—Turkey or other fowl having been well drawn, washed and dried with a towel, rubbed with salt and pepper, stuffed and sewn up, the legs and wings carefully skewered in place, put it in the oven, with the giblets, and about a quart of water in the pan. Bake until done, basting often, being careful not to burn it. The browning is done at the last. It will require three hours to roast a large turkey or a goose, and not much less for a brace of large ducks.

XV. Baked Ham.

The ham is first boiled. Very few persons know how to boil a ham. Wrap the ham in clean straw, or fill in around it in the pot with clean oat-straw. Add a clove or two of garlic (*not* a whole garlic), cloves, mace, allspice, thyme and pepper to the water in the pot. Add also a quart of cider and boil until done. If the water in which the ham is boiled is one-half old sound cider so much the better. Let it stand in the liquor until cool. If it is to be served without baking, skin and garnish with whole cloves stuck in the fat, and such other garnishing as may suit the taste. But a ham is better if baked after boiling.

How to Bake a Ham.—Skin the ham after boiling. Lay two flat pieces of wood in the bottom of the bake pan; lay the ham on them, and cover with a batter of flour and water spread equally. Bake two or three hours slowly, according to the size of the ham, remove the crust of batter, garnish and serve. It is excellent hot or cold, and all the better for having a half-pint of claret poured over it; or it may be eaten with a sauce of which wine is the basis.

XVI. Baked Beans.

The marrow beans are really best, but the small navy beans are generally used. Put them to soak early in the evening, change the water before going to bed, and again in the morning. Parboil for two hours, or until they are tender, but will not break up. Pour off nearly all the water. Place the beans in a bean-pot—a deep pan will do if unsoldered. Score a piece of salt pork. Sink it into the middle of the beans, so it is just level with the surface, and add a very little molasses. Bake six hours, raising the pork toward the last so it may be well browned.

To cook beans in the camp, after boiling, a hole is dug at the foot of the fire, filled with hot coals, the bean-pot is filled around with coals and covered with hot ashes, where it remains from supper-time in the evening until breakfast the following morning—about fourteen or fifteen hours. This, in fact, is the perfection of art in cooking beans for imparting a fine flavor.

XVII. Broiling and Frying.

In broiling and frying the same principle is carried out as in roasting, but it is somewhat different from that of baking. In baking, heat is applied to all sides of the meat during the whole operation. In roasting before the fire, heat is applied alternately to every side of the meat; the same thing is done in broiling and frying.

Frying comes nearer to baking than broiling does, when the frying is done in a skillet with little fat. In broiling the meat comes into direct contact with the heat of the fire and is altogether preferable, except for ham, pork, bacon and fish, which are generally fried. The thing to do, in both broiling and frying, is to have a strong, clear fire, without smoke—hard coal, charcoal, or the coals of hard wood, are indispensable for broiling.

Now that so many excellent and cheap implements for broiling are manufactured, so that the steak, cutlets, fish, etc., may be clasped between the leaves, thus saving the handling, frying is pretty much discontinued. In fact, in frying now, the articles

BROILER AND COVER.

are generally seethed in very hot fat, a preferable plan to frying, unless it be dry frying. If the fat is hot enough the meat will not absorb the fat, but come out exceedingly savory.

In broiling or frying, a fork should never be used. It pierces the meat and allows the juices to escape. With the modern broiler, turning with a fork is unnecessary; but every cook should have a pair of meat-tongs for turning and handling steaks, cutlets, etc., when necessary.

The Thickness for Broiling.—Beefsteaks should never be less than half an inch thick, and if a rich, juicy broil is desired the steak should be three-quarters of an inch thick. Pork and mutton chops, veal cutlets, and lamb chops should never be more than half an inch thick, and less is better. Salt pork, ham and bacon should be cut thin. Young chickens and other birds for broiling, should be cut down the back, pressed out, and pounded or broken down perfectly flat. Flat fish and all small fish are fried whole; round fish are slit down the back. No broiled or fried meat, except beef, must be rare enough to show the blood. Mutton is often liked slightly rare. All other meats must be thoroughly cooked through, especially fowl, pork and fish. All these meats are apt to contain germs, that unless destroyed by a seething heat, may be dangerous, as for instance, trichinæ in pork, fowls and fish.

Frying in Boiling Fat.—What we have said will fully cover the ground of boiling and dry frying. Frying, however, is nicely done in boiling fat, using enough to completely cover the article cooked. It must be very hot, so as to brown the substance properly. Meat or fish need not be entirely covered up if it be turned, but it is better to fry without turning. The fat—sweet drippings—may be used over and over again, but that used for fish should never again be used for other dishes, but the oil that meat has been fried in may be used for fish.

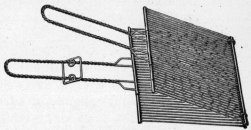

OYSTER BROILER.

Fried Oysters.—Select large, freshly shelled oysters, dry them in a towel, dip into egg that has been slightly whisked, and then roll in bread-crumbs or powdered cracker. Let them dry, and cook in boiling lard until a light brown; or they may be fried in a little butter by turning.

Oyster Fritters.—Make a stiff batter of eggs beaten, bread-crumbs, flour and milk, or cream; season with pepper and salt. Fry in a pan with butter or lard by dropping in a spoonful of batter, then an oyster or two according to size, and cover with more batter. Cook both sides brown by turning.

IMPROVED FRYING-PAN—HANDLE ALWAYS OVAL.

XVIII. Prepared Dishes Baked.

Oysters.—This is a nice breakfast dish. Separate the oysters from the liquor, and put some of them in a baking platter, or pan. Make a seasoning of grated bread-crumbs, pepper, salt, celery-seed and small bits of butter. Put this over the oysters, add some seasoning and again oysters until you have enough. Pour a glass of wine or sound cider over all, and add the liquor of the oysters, as much as may be necessary. Bake until hot throughout and serve immediately.

Meat Pot-Pie.—The old-fashioned way of making a pot-pie, and the only way in which the lower crust can be properly browned, is to use a thick round-bottomed pot over an open fire. The bottom and sides of the pot are lined with thick crust, and just before the pie is done the fire is increased until the crust is well browned. It cannot be made in a flat-bottomed or thin kettle. The modern way of making a pot-pie in the oven is to brown the top crust. It is the next best thing to a browned bottom crust. Chicken, game, beef and veal are the meats used.

XIX. Pastry for Meat Pies.

MANY prefer a crust that is firm enough to be raised around the meat. Care must be taken that the flour is dry, and it must be sifted. Boil water with a very little lard and an equal quantity of sweet drippings or butter. While hot, mix this with as much flour as you want, making the paste very smooth and stiff, by beating and kneading. When ready, place it in a cloth or under a pan until cold. To make the form, place the left hand on the lump of crust, and with the right keep working it up the back of the hand until the proper shape is obtained. It must be thick enough to support itself.

When worked into the desired form, the meat is put in with the necessary dressing, and the cover cemented on with the white of an egg, leaving vent holes for the escape of steam. You will have difficulty in forming the paste without practice, and in reality these raised pies are more fanciful than practical. Before putting into the oven the whole must be glazed with white of eggs.

Another Way.—Take a tin, half the height of the required pie, roll the paste of a proper thickness, cut out the top and bottom, and a long piece for the outside. Butter the dish, mould the side, lay on the white of an egg where the bottom is to join the sides, drawing it down over the bottom of the dish. Lay on the bottom, pressing all firm so it may not leak. Fill the pie, put on the cover, and pinch well together. The usual way is to press the paste into a buttered tin form, take the pie carefully out when firm enough, and again put it into the oven to brown.

Brioche Paste.—Seven fresh eggs are required for ten pounds of dried and warmed flour and one pound of sweet butter; also a little compressed yeast. Put the yeast into a portion of the flour, add warm water and mix to form the leaven. Set it to rise after making a slit in the top. The leaven being ready, take the remaining flour, make a hole in the middle, put in a saltspoonful of salt and the same quantity of powdered sugar, with a little water to melt it. Slightly whisk the eggs, break the butter small, and work the whole well together by kneading and spreading alternately. Spread again, lay the leaven evenly over all, and knead and work until the whole is evenly and thoroughly mixed. When finished, flour a towel, wrap the paste in it, and put in a cold place in hot weather, or a warm place in cold weather, to have it ready for the next day.

This makes the most delicate dumplings for soups or stews, cut in shapes and fried; it is nice with braised dishes. It makes the best case for lobster or other patties, and is an excellent side-dish, cut and fried in shapes. It may be boiled in cup shapes and served with asparagus, cut small and heaped on top with white sauce around it. In fact, there is an endless variety of uses to which the cook may put this paste; the only drawback is, it is troublesome to make.

Light Plain Paste.—Take one pound of flour, six ounces of lard and ten ounces of butter. Rub the lard into the flour, which must be thoroughly dry, work into a smooth paste with only a little water. Roll out thin, press the butter in a cloth to absorb the moisture, put the butter in the center of the paste, fold and press lightly down and roll very thin, dredging the board with a little flour. Fold in three laps and roll and set it in a cool place for a short time. Give it two more workings at intervals; fold again and it should be ready for use. This will make a good light paste for almost any purpose.

Crusts with Melted Fat.—These are made by pouring the melted fat into the flour and mixing until all is fine, using one egg to each pound of flour. The proportions are: Flour one pound, drippings of lard or butter, six to twelve ounces—according to how short it is wanted—and one egg.

Potato Crust.—An excellent crust is made with potatoes. Peel, boil and pass through a sieve, twelve potatoes. The other ingredients are: A gill of cream, two heaping tablespoonfuls of butter, a teaspoonful of salt, and enough flour to form the paste.

XX. Ingredients for Meat Pies.

For meat pies take a dozen slices of fried pork cut in pieces, together with the beef, veal or chicken, and stew all together in only water enough to form the gravy. Peel and slice potatoes and add them to the stew. Let the crust form the bottom, top and sides of the dish. Then form alternate layers of meat and crust, until the whole is finished, seasoning each layer of meat properly. Then pour on the liquor the meat was cooked in, until it just covers all; put on the top crust, and bake until the bottom crust is done and the top brown. If the liquor dries down, more must be added, or else water, in either case, boiling hot.

Pot-Pie of Fowl.—A pie may be made of chicken or wild fowl. The yolks of six hard-boiled eggs to each fowl may be stirred in.

Chicken Pie with Rice.—This may be made with or without crust. If without crust, line the dish with slices of boiled ham; cut up the boiled chicken, pour over it the gravy or melted butter, and fill in the interstices with boiled rice. Cover the top thickly with the same. Bake about three-quarters of an hour.

Giblet Pie with Oysters.—Take the giblets of two chickens or of a turkey; stew until nearly done, and cut into inch pieces. Line the pan with a rich paste, mix the giblets with a quart of oysters, adding liquor enough to make the pie juicy. Add flour or rolled cracker enough to thicken it somewhat; also butter, pepper and salt; cover with crust, and bake until the top is brown.

Fish and Oyster Pies.—Fish, eels and oysters are made into pies. The seasoning for these is generally high, and includes various spices, parsley, thyme, basil and the yolks of hard-boiled eggs. A modification by which these are baked in rich gravy or stock, is generally preferred by American palates. Any fish may be used, but the hard-fleshed fish are preferred.

XXI. Dishes of Eggs.

MANY dishes are made of eggs. Broken in water just simmering, they are called poached eggs.

Fried Eggs.—They are fried in hot lard or oil, first being broken into a dish and carefully turned into the frying-pan. While cooking, turn the hot fat over the tops of the eggs with an iron spoon to cook the tops, or turn them and fry both sides if this way is liked. They are nice boiled in the shell, three minutes by the watch, or somewhat less if wished very soft.

VEGETABLE OR EGG BOILERS.　　OMELET-PAN.　　STEAMER

Scrambled Eggs.—They are scrambled by breaking them into a hot frying-pan, containing only enough butter to thoroughly grease it, and constantly stirring the eggs until done. Of course, whatever the way of cooking, they must be properly seasoned with pepper and salt.

Baked Eggs.—Break the eggs carefully into a dish, so the yelks will not be broken. Turn them into a granite-ware pan well buttered, season with pepper and salt, and drop a small piece of butter on each egg. Set in a hot oven and bake until the whites are set.

Omelettes.—Eggs are not to be beaten for an omelette. If a soufflé is desired, they are beaten. For an omelette, they are stirred until the yolks and whites are properly mixed, one teaspoonful of cold water being used to each egg. The omelette takes its name from the flavoring used, parsley, ham, cheese, etc., either of which must be chopped fine; the material in which they are fried must be the sweetest butter. The butter being hot, pour the stirred eggs, and other material if used, into the omelette-pan, shaking the pan occasionally, as the mass sets, so it will not burn. If desired browned on both sides, turn; or when done, if fried only on one side, fold or roll together lightly, and garnish the top with bits of parsley.

XXII. Steamed Dishes.

OYSTERS, many vegetables, especially potatoes, puddings and various prepared dishes are steamed. This is simply utilizing the action of steam instead of hot water to break down the tissue and render the substance palatable. The articles are placed in a vessel with a perforated bottom, and fitting tight upon a pot. The lid must also be tight so as to allow some pressure of steam. An excellent form of steamer wherein the cover fits down upon a projection instead of over the side is shown in the illustration.

CHAPTER III.

SAUCES, SALADS, PICKLES AND CONDIMENTS.

I. SAUCES AND GRAVIES.—II. SALADS AND THEIR DRESSING.—III. VARIOUS MADE DISHES.—IV. PICKLES, CATSUPS AND CONDIMENTS.—V. LEAVES FOR FLAVORING.—VI. SOUR PICKLES—CUCUMBERS.—VII. CHOW-CHOW.—VIII. PICCALILLI.—IX. SWEET PICKLES.—X. CATSUPS.—XI. CONDIMENTS.—XII. FLAVORED VINEGAR.—XIII. STRAWBERRY ACID.

I. Sauces and Gravies.

THE French have a saying that the English have but one sauce—melted butter. This may have been measurably true once, but now the English and the Americans draw upon the products of every climate to please their palates. Our sauces are numerous, and it must be confessed, many of them are of little account; others are expensive and troublesome to prepare. The most costly and elaborate are now sold by grocers and purveyors, and are bought ready-made in sealed cans. However the sauce is prepared, the utensils must be clean, and a wooden spoon should be used for stirring. Melted butter, stock, bread sauce, white sauce and brown sauce are the bases of the principal sauces.

SOUP OR SAUCE STRAINER.

GRAVY STRAINER.

Melted Butter.—It is made in the relative proportions of two ounces of butter, two tablespoonfuls of water, and a little flour dredged in, prepared over a hot fire, and shaken back and forth. Another good way of making is to rub two tablespoonfuls of flour into a quarter of a pound of butter, adding five teaspoonfuls of water; set the sauce-pan containing it in a vessel of water kept boiling until it simmers.

Sauce for Fish and Fowl.—The melted butter sauce makes a good condiment for fish by adding hard-boiled eggs, chopped fine, and for boiled fowl, by adding chopped oysters when it is simmering.

Egg Sauce.—This may be made either with melted butter or with white sauce. Five or six hard-boiled eggs, cut into small slices, using only half the whites, are put into a sauce-pan, to half a pint of melted butter or white sauce, with a little cream, all poured onto the eggs hot.

Good White Sauce.—This is made by taking stock or the liquor in which the fish, flesh or fowl has been cooked, a little flour, pepper and salt. Turn in two beaten eggs, and let the whole come to a boil, stirring constantly. Or it may be made with one pint of milk, a small onion, a small head of celery and a little parsley, white pepper and salt, and two ounces of butter. The butter is melted in a sauce-pan; dredge the flour slowly in until mixed. Previously, the milk, herbs, pepper and salt, must have been cooked together in a "bain marie," or some vessel placed in another of boiling water. Stir the milk slowly in, then the well-whisked egg, stirring all the while. When it simmers it is done.

Another Good White Sauce.—Boil in a sauce-pan, half-pint of water, two cloves, fifteen pepper-corns and a blade of mace. Add two anchovies chopped fine, a quarter of a pound of butter, a little flour and a pint of cream. Let it boil three minutes, stirring it constantly.

Brown Sauce.—Put two ounces of butter into a stew-pan, with a quarter of a pound of lean bacon or ham cut fine, and two pounds of lean beef cut into strips. Add a little water, two cloves, pepper, salt and one bay-leaf. Set it over the fire, stirring constantly, until it is brown and rich. Then add two quarts of water, and when it boils, let it simmer slowly for an hour and a half. Strain through a sieve, and it is ready for use.

Sauce for Roast Meat.—A good sauce for roast meat may be made in ten minutes, with a quarter of a pint of water, the juice of a lemon strained, a sprig of parsley chopped fine, an ounce and a half of butter, all seasoned with white pepper and salt. Set the whole over the fire in a glazed sauce-pan and keep it there until it is just ready to boil; then serve. This may be varied by adding two tablespoonfuls of nasturtiums or capers. It may also be made with white vinegar, in place of the lemon; or the flavoring may be used with the gravy sauce usually served with the meat.

Wine Sauce, for Roast Game.—Take a pint and a half of jelly, three-quarters of a pound of butter, three tablespoonfuls of brown sugar, and half as much ground allspice, and a quart of port wine. Stew together until thick. This may be used for any roasted meats.

Onion Sauce.—Many people are fond of a sauce of onions, which are healthy, and to most palates agreeable. If it were not for the unpleasant odor they give the breath, onions would be universally used both as food and flavoring. A sauce is made by boiling the onions, until tender, in milk and water. Drain and chop the onions fine, adding pepper and salt. Pour drawn butter over them, and add milk or cream. When the whole comes to a boil, the sauce is ready.

Cold Meat Sauce.—A good sauce for cold meat is made by beating the yolks of three eggs, and adding a wineglassful of jelly cut up. The seasoning is made with a tablespoonful each of flour and mustard, softened with vinegar. Put the whole in a sauce-pan with a tablespoonful of butter and half a teacupful of vinegar. Boil, stirring constantly until thick. Any solid pickle, like cucumber, may be chopped fine, and stirred thoroughly in when the sauce is cold.

Another Sauce for Cold Meat.—Take equal quantities of ripe tomatoes and young okras; chop the okras fine, skin the tomatoes, and slice one onion. Stew all together very slowly until tender, and season with half a tablespoonful of butter and a little cayenne pepper and salt.

Sauce of Many Names.—A sauce that goes by the name of the flavoring added, as caper, mushroom, chopped cucumber, hard-boiled eggs, and various herbs, is made by mixing together two large tablespoonfuls of butter and a tablespoonful of flour; put into a sauce-pan, and add two cups of broth or water; set on the fire, and when thick add of the articles mentioned to suit the taste; salt; take from the fire, add the yolk of an egg, beaten, and serve. Thus you have cucumber, egg, herb, or mushroom sauce.

Sauce for Fowl.—Put half a pint of veal or chicken broth into a stew-pan, with a wineglassful of port wine, the juice of a lemon and the juice of an orange. Season with pepper and salt, boil for five minutes, pour over the fowls and serve.

II. Salads and their Dressing.

The value of a salad is said to be in the dressing. However this may be, most people like salads; and yet very few know how to prepare them. Salads may be called purely luxuries. They certainly are elegant additions to any table, and most appetizing.

Proverbial Salad.—The Spanish have a proverb that four persons are necessary to make a good salad: A spendthrift for oil, a miser for vinegar, a barrister for salt,

SALAD WASHER.

and a madman to stir it up. Vegetables for a salad must be fresh and crisp. Those kept fresh by soaking in water are always ruined. If they must be kept, lay them between folds of damp cloth. They should be young and well blanched. After washing, dry in a cloth before putting them in the salad bowl. They should be eaten soon after being prepared. Lettuce, cabbage, endive, celery, water-cress and cucumber are the principal vegetable ingredients for salads; beet-root, hard-boiled eggs and tomatoes are used for the garnishing. Mayonnaise sauce is necessary for all elaborate salads of meat or cooked vegetables. It is also a good foundation for all cold sauces.

Mayonnaise Sauce.—For half a pint of sauce, put the yolk of an egg in a basin, with half a tablespoonful of tarragon or other flavored vinegar, and a tablespoonful of pure vinegar, with a little salt and pepper. Mix these thoroughly with a wooden spoon. Then add oil, drop by drop, mixing thoroughly. Never add more until the first is well mixed. When about forty drops are mixed, the quantity added may be a teaspoonful at a time, until four ounces are added. Then taste, and add more vinegar, pepper and salt, if necessary. If you like it, a little eschalot, or onion and parsley, thoroughly mixed, may be added.

French Vegetable Salad.—Boil equal weights, separately, of the tender tips of asparagus, string beans, green peas, carrots and turnips. Dry them in a clean cloth, and when cold cut into small squares. These should be arranged on a dish, the beans in the bottom and center. Then, around them, in equal rows, the carrots, peas, turnips and, last, the asparagus. If there are vegetables enough, proceed as before, and over all sprinkle finely minced chervil, tarragon, burnet, chives and garden cress, all having been first blanched, strained, cooled and dried in a cloth. If you have not these, substitute others of a similar character. Serve with Mayonnaise sauce. If the vegetables are fresh, young and, of course, tender, it makes a delicious dish.

Salad of Meat, Fowl or Fish.—The cooked cold meat, chicken, game, fish or lobster, is to be cut into small scallops or pulled to pieces, and dipped into Mayonnaise sauce, and the lettuce well blanched as well as the endive. Prepare these and water-cress by washing and drying in a cloth. Break into pieces of an inch in length. Mince a sprig of chervil, two leaves of tarragon and a little sorrel. Peel and slice a fresh cucumber and a boiled red beet. Mix all these together thoroughly, make a foundation of the vegetables, then a layer of the fish, flesh or fowl, etc. So continue until you have the whole complete, saving some cucumber and beet-root for the outside of the dish. Garnish with hard-boiled eggs, properly cut; also, with jelly or olives. Serve with Mayonnaise sauce, from a boat or other suitable vessel.

Lettuce Salad.—Every vegetable salad should have a good paste. To make this requires care. Put the yolks of two boiled eggs into a dish with a teaspoonful of dry mustard and a tablespoonful of perfectly sweet olive oil, with enough pepper and salt to season. With a wooden spoon work this to a perfectly smooth paste. Then gradually add three tablespoonfuls more of oil, two of vinegar, and mix to the consistency of cream. Add two or three leaves of tarragon and a small eschalot, or one small white onion finely minced; also the whites of the two eggs cut in very small slices. Then add the lettuce and some water-cress, broken into inch pieces. When all is thoroughly mixed with the sauce, serve. We have given all the ingredients of a perfect lettuce salad. They may be all used or not, according to the taste. The French rub the dish in which it is sliced with garlic, but ours is not a garlic-eating nation.

Another Salad Dressing.—Here is a simple one. Take the yolks of four eggs, beaten, one teaspoonful of sugar, salt-spoonful of cayenne pepper, two teaspoonfuls of made mustard, six tablespoonfuls of salad oil and five of celery or other flavored vinegar, Mix all these thoroughly, put in a saucepan and boil three minutes with constant stirring. When cold pour it over the chicken or other salad.

Potato Salad.—Boil and mash the potatoes fine and smooth, and season well with butter, pepper, salt and a little cream. Use three hard-boiled eggs to each quart of potatoes. Chop the whites fine, and work the yellows smooth with mustard, a trifle of sugar, pepper and salt, according to taste, with only enough vinegar to moisten the whole. When thoroughly mixed, including the whites of the eggs, put a layer of the mashed potatoes in a flat-bottomed dish; drop the dressing in spots over the potatoes and so proceed until the whole is finished, with a layer of potatoes, nicely smoothed over the top, and so arranged that the salad will be crowning at the top. Then brown the whole in the oven and garnish.

Oyster Salad.—For a can of cove oysters, take a teacup half full of cream, a heaping tablespoonful of butter, a teacupful or less of vinegar, a teaspoonful of made mustard, cayenne or black pepper, and salt and sugar to suit. Whisk the eggs thoroughly; mix in the other ingredients; put in a saucepan; set this in a vessel of boiling water, and cook until the whole is thick. Use any flavoring you like, and some pounded cracker. Chop the oysters fine, pour over the dressing, mixing all together.

Cole Slaw.—To make cole slaw, take one heaping teaspoonful each of mustard and salt, two tablespoonfuls of cream, one tablespoonful of butter, three tablespoonfuls of sugar, two-thirds of a cup of vinegar, yolks of two eggs, well beaten; stir all together and set on the fire, stirring constantly until it thickens, then pour on chopped cabbage.

III. Various Made Dishes.

The talent of the cook may be employed in preparing many nice dishes from remnants that are too often thrown away. The stew-pan, the saucepan, the gridiron, the omelette-pan and the toasting-iron come in well here into play. These may be called dishes of taste and economy.

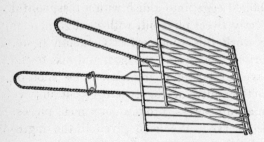

TOASTER AND LIGHT BROILER.

An Economical Dish.—Take the remnant of a cold boiled leg of mutton, or of a roast of beef; shave it into thin slices; season, and add, if you like it, an onion chopped fine, or a pinch of sweet herbs. Put this on a baking-dish, and pour over the gravy, if you have any; if not, a little water, butter and flour. Then take hot boiled potatoes; mash fine; add a little milk and salt, or butter, to soften them into a smooth paste, which lay over the meat. Put the dish in the oven, and bake a nice brown.

Potato Croquettes.—Peel, boil and mash a quart of potatoes, mix with them the yolks of four eggs and two ounces of milk; set on the fire, stir for two minutes, spread in a dish to get cold, or leave over night, if designed for breakfast, in which case, a little milk may be added to moisten their dryness; mix thoroughly, divide into tablespoon parts, shape them, roll in bread-crumbs, dip into beaten whites of eggs, roll in bread-crumbs again, and fry in hot fat. Take off when done, drain, dish and serve immediately. Shaped flat, they are "croquettes *a la duchesse.*"

Chicken Croquettes.—Take the remnants of cold chicken and chop the meat fine. Also chop an onion for each chicken. Fry the onion in a little butter, adding half a spoon of flour; stir a minute, then add the chopped chicken and a gill of broth, salt, pepper and nutmeg. Stir for two minutes. Put all back on the fire, stirring gently. Spread the mixture on a dish, and put it away to cool. When cold, stir the top well in with the rest, and if very dry, add a little broth. Divide into parts on the paste board of about a tablespoon to each. Have bread-crumbs on the board; make the parts round, dipping each one into beaten eggs, then into the bread-crumbs, and fry in hot lard. Serve hot. All kinds of meat may be made into croquettes after the same manner.

Fried Bread.—Cut dry bread into slices, dip it in water, and fry in hot lard or gravy, and butter. Fry until nicely browned, then pour cream and eggs well beaten over it. Let it fry to the bread and serve hot.

Relish for Breakfast or Lunch.—Take a quarter of a pound of good, fresh cheese; cut it up in thin slices and put in a spider, turning over it a large cupful of sweet milk; add a quarter of a teaspoonful of dry mustard, a dash of pepper, a little salt, and a piece of butter as large as a small egg; stir the mixture all the time. Take three Boston crackers or soda biscuits, finely powdered or rolled; sprinkle them in gradually; as soon as they are stirred in, turn the contents into a warm dish and serve.

Boiled Beans.—Put the dry beans in cold soft water, and let them soak three or four hours. Then put them in cold water—two quarts of water to one quart of beans, adding a tablespoonful of salt—bring them to a boil, and let them simmer until tender, say two, or two and a half hours. Pour the water away from them; let them stand by the side of the fire, with the lid of the saucepan partially off, to allow the beans to dry, then add an ounce of butter for every quart of beans, and seasoning of pepper and salt.

Side-Dish of Eggs.—Cut hard-boiled eggs in half, the long way. Take out the yolks and mix them with bread-crumbs, salt, pepper and butter. Put them back in the whites. Set the halved eggs in the pan, with the yolks up, and bake until the yolks, but not the whites, are browned.

Stewed Tripe.—Soak the cleaned tripe in salt water for several hours. Then boil until quite tender. When cool, cut into small strips, dredge thoroughly with flour, and cook in a stew-pan with butter until hot. Pour in half a pint of cream, stir until thickened, and serve.

Croquettes.—Meat, chicken or game croquettes are made as follows: Chop the meat fine, and with it allow three teaspoonfuls of chopped parsley to each pound of meat or fowl; add the same quantity of onions chopped fine if you like them. Add salt, pepper and mace to taste. Make a panada, with but little water, of half a pound of bread to each pound of meat. Butter the bread well. Break four eggs, add a grated nutmeg, and beat meat and eggs thoroughly. Then add the panada, mix, add three tablespoonfuls of cream, and work the whole thoroughly together. Roll into proper shapes, dip in white of egg, and then in bread-crumbs, and fry in boiling lard.

Salmon Croquettes.—Salmon or other fish is made into croquettes, by mixing the cooked fish, hot mashed potatoes, the yolks of eggs, and bread-crumbs or pulverized cracker together. Form them into tasty shapes, dip into egg, dredge them with cracker dust and fry in boiling lard.

Rice Croquettes.—Boil rice until it is thoroughly done, and dry. To every half-pound of rice, allow two tablespoonfuls of grated cheese, one teaspoonful of mace, and butter to moisten it. Chop six tablespoonfuls of the breast of boiled or roast chicken or turkey and the soft parts of six large oysters, a little parsley, a grated nutmeg, and the yellow rind of a lemon. Moisten with cream so that all may be thoroughly mixed. Then take a portion of the rice the size of an egg, flatten it and in the center put a dessert-spoonful of the mixture and close the rice around it; cover with the whisked yolk of egg, roll in pulverized cracker, brown in boiling lard and serve hot.

Sandwiches.—Chop cold boiled ham very fine, and mix it with the yolks of eggs (beaten), a little mustard and pepper, and spread on very thin slices of bread, buttered on the loaf; trim off the crust, and cut into squares.

Minced Liver.—Cut liver into small pieces and fry with salt pork; cut both into square bits, nearly cover with water, add pepper and a little lemon-juice; thicken the gravy with fine bread crumbs.

Fried Potatoes with Eggs.—Slice cold boiled potatoes and fry in butter until brown; beat up one egg for each person to be served, and stir them into the mess; do not leave them a moment on the fire after the eggs are in, for if they harden they are not good.

Macaroni and Cheese.—Boil macaroni until tender; butter the bottom of a pudding-dish, and put in a layer of macaroni, then a layer of grated cheese; season with butter, pepper and salt; then another layer of macaroni, and so on, finishing with a layer of cheese; cover with milk and bake forty minutes.

Parsnip Fritters.—Boil in salted water until very tender; then mash, seasoning with a little butter, pepper and salt, add a little flour and one or two eggs, well beaten; make into small balls or cakes and fry in hot lard.

Timbale of Potatoes.—Cook, drain, mash, and pass through a fine sieve two quarts of Irish potatoes; put this in a saucepan, with six ounces of butter, two whole eggs, the yolks of six eggs, salt, pepper, nutmeg and a little sugar; have a plain two-quart timbale mould, well buttered and sprinkled with fresh bread-crumbs; put the preparation in it, with a little more bread-crumbs, and bits of butter on the top; bake for half an hour in a moderately hot oven; before serving, pass the blade of a knife between the potatoes and the mould, turn over carefully, and in a few minutes take the mould off and serve.

PUDDING OR TIMBALE PAN.

Duchesse Potatoes.—Take eight large potatoes, boiled and mashed fine, one tablespoonful of butter, the yolks of two raw eggs, a little salt; stir all together over the fire, then set it away to cool. When quite cold, roll it on a board with flour to keep from sticking. Make it in cake or any form you wish. Take the white of the egg, beat with a little water, dip in the potato and roll in bread or cracker crumbs. Fry in hot lard.

Dried Beef Stewed.—Heat milk and water (about half of each), and thicken with a beaten egg and a little flour; when boiled add the beef, sliced as thin as possible, and remove from the fire at once; if the beef is very salt it will need freshening in a little hot water before going into the gravy, but if not it will season it without being freshened.

French Stew.—The French have a way of cooking tough meat, called "Daube," as follows: Season a thick steak with salt and pepper, and fry slowly in a little lard. Turn it often, so that both sides may be cooked alike and equally browned. When well browned, put in a good-sized porcelain-lined kettle, add a small quantity of water, half a sliced onion, some minced parsley and thyme, thicken with a spoonful of flour, cover close and leave it for an hour on the back of the stove where it may simmer slowly; after this has been done add a pound can of tomatoes, then let the "daube" cook about two hours, or until the meat is ready to fall to pieces.

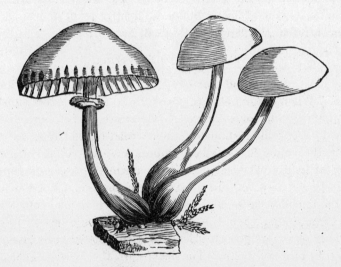

MUSHROOMS—GENUS AGARICUS.

Testing and Cooking Mushrooms.—Peel off the outer skin, break out the stem, and set the cap, top down, on a hot stove. In the spot where the stem formerly stood put a little salt, and, if desired, a small bit of butter. Scatter some salt over the gills. When the butter or salt melts, the cooking is done; and as soon as it is cool enough the fungus should be eaten, carefully saving the juice. Some fungi that do not seem particularly delicious when thus cooked, will, when slowly stewed with a little butter, and flour dredged in, with salt and pepper, make most delicious stews.

Agaricus campestris is the one generally raised, but all the mushrooms, cantharellus, marisimus, boletus, indeed all of the edible fungi named, will stew together, and form a dish that, alone or as an entrée, cannot be surpassed in delicacy of flavor and gastronomic satisfaction.

In testing new fungi, one eats a little of the cap with common salt to ascertain whether it tastes good, and whether it affects the fauces of the throat disagreeably; when a burning or stinging sensation accompanies or follows the swallowing, eat no more but take a copious dose of common salt, which generally neutralizes the poison. Some species, unpleasant or slightly injurious when raw, lose their harsh qualities in cooking; but as there are so many that are delicious, it is well to give up the doubtful kinds.

IV. Pickles, Catsup and Condiments.

PICKLES are any vegetables, fruits or other substance preserved in vinegar. Sweet pickles are preserved with vinegar, sugar and spices. Catsups are, in some sense, liquid pickles. The juices of the vegetables and vinegar compose the catsup with the addition of spices. A condiment is something to give zest to dishes, or to add to, or bring out flavor. Leaves are useful in flavoring; those of herbs are well known; many common varieties should be better known. Peach leaves and those of the laurel contain the virtue of bitter almonds, itself a form of prussic acid, a deadly poison when concentrated, but harmless in small quantities.

V. Leaves for Flavoring.

ONE of the most useful and harmless of all leaves for flavoring is that of the common syringa. When cucumbers are scarce these are a perfect substitute in salads or anything in which that flavor is desired. The taste is not only like that of cucumbers, but identical—a curious instance of the correlation of flavors in widely different families. Again, the young leaves of cucumbers have a striking likeness, in the way of flavor, to that of fruit. The same may be affirmed of carrot tops, which are as like carrots in taste as may be. In most gardens there is a waste of celery flavor in the sacrifice of the outside leaves and their partially blanched footstalks. Blanched celery is cut up into soup, when the outside leaves would flavor it equally well or better. The young leaves of gooseberries added to bottled fruit give a fresher flavor and greener color to pies and tarts. The leaves of the flowering currant give a sort of intermediate flavor between that of black currants and red. Orange, citron and lemon leaves impart a flavoring equal to that of the fruit and rind combined, and somewhat different from both. A few leaves added to pies, or boiled in the milk used to bake with rice, or formed into crusts or paste, impart an excellent flavor. In fact, leaves are not half so much used for seasoning purposes as they might be.

VI. Sour Pickles—Cucumbers.

As the cucumbers are gathered, pack them down in any suitable vessel with salt enough to make a strong brine, for which the cucumbers will usually furnish moisture enough; keep them under by placing over them a follower, fitting the inside of the barrel and weighted sufficiently to press firmly upon them. They may remain thus until wanted for the table. Then soak them in water until fresh, changing it as often as necessary. Pack them in jars, and pour over them scalding vinegar, seasoned to suit the taste with cloves and other spices, but if it is desired to have them of the fashionable and deleterious green color, they should be just scalded in a brass kettle, with water and a little alum to harden them. The verdigris which is formed during the act of scalding imparts the fashionable tint and the process is continued until the desired green color is reached; after which, pack as before directed, with hot vinegar. If this greening process is carried too far, the pickles become absolutely poisonous—sometimes fatally so. Sensible persons generally take their pickled cucumbers without this poison, and if not exposed to the light, ungreened pickles do not fade much. Cucumbers for pickles should never exceed over three inches in length. The smaller they are the better.

Quick Pickles.—Take small cucumbers of a uniform size, wash, put in a porcelain kettle, cover with cold water, add a little salt; set it on the stove, let it heat gradually and boil five minutes; then drain off all the water; add good vinegar; to one gallon of vinegar add one cup molasses, one tablespoonful cloves, and the same of cinnamon; let boil five minutes; remove to an earthen or stone dish; pour over them the hot vinegar; cover tight; when cold, they are ready for use.

Indian Pickle.—To each gallon of vinegar (cold) add half a pound of mustard, six ounces of turmeric, a handful of salt, and a little grated ginger; boil the vinegar and spices together, and let the mixture cool. Boil or scald the vegetables with vinegar, taking care to have among them a little garlic and onions; put them in your jar, and pour on the pickle. Afterward put in the jar a bag containing a quarter of a pound of ginger, one ounce of long red pepper pods, one of black pepper, one of cloves, and half an ounce of cayenne. If this is too hot for the taste, omit the cayenne to suit.

Pickled Red Cabbage.—Select sound red cabbages, and to each quart of vinegar, add half an ounce of ginger, well bruised, one ounce of whole black pepper, and, when liked, a little cayenne. Take off the outside leaves, cut in quarters, remove the stalks, and cut it across in very thin slices. Lay these on a dish, and strew them plentifully with salt, covering them with another dish. Let them remain for twenty-four hours, turn into a colander to drain, and, if necessary, wipe lightly with a clean, soft cloth. Put them in a jar; boil up the vinegar with spices in the above proportion, and when cold, pour it over the cabbage. It will be fit for use in a week,

COLANDER.

but if kept for a very long time, the cabbage is liable to get soft and to discolor. To be really nice and crisp, and of a good red color, it should be eaten almost immediately after it is made. A little bruised cochineal boiled with the vinegar, adds much to the appearance of this pickle. Tie a bladder over the top of the vessel in which it is kept, and keep in a dry place.

Pickles Without Vinegar.—In places where vinegar cannot be readily obtained, cucumbers and other articles for pickling may be made sour as follows: Take one quart of good alcohol to four quarts of water; put the cucumbers in fresh from the vines. Wipe them first with a wet cloth, or wash and drain them. Put in a warm place until fit for the table, then keep them in a cellar, or a cool place. They remain hard and green; and are always ready for the table.

Green Pickles, Natural.—Heat together one gallon of water to each two pounds of pure salt; pour this scalding hot over the washed cucumbers; at the end of three hours pour off and cover with scalding hot, sound cider vinegar. At the end of three days pour this off and scald the cucumbers in fresh vinegar. They will be naturally greened.

VII. Chow-Chow.

To each two quarts of small green cucumbers, or green tomatoes, and the same of cabbage, allow two dozen small onions and half a dozen green mango peppers. Sprinkle the onions or tomatoes with salt, also the cabbage, separately. At the end of six hours press out the water. Cut the onions in half, pour boiling water over them and let them stand for a little while. Cut the green peppers into inch-square pieces; and the cabbage and the tomatoes into pieces of suitable size. The tomatoes, if small, need not be cut. Mix all together. Then, to one teacupful of ground mustard add two cups of white mustard-seed, three tablespoonfuls of turmeric, three of celery-seed, one of mace, one of cayenne pepper and one of ground cinnamon, well mixed. Add boiling vinegar enough to cover. The vinegar should be sweetened with one pound of sugar to cover the whole pickle.

A Better Chow-Chow.—Take the white part of one head of cabbage, two medium heads of cauliflower, one quart of string beans, one quart of very small cucumbers, six roots of celeriac, six mango (sweet) peppers, one quart of small white onions and two quarts of green tomatoes. Boil each of these articles separately— except the cucumbers, which must be scalded in vinegar—until just done, but not soft. The cauliflower may be pulled apart, piece by piece, and the rest of the vegetables cut into suitable pieces, rather fine. If the cucumbers are not very small, they may also be cut; but if small, pack all the ingredients in a jar in regular layers, or mixed, so some of each may be taken out together. Then prepare the following pickle: Two gallons of strong cider vinegar, four ounces of mustard, four ounces white mustard-seed, a pot of French mustard, one ounce of cloves and two ounces of turmeric. Put the spices and vinegar into an enameled kettle, and when they come to a boil, pour over the vegetables. A pound of sugar and any spices liked may be added to the pickles, and any vegetables may be added to or omitted, if you keep the proportions correct. The vinegar must cover the whole completely, and it is a good plan to place a follower in the jar, to keep the chow-chow submerged. Chow-chow may be made by salting, draining, cutting into proper pieces and then scalding the vegetables in weak brine, packing and pouring over the dressing or pickle. Small martynias, nasturtiums, okra, Chili pepper, very young radish pods, or anything of the like kind add to the value—because to the variety—of chow-chow, but according to the addition of Chili pepper, omit the mango pepper.

Tomato Chow-Chow.—Slice one peck of green tomatoes, six green peppers, one onion, strewing a cup of fine salt over them. After standing one night, turn off the water. Put them in a preserving kettle, with vinegar enough to cover them, add one cup of sugar, one cup of grated horseradish, one tablespoon of whole cloves, one of ground cinnamon. Stew slowly until perfectly soft.

Imitation Chutney.—Indian chutney is a compound of mangoes, Chillies and lime-juice, with some portion of other native fruits, such as tamarinds, etc., the flavor being heightened by garlic. For family use the following recipe will be found suitable: Chillies, one pound to one and a half pounds; apples, one pound; red tamarinds, two pounds; sugar candy, one pound; fresh ginger-root, one pound; garlic, one-half pound to three-fourths of a pound; sultana raisins, one and a half pound; fine salt, one pound; distilled vinegar, five bottles. The Chillies are to be soaked for an hour in the vinegar, and the whole ground with a stone and muller to a paste.

Another recipe which may be depended upon for making an excellent chutney is as follows: One pound salt; one pound mustard-seed; one pound stoned raisins; one pound brown sugar; twelve ounces garlic; six ounces cayenne pepper; two quarts unripe gooseberries, and two quarts best vinegar. The mustard-seed should be gently dried and bruised, and the sugar made into a syrup with a pint of the vinegar; the gooseberries dried and boiled in a quart of the vinegar; the garlic to be well bruised in a mortar. When cold, gradually mix the whole in a mortar, and with the remaining vinegar thoroughly amalgamate them. To be tied down close; the longer kept the better.

VIII. Piccalilli.

This is simply mixed pickle, chopped. It may be fairly cooked, or 'better, only scalded. Here is a good recipe: To a peck of green tomatoes, sliced thin, add a pint of salt. Cover with cold water and let them stand twenty-four hours. Then chop fine one head of cabbage, six onions (or not as you please) and twelve sweet peppers. Cover with scalding vinegar and drain through a sieve. Pack the mixture in a jar, mix with vinegar enough to cover the whole; add a tablespoonful each of cloves and allspice, two ounces of white mustard-seed and a pint of molasses. Let the whole come to a boil and add the vegetables, either hot or cold, preferably cold.

Mixed Pickles.—Every suitable vegetable, even melon cut in pretty shapes, may be used. All small things, like cucumbers, onions, martynia, radish pods—all these should be small, and go in whole. Other things as cabbage, cauliflower, tomatoes, beans, etc., are sliced or cut in proper pieces. Put whatever you make your pickles of in strong brine for twenty-four hours. Drain three hours, place in a preserving kettle with eight ounces of white mustard-seed, a tablespoonful of ground black pepper and enough vinegar to just cover all, adding a saltspoonful of powdered alum. Let the whole come to a boil. Drain again, and when cold mix in half a pint of ground mustard, and cover with strong cider vinegar, adding turmeric enough to color if you like. These proportions are sufficient for three hundred small cucumbers, three heads of cabbage, the same of cauliflower, a quart of beans, the same of tomatoes, and the necessary horseradish, pepper, onions, etc.

IX. Sweet Pickles.

SWEET pickles are green, or nearly ripe fruits, prepared with spices, and the addition of a good deal of sugar to the vinegar. The general plan of preparation is alike with all, the spices and preparations of sugar being varied to meet extra acidity in the fruits, or to suit the special taste of individuals. Peaches and all soft, ripe fruits may be prepared as follows:

LIPPED PRESERVING AND PICKLING KETTLE.

Ripe Tomato and Fruit Pickles.—To seven pounds of fruit add three pounds of sugar and one quart of vinegar. Simmer them together for fifteen minutes, skim out the fruit, boil the syrup a few minutes longer, add cloves, cinnamon and other spices to suit the taste, and pour over while warm, preferably in glass jars that may be sealed. The same will apply to all mixed or other fancy pickles, since they look pretty if nicely arranged.

Sweet Ripe Pickles.—Take sound, ripe cucumbers, peel and remove the seeds, cut lengthwise into strips an inch wide. To three quarts of the pieces add three cups of vinegar and four of water; soak twenty-four hours, stirring once or twice. Put one quart of vinegar on the fire, add one pint of sugar, a little stick cinnamon and a teaspoonful of pimento tied in a bit of cloth; scald all together· add the cucumber and boil till soft.

Sweet Green Pickles.—The green tomatoes or other fruits should be sliced, and six large sliced onions added—if they are liked—for each peck of material. Sprinkle between the layers a teacupful of salt, and let them stand over night. Drain and boil in two quarts of water and one of vinegar; drain again. The fruit must then be boiled fifteen minutes in the following pickle: Two pounds brown sugar, four quarts strong vinegar, half a pound of green mustard, two tablespoonfuls each of cloves, ginger and cinnamon, to which have been added six sliced mango peppers.

X. Catsups.

ALL pickles, catsups, condiments or preserved fruits, must be kept in a cold (not freezing), dark place. Catsup is a semi-liquid condiment, and is made of tomatoes, green walnuts, mushrooms, or similar substances that will impart a pleasant flavor.

Tomato Catsup.—The tomatoes must be fresh and fully ripe. Scald them and press through a sieve that will retain all the seeds and skins. To each gallon thus prepared, when cold, add four tablespoonfuls of salt, three of ground mustard, two of black pepper, one of allspice, half as much cloves, and half as much cayenne pepper; also a pint of the strongest cider or wine vinegar. Simmer the whole together for four hours. Bottle and cork tight.

XI. Condiments.

Worcestershire Sauce.—Worcestershire sauce is generally adulterated. If made according to this formula, it will be good and pure: White wine vinegar, fifteen gallons; walnut catsup, ten gallons; Maderia wine, five gallons; mushroom catsup, ten gallons; table salt, twenty-five pounds; Canton soy, four gallons; powdered capsicum, two pounds; powdered allspice, one pound; powdered coriander seeds, one pound; cloves, mace and cinnamon, of each one-half pound; assafœtida, one-fourth pound, dissolved in one gallon of brandy. Boil twenty pounds of hog's liver in ten gallons of water, for twelve hours, renewing the water from time to time. Take out the liver, chop it, mix with water, and work it through a sieve. Mix with the sauce, and bottle for use. If less is wanted, use carefully estimated proportionate quantities of ingredients.

Imitation Worcestershire Sauce.—Here are two formulas, both excellent for making palatable sauces resembling the real Worcestershire: 1. White vinegar, one gallon; Canton soy, one pint; molasses, one pint; walnut catsup, one and one-half pints; table-salt four ounces; powdered capsicum, one ounce; allspice, one ounce; coriander, one-fourth ounce; cloves, one-half ounce; mace, one-half ounce; cinnamon, six drachms; assafœtida, one-fourth ounce, digested in four ounces of rum; mix.

2. Take port wine and mushroom catsup, of each one quart; walnut pickle, one pint; soy, one-half pint; pounded anchovies, one-half pound; fresh lemon peel, minced shallots and horseradish, each two ounces; allspice and black pepper, bruised, each one ounce (or currie powder one-fourth ounce), digest for fourteen days, strain and bottle.

How to Mix Mustard.—Mustard should always be mixed with water that has been boiled and allowed to cool. Put the mustard in a cup with a pinch of salt, and mix with it, gradually, sufficient boiled but cold water to make it 'drop from a spoon without being watery Stir and mix well and rub away all lumps with the back of a spoon. The mustard-pot should not be over half full, as mustard is better when freshly made.

Tomato Sauce.—Take any quantity of ripe tomatoes, put them into an earthen jar, and place them, covered over, in a hot oven till perfectly soft; then rub them through a fine sieve, to keep out the seeds and skins. To every quart of juice add a clove of garlic, or, if the flavor is preferred, two shallots, bruised, a quarter of an ounce of ginger, the same quantity of black pepper, and a tablespoonful of salt; boil for about twenty minutes, and bottle, cork down, and wax it at once. If liked, the juice of two lemons may be added to the above before boiling.

TINNED RIM KITCHEN SIEVE.

Chili Sauce.—Take nine large tomatoes, four large onions, four red peppers, or in the same proportion. Chop them together; then add four cups of vinegar, three tablespoonfuls of sugar, two of salt, two teaspoonfuls of cloves, the same of cinnamon, ground, of ginger, of allspice and of nutmeg. Boil one hour, and bottle for use.

XII. Flavored Vinegar.

Raspberry Vinegar.—Take six pounds of ripe raspberries, and pour on them four pints of the best vinegar. Leave them thus for four days, frequently stirring, but not mashing the fruit so as to bruise the seeds; then place a piece of clean fresh washed linen or flannel in a colander, and filter the vinegar; to each pint of juice add two pounds of loaf-sugar; put it into glazed jar or pan, which place in boiling water and keep there till the juice boils thick and syrupy. Let it become cold, then bottle it. The whole process should be carried on in a glazed kettle or earthen vessel. The same formula may be used with other small fruits.

Horseradish Vinegar.—Put into a jar, four ounces of grated horseradish, a teaspoonful of cayenne, two teaspoonfuls of salt and one tablespoonful of mustard; pour over them a quart of boiling vinegar, and set the jar, covered, by the fire for a fortnight; then boil up the vinegar, let it cool, strain through a jelly-bag, and bottle. It is an excellent relish for salads, cold meats, etc. The same means may be used for peppers, mushrooms, green walnuts and other articles, the flavor of which is desired.

XIII. Strawberry Acid.

Dissolve in a quart of spring water, two ounces of citric acid, and pour the solution on as many quite ripe strawberries—the wild fruit is preferable—stripped from their stalks, as it will just cover; in twenty-four hours drain the liquid closely from the fruit, and pour it over as many fresh strawberries as it will cover, keeping it in a cool place; the next day drain the liquid again entirely from the fruit, and boil it gently for three or four minutes, with its own weight of very fine sugar, which should be dissolved in the juice before it is placed over the fire. It should be boiled in an enameled pan. When perfectly cold, put it into small, dry bottles, closely corked for use, and store it in a cool place. It is one of the most delicate and deliciously-flavored preparations possible, and of a beautiful color. If allowed to remain longer in the preparation than forty-eight hours before it is boiled, it commences to ferment. In very hot weather, fermentation may take place inside of forty-eight hours.

CHAPTER IV.

BREAD-MAKING.

I. SELECTING THE FLOUR.—II. SOME THINGS TO BE REMEMBERED.—III. YEAST AND YEAST-MAKING.—IV. BREAD OF FINE FLOUR.—V. HEATING THE OVEN.—VI. MILK BREAD, POTATO BREAD AND CREAM BREAD.—VII. RYE BREAD.—VIII. GRAHAM BREAD.—IX. BOSTON BROWN BREAD.—X. VARIOUS RECIPES FOR BREAD.—XI. BISCUITS, ROLLS, GEMS, ETC.—XII. OATMEAL BREAKFAST CAKES—XIII. RUSKS AND ROLLS.

I. Selecting the Flour.

IN selecting flour, if it is a dead white, or bluish white, refuse it. If it has a yellowish tinge, it should be good. If, upon being squeezed in the hand, and then thrown against a smooth surface, it falls like powder, it is bad; if, on the contrary, it sticks, and what falls does not disintegrate, it should be good. In wetting and kneading the flour between the fingers, if it is sticky, refuse it. If it moulds kindly it is good. Squeeze some of the flour in the hand; if it retains the shape given it, it is a good indication. If flour stands the tests here given, there should be no difficulty in making good bread from it.

II. Some Things to be Remembered.

CERTAIN rules are necessary to be observed in making bread. In cold weather, the flour should not be chilled, and the sponge should be kept moderately warm. The kneading should be thorough; the bread should be baked with a uniform, rather quick heat; when just done (a splinter thrust in and coming out dry is a good test), remove at once and cover with a light cloth where it will not cool too quickly.

Experience is a good teacher, but with this, one must have good flour, good yeast, and care must be observed in the making. A pint of finely mashed potatoes to each loaf, will keep bread moist.

Rolls and Biscuit.—Rolls and biscuit should be baked quickly; they should be brushed with warm water before being put in the oven. If a glaze is desired, brush lightly with a mixture of milk and sugar. Baking powders should always be of the best, and when used, get the dough into the oven as quickly as possible after being moulded.

Gems, Fritters, etc.—The pans for what are called gems, should always be hot, and be well buttered before the gems are put in. Bake quickly.

In making fritters, use haste, but beat the batter thoroughly, and cook at once. Pancakes should also be well beaten, and if eggs are used; these should be beaten separately, and added the last thing.

Sponge.—To make sponge, sift the flour, and in the middle of it pour the yeast, mix thoroughly, adding lukewarm water, from time to time, as needed, so the whole will be like thick batter. Pour this slowly on flour. If made at night, work the first thing in the morning, using flour enough to make the dough of the proper consistency. Some persons mould once and bake, and others work the dough the second time. When risen, put in the oven at once, and bake an hour for ordinary sized loaves.

III. Yeast and Yeast-making.

WITHIN the last few years, grocers in cities and principal towns keep compressed yeast, in its natural and moist state. This is made from the superfluous yeast of breweries and distilleries, and also, by brewing, directly, as an article of commerce. Dried yeast or yeast-cakes are also universally sold. They are not always good. The working principle of all yeast is a plant—a microscopic fungus—and to the ordinary observer, yeast is simply a thick, creamy froth, which causes bread to rise, and beer and other liquids to ferment. If the fermentation goes too far, the bread or liquid becomes sour. The next stage is mold. The purer the yeast, the less liable are substances to run into acid fermentation and mold.

The recipes which we give will fully provide for all that is necessary for household purposes.

Hop Yeast and Yeast-Cakes.—To a handful of good hops, add a quart of water; boil until reduced one-half; strain, wipe out the vessel, return the liquid, and set over the fire again. While the hops are boiling, mix enough flour with a little cold water to make it as thick as stiff starch. When it begins to boil after returning it to the vessel, stir in the dissolved flour. Let it boil a few minutes, then set it to cool. When cool enough stir in some good yeast or a dissolved yeast-cake, and a tablespoonful or a little more of sugar. Pour into a jug, cork, and let it work. It is best to set it in something suitable, in case it should work and run over before you are aware, and you lose a good deal. After it works, cork tight, and keep in a cool place. To make the cakes, sift into a pan one-third of flour and two-thirds of meal, as fine as you can, that there may be as little meal as possible in your bread; pour in the yeast, and work well into a stiff dough. Make off into cakes the size of a small teacup, and about three-quarters of an inch thick; dry in the shade. From this start you may have perpetual yeast by taking off a piece of dough after it has risen for bread, and working it with your hands in water enough to make a thick batter, and then sifting in meal enough to make the cakes as before stated; or a piece of the risen dough will raise another batter of dough the same as the yeast.

Hop Yeast, No. 2.—One and one-half pounds of grated raw potato, one quart of boiling water in which a handful of hops have been boiled, one teacup of white sugar, one-half teacup of salt. When almost cold put a little good yeast to start it, say about half a pint. One pint of this yeast makes four good-sized loaves of excellent bread.

Potato Yeast.—Take five or six potatoes; grate fine. Then add two tablespoonfuls of sugar and one of salt. Take one quart of water and a handful of hops. Boil a few minutes, strain and stir into potatoes. Set on the stove and stir until thick. When cool add one cup of yeast.

Salt Hop Yeast.—Put a gallon of cold water on the fire; let it come to a boil, and then put into it eight good potatoes and boil until well done, when they must be mashed fine, together with one teacupful of salt and one of white sugar. Directly after taking out the potatoes, put a handful of hops into the water, and let it boil while you are preparing the potatoes. Mix these with the hop-water, which must be boiling hot when added, and unstrained. When nearly cold, add one cup of lively yeast, stir well and set aside to rise in a warm place for twenty-four hours. Then strain, bottle and cork it up tightly. Strain through a sieve so that as much potato as possible may pass through. Allow half a teacup of this yeast to a quart of good flour in cold weather; in the summer a less quantity is required. Keep the bread warm while rising.

Bread Without Yeast—Salt-Rising.—In the morning set a sponge in a pitcher, by taking one teaspoonful of sugar, two-thirds teaspoon of salt, one-half as much saleratus, and one coffee-cup of new milk. Pour on this one pint of boiling water; let it stand until it is only blood-warm, and stir in flour to make a stiff batter. Keep it warm by setting it in warm water, and in five hours it will be a foam—if not, stir in a little flour. Then mix soft in your bread-pan with about a pint of milk and water, salt, and cover lightly with flour. This will rise again in about an hour; then mix rather firm and put in your tins.

IV. Bread of Fine Flour.

In making any bread the mixer must know the strength of the yeast and the quality of the flour, for flour varies in its rising qualities just as yeast does in strength. The rule is that more yeast is required in cold weather than in warm, and closer watching of the sponge is required in warm than in cold weather. The "good old-fashioned way," when bread was only made once a week and baked in a brick oven, was to take from five to six gallons of flour, put it in the kneading-trough and make a hole in the center of the flour, into which was poured a pint or more of yeast —according to strength—well mixed with a pint of milk-warm water. With a spoon this was stirred into a smooth batter and sprinkled over with all the dry flour left. Then the plan was to cover with a cloth and in summer set it in a rather cool place, or in winter in a warm place. When the sponge has risen, scatter over it two tablespoonfuls of salt, add warm water by degrees as you mix all thoroughly together. Work and knead the whole until it will no longer stick to the hands. Cover the mass with flour and set it under a cloth, where it is warm, to rise. When risen, divide into suitable loaves, mould them lightly upon the pastry-board, place in floured tins or pans and bake as quickly as possible, but not to brown the crust too much; that is, the oven must be of such a temperature that a large loaf will bake in from one hour to one hour and a half. This is the best way to make bread where large quantities are baked, as, where many hands are kept. Then a brick oven is economical.

V. Heating the Oven.

To heat a brick oven requires judgment. The fuel should be hard wood, quite dry, and cut fine. Then the oven may be heated in an hour. The oven is cleaned by drawing out the coals and sweeping with a broom, having the brush on the handle similar to the teeth of a rake. If, upon throwing a little dry flour into the oven, it turns dark, the oven must be cooled a little. The bread once in the oven, shut and close tight.

VI. Milk Bread, Potato Bread and Cream Bread.

THESE are made with milk and water, instead of clear water, or with cream and water, if the bread is liked a trifle short. This also makes excellent rolls. Potato bread is made by adding a little warm mashed boiled potato to the sponge, the manipulation of the whole being otherwise identical with the operations for bread previously given.

Rice bread is also made by adding boiled rice, crushed fine, to the sponge. A little fine corn-meal is sometimes added to the sponge, under the supposition that it keeps the bread moist.

The intelligent housewife may modify her bread-making in many ways. In fact, hardly two persons make bread exactly alike.

VII. Rye Bread.

TAKE as much flour as you wish for one baking of bread; make a hole in the center of the flour and stir into it a teacupful of good hop yeast and a pint of new milk. Stir the batter a little stiffer than for griddle-cakes, salting the batter to the taste. Then cover closely, and let it rise over night. In the morning, add more new milk, and knead up your bread very stiff. Then make a hole through the center of the dough, and let it rise until it is even on the top. Your bread is then ready to put in the pans for the third rising, which will take about half an hour. Bake as you would wheat bread. It is also good if mixed with two-thirds wheat flour to one of rye.

VIII. Graham Bread.

MAKE a sponge as for white bread, as a Graham sponge is apt to sour, or use part of the sponge to make a soft dough. If liked, a little syrup or sugar may be added. Let it rise only once before putting into the pans. If you proceed as you do with white bread, you will not fail. If you wish all Graham bread, for a change, you can make the "gems" by stirring the Graham flour into cold water, so as to form a stiff batter, and bake in gem pans, in a quick oven; or puffs may be made by taking one cupful of sweet milk, one egg and one cupful of Graham flour, and bake as above.

IX. Boston Brown Bread.

The rule is two parts of Indian meal to one of rye. Wheat flour is sometimes used, and sometimes wheat and rye. To three quarts of mixed meal, add a gill of molasses (not glucose syrup), two teaspoonfuls of salt, one of saleratus and a half-teacupful of good yeast. Water, or better, skimmed milk enough may be used to make a very thick batter. Put in a baking-pan covered, and set in a warm place to rise. When it cracks on the top, smooth it over with the wet hand and place it in the warm oven (not hot) until risen. Then bake with a brisk heat for three or four hours; it may even take five or six hours if the loaf is large, for the oven must not be very hot at any time. When baked let it cool in the oven, and serve either warm or cold. It is good toasted, as an accompaniment to vegetables at dinner, and may be used with butter, soup, or the gravy of meat. At breakfast it is often eaten with butter and syrup.

KNEADING-PAN.

X. Various Recipes for Bread.

Potato Bread.—Take six good-sized potatoes, boil and mash very fine. Add three pints of boiling water. Stir flour in till it makes a stiff batter. When lukewarm, add yeast; set it in a moderately warm place. In the morning add the salt and knead in flour as stiff as you can. Set in a warm place to rise; knead again, adding as little flour as possible. Let it rise again, and then put it into your pans, making them half full. When the loaves have risen to the top of the pans, bake them to a nice rown.

Corn and Rice Bread.—Take one pint of well-boiled rice, one pint of corn-meal, one ounce of butter, two eggs, one pint of sweet milk, two teaspoonfuls of baking powder. Beat the eggs very light, then add the milk and melted butter, beat the rice until perfectly smooth, and add to the eggs and milk. Lastly add the corn-meal. Beat all together until very light.

Kentucky Corn Bread.—To one and a half pints of corn-meal, use a pint of buttermilk, one egg, a small teaspoonful of soda, one of salt, and one tablespoonful of lard or butter. Mix thoroughly and bake in a quick oven.

Corn Pone.—Mix thoroughly five teacupfuls of corn-meal, two of Graham flour, one of New Orleans molasses, and two teaspoonfuls of soda with one quart of buttermilk. Put in a tin or porcelain kettle, buttered, and flaring at the top; fasten the top securely, plunge into another kettle so it is submerged, fill with boiling water and boil hard for six hours; then slip it from the kettle to a pan and bake slowly for two hours.

Corn Dodger.—Take a scant quart of meal, one teaspoonful of soda, half as much salt, and a pint of buttermilk; mix well and bake in a moderately heated oven.

Western Corn Bread.—Take one pint of meal, pour boiling water on it, and stir until it is about as thick as griddle-cakes. Add one small tablespoonful of salt, about three-fourths of a cup of brown sugar or molases; then stir in wheat flour until it is quite stiff. When cool enough, add one-half cup of lively yeast. Put it in a warm place to rise, and when light mould it into a loaf with flour, mixing as soft as possible; when light so it fills the basin, bake in a moderate oven two hours.

CORN-CAKE PANS.

Buttermilk Bread.—Take one quart of buttermilk, set it over the fire until it is scalded (but not boiling), then stir in flour enough to make it thick as for griddle-cakes, and set it in a warm place to rise; when it is light, put in one-half teaspoonful of soda, and as much salt, stir or knead it again and let it rise. When light, knead and put into loaves, and when it has come up again, bake.

Steamed Corn Bread.—One and a half cups of sweet milk, a tablespoonful of molasses, half a teaspoonful of soda, and a pinch of salt. Make a thick batter of one-third flour and two-thirds corn-meal. Steam an hour and a half in a basin.

WOOD ROLLING-PIN.

XI. Biscuits, Rolls, Gems, etc.

Breakfast Biscuit.—Take a piece of risen bread dough, and work into it one beaten egg and a teaspoonful of butter; when all is thoroughly worked together, flour your hands and make it into balls the size of an egg; rub a tin over with cream, put in the biscuits and set in a quick oven for twenty minutes, and serve hot for breakfast. When eaten, break them open—to cut would make them heavy.

Thin Biscuits, or Notions.—Take one pint of flour, and make into dough as soft as can be rolled, with sweet milk, a saltspoonful of salt, and two ounces of butter. Roll into large, round cakes, and of wafer-like thickness. Stick well with a fork. In baking do not allow them to brown, but remove from the oven while they retain their whiteness, yet are crisp, and will melt in the mouth.

Raised Biscuit.—One pint of milk, one egg, one gill of butter, half-pint, or less, of sugar, two potatoes baked quite dry and mashed through a colander. Mix together over night, with rather less than half a pint of yeast, and flour in proportion. In the morning mould them by hand, with as little flour as possible. These quantities will make three dozen biscuits.

Breakfast Puffs.—Take two eggs well beaten, and stir into a pint of milk, a little salt, a piece of butter, and a pint and a half of flour. Beat the eggs, and stir the milk. Add the salt, melt the butter, and stir in. Then pour all into the flour, so as not to have it lumpy. Stir up thoroughly, and butter the cups into which the batter is poured, filling them two-thirds full. Eat with sauce.

Bread Cake.—Three cups of dough, very light; three cups of sugar, one cup of butter, three eggs, a nutmeg, raisins, one teaspoonful of pearlash dissolved in a little hot water. Rub the butter and sugar together, add the eggs and spice, and mix all thoroughly with the dough. Beat it well and pour into the pans. It will do to bake immediately, but the cake will be lighter if it stands a short time to rise, before putting it into the oven.

Graham Gems.—Take one quart of sweet milk. Stir in Graham flour until the batter is a little thicker than for griddle-cakes. Add salt. Bake in gem-pans in a quick oven. The gems are better if the batter is stirred up an hour before needed. The above will make gems as light as can be made with baking powder. If you wish them very nice, add one egg and a tablespoonful of sugar. Make the gem-pans very hot before the batter is put in.

Coffee Bread.—Take two teacups of hop-yeast dough, and add two eggs well beaten up, three tablespoonfuls of sugar, two of butter. Mix them well. Roll out and place on a buttered tin. Spread a little butter on the top, sprinkle with white sugar, add cinnamon. Set to rise and bake quite slowly. This is nice for lunch, dipped in hot coffee and cream.

XII. Oatmeal Breakfast Cake.

TAKE partly cooked oatmeal, sold by all grocers, add water enough to saturate it, and a very little salt. Pour it into a baking tin half an inch or three-quarters deep, shake it down level, and when this is done it should be so wet that two or three spoonfuls of water should run freely on the surface. Put it in a quick oven and bake twenty minutes. Eat warm. It will be as light and tender as the best "Johnny cake," unless you have wet it too much or baked it too long. If you have only the ordinary oatmeal, that requires long cooking, it may be partly cooked the night before. Scarcely any wholesome thing in the bread line can be prepared more readily. It can be made still thinner and baked quicker. It is good, either crisp or moist. For emergencies every housekeeper will find it convenient to be able to make the breakfast cake. Many use partly cooked oatmeal mixed with buckwheat, wheat or corn-meal mush for griddle-cakes. Take one-half pint of the porridge or the mush, diffuse it in one quart of water and add the wheat or buckwheat meal, sifting it in and stirring slowly.

XIII. Rusks and Rolls.

THREE-FOURTHS of a pound of sugar; one-half pound of butter; one pint of sweet milk; five eggs; three and a half pounds of flour. Beat the eggs very light. put milk, sugar and butter together over the fire till the butter is melted; when cooled, add one-half pint of yeast, then the eggs and flour. Mix quickly and set to rise. Mould by hand in round cakes, about half an inch in thickness. The cakes should be placed in the pan in a double layer—one cake on top of another.

BAKE-PANS FOR ROLLS.

Split Rolls.—One egg well beaten; one tablespoonful of sugar; one yeast-cake dissolved in a cup of warm milk; two teaspoonfuls of salt; flour enough to make a stiff batter; set to rise; when risen, work in a large spoonful of butter, and flour enough to roll; roll out an inch thick, spread over with butter or lard; fold in half; cut with biscuit-cutter, and let it rise and bake.

Cinnamon Rolls.—Take some of the dough you make bread of. Work in shortening and sugar. Then make a paste of butter, sugar and cinnamon. Roll your dough out thin, spread in this paste and roll up, putting it in your pans. Let them stand until they become light, and bake. After they are done, eat them with your coffee or tea, just as you like.

CHAPTER V.

PASTRY AND PUDDINGS.

I. DIGESTIBLE PASTRY.—II. PIES FOR DYSPEPTICS.—III. MINCE PIES.—IV. RHUBARB PIE.—V. SOME EVERY-DAY PIES.—VI. TARTS AND TART CRUSTS.—VII. FRUIT SHORT-CAKE.—VIII. PUDDINGS AND THEIR SAUCES.—IX. DEVONSHIRE CREAM.—X. ENGLISH PLUM PUDDING.—XI. OATMEAL PUDDING OR PORRIDGE.—XII. FOUR PUDDINGS OF POTATOES.—XIII. BROWN BETTY.—XIV. SOME GOOD PUDDINGS.—XV. DUMPLINGS.—XVI. A HEN'S NEST AND THE SAUCE.—XVII. FRUIT PUDDINGS.—XVIII. PUDDINGS OF GRAIN.—XIX. MISCELLANEOUS PUDDINGS.—XX. CUSTARDS AND CREAMS—FROZEN CUSTARD.

I. Digestible Pastry.

Fine Puff Paste.—To make fine puff paste take one pound of flour, half a pound of butter and half a pound of lard. Cut the hard butter and lard in thin shelly pieces through the cold sifted flour. Mix the whole with enough ice-water to make it roll easily. There must be no kneading, and the warm hands should come as little in contact with the dough as possible.

Plain Paste.—This may be made with rather less butter and lard than the above. Mix all together, roll out into thin sheets and fold over and over into a roll. Cut from the end of this for your crust, and roll out to a proper thickness.

German Puff Paste.—Take one pound of butter and one of flour. Mix the butter into one-half of the flour, using a knife; mix the remainder of the flour with the yolk of one egg and half a cup of milk; no salt; roll it out and divide it into four parts. Then divide the other portion in four parts; roll out one of the quarters without shortening; place one of the quarters, which has the butter in, on it, and fold over; then roll out; repeat this three times; do the same with the other quarters. This is enough for eight pies, for covers; the under crust can be made of the following: Three cups flour, one cup lard, one-half cup butter, one-half cup water; stir lightly with a knife.

II. Pies for Dyspeptics.

Some persons will eat pies, when they know they disagree with them. If so, let them stick to the pies mentioned in this section, and if they must eat the crust, let it be made by taking equal quantities of Graham and white flour, wet with thin sweet cream, bake in a hot oven, as common pie-crust. Or take a piece of bread dough, after it has risen, and roll in a small piece of butter; roll out as pie-crust.

Pumpkin Pie.—Stew, sift, add as much boiling milk as will make it about one-third thicker than for common pumpkin pie; sweeten with sugar or molasses, bake in a hot oven. Or add rolled cracker or flour to the sifted pumpkin; add milk to the thickness of common pumpkin pie. Squash and sweet-potato pies are made in the same way.

Peach Pie.—Take small juicy peaches; fill the pie-dish: sprinkle sugar, a little flour, a tablespoonful of water; cover, and bake one hour.

Cranberry Pie.—Stew the cranberries, strain through a sieve, add sugar; bake on under crust.

Apple Pie.—This is made in the same way as cranberry pie, with cream crust. None of these will hurt weak stomachs if moderately indulged in.

III. Mince Pies.

EVERYBODY likes mince pie. The mince-meat is generally made in quantities to last five or six weeks, and is used as wanted. If it is to be rich in fruit, add to the raisins (which should be stoned and chopped) a few Zante currants and some citron sliced thin. The following are good formulas for preparing the filling for the pies:

SCALLOPED PIE-PLATE.

Pies with Cider.—Three pounds of good beef, lean and fat together; nine quarts of green apples quartered; three pounds of good raisins; nine cups of good hard cider, or five cups of good vinegar and four cups of water; six pounds of sugar, or twelve cups pressed full and rounded; one and one-half cups suet cut fine, or the same of butter; one and a half ounces of cinnamon and three-fourths ounce of cloves ground together. Put all into a kettle and simmer until well heated through, then pack into a jar for use.

Pies without Cider.—Take four pounds of boiled lean beef; one-half pound suet; four ounces cinnamon; two ounces mace or nutmeg; one ounce cloves; four pounds raisins; one pint molasses; one quart brandy; sugar to make it very sweet. To the above add an equal weight, nearly twelve pounds, of tart apples, chopped fine. This will keep for months. Before baking, add a tablespoon of strong cider vinegar to each pie.

We prefer making the mince-meat with sound cider; or put in half the prescribed quantity of boiled cider, and just before bringing the pies to the table, cut around the top crust with a sharp knife, remove it, and pour equally over the filling, a tablespoonful of brandy, or a wineglassful of wine to each pie.

IV. Rhubarb Pie.

PREPARE the stalks by peeling off the thin, reddish skin, and cutting in half or three-quarter inch pieces, which spread evenly in your crust-lined tins. Sift on a little flour, to which add a bit of butter and a teacup of sugar, if for a large pie. However, when it is desirable to economize sugar, or when a very sharp, sour taste is not relished, a pinch of soda may be used to advantage, with less sugar, as it goes far toward neutralizing the acid. If you live in a new country without fruit, raise a good patch of rhubarb, save all your surplus, prepare as for use, and dry in the sun, as stove heat turns it dark colored. Soak and stew for winter use, with sugar and soda as above for pies. It makes, also, a nice sauce for tea. All tart fruit pies may be made in the same manner as directed for rhubarb pies, simply varying the proportions of sugar according to the fruit, and omitting the flour.

V. Some Every-Day Pies.

Good Lemon Pie.—Take one lemon, one cup of water, one cup of brown sugar, two tablespoonfuls of flour, five eggs, three tablespoonfuls of white sugar.

OBLONG PIE-PLATE

Grate the rind of the lemon; squeeze out the juice, put all together and add the water, brown sugar and flour, working the mass into a smooth paste. Beat the eggs and mix with the paste, saving the whites of three of them. Make two pies, baking without top crust. While these are baking, beat the whites of the three eggs saved for that purpose to a stiff froth, and stir in the white sugar. When the pies are done spread this frosting evenly over them and set again in the oven and brown slightly.

Lemon Pie without Lemon.—Take one-half a teaspoonful of tartaric acid dissolved in half a cupful of cold water, half a teaspoonful of extract of lemon, one of sugar, yolk of egg, one soda cracker. After dissolving the acid stir the yolk and sugar together, and mix with the acid and water, then the extract, then cracker crumbled in. Bake in crust as for custard pie, and cover with the white of an egg and brown. Many prefer this to pie made with the lemon.

Raisin Pie.—Take one pound of raisins, turn over them one quart of boiling water and boil one hour. Keep adding water, so there will be a quart when done. Grate the rind of one lemon into one cup of sugar, three spoonfuls of flour and one egg. Mix well together. Turn the raisins over the mixture, stirring the while. This makes three pies.

Peach Pie.—Line a dish with a good crust. Then place in it a single layer of peaches, cut in halves; sprinkle sugar over them, and pour on enough sweet cream to fill the dish, and bake. Use no upper crust.

"Homely" Pie.—Take one cup of molasses, one cup of good vinegar, one cup of water, one small spoonful of extract of lemon and a piece of butter the size of a hen's egg. Let it all come to a boil and thicken it with corn starch. This makes two pies. Don't put on a top crust, but lay strips of the paste on, as there is danger of its foaming over.

Cream Pie.—Beat two eggs well, in a coffee-cup of sugar and one of thick sour cream. Stir until thoroughly mixed. Add a teaspoonful of extract of lemon or vanilla. Bake with two crusts. This quantity will make two pies.

Apple Custard Pie.—Peel sour apples and stew until soft, and there is not much water left in them. Then rub through a colander. Beat three eggs for each pie to be baked; and put in at the rate of one cup of butter and one of sugar for three pies. Season with nutmeg.

Plain Pumpkin Pie.—Cut the pumpkin in pieces of convenient size to handle. Grate with a common grater, and add milk enough to make it a little thinner than common stewed pumpkin. To enough pumpkin for three pies, add one egg. Season with cinnamon. Bake a little longer than if the pumpkin was stewed.

Potato Pie.—Take one cupful of mashed potatoes, one cupful sugar, half a cupful of butter, half a cupful of sweet cream, two eggs, flavor to the taste with nutmeg. Bake in an under crust.

VI. Tarts and Tart-Crusts.

The Crust.—Tarts may have a rather firm crust if they are to be filled with watery or semi-liquid material. For tarts of this kind a crust may be made of one cup of lard, one tablespoonful of white sugar, white of an egg, three tablespoonfuls of cold water, and flour added to knead stiff.

For jellies, etc., only the finest puff paste should be used, as directed for pies. In fact, however, no tart-form can be really good unless it is made of the best puff paste.

SCALLOPED PATTY-PAN.

Strawberry Tarts.—For strawberry tarts the crust should be made into a puff paste. Then make a syrup of one pound of sugar and one teacupful of water; add a little white of an egg, put it into a kettle, let it boil, and skim it until only a foam arises; then put in a quart of berries, free from hulls and stems; let them boil until they look clear, and the syrup is quite thick. This, put into the puff paste, when warm, makes a most delicious tart for tea. This recipe is also good for tarts of other acid fruits.

VII. Fruit Short-Cake.

Strawberry Short-Cake.—Make a nice paste for the crust; roll out in thin cakes about the size of a breakfast-plate; put in a layer of strawberries with a light sprinkle of sugar, then another cake of dough, another layer of strawberries and sugar, with a top layer of dough; bake it slowly in an oven or stove, and eat for lunch or dessert, with sugar and butter sauce. This is the simple way to make strawberry short-cake; any other acid berry may be treated in the same manner. The usual way for other fruits is to make them into tarts; but try cherries, stoned and made into a short-cake, and after the recipe for Grandmother's Short-cake.

Grandmother's Strawberry Short-Cake.—Take a coffee-cup of cream or sour milk, beat into it a little salt and a small teaspoonful of soda, and before it stops foaming stir in enough flour to enable you to roll it out, but be sure not to get it very stiff. Roll into three circles, spread butter on top of each, and place one on top of the other. Bake until well done, then pull the three layers apart, butter one and cover with strawberries, then butter the second and lay, crust downwards, over the first; then another layer of strawberries, and cover with the third crust. Set in the oven a few minutes, and then, before serving hot, with cream, cover the top crust with large fresh strawberries. Before making the crust, stir into three pints of ripe, rich strawberries a coffee-cup of granulated sugar, and leave it covered over until the crust is done. If cream or sour milk is not plenty, use sweet milk, and sift into the flour two teaspoonfuls (scant) of baking powder, and as you roll out spread on three tablespoonfuls of ice-cold butter. Pounded ice is excellent eaten on top of a saucer of sugared berries. Wrap the ice in a clean, coarse towel, and pound fine with a potato-masher.

Pineapple Short-Cake.—A couple of hours before bringing the cake on the table, take a very ripe, finely-flavored pineapple, peel it, cut it as thin as wafers, and sprinkle sugar over it liberally; then cover it close. For the short-cake take sufficient flour for one pie-dish, of butter the size of a small egg, a tablespoonful or two of sugar, the yolk of an egg, two teaspoonfuls of baking powder, a very little salt, and milk enough to make a very soft dough. Do not knead the dough, but just barely mix it, and press it into the pie-plate. The baking powder and butter, sugar and salt, should be rubbed well through the flour, and the other ingredients then quickly added. When time to serve, split the cake, spread the prepared pineapple between the layers, and serve with nothing but sugar and sweet cream.

VIII. Puddings and their Sauces.

PUDDINGS are either steamed, baked or boiled, and are made with and without fruits. Bread, flour, rice, tapioca, corn starch, and various other gelatinous substances are used. Pudding sauces are made hard or liquid, and generally somewhat acid.

White Sauce.—White sauce may be made with the whites of two unbeaten eggs, and one cupful of white sugar, beaten together; add a teaspoonful of white wine vinegar or other light-colored vinegar; beat well; just before carrying to the table, add two-thirds of a cupful of cream and a tablespoonful of wine.

Wine Sauce.—This is made of three measures of sugar, one of butter and one of wine.

Boiled Pudding Sauce.—The following is good: Beat a coffee-cupful of sugar and one of butter thoroughly together. Then add a whisked egg. Mix well, place on the fire and stir until melted; add a tablespoonful of wine or brandy, and serve at once.

Plain Sauce.—To each wineglassful of thick paste, made of corn starch, add a teacupful of butter and one of sugar. Work these together, with the yolk of an egg, until thoroughly blended. Add the paste and the white of the egg beaten to a froth. Mix thoroughly, and add any flavoring you like.

IX. Devonshire Cream.

An excellent dish, to be eaten with puddings, some pies, fruits, etc., and one of the noted luxuries of the West of England, is "Devonshire cream," or "clotted cream." It is prepared as follows: From six to eight quarts of milk are strained into a thick earthenware pan or crock, which, when new, is prepared for use by standing in clear cold water for several days, and then scalded three or four times with skimmed milk. Tin pans may be used if they are scalded in hot bran and left to stand with the bran in them for twenty-four hours. The milk being strained into the pans is set in a cool room, from nine to fourteen hours, according to the temperature. It is then carefully moved to the top of the stove or range, or placed over a bright fire, not too near it, and slowly heated, so that at the end of half an hour the cream will have shrunken away from the sides of the pan and gathered into large wrinkles, the milk at the sides of the pan commencing to simmer. The pan is then carefully returned to the cool-room and left about ten hours. Then the cream is skimmed off. This cream is very delicious to use on fruit or preserves, and is esteemed a great luxury—selling for about the price per pound of the best butter.

X. English Plum Pudding.

The ingredients are: One and a half pounds suet, one and a half pounds of dry light brown sugar, one and a half pounds currants, washed and dried thoroughly, one and a half pounds raisins, four nutmegs sifted through a small tea-strainer and thoroughly mixed, so they will not be lumpy; one-quarter pound candied lemon-peel, one-quarter pound citron, a heaping spoonful of fine salt, mixed in the same way as the nutmegs, baker's bread enough to make a quantity equal in bulk to the suet. Use only the crumb of the loaf, rejecting the crust, (it will take nearly a loaf and a half of ordinary size) a half pint of flour, nine eggs beaten very light, and milk enough to wet the mixture. Chop the suet first, then add the bread-crumbs, sliced citron and peel, raisins and currants. Then sift the salt and nutmegs in, stirring thoroughly. Then add the sugar, and next sieve the flour in. Then pour in the eggs, mixing thoroughly as before. You only need sufficient milk to moderately moisten the pudding. Butter your tin basin well, put in your pudding, only leaving room for a stiff batter of flour and water which must be spread over the whole top of the pudding to exclude the air and water. Then take stout, unbleached cotton, tie it firmly over the top, round the rim of the basin, and bring the corners that hang down back again over the top, pinning them securely. Put the pudding into boiling water, tied in a pudding bag, and let it boil steadily at least ten hours. The best way is to make a pudding, in cool weather, two or three days before needed, and then put on again the day it is to be eaten, and boil three to four hours. Use cold sauce made of sugar, butter and wine, or hot brandy sauce.

XI. Oatmeal Pudding or Porridge.

Oatmeal mush, like corn-meal, requires long boiling to cook it fully; but it is

FARINA AND PORRIDGE BOILER.

now prepared by partly cooking at the manufactories, so that from five to fifteen minutes' boiling makes it ready for the table. It should be cooked in a porcelain or enameled pipkin. If the raw oatmeal is used, cook as directed: Have the pipkin two-thirds full of boiling water, into which put a half teaspoonful of salt. Into this drop the oatmeal with one hand, stirring with a wooden spatula held by the other. When it is the thickness of mush, cover it and set it where it will keep boiling slowly for an hour, beating it up occasionally to keep it well mixed, and free from lumps. Dish and eat it hot, with cold milk or cream. Butter and sugar melted upon it destroy its fine diuretic qualities, and make it really less palatable. Porridge, gruel, thin cakes and a sort of crackers, are the principal methods of using oatmeal. As a breakfast dish, the porridge made in the way described above has no superior. It stimulates the action of the liver, and, in conjunction with cranberries, eaten with a sauce, will restore a torpid liver to healthful activity, if used for the morning meal, to the exclusion of fried meats, broiled ham, and the like.

XII. Four Puddings of Potatoes.

1. Mix together twelve ounces of boiled mashed potatoes, one ounce suet, one ounce (one-sixteenth of a pint) of milk, and one ounce of cheese. The suet and cheese to be melted or chopped as fine as possible. Add as much hot water as will convert the whole into a tolerably stiff mass; then bake it for a short time in an earthen dish, either in front of the fire or in an oven.

DEEP PUDDING-PANS.

2. Twelve ounces of mashed potatoes, one ounce of milk, and one ounce of suet, with salt. Mix and bake as before.

3. Twelve ounces of mashed potatoes, one ounce of suet, one ounce of red herring, chopped fine or bruised in a mortar. Mix and bake.

4. Twelve ounces of mashed potatoes, one ounce of suet, and one ounce of hung beef, grated or chopped fine. Mix and bake.

XIII. Brown Betty.

One cup of bread-crumbs, two cups of chopped tart apples, half a cupful of sugar, one teaspoonful of cinnamon, and two teaspoonfuls of butter cut into small pieces. Butter a deep dish, and put a layer of the chopped apples at the bottom; sprinkle with sugar, a few bits of butter and cinnamon; cover with bread-crumbs, then more apple. Proceed in this order until the dish is full, having a layer of crumbs at the top. Cover closely, and steam three-quarters of an hour in a moderate oven;

then uncover and brown quickly. Eat warm with sugar and cream or sweet sauce. Serve in the dish in which it is baked.

XIV. Some Good Puddings.

Suet Pudding.—Take one teacupful of chopped suet, one of sour milk, one of molasses; also a teaspoonful of saleratus. Add flour to make it stiff. Use one teacupful of raisins, one of currants, one teaspoonful of each kind of spice, and three eggs. Boil three hours.

Steamed Pudding.—Take one cupful of suet, chopped very fine, and one cupful of molasses, one cupful of sour milk, one teaspoonful of cinnamon, half a teaspoonful of cloves, one teaspoonful of soda; beat up three eggs, one cupful of chopped raisins, four cupfuls of flour. Steam two hours. Eat with sugar and butter, or sugar and milk.

Quaker Pudding.—Lay slices of light bread, cut thin and spread with butter, in a pudding-dish, alternating the layers of bread with raisins until near the top. Beat five eggs up well, and add to them a quart of milk, salted and spiced according to taste. Pour this liquid over the contents of the dish. Bake the pudding half an hour, and eat with sweet sauce. It will be necessary to boil the raisins in a very little water so as to make them tender, and add the water with the rest.

Yorkshire Pudding.—Beat up four eggs, nine tablespoonfuls of flour, one pint of milk, and half a teaspoonful of salt. Put it in the pan under beef which is being roasted. Bake half an hour. Serve with the beef.

Hominy Pudding.—Prepare as for batter cakes, adding one egg to each pint, some cinnamon, a few raisins, sugar to suit the taste. Then bake just as you would rice pudding. A little butter or chopped suet can be added. Serve hot, with or without sauce.

"Every-Day Pudding."—Take half a loaf of stale bread soaked in a quart of milk; four eggs, four tablespoonfuls of flour; a little fruit, dried or fresh, is a great addition. Steam or boil three-fourths of an hour. Serve with the following sauce: Butter, sugar and water, thickened with a little corn-starch, and flavored with lemon extract or lemon juice and rind.

XV. Dumplings.

Oxford Dumplings.—Mix well together two ounces of grated bread, four ounces of currants, four ounces shred suet, a tablespoonful sifted sugar, a little allspice, and plenty of grated lemon-peel. Beat up well two eggs, add a little milk, and divide the mixture into five dumplings. Fry them in butter to a light-brown color, and serve them with wine sauce.

The following is one form of dumpling for apples or other fruit, and is easily prepared. The proportions are: One quart of flour, one egg, one teaspoonful of soda, one pint of buttermilk, or enough to mix your flour, and a little salt. Have ready plenty of boiling water. Roll your dough about three-fourths of an inch thick, cut out as for biscuit, and drop into the boiling water. Boil ten minutes. Do not let them remain after they are done, or they will fall, on taking them from the water. Have apple-sauce or other fruit previously prepared, and spread it on the dumplings in your plate and pour sweetened cream over it.

Apple Dumplings.—Pare and core as many apples as you wish and enclose them in puff paste, after putting half a clove and a little lemon-peel, and a trifle of mace, into the hole made by taking out the core. Wrap in bits of linen, or put into a net, and boil an hour. Before serving cut out a plug of the paste, put a teaspoonful of sugar and a little butter inside and replace the plug. Strew powdered sugar over all and serve with sauce.

XVI. A Hen's Nest, and the Sauce.

TAKE half a dozen eggs, make a hole at one end and empty the shells, fill them with blanc-mange; when stiff and cold take off the shells; pare lemon-rind very thin, boil in water until tender, then cut in thin strips to resemble straw, and preserve in sugar; fill a deep dish half full of jelly or cold custard, put the eggs in and lay the straws, nest-like, around them. To make sauce for the pudding, take one cupful of butter, one cupful of sugar, yolk of one egg; beat together and stir in one cup of boiling water. Let it come to a boil, and when ready for use, flavor to taste.

A nice dish to go with this is made by filling coffee-cups loosely with strawberries, and pouring over them Graham flour mush, or instead, thicken sweet boiling milk to a consistency which is thin enough to fill the interstices between the berries, and yet thick enough to be firm when cool. Turn out and serve up with cream and sugar.

XVII. Fruit Puddings.

MAKE a crust of Graham flour, sour cream, soda, and a pinch of salt. Pass the flour through a coarse sieve, so as to relieve it of the coarser bits of bran. For a family of six persons line a quart basin with the crust, a quarter of an inch thick. Fill the basin thus lined, with fruit—plums or peaches are best. Let the fruit be of the choicest variety. Cover the whole with a rather thick crust, and steam until the crust is thoroughly cooked. Serve with white sugar and thick, sweet cream. This has been called Queen of puddings, and can be eaten with a comparatively clear conscience.

Apple Tapioca Pudding.—One coffee-cupful tapioca, covered with three pints of cold water, and soaked over night. In the morning set it on the side of the range, or stove, stirring it often until it becomes transparent. If too thick, add more water, until it is as thin as good, clear starch. Stir in a small teaspoonful of salt. Pare and core, without breaking, as many good apples as will lie close on the bottom of a medium-sized pudding-dish. Fill the holes full of sugar, and a very little nutmeg and cinnamon; then pour over the tapioca, and bake slowly until the apples are soft and well done. To be eaten with hard sauce, which is made as follows: One cup sugar, two-thirds of a cup butter, beaten together until perfectly smooth and white.

Boiled Grape Pudding.—Pare rich, tart apples, and cut to the size of a chestnut by cutting each quarter in four pieces, and add an equal measure of grapes, say one pint of each, and stir into it two spoonfuls of wheat meal. Then make a scalded wheat-meal crust, roll to one-third of an inch thick, place in it the prepared fruit, close it over the fruit, sew up in a napkin, put into boiling water and boil an hour. Grape dumplings may be made with the same materials; wrapping up half a teacupful of the fruit in a crust, and, for convenience, placing it in a patty-pan, and setting in the steamer. Cook until the apples are rather soft. Serve warm with sauce.

Plain Apple Pudding.—Pare, quarter and core apples to fill a small dish rather more than half, and pour in water two inches deep. Make a crust of one pint of flour, one-half teaspoonful of salt, and baking powder enough to make it light. Add a level teaspoonful of lard, and flour enough to make a wet dough, and roll out quickly, put over the pudding-dish, and set on a hot stove. Cover tightly with a tin cover, on which put a flat-iron. The steam produced cooks the pudding quickly. Fifteen minutes will be found long enough. Serve hot, with hard sauce made of butter and sugar.

Sweet-Apple Pudding.—One pint of scalded milk, half a pint of Indian meal, one small teacupful of finely-chopped suet, two teaspoonfuls of salt, six sweet apples cut in small pieces, one great-spoon of molasses, half a teaspoonful of ginger, nutmeg or cinnamon—whichever is most desirable—two eggs well beaten, and half a teaspoonful of soda. Beat all well together, put into a pudding-mould, and boil two hours.

Dried-Peach Pudding.—Cut in small pieces one pint of dried peaches, wash them, and boil in just enough water to cover them. When they are tender, add two tablespoonfuls of brown sugar, and boil a few minutes longer, and then they will resemble cooked raisins. Make a stiff batter of three eggs, one tablespoonful of butter, one teacupful of sweet milk, one teaspoonful of soda and two of cream of tartar, sifted in a quart of flour. You may not have to use all the flour—just enough to make a stiff batter. Stir your peaches in the batter and bake in a buttered pan, and you will have a delicious pudding, which no one can tell from one made of raisins. Any other dried fruit may be used in the same way. Serve with butter sauce

Fig Pudding.—Take one pound of figs, six ounces of suet, three-quarters of a pound of flour, and milk. Chop the suet finely, mix it with the flour, and make into a smooth paste with milk. Roll it out about half an inch thick, cut the figs in small pieces, and stew them over the paste. Roll it up, make the ends secure, tie the pudding in a cloth, and boil from one and a half to two hours.

Cherry Pudding.—A nice pudding can be made by boiling one-half pint of rice half an hour in five times as much water, and pouring it boiling hot into one pint of wheat meal. Mix thoroughly, and place it in small spoonfuls in a nappy—a round earthen dish with flat bottom and sloping sides—interlaying it with a pint of cherries. Steam half or three-quarters of an hour. Serve warm, trimming it with melted sugar, or sweetened cherry-juice, or some other sweet sauce. This recipe can be used for such other small fruits in their season as will bear cooking enough to do the wheat-meal. Half an hour is the least that will answer for that purpose; three-quarters of an hour is better.

Apple Souffle.—Stew the apples and add a little grated lemon peel and juice, omitting butter; line the sides and bottom of a baking-dish with them. Make a boiled custard with one pint of milk and two eggs, flavoring with lemon and sweetening it to taste. Let it cool and then pour into the center of the dish. Beat the whites of two eggs to a stiff froth, spread them over the top; sprinkle white sugar all over them, and brown in the oven. The stewed apple should be about half an inch thick on the bottom and sides of the pudding-dish.

OVAL PUDDING-PAN.

Bird's Nest Pudding.—Take sour apples, peel, quarter and core enough to cover the bottom of a common square tin. Make a batter of one cup of buttermilk, one-half cupful of cream, two eggs, a little salt, one teaspoonful of soda, and flour enough to thicken about like fritters. Pour this over the apples and bake in a quick oven. Eat while hot, with butter or cream sauce.

XVIII. Puddings of Grain, etc.

Rice Pudding.—Rice pudding is eaten by everybody, even the most delicate. A good way to make it is as follows: In a quart bowl, take two eggs and two heaping tablespoonfuls of sugar, well beaten together; fill the bowl half full of cooked rice, bits of butter, and a handful of raisins; stir all well together, and then fill the bowl with new milk. After the ingredients are thoroughly mixed, bake in a hot oven half an hour. When set away to cool, take a spoon and stir it up, so as to mix in the melted butter on the top and the raisins in the bottom. Eat with cream, slightly sweetened. Season with nutmeg, or whatever you like.

Rice Pudding without Eggs.—A pudding without eggs can be made by taking one cup of rice to one-half gallon milk and one cup of sugar. Bake until the rice is done. Flavor to your taste.

Corn Pudding.—Take canned corn (in the season, green corn scraped from the cob) and add to one can of corn a quart of cold milk, three eggs well beaten, two tablespoonfuls of sugar and one teaspoonful of salt. If not sweet enough, add sugar; if too thick, more milk. Pour this into buttered dishes and bake. It is delicious for tea.

Rizena Pudding.—Rizena is a food preparation of rice. A pudding of this is made by mixing four large spoonfuls of rizena with half a pint of cold milk, and stir it into a quart of boiling milk until it boils again; then remove, stir in butter the size of an egg and a little salt; let it cool, and add four eggs, well beaten, two-thirds of a cup of white sugar, grated nutmeg, and half a wineglassful of brandy, or other flavoring if preferred; bake in a buttered dish twenty minutes. To be eaten hot, with sauce. It can hardly be said to be superior to rice.

XIX. Miscellaneous Puddings.

Floating Island.—Take the yolks of seven eggs to one quart of milk, one cupful of sugar, a little salt, and flavor with lemon. Beat all together, and set in a kettle of water other than the kettle it is boiled in. Beat the whites of the eggs to a stiff froth, and pile in heaps on top of the boiled milk, after it has been put in the float glasses. This will make twelve glasses full. They look very pretty set in a circle round a bouquet of flowers, in the center or at each end of the table.

Charlotte Russe.—Take a box of sparkling gelatine and pour on it a scant pint and a half of cold water; when it has stood ten minutes add the same quantity of boiling water, and stir until the gelatine is dissolved; stir in half a pound of white sugar; have ready six eggs, well beaten separately, and then together, and when the jelly is cool, but not congealed, beat it into the eggs; whip very lightly three pints of rich cream, flavored with vanilla or almond, or both, and when the eggs and jelly begin to congeal, beat it in as rapidly as possible, and pour the mixture in a bowl lined with lady-fingers or sponge-cake

CHARLOTTE RUSSE PAN.

Spice Pudding.—Take one cupful of butter, one cupful of molasses, and one cupful of sweet milk, three cupfuls of flour, one teaspoonful of ground cloves, one of cinnamon, half a teaspoonful of allspice, one teaspoonful of soda, one egg, and plenty of raisins. Steam three hours. A liquid sauce for spice pudding is made by taking six tablespoonfuls of sugar, four tablespoonfuls of butter, two tablespoonfuls of vinegar, one tablespoonful of flour, ten of boiling water, and a small lump of tartaric acid; flavor with lemon. Mix thoroughly and boil.

Delicate Pudding.—Take one quart of milk, and while boiling stir in one pint of sifted flour, six eggs, six tablespoonfuls of white sugar, one spoonful of butter, the grated peel and juice of two lemons. All the ingredients must be well beaten together before they are stirred into the milk. Stir one way without stopping for a minute or two, take it off, and turn into your pudding-dish. It is to be eaten cold, with sugar and cream if you like.

Orange Pudding.—Take four fair-sized oranges, peel, seed, and cut in small pieces. Add one cup of sugar, and let it stand. Into one cup of nearly boiling milk stir two tabespoonfuls of corn starch, mixed with a little water and the yolks of three eggs. When done, let it cool, and mix with the orange. Make a frosting of the whites of the eggs and half a cup of sugar. Spread over the top of the pudding, and put it into the oven for a few moments to brown.

Eve's Pudding.—Take half a pound of apples, half a pound of bread-crumbs, a pint of milk, half a pound of currants, six ounces of sugar, two eggs, and the grated rind of a lemon. Chop the apple small; add the bread-crumbs, currants, sugar and lemon-peel, then the eggs, well beaten; boil it three hours, in a buttered mould, and serve with sweet sauce.

Bachelor's Pudding.—Three eggs, well beaten, (the white of one beaten separately until firm enough to cut with a knife), two teacupfuls of milk, one teacupful of sugar, one soda cracker broken in six pieces, a slice of peeled orange laid on each piece and sprinkled with sugar; put them in the dish and they will float; bake in a very hot oven, and, when half done, put a spoonful of the white of beaten egg on each piece; return to the oven and bake five minutes, and you have a splendid dish.

Cocoanut Bread Pudding.—Boil one quart of milk submerged in a boiler. When hot, add a teacupful of grated cocoanut, and boil two hours. Add a cup of bread-crumbs, two eggs well beaten, and half a cup of sugar. Currants or raisins may be added. Boil one hour, and eat cold.

XX. Custards and Creams.

Frozen Custard.—This is a nice dish for dessert, and very easily prepared: Boil two quarts of rich milk. Beat eight eggs and a teacupful of sugar together, and after the milk has boiled, pour it over the eggs and sugar, stirring all the while. Pour the whole mixture into your kettle, and let it come to a boil, stirring it constantly. Then take it off the fire, and let it become cold. Flavor it with whatever essence you prefer. Then freeze it.

Chocolate Custard.—Scrape half a cake of good chocolate, and put it into a stew-pan, and moisten by degrees with a pint of warm milk and cream; when well dissolved, mix with the yolks of eggs, and finish the same as for other custards.

Bohemian Cream.—Take four ounces of any kind of fruit, stone it, and sweeten. Pass it through a sieve, adding one ounce and a half of melted or dissolved isinglass to each half pint of fruit. Mix well, then whip a pint of rich cream, and add the isinglass and fruit gradually to it. Pour all into a mould, set it on ice or where it is very cool, and when set, dip the mould a moment into water, and then turn it out ready for the table.

Whipped Cream.—Sweeten one pint of sweet cream, and add essence of lemon. Beat up the whites of four eggs until they are very light, adding them to the cream. Whip both together. As fast as the froth rises, skim it off, put in glasses, and continue until they are full.

CHAPTER VI.

CAKE-MAKING.

I. CAKE AN ECONOMICAL FOOD.—II. GENERAL RULES FOR MAKING CAKE.—III. ICING, GLAZING AND ORNAMENTING.—IV. RECIPES FOR FROSTING.—V. ORNAMENTING CAKE.—VI. SPECIAL PREPARATIONS.—VII. FRUIT CAKE, DARK.—VIII. RICH POUND-CAKE.—IX. MISCELLANEOUS CAKES.—X. MORE GOOD CAKES.—XI. GINGERBREAD AND OTHER "HOMELY" CAKES.—XII. THE HOUSE-WIFE'S TABLE OF EQUIVALENTS.

I. Cake an Economical Food.

WE have known persons begrudge their families cake, on the ground that it was expensive. This is a mistake. Milk, eggs, butter, flour and sugar are the ingredients of most cakes. The three first all farmers have, or should have plenty of, and sugar is no longer costly. As a condensed food, cake is cheaper than the best beefsteak, even in the country, and more than twice as nourishing. Eggs are more nourishing, pound for pound, than fresh meat, and a quart of good milk contains as much nourishment as a pound of fresh beef.

II. General Rules for Making Cake.

THERE are some general rules for making cake that must be observed:

1. The ingredients must be of the best, for the best are most economical.
2. Never allow butter to get oily before mixing it in the cake.
3. Always have an earthen or other enameled dish to mix and work the materials for cake. Tin, if not new, is apt to discolor the material. Remember that egg will tarnish even silver. Hence always use a clean wooden spoon.
4. As a rule, in mixing cake, first beat the sugar and butter together to a cream; then add the yolks of the eggs. If spices or liquors are used, these come in with the yolks of eggs; then comes milk; and last, the thoroughly whisked whites of the eggs and the flavor. If fruit is a portion, this is put in with the flour.

BEATING BOWL.

5. For small cakes the oven should be pretty hot; for larger cakes only moderately so. If a broom-straw, pushed through the thick part of the cake, comes out clean and free from dough, the cake is done.

6. When you take the cake from the oven, do not remove from the pans until it is somewhat cool—not sooner than fifteen minutes. When you take it from the pans, do not turn it over; set it down on a clean cloth, on its bottom, and cover with another clean cloth.

These directions have as many parts as an old-fashioned sermon. Fortunately they are not so long.

III. Icing, Glazing and Ornamenting.

1. A GLAZED shallow earthen dish should be used in making the icing.
2. Allow a full quarter of a pound, or more, of the finest white sugar to the white of each egg.
3. Lemon-juice and tartaric acid whiten the icing. If used, more sugar will be required.
4. Sprinkle the egg with part of the sugar, and beat, adding more sugar from time to time. If you use flavoring, add it last.
5. Dredge the cake thoroughly with flour after it is baked; then wipe it carefully before icing or frosting. It will then spread more kindly.
6. Put the frosting on in large spoonfuls. Begin in the center and spread with a thin-bladed knife or spatula, dipped from time to time in ice-water.
7. Let the frosting dry in a cool place.

IV. Recipes for Frosting.

TAKE the whites of eight eggs; beat to a stiff and perfect froth. Add pulverized white sugar, two pounds; starch, one tablespoonful; pulverized gum Arabic, one-half ounce, and the juice of a lemon. Sift sugar, starch and gum Arabic into the beaten eggs, and stir until perfectly firm.

Beat the white of an egg until you can turn the plate over without the egg running off, then add five heaping tablespoonfuls of pulverized sugar and one of starch. This quantity will frost one small cake. Flavor to taste.

Glazing.—Put into a porcelain or other glazed vessel, with a little water, the white of one egg well beaten, and stirred well into the water; let it boil, and whilst boiling, throw in a few drops of cold water. Then stir in a cupful of pounded sugar. This must boil to a foam, then be used; this makes a nice glace for cakes.

V. Ornamenting Cakes.

FOR figures or flowers, beat up two eggs, reserving a third (white) till the cake has become dry after icing. Then insert a clean glass syringe into the remainder, and direct as you choose over the iced cake. Dry again. Ripe fruit may be laid on the icing when about half dry, with a very pretty effect, such as berries, etc. Save a little icing out, dilute with rosewater, and put on when that first done is dry. It gives a smooth, gloss.

The ornamentation may be colored pink by mixing a very little carmine or strawberry juice in the egg. The yellow rind of lemons, put in a bag and squeezed hard into the icing will give a yellow. So will a little butter-color, or preparation of anatto. For raised figures formed of frosting, a cone of strong white paper, rolled, with a proper orifice at the bottom, answers well, since it may be held upright, and easily directed to make the desired figures.

VI. Special Preparations.

CHOCOLATE preparation is made as directed for other frosting, with the whites of two eggs, one and a half cupfuls of best white ground sugar, six tablespoonfuls of grated chocolate and two tablespoonfuls of vanilla. Spread between the layers and on top of the cake, and serve while fresh, or when not more than one day old.

Ice-Cream Icing for Cake.—This is used for white cake: Take two cups of white sugar boiled to a thick syrup; add three teaspoonfuls of vanilla, and when cool, the whites of three eggs beaten to a froth; flavor with two teaspoonfuls of citric acid.

OCTAGON CAKE-MOULD.

TURK'S-HEAD CAKE-MOULD.

VII. Fruit Cake—Dark.

THE quantity of fruit is according to how rich the cake is to be made. The proportions for a rich, dark cake may be: Two pounds of raisins (stoned), two pounds of currants, one pound of almonds (blanched), one pound of citron or candied peel and fruit, one pound of moist sugar, one pound of butter, one pound of flour, one dozen eggs, one teaspoonful of mace, one tablespoonful of cinnamon, one nutmeg, one wineglassful of brandy, and one of wine. The fruit should be cut up rather coarse, and the almonds in not more than three pieces. Roll the fruit in flour to separate it, reserving some almonds and citron for sticking in the top of the cake, but entirely out of sight. Beat the fruit into the eggs after they have been perfectly whisked; also the butter and sugar after they are creamed together. Let the rest of the flour be lightly stirred in just before putting the cake to rise. Put embers under it, and let it rise for three hours. Bake slowly for three hours, or until, by trying with a straw, you find it quite done. When taken from the oven, let the cake stand in the pan at least two hours, or if it is very large, leave it in a warm place all night. It will then be ready for frosting, and will keep indefinitely in a dark, cool place. The pan in which it is baked must be lined with buttered white paper. The white paper is also used for pound cake.

VIII. Rich Pound-Cake.

Take one pound each of white sugar, butter, and flour; ten eggs, a wineglassful of brandy, half a nutmeg, and a teaspoonful of vanilla or essence of lemon. Beat the sugar and butter to a cream, whisk the eggs to a froth, and beat all the ingredients together until perfectly light. Bake in a moderately heated oven an hour. Turn the cake out of the tin, invert it, and set the cake on the bottom to cool. Put on the frosting when cold.

Cocoanut Pound-Cake.—This is made with one pound of sugar, half a pound of butter, one teacupful of fresh milk, one pound of flour, one cocoanut grated, four eggs, the peel of half a lemon grated, or half a teaspoonful of essence of lemon, and a teaspoonful of carbonate of soda. Make as directed for pound-cake, but put in the cocoanut last. Bake in buttered tins, the cake-batter being put in an inch deep. The heat should be rather quick, and the cake is to be iced as directed for fruit cake.

SPONGE-CAKE PANS.

DEEP JELLY-CAKE PANS.

IX. Miscellaneous Cakes.

Roll Jelly Cake.—Take one cupful of white sugar, one-half teacupful of sweet milk, two eggs, one cupful of flour, two teaspoonfuls of cream of tartar, one-fourth of a teaspoonful saleratus, a pinch of salt, and such flavoring as you like. This will make two cakes in a square tin. Have the oven ready, put the cakes in, and while they are baking, get a cloth and the jelly ready on the table. As soon as they are baked, take them out and turn them one at a time on the cloth, spread quickly with jelly or marmalade, and roll up tightly in the cloth and lay them where they will cool. Handle them carefully or they may fall. Cut them with a sharp knife in slices.

Sponge Cake.—Take one pound of granulated sugar beaten with the yolks of ten eggs. Grate into this the yellow rind of two lemons, and add the juice of one; then beat the whites of the ten eggs separately, very light, and add the same, stirring lightly together. To this add three-fourths of a pound of flour, and stir lightly without beating. This will make three good-sized loaves. Care must be taken in baking not to put the pans in too hot an oven.

French Cream Cake.—Take three eggs, one cupful of sugar, one and a half cupfuls of flour, one teaspoonful of baking powder, and two tablespoonfuls of cold water. This is enough for two pans. Split the cakes while warm and spread the custard while hot between them.

To make the custard, boil nearly one pint of sweet milk. Take two tablespoonfuls of corn starch. Beat up with a little milk to this. Add two well-beaten eggs.

When the milk has boiled up stir this in slowly with nearly a teacupful of sugar. When almost done add half a cupful of butter and flavor to taste.

Delicate Cake.—Take one and one-half cupfuls of white sugar, half a cupful of butter; rub these to a cream. Add half a cupful of sweet milk, in which dissolve half a teaspoonful of soda, and two cupfuls of flour, in which rub one teaspoonful of cream of tartar; add a little salt and flavor with vanilla or lemon. Beat the whites of four eggs to a stiff froth, and add last. Bake slowly an hour in a moderate oven. This recipe will make a two-quart basin loaf, and if the proportions are followed exactly, a beautiful cake will be the result.

Marble Cake.—The white part is made with one-half cupful of white sugar, one-half cupful of butter, half a cupful of sweet milk; whites of four eggs, two and one-half cupfuls of flour, one teaspoonful of baking powder. Flavor with lemon.

For the spiced part take one cupful of brown sugar, one-half cupful of molasses, one-half cupful of butter, one cupful of sour milk. Take the yolks of five eggs, and the white of one egg, two cupfuls of flour, one teaspoonful of cinnamon, one of cloves, one nutmeg, one-half teaspoonful of soda.

Orange Cake.—Use one cupful of butter, one of sweet milk, two of sugar, two teaspoonfuls of baking powder, five eggs, reserving the whites of three, to be beaten to a stiff froth to go between the cakes. The remainder of the five eggs must go in the batter. Three and a half cupfuls of flour; grate two oranges (picking out the seeds and large pieces) into the batter. Take two cupfuls of pulverized sugar, beat with the reserved whites as frosting; then put between cakes as you would jelly cake.

Drop Cakes.—Take one pound of flour, half a pound of sugar, half a pound of butter and three eggs. Beat the butter and flour to a cream, beat the eggs separately, add the yolks and part of the flour, then the whites and the remainder of the flour. Stir in half a pound of currants, a quarter of a pound of citron, and a teaspoonful of mace or cinnamon. Drop with a spoon upon flat tins, and sift sugar over them.

Lady Fingers.—Beat the yolks of four eggs with a quarter of a pound of sugar until smooth and light; whisk the whites of the eggs and add to these, and sift in a quarter of a pound of flour. Make into a smooth paste, and lay it on buttered paper, in the size and shape the cakes are required. Bake quickly. While hot, press two of the cakes into one on the flat side.

Newport Cake.—Sift one quart of flour; add three eggs, three tablespoonfuls of white sugar, three of butter, two teaspoonfuls of cream tartar, one of soda, one cupful of sweet milk, or sufficient to make a stiff batter. Bake quick and eat warm or cold. This is a superior tea cake.

A Nice Cake.—Take two and one-half pounds of flour, one and one-fourth pounds of pulverized white sugar, ten ounces of fresh butter, five eggs, well beaten, one-eighth ounce of carbonate of ammonia, one pint of water; milk is better if you have it. Roll out, cut into cakes and bake. While yet hot, dredge over with coarse sugar.

X. More Good Cakes.

Children's Party Cakes.—Take three heaping tablespoonfuls of powdered sugar, two of butter, one of maizena or corn starch, one egg; put with this two cupfuls of flour, half a cupful of sweet milk, a teaspoonful of cream of tartar, half a teaspoonful of soda, a pinch of salt, and Zante currants. Roll this out in powdered sugar, cut the dough in strips, and twist round a thimble-sized pin. Sprinkle over this candied caraway-seeds, and bake in a brisk oven on flat tins. These are called children's party cakes, and also "goody" cakes.

Scotch Cake.—Take one pound of fine flour, a half-pound of fresh butter, a half-pound of finely sifted loaf sugar; mix well in a paste, roll out an inch thick in a square shape, pinch the edges so as to form small points; ornament with comfits and orange-peel chips; bake in a quick oven until of a pale lemon color.

Rice Cake.—Take about four ounces of rice flour, sift three ounces of wheat flour into it, add eight ounces of granulated sugar, the rind of a lemon grated fine, six eggs, using all the yolks, and but half the whites. Beat the whole together for about twenty minutes, and bake about three-quarters of an hour.

CAKE-CUTTER.

Tea Cake.—Break one egg into a teacup. Fill the cup with sweet milk. One cupful of sugar, one-half cupful of butter, a little nutmeg, one teaspoonful each of saleratus and cream of tartar. Flour to make it the consistency of common sponge cake.

Christmas Cake.—Two eggs, one-half cupful of butter, one cupful of molasses, one cupful of raisins, two cupfuls of flour, and various spices. Mix and bake in a rather brisk oven.

Taylor Cake.—Two and one-half cupfuls of flour, one and one-half cupfuls of sugar, one-half cupful of butter, one-half cupful of milk, one egg, one-half teaspoonful of soda, and with or without fruit.

Silver Cake.—Take one half coffee-cupful of butter, one and one-half cupful of sugar, two cupfuls flour, one-half cupful milk, one teaspoonful cream of tartar, one-half teaspoonful soda and the whites of eight eggs.

Gold Cake.—Use the same ingredients, and proceed in the same manner, only substituting the yolks of the eggs.

"Widow's Cake."—A palatable cake to be eaten as bread or rusks at tea is made with two cupfuls flour, one of meal, teaspoonful soda; one cupful molasses, two eggs, salt. Mix with warm milk. Bake in a quick oven.

Spice Cake.—One cupful of sugar, one cupful of sour milk, one cupful of raisins, one egg, a nutmeg, one teaspoonful of cloves, two teaspoonfuls of cinnamon, one teaspoonful of soda and three cupfuls of flour. Bake slowly but steadily until done.

Another.—One cupful butter, one of brown sugar, one and one-half of sour milk, one pint molasses, one tablespoonful saleratus, three eggs, cinnamon, cloves, allspice, nutmegs, citron, currants, raisins. Stir stiff with flour.

Hickory-Nut Cake.—Take a half cupful of butter, two cupfuls of sugar and four eggs beaten separately; then three cupfuls of flour, one-half cupful of sweet milk, two teaspoonfuls of baking powder, two cupfuls of hickory-nut meats cut fine, with one teaspoonful extract vanilla.

XI. Gingerbread and Other "Homely" Cakes.

Ginger Snaps.—One cupful of brown sugar, one cupful of molasses, one cupful of lard, two eggs, a small teacup half full of boiling water, two teaspoonfuls of ginger, two of cinnamon, a teaspoonful of saleratus. Roll thin, cut out, and bake in a quick oven.

Ginger Cookies.—One cupful of molasses, one-half cupful of sugar, two-thirds cupful of butter, one-half cupful of water, one egg, two teaspoonfuls of saleratus, one-half teaspoonful of alum, one teaspoonful of ginger, flour enough to roll out soft. Bake quick.

COOKIE-PANS.

Sponge Gingerbread.—Sift two teaspoonfuls of soda and a dessert-spoonful of ginger, in two cupfuls of molasses. Stir thoroughly, and add four well-beaten eggs, one cupful of butter, melted, one cupful of sour milk or buttermilk, in which is dissolved one teaspoonful of soda. Add flour until the whole is of the consistency of a pretty thick batter. Make into two loaves and bake.

Soft Gingerbread.—Take one cupful of molasses, one cupful of sugar, one cupful of sour milk, half a cupful of butter, five cupfuls of sifted flour, and somewhat more than half a teaspoonful of soda. Melt the butter in the molasses and sugar, with the soda, add the ginger, and, if you like, a little cloves, the sour milk and flour. The cake should be just stiff enough (a thick batter) to rise nicely in baking and not fall afterwards.

Ginger Cake.—Two cupfuls molasses, one cupful butter, one and a half cupfuls sour milk, three and a half cupfuls flour, three eggs, two teaspoonfuls saleratus, one tablespoonful ginger, one tablespoonful cinnamon, one tablespoonful cloves.

Cookies.—One cupful of white sugar, rolled fine, and mixed with a half-cupful of butter; a half cupful of sour cream, mixed with a half teaspoonful saleratus. Add two eggs thoroughly beaten. Season with caraway-seeds or nutmeg. Roll thin, sprinkle sugar on. Roll lightly once, cut them out in a circular shape and bake them in a quick oven.

Soda Cakes.—Take one quart of flour, one teaspoonful of soda, and one of cream of tartar, dissolved in hot water; one tablespoonful of lard and one of butter, rubbed into the flour; a little salt, mix soft with sour milk or buttermilk, and cut with a tin in round cakes; bake in a quick oven.

Short-Cake.—Mix with a pint of flour a lump of butter the size of an egg, rub up well with baking powder, or use two teaspoonfuls of cream of tartar in flour;

powder fine one teaspoonful saleratus. Add one cupful of cold water. Make a stiff batter; add flour if needed. Bake on tin for tea. If you use buttermilk you will not need cream of tartar, nor as much butter.

Custard Cake.—Two cupfuls powdered sugar, one-half cupful sweet milk, six tablespoonfuls melted butter, one teaspoonful baking powder, two and one-half cupfuls flour; bake as for jelly cake, and when cool, add the following custard: One pint milk, three eggs, sugar and flour to suit the taste, and prepare as for boiled custard.

Drop Johnnies.—One cupful sugar and two eggs well beaten together, one cupful cream, three cupfuls buttermilk, one large heaping teaspoonful of saleratus. Salt and spice to suit your taste. Thicken with flour to a stiff batter. Drop in hot fat, a spoonful at a time. Fry the same as fried cakes.

Virginia Apple Cake.—One cupful of bread dough, one and a half cupfuls of sugar. When ready, roll an inch thick, put it in a long pan, then slice good baking apples thin, and put smoothly over the dough; sprinkle sugar, butter and cinnamon over, and bake.

Pork Cake.—Chop one pound fat pork very fine. Stone and chop one pound raisins. Pour a pint of boiling water over the pork. Use one cupful of molasses, two of sugar, eight of flour, one tablespoonful ground cloves, one of cinnamon, one of saleratus, one egg—the white to be added last.

Yankee Doughnuts, Raised.—Heat a pint of milk just lukewarm, and stir into it a small cupful of melted lard, and sifted flour until it is a thick batter; add a small cupful of domestic yeast, and keep it warm until the batter is light; then work into it four beaten eggs, two cupfuls of sugar, rolled free from lumps, a teaspoonful of salt and two of cinnamon. When the whole is well mixed, knead in wheat flour until about as stiff as biscuit dough. Set it where it will keep warm until of a spongy lightness; then roll the dough out half an inch thick and cut it into cakes. Let them remain until light, then fry them in hot lard.

Fried Cakes.—Take four cupfuls of white sugar, four of buttermilk, one of butter, two eggs, two teaspoonfuls of soda. Season with cinnamon, mix quite hard, roll half an inch thick, cut in rings. They will fry much nicer than when twisted.

Griddle-Cakes.—To one quart of flour add one teaspoonful of cream of tartar and one three-fourths full of soda, mix with sour or butter milk, and bake on a griddle; season to taste. Buttermilk cakes made the same way, adding two eggs, are very nice.

Coffee Cake.—One teacupful of brown sugar, one of molasses, one of lukewarm strong coffee, one egg, one cupful of butter, one teaspoonful of soda, one pound of raisins. Use plenty of spice. This cake is much nicer for dipping in coffee if it is not cut until it is several days old.

XII. The Housewife's Table of Equivalents.

OFTEN in giving recipes—cup, wineglass, spoon, etc., are mentioned. It is the

CAKE-MAKING.

usual way in which ladies measure, and the majority of the recipes we have given were furnished by ladies who are excellent cooks. The following table is one of equivalents that will be found approximately correct.

Wheat flour—one pound is	One quart.
Indian meal—one pound two ounces are	One quart.
Butter, when soft—one pound is	One quart.
Loaf sugar, broken—one pound is	One quart.
White sugar, powdered—one pound one ounce are	One quart.
Best brown sugar—one pound two ounces are	One quart.
Ten eggs are	One pound.
Sixteen large tablespoonfuls are	One-half pint.
Eight large tablespoonfuls are	One gill.
Four large tablespoonfuls are	One-half gill.
Two gills are	One-half pint.
Two pints are	One quart.
Four quarts are	One gallon.
A common-sized tumbler holds	One-half pint.
A common-sized wineglass holds	One-half gill.
A large wineglass equal to	Two ounces.
A tablespoonful equal to	One-half ounce.
A teacup holds	One gill.
A large wineglass holds	One gill.
Forty drops are equal to	One teaspoonful
Four teaspoonfuls are equal to	One tablespoonful.

CHAPTER VII.

BEVERAGES, ICES AND CANDIES.

I. PURE WATER AS A BEVERAGE.—II. TEA AND COFFEE.—III. HOW TO MAKE TEA.—IV. THE TEA-MAKING OF VARIOUS PEOPLES.—V. A CUP OF COFFEE.—VI. CHOCOLATE.—VII. REFRESHING DRINKS.—VIII. SUMMER DRINKS.—IX. TOMATO BEER.—X. ICE CREAM AND WATER ICES.—XI. CANDY-MAKING.—XII. CANDIED FRUIT.

I. Pure Water as a Beverage.

NO person, now-a-days, can altogether get along without some beverage other than pure water. Not that the water drunk by man *is* pure; none of it is, for the minerals contained in the purest spring water, from the chemist's standpoint render it impure, but *not*, on this account, unhealthful. In fact, the lime, soda, magnesia, and other minerals of spring and well water, if it is in no way impregnated by leachings of the house or barn-yard, or uncontaminated with sewage, is more healthful than chemically pure water. But if impregnated with these last-named impurities, it is more deadly than the miasma of Roman marshes in the dog-days.

Impurities in Well-Water.—The water of a well may be bright, sparkling and most pleasant to the taste, and yet contain the deadly typhus and noxious germs, bringing diphtheria, meningitis, and other diseases that so mysteriously appear in neighborhoods apparently good in sanitary surroundings. How, then, do these germs reach a well, sunk in strong clay to a living stream of water, deep in gravel below? They come in by the surface water. The roots of the willows and most other trees go to water if they can. Every insect burrowing in the soil must have water, and they invariably burrow there, especially in great droughts, and seek the water of the well. Their burrows convey water from cesspools, house-drains, barn-yards, etc., for considerable distances, through the otherwise impervious clay. In digging or boring a well, the section down to, and partly into, the impervious clay should be larger than the rest, and strongly cemented with the best water lime at the back, and the stone or brick laid with the same material. But, after all, the only safe way to escape impurities is to boil all the water that is used for drinking. If the water is muddy, or has other mechanical impurities, it may be improved by filtering. But water is not, never has been, and never will be, the exclusive beverage

WATER-FILTER AND COOLER.

of civilized man. It is not so of even the most savage nations. Let us, therefore,

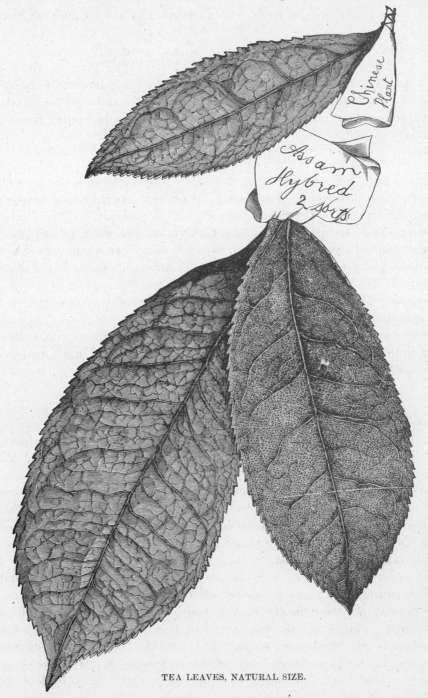

TEA LEAVES, NATURAL SIZE.

give some of the more innocent and pleasant of the artificial beverages.

II. Tea and Coffee.

TEA, to be good, must be fresh, unadulterated, and be kept dark, and away from the air.

Roasting Coffee.—Coffee when roasted, and especially when ground, loses its aroma. If roasted too fast or too much, it is little better than so much charcoal. Hence it should be roasted in an implement made for the purpose, or else in a closed vessel slowly and with constant motion. If a little butter and sugar be beaten together and added to the coffee after it becomes hot, it will assist in holding the aroma, the essential part of coffee.

III. How to Make Tea.

THE old-fashioned rule, and a good one, is a heaping teaspoonful for each person and an extra one for the pot. For, unlike the Chinese, we drink tea strong, and with milk or sugar, or both. Tea should be made with soft water. Filtered rain-water is good. One way is to scald a metal teapot, put in the tea, pour in half the required quantity of boiling water, cover the pot with a "cosey," (a quilted cover to slip over the pot to keep it hot), and at the end of ten or fifteen minutes add the other half of the water. It is then ready to be poured into hot teacups.

Serve by filling the cups half full. Then add more water to the pot and fill the cups, not too full.

Another way of making tea is to scald the pot, again fill it with boiling water, then put in the proper quantity of tea and let it stand, covered, until the leaves settle to the bottom of the pot, or about ten minutes.

IV. The Tea-Making of Various Peoples.

THE Chinaman puts his tea in a cup, and pours hot water upon it, and drinks the infusion of the leaves without addition. The Japanese triturates the leaves before putting them into the pot. In Morocco, they put green tea, a little tansy and a great deal of sugar in the teapot, and fill up with boiling water. In Bokhara, every man carries a small bag of tea about with him, a certain quantity of which he hands over to the booth-keeper whom he patronizes, who concocts the beverage for him. The Bokhariote tea-toper finds it as difficult to pass a tea-booth as our dram-drinker does to go by a whiskey-shop. His breakfast beverage is Schitschaj, that is, tea flavored with milk, cream or mutton fat, in which bread is soaked. During the daytime, sugarless green tea is drunk, with the accompaniment of cakes of flour and mutton suet. It is considered an inexcusable breach of manners to cool the hot cup of tea with the breath; but the difficulty is overcome by supporting the right elbow in the left hand and giving a circular movement to the cup. How long each kind of tea takes to draw is calculated to the second; and when the teapot is emptied it is passed round among the company for each tea-drinker to take up as many leaves as can be held between the thumb and finger—the leaves being esteemed by these people an especial dainty.

V. A Cup of Coffee.

SHERBADDIN, an old Arab author, asserts that the first man who drank coffee was a Mufti of Aden, who lived about A. D. 1500, or, as he puts it, in the ninth century of the Hegira. Even Arab authors should always leave room for a proviso. Perhaps some obscure person whose name has never come down to posterity, may have seen goats get "skittish" from eating the berries, as is related of a certain Dervish who is also credited with thus having discovered the virtue of *cahui*, as it was originally called. Coffee is good enough English, though an ex-alderman of Chicago is said to have spelled it without using a single correct letter, "kawphy." He did, however, get in two letters of the original name. He spells the name of his adopted city "Shecawgow."

COFFEE-ROASTER.

How to Make Coffee.—There are many ways of making it. It is brewed, boiled, filtered and generally baked, not roasted. We have shown that it should be roasted. Good coffee is made by taking freshly ground coffee (or if cold, warm the ground coffee), at the rate of four heaping tablespoonfuls for each three cups, on the principle of one for the pot. Scald the pot, put in the coffee, pour on boiling water, let it steep five minutes, strain, and then let it just boil up. If you have a filtering machine, patent digester, etc., use if you like.

When coffee is made it should be drunk at once. The cups should be hot, the cream thick, well stirred, and the sugar white. If you have bought whole coffee, of good quality; and if you have dried, roasted, and ground it yourself, there is no reason why you, the farmer's wife, with cream at home, should not have coffee of the best.

Artificial Cream for Coffee.—Beat well one egg, with one spoonful of sugar; pour a pint of scalding hot milk over this, stirring it briskly. Make it the night previous.

VI. Chocolate.

THE rule for chocolate is, two ounces of the cake, grated or thinly sliced, to each pint of boiling milk. Put the chocolate into a pot fitted with a "muller," pour on the boiling milk by degrees, mulling it as you proceed, over a slow heat, until it is hot and frothy. Or it may be frothed, fairly, with any of the modern whiskers for beating eggs.

When chocolate is used every day, a cake of chocolate is dissolved in a pint of boiling water by mulling it, but not on the fire. When mulled, set it on the fire until it boils up. It will keep ten days or more in a cool place. When used, mix in proper proportion with milk, and mull as heretofore directed.

VII. Refreshing Drinks.

Most persons drink too much, and, especially in hot weather, too much at a time. To drink a little slowly, is the way to quench thirst. Ice-water, especially, should be drunk sparingly. A most excellent substitute for it is pounded ice, taken in small lumps in the mouth, and allowed to dissolve upon the tongue. This will prove refreshing, and much more enduring in its effects.

To Make Lemonade.—Roll the lemons until they become soft. Grate the rinds, cut the lemons in slices, and squeeze them into a pitcher (a new clothes-pin will answer for a squeezer in lieu of something better); pour in the required quantity of water, and sweeten according to taste. The grated rinds, for the sake of their aroma, should be added to it. After mixing thoroughly, set the pitcher aside for half an hour; then strain the liquor through a jelly-strainer, and put in the ice.

Travelers may carry a box of lemon sugar, prepared from citric acid and sugar, a little of which in a glass of ice-water will furnish quite a refreshing drink, and one that oftentimes averts sick-headache and biliousness. Citric acid is obtained from the juice of lemons and limes.

Cherry Syrup.—Take six pounds of cherries, and bruise them; pour on a pint and a half of hot water, and boil for fifteen minutes; strain through a flannel bag, and add three pounds of sugar; boil half an hour or more, or until the liquid will sink to the bottom of a cup of water (try it with a teaspoonful of the liquid); then turn into jelly-cups, and cover with paper dipped in the white of an egg. A syrup may thus be prepared of any fruit.

To Prepare the Drink.—Put a spoonful of the jelly in a goblet of water, and let it stand about ten minutes; then stir it up, and fill it with pounded ice. Currants and raspberries made into "shrub," furnish a pleasant and cooling drink when mixed with ice-water. Pounded ice is also an agreeable addition to a saucer of strawberries, raspberries or currants. Pound it until it is almost as fine as snow, and spread it over the berries. With fruit it is also an excellent substitute for cream.

VIII. Summer Drinks.

Spruce Beer.—Allow an ounce of hops and a tablespoonful of ginger to a gallon of water. When well boiled strain it, and put in a pint of molasses and half an ounce or less of the essence of spruce; when cold add a teacupful of yeast; put it in a clean, tight cask (a jug will do), and let it ferment for a day or two; then bottle it for use—you will find it good after three days.

Beer of Sulphuric Acid.—Take of dilute sulphuric acid and concentrated infusion of orange-peel, each twelve drachms; syrup of orange-peel, five fluid ounces. This quantity is added to two imperial gallons of water. A large wineglassful is taken for a draught, mixed with more or less water according to taste. This beer is entirely harmless, even if taken in considerable quantities, and is refreshing in hot weather.

Cream of Tartar Beer.—Mix two ounces of cream of tartar, three pounds of brown sugar, three quarts of yeast. To be mixed and allowed to work. This makes ten gallons, and should be drunk as soon as worked. A strong syrup of pie-plant stalks makes an excellent beer prepared as above, but without the tartaric acid.

Beer of Various Fruits.—Have two quarts of water boiling, split six figs, and cut two apples into six or eight slices each; boil the whole together twenty minutes; pour the liquid into a basin to cool, and pass through a sieve, when it is ready for use. The figs and apples may be drained for eating with a little boiled rice. A delicious beverage may be made from currants, cherries or blackberries by this recipe.

Cream Beer.—Two and one-fourth pounds white sugar, two pounds of tartaric acid, and the juice of two lemons and three pints of water. Boil together five minutes. When nearly cold, add the whites of three eggs well beaten, half a cupful of flour well beaten, one-half ounce of wintergreen essence. Bottle and keep in a cool place. Use two tablespoonfuls of the syrup in a tumbler of ice-water. Add one-fourth teaspoonful of soda just at the moment you wish to drink, but shake the bottle of syrup before using. It is cool and refreshing.

IX. Tomato Beer.

GATHER the fruit, stem, wash and mash it; strain through a coarse linen bag, and to every gallon of the juice add three pounds of good brown sugar. Let it stand nine days, and then pour it off from the pulp which will settle in the bottom of the jar. Bottle it closely, and the longer you keep it the better it is. Take a pitcher that will hold as much as you want to use, fill it nearly full of fresh sweetened water, add a few drops of essence of lemon. To every gallon of sweetened water add a half-tumblerful of beer. This is a favorite drink in the Southern States of America, and is healthful.

Home-Made Bitters.— Take half an ounce of the yolk of fresh eggs carefully separated from the whites; half an ounce of gentian-root; one and a half drachms of orange-peel, and one pint of boiling water. Pour the water hot upon the ingredients mentioned, and let them steep in it for two hours; then strain, and bottle for use.

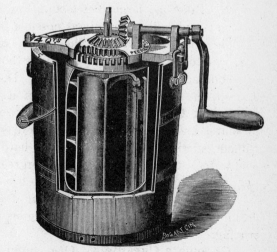

ICE CREAM FREEZER—INTERNAL ARRANGEMENT.

X. Ice Cream and Water Ices.

ICE cream is a preparation of milk or cream, with egg, sugar and flavoring, and frozen in an ice cream freezer. Water ices are the juices of fruits sweetened with sugars, syrup, and then frozen like ice cream. Ices are often made with fruit flavors chemically prepared. They should never be swallowed unless you know that they are made of the juices of real fruits. The ice cream of cheap restaurants, also, is often made of the most horrible compounds, including French clay and poisonous colorings. It is often, like the lemonade sold by circus-men, without a particle of what should constitute it—except the water—but much that should not be there. Ice cream and water ices are easily made by any family having ice.

Ice Cream.—In every quart of cream mix six ounces of crushed white sugar, and flavor with extract of vanilla, strawberry, pineapple, lemon or other flavor you may like. Add the white of an egg, frothed. Mix the whole together by thoroughly beating it, and stir in an ice cream freezer, until fully congealed.

Water Ice.—Water ices are made by making a syrup of white sugar of the proper sweetness. Then add the fruit-juice, the whites of eggs, dilute and freeze in the ice cream freezer. A few trials will enable you to make it to your taste. Try the syrup of a strength, first, sufficient to bear up a fresh egg, so that a section of the egg the size of a ten-cent piece shows above the surface, and you will soon learn how to vary it.

XI. Candy-Making.

Candy can, probably, be bought more cheaply than it can be made at home. Home-made candy, however, is pure. Candy sold at, or near, the price of sugar is not pure.

Molasses Candy.—We give two excellent recipes:

1. Take two cupfuls of molasses, one cupful of sugar, one tablespoonful of vinegar, and a piece of butter the size of a hickory-nut. Boil briskly twenty minutes, stirring all the time. When cool, pull until white.

2. Take one large coffee-cupful of molasses and two very large tablespoonfuls of sugar, and boil as rapidly as possible for twenty minutes. Try if it is brittle by dropping into cold water. When done, rub one-half teaspoonful soda smooth, and stir dry into the boiling candy. Mix it thoroughly and pour into buttered pans. Stir while boiling to keep it from burning. Do not pull. If you like pop-corn balls, pop it fresh, and stir into a part or whole of it.

Sugar Candy.—Six cupfuls sugar, one of vinegar, one of water, one spoonful of butter, and one teaspoonful of soda dissolved in a little hot water; boil all together without stirring. When it becomes hard, not brittle (test by dropping a little into cold water), flavor with lemon, wintergreen or peppermint, and turn out on buttered plates to cool. It is nice pulled, or left on the plate and cut in squares.

White Sugar Candy.—Two cupfuls of white sugar, half a teaspoonful of cream of tartar, a cupful of cold water, and teaspoonful of butter. Boil without stirring.

Chocolate Caramels.—One cupful of molasses, one cupful of brown sugar, one cupful of milk, one-half cupful of grated chocolate, butter the size of an egg. Boil half an hour.

Cocoanut Candy.—Two cupfuls of white sugar, one-fourth cupful of water; boil; put the pan in a larger pan of water, and stir until cool; when it begins to get somewhat stiff and cool enough, stir in the grated cocoanut, and stir until cold. Cut into cakes. The meats of any nuts, chopped or grated, may be used. The name of the candy coming from the nuts.

XII. Candied Fruit.

AFTER peaches, plums, citrons or quinces have been preserved, take the fruit from the syrup and drain it in a sieve. To a pound of loaf sugar put half a teacupful of water; when it is dissolved, set it over a moderate fire; when boiling-hot, put in the fruit; stir it continually until the sugar is candied about it; then take it upon a sieve, and dry it in a warm oven or before a fire. Repeat this two or three times, if you wish.

CHAPTER VIII.

PRESERVING, DRYING AND CANNING FRUIT.

I. OLD AND NEW WAYS OF PRESERVING.—II. CANNING FRUIT.—III. HOW TO PRESERVE FRUIT.—IV. CANNING WHOLE FRUIT—PEACHES.—V. CANNING TOMATOES.—VI. CANNING VEGETABLES.—VII. PRESERVING IN SUGAR.—VIII. MARMALADE.—IX. JAM OF APPLES AND OTHER FRUITS.—X. JELLIES.—XI. SYRUPS—BLACKBERRY, ETC.—XII. DRYING FRUITS.—XIII. MISCELLANEOUS RECIPES FOR PRESERVING.—XIV. BRANDY PEACHES AND OTHER BRANDIED FRUITS.

I. Old and New Ways of Preserving.

MANY persons, not much past middle age, can remember the time when preserving meant cooking the articles in sugar, pound for pound, making a conserve rather than a preserve. The other plan of preserving was to dry the fruit in the sun, having first cut it into thin strips, or other sections. Since then, the world has moved. Very little preserving, in the old-fashioned way, with sugar, "pound for pound," is now done. The fruits are either dried, put up in self-sealing cans or made into jelly.

Rules for Preserving.—A flannel bag is the best for straining jelly. If possible, avoid putting jelly in any stage in a metal vessel, unless silvered. For every pint of strained juice allow a pound of sugar. Granulated sugar is the best.

In all cases it is best to boil the juice fifteen minutes before adding the sugar, thus insuring the necessary evaporation, and avoiding the liability to burn it.

It is well also to heat the sugar before it is added, as in so doing the boiling process will not be interrupted.

All jelly should be made over a moderate fire, and be carefully watched and skimmed.

In making preserves, there must be no economy of time and care, and the fruit must be fresh.

Boil without covering, and very gently.

Jellies and jams must not be covered and put away until cold.

Marmalades require constant stirring.

In making jams, boil the fruit fifteen minutes before adding the sugar.

Mash the fruit before cooking.

Jellies.—In making jellies, from half to three-quarters of a pound of sugar is allowed for each pint, or pound, of strained juice; currants require a pound to a pint.

II. Canning Fruits.

IN canning fruits only enough sugar is used to suit the taste. One quarter of a pound to a pound of fruit is enough; but many use half a pound. None but the finest white sugar is to be used.

If put in glass, the cans must be kept in a perfectly dark place, and kept as cold as possible without freezing. The larger fruits, such as peaches, pears, etc., may be placed in a steamer, over a kettle of boiling water to cook. Then drop the fruit into a syrup of the right consistency, fill from there into the cans, pour over all the boiling syrup, and seal immediately.

Fruits, and How to Can.—Fruits for canning should be of the best quality, and not over ripe. Berries and all that kind of fruits are to be cooked in the syrup and then ladled into cans. The cans should always sit in a hot-water bath whilst being filled. Fit on the cover and pour the wax around the cover of the lid. The wax must cover every crevice. Set the cans carefully away, and in three or four days examine them to see that they are perfectly tight. Those that are not so must be reheated and rewaxed. Examine again in a week or ten days for signs of ferment.

If these directions are carefully observed, tin is as good as glass, provided it is clean and bright, and again thoroughly cleaned and dried the minute the fruit is taken out. Then, if put away in a perfectly dry place, the fruit will remain perfect for years.

III. How to Preserve Fruit.

We give a table of the time different fruits should be boiled, and the amount of sugar per quart, can or jar. Thus any person by observing the foregoing rules may can any of the articles named:

TIME OF BOILING FRUIT.		AMOUNT OF SUGAR TO A QUART CAN OR JAR.	
Cherries, moderately,	5 minutes.	For Cherries,	6 ounces.
Raspberries, moderately,	6 "	" Raspberries,	4 "
Blackberries, moderately,	6 "	" Blackberries,	6 "
Plums, moderately,	10 "	" Field blackberries,	6 "
Strawberries, moderately,	15 "	" Strawberries,	8 "
Whortleberries, moderately,	5 "	" Whortleberries,	4 "
Pie plant, sliced,	10 "	" Quinces,	10 "
Small sour pears, whole,	30 "	" Small sour pears whole,	8 "
Bartlett pears, in halves,	20 "	" Wild grapes,	8 "
Peaches, in halves,	8 "	" Peaches,	4 "
Peaches, whole,	15 "	" Bartlett pears,	6 "
Pineapples, sliced,	15 "	" Pineapples,	6 "
Siberian or crab-apples, whole,	25 "	" Siberian or crab-apples,	8 "
Sour apples, quartered,	10 "	" Pie plant,	1) "
Ripe currants,	6 "	" Plums,	8 "
Wild grapes,	10 "	" Sour apples,	6 "
Tomatoes,	20 "	" Ripe currants,	8 "

All stone fruits should be pitted, and pip fruits—apples, pears, etc., should be peeled and have the core removed. Berries are cooked in their natural state.

IV. Canning Whole Fruit—Peaches.

THE directions for canning peaches will serve for all fruits that are to retain their shape. Select fruit of firm and good quality. It is nonsense to suppose that inferior fruit is good enough to can. Pare and place in a steamer over boiling water. Put a dish under the fruit to catch the juice, afterwards to be strained and added to the syrup. Let them steam according directions in the table, or until they may be pierced with a broom straw. Make a syrup of the best sugar, have it boiling hot, dip the fruit into the syrup and put it into the cans or jars. Then pour over the boiling syrup to fill the vessel. Seal immediately over the steam.

When Glass is Used.—If glass is used the jars should be set in the water bath on straw or folded cloth, and come to a boiling heat gradually, or they will break. Another thing to be remembered, is, that syrup should be well skimmed before being poured over the fruit.

V. Canning Tomatoes.

POUR scalding water over tomatoes that are ripe, but not too ripe. Remove the skins, slice, cook in a porcelain-lined kettle, with a little salt, pour hot into the cans and seal. Those that are to be used during the winter may be put into wide-mouthed jugs of one gallon each, since in cold weather they will keep for some time after being opened.

VI. Canning Vegetables.

THE difficulty in keeping vegetables prevents their being canned in the country. They really should be sealed hermetically in a dense cloud of steam, and when boiling hot. The directions for canning corn will suffice for the kitchen. Cut, or better, pare and scrape the corn from the cob, and when it comes to a boil fill it into tin cans and solder hermetically—air tight. Puncture the top of each can with a small hole for the escape of steam. Set the cans in a vessel of water and boil hard for two hours. Then, while the steam is rising, drop a little solder over the hole. Keep in a cool place.

VII. Preserving in Sugar.

VERY little fruit is so preserved nowadays. This plan is generally used for rinds, like citron, melon, etc. The directions for watermelon will answer for all: After cutting your rind properly, boil it it clean water, with vine leaves between each layer; a piece of alum, the size of a hickory nut, is sufficient for a kettleful. After boiling it, put it into ice-water to cool; then repeat this a second time, each time putting it to cool; each time boiling one hour. Prepare the syrup with one and one-fourth pounds of sugar to each pound of fruit; green ginger boiled in the water you make your syrup with flavors it, or three lemons to six pounds of fruit. If the syrup thickens too fast, add a little water; the rind should be boiled in the water until clear and green.

VIII. Marmalade.

This is a kind of preserves that is much liked. Quince, peach and apple marmalade are the kinds mostly prepared, though any fruits may be used. Marmalade should be put away in jars covered with oiled paper, and made perfectly air-tight.

Quince Marmalade.—Select ripe yellow quinces, wash clean, pare and core them and cut them into small pieces. To each pound of quinces allow half a pound of white sugar; put the parings and cores into a kettle, with enough water to cover them, and boil slowly until quite soft. Then, having put the quinces with the sugar in a porcelain kettle, strain over them, through a cloth, the liquid from the parings, and cover; boil the whole over a clear fire until it becomes quite smooth and thick, keeping it covered, except when you are skimming it, and watching and stirring closely, to prevent sticking at the bottom. When cold, put in glass jars.

Peach Marmalade.—Allow three-quarters of a pound of sugar to a pound of fruit; boil the pits until the water is well flavored; peel and quarter the peaches, and add to the water boiling, half an hour before adding the sugar; stir constantly; boil an hour after adding the sugar.

Apple Marmalade.—Select four pounds of cooking apples; pare and core them, put them in an enameled saucepan with about a quart of sweet cider and two pounds of white sugar. Boil them until the fruit is quite soft. Squeeze it through a colander, and then through a sieve.

Strawberry Marmalade.—Pick ripe strawberries free from hull; to a pound of fruit put three-quarters of a pound of sugar; mash them together in a smooth mass; put them in a kettle over a gentle fire; stir with a wooden spoon, and cook until it is jelly-like and thick; cool a little, and if it thickens up like jelly, it is done. Then put in small jars or tumblers, and cover with paper as directed above.

Plum Marmalade.—Simmer the plums in water until they become soft, and then strain them and pass the pulp through a sieve. Put in a pan over a slow fire, together with an equal quantity of powdered loaf sugar; mix the whole well together, and let it simmer for some time until it becomes of the proper consistence. Then pour it into jelly-pots, and cover the surface with powdered loaf sugar.

Orange Marmalade.—Boil small oranges in water until they can be easily pierced with a straw, and then cut in quarters. Allow half a pound of sugar to a pound of fruit, and make a clear syrup; put in the fruit and cook over a slow fire until the fruit is clear; then stir in an ounce of isinglass and let it boil again; first take out the oranges and strain the jelly over them.

IX. Jam of Apples and Other Fruits.

The apples should be ripe and of the best eating sort. Pare and quarter, put into a pan with just water enough to cover them, and boil until they can be reduced to a mash. Then for each pound of the pared apples, a pound of sifted sugar is added, sprinkled over the boiling mixture. Boil and stir it well until reduced to a jam. Then put it into pots. The above is the simple way of making. To have it of the best possible clearness, make a thick syrup with three pounds of sugar to each pint of water and clarify it with an egg. Then add one pint of this syrup for every three pounds of apples, and boil the jam to a proper thickness. This recipe will answer for all the pip fruits. Sour stone fruits require up to half a pound of sugar to a pound of fruit, according to taste.

Gooseberry Jam.—Boil the fruit until perfectly tender, then add three-quarters of a pound of white sugar to every pound of fruit, and cook an hour.

Spiced Jam.—A nice relish for cold meats is as follows: Take five pounds of gooseberries, or other acid fruit, three pounds of sugar, well-cooked. Add a teaspoonful of salt, one of pepper, one of cloves, one of cinnamon, one of allspice and a little mace, if you like it; cook as above.

Strawberry or Raspberry Jam.—To one pound of berries allow one and one-quarter pounds of sugar; heat an earthen bowl hot on the stove, then remove it from the stove and put into it the berries and sugar, and beat them hard with a wooden spoon for as much as an hour and a half; do not cook at all, but put in jars with egg papers.

X. Jellies.

JELLY making is not difficult. It simply requires exactness and care.

1. The fruit for jellies requires to be ripe, but not dead ripe, for if much over-ripe the juice is not so rich.

2. The fruit must be gathered when dry; it must not be wet with rain and dew.

3. It should not lie long in bulk after being picked. Here is where those who raise their own fruit have the advantage, for any purpose, over those who buy their fruit: it is always fresh.

4. Weigh the fruit, if currants without removing the stems, and allow half a pound of loaf or pure granulated sugar to each pound of fruit

5. If the fruit requires rinsing do so quickly and let the fruit dry again.

6. Use only a porcelain-lined kettle for preserving. There is really no objection to a brass kettle but it must be thoroughly cleaned both before and after using.

FRUIT AND JELLY PRESS.

7. In making jelly, pound a portion of the fruit, to get some juice for the bottom of the kettle, to prevent burning. Then add the remainder and boil freely for twenty minutes or more, stirring often enough to prevent burning.

8. Let your straining-bag be three-cornered and of strong material and long, so it may be properly twisted to get pressure. Strain the juice into a porcelain-lined pan.

9. Return the liquid to the kettle, and when it has boiled up add the sugar, previously weighed.

10. When the sugar has thoroughly dissolved, the jelly will be done, and should be put into the jelly-glasses or forms. If you have properly observed the directions, it will coagulate upon the side of the dipper as it is taken out. So there can be no doubt of the result.

11. The general rules for making jellies may be stated as follows: In making jellies of apricots, quinces, peaches, apples or plums, peel, remove the stones or cores, cut in pieces, cover with water, and boil gently until well cooked; then strain the juice gently through a jelly bag, and add a half pint of sugar to a pint of juice. For berries, a pound of sugar to a pint of juice; boil until it ropes from the spoon, or from fifteen to twenty minutes. In making raspberry jelly use one-third currants and two-thirds raspberries.

12. To keep jellies from molding: Pulverize loaf sugar as fine as flour if possible, and cover the surface of the jelly with this to the depth of one-fourth of an inch. This will prevent mold, even if the jellies are kept for years.

By the rules we have given, jelly may be made of any of the fruits, but some fruits require strong pressure. In fact, the jelly bag may always be profitably twisted by means of a stick in the hands of two persons, the stick having been entangled in the end of the jelly bag.

Currant Jelly Without Cooking.—Press the juice from the currants, and strain it; to every pint put a pound of fine white sugar; mix them together until the sugar is dissolved; then put it into jars; seal them and expose them to a hot sun for two or three days.

Crab-Apple Jelly.—Select fresh, sound fruit, not more than fully ripe. Place one gallon of the fruit in an earthen or porcelain kettle and add one pint of water. Heat slowly until it boils. Continue to cook slowly until the fruit begins to come to pieces, then turn off the juice immediately, pressing the fruit gently back in the kettle as long as the liquor will run off clear. Then strain twice through a fine cloth strainer. Add one pound of the best white sugar for every pound of the juice. Boil ten or fifteen minutes. Skim carefully while boiling.

Grape Jelly.—Grapes for jelly should be used before they are ripe, or when just turning. Stem the grapes and slightly cook them; then strain and use a pint of sugar to a pint of juice. It makes the jelly of a light-red color, and much finer flavored than ripe grapes.

Fig Jelly.—Wash the figs, and add water enough to cover the fruit. Boil twenty minutes, strain, then add sugar, and boil ten to twenty minutes.

XI. Syrups.

Syrups are used principally for their medicinal qualities. Blackberry and elderberry are generally used, but all fruits have more or less cooling, anti-febrile qualities. The directions here given will apply as well to any other berries of which the extract can be gotten, as to blackberries.

Blackberry Syrup.—Make a simple syrup of a pound of sugar to each pint of water, and boil it until it is rich and thick. Then add to it as many pints of the expressed juice of ripe blackberries as there are pounds of sugar; put half a nutmeg, grated, to each quart of syrup; let it boil fifteen or twenty minutes; then add to it half a gill of fourth-proof brandy for each quart of syrup; set it by to become cold, then bottle it for use. A tablespoonful for a child, or a wineglassful for an adult, is a dose.

XII. Drying Fruits.

Fruits are dried in four ways:

1. By slicing thin, and exposing the natural fruit, spread on cloth or frames of silvered wire to the heat of the sun, taking them under cover in the event of rain, and at night. When dry they are placed loosely in paper bags and hung in a dry place.

2. The fruit is cured after slicing by being kept in a warm oven until dry.

3. They are cooked sufficiently to make them soft and then dried by fire heat.

4. They are also dried in dry-houses, more or less simple, by continuous fire heat. This plan is altogether the most economical when a considerable quantity is dried each year. In all fruit neighborhoods these houses may be found where the fruit is dried either on shares or at a given price per pound.

Pip fruits are pared, cored and sliced. Stone fruits may or may not have the stones removed. Peaches always should, and it is better to pare and slice them. Plums are generally halved, and cherries have the pits removed.

To Dry Currants.—Berries and other fruits are sometimes dried with sugar and heat. The directions for currants will also serve for other fruits: Take equal weights of stemmed currants and sugar. Let them boil together for one minute, then carefully skim the currants from the liquor, and spread on dishes to dry. Dry them in the oven. The remaining syrup may be used for jelly.

To Dry Plums.—Split ripe plums, take the stones from them, and lay them on plates or sieves to dry, in a warm oven or hot sun; take them in at sunset, and do not put them out again until the sun will be upon them; turn them that they may be done evenly; when perfectly dry, pack them in jars or boxes lined with paper, or keep them in jars; hang them in an airy place.

XIII. Miscellaneous Recipes for Preserving.

Preserved Plums Without the Skins.—Pour boiling water over large egg or other suitable plums; cover them until cold, then pull off the skins. Make a syrup of a pound of sugar and a teacupful of water for each pound of fruit, and pour it over; let them remain for a day or two, then drain off and boil again; skim it clear and pour it hot over the plums. Let them remain until the next day, then put them over the fire in the syrup; boil them very gently until clear; take them from the syrup with a skimmer into the pots or jars; boil the syrup until rich and thick; take off any scum which may arise, then let it cool and settle, and pour it over the plums.

Grapes Preserved with Honey.—Take seven pounds of sound grapes on the stems, the branches as perfect as possible, and pack them away snugly, without breaking, in a stone jar. Make a syrup of four pounds of honey and one pint of good vinegar, with cloves and cinnamon to suit, (about three ounces of each is the rule). Boil them well together for twenty minutes, and skim well, then turn boiling-hot water over the grapes and seal immediately. They will keep for years, if you wish. Apples, peaches and plums may be done in this way.

Figs of Tomatoes.—Pour boiling water over the tomatoes in order to remove the skins; then weigh them and place them in a stone jar, with as much sugar as you have tomatoes, and let them stand two days. Pour off the syrup and boil and skim it until no scum rises. Then pour it over the tomatoes, and let them stand two days, as before, then boil and skim again. After a third time they are fit to dry, if the weather is good; if not, let them stand in the syrup until drying weather. Then place on large earthen plates or dishes, and put them in the sun to dry, which will take about a week, after which, pack them down in small wooden boxes, with fine white sugar between each layer. Tomatoes prepared in this manner will keep for years.

Syrup of Lemons.—Clarify three pounds lump sugar; then pour into this, while at weak candy height and boiling, the juice of eighteen lemons, and the peel of three, grated. Boil together four minutes, strain through lawn and bottle. When cold, cork tight for use. This syrup is then ready for lemonade, punch, ices, jellies, etc.

XIV. Brandy Peaches and Other Brandied Fruits.

ALL the stone fruits, and also strawberries, raspberries and blackberries are sometimes preserved in brandy. Choke cherries, and other wild cherries, and any of the small fruits, are also preserved, by filling a vessel, that may be sealed tight, with the fruit, and filling up with brandy, or equal parts of strong proof spirits, and soft filtered water. The liquid is then used, properly diluted, as a warming medicine or stomachic. When preserved by heat and by the addition of brandy, the following recipe for peaches will apply to the preserving of all stone fruits with brandy as one of the preservative integers:

Recipe for Brandying.—One pound of sugar to each pound of fruit; boil the fruit until soft, make the syrup with as little water as possible. Take the peaches and lay separately on a dish, boil the syrup again until of the right consistency; put the peaches in the jar, then add one part brandy to two parts of syrup, stir and fill up the jar.

PART X.

DEPORTMENT AND SOCIETY.

SOCIAL FORMS AND CUSTOMS.

SELF HELP, RULES OF ETIQUETTE ETC.

DIRECTIONS FOR LETTER-WRITING, ETC.

COMPLETE SOCIAL GUIDE.

DEPORTMENT AND SOCIETY.

CHAPTER I.

PHILOSOPHY AND PRECEPTS OF ETIQUETTE.

I. THE PHILOSOPHY OF ETIQUETTE.—II. ETIQUETTE AN AID TO SUCCESS.—III. WHAT IT INCULCATES.—IV. ETIQUETTE OF DINING—HOW MANY TO INVITE.—V. DINNER COSTUMES.—VI. INFORMAL DINNERS.—VII. HOW TO RECEIVE GUESTS.—VIII. AT THE TABLE.—IX. HOW TO SERVE A DINNER.—X. FAMILY DINNERS.—XI. A FEW USEFUL HINTS.—XII. TABLE USAGES; WHAT TO DO AND WHAT TO AVOID.—XIII. WINES AT FORMAL AND OFFICIAL DINNERS.—XIV. SENSIBLE HINTS FOR DINNER GIVERS.—XV. AFTER DINNER.—XVI. BREAKFAST AND SUPPER. XVII. LUNCHEON—INVITATIONS AND SERVICE.—XVIII. ETIQUETTE OF DRESS AND CONVERSATION.—XIX. THE GOLDEN RULE.—XX. THINGS TO AVOID.—XXI. CALLS.—XXII. GENERAL ETIQUETTE OF CALLS.—XXIII. EVENING CALLS.—XXIV. VISITING CARDS.—XXV. NEW YEAR'S CALLS.

I. The Philosophy of Etiquette.

THERE is a philosophy in all the requirements of good breeding, whether in the etiquette of the table, the street, the call or in the discharge of other social duties and pleasures. The requirements which polite society demands of its votaries are not mere arbitrary rules, but will be found to be invariably the result of a careful study of the greatest good and pleasure of the greatest number. Take, for instance, a very gross and marked example: etiquette requires that the food shall be borne to the mouth on the fork and never on the knife. It is, evidently, most unclean, and, therefore, disagreeable, to see a person thrust a knife into his mouth, and exceedingly trying to delicate nerves to see him in continual danger of involuntarily enlarging his mouth by an awkward slip of the knife.

If you have ever eaten next to a left-handed person at a crowded table, you need not be told of the philosophy of the rule that every one should, at least, eat "right-handed."

What is true of these is also true of all the other demands of etiquette, and he is unwise, as well as boorish, who will not adapt himself to custom in such particulars after ascertaining what the usages of good society are.

II. Etiquette an Aid to Success.

SOME pretend to think that these observances are useless, but we venture the assertion that if two persons, with equal intellects and advantages of person and society, start in life, he who conforms to the decencies of life—for reasonable etiquette is nothing more—will advance himself with double the rapidity of the one who considers etiquette a bore, picks his teeth at the table, uses his napkin as a handkerchief, and his knife when he should use his fork.

III. What it Inculcates.

Of course, foppishness is not inculcated, but strict cleanliness and perfect propriety of demeanor are, and he who enters society should be willing to pay that amount of deference to the wishes of others. He who feels it too great a bother, should wisely resolve to keep himself in strict seclusion, and not subject himself to the observation and ridicule which will most certainly be excited by singularity and awkwardness of behavior.

So true is this, that in this age of the world the plain and ordinary rules of etiquette have become a necessary part of education, and we find deportment a branch of study taught in all finishing schools.

How many have suffered almost torture from bashfulness occasioned by ignorance of what are the correct things to be done in certain contingencies, or from an awkward unfamiliarity with the requirements of custom and society. It is to relieve all such embarassments that we here condense and present in their simplest form, the rules of etiquette pertaining to every social custom; making a brief but comprehensive and complete code of the laws of society.

IV. Etiquette of Dining—How Many to Invite.

Invitations to a formal dinner are usually issued from two to three days to a couple of weeks before it is to occur, and the card of invitation should be at once answered, either by an acceptance or a "regret," couched in proper and becoming terms. We give some forms of both in their proper places.

1. The success of the dinner depends upon the cook and caterer, but the truest pleasures of it,

"The feast of reason and the flow of soul,"

depend greatly upon the tact and judgment of the host and hostess.

2. Pleasant people must be invited, and a proper number brought together—neither too many nor too few. If there is a large crowd, acquaintance progresses more slowly, and it is more difficult to select in such numbers the persons best suited to cause a feeling of genial good humor, or, as the French aptly express it, *bonhommie*.

3. On the other hand, if the party is too small, it will lack the proper element of diversity, without which it is apt to sink into monotonous insipidity.

4. From six to fifteen (never thirteen, since the superstitions of some people concerning that number should be consulted) make a party combining all of the necessary advantages.

5. In regard to the number thirteen, it may not be generally known that, until the close of the Bourbon dynasty in France, there were certain persons who made a living by acting as diners-out, to fill up any parties which from accident or other causes were found, on sitting down to the table, to consist of this unlucky number. These *trieziemes*, as they were called, were men of culture, wit and accomplishments, and were always ready at a moment's notice to make up a party to the number of fourteen, receiving therefor a stipulated wage.

V. Dinner Costumes.

1. If an invitation to a formal dinner is accepted, it is expected that the guest will appear in the regulation "full dress," which consists of black waistcoat, trowsers and coat, with white gloves and tie.

2. The coat must be of the swallow-tail pattern. White waistcoats were long worn with this costume, but the indications are that they will be discarded even in summer.

3. At a dinner party a gentleman is allowed to wear richer jewelry than would be considered correct on any other occasion. But few, however, avail themselves of this latitude, and the only essential requirement is that the costume should be clean and well brushed.

4. Patent leather boots are never proper, but very low shoes of this material and black silk socks are fashionable.

5. The costume of a lady should be elaborate, fully as much so as for a ball, and she is expected to wear her richest jewelry. The dress should be of silk, or of some thin white fabric, and fashionably made. The hair, dressed; the gloves, delicate in color and perfect in fit; and the fan, either match or contrast with the color worn.

VI. Informal Dinners.

1. A SUNDAY dinner, or an informal invitation, does not require full dress either for lady or gentleman; though for either, dark, neat clothing should be worn.

2. The costume of the lady should be *demi toilette*.

3. For the gentleman, a frock-coat will do well in most parts of the United States.

4. It is very important to reach the house to which you have been invited at the proper time. If you call too early, your hostess may not be ready to receive; if too late, you commit the unpardonable rudeness of causing the dinner to lose its freshness, and the other guests to wait.

5. Ten to fifteen minutes before the appointed hour is about the proper time to reach the house.

6. The very sensible rule now obtains never to wait the dinner for a single guest, so if you are the solitary exception to promptness, it should cause you no annoyance to find the rest at dinner.

VII. How to Receive Guests.

THE hostess, before the arrival of the earliest, should make a last survey of dining-room and parlor to see that all is arranged in order, and then, accompanied by the grown members of her family, wait in the parlor to receive the guests.

1. The room must be neatly and tastefully arranged, well lighted, and in winter, well warmed.

2. The welcome should be pleasant and cordial, the lady advancing slightly to receive each guest as announced. A formal stiffness should be avoided, and should

either the dinner or the guests prove tardy, the defect should be atoned for by pleasant and diverting conversation, not a trace of annoyance being allowed to become visible.

3. If the dinner is ultra formal, a tray containing cards is handed around among the gentlemen, each card containing the name of a gentleman and that of the lady he is to escort in to dinner.

4. These cards are sometimes enclosed in envelopes and left on a tray in the hall, or handed to the gentlemen, by the servant, as he is announced.

5. In less formal parties the hostess or host pairs off the couples. Each guest should be introduced to those with whom he or she may be unacquainted.

VIII. At the Table.

1. On going in to the table the host leads the way with the most distinguished or eldest lady, the hostess follows last with the gentleman who is most entitled to be honored.

2. The host places his escort at his right, and this order is followed with all the guests; the ladies sitting always to the right of the gentlemen.

3. When all have reached the table, they stand at their designated places until the hostess seats herself; then the other ladies immediately follow, and their escorts then take seats.

4. There should be no observable difference in the time of the seating of the two sexes, but the lady having taken her chair, should be followed immediately by her escort.

5. In passing the various dishes, the servants begin at the lady upon the right of the host, ending with the hostess; while those serving the other side of the table begin with the gentleman seated to the right of the hostess and end with the host.

6. As soon as seated, guests remove their gloves, and lay over them their napkins.

7. The napkin, it is almost useless to say, is never to be used as a handkerchief, or tucked, as a child's bibb, into the collar of the coat or waist-coat.

8. Unless raw oysters are provided, soup is always the first course. If it is not relished, a sip or two may be taken, or it may, without any breach of etiquette, be left untasted.

9. Of soup, or of fish, which forms the second course, no one should ever take the second plate.

10. It is not necessary to wait until all are served before beginning to eat.

11. In a formal dinner all are helped by the servants, who places the portion before each.

12. Sauce should never be poured upon any article of food; if you wish it, the footman or servant helps you with a sauce-ladle.

13. Never call for any particular part of a dish, but being asked, do not hesitate to state your preference.

14. Never suffer your plate to be helped to a dish of which you do not know the nature; it is best to ask modestly and plainly what it is, as no one is supposed to be a *chef de cuisine*, and familiar with all dishes.

IX. How to Serve a Dinner.

EVERY dinner may not be grand, but no matter what its cost, it can be served in such a manner as to impress all the guests favorably. Table and room should be handsomely decorated, and if it is possible to obtain them, flowers should adorn both room and board. The glass should be brilliant, the silver and cutlery well polished, and cloth and napkins fresh and white. *Creme* or *ecru* cloths and napkins, which have began to come into favor, are used only for breakfast, luncheon or cold suppers; never at a formal dinner-table, with its broad glare of light.

Servants, Carving, etc.—1. Servants should be well trained, and everything go off smoothly, and without vexation or nervousness on the part of host or hostess.

2. The fashion denominated service *a la Russe*, but which is really of French origin, relieves the host of a very unpleasant duty; all of the carving being done by one of the servants, before the joints, roasts, etc., are brought to the table.

3. Where the *menu*, or bill of fare is in vogue, the guest is thus notified in advance what to expect, and those who have preferences are enabled to await their appearance.

Order of Dishes.—1. After soup comes the fish, then the *entrees*, or made dishes; and next comes the turkey, beef, lamb, or other *piece de resistance*. Where raw oysters are served they are placed, opened, but in one side of the shell, upon the plates before the guests enter the room, and on their removal the soup is served.

2. Game, puddings, jellies, etc., are next in order.

3. Soups are frequently placed on the table, the tureen before the lady, or, if there should be two kinds, one before the host and the other before the hostess, to serve.

4. If there are two soups, there should be two kinds of fish. If only one of each, the soup should be placed before the hostess; the fish, in its turn, before the host.

5. It is perfectly correct for the gentleman occupying the post of honor to relieve the hostess of helping the soup.

6. Side dishes should not be put upon the table, but should be handed around. After these have been removed and the plates changed, the fowls and meats are brought in, in the order mentioned.

7. Fowls are placed before the hostess, heavy meats before the host, to carve.

8. Game should be placed before the gentleman, the pudding before the lady of the house, to serve.

9. Cheese precedes the dessert, which is passed in the following order: First, ices; then fruits, etc. After which the servants leave the room.

X. Family Dinners.

1. At family dinners and chance invitations, there should be no attempt at show. The dinner is supposed to be in a great measure an impromptu affair, and nothing is more out of taste than to deluge a guest with apologies.

2. At affairs of this kind, serve to the guest or guests both soup and fish, fowl and meat.

3. In carving the fowl, give to each a piece of the white and black meats. Add to each plate dressing, and, if so requested, a small portion of the gravy.

4. The carver should stand up when carving, and should not hack up the meat into small fragments.

5. Vegetables, sauces, etc. should be passed quickly, yet quietly

6. If a clergyman is present he should be asked to say grace.

XI. A Few Useful Hints.

1. Never place together husband and wife, near relatives, or members of the same profession, as in such cases the almost invariable tendency of such neighbors is to "talk shop."

2. Always endeavor to have a nearly equal number of each sex, as these are always found to be the most pleasant parties.

3. Probably the poorest of all policies is to secure some lion or distinguished person for your party, hoping thus to make it a brilliant affair. Either your lion is talkative, thus boring and silencing the other guests, or he is moody, sullen and isolated in his grandeur, when he is sure to cast a gloom over all the others. In either case the dinner proves a failure.

4. Never permit wrangling, argument or heated discussion.

5. All politics and religion should be tabooed subjects, and if the guest should so far forget himself as to give way to contention, the host or hostess can, by the exhibition of a little tact and judgment, lead him off, by degrees, from his hobby, by dextrously asking his opinion of some other matter, or by engaging him on a different topic, thus allowing him time to cool off and see his blunder.

6. Neither host nor guest should look vexed or nervous at any blunder of the servants. Above all things, neither can afford to rebuke or criticise them.

7. Be punctual, not only in arriving, but also in being ready to rise with the other guests, and do not prolong your leave-taking unreasonably.

XII. Table Usages—What to Do and What to Avoid.

1. HAVING removed your gloves, if it is a formal dinner party and you have them on, and taken your napkin, sit perfectly erect and moderately close to the table.

2. The posture should not be stiff and constrained, but easy and natural, and should be maintained until you are served and begin eating.

3. When eating soup, hold a piece of bread in your left hand and your spoon in your right, and sip noiselessly from the side of the spoon near the end.

4. *Never put the point of the spoon into your mouth*.

5. Do not cut your food into bits ready for eating, as if you had but a limited time in which to eat, but cut off what you desire at the time and carry it to your mouth with your fork.

6. Never use a knife, *under any circumstances*, to convey food to the mouth. This is perhaps the most quickly noticeable, as well as the most disgusting, of all table blunders.

7. Remember that the napkin is not intended as a handkerchief, nor the handkerchief as a napkin.

8. No well-bred person will ever pick his teeth at the table; it would be fully as cleanly and decent to trim and clean the finger-nails

9. Neither will he use his handkerchief at the table, except in the most modest way and without the slightest noise.

10. While it is not necessary for any one to wait until others are helped, there should not be the slightest haste or awkwardness in eating.

11. Bread must neither be bitten nor cut; it should be broken.

12. At breakfast, or where such are provided, never drink from your saucer; wait until your tea, coffee, or chocolate, is cold enough to be taken from the cup. Never sip it from your spoon.

13. Some very eccentric persons raise the cup in the saucer by taking hold of the latter, while others bend down to the table and sip from the cup without raising it. These may be classed with those persons who use the napkin as a handkerchief, pick their teeth at the table and eat with their knives, and who sin through excessive ignorance of good breeding.

14. Place, at breakfast, your egg in the egg-cup small end downward, chip off a portion of the shell and season and eat by scooping from the shell.

15. Never hesitate about taking the last piece of anything passed to you; it is a poor compliment to your host to suppose he has not made ample provision.

16. Only among the Hottentots and Bushmen is smacking the lips, and making other unseemly noises while eating, considered correct.

17. Wine should be sipped slowly, not swallowed at a single draught.

18. Toasts and healths are out of fashion.

19. Hold the glass by the stem, not the bowl.

20. Port and sherry, not port wine and sherry wine, are correct terms in speaking of these beverages.

21. Vegetables should be passed and taken singly—two kinds should never appear on the waiter at once, though the plate may be helped to the two kinds which may accompany each course.

22. Plates should be changed after each meat and pastry.

23. Pork never figures among the dinner dishes.

24. Cut the meat upon your plate as cleanly as possible from the bones, but never hold the latter in your fingers to eat from.

25   Never use your own knife or fork to help yourself to butter, or to any dish that may be passed or placed near you.

26. A plate should never be overloaded.

27. Never play with knife, fork, spoon, glass or food, nor move about in your seat.

28. Never appear to be making a selection of any food passed to you—take the first that comes to hand.

29. Never talk while the mouth is full, and at no time monopolize the conversation; remember good listeners are always appreciated—never laugh nor talk loudly.

30. Never ask to be helped a second time to any dish; if it is passed to you unsolicited, you may help yourself.

31. Never tilt your chair, slouch around in it, nor lean your elbows on the table.

32. Never tilt plate, glass, nor dish, to drain the last morsel.

33. Do not thank, and above all things do not, as some would-be fashionables sometimes do, apologize to the waiters for troubling them—they are paid for their service and are merely performing their duties, for which no man expects thanks.

XIII. Wines at Formal and Official Dinners.

At a formal dinner wine is deemed necessary with each course. The best plan is to place upon the table, between each pair of guests, "caraffes," or open decanters of white glass, filled with the mild red wine, which is always the one most used. At each plate four or five wineglasses, of different shapes and sizes, are placed; for each wine has both bottles and glasses appropriate to it alone. With the raw oysters the servant fills the glasses with chablis, or other white wine. After the soup, sherry is served; with the fish, the white wine again; with the meats, champagne, or other sparkling white wine. After the meats and pastry a higher grade of claret (red wine from Bordeaux), Burgundy, port, or a liqueur may be given.

Red Wine Served Warm.—All red wines must be served warm; say as warm, or a little warmer, than the room in which you sit. Cold red wine has no flavor and is a barbarism. It is better to have it too warm than too cold, and if there is any danger of this, place the bottles in a tub of lukewarm water. White wine must be served cold; sparkling wines, very cold indeed. Never put ice in a wineglass; it is simply ridiculous, spoils the flavor of the wine, looks awkward, and shows want of knowlege of the world.

These Occasions Rare.—What has been said about the variety of wines applies only to formal dinners, in houses where the host is rich, and the servants can be relied upon to carry out a formal dinner without making it a burlesque. Such houses are rare in America, and the occasions rarer, except for people whose official position calls for this class of entertainment.

Native Wines.—There are now made in this country many wholesome wines. The use of these, by people accustomed to drink wine, not only helps to digest the food, but destroys the desire for strong spirits; a daily wine-drinker seldom cares for whiskey. A good native red-wine, which should be had at a cost of not more than twenty to thirty-five cents per quart, is thought by many to be an acquisition to the family dinner table, both in helping digestion and in promoting temperance among the growing members of the family, by destroying the taste for strong drink.

XIV. Sensible Hints to Dinner-Givers.

IF you ask friends to dine, do not try to provide anything very different from your own daily meal. If you do, mistakes will be made and the dinner be stiff. Have it of good material and well cooked, in ways well understood by you and your servant.

1. Never mix the courses. Let your soup be taken away before anything else is brought. If you have fish, the plates, knives and forks must be changed before the meats.

2. Never put a number of articles of food in the same plate. One vegetable, or, at the most, two, may be served.

3. Never put a number of discordant messes before a guest. A heaped-up plate and half a dozen little plates or saucers full of varied viands placed before one at the same time are nauseating and vulgar

4. Let your dinner be simple unless your servants are trained to serve elaborate courses, and can do so without a fault.

5. A good soup, a well-cooked and well served joint of meat, a fowl, and some fruit or cheese, with a bottle of good, sound, native red wine is far better than an attempt at a dinner in many courses, unless the latter is served and managed by an expert.

6. Never hire waiters for a dinner. Never borrow finery for your table. Both are vulgar shows. Serve what you have as well as you can.

7. Try to have your table service good every day. It costs nothing but a little time and that is well spent. Then, when you invite a friend, it will be much easier to have a dinner successful. You cannot dine every day in your shirt-sleeves, eat with your knife, and have your dinner served as if to fill a swill-barrel, and then, on occasions, be fine. Something will betray the daily custom.

XV. After Dinner.

1. IN English society, when the dinner is over, it is usual, at a signal from the hostess, for the ladies to rise and retire to the drawing-room, while the gentlemen remain to indulge in wine, politics, etc.

2. This habit, which has the advantage of giving to the sexes half an hour's time between a long and heavy dinner and the evening's entertainment, has never become popular in America, where the custom is for all to rise together, and adjourn to the drawing-room.

3. A cup of tea or coffee is handed around, after which the conversation becomes

general until the time for leave-taking. Music is appropriate and pleasant during this time.

4. Each guest should remain two or three hours after dinner.

5. Within a week after attending a party of this kind, each guest should make a call upon his hostess; to delay it beyond two weeks is inexcusable.

XVI. Breakfast and Supper.

Customs of Different Countries.—Breakfast is not so often made a meal of ceremony and invitation in America as in England; owing to the later hours which prevail in the latter country. On the Continent, especially in France and the southern countries of Europe, "breakfast" is a substantial meal of meat, wine, etc., taken at from eleven to twelve.

Do not therefore suppose the people are late risers; the custom is to take simply a cup of coffee and a bit of bread on rising, and to breakfast four or five hours afterwards. Many persons in the large cities, such as the brokers and others, often finish the usual day's work in this time.

Dinner, which differs in French households from breakfast more in having soup than in any other particular, is generally eaten at six, and is over in time to go to the theatre at eight. Parties begin in Paris at about midnight, or after the opera, and the doors of many of the large public balls open at that witching hour.

Many who rise early, take a nap after the breakfast, and, in Spain, this is done by all classes; shops are closed and the streets deserted for a couple of hours for the "siesta"—but then the Spaniard begins his work before day.

In England, family breakfasts are generally of cold meats, and the dinner comes at eight in the evening, the ladies having a cup of tea in the afternoon, at about five o'clock.

In the cold countries of the North, especially among the Scandinavians, more meals are taken—often as many as five, at which meat is served—and these do not correspond to our meals or those of Southern Europeans, enough even to bear the same names.

Indeed, the hours for eating must be regulated by the hours of employment. If one is in the whirl of fashionable life of a great city, and nightly out until four or five o'clock in the morning, the heavy meal of the day must come late, and a light supper at midnight becomes a necessity. If you are a hard-working farmer, living in the country and going to bed between nine and ten, the meal hours must correspond.

We often read of great changes in the dinner hour; that Henry the Eighth "dined" at ten in the morning. So he did. But until very modern times the heavy meal at close of day was called supper; they eat it in England to-day at nearly the same hour, and call it dinner. "What's in a name?"

Supper Parties.—Suppers are now given at balls and parties, after the opera, etc., and it is rare to invite a guest to supper alone. An invitation to a dancing or card party is generally understood to include a supper of some kind, although it is never mentioned. Such terms as "an oyster supper," "a champagne supper," etc., are never heard among decent people. Oysters and champagne may be given, but attention is not called to the fact any more than one would ask a guest to "a meat and claret dinner" when these were to be given. In fact, the terms mentioned should be restricted to the vulgar haunters of bars and billiard rooms.

Common Sense Hours.—Meals should be timed by common sense. It is probably more healthful to take a rather light breakfast, which, in a malarious country, should always include coffee. At noon a more substantial meal is in order, but as, in this country, several hours of hard work are to follow, it should not be too heavy, and a quarter or half an hour's rest after it is time well spent.

The dyspepsia so common in this country comes from taking, as a habit, more food than is necessary, and then working with head or body immediately afterwards. Digestion requires repose and most people eat far more than nature calls for.

The best time for a heavy dinner is after the hard work of the day is over and a couple or more of hours can be given to comfortable rest, reading, conversation or light amusement. Eat slowly, not too much at any one meal, take small pieces which can be easily masticated, and do not go directly from the table to violent exercise or severe brain work. Make your dinner (or evening meal, by whatever name you choose to call it,) a pleasant, social affair, which tempts you to linger over it; not a place to bolt, in haste, a certain amount of unmasticated food, and then fly from. Cultivate the beauties and the social aspects of the meal daily, and it will prove not only a delight, but a source of health as well as of civilization. Then your dinner parties to strangers will need only a little more care than the daily event, not a contrast which upsets the household.

XVII. Luncheon—Invitations and Service.

1. This is a strictly orthodox affair, and to provide a suitable luncheon is almost as great a test of one's catering powers as to triumph in a dinner. It is true that it is usually considered only a light repast, made up of elegant little trifles, but amongst fashionable people, the table is often dressed and garlanded as if for a ceremonious dinner, and a great variety of dishes are served.

To this affair, invitations are sent out, which may be autographic, or the visiting card, with date and hour added, will answer, thus:

Mrs. John Smythe.

Luncheon at 12½ Wednesday, Oct. 9th, '83.

How to Serve.—Some have the luncheon brought to the table in courses, but this adds a stiffness and formality not to be desired. The most pleasant way is to have the dishes upon the table, and to dispense with the aid of servants.

The luncheon at a bridal party is usually more formal; on such occasions it is customary to darken the room and light the gas or wax candles, by whose aid the feast is eaten.

Boquets presented to each guest with their napkin, and vases containing rich but not gaudy flowers are a great help to this meal.

The dress for this occasion is, of course, a street costume, but this should be fresh and elegant. The observances at luncheon are not rigidly formal, and, in fact, when well managed, this is one of the most delightfully pleasant and informal of all social meetings.

XVIII. Etiquette of Dress and Conversation.

1. CHESTERFIELD, accounting for the benefits that accrue from a polite demeanor rightly said, that but few possessed or were judges of science, art and grand achievements, but that all understand and appreciate grace, civility and politeness.

2. To be ignorant of the customs and usages of society will cause one to become constrained and bashful, and blunders and awkwardness are the result.

3. All this may easily be avoided by the study and acquirement of the few rules necessary to our guidance through all the shoals, rocks and quicksands of ignorance, awkwardness and ill-breeding. This being the case, is it not worth our while to make ourselves familiar with the canons of polite society, since it is with that class, if any, that we should desire to mingle.

XIX. The Golden Rule.

TRUE politeness is merely the practical observance, in small matters, of the "golden rule:" Not to offend the tastes of another; not to annoy him; not to place self before our neighbor, are the bases of all etiquette.

2. State your opinions plainly and mildly. Never talk loudly, nor make broad sweeping assertions.

3. Never offer to back up an opinion with a bet. Of course no gentleman will be guilty of the rudeness of an oath.

4. Always show a deference to age.

5. Never contradict any one flatly; always beg leave, smilingly, not sarcastically, to differ with them.

6. Never anticipate a slight, nor be ever ready to take one.

7. Above all, never give way to abusive argument or a quarrel.

8. Loud laughter and slang phrases are the wit and humor of the jockey and the clown. No lady or gentleman can afford to use them.

XX. Things to Avoid.

1. THE most despicable figure in society is that of the coarse, purse-proud man or woman, who depends solely upon money for standing and consideration. Next to these, if not in the same rank, is the vulgar creature who knows everything.

2. Never volunteer an opinion, nor try to monopolize the conversation.

3. It is not necessary to be foppish in order to be neat. The fop is as far at one extreme as the slouch is at the other.

4. Dress quietly, but let the material be rich; never dress loudly, and avoid much jewelry.

5. Never wear plated ornaments nor imitation gems.

6. Never whisper in company, nor attempt to monopolize the attention of a person.

7. Abstruse subjects, professional topics, religion and politics should be avoided. "The shop," as the English designate business affairs, should never enter into social conversation.

8. Indulge but seldom in quotation; never in inuendo, insinuation or punning.

9. Avoid all satire and sneering—the devil is painted always with a sneer upon his lips.

10. Never flatter, nor volunteer advice.

11. Never talk scandal.

12. Never laugh at your own jokes.

13. Never correct an error, misquotation nor other mistake of any one.

14. Never interrupt a conversation without good cause, and always apologize for so doing.

15. To inveigh against religion, or the nationality or sentiments of any one, is in the very worst of taste.

16. Sit or stand at your ease; avoid lolling, hitching about, playing with your chain or other part of your clothing.

17. Be cool, quiet and collected; avoid haste and worry.

18. The drawing-room comedian is the silliest of the silly. Buffoonery should be left to professional clowns.

19. Never exaggerate nor use highly-colored adjectives.

20. Never attempt to " show off."

21. Never bring in such sentences as " When I was in Rome," or " One day in Paris," etc.

22. Never make yourself the hero of the adventures you relate. It is homely but wise advice never to " blow your own bugle."

23. If your opinion is asked on some subject with which you are familiar, give it modestly, not as though it were infallible.

24. The practical joke is both low and cruel; no gentleman or lady would think of indulging in one.

25. Never use any foreign language, not understood by the company, unless there should be some one of that nation present who does not understand English.

26. Never, as it is termed, " take the word out of any one's mouth." Be patient, and in due time, no doubt, he who is speaking will find the word or phrase for which he is seeking.

27. Never utter a remark that you think may offend any other of the company.

28. Avoid all profanity and coarse language.

29. Avoid appealing to others to prove your assertions.

XXI. Calls.

1. Modern fashion declares a call made between noon and five o'clock, a morning call, though, in some cities, calls are still made as early as eleven o'clock.
2. In extreme cases, however, strict formality is not adhered to.
3. A formal call should not exceed fifteen or twenty minutes. These are always morning calls.
4. Evening calls are neither so short nor so formal.
5. A gentleman is expected to make a call: 1st. The day after escorting a lady to an entertainment—the call being to inquire after her health. 2d. When congratulations and condolences should be tendered, as after a marriage or a death. 3d. When he desires acknowledging hospitalities received elsewhere. 4th. When he desires to hand letters of introduction he may have received. 5th. Within a week after receiving an invitation to a house, even though it was not accepted. 6th. When a friend has returned from a long absence. 7th. When he desires to acknowledge any courtesy.

XXII. General Etiquette of Calls.

1. The gentleman retains his hat and gloves in his hand, during his call, which must be brief.
2. A friend should never be introduced without previous permission.
3. Ladies making a morning call, generally keep on their gloves, and also retain their parasols.
4. When callers retire, the hostess rings for a servant to see that they are attended to the door.
5. A hostess may retain any fancy work she may be engaged upon, but of course anything heavy is out of the question.
6. The callers should always be provided with cards, which should be sent up to insure accuracy in the name, and also to leave in case the lady of the house should not be at home.
7. When retiring after a call, do so in a gentle, graceful manner, not abruptly upon the entrance of other callers.

XXIII. Evening Calls.

1. An evening call should under no circumstances be made later than nine o'clock, and should not, under ordinary circumstances, exceed an hour.
2. If a gentleman's first call, he will retain hat and gloves. Intimate friends only should exceed an hours' stay.
3. On the second call, guests lay aside their hats and gloves on the invitation of the hostess; when, if solicited, they may spend the evening.
4. Calls should be returned within a week, especially if made by a stranger.
5. Those who settle in a new locality expect the first calls, which must be made as soon as their house is supposed to be in order.

6. On mere ceremonial calls, the lady usually leaves her own and husband's cards.

7. "Not at home" is the usual excuse alleged when it is inconvenient to receive a call.

8. A guest admitted must be seen, no matter how inconvenient.

9. An informal caller should not be detained while the hostess dons an elaborate toilet; they should be received in a morning dress.

10. No call should be so made as to come in conflict with any meal-hour of the person called on.

11. All customs bow not only to "great kings." but also to the visitor from a distance, who may not have the time to consult all of the ceremonies in calls, etc.

12. If any acquaintance has a visiting friend, you should call, and it is the duty of the acquaintance and friend to return the call.

13. Only during long protracted illness, may lady friends visit a gentleman.

14. On recovering from a spell of sickness, all calls that have been made should be returned. Leaving your card will answer.

15. After attending any entertainment at a house, leave your card there within a week. If unable to attend, call earlier to express regrets.

16. Immediately on hearing of a bereavement, leave your card, and call within a week.

17. A gentleman is received, by the hostess slightly rising and bowing; a lady, by her rising and advancing towards her.

18. When a lady retires, if there be no other guests, the hostess should attend her to the door.

19. If others are present, she may only be able to rise and bid her adieu.

20. A gentleman receives his friend by meeting him at the door, cordially shaking his hand, and assisting him with his overcoat, hat, etc.

21. During visits, strictly of ceremony, the gloves are not removed.

22. Be easy and natural while calling. Do not fidget, nor hitch about on your chair.

23. While waiting for the hostess, never try the piano, examine the cards, pictures, etc. Such curiosity is contemptible.

24. In leaving, make no excuse, such as: "Well, I must go." "What a time I've stayed," etc. Merely rise gracefully, say "good-bye," "good evening," or "good day," and quietly withdraw.

XXIV. Visiting Cards.

1. THE styles in these are legion. They should bear no titles except such as are intended to make them descriptive. "Miss," "Mrs.," "The Misses," are of course permissible; but "Prof.," "Hon.," "Esq.," are tabooed. It is even doubtful if a physician should use the "Dr.," or the "M. D.," on a visiting card bearing his name and initials.

2. Army and naval officers in service are allowed their titles on their cards, as Capt., etc.; though it is better to have only the name and the initials, "U. S. A.," or "U. S. N.," below the name.

3. The eldest girl of a family is "Miss Jones," "Miss Brown," or whatever the name may be; the others are called by their Christian names, with this title prefixed, as "Miss Mary," "Miss Bella," etc.

4. It is correct, when several sisters use a single card, to use "The Misses Holcomb," "The Misses Dye," etc.

XXV. New Year's Calls.

1. ARE generally made by two, or even more gentlemen together, and it is the occasion for renewing their acquaintance with lady friends.

2. Cards with emblematic designs, and "Happy New Year," etc., are left.

3. These calls should be very short.

4. Refreshments are almost invariably offered, but may be accepted or not.

5. Of course no gentleman will suffer himself to become intoxicated upon such an occasion.

6. Overcoats, gloves and hats should be removed in the hall, where a servant should be in waiting to assist the caller.

7. It is usual in city circles to send the card from the hall to the drawing-room, and to follow it after removing wraps.

8. Many ladies receive with friends either at their own homes or at those of their friends.

9 When receiving away from home, notification should be given.

10. Cards must be left with every lady receiving.

11. A gentleman introduced on this occasion is not privileged to call again without special invitation.

12. A New Year's call may be made as early as ten A. M., but never later than nine P. M.

13. The two or three days succeeding the New Year, ladies devote to calling among themselves, and hence they are often called "ladies' days."

14. For receiving calls, halls and rooms should be warm, and the reception room decorated. These rooms are usually darkened, and the gas lit.

CHAPTER II.

ETIQUETTE OF THE STREET, BALL, CHURCH, ETC.

I. STREET DEPORTMENT.—II. GENERAL RULES OF STREET DEPORTMENT.—III. SPECIAL RULES OF STREET DEPORTMENT.—IV. ETIQUETTE OF INTRODUCTIONS.—V. SALUTATIONS.—VI. RIDING AND DRIVING.—VII. BALL AND PARTY ETIQUETTE.—VIII. THE SUPPER, DRESSING ROOMS, ETC. IX. SOME GENERAL RULES OF PARTY ETIQUETTE.—X. EVENING PARTIES—THE CONVERSAZIONE· XI. CONCERTS, THEATRICALS, ETC.—XII. PARLOR LECTURES.—XIII. CHURCH ETIQUETTE.— XIV. ETIQUETTE OF VISITS.—XV. RULES FOR GENERAL GUIDANCE —XVI. ETIQUETTE OF THE FUNERAL.—XVII. ETIQUETTE OF THE CHRISTENING—GOD-FATHER AND GOD-MOTHER—PRESENTS, ETC.

I. Street Deportment.

THE first recognition should come from the lady.

2. Always raise the hat with the hand farthest from the person saluted.

3. Merely touch or but slightly raise the hat to a gentleman friend, unless of high rank, advanced years, a clergyman, a person in some manner distinguished, or accompanied by a lady.

4. A gentleman should never stop a lady in the street, though a lady may venture to stop a gentleman.

5. A gentleman should always carry a lady's packages and bundles, even the smallest, but should never volunteer to carry her parasol

6. He should never smoke while escorting a lady or speaking to one.

7. Gentlemen friends meeting should slightly raise the hat with the left hand, while the right hand is extended and shaken.

8. In shaking hands it is boorish to hold the hand for any length of time.

9. To give a violent jerk to the arm or to violently wring another's hand is very rude.

10. Only a dude or simpleton extends two fingers to be shaken.

11. Always give to those feebler, or more aged than yourself, or those of exalted position, the inner side of the walk.

12. Always accommodate your gait to that of a lady or an aged or infirm person.

13. Do not rush violently or swing the arms and body ungracefully in walking.

II. General Rules of Street Deportment.

1. A LADY is not expected to recognize a friend across a street.

2. Neither ladies nor gentlemen should stare about them or indulge in loud talk or laughter on the street or in a public conveyance.

3. Never call to a person across a street.

4. Never turn and look after a person. If you must see them again, it is better to turn back and go in the direction they are going.

5. No lady ever was, nor ever could be, guilty of the small and contemptible meanness of sneering at the dress of another, or of turning around to gaze superciliously at it, nor make uncomplimentary remarks about it. Only a fishwoman or a *parvenue* can condescend to such a thing.

6. Do not eat in the street; it can never be done gracefully.

7. Never nod to a person in a store; if you wish to speak to them, go into the store.

8. When accosted by a lady in the street, never show signs of impatience; let her intimate the termination of the interview by a slight bow.

9. Never attempt to force your way with a lady through a crowd. If politely requested, the throng will always make way.

10. Always introduce any friend who may be with you when stopped on the street. If spoken to and stopped by a lady, all of the party of gentlemen with whom you are walking should pause and raise their hats, and the one with whom you are side by side should be introduced.

11. Gentlemen should always uncover to ladies when spoken to on the street.

III. Special Rules of Street Deportment.

1. Never call a friend out from a party he may be with for a long talk. If necessary to talk with him, apologize to the others, and make your interview brief.

2. Never discuss private or personal matters in a crowd or on the street.

3. Any gentleman may offer a lady his umbrella in a storm, but if a stranger it should be pleasantly yet firmly declined. If an acquaintance, it may be accepted, but should be promptly returned.

4. In a 'bus or street car, a lady's fare should be passed. Most gentlemen will give their seats to ladies unable to obtain one, but no lady will accept without thanking the donor, who should bow in return.

5. Any stranger may assist a lady, an old person, or an invalid, who is in difficulties. This assistance is repaid with thanks, and is no basis for acquaintance.

6. If the way is clear, allow the lady to precede you; if any difficulty or danger is in the way, the gentleman should take the lead.

7. No gentleman after rendering assistance to a strange lady, will attempt to force his acquaintance upon her.

IV. Etiquette of Introduction.

1. Never introduce persons unless there is a mutual desire on their part for an acquaintance.

2. Never presume to advise an introduction by saying "you ought to know him or her," "I know you'd like each other," etc. This is silly and presumptuous.

3. Gentlemen are always introduced to ladies, and inferiors to superiors

4. If gentlemen or ladies are of equal rank, introduce the younger to the older.

5. "Mr. Smith, permit me to introduce to you my friend, Mr. Jones;" "Mr. Brown, allow me to present to you Mr. Johnson," are simple but sufficient formula for the introduction. After this formula is gone through with, the names are pronounced in a lower tone in a reversed order.

6. When introducing anyone to a lady, your bow should show more deference than in presenting a friend to a gentleman.

7. In introducing a number, as at a reception, it is only necessary to call the most honored name first, and join the others together as: "Allow me to present Mr. Smith, Mr. Jones, Mr. Brown, Mr. Johnson," at the same time indicating each with a slight bow. Pronounce names clearly to avoid mistakes.

8. Every introduction does not entitle one to a continuance of the acquaintance so formed. If either party desires, the acquaintance may be dropped.

9. Where the French fashion prevails, you are at liberty to address, without an introduction, any person you may meet socially at the house of an acquaintance. Being there is a sufficient guarantee of the respectability of each.

10. The slightest intimation from a lady that an introduction is not desirable, should suffice, as an explanation might prove embarrassing.

11. Ladies are given the privilege of dropping ball acquaintances, or those formed on any festive occasion.

12. It is permissible, in introducing a celebrity, to mention his distinction, as "Mr. Lowery, the artist."

13. You may drop acquaintances made in calls, unless the person should be a visitor to your friend from some other place, in which case he or she must be treated courteously during the stay.

14. A guest should be introduced to all callers.

15. A person should always be introduced by title as "Dr. Blank."

16. Two friends with different parties may stop for a short time and converse without introducing the other members of the company. Should any one be introduced, under such circumstances, recognition is not afterward obligatory.

17. If in the house of an acquaintance you are introduced to a person with whom you are at enmity, acknowledge the presentation courteously, but with reserve, as though an utter stranger.

18. Promiscuous introductions in large assemblies are not correct. Guests may introduce each other in large parties; in small ones it is the privilege of the host and hostess.

19. A married lady may shake hands with a gentleman when introduced, but a single lady should not.

20. No lady should dance with a gentleman to whom she has not been introduced.

21. Never introduce disagreeable or disreputable persons to any one.

22. Always raise your hat when introduced on the street.

23. You are entitled to call upon the President of the United States or the Governor of your own State at any public reception. In this case hand your card to the master of ceremonies. For a private interview it would be better to obtain the aid of some official, as Representative or Senator.

24. In calling upon the Governor of any State but your own, it would be best to carry letters from some well-known person.

25. In order to be presented to the Queen, in England, or at the court of other European sovereigns, it would be necessary to obtain the aid and advice of the resident Minister from the United States, at that court, who will give proper credentials and also inform you of the ceremonies and requirements necessary.

V. Salutation.

1. GENTLEMEN friends, meeting, bow and shake hands. A married lady may shake hands with a gentleman, or an old gentleman with a young lady.

2. In shaking hands do not embarrass yourself and others by waiting to draw off your glove; merely ask to be excused for not removing it. Never receive your friends in your own house with your gloves on.

3. If a lady gives no sign of recognition, a gentleman must pass without salutation.

4. Return the bow of any respectable person, male or female, whether acquainted or not, unless you know it to be some one seeking to force an acquaintance.

5. Ladies rarely find it necessary to stop a gentleman on the street; gentlemen never presume to stop a lady thus.

6. A young unmarried lady can have no pretext for speaking to any gentleman on the street, unless it be a near relative. A bow is sufficient for other acquaintances.

7. In making a bow, or lifting the hat, there should be an easy, graceful motion; only the dude, or the cad, affects angularity, and jerks off the hat with a rapid downward motion, similar to that of an organ-grinder's monkey.

8. A lady is required, in her own house, to extend her right hand to all guests.

9. On horseback, the lady bows but slightly to friends; the gentleman, holding reins and whip in the left hand, must raise his hat to ladies, and also, slightly, to gentlemen friends.

10. The gentleman precedes the lady going up stairs; the lady descends first.

11. In entering a room, a gentleman must carry his gloves, hat, etc., in his left hand, that his right may be free to offer to his friends.

12. Boisterous merriment, coarse conversation, loud talk and laughter, argument, anger and eccentricity, should be left to grooms and stable boys.

VI. Riding and Driving.

IN riding, we will suppose both lady and cavalier to be familiar with the exercise.

1. In this case, having an appointment to ride with a lady, the first care of the gentleman must be to see her safely mounted. It is usual to have the lady's horse sent by a groom some minutes before the appearance of the escort.

2. In mounting, the groom should stand at the head of the lady's horse; the escort at his shoulder, while we will suppose the lady at the left side of the animal, with her skirt held by her left hand, her right holding to the pommel of the saddle.

3. The escort stoops and holds out his left hand, in which the lady places her foot, and springs and is lifted to the saddle.

4. When the foot has been placed in the stirrup, her robe properly arranged and her seat firmly assured, the gentleman must lose no time in mounting, and then they start.

5. You ride on the lady's right side, never touching her rein, unless requested to curb her horse.

6. Accommodate the gait of your horse to that of hers. Be vigilant that no accident occurs.

7. In dismounting, the lady sees that her skirts are not held by the pommel of her saddle, and giving her left hand to the gentleman (who takes it in his right), places her foot in his left hand, and is gently assisted to the ground.

8. In the carriage, those most honored ride with their faces toward the horses; seats with back to the horses are for the less distinguished, the younger and servants.

9. Ladies enter first, but gentlemen leave the vehicle first.

10. In assisting ladies to enter or dismount, be careful that their dresses do not become soiled by the wheels, steps, etc. The place of the footman is to open and close doors, but not to assist the ladies.

11. Always drive close to the sidewalk, and then "cut" or turn the front wheels, so that there may be a larger space for ingress and egress.

12. In America, the driver sits to the right, and vehicles also turn to this direction to avoid others.

13. It is not only silly, but a breach of etiquette for a lady, when frightened, to grasp the arm of a gentleman driving.

VII. Ball and Party Etiquette.

1. For a ball or a ceremonious party, cards should be issued from ten days to three weeks in advance. If it is to be a large or brilliant affair, three weeks would be best.

2. Do not overcrowd your rooms, especially at a ball, and have as few "wall-flowers," or guests who do not dance, as possible.

3. The ball-room should be well lighted, but not too warm. The floor should be well waxed, or, when it is an impromptu affair, the carpets should be smoothly covered with sail-cloth or canvas.

4. The rooms should be tastefully ornamented. Flowers, foliage, plants, etc., being always in order. The music should be slightly elevated, and, if possible, handsomely screened off from the body of the room.

5. For a ball of any pretensions, there should be a programme of dances, so that the ladies may keep a list of their engagements.

6. The music, even in the most hastily gotten-up affairs, should never consist of the piano alone. A violin, cornet, or other instrument, at least, should be added.

7. In number, there should be at least eighteen dances, and never more than twenty-four.

8. There should always be a supper, or refreshments of some kind.

9. The pleasures of the evening usually begin with a lively march; then a quadrille, waltz, etc.

VIII. The Supper, Dressing-Rooms, etc.

1. THE supper is usually eaten standing, and ices are generally to be had, even after the supper is over.

2. Good beef tea, made strong, is admirable at balls, and should always be given in cold weather.

3. The meats, fowls, etc., are ready carved. Delicious salads should always be provided, and strong coffee. Wine may be given.

4. A dressing-room for ladies, and one for gentlemen, should be provided with all the necessaries for making the toilet.

5. When guests arrive, they should be met near the door of the reception room by the hostess, who should receive them cordially.

6. Other members of the family should busy themselves in introducing the guests, finding partners for those unprovided, and in other ways seeking to make the affair enjoyable to all.

7. A gentleman cannot refuse an introduction to a lady at a ball.

IX. General Rules of Party Etiquette.

1. IF unacquainted with a dance, a gentleman should never attempt it, unless invited by a lady to do so, and even then should acknowledge his ignorance.

2. The will of the gentleman must be completely subordinated to that of his partner, should she for any cause decline a dance, or having begun it, desire to retire, he must cheerfully acquiesce.

3. In conducting a lady to her place in a set, or to her seat after a dance, offer the arm respectfully.

4. A lady has the right to decline introductions at public balls. An introduction, even at a private ball, does not necessitate after-recognition.

5. Any excuse offered by a lady is valid.

6. A gentleman who has been declined as a partner should not ask any lady in hearing to dance in that set, but may go to another part of the room and do so. The reasons for this are obvious.

7. If a lady refuse a dance, for which she has no prior engagement, and no good excuse, a gentleman should not ask her again to honor him.

8. If a lady pleads only that she doesn't like the particular dance about to begin, the next may be asked for. Should she plead excessive fatigue, she should not dance again, as her inconsistency would have the appearance of falsehood.

9. Never attempt "fancy" steps in dancing, and do not dance too well—that is, with the air of a dancing master. Walk gracefully through quadrilles, and in waltzing, do so with a quiet grace unmarked by effort.

10. The formula: "May I have the pleasure?," "Allow me the pleasure," or, "Will you honor me?" is all that is necessary in asking for a dance.

11. It is the duty of the escort to attend the lady at the end of each dance, to hold her gloves, fan, boquet, etc., to see that she does not enter or cross the ball-room alone, and to provide her with partners.

12. Unmarried ladies should restrict themselves to two dances with any gentleman; more are apt to cause remark.

13. Never occupy the seat next to a lady you do not know; if you cannot obtain an introduction, do not embarrass her by taking that seat.

14. If a lady has no escort, and the hostess has made no arrangement for one to see her to supper, the gentleman who danced with her immediately preceding supper, will escort her to that refreshment. In entering the supper-room, do so slowly and gracefully; avoid all appearance of haste. Ladies should not remain more than ten or fifteen minutes at the table.

15. The escort, if invited to enter the lady's house, on the return from the ball, should, under ordinary circumstances, politely decline, but should call the next day.

X. Evening Parties—The Conversazione.

OF these social gatherings, *conversaziones* are of the most pleasant. They are intellectual gatherings, where amusement and instruction go hand in hand. The *conversazione* is usually given in honor of some distinguished guest, who is either a literary man, a warrior, explorer, or other celebrated personage. All the guests should be introduced, and the conversation should be general, and may be interspersed with music, dancing, etc. On the Continent of Europe, a recital of some interesting incident, the reading of a poem or essay, or the singing or execution of some brilliant piece of music, often forms part of the programme. The guests should be carefully selected for some distinction, as above suggested; in fact, the endeavor usually is to make them typical gatherings, in which there shall be a mutual interchange of ideas.

XI Concerts, Theatricals, etc.

A VERY pleasant way of passing the long winter evenings is the organization of neighborhood talent into amateur theatrical and concert companies. In these combinations, it is best to appoint a permanent stage manager, from whose casts or distribution of parts there shall be no appeal nor sulking.

1. Where there is no permanent manager, that duty usually falls upon the host or hostess of the house in which the party may be assembled.

2. Farces, one-act, or at most, two-act comedies and burlesques are the best selections, as they require little stage room, and not over one change of scene or costume.

3. The amateur company may be merely a neighborhood affair, or it may give representations in neighboring towns. It is not only a very pleasant, but a highly useful amusement, leading to a study and appreciation of the highest grade of literary productions, and educating a taste for the best and most classical works.

4. Light suppers should be provided at the places of meeting, and between the acts there may be intermissions for promenading, handing around cakes, ices, etc.

5. Loud bravoes, clapping the hands boisterously, and other rude methods of applause, are to be avoided.

6. A code of by-laws and a schedule of fines for non-attendance, failure to rehearse or to know parts, etc., should be adopted.

XII. Parlor Lectures.

The etiquette of tea parties, lawn parties, picnics, and out-of-door parties generally, is everywhere well known, but the parlor lecture is just coming into vogue in polite society, and certainly no better mode of spending an evening to intellectual advantage was ever devised.

1. Two plans obtain: First, the members of a community, or those of them, at least, who compose the more intellectual class, organize a society, any of the members of which are supposed to be in readiness to respond to an invitation to lecture before the society, upon some subject to be chosen by the lecturer

2. The society meets at the residence of one of its members every week, or if deemed better, every two weeks, and as there is a lecture at every meeting, there is room for considerable interchange of ideas, and every member in turn gives the results of his best thought and study.

3. A lady member may contribute a song, the rendition of a brilliant piece of music, or a drawing, painting or piece of art needlework for admiration and criticism. Recitations, equally with original efforts, are acceptable.

4. The second method is to invite from neighboring towns or cities lecturers eminent in some special line of research, and assemble the society in the parlor of one of its number to give him audience. The expense is trifling, when divided among a number, and the instruction and entertainment is beyond computation in dollars and cents.

5. At these entertainments it is usual to provide a table and a lamp with an argand shade for the lecturer, who sits at the table and reads from his notes, or delivers his lecture in a colloquial tone.

6. The lecturer, if employed especially for the occasion, should be treated just as any other guest; he has contributed his time and talent in exchange for your money, and occupies no menial position.

7. The lecture should begin at eight and not occupy more than two hours in its delivery. Music is not necessary at these entertainments.

8. The company should show the greatest respect, even should the subject prove uninteresting either in selection or delivery.

9. Whispering and all remarks will cease from the beginning of the lecture. At its close any questions pertaining to its subject may be propounded to the lecturer. No chairman or other presiding officer is necessary at these lectures.

XIII. Church Etiquette.

GOING to church is so general an English and American custom, that almost every one attends some place of worship. It is a matter of etiquette:

1. To arrive in time, so that the rest of the congregation will not be disturbed by a late entrance.
2. Never intrude into a pew without an invitation. To do so is a trespass upon private property.
3. On entering a strange church, advance a slight distance up the aisle, and wait until the usher, or some pew-owner, invites you to a seat.
4. In escorting a lady to a seat, walk beside her until the pew is reached, then permit her to enter, and follow.
5. If the services have not begun, grave and decorous conversation, in a low tone, is not improper; but there should be no whispering, giggling nor laughing.
6. When the services have begun, no remarks should be made.
7. If the forms, ritual or worship seem singular, there should be no merriment; remember that the Golden Rule is the basis of all etiquette; act as you would wish a stranger to act in the church of your choice.
8. If you should offer a seat in your pew to any one, do so in silence; a slight, graceful gesture will convey the invitation as plainly as words. A hymnal and prayer-book should be passed to the stranger. These should be open at the song or service for the day.
9. The hat, cane, umbrella, etc., should be taken into the pew and carefully placed so that neither cane nor umbrella shall fall and make a noise, which is apt to distract the attention of all.
10. When persons are entering, do not turn around to get a view of them.
11. Never bow across the church to any one after services have begun.
12. No matter how bored you may be by a long, dull and tiresome sermon, never yawn or leave the church. Your sole recourse should be in not again attending when the same minister officiates.
13. If obliged by necessity to leave church, do so as gently and noiselessly as possible, and never during prayers.
14. At the conclusion of a funeral service, wait until the relatives and nearest friends of the deceased have made their exit before leaving.
15. Only dudes and other vulgarians and simpletons gather at church entrances and in front of places of amusement to watch those who enter or come out. Such gaping, ill-bred curiosity only befits the boor, and no gentleman should be guilty of it.

XIV. Etiquette of Visits.

1. INFORMAL and general invitations should seldom, if ever, be accepted. If your society is desired, a specific invitation will be given, and this should always be in writing.

2. Never by hints, or in any other manner, seek an invitation. A letter to friends or relatives, telling them that you will come, if convenient, is the height of ill-breeding, since their only excuse to avoid an invitation, without seeming as ill-bred as yourself, must be sickness. No matter how great a bore a person may be, these self-invitations are usually honored, though always regretfully.

3. No one but a thoroughly stupid, selfish and vulgar person, will be guilty of writing a letter or making a remark, that will force such an invitation.

4. It is best to follow the usual course in these matters.

5. A written invitation should be given, to which a reply, also in writing, should be returned acknowledging, with thanks, the invitation, and naming a day upon which you will arrive.

6. Your host should then meet you at the depot and escort you to his house, where, after meeting the members of the family, you will be shown to the room that has been prepared for you. Here you will make your toilet, put on fresh clothing, and be ready to descend to the meal which is next announced.

7. If it is late at night when you arrive, or your journey has been a long or tiresome one, the hostess will suggest early retiring.

8. Breakfast will also be later than usual, that you may have the opportunity of getting plenty of rest. At breakfast it is usual and proper that the host should give you an idea of the daily routine, the hours for meals, etc.

XV. Rules for General Guidance.

1. YOUR friends, and also the friends of the family, will be invited to meet you while in the city, and you will be taken to church, concerts, theaters, and whatever amusements there may be.

2. Of course, you are not expected to adopt the religion of your host, nor he yours, while you are visiting him. Each may attend different churches; but it is his duty to accompany you to your church, and to call for you after services.

3. Ladies may assist each other in little household duties while on a visit; but in volunteering and accepting such light service, judgment must be used. The time of the guest, as well as the room allotted to him, must be sacred.

4. Should sudden illness occur in the family, you may either volunteer to assist in ministering to the sufferer, or, if you can be of no assistance, it is best to take an immediate leave. Especially is the latter best, if the disease should be contagious.

5. Make no demurrer to calls and amusements suggested by host or hostess, and never show that you are bored or dissatisfied if the calls should be upon dull people, or the amusements not the most entertaining.

6. If you have no opportunity to return hospitality, small presents to the hostess

and her children should be made, not as a payment, but as a complimentary testimonial of the pleasure you have received.

7. Never, to use a homely phrase, "outstay your welcome." Remember that your entertainers are not only put to additional expense while you are with them, but that their mode of life is disarranged and you are occupying a great deal of their time.

8. The announcement of your departure should be made, under ordinary circumstances, at least a day ahead.

9. While expressing regrets that you must leave, a well-bred host and hostess will not annoy you with solicitations to stay. In these matters it is taken for granted that the guest has a proper idea of the length of time he or she should remain.

XVI. Etiquette of the Funeral.

EVEN Death, the grim visitor that waits alike upon all, demands a special etiquette. When the loving circle of the home has been broken, and one of its members lies in the icy sleep of death, the grief of the remaining members demands that some dear friend shall step forward and make all the arrangements for the funeral.

1. This friend, especially if inexperienced, should call to his aid the undertaker who is to have charge of the funeral, and whose advice in regard to the ceremonial will be valuable.

2. If possible, announcements of the death and the time and place of burial should be made in the local papers, and an invitation extended to all friends to be present.

3. Invitations, written or printed, may also be sent out, of which forms will be given in the proper place.

4. It is usual for friends to call only to offer their services or leave cards while the funeral preparations are being made.

5. It is best that the friend in charge should receive all calls, so as to relieve the afflicted family from intrusion.

6. To prevent ordinary calls, as soon as the death occurs, crape should be affixed to the door or bell-knob or knocker.

7. If an old person, the crape should be black and tied with a black ribbon; if a child, white, tied with a white ribbon, and if a youthful person, or unmarried, black crape tied with white ribbon is usual.

8. The coffin should rest in the parlor, and it is here that the guests will assemble. Services may be held there, or at the church.

9. Guests must expect no attention from the members of the family. Some relative or chosen friend will receive them upon the sad occasion.

10. If there is to be a sermon in church, the coffin will be placed in front of the altar, and should be covered with a black cloth.

11. After the services an opportunity will be given to friends to take a last view of the dead. This should be done in solemn, decorous order; the congregation mov-

ing in one direction and passing by the coffin. The halt should be short. There should be no conversation.

12. When the services are concluded, the coffin will be borne to the hearse by pall-bearers, who have been selected from the nearest friends of the deceased.

13. The clergyman should now enter a carriage, and precede the hearse to the cemetery.

14. Immediately following the hearse is the carriage bearing the bereaved family; relatives and friends follow in the order of their relationship or friendship.

15. The friend in charge should see the family to their carriage, seat them, and close the door. He should also arrange the order of precedence, and attend to other matters.

16. At the gate of the cemetery, it is usual to dismount from the carriages and follow the coffin, on foot, to the grave.

17. Flowers used to decorate the coffin and the room, where the dead is lying, should be white. If an aged or married person, the ornament on the coffin is usually a cross; if young or unmarried, a wreath.

18. Societies of which deceased was a member may be notified through the papers to attend, or an invitation to the president of each society for its presence is sufficient.

19. If the death has occurred from any contagious disease, the fact should be stated in the funeral notice, and the invitation to attend omitted.

20. In England it is usual for the most deeply afflicted to remain at home, and in this country excess of grief often prevents their attendance at the grave.

21. Cards should be left for the bereaved family the week following the funeral, and in the second week brief calls may be made.

22. From these calls of condolence persons themselves lately afflicted, or in mourning, may be excused, since they might only renew their own grief.

XVII. Etiquette of the Christening.

1. When a child is to be baptized, near friends or relatives are chosen to act as sponsors or god-parents.

2. If it is the first child in the family, the preference is usually given to the grandfather on the father's side and the grandmother on the maternal side. Should this not be feasible, the other grandparents have the next preference.

3. For other children, and often for the first, other sponsors are chosen.

4. The child, clothed all in white, and held by its nurse, is carried into the church, or if at home, into the room, followed by its sponsors, side by side, but not arm in arm, and they in turn are followed by the parents; or the father alone, if the mother should be unable to attend.

5. When the question, "Who are the sponsors for this child?" is asked, the god-parents simply bow, and the ceremony proceeds. The nurse stands near the baptismal font, the child upon her left arm. Upon her right stands the god-father;

the god-mother, upon her left. The parents are next to them, and behind are the relatives and friends in their order of relationship.

6. Light refreshments may be passed after the ceremony, if performed in the house; if in church, the guests usually disperse to their homes.

7. God-parents are expected to make to baby as rich presents as are in keeping with their means, and they are supposed to look after its future welfare, both temporal and spiritual.

8. No one should volunteer as a god-parent, unless he or she knows that the offer will be gladly received.

9. It is sometimes appropriate for persons of great wealth or high station to so volunteer, when they have good cause to suppose that only the superiority of their position prevents their being invited.

CHAPTER III.

ETIQUETTE OF THE WEDDING, THE ROAD AND THE CAPITAL.

I. ETIQUETTE OF WEDDING ENGAGEMENTS.—II. THE WEDDING.—III. THE CEREMONY IN CHURCH. IV. WEDDING RECEPTIONS.—V. ETIQUETTE OF THE ROAD—TRAVELING.—VI. LADIES TRAVELING—THE ESCORT.—VII. GENERAL RULES FOR TRAVELING.—VIII. ETIQUETTE IN WASHINGTON.—IX. ETIQUETTE OF SHOPPING.—X. SPECIAL RULES OF DEPORTMENT.—XI. GEORGE WASHINGTON'S ONE HUNDRED RULES OF LIFE GOVERNMENT.

I. Etiquette of Wedding Engagements.

IT is now the vile and outrageous fashion to announce engagements in the society items of the press. This fashion is an European innovation, entirely unsuited to this country, and is of doubtful merit, except among people so great as to have their marriages a matter of public importance.

2. When an engagement has been formed, the matter should be promptly announced to the members of the two families, and if living sufficiently near, visits should be exchanged; the gentleman's family making the first call.

3. This call should, of course, be returned within a reasonable time.

4. Should the families reside some distance apart, a pleasant interchange of compliments, beginning as above, with the gentleman's family, may take place by letter.

5. Interchanges of presents between the engaged couple are not inappropriate; those of the lady usually consisting of articles of her own handiwork, as a watch-case, slippers, etc

6. The fore-finger of the left hand is the one upon which the engagement ring is worn; the wedding ring is placed upon the third finger (that next to the little finger) of the same hand.

7. An invitation to a lady known to be engaged should include her lover, and if it does not, she is justified in ignoring it.

8. When from any cause it becomes necessary to break an engagement, it should be done in a frank manner. Of course there must be grave cause to justify such a rupture.

9. There should be no absurd jealousies; neither should there be, upon either side, the remotest approach to the vulgar and odious practice of flirting with others. The behavior toward each other should be respectful and tender, but in it there should be nothing of love-sickness

II. The Wedding.

1. WEDDINGS are usually celebrated in the early days of the week, and rarely later than Thursday. The month of May is usually avoided, owing to a superstitious belief in its ill luck.

2. A marriage should never take place in Lent. June, July and August are the months usually chosen by fashionable people in Europe; in this country no particular months seem to have a preference. Forms for wedding invitations will be given among other forms.

3. A private wedding requires neither bridesmaids nor groomsmen; in church there may be any even number not exceeding eight of each.

4. Bridesmaids should be younger than the bride.

5. The principal decoration of the dresses should be flowers.

6. The ornaments of the bride should be few.

7. Plain white of rich fabric with garland and veil is the most appropriate costume.

8. Bridesmaids may wear more jewelry, but should not wear richer dresses than the bride.

9. The "best man," or nearest friend of the groom, should relieve him from all bother incident to the ceremony, and the tour which succeeds it, such as making arrangements, procuring tickets, checking baggage, settling bills, etc.

III. The Ceremony in Church.

1. When married in church, it is the duty of the bridegroom to send a carriage for the minister who is to perform the ceremony.

2. The front seats of the church are usually separated from the others by a white ribbon, and are reserved for the near relatives and friends of the bride and groom.

3. Ushers, wearing each a white rose, are on hand to show guests to seats.

4. The ushers attend the bridal party at the door of the church and escort them to the altar.

5. The procession is formed thus: the "best man" and chief bridesmaid lead the way to the altar, followed by the other attendants, in the order of their preference, and followed by the bridegroom with the bride's mother on his arm. Last comes the bride upon the arm of her father.

6. Arriving at the altar, the bride takes her position upon the left of the groom, the bridesmaids standing upon her left, slightly in the rear; the bridegroom's attendants standing to his right, also slightly to the rear. The father, who gives away the bride, stands just behind the young couple, and slightly in advance of the bride's mother.

7. If a ring is to be used, it is the duty of the chief bridesmaid to remove the glove of the bride.

8. The responses should be in a low tone, but clear and distinct.

9. After the ceremony, the first to approach and speak to the bride will be her parents; next, the parents of her husband; then relatives and friends in the order of nearness.

10. The bridegroom now gives his arm to the bride and moves toward the vestry, where he raises her veil and kisses her.

11. She may be also kissed by a few of her female friends and her relatives.

12. A wedding march is appropriate as the exit from the church begins, and as the carriages drive off the church bells should ring merrily.

IV. Wedding Receptions.

1. A SHORT reception is usually given at the bride's home. Those who call to congratulate the happy couple must first address the bride, unless acquainted only with the groom, who will present them.

2. At the wedding feast, the newly-married pair occupy the center of the table, the bride's father and mother sitting, as usual, at the head and foot of the table. The bride sits on the groom's right.

3. After retiring from this meal, the bride changes her dress, putting on her traveling costume.

4. This costume should be neat and quiet.

5. Occasionally the bride is wedded in her traveling dress, but this is seldom the case when the wedding is in church.

6. Upon the return of the bride, she will be assisted in her reception by her mother, sister, or some intimate friend.

7. These calls should be returned within a week.

V. Etiquette of the Road—Traveling.

EVEN in the bustle of railway and steamer travel there is a certain etiquette to be observed. Nothing more severely tests the politeness of a person than the disagreeable conditions that often attend modern journeying. The crowded car, the impatient throng at the ticket-office, and often the surly and ill-bred servants of railway corporations, try the patience and good breeding of the traveler. To be cool and careful, neither dilatory nor in a violent hurry, and to take matters with imperturbable serenity, mark the man of culture, being familiar with the various modes and miseries of travel.

1. The novice is apt to be too suspicious or too confiding.

2. Few questions should be asked, and when possible, always of railway porters, passenger directors and other public servants.

3. If compelled to ask a question of a stranger, do so in a polite manner, prefacing your question with, "Excuse me, sir!" or, "I beg your pardon, sir!"

4. To ask questions, especially needless ones, betrays the greatest verdancy, and the questioner is apt to be taken advantage of.

5. If with a friend who is familiar with a city, never appeal for any information about it to a stranger. Such conduct is an insult to the common sense or honesty of your friend.

6. Always, when possible, procure through tickets. Buy your tickets a day in advance, if possible, and thus save haste and confusion. The lower central berths in a sleeping-car are the best.

7. Never thrust your attentions on a lady traveler, but render any assistance in your power. Never attempt to force a conversation or an acquaintance in return for any service.

8. If appealed to, give your advice, but never intrude it.

9. Of course if you should hear false information given or an improper place recommended, you may in a genteel way inform the intended victim of the true state of the case.

10. Never be anxious to trumpet the praises of any hotel. If your opinion is asked, merely say that you put up at such a house and find it well kept, and a desirable stopping place. To a friend who is visiting a strange place, with which you are acquainted, you may recommend a hotel.

11. If requested, or if you see that a lady or gentleman is an invalid, timid, or disabled, you may offer to procure their tickets, check their baggage, etc,

VI. Ladies Traveling—The Escort.

1. If traveling with a lady, it is your duty to perform for her the services just mentioned.

2. When you reach a city, you should engage a carriage, and see her to her abode. If it is a hotel, escort her to the parlor, excuse yourself, go to the office and register her name, and get a desirable room for her.

3. Then escort her to her door and leave her, having mentioned the hour of the next meal, for which you should await her in the parlor, and to which you should accompany her.

4. It is neither your duty nor your privilege to pay for the tickets of a lady with whom you may be traveling. Small items, such as a dinner, 'bus and car fare, you may pay; but she will prefer to pay any considerable sum herself, and should be allowed to do so, as she thereby retains her self-respect and independence.

5. No lady can afford to accept money favors from any gentleman, save a very near relative.

6. No lady should accept promiscuous attentions when traveling; to do so is unwise and dangerous.

7. If any person seeks to intrude himself upon a lady, the proper reserve will generally cause him to resume his distance. Should this not be sufficient, an appeal to the conductor or other official will secure her from further annoyance.

8. Among ladies, it is customary to take turn about in paying street car fares, also for lunches, ices, etc.

9. In this matter, there should be a reciprocity; a lady who suffers her friend to pay all such little costs is wanting in breeding. Such conduct in a man would be called selfish vulgarity, and he would be promptly denounced as a "sponge."

VII. General Rules for Traveling.

1. In traveling, be considerate of others. Never occupy two seats while others are standing.

2. Make way promptly and pleasantly; not as if you were doing a favor, but only granting a just right.

3. Never keep your window open, if it annoys a lady, or even a gentleman who is an invalid, or weaker or much older than yourself.

4. You are not obliged to continue an acquaintance begun while traveling, though if mutually agreeable there is no reason why a friendship may not be formed thus.

5. Never permit a lady to stand while you are seated. An aged gentleman demands the same courtesy.

6. Never enter into disputes with either passengers or officials. Avoid all causes of contention. If any imposition is practiced, it may be remedied by an appeal to the higher officials of the company, or if they should refuse to act, by bringing the matter into court.

7. Avoid all games of chance with strangers, however innocent they may seem. It is only a very silly person who is ready to volunteer to assist in making up a game of cards with strangers. A gambler, able to protect himself against swindling, might do so with impunity, but a gentleman can not afford to take such chances.

8. Avoid any exhibitions of activity in getting on and off trains while in motion, or springing to a wharf before a steamer has fairly landed. The slightest slip may cause a loss of a limb or of life.

VIII. Etiquette in Washington.

The etiquette of the courts of Europe is firmly established by a code of enactment, as rigorously observed as is the civil, or penal code of the country.

1. Even at our national capital, a code has been agreed upon, to prevent the frequent clashing of rival claims. The order of public precedence has been fixed thus:

>*First*, The President and members of his family.
>*Second*, Heads of Departments.
>*Third*, Governors of States.
>*Fourth*, Justices of the Supreme Court.
>*Fifth*, Members of Congress.
>*Sixth*, The Diplomatic Corps.
>*Seventh*, Military and Naval Officers.
>*Eighth*, All others.

2. This is the order of reception on the First of January and the Fourth of July, when the public receptions of the President are held, beginning at noon.

3. On all other occasions the Supreme Judges rank next to the President and Vice-President.

4. The Vice-President pays a final call on the President on the assembling of Congress.

5. From all others he is entitled to the first call. These calls he may return in person or by card.

6. The members of the Supreme Bench call upon the President and Vice-President on the meeting of their court, which is in December. Also on New Year's day and the Fourth of July.

7. From all others they are entitled to the first call, and thus, socially, outrank all others, except the President and Vice-President.

8. Cabinet members must call first upon the President, Vice-President, the Supreme Judges, Senators, and the Speaker of the House.

9. All others must call first upon them.

10. Senators call first upon President and Vice-President, upon the Supreme Judges, and the Speaker of the House.

11. They are entitled to the first call from all others.

12. The Speaker of the House calls first upon the President, Vice-President, and Judges of the Supreme Court; but all others must call first upon him.

13. Members of the House of Representatives call upon the President, Vice-President, Supreme Judges, their Speaker, Senators, Cabinet Ministers, and Foreign Ministers.

14. Foreign Ministers call upon the President, Vice-President, Cabinet Officers, the Supreme Justices, and Speaker of the House. All others should call first upon them. Judges of the Court of Claims rank next in social order.

15. The wives of cabinet ministers hold receptions every Wednesday from two or three o'clock until half past five. Their houses are open to all. Refreshments, generally of a light character, as coffee, chocolate, tea, cakes, etc., are provided. Callers at their Wednesday receptions are entitled to two return calls; the first by the ladies of the family who have the official card of the minister; the second call is an invitation to an evening reception.

16. Cabinet officers are expected to entertain almost all governmental officials and members of the Diplomatic corps and their families at dinner parties. With other officials, as Senators, Representatives, etc., it is optional whether they entertain or not.

IX. Etiquette of Shopping.

It is impossible for the well-bred person to treat any one with wanton rudeness, and in fact a due amount of consideration for all is the strongest mark of good breeding.

1. Though a person may be employed for no other purpose than to wait upon all comers, yet he or she is entitled to polite and considerate treatment. It is only the rude and boorish who treat with scorn or roughness any of those whom they may deem their inferiors.

2. The truly gracious and refined see in an humble position still the greater need for courteous treatment, lest they may add further bitterness to a lot already sufficiently hard.

3. To go into a shop, ask the clerk to pull down a lot of goods, look them over, and cause him to lose his temper, and probably soil the goods, without any intention of buying, is neither fair nor honest. Such a proceeding is a theft of his time, and discredits him in the eyes of his employer, who thinks he should have made a sale.

4. Many well-bred ladies, who desire an article and cannot find it, reward the clerk's patience with a small purchase, if they can find at his counter anything they can use.

5. Should a stranger be examining goods, never touch them until she has finished her examination.

6. Never volunteer favorable or unfavorable opinions as to the merits of goods. If a friend asks an opinion, you may give it politely and mildly.

7. Never speak about goods you may have seen in other shops; if you do not like the goods exhibited, take your leave in a quiet way.

8. Some clerks seek to influence a sale by making the statement that Mrs. ——— (naming usually some wealthy or well-known person) has just ordered a pattern from it. This is exceedingly silly, and is a reflection upon the taste and judgment of the examiner, as well as an insult to her independence. Merchants should discourage such ignorance among their clerks.

9. Never haggle over prices. Deal only at stores where there is one price, and where every one is treated alike. No honest merchant has two prices for his goods; one for the unwary and ignorant, the other for the shrewd and haggling buyer.

10. The largest and oldest shops in cities are usually the best; as large and stable businesses can only be built up by honest dealing.

11. Treat all with whom you come in contact with courtesy, from the merchant to the cash-boy. Remember all of them have rendered you some equivalent for the money you may have spent.

X. Special Rules of Etiquette.

1. Be cool, quiet and self-possessed in all situations.

2. When you enter a room, bow to all therein. You can afterwards more particularly salute your friends.

3. Never go into company with soiled clothing: use no musk, and remove all offensive odors from clothes and person.

4. "Cleanliness is next to godliness," and is one of the cardinal points of good breeding.

5. Be courteous to all ladies, whatever may be their rank.

6. Gentlemen never cast slurs upon the softer sex, and he is churlish, as well as ill-bred, who maligns woman in general.

7. Shakspeare gives many excellent general rules for social government, amongst them: "Be thou familiar, but by no means vulgar," showing that even among friends intimacy should not degenerate into vulgar disregard of all conventionalities.

8. Beware of sudden familiarities.

9. Your dress should be of as rich material as you can afford. but not flashy. In cut and color it should be quiet and modest.

10. Be prompt in keeping engagements and punctual in meeting all obligations.

11. Avoid borrowing or lending. No man can be independent and but few honest when in debt.

12. In speaking of friends and acquaintances to others, no matter how intimate, give them the prefix of Mr., Miss or Mrs., as the case demands.

13. Avoid sneering and sarcasm.

14. Be not witty at the expense of another; no humor is permissible but that which is perfectly innocent.

15. Punning is a weak apology for wit, and should be eschewed.

16. Never look over anyone's shoulder while reading a book, paper or letter.

17. Never search through a card basket or an album unless invited.

18. Do not be ashamed to tender an apology, if in the wrong. Always accept one with gentle courtesy.

19. If a secret is intrusted to you, never reveal it; it is neither honorable nor honest to give away that which is not yours.

20. Exaggeration is foolish. If you must speak, speak the truth.

21. Never display any form of curiosity; it is a despicable trait of character to be curious about things that do not concern you.

22. Never flatter. A delicate compliment may be innocently offered and well received, but flattery is odious.

23. Do not whisper in society, and avoid signaling to friends in company.

24. Avoid the use of languages unknown to the generality of the company

25. Never be dogmatic, nor make dictatorial assertions.

26. In entering a house, even your own, always remove your hat, and do not be boisterous or restless.

27. It is better to have no associations than to have evil ones. Good books or good thoughts are better than evil companions.

28. Never back your opinions with an oath or a bet.

29. Avoid all profanity, loud talking and boisterous merriment.

30. At the breakfast table, politely salute all assembled, if it be the first time of meeting for that day. A cheerful "good morning" should be passed between the members of the home circle.

31. Of course, no gentleman will chew tobacco in a church, parlor, or in the presence of ladies.

32. Be natural. Avoid eccentricity and affectation.

33. Do not ape any one.

34. Your room is the place for making your toilet. Do not arrange your clothing in company.

35. In company avoid paring or cleaning your nails, picking your teeth, scratching your head, etc.

36. Be not egotistical nor pompous. These faults would cloud the most brilliant genius; how much more so mere ordinary mortals.

37. Volunteer your aid to any lady in distress, or to an invalid or aged person.

38. You cannot afford to let one beneath you in station exceed you in politeness. Be courteous to every one.

39. Boast of nothing; especially not of your wealth, since that is the least qualification of a gentleman.

40. A wife or husband should speak respectfully of each other, and should be mentioned as Mr. ———, or Mrs. ———.

41. Ostentation is silly and vulgar.

42. Never make your ailments or your troubles a topic of conversation, but treat sympathetically those who do.

43. Never contradict in a rude manner. Always point out a mistake with gentle courtesy.

44. Never soil or mark a book that has been lent to you. Return it in good order; and, if unavoidably injured, return it and a fresh copy also.

45. Never correct a person in grammar, deportment, or in a mistake that does not implicate you in a wrong.

46. Never remark upon the personal deformity or mental peculiarities of acquaintances.

47. Upon the street, the lady must first recognize the gentleman.

48. In dancing, gloves should always be worn.

49. You have no right to forget an engagement. To do so without a prompt and ample apology is equivalent to an insult.

50. A promise made must be carried out, if possible, at any cost.

51. No lady ever sneers at, or comments upon, the dress of another in the streets.

52. Avoid all slang and florid adjectives. The conversation, like the manners and the morals, should be quiet, chaste, and simple.

53. Learn to say "No," to all evil invitations and promptings; the true gentleman should be courageous as well as kind.

54. No amount of learning, wit and genius can atone for coarseness and ill-breeding.

55. Depend neither on wit, wealth, nor raiment for your status in society.

XI. George Washington's One Hundred Rules of Life Government.

BUT few men display, as did the "Father of his Country," the varied talents of the soldier, the statesman, the farmer, and the man of business, and if the code of self-government, which he is said to have prescribed to himself at the early age of thirteen, had anything to do with his success—and no doubt it did—it is certainly worthy of the deep consideration of all.

1. Every action in company ought to be some sign of respect to those present.

2. In the presence of others sing not to yourself with a humming noise, nor drum with your fingers or feet.

3. Speak not when others speak, sit not when others stand, and walk not when others stop.

4. Turn not your back to others, especially in speaking.

5. Be no flatterer; neither trifle with any one that does not delight in such familiarities.

6. Read no letters, books or papers, in company except when necessary; then ask to be excused.

7. Come not near the books or writing of any one so as to read them unasked.

8. Let your countenance be pleasant, but in serious matters somewhat grave.

9. Show not yourself glad at the misfortunes of another, though he were your enemy.

10. They that are in dignity or office have in all places precedency; but whilst they are young they ought to respect those that are their equals in birth or other qualities, though they have no public charge.

11. It is good manners to prefer those to whom we speak before ourselves, especially if they be above us—with whom in no sort should we take the lead.

12. Let your discourse with men of business be short and comprehensive.

13. In writing or speaking give to every one his due title, according to his degree and the custom of the place.

14. Strive not with your superiors in argument, but always submit your judgment to others with modesty.

15. When a man does all he can, though it succeeds not well, blame not him that did it.

16. It being necessary to advise or reprehend any one, consider whether it ought to be done in public or in private, presently or at some other time, also in what terms to do it.

17. In reproving any one, do it with no sign of choler, but with sweetness and mildness.

18. Mock not, nor jest at anything of importance.

19. Break no jests that are sharp and biting.

20. Laugh not at your own wit.

21. Wherein you reprove another be unblamable yourself, for example is more impressive than precept.

22. Use no vituperative language against any one.

23. Avoid all blasphemy.

24. Be not hasty to believe disparaging reports against any one.

25. Avoid all gossip and scandal.

26. In your dress be modest. Affect nothing singular or unusual.

27. Go to no extreme of fashion; be well but not gaudily dressed.

28. Play not the peacock, looking about on every side to see if you be well decked.

29. Never play with your dress in company, nor look at yourself to see if your clothes fit, or if they be awry.

30. Associate yourself with men of good quality if you esteem your own reputation.

31. It is better to be alone than in evil company.

32. Let your conversation be without malice or envy.

33. When angry, beware of haste; give reason time to resume her sway.

34. Do not urge any one to discover to you his secrets.

35. To reveal the secrets of another is base and dishonest.

36. Do not tell extravagant or marvelous stories.

37. Utter not base and frivolous things amongst grown or learned men.

38. Do not discourse on learned subjects to the ignorant, neither use obscure words or language in conversation with them.

39. Speak not of doleful things in time of mirth, nor at the table. Never speak of melancholy things at inappropriate times; of death and wounds; and if others mention them, change if you can the discourse.

40. If you must tell your dreams, do so only to intimate friends.

41. Break not a jest when none take pleasure in mirth.

42. Laugh not loudly, nor at all without occasion.

43. Do not talk loudly, nor exhibit a boisterous demeanor.

44. Deride no man's misfortunes, though there seem to be cause to do so; neither laugh at the calamity of any one.

45. Speak not injurious words, neither in jest nor in earnest; scoff at none, even though they give occasion.

46. Be not forward, but friendly and courteous.

47. Salute all who pay you that courtesy; hear and answer politely.

48. During a conversation affect not sad and pensive airs, or abstraction.

49. Neither detract from others nor be excessive in commending.

50. Go not where you are doubtful of a welcome.

51. Give no advice without being asked; then let it be brief.

52. When two are contending take not the part of either.

53. In indifferent matters, go with the majority.

54. Do not presume to correct the mistakes of others; that is the privilege of parents, masters and superiors.

55. Gaze not rudely on any one, neither note their deformities or peculiarities.

56. Do not use any foreign tongue in company, except to one ignorant of English.

57. Let your conversation be modest, and your language that of good society.

58. Speak plainly; do not drawl out your words, nor speak through your nose.

59. Treat solemn and sacred things with reverence.

60. Let your conversation indicate thought; silence is better than idle talk.

61. When another is speaking, be attentive. Should he hesitate for words, do not supply them. Never interrupt another while talking.

62. Select the proper time to talk upon any kind of business.

63. Never whisper in the company of others.

64. Make no odious comparisons.

65. Should you hear any one commended for any act, commend not another for the same or a greater action.

66. Be not curious to learn the affairs of others.

67. Never intrude yourself upon others that speak in private.

68. Undertake not what you cannot perform; make no promises you cannot fulfill.

69. Never attempt in an argument to bully others; give to every one perfect liberty in expressing himself, and always be willing to submit to the majority.

70. Be not tedious in discourse; make not many digressions, nor repeat the same tales.

71. Speak not ill of the absent; it is both cowardly and unjust.

72. Let all your pleasures be pure and manly.

73. Neither speak nor laugh when your superiors are talking; listen respectfully and without impatience.

73. Never be angry at the table; if annoyed, conceal your vexation, lest others, too, be made unhappy.

74. Jog not the table or desk at which another is reading or writing.

75. Lean not on any one, nor slap friends and acquaintances on the back or shoulder.

76. Affect not singularity in dress, manner or conversation.

77. Avoid many and extravagant adjectives.

78. Never look on when another is reading or writing.

79. Avoid sudden friendships.

80. Distrust those that protest vehemently.

81. Make no friendships with silly or evil persons.

82. Never seem to indorse any one that is disreputable.

83. It is best to avoid association with those who show any disrespect for old age.

84. Observe the customs of those older and wiser than yourself

85. Avoid becoming a borrower or lender of money.

86. Never do any action of which you have not well studied the consequences.

87. Be neither prodigal nor miserly; avoid both extremes.

88. A good listener is more esteemed by all than a good talker.
89. Avoid all vulgar ostentation; do nothing for show.
90. Be upright in all dealings.
91. Never be outdone in courtesy or politeness
92. Live temperately, but be not ascetic.
93. Avoid hypocrisy; never seem to be what you are not.
94. Avoid fanaticism and be not dictatorial nor too positive.
95. Never oppress nor deride those weaker, poorer or more ignorant than yourself.
96. Avoid all games of chance, especially with those who make a proposition of cards or dice.
97. Never attempt to make good an assertion with a wager.
98. Live not only honestly, but honorably; be chaste, moral and correct in all things.
99. Obey your parents in all things.
100. Revile not religion; when you speak of God, his works or attributes, do so reverently, and in church let your conduct be serious and solemn.

CHAPTER IV.

FORMS, LETTERS, FRENCH PHRASES, ETC.

I. WRITTEN INVITATIONS TO DINNER AND SOCIAL PARTIES.——II. OTHER INVITATIONS—EVENING PARTY.——III. ACCEPTANCES AND REGRETS.——IV. FRIENDLY INVITATIONS.——V. FRIENDLY ACCEPTANCES AND REGRETS.——VI. LETTERS OF INTRODUCTION.——VII. LETTERS OF RECOMMENDATION.——VIII. ASKING A LOAN, AND THE REPLY.——IX. DIRECTING A LETTER.——X. SUGGESTIONS FOR LETTER-WRITERS.——XI. STYLES OF CARDS.——XII. FRENCH WORDS AND PHRASES IN GENERAL USE.——XIII. TREATMENT OF CHILDREN.——XIV. SEVENTY-FIVE CARDINAL RULES OF ETIQUETTE.——XV. ALPHABET OF ETIQUETTE.

I. Forms of Written Invitation to Dinner and Social Parties.

Mr. and Mrs. E. H. Emory request the pleasure of Mr. and Mrs. C. H. Pierce's company, on Tuesday, Jan'y 9th, at five o'clock.

No. 27 Caroline Terrace.

Mr. William Gaw requests the pleasure of Mr. John A. Wheeler's company, at dinner, on Tuesday, April 1st, at 7½ o'clock.

Windsor Hotel.

INVITATION TO SOCIAL PARTY.

Mr. and Mrs. T. S. Strader request the pleasure of Mr. and Mrs. C. Adams' company, on Monday, Aug. 25th, from 8 to 12 o'clock.

No. 819 Boulevard Haussman.

Mr. and Mrs. Reuben Springer request the pleasure of Mr. and Mrs. Samuel Randall's company, on Wednesday, March 11th, at 8 o'clock.

85 East Ratcliffe Road. *Soirée Dansante.*

INVITATION TO MUSICAL PARTY.

Compliments of Mrs. Jno. H. Scudder, to Mr. and Mrs. J. J. Terry, and requests the pleasure of their company on next Friday evening, June 9th, to meet members of the Arion Society.

27 North Beaumont Ave.

II. Other Invitations—Evening Party.

Mrs. Ridgely Eveline DeVere requests the pleasure of the company of Mr. and Mrs. Jno. Estin Cook and family, to evening party, Tuesday, December 8th, at 8 o'clock.

Music and cards.

Mrs. Junius B. Allison requests the pleasure of Mr. and Mrs. Renick DeBar's company, on November 19th, at half-past nine o'clock.

No. 9 Benton Place.

INVITATION TO CONVERSAZIONE (ENGLISH).

Mrs. Brocton DeLisle-Monk requests the pleasure of the company of Mr. and Mrs. Lester Wallack and Miss Mary Anderson, on Tuesday, Sept. 18th, at 8 o'clock, to meet Mr. Henry Irving.

    58 Park Place, Chodmondely,
       North Brompton Road, E. S.

NOTE.—Invitations are often dated similarly to letters, thus:

                                       February 18th, 1883.

Mrs. Redway Benton requests the pleasure of Miss Bernice Plympton's company to an evening party, on Thursday next, at 8 o'clock.

    Benton Villa.

INVITATION TO CHILDREN'S PARTY.

Miss Pansie Abel requests pleasure of Miss Bertie Lindell's company on Monday evening, January 12th, from 5 to 10 o'clock.

    No. 18 Sedgwick St.

INVITATION TO A BALL.

Invitations to balls are almost invariably printed, and a form is hardly necessary here, since they can be supplied at any printing office. A written invitation should follow this, or a nearly similar form:

        The pleasure of your company is requested
                    at a Hop,
at Mrs. S. H. Allen's, on Wednesday evening, December 28th, 1883,
                    at 9 o'clock.

III. Acceptances and Regrets.

Mr. John Jones accepts, with pleasure, Mr. Robert Smith's kind invitation for Monday evening, Aug. 12th.

    Southern Hotel.
       Tuesday, Aug. 6th, 1883.

Mrs. Nicholas Longworth accepts the kind invitation of Mr. and Mrs. Redway Benton, for Thursday evening, May 12th.

    Caroline Terrace,
       Tuesday, May 5th, 1883.

REGRETS.

Mr. John Smith regrets that he cannot accept Mr. Edward Brown's polite invitation, for Tuesday evening.

    Buckingham Hotel,
       Friday, March 11th, 1883.

A previous engagement, sudden illness of any member of family, or any other valid excuse may be added. When an acceptance has to be revoked, the excuse should always be given, and should be a sufficiently weighty one.

IV. Friendly Invitations.

When persons are intimately acquainted, of course, less formal invitations may be extended. A few forms are here given:

TO DINNER.

John Corlew, Esqr. New York, May 20th, 1883.
    Friend Corlew:

        I have invited a few friends to a dinner, which is set for Wednesday, May 26th, at six o'clock, and it would give me great pleasure if you would form one of the number.

    Please let me know if you are at liberty to accept this invitation, and oblige

                Yours Very Truly,

Hoffman House. A. J. Thomas.

TO A PICNIC.

Miss Milly Adams.
    Dear Milly:

        We have made up a little party to go to "The Heights" on a picnic party, on the 16th. We anticipate quite an enjoyable time, which, on my part at least, will be heightened by your making one of the number. Will you do so? Please let me know, and oblige,

                Your Friend,

Honeysuckle Glen. Ruth Herndon.

FOR A DRIVE.

Miss Minnie Martin. St. Louis, Oct. 20th, 1883.
    My Dear Minnie:

        I am going for a drive in Forest Park to-morrow, and would be delighted if you would accompany me. Can you not do so? Mamma will act as chaperone. We start at two o'clock. Let me know if it will be convenient for you to go.

                Very Truly, Your Friend,

99 Lucas Place. Mamie Stevens.

V. Friendly Acceptances and Regrets.

The acceptances to these informal invitations require no greater amount of ceremony; the following answers would be suitable.

A. J. Thomas, Esqr. N. Y., May 21st, 1883.
    Friend Thomas:

        Your kind invitation was duly received, and it affords me great pleasure to be able to accept. In the meantime, believe me, in haste, but as ever,

                Your Friend,

Fifth Avenue Hotel. John Corlew.

Miss Ruth Herndon. Brunswick, Mo., June 12th, 1883.
    Dear Ruth:

        I received your note of the 18th with invitation, which it gives me great pleasure to be able to accept, so you may count me as one of the party. I don't doubt but that we will have a good time, though I fear I shall be able to contribute but little to your enjoyment, not from want of desire, however. You may certainly expect me, and until then, adieu.

                Yours, Ever,

Rose Hill. Milly Adams.

FORMS FOR REGRETS.

A regret should be worded somewhat in this manner:

Miss Mamie Stevens.   St. Louis, Oct. 20th, '83.
    Dear Mamie:

                     I rec'd your note with its kind invitation a few moments since. I am so sorry that a prior engagement compels me to decline. You don't know how much I regret my inability to accept. With kindest regards to your mother, believe me, dear Mamie,
                                               Ever Yours,
        Beaumont Flats.   *Minnie Martin.*

VI. Letters of Introduction.

A GENTLEMAN INTRODUCING A FRIEND.

Mr. John Buckmaster.   Boston, Oct. 15th, 1883.
    Friend Buckmaster:

                     This will introduce to you my friend, Joachin Miller, whom I am desirous of having you meet. Mr. Miller visits St. Louis on social and business matters, and anything you may be enabled to do that will add to his pleasure or forward his interests, will be properly appreciated, and on occasion reciprocated by
                                             Your Friend,
                                                       Wallace Overton.

A LADY INTRODUCING A LADY.

Miss Della Mansfield.   Denver, Nov. 1st, 1883.
    Dearest Della:

                     Allow me to take this occasion to introduce to you Miss Stella Ball, the bearer of this letter. I know you have heard me speak of her a hundred times, and believing that an acquaintance would confer mutual pleasure, I have urged her to call upon you, while in your city. Any attention bestowed upon her will be taken as a personal favor by
                                             Your Friend,
                                                     Stasia Mansfield.

VII. Letters of Recommendation.

GENERAL LETTER RECOMMENDING A SERVANT.

                                             New Orleans, Sept. 9th, '83.
    To Whom it May Concern:—*The bearer of this letter, Robert Hawkshaw, has been in my employ for the past three years, as groom and driver, and this is to certify that he has always proved himself honest and efficient. I therefore take pleasure in recommending him to any one desiring the services of a careful and competent coachman.*
                                                       John P. Rogers.
    98 Felicity Road.

SPECIAL LETTER RECOMMENDING A CLERK.

Fletcher, Ames & Co.,   Gunnison, Col., Oct. 12th, 1883.
    New York.

            Gentlemen:—*The bearer of this letter, Mr. Willard Hopkins, has been in our employ for the last two years, and we have found him efficient and honest. His large acquaintance throughout the West induces him to seek a more extended field of operations, hence, his visit to your city to obtain a position as traveling salesman.*

    While we are loath to part with Mr. Hopkins, yet we cheerfully recommend him as a first-class man for the position he seeks, and we will be glad to hear of his success in securing a good position.   *Very Truly,*
                                                         Jamison, Sells & Co.

FORMS, LETTERS, FRENCH PHRASES, ETC.

VIII. Asking a Loan and the Reply.

Chas. Jones, Esqr.   Chicago, Aug. 8th, 1883.
    Dear Sir:

        Having failed to make some collections I had regarded as certain, and not receiving remittances upon which I had counted, I would take it as a great favor if you would accommodate me with One Hundred Dollars until the 15th inst., when it will be promptly repaid.   Yours Truly,
                                                                              Herkimer Hudson.

ANSWER COMPLYING.

H. Hudson, Esqr.   Chicago, Aug. 9th, 1883.
    Dear Sir:

        Yours of yesterday (8th) rec'd, and in reply would say that though contrary to my usual custom, I herewith enclose am't requested ($100.) Please acknowledge receipt, and don't fail to return by the 15th, as I shall need it at that time.
                                                    Yrs., &c.,
                                                                Chas. Jones.

ANSWER REFUSING.

H. Hudson, Esqr.   Chicago, Aug. 9th, 1883.
    Dear Sir:

        Your favor of the 8th rec'd, and I regret to say that it is out of my power to accommodate you with am't requested. I am myself temporarily short, or should take pleasure in complying.   Yours Truly,
                                                                Chas. Jones.

IX. Directing Letters.

IN directing envelopes be careful to write plainly, and especially avoid any eccentricities, such as some silly persons indulge in. It is best in directing a letter for a small place, to put county as well as State; but both may be omitted from letters going to metropolitan cities, as Boston, St. Louis, New Orleans, etc.

The following form is best:

```
┌─────────────────────────────────────────────────┐
│  . . . . . . .                       ┌───────┐  │
│  . . . . . . .                       │       │  │
│  . . . . .                           │ STAMP │  │
│                                      │       │  │
│                                      └───────┘  │
│                                                 │
│              MR. WAT HERNDON,                   │
│                                                 │
│                       WASHINGTON,               │
│                                                 │
│                            Chariton Co.,        │
│                                                 │
│  . . . . . . .                         Mo.      │
│  . . . . . . .                                  │
│  . . . . .                                      │
└─────────────────────────────────────────────────┘
```

A return request may be written in the upper left-hand corner, as indicated by dotted lines. If it is desired to send "in care" of any one, the notification should be in lower left-hand corner, as indicated by dotted lines.

X. Suggestions for Letter-Writers.

A FEW suggestions in regard to writing and sending letters may not be amiss, and are herewith given:

1. It is best to read your letters carefully after having finished them, to see that you have omitted nothing of importance. In the haste of composition much may be overlooked.

2. Letters on business should be brief, and to the point. Remember that to a business man time is money.

3. Always write at least legibly. This will ensure your letters being read, which is not the case with all scrawls.

4. All letters, especially those relating to business, should be promptly answered.

5. Don't fail to copy all important letters that you may write.

6. It is best to give your address, town, county and State, in each letter. If living in a city, street and number will be sufficient.

7. It is best to use what the Government calls "request envelopes," that is, envelopes with the request printed on them that, "if not called for in —— days, they shall be returned to ——."

8. Never forget to date your letters. Often a great deal depends on the correct dating of a letter. Lives and fortunes have been lost by this slight omission.

9. By draft, P. O. order, the new postal-card orders, checks, or by express, are all better modes of sending money than by registered letter. The Government is not responsible for money lost in a registered letter.

10. Letters of any importance should be preserved. File them carefully, endorsing on back date of receipt, nature of contents and name of writer.

11. In writing, use moderately heavy paper. Black ink is best, though many now use only the purple inks, as they flow very freely and do not corrode steel pens.

12. In dating and beginning letters, the following is the correct form in which the writing should appear, though, of course, any other wording may be used:

Honolulu, Sandwich Islands,
Oct. 1st, 1883.

Maxwell B. Norton, Esqr.
    Dear Sir:
        Your favor of the 12th day of September is at hand, etc., etc.

XI. Styles of Cards.

VISITING CARDS.

MARY WALKER, M. D.	THE MISSES SIMPSON.
FOR A PROFESSIONAL LADY.	FOR SISTERS CALLING TOGETHER.
MRS. JNO. D. PERRY.	MISS MARY SMITH.
FOR MARRIED LADY.	FOR UNMARRIED LADY.
JAMES A. JACKSON.	WM. GIBSON, U. S. A.
FOR GENTLEMAN.	FOR ARMY OFFICER.
JNO. HARRINGTON, U. S. N.	MRS. CHAS. GREEN, MISS GREEN.
FOR NAVAL OFFICER.	FOR LADY AND DAUGHTER.

Visiting cards may have printed on them the day of week on which their owners receive, thus:

```
┌─────────────────────────────┐   ┌─────────────────────────────┐
│                             │   │                             │
│     Mrs. Wm. James,         │   │    Col. and Mrs. T. Green,  │
│                             │   │                             │
│              12 Berry Place.│   │             18 Glenwood Ave.│
│ Tuesdays.                   │   │ Thursdays.                  │
└─────────────────────────────┘   └─────────────────────────────┘
```

WITH DAY OF RECEPTION.

OTHER CARDS.

```
┌─────────────────────────────┐   ┌─────────────────────────────┐
│                             │   │   Mr. and Mrs. W. Manning,  │
│                             │   │          at home,           │
│    Mr. and Mrs. B. Jones,   │   │  Tuesday, November 9th, 1883,│
│                             │   │       3 to 6 o'clock.       │
│       Tuesday Evenings.     │   │                             │
│ 19 Block Place.             │   │ 8 Westminster Terrace.      │
└─────────────────────────────┘   └─────────────────────────────┘
```

A lady's visiting cards should not have street and number printed upon them—especially is this the case if the card belongs to an unmarried lady. If it is necessary to place the number on the card, as in case of a removal, visiting a stranger, etc., it may be neatly added with a pencil.

XII. French Words and Phrases in General Use.

There are many French phrases that have come into such general usage among society people that they have almost entirely displaced the corresponding English phrases. As it is important to know them, we herewith append a list of those most generally in use, with a translation of them, which will, no doubt, prove of service to many novices.

1. *Affaire d'amour*, ... A love affair.
2. *A la mode*, ... According to the fashion.
3. *Apropos*, ... To the purpose.
4. *Au contraire*, ... On the contrary.

5.	*Au fait,*	Correct.
6.	*Au revoir,* or, better, *A revoir,*	Until we meet again.
7.	*Bal masqué,*	Masked ball.
8.	*Blasé,*	Faded, satiated.
9.	*Billet-doux,*	Love-letter.
10.	*Bon jour,*	Good day.
11.	*Bon mot,*	A witty saying
12.	*Bon soir,*	Good night.
13.	*Bon ton,*	Good style, fashion.
14.	*Carte blanche,*	Full power (literally, a blank card).
15.	*Chacun a son gout,*	Each to his taste.
16.	*Château en Espagne,*	Air castles (literally, castles in Spain).
17.	*Chef d' Oeuvre,*	A masterpiece.
18.	*Chèr ami,*	Dear friend (male).
19.	*Chère amie,*	Dear friend (female),
20.	*Ci-devant,*	Former.
21.	*Comme il faut,*	Correct, as it should be.
22.	*Compagnon de voyage,*	Traveling comrade.
23.	*Costume de rigeuer,*	Full dress.
24.	*Coup d' Oeil,*	A glance (literally, a stroke of the eye).
25.	*Début,*	First appearance.
26.	*Dénouement,*	The sequel, disclosure.
27.	*Dot,*	Dowry.
28.	*Double entendre,*	Double meaning.
29.	*Eclat,*	Dash, brilliance.
30.	*Elite,*	Very select, choice
31.	*Embonpoint,*	Fatness.
32.	*Encore,*	Again.
33.	*Ennuie,*	Lassitude, weariness.
34.	*En regle,*	Regularly, properly.
35.	*Entente cordial*	Amicable relation.
36.	*Entrée,*	Entrance.
37.	*Entre nous,*	Between us.
38.	*E. P. (En personne),*	In person.
39.	*E. V. (En ville),*	In the town, or city.
40.	*Faux pas,*	False step.
41.	*Fête,*	An entertainment.
42.	*Fête champetre,*	Rural entertainment.
43.	*Haut ton,*	High fashion.
44.	*Jeu d'esprit,*	A witticism (literally, play of words).
45.	*Nom de plume,*	Assumed literary name.
46.	*Nous verrons,*	We shall see.
47.	*On dit,*	They say.
48.	*Outré,*	Garish, ridiculous.
49.	*Parvenu,*	A would-be fashionable
50.	*P. P. C. (Pour prendre congé),*	To take leave.
51.	*R. S. V. P. (Respondez s'il vous plait),*	Please answer.
52.	*Soi-disant,*	Self-styled.
53.	*Soirée,*	Evening entertainment.
54.	*Soirée Dansante,*	Dancing party, a ball.
55.	*Tête-a-tête,*	Private, face to face (literally, head to head).
56.	*Tout ensemble,*	General appearance, the whole.
57.	*Vis-a-Vis,*	Facing, opposite.

XIII. Treatment of Children.

1. Never swear or use coarse language in the presence of children.
2. Never lower their self-respect by calling them harsh names.
3. Be free to praise a child judiciously when deserving.
4. Never break promises made to children; teach truth by example
5. If necessary to chastise a child, do not do so brutally.
6. Do not expect of children the judgment and care of older persons.
7. It is cruel to keep children up late at night, or to waken them early in the morning. They require and should have more sleep than grown persons.
8. Make of your child a companion, counsel with it and listen to its sorrows and joys as to those of a friend.
9. Do not cruelly repel its love and drive it to other confidants.
10. Do not embitter with brutality and harshness the only portion of life that can ever be happy.
11. A child should be dressed respectably; to cause it to wear coarse or ill-fitting clothes is sure to degrade it.
12. Teach a child the value of money. Let it have small sums to expend, but require an account to be kept, and then show to it whether its purchases are wise or not.
13. Reason with your children and show them the evils of vice, intemperance, and other bad habits.
14. Teach them to be careful, cleanly, considerate, true and honest.
15. Do not overtask them mentally or physically
16. Give plenty of time for recreation, and encourage healthful out-of-door games and exercises.
17. Teach by precept and example the observances of etiquette. How to eat correctly, how to enter a room, how to salute a person, etc., should be a part of the child's daily training.

XIV. Seventy-five Cardinal Rules of Etiquette.

1. IN riding, driving or walking, pass to the right.
2. A gentleman should insist on carrying any packages a lady may have.
3. A true lady will always thank any one who accommodates her, as by giving up a seat in a car, opening a door, etc.
4. No gentleman will stand on a corner, before a theater, or at the door of a church, and stare at or make remarks upon ladies.
5. No gentleman or lady will be guilty of the vulgarity of flirting.
6. A gentleman who has rendered any assistance to a lady must, as soon as the service is over, bow respectfully, and pass on.
7. A lady in crossing a muddy street should raise her dress gracefully, with one hand, only to the top of her shoes.

8. It is allowable for a gentleman to offer his arms to two ladies, but no lady should take the arms of two gentlemen

9. Ladies are usually given the inside of the walk. When, in order to preserve this position, changes are to be made, the gentleman must pass behind the lady.

10. No gentleman will smoke in a parlor, nor is it allowable even on the street in the presence of a lady.

11. On very narrow crossings the lady should precede her escort, as then he is enabled to see that she gets across without accident.

12. No lady should wear a trailing dress upon the street.

13. In order the more effectually to protect a lady, a gentleman should give his left arm, and in cases of danger this should be done even if it places the lady upon the outer side of the walk.

14. A lady precedes her escort into a door or gate, down stairs and over a difficult crossing; the lady takes precedence in going down stairs.

15. A gentleman should always offer his arm to a lady in the evening. In the daytime it is not usual to do so.

16. If a lady is in distress, any gentleman, whether an acquaintance or not, should offer his assistance.

17. A lady should not go on the street at night unattended.

18. No gentleman ever makes sweeping innuendoes against the sex. To do so, argues an acquaintance only with the most disreputable.

19. Do not clean your nails or pick your teeth in company.

20. It is vulgar to indicate a person by pointing.

21. A practical joke is worse than vulgar: it is cruel.

22. Always give precedence to ladies, invalids and elderly persons.

23. Do not make promises that you do not intend to fulfill.

24. Never touch an acquaintance to call his attention, and never touch any one, however menial may be their position, with a cane, parasol or umbrella to secure their notice.

25. Only vulgarians touch pictures, statues, etc., with canes or parasols, to point them out, or to indicate any defect or beauty.

26. Every polite question deserves a civil answer.

27. You have no right to lend an article that you yourself have borrowed.

28. Never stand across the pavement, blocking it, and do not occupy a doorway or narrow hall to the annoyance and retarding of passers.

29. A gift made or a favor rendered should never be alluded to.

30. A gentleman or lady will not only not open the letter of another, but will not read one already opened; unless requested to do so.

31. Charity to street beggers is a doubtful method of relieving necessities.

32. Only a miserable cad or senseless dude ever boasts of his conquests amongst the fair sex.

33. Never betray a confidence.

34. Remember that "too much familiarity breeds contempt."
35. Rejoice not at the calamities of another, even though an enemy.
36. Be gentle and graceful in all things.
37. A present should never be undervalued, nor sent in hopes of a return.
38. Always close the door after you gently and without slamming.
39. Do not boast of your own exploits.
40. Never without necessity pass between two persons engaged in a conversation—should it become necessary, always apologize.
41. Apologize for passing between any one and the fire, and avoid doing so if possible.
42. Be punctual in keeping appointments.
43. Never attempt to monopolize the conversation.
44. Never assert anything with a great degree of certainty.
45. Do not call newly made acquaintances by their given names.
46. Do not laugh at your own wit.
47. Be not often the hero of your own legends.
48. In foreign countries boast not overmuch of your own.
49. It is vulgar to question children or servants about the affairs of their families.
50. Be neither rude to nor familiar with servants. Treat them humanly, yet with a proper degree of dignity.
51. It is neither just nor honest to punish children for faults in the commission of which you set them the example.
52. Treat even your enemies courteously under your own roof.
53. Never answer questions addressed to others, or even those that have been addressed generally to the company of which you are a member.
54. It is better to teach by precept than by example.
55. Avoid satire and sneering.
56. Wit that wounds should be carefully avoided.
57. Be kind to all; treat courteously those who ask and those who extend favors.
58. Be truthful. Lying is the vice of slaves.
59. Avoid hypocrisy, nor make any false pretences.
60. Invitations should be promptly answered.
61. Do not neglect calling upon friends.
62. If writing to another upon business (of benefit to yourself) do not fail to enclose postage stamp for reply.
63. Do not cross your legs, or extend your feet out into the aisles of cars or 'busses, as others will be troubled thereby.
64. An offered apology should always be accepted, though it is not necessary to again begin a friendship with the person offering it.
65. A lady should not accept expensive gifts, except from near relatives or an accepted lover.

66. Avoid all boisterous conduct, in laughing or talking.
67. All extremes should be avoided.
68. Flashy vestments and much jewelry should be avoided.
69. Do nothing that will interfere with the pleasures of others.
70. Do not scoff at the beliefs of others.
71. Be well dressed, but never over-dressed.
72. Treat children with gentleness and consideration. He who beats or abuses a child, degrades it.
73. Never indulge in profanity; it is a vice without pleasure or reward.
74. Never associate with any one whom you would not willingly introduce to your family.
75. Adopt, if not sinful or degrading, the customs of those amongst whom your lot is cast. Never be singular.

XV. Alphabet of Etiquette.

Avoid thou all evil, all rudeness, all haste,
Be gentle, be cheerful, be kindly, be chaste.
Consider the needs of the old and the weak;
Don't volunteer counsel, think twice ere you speak.
Ever think last of self, be not boastful or proud,
Fear scandal and gossip, let your talk be not loud.
Greet with equal politeness the high and the low,
Have a heart full of kindness, a soul pure as snow.
Injure none by a look, or a word or a tone;
Join not those that are evil; far better alone.
Keep promise and counsel, let your word be your bond,
Leave lying to slaves, of yourself be not fond.
Move gently; be modest in action and dress;
Never swear, never mock at another's distress.
Over dressing avoid, but at fashion don't sneer,
Pay due tribute to usage, but bend not to fear.
Quit all that is harmful to self or to others,
Remember this world is a wide band of brothers!
Shun the fool and the ruffian; the fop and the boor,
Take pleasure in helping the weak and the poor.
Use good language or none, all coarseness avoid,
Vulgarity's sinful, or with sin alloyed.
Wax ever in virtue, in grace and good will,
Xcelling in good and decreasing in ill.
Yonder sun be thy guidance in everything bright,
Zero marking thy standing in all that's not right.

---

*Composed by the late Lord D'Israeli, the English Premier, at the age of 12 years.

PART XI.

MISCELLANEOUS.

VALUABLE TABLES AND RECIPES.

FOODS, SPICES AND CONDIMENTS.

WEIGHTS, MEASURES, LEGAL FORMS, ETC.

MISCELLANEOUS.

CHAPTER I.

FOOD PRODUCTS OF COMMERCE.

I. FLOUR AND ITS MANUFACTURE.—II. RYE AND ITS PRODUCTS.—III. BARLEY AND ITS PRODUCTS.—IV. OATS AND THEIR PRODUCTS.—V. MAIZE AND ITS PRODUCTS.—VI. BEANS AND PEAS AND THEIR PRODUCTS.—VII. POTATOES AND POTATO PRODUCTS.—VIII. SAGE AND TAPIOCA.—IX. CHOCOLATE AND COCOA.—X. COFFEE.—XI. TEA.—XII. COTTON-SEED OIL.—XIII. SPICES AND THEIR ADULTERATION—PEPPER.—XIV. CINNAMON; HOW TO KNOW IF PURE.—XV. CLOVES AND ALLSPICE.—XVI. NUTMEGS AND MACE.—XVII GINGER AND ITS PREPARATION.—XVIII. CAPERS, TRUE AND SPURIOUS KINDS.—XIX. THE TAMARIND

I. Flour and Its Manufacture.

ORIGINALLY the word "flour" was used only to designate the bolted flour of wheat. Rye, barley, oats, buckwheat and corn were more coarsely ground, and their products were called meal. Improvements in milling have now so changed the nomenclature, that the products of wheat, rye and buckwheat are now also called flour. Corn-meal and oatmeal are the only meals, so called, in the United States, though, in some parts of Europe, and in Asia and Africa, barley is ground into meal. The grades of wheat flour are numerous under what is known as the "patent process," by which wheat, often being separated from its bran, is reduced by successive operations to the requisite fineness. It is not ground by crushing as formerly, but by granulating. Wheat and rye are separable into bran, shorts, middlings, fine, superfine and extra flour. There are several patent processes in successful use in the manufacture of flour, and the increased product, per bushel, has driven most of the ordinary mills to the wall; so that now the farmer often finds it cheaper to sell his wheat and buy his flour than to have it ground at country mills. The principal grades of flour are: Fine, Superfine, Family, and XX, XXX and XXXX extras; and every great mill has its own special brand of superior flour.

II. Rye and Its Products.

RYE flour is seldom used alone for bread, but is mixed with fine corn-meal or wheat flour. Thus used it makes a moist bread, from the quantity of gluten it contains. Rye and Indian bread is much used in cities for breakfast. It requires long baking, and hence is seldom prepared except by regular bakers. It was formerly much used ground, as a portion of the cut feed for horses. Its principal use now, in the United States, is in the manufacture of whiskey.

Rye is not generally sown as a field crop On sandy land considerable quantities are raised, the product being white and in every way better than that of mucky soils. The crop, however, is large, and our exports are considerable. The crop of 1881 was 20,704,950 bushels, being less than that of any year since 1876. In 1878 the largest crop of rye ever known in the country was produced, being 25,843,790 bushels. The average value of this grain in 1881 was 93.3 cents per bushel, the average yield 11.6 bushels per acre, and the average value $10.82 per acre. The average yield per acre for the last eleven years was 13.9 bushels and the average price for the same period was 72.2 cents per bushel.

III. Barley and its Products.

BARLEY is not used for bread in the United States, but it is an important and valuable crop, immense quantities being used in the manufacture of ale and beer. Pearled barley is manufactured to a limited extent. In this process the barley is first hulled and then rounded by the attrition of machinery. California, New York and Wisconsin produce more than half the annual crop. The crop for 1882 was 45,000,000 bushels—an average of 23.5 bushels per acre, a greater yield than any previous year, except in 1880, when the yield was 45,165,346 bushels, with an average of 24.5 bushels per acre. The average price for the last eleven years is 73.2 cents per bushel; the average yearly price ranging from 58.9 to 92.1 cents per bushel.

IV. Oats and Their Products.

OATS are every year more and more used as food in the shape of oatmeal. The grain is kiln-dried, hulled, and then broken or granulated in a peculiar mill. The best oatmeal is that prepared by partial cooking, again dried, so that its preparation for puddings (mush) requires only ten to fifteen minutes. The yield of oats in the United States in 1862 was 480,000,000 bushels, the largest yield ever known. Illinois, Iowa, New York, Wisconsin, Missouri, Pennsylvania, Ohio, Indiana and Kansas are the States of largest production. The average produce per acre for the last eleven years was 27.6 bushels per acre. Certain fields produced over one hundred bushels per acre.

V. Maize and Its Products.

THIS is the most important grain crop of the United States. As human food it ranks high, its various products thus used being meal, hominy, samp, hulled corn, farina and other preparations. Corn is also the great stock-feeding grain. The starch of commerce is manufactured from it. It is the principal agent in the production of alcohol, and many millions of bushels are now annually used in the manufacture of glucose (corn sugar). The Mississippi Valley is the great corn-producing region of the world, no other country seeming to have such combined capabilities of soil and climate. In 1882, the States north of Kentucky and west of Pennsylvania produced 1,250,000,000 of the total crop of 1,680,000,000 bushels.

VI. Beans and Peas and Their Products.

Beans and peas in their fully ripened and dry state form no insignificant article of the world's commerce and food supply. Beans are known commercially only in their natural state, the portion ground into bean-meal being insignificant. Peas, however, are found in a variety of forms, but principally as split peas (divested of the skin and halved), and as pea-meal. The trade in canned green peas is yearly assuming larger and larger proportions. The production of dried peas for shipment is confined entirely to the most northerly portions of the United States and to Canada, the ravages of the pea-beetle, or pea-weevil (*Brucus pisi*), incorrectly called pea-bugs, in the regions south of the latitude named, preventing the production of perfect ripened seeds. The great bulk of the split peas of commerce come from England and Scotland, whose climates are favorable to the crop. Two cuts are given showing the germination of the bean and the pea, as companion pictures, and as a lesson in botany. In the case of beans, the two lobes of the seed, in

GERMINATION OF THE BEAN.

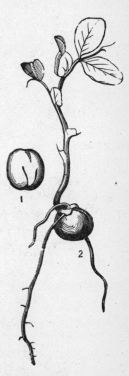

GERMINATION OF THE PEA.

germinating, are thrust above ground, thus forming the first leaves of the plant, and upon which it is supported until the roots are enabled to draw nourishment from the earth. In the case of peas, the shoot is thrust upwards, forming true leaves at once, the seed remaining below the surface.

The varieties of beans are innumerable, the sorts in greatest request for their green pods being colored, while those used for their dried seeds, as food, being the white varieties. The pea, (navy) bean, the marrow bean, and the Lima bean, are carried to all parts of the civilized world, and are considered indispensable to armies, navies and mariners generally. They are both exceedingly rich in flesh-forming material, as starch, gum and gluten, and deficient in oil.

VII. Potatoes and Their Products.

Potatoes are now almost exclusively used as human food. Formerly, large quantities of starch were made from them, and they were much used for distillation into spirits. Alcohol and starch are now made principally from corn. The average yearly production of potatoes in the United States in the past eleven years, was 135,491,019 bushels, the largest yield in one year being 181,626,400 bushels, and the smallest, 105,981,000 bushels.

New York is the great potato-growing State, giving 20,000,000 bushels annually; next comes Pennsylvania with about 9,000,000 bushels. Then come Michigan, Wisconsin, Iowa, Illinois, Minnesota, California and Ohio, the productions in these States being from more than 3,000,000 bushels annually, to over 7,000,000.

A true wild potato (*Salanum Fendliri*), supposed to be the original of the cultivated potato, has lately been found growing in abundance in the northern part of New Mexico, between Wingate and Fort Defiance. The tubers are half to three-quarters of an inch in diameter, and when boiled are esculent. They are eagerly sought by the squaws. If eaten in large quantities they produce griping, and hence the Navaho Indians eat magnesian earth with their meal of wild potatoes. The illustration shows the plant and tubers.

WILD POTATO OF NEW MEXICO. (*Salanum Fendliri*)

VIII. Sago and Tapioca.

Sago is a starchy farina, prepared from the pith of various palm trees. The arrow-root of Florida is a variety of sago prepared from indigenous plants, *Zamia integrifolia* and *Z. pumila*. Tapioca is prepared from a tropical plant, grown principally in Brazil—*Jatropa manihot* There are two varieties. The acidity of the roots is destroyed by roasting and washing. Good substitutes are prepared from the starch of corn.

IX. Chocolate and Cocoa.

BOTH of these are preparations of the nuts of the chocolate tree (*Theobroma cocoa*), the first from the meat, and the second from the cotyledons or nibs. The West Indies, Mexico, Central America and Brazil are the principal producing countries. The tree is inter-tropical, and there are several species. The beverage is nourishing, but does not possess the refreshing qualities of tea and coffee.

How Chocolate is Made.—In the regions where chocolate is cultivated, the ripe nuts are gathered and exported; or when manufactured at home, the fruits are split open, and the nuts (seeds) removed. The latter are then cleaned of the pulpy matter surrounding them, and subjected to a process of fermentation, for the purpose of developing their color, and when this process is completed they are dried in the sun and packed for transportation. The seeds are prepared for use by roasting in revolving metal cylinders, then bruising them to loosen the skins, which are removed by fanning. The cotyledons,

YOUNG PLANT OF ARABIAN COFFEE.

commonly called cacao-nibs, are separated and crushed and ground between heated rollers, which softens the oily matter, and reduces them to a uniform, pasty mass; this is then mixed with variable quantities of sugar and starch, to form the different kinds of cacao, or sweetened and flavored with vanilla or other substances for the formation of chocolate.

X. Coffee.

COFFEE and tea are articles of almost universal consumption among civilized peoples. Coffee is strictly an inter-tropical plant, and is scarcely hardy even in the warmest portions of Florida and California. The principal countries producing coffee are the West Indies, Central America and Brazil. The grades of so-called Mocha, Java, etc., are mostly produced in the localities above named from seed originally imported. Since the export from Oriental countries is comparatively insignificant, Brazil is the great coffee-producing country of the globe. Coffee owes its flavor both to the nature of the plants and to the soil upon which it grows. In this respect it is like tobacco, the peculiarity of soil and climate giving it flavor. The chief requirements of a coffee plantation are constant heat, and protection from the direct rays of the sun.

XI. Tea.

TEA is indigenous to the sub-tropical regions of China. It has been extensively naturalized in other countries, and will survive the winters in the United States as far north as Virginia. Its native country is not known, but Assam is the only country where it is found growing wild. It is injured by long sea voyages, even when the greatest care is observed, and hence, since the route to China and the other Oriental tea-producing countries has been opened by the railways to the Pacific, the fragrance is better preserved than heretofore, since the water voyage is materially lessened in point of time. Russia is said to receive the finest tea, outside of China, it being brought overland. In every part of the South where the shrub is hardy, sufficient should be produced for home consumption. The cost of labor in its preparation alone prevents its successful cultivation in the hill regions of the South.

XII. Cotton Seed Oil.

WITH the falling off in the cultivation of olives, various oils have been used for the adulteration of olive oil, or for entirely supplanting its use. One much used is cotton seed oil, of which vast quantities are now yearly exported to Europe to be returned to us, bottled and duly labeled *heuile d'olive*. It certainly would be better to use it in its original form without paying import dues, since it can be so cheaply made that it is largely used in the adulteration of lard, butter, etc.

Cotton-seed oil is a bland, pure oil, agreeable to the taste, and in no way injurious. The discovery of its valuable qualities has very much enhanced the profits of the cotton crop of the South, since the seeds have been utilized for the production of this oil as well as for the feeding of stock, instead of allowing them to rot

TEA IN VARIOUS STAGES OF MANUFACTURE.

on the ground for manure, as used to be the common practice in former years.

XIII. Spices and Their Adulteration—Pepper.

The pepper of commerce is of two kinds. Red, or cayenne pepper, is the product of our common bird or Chili pepper, an annual plant which may be ripened any where in the United States south of forty-five degrees, by starting the plants in a hot-bed, and transplanting. The Cayenne pepper of commerce is the fully ripened fruit pods, thoroughly dried and ground to powder.

Black and White Pepper.—The true pepper of commerce is produced in tropical climates, from the berry of a perennial climbing shrub, a native of Sumatra, but extensively cultivated in Java, Ceylon, and other tropical regions of Asia. The vine bears its seeds at three years old, on spikes. Black pepper is this seed. White pepper is the seed divested of its hull. The botanical name is *piper nigrum*.

Intoxicating Pepper.—The celebrated intoxicating pepper is the fruit of another species, *piper betel*, and is chewed by the natives of India prepared with lime, and wrapped in the leaves of the betel pepper (*chavica betel*.) One hundred million people are said to chew this intoxicating substance.

Adulterations.—Pepper is adulterated with ground rice, mustard, sweepings of warerooms, etc., and even the berries are counterfeited by a mixture of oil cake, clay and cayenne. The only safety in buying ground pepper is in the integrity of the firms who manufacture it.

XIV. Cinnamon—How to Know It Pure.

The botanical name of the true cinnamon is *Cinnamomum Zeylanicum*, (a species of the laurel family) a tropical tree reaching a height of thirty feet, and cultivated in many countries. Ceylon has long been noted for the excellence of its cinnamon, but commerce has been largely supplied from the West Indies and South America, and there is much inferior bark sold as the genuine article. It is prepared by stripping the bark from the branches, when it naturally rolls up into quills, the smaller of which are introduced into the larger, and then dried in the sun. Good cinnamon is known by the thinness of the bark; as a rule, the thinner and more pliable, the finer the quality. When it is broken the fracture is splintery. It is largely used as a condiment for its pleasant flavor, and its astringent and cordial properties give it a value as a medicine.

Cassia.—This is the bark of *Cinnamomum cassia*, a tree growing forty to fifty feet in height, cultivated to a considerable extent in China, the Philippine Islands, the western coast of Africa, and in Brazil. The China cassia is considered superior in perfume and flavor to any spice of its class. This bark resembles the true cinnamon, but is thicker, coarser, and not so delicate in flavor, but being cheaper is frequently used to adulterate the true article. For confectionary purposes this affords a stronger flavor than cinnamon, and is therefore preferred. The bark is collected and prepared as for cinnamon. Cassia bark is distinguished from cinnamon by being more brittle, and of less fibrous texture; it is not so pungent, and has more of a mucilaginous or gelatinous quality.

XVI. Cloves and Allspice.

Cloves.—Cloves are the flower-buds of a species of myrtle, (*Caryophyllus aromaticus*) a small evergreen tree, native to the Malaccas, but cultivated in various tropical countries, especially the East and West Indies. The flower buds are collected before they expand, are cleaned, dried, and darkened by smoking them over a wood fire. All parts of the plant are aromatic; the flower buds only are used in commerce.

Allspice.—This is of two kinds, Jamaica pepper, the fruit of a species of myrtle (*M. pimenta*), and allspice-pimento, the fruit of *Eugenia pimento*. It takes the name allspice from combining the flavor of various spices, as cinnamon, cloves and nutmeg. The allspice-pimento tree is the species generally cultivated, and is planted in orchard-like rows, forming high beautiful trees. The berries are about the size of a small pea, of a dark color, and, as seen in commerce, are surmounted by the remains of the calyx. They are prepared by being gathered before they are fully ripe, and then dried in the sun, when they acquire that reddish-brown tint which makes them marketable.

XVI. Nutmegs and Mace.

The nutmeg is the kernel of a small tropical tree named *Myristica Moschata*. The leaves are aromatic, and the fruit is very much like a peach, having a longitudinal groove on one side, and bursting into two pieces, when the inclosed seed, covered by a false aril which constitutes the substance known as the mace, is exposed. The seed itself has a thick outer shell which may be removed when dry, and which incloses the nucleus of the seed, the nutmeg of commerce. The fruit is gathered at various seasons as it attains maturity. The mace, or covering, of a saffron or orange color, is used in the same manner as the nutmeg.

XVII. Ginger and Its Preparation.

There are a number of East Indian plants that are used in the place of ginger, the principal ones of which are *Curcuma amada*, *aromatica* and *zedoria*. The true ginger of commerce is the root-stocks (*rhizomes*) of *Zingiber officinale*, a plant much cultivated both in the East and West Indies, as well as in South America, Africa and China. The rhizome, or woody root-stock, which forms the ginger, is dug up when of sufficient size, cleaned, scraped and dried, and in this state is called uncoated ginger; but when the outer skin is not removed from the root-stocks it is called coated, and presents a dirty-brown appearance. Independent of this difference in color, which is in the mode of preparation, it is supposed that there are two varieties of the species, one producing white, and the other dark-colored ginger. The darker kinds are sometimes bleached by exposure to the fumes of chloride of lime, or burning sulphur. Ginger when broken across shows a number of small fibers imbedded in floury tissue. Its well-known hot, pungent taste is due to the presence of a volatile oil; it also contains a large quantity of starch and yellow coloring matter, inclosed in large cells.

Adulterations.—Ground ginger is largely adulterated with starch, wheat-flour, ground rice, mustard, husks, etc., in various proportions. In a young state the rhizomes are tender, fleshy, and mildly aromatic, at which time, preserved in syrup they form the conserve known as preserved ginger. West India gingers are preferred to those from the East Indies.

XVIII. Capers—True and Spurious Kinds.

THE true caper of commerce is known botanically as *Capparis spinosa*, a creeping plant, a native of the south of Europe. The flower-buds, and in some parts of Italy the unripe fruit, are pickled in vinegar, and form what are known as capers. An African species, *C. sodada*, furnishes berries with a pepper-like, pungent taste, and when dried are used as food. The flower-buds of *Zygophyllum fabago*, a native of the Cape of Good Hope, are used instead of capers, or substituted for that condiment. *Z. coccineum* has aromatic seeds, which are used by the natives in place of pepper. These and several other species are possessed of vermifuge properties. The leaves of *Z. simplex* are used for diseases of the eye. The smell of this plant is said to be so detestable that animals will not eat the foliage.

XIX. The Tamarinds.

TAMARINDS are the fruit pods of one of the handsomest of tropical trees, the *Tamarindus Indica* having beautifully pinnated foliage. There are varieties of this tree, distinguishable chiefly by the size of the pods. The pods vary in length from three to six inches, and are slightly curved. They consist of a brittle shell, inclosing a soft, acid, brown pulp, traversed by strong woody fibers; the seeds are again immediately invested by a thin membranous covering. They owe their grateful acidity to the presence of citric, tartaric and other vegetable acids. Tamarinds form an important ingredient in the cookery of Eastern nations, as in the curries of the East Indies.

CHAPTER II.

LAW, COMMERCIAL AND OTHER FORMS.

I. INDENTURE OF APPRENTICESHIP OR FOR SERVICE.—II. ARREARS OF PAY AND BOUNTY.—III. FORMS FOR BOUNTY LAND.—IV. AGREEMENTS AND CONTRACTS.—V. WARRANTY DEED.—VI. MORTGAGE OF PERSONAL PROPERTY.—VII. BILLS OF SALE.—VIII. CERTIFICATES, RELEASES AND DISCHARGES.—IX. POWERS OF ATTORNEY.—X. REVOCATION OF POWER OF ATTORNEY.—XI. PROXY REVOKING ALL PREVIOUS PROXIES.

I. Indenture of Apprenticeship or for Service.

AN indenture must be made with the consent of the father, mother, or guardian; it must state whether it be as an apprentice, clerk, or servant. The term of years of service, age, sex, compensation if any, and when and how to be paid.

If with the consent of the mother, the father being not capable of giving consent, or if a guardian, the matter enclosed in [ ] in the indenture below must be omitted, and the following inserted:

INDENTURE WITH CONSENT OF MOTHER.

THIS INDENTURE, made this —— day of ——— 18—, witnesseth: That ———, of the town of ——— in the county of ———, and State of ———, now aged —— years, with the consent of ———, his mother, his father having abandoned and neglected to provide for his family, [*or*, being dead, *or*, being insane and not in a legal capacity to give his consent], as hereon certified by a justice of the peace of the said town, said consent being also hereon endorsed,

INDENTURE WITH CONSENT OF FATHER.

[THIS INDENTURE, made this —— day of ———, 18—, witnesseth: That ———, of the town of ———, in the county of ———, and State of ———, now aged ———, with the consent of ———, his father, hereon indorsed,] does hereby, of his (or her) own free will, bind himself (or herself) to serve ———, of the town of ———, in the county of ———, and State of ———, as —— in the ——— (*trade, profession or employment,*) and to learn the said trade, profession, or employment, until the said ——— shall have attained the age of —— years, which will be on the —— day of ———, in the year 18—, during all which time the said apprentice shall serve the said master faithfully, honestly, and industriously, his secrets keep, and lawful commands everywhere readily obey; at all times protect and preserve the goods and property of the said master; and not suffer or allow any to be injured or wasted. He (or she) shall not buy, sell, or traffic with his (or her) own goods or the goods of others, nor be absent from the said master's service, day or night, without leave; but in all things behave as a faithful apprentice ought to do, during the said term. And the said master shall clothe and provide for the said ——— in sickness and in health, and supply him (or her) with suitable food and clothing; and shall use and employ the utmost of his endeavors to teach, or cause the said apprentice to be taught or instructed in, the art, trade or mystery of ———; and also cause the said apprentice, within such term, to be instructed to read and write (and in the general rules of arithmetic); and at the expiration of the service, give the said apprentice $ ———.

And the said ———, acknowledges that he has received, with the said ———, from the said ———, the sum of ——— dollars, as a compensation for his instruction, as above mentioned.

(If wages are to be paid for the service of the apprentice, insert:) And the said ———— further agrees to pay to the said ———— the following sums of money—viz., for the first year of —— service, —— dollars; for the second year of —— service, —— dollars; and for every subsequent year, until the expiration of —— term of service, —— dollars; which said payments are to be made on the —— day of —— in each year.

AND for the true performance of all and singular the covenants and agreements aforesaid, the said parties bind themselves, each unto the other, firmly by these presents.

IN WITNESS WHEREOF, the parties aforesaid have hereunto set their hands and seals, this —— day of ——, A. D. 18—.

(Signature of Apprentice.) [SEAL.]
(Signature of Master.) [SEAL.]

CONSENT OF PARENT OR GUARDIAN.

I do hereby consent to, and approve of, the binding of my ——, as in the above —— indenture mentioned. Dated the —— day of ——, in the year 18—.

(Signature.)

CONSENT OF THE CHILD.

I hereby consent to the foregoing indenture, and agree to conform to the terms thereof, in all things on my part to be performed. Dated the —— day of ——, in the year 18—.

(Signature.)

Complaint of a Master to a Magistrate.—If an apprentice does wrong in any manner, thereby causing loss to the employer, then the following complaint may be made:

To ————, a Justice of the Peace, etc.:

I, —— of ——, in said ——, machinist, hereby make complaint, that ——, an apprentice lawfully indentured to me, and whose term of service is still unexpired, with whom I have not received, nor am I entitled to receive, any sum of money in compensation for his instruction (or as the facts may be), refuses to serve me and conducts himself in a disorderly and improper manner, in this, to-wit: (state the wrong doing), and utterly refuses to perform the conditions of said indenture, as required by law. Dated the —— day of ——, A. D. 18—.

State of ———— } ss.
County of ————

————, the person named in the foregoing complaint, being duly sworn, deposes and says, that the facts and circumstances stated and set forth in the said complaint are true.

Before me, this —— day of ——, A.D. 18—.

————, Justice of the Peace.

Discharge of Apprentice, etc.—In case the master abuses the apprentice, or fails to perform his obligations, the courts will give redress as follows:

State of ———— } ss.
County of ————

Complaint on oath having been made to the undersigned, Justice of the Peace in and for said county, upon oath by ——, apprentice of —— of ——, in said county, machinist, that the said —— to whom said —— is bound by indentures of apprenticeship, the term of service in which has not yet expired, had cruelly beat, etc. (as in complaint and summons), and the said ——, by virtue of our summons thereupon issued, having been brought before us, and upon due examination of the parties and of the evidence adduced by them, it satisfactorily appearing to us that the said ——— is guilty of the matters charged against him as aforesaid; now therefore, we do hereby discharge the said——— from the service of the said ———, any thing in his indentures of apprenticeship aforesaid to the contrary notwithstanding.

Given under our hands and seal this ——— day of ———, A. D. 18—.

———[L. S.]
———[L. S.]

Justices of the Peace for said———

II. Arrears of Pay and Bounty.

IN all applications for arrears of army pay and bounty, the application is sworn to, and must mention the name and title of the person administering the oath, the rank of the applicant, the company designated, by its letter, name of captain, name of regiment, and the branch of the service. We give the first form filled out with fictitious names and statements. The other forms are in blank.

APPLICATION OF INVALID FOR ARREARS OF PAY OR BOUNTY.

STATE OF ———
County of ——— } ss.

On this——day of———18—, personally appeared before me, John Doe, a justice of the peace, in and for the county and State aforesaid, Richard Roe, of———, in the county of———, and State of——— who being duly sworn, declares that his age is——years; that he is the same Richard Roe who was a private in Company A, commanded by James Arnold, in the tenth Regiment of Illinois volunteers, cavalry, who was honorably discharged from the service of the United States at———, in the State of———, on or about the——day of———18—, by reason of amputation of the left leg.

This declaration is made to recover all arrears of pay and other allowances due said Richard Roe from the United States, and the bounty provided by the——section of the act of Congress, approved *(date of approval).* AND HE HEREBY APPOINTS Thomas Bliss, of———, as his lawful attorney, and authorizes him to present and prosecute this claim, and to receive and receipt for any orders or moneys that may be issued or paid in satisfaction thereof. The post-office address of the claimant is Clarendon, Coles county, Illinois.

[*Signature of claimant.*]

ALSO personally appeared before me, Edwin Wright and Able Strong, of the county of ——— and State of ———, to me well known as credible persons, who, being duly sworn, declare that they have been for ——— years acquainted with the above-named Richard Roe, who was a private in Company A, of the 10th Regiment of Illinois Volunteers, and know that the applicant is the identical person he represents himself to be; that they saw him sign the foregoing declaration; and that they have no interest whatever in this application. (Signatures of witnesses.)

Sworn to and subscribed before me, this———day of———, 18—; and I hereby certify that I have no interest, direct or indirect, in the prosecution of this claim.

(Signature and title of Magistrate.)

WIDOW'S APPLICATION FOR ARREARS OF PAY AND BOUNTY.

STATE OF———
County of——— } ss.

On this———day of———, 18—, personally appeared before me ——— in and for the county and State aforesaid, ———, of———, in the county of ———, and State of ———, who, being duly sworn, declares that her age is —— years; that she is the widow of ———, late of the county of——— and State of———, who was a ——— in Company ———, commanded by ———, in the ——— Regiment of ———, who died in the service of the United States at———, in the State of ———, on or about the

—— day of——, 18—; that her maiden name was ——, and that she was married to said ——, deceased, on or about the —— day of——, 18—, at ——, in the State of——, by ——; that she has remained a widow since the decease of her said husband, and knows there is no record evidence of said marriage (or that there is record evidence of said marriage, to wit: stating the evidence).

This declaration is made to recover all arears of pay and other allowances due said deceased from the United States, and the bounty provided by the 6th section of the act of Congress, approved July 22, 1861. And she hereby appoints ——, of——, as her lawful attorney, and authorizes him to present and prosecute this claim, and to receive and receipt for any orders or moneys that may be issued or paid in satisfaction thereof. Her post-office address is as follows:

(Signature of claimant.)

Also personally appeared before me, —— and ——, of the county of ——, and State of ——, to me well known as credible persons, who, being duly sworn, declare that they have been for—— years acquainted with the above-named applicant, and with said —— deceased, who was a —— in Company ——, of the —— Regiment of ——, and know that said deceased recognized said applicant as his lawful wife, and that she was so recognized by the community in which she resided; and that they have no interest whatever in this application. (Signatures of witnesses.)

Sworn to, etc.

CHILDREN'S APPLICATION FOR ARREARS OF PAY AND BOUNTY.

STATE OF——
County of—— } ss.

On this——day of——, 18—, personally appeared before me, a——, in and for the county and State aforesaid, ——, of——, in the State of——, aged——years (and——of——, in the State of——, aged—— years), who, being duly sworn, declare that the above-named persons are of the age stated, and are the legitimate children of——, late of ——, in the county of ——, and State of ——, who was a —— in Company ——, commanded by —— of the —— Regiment of ——, who died, or was killed, in the service of the United States at——, on or about the——day of ——, 18—. That their mother's name was——, and that she is dead.

This application is made to recover all arrears of pay and other allowances due the deceased from the United States, and the bounty provided by the 6th section of the act of Congress, approved July 22, 1861. AND THE APPLICANT HEREBY APPOINT——, of——, their lawful attorney, and authorize him to present and prosecute this claim, and to recover and receipt for any orders or moneys that may be issued or paid in satisfaction thereof. Their post-office address is as follows:

(Signatures of claimants.)

ALSO personally appeared before me, —— and —— of the county of——, and State of——, to me well known as credible persons, who, being duly sworn, declare that they have been for—— years acquainted with the above-named applicants, and with said——, deceased, who was a—— in Company ——, of the —— Regiment of ——, and know the above-named children to be the legitimate children of said deceased; and that the deponents have no interest whatever in this application.

Sworn to, etc. (Signatures of witnesses.)

The widow and children of person entitled to arrears of pay or bounty being dead, the fathers, mothers, brothers and sisters of the dead soldier may apply, changing the form to suit the respective case.

III. Forms for Bounty Lands.

LAND is given for service in the army of the United States and in the volunteer service under certain conditions. The widow of the deceased soldier may apply, and also for a second warrant, the first not having been received, only changing the words to suit the circumstances. Every detail must be stated as to discharge, disability, etc.

LAW, COMMERCIAL AND OTHER FORMS.

APPLICATION FOR BOUNTY LAND BY ONE NEVER BEFORE APPLYING.

STATE OF ——
County of ——

On this —— day of ——, A. D. one thousand eight hundred and ——, personally appeared before me, a justice of the peace, within and for the county and State aforesaid, ——, aged —— years, a resident of ——, in the State of ——, who, being duly sworn according to law, declares that he is the identical —— who was a —— in the company commanded by Captain ——, in the —— Regiment of ——, commanded by ——, in the war with ——; that he enlisted at ——, on or about the —— day of ——, A. D. ——, for the term of ——, and continued in actual service in said war for the term of ——, and was honorably discharged at ——, on the —— day of ——, A. D. ——, by reason of —— on the —— day of —— 18—, at ——, in the State of ——, while in the service aforesaid, and in the line of his duty. (A) He makes this declaration for the purpose of obtaining the bounty land to which he may be entitled under the act approved ——. He also declares that he has not received a warrant for bounty land under this or any other act of Congress, nor made any other application therefor.

(Signature of the claimant.)

OATH TO IDENTITY.

We, —— and ——, residents of ——, in the State of ——, upon our oaths, declare that the foregoing declaration was signed and acknowledged by ——, in our presence, and that we know personally (or we believe, from the appearance and statements of the applicant, setting forth any further grounds of belief the witness may have) that he is the identical person he represents himself to be.

(Signatures of witnesses.)

MAGISTRATE'S CERTIFICATE.

The foregoing declaration and affidavit were sworn to and subscribed before me on the day and year above written; and I certify that I know the affiants to be credible persons; that the claimant is the person he represents himself to be [or, that I believe, from the appearance and statements of the claimant, and from the facts that ——] (set forth the grounds of belief that he is the person he represents himself to be), and that I have no interest, direct or indirect, in the prosecution of this claim.

(Signature of magistrate.)

CERTIFICATE OF OFFICIAL CHARACTER AND SIGNATURE.

STATE OF ——
County of —— } ss.

I ——, clerk of the —— Court, a court of record of said county, do hereby certify that ——, the person subscribing the foregoing certificate and affidavit, and before whom the same was made, was, on the —— day of ——, 18—, therein mentioned, a —— in and for said county, duly authorized to administer oaths and affirmations for general purposes, and that I am well acquainted with the handwriting of the said ——, and verily believe that the name of ——, subscribed to the said certificate, is his proper and genuine signature.

IN TESTIMONY WHEREOF, I have hereunto set my hand, and affixed the seal of said court, this —— day of ——, 18—.

——, Clerk of —— Court.

(Seal of court.)

APPLICATION FOR A SECOND WARRANT.

[Proceed as in Application for Bounty Land, to (A), and add.] And that he has heretofore made application for bounty land under the act of September 28, 1850 (or other act, as the case may be), and received a land warrant, No. ——, for —— acres.

He makes this declaration for the purpose of obtaining the additional bounty land to which he may be entitled under the act approved the 3d day of March, 1855. He also declares that he has never applied for nor received, under this or any other act of Congress, any bounty land-warrant except the one above mentioned.

(Signature of the claimant.)

IV. Agreements and Contracts.

An agreement is a promise to fulfill whatever may be stipulated in the writing. A contract is an agreement to perform certain acts. A breach of contract entitles the party aggrieved to remedy and damages at law, and if fraud is proved, to criminal damages.

If there is no loss from breach of contract, the plaintiff is entitled to nominal damages and costs.

Failure to deliver property according to contract entitles the plaintiff to the value of the property at the time and place at which it should have been delivered.

The damages recoverable on contract to deliver goods on demand is their value at the time of the demand.

If land is not conveyed according to covenant, the plaintiff is entitled to the value of the land at the time it was to have been conveyed.

In the case of loss of goods by a common carrier (railway company, express, etc.), the damage would be the wholesale price of the articles at the place of delivery, less the freight.

In all agreements and contracts the agreement or matter between the parties must be specifically stated, whether it is something to do or to pay, whether in labor, money or otherwise.

A GENERAL FORM OF AGREEMENT.

This Agreement, made this——day of———, one thousand eight hundred and———, between John Doe, of the——of——, in the county of——and State of———, of the first part, and Richard Roe, of the ——of——, in the county of———, State of———, of the second part, witnesseth: That the said John Doe, in consideration of the covenants on the part of the party of the second part hereinafter contained, doth covenant and agree to and with the said Richard Roe, that———; and the said Richard Roe, in consideration of the covenants on the part of the party of the first part, doth covenant and agree to and with the said John Doe, that———

In Witness Whereof, we have hereunto set our hands and seals, the day and year first above written.

Signed, sealed and delivered in presence of
    Philip Prim,
    James Johns.

JOHN DOE, [SEAL.]
RICHARD ROE. [SEAL.]

AGREEMENT FOR THE SALE OF A HORSE.

This Agreement, made this——day of———, in the year of our Lord one thousand eight hundred and———, between John Doe, party of the first party, and Richard Roe, party of the second part, witnesseth: That the said John Doe hereby agrees to sell to the said Richard Roe, his——horse, with a white star in the forehead, and black mane and tail, and to warrant the said horse to be well broken, to be kind and gentle, both under the saddle and in single and double harness, to be sound in every respect and free from vice, for the sum of——dollars, to be paid by the said Richard Roe, on the——day of ———next.

In consideration whereof, the said Richard Roe agrees to purchase the said horse, and to pay therefor to the said John Doe the sum of——dollars, on the——day of———next

In Witness Whereof, we have hereunto set our hands and seals the day and year first above written.

Signed, sealed and delivered in presence of
    Philip Prim,
    James Johns.

JOHN DOE, [SEAL.]
RICHARD ROE. [SEAL.]

AGREEMENT FOR WARRANTY DEED.

ARTICLES OF AGREEMENT, made this——day of——in the year of our Lord one thousand eight hundred and——, between——, party of the first part, and——, party of the second part, witnesseth: That said party of the first part hereby covenants and agrees, that if the party of the second part shall first make the payment and perform the covenants hereinafter mentioned on——part, to be made and performed, the said party of the first part will convey and assure to the party of the second part, in fee simple, clear of all incumbrances whatever, by a good and sufficient warranty deed, the following lot, piece, or parcel of ground, viz.:—— And the said party of the second part hereby covenants and agrees to pay to said party of the first part, the sum of——dollars, in the manner following:——dollars, cash in hand paid, the receipt whereof is hereby acknowledged, and the balance——with interest at the rate of——per centum per annum, payable——annually, on the whole sum remaining from time to time unpaid, and to pay all taxes, assessments, or impositions that may be legally levied or imposed upon said land, subsequent to the year 18— And in case of the failure of the said party of the second part to make either of the payments, or perform any of the covenants on——part hereby made and entered into, this contract shall, at the option of the party of the first part, be forfeited and determined, and the party of the second part shall forfeit all payments made by——on this contract, and such payments shall be retained by the said party of the first part in full satisfaction and in liquidation of all damages by——sustained, and——shall have the right to re-enter and take possession of the premises aforesaid.

It is mutually agreed that all the covenants and agreements herein contained, shall extend to and be obligatory upon the heirs, executors, administrators and assigns of the respective parties.

IN WITNESS WHEREOF. The parties to these presents have hereunto set their hands and seals the day and year first above written.

Signed, sealed and delivered in presence of——   [SIGNATURE.]   [SEAL.]
                                                    [SIGNATURE.]   [SEAL.]

V. Warranty-Deed.

THIS INDENTURE, made this——day of——, in the year one thousand eight huundred and——, between——, of the city of——, and State of——, merchant, and——, his wife, of the first part, and——, of ——, in the said county, farmer, of the second part, WITNESSETH, that the said parties of the first part, in consideration of the sum of——dollars, lawful money of the United States, to them in hand paid by the said party of the second part, at or before the ensealing and delivery of these presents, the receipt whereof is hereby acknowledged, and the said party of the second part, his executors and administrators, forever released and discharged from the same, by these presents, have granted, bargained, sold, aliened, remised, released, conveyed and confirmed, and by these presents do grant, bargain, sell, alien, remise, release, convey and confirm unto the said party of the second part, and to his heirs and assigns, forever, all [*here insert description*] together with all and singular the tenements, hereditaments and appurtenances thereunto belonging or in any wise appertaining; and the reversion and reversions, remainder and remainders, rents, issues and profits thereof; and also all the estate, right, title, interest, dower and right of dower, property, possession, claim and demand whatsoever, both in law and in equity, of the said parties of the first part, of, in and to the above-granted premises and every part and parcel thereof, with the appurtenances. TO HAVE AND TO HOLD the above mentioned and described premises, with the appurtenances and every part thereof, to the said party of the second part, his heirs and assigns, forever. And the said——and his heirs, the above-described and hereby granted and released premises, and every part and parcel thereof, with the appurtenances, unto the said party of the second part, his heirs and assigns, against the said parties of the first part, and their heirs, and against all and every person and persons whomsoever, lawfully claiming or to claim the same or any part thereof, shall and will warrant, and by these presents forever defend.

IN WITNESS WHEREOF, the said parties of the first part have hereunto set their hands and seals the day and year first above written.

Signed, sealed and delivered in presence of ——
         [*Signature of witness.*]   [*Signatures and seals.*]

We give this warranty deed to show the general form, so that the reader may be familiar with such papers. All important documents should be drawn up by a lawyer, magistrate or notary public, when the amount is considerable.

VI. Mortgage of Personal Property.

In mortgaging personal property all goods and chattels should be named. If there are too many for the printed form, then say "goods and chattels mentioned in the schedule hereto annexed." The form for mortgage of personal property is as follows:

FORM OF CHATTEL MORTGAGE.

I——, of——, in consideration of——dollars to me paid by——of——, convey to the said—— the following personal property to-wit:——and now in the——, in the town of——aforesaid.

To hold the aforegranted goods and chattels, to the said——and his assigns forever.

And I covenant, that I am the lawful owner of said goods and chattels, and have good right to dispose of the same in the manner aforesaid.

Provided, nevertheless, that if the said——pay to the said——or his assigns the sum of——dollars in——from date, with interest on said sum at the rate of——per cent per annum, payable——, then this deed, as also a certain note of even date with these presents, given by said——to said——or order, to pay the said sum and interest at the times aforesaid, shall be void.

In Witness Whereof, I hereto set my hand and seal, this——day of——, in the year of our Lord one thousand eight hundred and——

Executed and delivered in presence of——　　　　　　　　　—— [SEAL.]

CHATTEL MORTGAGE, WITH POWER OF SALE.

Know All Men by These Presents, That I——, of——, in the county of——, and State of——, in consideration of——dollars, to me paid by——of the town——of——in the county of——and State of ——, the receipt whereof is hereby acknowledged, do hereby grant, bargain and sell unto the said——, and his assigns, forever, the following goods and chattels, to-wit:——

To Have and to Hold, All and singular, the said goods and chattels unto the mortgagee herein, and his assigns, to their sole use and behoof forever. And the mortgagor herein, for himself and for his heirs, executors and administrators, does hereby covenant to and with the said mortgagee and his assigns, that said mortgagor is lawfully possessed of the said goods and chattels, as of his own property; that the same are free from all incumbrances, and that he will warrant and defend the same to him, the said mortgagee and his assigns, against the lawful claims and demands of all persons.

Provided, Nevertheless, that if the said mortgagor shall pay to the mortgagee, on the——day of——in the year——the sum of——dollars, then this mortgage is to be void, otherwise to remain in full force and effect.

And Provided Further, that until default be made by the said mortgagor in the performance of the condition aforesaid, it shall and may be lawful for him to retain the possession of the said goods and chattels, and to use and enjoy the same; but if the same or any part thereof shall be attached or claimed by any other person or persons at any time before payment, or the said mortgagor, or any person or persons whatever, upon any pretence, shall attempt to carry off, conceal, make way with, sell, or in any manner dispose of the same or any part thereof, without the authority and permission of the said mortgagee or his executors, administrators or assigns, in writing expressed, then it shall and may be lawful for the said mortgagee, with or without assistance, or his agent or attorney, or his executors, administrators or assigns, to take possession of said goods and chattels, by entering upon any premises wherever the same may be, whether in this County or State, or elsewhere, to and for the use of said mortgagee or his assigns. And if the moneys hereby secured, or the matters to be done or performed, as above specified, are not duly paid, done or performed at the time and according to the conditions above set forth, then the said mortgagee, or his attorney or agent, or his executors, administrators or assigns, may by virtue hereof, and without any suit or process, immediately enter and take possession of said goods and chattels, and sell and dispose of the same at public or private sale, and after satisfying the amount due, and all expenses, the surplus, if any remain, shall be paid over to said mortgagor or his assigns. The exhibition of this mortgage shall be sufficient proof that any person claiming to act for the mortgagee, is duly made, constituted and appointed agent and attorney to do whatever is above authorized.

IN WITNESS WHEREOF, The said mortgagor has hereunto set his hand and seal this—— day of ——, in the year of our Lord one thousand eight hundred and ——

Executed and delivered in presence of ——

———— [SEAL.]
———— [SEAL.]

State of ————
———— County. } ss.

This mortgage was acknowledged before me, by —— (the mortgagor), this —— day of —— A. D. 18—

The mortgagor is the person who borrows money on the mortgage. The mortgagee is the person who lends it.

VII. Bills of Sale.

IN giving bills of sale every article is to be specified, and warranties of every kind are to be inserted, as to kind, quality, etc.

BILL OF SALE OF PERSONAL PROPERTY.

KNOW ALL MEN BY THESE PRESENTS, That I ——, in the County of ——, for and in consideration of the sum of —— to —— in hand well and truly paid, at or before signing, sealing, and delivery of these presents by ——, the receipt whereof I, the said ——, do hereby acknowledge, have granted, bargained, and sold, and by these presents do grant, bargain, and sell unto the said ——, the following articles of personal property, to wit: ——

TO HAVE AND TO HOLD the said granted and bargained goods and chattels, unto the said —— heirs, executors, administrators and assigns, to —— only proper use, benefit, and behoof forever, and —— the said —— does vouch himself to be the true and lawful owner of the goods and effects hereby sold, and to have in himself full power, good right, and lawful authority to dispose of the said —— in manner as aforesaid, and I do, for myself, my heirs, executors and administrators, hereby covenant and agree to warrant and defend the title of said goods and chattels hereby sold unto the said —— heirs, executors, and administrators.

Executed and delivered in the presence of ——

[SIGNATURE.] [SEAL.]

BILL OF SALE WITH WARRANTY.

I, ——, in consideration of —— dollars to me paid before the delivery hereof by —— of ——, the receipt whereof is hereby acknowledged, have sold, and by these presents do convey to said —— and his assigns, the following articles of personal property: ——. And I hereby agree with said —— to warrant and defend the title of said goods and chattels hereby sold, to him and his assigns against all and every person. And I hereby warrant the said —— to be in perfect condition. (*Add any other warranty as to quality or otherwise as may be desired.*)

Witness my hand this —— day of ——, A. D. 18——.

Executed and delivered in presence of —— (SIGNATURE) (SEAL.)

VIII. Certificates, Releases and Discharges.

A CERTIFICATE for work or labor must be recorded. The following is the form. If the oath is objected to the person may affirm:

MECHANIC'S CERTIFICATE.

I, ——, of ——, hereby claim a lien upon the estate situated (here describe the premises); to secure payment of —— dollars and — cents, for wages due me, after deducting all just credits, for work done and performed in building (if there has been altering, repairing, or furnishing materials, etc., state them) said premises, according to the following bill:

(Here insert the bill.)

——, of ——, is owner of said premises, and ——, of ——, the contractor. under whom the work was performed.

(SIGNATURE.)

State of ——, } ss. ——, ——, 18—.
County of ——,

Personally appeared the above named ——, and made oath (or affirmed) that the foregoing certificate by him subscribed is true.

Before me,

——, Justice of the Peace.

RELEASE AND DISCHARGE OF A MECHANIC'S LIEN.

I DO HEREBY CERTIFY, That a certain mechanic's lien, filed in the office of the clerk of the —— county of ——, the —— day of ——, one thousand eight hundred and ——, at —— o'clock in the —— noon, in favor of —— claimant against the building and lot, —— situated —— side of —— street, and known as No. ——, in said street, whereof —— is owner, and —— is contractor, is discharged.

(SIGNATURE.)

—— ss. On the —— day of ——, one thousand eight hundred and ——, before me came ——, who is known to me to be the individual described in, and who executed the above certificate, and acknowledged that he executed the same.

(SIGNATURE.)   (SEAL.)

IX. Powers of Attorney.

A POWER of attorney must name the person making the power, the person named as attorney, and must prescribe the specific things to be done and the revocations must state the same. The form is as follows:

POWER OF ATTORNEY, SHORT FORM.

KNOW ALL MEN BY THESE PRESENTS, That I ——, have made, constituted, and appointed, and by these presents do make, constitute, and appoint——, my true and lawful attorney, for me and in my name, place, and stead to——, giving and granting unto my said attorney full power and authority to do and perform all and every act and thing whatsoever requisite and necessary to be done in and about the premises, as fully, to all intents and purposes, as I might or could do if personally present, with full power of substitution and revocation; hereby ratifying and confirming all that my said attorney or his substitute shall lawfully do or cause to be done by virtue thereof.

IN WITNESS WHEREOF, I have hereunto set my hand and seal, the —— day of —— in the year one thousand eight hundred and ——.

Executed and delivered in the presence of ——   (SIGNATURE.)   (SEAL.)

PROXY, OR POWER OF ATTORNEY TO VOTE.

KNOW ALL MEN BY THESE PRESENTS, That I, ——, of ——, do hereby appoint —— to be my substitute and proxy for me, and in my name and behalf to vote at any election of directors or other officers, and at any meeting of the stockholders of the ——, as fully as I might or could were I personally present.

IN WITNESS WHEREOF, I have hereunto set my hand and seal, this —— day of —— 18 ——

Witnesses present,)   (SIGNATURE.)   (SEAL.)

X. Revocation of Power of Attorney.

WHEREAS, I, —— of the —— of —— in the county of —— and State of ——, by my certain power of attorney, bearing date the —— day of ——, in the year one thousand eight hundred and ——, did appoint

———, of the———, my true and lawful attorney, for me and in my name, to (here set out what he was authorized to do, using the precise language of the power of attorney originally given him), as by the said power of attorney, reference thereunto being had, will more fully appear:

THEREFORE, KNOW ALL MEN BY THESE PRESENTS, That I———, aforesaid, have countermanded and revoked, and by these presents do countermand and revoke the said power of attorney and all power and authority thereby given to the said———

IN WITNESS WHEREOF, I have hereunto set my hand and seal, this———day of———, one thousand eight hundred and———

Sealed and delivered in presence of———

[SIGNATURE.]   [SEAL.]

XI. Proxy, Revoking all Previous Proxies.

KNOW ALL MEN BY THESE PRESENTS, That I, the undersigned, stockholder in the———do hereby appoint———my true and lawful attorney, with power of substitution, for me and in my name to vote at the meeting of the stockholders in said company, to be held at———, or at any adjournment thereof, with all the powers I should possess if personally present, hereby revoking all previous proxies.

18———

(*Witness.*)

[SIGNATURE.]

CHAPTER III.

TABLES OF WEIGHTS, DIMENSIONS, STRENGTH, GRAVITY, ETC.

I. TABLES OF WEIGHTS.—II. THE METRIC SYSTEM.—III. THE METRIC SYSTEM COMPARED WITH OUR OWN.—IV. TABLES RELATING TO MONEY.—V. FOREIGN EXCHANGE.—VI. SPECIFIC GRAVITY.—VII. EARTHS AND SOILS.—VIII. COHESION OF MATERIALS.—IX. STRENGTH OF COMMON ROPES.—X. HUMAN FORCE.—XI. HEAT AND ITS EFFECTS.—XII. CAPACITY OF SOILS FOR HEAT.—XIII. RADIATING POWER, ABSORPTION AND EVAPORATION.—XIV. TEMPERATURES REQUIRED BY PLANTS.—XV. TEMPERATURES OF GERMINATION.—XVI. CONTRASTS BETWEEN ANIMAL AND PLANT LIFE.—XVII. THERMOMETERS.—XVIII. DIMENSIONS AND CONTENTS OF FIELDS, GRANARIES, CORN-CRIBS, ETC.—XIX. RAINFALL IN THE UNITED STATES.—XX. FORCE AND VELOCITY.—XXI. WEIGHT OF AGRICULTURAL PRODUCTS.

I. Tables of Weights.

IN the dawn of our modern civilization we find many odd weights and measures, indicating a notable want of scales or implements. For many of these, rather fantastic origins have been given. We know that Charlemagne, when Emperor of what was nearly the whole known world, finding in every country arbitrary measures of distance, struck his huge foot to earth, and ordered that its length should be the sole standard for the world. For the rest the following are the accepted legends: It is thus said that the English standard, the "grain," was originally derived from the average weight of a grain of barley. The inch was determined from the length of three barley corns, round and dry. The weight of the English penny, by act of Henry III., in 1266, was to be equal to that of thirty-two grains of wheat, taken from the middle of the ear and well dried. Among the nations of the East, we have the "finger's length," from that of the digit, or second joint of the fore-finger, the finger's breadth, the palm, the hand, the span, the cubit or length of the fore-arm, the stretch of the arms, length of the foot, the step or pace, the stone, pack, etc.

Below we give a table showing these measures, now only used for especial purposes:

A sack of wool is 22 stone, 14 pounds to the stone, or 308 pounds.
A pack of wool is 17 stone 2 pounds, or 240 pounds, considered a pack-load for a horse.
A truss of hay is, new, 60 pounds; old, 50 pounds; straw, 40 pounds.
A load of hay is 36 trusses; a bale of hay is 300 pounds; a bale of cotton, 400 pounds; a sack of Sea Island cotton, 300 pounds.
In England, a firkin of butter is 56 pounds. In the United States, a firkin of butter is 50 pounds. Double firkins, 100 pounds.

    196 pounds . . . . make a barrel of flour.
    200 " . . . . "   " beef, pork or fish.
    280 " . . . . "   " salt.

A fathom is 6 feet; 880 fathoms, 1 mile. A ship's cable is a chain, 120 fathoms, or 720 feet long.

A hair's breadth is one forty-eighth part of an inch.

A knot, or geographical mile is one-sixtieth of a degree; 3 knots make a marine league; 60 knots, or $69\frac{1}{2}$ statute miles, 1 degree.

The following tables are generally recognized:

DISTANCE.

3 inches . . . . make 1 palm.	21.8 inches . . make 1 Bible cubit.
4 " . . . . " 1 hand.	$2\frac{1}{2}$ feet . . . " 1 military pace.
6 " . . . . " 1 span.	3 " . . . " 1 common pace.
18 " . . . . " 1 cubit.	3.28 " . . . " 1 metre.

WEIGHT.

3 pounds make 1 stone butcher's meat.	$6\frac{1}{2}$ tods . . make 1 wey of wool.
7 " " 1 clove.	2 weys . . " 1 sack "
2 cloves " 1 stone common articles.	12 sacks . . " 1 last "
2 stone " 1 tod of wool.	240 pounds . " 1 pack "

CLOTH MEASURE.

$2\frac{1}{2}$ inches . . . make 1 nail.	5 quarters . . make 1 English ell.
4 nails . . . " 1 quarter.	6 " . . " 1 French ell.
4 quarters . . " 1 yard.	$4\frac{2}{15}$ " . . " 1 Scotch ell.
3 " . . " 1 Flemish ell.	

DRY MEASURE.

2 quarts . . . . . make 1 pottle.	5 quarters . . . . . make 1 load.
2 bushels . . . . . " 1 strike.	3 bushels . . . . " 1 sack.
2 strikes . . . . . " 1 coom.	36 " . . . . " 1 chaldron.
2 cooms . . . . . " 1 quarter.	

WINE MEASURE.

18 United States gallons . make 1 runlet.	63 United States gallons . make 1 hogshead.
25 English gallons, or ⎫ " 1 tierce.	2 hogsheads . . . . . " 1 pipe.
42 United States gallons ⎭	2 pipes . . . . " 1 tun.
2 tierces . . . . . " 1 puncheon.	$7\frac{1}{2}$ English gallons . . . " 1 firkin of beer.
$52\frac{1}{4}$ English gallons . . . " 1 hogshead.	4 firkins . . . . . . " 1 barrel.

II. The Metric System.

English Standards.—England and America yet hold to their old measures of weight, length, area and volume. These were taken as they were found, but have been definitely fixed. For example, there is a brass rod kept in one of the public offices in London, which, at the temperature of 62 degrees Fahr., gives, between two gold studs, the legal yard. For fear this might, in time, be lost, a commission, appointed for the purpose, has given the following formula for its recovery: "The length of a pendulum vibrating seconds, in vacuo, on the level of the sea in London is, when the thermometer stands at 62° Fahr., 39.13929 inches." The French metre is 39.37079 inches, The standard pound is also fixed by a brazen weight kept at the same place, and in case of loss, can be thus recovered: "The weight of a cubic inch of distilled water, at 62° Fahr., is 252.724 grains."

The Metre Explained.—In France, all this has been simplified, and a decimal system of weights and measures adopted. This is an enormous advantage to any people, and, as we have the decimal system in our money, so should we have it for all purposes. The French have taken for their standard, the metre, or the 1-10,000,000 part of the quadrant of the meredian of Paris, as measured by Delambre and Mechain:

1. The METRE, the unit of length, is the basis of all the other metric measures.
2. The ARE, the unit of land measure, is the square of ten metres.
3. The LITRE, the unit of measure of capacity, both liquid and dry, is the cube of the tenth part of a metre.
4. The STERE, the unit of solid or cubic measure, is equal to one cubic metre.
5. The GRAM, the unit of measures of weights, is the weight, in vacuo of the quantity of distilled water at a certain temperature which would be contained in a vessel whose inside measure equals a cubic centimetre.
6. The FRANC, the unit of metric money, is equal to 19.3 cents, and weighs five grains.

Multiples of Metric Measure.—The multiples are taken from the Greek, and are: deka, or ten; hecto, hundred, and kilo, a thousand.

Sub-Multiples.—These are from the Latin, and are: deci, a tenth; centi, hundredth, and milli, thousandth. Thus with the metre, we should have: The metre equals 1 metre; decametre, equals 10 metres; hectometre equals 100 metres, and kilometre equals 1,000 metres. Reversing this we should reduce the metre thus by sub-multiples. The metre equals 1; decimetre equals one-tenth of a metre; centimetre equals one hundredth of a metre, and millimetre equals one thousandth of a metre. The same system of enumerations applies to the others, as litre, decalitre and decilitre. The stere is a cubic metre, and the litre a cubic decimetre. The gram, would, raised by ten, be 1 decagram, and divided by ten, be 1 decigram. In money no multiplying prefixes are used: *e. g.* the tenth of a frank is the decime, etc.

Abbreviations.—The following are the metric terms and the abbreviations, representing the several terms:

Kilogram or kilo,	K. or Kg.	Milligram,	Mg.
Kilometre,	Km.	Millimetre,	Mm.
Litre,	L.	Stere,	St.
Metre,	M.		Etc.

III. The Metric System Compared With Our Own.

LONG MEASURE—ENGLISH.

3 lines make 1 inch.
12 inches make 1 foot.
3 feet make 1 yard.

5½ yards make 1 rod or pole.
40 rods make 1 furlong.
8 furlongs 1 mile.

COMPARATIVE SCALE.

MILE.		FURLONG.		ROD.		YARD.		FEET.		INCH.
1	equals	8	equal	320	equal	1760	equal	5280	equal	63360
		1	"	40	"	220	"	660	"	7920
				1	"	5½	"	16½	"	198
						1	"	3	"	36
								1	"	12

LONG MEASURE—METRIC.

1 millimetre,	equals	0.039 inch.
10 millimetres equals 1 centimetre,	"	0.3937 "
10 centimetres " 1 decimetre,	"	3.937 inches.
10 decimetres " 1 metre,	"	39.37 "
10 metres " 1 decametre,	"	32.80 feet.
10 decametres' " 1 hectòmetre,	"	328.01 "
10 hectometres " 1 kilometre,	"	3,280.10 "
10 kilometres " 1 myriametre,	"	6.2137 miles.

CLOTH MEASURE.

This is by the same standard as long measure, thus:

2 sixteenths . . . equal 1 eighth.	2 quarters	equal 1 half.
2 eighths . . . . " 1 quarter.	4 quarters	" 1 yard.

SURVEYOR'S LONG MEASURE.

25 links . . make 1 rod.   4 rods . . make 1 chain.   80 chains . . make 1 mile.

SQUARE MEASURE.

144 square inches . make 1 square foot.	40 square rods make 1 rood, or quarter acre.
9 " feet . . " 1 " yard.	4 roods . . " 1 acre.
30¼ " yards . " 1 " rod.	640 acres . . " 1 square mile or section.

SURVEYORS' SQUARE MEASURE.

625 square links make 1 square rod, . . sq. rd.	640 acres makes 1 square mile, . . . sq. mi.
16 " rods " 1 " chain, . sq. ch.	36 square miles (six miles square) make
10 " chains " 1 acre, ... A.	1 township, ... Tp

COMPARATIVE SCALE.

A.	R.	RODS.	SQUARE YARDS.	SQUARE FEET.	SQUARE INCH.
1 equals 4	equal	160 equal	4,840 equal	43,560 equal	6,272,640
1	"	40 "	1,210 "	10,890 "	1,568,160
		1 "	30¼ "	272¼ "	39,204
			1 "	9 "	1,296
				1 "	144

METRIC SQUARE MEASURE AND EQUIVALENTS.

1 square centimetre	equals	1.155 square inches.
100 square centimetres equal 1 square decimetre	"	115.50 square inches.
100 square decimetres " 1 square metre (centare) . .	"	1.196 square yards.
100 centares " 1 are	"	119.6 square yards.
100 ares " 1 hectare	"	2.471 acres.
00 hectares " 1 square kilometre	"	3.861 square miles.
1 square mile	"	258.99 hectares.
1 square acre	"	40.47 ares.
1 square rood	"	10.12 ares.
1 square rood	"	25.29 square metres.
1 square yard	"	0.836 square metre.
1 square foot	"	.093 square metre.
1 square inch	"	6.45 square centimetre.

CUBIC OR SOLID MEASURE.

1728	cubic inches	make 1 cubic foot.
27	cubic feet	" 1 cubic yard.
40	cubic feet of round timber or ⎫	" 1 ton or load.
50	cubic feet of hewn timber ⎭	
8	cubic feet	" 1 cord foot.
16	cord feet or ⎫	" 1 cord of wood.
128	cubic feet ⎭	
24¾	cubic feet	" 1 perch of stone, or masonry.

METRIC CUBIC MEASURE AND EQUIVALENTS.

1000 cubic centimetres equal 1 cubic decimetre or litre . . equal 0.308 cubic foot.
1000 cubic decimetres " 1 cubic metre or stere . . . " 1.308 cubic yards.
1 cubic foot . . . . " 28,315.31 cubic centimetres.
1 cubic inch . . . . " 16.386 cubic centimetres.

ENGLISH AND METRIC EQUIVALENTS.

1 cubic inch	equals	16.387	cubic centimetres.	1 litre	equals	⎧ 1.0567 quarts, liquid meas.
						⎩ .928 quart, dry measure.
1 cubic foot	"	⎧ 28.34	litres.	1 hecto-	"	⎧ 2.837 bushels, dry measure.
		⎩ .0283	steres.	litre		⎩ 26.417 gallons, liquid meas.
1 cubic yard	"	.76531	steres.	1 kilo-		⎧ 35.316 cubic feet.
1 cord	"	3.6281	steres.	litre		
1 fluid ounce	"	.02958	litres.	1 cubic	equal	1.308 cubic yards.
1 gallon	"	3.786	litres.	metre		264.17 gallons, liquid meas.
1 bushel	"	35.24	litres.	1 stere		⎩ .2759 cord.

DRY MEASURE.

The standard is the Winchester bushel, which contains 2150.42 cubic inches, or 77.627 lbs. avoirdupois of distilled water at its maximum density. Its dimensions are 18½ inches diameter inside, 19½ inches outside, and 8 inches deep.

TABLE OF DRY MEASURE.

2 pints (pt.) . . . make 1 quart (qt.)   4 pecks . . . . make 1 bushel (bu.)
8 quarts . . . . . " 1 peck (pk.)   36 bushels . . . " 1 chaldron (cald.)

COMPARATIVE SCALE.

CALD.		BU.		PKS.		QTS.		PTS.
1	equals	36	equal	144	equal	1152	equal	2304
		1	"	4	"	32	"	64
				1	"	8	"	10
						1	"	2

METRIC DRY MEASURE AND EQUIVALENTS.

1 millilitre or cubic centimetre equals 0.061 cubic inch.
10 millilitres or cubic centimetres equal 1 centilitre, equals 0.6102 cubic inch.
10 centilitres equal 1 decilitre, equals 6.1022 cubic inches.
10 decilitres equal 1 litre, equals 0.908 quart.
10 litres equal 1 decalitre, equals 9.08 quarts.

10 decalitres equal 1 hectolitre, equals, 2 bushels, 3.35 pecks.
10 hectolitres equal 1 kilolitre or cubic metre, equals 1.308 cubic yard.
1 bushel equals 35.237 litres.
1 peck equals 8.809 litres.
1 quart equals 1.101 litre.

LIQUID OR WINE MEASURE

The wine gallon, English and American, are the same. The beer, the ale or beer gallon of the United States contains 4.62 litres metric.

TABLE.

gills make 1 pint, ... pt.	31½ gallons make 1 barrel, ...	bbl.
pints " 1 quart . . . . . . qt.	2 barrels } make 1 hogshead . .	hh.
4 quarts " 1 gallon, ... gal.	63 gallons }	

COMPARATIVE SCALE.

HOGSHEAD.		BARRELS.		GALLONS.		QUARTS.		PINTS.		GILLS.
1	equals	2	equal	63	equal	252	equal	504	equal	2016
		1	"	31½	"	126	"	252	"	1008
				1	"	4	"	8	"	32
						1	"	2	"	8
								1	"	4

LIQUID MEASURE—APOTHECARIES.

60 minims . . . . . 1 fluid drachm.	16 fluid ounces . . . . . 1 fluid pint.
8 fluid drachms . . . 1 fluid ounce.	8 pints . . . . 1 gallon.

Dry and liquid measures are computed alike by the metric tables. The equivalents are:

METRIC LIQUID MEASURE AND EQUIVALENTS.

1 millilitre . . . . . equals 0.27 fluid dram.	1 U. S. gallon . equals 3.785 litres.
1 centilitre . . . . . " 0.338 fluid ounce.	1 quart . . . . . " 0.946 litre.
1 decilitre . . . . . " 0.845 gill.	1 pint . . . . " 0.473 litre.
1 litre . . . . . . " 1.0567 quart.	1 gill . . . . " 0.118 litre.
1 decalitre . . . . . " 2.6417 gallons.	1 fluid ounce . . " 29.57 cubic centimetres.
1 hectolitre . . . . " 26.417 gallons.	1 fluid dram . . " 3.69 cubic centimetres.
1 kilolitre . . . . . " 264.17 gallons.	1 minim . . . " 0.0616 cubic centimetres.
1 imperial gallon . . " 4.543 litres.	

Avoirdupois Weight.—The ounce and pound avoirdupois differ in weight from those of the same denomination in Troy and Apothecaries' weight, though the Troy grain and Apothecaries' grain are the same. Troy weights are used in weighing precious metals and stones—Apothecaries' weight in preparing medicines.

AVOIRDUPOIS WEIGHT.

16 drams (dr.) . . . . equal	1 ounce, ...	oz.
16 ounces . . . . "	1 pound, ...	lb.
25 pounds . . . . "	1 quarter, ...	qr.
4 quarters . . . . "	1 hundredweight, ...	cwt.
20 hundredweight . . . . "	1 ton, ...	t.
100 pounds . . . . "	1 cental, ...	c.

SCALE OF COMPARISON.

TON.	HUNDREDWEIGHT.	QUARTER.	POUND.	OUNCE.	DRAM.
1 equals	20 equal	80 equal	2,000 equal	32,000 equal	512,000
	1 "	4 "	100 "	,000 "	25,600
		1 "	25 "	400 "	6,400
			1 "	16 "	256
				1 "	16

METRIC AVOIRDUPOIS WEIGHT.

1 milligram	. . . .	equals	0.0154 grain.
10 milligrams	. . . equal 1 centigram	"	0.1543 "
10 centigrams	. . . . " 1 decigram	"	1.5432 "
10 decigrams	. . . . " 1 gram	"	15.432 grains.
10 grams	. . . . . . " 1 decagram	"	0.3527 ounce.
10 decagrams	. . . . " 1 hectogram	"	3.5274 ounces.
10 hectograms	. . . " 1 kilogram or kilo . .	"	2.2046 pounds.
10 kilograms	. . . . " 1 myriagram . . .	"	22.046 "
10 myriagrams	. . . " 1 quintal	"	220.46 '
10 quintals	. . . . . " 1 tonneau	"	2204.6 "

COMPARISON:

1 ton (2,000 lbs.) . . equals 907.18 kilos.	1 ounce . . . . . equals 28.35 grams.
1 hundred-weight . . " 45.36 "	1 dram . . . . . " 1.772 gram.
1 pound . . . . " 0.454 kilo.	

TROY WEIGHT.

24 grains (gr.) make 1 pennyweight, dwt.   12 ounces make 1 pound, ... lb.
20 pennyweights " 1 ounce, . . oz.   3⅕ grains " 1 carat (diamond wt.), k.

COMPARISON:

POUND.	OUNCES.	PENNYWEIGHTS.	GRAINS.
1 equals	12 equal	240 equal	5760
	1 "	20 "	480
		1 "	24
		1k. "	3⅕

METRIC EQUIVALENTS.

1 pound . . . equals 373.24 grams.   1 pennyweight . . . equals 1.55 gram.
1 ounce. . . . . " 31.102 "   1 grain . . . . . . " 0.065 "

APOTHECARIES' WEIGHT.

20 grains (gr.) make 1 scruple, . . sc.   8 drams . . make 1 ounce, . . oz.
3 scruples . " 1 dram . . dr.   12 ounces . . " 1 pound, . . lb.

COMPARISON.

POUND.	OUNCES.	DRAMS.	SCRUPLES.	GRAINS.
1 equal	12 equal	96 equal	288 equal	5760
	1 "	8 "	24 "	480
		1 "	3 "	60
			1 "	20

TABLES OF WEIGHTS, DIMENSIONS, STRENGTH, GRAVITY, ETC.

METRIC EQUIVALENTS.

1 pound . . . . equals 373.24 grams. 1 scruple . . . . . equals 1.29 gram.
1 ounce . . . . . " 31.102 " 1 grain . . . . . . " 0.06 "
1 dram . . . . . " 3.88 "

COMPARATIVE TABLES OF MEASURES AND WEIGHTS OF CAPACITY.

1 gallon or 4 quarts wine measure	contains 231	cubic inches.
½ peck or 4 quarts dry measure	" 268.8	cubic inches.
1 gallon or 4 quarts beer measure	" 282	cubic inches.
1 bushel dry measure	" 2150⅓	cubic inches.
1 United States bushel	" 2150.42	cubic inches.
1 English bushel	" 2218.19	cubic inches.
1 United States gallon	" 231	cubic inches.
1 English gallon	" 277.26	cubic inches.
1 French litre	" 61.533	cubic inches
1 United States pound troy	equals 5760	grains troy.
1 English pound troy	" 5760	grains troy.
1 pound apothecaries'	" 5760	grains troy.
1 United States or English pound avoirdupois	" 7000	grains troy.
144 pounds avoirdupois	" 175	pounds troy.
1 French gram	" 15.433	grains troy.
1 United States or English yard	" 36	inches.
1 French metre	" 39.368	inches.
1 French are	" 119.664	square yards.

IV. Tables Relating to Money.

MONEY of the United States is computed by the decimal system, as follows:

```
10 mills (m.) 1 cent, ct.
10 cents     1 dime, d.    100 mills.
10 dimes     1 dollar, $   1000 mills  100 cents.
10 dollars   1 eagle E     10000 mills 1000 cents 100 dimes.
             1 eagle (gold) weighs 258   troy grains.
             1 dollar (silver)  "  412.5 troy grains.
             1 cent (copper)    "  168   troy grains.
             23.2 grains of pure gold equal $1.00.
```

FRENCH, AMERICAN AND ENGLISH MONEY COMPARED.

The following table will show the relative values. The franc, dollar and pound sterling being the units in the several countries.

FRANCS.		DOLLARS.		POUNDS STERLING.		SHILLINGS.		PENCE.
1	equal	0.1930	equal	0.03968	equal	0.7936	equal	9.523
5	"	0.9648	"	0.19840	"	3.968	"	47.61
5.1826	"	1	"	0.2056	"	4.11	"	49
25.913	"	5	"	1.0280	"	20.56	"	247
25.20	"	4.863	"	1	"	20	"	240
126	"	24.315	"	5	"	100	"	1200

V. Foreign Exchange.

The value of the standard coins of foreign countries is given in the following table:

Country.	Monetary Unit.	Standard.	Value in U. S. Money.	Standard Coin.
Austria,	Florin,	Silver,	$ .413	
Belgium,	Franc,	Gold and silver,	.193	5, 10 and 20 francs.
Bolivia,	Boliviano,	Silver,	.836	Boliviano.
Brazil,	Milreis of 1000 reis,	Gold,	.545	
British Possessions in N. A.,	Dollar,	Gold,	1.00	
Central America,	Peso,	Silver,	.836	Peso.
Chili,	Peso,	Gold,	.912	Condor, doubloon and escudo.
Denmark,	Crown,	Gold,	.268	10 and 20 crowns.
Ecuador,	Peso,	Silver,	.836	Peso.
Egypt,	Pound of 100 Piasters,	Gold,	4.974	5, 10, 25 and 50 piasters.
France,	Franc,	Gold and silver,	.193	5, 10 and 20 francs.
Great Britain,	Pound sterling,	Gold,	4.866⅔	½ sovereign and sovereign.
Greece,	Drachma,	Gold and silver,	.193	5, 10, 20, 50 and 100 drachmas.
German Empire,	Mark,	Gold,	.238	5, 10 and 20 marks.
India,	Rupee of 16 annas,	Silver,	.397	
Italy,	Lira,	Gold and silver,	.193	5, 10, 20, 50 and 100 lire.
Japan,	Yen (gold),	Gold and silver,	.997	1, 2, 5, 10 and 20 yen.
Liberia,	Dollar,	Gold,	1.00	
Mexico,	Dollar,	Silver,	.909	Peso or dollar, 5, 10, 25 and 50 centavo.
Netherlands,	Florin,	Gold and silver,	.402	
Norway,	Crown,	Gold,	.268	10 and 20 crowns.
Peru,	Sol,	Silver,	.836	Sol.
Portugal,	Milreis of 1000 reis,	Gold,	1.08	2, 5 and 10 milreis.
Russia,	Rouble of 100 copecks,	Silver,	.669	¼, ½ and 1 rouble.
Sandwich Islands,	Dollar,	Gold,	1.00	
Spain,	Peseta of 100 centimes,	Gold and silver,	.193	5, 10, 20, 50 and 100 pesetas.
Sweden,	Crown,	Gold,	.268	10 and 20 crowns.
Switzerland,	Franc,	Gold and silver,	.193	5, 10 and 20 francs.
Tripoli,	Mahbub of 20 piasters,	Silver,	.748	
Turkey,	Piaster,	Gold,	.044	25, 50, 100, 250 and 500 piasters.
United States of Colombia,	Peso,	Silver,	.836	Peso.

VI. Specific Gravity.

SPECIFIC GRAVITY OF METALS.

Name.	Sp. Grav.	Name.	Sp. Grav.
Antimony	6.712	Iron, cast	7.207
Arsenic	5.763	Iron, bars	7.778
Bismuth	9.823	Lead	11.352
Brass	7.820	Mercury	13.598
Bronze	8.700	Nickel	8.275
Copper	8.788	Platinum	22.069
Copper Wire	8.878	Silver	10.477
Gold, pure	19.258	Steel	7.833
Gold, 22 carat	17.486	Tin	7.291
Gold, 20 carat	15.709	Zinc	6.861

SPECIFIC GRAVITY OF ROCK AND EARTH.

NAME.	SP. GRAV.	NAME.	SP. GRAV.
Alabaster	2.730	Emory	4.000
Amber	1.078	Flint	2.590
Asbestos	3.073	Glass	2.930
Borax	1.714	Granite	2.625
Brick	1.900	Grindstone	2.143
Chalk	2.784	Gypsum	2.168
Charcoal	.441	Ivory	1.822
Coral	2.700	Limestone	3.180
Coal, bituminous	1.270	Lime, quick	.804
Coal, anthracite	1.556	Manganese	7.000
Diamond	3.521	Marble, parian	2.838

SPECIFIC GRAVITY OF THOROUGHLY DRY WOOD.

NAME.	SP. GRAV.	NAME.	SP. GRAV.
Apple	.793	Logwood	.919
Alder	.800	Mahogany	1.063
Ash	.845	Maple	.750
Beech	.852	Mulberry	.896
Box	1.231	Orange	.705
Campeachy	.913	Pine, yellow	.660
Cherry	.715	Pine, white	.554
Cocoa	1.040	Pear	.661
Cork	.240	Plum	.785
Cypress	.644	Quince	.705
Ebony	1.331	Sassafras	.482
Elder	.695	Walnut	.671
Elm	.671	Willow	.585
Fir, yellow	.657	Yew	.798
Fir, white	.669	Hickory	.838
Lignum-vitæ	1.333	Poplar	.383
Live Oak	1.120	Poplar, white	.529

SPECIFIC GRAVITY OF MISCELLANEOUS ARTICLES.

NAME.	SP. GRAV.	NAME.	SP. GRAV.
Beeswax,	.96	Oil, whale,	.92
Butter,	.94	Oil, turpentine,	.87
Honey,	1.45	Sea water,	1.02
Lard,	.94	Sugar,	1.60
Milk,	1.03	Tallow,	.93
Oil, linseed,	.94	Vinegar,	1.01 to 1.08

The specific gravity of a substance not aeriform is determined by its relative weight to an equal volume of distilled water at a temperature of 60 degrees Fahrenheit, the barometer being 30 inches. To find the weight of a cubic foot of metal, etc., remove the decimal point representing the specific gravity three places to the right and the result is the weight in ounces. To reduce to pounds, divide for precious metals by twelve and all other substances by sixteen. Thus antimony's specific gravity is 6.712. Weight 6,712 ounces per cubic foot; 6,712÷16=419 pounds 8 ounces.

VII. Earth and Soils.

NAME OF EARTH.	SP. GRAV.	WEIGHT OF A CUBIC FOOT, IN LBS.	
		DRY.	WET.
Calcareous sand,	2.722	113.6	141.3
Silicious sand,	2.653	111.3	136.1
Gypsum powder,	2.331	91.9	127.6
Sandy clay,	2.601	97.8	129.7
Loamy clay,	2.581	88.5	124.1
Stiff clay or brick earth,	2.560	80.3	119.6
Pure gray clay,	2.553	75.2	115.8
Pipe clay,	2.440	47.9	102.1
Fine carbonate of lime (chalk),	2.468	53.7	103.5
Garden mold,	2.332	68.7	102.7
Arable soil,	2.401	84.5	119.1
Fine slaty marl,	2.631	112.0	140.3

WEIGHT OF VARIOUS SUBSTANCES PER CUBIC FOOT.

NAME.	POUNDS.	NAME.	POUNDS.
Air,	0.0753	Portland stone,	157.5
Cork,	15	Clay and stones,	160
Fir,	34.375	Crown glass,	180.75
Tallow,	59	Mason's work,	205
Distilled water,	62.5	Cast iron,	450.45
Mahogany,	66.4	Copper,	486.75
Oak,	73.15	Steel,	489.8
Loose earth or sand,	95	Pure silver,	654.8
Common soil,	124	Lead,	709.5
Brick,	125	Pure gold,	1203.625
Strong soil,	127	Platina,	1218.75
Clay,	135		

BULK OF A TON OF DIFFERENT SUBSTANCES.

Twenty-three cubic feet of sand, eighteen cubic feet of earth, or seventeen cubic feet of clay, make a ton. Eighteen cubic feet of gravel or earth before digging, make twenty-seven cubic feet when dug; or the bulk is increased as three to two.

VIII. Cohesion of Materials.

THE force which binds similar particles together is called cohesion, and the measure of this cohesion is the strain which it will bear. The two following tables show the pounds of force necessary to rend a prism an inch square, and the length of prism necessary to tear it apart by its own weight.

WOODS.

NAME.	POUNDS.	FEET.	NAME.	POUNDS.	FEET.
Teak,	12,915	36,049	Elm,	9,720	39,050
Oak,	11,880	32,900	Memel fir,	9,520	40,500
Sycamore,	9,630	35,800	Norway fir,	12,346	55,500
Beech,	12,225	38,940	Larch,	12,240	42,160
Ash,	14,130	42,080			

TABLES OF WEIGHTS, DIMENSIONS, STRENGTH, GRAVITY, ETC.

METALS.

NAME.	POUNDS.	FEET.	NAME.	POUNDS.	FEET.
Cast Steel,	134,256	39,455	Cast copper,	19,072	5,093
Swedish malleable iron,	72,064	19,740	Yellow brass,	17,958	5,180
English,	55,872	19,740	Cast tin,	4,736	1,496
Cast iron,	19,096	6,110	Cast lead,	1,824	348

METALS, THEIR GRAVITY AND MELTING POINTS.

NAMES OF METALS.	SPECIFIC GRAVITY.	MELTING POINTS.	NAMES OF METALS.	SPECIFIC GRAVITY.	MELTING POINTS.
		Fahrenheit.			Fahrenheit.
1. Gold,	19.25	2016°	21. Chromium,		
2. Silver,	10.47	1873	22. Columbium,		
3. Iron,	7.78	2800? Smith's forge.	23. Palladium,	11.50	oxyhydrogen blowpipe.
			24. Rhodium,		
4. Copper,	8.89	1996	25. Iridium,		
5. Mercury,	13.56	—39	26. Osmium,		
6. Lead,	11.35	612	27. Cerium.		
7. Tin,	7.29	442	28. Potassium,	0.86	136
8. Antimony,	6 70		29. Sodium,	0.97	190
9. Bismuth,	9.80	497	30. Barium,		
10. Zinc,	7.00	773	31. Strontium,		
11. Arsenic,	5.88		32. Calcium,		
12. Cobalt,	8.53	2810?	33. Cadmium,	8.60	442
13. Platinum,	20.98	oxyhydrogen blowpipe.	34. Lithium,		
			35. Silicium,		
14. Nickel,	8.27	2810? Smith's forge.	36. Zirconium,		
15. Manganese,	6.85		37. Aluminum,		
16. Tungsten,	17.60		38. Glucinum,		
17. Tellurium,	6 11	620?	39. Yttrium,		
18. Molybdenum,	7.40	oxyhydrogen blowpipe.	40. Thorium,		
19. Uranium,	9.00		41. Magnesium,		
20. Titanium,	5.30		42. Vanadium,		

IX. Strength of Common Ropes.

The following shows the breaking weight and also the safe weight which may be borne by common ropes:

ROPE.	BREAKING WEIGHT.	BORNE WITH SAFETY.
One-eighth inch diameter	78 lbs.	31 lbs.
One-fourth inch "	314	125
One-half inch "	1,250	500
One-inch "	5,000	2,000
One and a fourth inch "	7,500	3,000
One and a half inch "	12,500	4,500

Experiments some years since by the British admiralty showed a wire rope two inches in circumference to be as strong as a hemp rope five inches in circumference, and either would bear seven tons just before breaking. A wire rope three inches in circumference was equal to one of hemp eight inches, and bore thirteen tons. One four inches in circumference was equal to hemp ten inches, and sustained twenty-one tons. We may add it is never safe to subject any rope or cable to more than half its ultimate strength, even when entirely free from swaying. The larger the rope, the less it will bear in proportion. A hemp rope from half an inch to an inch in diameter, will support 8,700 pounds for each square inch of section; from one to three inches in diameter, 6,800 pounds for each square inch; if from five to seven inches, it will bear only 4,800 pounds per square inch. Manila rope will only bear about one-half the strain that the best hemp rope will.

X. Human Force.

The proportional force between the human hands on the tool and the force exerted by the tool are given respectively in the first and second columns following:

	HAND.	TOOL.
Drawing knife,	100 pounds,	100 pounds.
Large auger,	100 "	about 800 "
Screw-driver, one hand,	84 "	" 250 "
Bench vice handle,	72 "	" 1000 "
Windlass, one hand,	60 "	180 to 700 "
Handsaw,	36 "	36 "
Brace bit,	16 "	150 to 700 "
Button screw, thumb and finger,	14 "	14 to 70 "

XI. Heat and Its Effects.

The following tables show various temperatures (Fahrenheit), and the effects of heat on various substances:

FUSING POINTS.

		DEGREES.			DEGREES.
Gold	melts	2590	Antimony	melts	951
Silver	"	1250	Bismuth	"	476
Copper	"	2548	Cadium	"	600
Wrought iron	"	3980	Steel	"	2500
Cast iron	"	3479	Lead	"	504
Glass	"	2377	Tin	"	424
Brass	"	1900	Zinc	"	740

TABLE OF HEATS.

	DEGREES.
Furnace under steam boiler,	1100
Common fire,	270
Iron bright red, in dark,	752
Iron red-hot, in twilight,	884
Heat, cherry red,	1500
Heat, bright red,	1860
Heat, white,	2900
Heat, visible by day,	1077
Heat of air furnace,	3300
Heat of human blood,	98
Heat snow and salt, equal parts,	0
Highest natural heat in shade (Egypt),	117
Greatest natural cold (below zero),	65
Greatest Artificial cold (below zero),	160
Ice melts	32

BOILING POINTS.

Water in vacuo	boils at	98 deg.	Ether	boils at	95 deg.
Water in vacuum pan	"	114 "	Ether, nitrous	"	57 "
Water in open air	"	212 "	Iodine	"	347 "
Alcohol	"	173 "	Mercury	"	602 "
Bromine	"	145 "	Olive oil	"	600 "

FREEZING POINTS.

Bromine	Freezes at	4 deg.	Oil olive	Freezes at	60 deg.
Mercury—below zero	"	39 "	Oil rose	"	60 "
Oil anise	"	50 "	Water	"	32 "

HEATING POWER OF FUEL.—The celebrated experiments of Marcus Bull, of Philadelphia, showed that one cord of hickory wood and one ton of anthracite coal were equal in heating power. The following is his table of quantities required for a given amount of heat:

Hickory,	4	cords.	Pitch pine,	9¼	cords.
White oak,	4¾	"	White pine,	9⅕	"
Hard maple,	6⅔	"	Anthracite coal,	4	tons.
Soft maple,	7⅕	"			

XII. Capacity of Soils for Heat.

SHUBLER, a learned German, heated a given quantity of soil to 145 degrees Fahr., placed a thermometer in it and observed the time it required to cool down to 70 degrees, the atmospheric temperature being 31 degrees Fahr. The following table gives the results of his experiments. The first column shows the time required for cooling, the second the relative power of retaining heat—the relative capacity of soils for holding heat being, sand, loam, clay and humus soils.

Lime sand,	Time of Cooling,	3 hours, 30 Minutes.	Capacity for Heat,		100
Quartz sand,	"	2 " 27 "	"	"	95.6
Clay loam,	"	2 " 30 "	"	"	71.8
Clay plow land,	"	2 " 27 "	"	"	70.1
Heavy clay,	"	2 " 24 "	"	"	68.4
Pure gray clay,	"	2 " 19 "	"	"	66.7
Garden earth,	"	2 " 16 "	"	"	64.8
Humus,	"	1 " 43 "	"	"	49.0

The absorption of heat from the sun, with the thermometer at 77, was found by Becquerel to be (moist earth heats slowest because the heat is constantly being dissipated in evaporating the moisture) as follows:

KINDS OF EARTH.	MAXIMUM TEMPERATURE, TOP LAYER.	
	Moist Earth.	Dry Earth.
	Degrees.	Degrees.
Silicious sand, yellowish gray,	99.05	112.55
Calcareous sand, whitish gray,	99.10	112.10
Argillaceous earth, yellowish gray,	99.28	112.32
Calcareous earth, white,	96.13	109.40
Mold, blackish gray.	103.55	117.27
Garden earth, blackish gray,	99.50	113.45

It will be seen that soils, under the action of continued heat, accumulate it, so as to become much warmer than the superjacent air.

XIII. Radiating Power, Absorption and Evaporation.

THE radiating power of substances have very much to do with the deposit of dew in warm weather, and frost in cold weather. That is: the stronger the radiating power the cooler the substance becomes. The following table shows the relative power of various substances, the first named being 100.

Lampblack,	100	Poplar sawdust,	99
Grasses,	103	Varnish,	97
Silicious sand,	103	Glass,	93
Leaves of the elm and the poplar,	101	Vegetable earth,	92

Absorption of Water.—The quantity of water a soil will absorb and retain before allowing it to run away, was thus determined by Schubler:

A cubic foot of
- Silicious sand . . . held of water 27.3 lbs.
- Calcareous sand . . . " " 31.8 "
- Sandy clay . . . . " " 38.8 "
- Loamy clay . . . . " " 41.4 "

A cubic foot of
- Stiff clay, or brick earth held of water 45.4 lbs.
- Arable soil . . . . " " 46.8 "
- Garden mold . . . " " 48.4 "

Of 200 parts of each earth exposed for four hours, on a thin surface in a closed room, at 65¾ Fahr., there was an evaporation of absorbed water as follows:

- Silicious sand . . . . . . lost 88.4 parts in 100 parts of absorbed water.
- Calcareous sand . . . . " 75.9 " " " " "
- Sand clay . . . . " 52.0 " " " " "
- Loamy clay . . . . " 45.7 " " " " "
- Stiff clay . . . . " 35.9 " " " " "
- Arable . . . . " 32.0 " " " " "
- Garden mold . . . . " 24.3 " " " " "

XIV. Temperatures Required by Plants.

THE reason that the West is so prolific in plant life, is the great heat of the summers. One can not grow pineapples because they require a full year to perfect themselves, but one *can* grow melons that only require one degree of heat less, as they ripen from seed in three to four months. The following table shows the maximum and minimum temperatures, in degrees Fahr., for the growth of various products:

NAME.	MAXIMUM.	MINIMUM.	NAME.	MAXIMUM.	MINIMUM.
Chocolate bean,	82° F.	73°	Pineapple,	82°	68°
Banana,	"	64	Melon,	82	67
Indigo,	"	71	Coffee,	79	74
Sugar-cane,	"	71	Wheat,	74	44
Cocoanut,	"	78	Barley,	74	59
Palm,	"	78	Potatoes,	75	52
Tobacco,	"	65	Flax,	74	54
Maize,	"	59	Apple,	72	59
French beans (haricots),	"	59	Oak,	67	61
Rice,	"	75			

Ripening of Plants.—The more steady the heat the less the number of days between germination and ripening. Hence we see why plants mature in one locality, sheltered, and are caught by frost in another, exposed. Fahrenheit being the standard, wheat, with a mean temperature of 59 degrees, requires 137 days to mature; with a temperature of 56 degrees, 160 days; with a temperature of 76 degrees, 92 days. In other words, the lower the mean temperature of the climate, the longer the crop must be in the ground before harvesting.

XV. Temperature of Germination.

	LOWEST TEMPERATURE.	HIGHEST TEMPERATURE.	TEMPERATURE OF MOST RAPID GERMINATION.
Wheat,	41 degrees Fahr.	104 degrees Fahr.	84 degrees Fahr.
Barley,	41 " "	104 " "	84 " "
Pea,	44 5 " "	102 " "	84 " "
Maize,	48 " "	115 " "	93 " "
Scarlet bean,	49 " "	111 " "	79 " "
Squash,	54 " "	115 " "	93 " "

XVI. Contrasts Between Animal and Vegetable Life.

Dumas and Boussingault give the following as the chemical and physiological balance of organic nature.

AN ANIMAL IS	A VEGETABLE IS
An Apparatus of Combustion;	An Apparatus of Reduction;
Possesses the faculty of Locomotion;	Is fixed;
Burns Carbon,	Reduces Carbon,
Hydrogen,	Hydrogen,
Ammonium,	Ammonium,
Exhales Carbonic acid,	Fixes Carbonic acid,
Water,	Water,
Oxide of Ammonium,	Oxide of Ammonium,
Nitrogen;	Nitrogen;
Consumes Oxygen,	Produces Oxygen,
Neutral nitrogenized matters,	Neutral nitrogenized matters,
Fatty matters,	Fatty matters,
Amylaceous matters, sugars, gums;	Amylaceous matters, sugars, gums;
Produces Heat,	Absorbs Heat,
Electricity;	Abstracts Electricity;
Restores its elements to the air,	Derives its elements from the air,
or to the earth,	or from the earth;
Transforms organized matters	Transforms mineral matters into
into mineral matters.	organized matters.

XVII. Thermometers.

A THERMOMETER is an instrument for measuring heat. Fahrenheit's scale, in which zero is the cold produced by a mixture of salt and snow, or 32 degrees below the freezing point of water, is the one in common use. The boiling point of water is 212 degrees, Fahr. Reaumer's and the Centigrade thermometers are used for scientific purposes. The standard points compare thus: 212 degrees Fahr., equal 100 degrees

Centigrade, and 88 degrees Reaumer. Thirty-two degrees of Fahr., equal 0 Centigrade, and 0 Reaumer. The following table will show the correspondence between these several thermometers:

REAUM.	CENT.	FAHR.	REAUM.	CENT.	FAHR.	REAUM.	CENT.	FAHR.	REAUM.	CENT.	FAHR.
80	100.	212.	54	67.5	153.5	29	36.25	97.25	4	5.	41.
79	98.75	209.75	53	66.25	151.25	28	35.	95.	3	3.75	38.75
78	97.5	207.5	52	65.	149.	27	33.75	92.75	2	2.5	36.5
77	96.25	205.25	51	63.75	146.85	26	32.5	90.5	1	1.25	34.25
76	95.	203.	50	62.5	144.5	25	31.25	88.25	0	0.	32.
75	93.75	200.75	49	61.25	142.25	24	30.	86.	1	1.25	29.75
74	92.5	198.5	48	60.	140.	23	28.75	83.75	2	2.5	27.5
73	91.25	196.25	47	58.75	137.75	22	27.5	81.5	3	3.75	25.25
72	90.	194.	46	57.5	135.5	21	26.25	79.25	4	5.	23.
71	88.75	191.75	45	56.25	133.25	20	25.	77.	5	6.25	20.75
70	87.5	189.5	44	55.	131.	19	23.75	74.75	6	7.5	18.5
69	86.25	187.25	43	53.75	128.75	18	22.5	72.5	7	8.75	16.25
68	85.	185.	42	52.5	126.5	17	21.25	70.25	8	10.	14.
67	83.75	182.75	41	51.25	124.25	16	20.	68.	9	11.25	11.75
66	82.5	180.5	40	50.	122.	15	18.75	65.75	10	12.5	9.5
65	81.25	178.25	39	48.75	119.75	14	17.5	63.5	11	13.75	7.25
64	80.	176.	38	47.5	117.5	13	16.25	61.25	12	15.	5.
63	78.75	173.75	37	46.25	115.25	12	15.	59.	13	16.25	2.75
62	77.5	171.5	36	45.	113.	11	13.75	56.75	14	17.5	0.5
61	76.25	159.25	35	43.75	110.75	10	12.5	54.5	15	18.75	1.75
60	75.	167.	34	42.5	108.5	9	11.25	52.25	16	20.	4.
59	73.75	164.75	33	41.25	106.25	8	10.	50.	17	21.25	6.25
58	72.5	162.5	32	40.	104.	7	8.75	47.75	18	22.5	8.5
57	71.25	160.25	31	38.75	101.75	6	7.5	45.5	19	23.75	10.75
56	70.	158.	30	37.5	99.	5	6.25	43.25	20	25.	13.
55	68.75	155.75									

XVII. Dimensions and Contents of Fields, Granaries, Corn-Cribs, etc.

FARMERS often wish to know the contents of a field or lot, crib, bin, cistern, etc. To find the number of acres in any square or rectangular field, multiply the length in rods and breadth in rods together, and divide by 160; or, multiply the length in feet by breadth in feet and divide by 43,560, the number of square feet in an acre. Thus:

    10 rods by 16 rods . . . make 1 acre.   25 feet by 100 feet . . make .0574 acre.
      8 rods by 20 rods . . . " 1 acre.   25 feet by 110 feet . . " .0631 acre.
      5 rods by 32 rods . . . " 1 acre.   25 feet by 120 feet . . " .0688 acre.
      4 rods by 40 rods . . . " 1 acre.   25 feet by 125 feet . . " .0717 acre.
      5 yards by 968 rods . . . " 1 acre.   25 feet by 150 feet . . " .109 acre.
    10 yards by 484 yards . . " 1 acre.   2178 square feet . . . " .05 acre.
    20 yards by 242 yards . . " 1 acre.   4356 square feet . . . " .10 acre.
    40 yards by 121 yards . . " 1 acre.   6534 square feet . . . " .15 acre.
    80 yards by 60½ yards . . " 1 acre.   8712 square feet . . . " .20 acre.
    70 yards by 69½ yards . " 1 acre.   10890 square feet . . . " .25 acre.
   220 feet by 198 feet . . . " 1 acre.   13068 square feet . . . " .30 acre.
   440 feet by 99 feet . . . " 1 acre.   15246 square feet . . . " .35 acre.
   110 feet by 369 feet . . . " 1 acre.   17424 square feet . . . " .40 acre.
    60 feet by 726 feet . . . " 1 acre.   19603 square feet . . . " .45 acre.
   120 feet by 363 feet . . ' 1 acre.   21780 square feet . . . " .50 acre.
   240 feet by 181½ feet . . " 1 acre.   32670 square feet . . . " .75 acre.
   200 feet by 108.9 feet . . " ½ acre.   31848 square feet . . . " .80 acre.
   100 feet by 145.2 feet . . " ⅓ acre.   43560 square feet . . . " 1. acre.
   100 feet by 108.9 feet . . " ¼ acre.

TABLES OF WEIGHTS, DIMENSIONS, STRENGTH, GRAVITY, ETC.   999

Never make the mistake of supposing that *square feet* and *feet square* are the same; one foot, yard, rod or mile, etc., square; or one square foot, yard, rod, mile, etc., are the same, but when you leave the unit, the difference increases with the square of the surface, thus:

FRACTIONS OF AN ACRE.	SQUARE FEET.	FEET SQUARE.	FRACTIONS OF AN ACRE.	SQUARE FEET.	FEET SQUARE.
1-16	2722½	52½	½	21780	147½
⅛	5445	73¾	1	43560	208¼
¼	10890	104½	2	87120	295¼
⅞	14520	120½			

Contents of Cribs.—In the West, and wherever dent corn is raised, three heaping half-bushels are roughly estimated to make a bushel of shelled corn of fifty-six pounds. In reality sixty-eight pounds of ears of sound dent corn, well dried in the crib, will do so. Four heaping half-bushels of flint corn is roughly allowed for a bushel. One rule for finding the contents is to multiply the length, breadth and height together, in feet, to obtain the cubic feet; multiply this product by four, strike off the right-hand figure, and the result will be shelled bushels, nearly. This is on the basis of four half-bushels, per bushel, of shelled corn.

For Dent Corn.—When the crib is flared both ways, multiply half the sum of the bottom breadths in feet by the perpendicular height in feet, and the same again by the length in feet; multiply the last product by .63 for heaped bushels of ears, and by .42 for the number of bushels in shelled corn. This rule is based on the generally accepted estimate that three heaped half-bushels of ears, or four even full, form one of shelled corn.

Length,	10	11	12	13	14	15	16	18	20	22	24	26	28	30
Breadth in feet, 3	135	149	162	175	189	202	216	243	270	297	324	351	378	405
" " 3½	158	173	189	205	221	236	258	284	315	347	378	410	451	473
" " 4	180	198	216	234	252	270	288	324	360	396	432	468	504	540
" " 4½	203	223	243	263	283	304	324	365	405	446	448	527	567	608
" " 5	225	248	270	292	315	337	360	405	450	495	540	585	630	675
" " 5½	248	272	297	322	347	371	396	446	495	545	594	644	693	743
" " 6	270	297	324	351	378	405	432	486	540	594	648	702	756	810
" " 6½	293	322	351	380	410	439	468	527	585	644	702	761	819	878
" " 7	315	347	378	409	441	472	504	567	630	693	756	819	882	945
" " 7½	338	371	405	439	473	506	540	608	675	743	810	878	945	1013
" " 8	360	396	432	468	504	540	576	648	720	792	864	936	1008	1080
" " 8½	383	421	459	497	536	574	612	689	765	842	918	995	1071	1148
" " 9	405	446	486	526	567	607	648	729	810	891	972	1053	1134	1215
" " 10	450	495	540	585	589	675	720	810	900	990	1080	1170	1260	1350
" " 11	495	545	594	643	693	742	792	891	990	1089	1188	1287	1386	1485
" " 12	540	594	648	702	756	810	864	972	1080	1188	1296	1404	1512	1620

The above table is based on the supposition that the crib is ten feet high, that the breadth is the average breadth, or, that half way from the bottom to the top, if flared and on the basis of 3,840 cubic inches of ears to the bushel of shelled corn.

Then a crib five feet wide, ten feet high and thirty feet long will contain 675 bushels of shelled corn. Look on the left side for the width of the crib and follow the line along until you come under the length of the crib, and the number on that line will be the bushels.

Contents of Granaries.—To find the contents of granaries, multiply length, breadth and height together, to get the cubic feet. Divide this by 56, and multiply by 45, and the result will be struck measure. The following table will give the capacities of grain bins, etc., ten feet high.

Width in feet.	Bin 6 feet Long.	Bin 7 feet Long.	Bin 8 feet Long.	Bin 9 feet Long.	Bin 10 feet Long.	Bin 11 feet Long.	Bin 12 feet Long.	Bin 13 feet Long.	Bin 14 feet Long.	Bin 15 feet Long.	Bin 16 feet Long.	Bin 20 feet Long.	Bin 22 feet Long.
	Bush.	Bush.	Bush.	Bush.	Bush.	Bush.	Bush.	Bush.	Bush.	Bush.	Bush.	Bush.	Bush.
3	145	169	192	217	241	265	289	313	338	362	386	482	530
4	193	225	257	289	321	354	386	418	450	482	514	643	708
5	241	282	321	362	402	442	482	522	563	603	643	804	884
6	290	338	386	434	482	530	579	627	675	723	771	964	1060
7	338	394	450	500	563	619	675	731	788	844	900	1125	1238
8	386	450	514	579	643	707	771	836	900	964	1029	1286	1414
9	434	507	579	651	723	796	868	940	1013	1085	1157	1446	1592
10	482	563	643	723	804	884	964	1045	1125	1205	1286	1607	1768
11	531	619	707	796	884	972	1061	1149	1238	1326	1414	1768	1944
12	579	675	771	868	964	1061	1157	1254	1350	1446	1542	1920	2122

The Standard Bushel.—A standard bushel is a measure eight inches deep and eighteen and a half inches inside diameter, containing 2,150 cubic inches. The heaped bushel requires six inches in the height of the cone above the top of the struck bushel, and contains 2,748 cubic inches in all. From these figures may be calculated the contents of cribs or granaries.

Contents of Cisterns.—Thirty-six inches of rain per year will yield seventy-two barrels of water for each ten feet square (100 square feet) of roof. Thus a 30x40 barn may supply two barrels per day throughout the year. Hence in dry countries—that is, countries where heavy rains are succeeded by long droughts—the cisterns must be larger than in those countries where rains are more constant, or, in countries having an average fall of water. When the water is to be used daily a 30x40 barn should have a cistern ten feet in diameter and nine feet deep; this will hold 168 barrels. But, if to be drawn from only in time of drought, it should be three times this capacity.

To determine the contents of a circular cistern of equal size at top and bottom, find the depth and diameter in inches; square the diameter and multiply the square by the decimal .0034, which will find the quantity of gallons for one inch in depth. Multiply this by the depth in inches, and divide by thirty-one and a half, and the result will be the number of barrels the cistern will hold. The following table shows the number of barrels of liquid the following diameters will hold, for each twelve inches in depth:

5 feet diameter, capacity per foot, in depth, ... 4.66 barrels.
6 " " " " " . . . . 6.71 "
7 " " " " " . . . . 9.13 "
8 " " " " " . . . . 11.93 "
9 " " " " " . . . . 15.10 "
10 " " " " " . . . . 18.65 "

Contents of a Square Cistern.—To find the contents of a square cistern, multiply the length by the breadth, and multiply the result by 1,728 and divide by 231. The result will be the number of gallons for each foot in depth. The following table shows the barrels for the sizes named for square cisterns:

5 feet by 5 feet has capacity per foot in depth of . . . . . . 5.92 barrels.
6 " 6 " " " " " . . . . 8.54 "
7 " 7 " " " " " . . . . 11.63 "
8 " 8 " " " " " . . . . 15.19 "
9 " 9 " " " " " . . . . 19.39 "
10 " 10 " " " " " . . . . 23.74 "

XIX. Rainfall in the United States.

The following tables of rainfall will be valuable. They are: First, The rainfall in some principal cities of the Atlantic seaboard and the Mississippi Valley; second, that of the Pacific slope climates, and also the rainfall of principal points on the plains.

AMERICAN ATLANTIC CLIMATES—INCHES OF RAIN.

STATIONS.	SPRING.	SUMMER.	AUTUMN.	WINTER.	TOTAL.
Cincinnati,	11.9	14.2	10.0	11.3	47.5
Cleveland,	9.1	11.6	8.9	6.9	27.4
Ann Arbor,	7.3	11.2	7.0	3.1	28.6
Pittsburg,	9.5	12.3	7.6	7.4	36.8
St. Louis,	12.7	14.6	8.7	7.0	42.5
Nashville,	14.1	14.0	12.3	12.4	52.8

AMERICAN PACIFIC CLIMATES—INCHES OF RAIN.

STATIONS.	SPRING.	SUMMER.	AUTUMN.	WINTER.	TOTAL.
CALIFORNIA.					
Sacramento,	3.3	0.1	3.2	6.9	13.5
San Francisco,	4.6	0.7	3.7	8.8	17.8
Los Angeles,	2.5	0.1	1.6	5.5	9.7
NEW MEXICO.					
El Paso,	0.6	6.6	4.9	0.3	12.4
Albuquerque,	0.6	5.6	1.2	1.0	8.4

The following table gives the average precipitation of rain and snow reduced to water, for the number of years mentioned above the proper column, in inches and hundredths, at various places on the plains; also the average for each month in the year; also altitude, latitude and longitude, as observed by officers of the United States Government. It will be found valuable in many respects in agriculture, and especially so as showing the average available water of the dry districts of the plains regions.

TABLE OF RAINFALL.

Stations.	Latitude.	Longitude.	Altitude.	Years.	January.	February.	March.	April.	May.	June.	July.	August.	September.	October.	November.	December.	Spring.	Summer.	Autumn.	Winter.	Year.
	° ′	° ′	Feet																		
Fort Towson, Arizona, . .	34 0	95 33	300	14	3.19	2.99	4.38	5.33	5.84	5.78	4.62	3.96	3.41	4.59	4.23	2.84	15.55	14.36	12.23	8.94	51.08
Fort Arbuckle, Ind. Ter., .	34 27	97 09	1,000	9¾	0.89	2.98	1.12	2.39	4.46	3.16	3.13	4.12	3.38	2.08	2.97	1.57	7.97	10.83	8.43	5.44	32.69
Fort Belknap, Texas, . . .	33 03	98 48	1,600	5⅚	0.47	2.29	1.32	0.88	4.21	3.98	2.49	3.97	2.77	2.92	2.65	1.10	6.41	9.44	8.34	3.86	28.05
Fort Scott, Kansas, . . .	37 45	94 35	1,000	10¼	1.92	1.18	1.79	3.70	7.08	8.13	4.55	3.69	2.30	2.66	3.46	1.69	12.57	16.37	8.42	4.79	42.15
Council Grove, " . . .	38 40	96 30	1,200	6	1.17	2.31	1.81	3.16	5.14	5.33	5.19	6.39	4.59	3.11	1.75	1.14	10.11	16.91	9.45	4.62	41.09
Manhattan, " . . .	39 15	96 40	1,100	9	0.52	1.63	1.36	3.10	2.37	4.86	4.90	3.52	3.35	1.69	1.52	1.06	8.50	13.08	6.56	3.21	29.69
Fort Riley, " . . .	39 03	96 35	1,300	14	0.77	1.01	0.75	1.74	3.01	3.93	3.00	3.22	3.09	1.21	1.15	0.74	6.83	10.15	5.45	2.52	23.62
Fort Atkinson, ...	37 47	100 14	2,330	1	0.04	0.49	0.96	3.38	9.34	4.35	3.00	2.80	3.85	6.81	1.39	1.60	13.68	10.15	12.05	2.13	38.01
Bellevue, Nebraska, . . .	41 20	95 57	1,250	8	0.89	1.57	0.79	1.92	3.76	3.58	3.91	3.82	3.18	1.76	1.36	2.10	6.47	11.31	6.30	4.56	28.64
Omaha Mission, " . . .	42 05	96 10	1,500	6½	1.03	0.84	1.42	2.56	4.42	3.00	4.42	3.57	3.06	1.68	1.67	1.13	8.40	10.99	6.41	3.00	28.80
Fort Kearney, " . . .	40 38	98 57	2,360	14¼	0.59	0.43	1.25	2.26	4.30	3.69	4.74	2.70	2.29	1.57	0.97	0.46	7.81	11.13	4.83	1.48	25.25
Fort Randall, Dakota, . . .	43 01	98 12	1,456	8⅚	0.49	0.42	0.99	1.10	2.67	2.30	1.78	2.56	2.43	1.09	0.47	0.30	4.76	6.64	3.90	1.21	16.51
Fort Laramie, Wyoming, .	42 12	104 31	4,519	12½	0.61	0.46	0.84	1.06	3.74	1.90	1.63	1.37	1.17	0.97	0.84	0.57	5.64	4.90	2.98	1.64	15.16
Denver, Colorado, . .	39 40	105 10	5,500	2	0.81	0.97	1.26	2.80	1.45	0.29	0.51	0.20	2.85	0.54	1.82	0.75	5.51	1.00	5.21	2.53	14.25
Fort Lyon, " . .	38 08	102 50	4,000	1	0.32	0.12	0.16	2.09	4.84	1.40	2.53	0.37	0.04	.	0.07	0.15	7.09	4.30	0.11	0.59	12.09
Golden City, " . .	39 45	105 20	5,240	1	.	1.00	1.40	2.80	5.40	2.10	2.37	0.45	2.20	.	.	.	9.60	4.92	. . .	.	. .
Fort Massachusetts, " . .	37 32	105 23	8,365	5	3.34	0.86	0.61	1.15	1.19	0.71	2.01	2.84	1.80	0.87	3.61	1.07	2.95	5.56	6.28	2.27	17.06
Great Salt Lake City, Utah.	40 46	112 06	4,320	5⅔	1.68	2.25	2.47	1.39	1.34	1.49	2.54	1.01	1.37	2.20	2.35	3.76	5.20	5.04	5.92	7.69	23.85
Camp Douglas, "	40 39	111 42	4,800	3¼	2.77	1.53	2.87	1.11	1.58	0.60	0.84	0.60	0.99	1.79	1.71	4.18	5.76	2.04	4.49	8.48	20.57
Fort Defiance, Arizona, . .	35 43	109 10	6.500	8½	0.98	0.70	0.84	0.67	0.52	0.74	2.44	2.73	1.86	0.70	1.16	0.87	2.03	5.91	3.72	2.05	14.21
Fort Ruby, Nevada . . .	40 01	115 35	5,922	1¾	2.24	0.23	1.63	1.57	2.04	0.59	0.59	1.82	0.53	3.33	0.94	1.76	5.24	3.00	4.80	4.23	17.27
Deer Lodge City, Montana, .	46 15	112 30	6,000	2¾	1.45	1.81	1.64	1.40	2.28	1.97	0.58	0.58	0.86	0.48	1.38	0.49	5.32	3.13	2.72	3.75	14.92
Helena, " .	46 45	111 50	4,150	1⅓	2.12	0.43	1.15	1.80	4.30	3.50	0.07	0.20	1.80	2.61	0.50	1.00	7.25	3.77	4.91	3.55	19.48
Fort Benton, " .	47 52	110 40	2,674	1½	1.01	0.37	1.11	1.81	2.86	1.14	4.62	0.61	1.82	.	.	1.30	5.78	6.37	5.00	2.68	19.83
Fort Sully, Dakota, ...	44 39	100 40	1,491	½	.	.	.	.	2.98	2.34	6.48	1.53	0.21	.	.	.	. .	10.35	. . .	.	20.00
Cheyenne, Wyoming, . . .	41 12	104 42	6,000	1	0.02	0.27	6.38	1.61	1.99	1.84	3.90	2.05	1.03	.	.	.	3.98	7.79	. . .	.	16.00
Corinne, Utah, ...	41 30	112 18	5,000	1	0.70	2.42	0.55	1.43	2.66	0.47	0.11	1.04	0.14	0.35	3.22	4.04	4.64	1.62	3.71	7.16	17.13

XX. Force and Velocity.

Of the Wind.—According to Burnell's Hydraulic Enquirer, the velocity of the wind and its pressure are as follows:

CHARACTER OF WIND.	Velocity per Second.		Effect per Yard Square.
	Ft.	In.	Pounds.
Light breeze, hardly perceptible,	1	8	0.04989
Gentle breeze,	3	4	0.19756
Light wind,	6	8	0.79130
Rather strong wind, best for sailing,	18	0	6.06996
Strong wind,	33	0	20.06690
Very strong wind,	66	0	80.26760
Tempest or storm,	70	0	101.62790
Great storm,	90	0	146.34430
Hurricane,	118	0	260.05670
Hurricane able to tear up trees, etc., etc.,	150	0	406.51180

Of Water in Tile Drains.—An acre of land, in a wet time, contains about 1,000 spare hogsheads of water. An underdrain will carry off from a strip of land about two rods wide, and one eighty rods long will drain an acre. The following table will show the size of the tile required to drain an acre in two days time, (the longest admissible,) at different rates of descent; or the size for any larger area:

Diameter of Bore.	Rate of Descent.	Velocity of current.	Hogsheads discharged.
2 inches,	1 foot in 100	22 inches per second.	400 in 24 hours.
2 inches.	1 foot in 50	32 inches "	560 "
2 inches.	1 foot in 20	51 inches "	900 "
2 inches.	1 foot in 10	73 inches "	1290 "
3 inches.	1 foot in 100	27 inches "	1170 "
3 inches.	1 foot in 50	38 inches "	1640 "
3 inches.	1 foot in 20	67 inches "	3100 "
3 inches.	1 foot in 10	84 inches "	3600 "
4 inches.	1 foot in 100	32 inches "	2500 "
4 inches.	1 foot in 50	45 inches "	3500 "
4 inches.	1 foot in 20	72 inches "	5600 "
4 inches.	1 foot in 10	100 inches "	7800 "

A deduction of one-third to one-half must be made for the roughness of the tile or imperfection in laying. The drains must be of some length to give the water velocity, and these numbers do not, therefore, apply to very short drains, and in computing capacity of tiles, the head and fall, is of fully as much an integer as the size of the pipe.

Strength of Horses.—The following table from Tredgold shows the average greatest velocity which a horse can travel, according to time consumed in travel:

Time of March in Hours	1	2	3	4	5	6	7	8	9	10
Greatest velocity per hour in miles	14.7	10.4	8.5	7.3	6.6	6.0	5.5	5.2	4.9	4.6

PLOWING.

Breadth of Furrow-slice or Cultivator.	Space Traveled in plowing an acre.	Extent plowed per day, at the rate of		Breadth of Furrow-slice or cultivator.	Space traveled in plowing an acre.	Extent plowed per day. at the rate of	
Inches.	Miles.	18 Miles. Acres.	16 Miles.	Inches.	Miles.	18 Miles. Acres.	16 Miles.
7	14⅛	1¼	1⅛	46	2 1-6	8½	7 2-5
8	12¼	1½	1¼	47	2 1-10	8	7 3-5
9	11	1 3-5	1½	48	2 1-12	8¾	7¾
10	9 9-10	1 4-5	1 3-5	49	2	8 9-10	7 9-10
11	9	2	1¾	50	2	9 9-10	8 1-10
12	8¼	2 1-5	1 9-10	51	1 9-10	9 1-5	8¼
13	7½	2⅓	2 1-10	52	1 9-10	9½	8 2-5
14	7	2½	2¼	53	1 9-10	9¾	8½
15	6½	2¾	2 2-5	54	1 4-5	9 4-5	8 9-10
16	6 1-6	2 9-10	2 3-5	55	1 4-5	10	8
17	5¾	3 1-10	2¾	56	1¾	10¼	9
18	5½	3¼	2 9-10	57	1¾	10 2-5	9 1-5
19	5¼	3½	3 1-10	58	1 7-10	10 3-5	9⅓
20	4 9-10	3 3-5	3¼	59	1 7-10	10¾	9½
21	4 7-10	3 4-5	3⅓	60	1 3-5	10 9-10	9 7-10
22	4½	4	3½	61	1 3-5	11 1-5	9 4-5
23	4¼	4 1-5	3 7-10	62	1 3-5	11⅓	10
24	4	4⅓	3 9-10	63	1 3-5	11½	10 1-5
25	4	4½	4	64	1½	11 7-10	10⅓
26	3 4-5	4¾	4 1-5	65	1½	11 4-5	10½
27	3 3-5	4 9-10	4⅓	66	1½	12	10 3-5
28	3½	5⅛	4½	67	1½	12¼	10 4-5
29	3½	5¼	4 3-5	68	1½	12 2-5	11
30	3⅓	5¾	4 4-5	69	1 2-5	12 3-5	11⅛
31	3 1-5	5	5	70	1 2-5	12¾	11⅓
32	3 1-10	5 4-5	5¼	71	1 2-5	12 9-10	11½
33	3	6	5⅓	72	1 2-5	13⅛	11 3-5
34	2 9-10	6 1-5	5½	73	1⅓	13⅓	11 4-5
35	2 4-5	6⅓	5 3-5	74	1½	13½	12
36	2¾	6½	5 4-5	75	1⅓	13 3-5	12⅛
37	2⅔	6¼	6	76	1 3-10	13 4-5	12¼
38	2 3-5	6 9-10	6⅛	77	1 3-10	14	12½
39	2½	7½	6⅓	78	1¼	14¼	12 3-5
40	2½	7⅓	6½	79	1¼	14 2-5	12¾
41	2 2-5	7¾	6¾	80	1¼	14 3-5	12 9-10
42	2⅓	7	6 2-3	81	1 1-5	14¾	13 1-10
43	2 3-10	7 4-5	7	82	1 1-5	15	13¼
44	2	8	7 1-10	83	1 1-5	15⅛	13 2-5
45	2 1-5	8 16	7¼	84	1 1-6	15⅓	13 3-5

XXI. Weight of Agricultural Products, etc.

THE following very complete table, as compiled by the well-known agricultural engineer, Gen. E. Waring, shows the weight of the bushel of agricultural products, etc., as established by law in the United States, Territories, and British Provinces, compared with the most recent enactments:

TABLES OF WEIGHTS, DIMENSIONS, STRENGTH, GRAVITY, ETC.

	California.	Canada.	Connecticut.	Dakota.	Delaware.	Illinois.	Indiana.	Iowa.	Kansas.	Kentucky.	Louisiana.	Maine.	Maryland.	Massachusetts.	Michigan.	Minnesota.	Missouri.	Nebraska.	Nevada.	New Brunswick.	New Hampshire.	New Jersey.	New York.	Nova Scotia.	Ohio.	Oregon.	Pennsylvania.	Rhode Island.	Vermont.	Washington Ter.	Wisconsin.
Apples,	.	.	.	.	.	.	.	.	.	.	.	.	.	.	.	.	.	.	.	.	.	.	.	.	45	.	.	.	45	.	.
Barley,	50	48	48	48	.	48	48	48	48	48	32	48	.	48	48	48	48	48	50	50	.	48	48	†52	48	46	47	.	48	45	48
Bran,	.	.	.	.	.	20	.	20	20	20	.	.	.	.	.	.	20	20	.	.	.	.	.	.	.	.	.	.	.	.	.
Broom corn seed,	.	.	.	.	.	.	.	30	.	.	.	.	.	.	.	.	.	.	.	.	.	.	.	.	.	.	.	.	.	.	.
Buckwheat,	40	48	48	42	.	52	50	52	50	52	.	48	.	48	48	42	52	52	40	50	.	50	48	.	50	42	48	.	46	42	42
Carrots,	.	60	55	.	.	.	.	.	.	.	.	50	.	.	.	.	.	.	.	56	.	.	.	.	.	.	.	50	.	50	.
Castor beans,	.	40	.	.	.	46	46	46	46	.	.	.	.	46	.	46	46	.	.	.	.	.	.	.	.	.	.	.	.	.	.
Coal, mineral,	.	.	.	.	.	80	*80	80	80	.	.	.	.	.	80	.	80	80	.	.	.	.	.	.	‡80	.	.	.	.	.	.
Corn, shelled,	52	56	56	56	56	56	56	56	56	56	56	56	.	56	56	56	56	56	52	60	56	56	58	58	56	56	56	.	56	56	56
Corn, in ear,	.	.	.	72	.	70	68	70	70	.	.	.	.	70	.	.	70	.	.	.	.	.	70	.	.	.	.	.	.	.	.
Corn meal,	.	.	50	.	.	48	50	.	.	50	.	.	.	50	50	.	.	50	.	.	50	.	.	.	60	.	50	.	.	.	.
Cranberries,	.	.	.	.	.	.	.	.	.	.	.	.	.	40	.	.	.	.	.	.	.	.	.	.	.	.	.	.	.	.	.
Dried apples,	.	22	.	.	.	24	25	24	24	.	.	.	.	22	28	24	24	.	.	.	.	.	.	.	25	28	.	.	.	28	28
Dried peaches,	.	33	.	.	.	33	33	33	33	.	.	.	.	28	28	33	33	.	.	.	.	.	.	.	33	28	.	.	.	28	28
Dried plums,	.	.	.	.	.	.	.	.	.	.	.	.	.	28	.	.	.	.	.	.	.	.	.	.	.	.	.	.	.	.	.
Flax seed,	.	50	.	.	.	56	.	56	56	56	.	.	.	56	.	56	56	.	.	.	55	55	.	.	56	.	.	.	.	.	56
Grass seed, blue,	.	14	.	.	.	14	14	14	.	14	.	.	.	14	.	14	14	.	.	.	.	.	.	.	.	.	.	.	.	.	.
" clover,	.	60	.	60	.	60	60	60	60	60	.	.	.	60	60	60	60	.	.	.	64	60	.	.	62	60	62	.	60	60	60
" Hungarian	.	.	.	.	.	.	.	45	.	.	.	.	.	50	.	60	.	.	.	.	.	.	.	.	50	.	.	.	.	.	.
" millet,	.	.	.	.	.	.	.	45	.	.	.	.	.	50	.	85	.	.	.	.	.	.	.	.	50	.	.	.	.	.	.
" orchard,	.	.	.	.	.	.	.	.	.	.	.	.	.	14	.	.	.	.	.	.	.	.	.	.	.	.	.	.	.	.	.
" red top,	.	.	.	.	.	.	.	.	.	.	.	.	.	14	.	.	.	.	.	.	.	.	.	.	.	.	.	.	.	.	.
" timothy,	.	48	.	42	.	45	45	45	45	45	.	.	.	45	.	45	45	.	40	.	.	44	.	.	45	.	.	.	42	40	46
Hair, plastering,	.	.	.	.	.	8	.	.	.	.	.	.	.	.	.	.	.	.	.	.	.	8	.	.	.	.	.	.	.	.	.
Hemp seed,	.	44	.	.	.	44	44	44	44	44	.	.	.	.	44	.	44	44	.	.	.	.	.	.	44	.	.	.	.	.	.
Lime, unslacked,	.	.	.	.	.	80	.	80	.	.	.	.	.	.	.	.	80	.	.	.	.	.	.	.	.	.	.	.	.	.	.
Malt,	.	36	.	.	.	38	.	.	.	.	.	.	.	.	.	.	30	.	.	.	.	39	34	.	.	.	.	.	.	.	.
Mangel-wurzel,	.	.	60	.	.	.	.	.	.	.	60	.	.	.	.	.	.	.	.	.	.	.	.	.	.	.	50	.	.	.	.
Oats,	32	34	32	32	.	32	.	33	32	33¼	32	30	.	32	32	32	35	34	32	36	30	30	32	34	33	36	30	.	30	35	32
Onions,	.	60	50	.	.	57	49	57	57	57	.	52	.	52	54	.	57	57	.	56	.	.	.	.	.	.	.	.	50	.	50
Onions, top,	.	.	.	.	.	.	.	.	.	.	.	.	.	.	.	.	.	.	.	25	.	.	.	.	.	.	.	.	.	.	.
Osage-orange seed,	.	.	.	.	.	.	.	32	.	.	.	.	.	.	.	33	.	32	.	.	.	.	.	.	.	.	.	.	.	.	.
Parsnips,	.	60	45	.	.	.	.	.	.	.	.	.	.	.	.	.	.	.	.	.	.	60	.	.	60	.	50	.	.	.	.
Pears,	.	.	.	.	.	.	.	.	.	.	.	.	.	.	.	.	.	.	.	.	.	.	.	.	45	.	.	.	45	.	.
Peas,	.	60	60	.	.	.	.	.	.	.	60	.	.	60	.	60	.	.	.	.	60	.	.	.	.	.	.	.	60	60	.
Potatoes, Irish,	.	60	60	60	.	60	60	60	60	56	.	56	.	60	60	.	60	60	.	56	60	60	60	.	60	60	56	60	.	60	60
Potatoes, sweet,	.	.	.	.	.	55	.	46	.	.	.	56	.	.	50	.	.	.	.	.	.	50	.	.	.	.	.	.	.	.	.
Rutabaga,	.	.	60	.	.	.	.	.	.	.	60	.	.	.	.	.	.	.	.	56	.	.	.	.	.	.	50	.	50	.	.
Rye,	54	56	56	56	.	56	56	56	56	56	32	56	.	56	56	56	56	56	54	56	56	56	56	56	56	56	56	.	56	56	56
Salt, coarse,	.	.	.	.	.	.	.	.	.	.	.	.	.	.	.	.	.	.	.	.	.	.	.	.	.	.	85	.	.	.	.
Salt, fine,	.	.	.	.	.	55	.	.	.	.	.	.	.	.	.	.	.	.	.	.	.	.	.	.	.	.	62	.	.	.	.
Salt, ground,	.	.	.	.	.	.	.	.	.	.	.	.	.	.	.	.	.	.	.	.	.	.	.	.	.	.	70	.	.	.	.
Sand,	.	.	.	.	.	.	.	130	.	.	.	.	.	.	.	.	.	.	.	.	.	.	.	.	.	.	.	.	.	.	.
Sugar beets,	.	60	60	.	.	.	.	.	.	.	60	.	.	.	.	.	56	.	.	.	.	.	.	.	.	.	50	.	50	.	.
Turnips,	.	60	50	.	.	55	.	.	.	.	50	.	.	58	.	55	56	.	.	.	.	.	.	.	.	.	50	.	50	.	.
Wheat,	60	60	60	60	60	60	60	60	60	60	60	60	.	60	60	60	60	60	60	60	60	60	60	60	60	60	60	.	60	60	60
White beans,	.	60	60	60	.	60	60	60	60	60	.	64	.	60	.	60	60	.	.	60	.	62	.	60	.	.	.	.	60	60	.

* Mined within the State, 70 lbs.; without the State, 80 lbs. † Foreign—Barley produced in the Province, 48 lbs.
‡ Bituminous—Cannel coal, 70 lbs.

"Salt" in Canada is 56 pounds to the bushel; in Illinois, Indiana, Iowa, Kansas, Kentucky, Missouri and Nebraska, it is 50 pounds to the bushel. In Michigan, "Michigan Salt" is 56 pounds to the bushel. In Massachusetts, "Salt" is 70 pounds to the bushel.

Coal in Kentucky is 76 pounds per bushel, *except* Wheeling coal, which is 84, and Kentucky River, which is 78 pounds per bushel, and Adrian Branch, or Cumberland River coal, which is 72 pounds per bushel. Cotton seed is 33 pounds to the bushel in Missouri.

Sorghum seed is 30 pounds to the bushel in Iowa and Nebraska. Strained honey is 12 pounds to the gallon in Nebraska.

To reduce cubic feet to bushels, struck measure, divide the cubic feet by 56 and multiply by 45.

CHAPTER IV

TABLES AND DIAGRAMS OF PRACTICAL VALUE.

I. SEEDS AND PLANTS TO CROP AN ACRE.—II. VEGETABLE SEEDS TO SOW 100 YARDS OF DRILLS.—III. PLANTS PER ACRE AT VARIOUS DISTANCES.—IV. VITALITY OF SEEDS.—V. PLANTS PER SQUARE ROD OF GROUND.—VI. FORETELLING THE WEATHER.—VII. COMPARISON OF CROPS IN GREAT BRITAIN AND THE UNITED STATES.—VIII. IMPROVED AND UNIMPROVED LANDS IN THE STATES AND TERRITORIES.— IX. FOREST AREAS—EUROPE AND UNITED STATES.— X. SURVEYED AND APPROPRIATED LANDS IN STATES AND TERRITORIES.—XI. TABLES OF NUTRITIVE EQUIVALENTS, ETC.— XII. TABLE SHOWING PRICES PER POUND.— XIII. TABLE OF INTEREST AT SIX PER CENT.—XIV. GROWTH OF MONEY AT INTEREST.—XV. MEAN DURATION OF LIFE.—XVI. MORTALITY RATES.—XVII. HOW TO CALCULATE SALARIES AND WAGES.—XVIII. THE EARTH'S AREA AND POPULATION.—XIX. THE WORLD'S COMMERCE.—XX. PAY OF THE PRINCIPAL OFFICERS OF THE UNITED STATES.—XXI. PUBLIC DEBT OF THE UNITED STATES.—XXII. THE UNITED STATES AND TERRITORIES.—XXIII. DIAGRAMS GIVING VALUABLE STATISTICS.

I. Seeds and Plants to Crop an Acre.

Asparagus in 12-inch drills,	16	quarts.	Flax, broadcast,	2	bushels.
Asparagus plants at 4x1½ feet,	8000		Grass, timothy, with clover,	6	quarts.
Barley,	2½	bushels.	Grass, timothy, without clover,	10	quarts.
Beans, Bush, in drills at 2½ feet,	1½	bushels.	Grass, orchard,	25	pounds.
Beans, pole, Lima, at 4x4 feet,	20	quarts.	Grass, red top, or herds,	20	pounds.
Beans, Carolina, prolific, etc. at 4x3,	10	quarts.	Grass, blue,	28	pounds.
Beets and mangolds in drills at 2½ ft.	8	pounds.	Grass, rye,	20	pounds.
Broom corn in drills,	12	pounds.	Grass, millet,	32	quarts.
Cabbage, sown in outdoor beds, for transplanting.	10	ounces.	Hemp, broadcast,	1¼	bushel.
			Kale, German greens,	3	pounds.
Cabbage, sown in frames,	4	ounces.	Lettuce, in rows at 2½ feet,	3	pounds.
Carrot, in drills at 2½ feet,	3	pounds.	Leek, in rows at 2½ feet,	3	pounds.
Celery. seed,	8	ounces.	Lawn grass,	35	pounds.
Celery, plants, at 4½ feet,	25000		Melons, water, in hills 8x8 feet,	3	pounds.
Clover, white Dutch,	12	pounds.	Melons, citron, in hills 4x4 feet,	2	pounds.
Clover, Lucerne,	10	pounds.	Oats,	2	bushels.
Clover, Alsike,	12	pounds.	Okra, in drills 2½x¼ feet,	20	pounds.
Clover, large red, with timothy,	12	pounds.	Onion, in beds, for sets,	35	pounds.
Clover, large red, without timothy,	16	pounds.	Onion, in rows, to make large bulbs,	5	pounds.
Corn, sugar,	9	quarts.	Parsnip, in drills at 2½ feet,	5	pounds.
Corn, field,	7	quarts.	Pepper, plants 2¼x1 feet,	17500	
Corn, salad, in drills at 10 inches, large seed,	25	pounds.	Pumpkin, in hills 8x8 feet,	2	quarts.
			Parsley, in drills at 2 feet,	4	pounds.
Cucumber, in hills,	2	quarts.	Peas, in drills, short varieties,	2	bushels.
Cucumber, in drills,	3	quarts.	Peas, in drills, tall varieties,	1 to 1½	bushels.
Egg-plants, plants 3x2 feet,	4	ounces.	Peas, broadcast,	3	bushels.
Endive. in drills at 2½ feet,	3	pounds.	Potatoes,	8	bushels.

TABLES AND DIAGRAMS OF PRACTICAL VALUE.

Radish, in drills 2 feet,	8	pounds.	Turnips, in drills at 2 feet,	2	pounds.
Rye, broadcast,	2	bushels.	Turnips, broadcast,	2½	pounds.
Rye, drilled,	1½	bushels.	Tomatoes, in frame,	3	ounces.
Salsify, in drills at 2½ feet,	10	pounds.	Tomatoes, seed, in hills, 3x3	8	ounces.
Spinach, broadcast,	30	pounds.	Tomatoes, plants,	3800	
Squash, Bush, in hills 4x4 feet,	3	pounds.	Wheat, in drills,	1¼	bushels.
Squash, running, in hills 8x8 feet,	2	pounds.	Wheat, broadcast,	2	bushels.
Sorghum,	4	quarts.			

II. Vegetable Seeds to Sow 100 Yards of Drill.

Asparagus,	8	ounces.	Lettuce,	2	ounces.
Beans, Bush,	3	quarts.	Melon, water,	2	ounces.
Beans, Lima,	3	pints.	Melon, citron,	1	ounce.
Beans, pole,	1	pint.	Mustard,	4	ounces.
Beet,	4	ounces.	Okra,	12	ounces.
Broccoli,	½	ounce.	Onions, for large bulbs,	2	ounces.
Brussels sprouts,	½	ounce.	Onions, for sets,	6	ounces.
Cabbage,	1	ounce.	Parsley,	2	ounces.
Carrot,	3	ounces.	Peas,	3	quarts.
Cauliflower,	½	ounce.	Pepper,	½	ounce.
Celery,	3	ounces.	Pumpkin,	2	ounces.
Collards.	½	ounce.	Radish,	6	ounces.
Corn,	1	pint.	Rhubarb,	4	ounces.
Cress,	4	ounces.	Salsify,	4	ounces.
Cucumber,	4	ounces.	Spinach,	6	ounces.
Egg-plant,	½	ounce.	Squash,	3	ounces.
Endive,	2	ounces.	Tomato,	1	ounce.
Leek,	2	ounces.	Turnips,	3	ounces.

These quotations are far too large to stand, but you must have more than enough in order to provide for contingencies.

TABLES OF CIRCULAR MEASURE, TIME AND LONGITUDE.

CIRCULAR MEASURE.		MEASURES OF TIME.	
60 seconds,	1 minute.	60 seconds,	1 minute.
60 minutes,	1 degree.	60 minutes,	1 hour.
360 degrees.	1 circle.	24 hours,	1 day.
30 degrees,	1 sign of zodiac.	7 days,	1 week.
12 signs,	1 zodiac circle.	28 days,	1 lunar month.
360 degrees, the circumference of the earth.		28, 29, 30 or 31 days,	1 calendar month.
24,899 statute miles, circumference of the earth at the equator.		12 calendar months,	1 year.
69.124 statute miles, 1 degree of the equator.		365.25 days,	1 common year.
1.1527 statute miles, 1 geographic mile.		366 days,	1 leap year,
60 geographic miles, 1 degree.			

LONGITUDE AND TIME COMPARED.

LONGITUDE.	TIME.	LONGITUDE.	TIME.
1 second,	.0666 second.	1 degree,	4 minutes.
1 minute,	4 seconds.	360 degrees,	1 day.
15 minutes,	1 minute.		

Add difference of time for places east, and subtract for places west, of the given place.

III. Plants Per Acre at Various Distances.

The following table shows the number of plants required per acre, for the given feet and inches:

Ft. In.	Ft. In.	Plants.	Ft. In.	Ft. In.	Plants.	Ft. In.	Ft. In.	Plants.	Ft. In.	Ft. In.	Plants.
1 0	by 1 0	43360	3 6	by 2 3	5531	4 9	by 1 0	9170	6 0	by 5 0	1452
1 3	" 1 0	34848	" "	" 2 6	4978	" "	" 1 3	7336	" "	" 5 6	1320
" "	" 1 3	27878	" "	" 2 9	4525	" "	" 1 6	6113	" "	" 5 9	1262
1 6	" 1 0	29040	" "	" 3 0	4148	" "	" 1 9	5248	" "	" 6 0	1210
" "	" 1 3	23232	" "	" 3 3	3829	" "	" 2 0	4585	6 6	" 1 0	6701
" "	" 1 6	19369	" "	" 3 6	3555	" "	" 2 3	4075	" "	" 1 6	4467
1 9	" 1 0	24454	3 9	" 1 0	14616	" "	" 2 6	3668	" "	" 2 0	3350
" "	" 1 3	19913	" "	" 1 3	9272	" "	" 2 9	3334	" "	" 2 6	2680
" "	" 1 6	16594	" "	" 1 6	7744	" "	" 3 0	3056	" "	" 3 0	2233
" "	" 1 9	14223	" "	" 1 9	6637	" "	" 3 3	2821	" "	" 3 6	1914
2 0	" 1 0	21780	" "	" 2 0	5808	" "	" 3 6	2620	" "	" 4 0	1675
" "	" 1 3	17424	" "	" 2 3	5162	" "	" 3 9	2445	" "	" 4 6	1489
" "	" 1 6	14520	" "	" 2 6	4646	" "	" 4 0	2292	" "	" 5 0	1340
" "	" 1 9	12445	" "	" 2 9	4224	" "	" 4 3	2157	" "	" 5 6	1218
" "	" 2 0	10890	" "	" 3 0	3872	" "	" 4 6	2037	" "	" 6 0	1116
2 3	" 1 0	19260	" "	" 3 3	3574	" "	" 4 9	1930	" "	" 6 6	1031
" "	" 1 3	15488	" "	" 3 6	3318	5 0	" 1 0	8712	7 0	" 1 0	6222
" "	" 1 6	12905	" "	" 3 9	3097	" "	" 1 3	6969	" "	" 1 6	4148
" "	" 1 9	11062	4 0	" 1 0	10890	" "	" 1 6	5808	" "	" 2 0	3111
" "	" 2 0	9680	" "	" 1 3	8712	" "	" 1 9	4078	" "	" 2 6	2489
" "	" 2 3	8604	" "	" 1 6	7260	" "	" 2 0	4336	" "	" 3 0	2074
2 6	" 1 0	17424	" "	" 1 9	6222	" "	" 2 3	3874	" "	" 3 6	1777
" "	" 1 3	13939	" "	" 2 0	5445	" "	" 2 6	3484	" "	" 4 0	1555
" "	" 1 6	11616	" "	" 2 3	4840	" "	" 2 9	3168	" "	" 4 6	1382
" "	" 1 9	9956	" "	" 2 6	4356	" "	" 3 0	2904	" "	" 5 0	1244
" "	" 2 0	8742	" "	" 2 9	3960	" "	" 3 3	2680	" "	" 5 6	1131
" "	" 2 3	7740	" "	" 3 0	3630	" "	" 3 6	2489	" "	" 6 0	1037
" "	" 2 6	6960	" "	" 3 3	3350	" "	" 3 9	2323	" "	" 6 6	957
2 9	" 1 0	15810	" "	" 3 6	3111	" "	" 4 0	2178	" "	" 7 0	888
" "	" 1 3	12670	" "	" 3 9	2904	" "	" 4 3	2049	8 0	" 3 0	1815
" "	" 1 6	10560	" "	" 4 0	2722	" "	" 4 6	1936	" "	" 4 0	1361
" "	" 1 9	9054	4 3	" 1 0	10249	" "	" 4 9	1834	" "	" 5 0	1089
" "	" 2 0	7920	" "	" 1 3	8199	" "	" 5 0	1742	" "	" 6 0	905
" "	" 2 3	7040	" "	" 1 6	6832	5 6	" 1 0	7910	" "	" 7 0	777
" "	" 2 6	6336	" "	" 1 9	5856	" "	" 1 3	6336	" "	" 8 0	680
" "	" 2 9	5760	" "	" 2 0	5124	" "	" 1 6	5280	9 0	" 5 0	968
3 0	" 1 0	14520	" "	" 2 3	4555	" "	" 1 9	4525	" "	" 6 0	806
" "	" 1 3	11616	" "	" 2 6	4099	" "	" 2 0	3960	" "	" 7 0	691
" "	" 1 6	9680	" "	" 2 9	3727	" "	" 2 3	3520	" "	" 8 0	605
" "	" 1 9	8297	" "	" 3 0	3416	" "	" 2 6	3168	" "	" 9 0	537
" "	" 2 0	7260	" "	" 3 3	3153	" "	" 2 9	2886	10 0	" 10 0	435
" "	" 2 3	6453	" "	" 3 6	2914	" "	" 3 0	2640	11 0	" 11 0	360
" "	" 2 6	5808	" "	" 3 9	2733	" "	" 3 3	2436	12 0	" 12 0	302
" "	" 2 9	5289	" "	" 4 0	2562	" "	" 3 6	2262	13 0	" 13 0	257
" "	" 3 0	4840	" "	" 4 3	2411	" "	" 3 9	2112	14 0	" 14 0	222
3 3	" 1 0	13403	4 6	" 1 0	9680	" "	" 4 0	1980	15 0	" 15 0	193
" "	" 1 3	10722	" "	" 1 3	7744	" "	" 4 3	1863	16 0	" 16 0	170
" "	" 1 6	8935	" "	" 1 6	6453	" "	" 4 6	1760	17 0	" 17 0	150
" "	" 1 9	7658	" "	" 1 9	5531	" "	" 4 9	1667	18 0	" 18 0	134
" "	" 2 0	6701	" "	" 2 0	4840	" "	" 5 0	1584	19 0	" 19 0	120
" "	" 2 3	5956	" "	" 2 3	4302	" "	" 5 3	1508	20 0	" 20 0	108
" "	" 2 6	5361	" "	" 2 6	3872	" "	" 5 6	1417	24 0	" 24 0	75
" "	" 2 9	4873	" "	" 2 9	3520	6 0	" 1 0	7260	25 0	" 25 0	69
" "	" 3 0	4818	" "	" 3 0	3226	" "	" 1 6	4840	27 0	" 27 0	59
" "	" 3 3	4124	" "	" 3 3	2978	" "	" 2 0	3630	30 0	" 30 0	48
3 6	" 1 0	12445	" "	" 3 6	2765	" "	" 2 6	2904	40 0	" 40 0	27
" "	" 1 3	9956	" "	" 3 9	2581	" "	" 3 0	2420	50 0	" 50 0	17
" "	" 1 6	8297	" "	" 4 0	2420	" "	" 3 6	2074	60 0	" 60 0	12
" "	" 1 9	7111	" "	" 4 3	2277	" "	" 4 0	1815	66 0	" 66 0	10
" "	" 2 0	6222	" "	" 4 6	2151	" "	" 4 6	1613			

IV. Vitality of Seeds.

The following table shows the length of time various seeds will retain their vitality or germinating power:

SEEDS OF VEGETABLES.	YEARS.	SEEDS OF VEGETABLES.	YEARS.
Artichoke,	5 to 6	Parsley,	2 to 3
Asparagus,	2 to 3	Parsnip.	2 to 3
Beans, all kinds,	2 to 3	Pea,	5 to 6
Beet,	3 to 4	Pumpkin,	8 to 10
Broccoli,	5 to 6	Rhubarb,	3 to 4
Carrot,	2 to 3	Squash,	8 to 10
Cress,	3 to 4	Lettuce,	3 to 4
Corn kept on the cob,	2 to 3	Melon,	8 to 10
Cucumber,	8 to 10	Mustard,	3 to 4
Egg plant,	1 to 2	Okra,	3 to 4
Endive,	5 to 6	Spinach,	3 to 4
Leek,	2 to 3	Tomato,	2 to 3
Cauliflower,	5 to 6	Turnip,	5 to 6
Celery,	2 to 3	Pepper,	2 to 3
Chervil,	2 to 3	Radish,	4 to 5
Corn salad,	2 to 3	Salsify,	2 to 3
Onion,	2 to 3		

SEEDS OF HERBS.	YEARS.	SEEDS OF HERBS.	YEARS.
Anise,	3 to 4	Hyssop,	3 to 4
Balm,	2 to 3	Lavender,	2 to 3
Basil,	2 to 3	Sweet marjoram,	2 to 3
Caraway,	2	Summer savory,	1 to 2
Coriander,	1	Sage,	2 to 3
Dill,	2 to 3	Thyme,	2 to 3
Fennel,	2 to 3	Wormwood,	2 to 3

V. Plants per Square Rod of Ground.

A ROD has 272½ square feet, or 39,204 square inches. It is well to know how many plants small plats will contain. The following table will show this per square rod. To find any given number divide the square of the distance apart into the square feet or inches in the plat; as, 3 x 3 inches = 9 inches, and this, divided into 39,204, gives 4,356, the number of plants, three inches apart, to the square rod.

Trees or Plants.	Inches over.	Number of Inches apart.	Square Inches to each.	Trees or Plants.	Inches over.	Number of Inches apart.	Square Inches to each.
2450	4	4 by 4	16	612	36	8 by 8	64
1960	..	5 " 4	20	490	4	10 " 8	80
1633	12	6 " 4	24	392	4	10 " 10	100
1069	..	6 " 6	36	272	36	12 " 12	144
816	36	8 " 6	48	261	54	15 " 10	150

Distances Apart for Trees, Shrubs, etc.—Forest trees should be 3 x 3 feet when first set out from the nursery, and ultimately thinned to forty-eight feet for the very largest, and so down to twenty-four feet for the smaller. Fruits may be set as in schedule on next page.

	FEET.		FEET.
Apples, standard,	25 to 33	Plums, standard,	15
Apples, dwarf,	5 to 8	Plums, dwarf,	8 to 10
Pears, standard,	20	Quinces,	6 to 8
Pears, dwarf,	8 to 10	Grapes,	10 to 12
Peaches, headed back,	12	Gooseberries and currants,	4
Cherries, standard,	20	Raspberries,	4
Cherries, dwarf,	8 to 10	Blackberries,	6 to 8

VI. Foretelling the Weather.

The many prognostics for foretelling weather a long time ahead are untrustworthy. The following humorous and true prognostics for twenty-four hours ahead, by the celebrated Dr. Jenner, were written over a century ago, and comprise about all that is known to-day:

> The hollow winds begin to blow,
> The clouds look black, the glass is low;
> The soot falls down, the spaniels sleep,
> And spiders from their cobwebs peep.
> Last night the sun went pale to bed,
> The moon in halos hid her head;
> The boding shepherd heaves a sigh,
> For, see, a rainbow spans the sky;
> The walls are damp, the ditches smell,
> Closed is the pink-eyed pimpernel.
> Hark! how the chairs and tables crack,
> Old Betty's joints are on the rack;
> Loud quack the ducks, the peacocks cry;
> The distant hills are looking nigh.
> How restless are the snorting swine,
> The busy flies disturb the kine;
> Low o'er the grass the swallow wings;
> The cricket, too, how sharp he sings;
> Puss, on the hearth, with velvet paws,
> Sits, wiping o'er her whisker'd jaws.
>
> Through the clear stream the fishes rise,
> And nimbly catch th' incautious flies;
> The glow-worms, numerous and bright,
> Illum'd the dewy dell last night.
> At dusk the squalid toad was seen,
> Hopping and crawling o'er the green;
> The whirling wind the dust obeys,
> And in the rapid eddy plays;
> The frog has chang'd his yellow vest,
> And in a russet coat is dressed.
> Though June, the air is cold and still;
> The blackbird's mellow voice is shrill,
> My dog, so alter'd is his taste,
> Quits mutton bones, on grass to feast;
> And see, yon rooks, how odd their flight,
> They imitate the gliding kite,
> And seem precipitate to fall—
> As if they felt the piercing ball.
> 'Twill surely rain, I see with sorrow,
> Our jaunt must be put off to-morrow.

VII. Comparison of Crops in Great Britain and the United States.

GREAT BRITAIN, 1878.		UNITED STATES, 1878.	
Total acreage in crops, fallow, and hay,	47,327,000	Total acreage in crops, fallow, and hay,	179,000,000
Wheat (acres),	3,382,000	Wheat (acres),	32,000,000
Oats (acres),	4,124,000	Oats (acres),	13,176,000
Barley (acres),	2,723,000	Barley (acres),	1,790,000
Total cereals, including peas and beans,	11,030,000	Total of cereals, including corn,	97,960,000
Potatoes,	1,365,000	Potatoes,	1,776,000
Turnips and other green crops,	3,400,000	Cotton, sugar, and tobacco,	13,000,000

LIVE STOCK.

Horses,	1,927,000	Horses,	10,611,000
Cattle,	9,761,000	Cattle,	31,850,000
Sheep,	32,571,000	Sheep,	36,575,000
Hogs,	3,768,000	Hogs,	33,134,000
Total live stock,	48,027,000	Total live stock,	112,170,000

VIII. Improved and Unimproved Lands in the States and Territories.

States and Territories.	Acres of Improved Land in Farms.		Acres of Unimproved Land in Farms.	
	1850.	1860.	1850.	1860.
Maine,	2,039,596	2,677,216	2,515,797	3,023,539
New Hampshire,	2,251,488	2,367,039	1,140,926	1,377,591
Vermont,	2,601.409	2,758,443	1,524,413	1,402,396
Massachusetts,	2,133,436	2,155,512	1,222,576	1,183,212
Rhode Island,	356,487	329,884	197,451	189,814
Connecticut,	1,768,178	1,839,808	615,701	673,457
New York,	12,408,964	14,376.397	6,710,120	6,616,553
New Jersey,	1,767,991	1,944,445	984,955	1,039,086
Pennsylvania,	8,628,619	10,463,306	6,294,728	6,548,847
Delaware,	580,862	637,065	375,282	367,230
Maryland,	2,797,905	3,002,269	1,836,445	1,833,306
District of Columbia,	16,267	16,267	11,187	16,789
Virginia,	10,360,135	11,435,954	15,792,176	19,578,946
North Carolina,	5,453,975	6,517,284	15,543,008	17,245,685
South Carolina,	4,072,651	4,572,060	12,145,049	11,623,860
Georgia,	6,378,479	8,062,758	16,442,900	18,587,732
Florida,	349,049	676,464	1,246,240	2,273,008
Alabama,	4,435,614	6,462,987	7,702,067	12,687,913
Mississippi,	3,444,358	5,150,008	7,046,061	11,703,556
Louisiana,	1,590,025	2,734,901	3,399,018	6,765,879
Texas,	643,976	2,649,207	10,852,363	20,486,990
Arkansas,	781,530	1,933,036	1,816,684	7,609,938
Tennessee,	5,175,173	6,897,974	13,808,849	13,457,960
Kentucky,	5,968,270	7,644,217	10,981,478	11,519,059
Ohio,	9,851,493	12,665,587	8,146,000	8,075,551
Michigan,	1,929,110	3,419,861	2,454,780	3,511,581
Indiana,	5,046,543	8,161,717	7,746,879	8,154,059
Illinois,	5,039,545	13,251,473	6,997,867	7,993,557
Wisconsin,	1,045,499	3,746,036	1,931,159	4,153,134
Minnesota,	5,035	554,397	23,846	2,222,734
Iowa,	824,682	3,780,253	1,911,382	5,649,136
Missouri,	2,938,425	6,246,871	6,794,245	13,737,939
Kansas,		372,835		1,284,626
California,	32,454	2,430,882	3,861,531	6,533,858
Oregon,	132,857	895,375	299,951	5,316,817
Washington,				300,897
Utah,	16,333	16,333	30,516	58,898
New Mexico,	166,201	166,201	124,370	1,177,055
Nebraska,				501,720
Dakota,				24,333

IX. Forest Areas, Europe and United States.

EUROPEAN COUNTRIES.	Per cent.	Acres per head of population.	EUROPEAN COUNTRIES.	Per cent.	Acres per head of population.
Norway,	66	24.61	Sardinia,	12.29	0.223
Sweden,	60	8.55	Naples,	9.43	0.138
Russia,	30.90	4.28	Holland,	7.10	0.12
Germany,	26.58	0.663	Spain,	5.52	0.291
Belgium,	18.52	0.186	Denmark,	5.50	0.22
France,	16.79	0.376	Great Britain,	5	0.1
Switzerland,	15	0.396	Portugal,	4.40	1.182

Percentage in Farms.—In the following table, the first column shows the percentage of areas in farms; the second column the area of forest:

UNITED STATES.	Percentage in farms.	Total percentage forest.	UNITED STATES.	Percentage in farms.	Total percentage forest.
Maine,	38.1	46.9	Michigan,	40.7	47.1
New Hampshire,	29	37.2	Indiana,	39.6	34.8
Vermont,	30.6	36.5	Illinois,	19.6	16.9
Massachusetts,	25.8	29.2	Wisconsin,	29.3	20.9
Rhode Island,	33.7	24.2	Minnesota,	20.6	17.1
Connecticut,	24.4	21.2	Iowa,	16.2	14.1
New York,	25.5	27.6	Missouri,	41.3	45.4
New Jersey,	24	28.1	Kansas.	11.2	5.6
Pennsylvania,	31.9	38.9	Nebraska,	10.2	5.2
Delaware,	28	29.2	California,	4.1	7.9
Maryland,	31.8	38.4	Oregon,	31.8	25.2
Virginia,	45.7	49.4	Nevada,	6.4	5
North Carolina,	60.6	64.2	Colorado,	3.5	10
South Carolina,	53.2	60.6	**TERRITORIES.**		
Georgia,	54.6	60.2			
Florida,	60	50.6	Utah,	0.1	10
Alabama,	56	63.5	New Mexico,	12.7	6
Mississippi,	60.6	65.9	Washington,	44.8	33
Louisiana,	56.9	59.1	Dakota,	7.4	3
Texas,	41.6	26.7	Montana,	0.8	16
Arkansas,	51.4	58	Idaho,	9.6	15
Tennessee,	55	59.9	Arizona,		6
West Virginia,	51.1	54.9	Wyoming,	0.8	8
Kentucky,	48.9	49.1	Indian,		8
Ohio,	31.7	28.4	Alaska,		30

X. Surveyed and Appropriated Lands in States and Territories.

States and Territories.	Area in Acres.	Acres Surveyed.	Acres Appropriated.
California,	120,947,840	38,805,776	20,877,602
Dakota Territory,	96,596,128	13,863,913	5,835,604
Montana Territory,	92,016,640	6,784,481	5,179,821
New Mexico Territory,	77,568,640	5,486,185	6,864,082
Arizona Territory,	72,906,240	3,135,753	4,050,350
Nevada,	71,737,600	8,198,194	4,669,383
Colorado,	66,880,000	15,683,086	4,303,329
Wyoming Territory,	62,645,068	4,748,841	3,480,281
Oregon,	60,975,300	15,255,617	9,515,744
Idaho Territory,	55,228,160	4,014,953	3,102,407
Utah Territory,	54,065,042	5,984,792	5,315,086
Minnesota,	53,459,840	35,897,912	19,516,340
Kansas,	52,043,520	45,770,685	10,544,439
Nebraska,	48,636,800	32,372,410	8,869,943
Washington Territory,	44,796,160	10,190,046	3,556,967
Indian Territory,	44,154,240	22,832,725	
Missouri,	41,284,000	41,284,000	40,549,368
Florida,	37,931,520	29,345,870	20,643,611
Michigan,	36,128,640	36,128,640	32,468,110
Illinois,	35,462,400	35,462,400	35,462,400
Iowa,	35,228,800	35,228,800	34,036,220
Wisconsin,	34,511,360	34,511,360	26,118,729
Alabama,	34,462,080	34,462,080	28,522,448
Arkansas,	34,406,720	38,046,720	22,463,872
Mississippi,	30,179,840	30,179,840	25,531,387
Louisiana,	26,461,440	23,909,253	20,033,897
Ohio,	25,576,980	25,576,960	25,576,960

XI. Tables of Nutritive Equivalents, etc.

KINDS OF FOOD.	Standard water per cent.	Nitrogen per cent—dry.	Nitrogen per cent—not dried.	Standard.
Ordinary natural meadow hay,	11.0	1.34	1.15	100
Do. of fine quality,	14.0	1.50	1.30	98
Do. select,	18.8	2.40	2.00	58
Do. freed from woody stems,	14.0	2.44	2.10	55
Lucerne hay,	16.6	1.66	1.38	83
Red clover hay, second year's growth,	10.1	1.70	1.54	75
Red clover cut in flower, green, do.,	76.0	...	0.64	311
New wheat straw,	26.0	0.36	0.27	426
Old wheat straw,	8.5	0.53	0.49	235
Do. do. lower parts of the stalk,	5.3	0.43	0.41	280
Do. do. upper parts of the stalk and ear,	9.4	1.42	1.33	86
New rye straw,	18.7	0.30	0.24	479
Old do.,	12.6	0.50	0.42	250
Oat straw,	21.0	0.36	0.30	383
Barley do.,	11.0	0.30	0.25	460
Pea do.,	8.5	1.95	1.79	64
Millet do.,	19.0	0.96	0.78	147
Buckwheat do.,	11.6	0.54	0.48	240
Drum cabbage,	92.3	3.70	0.28	411
Swedish turnip,	91.0	1.83	0.17	676
Turnip,	92.5	1.70	0.13	885
Field beet,	87.8	1.70	0.21	548
Do. white Silesian,	85.6	1.43	0.18	669
Carrots,	87.6	2.40	0.30	382
Jerusalem artichokes,	79.2	1.60	0.30	348
Do.,	75.5	2.20	0.42	274
Potatoes,	65.9	1.50	0.36	319
Field beans,	7.9	5.50	5.11	...
White peas,	8.6	4.20	3.84	...
New Indian corn,	18.0	2.00	1.64	...
Buckwheat,	12.5	2.40	2.10	...
Barley,	13.2	2.02	1.76	...
Barley-meal,	13.0	2.46	2.14	...
Wheat,	10.5	2.33	2.09	...
Do. from highly manured soil,	16.6	3.18	2.65	...
Recent bran,	37.1	2.18	1.36	...
Wheat husk or chaff,	7.6	0.94	0.85	...
Linseed cake,	13.4	6 00	5.20	...
Colza do.,	10.5	5.50	4.92	...
Madia do.,	6.5	5.93	5.51	...
Hemp do.,	5.0	4.78	4.21	...
Poppy do.,	6.8	5.70	5.36	...
Nut do.,	6.0	5.59	5.24	...
Beech mast do.,	6.2	3.53	3 31	...
Arachis (Pindars) do.,	6.6	8.89	8.33	...

Proximate Principles.—The following tables, after M. Payen, show the amount of proximate principles of maize as compared with the other cereal grains:

100 PARTS OF	STARCH.	GLUTEN AND OTHER AZOTIZED MATTER.	DEXTRINE, GLUCOSE, ETC	FATTY MATTERS.	CELLULOSE.	MINERAL MATTER AND SALTS.
Wheat,	58.12	22.75	9.50	2.61	4.00	3.02
Rye,	65.65	13.50	12.00	2.15	4.10	2.60
Barley,	65.43	13.96	10.00	2.76	4.75	3.10
Oats,	60.54	14.38	9.25	5.50	7.06	3.25
Maize,	67.55	12.50	4.00	8.80	5.90	1.25
Rice,	89.15	7.05	1.00	.80	3.00	.90

PROXIMATE PRINCIPLES OF CLOVER AND GRASS.

Green State.	Red Clover.	White Clover.	Lucerne.	Dry State.	Red Clover.	White Clover.	Lucerne.
Water	76.0	80.0	75.0	Flesh formers	22.55	18.76	12.76
Starch	1.4	1.0	2.2	Fat formers	44.00	40.00	38.00
Wood fiber	13.9	11.5	14.3	Accessories	24.00	30.00	36.00
Sugar	2.1	1.5	0.8	Mineral matter	9.45	11.25	13.24
Albumen	2.0	1.5	1.9				
Extractive matter and gum	3.5	3.4	4.4				
Fatty matter	0.1	0.2	0.6				
Phosphate of lime	1.0	0.9	0.8				

Or, economically:					Grass	Meadow Hay.
Green State.	Red Clover.	White Clover.	Lucerne.	Water	68.23	14.61
				Flesh-producing or nitrogenized substances	4.86	8.44
Water	76.0	80.0	75.0	Fat-producing or non-nitrogenized substances	11.45	43.63
Flesh formers	2.0	1.5	1.9	Woody fiber	12.60	27.16
Fat formers	3.6	2.7	3.6	Ash	2.86	6.16
Accessories	17.4	14.9	18.7			
Mineral Water	1.0	0.9	0.8		100.00	100.00

Nutritive Value of Various Foods.—The following table shows the nutritive value of several substances. First, from analysis, or theoretically; and second, according to the average of several different experiments; the figures giving the quantity, in pounds, to be taken of each kind to be equal to any other.

COMPARATIVE VALUE OF FOODS BY ANALYSIS AND EXPERIMENT.

	Value by Analysis.	Value by Experiment.		Value by Analysis.	Value by Experiment.
Good hay,	100	100	Beans,	29	46
Red clover hay (well cured),	77	95	Peas,	30	44
			Indian corn,	70	56
Rye straw,	502	355	Barley,	65	51
Oat straw,	364	220	Rye,	58	49
Ruta-bagas,	676	262	Oats,	60	59
Field beets,	391	346	Buckwheat,	74	64
Carrots,	412	280	Wheat,	47	43
Potatoes,	324	195	Linseed oil-cake,	22	64

XII. Table Showing Prices per Pound.

Any commodity which is sold by the ton may be readily computed in smaller quantities by the table on opposite page. For example, if the price per ton is $10, the price of 50 pounds will be twenty-five cents, and the price of 3,900 pounds will be $19.50. Any desired combination may be had by adding two or more quantities together. The pounds are in the left-hand column, the price per ton at the top of each column, and the price for given numbers of pounds opposite the column of quantities.

TABLES AND DIAGRAMS OF PRACTICAL VALUE.

Lbs.	25c.	50c.	$1	$2	$3	$5	$6	$7	$8	$9	$10	$11	$12	$13	$14	$15	$16	$17	$19	$20	$25	$30	$40
	c.	c.	c.	$ c.	$ c.	$ c.	$ c.	$ c.	$ c.	$ c.	$ c.	$ c.	$ c.	$ c.	$ c.	$ c.	$ c.	$ c.	$ c.	$ c.	$ c.	$ c.	$ c.
3	.	.	.	.	.	.01	.01	.01	.01	.01	.02	.02	.02	.02	.02	.02	.02	.03	.03	.03	.04	.05	.06
7	.	.	.	.01	.01	.02	.02	.02	.03	.03	.04	.04	.04	.05	.05	.05	.06	.06	.07	.07	.09	.10	.14
10	.	.01	.01	.01	.02	.03	.03	.04	.04	.05	.05	.06	.06	.07	.07	.08	.08	.09	.10	.10	.13	.16	.20
20	.01	.01	.02	.02	.03	.05	.06	.07	.08	.09	.10	.11	.12	.13	.14	.15	.16	.17	.19	.20	.25	.38	.40
30	.01	.01	.03	.03	.05	.08	.09	.11	.12	.14	.15	.17	.18	.20	.21	.23	.24	.26	.29	.30	.38	.46	.60
50	.02	.02	.04	.05	.08	.13	.15	.18	.20	.23	.25	.28	.30	.33	.35	.38	.40	.43	.48	.50	.63	.74	1.00
70	.01	.02	.04	.07	.11	.18	.21	.25	.28	.32	.35	.39	.44	.46	.49	.53	.56	.60	.67	.70	.88	1.06	1.40
80	.02	.02	.04	.08	.12	.20	.24	.28	.32	.36	.40	.44	.48	.52	.56	.60	.64	.68	.76	.80	1.00	1.20	1.60
90	.02	.02	.05	.09	.14	.23	.27	.32	.36	.41	.45	.50	.52	.59	.63	.68	.72	.77	.86	.90	1.13	1.36	1.80
100	.01	.03	.05	.10	.15	.25	.30	.35	.40	.45	.50	.55	.60	.65	.70	.75	.80	.85	.95	1.00	1.25	1.50	2.00
200	.03	.05	.10	.20	.30	.50	.60	.70	.80	.90	1.00	1.10	1.20	1.30	1.40	1.50	1.60	1.70	1.90	2.00	2.50	3.00	4.00
300	.04	.08	.15	.30	.45	.75	.90	1.05	1.20	1.35	1.50	1.65	1.80	1.95	2.10	2.25	2.40	2.55	2.85	3.00	3.75	4.50	6.00
400	.05	.10	.20	.40	.60	1.00	1.20	1.40	1.60	1.80	2.00	2.20	2.40	2.60	2.80	3.00	3.20	3.40	3.80	4.00	5.00	6.00	8.00
500	.06	.13	.25	.50	.75	1.25	1.50	1.75	2.00	2.25	2.50	2.75	3.00	3.25	3.50	3.75	4.00	4.25	4.75	5.00	6.25	7.50	10.00
600	.08	.15	.30	.60	.90	1.50	1.80	2.10	2.40	2.70	3.00	3.60	3.60	3.90	4.20	4.50	4.80	5.10	5.70	6.00	7.50	9.00	12.00
700	.09	.18	.35	.70	1.05	1.75	2.10	2.45	2.80	3.15	3.50	3.85	4.20	4.55	4.90	5.25	5.60	5.95	6.65	7.00	8.75	10.50	14.00
800	.10	.20	.40	.80	1.20	2.00	2.40	2.80	3.20	3.60	4.00	4.40	4.80	5.20	5.60	6.00	6.40	6.80	7.60	8.00	10.00	12.00	16.00
900	.11	.23	.45	.90	1.35	2.25	2.70	3.15	3.60	4.05	4.50	4.95	5.40	5.85	6.30	6.75	7.20	7.65	8.55	9.00	11.25	13.50	18.00
1000	.13	.25	.50	1.00	1.50	2.50	3.00	3.50	4.00	4.50	5.00	5.50	6.00	6.50	7.00	7.50	8.00	8.50	9.50	10.00	12.50	15.00	20.00
1100	.14	.28	.55	1.10	1.65	2.75	3.30	3.85	4.40	4.95	5.50	6.05	6.60	7.15	7.70	8.25	8.80	9.35	10.45	11.00	13.75	16.50	22.00
1200	.15	.30	.60	1.20	1.80	3.00	3.60	4.20	4.80	5.40	6.00	6.60	7.20	7.80	8.40	9.00	9.60	10.20	11.40	12.00	15.00	18.00	24.00
1300	.16	.33	.65	1.30	1.95	3.25	3.90	4.55	5.20	5.85	6.50	7.15	7.80	8.45	9.10	9.75	10.40	11.05	12.35	13.00	16.25	19.50	26.00
1400	.18	.35	.70	1.40	2.10	3.50	4.20	4.90	5.60	6.30	7.00	7.70	8.40	9.10	9.80	10.50	11.20	11.90	13.30	14.00	17.50	21.00	28.00
1500	.19	.38	.75	1.50	2.25	3.75	4.50	5.25	6.00	6.75	7.50	8.25	9.00	9.75	10.50	11.25	12.00	12.75	14.25	15.00	18.75	22.50	30.00
1600	.20	.40	.80	1.60	2.40	4.00	4.80	5.60	6.40	7.20	8.00	8.80	9.60	10.40	11.20	12.00	12.80	13.60	15.20	16.00	20.00	24.00	32.00
1700	.21	.43	.85	1.70	2.55	4.25	5.10	5.95	6.80	7.65	8.50	9.35	10.20	11.05	11 90	12.75	13.60	14.15	16.15	17.00	21.25	25.50	34.00
1800	.23	.45	.90	1.80	2.70	4.50	5.40	6.30	7.20	8.10	9.00	9.90	10.80	11.70	12.60	13.50	14.40	15.30	17.10	18.00	22.50	27.00	36.00
1900	.24	.48	.95	1.90	2.85	4.75	5 70	6.65	7.60	8.55	9.50	10.45	11.40	12.35	13.30	14.25	15.20	16.15	18.05	19.00	23.75	28.50	38.00

XIII. Table of Interest at Six Per Cent.

In fractions of half a cent or more, one cent is taken. If less, nothing is taken. Where cents form part of the principal, if they amount to half a dollar, or upward, the discount is taken as for a dollar; when less than half a dollar, they are disregarded. The left-hand column shows the time, and the columns on the top of the table show the amount. Under the column of amount and opposite the time column, will be found the interest. To find the interest at any other rate, multiply the interest found in the column by the desired rate of interest, and divide by 6. Thus, $60, the interest of $1000 for one year, multiplied by 8, gives $480; divided by 6 gives $80, the interest of $1000 for one year at 8 per cent.

Time.	$1	$2	$3	$4	$5	$6	$7	$8	$9	$10	$20	$30	$40	$50	$60	$70	$80	$90	$100	$1000
1 Day,	0	0	0	0	0	0	0	0	0	0	0	1	1	1	1	1	1	2	2	17
2 Days,	0	0	0	0	0	0	0	0	0	0	1	1	1	2	2	2	3	3	3	33
3 "	0	0	0	0	0	0	0	0	0	1	1	2	2	3	3	4	4	5	5	50
4 "	0	0	0	0	0	0	0	1	1	1	1	2	3	3	4	5	5	6	7	67
5 "	0	0	0	0	0	1	1	1	1	1	2	3	3	4	5	6	7	8	8	83
6 "	0	0	0	0	1	1	1	1	1	1	2	3	4	5	6	7	8	9	10	1 00
7 "	0	0	0	0	1	1	1	1	1	1	2	4	5	6	7	8	9	11	12	1 17
8 "	0	0	0	1	1	1	1	1	1	1	3	4	5	7	8	9	11	12	13	1 33
9 "	0	0	0	1	1	1	1	1	1	2	3	5	6	8	9	11	12	14	15	1 50
10 "	0	0	1	1	1	1	1	1	2	2	3	5	7	8	10	12	13	15	17	1 67
11 "	0	0	1	1	1	1	1	2	2	2	4	6	7	9	11	13	15	17	18	1 83
12 "	0	0	1	1	1	1	1	2	2	2	4	6	8	10	12	14	16	18	20	2 00
13 "	0	0	1	1	1	1	1	2	2	2	4	7	9	11	13	15	17	20	22	2 17
14 "	0	0	1	1	1	1	2	2	2	2	5	7	9	12	14	16	19	21	23	2 33
15 "	0	1	1	1	1	2	2	2	2	3	5	8	10	13	15	18	20	23	25	2 50
16 "	0	1	1	1	1	2	2	2	2	3	5	8	11	13	16	19	21	24	27	2 67
17 "	0	1	1	1	1	2	2	2	3	3	6	9	11	14	17	20	23	26	28	2 83
18 "	0	1	1	1	2	2	2	2	3	3	6	9	12	15	18	21	24	27	30	3 00
19 "	0	1	1	1	2	2	2	3	3	3	6	10	13	16	19	22	25	29	32	3 17
20 "	0	1	1	1	2	2	2	3	3	3	7	10	13	17	20	23	27	30	33	3 33
21 "	0	1	1	1	2	2	2	3	3	4	7	11	14	18	21	25	28	32	35	3 50
22 "	0	1	1	1	2	2	3	3	3	4	7	11	15	18	22	26	29	33	37	3 67
23 "	0	1	1	2	2	2	3	3	3	4	8	12	15	19	23	27	31	35	38	3 83
24 "	0	1	1	2	2	2	3	3	4	4	8	12	16	20	24	28	32	36	40	4 00
25 "	0	1	1	2	2	3	3	3	4	4	8	13	17	21	25	29	33	38	42	4 17
26 "	0	1	1	2	2	3	3	3	4	4	9	13	17	22	26	30	35	39	43	4 33
27 "	0	1	1	2	2	3	3	4	4	5	9	14	18	23	27	32	36	41	45	4 50
28 "	0	1	1	2	2	3	3	4	4	5	9	14	19	23	28	33	37	42	47	4 67
29 "	0	1	1	2	2	3	3	4	4	5	10	15	19	24	29	34	29	44	48	4 83
1 Mo.,	1	1	2	2	3	3	4	4	5	5	10	15	20	25	30	35	40	45	50	5 00
2 Mos.,	1	2	3	4	5	6	7	8	9	10	20	30	40	50	60	70	80	90	1 00	10 00
3 "	2	3	5	6	8	9	11	12	14	15	30	45	60	75	90	1 05	1 20	1 35	1 50	15 00
4 "	2	4	6	8	10	12	14	16	18	20	40	60	80	1 00	1 20	1 40	1 60	1 80	2 00	20 00
5 "	3	5	8	10	13	15	18	20	23	25	50	75	1 00	1 25	1 50	1 75	2 00	2 25	2 50	25 00
6 "	3	6	9	12	15	18	21	24	27	30	60	90	1 20	1 50	1 80	2 10	2 40	2 70	3 00	30 00
7 "	4	7	11	14	18	21	25	28	32	35	70	1 05	1 40	1 75	2 10	2 45	2 80	3 15	3 50	35 00
8 "	4	8	12	16	20	24	28	32	36	40	80	1 20	1 60	2 00	2 40	2 80	3 20	3 60	4 00	40 00
9 "	5	9	14	18	23	27	32	36	41	45	90	1 35	1 80	2 25	2 70	3 15	3 60	4 05	4 50	45 00
10 "	5	10	16	20	25	30	35	40	45	50	1 00	1 50	2 00	2 50	3 00	3 50	4 00	4 50	5 00	50 00
11 "	6	11	17	22	28	33	39	44	50	55	1 10	1 65	2 20	2 75	3 30	3 85	4 40	4 95	5 50	55 00
1 Year,	6	12	18	24	30	36	42	48	54	60	1 20	1 80	2 40	3 00	3 60	4 20	4 80	5 40	6 00	60 00

TABLES AND DIAGRAMS OF PRACTICAL VALUE.

XIV. Growth of Money at Interest.

As properly following the interest table, we give two companion tables, one showing the growth of one dollar for 100 years, at various rates of interest, the interest being annually added to the principal. It will be seen that one dollar invested for 100 years at 24 per cent would about cancel the national debt. The other shows the value of daily savings, at compound interest. It will be seen that the insignificant sum of 2¾ cents per day amounts to $10 per year, and if this sum is saved daily from the age of twenty-one years to the age of seventy, the snug sum of $2,900 is reached. Fifty-five cents a day saved, in ten years reaches $2,600; money enough to buy a snug home, or an eighty-acre farm.

GROWTH OF MONEY AT COMPOUND INTEREST.

One dollar, 100 years at 1 per cent,	$ 2¾
One dollar, 100 years at 2 "	7¼
One dollar, 100 years at 3 "	19¼
One dollar, 100 years at 4 "	50½
One dollar, 100 years at 5 "	131½
One dollar, 100 years at 6 "	340
One dollar, 100 years at 7 "	868
One dollar, 100 years at 8 "	2,203
One dollar, 100 years at 9 "	5,513
One dollar, 100 years at 10 "	13,809
One dollar, 100 years at 12 "	84,675
One dollar, 100 years at 15 "	1,174,405
One dollar, 100 years at 18 "	15,145,000
One dollar, 100 years at 24 "	2,551,799,404

DAILY SAVINGS AT COMPOUND INTEREST.

Cents per Day.	Per Year.	Ten Years.	Fifty Years.
2¾	$ 10	$ 130	$ 2,900
5¼	20	260	5,800
11	40	520	11,600
27½	100	1,300	29,000
55	200	2,600	58,000
$1.10	400	5,200	116,000
1.37	500	6,500	145,000

By the above table it appears that if a mechanic or clerk saves 2¾ cents per day from the time he is twenty-one until he is seventy, the total with interest will amount to $2,900, and a daily saving of 27½ cents reaches the important sum of $29,000.

XV. Mean Duration of Life.

Comparative age.	Carlisle Experience.	Northampton Experience.	Comparative age.	Carlisle Experience.	Northampton Experience.	Comparative age.	Carlisle Experience.	Northampton Experience.
0	38.72	25.18	27	36.41	29.82	54	18.28	16.06
1	44.68	32.74	28	35.69	29.30	55	17.58	15.58
2	47.55	37.79	29	35.00	28.79	56	16.89	15.10
3	49.82	39.55	30	34.34	28.27	57	16.21	14.63
4	50.76	40.58	31	33.68	27.76	58	15.55	14.15
5	51.25	40.84	32	33.03	27.24	59	14.92	13.68
6	51.17	41.07	33	32.36	26.72	60	14.34	13.21
7	50.80	41.03	34	31.68	26.20	61	13.82	12.75
8	50.24	40.79	35	31.00	25.68	62	13.31	12.28
9	49.57	40.36	36	30.32	25.16	63	12.81	11.81
10	48.82	39.78	37	29.64	24.64	64	12.30	11.35
11	48.04	39.14	38	28.96	24.12	65	11.79	10.88
12	47.27	38.49	39	28.28	23.60	66	11.27	10.42
13	46.51	37.83	40	27.61	23.08	67	10.75	9.96
14	45.75	37.17	41	26.97	22.56	68	10.23	9.50
15	45.00	36.51	42	26.34	22.04	69	9.70	9.05
16	44.27	35.85	43	25.71	21.54	70	9.18	8.60
17	43.57	35.20	44	25.09	21.03	71	8.65	8.17
18	42.89	34.58	45	24.46	20.52	72	8.16	7.74
19	42.17	33.99	46	23.82	20.02	73	7.72	7.33
20	41.46	33.43	47	23.17	19.51	74	7.33	6.92
21	40.75	32.90	48	22.50	19.00	75	7.01	6.54
22	40.04	32.39	49	21.81	18.49	76	6.69	6.18
23	39.31	31.38	50	21.11	17.99	77	6.40	5.83
24	38.59	31.36	51	20.39	17.50	78	6.12	5.48
25	37.86	30.85	52	19.68	17.02	79	5.80	5.11
26	37.14	30.33	53	18.97	16.54	80	5.51	4.75

EITHER the Carlisle or Northampton tables are approved standards, and admitted by the courts as a basis for computing the value for life-estates, or losses resulting from injury in suits to recover damages. In the West, the Carlisle tables seem to be in favor; in the East and South, the Northampton tables are used. The Carlisle show the probable lease of life at birth to be 38.72 years; at 80 years, 5.51 years. We give on preceding page a table of probabilities from birth to eighty years.

XVI. Mortality Rates.

The following figures are from the Carlisle tables showing the annual average deaths and the number alive in 10,000 individuals from birth up to the age of 104 years. Thus 1,539 persons out of every 10,000 die the first year after birth and 8,461 are left; at 104 only one is left, and that year this person dies:

YEAR.	No. Alive.	Deaths.	YEAR.	No. Alive.	Deaths.	YEAR.	No. Alive.	Deaths.	YEAR.	No. Alive.	Deaths.
At Birth.	10000	1539	27 ..	5793	45	54 ..	4143	70	81 ..	837	112
1 ..	8461	682	28 ..	5748	50	55 ..	4073	73	82 ..	725	102
2 ..	7779	505	29 ..	5698	56	56 ..	4000	76	83 ..	623	94
3 ..	7274	276	30 ..	5642	57	57 ..	3924	82	84 ..	529	84
4 ..	6998	201	31 ..	5585	57	58 ..	3842	93	85 ..	445	78
5 ..	6797	121	32 ..	5528	56	59 ..	3749	106	86 ..	367	71
6 ..	6676	82	33 ..	5472	55	60 ..	3633	122	87 ..	296	64
7 ..	6594	58	34 ..	5417	55	61 ..	3521	126	88 ..	232	51
8 ..	6536	43	35 ..	5362	55	62 ..	3395	127	89 ..	181	39
9 ..	6493	33	36 ..	5307	56	63 ..	3268	125	90 ..	142	37
10 ..	6460	29	37 ..	5251	57	64 ..	3143	125	91 ..	105	30
11 ..	6431	31	38 ..	5194	58	65 ..	3018	124	92 ..	75	21
12 ..	6400	32	39 ..	5136	61	66 ..	2894	123	93 ..	54	14
13 ..	6368	33	40 ..	5075	66	67 ..	2771	123	94 ..	40	10
14 ..	6335	35	41 ..	5009	69	68 ..	2648	123	95 ..	30	7
15 ..	6300	39	42 ..	4940	71	69 ..	2525	124	96 ..	23	5
16 ..	6261	42	43 ..	4869	71	70 ..	2401	124	97 ..	18	4
17 ..	6219	43	44 ..	4798	71	71 ..	2277	134	98 ..	14	3
18 ..	6176	43	45 ..	4727	70	72 ..	2143	146	99 ..	11	2
19 ..	6133	43	46 ..	4657	69	73 ..	1997	156	100 ..	9	2
20 ..	6090	43	47 ..	4588	67	74 ..	1841	166	101 ..	7	2
21 ..	6047	42	48 ..	4521	63	75 ..	1675	160	102 ..	5	2
22 ..	6005	42	49 ..	4458	61	76 ..	1515	156	103 ..	3	2
23 ..	5963	42	50 ..	4397	69	77 ..	1359	146	104 ..	1	1
24 ..	5921	42	51 ..	4338	62	78 ..	1213	132			
25 ..	5879	43	52 ..	4276	65	79 ..	1081	128			
26 ..	5836	43	53 ..	4211	68	80 ..	953	116			

XVII. How to Calculate Salaries and Wages.

The computation of wages or salaries for short periods is vexatious to many. The subjoined table will show at a glance the wages per day, week and month of the sums given. To find the monthly, weekly or daily wages for annual wages not given in the table, find the rate for figures that are given, and add, subtract or multiply proportionately. Thus, suppose you wish to know the wages per day corresponding to $1,265 per year. The wages for a year are $1,250; per month is $104.17; per

TABLES AND DIAGRAMS OF PRACTICAL VALUE.

week, $23.29; per day, $3.42. Add to this the half of the wages of $30.00, and we have $1.25 per month, 29 cents per week, and 8 cents per day. Hence the wages of $1,265 per year would be: per month, $105.42; per week, $23.58, and per day, $3.50. The left-hand columns give the salaries per year, and the figures opposite, the wages per month, week and day:

Per Year.	Per Month.	Per Week.	Per Day.	Per Year.	Per Month.	Per Week.	Per Day.	Per Year.	Per Month.	Per Week.	Per Day.
$	$	$	$	$	$	$	$	$	$	$	$
20 is	1 67	.38	.05	195	16.25	3.74	.53	450	37.50	8.63	1.23
25	2.08	.48	.07	200	16.57	3.84	.55	475	39.58	9.11	1.30
30	2.50	.58	.08	205	17.08	3.93	.56	500	41.67	9.59	1.37
35	2.92	.67	.10	210	17.50	4.03	.58	525	43.75	10.07	1.44
40	3.33	.77	.11	215	17.92	4.12	.59	550	45.83	10.55	1.51
45	3.75	.86	.12	220	18.33	4.22	.60	575	47.92	11.03	1.58
50	4.17	.96	.14	225	18.75	4.34	.62	600	50.00	11.51	1.64
55	4.58	1.06	.15	230	19.17	4.41	.63	625	52.08	11.99	1.71
60	5.00	1.15	.16	235	19.58	4.51	.64	650	54.17	12.47	1.78
65	5.42	1.25	.18	240	20.00	4.60	.66	675	56.25	12.95	1.85
70	5.83	1.34	.19	245	20 42	4.70	.67	700	58.33	13.42	1.92
75	6.25	1.44	.21	250	20.83	4.79	.69	725	60 42	13.90	1.99
80	6.67	1.56	.22	255	21.25	4.89	.70	750	62.50	14.38	2.05
85	7.08	1.63	.23	260	21.67	4.99	.71	775	64.58	14.86	2.12
90	7.50	1.73	.25	265	22.08	5.08	.73	800	66.67	15.34	2.19
95	7.92	1.82	.26	270	22.50	5.18	.74	825	68.75	15.82	2.26
100	8.33	1.92	.27	275	22.92	5.27	.75	850	70.83	16.30	2.33
105	8.75	2.01	.29	280	23.33	5.37	.77	875	72.92	16.78	2.40
110	9.17	2.11	.30	285	23.75	5.47	.78	900	75.00	17.26	2.47
115	9.58	2.21	.32	290	24.17	5.56	.79	925	77.08	17.74	2.53
120	10.00	2.30	.33	295	24.58	5.66	.81	950	79.17	18.22	2.60
125	10.42	2.40	.34	300	25.00	5.75	.82	975	81.25	18.70	2.67
130	10.83	2.49	.36	310	25.83	5.95	.85	1000	83.33	19.18	2.74
135	11.25	2.59	.37	320	26.67	6.14	.88	1050	87.50	20.14	2.88
140	11.67	2.69	.38	325	27.08	6.23	.89	1100	91.67	21.10	3.01
145	12.08	2.78	.40	330	27.50	6.33	.90	1150	95.83	22.06	3.15
150	12.50	2.88	.41	340	28.33	6.52	.93	1200	100.00	23.01	3.29
155	12.92	2.97	.42	350	29.17	6.71	.96	1250	104 17	23.29	3.42
160	13.33	3.07	.44	360	30.00	6.90	.99	1300	108.33	24.93	3.56
165	13.75	3.16	.45	370	30.83	7.10	1.01	1350	112.50	25.89	3.70
170	14.17	3.26	.47	375	31.25	7.19	1.03	1400	116.67	26.85	3.84
175	14.58	3.36	.48	380	31.67	7.29	1.04	1450	120.84	27.80	3.98
180	15.00	3.45	.49	390	32.50	7.48	1.07	1500	125.00	28.77	4.11
185	15.42	3.55	.51	400	33.33	7.67	1.10	1600	133.35	30.68	4.38
190	15.83	3.64	.52	425	35.42	8.15	1.16	. . . .	. . . .	. . . .	. . .

XVIII. The Earth's Area and Population.

Divisions.	Area.	Population.	Population to Square Mile.
America,	14,700,000	88,061,148	6
Europe,	3,800,000	296,713,500	80
Asia,	15,000,000	699,863,000	46
Africa,	10,800,000	67,414,000	6
Oceanica,	4,500,000	25,924,000	6
TOTAL,	48,800,000	1,177,975,648	24

CLASSIFICATION OF THE POPULATION.

RACES.		RELIGIONS.	
Whites,	550,000,000	Pagans,	676,000,000
Mongolian,	550,000,000	Christians,	220,000,000
Black,	173,000,000	Mohammedans,	140,000,000
Copper-colored,	12,000,000	Jews,	14,000,000

Christians are divided as follows:

ROMAN CATHOLICS.	EASTERN OR GREEK CHURCH.	PROTESTANTS
170,000,000.	60,000,000.	90,000,000.

The Greek Church and the Church of Rome were originally one. The first differences occurred in A. D. 482; the second in 732. The most bitter feuds were carried on, from time to time, between the two churches. The conquest of Constantinople, in 1453, increased the hostility to the Church of Rome. As late as 1848, Pope Pius IX, by an encyclical letter, again invited the entire Eastern Church to a corporate union with the Church of Rome. The invitation was rejected by the Greek Church.

XIX. The World's Commerce.

Countries.	Population.	Commerce.	Imports.	Exports.
Europe,	289,000,000	$9,976,000,000	$5,650,400,000	$4,336,200,000
America,	84,840,000	2,140,000,000	972,800,000	1,167,200,000
Asia,	806,700,000	1,131,000,000	489,000,000	641,600,000
Australasia,	1,800,000	462,000,000	237,800,000	224,400,000
Africa,	80,000,000	291,000,000	134,400,000	156,600,000
TOTAL,	1,262,340,000	$14,000,000,000	$7,747,400,000	$6,526,000,000

IMMIGRATION INTO THE UNITED STATES.

Year.	Total Immigrants.	Year.	Total Immigrants.	Year.	Total Immigrants.	Year.	Total Immigrants.
1820,	8,385	1836,	76,242	1852,	371,603	1868,	282,189
1821,	9,127	1837,	79,340	1853,	368,645	1869,	352,768
1822,	6,911	1838,	38,914	1854,	427.833	1870,	387.203
1823,	6,354	1839,	68,069	1855,	200.877	1871,	321,350
1824,	7,912	1840,	84,066	1856,	195,857	1872,	404,806
1825,	10,199	1841,	80,289	1857,	246,945	1873,	459,803
1826,	10,837	1842,	104,565	1858,	119,501	1874,	313,339
1827,	18,875	1843,	52,496	1859,	118,616	1875,	227,498
1828,	27,382	1844,	78,615	1860,	150,237	1876,	169,986
1829,	22,520	1845,	114.371	1861,	89,724	1877,	141,857
1830,	23,322	1846,	154,416	1862,	89,007	1878,	138,469
1831,	22,633	1847,	234.968	1863,	174,524	1879,	177,826
1832,	60,482	1848,	226,557	1864,	193,195	1880,	457,257
1833,	58,640	1849,	297,024	1865,	247,453	1881,	669,431
1834,	65,365	1850,	369,980	1866,	167,757		
1835,	45,374	1851,	379,466	1867,	298,967		

TOTAL, 10,808,189

CHINESE IMMIGRATION INTO THE UNITED STATES.

Year.	No.	Year.	No.	Year.	No.	Year.	No.
1855,	3,526	1862,	3,633	1869,	14,902	1876,	16,879
1856,	4,733	1863,	7,214	1870,	11,943	1877,	10,379
1857,	5,944	1864,	2,795	1871,	6,039	1878,	8,468
1858,	5,128	1865,	2,942	1872,	10,642	1879,	9,189
1859,	3,457	1866,	2,385	1873,	18,154	1880,	7,011
1860,	5,467	1867,	3,863	1874,	16,651	1881,	13,704
1861,	7,518	1868,	10,684	1875,	19,033		

TOTAL, 232,283

XX. Pay of the Principal Officers of the United States.

President,	$50,000
Vice-President,	10,000
Cabinet Ministers,	10,000
Chief Justice Supreme Court,	10,500
Justices of the Supreme Court,	10,000
Senators and Representatives in Congress, with Mileage,	5,000
Speaker House of Representatives, with Mileage,	10,000
Secretary of Senate,	5,000
Clerk House of Representatives,	5,000
Assistant Secretaries of Departments,	6,000
Heads of Bureaus,	4,500
Superintendent Coast Survey,	6,000
Judges District of Columbia,	3,000
Secretary Smithsonian Institution,	4,000
Ministers Plenipotentiary to Great Britain, France, Germany, Russia, each,	17,500
Ministers Plenipotentiary to Spain, Austria, China, Italy, Mexico, Brazil and Japan, each	12,000
Ministers Resident and Plenipotentiary to Chili, Peru, Uruguay, Guatemala, Costa Rica, Honduras, Nicaragua and San Salvador, each,	10,000
Ministers Resident to Portugal, Belgium, Netherlands, Denmark, Sweden and Norway, Switzerland, Turkey, Hawaiian Islands, Hayti, Columbia, Venezuela, Ecuador, Argentine Republic, Paraguay, Bolivia and Greece, each,	7,500
Interpreter and Secretary of Legation to China,	5,000
Dragoman and Secretary of Legation to Turkey,	3,000
Consul-General to Cairo,	4,000
Consul-General to London, Paris, Havana and Rio Janeiro,	6,000
Consul-General to Calcutta and Shanghai,	5,000
Consul-General to Melbourne,	4,500
Consul-General to Kanagawa, Montreal and Berlin,	4,000
Consul-General to Vienna, Frankfort, Rome and Constantinople, each,	3,000
Consul-General to Turkey and Egypt,	3,500
Consul-General to St. Petersburg and Mexico,	2,000
Consul-General to Liverpool,	6,000
Secretaries of Legation, average,	2,000
Consuls, average,	5,000

Of officers of the line, there are, one General, salary, $13,000; one Lieutenant-General, salary, $11,000; three Major-Generals with a salary each of $7,500, and six Brigadier-Generals, with a salary each of $5,500. Of the staff officers, there are twenty-nine Aids-de-Camp, six of them, with the pay of a Colonel, are Aids-de-Camp to the General of the army; two of them, with the pay of a Lieutenant-Colonel, are Aids-de-Camp to Lieutenant-General; eight of them, $200 in addition to pay in line, are Aids-de-Camp to Major-Generals; thirteen, $150 in addition to pay in line, are Aids-de-Camp to Brigadier-Generals.

REGIMENTAL AND COMPANY OFFICERS.

Infantry	No.	Salary.	Cavalry.	No	Salary.	Artillery.	No.	Salary.
Colonels,	25	$3,500	Colonels,	10	$3.500	Colonels,	5	$3,500
Lieut-Colonels,	25	3,000	Lieut.Colonels,	10	3.000	Lieut-Colonels,	5	3,000
Majors,	25	2,500	Majors,	30	2,500	Majors,	15	2,500
Captains,	250	1,800	Captains,	120	2,000	Captains,	60	2,000
Adjutants,	25	1,800	Adjutants,	10	1,800	Adjutants,	5	1,800
Reg. Qrs.	25	1,800	Reg. Qrs.	10	1,800	Reg. Qrs.,	5	1,800
1st Lieutenants,	250	1,500	1st Lieutenants,	120	1,600	1st Lieutenants.	120	1,600
2d Lieutenants,	250	1,400	2d Lieutenants,	120	1,500	2d Lieutenants,	65	1,500
Chaplains,	2	1,500	Chaplains,	2	1,500			

OFFICERS OF THE UNITED STATES NAVY IN ACTIVE SERVICE.

Line—At Sea.	No.	Salary.	Staff.	No.	Salary.
Admiral,	1	$13,000	Medical Directors,	15	$2,800 to $4,400
Vice-Admiral,	1	9,000	Pay Directors,	13	2,800 to 4,200
Rear Admirals,	11	6.000	Chief Engineers,	70	2,800 to 4,200
Commodores,	25	5,000	Surgeons,	50	2,800 to 4,200
Captains,	50	4,500	Paymasters,	50	2,800 to 4,200
Commanders,	90	3,500	Passed or Asst. Surgeons,	100	1,900 to 2,200
Lieut.-Commanders,	80	$2,800 to 3,000	Passed or Asst. Paymasters,	30	2,000 to 2,200
Lieutenants,	280	2,400 to 2,600	Chaplains,	24	2,500 to 2,800
Masters,	100	1,800 to 2,000	Naval Constructors,	11	3,200 to 4,200
Ensigns,	100	1,200 to 1,400	Asst. Constructors,	5	2,000 to 2,600
Midshipmen.	40	1,000	Profs. Mathematics,	12	2,400 to 3,500
Cadet Midshipmen,	334	500 to 950	Civil Engineers,	9	2,400 to 3,500
Mates,	42	900	Cadet Engineers,		500 to 1,000

OFFICERS OF THE MARINE CORPS IN ACTIVE SERVICE.

1 Colonel Commander,	Salary,	$3,500
1 Colonel,	"	3,500
2 Lieutenant Colonels,	"	3,000
1 Major.	"	2,500
18 Captains,	Salary,	$1,800
30 First Lieutenants,	"	1,500
20 Second Lieutenants,	"	1,400

XXI. Public Debt of the United States—Ninety Years.

Year	Debt	Year	Debt	Year	Debt	Year	Debt
1791	$75,463,476 52	1814	$ 81,487,846 24	1837	$ 3,308,124 07	1860	$ 64,842,287 88
1792	77,227,924 66	1815	99,833,660 15	1838	10,434,221 14	1861	90,580,873 72
1793	80,352,634 04	1816	127,334,933 74	1839	3,573,343 82	1862	524,176,412 13
1794	78,427,404 77	1817	123,491,965 16	1840	5,250,875 54	1863	1,119,772,138 63
1795	80,747,587 39	1818	103,466,633 83	1841	13,594,480 73	1864	1,815,784,370 57
1796	83,762,172 07	1819	95,529,648 28	1842	20,601,226 28	1865	2,680,647,869 74
1797	82,064,479 33	1820	91,015,566 15	1843	32,742,922 00	1866	2,773,236,173 69
1798	79,228,529 12	1821	89,987,427 66	1844	23,461,652 50	1867	2,678,126,103 87
1799	78,408,669 77	1822	93,546,676 98	1845	15,925,303 01	1868	2,611,687,851 19
1800	82,976,294 35	1823	90,875,877 28	1846	15,550,202 97	1869	2,588,452,213 94
1801	83,038,050 80	1824	90,269,777 77	1847	38,826,534 77	1870	2,480,672,427 81
1802	86,712,632 25	1825	83,788,432 71	1848	47,044,862 23	1871	2,353,211,332 32
1803	77,054,686 30	1826	81,054,059 99	1849	63,061,858 69	1872	2,253,251,328 78
1804	86,427,120 88	1827	73,987,357 20	1850	63,452,773 55	1873	2,234,482,993 20
1805	82,312,150 50	1828	67,475,043 87	1851	68,304,796 02	1874	2,251,690,468 43
1806	75,723,270 66	1829	58,421,413 67	1852	66,199,341 71	1875	2,232,284,531 95
1807	69,218,398 64	1830	48,565,406 50	1853	59,803,117 70	1876	2,180,395,067 15
1808	65,196,317 97	1831	39,123,191 68	1854	42,242,222 42	1877	2,205,301,392 10
1809	57,023,192 09	1832	24,322,235 18	1855	35,586,858 56	1878	2,256,205,892 53
1810	53,173,217 52	1833	7,001,698 83	1856	31,972,537 90	1879	2,245,495,072 04
1811	48,005,587 76	1834	4,760,082 08	1857	28,699,831 85	1880	2,120,415,370 63
1812	45,209,737 90	1835	37,513 05	1858	44,911,881 03	1881	2,069,013,569 58
1813	55,962,827 57	1836	336,957 83	1859	58,496,837 88	1882	1,918,312,994 03

XXII. The United States and Territories.

The dates of admission or organization of the several States and Territories from 1800 to 1880 are given below. It will be interesting as showing the relative growth:

STATES AND TERRITORIES.	DATE.	AREA IN SQUARE MILES.	TOTAL POPULATION.				
			1800.	1820.	1840.	1860.	1880.
Maine,	1820	31,765	151,719	298,335	501,793	628,279	648,945
New Hampshire,	1788	9,280	183,762	244,161	284,574	326,073	347,784
Vermont,	1791	9,056	154,465	235,764	291,948	315,098	332,286
Massachusetts,	1788	7,800	423,245	523,287	737,699	1,231,066	1,783,086
Rhode Island,	1790	1,046	69,122	83,059	108,830	174,620	276,528
Connecticut,	1788	4,730	251,002	275,202	309,978	460,147	622,683
New York,	1788	50,519	586,756	1,372,812	2,428,291	3,880,735	5,083,173
New Jersey,	1787	8,320	211,949	277,575	373,306	673,035	1,130,892
Pennsylvania,	1787	46,000	602,361	1,049,458	1,724,033	2,906,115	4,282,738
Delaware,	1787	2,120	64,273	72,749	78,085	112,216	146,654
Maryland,	1788	11,124	341,548	407,350	470,019	687,049	935,139
District of Columbia,	1790	60	14,093	33,039	43,712	75,080	177,638
Virginia (inc. W. Va.),	1788	61,352	880,200	1,065,379	1,239,797	1,596,318	1,512,203
West Virginia,	1863	23,000	. . . .	. . . .	. . . .	. . . .	618,193
North Carolina,	1789	45,000	478,103	638,829	753,419	992,622	1,400,000
South Carolina,	1788	30,213	345,591	502,741	594,398	703,708	995,706
Georgia,	1788	58,000	162,101	340,987	691,392	1,057,286	1,538,983
Florida,	1845	59,268	. . . .	. . . .	54,477	140,425	266,566
Alabama,	1819	50,722	. . . .	127,901	590,756	964,201	1,262,344
Mississippi,	1817	47,156	8,850	75,448	375,651	791,305	1,131,899
Louisiana,	1812	41,255	. . . .	153,407	352,411	708,002	940,263
Texas,	1845	237,504	. . . .	. . . .	. . . .	604,215	1,597,509
Arkansas,	1836	52,198	. . . .	14,273	97,574	435,450	802,564
Tennessee,	1796	45,600	105,602	422,813	829,210	1,109,801	1,542,463
Kentucky,	1792	37,680	220,955	564,317	779,828	1,155,684	1,648,599
Ohio,	1802	39,964	45,365	581,434	1,519,467	2,339,502	3,197,794
Michigan,	1837	56,243	. . . .	8,896	212,267	749,113	1,634,096
Indiana,	1816	33,809	4,875	147,178	685,866	1,350,428	1,978,358
Illinois,	1818	55,405	. . . .	55,211	476,183	1,711,951	3,078,636
Wisconsin,	1848	53,924	. . . .	. . . .	30,945	775,881	1,315,386
Minnesota,	1858	81,259	. . . .	. . . .	. . . .	173,855	780,807
Iowa,	1846	50,914	. . . .	. . . .	43,112	674,948	1,624,463
Missouri,	1821	67,380	. . . .	66,586	383,702	1,182,012	2,169,091
Kansas,	1861	78,418	. . . .	. . . .	. . . .	107,206	995,335
California,	1850	155,500	. . . .	. . . .	. . . .	379,994	864,686
Oregon,	1859	80,000	. . . .	. . . .	. . . .	52,465	174,767
Nebraska,	1867	75,995	. . . .	. . . .	. . . .	28,841	452,432
Nevada,	1864	81,539	. . . .	. . . .	. . . .	6,857	62,265
Colorado,	1875	104,500	. . . .	. . . .	. . . .	34,277	174,649
Washington Territory,	1853	Estimated about 800,000.	. . . .	. . . .	. . . .	11,594	75,120
Utah "	1850		. . . .	. . . .	6,100 persons in U. S. Navy.	40,273	143,907
New Mexico "	1850		. . . .	. . . .		93,516	118,430
Dakota "	1861		. . . .	. . . .		4,837	134,502
Idaho "	1867		. . . .	. . . .		. . . .	32,611
Montana "	1867		. . . .	. . . .		. . . .	39,157
Wyoming "	1868		. . . .	. . . .		. . . .	20,788
Arizona "	1867		. . . .	. . . .		. . . .	40,441
Total,			5,305,937	9,638,191	17,069,453	31,445,080	50,152,559

XXIII. Diagrams Giving Valuable Statistics.

The following statistical diagrams show Production of Corn and Wheat; Export and Production of Cotton; Average Wages; Miles of Railroad in Operation, and Aggregate Tons of Freight Moved, and will be found valuable for reference.

Diagram showing the production of Corn for the year 1849-1859-1869-1879 by the principal Corn producing States.

TABLES AND DIAGRAMS OF PRACTICAL VALUE.   1025

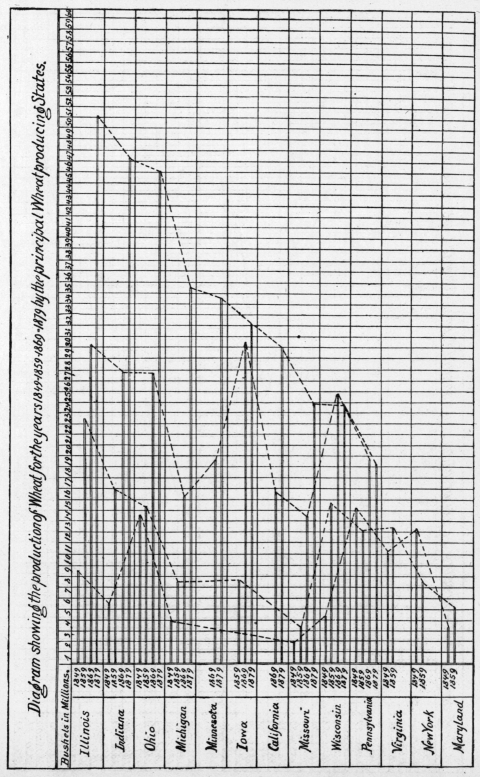

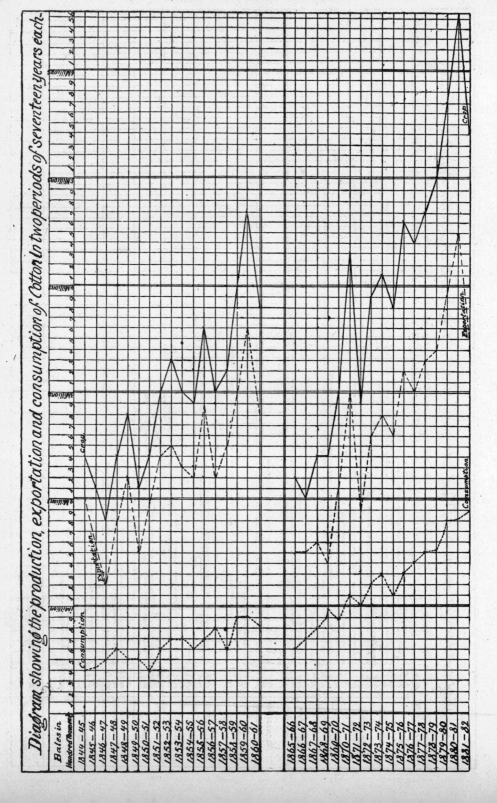

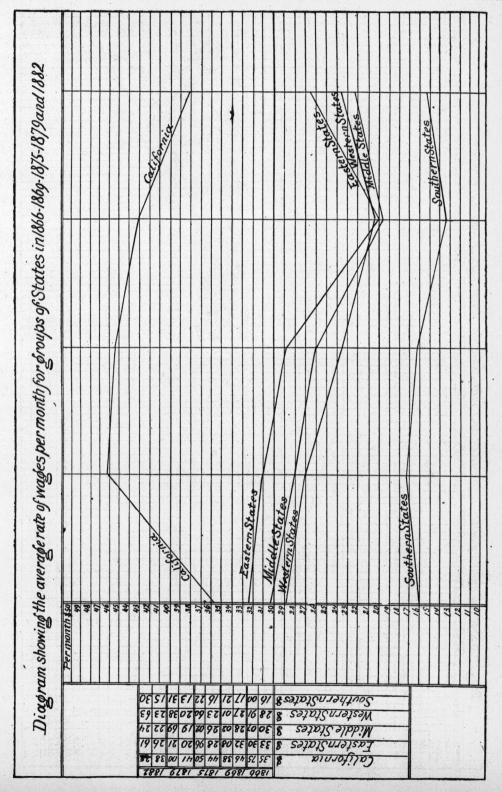

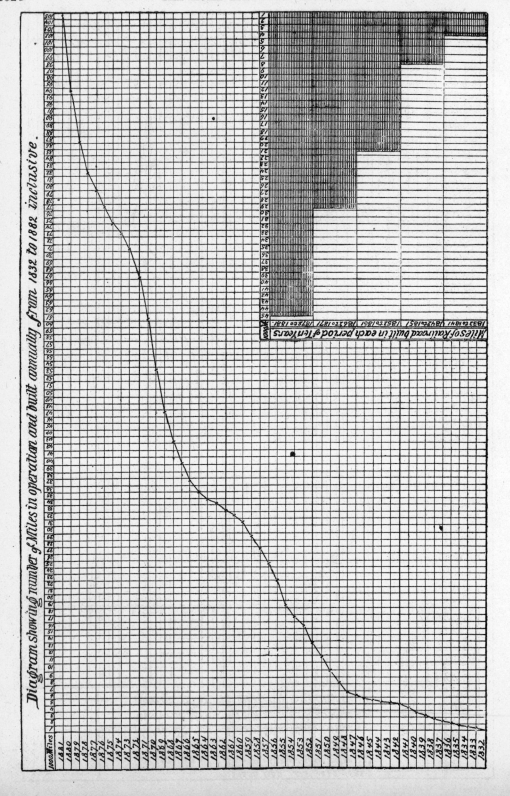

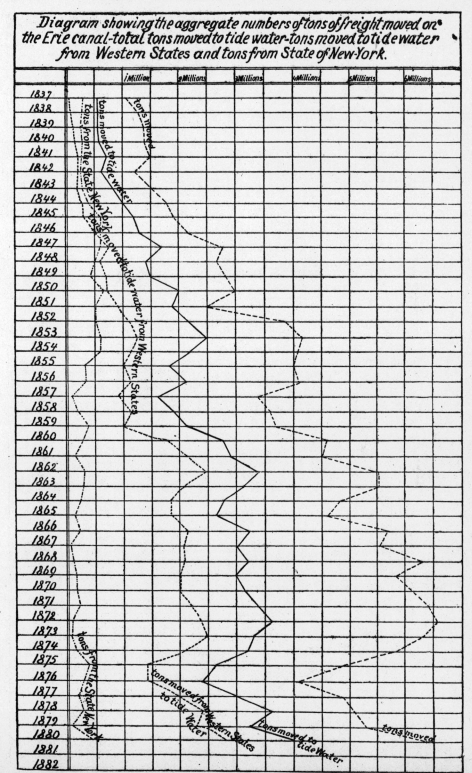

ALPHABETICAL AND ANALYTICAL INDEX

—TO—

THE HOME AND FARM MANUAL.

	PAGE.
Absorbing power of soils,	86
Absorption of oxygen by soil,	86
" by various substances,	996
Abuses in courts,	674
Acceptance of invitation,	946
Accurate knowledge, value of	142
Acid stains, to remove,	788
Activity of youth,	39
Adapting crops to localities,	286
" " soils,	286
Addressing a letter,	949
Admission or organization of the several States and Territories,	1023
Adorning the home,	40
Advantage of soiling,	172
After dinner, etiquette of	911
" " rules for	911
Agreement between landlord and tenant,	713
" general form of	976
" for building,	382
Agreements and contracts,	976
Agricultural alphabet,	151
" ideals, ten	83
" principles and practice,	75
" products, weight of	1005
" works of Mago,	48
Agriculture and the ancients,	44
" and science,	84
" defined,	47
" history of	47
" of Arizona Indians,	46
" of the Indians,	44
" progressive,	84
" mediæval and modern,	59
" the basis of wealth,	42
A la mode beef,	830
Alaska lands,	672
Albumen of meat,	818
Alcohol pickles,	848
Aleck and Sarah's ornamental work,	742
Aleck's quilting frame,	745
Alfalfa, soil and cultivation for	164
" or luzerne,	163
Aligning fence,	303
Allspice, varieties of	969
Alphabetical list, sick-room remedies,	771–774
Alphabet of agriculture,	151
" of etiquette,	957
Alsike or Swedish clover,	163
Alteration of grades,	347
Always take a receipt,	710
Amuse children,	762
Analysis of the soil unnecessary,	75

	PAGE.
Ancient and modern dress,	754
" crop yields,	58
" cultivation precise,	51
" farms and implements,	50
" farms small,	51
" manner of seeding,	58
" manure making,	56
" plowing superficial,	51
" plows and plowing,	57
" reaping machines,	55
" Roman implements,	51
" whims and vegetables,	753
" writers on draining,	325
Animal and vegetable life, contrasts,	997
" husbandry, definition of	47
Animals in open highway, liability for	661
" trespassing,	662
Antique crops,	54
Ants, to drive away,	784
Apothecaries' measure, liquid,	987
" weights, comparative scale,	988
Apparatus for hatching fish,	565–566
Apple and other fruit waters,	770
" cake, Virginia,	882
" custard pie,	865
" dumplings,	870
" jam,	895
" marmalade,	895
" pie,	863
" pudding, plain,	871
" souffle,	872
" tapioca pudding,	870
Apples, Canada reinette,	453
" Early Joe,	455
" Grimes golden,	456
" Higby sweet,	456
" Summer rose,	456
" Tetofsky,	454
" cultivation and varieties,	453–458
" distance apart of trees,	449
" pruning,	453
" some good ones described,	453–457
" valuable, illustrated,	450–457
Application for bounty lands,	975
" of manure,	56
Applications, various, for arrears of pay,	973
Apprentice, complaint of	972
Apprenticeship, indenture of	971
Arable lands, ancient,	52
Arbitration,	717
" form of submission,	717
Arbitrators, award of	717
Arboriculture, definition of	47

(1031)

	PAGE.
Arbor-vitæ hedge,	320
Architect, consultation with	383
Architecture, glossary of terms,	384–392
"　　rural,	361
Area of countries,	1019
"　timber for the farm,	570
Arizona Indians,	46
Arrangement of furniture,	735
"　　and care of tools,	427–431
"　　of privies,	411
Arrears of pay and bounty, application for	973
"　　"　various forms for.	973
Artichoke, cultivation of	177
"　Jerusalem,	177
Artificial cross fertilization,	98
"　fish breeding,	562
"　hatching of fish,	564
"　spawning of fish,	563
"　timber and lumber,	570
Art of grafting,	482
"　plowing,	63
Asking a favor, letter,	949
Assignment of mortgage,	715
A stitch in time,	756
Attorney, power of	714
Autumn roses,	516
Avoirdupois, scale of comparison,	988
"　　weight,	987
Award of arbitrators,	717
Bachelor's pudding,	874
Back furrows,	65
Bacon, how to know good,	810
Bad seasons, and rotation,	79
"　training and idleness,	39
Baked rice and apples,	769
"　eggs,	836
"　meats,	831
"　oysters,	833
Balance gate,	309
Ball etiquette,	923
"　invitation to	946
Balloon frame, barn,	441
"　frames, how to build,	439
"　frames strongest,	439–440
Bands for corn shocks,	116
Bank bills,	718
"　notes, engraving of	718
Barberry hedge,	321
Bark lice,	597–598
"　to destroy,	584
Barley and its products,	962
"　"　cultivation,	102
"　"　new varieties,	104
"　"　table of best soil,	103
"　"　harvesting,	103
"　"　seed per acre,	103
"　"　when sown,	103
Barriers, trees for	322–323
Bars and gates,	307
Barrel coop,	408
Bark summer house,	406
Barn basement,	417
"　fixtures,	414
"　farm, main floor,	373
"　for cattle,	372
"　for horses and cows,	416
"　horse, ground floor,	373

	PAGE.
Barn machinery,	415
"　with shed,	417
Barns, stables and corn cribs,	414
Basement of barn,	417
Basting meat,	829
Bathing infants,	762
"　proper time for	763
Beans, products of	963
Bean soup,	823
Beautifying the home,	725
Bed-bug,	610
Bed-bugs, to kill,	784
Bedding for the household,	726
Bed-rooms,	731
Beech, weeping,	551
Beef tea, to make,	767
"　a la mode,	830
"　boiled,	825
"　killing on the farm,	813
"　soup, clear,	821
"　stew,	827
Beefsteak stew with jelly,	827
Beer of sulphuric acid,	888
Beet sugar in the United States,	247
"　"　of Europe,	246
Beginning a dairy,	287
"　market garden,	494
Belts of wheat and corn,	90
Beneficial insects,	647–650
"　"　illustrated,	586–587
Bent grasses,	159
Best cotton States,	197
Biennial flowers, select,	513
Bill-bug, corn, preventives from	627
Bill of sale, personal property,	979
"　"　with warranty,	979
Bills of sale,	979
"　"　why given,	716
Bird friends,	652
Birds, classified by their food,	654
"　climbers,	654
"　deserving extermination,	657
"　fruit eaters,	653
"　injudicious destruction of	655
"　in economy of nature,	651–655
"　M. Provost on	657
"　natural enemies of insects,	655
"　of doubtful utility,	656
"　predacious,	653
"　the farmer should foster,	656
"　to be fostered,	656
"　versus insects,	579
"　what to kill,	652
"　which render best service,	657
Bird's-nest pudding,	872
Birch, weeping,	550
Biscuits, raised,	859
Bitters, home-made,	889
Black pepper,	968
Blackbirds,	654
Blackberries, care of	472
Black walnut coloring,	803
Blank indorsements,	706
Blanc mange,	769
Blankets, to cleanse,	787
Bleaching cotton goods,	789
"　linens,	789
Blight in pears,	457

ALPHABETICAL AND ANALYTICAL INDEX.

	PAGE.
Blind drains,	327
Blood-sucking cone-nose,	612
Board fence, how to build,	303
Bohemian cream,	874
Boiled beans,	843
" dinner,	826
" dishes,	825–826
" fowl,	825
" grape pudding,	871
" lamb,	826
" meats, sauce for	826
" and stewed dishes,	819
Boiling, famous cook on	819
" maple sap,	266
" sorghum juice,	264
" things to know in	819
" points of fluids,	995
" fish,	767
Bonds, why given,	716
Book-destroying insects,	786
Book farming,	44
Books, value of	44
Boston brown bread,	858
Boundaries of farm,	669
Bounty land, oath to identity,	975
" " various forms for	975
Brandied fruits,	899
Bread and bread-making,	856–860
" of fine flour,	856
" without yeast,	856
Breaded chops and cutlets,	767
Breakfast,	751
" and supper,	912
" biscuit,	859
" cake, oatmeal,	860
" puffs,	860
" relish,	843
Breaking hemp for market,	206
Brioche paste,	834
Brick, unburned, to make,	376
Bride at home, the	36
British agriculture, early	59
" farm tools, Norman,	59
Broiling fowl and game,	768
" and frying,	831
Brome or rescue grass,	150
Brown Betty,	868
" dye,	802
" sauce,	839
Brooms, care of	792
Broths, to make,	767
Brushes and combs,	760
" to clean,	760
Buckwheat, best soil,	108
" cultivation of	108
" seed per acre,	108
" time of sowing,	108
" time to harvest,	108
Budding and grafting,	481–489
" care after,	489
" materials necessary for	487
" spring,	488
" terminal,	488
" the art of	487
" time for	486
Buds, loosing strings,	489
" protection in winter,	489
" to prepare,	487

	PAGE.
Buffalo grass,	145
Bugs, general means for destroying,	615
Building happy homes,	33
Build according to means,	361
Building a log house,	282
" a wall,	377
" board fence,	303
" details, judgment in	370
" material,	376
" stone wall,	438
Building-bee.	283
uildings and fences, relative cost of	298
" should be substantial,	371
" specification for	377
Bulking tobacco,	239
Bulbs, summer flowering,	514
Bunching hemp,	206
Bushels in cribs by height and breadth,	999
Business, characters used in	721
" law, points on	719
" maxims,	719
" transactions, law forms on	703–721
Buttermilk bread,	859
Buy of responsible dealers,	704
Buying a farm, things to consider,	290–292
Cabbage soup,	822
Cake and cake-making,	875–882
" economical as food,	875
Cake-making, general rules for	875–876
Cakes, ornamenting,	876
Calf figured for cutting,	813
Calves and their best parts,	812
Calf's-foot jelly,	768
California game laws,	678
Calls, etiquette of	916
Calling on New Year,	918
Cancelling notes,	710
Candles, home-made,	790
Candied fruits,	891
Candy-maker's art,	890
Cannibal bugs,	611–614
Cane and other sugar compared,	246
" sugar,	248
" sugars of the world,	246
Canning fruits,	893
" fruits, table of boiling.	893
" fruits, table of sugar per quart,	893
" tomatoes,	894
" vegetables,	894
" whole fruit, peaches,	894
Capabilities of soil,	84
Capacity of soils for water,	348
Capers, true and false,	970
Cards, "At Home,"	952
" reception,	952
Cardinal rules of etiquette,	954
Careful farmer's barn,	275
Care after budding,	489
" and arrangement of tools,	427–431
" of blankets and furs,	794
" of brooms,	792
" of children,	762
" of combs and brushes,	760
" of the hair,	759
" of trees against insects,	583
" of hop-yard,	227
" of hops in crop years,	227

	PAGE.
Care of hops when dried,	230
" silk-worms,	217
Carnivorous bugs,	611–614
Carp breeding,	567
Carpet bedding, plants for	541
" rags, coloring green,	804
" " to color,	804
Carpets and bedding,	726
" laying them,	733
" to brighten,	796
" to clean,	794
Carriage house and farm stable,	368
" " stable,	414
Carrot, red altringham,	177
Carrots, Belgian,	175
Carving at table,	748
" by servants,	907
" fish,	751
" four-footed game,	749
" small birds,	750
" turkey,	749
Cassia, where produced,	968
Catalpa,	530
" sphinx, remedies for	637
Cato on agriculture,	49
Catsups, ",	851
Cattle barn,	372
" feeding-barn,	416
Cellar economy,	732
Cellaring and pitting roots,	176
Cereals described,	89
Certificate, mechanics',	979
Chalcis fly,	649
Challoner's level,	344
Chamomile tea,	770
Character of fruits,	465
Characters used in business,	721
Charlotte Russe,	873
Chattel mortgage, form of	978
" " with power of sale,	978
" security on lease,	294
Cheerfulness in sick-room,	764
Chemistry of bread-making,	817
" of the kitchen,	817
Cherry pudding,	871
" syrup,	888
" (cerasus),	528
Cherries, good sorts for the East,	467–472
" for the West,	468
" their cultivation,	467
Cherokee rose hedge,	321
Chicken croquettes,	842
" coops,	408
" enclosure,	409
" houses,	409
" pot-pie,	835
Childhood's democracy,	41
" sports,	41
Children, care of	762
" duty to	739
" treatment of	954
" trespassing while berrying, etc.,	662
Children's application for arrears of pay and bounty,	974
Children's party cake,	880
" " invitation to	946
Chili sauce,	852
Chinch bug, remedies for	607, 616

	PAGE.
Chinese plow, ancient	48
Chinese wistaria,	527
Chinking the house,	284
Chocolate,	887
" caramels,	890
" custard,	874
" preparation for cake,	877
" and cocoa,	965
" how made,	965
Choice parts of beef,	814
Chow-chow,	848
Chronicles of Columella,	50
Christian population of the earth,	1020
Christmas cake,	880
Christening, etiquette at	930
" how to conduct,	930–931
Church architecture,	403
" etiquette, rules for	927
Chutney, imitation,	849
Cicero and agriculture,	49
Cinnamon, how to know when pure,	968
" rolls,	861
Circular measure,	1007
Cisterns,	732
" contents of	1000
Claim of land, selection of	276
Clam soup,	824
Clarifying stock in soup,	828
Classification of insects,	588
Clay as a deodorizer,	775
" in soils, per cent of	85
Cleaning grain, ancient,	55
" house,	735, 794
" kitchen utensils,	816
" silver,	791
" trees of insects,	583
Cleanliness in the kitchen,	816
Cleansing the hair,	798
Clearing a timbered farm,	282
Cleft grafting,	482–484
Clematis,	526
Climate for cotton,	197
Climbing and trailing shrubs,	525–527
" shrubs, clematis,	526
" " honeysuckle,	527
" " ivy,	527
" " moon seed,	527
" " silk vine,	527
" " wistaria,	527
" birds,	654
Cloth measure,	985
" and fur moths,	785
Clothe according to means,	755
Clothes, care of	760
" conserve heat,	755
Clothes-pins, to preserve,	788
Clothing of children,	756
Clotted cream,	867
Clover-root borer, remedy for,	618
Clover-stem borer,	617
Clover-leaf beetle, "	619
" midge,	618
Clover as related to husbandry,	160
" Japan, on bush,	164
" Mexican,	165
" seed crop,	161
" valuable varieties of	161
" white	162

	PAGE.
Clovers as soiling crops,	170
" for the South,	163
Cloves,	969
Coal in sick-rooms,	782
Cockroaches, to kill	784
Cocoa and chocolate,	965
Cocoanut bread pudding,	874
" candy,	891
" pound-cake,	878
Codicil to will,	714
Coffee, a cup of	887
" bread,	860
" cake,	882
" grades of	966
" how to make,	887
Coin, counterfeit,	719
Cohesion of materials,	992
Cold and warm effects in color,	757
" baths,	763
" frame, planting in,	496
" meat sauce,	839
Cold-made soap,	790
Cole slaw,	842
Color and combination,	757
" to restore,	788
Colored washes for walls,	735
Coloring black,	802
" blue,	801
" carpet rags,	804
" " " blue or yellow,	805
" " " green or red,	805
" cotton,	801
" fabrics,	800
" green,	801
" scarlet and pink,	802
" with butternut bark,	803
" " walnut bark,	803
" yellow,	800
Columella and agriculture,	50
Comfort in the homestead,	275
Comforts of to-day,	42
Commerce of the earth,	1020
Commercial fertilizers—ancient,	56
Common sense of etiquette,	903
Compacting soil,	69
Comparison of proximate principles,	1013
Comparative tables of measures and weights,	989
" values of sorghum (tables),	256
Complaint of apprentice,	972
" of a master	972
Complete dairy barn,	415
Composition of soils,	76
Complying with favor asked,	949
Component parts of meat,	817
Compound fence,	306
Comstock's retinia,	639
Concerts, private,	925
Conclusions on sorghum sugar-making,	266
Condensed information, value of	780
Conditional indorsement,	706
Condiments,	851
Connecticut game laws,	676
Connecting laterals with mains,	349
Consent to indenture,	972
Consulting an architect,	383
Contents and dimensions of fields,	998
" of cisterns,	1000
" of granaries.	1000

	PAGE.
Contracting for the crop,	289
Contract for building,	381–382
Contracts and agreements,	976
Contrast between homes,	33
Contributions on household economy,	780
Convalescent cookery,	767
Convenient cottage, ground plan,	398
" " second floor,	398
Conversation, etiquette of	914
Conversazione, invitation to	946
" the	925
Conveyor's right in rights of way,	664
Cookies,	881
Cooking, general rules,	765
" mushrooms,	845
" for convalescents,	767–770
" for invalids,	765
" vegetables,	766
Cooling pork,	809
" and heating of soil,	995
Cooling-room for hops,	230
Copperas for cess-pools, etc.,	775
Corn and rice bread,	858
" and wheat belts,	90
" beef, boiled,	825
" bill-bug, preventives for	627
" blossom,	134
" borer, smaller	625
" " " preventives for,	625
" bread, steamed,	859
" crios,	414–422
" " contents of	999
" " with driveway	424
" crop, cost of	117
" " of the United States,	109
" cutting and shocking,	116
" depth to cultivate,	114
" dodger,	859
" early cultivation,	112
" for soiling,	169
" germinating temperature,	117
" houses,	422
" how often to cultivate,	114
" manures for	110
" planters,	72
" planting,	112
" pone,	858
" prize crops,	110
" production—table,	90
" pudding,	872
" results of bad cultivation,	109
" shocking around tables,	116
" soup,	822
" to increase the average,	109
" to prepare the soil,	111
" varieties of	119
" worm, preventives for,	628
" yellow dent,	111–119
Corn-horse for shocking,	116
Corn-shock binder,	117
Cost and profit of drainage,	332
" of growing hops,	223
" of farm fences,	299
" of fence per rod,	299
Costumes at dinner.	905
Cottage, farm or suburban,	365
" for farm hand,	394
" of pioneer, cost of	272

Cottages and farm houses,	393
Cotton a child of the sun,	197
" antiquity of	194
" burying the stalks,	199
" climate, the	197
" crop by States—table,	196
" crops, table of	196
" cultivation, improved implements for	199
" " of	198
" destroying insects on	201, 628–631
" family, the	194
" first cultivation,	200
" history and cultivation,	194
" " of, United States,	195
" importance of	196
" long staple,	195
" plowing and fertilizing,	200
" preparation of soil for	199
" second cultivation,	200
" seed, products of	966
" soils,	195
" species of	194
" States, best	197
" uplands,	195
Cotton-boll and corn-worm, natural enemies of	628
" worm,	628
" " preventives for,	628
Cotton-plant louse,	596
Cotton-worm,	629–634
" killing, illustrated,	201
" poisons for.	631
" remedies for	629
" spraying poisons on,	630
Country home, comfortable,	364
" " ideal,	35
Counterfeit bills, to detect,	718
Court plaster,	798
Covering wheat proper depth,	94
Cow's heel jelly,	769
Cow-pea for forage,	170
Crab-apple jelly,	897
" (pyrus),	530
Cranberry bog, leveling,	477
" bogs, dykes for	477
" " sanding,	477
" cultivation of	477–480
" pie,	863
" situations for	477
Cranberries, curing and packing,	480
Cream, artificial for coffee,	887
" beer,	889
" bread,	857
" cake, French,	878
" clotted,	867
" pie,	864
Cream of tartar beer,	888
Cresinus' defence,	51
Crib of poles,	422
Cribs, to find contents, dent corn,	999
" " " flint corn,	999
Crop, contracting for	289
" second year's,	279
Crops and rotation,	78
" diversity of	279
" export of	105
" Great Britain and United States,	1010
" indicated by soil,	286
" of grass seed,	80
Crops, pulled by hand,	55
" Roman,	53
" to localities, adaptation of	287
" to raise,	277
" when to hold,	290
" when to sell,	289
Cross fertilization, artificial,	98
" " diagram,	99
Croquettes,	843
Crows,	653
Crust for tarts,	865
" with melted fat,	835
Cubic feet per ton,	992
" or solid measure,	986
Cucumbers, to pickle,	847
Cultivating garden, economy in	503
" orchards against insects,	582
" the hedge,	318
" tobacco, South,	232
Cultivatable fishes,	559
Cultivation and care of tobacco,	238
" of barley,	102
" of cherries,	467
" of cotton,	198
" of flowers,	508
" of hemp,	204
" of hop-yard,	227
" of jute,	208
" of oats,	105
" of peanuts,	242
" of ramie plants,	209
" of rice,	125
" of sorghum,	249–252
" of sugar-cane,	249
" of the vine,	474
" " in the moon " theory,	505
Cultivator, one-horse,	71
" straddle row,	70
" walking,	71
Curing and packing cranberries,	480
" tobacco,	238
Curious dishes,	752
Currant and moss jelly,	769
Curtains for windows,	728
Currant jelly without cooking,	897
Currants, care of	472
Curves, laying out,	539
" of walks,	540
Custard cake,	882
Custards and creams,	874
Cutting and handling sorghum-cane,	250
" and saving scions,	485
" and shocking corn,	116
" hemp,	205
" scions, time for	489
" tobacco,	238
" up a hog,	810
" up a lamb,	812
" up an ox,	814
" up a sheep,	811
Cuttings, propagation by	490
Cutlets, breaded,	767
Cypress, Lawson's,	547
Dactylis (orchard grass),	159
Daily savings, growth of	1017
Dairy barn, complete,	415
" districts and soiling,	167

ALPHABETICAL AND ANALYTICAL INDEX.

	PAGE.
Dairy fixtures,	289
" house,	287
" to start,	287
Dakota public lands,	672
Damper in stove,	781
Dancing, rules for	925
Daniel Webster's plow,	62
Danvers carrot,	176
Dating and beginning a letter,	950
Deadening timber,	285
Dealers, liability of	667
Deciduous hedges, care of	322
' trees for landscape effect,	547
Decoloring sorghum sugar,	265
Decorating dining-room,	746
" the parlor,	740
" the table,	746
Decorum in children,	739
Deeds, erasures in	716
' interlining,	716
" warranty,	715
Defences defeating payment,	709
Definition of mercantile terms,	720–721
Delaware game laws,	677
Delicate cake,	879
" pudding,	873
Democracy of childhood,	41
Dent corn,	119
Dentzia,	519
Deoderizers,	775
Deportment in the street,	919
Depth for germinating of grass seed,	152
" to cultivate corn,	112
Description of cereals,	89
" of nectarines,	467
Desert land act,	672
Design for village lot,	537
Destroying cotton insects,	201
Details, must be watched,	274
Detecting counterfeit money,	718
Devonshire cream	867
Device for moving trees,	540
Diagrams of balloon frames,	439–440
" of flower beds,	540–541
Diagram, average tons moved, Erie canal, etc., series of years,	1029
Diagram, miles of railways built and operated, series of years,	1028
Diagram of average wages, groups of States,	1027
" production of corn, series of years,	1024
" " wheat, series of years,	1025
" relating to cotton, series of years,	1026
" cross fertilization,	99
Difficulties in pear culture,	457
Digestibility of some animal foods,	766
Dinner,	752
" costumes, rules for	905
" givers, hints to	911
" how many to invite,	904
" how to serve,	907
" order of dishes at	907
" useful hints for	908
" family,	908
Dinner-table, well-arranged,	747
Dining-room decoration,	728
" and dining-service,	746
" furniture,	728, 746
Directing youthful sports,	42

	PAGE
Directing the talents of youth,	38
Disinfection,	775
Disinfectants of high authority,	778
" how to use,	778
" table of relative values,	777
Disinfecting clothing,	775
" the sick-room,	775, 778–779
District school-house,	403
Dishes, order of, at dinner,	907
Ditch fence,	306
Diversity of crops,	279
Division fences, how built,	663
" of insects by food,	593
Divisions of agriculture,	47
" of horticulture,	47
Dog laws of the several States,	679–682
Dogwood,	517
Doughnuts, Yankee, raised,	882
Doctoring bruised furniture,	795
Doctor's bills, to save,	761
Dolly and Tom's household ornaments,	742
Doses and their graduation,	775
Double-braced gate,	312
Drab and green coloring for carpet rags,	804
Drafts explained,	710
Drainage, ancient writers on	325
" and fences,	337
" by French monks,	327
" cost and profit of	332
" defined,	326
" experience in Illinois,	332
" fixing gradients,	342
" from the house,	374
' furrows,	65
" how to know when necessary,	356
" importance of, to stockmen,	357
" indestructible,	328
" John Johnston's testimony,	329
" lands requiring,	355
" laying out the work,	341
" map and pond,	336
" of the Greeks,	325
" right and wrong way,	333
" tools,	342, 345, 350
" various means of	338
" Versailles experiments,	358
Drainer's level,	342
" spades,	345
Draining a whole farm,	354
" a field,	350
" and drainer's art,	324
" antiquity of	324
" into ponds,	335
" in Indiana,	331
" in Michigan,	331
" orchard soils,	450
" S. F. Woolley's experience,	330
" when it pays,	351
Drains, form of	341
" leveling the bottom,	344
" open,	327
" protecting the banks,	343
" slab and pole,	340
" stock water from	334
" stone laid,	339
" surface leveling,	346
" grades, alteration of	347
" and springs,	328

	PAGE.
Drawing-room, the	737
Dress, ancient and modern,	754
" children comfortably,	757
" conserves heat,	755
" its true purpose,	754
" true taste in	754
Dressing hemp fiber,	206
Dried beef, stewed,	845
" peach pudding,	871
Drilling wheat, advantages of	94
Drills for seeding,	72
Drinks, refreshing,	888
Drives and walks, forming curves,	539
" keeping in order,	540
Driving etiquette,	922
" in plowing,	64
" out flies,	785
" three abreast,	278
Drop cakes,	879
" Johnnies,	882
Dry-house and hot-bed,	410
Drying currants,	898
" hops,	228
" plums,	898
" tobacco,	239
Dry-kiln for hops,	229
Dry measure, comparative scale,	986
" soil grasses,	142
" surface may need draining,	330
Duchesse potatoes,	844
Duck enclosure,	409
Due-bills, forms of	708
Dumplings,	869–870
" for soup,	834
Duration of bath,	763
" of life, mean,	1017
Dusting furniture,	736
Duties of parents morally,	39
Dyeing brown with catechu,	802
Dyer's art, the	800
Dyes for dresses,	800
Dyking cranberry bogs,	477
Eagle,	653
Early rhubarb, forcing,	500
Earth closet,	412, 775
" silos,	180
Earth's Christian population,	1020
" population, classification of	1020
" specific gravity and weight,	992
Eastern rotation,	80
Eating, rules for	903, 908–910
Economy dish,	842
" in cultivation,	503
" in coal,	782
" in fuel,	781
" in the kitchen,	781
" of fertilization,	77
" of the garden,	491
Educate to a purpose,	37
Eel soup,	824
Effervescent drink,	771
Egg sauce,	836
Eggs, dishes of	836
Elaborate rotation,	79
Elderberry wine,	798
Elruge nectarines,	467
English gothic cottage,	400

	PAGE.
English gothic cottage, ground plan,	401
" and metric equivalents,	986
" long measure,	985
" standards of measure,	983
Ensilage and silos,	178
" condensed facts on	186–191
" cost of, detailed account,	192
" crops,	187
" definition of	178
" effects on dairy products,	189
" facts in feeding,	189–190
" father of	180
" feeding, practical conclusions,	193
" value of	182
" history of	178–182
" in the United States,	183
" " West in 1870,	178
" long known in Europe,	178
" materials for	181
" plants, yield of	183
" pressure of	187
" rations,	190
" " for cows,	185
" should not ferment,	181
" value of	182
Entertaining visitors,	738
Enticing laborers from work,	668
Entomology on the farm,	579
" practical value of	579
Epax apicaulis,	649
Equivalents, the housewife's,	883
Erecting rough buildings,	435
Establishing a hop-yard,	224
Estray laws,	683
Etiquette of introductions,	920
" of shopping,	937
" of travel, for ladies,	935
" of the road,	934
" of visiting,	928
" philosophy of	903
" special rules of	938
" of the parlor,	737
" of the table,	747, 903–912
" alphabet of	957
" and success in life,	903
" at balls and parties,	923
" at church,	927
" at funerals,	929
" at luncheon,	913
" in dining,	904
" in Washington,	936
" of calls,	916
" of courses at dinner,	907
" of dress,	914
European rotation,	82
Evaporation from various substances,	996
Everlasting flowers,	533
Evening calls,	916
" parties,	925
" party, invitation to	945
Every-day pies,	864–865
" " pudding,	869
Eve's pudding,	873
Example before children,	954
Exercise of children,	764
" of taste at home,	40
Expelling flies from rooms,	785
Experimental patches of grass,	135

ALPHABETICAL AND ANALYTICAL INDEX.

	PAGE.
Experiments, feeding ensilage,	193
" with ensilage,	190–191
Explanation of drafts,	710
Exports of food crops,	105
Face, to remove spots, etc., on	797
False hair, to curl,	797
Fall and spring plowing,	66
" plowing for root crops,	174
Fallow crops, Roman,	54
Family government,	35
" dinners,	908
Farm and forest area, United States,	1012
" a garden	296
" barn, main floor,	373
" buildings, grouping,	414–415
" cottage,	365
" " ground plan,	367
" " second floor,	367
" fences, cost of	299
" gate, southern,	312
" house toilet,	758
" " Italian style,	398
" " simple plans,	364
" houses and cottages,	393
" " substantial,	361
" how to buy,	290
" ice-house.	411
" garden and orchard,	445
" lease,	712
" leasing,	292–294
" map of	296
" commencing a	277
" orchard and garden,	445
" paint shop,	434
" park, view of	368
" situation of, important,	291
" to lay out,	295
" to select,	290
" stable and carriage-house,	368
" workshop,	426
" what is it?	668
" implements, modern,	61
" " kept in order,	429
" sons and daughters on	37
Farmer and gardener,	492
" unthrift,	275
Farmer's animals, law governing,	661
" wife, the	36
Farming, attention to details,	273
" fancy, ancient,	51
" pioneer,	271
Fat and fiber of meat,	818
Feeding barn for cattle,	416
" boxes,	418
" green fodder,	169
" rack for sheep,	420
" silk-worms,	215
" shelves for silk-worms,	216
" value of ensilage,	182
Feet, square, and square feet in areas,	999
Fence, aligning,	303
" capped and battened,	304
" compound,	306
" finishing,	303
" how to build,	300
" laws in general,	696
" " in the South,	697–698

	PAGE.
Fence laws Middle States,	697
" " of the several States,	695–702
" " New England States,	696–697
" " Pacific Slope,	702
" " Western States,	699–702
" per rod, cost of,	299
" portable,	304
" post-and-rail,	301
" setting the posts,	303
" sod and ditch,	306
" straight with stakes,	301
" wire,	305
Fences and buildings, relative cost of	298
" cost of	299
Fencing hillsides,	307
" vs. soiling,	168
Fermentation in ensilage,	181
" in the silo, effects of	184
Fertility, practical test of	77
Fertilizers, economy of	77
Fertilizing cotton soils,	200
Field birds, to carve,	750
" cultivation of sweet potatoes,	244
" drainage illustrated,	335
" drained,	351
" husbandry, definition of	47
Fields, dimensions and contents of,	998
Fig jelly,	897
" pudding,	871
Figs of tomatoes,	899
Filing implements,	432
Filtering and liming sorghum juice,	264
Fine flour bread,	856
Finch-family,	654
Finished sugar, tests for	268
First furrows,	64
" pastures,	280
" steel plow,	61
Fish and fowl sauce,	838
" " game laws,	675–679
" " oyster pies,	835
" breeding, artificial,	562
" chowder,	825
" culture, mistakes in	558
" hatching, artificial,	564
" " boxes,	565
" of New England, spawning,	561
" of the West, spawning,	560
" on the farm,	557
" pond, to form,	568
" soup,	823
" spawning,	563
" transportation, rules for	562
" time required to cook,	767
" to carve,	751
Fishes for cultivation,	559
Fixtures for the dairy,	289
Flat furrows, turning,	279
Flavor gives name to a sauce,	840
Flavoring, leaves for	846
Flax and its cultivation,	201
" and hemp, conclusions on	207
" fiber,	203
" harvesting of	204
" soil, preparation of	203
" soils,	203
" straw, dew and water rotting,	204
" threshing,	204

	PAGE.
Flaxseed,	201
" quantity to sow,	203
" selection of	203
Flies, to destroy,	784
" to drive out,	785
Flint corn,	121
Floating island,	873
Floriculture,	47
Flour, its manufacture,	961
" original meaning of	961
Flower beds, diagrams of	540–541
" " laying out,	541
" " planting,	541
" garden,	507
" " Lincoln Park, Chicago,	542
Flowers, everlasting,	533
" not transplantable,	509
" perennial, select list of	513
" select list of	510–511
" succession of	509
" to cultivate,	508
Flowering of grasses illustrated,	136–141
" plants and vines,	514
" shrubs,	517
" trees,	528
Flood-gates,	314
Flooding the rice crop,	126, 128
Fluids, freezing points of	995
" boiling points of	995
Food crops, export of	105
" of silk-worm,	221–222
Foods and digestion,	765
" by analysis and experiment,	1014
" for convalescents,	767
" for invalids,	765
" nutritive value, for animals,	1014
Forage of cow-peas,	170
" root crops for	173
Force in using tools,	994
Foreign exchange, table of	990
Forcing plants, cold frame,	495
" " in spring,	497
" " troughs for	494
" rhubarb,	500
Forest, the poetry of	574
" and farm areas, United States,	1011
" Europe and United States compared,	1011
Foretelling the weather,	1010
Foreclosure of mortgage,	715
Formal dinners, wines at	910
Form of award of arbitrators,	717
" of bond,	716
" of check,	705
" of will,	713
Forms for bounty lands,	974
" of bills of sale,	979
" of fruits explained,	463–465
" of indorsements,	706
" of notes,	707
" for remitting funds,	711
Fowl pot-pie,	835
Fowls trespassing,	663
Frame, cold,	495
Fraud in selling a farm,	669
Fraudulent money,	718
Freckles, to remove,	797
Frederick the Great and sparrows,	657
Free farm, how to get,	669

	PAGE.
Freezing point of fluids,	995
French, American and English money compared,	989
French polish,	796
" phrases and their meaning,	952–953
" stew,	845
Fried bread,	843
" cakes,	882
" eggs,	836
" oysters,	833
" potatoes with eggs,	844
Friendly acceptances,	947
" regrets,	947
Frosting, recipes for	876
Frozen custard,	874
Fruit beers,	889
" cake, dark,	877
" eating birds,	653
" growing,	47
" protection of from insects,	584
" sugar,	248
" waters,	770
Fruits, character of	465
" forms explained,	463–465
" in home orchard,	445
" how long to boil,	893
" how to can,	893
" how preserved,	892–895
" size of	465
" recipe for brandying,	899
" to dry,	898
Frying in boiling fat,	833
Fuel in the kitchen,	815
Fuller's rose beetle, remedy for	644
Fumigating clothing and rooms,	779
Funeral, etiquette of	929
Furnishing the house,	726
Furniture arrangement of	735
" bruises, to take out,	795
" of dining-room,	746
" of the kitchen,	815–816
" to clean,	796
Furrows, back,	65
" for draining,	65
Furs, to preserve,	786
Fusing temperatures of metals,	994
Gama grass,	145
Game laws in old States,	678
" in new States,	678
Gang plow, stubble,	64
Garden and orchard,	445–448
" cultivation of sweet potatoes,	245
" lessons,	42
" farm,	296
" pot herbs,	502
" small fruits in	502
" structures,	405–407
Gardening by farmers,	492
" green-houses and	500
" in spring,	501
" pricking out plants,	499
" starting the hot-bed,	498
" tact,	502
" the moon theory,	505
" transplanting in hot-bed,	499
Garments, mending,	755
" next the skin,	759

ALPHABETICAL AND ANALYTICAL INDEX.

	PAGE.
Gate, balance,	309
" double-braced,	312
" flood and water,	314
" rollers for	311
" self-closing,	310
" slide and swing,	308
" southern,	312
" strap-hinged,	311
" stream, with footway,	314
Gates and bars,	307
" ornamental,	313
" sagging, to prevent,	313
Gathering pears,	457
" root crops,	176
Gelatine of meat,	818
Gems, fritters, etc.	854
" graham,	860
Genera, species and varieties of grass,	155
Gentle firmness with children,	762
Genus,	158
German puff paste,	862
Germinating temperatures.	997
Giblet pie with oysters,	835
Gingerbread and other "homely" cakes,	881
Ginger cake.	881
" cookies,	881
" how to know,	969
" snaps.	881
" species producing it,	969
Glass in canning, caution,	894
Glazing for cakes,	876
Glossary of architectural terms,	384–392
Glue, recipes for	799
Glycerine ointment,	798
Going to bed,	738
" table,	906
Gold cake,	880
" chains, to clean,	797
Golden bell (*Forsythia*),	519
" rules in etiquette,	914
Gooseberries,	472
Gooseberry jam,	896
Gothic cottage, English,	400
Goubers or peanuts,	241
Grade, altering,	347
Gradients of drains,	342
Grading the ditch,	343
" the necessary fall,	348
" the proper slope,	343
Grafts and cuttings harbor insect eggs,	580
Grafter's art,	482
Grafting and budding,	481–490
" by approach,	483
" directions for	482
" old orchards,	484
" on the farm,	482
" the grape,	489
" tools,	483
" wax,	482, 485
" " to make,	485
Graham bread,	857
Grain drills,	72
" general conclusions,	98
" pedigree	97
" plant-louse,	596
" shocking,	96
Granaries,	422
" contents of	1000

	PAGE.
Granary, rat-proof,	422
Granivorous birds,	655
Grandmother's strawberry short-cake,	865
Grape, grafting,	489
" jelly,	897
" phyloxera,	641
" sugar,	248
Grapes, American species of	473
" for farmers,	474
" preserved with honey,	899
Grape-vine borers,	641
" flea-beetle,	642
Grass blossoms,	134
" family,	158
" genera, species and varieties of	155
" how to know it,	133
" is king,	132
" on timbered farms,	285
" scientific classification,	133
" seeds, crops of	80
" value of	132
Grasses, bent,	159
" disappearance of native,	146
" experimental patches of	135
" favorite pasture,	158
" flowering, illustrated,	136–141
" for arable loams,	153
" for dry soils,	142, 154
" for hay and pasture,	143
" for meadow soils,	152
" for moist soils,	142
" for overflowed lands,	152
" for the South,	146–150
" for various regions,	160
" for wet, undrained soils,	153
" illustrated,	134–160
" list of good	143
" special for Alabama,	160
" " California,	160
" " Florida,	160
" " Georgia,	160
" " Idaho and Montana,	160
" " Louisiana,	160
" " Mississippi,	160
" " Texas,	160
" " Washington Territory,	160
" " Oregon,	160
" value of interchange of	166
" Woburn tables	156–157
Grasshopper or locust,	634
Grass-worm, remedies for	626
Gray tree-bug,	602
Grease-spots on wood,	784
" " taken from carpets,	796
Grease, to remove,	788
Grecian agriculture,	48
Grecian drainage,	325
Greek implements, ancient,	48
Greenback paper,	718
Green coloring for cotton,	801
Green fodder, to feed,	169
Green tree-bug,	603
Griddle-cake, flour,	882
Grinding soils, implement for	69
Ground beetles,	649
" plan of barn and sheds,	417
" " cottage,	367
" " of crib and granary,	424

	PAGE.
Ground plan of English cottage,	401
" " of granary,	424
" " of hog barn,	421
" " sheep barn and yards,	420
" floor of horse barn,	373
Grouping farm buildings,	414–415
Grove, how to plant,	280
Groves and wind-breaks,	280
Growth of timber,	571
Gruel of groats,	769
Gruels, to make,	769
Guard against swindlers,	703
Guest rooms,	737
Guests, behavior of	738
" how to receive,	905
" of the family,	737
Guinea grass,	149
Hair stimulant,	797
" wash,	798
Ham, how to carve,	748
" how to bake,	831
Hanging pork,	809–810
Hard and soft ground crops,	83
" " soils,	286
Hardening tallow,	791
" plants,	495
Hardiness of the pear,	457
Harlequin cabbage bug,	602
Harrow, double,	68
" rotary,	68
Harrowing corn,	112
Harvesting, ancient,	55
" barley,	103
" flax,	204
" hemp,	205
" jute,	208
" machinery,	73
" peanuts,	242
" potatoes,	505
" rice,	129
" root crops,	175
" wheat,	96
Hatching fish,	564
" silk-worm eggs,	215
Hawks,	653
Hay and pasture grasses,	143
" rack, to make,	436
Heat and its effects,	994
" and ripening of plants,	997
" table of degrees of	994
Heating the oven,	857
Heavy stones, to move,	437
" wooded pine,	547
Hedge and ditch fence,	306
" arbor vitæ,	320
" barberry,	321
" Cherokee rose,	321
" cultivation of	318
" deciduous, care of	322
" hemlock,	320
" Japan quince,	321
" laying down,	318
" locust,	317
" **Norway spruce**,	319
" Osage orange,	319
" plants,	317–321
" to plant,	317

	PAGE.
Hedge privet,	321
" row, to prepare,	317
" trimming of	318
" white thorn,	321
Hedges, ornamental,	319
" poetry of	316
" use of	316
Heeling in trees,	451
Helps in washing,	787
Hemlock hedge,	320
Hemp and flax, conclusions on	207
" and its cultivation,	204
" for its lint,	205
" harvesting of	205
" male and female plants of	204
" preparing for market,	206
" sexes of plants,	205
Hen's-nest pudding,	870
Hickory-bark coloring,	803
Hickory-nut cake,	881
Hiding the seed corn,	45
Hidden drains,	327
Hillside cottage,	372
" fence,	307
Hints for dinner-givers,	911
" for dinner use.	908
Hired help—specific wages,	668
Hitching three abreast,	278
Hog barn with wings,	421
Hog figured for cutting,	810
Home and its adornment,	40
" and its charms,	33
" and the children,	35
" and the husband,	34
" and the wife,	34
" comforts to-day,	42
" made beautiful,	725
" of the pioneer	38
" orchard, arrangement of	445–448
Home-like cottage,	399
Homely pie,	864
Homestead, comfort in,	271
" improving the	361
" its improvement,	41
" how to acquire,	670
" law,	670
" in after life,	39
Hominy pudding,	869
Honeysuckle (*lonicera*),	527
Hop crop, preparing for	224
" tea,	770
" yeast,	855
Hops after drying,	230
" care in crop years,	227
" cooling room for,	230
" cost of raising,	223
" drying by hot air,	229
" kiln-drying,	228
" male and female plants,	226
" new plantation, cultivation of	226
" picking,	227
" planting the sets,	226
" proper situation and soil,	226
" the dry-kiln,	229
" trenching for	225
Hop-growing in America,	223
Hop-kiln, management of	229
Hop-yard,	224

ALPHABETICAL AND ANALYTICAL INDEX.

	PAGE.
Hop-yard, care of	227
Horse, agreement for sale of	976
" and cow barn,	416
" barn, ground floor of	373
" chestnuts (æsculus),	528
" hoe,	71
Horse-radish vinegar,	852
Horses, strength of	1003
Horticulture, divisions of	47
Hot-bed, materials for	497
" pressure of	498
" pricking out plants,	499
" soil for	498
" to make,	498
" water and ventilation,	500
" winter forcing,	499
Hot-beds and dry-house,	408
Hotch-potch,	828
House cisterns,	732
" cleaning,	735, 794
" drainage,	374
" in Italian style,	399
" how to build,	369
" ventilation,	374
Household art and taste,	725
" hints,	795
" recipes,	780
Housewife's equivalents,	882
Human force applied,	994
Hunters trespassing,	662
Hygiene of bedding,	726
Husbandry and clover,	160
" animal,	47
" field,	47
Husking corn,	115
" " from the hill,	115
Hybrids explained,	158
Ice cream,	890
" and water ices,	889
" icing for cake,	877
Ice-house and preservatory,	373
" on the farm,	410
Iceland moss jelly,	769
Idleness from bad training,	39
Ignorance vs. intelligence,	84
Icing, glazing and ornamenting,	876
Illinois experience in drainage,	332
Illustrations of wheat,	93
Imitation chutney, best,	849
" Worcestershire sauce,	851
Immigration into the United States,	1020
Implements for draining,	342–345, 350
" of cultivation,	70
Important business rules,	704
" papers, to draw,	712
" points in root culture,	174
" on sorghum,	254
Improved land, United States,	1011
" implements for cotton crop,	199
" implements of to-day,	61–73
Improving the farm,	272, 285
" the homestead,	361
" the homestead,	41
" timbered farm,	285
Impurities in water, tests for	777
Indelible ink,	799
" " for brushes,	799

	PAGE.
Indenture of apprenticeship,	971
Indian corn, ancient and modern,	45
" " (see corn).	
" " corn, products of	962
" method of saving corn,	46
" pickles,	847
Indorsements, form of	706
Infants and their food,	762
Informal dinners, rules for	905
Ingenious and useful ornaments,	744
Injury to stock through trespass,	663
Ink on bank bills,	718
" not injured by freezing,	799
" spots on marble,	784
" for zinc labels,	799
Inks, home-made,	798
Insect nomenclature,	590–592
" " abdomen,	592
" " head,	590
" " legs,	592
" " thorax,	591
" " wings,	591
Insect borers, in soils,	348
" transformations,	590
" variation,	591
Insectivorous birds,	654
Insects and the nursery, second year,	581
" changing food with age,	593
" classification of	588
" destroying books,	786
" " grass,	617
" division of, by food,	593
" eggs on grafts and cuttings,	580
" eyes of	589
" hand-picking for	581
" herbivarous,	593
" injuring clover,	617–620
" " coniferous trees,	639–641
" " fruit trees,	642–645
" " plants,	645–647
" " the grape,	641–642
" " trees,	637–644
" in the soil,	580
" injurious,	594
" muscles of	589
" natural economy of	650
" nervous system of	589
" noxious,	594
" study of	585–588
" vs. birds,	579
Irregular areas, to plow	73
Interest at any rate per cent, to find,	1016
" money, growth of	1017
" table, six per cent,	1016
Intoxicating varieties of pepper,	968
Introduction, letters of	948
Introductions, etiquette of	920–922
Invalid cooking, rules for	765
" drinks,	770
Invitation to dinner, form for	945
" for a drive,	947
Invitations, friendly,	947
Irish stew,	826
Iron rust, to remove,	790
Italian house, with tower,	398
Ivory-nut plant,	555
Ivory, to whiten,	797
Ivy (Hedera),	527

	PAGE.
Jam of fruits,	895–896
Japan clover (*Lespedeza*),	164
" quince hedge,	321
" " (*Cydonia*),	518
Japanese silk-worm eggs best,	214
Jellies, rules for	897
Jelly-making, rules for	896–897
Jerusalem artichoke,	177
Judgment notes,	707
" in building details,	370
June roses,	517
Juniper, web-worm,	641
Jute, cultivation of	207–208
" harvesting,	208
" preparing the fiber,	208
" soil and climate for	208
" species of	207
Kalsomine, to prepare,	793
Kalsomining,	734, 793
Kentuckian's bill of fare,	761
Kentucky corn-bread,	858
Kerosene, care of	786
" test for	786
Killing beef on the farm,	813
" silk-worms in cocoons,	219
Kitchen, arrangement of	815
" chemistry,	817
" floor,	815
" furniture,	731
" gardening,	499
" utensils, list of	816
" walls and ceiling,	815
Klippart's twelve propositions on draining,	357
Lady fingers,	879
" birds,	647
" " California varieties of	647
Ladies traveling, rules for the escort,	935
" dress,	757
Lamb, age for killing,	812
" to cut up,	812
Land acquired under all acts,	671
" areas and contents,	998
" bugs,	604
" game birds,	654
" in United States,	1011
Landlord's agreement,	713
" certificate of lease,	294
Lands, how described,	669
" in each State and Territory,	1012
" public, how obtained,	699
" requiring drainage,	355
" yet open to settlement,	672
Landscape, deciduous trees,	547–555
" " " ash,	549
" " " beech,	550
" " " birch,	550
" " " elm,	555
" " " linden,	553
" " " maple,	548
" " " persimmon,	532, 556
" " " sycamore,	548
" " " yellowwood,	554
" gardening,	40, 535–556
" " in villages,	536
" gardener's art,	535
" effects,	545

	PAGE.
Landscape effect, Lincoln Park,	543
" planting trees for	546
" trees, evergreen,	546–547
" " maples,	548
Lard candles,	791
Lark,	654
Laterals connected with mains,	349
Law forms,	971
" " for business matters,	703–721
" governing farm animals,	661
" relating to specific work,	668
Laws farmers should study,	674
" relating to dogs,	679–685
" " to stock and estrays.	683
Layering the vine,	490
Laying down carpets,	733
" out a farm,	295
" out an orchard,	449
" out curves of walks and drives,	539
" out flower beds,	541
Lead in water, test for	777
Leaf-hoppers,	623
" remedies for	623
Leaf-rollers, remedies for	643
Lease, short form,	712
" and certificates, forms for	293
" chattel security,	294
Leasing a farm,	292
Leaves for flavoring,	846
Lebia grandis,	649–650
Lemonade, to make	888
Lemon pie without lemon,	864
" " good,	864
Lemons, syrup of	899
Letter writers, suggestions for	950
Letters, addressing,	949
" copies should be kept,	704
" dating,	950
" of introduction,	738, 948
" of recommendation,	948
Lettuce salad,	841
Leveling cranberry bogs,	477
" drains from the surface,	346
" the bottom of drains,	344
Liability from act on own land injuring others,	667
" of dealers,	667
Library, the	737
Lice, plant- (*aphides*)	595–597
Licensed dogs,	662
Life government, Washington's rules,	940
" mean duration of	1017
Light, plain paste.	835
Lime and sulphur for insects,	581
" tree winter moth, remedies for	643
Lincoln Park, Chicago, flower garden,	542
Linden or lime,	553
Linseed tea,	770
Liquid or wine measure,	987
Little-lined plant-bugs,	609
Live stock, Great Britain and United States,	1010
Loam and dry soil grasses.	152
Local markets for vegetables,	494
Locust hedge,	317
Locust, American,	634
" destroying the eggs,	635
" killing young.	635
" map of infested districts,	632–633
" not *cicada*,	634

	PAGE.
Locust or grasshopper, remedies for	634
" or hateful grasshopper,	634
" the Riley gatherer,	635
" trapping,	635
Log-house, chinking,	284
" carrying up the sides,	283
" laying the foundation,	282
" the fireplace,	284
" the roof,	283
Looking glasses, to clean,	795
Longitude and time compared,	1007
Long measure, surveyor's,	985
" staple cotton,	195
Lucerne (*Alfalfa*),	163
Luncheon, how to serve,	913
" invitation for	913
Lyme grasses,	146
Macaroni and cheese,	844
Mace and nutmegs,	969
Magnolia,	528
Mago on working cattle,	49
Maine game laws,	675
Mammoth red clover,	161
Management of hop-kiln,	229
" of rice fields,	126
Mangel-wurzel,	175
Manured tobacco, south,	232
Manures, ancient commercial,	56
" preparation of	56
Manuring by the ancients,	55
" by sheep, south,	81
Maize and its products,	962
" (see corn.)	
Map of the farm,	296
Maple sap, how long to boil,	268
" sugar,	266
" " boiling sap,	266
" " buckets for sap,	266
" " gathering tubs,	267
" " storing tubs,	267
" " sugaring off,	268
" " tapping the trees,	267
" " when done,	268
Marble, to clean,	783
" cake,	879
Marketing silk-worm cocoons or eggs,	220–221
Market garden,	499
" " beginning one,	494
" preparing vegetables for	503
" vegetables, what to grow,	501
Market prices,	289
Markets for vegetables,	494
Marriage procession,	933
" settlement,	369
" in church, ceremony,	933
Married women, obligation for	712
Masonry and masons' work,	377
Massachusetts game law,	675
Massing trees,	539
Materials for budding,	487
" for building,	376
Mayonnaise sauce,	840
May beetle, remedy for	648
Maxims for business,	719
Meadow, a rich,	53
Meadow grasses, tables of	152
" " summary of	154

	PAGE.
Meadow-sweet (*spirea*),	524
Meadows, Roman, irrigation of	52
" seeding,	150
Meals timed by common sense,	913
Measures of time,	1007
Meat and onion stew,	827
" and vegetables,	818
" jellies,	768
" pies, crust for	834
" " pastry for	834
" pot-pie,	834
" thickness of for broiling,	832
" to roast,	829
Mechanics on the farm,	426–441
" lien,	980
" tools,	427
Medicinal teas,	770
Melon worm, remedy for	645
Melted butter,	838
Mending clothes,	755
Merino, to clean,	789
Mercantile terms, definition of	720–721
Metals, gravity of	993
" melting points of	993
Meter explained,	984
Metric and English system compared,	984
" avoirdupois comparison,	988
" " weight,	988
" cubic measure and equivalents,	986
" dry " "	986
" equivalents, apothecaries' weight,	989
" " Troy weight,	988
" liquid measure and equivalents,	987
" long measure,	985
" measure, multiples of	984
" square measure and equivalents,	985
" system,	983
" terms, abbreviations of	984
Mexican clover (*Richardsonia*),	165
Michigan, draining in	331
" game laws,	678
Middle States, fence laws,	697
" " stock laws,	686–687
Milk bread,	857
Millet and Hungarian grass,	170–171
Minced-liver,	844
Mince-meat,	863
Mince-pies, with and without cider,	863
Minnesota game laws,	678
Mistakes in fish culture,	558
Mixed horticulture,	47
Mixed pickles,	849
Mock orange (*Syringa*)	522
Model silos, cost of	192
Modern farm implements,	61
Modified feeding floor,	417
Moist soil grasses,	142
Molasses candy,	890
Money at interest, growth of	1017
" by express,	711
" by letter,	711
" tables,	989
" spurious,	719
Moon-seed (*Menispermum*),	527
Moral duties of parents,	39
Mordants for dyes,	800
Mortality rates,	1018
Mortgage, short form,	715

Mortising, gauge for	302
" posts,	302
Moss-roses,	514
Moss and currant jelly,	769
Moths, in clothes and furs,	785
" night-flying, to kill,	622
" to eradicate,	786
Moulting of silk-worms,	217
Mound builders, the	46
Mushrooms, edible,	845
" testing,	845
Musical party, invitation to	945
Mustard, to mix	852
Mutton and lamb,	811
" boiled,	826
" broth,	822
" to carve,	749
" when best,	811
Names of the parts of the vine,	476
Napkins, to fold,	753
" use of	753
Native grasses, disappearance of	146
" western grasses,	144
" wines at dinner,	910
Nature the guide,	761
Nature's classic halls,	39
Nectarines described,	467
" difficulties of cultivation,	467
Needed reforms in farm law,	674
New England States fence laws,	696–697
" " " stock laws,	683–686
New Hampshire game laws,	675
New Jersey " "	677
New York " "	676
Newport cake,	879
New Year's calls,	918
Night-flying moths, to kill,	622
Norman farm implements,	59
Norway spruce hedge,	319
Note payable to bearer,	707
" with surety,	707
Notes, forms of	707
" jointly and severally,	707
" remarks on	709
Notice against trespassing,	662
" to quit,	713
Nursery and insects, second year's cultivation,	582
Nursery bathing,	762
" the	47
" trees, destroying insects on	579
" " proper forms,	452
Nursing weak patients,	765
Nutmegs and mace,	969
Nutritive equivalents, table of	1013
Oatmeal breakfast cake,	860
" gruel,	770
" porridge,	867
" pudding,	867
Oats and their cultivation,	105
" and their products,	962
" best soils for	107
" seed per acre,	107
" species of	106
" threshing,	107
" time of harvest,	107
" time of sowing,	106

Oats, valuable varieties,	106
" when to sow,	106
Obligation of married women,	712
Obstructions in roads,	665
Officers of the United States, pay of	1021
Ohio game laws,	677
" farmer on draining,	330
Oil-cloths, to lay down,	797
Okra soup,	822
Old house, remodeling,	364
Omelets, to make,	836
Omnivorous birds,	654
Onion sauce,	839
Oriole,	654
Ornamental cottage,	396
" " ground plan,	396
" gates,	313
" grasses,	533
" hedges,	319
" " planting,	322
" trellises,	534
" trees,	538–556
Ornamentation, modest,	740
Ornamenting cakes,	876
Open drains,	327
" " proper form,	333
Orange-leaf notcher, remedy for	644
Orange cake,	879
" marmalade,	895
" scale insects, remedies for	598
" pudding,	873
Orangeade,	770
Orchard and garden,	445
" cultivation destroys insects,	580
" digging the holes,	449
" fruits, picking and packing,	472
" how to prepare for	448
" laying out and planting,	449
" peach,	466
" pear,	457
" planting of	445
" root pruning,	450
" the	281
" trees to plant,	281
" " transplanting,	281
" " when to buy,	451
" " when to plant,	452
Orchards, grafting old	484
Orders, forms of	708
Orkney Islands plowing,	59
Osage orange hedge,	319
" " sphinx,	639
Osmazone of meat,	818
Outline for specifications for buildings,	377
Out-houses,	407–413
Oven, heating the	857
Overworked wife, the	36
Owls,	653
Oxford dumplings,	869
Ox-gall and lye for cleaning marble,	783
Ox, how to cut up,	814
Oxygen, absorption of by soil,	86
Oyster fritters,	833
" pies,	835
" salad,	842
" soup, plain,	824
" " rich,	824
" stew,	824

ALPHABETICAL AND ANALYTICAL INDEX. 1047

	PAGE.
Pacific States fence laws,	702
Packing orchard fruits,	472
Paint, inside, recipe for	793
" shop of the farm,	434
Painting,	734
" time for outside work,	793
Paintings and engravings,	727
Palladius and agriculture.	50
Pan fish, to serve,	751
Panicum varieties of	149
Papering walls,	792
Parlor decorations,	740
" etiquette,	737
" lectures, rules for	926
" plants,	728
Parsnip fritters,	844
Party etiquette,	923–924
Paste that will keep,	800
Pasture grasses,	132, 166
Pastures, ancient,	52
" first,	280
" permanent,	280
Pasturing and soiling compared,	167
Pastry, to make,	862
Pay of officers of the United States,	1021
Peach culture,	466
" marmalade,	895
" orchard,	466
" pie,	863–864
" yellows in	466
Peaches, curl in	466
" illustrated,	463–465
" planting,	466
" some good varieties,	466–467
Peas, products of	963
Pea-soup,	823
Peanuts, cultivation of	241
" harvesting,	242
" or goubers,	241
" shocking,	242
" save good seed,	243
Pears, varieties illustrated,	457–462
" bonne du puits ansault,	458
" Dix,	460
" Frederic Clapp,	458
" gathering,	457
" Howell;	458
" little Marguerite,	462
" on quince stock,	457
" paradise d'Automne,	462
" uncertainty of	457
" varieties and cultivation,	457
Pedigree grain,	97
Peep holes over drains,	347
Pennsylvania game laws,	677
Pepper and its adulterations,	968
Perennial flowers, select list of	513
Performance of obligations,	381
Permanent pastures,	280
Perpetual roses,	515
Persimmon (*diaspyras*),	532
Philosophy of etiquette,	903
Phosphate and potash crops,	83
Piccalilli,	849
Pickled red cabbage,	847
Pickles, mixed,	849
" sweet,	850
" without vinegar,	848

	PAGE.
Picking and curing cranberries,	480
" and packing orchard fruits,	472
" hops,	227
Picnic, invitation to	947
Pictures for the dining-room,	730
Pies and pie-making,	862–865
" for dyspeptics,	862
Pineapple short-cake,	866
Pink color for woolens,	802
Pine-tree borer,	639
Pioneer cottage, cost of	272
" " to build,	273
" farming,	271
Pioneer's plow, a	61
" rude home,	38
Pit silos, difficulties of	180
Pitting and cellaring roots,	176
Plain farm-house,	393
" oyster soup,	824
" paste,	862
" sauce,	866
Plan for secluded grounds,	538
" of plain house, two diagrams,	394
" of rural grounds,	402
Plank soil grinder,	69
Plantations of timber,	280
Planting corn,	112
" cranberries,	478
" flower beds,	541
" "in the moon,"	505
" ornamental hedge,	321
" ramie,	209
" sweet potato slips,	244
" the hedge,	317
" the hop-yard,	226
" the orchard,	449
" village lots,	538
Plant bugs,	601–607
" " classification of	601
" " remedies for	607–608
Plant-lice,	595–597
" remedies for	597
Plants and seeds per acre, to crop,	1006
" and vines, flowering,	514
" for carpet bedding,	541
" for hedging,	317–321
" for the dining-room,	729
" heat for ripening,	997
" per acre, various distances,	1008
" per square rod,	1009
" temperatures proper for	996
" to raise, number of	495
" value of interchange,	166
" wet weather,	356–358
Plashing the hedge,	318
Play hours for youth,	38
Pliny's agricultural writings,	50
Plover,	654
Plow and pasture land,	291
" attachments for trash,	63
" of Jethro Wood,	61
" of to-day,	62
" scooter,	201
" Webster's,	62
" with chain,	63
Plowing, art of	63
" and fertilizing cotton soils,	200
" by steam,	60

	PAGE.
Plowing, irregular areas,	73
" of the Romans,	57
" subsoil,	66
" sward,	65
" table of,	1004
" trench,	67
" without dead furrows,	74
Plows, ancient,	57
" first steel,	61
Plum, flowering,	523
" marmalade,	895
" pudding, English,	867
Plums, preserved without the skins,	898
Poetry of hedges,	316
" of the forest,	574
Points on business law,	719
Poisons and their antidotes,	777–778
Polish for collars, cuffs, etc.,	789
Polishing starched linen,	788
Pomology, definition of	47
Pond and river fish, families of	560–561
Pond fish of the West,	560
Ponds and drains,	335
" breeding fish in	567
" to make from drainage,	335
Population of countries,	1019
" square mile of countries,	1019
Pork cake,	882
Portable fence of wire,	307
" fences,	304–307
Post-and-rail fence,	301
Posts, fence, best,	301
" " mortising,	302
" " setting the	303
" to split,	302
Potash and phosphate crops,	83
Potato bread,	857–858
" croquettes,	842
" crust,	835
" pie,	865
" salad,	841
Potato-growing States,	964
Potatoes, how to raise,	504
" products of	963
" to harvest,	505
" varieties illustrated,	506
" when to plant,	505
Pot herbs in the garden,	502
Pot-pie, meat,	834
" chicken,	835
Poultry houses,	409
Pound cake, rich,	878
Pounds, per bushel, of grass seeds,	152
Power of attorney, proxy to vote,	980
" " revocation of	980
" " short form,	980
Powers of attorney, what for	714
Practical conclusions, ensilage feeding,	193
" test of fertility,	77
" aims,	43
Prairie breaking,	65
Precedence, rules of	936
Predatory birds.	585
" insects,	585
Preëmption law,	670
Prehistoric people,	46
Preparation of ginger,	969
Prepared dishes baked,	833

	PAGE.
Preparing cotton soils,	199
" hop-yard,	225
" jute fiber,	208
" wheat soils,	92
Preserving in sugar,	894
" pork,	809–810
" rules for	892
" scions,	485
Preservatory and ice-house,	373
Preservation of health,	761
Prices per pound and ton, table of	1015
Principles and practice in agriculture,	75
Printing and agriculture,	44
Primary school-house,	404
Private weddings, how conducted,	933
Privet hedge,	321
Privies without smell,	411
Products of barley,	962
" of beans,	963
" of maize,	962
" of oats,	962
" of peas,	963
" of potatoes,	963
" of rye,	961
Production of silk-worm eggs,	213
Profitable silk regions,	212
Programmes at dances,	923
Promiscuous introductions,	921
Promises inviolable,	704
Propagation of vines by cuttings,	490
Propagating pit,	408
Property may not be left in road,	665
Proportion of ears to shelled corn,	999
Proportionate doses,	775
Protection, trees for	322–323
Protecting cribs from rats,	425
" useful birds,	679
" " insects,	588
" buds in winter,	489
Protectors for trees,	539
Protests of notes, drafts, etc.	710
Proverbial salad,	840
Providing for children,	369
Proximate principles, grains,	1013
" " grass,	1014
Proxies,	981
Pruning apple-trees,	453
Public debt, United States, ninety years,	1022
" land, how divided,	670
" lands, how acquired,	669–673
" " how disposed of,	669
" roadways and public rights,	664
Puddings,	769, 866
" and their sauces,	866–869
" fruit,	870–872
" grain, etc.	872–874
" potato,	868
Puff paste, fine,	862
Puffs, breakfast,	860
Pulse family, importance of	165
Pumpkin pie,	862
" " plain,	865
Pure water as a beverage,	884
Putty, to soften,	794
Qualified indorsement,	706
Quaker pudding,	869
Quick pickles,	847

ALPHABETICAL AND ANALYTICAL INDEX.

	PAGE.
Quilting frame,	745
Quince, Japan,	518
" stocks for pears,	457
" marmalade,	895
Radiating, various substances,	996
Rainfall on the plains,	1002
" United States,	1001
Rails, sharpening,	302
Railway fences and trespass,	663
" right of way,	664
Railways running through farms,	664
Raising-bee,	283
Raising fish in ponds,	567
" hemp for lint,	205
" hemp-seed,	204
" potatoes,	504
" tobacco, South,	232
Raisin pie,	864
Ramie a perennial plant,	210
" fiber,	209
" in the United States,	209
" soil for and planting,	209
Rapacious birds,	653
Raspberry jam,	896
" syrup,	897
" vinegar,	852
Raspberries, care of	472
Rat-proof cribs,	423
Rates of mortality,	1018
Raw beef tea, to make	767
Reaping machines, ancient,	55
Receipts, forms of	708
" importance of taking	710
" in full,	710
Recipe for brandying fruits,	899
" for hard soap,	790
" for whitewash,	734
Recipes for household,	780–805
Recommendation, general letter of	948
" special letter of	948
Receiving calls,	918
Reception cards,	952
Receptions, wedding,	934
Red bug (cotton stainer),	608
" remedy for	608
Red Altringham carrot,	177
" cabbage, pickled,	847
" clovers, varieties of	161–162
" coloring for cotton,	801
" wine, how served,	910
Reeling silk,	220
Refining sorghum sugar,	265
Refusing favor asked,	949
Regrets at non-acceptance,	946
" forms for	948
Release and discharge, mechanics' lien,	980
" of mortgage,	715
Religions of the earth, census of	1020
Relish for lunch,	843
Remedies for the sick,	771–774
Remittances, charges on	712
" how made,	711
Remodeled farm-house.	364
" ground plan,	365
Remodeling the old house,	363
Renovating carpets,	736
Rent of farm, security for	295

	PAGE.
Repairing garments,	755
" implements,	433
" on the farm,	428
Re-plowing,	66
Reproduction by grafting and budding,	481
" from cuttings,	481
" from seedlings,	481
Rescue (brome) grass,	150
Restoring furniture,	796
Restrictive indorsement,	706
Revocation of power of attorney,	980
Rhode Island game laws,	675
Rhodotypus,	525
Rhubarb pies,	863
Rice and its cultivation,	123–129
" croquettes,	844
" cake,	880
" fields, management of	126
" harvesting and threshing,	129
" hulling,	130
" in the Mississippi delta,	130
" maggot,	128
" origin of	123
" pudding without eggs,	872
" seed and seeding,	127
" stalk-borer,	625
" swamp species,	123
" to boil,	826
" upland, northern limit,	123
" water with raisins,	770
" wild, description of	123
Riding and driving etiquette,	922
Right and wrong way of draining,	333
" of way by railways,	664
" " by sale,	666
" " by use,	666
" " over other's lands,	665
Rights over others' lands, how acquired,	666
Ripe cucumber pickles,	850
" fruit pickles,	850
" meat, meaning of	810
" tomato pickles,	850
River and pond fish, families,	560–561
" fish of the West,	560
Rizena pudding,	872
Road officers' rights,	664
" rights, limits to	666
Roads are for the public,	665
" keeping in order,	540
" when obstructed,	665
Roast beef, to carve,	748
" " how to prepare,	830
" fowl,	830
" pig, to carve	749
" turkey,	830
Roasted meat,	829
Roasting and boiling, time required for	767
" coffee,	886
" meats,	768
Rocking-chair and its covering,	741
Rocky Mountain locust,	634
Roller, iron,	69
Roll jelly-cake,	878
Rolls and biscuits,	854
Rollers for gates,	311
Roman agriculture,	49
" barley,	53
" crops,	54

	PAGE.
Roman manures,	55
" plowing,	57
" rotation,	53
" seeding,	54, 58
" threshing,	55
" wheat,	53
Roof, putting on	283
" to shingle,	436
Root crops, clean, rich land necessary,	174
" " cultivating,	174
" " fall plowing necessary for	174
" " for forage,	173
" " harvesting,	175
" " preparation for	174
" " cost of raising per acre,	174
" " singling or separating plants,	175
" " weeding,	175
" grafting,	482
" lice,	597
" pruning,	450
Ropes, strength of	993
Roses, autumn,	516
" June,	517
" moss,	514
" perpetual,	515
" tea,	515
Rotting flax straw,	204
" hemp,	206
Rough buildings, to erect,	435
" land,	276
Round cisterns, contents of	1000
Rubbish harbors insects,	580
Rotation, a simple,	78
" a southern,	81
" and bad seasons,	79
" an eastern,	80
" European,	82
" elaborate,	79
" and crops,	78
" system in	80
" substitution in	82
" table of	79
Rotary harrow,	68
Rules for conduct of funerals,	929
" for evening calls,	916
" for invalid cooking,	765
" for general guidance,	928
" for informal dinners,	905
" in making jellies,	896
" for parlor lectures,	926
" for preserving,	892
" for success of the dinner,	904
" for tobacco growers,	241
" of conduct in traveling,	934
" of etiquette, seventy-five cardinal	954–957
" of table usage,	908–910
" of guidance in business,	704
" relating to banking,	705
Rural architecture,	361
" buildings,	393
" grounds plan of	401
" home,	397
" " ground plan,	397
Rustic seats,	406
Rusks and rolls,	861
Rye and its cultivation,	101
" and its products,	961
" and Indian bread,	858

	PAGE.
Rye, best soil,	102
" bread,	857
" harvest time,	102
" Montana spring,	101
" seed per acre,	102
" sowing time,	102
" uses of in United States,	101
Saddle grafting,	482
Sagging gates, to prevent,	313
Sago and its preparation,	964
" whence derived,	964
Salaries, to calculate,	1018
Salad dressing, simple,	841
" of meat, fowl and fish,	841
Salads and their dressing,	840–842
Salmon croquettes,	844
Salt hop yeast,	856
Salt rising,	856
Sandwiches,	844
Salutations,	922
" recognition of	922
Sand and cranberry growing,	477
" in soils, per cent,	85
Sap buckets and tubs,	266
" frothing over,	268
Sashes, to tighten,	796
Sauce for boiled meats,	826
" for boiled pudding,	866
" for cold meat,	839
" for fish and fowl,	838
" for fowl,	840
" for hen's-nest pudding,	870
" for roast meat,	839
" of many names,	840
" white,	839, 866
Sauces and gravies,	838–840
" for puddings,	866
Savage agriculture,	44
Scale insects,	597–601
" California, remedies for	601
Scarlet dye for woolens,	802
Science in agriculture,	84
Scions, cutting and saving,	485
" selection of	489
" time to cut,	489
" to keep,	489
School-house architecture,	402
" interior,	404
" primary,	404
Scooter plow,	201
Scotch cake,	880
Scrambled eggs,	836
Scraper for roads,	540
Secluded grounds, plan for	538
Second year's crop,	279
Section of under-drain,	338
Security for rent,	295
Seed bed for tobacco, South,	232
" " of sweet potatoes,	244
" corn, test,	117
" crop of clover,	161
" husbandry,	47
" importance of good,	97
" improvement of, by selection,	98
" peanuts,	243
" potatoes, how to cut,	504
" saved in caves,	45

	PAGE.
Seeding and cultivation, Roman,	58
" for hay and pasture,	151
" machines,	72
" meadows,	150
Seeds and plants, per acre, to crop,	1006
" per ounce of grasses,	152
" to sow 100 yards of drill,	1007
" vitality of	1009
" value of interchange of	166
Select biennial flowers,	513
" list of flowers,	510–511
Selecting a claim of land,	276
" a farm,	290
" flour,	854
" the site,	277
Selling the crop,	289
Self-closing gates,	309–310
Servants and parlor service,	739
Servants' carving,	907
Service of the table,	751
Setting posts	303
Settler's first home,	41
Settling a new country,	372
Seventeen household facts,	795
Seventy-five rules of etiquette,	954–957
Shade trees for villages,	537
" " to save	282
"Shall I move the barn?"	274
Sharpening tools,	432
Sheds and barns for sheep,	418
Sheep barn and sheds,	418
" barn and yards,	419
" dipping box,	419, 420
" figured for cutting up,	811
" how to cut up,	811
" rack,	420
Sheltering groves,	572–576
Sherbet,	771
Shingling a roof,	436
Shocked corn, field of	115
Shocking corn around tables,	116
" green-cut forage,	179
" grain,	96
" peanuts,	242
Shopping etiquette,	937
Short-cake,	881
" fruit,	865
Short form, bill of sale,	716
" " of lease,	712
" " of mortgage,	715
" " warranty deed,	715
Shrubs, flowering,	517
Sick-room, disinfecting,	775
Side-dish of eggs,	843
Side of beef, to cut up,	814
Sight drafts,	710
Signing by mark,	711
Signs used in business,	721
Silk and silk-worms,	211
" reeling the	220
" vine (*periploca*),	527
Silk-reeling machine,	219
Silk-producing insects,	212
Silk-worm cocoons, to gather	219
" eggs,	213
" " hatching,	214
" " heat and moisture for	215
" " keeping,	214

	PAGE.
Silk-worms, care of	217
" feeding,	215
" " shelves,	216
" food of	221–222
" how to kill,	219
" Japanese eggs best,	214
" keeping,	214
" marketing cocoons and eggs,	220
" moulting of	213, 217
" regions,	212
" stages of	213
" varieties of	214
" ventilation of building,	215
" winding frames,	218
Silver bell (*halesia*),	519
Silt wells,	347
Silo, capacity,	187
" condition when opened,	189
" cost of	187
" " filling,	188
" covering,	187
" crops for	187
" the foundation,	185
" the superstructure,	186
" how to build,	185
Silos and ensilage,	178–193
" air-tight, modern,	179
" best form,	181
" condensed facts on	186–191
" cost of perfect,	192
" Crevat's experiments,	180
" fermentation in	184
" of earth,	180
" pit, illustrated,	179
" practical experience and results,	186
" size of	185
Silver cake,	880
" to clean,	791
Simple dishes,	769
" farm-house plans,	364
" poisons and their antidotes,	777
" sick-room remedies,	771
" slide-gate,	310
Simplifying law,	674
Sinks and wallows,	352
Site, selection of	277
Situation of the farm important,	291
Size of fruits,	465
Skim and trench plowing,	67
Slab and pole drains,	340
Sleeping-rooms,	731
Slide and swing gate,	308
" gate, self-closing,	309
Sloughs, soaks and springs,	353
Small fruits and their care,	472
" grains, ancient,	53
Smaller cornstalk borer,	625
Smoke-house, brick,	413
" wood,	413
Smoke-houses,	419
Snout beetles, remedies for	645
Snowball (*viburnum*),	525
Soap, home-made,	790
Social party, invitation to	945
Society, rules for guidance,	914–915
Sod and ditch fence,	306
" " " to build,	306
" fence, banking,	306

	PAGE.
Soda cakes,	881
Soft and hard soils,	286
" and hard ground crops,	83
" gingerbread,	881
" soap, to make	790
Soil, analysis unnecessary,	75
" and climate for jute,	208
" and cultivation for alfalfa,	164
" capabilities,	84
" for hot-bed,	498
" for sweet potatoes,	244
" grinder,	69
" proper for ramie,	209
" to test	277
" vegetation an index to	276
Soiling, advantages of	172–173
" compared with pasturing,	167
" crops, brown dhoura,	171
" crops, common millet,	171
" crops, German millet,	171
" crop, how to raise,	168
" crops, Hungarian,	170
" crops, pearl millet,	171
" crops, prickly comfrey,	172
" fodder and root crops,	167
" for dairy districts,	167
" results of, in Scotland,	173
" vs. fencing,	168
" with clover,	173
" with corn and sorghum,	169
Soils, absorbing power of	86, 996
" absorption of oxygen,	86
" and situations for tobacco,	231
" capacity of, for heat,	995
" capacity of, for water,	348
" crops, adaptation of,	286
" do not wear out,	76
" evaporation of	996
" for cotton,	195
" for flax,	203
" for wheat,	92
" organic matter of	996
" radiating power of	996
" temperature of	995
" unhealthful,	276
" valuable,	276
" weight of	85
Soiled beds and mattresses,	794
Soldier beetles,	648
" bug,	648
Solid or cubic measure,	986
Sons and daughters on the farm,	37
Sorghum, boiling in vacuum pan,	265
" boiling the juice,	264
" granulation,	264
" cane, cutting and handling,	250
" crystalizing and draining,	265
" cultivation of,	249–250
" for soiling,	169
" general conclusions on,	266
" important points on,	254
" juice, filtering and liming,	264
" " specific gravity,	250–253
" manufacture of	264
" real tests of value,	255
" sugar, decoloring,	265
" " produced,	247
" " refining,	265

	PAGE.
Sorghum sugar, whitening,	265
" valuable varieties illustrated,	257–263
" web-worm,	624
" when to cut,	250
Soup or stock, to clarify,	828
" should be simmered,	819
" vegetable	821
Soups,	821–824
" to color,	829
Sour pickles,	846–848
Southern farm-gate,	311
" grass-worm,	626
" States, fence laws,	697–698
Sowers, broadcast,	72
Sowing and cultivating root crops,	174
Spanish proverb on salads,	840
Spawning bed, artificial,	568
Special forms of notes,	707
" grass and other crops,	131
Species of grass,	158
" of American grapes,	473
" of jute,	207
" of oats,	106
" of wheat,	91
Specifications, cost of model silos,	192
" for building,	377–381
Specific gravity, how determined,	991
" " of sorghum juice,	250–253
" " of dry wood,	991
" " of earths,	991
" " of rock,	991
" " miscellaneous,	991
" " of metals, table of	990
Spice cake,	880
" pudding,	873
Spiced jam,	896
Spined tree-bugs,	603
Spirea (meadow sweet),	524
Spireas ten good,	524–525
Split rolls,	861
Splitting posts,	302
Sponge cake,	878
" for bread,	855
" gingerbread,	881
Sports of childhood,	41
Spring budding,	488
" forcing,	497
" garden work,	501
" work on tobacco, South,	233
Springs and drainage,	328
" soaks and sloughs,	353
Spruce beer,	888
Spurious coin,	719
Square cottage, first floor,	395
" " second floor,	395
" summer-house,	406
" cisterns, contents of	1001
" feet, and feet square, in areas,	999
" measure,	985
" " surveyor's,	985
" " comparative scale,	985
Squash bug,	605
Stable and carriage-house for farm,	368
" floor,	417
Stables and corn cribs,	414
Stalk-borer, rice,	625
" " preventives for	626
Standard bushel,	1000

Standards of measure,	983
Starch for lawns, etc.	789
Starching and ironing,	788
Starting a dairy,	287
" a grove,	280
" an orchard,	281
" the hot-bed,	497
" tobacco plants,	235
State laws on fencing,	695–702
States best for tobacco,	232
Statute defining fences,	663
Steam plowing,	60
Steamed dishes,	837
" pudding,	869
Stewed tripe,	843
Stewing,	820, 826–828
Stirring and stubble plow,	66
Stock killed on railway lines,	664
" laws,	683
" " Middle States,	686–687
" " New England States,	683–686
" " Southern States,	687–690
" " Western States,	690–695
Stockmen and drainage,	357
Stock water from drains,	334
Stock, how to make,	828
" or soup, to clarify,	828
Stone fences, how to build,	438
" heavy, to move,	437
" laid drains,	339
Stoves, care of	783
Straight fence, with stakes,	301
Strap-hinged gate,	311
Strawberry acid,	853
" jam,	896
" marmalade,	895
" short-cake,	865
" tarts,	865
Strawberries, care of	472
Stream gate and footway,	314
Street deportment,	919
" " special rules for	920
Strength of horses,	1003
" of ropes,	993
Stringing wire for fence,	305
Stripping tobacco,	239
Stubble plow, gang,	64
Study and play,	764
" the face of bank-bills,	718
Studying taste in ornamentals,	536
Subsoil plowing,	66
Suburban cottage,	365
" ornamental cottage,	396
Successful farming,	273
Suet pudding,	869
Sugar, beet, in United States,	247
" candy,	890
" to a quart of fruit,	893
" plants, our two greatest,	247
Sugars, comparison of	248–249
" the principal,	248
Sugar-cane and sugar-making,	246
" " beetle,	625
" " cultivation of	249
Sugar-making on the farm,	264
Suggestions on letter-writing,	950
Sulphur and lime for insects,	581
Summary of meadow grasses,	154
Summer bathing,	763
" drinks,	888–889
" flowering bulbs,	514
" house of bark,	406
" " elegant,	407
" " rustic,	406
" " wigwam,	405
Supply of water,	374
" " for house,	732
Supper and breakfast,	912
" parties,	912
" room, the	924
Surveyed lands, United States,	673
Sward plowing,	65
Swedish clover,	162
Sweeping, and care of brooms,	792
" carpets,	736
Sweet apple pudding,	871
" green pickles,	851
" pickles,	850–851
" potato root-borer and remedy,	645
Sweet potatoes, field cultivation of	244
" " garden cultivation of	245
" " planting the sets,	244
" " soils,	244
" " starting the plants,	244
" " to keep during winter,	245
Swimming birds,	654
Swindling by solicitors,	703
" through notes, etc.	703
Syringa (*Philadelphus*),	522
Syrup of lemons,	899
Syrups,	897–898
Table decorations,	746
" etiquette,	747
" service,	751
" usages at	908
Table of absorbing power of soils,	86
" absorption of oxygen by soil,	86
" cloth measure,	983
" colors,	758
" comparative values of sorghum,	256
" corn product,	90
" cost of corn crop,	118
" cotton crop by States,	196
" " crops,	196
" crushing force of metals,	993
" distance,	983
" dry measure,	983, 986
" earth's area and population,	1019
" equivalents for cooking,	883
" export of food crops,	105
" foods,	765
" foreign exchange,	990
" germinating depths for wheat,	95
" grass seed, pounds per bushel, seeds per ounce, depth for germination,	152
" grasses for special soils,	152
" " Woburn experiments,	156–157
" improvement in seed wheat,	98
" " wheat ears,	98
" plowing, acres,	1004
" price per pound,	1015
" proportionate doses,	775
" relative corn crop,	91
" rending force, wood,	992
" rye, seed, harvest, soils, etc.,	102

	PAGE.
Table of seeding, harvest, etc., of wheat,	95
" soils, seeding, etc., of barley,	103
" specific gravities,	990
" time of boiling fruit,	893
" " digestion,	766
" United States money,	989
" unusual weights,	982
" valuable western grasses,	144
" wages and salaries,	1019
" water discharged from tile,	349
" wet land plants,	358
" weight of soils,	85
" weights,	982
" wine measure,	983, 987
Tachina fly,	649, 650
Taking up public lands,	669–673
Tamarinds,	970
Tapioca, varieties and use,	964
Tarts and tart crusts,	865
Taste in building,	370
" in ladies' dress,	757
" in pictures, etc.,	730
Taylor cake,	880
Tea "	880
" how to keep,	886
" how to make,	886
" roses,	515
" where growing wild,	966
" where indigenous,	966
Tea-making of various peoples,	886
Teams, three-horse,	278
Temperature of germination,	997
" of soils,	995
" required by plants,	996
Tenant's certificate of lease,	294
" agreement,	713
" notice of quitting,	713
Ten ideals in agriculture,	83
Tepid baths, fresh and salt water,	763
Terminal budding,	488
Terminology of the vine,	476
Terraces,	539
Test for copper in water,	777
" for iron in water,	777
" for impurities in water,	777
" of fertility, practical,	77
" the soil,	291
Testing pork,	809–810
" soils,	277
" the value of species,	135
Tests for finishing sugar,	268
" of value in working sorghum,	255
Textile crops and fibers,	194
Theatricals, private,	925
Thermometers compared,	998
" relations between,	997
Things to avoid in society,	914
Thorn (*Cratægus*),	529
Three horses abreast, driving,	278
" " " hitching,	278
Threshing, ancient,	55
" barley,	103
" flax,	204
" rice,	129
Thrift vs. unthrift,	275
"Thy free, fair homes,"	33
Tiger beetles,	648
Tile, connections, small with large,	349

	PAGE.
Tile, different kinds of	341
" drainage, practical men on	329
" drains, velocity of water in	1003
" origin of	329
" water carried by	347
" " discharged from	349
" velocity of water in	349
Timbale of potatoes,	844
Timber culture act,	671
" climatic effects of	571
" deadening,	285
" valuable, how saved,	282
" fuel from	572
" growth of	571
" planting,	570–576
" plantations,	280
" protection and fuel,	572
" to plant,	571
" various uses of	572
" when a nuisance,	573
Timbered farm, to clear,	282
Time for meals,	913
" required to cook vegetables,	766
" " to roast or boil,	767
Tin covers, to clean,	784
Tobacco, belt of	231
" bulking,	239
" cultivation of, South,	232–234
" cutting,	238
" growers, twelve rules for	240
" growing, South,	232
" house,	239
" magnitude of the crop,	231
" situations for	231
" South, manuring,	232
" " seed-bed,	235
" States,	232
" spring cultivation, South,	233
" stripping,	239
" transplanting, North,	236–237
" transplanting, South,	233
" proper way to transplant,	237
" ventilation in drying,	239
" worming,	234
" raising the plants,	235
Tomato beer,	889
" catsup	851
" chow-chow,	849
" figs,	899
" or cotton worm,	628
" sauce,	852
Toilet recipes,	797
Toilet-room and bath,	758
" for guests,	924
" for farm-hands,	758
Tool-house,	427–431
Tools, how to use,	432–433
" ready for service,	273
Trade for cash,	704
Trailing and climbing shrubs,	525–527
Training grapevines,	475–476
" the vine to stakes,	475
Transplanting flowers,	509
" orchard trees,	281
Transporting fish, rules for	562
Trash, cleaning from plows,	63
Traveling etiquette,	934
Treatment of children,	954

ALPHABETICAL AND ANALYTICAL INDEX.

	PAGE.
Tree planting,	573
" producing tamarinds	970
" protectors,	539
Trees about homestead,	576
" flowering,	528
" for barriers,	322–323
" for village grounds,	537
" for protection,	322–323
" for shade,	282
" heeling in	451
" of California,	576
" shrubs, etc., distances apart for	1009–1010
" varieties to plant,	453
" when to buy,	451
" when to plant,	452
Trellises, ornamental,	534
Trellising the vine,	476
Trench plowing,	67
Trenching deep impracticable,	67
" the soil for hops,	225
Trespass from highway to fields,	663
" upon property,	662
Trimming the hedge,	318
Troughs for forcing plants,	494
Troy weight, comparison.	988
" table of	988
True and spurious capers,	970
" taste in dress,	754
Turning flat furrows,	279
Turkey and fowl, to roast,	830–831
" boiled,	825
" steamed,	825
Turning first furrows,	64
Unburned brick, to make	376
Underdraining,	326
" twelve propositions,	357
Underdrain, section of	338
Underdrains, formation of	337
Unfermented wine,	798
Unimproved land, United States,	1011
United States and Great Britain, crop comparison,	1010
Unsurveyed area, United States,	672
Upland cotton,	195
" rice,	123
" " cultivation,	125
Upper floor, farm cottage,	367
Vagabond crambus,	623
" remedies for	623
Varro's agricultural writings,	49
Veal, boiled,	826
" to cut up,	813
" to carve,	748
Vegetable and animal life, contrasts of	997
" garden, economy of	494
" " width of rows,	491
" gardening,	42
" salad, French,	841
" soup,	821
Vegetables, ancient,	753
" and fruits, necessity of	445
" and human names,	752
" for market,	501
" illustrated,	496–506
" preparing for market,	503
" that will sell,	502

	PAGE.
Vegetables, winter forcing of	499
" time required to cook,	766
" with meat,	818
Vegetation an index to soil,	276
Velocity and effect of wind,	1003
" of water in tiles,	349
" " in tile drains,	1003
Venison, roast saddle of	830
Ventilation in drying tobacco,	239
" of houses,	374
Verminous insects,	784
" animals and insects,	579
Vermont game laws,	675
Versailles experiments on drainage,	358
Vicious animals must be restrained,	661
Village lot, design for	537
" lots, planting,	538
" home,	34
Vinegar, flavored,	852
Vine, alternate system,	475
" and plants, flowering,	514
" cultivation of	471–475
" names of parts of	476
" covering in winter,	475
" layering,	490
" pinching and pruning,	475
" training to stakes,	475
Vineyard, the	473–476
Virgil on agriculture,	49
Virginia fence,	299–300
Visiting cards, styles for	951
" " titles, etc.,	917
" etiquette, rules for	928
" introductions at	916
Visitors, entertainment of	738
Vitality of seeds,	1009
Wagon jacks,	419
Wages, to calculate,	1018
Wallows and sinks,	352
Walls, how to paper,	792
Wall, to build,	377
Wardian case,	745
Warranty deed, agreement for	977
" " form of	977
" deeds,	715
Wash for face,	797
Washing dishes,	781
" fluid,	788
" fine fabrics,	787
" rice before boiling,	826
" summer suits,	787
" windows,	794
Washington's one hundred rules,	941
Waste paper, uses of	783
Water and drainage rights,	667
" birds,	654
" bugs,	614–615
" closets,	411
" gates,	314
" ice,	890
" lilies,	534
" meadows, ancient,	52
" plants,	533–534
" rice,	124
" supply,	373
" " for house,	732
Water-proof glue,	799

	PAGE.
Wax for grafting, to make,	485
Waxing grafts,	483
Weather, foretelling the	1010
Wedding engagements, etiquette of	932
" invitations,	932
" when celebrated,	932
Weigelia (*diervilla*),	519
Weight of various substances, cubic foot,	992
" per bushel, of products,	1005
Weights used for special purposes,	982
Well-water, importance of pure,	733
" impurities in	884
Western corn bread,	859
" " crib,	425
" indigenous grasses,	144
" States, fence laws,	699–702
" " stock laws,	690–695
Wet weather plants,	356
Wheat and corn belts,	90
" best soil,	95
" bushels of seed per acre,	95
" conclusion on improvement,	98
" drilling, advantage of	94
" depth of covering,	94
" favorite varieties,	91
" harvest time,	95
" illustrations of	93, 97, 100
" isosoma,	623–624
" reputable old varieties,	99
" soils, best, West,	92
" soils, preparation of	92
" time of sowing,	95
" varieties of, ancient,	53
" with and without fertilizer,	77–78
Whipped cream,	874
White clover,	162
" dent corn	120
" grub or May beetle,	647
" pepper,	968
" sauce,	839, 866
" sugar candy,	890
" spruce,	547
" thorn hedge,	320
Whitening sorghum sugar,	265
Whitewashed walls, to paper	793
Whitewashing,	734
White-wine whey,	770

	PAGE.
Wicket coop,	408
"Widow's cake,"	880
Wigwam summer-house,	405
Wild potato of New Mexico,	964
" rice,	123
Will, codicil to	714
" how to draw	713
Wind-breaks and groves,	280
Window curtains,	728
" sashes, to fasten,	796
" gardening,	40
Wind, velocity and effect of	1003
Winding frames for silk-worms,	218
Wine measure, comparative scale,	987
" or liquid measure,	987
" sauce,	866
" " for roast game,	839
Wines at official dinners,	910
" home-made,	798
Winnowing grain, ancient,	55
Winter bathing,	763
" forcing of vegetables,	499
Wire fence, bracing,	306
" stringing the wire,	305
Wisconsin game laws,	678
Woburn experiments on grasses,	154
" tables of grasses,	156–157
Wood, cement for	799
Woolley, S. J., on draining,	330
Woolly apple-tree blight,	597
Worcestershire sauce,	851
Working cattle,	49
Worm or Virginia fence,	299
Worming tobacco,	234, 238
Workshop on the farm,	426
Yeast and yeast-making,	855
Yeast cakes,	855
" of hops,	855
Yellow and blue color for carpet rags,	804
" color for cotton,	801
" dent corn,	120
Yellowwood (*virgilia*),	554
Yorkshire pudding,	869
Youth, the precocious,	39
Youthful activity,	39
" sports,	41